In accordance with the latest syllabus prescribed by the
Central Board of Secondary Education, New Delhi

MOST LIKELY CBSE QUESTION BANK

CLASS X

MATHEMATICS (Standard)

Category Wise & Chapter Wise

By

A PANEL OF AUTHORS

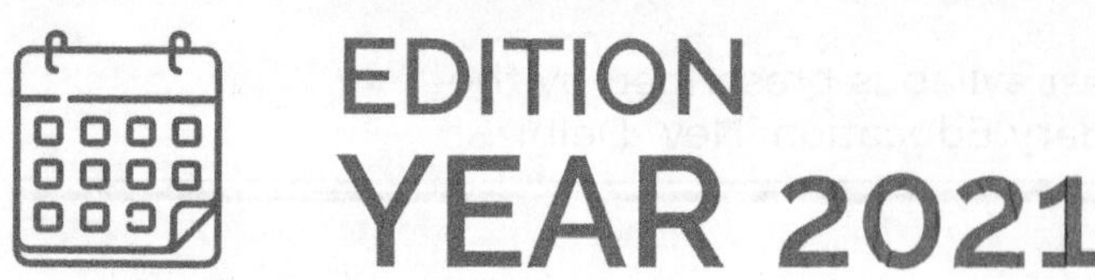

DISCLAIMER

With the ambition of providing standard academic resources, we have exercised extreme care in publishing the content. In case of any discrepancies in the matter, we request readers to excuse the unintentional lapse and not hold us liable for the same. Suggestions are always welcome.

ISBN : 978-93-91184-18-6

PRICE : ₹ 499.00

PUBLISHED BY

GURUKUL BOOKS & PACKAGING
(Unit of Oswal Printers & Publishers Pvt. Ltd.)

 1/12, Sahitya Kunj, M.G. Road, Agra - 282002

 (0562) 2527771-4, +91 7534077222

 info@oswalpublishers.in

 www.oswalpublishers.com

The cover of this book has been designed using resources from Freepik.com

Printed At Upkar Printing Unit, Agra

PREFACE

Gurukul's CBSE Most Likely Question Bank series is up-to-date with the latest syllabus given by the Central Board of Secondary Education.

This title highlights the knowledge-based and skill-based goals of the Bloom's Taxonomy by acquainting the students with relevant facts and concepts. It also teaches them ways to apply their subjective knowledge to get their best academic results.

In this book, the questions are arranged section-wise so that the students can revise the whole syllabus in less time and develop the ability of prioritising and categorising topics for effective learning. It covers all probable types of questions that can be asked in the exams. The solutions and explanations have been prepared by eminent subject experts. They follow the standard marking scheme of the CBSE Board. Additionally, questions from compartment paper, foreign paper and latest board paper have also been incorporated in the chapters.

The series is an attempt to instil confidence in students to face the board examination. The language used is simple, to the point and questions cover all the important topics as per the weightage given to them by the board.

We hope this book will be a valuable asset for the students. All suggestions towards improving the series are welcome and would be incorporated in the future editions.

Note : Questions marked with :
'★' are board exam questions from previous years
'★★' are frequently asked board exam questions

—Publisher

HOW TO PRACTICE THE 3R'S AT HOME?

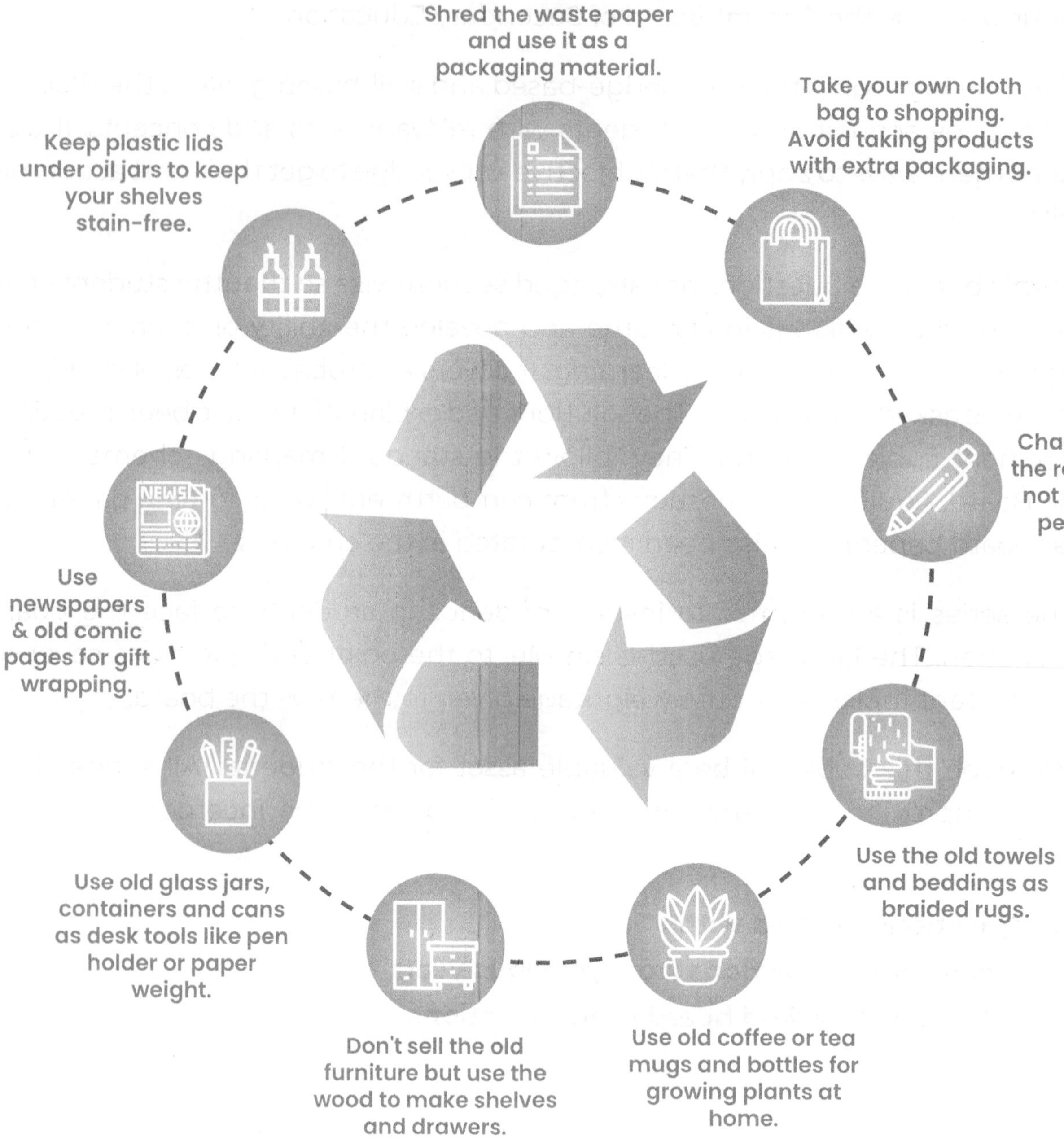

Shred the waste paper and use it as a packaging material.

Take your own cloth bag to shopping. Avoid taking products with extra packaging.

Change the refills not the pen.

Use the old towels and beddings as braided rugs.

Use old coffee or tea mugs and bottles for growing plants at home.

Don't sell the old furniture but use the wood to make shelves and drawers.

Use old glass jars, containers and cans as desk tools like pen holder or paper weight.

Use newspapers & old comic pages for gift wrapping.

Keep plastic lids under oil jars to keep your shelves stain-free.

CONTENTS

Note: Questions marked with :
'★' are board exam questions from previous years
'★★' are frequently asked board exam questions

COURSE STRUCTURE

Units	Unit Name	Marks
I	NUMBER SYSTEMS	06
II	ALGEBRA	20
III	COORDINATE GEOMETRY	06
IV	GEOMETRY	15
V	TRIGONOMETRY	12
VI	MENSURATION	10
VII	STATISTICS & PROBABILITY	11
	TOTAL	**80**

UNIT I : NUMBER SYSTEMS

1. **REAL NUMBERS** **(15) Periods**

   Euclid's division lemma, Fundamental Theorem of Arithmetic - statements after reviewing work done earlier and after illustrating and motivating through examples, Proofs of irrationality of $\sqrt{2}, \sqrt{3}, \sqrt{5}$.

   Decimal representation of rational numbers in terms of terminating/non-terminating recurring decimals.

UNIT II : ALGEBRA

1. **POLYNOMIALS** **(7) Periods**

   Zeros of a polynomial. Relationship between zeros and coefficients of quadratic polynomials. Statement and simple problems on division algorithm for polynomials with real coefficients.

2. **PAIR OF LINEAR EQUATIONS IN TWO VARIABLES** **(15) Periods**

   Pair of linear equations in two variables and graphical method of their solution, consistency/inconsistency.

   Algebraic conditions for number of solutions. Solution of a pair of linear equations in two variables algebraically- by substitution, by elimination, and by cross multiplication method. Simple situational problems. Simple problems on equations reducible to linear equations.

3. **QUADRATIC EQUATIONS** **(15) Periods**

   Standard form of a quadratic equation $ax^2 + bx + c = 0$, (a ≠ 0). Solutions of quadratic equations (only real roots) by factorization and by using quadratic formula. Relationship between discriminant and nature of roots.

   Situational problems based on quadratic equations related to day to day activities to be incorporated.

4. **ARITHMETIC PROGRESSIONS** **(8) Periods**

   Motivation for studying Arithmetic Progression Derivation of the nth term and sum of the first n terms of A.P. and their application in solving daily life problems.

UNIT III : COORDINATE GEOMETRY

1. **LINES (IN TWO-DIMENSIONS)** **(14) Periods**

   Review : Concepts of coordinate geometry, graphs of linear equations. Distance formula. Section formula (internal division). Area of a triangle.

UNIT IV : GEOMETRY

1. **TRIANGLES** **(15) Periods**

   Definitions, examples, counter examples of similar triangles.

1. (Prove) If a line is drawn parallel to one side of a triangle to intersect the other two sides in distinct points, the other two sides are divided in the same ratio.
2. (Motivate) If a line divides two sides of a triangle in the same ratio, the line is parallel to the third side.
3. (Motivate) If in two triangles, the corresponding angles are equal, their corresponding sides are proportional and the triangles are similar.
4. (Motivate) If the corresponding sides of two triangles are proportional, their corresponding angles are equal and the two triangles are similar.

5. (Motivate) If one angle of a triangle is equal to one angle of another triangle and the sides including these angles are proportional, the two triangles are similar.
6. (Motivate) If a perpendicular is drawn from the vertex of the right angle of a right triangle to the hypotenuse, the triangles on each side of the perpendicular are similar to the whole triangle and to each other.
7. (Prove) The ratio of the areas of two similar triangles is equal to the ratio of the squares of their corresponding sides.
8. (Prove) In a right triangle, the square on the hypotenuse is equal to the sum of the squares on the other two sides.
9. (Prove) In a triangle, if the square on one side is equal to sum of the squares on the other two sides, the angle opposite to the first side is a right angle.

2. CIRCLES **(8) Periods**

Tangent to a circle at, point of contact

1. (Prove) The tangent at any point of a circle is perpendicular to the radius through the point of contact.
2. (Prove) The lengths of tangents drawn from an external point to a circle are equal.
3. (Motivate) Alternative Segment theorem : If a chord is drawn through the point of contact of tangent to a circle, then the angles made by the chord with the tangent are respectively equal to the angles subtended by the chord in the alternate segments.

3. CONSTRUCTIONS **(8) Periods**

1. Division of a line segment in a given ratio (internally).
2. Tangents to a circle from a point outside it.
3. Construction of a triangle similar to a given triangle.

UNIT V : TRIGONOMETRY

1. INTRODUCTION TO TRIGONOMETRY **(10) Periods**

Trigonometric ratios of an acute angle of a right-angled triangle. Proof of their existence (well defined); motivate the ratios whichever are defined at 0° and 90°. Values of the trigonometric ratios of 30°, 45° and 60°. Relationships between the ratios.

2. TRIGONOMETRIC IDENTITIES **(15) Periods**

Proof and applications of the identity $\sin^2 A + \cos^2 A = 1$. Only simple identities to be given. Trigonometric ratios of complementary angles.

3. HEIGHTS AND DISTANCES : Angle of elevation, Angle of Depression. **(8) Periods**

Simple problems on heights and distances. Problems should not involve more than two right triangles. Angles of elevation/depression should be only 30°, 45° and 60°.

UNIT VI : MENSURATION

1. AREAS RELATED TO CIRCLES **(12) Periods**

Motivate the area of a circle; area of sectors and segments of a circle. Problems based on areas and perimeter/circumference of the above said plane figures. (In calculating area of segment of a circle, problems should be restricted to central angle of 60°, 90° and 120° only. Plane figures involving triangles, simple quadrilaterals and circle should be taken).

2. SURFACE AREAS AND VOLUMES **(12) Periods**

1. Surface areas and volumes of combinations of any two of the following : cubes, cuboids, spheres, hemispheres and right circular cylinders/cones. Frustum of a cone.
2. Problems involving converting one type of metallic solid into another and other mixed problems. (Problems with combination of not more than two different solids be taken).

UNIT VII : STATISTICS AND PROBABILITY

1. STATISTICS **(18) Periods**

Mean, median and mode of grouped data (bimodal situation to be avoided). Cumulative frequency graph.

2. PROBABILITY **(10) Periods**

Classical definition of probability. Simple problems on finding the probability of an event.

QUESTION PAPER DESIGN
CLASS X
Subject : Mathematics (041)

Time : 3 Hours **Max. Marks : 80**

S. No.	Typology of Questions	Total Marks	% Weightage (approx)
1.	**Remembering :** Exhibit memory of previously learned material by recalling facts, terms, basic concepts, and answers. **Understanding :** Demonstrate understanding of facts and ideas by organizing, comparing, translating, interpreting, giving descriptions, and stating main ideas.	43	54
2.	**Applying :** Solve problems to new situations by applying acquired knowledge, facts, techniques and rules in a different way.	19	24
3.	**Analyzing :** Examine and break information into parts by identifying motives or causes. Make inferences and find evidence to support generalizations. **Evaluating :** Present and defend opinions by making judgements about information, validity of ideas, or quality of work based on a set of criteria. **Creating :** Compile information together in a different way by combining elements in a new pattern or proposing alternative solutions.	18	22
	Total	80	100

INTERNAL ASSESSMENT	20 MARKS
Pen Paper Test and Multiple Assessment (5 + 5)	10 Marks
Portfolio	05 Marks
Lab Practical (Lab activities to be done from the prescribed books)	05 Marks

Basic Concepts

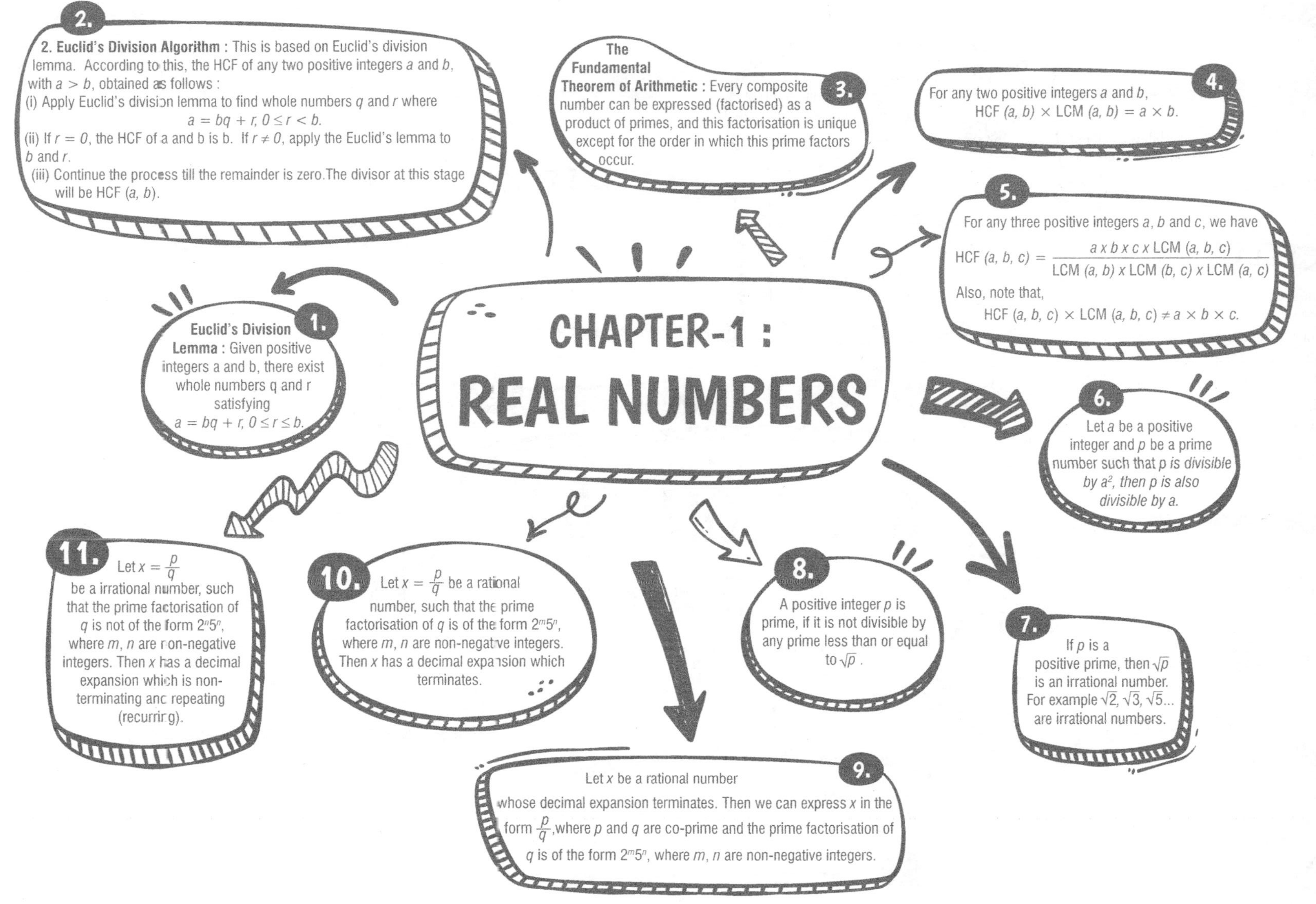

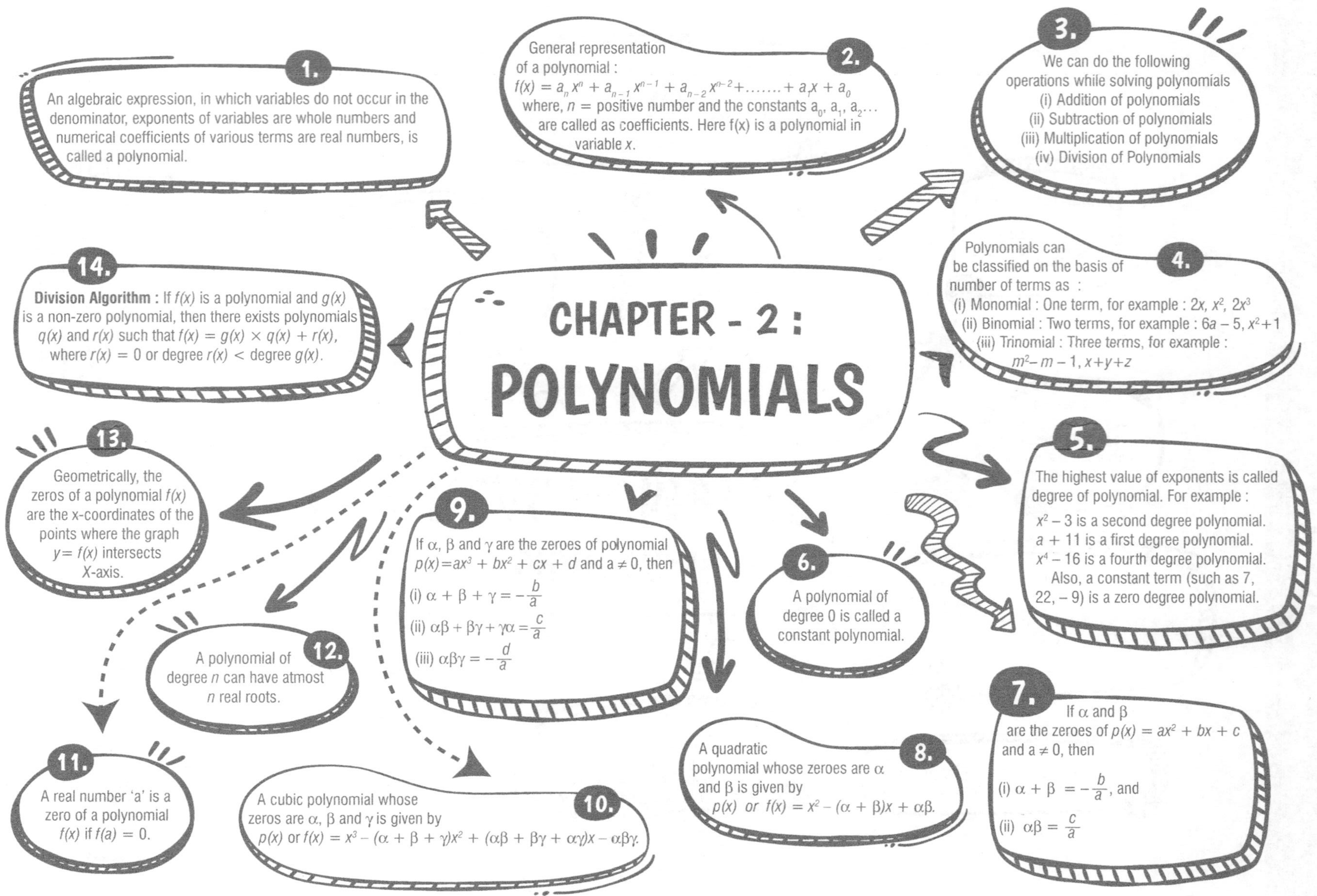

CHAPTER - 2 :
POLYNOMIALS
1. An algebraic expression, in which variables do not occur in the denominator, exponents of variables are whole numbers and numerical coefficients of various terms are real numbers, is called a polynomial.
2. General representation of a polynomial : $f(x) = a_n x^n + a_{n-1} x^{n-1} + a_{n-2} x^{n-2} + \ldots\ldots + a_1 x + a_0$ where, n = positive number and the constants $a_0, a_1, a_2 \ldots$ are called as coefficients. Here f(x) is a polynomial in variable x.
3. We can do the following operations while solving polynomials
(i) Addition of polynomials
(ii) Subtraction of polynomials
(iii) Multiplication of polynomials
(iv) Division of Polynomials
4. Polynomials can be classified on the basis of number of terms as :
(i) Monomial : One term, for example : $2x$, x^2, $2x^3$
(ii) Binomial : Two terms, for example : $6a - 5$, x^2+1
(iii) Trinomial : Three terms, for example : $m^2 - m - 1$, $x+y+z$
5. The highest value of exponents is called degree of polynomial. For example :
$x^2 - 3$ is a second degree polynomial.
$a + 11$ is a first degree polynomial.
$x^4 - 16$ is a fourth degree polynomial.
Also, a constant term (such as 7, 22, – 9) is a zero degree polynomial.
6. A polynomial of degree 0 is called a constant polynomial.
7. If α and β are the zeroes of $p(x) = ax^2 + bx + c$ and $a \neq 0$, then
(i) $\alpha + \beta = -\frac{b}{a}$, and
(ii) $\alpha\beta = \frac{c}{a}$
8. A quadratic polynomial whose zeroes are α and β is given by $p(x)$ or $f(x) = x^2 - (\alpha + \beta)x + \alpha\beta$.
9. If α, β and γ are the zeroes of polynomial $p(x) = ax^3 + bx^2 + cx + d$ and $a \neq 0$, then
(i) $\alpha + \beta + \gamma = -\frac{b}{a}$
(ii) $\alpha\beta + \beta\gamma + \gamma\alpha = \frac{c}{a}$
(iii) $\alpha\beta\gamma = -\frac{d}{a}$
10. A cubic polynomial whose zeros are α, β and γ is given by $p(x)$ or $f(x) = x^3 - (\alpha + \beta + \gamma)x^2 + (\alpha\beta + \beta\gamma + \alpha\gamma)x - \alpha\beta\gamma$.
11. A real number 'a' is a zero of a polynomial $f(x)$ if $f(a) = 0$.
12. A polynomial of degree n can have atmost n real roots.
13. Geometrically, the zeros of a polynomial $f(x)$ are the x-coordinates of the points where the graph $y = f(x)$ intersects X-axis.
14. **Division Algorithm :** If $f(x)$ is a polynomial and $g(x)$ is a non-zero polynomial, then there exists polynomials $q(x)$ and $r(x)$ such that $f(x) = g(x) \times q(x) + r(x)$, where $r(x) = 0$ or degree $r(x) <$ degree $g(x)$.

CHAPTER - 3 : PAIR OF LINEAR EQUATIONS IN TWO VARIABLES

1. A pair of linear equations in two variables x and y, can be algebraically represented as :

$$ax + by + c = 0 \text{ and } px + qy + r = 0.$$

Here a, b, c, p, q and r are real numbers and $a^2 + b^2 \neq 0$ and $p^2 + q^2 \neq 0$.

2. A pair of linear equations $ax + by + c = 0$ and $px + qy + r = 0$ graphically or geometrically represents a pair of straight lines which are :

(i) intersecting, if $\frac{a}{p} \neq \frac{b}{q}$

(ii) parallel, if $\frac{a}{p} = \frac{b}{q} \neq \frac{c}{r}$

(iii) coincident, if $\frac{a}{p} = \frac{b}{q} = \frac{c}{r}$

3. There are two methods of solving a pair of linear equations :

(i) Graphical method, and

(ii) Algebraic method.

4. While solving a pair of linear equations by the graphical method, we have to first draw the lines represented by them on the graph.

(i) If the resultant line intersect, then the pair is said to be consistent and the coordinates of the point provide an unique solution.

(ii) If the lines are parallel, then they do not have any solutions and are called inconsistent.

(iii) If the lines are coincident, then there are infinite numbers of solutions and they are called consistent but having infinite solutions.

5. A pair of linear equations can be solved by the algebraic method in three ways :

(i) By substitution method,

(ii) By elimination method, and

(iii) By cross-multiplication method.

6. For a pair of linear equations in two variables $ax + by + c = 0$ and $px + qy + r = 0$

(i) If, $\frac{a}{p} \neq \frac{b}{q}$ then the pair of linear equations represents a consistent and unique solution;

(ii) If, $\frac{a}{p} = \frac{b}{q} \neq \frac{c}{r}$ then the pair of linear equations represents an inconsistent solution; and

(iii) If, $\frac{a}{p} = \frac{b}{q} = \frac{c}{r}$ then the pair of linear equations represents consistent but infinite solutions.

CHAPTER - 4 : QUADRATIC EQUATIONS

1. A polynomial of degree 2 is called a quadratic polynomial. Its general form is $ax^2 + bx + c$, where a, b and c are real numbers such that $a \neq 0$ and x is a variable quantity.

2. If $p(x) = ax^2 + bx + c, a \neq 0$ is a quadratic polynomial and α is a real number, then $p(\alpha) = a\alpha^2 + b\alpha + c$ is known as the value of $p(x)$ at $x = \alpha$.

3. A real number α is said to be a zero or root of the quadratic polynomial $p(x) = ax^2 + bx + c$, if $p(\alpha) = 0$ *i.e.* $a\alpha^2 + b\alpha + c = 0$ where $a \neq 0$ and a, b and c are real numbers.

4. If $p(x) = ax^2 + bx + c$ is a quadratic polynomial, then $p(x) = 0$, *i.e.*, $ax^2 + bx + c = 0$, $a \neq 0$ is called a quadratic equation.

5. The nature of the roots of the quadratic equation $ax^2 + bx + c = 0$ depends upon the value of $D = b^2 - 4ac$ where D is the discriminant of the quadratic equation.

6. The roots of a quadratic equation $ax^2 + bx + c = 0$, where $a \neq 0$, can be found by using the formula $x = \dfrac{-b \pm \sqrt{b^2 - 4ac}}{2a}$, provided that $b^2 - 4ac \geq 0$.

7. The quadratic equation $ax^2 + bx + c = 0, a \neq 0$ has :
   (i) two distinct real roots, if $D = b^2 - 4ac > 0$
   (ii) two equal real roots, if $D = b^2 - 4ac = 0$
   (iii) imaginary or no real roots, if $D = b^2 - 4ac < 0$

8. If $ax^2 + bx + c, a \neq 0$ is factorizable into a product of two linear factors, then the roots of the quadratic equation $ax^2 + bx + c = 0$ can be found by equating each factor to zero.

CHAPTER - 5 : ARITHMETIC PROGRESSIONS

1. **Sequence** : A sequence is an arrangement of numbers or objects in a definite order.

2. A sequence $a_1, a_2, a_3, a_4, \ldots, a_n$ is called an Arithmetic Progression, if there exists a constant d such that $a_2 - a_1 = d$, $a_3 - a_2 = d$, $a_4 - a_3 = d, \ldots, a_{n+1} - a_n = d$ and so on. The constant d is called the common difference.

3. A sequence $a_1, a_2, a_3, a_4, \ldots, a_n$ is called an Arithmetic Progression, if $a_{n+1} - a_n$ is independent of n.

4. A sequence $a_1, a_2, a_3, \ldots, a_n$ is an A.P., if and only if its nth term a_n is a linear expression in n and is such that the coefficient of n is the common difference.

5. The nth term a_n of an A.P. with the first term a and common difference d is given by $a_n = a + (n - 1)d$.

6. If there are m terms of an A.P. with the first term as a and the common difference as d then, n^{th} term from the end $= (m - n + 1)^{th}$ term from the beginning $= a + (m - n)\,d$.
Also, n^{th} term from the end = where l denotes the last term.

7. The sum of n terms of an A.P. with the first term a and common difference d is given by

$$S_n = \frac{n}{2}\{2a + (n - 1)d\},$$

or, $S_n = \frac{n}{2}\{a + l\}$, where l = last term $= a + (n - 1)d$

8. If the ratio of the sums of n terms of two AFs is given, then to find the ratio of their n^{th} terms, we replace n by $(2n - 1)$ in the ratio of the sums of n terms.

9. A sequence is an AP if and only if the sum of its n terms is of the form $An^2 + Bn$, where A and B are constants and the common difference is $2A$ in such a case.

CHAPTER - 6 : COORDINATE GEOMETRY

1. **Abscissa and Ordinate :** From a given point, the abscissa and the ordinate are the respective distances from the X-axis and the Y-axis respectively.

2. The coordinates of a point on the X-axis are always $(x, 0)$ and on the Y-axis, are always $(0, y)$.

3. The distance between two points, $A(x_1, y_1)$ and $B(x_2, y_2)$ is given by the formula :

$$AB \text{ or } |AB| = \sqrt{(x_2 - x_1)^2 + (y_2 - y_1)^2}$$

4. Distance of a point $A(x, y)$ from the origin $O(0, 0)$ is given by

$$OA \text{ or } |OA| = \sqrt{x^2 + y^2}$$

5. **Section Formula :** If $C(x, y)$ divides the line joining $A(x_1, y_1)$ and $B(x_2, y_2)$ internally in the ratio $m : n$, then coordinates of $C(x, y)$ are

$$(x, y) = \left(\frac{mx_2 + nx_1}{m+n}, \frac{my_2 + ny_1}{m+n}\right)$$

If the division occurs externally, then

$$(x, y) = \left(\frac{mx_2 - nx_1}{m-n}, \frac{my_2 - ny_1}{m-n}\right)$$

6. **Mid-point :** The coordinates of the mid-point C, of a line segment joining points $A(x_1, y_1)$ and $B(x_2, y_2)$ (when $m = n$) are given by :

$$\left(\frac{x_1 + x_2}{2}, \frac{y_1 + y_2}{2}\right)$$

7. **Centroid of a Triangle :** The coordinates of the centroid O of a triangle whose vertices are $A(x_1, y_1)$, $B(x_2, y_2)$ and $C(x_3, y_3)$, are given by :

$$\left(\frac{x_1 + x_2 + x_3}{3}, \frac{y_1 + y_2 + y_3}{2}\right)$$

8. The equation of a line parallel to the X-axis is $y = b$ and parallel to the Y-axis is $x = a$.

9. **Area of a Triangle :** The area of a triangle whose vertices are $A(x_1, y_1)$, $B(x_2, y_2)$ and $C(x_3, y_3)$ is given by :

$$\frac{1}{2} \left| x_1(y_2 - y_3) + x_2(y_3 - y_1) + x_3(y_1 - y_2) \right|$$

10. If the points $A(x_1, y_1)$, $B(x_2, y_2)$ and $C(x_3, y_3)$ are collinear, then $x_1(y_2 - y_3) + x_2(y_3 - y_1) + x_3(y_1 - y_2) = 0$.

CHAPTER - 7 : TRIANGLES

1. **Similar Figures :** These are two geometrical figures having the same shape but not necessarily same size.

2. **Congruent Figures :** Two figures that are similar in shape and size, are congruent to each other.

3. All congruent figures are similar but converse is not true.

4. **Similar Polygons :** Two polygons having same number of sides are said to be similar if their corresponding angles are equal and the sides corresponding to those angles are proportional.

5. **Thales' Theorem or Basic Proportionality Theorem :** If a line joining two sides of a triangle at distinct points (not necessarily the mid-points of both the sides) is parallel to the third side, then it divides both the sides in equal proportions.
The Converse of Thales' Theorem : If a line divides any two sides of a triangle in the same ratio, then the line must be parallel to the third side.

6. **Mid-Point Theorem :** The line joining the mid-points of two sides of a triangle is parallel to the third side and its length is equal to half of the length of the third side.

7. **Angle-Bisector Theorem :** The internal bisector of an angle of a triangle divides the opposite side internally into two parts that are proportional to the other two sides of the triangle.

8. The external bisector of an angle of a triangle divides the opposite side externally in the ratio of the sides containing the angle.

9. The internal bisector of an angle of a triangle divides the opposite side internally in the ratio of the sides containing the angle.

10. Two triangles are considered to be similar if their corresponding angles are equal and their corresponding sides are proportional.

11. **AAA Similarity :** If in two triangles, the corresponding angles are equal, then their corresponding sides are proportional and hence the triangles are similar.

12. **AA Similarity :** If two angles of a triangle are equal to two angles of another triangle, then the two triangles are similar.

13. **SSS Similarity :** If the corresponding sides of two triangles are proportional, then the angles containing them are also equal and the two triangles are similar.

14. **SAS Similarity :** If one angle of a triangle is equal to one angle of another triangle and the sides including the equal angles are proportional, then the two triangles are similar.

15. If a perpendicular is dropped from the right-angled vertex of a right-angled triangle on to the hypotenuse, then the resultant triangles are similar to each other as well as to the triangle containing them.

16. The diagonals of a trapezium divide each other proportionally.

17. If the diagonals of a quadrilateral divide each other proportionally, then it is a trapezium.

18. If two triangles are equiangular, then the ratio of their corresponding sides is the same as the ratio of their corresponding medians and the ratio of their corresponding angle-bisector segments is same as the ratio of the corresponding sides.

19. The ratio of the areas of two similar triangles is equal to the ratio of the squares of any two corresponding sides, the squares of the corresponding altitudes, the squares of the corresponding medians and the squares of the corresponding angle-bisector segments.

20. If two similar triangles have equal areas, then the two triangles are congruent.

21. **Pythagoras Theorem :** The square of the hypotenuse of a right-angled triangle is equal to the sum of the squares of the other two sides that contain the right-angle.

22. **Converse of Pythagoras Theorem :** In a triangle, if the square of the longest side is equal to the sum of the squares of the other two sides, then the angle opposite to the longest side is a right-angle.

23. In any triangle, the sum of the squares of any two sides is equal to twice the square of half of the third side and twice the square of the median which bisects the third side.

24. Three times the sum of the square of the sides of a triangle is equal to four times the sum of the squares of the medians of the triangle.

25. Three times the square of any side of an equilateral triangle is equal to four times the square of its altitude.

CHAPTER - 8 : CIRCLES

1. A tangent to a circle is a line that touches the circle at only one point. There is only one tangent at a point on a circle.
2. From a point lying outside a circle, two and only two tangents can be drawn to it.
3. The lengths of the two tangents drawn from an external point to a circle are equal.
4. A tangent to a circle is perpendicular to the radius through the point of contact.
5. A line which intersects a circle at two distinct points is called secant of the circle.

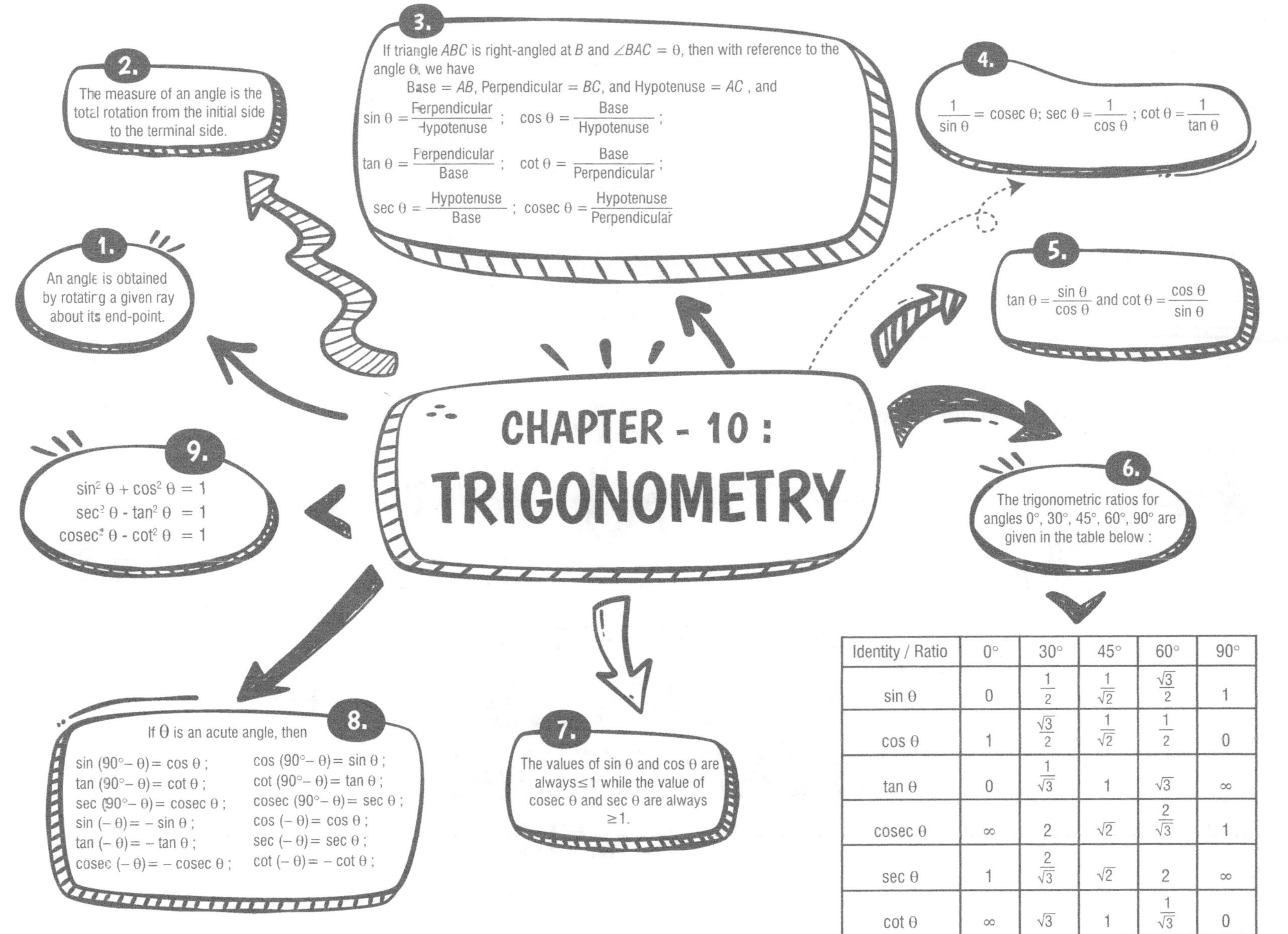

Identity / Ratio	0°	30°	45°	60°	90°
sin θ	0	$\frac{1}{2}$	$\frac{1}{\sqrt{2}}$	$\frac{\sqrt{3}}{2}$	1
cos θ	1	$\frac{\sqrt{3}}{2}$	$\frac{1}{\sqrt{2}}$	$\frac{1}{2}$	0
tan θ	0	$\frac{1}{\sqrt{3}}$	1	$\sqrt{3}$	∞
cosec θ	∞	2	$\sqrt{2}$	$\frac{2}{\sqrt{3}}$	1
sec θ	1	$\frac{2}{\sqrt{3}}$	$\sqrt{2}$	2	∞
cot θ	∞	$\sqrt{3}$	1	$\frac{1}{\sqrt{3}}$	0

CHAPTER - 11 : APPLICATIONS OF TRIGNOMETRY

1. The line joining the eyes of the observer and the object is called the line of sight.

2. The angle formed by the line of sight with the horizontal when the object is above the horizontal level is called the angle of elevation.

3. The angle formed by the line of sight with the horizontal when the object is below the horizontal level is called the angle of depression.

CHAPTER - 12 : AREAS RELATED TO CIRCLES

1. Distance moved by a wheel in 1 rotation = Circumference of the wheel.

2. Number of rotation by a wheel in 1 minute = Distance moved by wheel in 1 minute / Circumference of wheel

3. For a circle having radius r,

(i) Diameter = $2r$

(ii) Circumference = $2\pi r$

(iii) Area = πr^2

(iv) Area of semi-circle = $\frac{\pi r^2}{2}$

(v) Area of a quadrant = $\frac{\pi r^2}{4}$

(vi) Perimeter of semi-circle = $(\pi r + 2r)$

4. If R and r are the radii of two concentric circles such that $R > r$ then,

The area enclosed between the two circles $= \pi R^2 - \pi r^2$
$= \pi(R^2 - r^2)$

5. A segment of a circle is the region bounded by a chord and the arc subtended by the chord.

6. If a sector of a circle of radius r contains an angle of $\theta°$, then

(i) Length of the arc of the sector $= \frac{\theta}{360°} \times 2\pi r = \frac{\theta}{360°} \times$ (Circumference of the circle)

(ii) Perimeter of the sector $= 2r + \frac{\theta}{360°} \times 2\pi r$

(iii) Area of the sector $= \frac{\theta}{360°} \times \pi r^2 = \frac{\theta}{360°} \times$ (Area of the circle)

(iv) Area of the minor segment = Area of the corresponding sector – Area of the corresponding triangle

$$= \frac{\theta}{360°} \times \pi r^2 - r^2 \sin\frac{\theta}{2}\cos\frac{\theta}{2}$$

$$= \left\{\frac{\theta}{360°} \times \pi - \sin\frac{\theta}{2}\cos\frac{\theta}{2}\right\} r^2$$

$$= \left\{\frac{\theta}{360°} \times \pi - \frac{1}{2}\sin\theta\right\} r^2$$

(iv) Area of the major segment = Area of the circle – Area of the minor segment

CHAPTER-13 : SURFACE AREAS AND VOLUMES

1. Cuboid : In a cuboid, let l be the length, b be the breadth and h be the height, then

(i) Total surface area $= 2(lb + bh + lh)$ sq. units

(ii) Volume $= l \times b \times h$ cubic units

(iii) Diagonal of a cuboid $= \sqrt{l^2 + b^2 + h^2}$ units

(iv) Area of the four walls $= 2(l + b)h$ sq. units

2. Cube : If the side of a cube is 'a' units, then

(i) Total surface area $= 6a^2$ sq. units

(ii) Volume $= a^3$ cubic units

(iii) Diagonal of a cube $= \sqrt{3}a$ units

3. Right-Circular Cylinder : For a right-circular cylinder of base r and height h,

(i) Area of each end $= \pi r^2$ sq. units

(ii) Curved surface area $= 2\pi rh$ sq. units

(iii) Total surface area $= 2\pi r(h+r)$ sq. units

(iv) Volume $= \pi r^2 h$ cubic units

4. Hollow Cylinder : Let R and r be the external radius and the internal radius, respectively and h be the height of hollow cylinder, then

(i) Area of each end $= \pi(R^2 - r^2)$ sq. units

(ii) Curved surface area $= 2\pi(R + r)h$ sq. units

(iii) Total surface area $= 2\pi(R + r)(R + h - r)$sq. units

(iv) Volume $= \pi(R^2 - r^2)h$ cubic units

5. Right-Circular Cone : If r, h and l denote the radius of the base, height and slant height of a cone, respectively, then

(i) $l^2 = r^2 + h^2$

(ii) Curved surface area $= \pi rl$ sq. units

(iii) Total surface area $= \pi r^2 + \pi rl$ sq. units

(iv) Volume $= \frac{1}{3}\pi r^2 h$ cubic units

6. Sphere : For a sphere of radius r,

(i) Surface area $= 4\pi r^2$ sq. units

(ii) Volume $= \frac{4}{3}\pi r^3$ cubic units

7. Hemisphere :

For a hemisphere of radius r,

(i) Curved surface area $= 2\pi r^2$ sq. units

(ii) Total surface area $= 3\pi r^2$ sq. units

(iii) Volume $= \frac{2}{3}\pi r^3$ cubic units

8. Spherical Shell : For a spherical shell of outer radius R and inner radius r,

(i) Surface area $= 4\pi(R^2 - r^2)$ sq. units

(ii) Volume of material $= \frac{4}{3}\pi(R^3 - r^3)$ cubic units

9. Frustum of a Cone : If h, l, r_1 and r_2 denote the height, slant height and radii of the bases of a frustum of a cone, respectively, then

(i) Volume $= \frac{\pi}{3}(r_1^2 + r_1 r_2 + r_2^2)$ h cubic units

(ii) Lateral surface area $= \pi(r_1 + r_2)$ l sq. units

(iii) Total surface area $= \pi\{(r_1 + r_2)l + (r_1^2 + r_2^2)\}$ sq. units

(iv) Slant height of the frustum $= \sqrt{h^2 + (r_1 - r_2)^2}$

(v) Height of the cone of which the frustum is a part $= \frac{hr_1}{r_1 - r_1}$

(vi) Slant height of the cone of which the frustum is a part $= \frac{lr_1}{r_1 - r_1}$

CHAPTER-14 : STATISTICS

1. The three measures of central value are Mean, Median and Mode.

2. Mean is computed by the Direct method, Short-cut method and Step-deviation method.
For each of the three methods, the class mark x_i can be found by the following given formula :

$$x = \frac{1}{2}(\text{lower limit} + \text{upper limit})$$

3. If a variable X takes the values $x_1, x_2, x_3, \ldots, x_n$ with the corresponding frequencies being $f_1, f_2, \ldots, f_n$ respectively, then the arithmetic mean of these values is given by

(i) Mean $= \frac{\Sigma(f_i \times x_i)}{\Sigma f_i}$ [Direct Method]

(ii) Mean $= A + \frac{\Sigma f_i d_i}{\Sigma f_i}$ [Assumed Mean Method/Short-cut Method], where

$d_i = x_i - A\left[x_i = \textit{mid-value} = \frac{\text{lower limit} + \text{upper limit}}{2}\right]$

(iii) Mean $= A + \left\{h \times \frac{\Sigma(f_i \times u_i)}{\Sigma f_i}\right\}$ [Step-Deviation Method],

where h=class size and $u_i = \frac{x_i - A}{h}$

A=The Assumed Mean may be the class mark of the central class-interval.

4. The median of a distribution is the value of the variable which divides it into two equal parts. The median of a grouped or continuous frequency distribution may be computed by using the following formula :

$$\text{Median} = l + \frac{\frac{N}{2} - F}{f} \times h$$

where, l = lower limit of the median class*
f = frequency of the median class
h = width of the median class
F = cumulative frequency of the class preceding the median class

and $N = \sum_{i=1}^{n} f_i$

*To find the Median class, one must first find out the value of $\frac{N}{2}$, where N is the cumulative frequency of the given frequency distribution. The frequency of the class just greater than $\frac{N}{2}$ gives the Median class. The cumulative frequency (N) is given by the summation of all the frequencies in a grouped data.

5. Mode is the value of the variable with the maximum frequency. The mode of a continuous or grouped frequency distribution may be computed by using the following formula :

$$\text{Mode} = l + \frac{f - f_1}{2f - f_1 - f_2} \times h$$

where, l = lower limit of the modal class**
f = frequency of the modal class
h = width of the modal class
f_1 = frequency of the class preceding the modal class
f_2 = frequency of the class following the modal class

**In a frequency distribution, the Modal class is given by the class having the maximum frequency.

6. The three measures of the central value are connected by the following relation :
Mode = 3 Median – 2Mean

7. Ogive (s) are used to find the median of a frequency distribution.

8. If in a frequency distribution, the class-range is uneven, then they are adjusted by calculating the other class ranges in such a manner that they become multiples of the range with the smallest class-size. The frequencies are then adjusted by dividing the minimum class size by the class size and then multiplying with the frequency of the class. Such a frequency is known as Adjusted Frequency and is given by the following formula:

Adjusted Frequency=

$$\frac{\text{Minimum class size}}{\text{class size}} \times \text{Frequency of the class size}$$

CHAPTER-15 : PROBABILITY

1. **Experiment :** It is a process, resulting in some well-defined outcomes.

2. **Random Experiment :** In this type of experiment, all the outcomes are known in advance, but the outcome of any specific event is unpredictable.

3. **Outcome :** This is the result of an experiment.

4. **Sample Space :** The sample space is a set of all possible outcomes of a random experiment and is generally denoted by 'S'.

5. **Event :** An event is an occurrence.

6. In a random experiment, if there are n elementary events and m are favourable to an event E, then the probability P of happening of E, denoted by $P(E)$ or simply P, is defined as the ratio $\frac{m}{n}$.

   It is thus, a concept which measures, numerically, the degree of certainty of the occurrence of events.

   $$P(E) = \frac{\text{Favourable Outcomes}}{\text{Total Outcomes}}$$

7. The probability of a sure event is always 1.

8. The probability of an impossible event is always 0.

9. The sum of the probabilities of all the outcomes of elementary events is 1.

10. **Complementary Events :** For any event E, $P(E) + P(\overline{E}) = 1$, where $\overline{E}$ stands for not E, E and $\overline{E}$ are called complementary events.

11. For any event E associated to a random experiment, we have $0 \le P(E) \le 1$.

Multiple Choice Questions

| Set 1 |

Chapter 1. Real Numbers

Q. 1. Has the rational number $\frac{441}{2^2 5^7 7^2}$ a terminating or non-terminating decimal representation ?★

(i) Terminating **(ii) Non-terminating**

(iii) Neither **(iv) Not**

Solution : The rational number $\frac{441}{2^2 5^7 7^2}$ has a non-terminating decimal representation because

$$(2^2 5^7 7^2) \neq (2^m \times 5^n)$$

So the correct option is (ii).

Q. 2. The decimal expansion of the rational number $\frac{43}{2^4 5^3}$ will terminate after how many places of decimal ?★

(i) 1 **(ii) 2**

(iii) It is interminable **(iv) 4**

Solution : $\frac{43}{2^4 5^3} = \frac{43 \times 5}{2^4 \times 5^4}$

$$= \frac{215}{(2 \times 5)^4} = \frac{215}{10^4}$$

$$= 0.0215$$

Thus, the rational number terminates after 4 decimal places.

So the correct option is (iv).

Q. 3. Complete the missing entries in the following factor tree :★

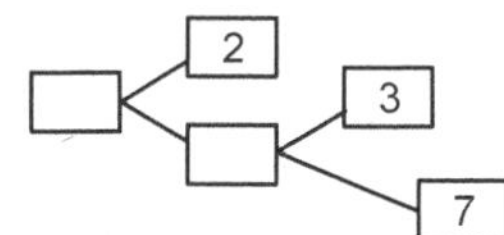

(i) 42 and 21 **(ii) 24 and 12**

(iii) 7 and 3 **(iv) 84 and 42**

Solution : The correct option is (i).

Q. 4. The sum of exponents of prime factors in the prime-factorisation of 196 is :★

(i) 3 **(ii) 4**

(iii) 5 **(iv) 2**

Solution : Prime factors of 196 = $2^2 \times 7^2$

So, sum of exponents = 2 + 2 = 4

So the correct option is (ii).

★ are board exam questions from previous years

Q. 5. Euclid's division Lemma states that for two positive integers a and b, there exists unique integer q and r satisfying $a = bq + r$, and★

(i) $0 < r < b$ **(ii) $0 < r \leq b$**

(iii) $0 \leq r < b$ **(iv) $0 \leq r \leq b$**

Solution : Euclid's division Lemma states that for two positive integers a and b, there exist unique integers q and r satisfying

$a = bq + r$, and $0 \leq r < b$.

So the correct option is (iii).

Q. 6. The total number of factors of a prime number is :★

(i) 1 **(ii) 0**

(iii) 2 **(iv) 3**

Solution : As we know that prime numbers are the numbers with two factors, 1 and the number itself.

∴ Total number of factors of a prime number will be 2.

So the correct option is (iii).

Q. 7. The HCF and the LCM of 12, 21, 15 respectively are :★

(i) 3, 140 **(ii) 12, 420**

(iii) 3, 420 **(iv) 420, 3**

Solution :

2	12
2	6
3	3
	1

3	21
7	7
	1

3	15
5	5
	1

∴ Prime factors of 12, 21 and 15 are

$$12 = 2 \times 2 \times 3 = 2^2 \times 3^1$$
$$21 = 3 \times 7 = 3^1 \times 7^1$$
$$15 = 3 \times 5 = 3^1 \times 5^1$$

⇒ HCF (12, 21, 15) = $3^1 = 3$

⇒ LCM (12, 21, 15) = $2^2 \times 3^1 \times 5^1 \times 7^1 = 420$

So the correct option is (iii).

Chapter 2. Polynomials

Q. 1. If $x = \dfrac{8 \pm \sqrt{(-8)^2 - 4 \times 3 \times 2}}{2 \times 3}$ **then the required polynomial is :**

(i) $3x^2 - 8x + 2 = 0$ (ii) $2x^2 - 8x - 2 = 0$

(iii) $3x^2 + 8x - 2 = 0$ (iv) $3x^2 + 8x + 2 = 0$

Solution : If $x = \dfrac{8 \pm \sqrt{(-8)^2 - 4 \times 3 \times 2}}{2 \times 3}$ then on comparing with $x = \dfrac{-b \pm \sqrt{b^2 - 4ac}}{2a}$ we have $b = -8$, $a = 3$ and $c = 2$. So the required polynomial is $3x^2 - 8x + 2 = 0$. So the correct option is (i).

Q. 2. Find the coefficient of x^0 **in** $x^2 + 3x + 2 = 0$.

(i) 3 (ii) – 3

(iii) 2 (iv) – 2

Solution : As $x^0 = 1$, so its coefficient is the constant term.

Hence the correct option is (iii).

Q. 3. In which condition will the polynomial $ax^2 + bx + c = 0$ **be a quadratic equation ?**

(i) $a \neq 0$ (ii) $a = b$

(iii) $a = c$ (iv) $a = 0$

Solution : The correct option is (i).

Q. 4. For which value of p **will the equation**

$$(p^2 - 1)x^2 + px + q = 0$$

not be a quadratic equation ?

(i) $p = 1$ (ii) $p = -1$

(iii) Both (i) and (ii) (iv) $p = 0$

Solution : The correct option is (iii).

Q. 5. Write the zeros of the polynomial★

$$f(x) = x^2 - x - 6.$$

(i) – 3, 2 (ii) – 3, – 2

(iii) 3, 2 (iv) 3, – 2

Solution : $f(x) = x^2 - x - 6$

$= x^2 - 3x + 2x - 6$

$= x(x - 3) + 2(x - 3)$

$= (x + 2)(x - 3)$

To find zeros put $f(x) = 0$

$\therefore \quad x = -2, 3$

So the correct option is (iv).

Q. 6. If $(x + a)$ **is a factor of** $f(x) = (2x^2 + 2ax + 5x + 10)$**, find** a.★

(i) 2 (ii) – 2

(iii) ± 2 (iv) 0

Solution : $f(x) = (2x^2 + 2ax + 5x + 10)$

$= 2x(x + a) + 5(x + 2)$

$\therefore \quad x + a = x + 2$

$\Rightarrow \quad a = 2$

So the correct option is (i).

Q. 7. For what value of k **is – 4 a zero of the polynomial** $f(x) = x^2 - x - (2k + 2)$ **?★**

(i) 6 (ii) – 6

(iii) 9 (iv) – 9

Solution : $f(x) = x^2 - x - (2k + 2)$

Substituting x for – 4, we have

$f(-4) = (-4)^2 - (-4) - (2k + 2)$

$= 16 + 4 - 2k - 2$

$= 18 - 2k = 0$

Thus, $18 - 2k = 0$

or $2k = 18$

or $k = 9$

So the correct option is (iii).

Q. 8. If α **and** β **are the zeros of a polynomial such that** $\alpha + \beta = -6$ **and** $\alpha\beta = -4$ **then write the polynomial.★**

(i) $x^2 - 6x - 4 = 0$ (ii) $x^2 + 6x - 4 = 0$

(iii) $x^2 + 6x + 4 = 0$ (iv) $x^2 - 6x + 4 = 0$

Solution : Since $\alpha + \beta = -6$ and $\alpha\beta = -4$

The polynomial is $f(x) = x^2 + 6x - 4 = 0$.

So the correct option is (ii).

Q. 9. If α **and** β **are the zeros of the polynomial** $f(y) = 2y^2 + 7y + 5$**, write the values of** $\alpha + \beta + \alpha\beta$.★

(i) 1 (ii) 0

(iii) – 2 (iv) – 1

Solution : $f(y) = 2y^2 + 7y + 5$

$$\alpha + \beta = -\frac{7}{2}, \ \alpha\beta = \frac{5}{2}$$

$$\therefore \quad \alpha + \beta + \alpha\beta = -\frac{7}{2} + \frac{5}{2} = -1$$

So the correct option is (iv).

Q. 10. For what value of k **is 3 a zero of the polynomial** $f(x) = 2x^2 + x + k$ **?★**

(i) $\sqrt{21}$ (ii) 21

(iii) $-\sqrt{21}$ (iv) – 21

Solution : $f(x) = 2x^2 + x + k$

Substituting 3 for x, we have

$f(3) = 2(3)^2 + (3) + k = 0$

$\Rightarrow \quad 18 + 3 + k = 0$

$\Rightarrow \quad k = -21$

So the correct option is (iv).

★ are board exam questions from previous years

Q. 11. For what value of k is -3 a zero of the polynomial $f(x) = x^2 + 11x + k$?★

(i) $\sqrt{24}$ (ii) $-\sqrt{24}$

(iii) -24 (iv) 24

Solution : $f(x) = x^2 + 11x + k$

Substituting -3 for x, we have

$$f(-3) = (-3)^2 + 11(-3) + k = 0$$

$$\Rightarrow \quad 9 - 33 + k = 0$$

$$\Rightarrow \quad k = 24$$

So the correct option is (iv).

Q. 12. For what value of k is -2 a zero of the polynomial $f(x) = 3x^2 + 4x + 2k$?★

(i) $\sqrt{2}$ (ii) $-\sqrt{2}$

(iii) -2 (iv) 2

Solution : $f(x) = 3x^2 + 4x + 2k$

Substituting -2 for x, we have

$$f(-2) = 3(-2)^2 + 4(-2) + 2k = 0$$

$$\Rightarrow \quad 12 - 8 + 2k = 0$$

$$\Rightarrow \quad k = -2$$

So the correct option is (iii).

Q. 13. If 1 is a root of the equations $ay^2 + ay + 3 = 0$ and $y^2 + y + b = 0$ then ab equals :★

(i) 3 (ii) $-\frac{7}{2}$

(iii) 6 (iv) -3

Solution : $f(y) = ay^2 + ay + 3 = 0$

Thus, $f(1) = a(1)^2 + a(1) + 3 = 0$

$$\Rightarrow \quad 2a + 3 = 0$$

$$\Rightarrow \quad a = -\frac{3}{2}$$

$$f(y) = y^2 + y + b = 0$$

$$\Rightarrow \quad f(1) = (1)^2 + 1 + b = 0$$

$$\Rightarrow \quad b = -2$$

Hence, $ab = \left(-\frac{3}{2}\right) \times (-2) = 3$

So the correct option is (i).

Q. 14. For which value of p will $(3p + 1)x^2 + 4x - 2 = 0$ not be a quadratic equation ?

(i) 0 (ii) $-\frac{1}{3}$

(iii) 1/3 (iv) none of the above

Solution : $(3p + 1)x^2 + 4x - 2 = 0$

$$\Rightarrow \quad x^2 + \frac{4}{3p+1}x - \frac{2}{3p+1} = 0$$

Put $3p + 1 = 0$

Thus, $p = -\frac{1}{3}$

Hence, $(3p + 1)x^2 + 4x - 2 = 0$ will not be a quadratic equation for $p = -\frac{1}{3}$.

So the correct option is (ii).

Q. 15. The roots of the equation $x^2 + x - p(p + 1) = 0$ where p is a constant are :★

(i) $p, p + 1$ (ii) $-p, p + 1$

(iii) $p, -(p + 1)$ (iv) $-p, -(p + 1)$

Solution : $x^2 + x - p(p + 1) = 0$

Let the roots be α and β.

Thus, $\alpha + \beta = -1$ and $\alpha\beta = -p(p + 1)$

$$\Rightarrow \quad \alpha = -(1 + \beta)$$

and $-(1 + \beta)\beta = -(p + 1)p$

Hence $\beta = p$ and $\alpha = -(p + 1)$

So the correct option is (iii).

Q. 16. The zeroes of the polynomial $x^2 - 3x - m(m + 3)$ are :★

(i) $m, m + 3$ (ii) $-m, m + 3$

(iii) $m, -(m + 3)$ (iv) $-m, -(m + 3)$

Solution : $x^2 - 3x - m(m + 3)$

$$= x^2 - (m + 3 - m)x - m(m + 3)$$

$$= x^2 - (m + 3)x + mx - m(m + 3)$$

$$= x\{x - (m + 3)\} + m\{x - (m + 3)\}$$

$$= (x + m)\{x - (m + 3)\}$$

$\therefore$ Its zeroes are $-m$, $(m + 3)$.

So the correct option is (ii).

Q. 17. If one of the zeroes of the quadratic polynomial $x^2 + 3x + k$ is 2, then the value k is :★

(i) 10 (ii) -10

(iii) -7 (iv) -2

Solution :

Let $p(x) = x^2 + 3x + k$

$\because$ $x = 2$ is a zero of $p(x)$

$\therefore$ $p(2) = 0$

$$\Rightarrow \quad (2)^2 + 3(2) + k = 0$$

$$\Rightarrow \quad 4 + 6 + k = 0$$

$$\Rightarrow \quad 10 + k = 0$$

$$\Rightarrow \quad k = -10$$

So the correct option is (ii).

Q. 18. The quadratic polynomial, the sum of whose zeroes is -5 and their product is 6, is :★

★ are board exam questions from previous years

(i) $x^2 + 5x + 6$ **(ii) $x^2 - 5x + 6$**

(iii) $x^2 - 5x - 6$ **(iv) $-x^2 + 5x + 6$**

Solution : Let the polynomial be

$$p(x) = k\,[x^2 - (\text{Sum of zeroes})\,x + \text{Product of zeroes}]$$

Sum of zeroes = – 5

Product of zeroes = 6

∴ Required polynomial is

$$p(x) = k[x^2 - (-5)x + (6)]$$
$$= k[x^2 + 5x + 6]$$

Let $k = 1$

$$\therefore \quad p(x) = x^2 + 5x + 6$$

So the correct option is (i).

Chapter 3. Pair of Linear Equations in Two Variables

Q. 1. The values of x and y in $2x + 3y = 2$ and $x - 2y = 8$ are :

(i) – 4, 2 **(ii) – 4, – 2**

(iii) 4, – 2 **(iv) 4, 2**

Solution : As per the question,

$$2x + 3y = 2 \quad \text{...(i)}$$
$$x - 2y = 8 \quad \text{...(ii)}$$

Multiplying equation (i) with 2 and equation (ii) with 3, we have

$$4x + 6y = 4 \quad \text{...(iii)}$$
and
$$3x - 6y = 24 \quad \text{...(iv)}$$

Thus, adding the equation (iii) & (iv), we get

$$7x = 28$$
$$\Rightarrow \quad x = 4$$

Putting $x = 4$ in equation (ii), we have

$$4 - 2y = 8$$
$$\Rightarrow \quad -2y = 4$$
$$\Rightarrow \quad y = -2$$

Hence, the values of x and y are 4, – 2 respectively.

So the correct option is (iii).

Q. 2. The values of x and y in $3x + 2y = 4$ and $2x - 3y = 7$ are :

(i) 2, – 1 **(ii) 2, 1**

(iii) – 2, – 1 **(iv) – 2, 1**

Solution : As per the question,

$$3x + 2y = 4 \quad \text{...(i)}$$
$$2x - 3y = 7 \quad \text{...(ii)}$$

Multiplying equation (i) with 3 and equation (ii) with 2, we have

$$9x + 6y = 12 \quad \text{...(iii)}$$
and
$$4x - 6y = 14 \quad \text{...(iv)}$$

Thus, adding the above two equations we get

$$13x = 26$$
$$\Rightarrow \quad x = 2$$

Replacing $x = 2$ in equation (i), we have

$$6 + 2y = 4$$
$$\Rightarrow \quad 2y = -2$$
$$\Rightarrow \quad y = -1$$

Hence, the values of x and y are 2, – 1 respectively.

So the correct option is (i).

Q. 3. The values of x and y in $x - y + 1 = 0$ and $3x + 2y - 12 = 0$ are :

(i) 2, – 3 **(ii) – 2, 3**

(iii) – 2, – 3 **(iv) 2, 3**

Solution : As per the question,

$$x - y = -1 \quad \text{...(i)}$$
$$3x + 2y = 12 \quad \text{...(ii)}$$

Multiplying equation (i) with 2

$$2x - 2y = -2 \quad \text{...(iii)}$$
and
$$3x + 2y = 12 \quad \text{...(iv)}$$

Thus, adding the above two equations we get

$$5x = 10$$
$$\Rightarrow \quad x = 2$$

Putting $x = 2$ in equation (i), we have

$$2 - y = -1$$
$$\Rightarrow \quad -y = -3$$
$$\Rightarrow \quad y = 3$$

Hence, the values of x and y are 2, 3 respectively.

So the correct option is (iv).

Q. 4. The values of x and y in $2x + 3y = 4$ and $3x - y = -5$ are :

(i) – 1, – 2 **(ii) – 1, 2**

(iii) 1, – 2 **(iv) 1, 2**

Solution : As per the question,

$$2x + 3y = 4 \quad \text{...(i)}$$
$$3x - y = -5 \quad \text{...(ii)}$$

Multiplying equation (ii) with 3, we have

$$2x + 3y = 4 \quad \text{...(iii)}$$
and
$$9x - 3y = -15 \quad \text{...(iv)}$$

Thus, adding the above two equations we get

$$11x = -11$$
$$\Rightarrow \quad x = -1$$

Putting $x = -1$ in equation (ii), we have

$$-3 - y = -5$$
$$\Rightarrow \quad -y = -2$$
$$\Rightarrow \quad y = 2$$

Hence, the values of x and y are – 1, 2 respectively.

So the correct option is (ii).

Q. 5. The values of x and y in $2x - 3y = 1$ and $3x - 4y = 1$ are :

(i) 1, – 1 **(ii) – 1, 1**
(iii) 1, 1 **(iv) – 1, – 1**

Solution : As per the question,

$$2x - 3y = 1 \quad \ldots(i)$$
$$3x - 4y = 1 \quad \ldots(ii)$$

Multiplying equation (i) with 3 and equation (ii) with 2, we have

$$6x - 9y = 3 \quad \ldots(iii)$$
and $$6x - 8y = 2 \quad \ldots(iv)$$

Thus, subtracting equation (iv) from equation (iii), we get

$$-y = 1$$
$$\Rightarrow \quad y = -1$$

Putting $y = -1$ in equation (i), we have

$$2x + 3 = 1$$
$$\Rightarrow \quad 2x = -2$$
$$\Rightarrow \quad x = -1$$

Hence, the values of x and y are – 1, – 1 respectively.
So the correct option is (iv).

Q. 6. The values of x and y in $2x + 3y = 8$ and $x - 2y + 3 = 0$ are :

(i) 1, 2 **(ii) – 1, – 2**
(iii) 1, – 2 **(iv) – 1, 2**

Solution : As per the question,

$$2x + 3y = 8 \quad \ldots(i)$$
$$x - 2y = -3 \quad \ldots(ii)$$

Multiplying equation (i) with 2 and equation (ii) with 3, we have

$$4x + 6y = 16 \quad \ldots(iii)$$
and $$3x - 6y = -9 \quad \ldots(iv)$$

Thus, adding the above two equations we get

$$7x = 7$$
$$\Rightarrow \quad x = 1$$

Putting $x = 1$ in equation (i), we have

$$2 + 3y = 8$$
$$\Rightarrow \quad 3y = 6$$
$$\Rightarrow \quad y = 2$$

Hence, the values of x and y are 1, 2 respectively.
So the correct option is (i).

Q. 7. The values of x and y in $2x - 5y + 4 = 0$ and $2x + y - 8 = 0$ are :

(i) 1, 2 **(ii) 3, 2**
(iii) 2, 1 **(iv) 2, 3**

Solution : As per the question,

$$2x - 5y + 4 = 0$$
$$\Rightarrow \quad 2x - 5y = -4 \quad \ldots(i)$$
and $$2x + y - 8 = 0$$
$$\Rightarrow \quad 2x + y = 8 \quad \ldots(ii)$$

Thus, subtracting equation (ii) from (i), we get

$$-6y = -12$$
$$\Rightarrow \quad y = 2$$

Replacing $y = 2$ in equation (ii), we have

$$2x + 2 = 8$$
$$\Rightarrow \quad 2x = 6$$
$$\Rightarrow \quad x = 3$$

Hence, the values of x and y are 3, 2 respectively.
So the correct option is (ii).

Q. 8. The values of x and y in $2x + y + 1 = 0$ and $2x - 3y + 8 = 0$ are :

(i) 1, 2 **(ii) $\frac{-11}{8}, \frac{7}{4}$**

(iii) $\frac{11}{8}, \frac{7}{4}$ **(iv) 2, 3**

Solution : As per the question,

$$2x + y + 1 = 0$$
$$\Rightarrow \quad 2x + y = -1 \quad \ldots(i)$$
and $$2x - 3y + 8 = 0$$
$$\Rightarrow \quad 2x - 3y = -8 \quad \ldots(ii)$$

Thus, subtracting equation (ii) from (i), we get

$$4y = 7$$
$$\Rightarrow \quad y = \frac{7}{4}$$

Replacing $y = \frac{7}{4}$ in equation (ii), we have

$$2x - 3\left(\frac{7}{4}\right) = -8$$
$$\Rightarrow \quad 2x = \frac{21}{4} - 8$$
$$\Rightarrow \quad 2x = \frac{21 - 32}{4} = -\frac{11}{4}$$
$$\Rightarrow \quad x = -\frac{11}{8}$$

Hence, the values of x and y are $-\frac{11}{8}, \frac{7}{4}$ respectively.

So the correct option is (iii).

Q. 9. The length and breadth of a room if its area is 120 m² and perimeter is 44 m are :

(i) 11 m and 2 m **(ii) 10 m and 2 m**
(iii) 12 m and 10 m **(iv) 12 m and 1 m**

Solution : Given,

Area = 120 m^2 and perimeter = 44 m

Let the length be l and the breadth be b.

Thus, $lb = 120$

and $2(l + b) = 44$

$\Rightarrow$ $l + b = 22$

$\Rightarrow$ $l = 22 - b$

$\Rightarrow$ $(22 - b)b = 120$

$\Rightarrow$ $b^2 - 22b + 120 = 0$

$\Rightarrow$ $b^2 - 12b - 10b + 120 = 0$

$\Rightarrow$ $b(b - 12) - 10(b - 12) = 0$

$\Rightarrow$ $(b - 12)(b - 10) = 0$

$\Rightarrow$ $b = 10$ m or 12 m

$\therefore$ $l = 12$ m or 10 m

So the correct option is (iii).

Q. 10. Find the marks obtained by Neha in two subjects, if their average is 75 and their product is 5600 :

(i) 75 and 75 **(ii) 70 and 80**

(iii) 56 and 100 **(iv) 73 and 77**

Solution : Let the marks obtained by Neha in the two subjects be x and y.

Hence, $xy = 5600$ and $\frac{x+y}{2} = 75$

Thus, $x + y = 150$

or $x = 150 - y$

and $xy = 5600$

$\Rightarrow$ $y(150 - y) = 5600$

$\Rightarrow$ $150y - y^2 = 5600$

$\Rightarrow$ $y^2 - 150y + 5600 = 0$

$\Rightarrow$ $y^2 - 80y - 70y + 5600 = 0$

$\Rightarrow$ $y(y - 80) - 70(y - 80) = 0$

$\Rightarrow$ $(y - 80)(y - 70) = 0$

$\Rightarrow$ $y = 70$ or 80

and $x = 80$ or 70

So the correct option is (ii).

Q. 11. The perimeter of an isosceles triangle is 65 cm and the unequal side is thrice as large as each of the equal sides. The lengths of the sides are :

(i) 13, 13, 39 **(ii) 39, 39, 26**

(iii) 13, 26, 26 **(iv) 13, 13, 26**

Solution : Given, Perimeter of an isosceles triangle = 65 cm

Let the sides of the triangle be a cm, a cm and b cm.

Now, $b = 3a$

Thus, $a + b + a = 65$

$\Rightarrow$ $a + 3a + a = 65$

$\Rightarrow$ $5a = 65$

$\Rightarrow$ $a = 13$ cm

$\therefore$ $b = 39$ cm

So the correct option is (i).

Q. 12. The length and breadth of a field if its area is 540 m^2 and perimeter is 96 m are :

(i) 36 m and 15 m **(ii) 18 m and 30 m**

(iii) 9 m and 60 m **(iv) 27 m and 27 m**

Solution : Given, area = 540 m^2 and perimeter = 96 m

Let the length and breadth be l m and b m respectively.

$\therefore$ $2(l + b) = 96$

$\Rightarrow$ $l + b = 48$ m

$\Rightarrow$ $l = (48 - b)$ m

and $lb = 540$

$\Rightarrow$ $(48 - b)b = 540$

$\Rightarrow$ $b^2 - 48b + 540 = 0$

$\Rightarrow$ $b^2 - 30b - 18b + 540 = 0$

$\Rightarrow$ $b(b - 30) - 18(b - 30) = 0$

$\Rightarrow$ $(b - 30)(b - 18) = 0$

$\Rightarrow$ $b = 18$ or 30 m

Hence, $l = 30$ m or 18 m

So the correct option is (ii).

Q. 13. Peter secured x marks in Mathematics and y marks in Physics. Find the value of x and y if $x - y = 2$ and $xy = 2600$.

(i) 48 and 50 **(ii) 52 and 54**

(iii) 50 and 52 **(iv) 51 and 53**

Solution : Given, $x - y = 2$ and $xy = 2600$

$\Rightarrow$ $x = y + 2$

So, $(y + 2)y = 2600$

$\Rightarrow$ $y^2 + 2y - 2600 = 0$

$\Rightarrow$ $y^2 + 52y - 50y - 2600 = 0$

$\Rightarrow$ $y(y + 52) - 50(y + 52) = 0$

$\Rightarrow$ $(y + 52)(y - 50) = 0$

$\Rightarrow$ $y = 50, -52$

But as marks cannot be negative, so $y = 50$

and $x = 2 + 50 = 52$

So the correct option is (iii).

Q. 14. In a two-digit number, the sum of the digits is 10 and the square of the tens digit when divided by the cube of the units digit equal to the value of the tens digit. Find the number.

(i) 82 **(ii) 73**

(iii) 91 **(iv) 64**

Solution : Let the digits of the number be x and y.

Thus, the number is $10x + y$.

Given, $x + y = 10$

and $\frac{x^2}{y^3} = x$

Thus, $x = y^3$

Now, $y^3 + y = 10$

$\Rightarrow y^3 + y - 10 = 0$

$\Rightarrow y^3 - 2y^2 + 2y^2 - 4y + 5y - 10 = 0$

$\Rightarrow y^2(y-2) + 2y(y-2) + 5(y-2) = 0$

$\Rightarrow (y-2)(y^2 + 2y + 5) = 0$

$\Rightarrow y = 2 \text{ and } \frac{-2 \pm \sqrt{2^2 - 4(1)(5)}}{2(1)}$

$\Rightarrow y = 2 \text{ and } \frac{-2 \pm \sqrt{4 - 20}}{2}$

$\Rightarrow y = 2 \text{ and } \frac{-2 \pm \sqrt{-16}}{2}$

Thus, y is taken as 2 because the other option is unreal.

$\Rightarrow x = y^3 = (2)^3 = 8$

$\therefore$ the number is $10(8) + 2 = 82$.

So the correct option is (i).

Q. 15. A number consists of 2 digits. The sum of the digits is 12 and the units digit, when divided by the tens digit gives the result as 3. Find the number.

(i) 84 **(ii) 48**

(iii) 93 **(iv) 39**

Solution : Let the number be $10x + y$

Given, $x + y = 12$ and $\frac{y}{x} = 3$

Thus, $y = 3x$

Hence, $x + 3x = 12$

$\Rightarrow 4x = 12$

$\Rightarrow x = 3$

and $y = 9$

So the number is $3(10) + 9 = 39$

and the correct option is (iv).

Q. 16. The value of k for which the system of linear equations $x + 2y = 3$, $5x + ky + 7 = 0$ is inconsistent is :★

(i) $-\frac{14}{3}$ **(ii) $\frac{2}{5}$**

(iii) 5 **(iv) 10**

★ are board exam questions from previous years

Solution : For the system of linear equations to be inconsistent, we have

$$\frac{a_1}{a_2} = \frac{b_1}{b_2} \neq \frac{c_1}{c_2}$$

Here $a_1 = 1, b_1 = 2, c_1 = -3$ and $a_2 = 5, b_2 = k, c_2 = 7$

So, $\frac{1}{5} = \frac{2}{k} \neq \frac{-3}{7}$

$\Rightarrow k = 10$

So the correct option is (iv).

Q. 17. The value of k for which the system of equations $x + y - 4 = 0$ and $2x + ky = 3$, has no solution, is :★

(i) – 2 **(ii) $\neq$ 2**

(iii) 3 **(iv) 2**

Solution : Given,

$$x + y - 4 = 0$$
$$2x + ky - 3 = 0$$

$a_1 = 1, b_1 = 1, c_1 = -4$

$a_2 = 2, b_2 = k, c_2 = -3$

For no solution, the condition is

$$\frac{a_1}{a_2} = \frac{b_1}{b_2} \neq \frac{c_1}{c_2}$$

$\Rightarrow \frac{1}{2} = \frac{1}{k} \neq \frac{-4}{-3}$

$\Rightarrow \frac{1}{2} = \frac{1}{k} \text{ and } \frac{1}{k} \neq \frac{4}{3}$

$\Rightarrow k = 2 \text{ and } k \neq \frac{3}{4}$

$\Rightarrow k = 2$

So the correct option is (iv).

Chapter 4. Quadratic Equations

Q. 1. If $y = 1$ is a common root of the equations $ay^2 + ay + 3 = 0$ and $y^2 + y + b = 0$, then ab equals :★

(i) 3 **(ii) $-\frac{7}{2}$**

(iii) 6 **(iv) – 3**

Solution : The given equations are

$$ay^2 + ay + 3 = 0 \text{ and } y^2 + y + b = 0$$

Substituting $y = 1$ in them we have

$$a(1)^2 + a(1) + 3 = 0 \text{ and } (1)^2 + (1) + b = 0$$

$$\Rightarrow \quad 2a + 3 = 0 \text{ and } 2 + b = 0$$

$$\Rightarrow \quad a = -\frac{3}{2} \text{ and } b = -2$$

Thus, $$ab = \left(-\frac{3}{2}\right)(-2) = 3$$

So the correct option is (i).

Q. 2. The roots of the equation $x^2 - 3x - 9 = 0$ are :★

(i) real and unequal

(ii) real and equal

(iii) roots are not equal

(iv) imaginary roots

Solution : $D = b^2 - 4ac$

$= (-3)^2 - 4(1)(-9)$

$= 9 + 36$

$= 45 > 0$

Thus, the roots are real and distinct.

So the correct option is (i).

Q. 3. The value of x in $x - \frac{18}{x} = 6$ is :★

(i) real and unequal

(ii) real and equal

(iii) roots are imaginary

(iv) roots are not equal

Solution : Given,

$$x - \frac{18}{x} = 6$$

or $x^2 - 6x - 18 = 0$

Thus, $D = b^2 - 4ac$

$= (-6)^2 - 4(-18)$

$= 36 + 72 = 108 > 0$

Thus, the roots are real and distinct.

So the correct option is (i).

Q. 4. The solution of a quadratic equation is as follows : $x = \frac{8 \pm \sqrt{(-8)^2 - 4(3)(2)}}{2(3)}$. Then the quadratic equation is :

(i) $3x^2 - 8x + 2 = 0$ **(ii) $2x^2 - 8x + 3 = 0$**

(iii) $3x^2 + 8x - 2 = 0$ **(iv) $3x^2 + 8x + 2 = 0$**

Solution : Given, $x = \frac{8 \pm \sqrt{(-8)^2 - 4(3)(2)}}{2(3)}$

Thus, $a = 3, b = -8$ and $c = 2$

The equation is given by $ax^2 + bx + c = 0$

or $3x^2 - 8x + 2 = 0$

So the correct option is (i).

Q. 5. The roots of the quadratic equation $x^2 - 0.04 = 0$ are :★

(a) ± 0.2 **(b) ± 0.02**

(c) 0.4 **(d) 2**

Solution : Given : $x^2 - 0.04 = 0$

$$\Rightarrow \quad x^2 - (0.2)^2 = 0$$

$$\Rightarrow \quad (x + 0.2)(x - 0.2) = 0$$

$$\Rightarrow \quad x = -0.2, 0.2$$

So the correct option is (i).

Chapter 5. Arithmetic Progressions

Q. 1. The first three terms of A.P. are $(3y - 1)$, $(3y + 5)$ and $(5y + 1)$. Then y equals :★

(i) –3 **(ii) 4**

(iii) 5 **(iv) 2**

Solution : The term of an A.P. are

$(3y - 1)$, $(3y + 5)$ and $(5y + 1)$

Thus, $d = (3y + 5) - (3y - 1) = (5y + 1) - (3y + 5)$

$$\Rightarrow \quad 6 = 2y - 4$$

$$\Rightarrow \quad 2y = 10$$

$$\Rightarrow \quad y = 5$$

So the correct option is (iii).

Q. 2. If k, $2k - 1$ and $2k + 1$ are three consecutive terms of an A.P., then the value of k is :★

(i) 2 **(ii) 3**

(iii) – 3 **(iv) 5**

Solution : The terms of an A.P. are

k, $(2k - 1)$ and $(2k + 1)$

Thus, $d = (2k - 1) - k = 2k + 1 - (2k - 1)$

$$\Rightarrow \quad 2k - 1 - k = 2k + 1 - 2k + 1$$

$$\Rightarrow \quad k - 1 = 2$$

$$\Rightarrow \quad k = 3$$

So the correct option is (ii).

Q. 3. The next terms of the A.P. $\sqrt{7}, \sqrt{28}, \sqrt{63}$ ... will be :★

(i) $\sqrt{70}$ **(ii) $\sqrt{84}$**

(iii) $\sqrt{97}$ **(iv) $\sqrt{112}$**

Solution : The given series is $\sqrt{7}, \sqrt{28}, \sqrt{63}$ ...

Thus, $a = \sqrt{7}$ and $d = \sqrt{28} - \sqrt{7} = \sqrt{63} - \sqrt{28}$

$$\Rightarrow \quad d = 2\sqrt{7} - \sqrt{7} = 3\sqrt{7} - 2\sqrt{7} = \sqrt{7}$$

Thus, $t_4 = a + (4 - 1)d$

$$\Rightarrow \quad t_4 = \sqrt{7} + (4-1)\sqrt{7}$$

$$\Rightarrow \quad t_4 = \sqrt{7} + 3\sqrt{7} = 4\sqrt{7}$$

$$= \sqrt{(16)7} = \sqrt{112}$$

So the correct option is (iv).

★ are board exam questions from previous years

Q. 4. The sum of first 20 odd natural numbers is :★

(i) 100 **(ii) 210**

(iii) 400 **(iv) 420**

Solution : The given A.P. is

1, 3, 5, 7, 9, 11, 13, 15, 17, 19, 21, 23, 25, 27, 29, 31, 33, 35, 37, 39

Thus $a = 1, d = 2$ and $n = 20$

We know that,

$$S_n = \frac{n}{2}\{2a + (n-1)d\}$$

$$= \frac{20}{2}\{2(1) + (20-1)2\}$$

$$= 10\{2 + 38\} = 400$$

So the correct option is (iii).

Q. 5. If the nth term of an A.P. is $(2n+1)$, then the sum of its first three terms is :★

(i) $6n+3$ **(ii) 15**

(iii) 12 **(iv) 21**

Solution : We know that,

$$t_n = a + (n-1)d = 2n + 1$$

$\Rightarrow \quad a + nd - d = 2n + 1$

$\Rightarrow \quad (a - d) + nd = 1 + 2n$

Thus, $a - d = 1$ and $d = 2$

So, $a = 2 + 1 = 3$

Then, $S_3 = \frac{3}{2}\{2(3) + (3-1)2\}$

$$= \frac{3}{2}\{6 + 4\} = 15$$

So the correct option is (ii).

Q. 6. The value of $t_{30} - t_{20}$ for the A.P. 2, 7, 12, 17, ... is :★

(i) 100 **(ii) 10**

(iii) 50 **(iv) 20**

Solution : $a = 2, d = 5$

So, $t_{30} = 2 + (30-1)5$

$= 2 + 145 = 147$

and $t_{20} = 2 + (20-1)5$

$= 2 + 95 = 97$

Thus, $t_{30} - t_{20} = 147 - 97 = 50$

So the correct option is (iii).

Q. 7. If the Common Difference of an A.P. is 3, then $t_{20} - t_{15}$ is :★

(i) 5 **(ii) 3**

(iii) 15 **(iv) 20**

Solution : Given,

$$d = 3$$

$$t_{20} = a + (20-1)3 = a + 57$$

$$t_{15} = a + (15-1)3 = a + 42$$

Thus, $t_{20} - t_{15} = a + 57 - (a + 42) = 15$

So the correct option is (iii).

Q. 8. If an A.P., if $d = -2$, $n = 5$ and $t_n = 0$, then the value of a is :★

(i) 10 **(ii) 5**

(iii) – 8 **(iv) 8**

Solution : Given, $d = -2$, $n = 5$ and $t_n = 0$

Thus, $t_5 = a + (5-1)(-2) = 0$

$\Rightarrow \quad a + (4)(-2) = 0$

$\Rightarrow \quad a - 8 = 0$

$\Rightarrow \quad a = 8$

So the correct option is (iv).

Q. 9. Write the value of a and d for the following A.P. – 5, – 1, 3, 7. **[NCERT]**

(i) – 5 and 4 **(ii) – 4 and 5**

(iii) 5 and 4 **(iv) – 5 and – 4**

Solution : The given A.P. is – 5, – 1, 3, 7

Thus, $a = -5$ and $d = -1 - (-5) = 4$

So the correct option is (i).

Q. 10. The first term of an A.P. is p and its common difference is q. Find the 10th term.★

(i) $p + 9q$ **(ii) $p + 11q$**

(iii) $p + 10q$ **(iv) $p + q$**

Solution : Given, $a = p$ and $d = q$

$\therefore \quad t_{10} = p + (10-1)(q)$

$= p + 9q$

So the correct option is (i).

Q. 11. For what value of p are $2p + 1$, 13, $5p - 3$, the three consecutive terms of an A.P. ?★

(i) 5 **(ii) 4**

(iii) 3 **(iv) 2**

Solution : For $2p + 1$, 13, $5p - 3$ to be consecutive the common difference should be the same.

Thus, $13 - 2p - 1 = 5p - 3 - 13$

$\Rightarrow \quad 12 - 2p = 5p - 16$

$\Rightarrow \quad 7p = 28$

$\Rightarrow \quad p = 4$

So the correct option is (ii).

★ are board exam questions from previous years

Q. 12. If $\frac{4}{5}$, a, 2 are three consecutive terms of an A.P., then find the value of a.★

(i) $\frac{5}{7}$ (ii) $\frac{5}{4}$

(iii) $\frac{1}{5}$ (iv) $\frac{7}{5}$

Solution : Since $\frac{4}{5}$, a, 2 are the three consecutive terms of an A.P., then

$$d = a - \frac{4}{5} = 2 - a$$

$$\Rightarrow 2a = 2 + \frac{4}{5}$$

$$\Rightarrow 10a = 10 + 4$$

$$\Rightarrow 10a = 14$$

$$\Rightarrow a = \frac{7}{5}$$

So the correct option is (iv).

Q. 13. If the sum of first p terms of A.P. is $ap^2 + bp$, find its common difference.★

(i) $\frac{2b}{p-1} + 2a$ (ii) $\frac{2a}{p-1} + 2b$

(iii) $\frac{2b}{a-1} + 2p$ (iv) $\frac{2a}{b-1} + 2p$

Solution : Let the first term be a and the common difference be d.

We know that,

$$S_p = \frac{p}{2}\{2a + (p-1)d\} = ap^2 + bp$$

$$\Rightarrow p\{2a + (p-1)d\} = 2p(ap + b)$$

$$\Rightarrow 2a + (p-1)d = 2ap + 2b$$

$$\Rightarrow (p-1)d = 2a(p-1) + 2b$$

$$\Rightarrow (p-1)(d-2a) = 2b$$

$$\Rightarrow d - 2a = \frac{2b}{p-1}$$

$$\Rightarrow d = \frac{2b}{p-1} + 2a$$

So the correct option is (i).

Q. 14. The common difference of the A.P.

$$\frac{1}{2q}, \frac{1-2q}{2q}, \frac{1-4q}{2q} \ldots \text{ is :★}$$

(i) – 1 (ii) 1

(iii) q (iv) $2q$

Solution : The given A.P. is $\frac{1}{2q}, \frac{1-2q}{2q}, \frac{1-4q}{2q}\ldots$

Thus, common difference

$$= \frac{1-2q}{2q} - \frac{1}{2q} = -\frac{2q}{2q} = -1$$

So the correct option is (i).

Q. 15. The common difference of the A.P.

$$\frac{1}{3}, \frac{1-3b}{3}, \frac{1-6b}{3}, \ldots \text{ is :★}$$

(i) $\frac{1}{3}$ (ii) $-\frac{1}{3}$

(iii) $-b$ (iv) b

Solution : The given A.P. is $\frac{1}{3}, \frac{1-3b}{3}, \frac{1-6b}{3}\ldots$

Thus, the common difference

$$= \frac{1-3b}{3} - \frac{1}{3} = -b$$

So the correct option is (iii).

Q. 16. The common difference of the A.P.

$$\frac{1}{2b}, \frac{1-6b}{2b}, \frac{1-12b}{2b}, \ldots \text{ is :★}$$

(i) $2b$ (ii) $-2b$

(iii) 3 (iv) – 3

Solution : The given is A.P. $\frac{1}{2b}, \frac{1-6b}{2b}, \frac{1-12b}{2b}\ldots$

Thus, the common difference

$$= \frac{1-6b}{2b} - \frac{1}{2b} = -3$$

So the correct option is (iv).

Q. 17. The common difference of the A.P. $\frac{1}{p}, \frac{1-p}{p}, \frac{1-2p}{p}$ .... is★

(i) 1 (ii) $\frac{1}{p}$

(iii) – 1 (iv) $-\frac{1}{p}$

Solution : We know,

Common difference, $d = a_2 - a_1$

$$= \frac{1-p}{p} - \frac{1}{p}$$

$$= \frac{1-p-1}{p}$$

$$= \frac{-p}{p}$$

$$= -1$$

So the correct option is (iii).

★ are board exam questions from previous years

Q. 18. The nth term of the A.P. $a, 3a, 5a,$ .... is :★

(i) na (ii) $(2n - 1)a$

(iii) $(2n + 1)a$ (iv) $2na$

Solution : In the A.P., $a = a$ and $d = 3a - a = 2a$

We know, n^{th} term of an A.P. is given by,

$$a_n = a + (n - 1)d$$

$$\Rightarrow \quad a_n = a + (n - 1)2a$$

$$\Rightarrow \quad a_n = a + 2an - 2a$$

$$\Rightarrow \quad a_n = 2an - a$$

$$\Rightarrow \quad a_n = a(2n - 1)$$

So the correct option is (ii).

Q. 19. The value of x for which $2x$, $(x + 10)$ and $(3x + 2)$ are the three consecutive terms of an A.P. is :★

(i) 6 (ii) – 6

(iii) 18 (iv) – 18

Solution : Let $a_1 = 2x$, $a_2 = (x + 10)$, $a_3 = (3x + 2)$

We know, $a_2 - a_1 = a_3 - a_2$

$$(x + 10) - 2x = (3x + 2) - (x + 10)$$

$$\Rightarrow \quad x + 10 - 2x = 3x + 2 - x - 10$$

$$\Rightarrow \quad -x + 10 = 2x - 8$$

$$\Rightarrow \quad 3x = 18$$

$$\Rightarrow \quad x = \frac{18}{3} = 6$$

$$\Rightarrow \quad x = 6$$

So the correct option is (i).

Q. 20. The first term of an A.P. is p and the common difference is q, then its 10th term is :★

(i) $q + 9p$ (ii) $p - 9q$

(iii) $p + 9q$ (iv) $2p + 9q$

Solution : We know that

$$a_n = a + (n - 1)d$$

Here $a = p$, $d = q$, $n = 10$

$\therefore$ 10th term, $a_{10} = p + (10 - 1)(q)$

$$\Rightarrow \quad a_{10} = p + 9q$$

So the correct option is (iii).

★ are board exam questions from previous years

Chapter 6. Coordinate Geometry

Q. 1. $ABCD$ is a rectangle whose vertices are $B(4, 0)$, $C(4, 3)$ and $D(0, 3)$. The length of one of its diagonals is :★

(i) 4 (ii) 5

(iii) 3 (iv) 25

Solution : In rectangle $ABCD$, $B(4, 0)$, $C(4, 3)$ and $D(0, 3)$ have these coordinates.

Thus, BD is the diagonal.

Hence, length of $BD = \sqrt{(4-0)^2 + (0-3)^2}$

$$= \sqrt{16+9} = 5$$

So the correct option is (ii).

Q. 2. If the points $A(x, 2)$, $B(-3, -4)$ and $C(7, -5)$ are collinear, then the value of x is :★

(i) – 63 (ii) 63

(iii) – 60 (iv) 60

Solution : Since $A(x, 2)$, $B(-3, -4)$ and $C(7, -5)$ are collinear.

$$\therefore \quad x_1(y_2 - y_3) + x_2(y_3 - y_1) + x_3(y_1 - y_2) = 0$$

$$\Rightarrow x[-4-(-5)] + (-3)[(-5)-2] + 7[2-(-4)] = 0$$

$$\Rightarrow \quad x + (-3)[-7] + 7[6] = 0$$

$$\Rightarrow \quad x + 21 + 42 = 0$$

$$\Rightarrow \quad x = -63$$

So the correct option is (i).

Q. 3. The perimeter of a triangle with vertices $A(0, 4)$, $B(0, 0)$ and $C(3, 0)$:★

(i) $7 + \sqrt{5}$ (ii) 5

(iii) 10 (iv) 12

Solution : Given, $A(0, 4)$, $B(0, 0)$ and $C(3, 0)$

Thus, Perimeter $= AB + BC + CA$

$$= [\sqrt{(0-0)^2 + (0-4)^2}$$

$$+ \sqrt{(3-0)^2 - (0-0)^2}$$

$$+ \sqrt{(3-0)^2 + (4-0)^2}]$$

$$= [\sqrt{16} + \sqrt{9} + \sqrt{9+16}]$$

$$= (4 + 3 + 5) \text{ units}$$

$$= 12 \text{ units}$$

So the correct option is (iv).

Q. 4. The area of the triangle in the given figure (in sq. units) is :★

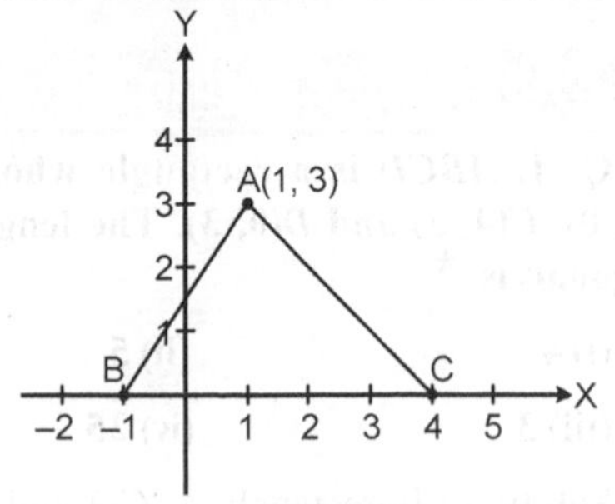

(i) 15 **(ii) 10**

(iii) 7.5 **(iv) 2.5**

Solution : Given, $A = (1, 3)$; $B = (-1, 0)$ and $C = (4, 0)$

Thus, Area of the ΔABC

$$= \frac{1}{2}[1(0-0) - 1(0-3) + 4(3-0)]$$

$$= \frac{1}{2}[3 + 12]$$

$$= 7.5 \text{ sq. units}$$

So the correct option is (iii).

Q. 5. If the point $C(k, 4)$ divides the join of points $A(2, 6)$ and $B(5, 1)$ in the ratio 2 : 3, then the value of k is :★

(i) 16 **(ii) $\frac{28}{5}$**

(iii) $\frac{16}{5}$ **(iv) $\frac{8}{5}$**

Solution : Given, $A(2, 6)$, $C(k, 4)$ and $B(5, 1)$ and $AC : BC = 2 : 3$

Thus, $$k = \frac{3\times 2 + 2\times 5}{2+3}$$

$$= \frac{6+10}{5} = \frac{16}{5}$$

So the correct option is (iii).

Q. 6. The distance of the point (– 3, 4) from the x-axis is :★

(i) 3 **(ii) – 3**

(iii) 4 **(iv) 5**

Solution : The given point is (– 3, 4).

Thus, the distance $= \sqrt{(-3-0)^2 + (4-0)^2}$

$= 4$ units

Hence the correct option is (iii).

Q. 7. In the given figure $P(5, -3)$ and $Q(3, y)$ are the points of trisection of the line segment joining $A(4, 7)$ and $B(1, -5)$. Then y equals :★

A(4, 7) P(5, –3) Q(3, y) B(1, –5)

(i) 2 **(ii) 4**

(iii) – 4 **(iv) $-\frac{5}{2}$**

★ are board exam questions from previous years

Solution : $$PQ = QB$$

$$\Rightarrow \sqrt{(5-3)^2 + (-3-y)^2} = \sqrt{(3-1)^2 + (y+5)^2}$$

$$\Rightarrow (2)^2 + (3+y)^2 = (2)^2 + (y+5)^2$$

$$\Rightarrow 4 + 9 + 6y + y^2 = 4 + 25 + 10y + y^2$$

$$\Rightarrow 6y + 13 = 29 + 10y$$

$$\Rightarrow 4y = -16$$

$$\Rightarrow y = -4$$

So the correct option is (iii).

Q. 8. If the coordinates of one end of a diameter of a circle are (2, 3) and the coordinates of the centre are (– 2, 5), then the coordinates of the other end of the diameter are :★

(i) (– 6, 7) **(ii) (6, – 7)**

(iii) (6, 7) **(iv) (– 6, – 7)**

Solution : Given the coordinates of the centre = (– 2, 5) and one end = (2, 3).

Thus, the centre is the mid-point of the two end points.

Now, let the coordinates of the other end be (x, y).

So, $$-2 = \frac{2+x}{2} \quad \text{and} \quad 5 = \frac{3+y}{2}$$

$$\Rightarrow -4 = 2 + x \quad \text{and} \quad 10 = 3 + y$$

$$\Rightarrow x = -6 \quad \text{and} \quad y = 7$$

Hence, the coordinates are (– 6, 7).

So the correct option is (i).

Q. 9. The coordinates of the point P dividing the line segment joining the points $A(1, 3)$ and $B(4, 6)$ in the ratio 2 : 1 are :★

(i) (2, 4) **(ii) (3, 5)**

(iii) (4, 2) **(iv) (5, 3)**

Solution : Given, $A = (1, 3)$; $B = (4, 6)$ and $AP : PB = 2 : 1$

Let the co-ordinates of point P be (x, y).

Thus $$x = \frac{2\times 4 + 1\times 1}{2+1} \text{ and } y = \frac{2\times 6 + 1\times 3}{2+1}$$

$$\Rightarrow x = \frac{8+1}{3} \text{ and } y = \frac{12+3}{3}$$

$$\Rightarrow x = 3 \text{ and } y = 5$$

$$\Rightarrow (x, y) = (3, 5)$$

So the correct option is (ii).

Q. 10. If $P\left(\frac{a}{2}, 4\right)$ is the mid-point of the line segment joining the points $A(-6, 5)$ and $B(-2, 3)$, then the value of a is :★

(i) – 8 **(ii) 3**

(iii) – 4 **(iv) 4**

Solution : Given, $P\left(\frac{a}{2}, 4\right)$ is the mid-point of the line segment joining the points $A(-6, 5)$ and $B(-2, 3)$.

Thus, $\frac{a}{2} = \frac{-6-2}{2}$

$= \frac{-8}{2} = -4$

or $a = -8$

So the correct option is (i).

Q. 11. If A and B are the points $(-6, 7)$ and $(-1, -5)$ respectively, then the distance $2AB$ is equal to :★

(i) 13 **(ii) 26**
(iii) 169 **(iv) 238**

Solution : Given, $A = (-6, 7)$ and $B = (-1, -5)$

Thus, $AB = \sqrt{(-6+1)^2 + (7+5)^2}$

$= \sqrt{(-5)^2 + (12)^2}$

$= \sqrt{25+144}$

$= \sqrt{169}$

$= 13$

Thus, $2AB = 26$ units

So the correct option is (ii).

Q. 12. The area (in sq. units) of the triangle formed by the points $A(a, 0)$, $O(0, 0)$ and $B(0, b)$ is :★

(i) ab **(ii) $\frac{1}{2}ab$**

(iii) $\frac{1}{2}a^2b^2$ **(iv) $\frac{1}{2}b^2$**

Solution : Given,

$A = (a, 0)$, $O = (0, 0)$ and $B = (0, b)$

Thus, Area of ΔABC

$= \frac{1}{2}[a(0-b) + 0(b-0) + 0(0-0)]$

$= -\frac{1}{2}ab$ sq. units

$= \frac{1}{2}ab$ sq. units

[as area cannot be negative]

So the correct option is (ii).

Q. 13. The point P which divides the line segment joining the points $A(2, -5)$ and $B(5, 2)$ in the ratio 2 : 3 lies in the quadrant :★

(i) I **(ii) II**
(iii) III **(iv) IV**

Solution : Let the coordinates of the point P be (x, y).

Given, $A = (2, -5)$, $B = (5, 2)$ and $m : n = 2 : 3$

Now, $P_x = \frac{2(5)+3(2)}{2+3}$ and $P_y = \frac{2(2)+3(-5)}{2+3}$

$\Rightarrow P_x = \frac{10+6}{5}$ and $P_y = \frac{4-15}{5}$

$\Rightarrow x = \frac{16}{5}$ and $y = -\frac{11}{5}$

$\Rightarrow P = \left(\frac{16}{5}, -\frac{11}{5}\right)$

Thus, P lies in Quadrant IV.

So the correct option is (iv).

Q. 14. The mid-point of segment AB is the point $P(0, 4)$. If the coordinates of B are $(-2, 3)$, then the coordinates of A are :★

(i) (2, 5) **(ii) (–2, –5)**
(iii) (2, 9) **(iv) (–2, 11)**

Solution : Given, $P = (0, 4)$, $B = (-2, 3)$ and $AP = PB$

Let $A = (x, y)$

Thus, $P_x = \frac{x-2}{2}$ and $P_y = \frac{y+3}{2}$

$\Rightarrow 0 = \frac{x-2}{2}$ and $4 = \frac{y+3}{2}$

$\Rightarrow x = 2$ and $y = 5$

Thus, $A = (2, 5)$

So the correct option is (i).

Q. 15. The point on the x-axis which is equidistant from the point $(-1, 0)$ and $(5, 0)$ is :★

(i) (0, 2) **(ii) (2, 0)**
(iii) (3, 0) **(iv) (0, 3)**

Solution : Given, $A = (-1, 0)$ and $B = (5, 0)$.

Let the point on the x-axis be $P(x, 0)$.

Thus, $AP = PB$

$\Rightarrow \sqrt{(x+1)^2 + (0-0)^2} = \sqrt{(5-x)^2 + (0-0)^2}$

$\Rightarrow (x+1)^2 = (5-x)^2$

$\Rightarrow x + 1 = 5 - x$

$\Rightarrow 2x = 4$

$\Rightarrow x = 2$

Thus, the point is (2, 0).

So the correct option is (ii).

Q. 16. The point P on x-axis equidistant from the points A(–1, 0) and B(5, 0) is :★

(i) (2, 0) **(ii) (0, 2)**
(iii) (3, 0) **(iv) (2, 2)**

Solution : Let P(h, 0) be the required point on X-axis.

So, according to the question

PA = PB

or $(PA)^2 = (PB)^2$

$\Rightarrow (h+1)^2 + (0-0)^2 = (h-5)^2 + (0-0)^2$

★ are board exam questions from previous years

$\Rightarrow \quad (h+1)^2 = (h-5)^2$

$\Rightarrow \quad h^2 + 2h + 1 = h^2 - 10h + 25$

$\Rightarrow \quad 12h = 24$

$\Rightarrow \quad h = 2$

So, the required point is (2, 0)

So the correct option is (i).

Q. 17. The co-ordinates of the point which is reflection of point (– 3, 5) in x-axis are :★

(i) (3, 5) (ii) (3, – 5)

(iii) (– 3, – 5) (iv) (– 3, 5)

Solution : The co-ordinates of the point which is reflection of point (– 3, 5) in x-axis is (– 3, – 5).

So the correct option is (iii).

Q. 18. If the point P(6, 2) divides the line segment joining A(6, 5) and B(4, y) in the ratio 3 : 1, then the value of y is :★

(i) 4 (ii) 3

(iii) 2 (iv) 1

Solution : Using section formula,

$$y\text{-coordinate of P} = \frac{3 \times y + 1 \times 5}{3+1}$$

$$\Rightarrow \quad 2 = \frac{3y+5}{4}$$

$\Rightarrow \quad 8 = 3y + 5$

$\Rightarrow \quad 3y = 3$

$\Rightarrow \quad y = 1$

So the correct option is (iv).

Q. 19. The distance between the points (a cos θ + b sin θ, 0) and (0, a sin θ – b cos θ), is :★

(i) $a^2 + b^2$ (ii) $a^2 - b^2$

(iii) $\sqrt{a^2+b^2}$ (iv) $\sqrt{a^2-b^2}$

Solution : The distance formula,

$$d = \sqrt{(x_2 - x_1)^2 + (y_2 - y_1)^2}$$

Here $x_1 = a\cos\theta + b\sin\theta$, $y_1 = 0$

$x_2 = 0$, $y_2 = a\sin\theta - b\cos\theta$

$$\therefore d = \sqrt{[0-(a\cos\theta + b\sin\theta)]^2 + [(a\sin\theta - b\cos\theta) - 0]^2}$$

$$= \sqrt{(a\cos\theta + b\sin\theta)^2 + (a\sin\theta - b\cos\theta)^2}$$

$$= \sqrt{\begin{aligned}&a^2\cos^2\theta + b^2\sin^2\theta + 2ab\sin\theta\cos\theta + a^2\sin^2\theta \\ &+ b^2\cos^2\theta - 2ab\sin\theta\cos\theta\end{aligned}}$$

$$= \sqrt{a^2(\sin^2\theta + \cos^2\theta) + b^2(\sin^2\theta + \cos^2\theta)}$$

$$= \sqrt{a^2 \times 1 + b^2 \times 1} \qquad \{\because \sin^2\theta + \cos^2\theta = 1\}$$

$$= \sqrt{a^2 + b^2}$$

So the correct option is (iii).

Q. 20. If the point P(k, 0) divides the line segment joining the points A(2, – 2) and B(– 7, 4) in the ratio 1 : 2, then the value of k is :★

(i) 1 (ii) 2

(iii) – 2 (iv) – 1

Solution :

(m) (n)
1 : 2

A (2, –2) (x_1, y_1) — P (k, 0) (x, y) — B (–7, 4) (x_2, y_2)

By section formula,

$$x = \frac{mx_2 + nx_1}{m+n}, \quad y = \frac{mx_2 + ny_1}{m+n}$$

Here $x = k$, $m = 1$, $n = 2$, $x_1 = 2$, $x_2 = -7$

∴ we get,

$$k = \frac{(1)(-7) + (2)(2)}{1+2}$$

$$\Rightarrow \quad k = \frac{-7+4}{3}$$

$$\Rightarrow \quad k = \frac{-3}{3}$$

$\Rightarrow \quad k = -1$

So the correct option is (iv).

Q. 21. The value of p, for which the points A(3, 1), B(5, p) and C(7, – 5) are collinear, is :★

(i) – 2 (ii) 2

(iii) – 1 (iv) 1

Solution : Given,

A(3, 1) $\Rightarrow x_1 = 3, y_1 = 1$

B(5, p) $\Rightarrow x_2 = 5, y_2 = p$

C(7, – 5) $\Rightarrow x_3 = 7, y_3 = -5$.

If A, B and C are collinear points, then, ar(ΔABC) = 0

$$\Rightarrow \frac{1}{2}[x_1(y_2 - y_3) + x_2(y_3 - y_1) + x_3(y_1 - y_2)] = 0$$

$$\Rightarrow \frac{1}{2}[3(p-(-5)) + 5(-5-1) + 7(1-p)] = 0$$

$$\Rightarrow \frac{1}{2}[3p + 15 - 25 - 5 + 7 - 7p] = 0$$

★ are board exam questions from previous years

$$\Rightarrow \quad [-4p - 8] = 0$$
$$\Rightarrow \quad -2p - 4 = 0$$
$$\Rightarrow \quad p = \frac{4}{-2} = -2$$
$$\Rightarrow \quad p = -2$$

So the correct option (i).

Chapter 7. Triangles

Q. 1. In a right triangle ABC, right-angled at B, BC = 12 cm and AB = 5 cm. The radius of the circle inscribed in the triangle (in cm) is :

(i) 4 **(ii) 3**

(iii) 2 **(iv) $\frac{1}{2}$**

Solution : Given, ΔABC is a right-angled at B.

Applying Pythagoras's Theorem,

$$AC^2 = BC^2 + AB^2$$
$$\Rightarrow \quad AC^2 = 12^2 + 5^2$$
$$= 144 + 25$$
$$= 169 = 13^2$$
$$\Rightarrow \quad AC = 13$$

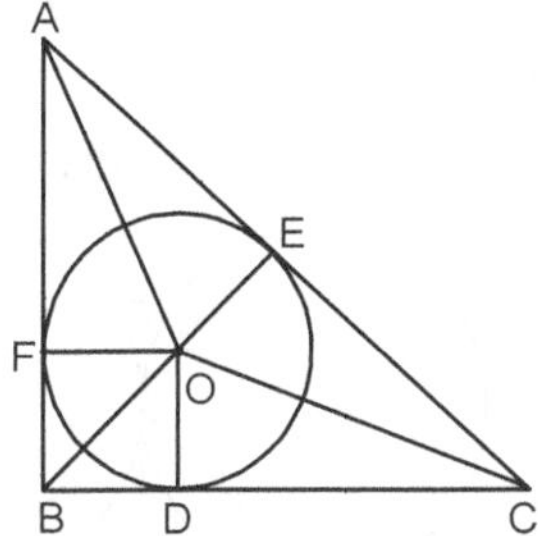

Let O be the centre of the circle inscribed within ΔABC such that OD, OE and OF are the altitude of ΔBOC, ΔCOA and ΔAOB and $OD = OE = OF = r$

Thus, Area of ΔABC = Area of ΔBOC + Area of ΔCOA + Area of ΔAOB

$$\Rightarrow \quad \frac{1}{2}AB \times BC = \frac{1}{2}OD \times BC + \frac{1}{2}OE \times AC + \frac{1}{2}OF \times AB$$
$$\Rightarrow \quad \frac{1}{2}(5 \times 12) = \frac{1}{2}r \times 12 + \frac{1}{2}r \times 13 + \frac{1}{2}r \times 5$$
$$\Rightarrow \quad 60 = 12r + 13r + 5r$$
$$\Rightarrow \quad 60 = 30r$$
$$\Rightarrow \quad r - 2$$

So the correct option is (iii).

Q. 2. A boy walks 12 m due east and 5 m due south. How far is he from the starting point ?

(i) 31 m **(ii) 26 m**

(iii) 62 m **(iv) 13 m**

★ **are board exam questions from previous years**

Solution : Applying Pythagoras' Theorem

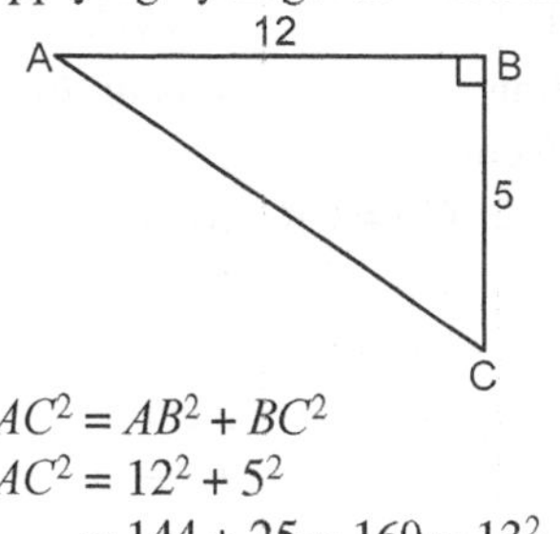

$$AC^2 = AB^2 + BC^2$$
$$\Rightarrow \quad AC^2 = 12^2 + 5^2$$
$$= 144 + 25 = 169 = 13^2$$
$$\Rightarrow \quad AC = 13 \text{ m}$$

So the correct option is (iv).

Q. 3. In ΔPQR, if PS is the internal bisector of $\angle P$ meeting QR at S and PQ = 15 cm, $QS = (3 + x)$ cm, $SR = (x - 3)$ cm and PR = 7 cm, then find the value of x.

(i) 2.85 cm **(ii) 8.25 cm**

(iii) 5.28 cm **(iv) 8.52 cm**

Solution : Since PS is the internal bisector of $\angle P$ and it meets QR at S

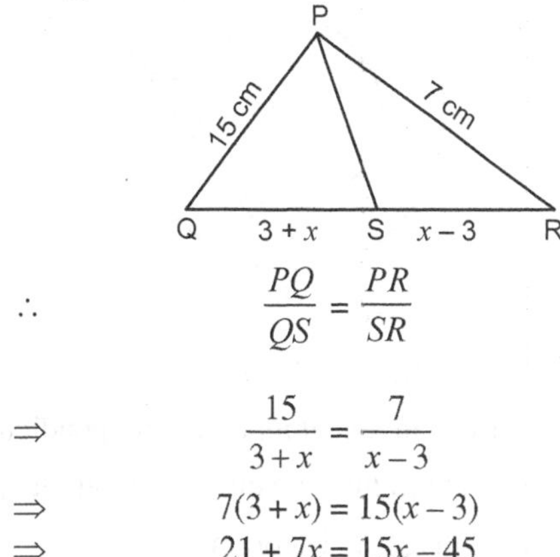

$$\therefore \quad \frac{PQ}{QS} = \frac{PR}{SR}$$
$$\Rightarrow \quad \frac{15}{3+x} = \frac{7}{x-3}$$
$$\Rightarrow \quad 7(3 + x) = 15(x - 3)$$
$$\Rightarrow \quad 21 + 7x = 15x - 45$$
$$\Rightarrow \quad 15x - 7x = 45 + 21$$
$$\Rightarrow \quad 8x = 66$$
$$\Rightarrow \quad 4x = 33$$
$$\Rightarrow \quad x = 8.25 \text{ cm}$$

So the correct option is (ii).

Chapter 8. Circles

Q. 1. In the given figure, a quadrilateral $ABCD$ is drawn to circumscribe a circle such that its sides AB, BC, CD and AD touch the circle at P, Q, R and S respectively. If $AB = x$ cm, BC = 7 cm, CR = 3 cm and AS = 5 cm, find x.★

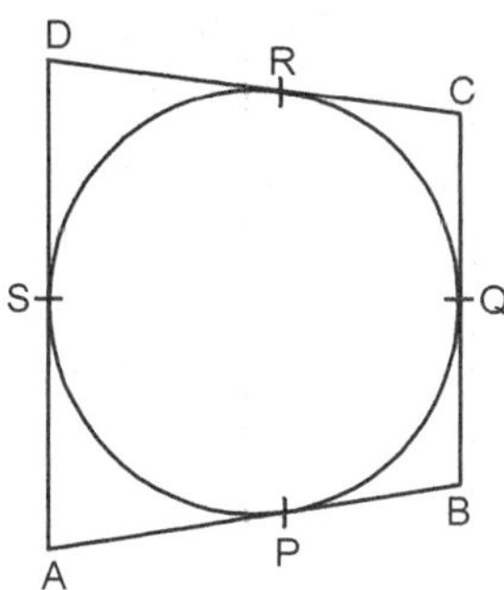

(i) 10 **(ii) 9**
(iii) 8 **(iv) 7**

Solution : Given, $AB = x$ cm, $BC = 7$ cm, $CR = 3$ cm and $AS = 5$ cm

Hence, $CQ = CR = 3$ cm [Tangents from C]

Thus, $BQ = BC - CQ = (7 - 3)$ cm $= 4$ cm

So, $PB = BQ = 4$ cm [Tangents from B]

Also, $AS = AP = 5$ cm [Tangents from A]

$\therefore$ $AB = AP + PB = (5 + 4)$ cm $= 9$ cm

So the correct option is (ii).

Q. 2. Two concentric circles have radii 5 cm and 3 cm. Find the length of the chord of the larger circle if it touches the smaller (tangent).★

(i) 4 **(ii) 5**
(iii) 8 **(iv) 10**

Solution : Given,

$OA = OB = 5$ cm; $OD = 3$ cm

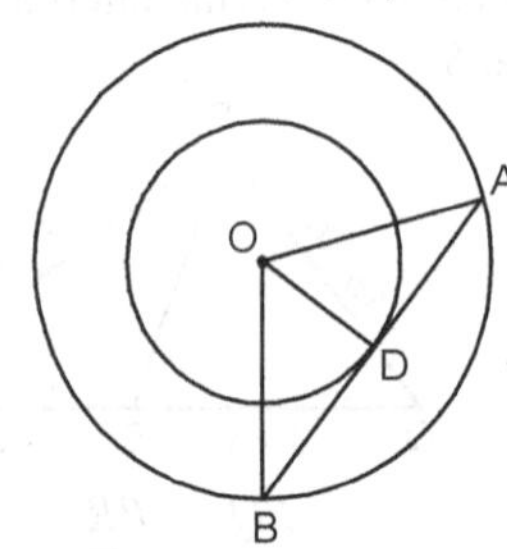

AB is a tangent at D.

So, $\angle ODB = \angle ODA = 90°$

[the tangent at a point is perpendicular to the radius of a circle through the point]

Applying Pythagoras Theorem

$OB^2 = BD^2 + OD^2$

$\Rightarrow$ $BD^2 = OB^2 - OD^2$

$\Rightarrow$ $BD^2 = (5)^2 - (3)^2$

$\Rightarrow$ $BD^2 = 25 - 9 = 16$

$\Rightarrow$ $BD = 4$ cm ...(i)

Also $OA^2 = AD^2 + OD^2$

$\Rightarrow$ $AD^2 = OA^2 - OD^2$

$\Rightarrow$ $AD^2 = (5)^2 - (3)^2$

$\Rightarrow$ $AD^2 = 25 - 9 = 16$

$\Rightarrow$ $AD = 4$ cm ...(i)

Hence, $AB = AD + BD = (4 + 4)$ cm $= 8$ cm

So the correct option is (iii).

Q. 3. A chord of a circle of radius 10 cm subtends a right angle at its centre. The length of the chord (in cm) is :★

(i) $5\sqrt{2}$ **(ii) $10\sqrt{2}$**
(iii) $\frac{5}{\sqrt{2}}$ **(iv) $10\sqrt{3}$**

Solution :

Given, $AO = OB = 10$ cm

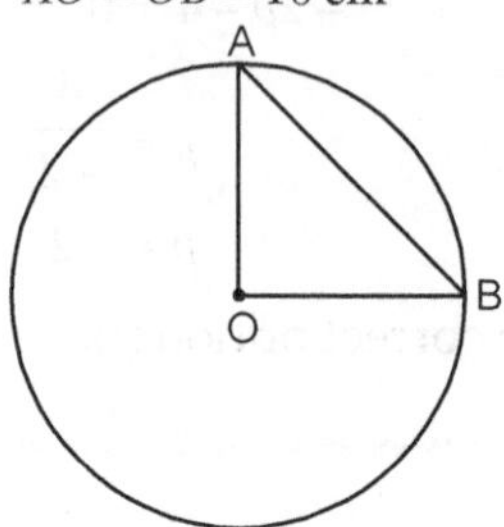

Thus, $AB^2 = AO^2 + OB^2$

[By Pythagoras Theorem]

$\Rightarrow$ $AB^2 = (10)^2 + (10)^2$

$\Rightarrow$ $AB^2 = 100 + 100$

$\Rightarrow$ $AB^2 = 200$

Thus $AB = 10\sqrt{2}$ cm

So the correct option is (ii).

Q. 4. Two circles touch each other externally at *P*. *AB* is a common tangent to the circles, touching them at *A* and *B*. The value of $\angle APB$ is :★

(i) 30° **(ii) 45°**
(iii) 60° **(iv) 90°**

Solution :

Let PO meet AB at O such that PO is a tangent.

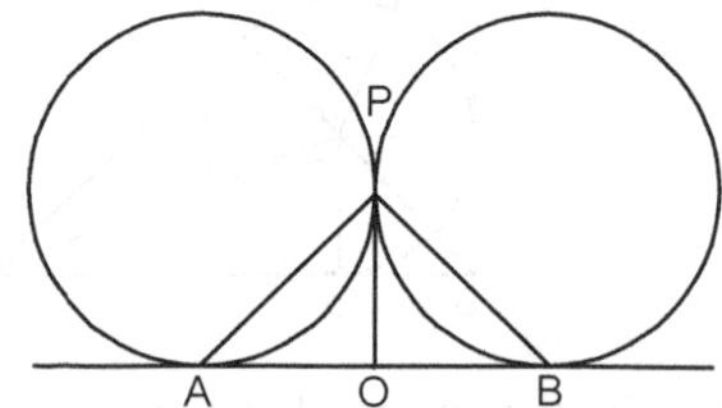

Thus $AO = PO = OB$ and $\angle AOP = \angle BOP = 90°$

As ΔAOP and ΔBOP are right-angled isosceles triangles.

So $\angle OAP = \angle OPA = \angle OBP = \angle OPB = 45°$

Thus $\angle APB = \angle APO + \angle OPB = 45° + 45° = 90°$

So the correct option is (iv).

Q. 5. In the given figure, *QR* is a common tangent to the given circles, touching externally at point *T*. The tangent at *T* meets *QR* at *P*. If *PT* = 3.8 cm, then the length of *QR* (in cm) is :★

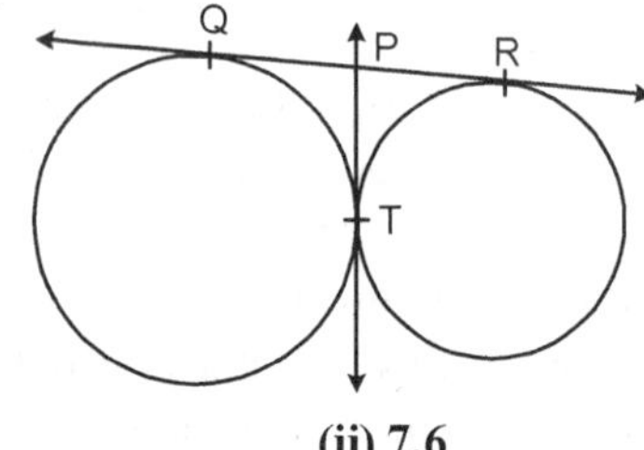

(i) 3.8 **(ii) 7.6**
(iii) 5.7 **(iv) 1.9**

★ are board exam questions from previous years

Solution :

Since PT intersects QR at P, it is the external point from two different tangents for the two circles are possible.

Thus, $PT = PQ$ and $PT = PR$

So, $PT = PQ = PR = 3.8$ cm

Thus, $QR = (3.8 + 3.8)$ cm $= 7.6$ cm

So the correct option is (ii).

Q. 6. In the given figure, PQ and PR are two tangents to a circle with centre O. If $\angle QPR = 46°$, then $\angle QOR$ equals :★

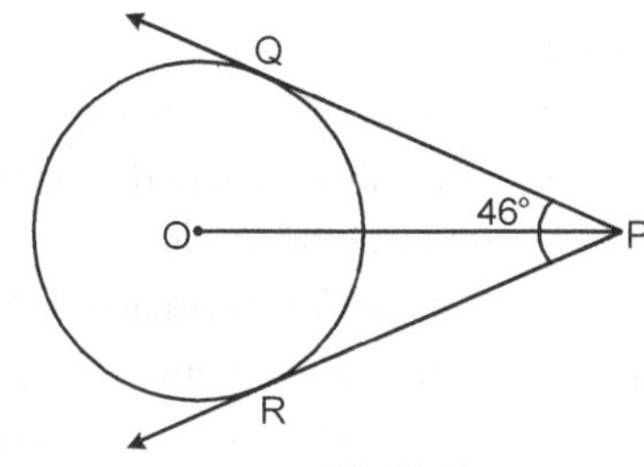

(i) 67° **(ii) 134°**

(iii) 44° **(iv) 46°**

Solution :

Join OP such that it bisects $\angle OPR$.

Thus, $\angle OPQ = \angle OPR = 23°$

So, $\angle QOP = \angle ROP = 180° - (90° + 23°)$

$= 180° - 113° = 67°$

Hence, $\angle QOR = \angle QOP + \angle ROP$

$= 67° + 67° = 134°$

So the correct option is (ii).

Q. 7. In the given figure, if $\angle AOD = 135°$, then $\angle BOC$ is equal to :★

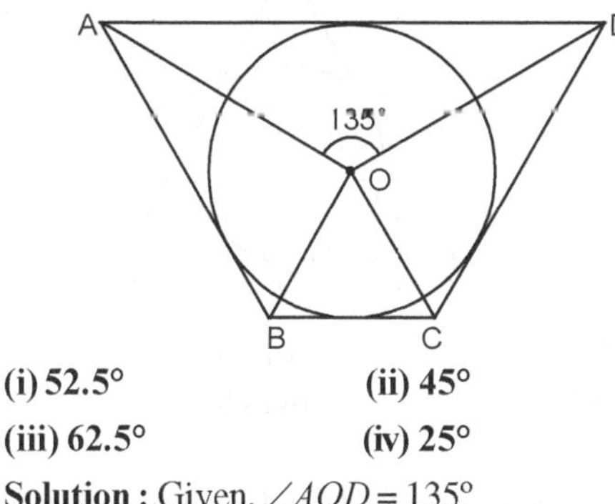

(i) 52.5° **(ii) 45°**

(iii) 62.5° **(iv) 25°**

Solution : Given, $\angle AOD = 135°$

$\therefore \quad \angle BOC = 180° - 135°$

$= 45°$

Angle subtende at the centre by a pair of opposite sides are supplementary.

So the correct option is (ii).

Q. 8. In the given figure, two circles touch each other at C and AB is a tangent to both circles. $\angle ACB$ equal to :★

★ are board exam questions from previous years

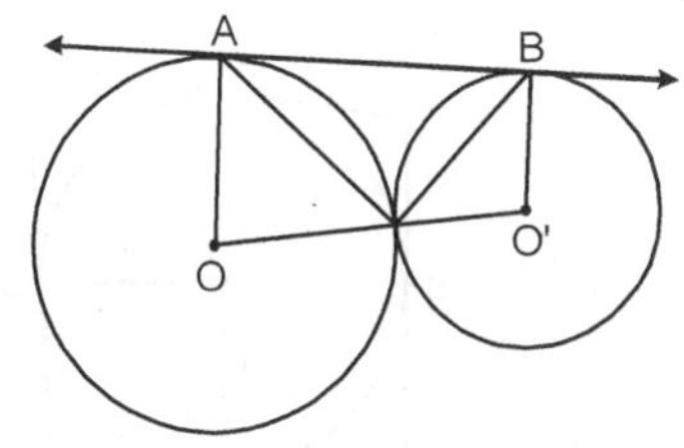

(i) 60° **(ii) 45°**

(iii) 120° **(iv) 90°**

Solution : Let AO and BO' be the two radii that are perpendicular to AB.

Thus AC bisects $\angle OAB$.

So, $\angle CAB = 45°$

Similarly $\angle CBA = 45°$

Thus, $\angle ACB = 180° - (45° + 45°)$

$= 180° - 90° = 90°$

So the correct option is (iv).

Q. 9. In the given figure, PA and PB are two tangents drawn from an external point P to a circle with centre C and radius 4 cm. If PA is perpendicular to PB, then the length of each tangent is :★

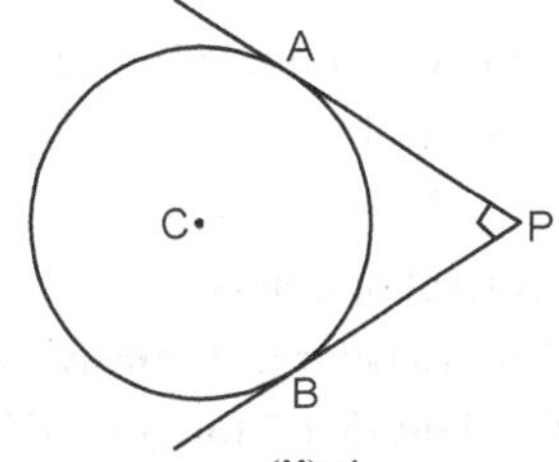

(i) 3 cm **(ii) 4 cm**

(iii) 5 cm **(iv) 6 cm**

Solution : Given,

$AP = BP$

[tagnents from external point P]

Join CA and CB such that they are perpendicular to AP and AB respectively.

Given $CA = CB = 4$ cm

$\angle ACB = 360° - [90° + 90° + 90°]$

or $\angle ACB = 90°$

As $APBC$ is a right-angled quadrilateral with its adjacent sides CA and CB being 4 cm.

So $AP = PB = 4$ cm

[opposite sides to adjacent sides of a right-angled quadrilateral]

So correct option is (ii).

Q. 10. In the given figure, a circle with centre O is inscribed in a quadrilateral $ABCD$ such that it touches the side BC, AB, AD and CD at points P, Q, R and S respectively. If $AB = 29$ cm, $AD = 23$ cm, $\angle B = 90°$ and $DS = 5$ cm, then the radius of the circle (in cm) is :★

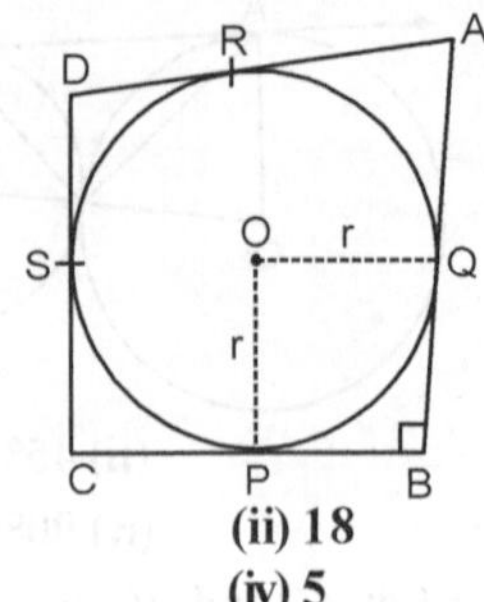

(i) 11 **(ii) 18**
(iii) 6 **(iv) 5**

Solution :

Given, $AB = 29$ cm, $AD = 23$ cm, $DS = 5$ cm and $\angle B = 90°$

Now, $DS = DR = 5$ cm [tangents from D]

So, $AR = AD - DR = (23 - 5)$ cm $= 18$ cm

Again $AR = AQ = 18$ cm [tangents from A]

So, $QB = AB - AQ = (29 - 18)$ cm $= 11$ cm

$\angle OPB = \angle OQB = 90°$

[OQ and OP are perpendicular to AB and BC respectively]

Thus, in the quadrilateral $OQBP$,

$$\angle POQ = 360° - (90° + 90° + 90°) = 90°$$

Hence, it is a right-angled quadrilateral with adjacent opposite sides equal.

$\therefore$ $OQ = PB = 11 \text{ cm} = r$

So the correct option is (i).

Q. 11. From a point Q, 13 cm away from the centre of a circle, the length of tangent PQ to the circle is 12 cm. The radius of the circle (in cm) is :★

(i) 25 **(ii) $\sqrt{313}$**
(iii) 5 **(iv) 1**

Solution : Given,

$PQ = 13$ cm and $OQ = 12$ cm

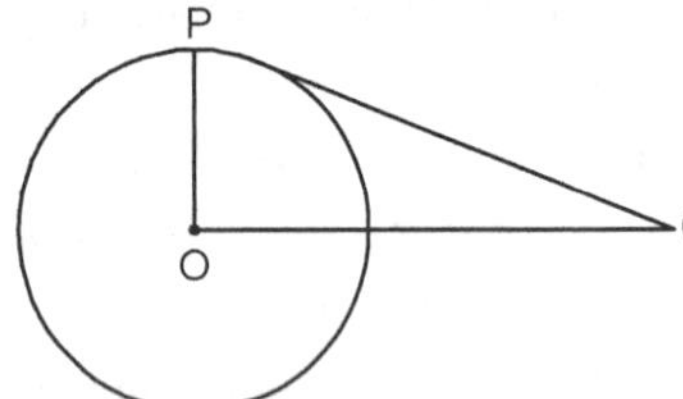

Thus, $r = PO = \sqrt{PQ^2 - OQ^2}$ [By Pythagoras Theorem]

$= \sqrt{(13)^2 - (12)^2}$

$= \sqrt{169 - 144}$

$= \sqrt{25} = 5$ cm

As length cannot be negative.

So the correct option is 5 cm.

Q. 12. In the given figure, AP, AQ and BC are tangents to the circle. If $AB = 5$ cm, $AC = 6$ cm and $BC = 4$ cm, then the length of AP (in cm) is :★

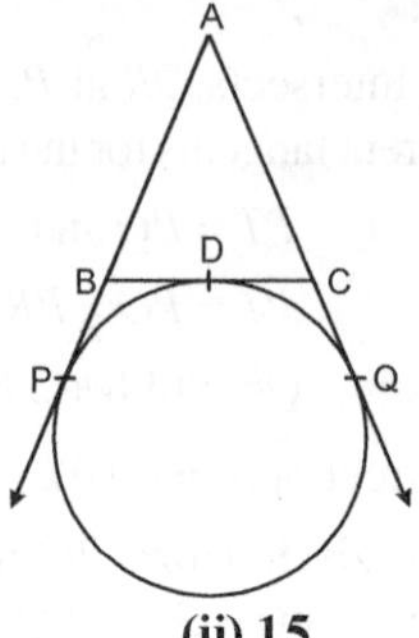

(i) 7.5 **(ii) 15**
(iii) 10 **(iv) 9**

Solution :

Given, $AB = 5$ cm, $AC = 6$ cm and $BC = 4$ cm

Let $PB = x$ cm, $CQ = y$ cm and D be the point where the tangent BC touches the circle.

So, $AP = (5 + x)$ cm and $AQ = (6 + y)$ cm

Also $PB = PD = x$ cm [tangents from B]

and $CQ = CD = y$ cm [tangents from C]

Now, $5 + x = 6 + y$ [$\because AP = AQ$]

$\Rightarrow$ $x - y = 1$ ...(i)

Again $x + y = 4$ [$\because BC = 4$] ...(ii)

Adding (i) and (ii),

$2x = 5$

or $x = 2.5$

Thus, $AP = (5 + x)$ cm $= 7.5$ cm

So the correct option is (i).

Q. 13. In the given figure, the sides AB, BC and CA of a triangle ABC touch a circle at P, Q and R respectively. If $PA = 4$ cm, $BP = 3$ cm and $AC = 11$ cm, then the length of BC (in cm) is :★

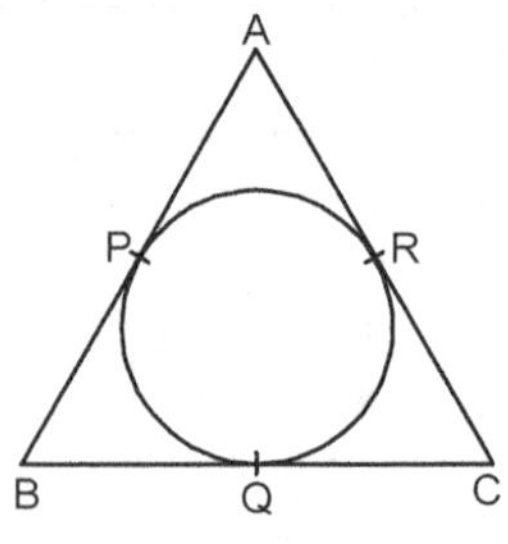

(i) 11 **(ii) 10**
(iii) 14 **(iv) 15**

Solution : Given, $PA = 4$ cm, $PB = 3$ cm and $AC = 11$ cm

Now, $BQ = BP = 3$ cm

[BQ and BP are tangents from B]

and $AP = AR = 4$ cm

[AP and AR are tangents from A]

Thus, $RC = AC - AR = (11 - 4)$ cm $= 7$ cm

Also $RC = CQ$

[RC and CQ are tangents from C]

So, $CQ = 7$ cm

★ are board exam questions from previous years

Hence, $BC = BQ + CQ = (3 + 7)$ cm $= 10$ cm

So the correct option is (ii).

Q. 14. In the given figure, *O* is the centre of the circle, *PQ* is a chord and *PT* is the tangent at *P*. If $\angle POQ = 70°$, then $\angle TPQ$ equals to :★

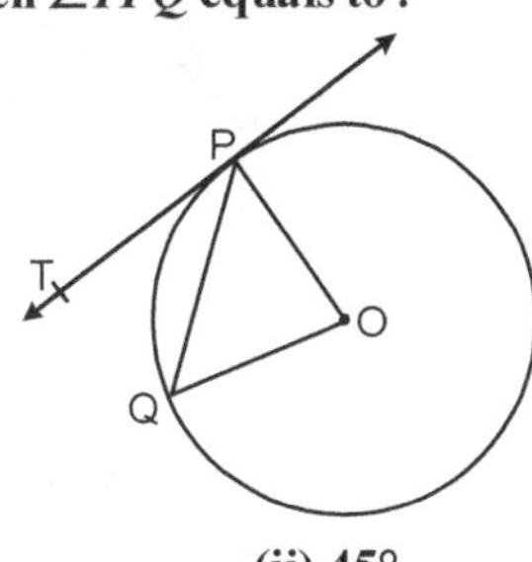

(i) 70° **(ii) 45°**
(iii) 90° **(iv) 35°**

Solution : Given,

$$\angle POQ = 70°$$

As *OP* and *OQ* are the radii of the circle so ΔPOQ is isosceles.

Thus, $\angle OPQ = \angle OQP$

Hence, $180° = \angle OPQ + \angle OQP + 70°$

$\Rightarrow$ $2\angle OPQ = 110°$

$\Rightarrow$ $\angle OPQ = 55°$

Also as *OP* is perpendicular to *PT*, so

$$\angle OPT = 90°$$

Thus, $\angle TPQ = \angle OPT - \angle OPQ$

or $\angle TPQ = 90° - 55° = 35°$

So the correct option is (iv).

Q. 15. In the given figure, *AB* and *AC* are tangents to the circle with centre *O* such that $\angle BAC = 40°$. Then $\angle BOC$ is equal to :★

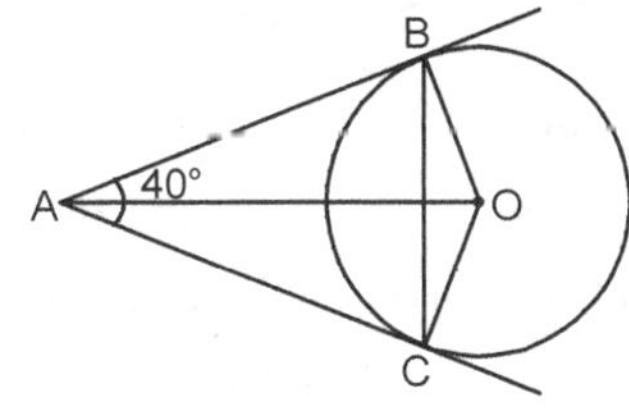

(i) 40° **(ii) 50°**
(iii) 140° **(iv) 150°**

Solution : Given, $AB = AC$, $OB = OC$ and $\angle A = 40°$

Join *AO* such that it is perpendicular to the chord *BC* and bisects $\angle A$.

Thus, $\angle BAO = \angle OAC = 20°$

Also, as *OB* and *OC* are perpendicular to *AB* and *AC* respectively.

$$\angle ABO = \angle ACO = 90°$$

So in ΔABO,

$$\angle BOA = 180° - (90° + 20°)$$

or $\angle BOA = 180° - 110° = 70°$

Similarly in ΔACO,

$$\angle COA = 180° - (90° + 20°)$$

or $\angle COA = 180° - 110° = 70°$

Hence, $\angle BOC = \angle BOC + \angle COA$

$= 70° + 70° = 140°$

So the correct option is (iii).

Q. 16. In the given figure, *O* is the centre of the circle, *AB* is the chord and *AT* is the tangent at *A*. If $\angle AOB = 100°$, then $\angle BAT$ is equal to :★

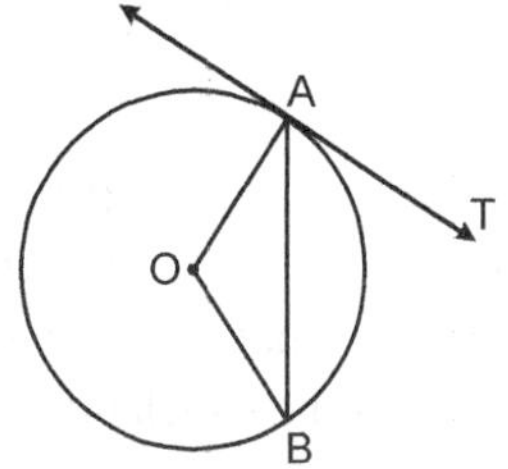

(i) 100° **(ii) 40°**
(iii) 50° **(iv) 90°**

Solution : Given,

$$\angle AOB = 100° \text{ and } OA = OB$$

[both are the radii of the circle with centre *O*]

Thus, ΔAOB is an isosceles triangle and

$$\angle OAB = \angle OBA$$

[base angles of an isosceles triangle]

Thus $\angle OAB + \angle OBA = 180° - \angle AOB$

$\Rightarrow$ $2\angle OBA = 180° - 100°$

$\Rightarrow$ $\angle OBA = 40° = \angle OAB$

As *OA* is perpendicular at *AT*,

$$\angle OAT = 90°$$

Hence, $\angle BAT = 90° - 40° = 50°$

So the correct option is (iii).

Q. 17. In the given figure, *PA* and *PB* are tangents to the circle with centre *O*. If $\angle APB = 60°$ then $\angle OAB$ is :★

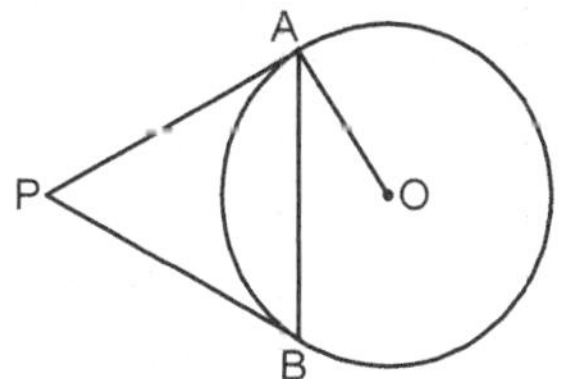

(i) 30° **(ii) 60°**
(iii) 90° **(iv) 15°**

Solution : Given,

$$\angle APB = 60°, AP = PB$$

[tangents from *P*]

In ΔPAB, $PA = PB$ [given]

So, ΔPAB is isosceles and $\angle PAB = \angle PBA$

[base angles of isosceles triangles]

Thus, $\angle PAB + \angle PBA = 180° - \angle APB$

$\Rightarrow$ $2\angle PAB = 180° - 60° = 120°$

$\Rightarrow$ $\angle PAB = 60° = \angle PBA$

Now, $\angle PAO = 90°$ [*OA* perpendicular to *AP*]

Thus, $\angle OAB = \angle PAO - \angle PAB$

★ are board exam questions from previous years

$= 90° - 60° = 30°$

So the correct option is (i).

Q. 18. In the given figure, point P is 26 cm away from the centre O of a circle and the length PT of the tangent drawn from P to the circle is 24 cm. Then the radius of the circle (in cm) is :★

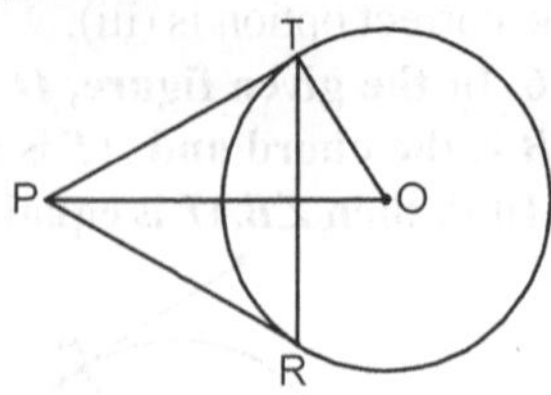

(i) 25 **(ii) 26**
(iii) 24 **(iv) 10**

Solution : Given, $OP = 26$ cm and $PT = 24$ cm

Let the radius be r such that

$OT = r$

Applying Pythagoras theorem

$OP^2 = PT^2 + OT^2$

$\Rightarrow OT^2 = OP^2 - PT^2$

$\Rightarrow r^2 = (26)^2 - (24)^2$

$\Rightarrow r^2 = 676 - 576$

$\Rightarrow r^2 = 100$

$\Rightarrow r = 10$ cm

[as radius cannot be negative]

So the correct option is (iv).

Q. 19. In the given figure, TP and QT are two tangents to a circle with centre O such that $\angle POQ = 110°$. Then $\angle PTQ$ is equal to :★

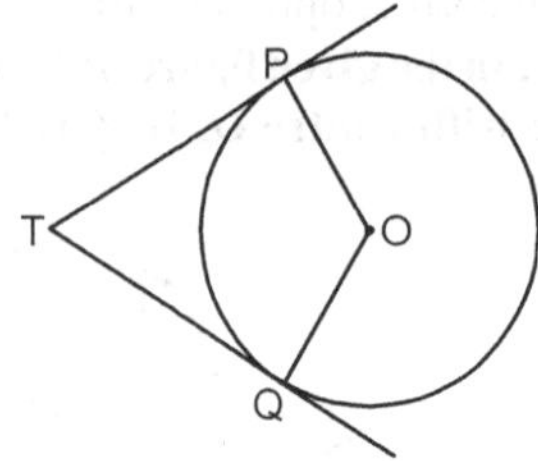

(i) 55° **(ii) 70°**
(iii) 110° **(iv) 90°**

Solution : Given,

$\angle POQ = 110°$

Now, $TQ = TP$ [both are tangents from T]

Join TO such that it bisects both $\angle PTQ$ and $\angle POQ$

Thus, $\angle TOP = 55°$ and $\angle OPT = 90°$

[OP is perpendicular to PT]

Hence, $\angle PTO = 180° - (90° + 55°)$

$= 180° - 145° = 35°$

Similarly $\angle QTO = 35°$

Hence, $\angle PTQ = (35° + 35°) = 70°$

So the correct option is (ii).

★ are board exam questions from previous years

Q. 20. In the given figure, ΔABC is circumscribing a circle. $AR = 4$ cm, $BR = 3$ cm and $AC = 11$ cm. The length of BC is :★

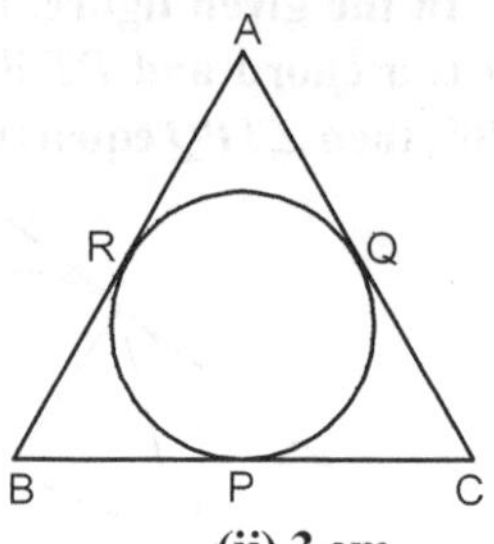

(i) 10 cm **(ii) 3 cm**
(iii) 4 cm **(iv) 11 cm**

Solution : Given, $AR = 4$ cm, $BR = 3$ cm and $AC = 11$ cm

Now, $BR = BP$ [tangents from B]

Thus, $BP = 3$ cm

Similarly $AR = AQ = 4$ cm [tangents from A]

Thus, $QC = AC - AQ = (11 - 4)$ cm $= 7$ cm

Also $QC = PC = 7$ cm [tangents from C]

Now, $BC = BP + PC = (7 + 3)$ cm $= 10$ cm

So the correct option is (i).

Q. 21. In the given figure, CP and CQ are tangents from an external point C to a circle with centre O. AB is another tangent that touches the circle at R. If $CP = 11$ cm and $BR = 4$ cm then BC is :★

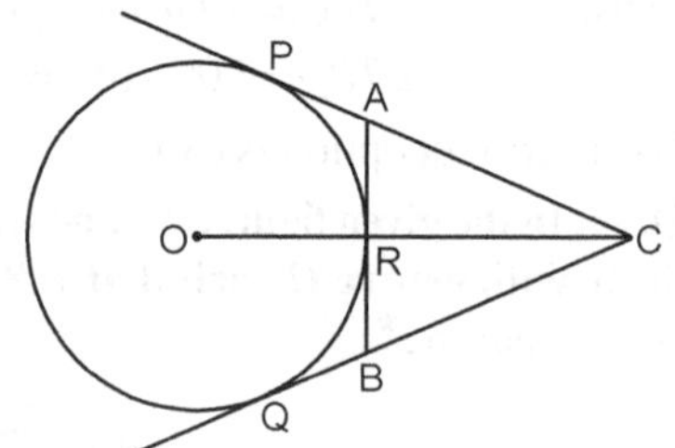

(i) 7 cm **(ii) 11 cm**
(iii) 4 cm **(iv) 10 cm**

Solution : Given, $CP = 11$ cm and $BR = 4$ cm

Now, $CP = CQ$ [tangents from C]

Thus, $CQ = 11$ cm

Again $BR = BQ$ [tangents from B]

Thus, $BQ = 4$ cm

Hence, $CQ = BQ + BC$

$\Rightarrow BC = CQ - BQ$

$\Rightarrow BC = (11 - 4)$ cm $= 7$ cm

So the correct option is (i).

Q. 22. In the given figure, a circle touches the DF of ΔEDF at H and touches ED and EF produced at K and M respectively. If $EK = 9$ cm, then the perimeter of ΔEDF is :★

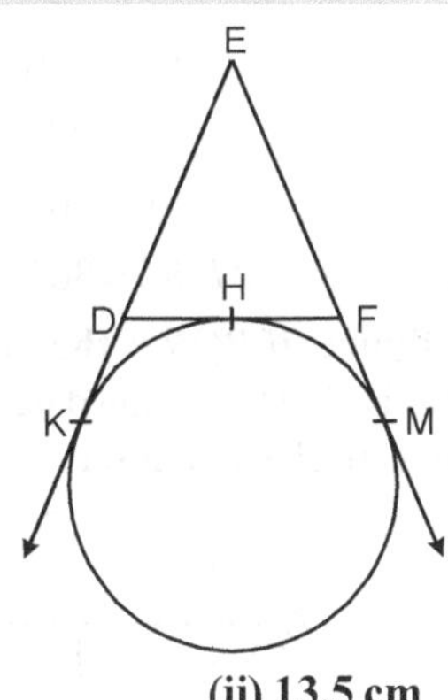

(i) 18 cm **(ii) 13.5 cm**
(iii) 12 cm **(iv) 9 cm**

Solution : Given,

$EK = 9$ cm

Now, $EK = EM = 9$ cm [tangents from E]

Let $DK = x$ cm and $FM = y$ cm

Now, $DH = DK = x$ cm [tangents from D]

and $FM = FH = y$ cm [tangents from F]

Thus, $DF = DH + HF = (x + y)$ cm

Also $ED = (9 - x)$ cm

and $EF = (9 - y)$ cm

So perimeter of ΔEDF

$= ED + DF + EF$

$= (9 - x) + (x + y) + (9 - y)$ cm

$= 18$ cm

So the correct option is (i).

Q. 23. In the given figure, *DE* and *DF* are tangents from the external point *D* to a circle with centre *A*. If *DE* = 5 cm and *DE* is perpendicular to *DF*, then the radius of the circle is :★

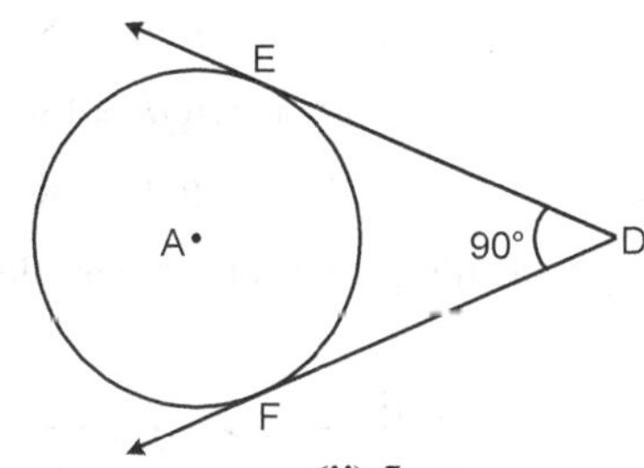

(i) 3 cm **(ii) 5 cm**
(iii) 4 cm **(iv) 6 cm**

Solution : Given,

$DE = 5$ cm

Now, $\angle AED = 90°$ [AE perpendicular to ED]

$\angle AFD = 90°$ [AF perpendicular to FD]

Thus, $EAF = 360° - (\angle AED + \angle AFD + \angle EDF)$

$\Rightarrow$ $EAF = 360° - (90° + 90° + 90°)$

$\Rightarrow$ $EAF = 360° - 270°$

$\Rightarrow$ $EAF = 90°$

Hence, $EDFA$ is a right-angled quadrilateral with $FD = AE$ and $AF = ED$

[opposite sides of a right-angled quadrilateral]

But $ED = DF$ [tangents from D]

Hence, $EDFA$ is a square and $AE = AF$ are the radii of the circle.

$\therefore$ the radius of the circle is 5 cm

[as $FD = AE$ and $AF = ED = 5$ cm]

So the correct option is (ii).

Q. 24. In the given figure, a circle with centre *O* is inscribed in a quadrilateral *ABCD* such that it touches sides *BC*, *AB*, *AD* and *CD* at points *P*, *Q*, *R* and *S* respectively. If *AB* = 29 cm, *AD* = 23 cm, ∠*B* = 90° and *DS* = 5 cm, then the radius of the circle (in cm) is :★

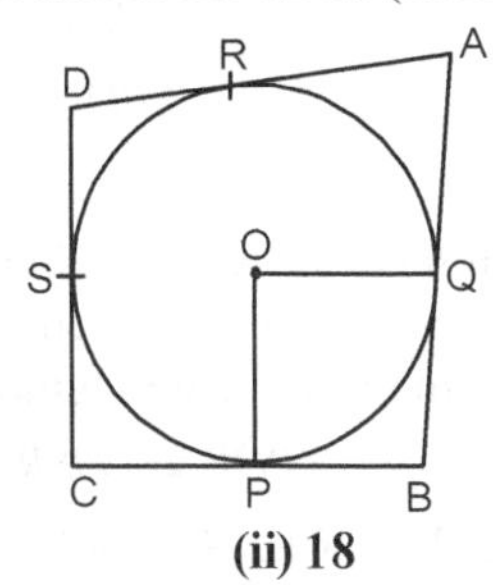

(i) 11 **(ii) 18**
(iii) 6 **(iv) 15**

Solution : Given,

$AB = 29$ cm, $AD = 23$ cm and $DS = 5$ cm

Now, $DR = DS = 5$ cm [tangents from D]

Thus, $AR = AD - DR = (23 - 5)$ cm $= 18$ cm

and $AQ = AR = 18$ cm [tangents from A]

Hence, $QB = AB - AQ = (29 - 18)$ cm $= 11$ cm

Now, in quadrilateral $POQB$,

$\angle OQB = \angle OPB = 90°$

[OQ and OP are perpendicular to AB and BC]

Now, $\angle POQ = 360° - (\angle OPB + \angle PBQ + \angle QOP)$

$= 360° - 270° = 90$

Hence, $QOPB$ is a right-angled quadrilateral.

Also $BQ = PB$ [tangents from B]

Hence, $OQ = OP = 11$ cm

[opposite sides of PB and OP respectively in right-angled quadrilaterals]

So the correct option is (i).

25. In the following figure, tangents *PQ* and *PR* are draw to a circle such that ∠*RPQ* = 50°.

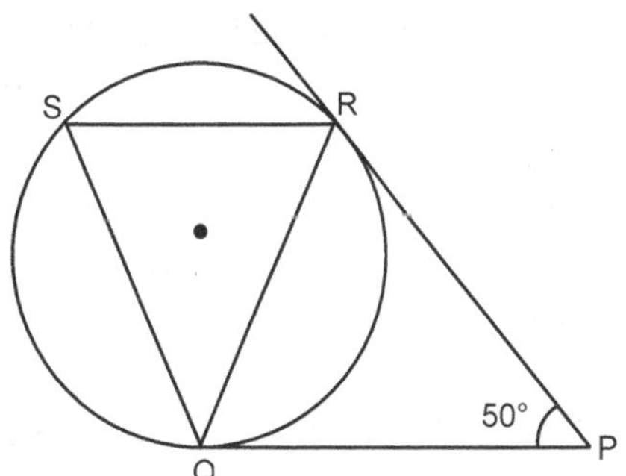

★ are board exam questions from previous years

(a) 50° **(b) 60°**

(c) 90° **(d) 45°**

Solution. (a)

Explanation : Since $PQ = PR$

Therefore $\angle PQR = \angle QRP$

Therefore in ΔPQR, we have

$$\angle PQR + \angle QRP + \angle QPR = 180^\circ$$

$$2\angle PQR + 50^\circ = 180^\circ$$

$$\angle PQR = 65^\circ$$

Now $SR \parallel QP$

So, $\angle SRQ = \angle RQP = 65^\circ$ (Alternate angles)

According to the Alternate Segment Theorem which states that the angle beteween chord and tangent is equal to the angle in the alternate segment, we have

$$\angle PQR = \angle QSR = 65^\circ$$

$$\angle Q + \angle R + \angle S = 180^\circ$$

$$\angle Q = 180^\circ - 65^\circ + 65^\circ$$

$$\angle Q = 50^\circ.$$

26. In the figure, *AB* is a diameter and *AC* is a chord of a circle centered at *O*. If the tangent at *C* intersects *AB* produced at *D* such that *BC* = *BD*, then $\angle BAC$ =

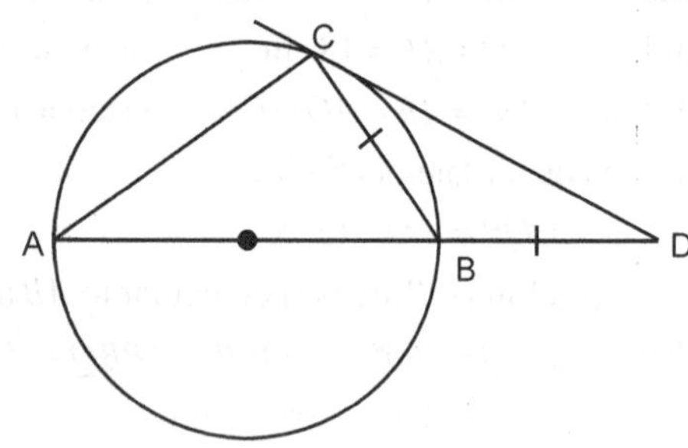

(a) 60° **(b) 30°**

(c) 90° **(d) 45°**

Solution. (b)

Explanation : Since

$BC = BD$ (given)

$\therefore$ $\angle BCD = \angle BDC = x$ (say)

(angles opposite to equal side of a Δ)

Now, $\angle ABC = \angle BCD + \angle BDC$

(exterior angle is equal to sum of opposite interior angles)

$\Rightarrow$ $\angle ABC = x + x = 2x$

Also, $\angle BAC = \angle BCD = x$

(alternate segment theorem)

And, $\angle ACB = 90^\circ$

(angle in a semicircle)

Now, $\angle ABC + \angle BAC + \angle ACB = 180^\circ$

(sum of angles of a Δ)

$\Rightarrow$ $2x + x + 90^\circ = 180^\circ$

$\Rightarrow$ $3x = 90^\circ$

$\Rightarrow$ $x = 30^\circ$

Hence $\angle BAC = 30^\circ.$

27. In figure, if *PQR* is the tangent to a circle at *Q* whose centre is *O*, *AB* is a chord parallel to *PR* and $\angle BQR$ = 50°, then $\angle AQB$ is equal to:

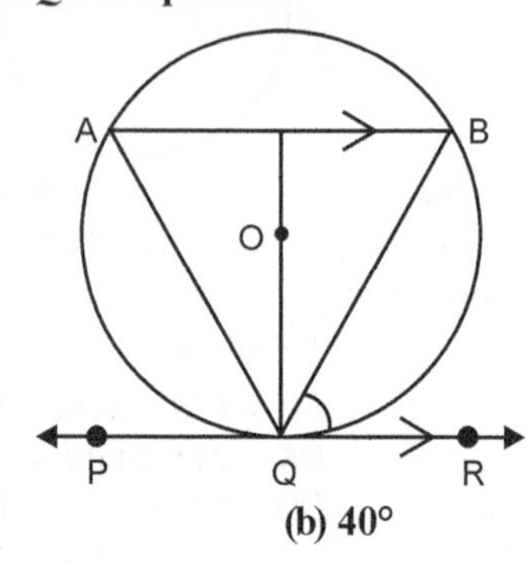

(a) 20° **(b) 40°**

(c) 35° **(d) 45°**

Solution. (b)

Explaination : $AB \parallel PR$

$\therefore$ $\angle ABQ = \angle BQR$

[Alternate interior angles]

$\Rightarrow$ $\angle ABQ = 50^\circ$

Also, $\angle BQR = \angle BAQ$

[Angles in alternate segment]

$\Rightarrow$ $\angle BAQ = 50^\circ$

In ΔAQB,

$$\angle BAQ + \angle ABQ + \angle AQB = 180^\circ$$

$\Rightarrow$ $50^\circ + 50^\circ + \angle AQB = 180^\circ$

$\Rightarrow$ $\angle AQB = 180^\circ - 100^\circ = 80^\circ.$

28. In the diagram shown below, the value of $\angle QPS$ is:

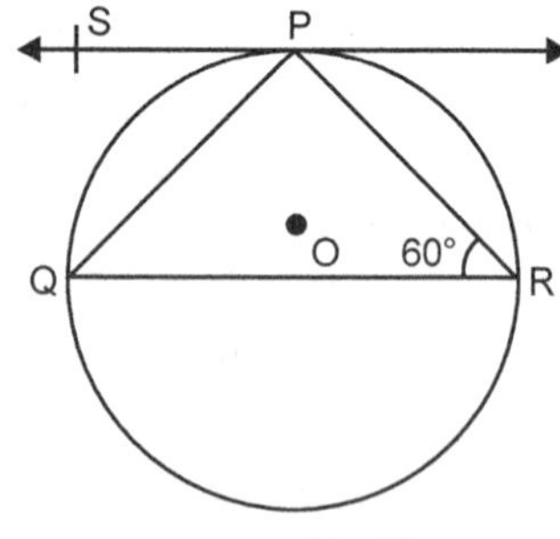

(a) 50° **(b) 45°**

(c) 60° **(d) 30°**

Solution. (c)

Explaination : By alternate segment theorem,

$$\angle QPS = \angle QRP = 60^\circ$$

$\therefore$ $\angle QPS = 60^\circ.$

Chapter 10. Introduction to Trigonometry

Q. 1. $\dfrac{2\tan 30°}{1+\tan^2 30°}$ is equal to : **[NCERT]**

(i) sin 60° **(ii) cos 60°**
(iii) tan 60° **(iv) sin 30°**

Solution : Consider,

$$\frac{2\tan 30°}{1+\tan^2 30°} = \frac{2}{\sqrt{3}\left(1+\frac{1}{3}\right)}$$

$$= \frac{2\times 3}{\sqrt{3}\times 4}$$

$$= \frac{\sqrt{3}}{2}$$

$$= \sin 60°$$

So the correct option is (i).

Q. 2. $\dfrac{1-\tan^2 45°}{1+\tan^2 45°}$ is equal to : **[NCERT]**

(i) tan 90° **(ii) 1**
(iii) sin 45° **(iv) sin 0°**

Solution : Given,

$$\frac{1-\tan^2 45°}{1+\tan^2 45°} = \frac{1-1}{1+1} = 0 = \sin 0°$$

So the correct option is (iv).

Q. 3. $\sin 2A = 2\sin A$ is true when $A =$ **[NCERT]**

(i) 0° **(ii) 30°**
(iii) 45° **(iv) 60°**

Solution : $\sin 2A = 2\sin A$

$\Rightarrow$ $2\sin A\cos A = 2\sin A$

$\Rightarrow$ $\cos A = 1$

$\Rightarrow$ $\cos A = 1 = \cos 0°$

$\Rightarrow$ $A = 0°$

So the correct option is (i).

Q. 4. $\dfrac{2\tan 30°}{1-\tan^2 30°}$ is equal to : **[NCERT]**

(i) cos 60° **(ii) sin 60°**
(iii) tan 60° **(iv) sin 30°**

Solution : Given,

$$\frac{2\tan 30°}{1-\tan^2 30°} = \frac{2}{\sqrt{3}\left(1-\frac{1}{3}\right)}$$

$$= \frac{2\times 3}{\sqrt{3}\times 2} = \sqrt{3} = \tan 60°$$

So the correct option is (iii).

Q. 5. In the given figure, if $AD = 4$ cm, $BD = 3$ cm and $CB = 12$ cm, find cot θ. **[CBSE 2008]**

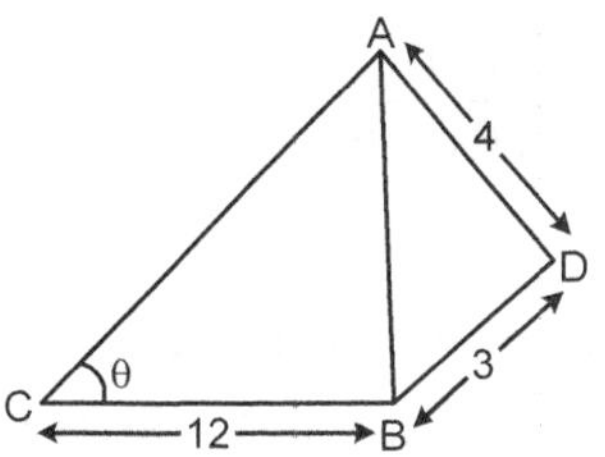

(i) $\dfrac{12}{5}$ **(ii) $\dfrac{5}{12}$**
(iii) $\dfrac{13}{12}$ **(iv) $\dfrac{12}{13}$**

Solution : Given, $AD = 4$ cm, $BD = 3$ cm and $CB = 12$ cm

Thus, $AB^2 = AD^2 + BD^2$ [Pythagoras Theorem]

$\Rightarrow$ $AB^2 = (4)^2 + (3)^2 = 16 + 9 = 25 = 5^2$

$\Rightarrow$ $AB = 5$ cm

Thus, $\cot\theta = \dfrac{\text{Base}}{\text{Perpendicular}} = \dfrac{12}{5}$

So the correct option is (i).

Q. 6. $9\sec^2 A - 9\tan^2 A$ is equal to : **[NCERT]**

(i) 1 **(ii) 9**
(iii) 8 **(d) 0**

Solution : Given,

$9\sec^2 A - 9\tan^2 A$

$$= 9(\sec^2 A - \tan^2 A)$$
$$= 9(1)$$
$$= 9$$

So the correct option is (ii).

Q. 7. $(1 + \tan\theta + \sec\theta)(1 + \cot\theta - \text{cosec}\,\theta) =$ **[NCERT]**

(i) 0 **(ii) 1**
(iii) 2 **(iv) – 1**

Solution : Given,

$(1 + \tan\theta + \sec\theta)(1 + \cot\theta - \text{cosec}\,\theta)$

$= [1 + \tan\theta + \sec\theta + \cot\theta + \tan\theta\cot\theta + \sec\theta\cot\theta - \text{cosec}\,\theta - \tan\theta\,\text{cosec}\,\theta - \sec\theta\,\text{cosec}\,\theta]$

$= [1 + \tan\theta + \sec\theta + \cot\theta + 1 + \text{cosec}\,\theta - \text{cosec}\,\theta - \sec\theta - \sec\theta\,\text{cosec}\,\theta]$

$= [2 + \tan\theta + \cot\theta - \sec\theta\,\text{cosec}\,\theta]$

$$= 2 + \frac{\sin\theta}{\cos\theta} + \frac{\cos\theta}{\sin\theta} - \frac{1}{\sin\theta\cos\theta}$$

$$= 2 + \frac{\sin^2\theta + \cos^2\theta - 1}{\sin\theta\cos\theta}$$

$$= 2 + \frac{1-1}{\sin\theta\cos\theta}$$

$$= 2$$

So the correct option is (iii).

Q. 8. (sec *A* + tan *A*) (1 – sin *A*) = **[NCERT]**

(i) sec *A* **(ii) sin *A***

(iii) cosec *A* **(iv) cos *A***

Solution : Given,

(sec *A* + tan *A*) (1 – sin *A*)

$$= \left(\frac{1}{\cos A} + \frac{\sin A}{\cos A}\right)(1 - \sin A)$$

$$= \left(\frac{1+\sin A}{\cos A}\right)(1 - \sin A)$$

$$= \left(\frac{1-\sin^2 A}{\cos A}\right)$$

$$= \frac{\cos^2 A}{\cos A}$$

$$= \cos A$$

So the correct option is (iv).

Q. 9. $\frac{1+\tan^2 A}{1+\cot^2 A}$ equal to : [NCERT]

(i) $\sec^2 A$ **(ii) – 1**

(iii) $\cot^2 A$ **(iv) $\tan^2 A$**

Solution : Given,

$$\frac{1+\tan^2 A}{1+\cot^2 A} = \frac{(\cos^2 A + \sin^2 A)\sin^2 A}{(\cos^2 A + \sin^2 A)\cos^2 A}$$

$$= \tan^2 A$$

So the correct option is (iv).

Chapter 11. Applications of Trigonometry

Q. 1. If the height of a vertical pole is $\sqrt{3}$ times the length of its shadow on the ground, then the angle of elevation of the Sun at that time is :★

(i) 30° **(ii) 60°**

(iii) 45° **(iv) 75°**

★ are board exam questions from previous years

Solution : Let the length of the shadow be *x* ft.

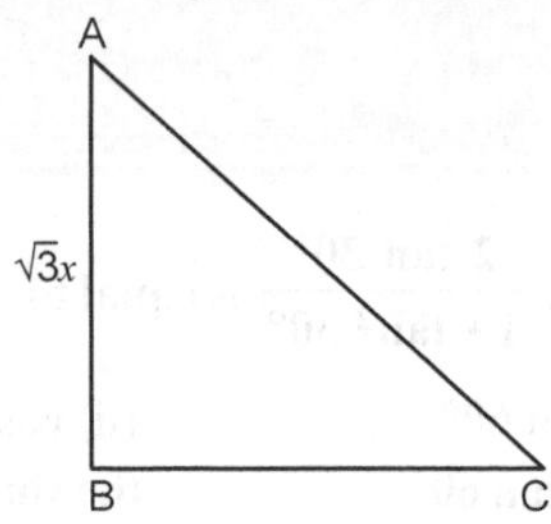

Thus, the height of the vertical pole is $\sqrt{3}x$ ft.

$$\tan\theta = \frac{\text{Perpendicular}}{\text{Base}}$$

$$\tan\theta = \frac{\sqrt{3}x}{x} = \sqrt{3} = \tan 60^\circ$$

or $\theta = 60^\circ$

Thus, the angle of elevation is 60°.

So the correct option is (ii).

Q. 2. From the top of a cliff 20 m high, the angle of elevation of the top of a tower is found to be equal to the angle of depression of the foot of the tower. The height of the tower is :★

(i) 20 m **(ii) 40 m**

(iii) 60 m **(iv) 80 m**

Solution : Let the angle of elevation = angle of depression = x°

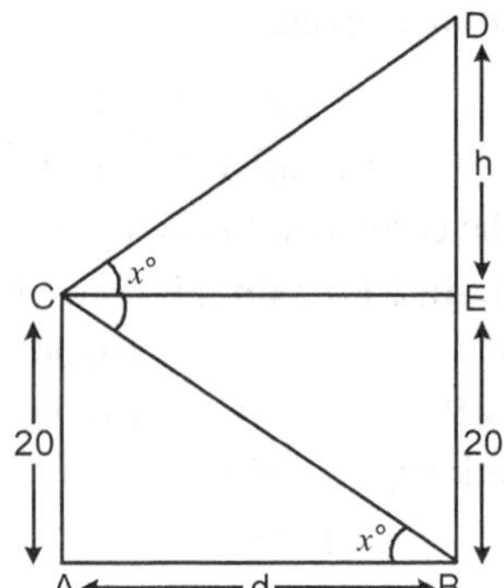

Also, let the distance between the tower (*BD*) and the cliff (*AC*) be *d* m.

Thus, in ΔABC, $\tan x = \frac{20}{d}$ ...(i)

Also, in ΔCDE, $\tan x = \frac{h}{d}$ ...(ii)

From equation (i) and (ii)

$$\frac{h}{d} = \frac{20}{d}$$

$\Rightarrow$ $h = 20$ m

$\Rightarrow$ Height of the tower is 20 + 20 = 40 m.

So the correct option is (ii).

Q. 3. A ladder makes an angle of 60° with the ground when placed against a wall. If the foot of the ladder is 2 m away from the wall, then the length of the ladder, in metres is :★

(i) $\frac{2}{\sqrt{3}}$ **(ii)** $2\sqrt{3}$

(iii) $2\sqrt{2}$ **(iv) 4**

Solution : Given, the distance of the ladder AC from the base of the wall AB is 2 m.

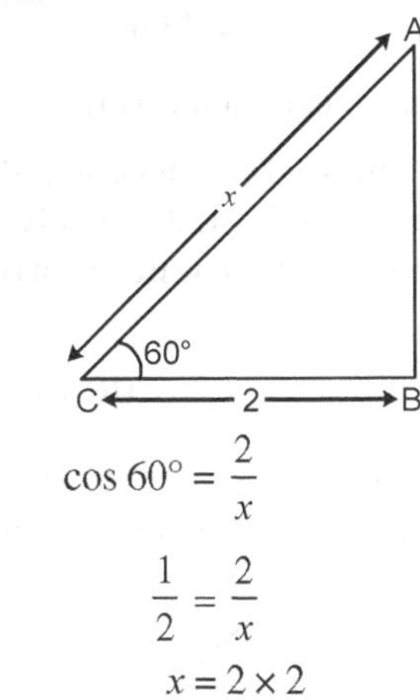

So, $\cos 60° = \frac{2}{x}$

$$\frac{1}{2} = \frac{2}{x}$$

$\Rightarrow$ $x = 2 \times 2$

The height of the ladder = 4 m

So the correct option is (iv).

Q. 4. The angle of depression of a car standing on the ground from the top of a 75 m high tower is 30°. The distance of the car from the base of the tower (in m) is :★

(i) $25\sqrt{3}$ **(ii)** $50\sqrt{3}$

(iii) $75\sqrt{3}$ **(iv) 150**

Solution : Let the distance of the car from the base of the tower be x m.

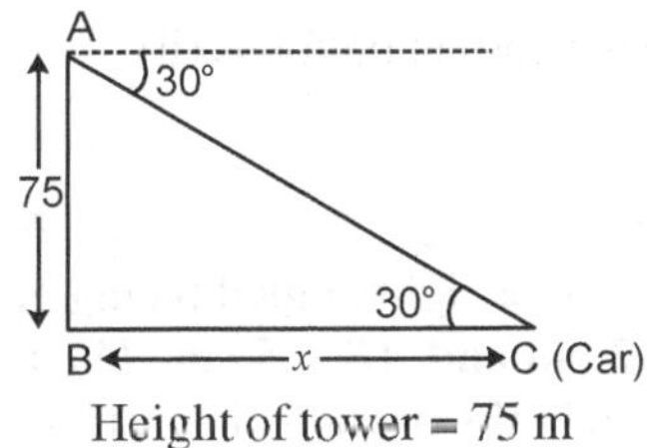

Height of tower = 75 m

$$\tan 30° = \frac{\text{Perpendicular}}{\text{Base}}$$

$\Rightarrow$ $\frac{1}{\sqrt{3}} = \frac{75}{x}$

$\Rightarrow$ $x = 75\sqrt{3}$

Thus, distance of the parked car from the base of the tower is $25\sqrt{3}$ and the correct option is (iii).

Q. 5. The angle of depression of a car parked on the road from the top of a 150 m high tower is 30°. The distance of the car from the tower (in metres) is :★

(i) $50\sqrt{3}$ **(ii)** $150\sqrt{3}$

(iii) $150\sqrt{2}$ **(iv) 75**

Solution : Let the distance of the parked car from the base of the tower be x m.

★ are board exam questions from previous years

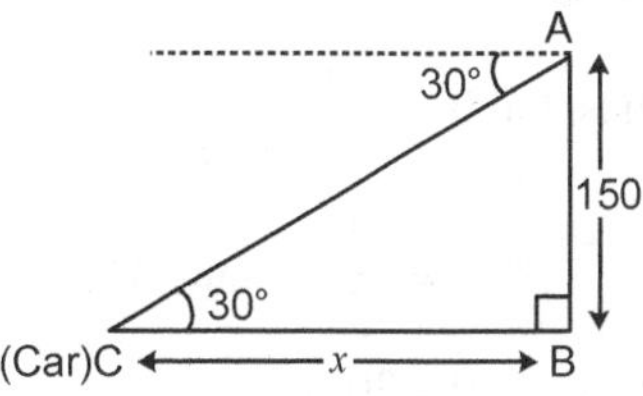

Height of tower = 150 m

$$\tan 30° = \frac{\text{Perpendicular}}{\text{Base}}$$

$\Rightarrow$ $\frac{1}{\sqrt{3}} = \frac{150}{x}$

$\Rightarrow$ $x = 150\sqrt{3}$

Thus, distance of the parked car from the base of the tower is $150\sqrt{3}$ and the correct option is (ii).

Q. 6. A kite is flying at a height of 30 m from the ground. The length of a string from the kite to the ground is 60 m. Assuming that there is no slack in the string, the angle of elevation of the kite at the ground is :★

(i) 45° **(ii) 30°**

(iii) 60° **(iv) 90°**

Solution : Given, perpendicular height = 30 m and hypotenuse = 60 m.

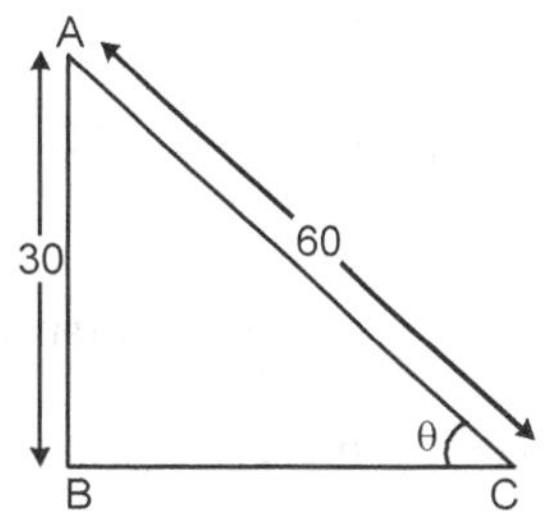

So, $\sin\theta = \frac{\text{Perpendicular}}{\text{Hypotenuse}} = \frac{30}{60} = \frac{1}{2} = \sin 30°$

Thus, $\theta = 30°$

So the correct option is (ii).

Q. 7. The length of shadow of a tower on the plane ground is $\sqrt{3}$ times the height of the tower. The angle of elevation of sun is :

(i) 45° **(ii) 30°**

(iii) 60° **(iv) 90°**

Solution : Let the height of the tower be h m.

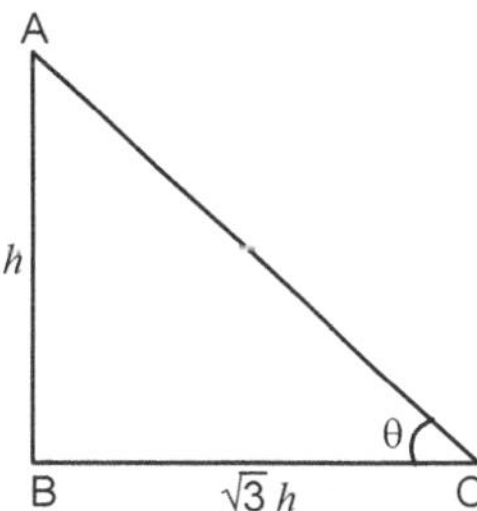

So, the length of the shadow will be $\sqrt{3}h$ m.

Thus, $\tan\theta = \frac{\text{Perpendicular}}{\text{Base}} = \frac{h}{\sqrt{3}h}$

$= \frac{1}{\sqrt{3}} = \tan 30^\circ$

Thus, $\theta = 30^\circ$

So, the correct option is (ii).

Q. 8. At some point of time in the day, the length of the shadow of a tower is equal to its height. Then the sun's altitude at that time is :★

(i) 30° **(ii) 60°**

(iii) 90° **(iv) 45°**

Solution : Let the height of the tower be h m and the length of the shadow be l m.

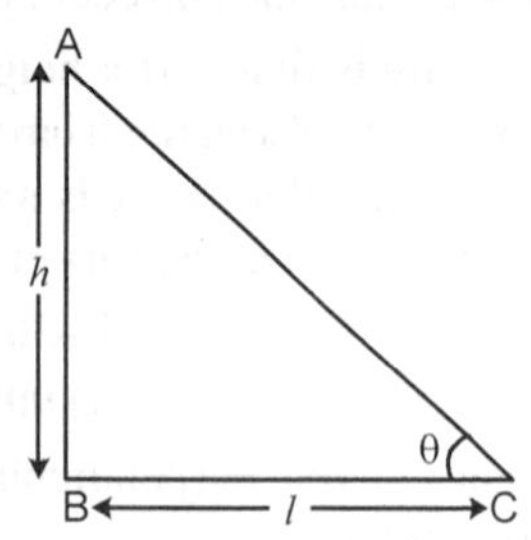

Now, $h = l$

Thus, $\tan\theta = \frac{\text{Perpendicular}}{\text{Base}}$

$= \frac{h}{l} = 1 = \tan 45^\circ$

Hence, $\theta = 45^\circ$

So the correct option is (iv).

Q. 9. A tower stands vertically on the ground. From a point on the ground which is 25 m away from the foot of the tower, the angle of elevation of the top of the tower is found to be 45°. Then the height (in metres) of the tower is :★

(i) $25\sqrt{2}$ **(ii) $25\sqrt{3}$**

(iii) 25 **(iv) 12·5**

Solution : Let the height of the tower be H m.

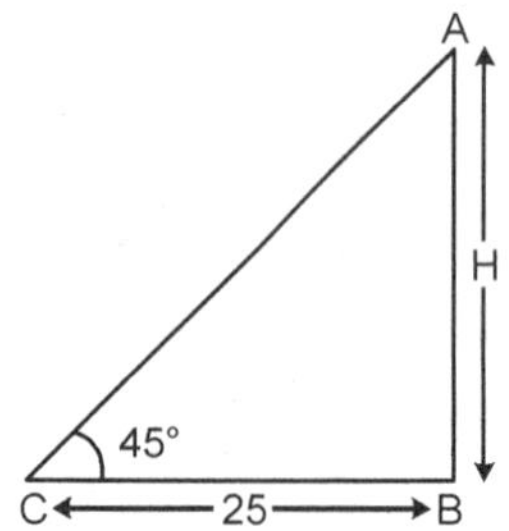

Thus, $\tan 45^\circ = \frac{\text{Perpendicular}}{\text{Base}}$

or $\frac{H}{25} = 1$

Hence, $H = 25$ m

So the correct option is (iii).

Q. 10. The angle of elevation of the top of a tower from a point on the ground which is 30 m away from the foot of the tower is 45°. The height of the tower (in metres) is :★

(i) 15 **(ii) 30**

(iii) $30\sqrt{3}$ **(iv) $10\sqrt{3}$**

Solution : Let the height of the tower be H.

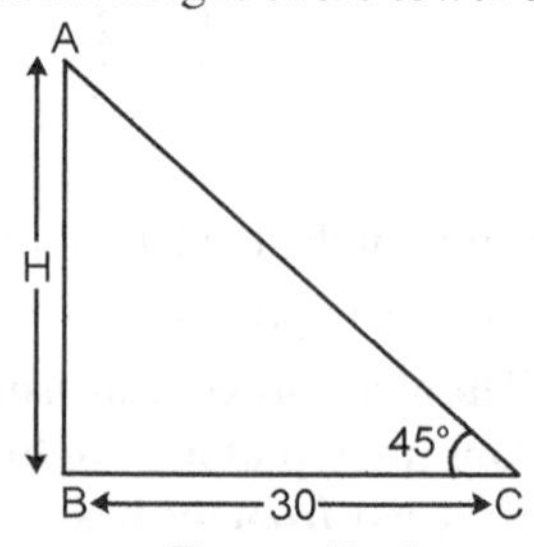

Thus, $\tan 45^\circ = \frac{\text{Perpendicular}}{\text{Base}}$

or $\frac{H}{30} = 1$

Hence, $H = 30$ m

So the correct option is (ii).

Chapter 12. Areas related to Circles

Q. 1. In a right angled triangle, right-angled at B, BC = 12 cm and AB = 5 cm. The radius of the circle inscribed in the triangle (in cm) is :★

(i) 4 **(ii) 3**

(iii) 2 **(iv) 1**

Solution : Given, BC = 12 cm and AB = 5 cm

Let $EO = DO = FO = r$ cm

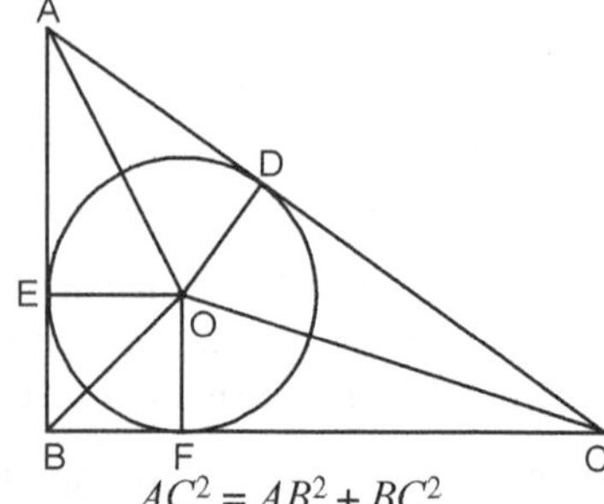

$AC^2 = AB^2 + BC^2$

[By Pythagoras Theorem]

$\Rightarrow AC^2 = (5)^2 + (12)^2$

$\Rightarrow AC^2 = 25 + 144$

$\Rightarrow AC^2 = 169$

★ are board exam questions from previous years

$\Rightarrow \quad AC^2 = (13)^2$

$\Rightarrow \quad AC = 13$

Now, Area of $\Delta ABC = \frac{1}{2} \times 12 \times 5 = 30 \text{ cm}^2$

But Area of ΔABC = Area of ΔBOC + Area of ΔCOA + Area of ΔBOA

$$\Rightarrow \quad 30 = \frac{1}{2} \times 12 \times r + \frac{1}{2} \times 13 \times r + \frac{1}{2} \times 5 \times r$$

$$\Rightarrow \quad \frac{r}{2}[12 + 13 + 5] = 30$$

$$\Rightarrow \quad \frac{r}{2}[30] = 30$$

$$\Rightarrow \quad r = 2 \text{ cm}$$

So the correct option is (iii).

Q. 2. If the difference between the circumference and the radius of a circle is 37 cm, then using $\pi = \frac{22}{7}$, the radius of the circle (in cm) is :★

(i) 154 **(ii) 44**

(iii) 14 **(iv) 7**

Solution : Let the radius be r cm.

Thus, circumference = $2\pi r$ cm

Now, $2\pi r - r = 37$

$$\Rightarrow \quad r\left(\frac{44}{7} - 1\right) = 37$$

$$\Rightarrow \quad r\left(\frac{44 - 7}{7}\right) = 37$$

$$\Rightarrow \quad r = \frac{7 \times 37}{37} = 7 \text{ cm}$$

So the correct option is (iv).

Q. 3. The circumference of a circle is 22 cm. The area of its quadrant (in cm²) is :★

(i) $\frac{77}{2}$ **(ii) $\frac{77}{4}$**

(iii) $\frac{77}{8}$ **(iv) $\frac{77}{16}$**

Solution : Given, circumference = $2\pi r$ = 22 cm

or $\quad r = \frac{11}{\pi}$ cm

$\therefore$ Area of the quadrant

$$= \frac{\pi r^2}{4} = \frac{\pi}{4} \times \frac{11}{\pi} \times \frac{11}{\pi}$$

$$= \frac{11 \times 11 \times 7}{4 \times 22}$$

$$= \frac{77}{8} \text{ cm}^2$$

So the correct option is (iii).

Q. 4. If the area of a circle is equal to sum of the areas of two circles of diameters 10 cm and 24 cm, then the diameter of the larger circle (in cm) is :★

(i) 34 **(ii) 26**

(iii) 17 **(iv) 14**

Solution : Given, the diameters of the two smaller circles = 10 cm and 24 cm.

Thus, radii of the two smaller circles = 5 cm and 12 cm

Let the radius of the larger cricle be R'.

Now, Area of the larger circle

= Sum of the area of the two smaller circles

$= \pi(R^2 + r^2)$

$= \pi[(12)^2 + (5)^2]$

$= \pi[144 + 25]$

$= \pi[169]$

$= \pi(13)^2$

$$\Rightarrow \quad \pi(R')^2 = \pi(13)^2$$

$$\Rightarrow \quad (R')^2 = (13)^2$$

$$\Rightarrow \quad R' = 13 \text{ cm}$$

Hence, diameter of the larger circle = 26 cm

So the correct option is (ii).

Q. 5. The perimeter of a square circumscribing a circle of radius a cm is :★

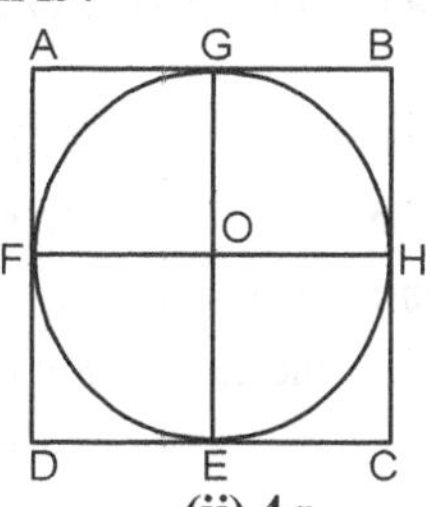

(i) $8a$ **(ii) $4a$**

(iii) $2a$ **(iv) $16a$**

Solution :

Let $ABCD$ be the square that inscribes a circle with centre O, touching it at the points E, F, G and H.

Given, the radius of the circle = a cm

Thus, $\quad OE = OF = OG = OH = a$ cm

Hence, $\quad OG + OE = AD = BC$ and $OF + OH = AB = DC$

So each side of the square equals $2a$ cm

Hence, perimeter of the square

$= 4(2a) = 8a$ cm

So the correct option is (i).

Q. 6. The radii of two circles are 4 cm and 3 cm respectively. The diameter of the circle having an area equal to the sum of the two circles (in cm) is :★

(i) 5 **(ii) 7**

(iii) 10 **(iv) 14**

★ are board exam questions from previous years

Solution :

Given, $r_1 = 4$ cm and $r_2 = 3$ cm

Let the radius of the larger circle = R

Thus, $\pi R^2 = \pi[(r_1)^2 + (r_2)^2]$

$\Rightarrow$ $R^2 = [(4)^2 + (3)^2]$

$\Rightarrow$ $R^2 = [16 + 9]$

$\Rightarrow$ $R^2 = 25$

$\Rightarrow$ $R = 5$ cm

[as radius cannot be negative]

Thus, the diameter of the larger circle = 10 cm

So the correct option is (iii).

Q. 7. If the area of a circle is numerically equal to twice its circumference then the diameter of the circle is :★

(i) 4 units **(ii) π units**

(iii) 8 units **(iv) 2 units**

Solution : Given, the area of a circle is numerically equal to twice its circumference.

Let the radius of the circle be r.

Thus, $\pi r^2 = 2(2\pi r)$

or $r = 4$ units

So the correct option is (i).

Q. 8. If π is taken as $\frac{22}{7}$, the distance (in meters) covered by a wheel of diameter 35 cm, in one revolution is :★

(i) 2.2 **(ii) 1.1**

(iii) 9.625 **(iv) 96.25**

Solution : Given, diameter = 35 cm

Now, one revolution = circumference

$= \pi d$

$= \frac{22 \times 35}{7}$

$= 110$ cm

$= 1.1$ m

So the correct option is (ii).

Chapter 13. Surface Areas and Volumes

Q. 1. A rectangular sheet of paper 40 cm × 22 cm is rolled to form a hollow cylinder of height 40 cm. The radius of the cylinder (in cm) is :★

(i) 3.5 **(ii) 7**

(iii) $\frac{80}{7}$ **(iv) 5**

Solution : Given, Area of the sheet of paper

$= (40 \times 22)$ cm^2

$= 880$ cm^2

Height of the cylinder = 40 cm

Let r be the readius of the cylinder

Now, Curved Surface Area of the cylinder = $2\pi rh$

Thus, $2\left(\frac{22}{7}\right)40r = 40 \times 22$

$\Rightarrow$ $2\left(\frac{1}{7}\right)r = 1 \times 1$

$\Rightarrow$ $r = \frac{7}{2}$ cm $= 3.5$ cm

So the correct option is (i).

Q. 2. The number of solid spheres each of diameter 6 cm that can be made by melting a solid metal cylinder of height 45 cm and diameter 4 cm are :★

(i) 3 **(ii) 5**

(iii) 4 **(iv) 6**

Solution : Given, height of cylinder = 45 cm

Diameter of cylinder = 4 cm

Diameter of each sphere = 6 cm

Thus, radius of cylinder = 2 cm

And radius of each sphere = 3 cm

$\therefore$ Number of spheres = $\frac{\text{Volume of cylinder}}{\text{Volume of each sphere}}$

$= \frac{\pi(2)^2 45}{\frac{4}{3}\pi(3)^3}$

$= \frac{45}{(3)^2}$

$= 5$

So the correct option is (ii).

Q. 3. The surface area of two spheres are in the ratio 16 : 9. The ratios of their volumes is :★

(i) 64 : 27 **(ii) 16 : 9**

(iii) 4 : 3 **(iv) $16^3 : 9^3$**

Solution : Let the radius of the larger sphere be R and the smaller be r.

Given, $4\pi R^2 : 4\pi r^2 = 16 : 9$

$\Rightarrow$ $R^2 : r^2 = 4^2 : 3^2$

$\Rightarrow$ $R : r = 4 : 3$

Thus, $\frac{4\pi}{3}R^3 : \frac{4\pi}{3}r^3 = 4^3 : 3^3$

or $\frac{4\pi}{3}R^3 : \frac{4\pi}{3}r^3 = 64 : 27$

So the correct option is (i).

Q. 4. A solid right-circular cone is cut into two parts at the middle of its height by a plane parallel to its base. The ratio of the volume of the smaller cone to the whole cone is :★

(i) 1 : 2 **(ii) 1 : 4**

(iii) 1 : 6 **(iv) 1 : 8**

★ are board exam questions from previous years

Solution : Let the height of the bigger cone be $2x$ cm.

Thus, height of the smaller cone = x cm

Also let the radius of the larger cone be $2r$ cm.

Thus, radius of the smaller cone

$= r$ cm [mid-point theorem]

Now, volume of the large cone

$$= \frac{1}{3}\pi(2r)^2(2x)$$

$$= \frac{8\pi}{3}r^2x$$

and Volume of the smaller cone

$$= \frac{1}{3}\pi(r)^2(x)$$

$$= \frac{\pi}{3}r^2x$$

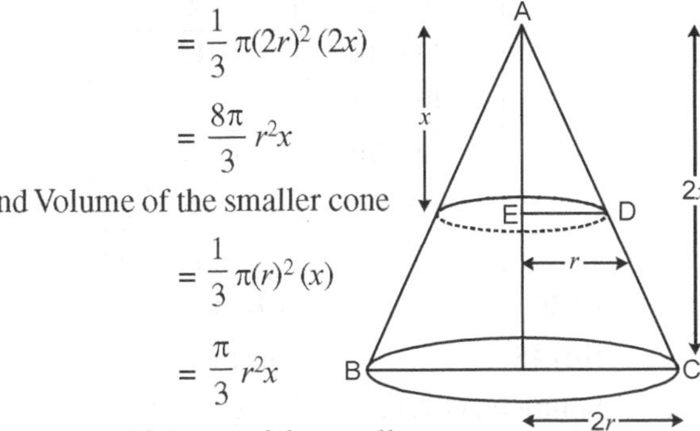

Thus, $\frac{\text{Volume of the smaller cone}}{\text{Volume of the larger cone}}$

$$= \frac{\frac{\pi}{3}r^2x}{\frac{8\pi}{3}r^2x} = \frac{1}{8}$$

So the correct option is (iv).

Q. 5. If the radius of the base of a right-circular cylinder is halved keeping the height same then the ratio of the volume of the cylinder thus obtained to the volume of the original cylinder is :★

(i) 1 : 2 **(ii) 2 : 1**

(iii) 1 : 4 **(iv) 4 : 1**

Solution : Let the original radius of the base be $2r$ cm and the height be h.

Thus, the new radius of the base = r cm

Now, Original volume = $\pi(2r)^2h$

New volume = $\pi(r)^2h$

Thus, $\frac{\text{Original volume}}{\text{New volume}} = \frac{\pi 4r^2h}{\pi r^2h}$

$$= \frac{4}{1}$$

$\Rightarrow$ $\frac{\text{New volume}}{\text{Original volume}} = \frac{1}{4}$

So the correct option is (iii).

Q. 6. The radius of the largest circular cone that can be cut out from a cube of edge 4.2 cm is :★

(i) 4.2 **(ii) 2.1**

(iii) 8.4 **(iv) 1.05**

Solution : Given, the edge of cube = 4.2 cm

Now, the largest circular cone will have diameter equal to the measure of the edge.

★ are board exam questions from previous years

Thus, diameter = 4.2 cm

Hence, radius = 2.1 cm

So the correct option is (ii).

Q. 7. A sphere of diameter 18 cm is dropped into a cylindrical vessel of diameter 36 cm partly filled with water. If the sphere is completely submerged then the water level rises (in cm) by :★

(i) 3 **(ii) 4**

(iii) 5 **(iv) 6**

Solution : Given,

Diameter of sphere = 18 cm

Diameter of cylinder = 36 cm

Let the increase in height of water in the cylinder = h cm

Radius of sphere = 9 cm

Radius of cylinder = 18 cm

Thus, volume of sphere = $\frac{4}{3}\pi(9)^3$ cm^3

$$= \frac{4}{3}\pi(81)(9) \text{ cm}^3$$

$$= (81)(12)\pi \text{ cm}^3$$

Thus, increase in volume of water level in the cylinder = 243π cm^3

Hence, $\pi(18)^2h = (81)(12)\pi$

$\Rightarrow$ $4h = 12$

$\Rightarrow$ $h = 3$ cm

So the correct option is (i).

Q. 8. A solid is hemispherical at the bottom and conical (of same radius) above it. If the surface area of the two are equal then the ratio of the radius and the slant height of the conical part is :★

(i) 2 : 1 **(ii) 1 : 2**

(iii) 1 : 4 **(iv) 4 : 1**

Solution : Given, radius of hemispehre = radius of cone

Also, curved surface area of hemisphere = Curved surface area of cone

$\Rightarrow$ $2\pi r^2 = \pi rl$ [l = slant height of the cone]

$\Rightarrow$ $2r = l$

$\Rightarrow$ $r : l = 1 : 2$

So the correct option is (ii).

Chapter 14. Statistics

Q. 1. Which measure of central tendency is given by the x-coordinate of the point of intersection of the 'more than ogive' and 'less than ogive' ?★

(i) Mean **(ii) Median**

(iii) Mode **(iv) All three**

Solution : The correct option is (ii).

Q. 2. Find the Class marks of class 10 – 25 and 35 – 55.★

(i) 1.75 and 45 **(ii) 17.5 and 4.5**

(iii) 1.75 and 4.5 **(iv) 17.5 and 45**

Solution : We derive the class mark x_i by the following formula :

$$x_i = \frac{1}{2}(\text{lower limit} + \text{upper limit})$$

Thus, $$x_i = \frac{1}{2}(10+25)$$

$$= \frac{1}{2}(35) = 17.5$$

and $$x_{ii} = \frac{1}{2}(\text{lower limit} + \text{upper limit})$$

Thus, $$x_{ii} = \frac{1}{2}(35+55)$$

$$= \frac{1}{2}(90) = 45$$

So the correct option is (iv).

Q. 3. Write down the Median Class of the following Frequency Distribution :★

Class Interval	Frequency
0 – 10	4
10 – 20	4
20 – 30	8
30 – 40	10
40 – 50	12
50 – 60	8
60 – 70	4

(i) 20 – 30 **(ii) 30 – 40**

(iii) 40 – 50 **(iv) 50 – 60**

Solution :

Class Interval	Frequency (f_i)	Cumulative Frequency
0 – 10	4	4
10 – 20	4	8
20 – 30	8	16
30 – 40	10	26
40 – 50	12	38
50 – 60	8	46
60 – 70	4	50
	$N = \Sigma f_i = 50$	

Thus, $$\frac{N}{2} = 25$$

The cumulative frequency just above 25 is 26.

★ **are board exam questions from previous years**

Hence, the Median Class is 30 – 40.

So the correct option is (ii).

Q. 4. Calculate the value of p from the following data :

Class	Frequency
0 – 20	8
20 – 40	15
40 – 60	p
60 – 80	12
80 – 100	5
	$N = \Sigma f_i = 60$

(i) 20 **(ii) 30**

(iii) 45 **(iv) 50**

Solution : Given, $N = \Sigma f_i = 60$

$\Rightarrow \quad 8 + 15 + p + 12 + 5 = 60$

$\Rightarrow \quad 40 + p = 60$

$\Rightarrow \quad p = 20$

So the correct option is (i).

Q. 5. In an inclusive series :

(i) The lower class boundary is same as the upper class boundary of the previous class.

(ii) The upper class boundary is same as the lower class boundary of the next class.

(iii) Both the lower and upper class boundaries are the same.

(iv) The lower and upper class boundaries are contained within the class and do not intersect with either the upper boundary of the previous class or the lower boundary of the next class.

Solution : (iv) The lower and upper class boundaries are contained within the class and do not intersect with either the upper boundary of the previous class or the lower boundary of the next class.

Chapter 15. Probability

Q. 1. In a family of three children, the probability of having at least one boy is :★

(i) $\frac{7}{8}$ **(ii) $\frac{1}{8}$**

(iii) $\frac{5}{8}$ **(iv) $\frac{3}{4}$**

Solution : Total number of children = 3

All possible outcomes = *GGG, GGB, GBB, GBG, BBB, BGG, BGB, BBG*

Let the event of getting at least one boy be *E*.

Then, the favourable outcomes are

GGB, GBB, GBG, BBB, BBG, BGG, BGB

Hence, number of favourable outcomes = 7

$\therefore \quad P(E) = \frac{7}{8}$

So the correct option is (i).

Q. 2. The probability that a number selected at random from the numbers 1, 2, 3, ..., 15 is a multiple of 4 is :★

(i) $\frac{4}{15}$ (ii) $\frac{2}{15}$

(iii) $\frac{1}{5}$ (iv) $\frac{1}{3}$

Solution : Total number = 15

Let the event of getting a number multiple of 4 be E.
Then, the favourable outcomes are 4, 8, 12
Hence, number of favourable outcomes = 3

$\therefore \quad P(E) = \frac{3}{15} = \frac{1}{5}$

So the correct option is (iii).

Q. 3. If two dice are rolled together, the probability of getting an even number on both the dice is :★

(i) $\frac{1}{36}$ (ii) $\frac{1}{2}$

(iii) $\frac{1}{6}$ (iv) $\frac{1}{4}$

Solution : Total number of possible outcomes = (1, 1); (1, 2); (1, 3); (1, 4); (1, 5); (1, 6); (2, 1); (2, 2); (2, 3); (2, 4); (2, 5); (2, 6); (3, 1); (3, 2); (3, 3); (3, 4); (3, 5); (3, 6); (4, 1); (4, 2); (4, 3); (4, 4); (4, 5); (4, 6); (5, 1); (5, 2); (5, 3); (5, 4); (5, 5); (5, 6); (6, 1); (6, 2); (6, 3); (6, 4); (6, 5), (6, 6) = 36

Thus, the favourable outcomes are (2, 2), (2, 4), (2, 6), (4, 2), (4, 4), (4, 6), (6, 2), (6, 4), (6, 6)

Hence, the total number of favourable outcomes = 9

$\therefore \quad P(E) = \frac{9}{36} = \frac{1}{4}$

So the correct option is (iv).

Q. 4. A number is selected at random from numbers 1 to 30. The probability that it is a prime number is :★

(i) $\frac{2}{3}$ (ii) $\frac{1}{6}$

(iii) $\frac{1}{3}$ (iv) $\frac{11}{30}$

Solution : Total number = 30

Let the event of getting prime number be E.

Then, the favourable outcomes are 2, 3, 5, 7, 11, 13, 17, 19, 23, 29.

Hence, number of favourable outcomes = 10

$\therefore \quad P(E) = \frac{10}{30} = \frac{1}{3}$

So the correct option is (iii).

★ are board exam questions from previous years

Q. 5. Two different coins are tossed simultaneously. The probability of getting at least one head is :★

(i) $\frac{1}{4}$ (ii) $\frac{1}{8}$

(iii) $\frac{3}{4}$ (iv) $\frac{7}{8}$

Solution : Total number of possible outcomes
= $HH, HT, TH, TT = 4$

Thus, the favourable outcomes are HH, HT, TH
Hence, the total number of favourable outcomes = 3

$\therefore \quad P(E) = \frac{3}{4}$

So the correct option is (iii).

Q. 6. A bag contains cards numbered 1 to 25. A card is drawn at random from the bag. The probability that the number on the card is divisible by both 2 and 3 is :★

(i) $\frac{1}{5}$ (ii) $\frac{3}{25}$

(iii) $\frac{4}{25}$ (iv) $\frac{2}{25}$

Solution : Total number = 25

Let the event of getting a number divisible by 2 or 3 be E.

Then, the favourable outcomes are 6, 12, 18, 24
Hence, number of favourable outcomes = 4

$\therefore \quad P(E) = \frac{4}{25}$

So the correct option is (iii).

Q. 7. The probability of getting an even number when a die is thrown once is :★

(i) $\frac{1}{2}$ (ii) $\frac{1}{3}$

(iii) $\frac{1}{6}$ (iv) $\frac{5}{6}$

Solution : Total number of possible outcomes
= 1, 2, 3, 4, 5, 6

Thus, the favourable outcomes are 2, 4, 6
Hence, the total number of favourable outcomes = 3

$\therefore \quad P(E) = \frac{3}{6} = \frac{1}{2}$

So the correct option is (i).

Q. 8. A box contains 90 discs, numbered from 1 to 90. If one disc is drawn at random from the box, the probability that it bears a prime number less than 23 is :★

(i) $\frac{7}{90}$ (ii) $\frac{10}{90}$

(iii) $\frac{4}{45}$ (iv) $\frac{9}{89}$

Solution : Total number of possible outcomes = 90

Thus, the favourable outcomes are 2, 3, 5, 7, 11, 13, 17, 19

Hence, the total number of favourable outcomes = 8

$$\therefore \quad P(E) = \frac{8}{90} = \frac{4}{45}$$

So the correct option is (iii).

Q. 9. If $P(A)$ denotes the probability of an event A, then :★

(i) $P(A) < 0$ **(ii) $P(A) > 1$**

(iii) $0 \leq P(A) \leq 1$ **(iv) $-1 \leq P(A) \leq 1$**

Solution : The probability of a sure event is always 1.

The probability of an impossible event is always 0.

The sum of the probabilities of all the outcomes of elementary events is 1.

Thus, $P(A)$ cannot be less than 0 and more than 1.

So, the correct option is (iii).

Q. 10. One ticket is drawn at random from a bag containing tickets numbered 1 to 40. The probability that the selected ticket has a number that is a multiple of 7 is :★

(i) $\frac{1}{7}$ **(ii) $\frac{1}{8}$**

(iii) $\frac{1}{5}$ **(iv) $\frac{7}{40}$**

Solution : The total number of possible outcomes = 40

The favourable outcomes are 7, 14, 21, 28, 35

Hence, the number of favourable outcomes = 5

$$\text{Thus,} \quad P(E) = \frac{5}{40} = \frac{1}{8}$$

So the correct option is (ii).

Q. 11. Two dice are thrown together. The probability of getting the same number on both the dice is :★

(i) $\frac{1}{2}$ **(ii) $\frac{1}{3}$**

(iii) $\frac{1}{6}$ **(iv) $\frac{1}{12}$**

Solution : Total number of possible outcomes = (1, 1); (1, 2); (1, 3); (1, 4); (1, 5); (1, 6); (2, 1); (2, 2); (2, 3); (2, 4); (2, 5); (2, 6); (3, 1); (3, 2); (3, 3); (3, 4); (3, 5); (3, 6); (4, 1); (4, 2); (4, 3); (4, 4); (4, 5); (4, 6); (5, 1); (5, 2); (5, 3); (5, 4); (5, 5); (5, 6); (6, 1); (6, 2); (6, 3); (6, 4); (6, 5), (6, 6) = 36

Thus the favourable outcomes are (1, 1), (2, 2), (3, 3), (4, 4), (5, 5), (6, 6).

Hence, the total number of favourable outcomes = 6

$$\therefore \quad P(E) = \frac{6}{36} = \frac{1}{6}$$

So the correct option is (iii).

★ are board exam questions from previous years

Q. 12. Cards bearing numbers 2, 3, 4, ..., 11 are kept in a bag. A card is drawn at random from it. The probability of getting a card with a prime number is :★

(i) $\frac{1}{2}$ **(ii) $\frac{2}{5}$**

(iii) $\frac{3}{10}$ **(iv) $\frac{5}{9}$**

Solution : Total number of possible outcomes = 10

Thus, the favourable outcomes are 2, 3, 5, 7, 11

Hence, the total number of favourable outcomes = 5

$$\therefore \quad P(E) = \frac{5}{10} = \frac{1}{2}$$

So the correct option is (i).

Q. 13. The probability of throwing a number greater than 2 with a single die is :★

(i) $\frac{2}{3}$ **(ii) $\frac{5}{6}$**

(iii) $\frac{1}{3}$ **(iv) $\frac{2}{5}$**

Solution : Total number of possible outcomes = 6

Thus, the favourable outcomes = 3, 4, 5, 6

Hence, the total number of favourable outcomes = 4

$$\therefore \quad P(E) = \frac{4}{6} = \frac{2}{3}$$

So the correct option is (i).

Q. 14. A card is drawn from a well-shuffled deck of 52 playing cards. The probability that the card will not be an ace is :★

(i) $\frac{1}{13}$ **(ii) $\frac{1}{4}$**

(iii) $\frac{12}{13}$ **(iv) $\frac{3}{4}$**

Solution : Total number of possible outcomes = 52

Number of aces in the pack = 4

Thus, the probability of not drawing an ace

$$\therefore \quad P(E) = \frac{52-4}{52} = \frac{12}{13}$$

So the correct option is (iii).

Q. 15. Which of the following cannot be the probability of an event :★

(i) 1.5 **(ii) $\frac{3}{5}$**

(iii) 25% **(iv) 0.3**

Solution : The sum of the probabilities of all the outcomes of elementary events is 1.

Thus, $P(A)$ cannot be less than 0 and more than 1.

So, the correct option is (i).

Q. 16. A die thrown once. What is the probability of getting a number less than 3 ?★

(i) $\frac{1}{2}$ (ii) $\frac{1}{3}$

(iii) $\frac{1}{6}$ (iv) $\frac{1}{4}$

Solution : The possible outcomes are 1, 2, 3, 4, 5, 6

The sum of the possible outcomes = 6

The favourable outcomes are 1, 2

The sum of the favourable outcomes = 2

$$\therefore \quad P(E) = \frac{2}{6} = \frac{1}{3}$$

So the correct option is (ii).

Q. 17. Cards bearing numbers 3 to 20 are placed in a bag and mixed thoroughly. A card is taken out of the bag at random. What is the probability that the number on the card taken out is an even number ?★

(i) $\frac{9}{17}$ (ii) $\frac{1}{2}$

(iii) $\frac{5}{9}$ (iv) $\frac{7}{18}$

Solution : Total number of possible outcomes = 18

The favourable outcomes are 4, 6, 8, 10, 12, 14, 16, 18, 20

Sum of the favourable outcomes = 9

$$\therefore \quad P(E) = \frac{9}{18} = \frac{1}{2}$$

So the correct option is (ii).

Q. 18. A bag contains 4 red and 6 black balls. A ball is taken out of the bag at random. What is the probability of getting a black ball ?★

(i) $\frac{2}{5}$ (ii) $\frac{3}{5}$

(iii) $\frac{1}{10}$ (iv) none of these

Solution : Total number of possible outcomes = 10

The total number of favourable outcomes = 6

$$\therefore \quad P(E) = \frac{6}{10} = \frac{3}{5}$$

So the correct option is (ii).

Q. 19. A card is drawn out from a well-shuffled deck of 52 cards. What is the probability of getting a black king ?★

(i) $\frac{1}{13}$ (ii) $\frac{1}{26}$

(iii) $\frac{2}{35}$ (iv) none of these

★ are board exam questions from previous years

Solution : Total number of possible outcomes = 52

The total number of favourable outcomes = 2

$$\therefore \quad P(E) = \frac{2}{52} = \frac{1}{26}$$

So the correct option is (ii).

Q. 20. Two friends were born in the year 2000. What is the probability that they have the same birthday ?★

(i) $\frac{1}{365}$ (ii) $\frac{1}{366}$

(iii) $\frac{2}{365}$ (iv) $\frac{1}{183}$

Solution : Since year 2000 is a leap year.

Total possible cases = 366

Favourable case = 1

$$\therefore \quad P(E) = \frac{1}{366}$$

So the correct option is (ii).

Q. 21. What is the probability that two friends have different birthdays ?★

(i) $\frac{1}{365}$ (ii) $\frac{2}{365}$

(iii) $\frac{364}{365}$ (iv) $\frac{363}{365}$

Solution : Total number of possible outcomes

$= 365 \times 365$

The total number of favourable outcomes

$= 365 \times (365 - 1) = 365 \times 364$

$$\therefore \quad P(E) = \frac{365 \times 364}{365 \times 365} = \frac{364}{365}$$

So the correct option is (iii).

Q. 22. A die is thrown once. The probability of getting a prime number is :★

(i) $\frac{2}{3}$ (ii) $\frac{1}{3}$

(iii) $\frac{1}{2}$ (iv) $\frac{1}{6}$

Solution : The maximum number of possible outcomes = 6

The favourable outcomes = 2, 3, 5

Total number of the favoruable outcomes = 3

$$\therefore \quad P(E) = \frac{3}{6} = \frac{1}{2}$$

So the correct option is (iii).

●●

Very Short Questions | Set 2 |

Chapter 1. Real Numbers

Q. 1. Has the rational number $\frac{441}{2^2 5^7 7^2}$ a terminating or non-terminating decimal representation ?

Solution : The rational number

$$\frac{441}{2^2 5^7 7^2} = \frac{7^2 \times 3^2}{2^2 5^7 \times 7^2} = \frac{3^2}{2^2 \times 5^7}$$

has a terminating decimal representation because

$(2^2 5^7 7^2) = (2^m \times 5^n)$ **Ans.**

Q. 2. The decimal expansion of the rational number $\frac{43}{2^4 5^3}$ will terminate after how many places of decimal ?

Solution : Given number is

$$\frac{43}{2^4 5^3} = \frac{43 \times 5}{2^4 \times 5^4}$$

$$= \frac{215}{(2 \times 5)^4} = \frac{215}{10^4}$$

$$= 0.0215$$

Thus, the rational number terminates after 4 decimal places. **Ans.**

Q. 3. Express 23150 as product of its prime factors. Is it unique ?

Solution : Prime factors of $23150 = 2 \times 5 \times 5 \times 463$. As per the fundamental theorem of Arithmetic, every number has a unique factorisation.

2	23150
5	11575
5	2315
403	453
	1

Q. 4. State whether the real number 52.0521 is rational or not. If it is rational, express it in the form $\frac{p}{q}$, where p, q are co-prime integers and $q \neq 0$. What can you say about prime factorisation of q ?

Solution : We have,

$$52.0521 = \frac{520521}{10000}$$

Yes, it is rational number.

where $q = 10000 = 2^4 \times 5^4$

The given decimal expression is a terminating decimal as the factors of q consist only 2 and 5. **Ans.**

★ are board exam questions from previous years

Q. 5. Find the largest number which divides 70 and 125 leaving remainder 5 and 8 respectively.

Solution : It is given that on dividing 70 by the required number, there is a remainder 5. This means that $70 - 5 = 65$ is exactly divisible by the required number.

Similarly, $125 - 8 = 117$ is also exactly divisible by required number.

Now, Required number = HCF (65, 117)

5	65
	13

3	117
3	39
	13

$\because \quad 65 = 5 \times 13$

and $\quad 117 = 3^2 \times 13$

$\therefore \quad$ HCF (65, 117) = 13

$\therefore$ Required number = 13 **Ans.**

Q. 6. If HCF (336, 54) = 6, find LCM (336, 54).★

Solution : Given, HCF (336, 54) = 6

We know,

HCF × LCM = one number × other number

$\Rightarrow \quad 6 \times \text{LCM} = 336 \times 54$

$$\Rightarrow \quad \text{LCM} = \frac{336 \times 54}{6}$$

$$= 336 \times 9$$

$$= 3024$$ **Ans.**

Q. 7. Write whether $\frac{2\sqrt{45} + 3\sqrt{20}}{2\sqrt{5}}$ on simplification gives a rational or an irrational number.

Solution : Given,

$$\frac{2\sqrt{45} + 3\sqrt{20}}{2\sqrt{5}} = \frac{6\sqrt{5} + 6\sqrt{5}}{2\sqrt{5}}$$

$$= \frac{12\sqrt{5}}{2\sqrt{5}} = 6$$

(a rational number)

Thus, on simplification $\frac{2\sqrt{45} + 3\sqrt{20}}{2\sqrt{5}}$ gives a rational number. **Ans.**

Q. 8. 96 Defective pens are accidentally mixed with 105 good pens. What is LCM of 96 and 105 ?

Solution : We have,

$$96 = 2^5 \times 3$$

$$105 = 3 \times 7 \times 5$$

$\therefore \quad \text{LCM}(96, 105) = 2^5 \times 3 \times 7 \times 5$

$= 3360$ **Ans.**

Q. 9. Atul, Ravi and Tarun go for a morning walk. They step off together and their steps measure 40 cm, 42 cm and 45 cm, respectively. What is the minimum distance each should walk so that each can cover the same distance in complete steps ?

Solution : We have,

$$40 = 2^3 \times 5$$
$$42 = 2 \times 3 \times 7$$
$$45 = 3^2 \times 5$$

$\therefore$ LCM of 40, 42 and 45 $= 2^3 \times 3^2 \times 5 \times 7$

$= 2520$

So, after the walk of 2520 cm each can cover the same distance in complete steps. **Ans.**

Q. 10. There is a circular path around a sports field, Priya takes 18 min. to drive one round of the field, while Ravish takes 12 min. for the same. Suppose they both start at the same point and at the same time and go in the same direction. After how many minutes will they meet again at the starting point?

Solution : No. of minutes after which they will meet again at the starting point

= L.C.M. of 18 and 12

$= 2^2 \times 3^2$

= 36.

Thus, Priya and Ravish will meet again at the starting point after 36 min. **Ans.**

Q. 11. Without actually performing the long division, state whether the following rational numbers will have terminating decimal expansions or non-terminating repeating decimal expansion. Also find the number of places of decimals after which the decimal expansion terminates.

(i) $\frac{17}{8}$ **[NCERT]** **(ii)** $\frac{64}{455}$ **[NCERT]**

(iii) $\frac{29}{243}$ **[NCERT]** **(iv)** $\frac{15}{1600}$ **[NCERT]**

(v) $\frac{13}{3125}$ **[NCERT]** **(vi)** $\frac{23}{2^3 5^2}$ **[NCERT]**

(vii) $\frac{35}{50}$ **[NCERT]** **(viii)** $\frac{77}{210}$ **[NCERT]**

(ix) $\frac{129}{2^2 5^7}$ **[NCERT]**

Solution : (i) $\frac{17}{8} = \frac{17}{2^3 5^0}$

Since, the denominator is of the form $2^m \times 5^n$, where m and n are non-negative integers, hence, $\frac{17}{8}$ has a terminating decimal expansion and it terminates after 3 decimal places.

(ii) $\frac{64}{455} = \frac{64}{5 \times 7 \times 13}$

Clearly, the denominator is not of the form $2^m \times 5^n$.

So, the decimal expansion of $\frac{64}{455}$ is non-terminating repeating.

(iii) $\frac{29}{243} = \frac{29}{3^5}$

Clearly, the denominator is not of the form $2^m \times 5^n$.

So, the decimal expansion of $\frac{29}{243}$ is non-terminating repeating.

(iv) $\frac{15}{1600} = \frac{3}{320} = \frac{3}{2^6 5^1}$

Since, the denominator is of the form $2^m \times 5^n$, where m and n are non-negative integers, hence, $\frac{15}{1600}$ has a terminating decimal expansion and it terminates after 6 decimal places.

(v) $\frac{13}{3125} = \frac{13}{2^0 5^5}$

Since, the denominator is of the form $2^m \times 5^n$, where m and n are non-negative integers, hence, $\frac{13}{3125}$ has a terminating decimal expansion and it terminates after 5 decimal places.

(vi) $\frac{23}{2^3 5^2}$

Since, the denominator is of the form $2^m \times 5^n$, where m and n are non-negative integers, hence, $\frac{23}{2^3 5^2}$ has a terminating decimal expansion and terminates so after 3 decimal places.

(vii) $\frac{35}{50} = \frac{7}{10} = \frac{7}{2^1 5^1}$

Since, the denominator is of the form $2^m \times 5^n$, where m and n are non-negative integers, hence, $\frac{35}{50}$ has a terminating decimal expansion and it terminates after 1 decimal place.

(viii) $\frac{77}{210} = \frac{11}{2 \times 5 \times 3}$

Clearly, the denominator is not of the form $2^m \times 5^n$.

So, the decimal expansion of $\frac{77}{210}$ is non-terminating rcpcating.

(ix) $\frac{129}{2^2 5^7}$

Since, the denominator is of the form $2^m \times 5^n$, where m and n are non-negative integers, hence, $\frac{129}{2^2 5^7}$ has a terminating decimal expansion and it terminates after 7 decimal place.

Q. 12. Explain why $(17 \times 5 \times 11 \times 3 \times 2 + 2 \times 11)$ is a composite number ?★

Solution : $17 \times 5 \times 11 \times 3 \times 2 + 2 \times 11$

$$= 17 \times 5 \times 3 \times 22 + 22$$
$$= 22\,(17 \times 5 \times 3 + 1)$$
$$= 22\,(255 + 1)$$
$$= 2 \times 11 \times 256 \quad \text{...(i)}$$

Equation (i) is divisible by 2, 11 and 256, which means it has more than 2 prime factors.

$\therefore$ $(17 \times 5 \times 11 \times 3 \times 2 + 2 \times 11)$ is a composite number. **Ans.**

Q. 13. Show that $3\sqrt{7}$ is an irrational number.

Solution : Let us assume, to the contrary, that $3\sqrt{7}$ is rational.

That is, we can find co-prime a and b $(b \neq 0)$ such that $3\sqrt{7} = \frac{a}{b}$

On rearranging, we get $\sqrt{7} = \frac{a}{3b}$

Since 3, a and b are integers, therefore, $\frac{a}{3b}$ can be written in the form of $\frac{p}{q}$, so $\frac{a}{3b}$ is rational, and so $\sqrt{7}$ is rational.

But this contradicts the fact that $\sqrt{7}$ is irrational.

So, we conclude that $3\sqrt{7}$ is irrational.

Hence Proved.

Q. 14. Find a rational number between $\sqrt{2}$ and $\sqrt{3}$.★

Solution : As $\sqrt{2} = 1{\cdot}414....$

$\sqrt{3} = 1{\cdot}732....$

So, a rational number between $\sqrt{2}$ and $\sqrt{3}$ is 1·5 or we can take any number between 1·414 and 1·732. **Ans.**

Q. 15. 50 people work in a cooperative society. They all use their own conveyance. 20 people use their scooters, 12 go by their cars, 16 go by public transport and 2 use bicycle. Find H.C.F. of 20, 16, 12 and 2.

Solution :
$$20 = 2^2 \times 5$$
$$12 = 2^2 \times 3$$
$$16 = 2^4$$
$$2 = 2^1$$
$$\therefore \quad \text{H.C.F.} = 2$$ **Ans.**

★ are board exam questions from previous years

Q. 16. Two positive integers a and b can be written as $a = x^3y^2$ and $b = xy^3$. x, y are prime numbers. Find LCM (a, b).★

Solution. Given, $a = x^3y^2$ and $b = xy^3$

$\Rightarrow$ L.C.M (a, b) = Product of the greatest power of each prime factors

$= x^3y^3$ **Ans.**

Q. 17. Express the number $0.\overline{3178}$ in the form of rational number $\frac{a}{b}$.

Solution : Let $x = 0.31783178....$ ...(i)

Multiplying by 10000, we get

$10000\,x = 3178.31783178.....$ ...(ii)

Subtracting (i) from (ii), we get

$$10000\,x = 3178.3178....$$
$$x = 0.3178......$$
$$- \quad -$$
$$999x = 3178$$

$\Rightarrow$ $x = \frac{3178}{999}$ **Ans.**

Chapter 2. Polynomials

Q. 1. If $x = \frac{8 \pm \sqrt{(-8)^2 - 4 \times 3 \times 2}}{2 \times 3}$ then find the required polynomial.

Solution : We have, $x = \frac{8 \pm \sqrt{(-8)^2 - 4 \times 3 \times 2}}{2 \times 3}$

Then, on comparing with $x = \frac{-b \pm \sqrt{b^2 - 4ac}}{2a}$,

we get $b = -8$, $a = 3$ and $c = 2$.

So the required polynomial is

$3x^2 - 8x + 2 = 0.$ **Ans.**

Q. 2. Find the coefficient of x^0 in $x^2 + 3x + 2 = 0$.

Solution : As $x^0 = 1$, so its coefficient is the constant term. Hence, the coefficient of x^0 is 2.

Q. 3. Find a quadratic polynomial, the sum and product of whose zeros are 0 and $-\sqrt{2}$ respectively.

Solution : Quadratic polynomial is given as,

$\Rightarrow x^2$ – (Sum of zeros) x + (Product of zeros)

$\Rightarrow$ $x^2 - (0)\,x + (-\sqrt{2})$

$x^2 - \sqrt{2}$ **Ans.**

Q. 4. Write the zeros of the polynomial

$$f(x) = x^2 - x - 6.$$

Solution : Given polynomial is,

$$\begin{aligned} f(x) &= x^2 - x - 6 \\ &= x^2 - 3x + 2x - 6 \\ &= x(x-3) + 2(x-3) \\ &= (x+2)(x-3) \end{aligned}$$

To find zeros, put $f(x) = 0$

$\therefore \quad x = -2, 3.$ **Ans.**

Q. 5. If $(x + a)$ is a factor of $f(x) = (2x^2 + 2ax + 5x + 10)$, find a.

Solution : We have,

$$f(x) = 2x^2 + 2ax + 5x + 10$$

Since $(x + a)$ is a factor of $f(x)$,

Then put $\quad (x + a) = 0$

$\Rightarrow \quad x = -a$

Put $x = -a$ in $f(x)$ and put $f(-a) = 0$

$\Rightarrow \quad 2(-a)^2 + 2a(-a) + 5(-a) + 10 = 0$

$\Rightarrow \quad 2a^2 - 2a^2 - 5a + 10 = 0$

$\Rightarrow \quad 5a = 10$

$\Rightarrow \quad a = 2$ **Ans.**

Q. 6. For what value of k is -4 a zero of the polynomial $f(x) = x^2 - x - (2k + 2)$?

Solution : Given polynomial is,

$$f(x) = x^2 - x - (2k + 2)$$

$\because -4$ is a zero of $f(x)$

$\therefore \quad f(-4) = 0$

$\Rightarrow \quad (-4)^2 - (-4) - (2k + 2) = 0$

$\Rightarrow \quad 16 + 4 - 2k - 2 = 0$

$\Rightarrow \quad 18 - 2k = 0$

$\Rightarrow \quad 2k = 18$

$\Rightarrow \quad k = 9$ **Ans.**

Q. 7. If α and β are the zeros of the polynomial $f(y) = 2y^2 + 7y + 5$, write the values of $\alpha + \beta + \alpha\beta$.

Solution : Given polynomial is,

$$f(y) = 2y^2 + 7y + 5$$

$\because \quad \alpha + \beta = -\dfrac{7}{2}, \ \alpha\beta = \dfrac{5}{2}$

$\therefore \quad \alpha + \beta + \alpha\beta = -\dfrac{7}{2} + \dfrac{5}{2} = -1$ **Ans.**

Q. 8. For what value of k is 3 a zero of the polynomial $f(x) = 2x^2 + x + k$?

Solution : We have, $f(x) = 2x^2 + x + k$

Substituting 3 for x, we get

$$f(3) = 2(3)^2 + (3) + k = 0$$

$\Rightarrow \quad 18 + 3 + k = 0$

$\Rightarrow \quad k = -21$ **Ans.**

Q. 9. For what value of k is -3 a zero of the polynomial $f(x) = x^2 + 11x + k$?

Solution : We have, $f(x) = x^2 + 11x + k$

Substituting -3 for x, we get

$$f(-3) = (-3)^2 + 11(-3) + k = 0$$

$\Rightarrow \quad 9 - 33 + k = 0$

$\Rightarrow \quad k = 24$ **Ans.**

Q. 10. For what value of k is -2 a zero of the polynomial $f(x) = 3x^2 + 4x + 2k$?

Solution : Given polynomial is,

$$f(x) = 3x^2 + 4x + 2k$$

$\because -2$ is a zero of $f(x)$

$\therefore \quad f(-2) = 0$

$\Rightarrow \quad 3(-2)^2 + 4(-2) + 2k = 0$

$\Rightarrow \quad 12 - 8 + 2k = 0$

$\Rightarrow \quad 2k = -2$

$\Rightarrow \quad k = -2$ **Ans.**

Q. 11. Find the zeros of the polynomial $x^2 + x - p(p + 1) = 0$, where p is a constant.

Solution : $\quad x^2 + x - p(p + 1) = 0$

Let the zeros be α and β.

Thus, $\quad \alpha + \beta = -1 \quad$ ...(i)

and $\quad \alpha\beta = -p(p + 1) \quad$ ...(ii)

From equation (i), we have

$$\alpha = -(1 + \beta)$$

Substituting the value of α in eq. (ii), we get

$$-(1 + \beta)\beta = -(p + 1)p$$

Hence $\quad \beta = p$ and $\alpha = -(p + 1)$ **Ans.**

Q. 12. If $x + k$ is the G.C.D. of $x^2 - 2x - 15$ and $x^3 + 27$, then find the value of k.

Solution : We have

$$\begin{aligned} x^3 + 27 &= x^3 + 3^3 \\ &= (x+3)^3 - 9x(x+3) \\ &= (x+3)[(x+3)^2 - 9x] \\ &= (x+3)(x^2 - 3x + 9) \\ x^2 - 2x - 15 &= x^2 - 5x + 3x - 15 \\ &= x(x-5) + 3(x-5) \\ &= (x+3)(x-5) \end{aligned}$$

$\therefore$ $(x + 3)$ is the G.C.D. of both the equations.

So $\quad x + k = x + 3$

or $\quad k = 3.$ **Ans.**

Q. 13. If $(x + a)$ is a factor of $2x^2 + 2ax + 5x + 10$, find a.

Solution : Given,

$$f(x) = 2x^2 + 2ax + 5x + 10 = 0$$
$$= 2x(x + a) + 5(x + 2) = 0$$

Thus $\quad x + a = x + 2$

or $\quad a = 2.$ **Ans.**

Q. 14. Show that $x = -3$ is a solution of $x^2 + 6x + 9 = 0$.

Solution : Let,

$$f(x) = x^2 + 6x + 9 = 0$$
$$= x^2 + 3x + 3x + 9 = 0$$
$$= x(x + 3) + 3(x + 3) = 0$$

$\Rightarrow \quad (x + 3)^2 = 0$

$\Rightarrow \quad x + 3 = 0$

$\Rightarrow \quad x = -3.$ **Hence Proved.**

Q. 15. If α, β are the zeros of a polynomial such that $\alpha + \beta = 6$ and $\alpha\beta = 4$, then write down the polynomial.

Solution : Let the required polynomial be $f(x)$.

So $\quad f(x) = x^2 - (\alpha + \beta)x + \alpha\beta = 0$

or $\quad x^2 - 6x + 4 = 0$

(substituting the values of $\alpha + \beta$ and $\alpha\beta$)

Thus, $x^2 - 6x + 4 = 0$ is the required polynomial.

Ans.

Q. 16. Sum of zeros of the polynomial $2x^2 - 4x + 5$ is 4. Navdeep at once said "it is false"

Do you agree with Navdeep ? Justify.

Solution : Yes

Here $a = 2$, $b = -4$, $c = 5$

$$\text{Sum of zeros} = \frac{-b}{a} = \frac{-(-4)}{2} = \frac{4}{2} = 2$$

$\therefore$ Navdeep is right. **Ans.**

Q. 17. Find the value of p so that the polynomial $px(x - 3) + 9 = 0$ has two equal roots.

Solution : Given,

$$px(x - 3) + 9 = 0$$

or $\quad px^2 - 3px + 9 = 0$

Let α, β be the zeros of the polynomial.

$\therefore \quad \alpha + \beta = 3$ and $\alpha\beta = \dfrac{9}{p}$

Also $\quad \alpha = \beta$

$\therefore \quad 2\alpha = 3$

or $\quad \alpha = \beta = \dfrac{3}{2}$

$\therefore \quad \dfrac{9}{4} = \dfrac{9}{p}$

$\Rightarrow \quad p = 4$

Hence, the zeros of the polynomial $px(x - 3) + 9 = 0$ will be equal when $p = 4$. **Ans.**

OR

$$px\,(x - 3) + 9 = 0$$
$$px^2 - 3px + 9 = 0$$

Comparing the equation $ax^2 + bx + c = 0$, then $a = p$, $b = -3p$, $c = 9$.

For equal roots

$$b^2 - 4ac = 0$$
$$(-3p)^2 - 4(p)(9) = 0$$
$$9p^2 - 36p = 0$$
$$9p\,(p - 4) = 0$$

If $p - 4 = 0 \Rightarrow p = 4$

$\therefore \quad p = 0$ and $p = 4$

But $p = 0$ does not satisfy the equation

$\therefore \quad p = 4.$ **Ans.**

Q. 18. For what value of k, -4 is a zero of the polynomial $x^2 - x - (2k + 2)$?

Solution : Given,

$$f(x) = x^2 - x - (2k + 2)$$

If -4 is the zero of the polynomial $f(x)$,

Then $\quad f(-4) = 0$

$$(-4)^2 - (-4) - (2k + 2) = 0$$
$$16 + 4 - (2k + 2) = 0$$
$$20 - (2k + 2) = 0$$
$$18 - 2k = 0$$
$$2k = 18$$
$$k = 9.$$ **Ans.**

Q. 19. If the product of the zeros of a polynomial $(ax^2 - 6x - 6)$ is 4, find the value of a.

Solution : Given polynomial is, $ax^2 - 6x - 6$ Let the zeros of the polynomial be α and β.

$\therefore \quad \alpha + \beta = \dfrac{6}{a}$

and $\quad \alpha\beta = -\dfrac{6}{a}$

But it is given that $\alpha\beta = 4$.

$\therefore \quad \alpha\beta = -\dfrac{6}{a} = 4$

or $\quad a = -\dfrac{3}{2}.$ **Ans.**

Q. 20. Write the polynomial, the product and sum of whose zeros are $-\frac{9}{2}$ and $-\frac{3}{2}$.

Solution : Let the zeros of the polynomial be α and β.

$$\therefore \quad \alpha+\beta = -\frac{3}{2}$$

$$\text{and} \quad \alpha\beta = -\frac{9}{2}$$

We know, the polynomial is given as,

$$x^2-(\alpha+\beta)x+\alpha\beta = 0$$

$$\Rightarrow \quad x^2+\frac{3}{2}x-\frac{9}{2} = 0$$

$$\Rightarrow \quad 2x^2+3x-9 = 0.$$ **Ans.**

Q. 21. A group consists of 12 honest people and 8 dishonest people. Write a quadratic polynomial whose roots are equal to number of honest people and number of dishonest people.

Solution : Number of honest people = 12

Number of dishonest people = 8

$$\therefore \quad \text{Sum of roots} = 12+8 = 20$$

$$\text{Product of roots} = 12\times 8 = 96$$

Quadratic polynomial $= x^2-20x+96$ **Ans.**

Q. 22. Manish engages a labour to get some repair work. Charges to be paid for this work are zeros of the polynomial x^2 – 300x + 22500. Find zeros of this polynomial.

Solution : We have,

$$x^2-300x+22500 = 0$$

$$\Rightarrow \quad x^2-150x-150x+22500 = 0$$

$$\Rightarrow \quad x(x-150)-150(x-150) = 0$$

$$\Rightarrow \quad (x-150)(x-150) = 0$$

$$\therefore \quad x = 150$$ **Ans.**

Q. 23. Given $f(x+1) = 3x+5$, evaluate $f(-2)$ and $f(x)$.

Solution : Given,

$$f(x+1) = 3x+5$$

$$\therefore \quad f(x) = f(x+1-1) = 3(x-1)+5 = 3x-3+5 = 3x-2$$

$$\text{and} \quad f(-2) = 3(-2)+5 = -6+5 = -1.$$ **Ans.**

Q. 24. If $f(x) = \dfrac{x}{x^2+1}$ then find :

(i) $f\left(\dfrac{1}{x}\right)$ and (ii) $f(x-1)$.

Solution : Given,

$$f(x) = \frac{x}{x^2+1}$$

(i) Thus,
$$f\left(\frac{1}{x}\right) = \frac{\frac{1}{x}}{\left(\frac{1}{x}\right)^2+1} = \frac{x^2}{x(x^2+1)} = \frac{x}{x^2+1}$$

(ii)
$$f(x-1) = \frac{x-1}{(x-1)^2+1} = \frac{x-1}{x^2-2x+2}$$ **Ans.**

Q. 25. Find the remainder when

$$f(x) = (2x^3-3x^2+7x-8)$$

is divided by $g(x) = (x-1)$.

Soution : Given, $f(x) = 2x^3-3x^2+7x-8$

and $g(x) = x-1$

Put $x-1 = 0 \Rightarrow x = 1$

$$\text{Remainder} = f(1) = 2(1)^3-3(1)^2+7(1)-8 = 2-3+7-8 = -2$$

Hence, the remainder is –2. **Ans.**

Q. 26. Show that $g(x) = (x-3)$ is a factor of

$$f(x) = (x^3-7x^2+15x-9).$$

Solution : Given, $f(x) = x^3-7x^2+15x-9$

and $g(x) = x-3$

Put $x-3 = 0 \Rightarrow x = 3$

$$\text{Remainder} = f(3) = (3)^3-7(3)^2+15(3)-9 = 27-63+45-9 = 72-72 = 0$$

Hence, $(x-3)$ is a factor of given polynomial of $f(x)$.

Hence Proved.

Q. 27. Find the H.C.F. of $f(x) = 3x^3 - 27x^2 + 60x$ and $g(x) = x^2 - 16$.

Solution : Given,

$$f(x) = 3x^3 - 27x^2 + 60x$$

and $g(x) = x^2 - 16$

Now
$$\begin{aligned} f(x) &= 3x^3 - 27x^2 + 60x \\ &= 3x(x^2 - 9x + 20) \\ &= 3x(x - 5x - 4x + 20) \\ &= 3x\{x(x-5) - 4(x-5)\} \\ &= 3x(x-5)(x-4) \end{aligned}$$

and
$$\begin{aligned} g(x) &= x^2 - 16 \\ &= (x-4)(x+4) \end{aligned}$$

$\therefore$ H.C.F. $= x - 4$. **Ans.**

Q. 28. Find the value of k if $(x - 2)$ is a factor of $f(x) = x^3 + 2x^2 - kx + 10$. Also determine whether $(x + 5)$ is also a factor.

Solution : Given,

$$f(x) = x^3 + 2x^2 - kx + 10$$

and $g(x) = x - 2$

Put $g(x) = x - 2 = 0$

or $x = 2$

Replacing x by 2 in $f(x) = x^3 + 2x^2 - kx + 10$, we get

$$(2)^3 + 2(2)^2 - 2k + 10 = 0$$

$\Rightarrow$ $8 + 8 - 2k + 10 = 0$

$\Rightarrow$ $-2k + 26 = 0$

$\Rightarrow$ $2k = 26$

$\Rightarrow$ $k = 13$

Thus $f(x) = x^3 + 2x^2 - 13x + 10$

Now, put $x + 5 = 0$

or $x = -5$

Replacing x with -5 in $f(x)$, we get

$$\begin{aligned} f(-5) &= (-5)^3 + 2(-5)^2 - 13(-5) + 10 \\ &= -125 + 50 + 65 + 10 \\ &= -125 + 125 = 0 \end{aligned}$$

Thus, $(x + 5)$ is also a factor of

$$f(x) = x^3 + 2x^2 - 13x + 10.$$ **Ans.**

Q. 29. Find the factors of the following polynomial

$$f(x) = 2x^3 + x^2 - 13x + 6.$$

Solution : Given,

$$\begin{aligned} f(x) &= 2x^3 + x^2 - 13x + 6 \\ &= 2x^3 - x^2 + 2x^2 - x - 12x + 6 \\ &= x^2(2x-1) + x(2x-1) - 6(2x-1) \\ &= (2x-1)(x^2 + x - 6) \\ &= (2x-1)(x^2 + 3x - 2x - 6) \\ &= (2x-1)\{x(x+3) - 2(x+3)\} \\ &= (2x-1)(x+3)(x-2) \end{aligned}$$

Hence, the zeros of $f(x)$ are -3, 2 and $\frac{1}{2}$. **Ans.**

Q. 30. Find the factors of the polynomial

$$f(x) = 3x^3 + 2x^2 - 19x + 6.$$

Solution : Given,

$$\begin{aligned} f(x) &= 3x^3 + 2x^2 - 19x + 6 \\ &= 3x^3 - x^2 + 3x^2 - x - 18x + 6 \\ &= x^2(3x-1) + x(3x-1) - 6(3x-1) \\ &= (3x-1)(x^2 + x - 6) \\ &= (3x-1)(x^2 + 3x - 2x - 6) \\ &= (3x-1)\{x(x+3) - 2(x+3)\} \\ &= (3x-1)(x-2)(x+3) \end{aligned}$$

$\therefore$ The zeros of $f(x) = 3x^3 + 2x^2 - 19x + 6$ are $2, -3, \frac{1}{3}$.

Ans.

Q. 31. If two zeros of the polynomial $x^3 - 4x^2 - 3x + 12$ are $\sqrt{3}$ and $-\sqrt{3}$, then find the third zero.

Solution : Let

$$\begin{aligned} f(x) &= x^3 - 4x^2 - 3x + 12 \\ &= x^3 - 3x - 4x^2 + 12 \\ &= x(x^2 - 3) - 4(x^2 - 3) \\ &= (x^2 - 3)(x - 4) \end{aligned}$$

As $\sqrt{3}$ and $-\sqrt{3}$ are the zeros of $f(x)$, therefore, $(x - \sqrt{3})$ and $(x + \sqrt{3})$ are its factors. So $(x - \sqrt{3})(x + \sqrt{3})$ is a factor of $f(x)$.

Let $(x - \sqrt{3})(x + \sqrt{3}) = g(x)$

or $x^2 - 3 = g(x)$

So, $$\frac{f(x)}{g(x)} = \frac{(x^2 - 3)(x - 4)}{(x^2 - 3)} = x - 4$$

Thus $x - 4 = 0$

or $x = 4$

Hence, the third zero of the polynomial is 4. **Ans.**

Q. 32. If $x = \frac{2}{3}$ and $x = -3$ are zeros of the polynomial $ax^2 + 7x + b = 0$, find the value of a and b.

Solution : The given polynomial is,

$$p(x) = ax^2 + 7x + b$$

$\because$ $x = \frac{2}{3}$ and $x = -3$ are the roots of the given polynomial

$$\therefore \quad p\left(\frac{2}{3}\right) = a\left(\frac{2}{3}\right)^2 + 7\left(\frac{2}{3}\right) + b = 0$$

$$\Rightarrow \quad \frac{4a}{9}+\frac{14}{3}+b=0$$

$$\Rightarrow \quad 4a+42+9b=0 \quad \text{...(i)}$$

and $\quad p(-3) = a(-3)^2+7(-3)+b=0$

$$\Rightarrow \quad 9a-21+b=0 \quad \text{...(ii)}$$

Multiply equation (ii) by 9 and then subtracting it with equation (i),

$$\begin{array}{r} 4a+\ 42+9b=0 \\ 81a-189+9b=0 \\ -\quad +\quad - \\ \hline -77a+231=0 \end{array}$$

$$\Rightarrow \quad a=\frac{231}{77}=3$$

Putting $a = 3$ in equation (ii), we get

$$9(3)-21+b=0$$

$$b=-6$$

$\therefore \quad a=3,\ b=-6$ **Ans.**

Q. 33. Find the value of k so that the polynomial equation $x^2-4kx+k=0$ has equal roots.

Solution : Let α and β be the roots of the polynomial equation $x^2-4kx+k=0$

So $\quad \alpha+\beta=4k \quad$ ...(i)

and $\quad \alpha\beta=k \quad$ ...(ii)

Also $\quad \alpha=\beta \quad$ (Given)

$\therefore$ From equation (i),

$$2\alpha=4k$$

$$\Rightarrow \quad \alpha=2k \quad \text{...(iii)}$$

and from equation (ii)

$$\alpha\cdot\alpha=k$$

$$\Rightarrow \quad \alpha^2=k \quad \text{...(iv)}$$

From equations (iii) and (iv),

$$\alpha=2k=2\alpha^2$$

$$\Rightarrow \quad 2\alpha=1$$

$$\therefore \quad \alpha=\beta=2k=\frac{1}{2}$$

Thus, $\quad k=\frac{1}{4}$

Hence, the two roots of the polynomial $x^2-4k+k=0$ would be equal when $k=\frac{1}{4}$. **Ans.**

Chapter 3. Pair of Linear Equations in Two Variables

Q. 1. Find the values of x and y in $2x+3y=2$ and $x-2y=8$.

Solution : Given equations are,

$$2x+3y=2 \quad \text{...(i)}$$

$$x-2y=8 \quad \text{...(ii)}$$

Multiplying equation (i) with 2 and equation (ii) with 3, we get

$$4x+6y=4 \quad \text{...(iii)}$$

and $\quad 3x-6y=24 \quad$ ...(iv)

Thus, adding the equations (iii) and (iv), we get

$$7x=28$$

$$\Rightarrow \quad x=4$$

Replacing $x=4$ in equation (ii), we get

$$4-2y=8$$

$$\Rightarrow \quad -2y=4$$

$$\Rightarrow \quad y=-2$$

Hence, the values of x and y are 4, – 2 respectively. **Ans.**

Q. 2. Find the values of x and y in $3x+2y=4$ and $2x-3y=7$.

Solution : Given equations are,

$$3x+2y=4 \quad \text{...(i)}$$

$$2x-3y=7 \quad \text{...(ii)}$$

Multiplying equation (i) with 3 and equation (ii) with 2, we get

$$9x+6y=12 \quad \text{...(iii)}$$

and $\quad 4x-6y=14 \quad$ ...(iv)

Thus, adding the above two equations, we get

$$13x=26$$

$$\Rightarrow \quad x=2$$

Replacing $x=2$ in equation (i), we get

$$3(2)+2y=4$$

$$\Rightarrow \quad 6+2y=4$$

$$\Rightarrow \quad 2y=-2$$

$$\Rightarrow \quad y=-1$$

Hence, the values of x and y are 2, – 1 respectively. **Ans.**

Q. 3. Find the values of x and y in $2x+3y=4$ and $3x-y=-5$.

Solution : Given equations are,

$$2x+3y=4 \quad \text{...(i)}$$

$$3x-y=-5 \quad \text{...(ii)}$$

Multiplying equation (ii) with 3, we get

$$9x-3y=-15 \quad \text{...(iii)}$$

Thus, adding equations (i) and (iii), we get

$$11x=-11$$

$$\Rightarrow \quad x=-1$$

Replacing $x=-1$ in equation (ii), we get

$$-3-y=-5$$

$$\Rightarrow \quad -y=-2$$

$\Rightarrow \quad y = 2$

Hence, the values of x and y are -1, 2 respectively.

Ans.

Q. 4. Find the values of x and y in $2x + 3y = 8$ and $x - 2y + 3 = 0$.

Solution : Given equations are,

$$2x + 3y = 8 \quad \ldots(i)$$

$$x - 2y = -3 \quad \ldots(ii)$$

Multiplying equation (i) with 2 and equation (ii) with 3, we get

$$4x + 6y = 16 \quad \ldots(iii)$$

and $$3x - 6y = -9 \quad \ldots(iv)$$

Thus, adding the above two equations, we get

$$7x = 7$$

$\Rightarrow \quad x = 1$

Replacing $x = 1$ in equation (i), we get

$$2 + 3y = 8$$

$\Rightarrow \quad 3y = 6$

$\Rightarrow \quad y = 2$

Hence, the values of x and y are 1, 2 respectively.

Ans.

Q. 5. Find the values of x and y in $2x - 5y + 4 = 0$ and $2x + y - 8 = 0$.

Solution : Given equations are,

$$2x - 5y + 4 = 0$$

$\Rightarrow \quad 2x - 5y = -4 \quad \ldots(i)$

and $\quad 2x + y - 8 = 0$

$\Rightarrow \quad 2x + y = 8 \quad \ldots(ii)$

Thus, subtracting equation (ii) from (i), we get

$$-6y = -12$$

$\Rightarrow \quad y = 2$

Replacing $y = 2$ in equation (ii), we get

$$2x + 2 = 8$$

$\Rightarrow \quad 2x = 6$

$\Rightarrow \quad x = 3$

Hence, the values of x and y are 3, 2 respectively.

Ans.

Q. 6. Given the linear equation $x - 2y - 6 = 0$, write another linear equation in these two variables, such that the geometrical representation of the pair so formed is :

(i) coincident lines (ii) intersection lines

Solution : (i) Given, $x - 2y - 6 = 0$

For lines to be coincident,

$$\frac{a_1}{a_2} = \frac{b_1}{b_2} = \frac{c_1}{c_2}$$

Thus, one possible option will be

$$2x - 4y - 12 = 0$$

Here, $\quad a_1 = 1, b_1 = -2, c_1 = -6.$

$\quad a_2 = 2, b_2 = -4, c_2 = -12.$

$$\frac{a_1}{a_2} = \frac{1}{2}; \frac{b_1}{b_2} = \frac{-2}{-4} = \frac{1}{2}; \frac{c_1}{c_2} = \frac{-6}{-12} = \frac{1}{2}$$

$$\frac{a_1}{a_2} = \frac{b_1}{b_2} = \frac{c_1}{c_2}$$

So, it is showing coincident lines. **Ans.**

(ii) Given, $x - 2y - 6 = 0$

For intersecting lines

$$\frac{a_1}{a_2} \neq \frac{b_1}{b_2}$$

Thus, one possible option will be

$$2x - 7y - 13 = 0$$

Here, $\quad a_1 = 1, b_1 = -2, c_1 = -6$

$\quad a_2 = 2, b_2 = -7, c_2 = -13$

$$\frac{a_1}{a_2} = \frac{1}{2}; \frac{b_1}{b_2} = \frac{-2}{-7} = \frac{2}{7}$$

and $$\frac{a_1}{a_2} \neq \frac{b_1}{b_2}$$

So, it is representing intersecting lines. **Ans.**

Q. 7. Peter secured x marks in Mathematics and y marks in Physics. Find the value of x and y, if $x - y = 2$ and $xy = 2600$.

Solution : Given, $\quad x - y = 2$

$\Rightarrow \quad x = y + 2 \quad \ldots(i)$

and $\quad xy = 2600 \quad \ldots(ii)$

Substuting the value of x from equation (i) to equation (ii), we get

$$(y + 2)y = 2600$$

$\Rightarrow \quad y^2 + 2y - 2600 = 0$

$\Rightarrow \quad y^2 + 52y - 50y - 2600 = 0$

$\Rightarrow \quad y(y + 52) - 50(y + 52) = 0$

$\Rightarrow \quad (y + 52)(y - 50) = 0$

$\Rightarrow \quad y = 50, -52$

But as marks cannot be negative, so $y = 50$ and from equation (i)

$$x = 2 + 50 = 52$$

$\therefore$ Marks secured in Mathematics = 52 and in Physics = 50 **Ans.**

Q. 8. In a two-digit number, the sum of the digits is 10 and the square of the tens digit when divided by the cube of the units digit is equal to the value of the tens digit. Find the number.

Solution : Let the digits of the number be x and y.

Thus, the number is $10x + y$.

According to the question,

$$x + y = 10 \quad \text{...(i)}$$

and $\dfrac{x^2}{y^3} = x$

$\Rightarrow \quad x = y^3 \quad$ ...(ii)

Now, $\quad y^3 + y = 10 \quad$ [From (i)]

$\Rightarrow \quad y^3 + y - 10 = 0$

$\Rightarrow \quad y^3 - 2y^2 + 2y^2 - 4y + 5y - 10 = 0$

$\Rightarrow \quad y^2(y-2) + 2y(y-2) + 5(y-2) = 0$

$\Rightarrow \quad (y-2)(y^2 + 2y + 5) = 0$

$\Rightarrow \quad y = 2$ and $\dfrac{-2 \pm \sqrt{2^2 - 4(1)(5)}}{2(1)}$

$\Rightarrow \quad y = 2$ and $\dfrac{-2 \pm \sqrt{4-20}}{2}$

$\Rightarrow \quad y = 2$ and $\dfrac{-2 \pm \sqrt{-16}}{2}$

Thus, y is taken as 2 because the other option is unreal.

Now, replacing $y = 2$ in eq. (ii), we get

$$x = y^3$$
$$= (2)^3$$
$$= 8$$

So the number is $10(8) + 2 = 82$. **Ans.**

Q. 9. A number consists of 2 digits. The sum of the digits is 12 and the units digit, when divided by the tens digit gives the result as 3. Find the number.

Solution : Let the number be $10x + y$

Given, $\quad x + y = 12 \quad$ ...(i)

and $\quad \dfrac{y}{x} = 3$

$\Rightarrow \quad y = 3x \quad$ ...(ii)

Hence, $\quad x + 3x = 12 \quad$ [From (i)]

$\Rightarrow \quad 4x = 12$

$\Rightarrow \quad x = 3$

and $\quad y = 9 \quad$ [From (ii)]

So the number is $3(10) + 9 = 39$ **Ans.**

Q. 10. The cost of 2 kg apples and 1 kg grapes was ₹ 160. After a month, the cost of 4 kg apples and 2 kg grapes was ₹ 300. Represent the situation algebraically. [NCERT]

Solution : Let the cost of 1 kg of apples be ₹ a and the cost of 1 kg of grapes be ₹ g.

Thus $\quad 2a + g = 160 \quad$ ...(i)

and $\quad 4a + 2g = 300$

or $\quad 2a + g = 150 \quad$ ...(ii)

Ans.

Q. 11. Cost of 2 pens and 3 pencils together is ₹ 40 and cost of 6 pens and 9 pencils together is ₹ 130. Express above statement in the form of Linear equations.

Solution : Let the cost of 1 pen be ₹ x and cost of 1 pencil be ₹ y

Then,

$$2x + 3y = 40$$

and $\quad 6x + 9y = 130 \quad$ **Ans.**

Q. 12. Speed of boat in downstream is 8 km per hour whereas speed in upstream is 2 km per hour. Find the time taken by boat in 16 km downstream and 8 km upstream.

Solution : Given, Speed of boat in downstream = 8 km/hour

Speed of boat in upstream = 2 km/hour

Time taken by boat in 16 km downstream

$$= \frac{16}{8} = 2 \text{ hours}$$

and, time taken by boat in 8 km upstream

$$= \frac{8}{2} = 4 \text{ hours.}$$ **Ans.**

Q. 13. Cost of a burger is ₹ 20 more than the cost of juice of one glass of orange. If cost of one burger and one glass of orange juice is ₹ 60. Find the cost of each.

Solution : Let the cost of burger be ₹ x and the cost of orange juice be ₹ y

Then $\quad x = 20 + y$

$\Rightarrow \quad x - y = 20 \quad$ ...(i)

and $\quad x + y = 60 \quad$ ...(ii)

On adding equations (i) and (ii), we get

$$2x = 80$$
$$x = 40$$

Replacing the value of x in eq. (ii), we get

$$40 + y = 60$$

$\Rightarrow \quad y = 20$

Thus, the cost of burger is ₹ 40 and orange juice is ₹ 20. **Ans.**

Q. 14. Solve the following linear equations algebraically : $(a \quad b)x + (a + b)y = a^2 - 2ab - b^2$ and $(a + b)(x + y) = a^2 + b^2$.

Solution : Given,

$(a - b)x + (a + b)y = a^2 - 2ab - b^2 \quad$ ...(i)

and $\quad (a + b)(x + y) = a^2 + b^2$

or $\quad (a + b)x + (a + b)y = a^2 + b^2 \quad$ ...(ii)

Subtracting (ii) from (i), we get

$$[(a-b)-(a+b)]x = a^2 - 2ab - b^2 - a^2 - b^2$$

$\Rightarrow \quad (a-b-a-b)x = -2b(a+b)$

$\Rightarrow \quad -2bx = -2b(a+b)$

$\Rightarrow \quad x = a+b$

Substituting $x = a + b$ in equation (i), we get

$$(a-b)(a+b) + (a+b)y = a^2 - 2ab - b^2$$

$\Rightarrow \quad a^2 - b^2 + (a+b)y = a^2 - 2ab - b^2$

$\Rightarrow \quad (a+b)y = -2ab$

$\Rightarrow \quad y = -\dfrac{2ab}{a+b}.$

Thus, $\quad x = a+b$

and $\quad y = \dfrac{-2ab}{a+b}$ **Ans.**

Q. 15. Solve for *x* and *y* :

$$mx - ny = m^2 + n^2 \text{ and } x + y = 2m.$$

Solution : Given,

$$mx - ny = m^2 + n^2 \quad \text{...(i)}$$

and $\quad x + y = 2m$

$\Rightarrow \quad nx + ny = 2mn \quad$ ...(ii)

[multiplying both sides by *n*]

Adding equations (i) and (ii), we get

$$mx + nx = m^2 + n^2 + 2mn$$

$\Rightarrow \quad (m+n)x = (m+n)^2$

$\Rightarrow \quad x = m+n$

Substituting $x = m + n$ in $x + y = 2m$, we get

$$m + n + y = 2m$$

$\Rightarrow \quad y = m - n$

Thus, $\quad x = m + n$ and $y = m - n$. **Ans.**

Q. 16. Solve for *x* and *y* :

$$\frac{bx}{a} - \frac{ay}{b} + a + b = 0 \text{ and } bx - ay + 2ab = 0.$$

Solution : Given,

$$\frac{bx}{a} - \frac{ay}{b} + a + b = 0$$

$\Rightarrow \quad b^2x - a^2y = -ab(a+b) \quad$ ...(i)

and $\quad bx - ay + 2ab = 0$

$\Rightarrow \quad abx - a^2y = -2a^2b \quad$ ...(ii)

Subtracting (ii) from (i), we get

$$b^2x - abx = -ab(a+b) - (-2a^2b)$$

$\Rightarrow \quad (b-a)bx = -a^2b - ab^2 + 2a^2b$

$\Rightarrow \quad (b-a)bx = a^2b - ab^2$

$\Rightarrow \quad (b-a)bx = ab(a-b)$

$\Rightarrow \quad (b-a)x = -a(b-a)$

$\Rightarrow \quad x = -a$

Substituting with $x = -a$ in equation (ii), we get

$$ab(-a) - a^2y = -2a^2b$$

$\Rightarrow \quad -a^2b - a^2y = -2a^2b$

$\Rightarrow \quad -a^2y = -a^2b$

$\Rightarrow \quad y = b$

Thus, $\quad x = -a$ and $y = b$. **Ans.**

Q. 17. Solve for *x* and *y* :

$$\frac{x}{a} + \frac{y}{b} = 2 \text{ and } ax - by = a^2 - b^2.$$

Solution : Given,

$$\frac{x}{a} + \frac{y}{b} = 2$$

or $\quad bx + ay = 2ab \quad$ ...(i)

and $\quad ax - by = a^2 - b^2 \quad$ ...(ii)

Multiplying (i) with *b* and (ii) with *a*, we have

$$b^2x + aby = 2ab^2 \quad \text{...(iii)}$$

$$a^2x - aby = a(a^2 - b^2) \quad \text{...(iv)}$$

Adding (iii) and (iv), we get

$$(a^2 + b^2)x = 2ab^2 + a(a^2 - b^2)$$

$\Rightarrow \quad (a^2 + b^2)x = a(2b^2 + a^2 - b^2)$

$\Rightarrow \quad (a^2 + b^2)x = a(a^2 + b^2)$

$\Rightarrow \quad x = a$

Substituting $x = a$ in (ii), we get

$$a^2 - by = a^2 - b^2$$

$\Rightarrow \quad -by = -b^2$

$\Rightarrow \quad y = b$

Thus, $\quad x = a$ and $y = b$. **Ans.**

Q. 18. Solve for *x* and *y* :

$$\sqrt{2}x - \sqrt{3}y = 0 \text{ and } \sqrt{3}x - \sqrt{8}y = 0.$$

Solution : Given,

$$\sqrt{2}x - \sqrt{3}y = 0 \quad \text{...(i)}$$

and $\quad \sqrt{3}x - \sqrt{8}y = 0$

$\Rightarrow \quad \sqrt{3}x - 2\sqrt{2}y = 0 \quad$ ...(ii)

Multiplying (i) with $2\sqrt{2}$ and (ii) with $\sqrt{3}$, we get

$$4x - 2\sqrt{6}y = 0 \quad \text{...(iii)}$$

and $\quad 3x - 2\sqrt{6}y = 0 \quad$ ...(iv)

Subtracting (iv) from (iii), we get

$$x = 0$$

Substituting $x = 0$ in (i), we get

$$0 - \sqrt{3}y = 0$$

or $\quad y = 0$

Thus $\quad x = y = 0$. **Ans.**

Q. 19. Solve for x and y :

$$\frac{2}{\sqrt{x}}+\frac{3}{\sqrt{y}}=2 \text{ and } \frac{4}{\sqrt{x}}-\frac{9}{\sqrt{y}}=-1. \quad \text{[NCERT]}$$

Solution : Given,

$$\frac{2}{\sqrt{x}}+\frac{3}{\sqrt{y}}=2 \text{ and } \frac{4}{\sqrt{x}}-\frac{9}{\sqrt{y}}=-1$$

Let $\frac{2}{\sqrt{x}}=u$ and $\frac{3}{\sqrt{y}}=v$

Thus, the equation becomes,

$$u+v=2 \quad \ldots\text{(i)}$$

and $$2u-3v=-1 \quad \ldots\text{(ii)}$$

Multiplying (i) with 3, we get

$$3u+3v=6 \quad \ldots\text{(iii)}$$

Adding (ii) and (iii), we get

$$5u=5$$

$\Rightarrow$ $u=1$

Thus $\frac{2}{\sqrt{x}}=1$

or $x=4$

Substituting $u=1$ in (i), we get

$$1+v=2$$

or $v=1$

Thus $\frac{3}{\sqrt{y}}=1$

or $y=9$

Thus $x=4$ and $y=9$. **Ans.**

Q. 20. Solve for x and y :

$$\frac{7x-2y}{xy}=5 \text{ and } \frac{8x+7y}{xy}=15. \quad \text{[NCERT]}$$

Solution : Given,

$$\frac{7x-2y}{xy}=5$$

or $$7x-2y=5xy \quad \ldots\text{(i)}$$

and $$\frac{8x+7y}{xy}=15$$

or $$8x+7y=15xy \quad \ldots\text{(ii)}$$

Multiplying (i) with 7 and (ii) with 2, we get

$$49x-14y=35xy \quad \ldots\text{(iii)}$$

$$16x+14y=30xy \quad \ldots\text{(iv)}$$

Adding (iii) and (iv), we get

$$65x=65xy$$

$\Rightarrow$ $y=1$

Substituting $y=1$ in (i), we get

$$7x-2=5x$$

$\Rightarrow$ $2x=2$

$\Rightarrow$ $x=1$

Thus $x=y=1$. **Ans.**

Q. 21. Solve for x and y :

$152x-378y=-74$ and $-378x+152y=-604$. **[NCERT]**

Solution : Given,

$$152x-378y=-74 \quad \ldots\text{(i)}$$

and $$-378x+152y=-604 \quad \ldots\text{(ii)}$$

Adding (i) and (ii), we get

$$-226x-226y=-678$$

$\Rightarrow$ $x+y=3 \quad \ldots$(iii)

Subtracting (ii) from (i), we get

$$530x-530y=530$$

$\Rightarrow$ $x-y=1 \quad \ldots$(iv)

Adding (iii) and (iv), we get

$$2x=4$$

$\Rightarrow$ $x=2$

Substituting $x=2$ in (iii), we get

$$2+y=3$$

$\Rightarrow$ $y=1$

Thus $x=2$ and $y=1$. **Ans.**

Q. 22. Solve by cross-multiplication :

$$\frac{ax}{b}-\frac{by}{a}=a+b \text{ and } ax-by=2ab.$$

Solution : Given, $\frac{ax}{b}-\frac{by}{a}=a+b$

or $$a^2x-b^2y-ab(a+b)=0 \quad \ldots\text{(i)}$$

and $$ax-by=2ab$$

or $$ax-by-2ab=0 \quad \ldots\text{(ii)}$$

Thus $$\frac{x}{[(-b^2)\times(-2ab)]-[ab^2(a+b)]}=\frac{y}{[-ba^2(a+b)]-[(a^2\times(-2ab))]}=\frac{1}{-(a^2b)-(-ab^2)}$$

$$\Rightarrow \frac{x}{2ab^3-a^2b^2-ab^3}=\frac{y}{-ba^3-b^2a^2+2a^3b}=\frac{1}{-ba^2+ab^2}$$

$$\Rightarrow \frac{x}{ab^3-a^2b^2}=\frac{y}{a^3b-a^2b^2}$$

$$= \frac{1}{ab^2 - a^2b}$$

$$\Rightarrow \quad \frac{x}{ab^2(b-a)} = \frac{y}{a^2b(a-b)}$$

$$= \frac{1}{ab(b-a)}$$

$$\Rightarrow \quad x = \frac{ab^2(b-a)}{ab(b-a)} = b$$

and $$y = \frac{a^2b(a-b)}{ab(b-a)}$$

$$= -\frac{a^2b(b-a)}{ab(b-a)} = -a$$

Thus $x = b, y = -a.$ **Ans.**

Q. 23. Solve by cross-multiplication :

$\boldsymbol{ax + by = c}$ **and** $\boldsymbol{bx + ay = 1 + c.}$ **[NCERT]**

Solution : Given, $ax + by = c$

$$\Rightarrow \quad ax + by - c = 0 \quad \text{...(i)}$$

and $bx + ay = 1 + c$

$$\Rightarrow \quad bx + ay - (1 + c) = 0 \quad \text{...(ii)}$$

$$\Rightarrow \frac{x}{-b(1+c)+ac} = \frac{y}{-bc+a(1+c)} = \frac{1}{(a\times a)-(b\times b)}$$

$$\Rightarrow \quad \frac{x}{-b-bc+ac} = \frac{y}{-bc+a+ac} = \frac{1}{a^2-b^2}$$

$$\Rightarrow \quad \frac{x}{c(a-b)-b} = \frac{y}{c(a-b)+a}$$

$$= \frac{1}{(a-b)(a+b)}$$

$$\Rightarrow \quad x = \frac{c(a-b)-b}{(a-b)(a+b)}$$

and $$y = \frac{c(a-b)+a}{(a-b)(a+b)}$$

$$\Rightarrow \quad x = \frac{c}{(a+b)} - \frac{b}{(a^2-b^2)}$$

and $$y = \frac{c}{(a+b)} + \frac{a}{(a^2-b^2)}.$$ **Ans.**

Q. 24. Solve by cross-multiplication :

$\boldsymbol{\frac{x}{a} = \frac{y}{b}}$ **and** $\boldsymbol{ax + by = a^2 + b^2.}$ **[NCERT]**

Solution : Given, $\frac{x}{a} = \frac{y}{b}$

$$\Rightarrow \quad bx - ay = 0 \quad \text{...(i)}$$

and $ax + by = a^2 + b^2$

$$\Rightarrow \quad ax + by - (a^2 + b^2) = 0 \quad \text{...(ii)}$$

Thus $$\frac{x}{a(a^2+b^2)} = \frac{y}{b(a^2+b^2)} = \frac{1}{a^2+b^2}$$

$$\Rightarrow \quad x = \frac{a(a^2+b^2)}{(a^2+b^2)} = a$$

and $$y = \frac{b(a^2+b^2)}{(a^2+b^2)} = b.$$ **Ans.**

Q. 25. Solve by cross-multiplication :

$\boldsymbol{6(ax + by) = 3a + 2b}$ **and** $\boldsymbol{6(bx - ay) = 3b - 2a.}$

Solution : Given,

$$6(ax + by) = 3a + 2b$$

or $$6ax + 6by - (3a + 2b) = 0 \quad \text{...(i)}$$

and $$6(bx - ay) = 3b - 2a$$

or $$6bx - 6ay - (3b - 2a) = 0 \quad \text{...(ii)}$$

Thus $$\frac{x}{-6b(3b-2a)-6a(3a+2b)}$$

$$= \frac{y}{-6b(3a+2b)+6a(3b-2a)}$$

$$= \frac{1}{-36a^2-36b^2}$$

$$\Rightarrow \quad \frac{x}{-18b^2+12ab-18a^2-12ab}$$

$$= \frac{y}{-18ab-12b^2+18ab-12a^2}$$

$$= \frac{1}{-36(a^2+b^2)}$$

$$\Rightarrow \quad \frac{x}{-18(a^2+b^2)} = \frac{y}{-12(a^2+b^2)}$$

$$= \frac{1}{-36(a^2+b^2)}$$

$$\Rightarrow \quad x = \frac{-18(a^2+b^2)}{-36(a^2+b^2)} = \frac{1}{2}$$

and $$y = \frac{-12(a^2+b^2)}{-36(a^2+b^2)} = \frac{1}{3}.$$ **Ans.**

Q. 26. Solve by cross-multiplication :

$$\boldsymbol{\frac{a^2}{x} - \frac{b^2}{y} = 0 \text{ and } \frac{a^2b}{x} + \frac{ab^2}{y} = a + b.}$$

Solution : Given, $\frac{a^2}{x} - \frac{b^2}{y} = 0$

and $$\frac{a^2b}{x} + \frac{ab^2}{y} = a + b$$

Let $$\frac{a^2}{x} = u \text{ and } \frac{b^2}{y} = v$$

∴ The equations become,

$$u - v = 0 \quad \text{...(i)}$$

and $$bu + av = a + b$$

or $$bu + av - (a + b) = 0 \quad \text{...(ii)}$$

Thus $$\frac{u}{a+b} = \frac{v}{a+b} = \frac{1}{a+b}$$

$\Rightarrow \quad u = 1 \text{ and } v = 1$

$\Rightarrow \quad \frac{a^2}{x} = 1 \text{ and } \frac{b^2}{y} = 1$

$\Rightarrow \quad x = a^2 \text{ and } y = b^2.$ **Ans.**

Q. 27. Determine the value of k, for which the given system of equations has infinitely many solutions :

$kx + 3y = k - 3$ and $12x + ky = k$.

[NCERT]

Solution : Given,

$$kx + 3y - (k - 3) = 0 \quad ...(i)$$

and $$12x + ky - k = 0 \quad ...(ii)$$

For this pair of equations to have infinite number of solutions, it must satisfy the condition of

$$\frac{k}{12} = \frac{3}{k} = \frac{k-3}{k}$$

$\Rightarrow \quad k^2 = 12(k - 3)$

$\Rightarrow \quad k^2 - 12k + 36 = 0$

$\Rightarrow \quad k^2 - 6k - 6k + 36 = 0$

$\Rightarrow \quad k(k - 6) - 6(k - 6) = 0$

$\Rightarrow \quad (k - 6)^2 = 0$

$\Rightarrow \quad k = 6$

Also $\quad k^2 = 36$

$\quad k = \pm 6$

Common value $\quad k = 6$

Thus, the pair of equations will have infinite number of solutions for $k = 6$. **Ans.**

Q. 28. Determine the value of k, for which the given system of equations has no solution :

$3x + y = 1$ and $(2k - 1)x + (k - 1)y = 2k + 1$.

[NCERT]

Solution : Given, $\quad 3x + y = 1 \quad ...(i)$

and $\quad (2k - 1)x + (k - 1)y = 2k + 1 \quad ...(ii)$

For this pair of equations not to have any solution, it must satisfy the condition of

$$\frac{3}{2k-1} = \frac{1}{k-1} \neq \frac{1}{2k+1}$$

$\Rightarrow \quad 3(k - 1) = 2k - 1$

$\Rightarrow \quad 3k - 3 - 2k + 1 = 0$

$\Rightarrow \quad k - 2 = 0$

$\Rightarrow \quad k = 2$

Thus, the pair of equations will have no solution for $k = 2$. **Ans.**

Q. 29. Determine the value of k for which the given system of equations has infinitely many solutions :

$x + (k + 1)y = 5$ and $(k + 1)x + 9y = 8k - 1$.

Solution : Given,

$$x + (k + 1)y - 5 = 0 \quad ...(i)$$

and $$(k + 1)x + 9y - (8k - 1) = 0 \quad ...(ii)$$

For this pair of equations to have infinite number of solutions, it must satisfy the condition of

$$\frac{1}{k+1} = \frac{k+1}{9} = \frac{5}{8k-1}$$

$\Rightarrow \quad 8k - 1 = 5(k + 1)$

$\Rightarrow \quad 8k - 1 = 5k + 5$

$\Rightarrow \quad 3k = 6$

$\Rightarrow \quad k = 2$

Also $\quad \frac{1}{k+1} = \frac{k+1}{9}$

$(k + 1)^2 = 9$

$k + 1 = \pm 3$

$k = 2 \text{ or } -4$

Common value $\quad k = 2$

Thus, the pair of equations will have infinite number of solutions for $k = 2$. **Ans.**

Q. 30. For what value of k will the system of equations $x + 2y = 5$ and $3x + ky - 15 = 0$ has (i) a unique solution and (ii) infinite solutions ?

Solution : Given, $\quad x + 2y - 5 = 0 \quad ...(i)$

and $\quad 3x + ky - 15 = 0 \quad ...(ii)$

(i) The system of equations will have a unique solution, if

$$\frac{1}{3} \neq \frac{2}{k}$$

or $\quad k \neq 6$ **Ans**

(ii) The system of equations will have infinite solutions, if

$$\frac{1}{3} = \frac{2}{k} = \frac{5}{15}$$

or $$\frac{1}{3} = \frac{2}{k} = \frac{1}{3}$$

Thus, $\quad k = 6.$ **Ans.**

Q. 31. Determine the value of α and β for which the given system of equations has infinitely many solutions :
$2x + 3y = 7$ and $2\alpha x + (\alpha + \beta)y = 28$.

Solution : Given, $2x + 3y - 7 = 0 \quad ...(i)$

and $\quad 2\alpha x + (\alpha + \beta)y - 28 = 0 \quad ...(ii)$

For this pair of equations to have infinite number of solution, it must satisfy the condition of

$$\frac{2}{2\alpha} = \frac{3}{\alpha+\beta} = \frac{7}{28}$$

$$\Rightarrow \quad \frac{1}{\alpha} = \frac{3}{\alpha+\beta} = \frac{1}{4}$$

$\Rightarrow \quad \alpha+\beta=12 \quad \ldots$(iii)

and $\quad \alpha=4$

From equation (iii),

$4+\beta=12$

$\Rightarrow \quad \beta=8$

Thus, the pair of equations will have infinite number of solutions for $\alpha=4$ and $\beta=8$. **Ans.**

Q. 32. For what value of k will the system of equations $(3k+1)x+3y-2=0$ and $(k^2+1)x+(k-2)y-5=0$ has no solution ?

Solution : Given, $(3k+1)x+3y-2=0 \quad \ldots$(i)

and $\quad (k^2+1)x+(k-2)y-5=0 \quad \ldots$(ii)

The system of equations will have no solution, if

$$\frac{3k+1}{k^2+1}=\frac{3}{k-2}\neq\frac{2}{5}$$

$\Rightarrow \quad (3k+1)(k-2)=3(k^2+1)$

$\Rightarrow \quad 3k^2+k-6k-2=3k^2+3$

$\Rightarrow \quad -5k=5$

$\Rightarrow \quad k=-1$

Clearly, for $k=-1$, $\dfrac{3}{k-2}\neq\dfrac{2}{5}$

Thus, for $k=-1$ the pair of equations will have no solution. **Ans.**

Q. 33. For what value of k will the system of equations $4x+ky+8=0$ and $2x+2y+2=0$ has a unique solution ? [NCERT]

Solution : The system of equations will have a unique solution, if

$$\frac{4}{2}\neq\frac{k}{2}$$

or $\quad k\neq 4$. **Ans.**

Q. 34. For what value of k will the system of equations $kx+3y=(2k+1)$ and $2(k+1)x+9y=(7k+1)$ has infinite solutions ?

Solution : Given, $kx+3y-(2k+1)=0 \quad \ldots$(i)

and $\quad 2(k+1)x+9y-(7k+1)=0 \quad \ldots$(ii)

For pair of equations to have infinite number of solutions, it must satisfy the condition of

$$\frac{k}{2(k+1)}=\frac{3}{9}=\frac{2k+1}{7k+1}$$

$\Rightarrow \quad \dfrac{3}{9}=\dfrac{2k+1}{7k+1}$

$21k+3=18k+9$

$3k=6$

$k=2$

Also $\quad 9k=6(k+1)$

$\Rightarrow \quad 9k-6k=6$

$\Rightarrow \quad 3k=6$

$\Rightarrow \quad k=2$

Thus, the pair of equations will have infinite number of solutions for $k=2$. **Ans.**

Q. 35. For what value of k will the system of equations $2x+(k-2)y=k$ and $6x+(2k-1)y=2k+5$ has infinite solutions ?

Solution : Given, $2x+(k-2)y-k=0 \quad \ldots$(i)

and $\quad 6x+(2k-1)y-(2k+5)=0 \quad \ldots$(ii)

For this pair of equations to have infinite number of solutions, it must satisfy the condition of

$$\frac{2}{6}=\frac{k-2}{2k-1}=\frac{k}{2k+5}$$

$2(2k-1)=6(k-2)$

$4k-2=6k-12$

$2k=10$

$k=5$

Also $\quad \dfrac{k-2}{2k-1}=\dfrac{k}{2k+5}$

$(k-2)(2k+5)=k(2k-1)$

$2k^2-4k+5k-10=2k^2-k$

$2k^2+k-10=2k^2-k$

$2k=10$

$k=5$

Thus, the pair of equations will have infinite number of solutions for $k=5$. **Ans.**

Q. 36. For what value of α will the system of equations $\alpha x+3y=\alpha-3$ and $12x+\alpha y=\alpha$ has no solution ?★

Solution : The system of equations will have no solution, if

$$\frac{\alpha}{12}=\frac{3}{\alpha}\neq\frac{\alpha-3}{\alpha}$$

$\Rightarrow \quad \alpha^2=36$

$\alpha=\pm 6$

Also $\quad 3\alpha\neq\alpha(\alpha-3)$

$3\alpha\neq\alpha^2-3\alpha$

$\alpha^2\neq 6\alpha$

$\alpha\neq 6$

Since $\alpha=6$ and $\alpha\neq 6$ is not possible, so for $\alpha=-6$, the pair of equations will have no solution. **Ans.**

★ are board exam questions from previous years

Q. 37. Determine the value of k for which the given system of equations have no solution :

$3x - y - 5 = 0$ and $6x - 2y + k = 0$ where $k \neq 0$.

Solution : Given, $3x - y - 5 = 0$ ...(i)

and $6x - 2y + k = 0$ ...(ii)

For this pair of equations not to have any solution, it must satisfy the condition of

$$\frac{3}{6} = \frac{1}{2} \neq -\frac{5}{k}$$

$\Rightarrow$ $k \neq -10$

Thus, the pair of equations will have no solution for $k \neq -10$. **Ans.**

Q. 38. The age of the father is twice the sum of the ages of his 2 children. After 20 years, his age will be equal to the sum of the ages of his children. Find the age of the father.

Solution : Let the sum of ages of the 2 children be x and the age of the father be y years.

According to the question,

$$y = 2x$$

$\Rightarrow$ $2x - y = 0$ ...(i)

and $x + 40 = y + 20$

$\Rightarrow$ $x - y = 20$ ...(ii)

Subtracting equations (i) and (ii)

$$2x - y = 0$$
$$x - y = -20$$
$$- \quad + \quad (-)$$
$$x = 20$$

$\therefore$ $y = 2 \times 20 = 40$

Thus, Father's age $= 40$ years **Ans.**

Chapter 4. Quadratic Equations

Q. 1. For which value of p will $(3p + 1)x^2 + 4x - 2 = 0$ not be a quadratic equation ?

Solution : $(3p + 1)x^2 + 4x - 2 = 0$

$\Rightarrow$ $x^2 + \frac{4}{3p+1}x - \frac{2}{3p+1} = 0$

Put $3p + 1 = 0$

Thus, $p = -\frac{1}{3}$

Hence, $(3p + 1)x^2 + 4x - 2 = 0$ will not be a quadratic equation for $p = -\frac{1}{3}$. **Ans.**

Q. 2. Find the marks obtained by Neha in two subjects, if their average is 75 and their product is 5600.

Solution : Let the marks obtained by Neha in the two subjects be x and y.

Hence, $xy = 5600$ and $\frac{x+y}{2} = 75$

Thus, $x + y = 150$

or $x = 150 - y$

and $xy = 5600$

$\Rightarrow$ $(150 - y)y = 5600$

$\Rightarrow$ $150y - y^2 = 5600$

$\Rightarrow$ $y^2 - 150y + 5600 = 0$

$\Rightarrow$ $y^2 - 80y - 70y + 5600 = 0$

$\Rightarrow$ $y(y - 80) - 70(y - 80) = 0$

$\Rightarrow$ $(y - 80)(y - 70) = 0$

$\Rightarrow$ $y = 70$ or 80

and $x = 80$ or 70 **Ans.**

Q. 3. The perimeter of an isosceles triangle is 65 cm and the unequal side is thrice as large as each of the equal sides. Find the lengths of the sides.

Solution : Given, Perimeter of an isosceles triangle $= 65$ cm

Let the sides of the triangle be a cm, a cm and b cm.

Then $b = 3a$

Also $a + b + a = 65$

$\Rightarrow$ $a + 3a + a = 65$

$\Rightarrow$ $5a = 65$

$\Rightarrow$ $a = 13$ cm

and $b = 39$ cm **Ans.**

Q. 4. If $y = 1$ is a common root of the equations $ay^2 + ay + 3 = 0$ and $y^2 + y + b = 0$, then find ab.

Solution : The given equations are

$$ay^2 + ay + 3 = 0 \text{ and } y^2 + y + b = 0$$

Substituting $y = 1$ in them, we have

$$a(1)^2 + a(1) + 3 = 0 \text{ and } (1)^2 + (1) + b = 0$$

$\Rightarrow$ $2a + 3 = 0$ and $2 + b = 0$

$\Rightarrow$ $a = -\frac{3}{2}$ and $b = -2$

Thus, $ab = \left(-\frac{3}{2}\right)(-2) = 3$ **Ans.**

Q. 5. Find the nature of roots of the quadratic equation $2x^2 - 4x + 3 = 0$.★

Solution : Given, $2x^2 - 4x + 3 = 0$

Comparing it with quadratic equation

$$ax^2 + bx + c = 0$$

★ are board exam questions from previous years

Here, $a = 2, b = -4$ and $c = 3$

$$\therefore \quad D = b^2 - 4ac = (-4)^2 - 4 \times (2)(3) = 16 - 24 = -8 < 0$$

Hence, $D < 0$ this shows that roots will be imaginary. **Ans.**

Q. 6. Is the roots of the equation $x^2 - 3x - 9 = 0$ are real and distinct?

Solution : Given, $x^2 - 3x - 9 = 0$

Now,
$$D = b^2 - 4ac = (-3)^2 - 4(1)(-9) = 9 + 36 = 45 > 0$$

Thus, the roots are real and distinct. **Ans.**

Q. 7. For what values of k, the roots of the equation $x^2 + 4x + k = 0$ are real ?★

Solution : The given equation is $x^2 + 4x + k = 0$

On comparing the given equation with $ax^2 + bx + c = 0$, we get

$a = 1, b = 4$ and $c = k$

$a = 1, b = 4$ and $c = k$

For real roots, $D \geq 0$

or $b^2 - 4ac \geq 0$

$16 - 4k \geq 0$

or $k \leq 4$

$\therefore$ For $k \leq 4$, equation $x^2 + 4x + k$ will have real roots. **Ans.**

Q. 8. Is the value of x in $x - \frac{18}{x} = 6$ is imaginary ?

Solution : Given,

$$x - \frac{18}{x} = 6$$

or $x^2 - 6x - 18 = 0$

Now,
$$D = b^2 - 4ac = (-6)^2 - 4(-18) = 36 + 72 = 108 > 0$$

Thus, the roots are real and distinct.

So the value of x is not imaginary. **Ans.**

Q. 9. The solution of a quadratic equation is as follows :

$$x = \frac{8 \pm \sqrt{(-8)^2 - 4(3)(2)}}{2(3)}.$$

Then find the quadratic equation.

Solution : Given, $x = \dfrac{8 \pm \sqrt{(-8)^2 - 4(3)(2)}}{2(3)}$

Thus $a = 3, b = -8$ and $c = 2$

The equation is given by $ax^2 + bx + c = 0$

Hence $3x^2 - 8x + 2 = 0$ **Ans.**

Q. 10. Solve :

$$\sqrt{3}x^2 - 2\sqrt{2}x - 2\sqrt{3} = 0.$$

Solution : Given, $\sqrt{3}x^2 - 2\sqrt{2}x - 2\sqrt{3} = 0$

$$\Rightarrow \quad x = \frac{2\sqrt{2} \pm \sqrt{(-2\sqrt{2})^2 - 4(\sqrt{3})(-2\sqrt{3})}}{2(\sqrt{3})}$$

$$\left[\text{applying the formula : } x = \frac{-b \pm \sqrt{b^2 - 4ac}}{2a}\right]$$

$$\Rightarrow \quad x = \frac{2\sqrt{2} \pm \sqrt{8 + 24}}{2\sqrt{3}} = \frac{2\sqrt{2} \pm \sqrt{32}}{2\sqrt{3}}$$

$$= \frac{\sqrt{2} \pm 2\sqrt{2}}{\sqrt{3}} = \sqrt{6} \text{ and } -\sqrt{\frac{2}{3}}.$$ **Ans.**

Q. 11. Solve :

$$4\sqrt{3}x^2 - 5x - 2\sqrt{3} = 0.$$

Solution : Given,

$$4\sqrt{3}x^2 - 5x - 2\sqrt{3} = 0$$

$$\Rightarrow \quad 4\sqrt{3}x^2 - 8x + 3x - 2\sqrt{3} = 0$$

$$\Rightarrow \quad 4x(\sqrt{3}x - 2) - \sqrt{3}(\sqrt{3}x - 2) = 0$$

$$\Rightarrow \quad (\sqrt{3}x - 2)(4x - \sqrt{3}) = 0$$

$$\Rightarrow \quad x = \frac{2}{\sqrt{3}} \text{ and } \frac{\sqrt{3}}{4}.$$

Ans.

Q. 12. Solve : $abx^2 + (b^2 - ac)x - bc = 0$.

Solution : Given,

$$abx^2 + (b^2 - ac)x - bc = 0$$

$$\Rightarrow \quad abx^2 + b^2x - acx - bc = 0$$

$$\Rightarrow \quad bx(ax + b) - c(ax + b) = 0$$

$$\Rightarrow \quad (ax + b)(bx - c) = 0$$

$$\Rightarrow \quad x = -\frac{b}{a} \text{ and } \frac{c}{b}.$$

Ans.

Q. 13. Solve for x : $\sqrt{6x + 7} - (2x - 7) = 0$

Solution : We have,

$$\sqrt{6x + 7} - (2x - 7) = 0$$

$$\sqrt{6x + 7} = (2x - 7)$$

Squaring both sides

$$\left(\sqrt{6x + 7}\right)^2 = (2x - 7)^2$$

$$\Rightarrow \quad 6x + 7 = 4x^2 + 49 - 28x$$

$$\Rightarrow \quad 4x^2 + 42 - 34x = 0$$

$$\Rightarrow \quad 2x^2 - 17x + 21 = 0$$

★ are board exam questions from previous years

$\Rightarrow \quad 2x^2 - 14x - 3x + 21 = 0$

$\Rightarrow \quad 2x(x-7) - 3(x-7) = 0$

$\Rightarrow \quad (2x-3)(x-7) = 0$

$\Rightarrow \quad x = \frac{3}{2}$ or 7

$\therefore$ $x = 7$ (as $x = 3/2$ doesn't satisfy the given equation)

Ans.

Q. 14. Solve : $12abx^2 - (9a^2 - 8b^2)x - 6ab = 0$.

Solution : Given,

$$12abx^2 - (9a^2 - 8b^2)x - 6ab = 0$$

$\Rightarrow \quad 12abx^2 - 9a^2x + 8b^2x - 6ab = 0$

$\Rightarrow \quad 3ax(4bx - 3a) + 2b(4bx - 3a) = 0$

$\Rightarrow \quad (3ax + 2b)(4bx - 3a) = 0$

$\Rightarrow \quad x = -\frac{2b}{3a}$ and $\frac{3a}{4b}$. **Ans.**

Q. 15. Solve : $p^2x^2 + (p^2 - q^2)x - q^2 = 0$.

Solution : Given,

$$p^2x^2 + (p^2 - q^2)x - q^2 = 0$$

$\Rightarrow \quad p^2x^2 + p^2x - q^2x - q^2 = 0$

$\Rightarrow \quad p^2x(x+1) - q^2(x+1) = 0$

$\Rightarrow \quad (p^2x - q^2)(x+1) = 0$

$\Rightarrow \quad x = \left(\frac{q}{p}\right)^2$ and -1. **Ans.**

Q. 16. Solve for x : $\sqrt{2x+9} + x = 13$

Solution : We have,

$$\sqrt{2x+9} + x = 13$$

$$\sqrt{2x+9} = 13 - x$$

Squaring both sides

$$\left(\sqrt{2x+9}\right)^2 = (13-x)^2$$

$$2x + 9 = 169 + x^2 - 26x$$

$$x^2 - 28x + 160 = 0$$

$$x^2 - 20x - 8x + 160 = 0$$

$$x(x-20) - 8(x-20) = 0$$

$$(x-8)(x-20) = 0$$

$$x = 20 \text{ or } 8$$

$\therefore$ $x = 8$ (as $x = 20$ doesn't satisfy the given equation).

Ans.

Q. 17. If the roots of the equation

$$(a-b)x^2 + (b-c)x + (c-a) = 0$$

are equal, prove that $b + c = 2a$.

Solution : Given, $(a-b)x^2 + (b-c)x + (c-a) = 0$

Comparing with $Ax^2 + Bx + C = 0$, we get

$$A = a - b,\ B = b - c \text{ and } C = c - a$$

Since, the roots are equal

$\therefore \quad D = 0$

$\Rightarrow \quad B^2 - 4AC = 0$

$\Rightarrow (b-c)^2 - 4(a-b)(c-a) = 0$

$\Rightarrow b^2 + c^2 - 2bc - 4(ac - bc - a^2 + ab) = 0$

$\Rightarrow b^2 + c^2 - 2bc - 4ac + 4bc + 4a^2 - 4ab = 0$

$\Rightarrow \quad b^2 + c^2 + 4a^2 + 2bc - 4ac - 4ab = 0$

$\Rightarrow \quad (b + c - 2a)^2 = 0$

$\Rightarrow \quad b + c - 2a = 0$

$\Rightarrow \quad b + c = 2a.$

Hence Proved.

Q. 18. Find the value of k for which the roots of $9x^2 + 8kx + 16 = 0$ are real and equal.

Solution : Given,

$$9x^2 + 8kx + 16 = 0$$

Since, the roots are real and equal,

$\therefore \quad D = (8k)^2 - 4(9)(16) = 0$

$\Rightarrow \quad 64k^2 = 64(9)$

$\Rightarrow \quad k^2 = 9$

$\Rightarrow \quad k = 3$ and -3. **Ans.**

Q. 19. Find the value of k for which the equation $x^2 + k(2x + k - 1) + 2 = 0$ has real and equal roots.

Solution : Given equation is,

$$x^2 + k(2x + k - 1) + 2 = 0$$

$\Rightarrow \quad x^2 + 2kx + k(k-1) + 2 = 0$

Here $\quad a = 1$, $b = 2k$ and $c = k(k-1) + 2$

For real and equal roots

$$b^2 - 4ac = 0$$

$\Rightarrow (2k)^2 - 4.1.[(k(k-1) + 2)] = 0$

$\Rightarrow \quad 4k^2 - 4(k^2 - k + 2) = 0$

$\Rightarrow \quad 4k^2 - 4k^2 + 4k - 8 = 0$

$\Rightarrow \quad 4k = 8$

$\Rightarrow \quad k = \frac{8}{4} = 2$ **Ans.**

Q. 20. Find the value of k for which the roots of $3x^2 + 2kx + 27 = 0$ are real and equal.

Solution : Given,

$$3x^2 + 2kx + 27 = 0$$

Since, the roots are real and equal,

$\therefore \quad D = (2k)^2 - 4(3)(27) = 0$

$\Rightarrow \quad (2k)^2 = 4(81)$

$\Rightarrow \quad 4k^2 = 4(81)$

$\Rightarrow \quad k^2 = 81$

$\Rightarrow \quad k = 9$ and -9. **Ans.**

Q. 21. A teacher on attempting to arrange the students for mass drill in the form of a solid square found that 24 students were left. When he increased the square by one row and one column, he was short of 25 students. Find the number of students.

Solution : Let the number of rows be x and the number of students in each row also be x.

Thus $\quad x^2 + 24 = (x+1)^2 - 25$

$\Rightarrow \quad (x+1)^2 - x^2 = 24 + 25$

$\Rightarrow \quad (x+1-x)(x+1+x) = 49$

$\Rightarrow \quad 2x + 1 = 49$

$\Rightarrow \quad 2x = 48$

$\Rightarrow \quad x = 24$

Thus, the total number of students $= x^2 + 24$

$= (24)^2 + 24 = 24(24+1)$

$= 24 \times 25 = 600$. **Ans.**

Q. 22. If one of the roots of the quadratic equation $2x^2 + kx - 6 = 0$ is 2, find the value of k. Also, find the other root.

Solution : Given,

$$2x^2 + kx - 6 = 0$$

As 2 is one of the roots, therefore, substituting 2 for x, we get

$$2(2)^2 + 2k - 6 = 0$$

$\Rightarrow \quad 8 + 2k - 6 = 0$

$\Rightarrow \quad 2k + 2 = 0$

$\Rightarrow \quad k = -1$

Substituting -1 for k, we get

$$2x^2 - x - 6 = 0$$

$\Rightarrow \quad 2x^2 - 4x + 3x - 6 = 0$

$\Rightarrow \quad 2x(x-2) + 3(x-2) = 0$

$\Rightarrow \quad (x-2)(2x+3) = 0$

$\Rightarrow \quad x = 2$ and $-\dfrac{3}{2}$

Thus, the other root is $-\dfrac{3}{2}$. **Ans.**

Q. 23. Solve : $4x^2 - 4a^2x + (a^4 - b^4) = 0$.

Solution : Given,

$$4x^2 - 4a^2x + (a^4 - b^4) = 0$$

$\Rightarrow \quad 4x^2 - 4a^2x + (a^2 - b^2)(a^2 + b^2) = 0$

$\Rightarrow 4x^2 - 2(a^2 - b^2)x - 2(a^2 + b^2)x + (a^2 - b^2)(a^2 + b^2) = 0$

$\Rightarrow 2x[2x - (a^2 - b^2)] - (a^2 + b^2)[2x - (a^2 - b^2)] = 0$

$\Rightarrow \quad [2x - (a^2 - b^2)][2x - (a^2 + b^2)] = 0$

$\Rightarrow \quad x = \dfrac{a^2 - b^2}{2}$ and $\dfrac{a^2 + b^2}{2}$. **Ans.**

Q. 24. Solve : $\sqrt{2}x^2 + 7x + 5\sqrt{2} = 0$. **[NCERT]**

Solution : Given,

$$\sqrt{2}x^2 + 7x + 5\sqrt{2} = 0$$

$\Rightarrow \quad \sqrt{2}x^2 + 5x + 2x + 5\sqrt{2} = 0$

$\Rightarrow \quad x(\sqrt{2}x + 5) + \sqrt{2}(\sqrt{2}x + 5) = 0$

$\Rightarrow \quad (\sqrt{2}x + 5)(x + \sqrt{2}) = 0$

$\Rightarrow \quad x = -\dfrac{5}{\sqrt{2}}$ and $-\sqrt{2}$. **Ans.**

Q. 25. Find the value of p, for which one root of the quadratic equation $px^2 - 14x + 8 = 0$ is 6 times the other.

Solution : Given equation is,

$$px^2 - 14x + 8 = 0$$

Let $\quad$ one root $= \alpha$

Then $\quad$ other root $= 6\alpha$

Sum of roots $= -\dfrac{b}{a}$

$\Rightarrow \quad \alpha + 6\alpha = \dfrac{-(-14)}{p}$

$\Rightarrow \quad 7\alpha = \dfrac{14}{p}$;

$\Rightarrow \quad \alpha = \dfrac{14}{p \times 7}$

or $\quad \alpha = \dfrac{2}{p}$ ...(i)

Product of roots $= \dfrac{c}{a}$

$\Rightarrow \quad (\alpha)(6\alpha) = \dfrac{8}{p}$

$\Rightarrow \quad 6\alpha^2 = \dfrac{8}{p}$ ...(ii)

Putting value of α from equation (i) to equation (ii), we get

$$6\left(\frac{2}{p}\right)^2 = \frac{8}{p}$$

$$\Rightarrow \quad 6 \times \frac{4}{p^2} = \frac{8}{p}$$

$$\Rightarrow \quad 24p = 8p^2$$

$$\Rightarrow \quad 8p^2 - 24p = 0$$

$$\Rightarrow \quad 8p(p-3) = 0$$

Either $8p = 0 \Rightarrow p = 0$

or $p - 3 = 0 \Rightarrow p = 3$

$\therefore p = 3$ (as $p = 0$ doesn't satisfy the given equation) **Ans.**

Q. 26. Find the value of k for which the equation $(k+1)x^2 - 2(k-1)x + 1 = 0$ has real and equal roots.

Solution : Given,

$$(k+1)x^2 - 2(k-1)x + 1 = 0$$

Since, the equation has real and equal roots,

$$\therefore \quad D = [2(k-1)]^2 - 4(k+1)(1) = 0$$

$$\Rightarrow \quad 4k^2 + 4 - 8k - 4k - 4 = 0$$

$$\Rightarrow \quad 4k^2 - 12k = 0$$

$$\Rightarrow \quad 4k(k-3) = 0$$

$$\Rightarrow \quad k = 0 \text{ or } 3$$ **Ans.**

Q. 27. If -5 is a root of the equation $2x^2 + px - 15 = 0$ and the quadratic equation $p(x^2 + x) + k = 0$ has equal roots, find the value of k.

Solution : As -5 is a root of the equation $2x^2 + px - 15 = 0$,

$$\therefore \quad 2(-5)^2 + p(-5) - 15 = 0$$

$$\Rightarrow \quad 2(25) - 5p - 15 = 0$$

$$\Rightarrow \quad 5p = 50 - 15$$

$$\Rightarrow \quad 5p = 35$$

$$\Rightarrow \quad p = 7$$

Now substituting $p = 7$ in $p(x^2 + x) + k = 0$, we have

$$7(x^2 + x) + k = 0$$

$$\Rightarrow \quad 7x^2 + 7x + k = 0$$

Since, the above equation has equal roots,

$$\therefore \quad D = (7)^2 - 4(7)(k) = 0$$

$$\Rightarrow \quad 49 - 28k = 0$$

$$\Rightarrow \quad 28k = 49$$

$$\Rightarrow \quad 4k = 7$$

$$k = \frac{7}{4}$$

Thus, the equation will have real roots, if $k = \frac{7}{4}$. **Ans.**

Q. 28. Find the value of k for which the equation $x^2 + 2(k+1)x + k^2 = 0$ has real and equal roots.

Solution : Given,

$$x^2 + 2(k+1)x + k^2 = 0$$

Since, the equation has real and equal roots,

$$\therefore \quad D = [2(k+1)]^2 - 4(1)(k)^2 = 0$$

$$\Rightarrow \quad 4(k^2 + 1 + 2k) - 4k^2 = 0$$

$$\Rightarrow \quad 4k^2 + 4 + 8k - 4k^2 = 0$$

$$\Rightarrow \quad 8k = -4$$

$$\Rightarrow \quad k = -\frac{4}{8}$$

$$\Rightarrow \quad k = -\frac{1}{2}$$

Thus, the equation will have real roots, if $k = -\frac{1}{2}$ **Ans.**

Q. 29. Find the value of k for which the roots of the equation $3x^2 - 10x + k = 0$ are reciprocal of each other.★

Solution : The given equation is $3x^2 - 10x + k = 0$

On comparing it with $ax^2 + bx + c = 0$, we get

$$a = 3, b = -10, c = k$$

Let the roots of the equation are α and $\frac{1}{\alpha}$

$\because$ Product of the roots $= \frac{c}{a}$

$$\therefore \quad \alpha . \frac{1}{\alpha} = \frac{k}{3}$$

or $k = 3$ **Ans.**

Q. 30. Find the value of k for which the equation $x^2 - 4kx + k = 0$ has real and equal roots.

Solution : Given,

$$x^2 - 4kx + k = 0$$

Since, the equation has real and equal roots,

$$\therefore \quad D = (-4k)^2 - 4(1)(k) = 0$$

$$\Rightarrow \quad 16k^2 - 4k = 0$$

$$\Rightarrow \quad 4k(4k - 1) = 0$$

$$\Rightarrow \quad k = 0, \frac{1}{4}$$

Thus, the equation will have real roots, if $k = 0, \frac{1}{4}$. **Ans.**

Q. 31. Find the value of k for which the equation $kx(x - 2\sqrt{5}) + 10 = 0$ has real and equal roots.

Solution : Given,

$$kx(x - 2\sqrt{5}) + 10 = 0$$

★ are board exam questions from previous years

$\Rightarrow \quad kx^2 - 2\sqrt{5}kx + 10 = 0$

Since, the equation has real and equal roots,

$\therefore \quad D = (-2\sqrt{5}k)^2 - 4(10)(k) = 0$

$\Rightarrow \quad 20k^2 - 40k = 0$

$\Rightarrow \quad 20k(k-2) = 0$

$\Rightarrow \quad k = 0, 2$

But $k = 0$ is not possible.

Thus, the equation will have equal roots, if $k = 2$.

Ans.

Q. 32. If the quadratic equation $px^2 - 2\sqrt{5}\,px + 15 = 0$, has two equal roots then find the value of p.

Solution : The given quadratic equation is,

$$px^2 - 2\sqrt{5}\,px + 15 = 0$$

This is of the form

$$ax^2 + bx + c = 0$$

where, $a = p, b = -2\sqrt{5}\,p, c = 15$

We know, $D = b^2 - 4ac$

$$= \left(-2\sqrt{5}p\right)^2 - 4 \times p \times 15$$

$$= 20p^2 - 60p$$

$$= 20p\,(p-3)$$

For real and equal roots, we must have

$$D = 0 \Rightarrow 20p\,(p-3) = 0$$

$$\Rightarrow p = 0, p = 3$$

$p = 0$, is not possible as whole equation will be zero.

Hence, 3 is the required value of p. **Ans.**

Q. 33. Find two consecutive numbers whose squares have the sum 85.

Solution : Let the two numbers be x and $(x + 1)$.

Thus, $\quad x^2 + (x+1)^2 = 85$

$\Rightarrow \quad x^2 + x^2 + 2x + 1 = 85$

$\Rightarrow \quad 2x^2 + 2x - 84 = 0$

$\Rightarrow \quad x^2 + x - 42 = 0$

$\Rightarrow \quad x^2 + 7x - 6x - 42 = 0$

$\Rightarrow \quad x(x+7) - 6(x+7) = 0$

$\Rightarrow \quad (x+7)(x-6) = 0$

$\Rightarrow \quad x = 6, -7$

Thus, the two numbers are either 6 and 7 or – 7 and – 6. **Ans.**

Q. 34. A peacock is sitting on the top of a pillar which is 9 m high. From a point, 27 m away from the bottom of a pillar, a snake is coming to its hole at the base of the pillar. Seeing the snake, the peacock pounces on it. If their speeds are equal, at what distance from the hole is the snake caught ?

Solution : Let the distance, where the snake is caught from the base of the pillar be x m. So the distance travelled by the snake before being caught is $(27 - x)$ m.

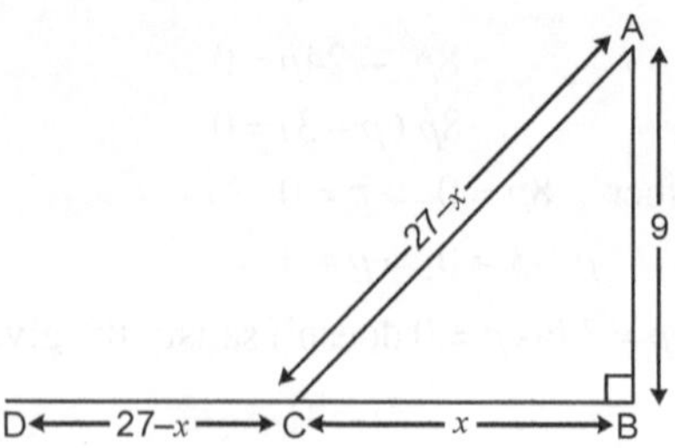

As the speed of the peacock and the snake is the same, so the distance travelled by the two is also same. Thus, the peacock travels $(27 - x)$ m from the top of the tower to catch the snake. Hence $(27 - x)$ is the hypotenuse. Applying Pythagoras theorem,

$$(9)^2 + x^2 = (27 - x)^2$$

$\Rightarrow \quad x^2 + 81 = 729 + x^2 - 54x$

$\Rightarrow \quad 54x = 729 - 81$

$\Rightarrow \quad 54x = 648$

$\Rightarrow \quad x = 12$ m

Hence, the snake is caught 12 m from the pillar. **Ans.**

Q. 35. Show that $x = -3$ is a solution of $x^2 + 6x + 9 = 0$.

Solution : The given equation is

$$x^2 + 6x + 9 = 0$$

Substituting $x = -3$, we have

$$\text{L.H.S.} = (-3)^2 + 6(-3) + 9$$

$$= 9 - 18 + 9$$

$$= 0 = \text{R.H.S.}$$

Thus, $x = -3$ is a solution of $x^2 + 6x + 9 = 0$.

Hence Proved.

Q. 36. Show that $x = -2$ is a solution of $3x^2 + 13x + 14 = 0$.

Solution : The given equation is,

$$3x^2 + 13x + 14 = 0$$

Substituting $x = -2$, we have

$$\text{L.H.S.} = 3(-2)^2 + 13(-2) + 14$$

$$= 12 - 26 + 14$$

$$= 0 = \text{R.H.S.}$$

Thus, $x = -2$ is a solution of $3x^2 + 13x + 14 = 0$.

Hence Proved.

Q. 37. Find the discriminant of the quadratic equation $3\sqrt{3}x^2 + 10x + \sqrt{3} = 0$.

Solution : The given equation is

$$3\sqrt{3}x^2 + 10x + \sqrt{3} = 0$$

Thus $D = (10)^2 - 4(3\sqrt{3})(\sqrt{3})$

or $D = 100 - 36$

$= 64$

Hence the discriminant of the quadratic equation $3\sqrt{3}x^2 + 10x + \sqrt{3} = 0$ is 64. **Ans.**

Chapter 5. Arithmetic Progressions

Q. 1. The first three terms of A.P. are (3*y* – 1), (3*y* + 5) and (5*y* + 1). Then find *y*.

Solution : The terms of an A.P. are

$(3y-1)$, $(3y+5)$ and $(5y+1)$

Thus, $d = (3y+5) - (3y-1) = (5y+1) - (3y+5)$

$\Rightarrow \quad 6 = 2y - 4$

$\Rightarrow \quad 2y = 10$

$\Rightarrow \quad y = 5.$ **Ans.**

Q. 2. If *k*, 2*k* – 1 and 2*k* + 1 are three consecutive terms of an A.P., then find the value of *k*.

Solution : The terms of an A.P. are

k, $(2k-1)$ and $(2k+1)$

Thus, $d = (2k-1) - k = 2k + 1 - (2k-1)$

$\Rightarrow \quad 2k - 1 - k = 2k + 1 - 2k + 1$

$\Rightarrow \quad k - 1 = 2$

$\Rightarrow \quad k = 3.$ **Ans.**

Q. 3. Find the next term of the A.P.

$\sqrt{7}, \sqrt{28}, \sqrt{63} \ldots$

Solution : The given series is $\sqrt{7}, \sqrt{28}, \sqrt{63} \ldots$

Thus, $a = \sqrt{7}$ and $d = \sqrt{28} - \sqrt{7} = \sqrt{63} - \sqrt{28}$

$\Rightarrow \quad d = 2\sqrt{7} - \sqrt{7} = 3\sqrt{7} - 2\sqrt{7} = \sqrt{7}$

Thus $t_4 = a + (4-1)d$

$\Rightarrow \quad t_4 = \sqrt{7} + (4-1)\sqrt{7}$

$\Rightarrow \quad t_4 = \sqrt{7} + 3\sqrt{7} = 4\sqrt{7}$

$= \sqrt{(16)7} = \sqrt{112}$ **Ans.**

Q. 4. Find the sum of first 20 odd natural numbers.

Solution : First 20 odd natural numbers are :

1, 3, 5, 7, 9, 11, 13, 15, 17, 19, 21, 23, 25, 27, 29, 31, 33, 35, 37, 39.

This is an A..P.

with $a = 1, d = 2$ and $n = 20$

We know that,

$$S_n = \frac{n}{2}\{2a + (n-1)d\}$$

$$= \frac{20}{2}\{2(1) + (20-1)2\}$$

$= 10\{2 + 38\} = 400$ **Ans.**

Q. 5. If the nth term of an A.P. is (2*n* + 1), then find the sum of its first three terms.

Solution : Here,

$a_n = 2n + 1$

$a_1 = 2(1) + 1 = 3$

$a_2 = 2(2) + 1 = 5$

$a_3 = 2(3) + 1 = 7$

$\therefore \quad a_1 + a_2 + a_3 = 3 + 5 + 7 = 15$ **Ans.**

Q. 6. Find the value of $t_{30} - t_{20}$ for the A.P. 2, 7, 12, 17, ...

Solution : Given A.P. is 2, 7, 12, 17, ...

Here, $a = 2, d = 7 - 2 = 5$

So $t_{30} = 2 + (30-1)5$

$= 2 + 145 = 147$

and $t_{20} = 2 + (20-1)5$

$= 2 + 95 = 97$

Thus $t_{30} - t_{20} = 147 - 97 = 50$ **Ans.**

Q. 7. If the common difference of an A.P. is 3, then find $t_{20} - t_{15}$.

Solution : Given,

$d = 3$

$\therefore \quad t_{20} = a + (20-1)3 = a + 57$

and $t_{15} = a + (15-1)3 = a + 42$

Thus $t_{20} - t_{15} = a + 57 - (a + 42) = 15.$ **Ans.**

Q. 8. In an A.P., if $d = -2$, $n = 5$ and $t_n = 0$, then find the value of *a*.

Solution : Given, $d = -2$, $n = 5$ and $t_n = 0$

We know that $t_n = a\,(n-1)\,d$

Thus $t_5 = a + (5-1)(-2) = 0$

$\Rightarrow \quad a + (4)(-2) = 0$

$\Rightarrow \quad a - 8 = 0$

$\Rightarrow \quad a = 8.$ **Ans.**

Q. 9. Write the value of *a* and *d* for the following A.P. – 5, – 1, 3, 7. [NCERT]

Solution : The given A.P. is – 5, – 1, 3, 7

Thus $a = -5$ and $d = -1 - (-5) = 4$ **Ans.**

Q. 10. For what value of *k* will *k* + 9, 2*k* – 1 and 2*k* + 7 are the consecutive terms of an A.P. ?

Solution : We have, $k + 9$, $2k - 1$ and $2k + 7$ as consecutive terms of an A.P.

Thus, $2(2k - 1) = k + 9 + 2k + 7$

[if a, b and c are in A.P. then $2b = a + c$]

$$4k - 2 = 3k + 16$$

$$k = 18$$ **Ans.**

Q. 11. Find the 9th term from the end (towards the first term) of the A.P. 5, 9, 13, ..., 185.

Solution : Given, A.P. is 5, 9, 13, ..., 185

Here, $l = 185$ and $d = 5 - 9 = 9 - 13 = -4$

Then, $a_9 = l + (n - 1)d$

$= 185 + (9 - 1)(-4)$

$= 185 + 8(-4)$

$\therefore$ $a_9 = 153$ **Ans.**

Q. 12. The first term of an A.P. is p and its common difference is q. Find the 10th term.

Solution : Given, $a = p$ and $d = q$

We know that

$$t_n = a + (n - 1)d$$

$\therefore$ $t_{10} = p + (10 - 1)(q)$

$= p + 9q.$ **Ans.**

Q. 13. For what value of p are $2p + 1$, 13, $5p - 3$ are the three consecutive terms of an A.P. ?

Solution : For $2p + 1$, 13, $5p - 3$ to be consecutive, the common difference should be the same.

Thus $13 - (2p + 1) = 5p - 3 - 13$

$\Rightarrow$ $12 - 2p = 5p - 16$

$\Rightarrow$ $7p = 28$

$\Rightarrow$ $p = 4.$ **Ans.**

Q. 14. If the sum of first p terms of A.P. is $ap^2 + bp$, find its common difference.

Solution : Given,

$$S_p = ap^2 + bp$$

$\therefore$ $S_1 = a(1)^2 + b(1)$

$= a + b$

and $S_2 = a(2)^2 + b(2)$

$= 4a + 2b$

Now, we know that

$$a_1 = S_1 = a + b$$

$$a_2 = S_2 - S_1$$

$= 4a + 2b - a - b$

$= 3a + b$

Now, $d = a_2 - a_1$

$= 3a + b - a - b$

$= 2a$

Hence the common difference is $2a$. **Ans.**

★ **are board exam questions from previous years**

Q. 15. Find the common difference of the A.P.

$$\frac{1}{2q}, \frac{1-2q}{2q}, \frac{1-4q}{2q} \ldots$$

Solution : The given A.P. is $\frac{1}{2q}, \frac{1-2q}{2q}, \frac{1-4q}{2q} \ldots$

Thus, common difference

$$= \frac{1-2q}{2q} - \frac{1}{2q}$$

$$= -\frac{2q}{2q} = -1$$

So, the common difference of AP is -1. **Ans.**

Q. 16. Find the common difference of the A.P.

$$\frac{1}{2b}, \frac{1-6b}{2b}, \frac{1-12b}{2b}, \ldots$$

Solution : The given is A.P. $\frac{1}{2b}, \frac{1-6b}{2b}, \frac{1-12b}{2b} \ldots$

Thus, the common difference

$$= \frac{1-6b}{2b} - \frac{1}{2b} = -3$$

So the common difference of given A.P. is -3.

Ans.

Q. 17. Find the common difference of the Arithmetic Progression (A.P.)★

$$\frac{1}{a}, \frac{3-a}{3a}, \frac{3-2a}{3a}, \ldots (a \neq 0).$$

Sol. Given, A.P. is $\frac{1}{a}, \frac{3-a}{3a}, \frac{3-2a}{3a}, \ldots.$

$$d = \frac{3-a}{3a} - \frac{1}{a}$$

$$= \frac{3-a-3}{3a}$$

$$= \frac{-a}{3a} = \frac{-1}{3}$$ **Ans.**

Q. 18. What is the common difference of an A.P. in which $a_{21} - a_7 = 84$?

Solution : Given, $a_{21} - a_7 = 84$

$\Rightarrow$ $(a + 20d) - (a + 6d) = 84$

$\Rightarrow$ $a + 20d - a - 6d = 84$

$\Rightarrow$ $20d - 6d = 84$

$\Rightarrow$ $14d = 84$

$\Rightarrow$ $d = \frac{84}{14} = 6$

Hence, common difference = 6. **Ans.**

Q. 19. The first and the last terms of an A.P. are 5 and 45 respectively. If the sum of all the terms is 400, find the common difference.

Solution : Given,

$$S_n = 400, a = 5 \text{ and } l = 45$$

We know $S_n = \frac{n}{2}[a+l]$

$\Rightarrow$ $400 = \frac{n}{2}[5+45]$

$\Rightarrow$ $n = \frac{400\times 2}{50} = 16$

Thus, applying $t_n = a + (n-1)d$, we have

$$45 = 5 + (16-1)d$$

$\Rightarrow$ $15d = 40$

$\Rightarrow$ $3d = 8$

$\Rightarrow$ $d = \frac{8}{3}$

Thus, the common difference is $\frac{8}{3}$. **Ans.**

Q. 20. Find how many integers between 200 and 500 are divisible by 8.

Solution : Smallest number divisible by 8 in the given range = 208

Last number divisible by 8 in range = 496

So, $a = 208$, $d = 8$, $n = ?$, $T_n = 496$

We know $T_n = a + (n-1)d$

$$496 = 208 + (n-1)\,8$$

or $8n + 208 - 8 = 496$

or $8n = 496 - 200$

$\Rightarrow$ $8n = 296$

$\Rightarrow$ $n = \frac{296}{8} = 37$

So number of terms between 200 and 500 divisible by 8 are 37. **Ans.**

Q. 21. Find the number of natural numbers between 101 and 999 which are divisible by both 2 and 5.

Solution : As the numbers are divisible by both 2 and 5, so their last digits will always be 0. Hence the series is 110, 120, 130, ..., 990.

Here $a = 110$, $d = 10$ and $t_n = 990$

We know that,

$$t_n = a + (n-1)d$$

$\Rightarrow$ $990 = 110 + (n-1)10$

$\Rightarrow$ $(n-1)10 = 880$

$\Rightarrow$ $n - 1 = 88$

$\Rightarrow$ $n = 89$

Thus, the number of natural numbers between 101 and 999 which are divisible by both 2 and 5 is 89. **Ans.**

Q. 22. The sum of the first n terms of an A.P. is $3n^2 + 6n$. Find the nth term of this A.P.

Solution : We have,

$$S_n = 3n^2 + 6n$$

$\therefore$ $S_{n-1} = 3(n-1)^2 + 6(n-1)$

Now, $t_n = S_n - S_{n-1}$

$= 3n^2 + 6n - 3(n-1)^2 - 6(n-1)$

$= 3n^2 + 6n - 3n^2 + 6n - 3 - 6n + 6$

$t_n = 6n + 3$ **Ans.**

OR

We know that

$$S_n = \frac{n}{2}\{2a + (n-1)d\}$$

Also, $S_n = 3n^2 + 6n$ (Given)

$\Rightarrow \frac{n}{2}[2a+(n-1)d] = 3n^2 + 6n$

$\Rightarrow$ $[2a + (n-1)d] = 6(n+2)$

$\Rightarrow$ $2a + dn - d = 6n + 12$

$\Rightarrow$ $dn + (2a - d) = 6n + 12$

$\therefore$ $d = 6$

and $2a - d = 12$

$\Rightarrow$ $2a - 6 = 12$

$\Rightarrow$ $2a = 18$

$\Rightarrow$ $a = 9$

Thus, $t_n = 9 + (n-1)6 = 9 + 6n - 6 = 6n + 3$. **Ans.**

Q. 23. The sum of the first n terms of an A.P. is $4n^2 + 2n$. Find the nth term of this A.P.

Solution : Given,

$$S_n = 4n^2 + 2n$$

Also, we know,

$$S_n = \frac{n}{2}[2a+(n-1)d]$$

$\therefore$ $\frac{n}{2}[2a+(n-1)d] = 4n^2 + 2n$

$\Rightarrow$ $2a + dn - d = 8n + 4$

$\Rightarrow$ $dn + (2a - d) = 8n + 4$

$\therefore$ $d = 8$

and $2a - d = 4$

$\Rightarrow$ $2a - 8 = 4$

$\Rightarrow \quad 2a = 12$

$\Rightarrow \quad a = 6$

Thus, $t_n = 6 + (n-1)8$

$= 6 + 8n - 8$

$= 8n - 2$ **Ans.**

Q. 24. If the numbers $2n-1, 3n+2$ and $6n-1$ are in A.P., find n and hence find the numbers.

Solution : The terms of an A.P. are $(2n-1)$, $(3n+2)$ and $(6n-1)$.

Thus, $d = (3n+2) - (2n-1) = (6n-1) - (3n+2)$

$\Rightarrow \quad n + 3 = 3n - 3$

$\Rightarrow \quad 2n = 6$

$\Rightarrow \quad n = 3$

Hence, the numbers are

$= 2(3) - 1, 3(3) + 2$ and $6(3) - 1$

$= 5, 11$ and 17. **Ans.**

Q. 25. Find the sum of all three-digit natural numbers which are multiples of 11.

Solution : The given A.P. series is

110, 121, 132, 143, ..., 990

Here, $a = 110, d = 11$ and $t_n = 990$

We know that

$t_n = a + (n-1)d$

$\Rightarrow \quad 990 = 110 + (n-1)11$

$\Rightarrow \quad 90 = 10 + n - 1$

$\Rightarrow \quad n = 81$

Now

$$S_n = \frac{n}{2}(a+l)$$

$$\Rightarrow \quad S_{81} = \frac{81}{2}(110+990)$$

$$= \frac{81}{2}(1100)$$

$= 81 \times 550$

$= 44550$. **Ans.**

Q. 26. How many two-digit numbers are divisible by 3 ?

Solution : Two-digit numbers divisible by 3 are 12, 15, 18, 21, ..., 99.

This is an A.P.,

$a = 12, d = 3$ and $t_n = 99$

We know that

with, $t_n = a + (n-1)d$

$\Rightarrow \quad 99 = 12 + (n-1)3$

$\Rightarrow \quad (n-1)3 = 87$

$\Rightarrow \quad n - 1 = 29$

$\Rightarrow \quad n = 30$. **Ans.**

Q. 27. Find an A.P. whose fourth term is 9 and the sum of its sixth term and thirteenth term is 40.

Solution : Let the first term be a and the common difference be d.

Thus, $t_4 = \{a + (4-1)d\} = 9$ (Given)

$\Rightarrow \quad a + 3d = 9$ ...(i)

Also $t_6 + t_{13} = 40$ (Given)

$\Rightarrow \quad a + 5d + a + 12d = 40$

$\Rightarrow \quad 2a + 17d = 40$ ...(ii)

Multiplying (i) with 2 and then subtracting from (ii), we have

$11d = 22$

$\Rightarrow \quad d = 2$

Thus, $a = 9 - 3(2) = 3$

Thus, the A.P. is 3, 5, 7, 9, 11, ... **Ans.**

Q. 28. Find whether – 150 is a term of the A.P. 17, 12, 7, 2, ..., ?

Solution : Given A.P. is 17, 12, 7, 2, ...

Here, $a = 17, d = 12 - 17 = -5$

Let – 150 be the nth term.

Thus $t_n = 17 + (n-1)(-5)$

$\Rightarrow \quad 17 - 5n + 5 = -150$

$\Rightarrow \quad -5n = -150 - 22$

$\Rightarrow \quad 5n = 172$

$$\Rightarrow \quad n = \frac{172}{5} = 34.4$$

Thus, – 150 is not a term of this A.P. as n is not a natural number. **Ans.**

Q. 29. Which term of the A.P. 5, 9, 13, 17, ... is 81 ? Also, find the sum.

Solution : From the given A.P, we have

$a = 5$ and $d = 9 - 5 = 4$

Let 81 be the nth term of the A.P.

Thus, $t_n = 5 + (n-1)(4) = 81$

$\Rightarrow \quad 5 + 4n - 4 = 81$

$\Rightarrow \quad 4n = 80$

$\Rightarrow \quad n = 20$

Thus, 81 is the 20th term of the A.P.

Hence $$S_{20} = \frac{20}{2}\{2(5) + (20-1)4\}$$

$= 20(5 + 38)$

$= 20 \times 43$

$= 860.$ **Ans.**

Q. 30. Which term of the A.P. 3, 15, 27, 39, ... will be 120 more than its 21st term ?

Solution : From the given A.P., we have

$a = 3, d = 12$ and $n = 21$

Thus, $t_{21} = 3 + (21 - 1)(12)$

$= 3 + 240 = 243$

Thus, the number 120 more than its 21st term is $243 + 120 = 363$

Let 363 be the mth term.

Hence $t_m = 3 + (m - 1)(12) = 363$

$\Rightarrow 3 + 12m - 12 = 363$

$\Rightarrow 12m - 9 = 363$

$\Rightarrow 12m = 372$

$\Rightarrow m = 31$

Thus, the 31st term will be 120 more than the 21st term. **Ans.**

Q. 31. Find the value of the middle term of the following A.P., 10, 7, 4, ..., (– 62).

Solution : From the given A.P., we have

$a = 10, d = -3$ and $t_n = -62$

Thus $t_n = 10 + (n - 1)(-3) = -62$

$\Rightarrow 10 - 3n + 3 = -62$

$\Rightarrow -3n = -75$

$\Rightarrow n = 25$

Thus, the middle term would be 13.

Hence $t_{13} = 10 + (13 - 1)(-3)$

$= 10 - 36 = -26.$ **Ans.**

Q. 32. Find the middle term of the A.P. 6, 13, 20, ... 216.

Solution : Given A.P. is 6, 13, 20,............, 216

Here, $a = 6, d = 13 - 6 = 20 - 13 = 7$

Let n be the number of terms.

Then $T_n = a + (n - 1)d$

$216 = 6 + (n - 1)7$

$216 = 6 + 7n - 7$

$217 = 7n$

$\therefore n = 31$

So middle term is $\left(\frac{n+1}{2}\right)^{th}$ term *i.e.*, 16th term

$\therefore T_{16} = 6 + (16 - 1)7$

$= 6 + 15 \times 7$

$\Rightarrow T_{16} = 111$

$\therefore$ Middle term of the A.P. is 111. **Ans.**

Q. 33. The 8th term of an A.P. is zero. Prove that the 38th term is triple its 18th term.

Solution : Let the first term of the A.P. be a and the common difference be d.

Thus $t_8 = a + (8 - 1)d = 0$ [Given]

or $a + 7d = 0$ ...(i)

Also $t_{18} = a + (18 - 1)d$

$= a + 17d$

$= (a + 7d) + 10d = 10d$ [from (i)]

and $t_{38} = a + (38 - 1)d$

$= a + 37d$

$= (a + 7d) + 30d$

$= 30d$ [from (i)]

$= 3(10d)$

$= 3t_{18}.$ **Hence Proved.**

Q. 34. If m times the mth term of an A.P. is equal to n times the nth term and $m \neq n$, then show that its $(m + n)^{th}$ term is zero.

Solution : Let the first term be a and the common difference be d.

Hence $t_m = a + (m - 1)d$ and $t_n = a + (n - 1)d$

Now $m\{a + (m - 1)d\} = n\{a + (n - 1)d\}$

$\Rightarrow ma + dm^2 - dm = na + dn^2 - dn$

$\Rightarrow (m - n)a + (m - n)(m + n)d - (m - n)d = 0$

$\Rightarrow a + (m + n)d - d = 0$

$\Rightarrow a + \{(m + n) - 1\}d = 0$

$\Rightarrow t_{m+n} = 0.$ **Hence Proved.**

Q. 35. For what value of n, are the nth terms of two A.Ps. 63, 65, 67,....... and 3, 10, 17,...... equal ?

Solution : 1st A.P. is 63, 65, 67, ...

Here, $a = 63, d = 65 - 63 = 2$

$\therefore a_n = a + (n - 1)d$

$= 63 + (n - 1)(2)$

$= 63 + 2n - 2 = 61 + 2n$

2nd A.P. is 3, 10, 17....

Here, $a = 3, d = 10 - 3 = 7$

$\therefore a_n = a + (n - 1)d$

$= 3 + (n - 1)(7)$

$= 3 + 7n - 7 = 7n - 4$

According to question,

$$61 + 2n = 7n - 4$$
$$61 + 4 = 7n - 2n$$
$$65 = 5n$$
$$n = \frac{65}{5} = 13$$
$$n = 13$$

Hence, 13th term of both A.P. is equal. **Ans.**

Q. 36. Which term of the A.P. 3, 8, 13, 18, ... will be 55 more than its 20th term ?

Solution : Given, $a = 3$, $d = 5$

Thus, $t_{20} = 3 + (20 - 1)5$
$= 3 + 95 = 98$

So $98 + 55 = 153$

Let the required term be n.

So $t_n = 3 + (n - 1)5 = 153$

$\Rightarrow$ $(n - 1)5 = 150$

$\Rightarrow$ $n - 1 = 30$

$\Rightarrow$ $n = 31.$ **Ans.**

Q. 37. Which term of the A.P. 5, 15, 25, ... will be 130 more than its 31st term ?

Solution : Given,

$a = 5, d = 10$

Thus, $t_{31} = 5 + (31 - 1)10$
$= 5 + 300$
$= 305$

So $305 + 130 = 435$

Let the required term be n.

So $t_n = 5 + (n - 1)10 = 435$

$\Rightarrow$ $(n - 1)10 = 430$

$\Rightarrow$ $n - 1 = 43$

$\Rightarrow$ $n = 44.$ **Ans.**

Q. 38. Find the 6th term from the end of the A.P. 17, 14, 11, ..., – 40.

Solution : Given, $a = 17, d = -3$ and $t_n = -40$

Now taking from the end

$a = -40, d = 3$ and $t_n = 17$

Hence, the 6th term from the end is

$t_6 = \{-40 + (6 - 1)3\}$
$= -40 + 15$
$= -25$

Thus, the 6th term from the end is – 25. **Ans.**

Q. 39. Find the 21st term of the A.P. ★

$$-4\frac{1}{2}, -3, -1\frac{1}{2}, \ldots$$

Solution : Given, , $-4\frac{1}{2}, -3. -1\frac{1}{2}, \ldots.$

$\Rightarrow$ $-\frac{9}{2}, -3, -\frac{3}{2}, \ldots.$

Here $a = -\frac{9}{2}$, $d = -3-\left(-\frac{9}{2}\right) = -3+\frac{9}{2}$

$= \frac{-6+9}{2} = \frac{3}{2}$

$\because$ $T_n = a + (n - 1)\,d$

$T_{21} = -\frac{9}{2}+(21-1)\frac{3}{2}$

$T_{21} = -\frac{9}{2}+20\times\frac{3}{2}$

$T21 = -\frac{9}{2}+30$

$T21 = \frac{-9+60}{2} = \frac{51}{2} = 25\frac{1}{2}$ **Ans.**

Q. 40. Determine k such that $(3k - 2)$, $(4k - 6)$ and $(k + 2)$ are three consecutive terms of an A.P. .

Solution : The terms of an A.P. are $(3k - 2)$, $(4k - 6)$ and $(k + 2)$.

Thus, $d = (4k - 6) - (3k - 2)$
$= (k + 2) - (4k - 6)$

$\Rightarrow$ $k - 4 = -3k + 8$

$\Rightarrow$ $4k = 12$

$\Rightarrow$ $k = 3.$ **Ans.**

Q. 41. Write the value of x for which $(x + 2)$, $2x$ and $(2x + 3)$ are three consecutive terms of an A.P. .

Solution : The terms of an A.P. are $(x + 2)$, $2x$ and $(2x + 3)$.

Thus, $d = 2x - (x + 2)$
$= (2x + 3) - 2x$

$\Rightarrow$ $x - 2 = 3$

$\Rightarrow$ $x = 5.$ **Ans.**

Q. 42. Find the sum of 25 + 28 + 31 + ... + 100.

Solution : Here, $a = 25$, $d = 3$ and $t_n = 100$

Now, $t_n = 25 + (n - 1)3 = 100$

$\Rightarrow$ $3(n - 1) = 75$

★ are board exam questions from previous years

$\Rightarrow \quad n - 1 = 25$

$\Rightarrow \quad n = 26$

$\therefore \quad S_n = \frac{26}{2}\{2(25) + (26 - 1)3\}$

$= 13[50 + 75]$

$= 13 \times 125$

$= 1625.$ **Ans.**

Q. 43. If the nth term of an A.P. is $2n + 1$, find the sum of the first n terms the A.P.

Solution : Given,

$t_n = 2n + 1$

$\Rightarrow \quad t_1 = 2 \times 1 + 1 = 3$

$\Rightarrow \quad t_2 = 2 \times 2 + 1 = 5$

Thus, $a = 3$ and $d = t_2 - t_1 = 2$

Hence, $S_n = \frac{n}{2}\{2(3) + (n - 1)(2)\}$

$= n\{3 + n - 1\}$

$= n(2 + n).$ **Ans.**

Q. 44. How many terms of the A.P. 65, 60, 55,.... be taken so that their sum is zero ?

Solution : Given, A.P. is 65, 60, 55......

there, $a = 65$, $d = 60 - 65 = 55 - 60 = -5$

Now, $S_n = 0$

$\therefore \quad S_n = \frac{n}{2}[2a + (n-1)d] = 0$

$\Rightarrow \quad \frac{n}{2}[2(65) + (n-1)(-5)] = 0$

$\Rightarrow \quad 130 - 5n + 5 = 0$

$\Rightarrow \quad 135 - 5n = 0$

$\Rightarrow \quad 5n = 135$

$\therefore \quad n = 27$

Hence, the number of terms are 27. **Ans.**

Q. 45. Find the sum of the first hundred even natural numbers, which are divisible by 5.

Solution : First hundred even natural numbers divisible by 5 are 10, 20, 30, 40, ...

This is an A.P.

So $a = 10$, $d = 10$ and $n = 100$

$S_{100} = \frac{100}{2}\{2(10) + (100 - 1)(10)\}$

$= 50\{20 + 990\}$

$= 50\{1010\}$

$= 50500.$ **Ans.**

Q. 46. The sum of the first n terms of an A.P. is $2n^2 + 5n$. Find the nth term of this A.P.

Solution : Given, $S_n = 2n^2 + 5n$

$\therefore \quad S_{n-1} = 2(n-1)^2 + 5(n - 1)$

Now $t_n = S_n - S_{n-1}$

$= 2n^2 + 5n - 2(n - 1)^2 - 5(n - 1)$

$= 2n^2 + 5n - 2n^2 + 4n - 2 - 5n + 5$

$= 4n + 3$ **Ans.**

Q. 47. The sum of the first n terms of an A.P. is $3n^2 - n$. Find the nth term, first term and the common difference of this A.P.

Solution : Given, $S_n = 3n^2 - n$

$\therefore \quad S_{n-1} = 3(n - 1)^2 - (n - 1)$

Now, $t_n = S_n - S_{n-1}$

$t_n = 3n^2 - n - 3(n - 1)^2 + n - 1$

$= 3n^2 - n - 3n^2 + 6n - 3 + n - 1$

$= 6n - 4$

$\therefore \quad a = t_1 = 6 - 4 = 2$

$t_2 = 6 \times 2 - 4$

$= 12 - 4 = 8$

and $d = t_2 - t_1 = -2 = 6$

Hence n^{th} term $= 6n - 4$, $a = 2$ and $d = 6$. **Ans.**

Q. 48. The first and last terms of an A.P. are 4 and 81 respectively. If its common difference is 7, how many terms are there and what is their sum ?

Solution : Given, $a = 4$, $d = 7$ and $t_n = 81$

We know $t_n = a + (n - 1)d$

$\Rightarrow \quad 4 + (n - 1)7 = 81$

$\Rightarrow \quad 7n - 3 = 81$

$\Rightarrow \quad 7n = 84$

$\Rightarrow \quad n = 12$

Thus, $S_{12} = \frac{12}{2}\{2(4) + (12 - 1)7\}$

$= 6\{8 + 77\}$

$= 6\{85\} = 510.$ **Ans.**

Q. 49. If $(2p - 1)$, 7 and $3p$ are three consecutive terms of an A.P., then find the value of p.

Solution : The terms of an A.P. are $(2p - 1)$, 7 and $3p$.

Thus, $7 - (2p - 1) = 3p - 7$

$\Rightarrow \quad 7 - 2p + 1 = 3p - 7$

$\Rightarrow \quad 8 - 2p = 3p - 7$

$\Rightarrow \quad 5p = 15$

$\Rightarrow \quad p = 3.$ **Ans.**

Q. 50. Which term of the progression 20, $19\frac{3}{4}$, $18\frac{1}{2}, 17\frac{3}{4}$ ..... is the first negative term ?

Solution : Given, A.P. is 20, $19\frac{1}{4}, 18\frac{1}{2}, 17\frac{1}{4}$,.....

$= 20, \frac{77}{4}, \frac{37}{2}, \frac{71}{4}, \ldots\ldots$

Here, $a = 20, d = \frac{77}{4} - 20 = \frac{77-80}{4} = \frac{-3}{4}$

Let a_n be its first negative term

Then $\quad a_n + (n-1)d < 0.$

$\Rightarrow 20 + (n-1)\left(-\frac{3}{4}\right) < 0$

$\Rightarrow \quad 20 - \frac{3}{4}n + \frac{3}{4} < 0$

$\Rightarrow \quad 20 + \frac{3}{4} < \frac{3}{4}n$

$\Rightarrow \quad \frac{83}{4} < \frac{3}{4}n$

$\Rightarrow \quad n > \frac{83}{4} \times \frac{4}{3}$

$\Rightarrow \quad n > \frac{83}{3} = 27.66...$

28th term will be first negative term of given A.P.

Ans.

Q. 51. Which term of the A.P. 8, 14, 20, 26, ..., will be 72 more than its 41st term ?

Solution : Given, AP is 8, 14, 20, 26,.....

Here, $\quad a = 8, d = 14 - 8 = 6$

Let $\quad a_n = a_{41} + 72$

$\Rightarrow \quad a + (n-1)d = a + 40d + 72$

$\Rightarrow \quad (n-1)\, 6 = 40 \times 6 + 72$

$= 240 + 72 = 312$

$\Rightarrow \quad n - 1 = \frac{312}{6} = 52$

$\Rightarrow \quad n = 52 + 1$

$= 53^{rd}$ term. **Ans.**

Q. 52. The nth term of an A.P. is $(7 - 4n)$. Find its common difference.

Solution : Given, $t_n = 7 - 4n$

$\therefore \quad t_{n-1} = 7 - 4(n-1)$

$= 7 - 4n + 4$

$= 11 - 4n$

Now, $\quad d = t_n - t_{n-1}$

$\Rightarrow \quad d = 7 - 4n - (11 - 4n)$

$\Rightarrow \quad d = -4.$ **Ans.**

OR

Given, $t_n = (7 - 4n)$.

Let the first term be a and the common difference be d.

Then $\quad t_n = a + (n-1)d = (7 - 4n)$

$\Rightarrow \quad a + (n-1)d\} = 7 - 4n$

$\Rightarrow \quad a + dn - d = 7 - 4n$

$\Rightarrow \quad (a - d) + dn = 7 - 4n$

$\Rightarrow \quad d = -4.$ **Ans.**

Q. 53. The first term of an A.P. is p and its common difference is q. Find its 10th term.

Solution : Given, $a = p$ and $d = q$

So $\quad t_n = \{p + (10-1)q\}$

$\Rightarrow \quad t_{10} = p + 9q.$ **Ans.**

Q. 54. Write the next term of the A.P.

$$\sqrt{8}, \sqrt{18}, \sqrt{32}, \ldots$$

Solution : The given A.P. is $\sqrt{8}, \sqrt{18}, \sqrt{32}, \ldots$

Thus, $\quad a = \sqrt{8}$

and $\quad d = \sqrt{18} - \sqrt{8}$

$= 3\sqrt{2} - 2\sqrt{2} = \sqrt{2}$

Hence, $\quad t_4 = \sqrt{8} + (4-1)\sqrt{2}$

$= 2\sqrt{2} + 3\sqrt{2}$

$= 5\sqrt{2}$

$= \sqrt{50}$

Thus, the next term is $\sqrt{50}$. **Ans.**

Q. 55. Which term of the A.P. 21, 18, 15, ... is zero ?

Solution : The given A.P. is 21, 18, 15, ...

Thus, $\quad a = 21$ and $d = 18 - 21 = -3$

Let the nth term be zero.

Hence, $\quad t_n = 21 - (n-1)3 = 0$

$\Rightarrow \quad 21 - 3n + 3 = 0$

$\Rightarrow \quad 24 - 3n = 0$

$\Rightarrow \quad 3n = 24$

$\Rightarrow \quad n = 8.$ **Ans.**

Q. 56. Find the number of all three-digit natural number disivible by 9.

Solution : Three digit natural numbers divisible by 9 are 108, 117, 126, 135, ..., 999.

The above series is an A.P.

with, $a = 108, d = 9$ and $t_n = 999$

Thus, $t_n = 108 + (n-1)9 = 999$

$\Rightarrow \quad 108 + 9n - 9 = 999$

$\Rightarrow \quad 9n = 999 - 99$

$\Rightarrow \quad 9n = 900$

$\Rightarrow \quad n = 100$

Thus, the number of three-digit natural numbers divisible by 9 are 100. **Ans.**

Q. 57. The 4th term of an A.P. is zero. Prove that the 25th term of the A.P. is three times its 11th term.

Solution : We know that

$$T_n = a + (n-1)d$$

Given, $T_4 = a + (4-1)d = 0$

$\Rightarrow \quad a + 3d = 0$

$\Rightarrow \quad a = -3d$ ...(i)

Now, $T_{25} = a + (25-1)d$

$= a + 24d$

$= (-3d) + 24d$

$= 21d$

and $T_{11} = a + (11-1)d$

$= a + 10d$

then, $3T_{11} = 3(a + 10d)$

$= 3a + 30d$

$= 3(-3d) + 30d$ [From (i)]

$= 30d - 9d = 21d = T_{25}$

$\therefore \quad 3T_{11} = T_{25}$ **Hence Proved.**

Q. 58. How many terms of the A.P. 18, 16, 14,... be taken so that their sum is zero ?

Solution : Given, A.P. is 18, 16, 14,...

there, $a = 18, d = 16 - 18 = 14 - 16 = -2$

Now, $S_n = 0$

$\therefore \quad S_n = \frac{n}{2}[2a + (n-1)d] = 0$

$\frac{n}{2}[2 \times 18 + (n-1)(-2)] = 0$

$36 - 2n + 2 = 0$

$2n = 38$

$n = 19$

Hence, the number of terms are 19. **Ans.**

Q. 59. If in an A.P., $S_5 + S_7 = 167$ and $S_{10} = 235$, then find the A.P., where S_n denotes the sum of its first n terms.

Solution : Given, $S_5 + S_7 = 167$

$\Rightarrow \quad \frac{5}{2}(2a + 4d) + \frac{7}{2}(2a + 6d) = 167$

$\Rightarrow \quad \frac{5}{2} \times 2(a + 2d) + \frac{7}{2} \times 2(a + 3d) = 167$

$\Rightarrow \quad 5a + 10d + 7a + 21d = 167$

$\Rightarrow \quad 12a + 31a = 167$ ...(i)

Also, $S_{10} = 235$

$\Rightarrow \quad \frac{10}{2}(2a + 9d) = 235$

$\Rightarrow \quad 10a + 45d = 235$

$\Rightarrow \quad 2a + 9d = 47$ ...(ii)

On multiplying equation (ii) by 6, we get

$12a + 54d = 282$ ...(iii)

On subtracting equation (i) from (iii), we get

$12a + 54d = 282$

$12a + 31d = 167$

$- \quad - \quad -$

$23d = 115$

$\Rightarrow \quad d = 5$

Substituting value of d in equation (i), we get

$12a + 31 \times 5 = 167$

$12a + 155 = 167$

$\Rightarrow \quad 12a = 12$

$\Rightarrow \quad a = 1$

Hence, A.P. is 1, 6, 11.... **Ans.**

Chapter 6. Coordinate Geometry

Q. 1. *ABCD* is a rectangle whose vertices are *B*(4, 0), *C*(4, 3) and *D*(0, 3). Find the length of one of its diagonals.

Solution : In rectangle *ABCD*, *B*(4, 0), *C*(4, 3) and *D*(0, 3) are the coordinates.

Thus, *BD* is the diagonal.

Hence length of $BD = \sqrt{(4-0)^2 + (0-3)^2}$

$= \sqrt{16+9}$

$= \sqrt{25} = 5$ units **Ans.**

Q. 2. If the points $A(x, 2)$, $B(-3, -4)$ and $C(7, -5)$ are collinear, then find the value of x.

Solution : Since $A(x, 2)$, $B(-3, -4)$ and $C(7, -5)$ are collinear,

$$\therefore \quad x_1(y_2 - y_3) + x_2(y_3 - y_1) + x_3(y_1 - y_2) = 0$$

$$\Rightarrow x[-4-(-5)] + (-3)[(-5)-2] + 7[2-(-4)] = 0$$

$$\Rightarrow \quad x + (-3)[-7] + 7[6] = 0$$

$$\Rightarrow \quad x + 21 + 42 = 0$$

$$\Rightarrow \quad x = -63$$

Ans.

Q. 3. Find the perimeter of a triangle with vertices $A(0, 4)$, $B(0, 0)$ and $C(3, 0)$.

Solution : Given, $A(0, 4)$, $B(0, 0)$ and $C(3, 0)$

Thus, Perimeter $= AB + BC + CA$

$$= [\sqrt{(0-0)^2 + (0-4)^2} + \sqrt{(3-0)^2 + (0-0)^2} + \sqrt{(3-0)^2 + (4-0)^2}]$$

$$= [\sqrt{16} + \sqrt{9} + \sqrt{9+16}]$$

$= (4 + 3 + 5)$ units

$= 12$ units. **Ans.**

Q. 4. Find the area of the triangle in the given figure (in sq. units).

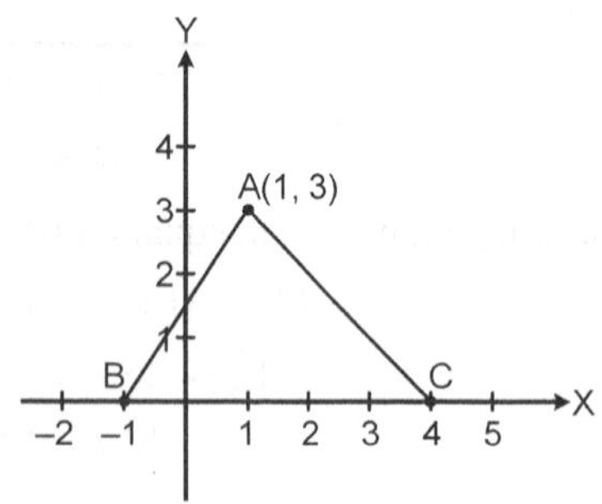

Solution : Given, $A = (1, 3)$, $B = (-1, 0)$ and $C = (4, 0)$.

Thus, area of the ΔABC

$$= \frac{1}{2}|1(0-0) - 1(0-3) + 4(3-0)|$$

$$= \frac{1}{2}|3 + 12|$$

$= 7.5$ sq. units. **Ans.**

Q. 5. If the point $C(k, 4)$ divides the join of points $A(2, 6)$ and $B(5, 1)$ in the ratio 2 : 3, then find the value of k.

Solution : Given, $A(2, 6)$, $C(k, 4)$ and $B(5, 1)$ and $AC : BC = 2 : 3$

Thus, using section formula,

$$k = \frac{3\times2+2\times5}{2+3}$$

$$= \frac{6+10}{5}$$

$$= \frac{16}{5}$$ **Ans.**

Q. 6. Find the distance of the point $(-3, 4)$ from the origin.

Solution : The given point is $(-3, 4)$.

We know that,

$$\text{Distance} = \sqrt{(x_1 - x_2)^2 + (y_1 - y_2)^2}$$

$$\text{Thus, the distance} = \sqrt{(-3-0)^2 + (4-0)^2}$$

$= 4$ units. **Ans.**

Q. 7. In the given figure, $P(5, -3)$ and $Q(3, y)$ are the points of trisection of the line segment joining $A(4, 7)$ and $B(1, -5)$. Then find y.

A(4, 7)   P(5, –3)   Q(3, y)   B(1, –5)

Solution : Since P and Q are the points of trisection, thus,

$$PQ = QB$$

$$\Rightarrow \sqrt{(5-3)^2 + (-3-y)^2} = \sqrt{(3-1)^2 + (y+5)^2}$$

$$\Rightarrow \quad (2)^2 + (3+y)^2 = (2)^2 + (y+5)^2$$

$$\Rightarrow \quad 4 + 9 + 6y + y^2 = 4 + 25 + 10y + y^2$$

$$\Rightarrow \quad 6y + 13 = 29 + 10y$$

$$\Rightarrow \quad 4y = -16$$

$$\Rightarrow \quad y = -4.$$ **Ans.**

Q. 8. If the coordinates of one end of a diameter of a circle are (2, 3) and the coordinates of the centre are (– 2, 5), then find the coordinates of the other end of the diameter.

Solution : Given, the coordinates of the centre = $(-2, 5)$ and one end = $(2, 3)$.

Let the coordinates of the other end be (x, y)

Since, the centre is the mid-point of the two end points

Thus, by mid point formula,

$$-2 = \frac{2+x}{2} \quad \text{and} \quad 5 = \frac{3+y}{2}$$

$$\Rightarrow \quad -4 = 2 + x \quad \text{and} \quad 10 = 3 + y$$

$$\Rightarrow \quad x = -6 \quad \text{and} \quad y = 7$$

Hence, the coordinates are $(-6, 7)$. **Ans.**

Q. 9. Find the coordinates of point A, where AB is diameter of a circle whose centre is (2, – 3) and B is the point (1, 4).★

Solution : Let the co-ordinates of point A be (x, y) and point O (2, – 3) be point the centre, then

By mid-point formula.

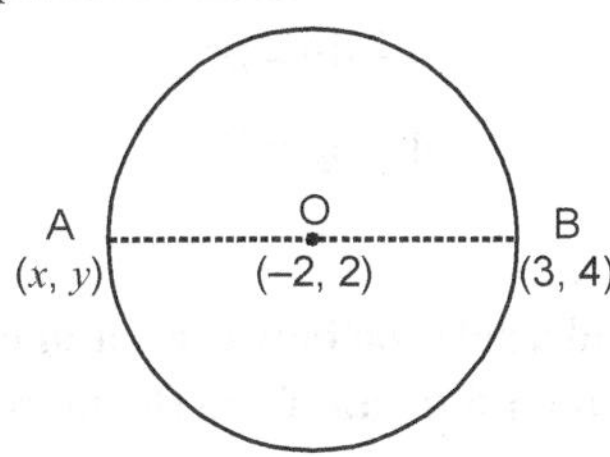

$$\frac{x+1}{2} = 2 \quad \text{and} \quad \frac{y+4}{2} = -3$$

or $\quad x = 4 - 1$ and $\quad y = -6 - 4$

$x = 3 \qquad y = -10$

∴ The co-ordinates of point A are (3, – 10) **Ans.**

Q. 10. Find the coordinates of the point P dividing the line segment joining the points $A(1, 3)$ and $B(4, 6)$ in the ratio 2 : 1.

Solution : Given, $A = (1, 3)$; $B = (4, 6)$ and $AP : PB = 2 : 1$

Let the coordinates of point P be (x, y).

Then, by section formula

$$x = \frac{2 \times 4 + 1 \times 1}{2+1} \text{ and } y = \frac{2 \times 6 + 1 \times 3}{2+1}$$

$$\Rightarrow \quad x = \frac{8+1}{3} \text{ and } y = \frac{12+3}{3}$$

$$\Rightarrow \quad x = 3 \text{ and } y = 5$$

$$\Rightarrow \quad (x, y) = (3, 5).$$ **Ans.**

Q. 11. If $P\left(\frac{a}{2}, 4\right)$ is the mid-point of the line segment joining the points $A(-6, 5)$ and $B(-2, 3)$, then find the value of a.

Solution : Given, $P\left(\frac{a}{2}, 4\right)$ is the mid-point of the line segment joining the points $A(-6, 5)$ and $B(-2, 3)$.

Thus $\quad \frac{a}{2} = \frac{-6-2}{2}$

$$= \frac{-8}{2} = -4$$

or $\quad a = -8.$ **Ans.**

Q. 12. If A and B are the points (– 6, 7) and (– 1, – 5) respectively, then find the distance $2AB$.

Solution : Given, $A = (-6, 7)$ and $B = (-1, -5)$.

By distance formula

$$AB = \sqrt{(-6+1)^2 + (7+5)^2}$$
$$= \sqrt{(-5)^2 + (12)^2}$$
$$= \sqrt{25 + 144}$$
$$= \sqrt{169}$$
$$= 13$$

Thus, $\quad 2AB = 2 \times 13 = 26$ units. **Ans.**

Q. 13. Find the area (in sq. units) of the triangle formed by the points $A(a, 0)$, $O(0, 0)$ and $B(0, b)$.

Solution : Given,

$$A = (a, 0),\ O = (0, 0) \text{ and } B = (0, b)$$

Thus, area of ΔABC

$$= \frac{1}{2}|a(0-b) + 0(b-0) + 0(0-0)|$$

$$= \frac{1}{2}|-ab|$$

$$= \frac{1}{2}ab \text{ sq. units.}$$ **Ans.**

Q. 14. The point P which divides the line segment joining the points $A(2, -5)$ and $B(5, 2)$ in the ratio 2 : 3 lies in which quadrant ?

Solution : Let the coordinates of the point P be (x, y).

Given, $\quad A = (2, -5)$, $B = (5, 2)$ and $m : n = 2 : 3$

Now, $$P_x = \frac{2(5) + 3(2)}{2+3} \text{ and } P_y = \frac{2(2) + 3(-5)}{2+3}$$

$$\Rightarrow \quad P_x = \frac{10+6}{5} \text{ and } P_y = \frac{4-15}{5}$$

$$\Rightarrow \quad x = \frac{16}{5} \text{ and } y = -\frac{11}{5}$$

$$\Rightarrow \quad P = \left(\frac{16}{5}, -\frac{11}{5}\right)$$

Thus, P lies in Quadrant IV. **Ans.**

Q. 15. The mid-point of segment AB is the point $P(0, 4)$. If the coordinates of B are $(-2, 3)$, then find the coordinates of A.

Solution : Given, $P = (0, 4)$, $B = (-2, 3)$ and $AP = PB$

Let $\quad A = (x, y)$

Thus $$P_x = \frac{x-2}{2} \text{ and } P_y = \frac{y+3}{2}$$

$$\Rightarrow \quad 0 = \frac{x-2}{2} \text{ and } 4 = \frac{y+3}{2}$$

★ are board exam questions from previous years

$\Rightarrow \quad x = 2 \text{ and } y = 5$

Thus $\quad A = (2, 5)$. **Ans.**

Q. 16. Find the point on the X-axis which is equidistant from the point $(-1, 0)$ and $(5, 0)$.

Solution : Given, $A = (-1, 0)$ and $B = (5, 0)$.

Let the point on the X-axis be $P(x, 0)$.

It is given that $\quad AP = PB$

$$\Rightarrow \sqrt{(x+1)^2+(0-0)^2} = \sqrt{(5-x)^2+(0-0)^2}$$

$$\Rightarrow (x+1)^2 = (5-x)^2$$

$$\Rightarrow x + 1 = 5 - x$$

$$\Rightarrow 2x = 4$$

$$\Rightarrow x = 2$$

Thus, the point is (2, 0). **Ans.**

Q. 17. Find the coordinates of a point A, where AB is a diameter of the circle with centre (– 2, 2) and B is the point with coordinates (3, 4).★

Solution. By mid-point formula

$$\frac{x+3}{2} = -2$$

$$x = -4-3$$

$$x = -7$$

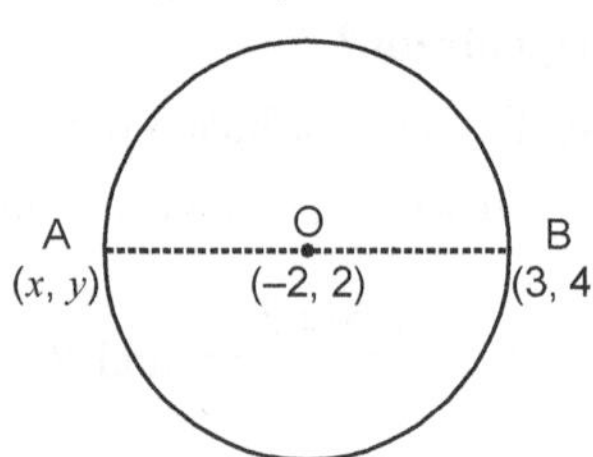

and $\quad \dfrac{y+4}{2} = 2$

$$y = 0$$

∴ Co-ordinates of point A are (– 7, 0). **Ans.**

Q. 18. Find the value of k if the point $P(2, 4)$ is equidistant from the points $A(5, k)$ and $B(k, 7)$.

Solution : Given, $A = (5, k)$, $B = (k, 7)$ and $P = (2, 4)$

Also, $\quad AP = PB$

$$\Rightarrow \sqrt{(5-2)^2+(k-4)^2} = \sqrt{(2-k)^2+(7-4)^2}$$

$$\Rightarrow (3)^2 + (k-4)^2 = (2-k)^2 + (3)^2$$

$$\Rightarrow 9 + (k-4)^2 = (2-k)^2 + 9$$

$$\Rightarrow k - 4 = 2 - k$$

$$\Rightarrow 2k = 6$$

$$\Rightarrow k = 3.$$ **Ans.**

Q. 19. If a point $A(0, 2)$ is equidistant from the point $B(3, p)$ and $C(p, 5)$ then find the value of p.

★ are board exam questions from previous years

Solution : Given, $\quad AB = AC$

$$\Rightarrow \sqrt{(3-0)^2+(p-2)^2} = \sqrt{(0-p)^2+(2-5)^2}$$

$$\Rightarrow 9 + (p-2)^2 = (-p)^2 + 9$$

$$\Rightarrow (p-2)^2 = p^2$$

$$\Rightarrow p^2 + 4 - 4p = p^2$$

$$\Rightarrow 4 - 4p = 0$$

$$\Rightarrow p = 1.$$ **Ans.**

Q. 20. Find a relation between x and y such that the points $P(x, y)$ is equidistant from the points $A(1, 4)$ and $B(-1, 2)$.

Solution : Given $A = (1, 4)$, $B = (-1, 2)$ and $P = (x, y)$

Also, $\quad AP = PB$

$$\Rightarrow \sqrt{(1-x)^2+(4-y)^2} = \sqrt{(x+1)^2+(y-2)^2}$$

$$\Rightarrow (1-x)^2 + (4-y)^2 = (x+1)^2 + (y-2)^2$$

$$\Rightarrow (1-x)^2 - (1+x)^2 = (y-2)^2 - (4-y)^2$$

$$\Rightarrow [(1-x+1+x)(1-x-1-x)]$$
$$= [(y-2+4-y)(y-2-4+y)]$$

$$\Rightarrow 2(-2x) = [(2)(2y-6)]$$

$$\Rightarrow -4x = 4(y-3)$$

$$\Rightarrow -x = y - 3$$

$$\Rightarrow x + y = 3.$$ **Ans.**

Q. 21. Find the value of x for which the distance between the points $P(x, 4)$ and $Q(9, 10)$ is 10 units.

Solution : Given, $P = (x, 4)$, $Q = (9, 10)$ and $PQ = 10$ units

Now $\quad 10 = \sqrt{(x-9)^2+(4-10)^2}$

$$100 = (x-9)^2 + (-6)^2$$

$$\Rightarrow 100 = x^2 + 81 - 18x + 36$$

$$\Rightarrow x^2 - 18x + 17 = 0$$

$$\Rightarrow x^2 - 17x - x + 17 = 0$$

$$\Rightarrow x(x-17) - 1(x-17) = 0$$

$$\Rightarrow (x-17)(x-1) = 0$$

$$\Rightarrow x = 1, 17$$

Thus, the value of x is 1 or 17. **Ans.**

Q. 22. Find the point on the X-axis which is equidistant from the points $(5, -2)$ and $(-3, 2)$.

Solution : As the point is on X-axis, let its coordinates be $(x, 0)$.

Now, $\sqrt{(5-x)^2+(-2-0)^2}$

$$= \sqrt{(-3-x)^2+(2-0)^2}$$

$$\Rightarrow (5-x)^2 + 4 = (-3-x)^2 + 4$$

$$\Rightarrow 25 - 10x + x^2 = 9 + 6x + x^2$$

$\Rightarrow \quad 16x = 16$

$\Rightarrow \quad x = 1$

Thus, the point is (1, 0). **Ans.**

Q. 23. If the point $P(x, y)$ is equidistant from the points $A(5, 1)$ and $B(-1, 5)$, prove that $3x = 2y$.

Solution : Given, $A = (5, 1)$, $B = (-1, 5)$ and $PA = PB$

Thus $\sqrt{(5-x)^2+(1-y)^2} = \sqrt{(-1-x)^2+(5-y)^2}$

$\Rightarrow 25 - 10x + x^2 + 1 - 2y + y^2$

$= 1 + 2x + x^2 + 25 - 10y + y^2$

$\Rightarrow \quad 26 - 10x - 2y = 26 + 2x - 10y$

$\Rightarrow \quad 2x + 10x = 10y - 2y$

$\Rightarrow \quad 12x = 8y$

$\Rightarrow \quad 3x = 2y.$ **Hence Proved.**

Q. 24. Find the coordinates of the point which is equidistant from the three given points $A(5, 1)$, $B(-3, -7)$ and $C(7, -1)$.

Solution : Given,

$A = (5, 1), B = (-3, -7)$ and $C = (7, -1)$

Let the required point be $P(x, y)$

Thus $\quad PA = PB = PC$

$\Rightarrow \sqrt{(5-x)^2+(1-y)^2} = \sqrt{(-3-x)^2+(-7-y)^2}$

$= \sqrt{(7-x)^2+(-1-y)^2}$

$\Rightarrow 25 - 10x + x^2 + 1 - 2y + y^2$

$= 9 + 6x + x^2 + 49 + 14y + y^2$

$= 49 - 14x + x^2 + 1 + 2y + y^2$

$\Rightarrow \quad 26 - 10x - 2y = 58 + 6x + 14y$

$= 50 - 14x + 2y$

$\Rightarrow \quad 26 - 10x - 2y = 58 + 6x + 14y$

and $\quad 26 - 10x - 2y = 50 - 14x + 2y$

$\Rightarrow \quad 16x + 16y = -32$ and $4x - 4y = 24$

$\Rightarrow \quad x + y = -2$ and $x - y = 6$

On adding, we get

$2x = 4 \Rightarrow x = 2$

and $\quad y = -4$

Thus, $\quad P = (2, -4).$ **Ans.**

Q. 25. If $A(6, -1)$, $B(1, 3)$ and $C(k, 8)$ are three points such that $AB = BC$, find the value of k.

Solution : Given, $A = (6, -1)$, $B(1, 3)$ and $C(k, 8)$ and $AB = BC$

Thus, $\quad AB = \sqrt{(6-1)^2+(-1-3)^2}$

$= \sqrt{(5)^2+(-4)^2} = \sqrt{41}$ units

Also $\quad BC = \sqrt{(1-k)^2+(3-8)^2}$

$= \sqrt{(1-k)^2+(-5)^2}$

$\because \quad AB = BC$

$\therefore \sqrt{(1-k)^2+(-5)^2} = \sqrt{41}$

$\Rightarrow \quad (1-k)^2 + 25 = 41$

$\Rightarrow \quad (1-k)^2 = 41 - 25$

$\Rightarrow \quad (1-k)^2 = 16$

$\Rightarrow \quad 1 - k = \pm 4$

$\Rightarrow \quad k = 5$ or -3

Hence, the value of k is 5 or -3. **Ans.**

Q. 26. Find the point on the Y-axis which is equidistant from the points $(-2, 5)$ and $(-2, 9)$.

Solution : As the point is on Y-axis, let its coordinates by $(0, y)$

Now, $\sqrt{(-2-0)^2+(5-y)^2} = \sqrt{(-2-0)^2+(y-9)^2}$

$\Rightarrow \quad 4 + (5-y)^2 = 4 + (y-9)^2$

$\Rightarrow \quad 5 - y = y - 9$

$\Rightarrow \quad 2y = 14$

$\Rightarrow \quad y = 7$

Thus, the point is (0, 7). **Ans.**

Q. 27. If the point $P(x, y)$ is equidistant from the points $A(6, -1)$ and $B(2, 3)$, show that $x - y = 3$.

Solution : Given $A = (6, -1)$, $B = (2, 3)$ and $PA = PB$

Let the point P be (x, y).

Thus $\sqrt{(x-6)^2+(y+1)^2} = \sqrt{(x-2)^2+(y-3)^2}$

$\Rightarrow 36 - 12x + x^2 + 1 + 2y + y^2$

$= 4 - 4x + x^2 + 9 - 6y + y^2$

$\Rightarrow \quad 37 - 12x + 2y = 13 - 4x - 6y$

$\Rightarrow \quad 12x - 4x = 6y + 2y + 24$

$\Rightarrow \quad 8x = 8y + 24$

$\Rightarrow \quad x - y = 3.$ **Hence Proved.**

Q. 28. Write the coordinates of a point P on x-axis which is equidistant from the point A$(-2, 0)$ and B$(6, 0)$.★

Solution : Let coordinates of P on x-axis is $(x, 0)$

Given, A$(-2, 0)$ and B$(6, 0)$

Here, $\quad$ PA = PB

$\therefore \quad \sqrt{(x+2)^2+(0-0)^2} = \sqrt{(x-6)^2+(0-0)^2}$

$\Rightarrow \quad \sqrt{(x+2)^2} = \sqrt{(x-6)^2}$

On squaring both sides, we get

$(x+2)^2 = (x-6)^2$

$x^2 + 4 + 4x = x^2 + 36 - 12x$

$\Rightarrow \quad 4 + 4x = 36 - 12x$

$\Rightarrow \quad 16x = 32$

★ are board exam questions from previous years

$\Rightarrow \quad x = 2$

Co-ordinates of P are (2, 0) **Ans.**

Q. 29. Find the coordinates of the point equidistant from the three given points $A(5,3)$, $B(5,-5)$ and $C(1,-5)$.

Solution : Given, $A = (5, 3)$, $B = (5, -5)$ and $C = (1, -5)$

Let the required point be $P(x, y)$.

Thus, $\quad PA = PB = PC$

$$\sqrt{(5-x)^2+(3-y)^2} = \sqrt{(5-x)^2+(-5-y)^2}$$

$$= \sqrt{(x-1)^2+(y+5)^2}$$

Now $\quad (3-y)^2 = (5+y)^2$

$\Rightarrow \quad 3 - y = 5 + y$

$\Rightarrow \quad 2y = -2$

$\Rightarrow \quad y = -1$

Also, $(5-x)^2 + (3+1)^2 = (x-1)^2 + (-1+5)^2$

$(5-x)^2 + 4^2 = (x-1)^2 + 4^2$

and $\quad (5-x)^2 = (x-1)^2$

$\Rightarrow \quad 5 - x = x - 1$

$\Rightarrow \quad 2x = 6$

$\Rightarrow \quad x = 3$

Hence $\quad P = (3, -1)$. **Ans.**

Q. 30. By the distance formula show that the points (1, – 1), (5, 2) and (9, 5) are collinear.

Solution : Let $A(1, -1)$, $B(5, 2)$ and (9, 5) be the given points.

Then by distance formula

$$AB = \sqrt{(5-1)^2+(2+1)^2}$$

$$= \sqrt{16+9} = 5$$

$$BC = \sqrt{(5-9)^2+(2-5)^2}$$

$$= \sqrt{16+9} = 5$$

$$AC = \sqrt{(1-9)^2+(-1-5)^2}$$

$$= \sqrt{64+36}$$

$$= 10$$

Clearly, $\quad AC = AB + BC$

Hence A, B, C are collinear points. **Hence Proved.**

Q. 31. If $A(-2, 4)$, $B(0, 0)$ and $C(-4, -2)$ are the vertices of a ΔABC, then find the length of the median through the vertex A.

Solution : Given, $A = (-2, 4)$, $B = (0, 0)$ and $C = (-4, -2)$

Let AD be the median on BC through the vertex A.

Thus $\quad BD = DC$

So $\quad D = \left(\dfrac{0-4}{2}, \dfrac{0-2}{2}\right)$

$\Rightarrow \quad D = (-2, -1)$

Thus $\quad AD = \sqrt{(-2+2)^2+(-1-4)^2}$

$= 5$ units

Hence, the length of the median is 5 units. **Ans.**

Q. 32. Two vertices of ΔABC are $A(6, 4)$ and $B(-2, 2)$ and its centroid is $G(3, 4)$. Find the coordinates of the third vertex C.

Solution : Given, $A = (6, 4)$ and $B = (-2, 2)$ and $G = (3, 4)$

Let the coordinates of C be (x, y).

From the formula for the coordinates of the centroid of a triangle,

We have, $\quad G_x = \dfrac{x_1+x_2+x_3}{3}$ and $G_y = \dfrac{y_1+y_2+y_3}{3}$

$\Rightarrow \quad 3 = \dfrac{6-2+x}{3}$ and $4 = \dfrac{4+2+y}{3}$

$\Rightarrow \quad x = 9 - 4$ and $y = 12 - 6$

$\Rightarrow \quad x = 5$ and $y = 6$

Thus $\quad C = (x, y) = (5, 6)$. **Ans.**

Q. 33. The line segment joining the points $A(3, -4)$ and $B(1, 2)$ is trisected at the points $P(p, -2)$ and $Q\left(\dfrac{5}{3}, q\right)$. Find the value of p and q.

Solution : Given $A = (3, -4)$, $B = (1, 2)$, $P = (p, -2)$ and $Q = \left(\dfrac{5}{3}, q\right)$

A(3, –4) P(p, –2) $Q\left(\frac{5}{3}, q\right)$ B(1, 2)

Now, $\quad AP = PQ = QB$

Thus $\quad AP : PB = 1 : 2$ and $AQ : QB = 2 : 1$

$\therefore \quad p = \dfrac{1(1)+2(3)}{2+1}$

and $\quad q = \dfrac{2(2)+1(-4)}{2+1}$

$\therefore \quad p = \dfrac{7}{3}$ and $q = 0$. **Ans.**

Q. 34. If $(2, p)$ is the mid-point of the line segment joining the points $A(6, -5)$ and $B(-2, 11)$, find the value of p.

Solution : Given, $A = (6, -5)$ and $B = (-2, 11)$

By mid-point formula

$$p = \frac{11-5}{2} = \frac{6}{2} = 3.$$ **Ans.**

Q. 35. Let P and Q be the points of trisection of the line segment joining the points $A(2, -2)$ and $B(-7, 4)$ such that P is nearer to A. Find the coordinates of P and Q.

Solution : Since, P and Q are the points of trisection of AB, thus P divides AB in the ratio 1 : 2.

P Q
A(2, – 2) B(– 7, 4)

$$\therefore \text{ Coordinates of } P = \left(\frac{1(-7)+2(2)}{1+2}, \frac{1(4)+2(-2)}{1+2}\right)$$

$$= \left(\frac{-3}{3}, \frac{0}{3}\right) = (-1, 0)$$

and Q is the mid-point of PB

$$\therefore \text{Coordinates of } Q = \left(\frac{-1+(-7)}{2}, \frac{0+4}{2}\right)$$

$$= (-4, 2)$$

So, $P = (-1, 0), Q = (-4, 2)$ **Ans.**

Q. 36. Find the ratio in which the point $P\left(\frac{3}{4}, \frac{5}{12}\right)$ divides the line segment joining the points $A\left(\frac{1}{2}, \frac{3}{2}\right)$ and $B(2, -5)$.

Solution : Let point $P\left(\frac{3}{4}, \frac{5}{12}\right)$ divides the line AB in ratio $k : 1$.

$P\left(\frac{3}{4}, \frac{5}{12}\right)$
$A\left(\frac{1}{2}, \frac{3}{2}\right)$ k 1 B(2, – 5)

Then, by section formula,
Coordinates of P are

$$\frac{2k + \frac{1}{2}}{k+1} = \frac{3}{4}$$

and $$\frac{-5k + \frac{3}{2}}{k+1} = \frac{5}{12}$$

$\Rightarrow$ $8k + 2 = 3k + 3$

and $-60k + 18 = 5k + 5$

$\Rightarrow$ $8k - 3k = 3 - 2$

and $65k = 18 - 5$

$\Rightarrow$ $5k = 1$

and $65k = 13 \Rightarrow k = \frac{1}{5}$ in each case.

Hence, the required ratio is 1 : 5. **Ans.**

Q. 37. Find the ratio in which the point $P(-6, a)$ divides the line segment joining the points $A(-3, -1)$ and $B(-8, 9)$. Also, find the value of a.

Solution : Given, $A = (-3, -1)$, $B = (-8, 9)$ and $P = (-6, a)$

Let the required ratio be $m : n$

m n
A P B
(–3, –1) (–6, a) (–8, 9)

Now, $$-6 = \frac{n(-3) + m(-8)}{m+n}$$

$\Rightarrow$ $-6(m + n) = -3n - 8m$

$\Rightarrow$ $-6m - 6n = -3n - 8m$

$\Rightarrow$ $2m = 3n$

$\Rightarrow$ $m : n = 3 : 2$

Now, $$a = \frac{2(-1) + 3(9)}{2+3}$$

$$= \frac{25}{5} = 5.$$ **Ans.**

Q. 38. Find the ratio in which the point $P(-3, k)$ divides the line segment joining the points $A(-5, -4)$ and $B(-2, 3)$. Also find the value of k.

Solution : Given, $A = (-5, -4)$, $B = (-2, 3)$ and $P = (-3, k)$

Let the required ratio be $m : n$.

Now, $$-3 = \frac{n(-5) + m(-2)}{m+n}$$

$\Rightarrow$ $-3(m + n) = -5n - 2m$

$\Rightarrow$ $-3m - 3n = -5n - 2m$

$\Rightarrow$ $-m = -2n$

$\Rightarrow$ $m : n = 2 : 1$

Now, $$k = \frac{1(-4) + 2(3)}{1+2}$$

$$= \frac{2}{3}.$$ **Ans.**

Q. 39. Find the centroid of ΔABC whose vertices are $A(-1, 0)$, $B(5, -2)$ and $C(8, 2)$.

Solution : Given, $A = (-1, 0)$, $B = (5, -2)$ and $C = (8, 2)$

Let the centroid be G (x, y).

Thus $$(x, y) = \left(\frac{-1+5+8}{3}, \frac{0-2+2}{3}\right)$$

$$= \left(\frac{12}{3}, \frac{0}{3}\right)$$

$$= (4, 0).$$ **Ans.**

Q. 40. Find the third vertex of a ΔABC if two of its vertices are $B(-3, 1)$ and $C(0, -2)$ and its centroid is at the origin.

Solution : Given, $B = (-3, 1)$, $C = (0, -2)$ and $O = (0, 0)$

Let the coordinates of A be (x, y).

Thus, $(0, 0) = \left(\dfrac{-3+0+x}{3}, \dfrac{1-2+y}{3}\right)$

$\Rightarrow$ $x - 3 = 0$ and $y - 1 = 0$

$\Rightarrow$ $x = 3$ and $y = 1$

So the coordinates of $A = (3, 1)$. **Ans.**

Q. 41. Find the area of the quadrilateral $ABCD$ whose vertices are $A(-4, -2)$, $B(-3, -5)$, $C(3, -2)$ and $D(2, 3)$.

Solution : Given, $A = (-4, -2)$, $B = (-3, -5)$, $C = (3, -2)$ and $D = (2, 3)$

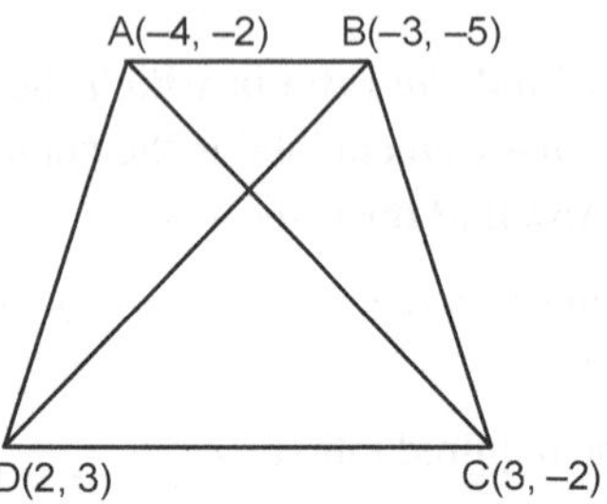

Area of $ABCD$ = Area of ΔABC + Area of ΔADC

$$= \frac{1}{2}|-4(-5+2) - 3(-2+2) + 3(-2+5)|$$

$$+ \frac{1}{2}|-4(3+2) + 2(-2+2) + 3(-2-3)|$$

$$= \frac{1}{2}|12 + 0 + 9| + \frac{1}{2}|-20 + 0 - 15|$$

$$= \frac{1}{2}|19| + \frac{1}{2}|-35|$$

$$= \frac{19}{2} + \frac{35}{2}$$

= 28 sq. units. **Ans.**

Q. 42. Find the value of k if the points $P(5, 4)$, $Q(7, k)$ and $R(9, -2)$ are collinear.

Solution : Given, $P(5, 4)$, $Q(7, k)$ and $R(9, -2)$ are collinear.

So, $|x_1(y_2 - y_3) + x_2(y_3 - y_1) + x_3(y_1 - y_2)| = 0$

Thus, $5(k + 2) + 7(-2 - 4) + 9(4 - k) = 0$

$\Rightarrow$ $5k + 10 - 42 + 36 - 9k = 0$

$\Rightarrow$ $-4k + 4 = 0$

$\Rightarrow$ $4k = 4$

$\Rightarrow$ $k = 1$

Hence, the value of $k = 1$. **Ans.**

Q. 43. Find the area of ΔABC, whose vertices are $A(-5, 7)$, $B(-4, -5)$ and $C(4, 5)$.★

Solution : Area of ΔABC

$$= \frac{1}{2}|-5(-5-5) - 4(5-7) + 4(7+5)|$$

$$= \frac{1}{2}|50 + 8 + 48|$$

= 53 sq. units. **Ans.**

Q. 44. Find the area of ΔABC whose vertices are $A(3, 8)$, $B(-4, 2)$ and $C(5, -1)$.

Solution : Area of ΔABC

$$= \frac{1}{2}|3(2+1) - 4(-1-8) + 5(8-2)|$$

$$= \frac{1}{2}|9 + 36 + 30|$$

= 37.5 sq. units. **Ans.**

Q. 45. Find the area of the quadrilateral $ABCD$ whose vertices are $A(1, 0)$, $B(5, 3)$, $C(2, 7)$ and $D(-2, 4)$.

Solution : Given, $A(1, 0)$, $B(5, 3)$, $C(2, 7)$ and $D(-2, 4)$

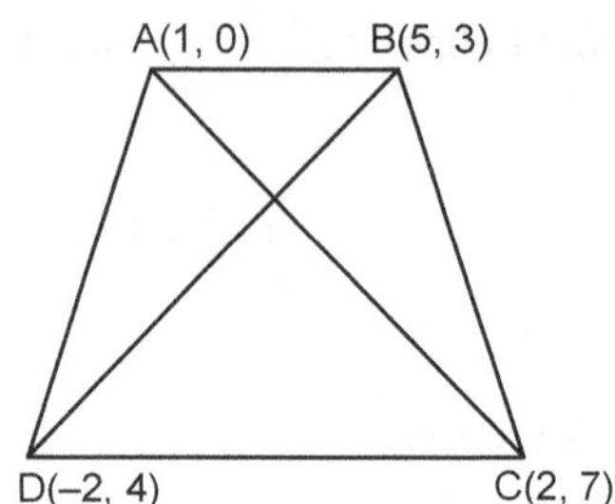

Area of $ABCD$ = Area of ΔABC + Area of ΔADC

$$= \frac{1}{2}|1(3-7) + 5(7-0) + 2(0-3)|$$

$$+ \frac{1}{2}|1(7-4) + 2(4-0) - 2(0-7)|$$

$$= \frac{1}{2}|-4 + 35 - 6| + \frac{1}{2}|3 + 8 + 14|$$

$$= \frac{1}{2}|25 + 25| = 25 \text{ sq. units}$$ **Ans.**

Q. 46. Find the value of p for which $A(3, 2)$, $B(4, p)$ and $C(5, 3)$ are collinear.

Solution : Given, $A(3, 2)$, $B(4, p)$ and $C(5, 3)$ are collinear

★ are board exam questions from previous years

So, $|x_1(y_2 - y_3) + x_2(y_3 - y_1) + x_3(y_3 - y_2)| = 0$

Thus, $3(p-3) + 4(3-2) + 5(2-p) = 0$

$\Rightarrow \quad 3p - 9 + 4 + 10 - 5p = 0$

$\Rightarrow \quad -2p + 5 = 0$

$\Rightarrow \quad 2p = 5$

$\Rightarrow \quad p = \frac{5}{2}$. **Ans.**

Q. 47. Find the distance between the points $\left(-\frac{8}{5}, 2\right)$ and $\left(\frac{2}{5}, 2\right)$.

Solution : Let $A = \left(-\frac{8}{5}, 2\right)$ and $B = \left(\frac{2}{5}, 2\right)$

Then, distance $= \sqrt{(x_2 - x_1)^2 + (y_2 - y_1)^2}$

Thus, $AB = \sqrt{\left(\frac{2}{5} + \frac{8}{5}\right)^2 + (2-2)^2}$

$= \frac{10}{5}$

$= 2$ units **Ans.**

Q. 48. A line intersects the *Y*-axis and *X*-axis at the points *P* and *Q* respectively. If (2, – 5) is the mid-point of *PQ*, then find the coordinates of *P* and *Q*.

Solution : Let the coordinate of P be $(0, y)$ and coordinates of Q be $(x, 0)$

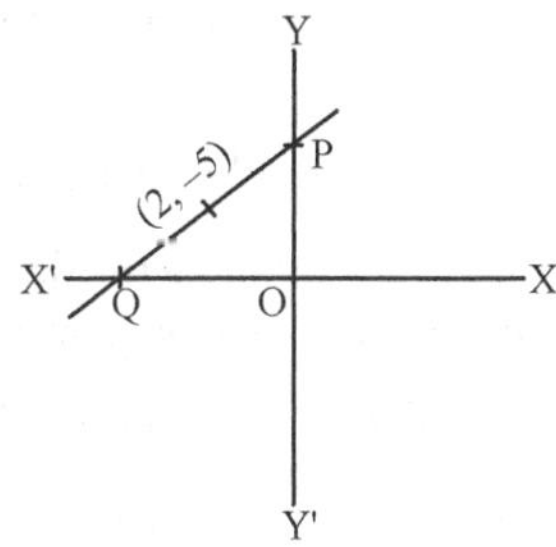

Since, mid-point is (2, –5)

$\therefore \left(\frac{x_1 + x_2}{2}, \frac{y_1 + y_2}{2}\right) = (2, -5)$

$\Rightarrow \left(\frac{x+0}{2}, \frac{0+y}{2}\right) = (2, -5)$

$\Rightarrow \quad \frac{x}{2} = 2; \quad \frac{y}{2} = -5$

$\Rightarrow \quad x = 4; \quad y = -10$

Coordinates of P (0, –10)

Coordinates of Q (4, 0) **Ans.**

★ **are board exam questions from previous years**

Q. 49. $A(-4, -2)$, $B(-3, -5)$, $C(3, -2)$ and $D(2, k)$ are the vertices of a quadrilateral *ABCD*. Find the value of *k* if the area of the quadrilateral is 28 sq. units.★

Solution. Given, $A = (-4, -2)$, $B = (-3, -5)$, $C = (3, -2)$ and $D = (2, k)$

Also, Area of $ABCD = 28$ sq. units

Now, Area of $ABCD$

$=$ Area of ΔABC + Area of ΔACD

A(–4, –2) B(–3, –5)

D(2, k) C(3, –2)

$\Rightarrow 28 = \frac{1}{2}|-4(-5+2) - 3(-2+2) + 3(-2+5)|$

$+ \frac{1}{2}|-4(-2-k) + 3(k+2) + 2(-2+2)|$

$\Rightarrow \quad 56 = 12 + 9 + 8 + 4k + 3k + 6$

$\Rightarrow \quad 7k = 56 - 35$

$\Rightarrow \quad 7k = 21$

$\Rightarrow \quad k = 3.$ **Ans.**

Q. 50. $A(3, 2)$ and $B(-2, 1)$ are two vertices of a triangle *ABC* whose centroid *G* has the coordinates $\left(\frac{5}{3}, -\frac{1}{3}\right)$. Find the coordinates of the third vertex of the triangle.

Solution : Given, $A = (3, 2)$, $B = (-2, 1)$ and centroid $G = \left(\frac{5}{3}, -\frac{1}{3}\right)$

Let the coordinates of the third vertex be $C(x, y)$.

Thus $\left(\frac{5}{3}, -\frac{1}{3}\right) = \left(\frac{x+3-2}{3}, \frac{y+1+2}{3}\right)$

$\Rightarrow \quad (5, -1) = (x+1, y+3)$

$\Rightarrow \quad (x, y) = (4, -4)$

Hence, the coordinates of the third vertex are (4, – 4).

Ans.

Q. 51. If the area of ΔABC formed by $A(x, y)$, $B(1, 2)$ and $C(2, 1)$ is 6 sq. units, then prove that $x + y = 15$.

Solution : Given, $A = (x, y)$, $B = (1, 2)$, $C = (2, 1)$ and area of $\Delta ABC = 6$ sq. units.

Thus, area of ΔABC

$= \frac{1}{2}|x(2-1) + 1(1-y) + 2(y-2)|$

$\Rightarrow \quad 6 = \frac{1}{2}|x + 1 - y + 2y - 4|$

$\Rightarrow \quad 12 = x + y - 3$

$\Rightarrow \quad x + y = 15.$ **Hence Proved.**

Q. 52. The *x*-coordinate of a point *P* is twice its *y*-coordinate. If *P* is equidistant from *Q*(2, – 5) and *R*(– 3, 6), find the coordinates of *P*.

Solution : Let the coordinates of point P be $(2y, y)$. Since, P is equidistant from Q and R

$\therefore \quad PQ = PR$

$\Rightarrow \quad \sqrt{(2y-2)^2 + (y+5)^2} = \sqrt{(2y+3)^2 + (y-6)^2}$

$\Rightarrow \quad (2y-2)^2 + (y+5)^2 = (2y+3)^2 + (y-6)^2$

$\Rightarrow 4y^2 + 4 - 8y + y^2 + 25 + 10y$

$= 4y^2 + 9 + 12y + y^2 + 36 - 12y$

$\Rightarrow \quad 2y + 29 = 45$

$\Rightarrow \quad 2y = 45 - 29$

$\Rightarrow \quad y = \frac{16}{2} = 8$

Hence, the coordinates of point P are (16, 8). **Ans.**

Q. 53. If (*x*, *y*) is on the line joining the points (1, – 3) and (– 4, 2), prove that $x + y + 2 = 0$.

Solution : Let $A = (1, -3)$ and $B = (-4, 2)$. Also, let $P = (x, y)$ is on the line joining AB. Thus A, B and P are collinear.

So, $x_1(y_2 - y_3) + x_2(y_3 - y_1) + x_3(y_1 - y_2) = 0$

Hence, $1(2 - y) - 4(y + 3) + x(-3 - 2) = 0$

$\Rightarrow \quad 2 - y - 4y - 12 - 5x = 0$

$\Rightarrow \quad -5y - 10 - 5x = 0$

$\Rightarrow \quad x + y + 2 = 0.$

Hence Proved.

Q. 54. Find the value of *k* if points *A*(7, – 2), *B*(5, 1) and *C*(3, 2*k*) are collinear.

Solution : Given, $A = (7, -2)$, $B = (5, 1)$ and $C = (3, 2k)$

So $x_1(y_2 - y_3) + x_2(y_3 - y_1) + x_3(y_1 - y_2) = 0$

Thus, $7(1 - 2k) + 5(2k + 2) + 3(-2 - 1) = 0$

$\Rightarrow \quad 7 - 14k + 10k + 10 - 9 = 0$

$\Rightarrow \quad -4k + 8 = 0$

$\Rightarrow \quad -4k = -8$

$\Rightarrow \quad k = 2.$ **Ans.**

Q. 55. If the point *P*(*m*, 3) lies on the line segment joining the points $A\left(-\frac{2}{5}, 6\right)$ and *B* (2, 8), find the value of *m*.

Solution : Given, $A = \left(-\frac{2}{5}, 6\right)$ and $B = (2, 8)$. Also $P = (m, 3)$ is on the line joining AB. Thus A, B and P are collinear.

So, $x_1(y_2 - y_3) + x_2(y_3 - y_1) + x_3(y_1 - y_2) = 0$

Thus, $-\frac{2}{5}(8 - 3) + 2(3 - 6) + m(6 - 8) = 0$

$\Rightarrow \quad -2 - 6 - 2m = 0$

$\Rightarrow \quad -2m = 8$

$\Rightarrow \quad m = -4.$

Ans.

Q. 56. Find the relation between *x* and *y* if the points *A*(*x*, *y*), *B*(– 5, 7) and *C*(– 4, 5) are collinear.

Solution : Given that the points $A(x, y)$, $B(-5, 7)$ and $C(-4, 5)$ are collinear.

So, the area formed by the vertices = 0.

Therefore,

$\frac{1}{2}[x_1(y_2 - y_3) + x_2(y_3 - y_1) + x_3(y_1 - y_2)] = 0$

$\Rightarrow \quad \frac{1}{2}[x(7 - 5) - 5(5 - y) - 4(y - 7)] = 0$

$\Rightarrow \quad \frac{1}{2}[x(2) - 5(5 - y) - 4(y - 7)] = 0$

$\Rightarrow \quad 2x - 25 + 5y - 4y + 28 = 0$

$2x + y + 3 = 0$

$y = -2x - 3$

which is the required relation between x and y. **Ans.**

Q. 57. If *R*(*x*, *y*) is a point on the line segment joining the points *P*(*a*, *b*) and *Q*(*b*, *a*), then prove that $x + y = a + b$.

Solution : Given, $P = (a, b)$ and $Q = (b, a)$ and $R = (x, y)$

So, $x_1(y_2 - y_3) + x_2(y_3 - y_1) + x_3(y_1 - y_2) = 0$

Thus, $x(b - a) + a(a - y) + b(y - b) = 0$

$\Rightarrow \quad x(b - a) + a^2 - ay + by - b^2 = 0$

$\Rightarrow \quad x(b - a) + y(b - a) - (b - a)(b + a) = 0$

$\Rightarrow \quad (b - a)[x + y - (a + b)] = 0$

But $a \neq b$ as for $a = b$, point P and Q will be same.

$\Rightarrow \quad x + y = b + a.$

Hence Proved.

Q. 58. If the vertices of a triangle are (1, – 3), (4, *p*) and (– 9, 7) and its area is 15 sq. units then find the value of *p*.

Solution : Let, $A = (1, -3)$, $B = (4, p)$, $C = (-9, 7)$ and area of ΔABC = 15 sq. units.

Area of $\Delta ABC = \frac{1}{2}|x_1(y_2 - y_3) + x_2(y_3 - y_1)$

$+ x_3(y_1 - y_2)|$

Thus, $15 = \frac{1}{2}|1(p-7) + 4(7+3) - 9(-3-p)|$

$\Rightarrow \quad 30 = p - 7 + 40 + 27 + 9p$

$\Rightarrow \quad 30 = 10p + 60$

$\Rightarrow \quad 10p = -30$

$\Rightarrow \quad p = -3.$ **Ans.**

Q. 59. What is the distance between the points $A(c, 0)$ and $B(0, -c)$?

Solution : Given, $A = (c, 0)$ and $B = (0, -c)$

Distance, $= \sqrt{(x_2 - x_1)^2 + (y_2 - y_1)^2}$

Thus, $AB = \sqrt{(c-0)^2 + (0+c)^2}$

$= \sqrt{(c)^2 + (c)^2}$

$= \sqrt{2}c$ units. **Ans.**

Q. 60. If the point $P(2, 6)$ is the mid-point of the line segment joining $A(6, 5)$ and $B(4, y)$, find the value of y.

Solution : Given, $P = (2, 6)$ is the mid-point of the line segment joining $A = (6, 5)$ and $B = (4, y)$.

$\therefore$ By mid-point formula

$$6 = \frac{5+y}{2}$$

$\Rightarrow \quad 5 + y = 12$

$\Rightarrow \quad y = 7.$ **Ans.**

Q. 61. If the distance between the points (3, 0) and $(0, y)$ is 5 units and y is positive then what is the value of y ?

Solution : Given, the distance between the points (3, 0) and $(0, y)$ is 5 units.

Distance $= \sqrt{(x_2 - x_1)^2 + (y_2 - y_1)^2}$

Thus, $5 = \sqrt{(3-0)^2 + (0-y)^2}$

$= \sqrt{9 + y^2}$

$\Rightarrow \quad 9 + y^2 = 25$

$\Rightarrow \quad y^2 = 16$

$\Rightarrow \quad y^2 = 4^2$

$\Rightarrow \quad y = \pm 4$

But as y is positive so the value of $y = 4$. **Ans.**

Q. 62. Prove that the points (3, 0), (6, 4) and (– 1, 3) are the vertices of a right angled isosceles triangle.

Solution : Let $A(3, 0)$, $B(6, 4)$ and $C(-1, 3)$ be the vertices of a triangle ABC.

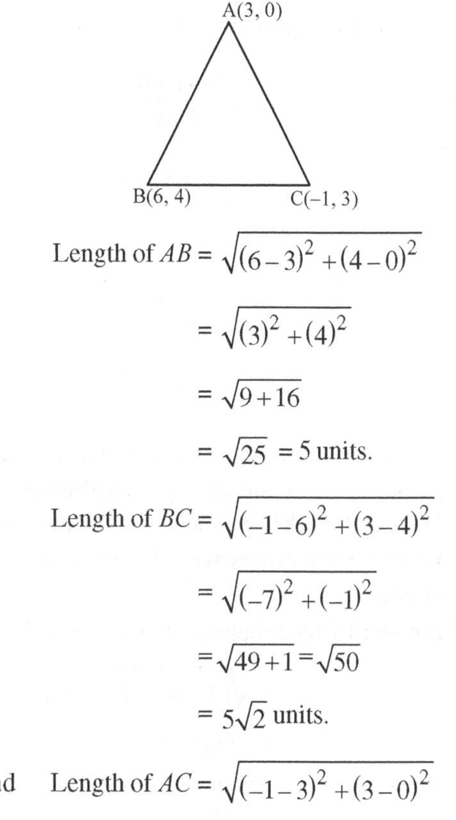

Length of $AB = \sqrt{(6-3)^2 + (4-0)^2}$

$= \sqrt{(3)^2 + (4)^2}$

$= \sqrt{9+16}$

$= \sqrt{25} = 5$ units.

Length of $BC = \sqrt{(-1-6)^2 + (3-4)^2}$

$= \sqrt{(-7)^2 + (-1)^2}$

$= \sqrt{49+1} = \sqrt{50}$

$= 5\sqrt{2}$ units.

and Length of $AC = \sqrt{(-1-3)^2 + (3-0)^2}$

$= \sqrt{(-4)^2 + (3)^2}$

$= \sqrt{16+9}$

$= \sqrt{25}$

$= 5$ units.

$\because \quad AB = AC$

and $(AB)^2 + (AC)^2 = (BC)^2$

Hence, ΔABC is an isosceles, right angled triangle.

Hence Proved.

Q. 63. Find the ratio in which Y-axis divides the line segment joining the points $A(5, -6)$ and $B(-1, -4)$. Also find the coordinates of the point of division.

Solution : Let the required ratio be $k : 1$ and point on Y-axis be $(0, y)$

k 1
A P B
(5, –6) (0, y) (–1, –4)

$\therefore \quad AP : PB = k : 1$

Then, by section formula

$$\frac{(-1)k + 5}{k+1} = 0$$

$$\frac{5-k}{k+1} = 0$$

$$5 - k = 0$$

$$\Rightarrow \quad k = 5$$

Hence, required ratio is 5 : 1

$$\therefore \quad y = \frac{-4k-6}{k+1}$$

$$y = \frac{-4(5)-6}{5+1}$$

$$\therefore \quad y = \frac{-26}{6} = -\frac{13}{3}$$

Hence, point on Y-axis is $\left(0, \frac{-13}{3}\right)$. **Ans.**

Q. 64. The coordinates of houses of Sonu and Labhoo are (7, 3) and (4, 3) respectively. Coordinates of their school is (2, 2). If both leave their house at the same time in the morning and also reach school in time then who travels faster ?

Solution : Distance between Sonu's house and school

$$= \sqrt{(2-7)^2 + (2-3)^2}$$

$$= \sqrt{26} \text{ units}$$

Distance between Labhoo's house and school

$$= \sqrt{(2-4)^2 + (2-3)^2}$$

$$= \sqrt{5} \text{ units}$$

$\because$ Distance of Sonu's house from school is more

$\therefore$ Sonu travels faster. **Ans.**

Chapter 7. Triangles

Q. 1. A farmer has a field in the form of triangle. He wants to give his land to his 4 sons equally. Explain how he will separate his field.

Solution : As the farmer wants to divide the field into equal parts, so he can divide it as shown in this figure.

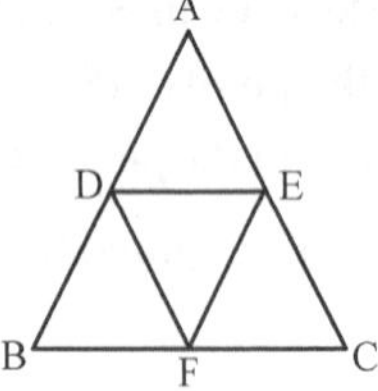

where D, E and F are mid-points of sides AB, BC and CA respectively and

$$\Delta ADE \cong \Delta DBF \cong \Delta DEF \cong \Delta EFC$$

So, he will give piece in the form of triangles ΔADF, ΔBDF, ΔCEF and ΔDEF. **Ans.**

Q. 2. In an isosceles ΔABC right angled at B, prove that $AC^2 = 2AB^2$.

Solution : Given, In ΔABC, $AB = BC$ [as triangle is isosceles]

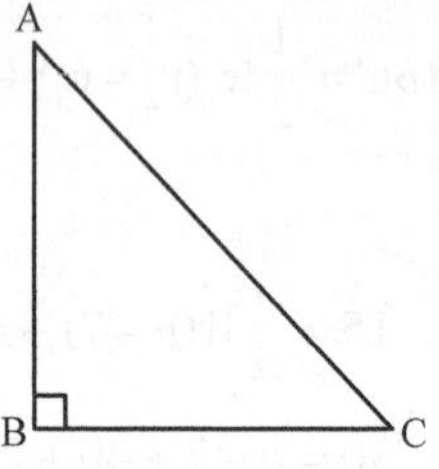

Now,

In ΔABC by Pythagoras theorem,

$$AC^2 = AB^2 + BC^2$$

$$AC^2 = AB^2 + AB^2 \quad [\because AB = BC]$$

$$AC^2 = 2AB^2$$ **Hence Proved.**

Q. 3. In Figure, ABC is an isosceles triangle right angled at C with AC = 4 cm. Find the length of AB. ★

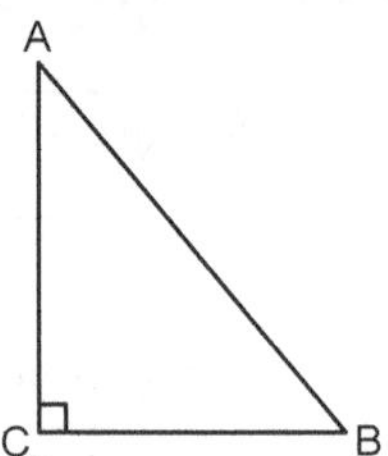

Solution : Given, $\angle C = 90°$ and $AC = 4$ cm, $AB = ?$

$\because$ ΔABC is an isosceles triangle so,

$$BC = AC = 4 \text{ cm}$$

On applying Phythagoras theorem, we have

$$AB^2 = AC^2 + BC^2$$

$$\Rightarrow \quad AB^2 = AC^2 + AC^2 \quad (\because BC = AC)$$

$$= 4^2 + 4^2$$

$$= 16 + 16 = 32$$

$$\Rightarrow \quad AB = \sqrt{32}$$

$$= 4\sqrt{2} \text{ cm}$$

Q. 4. A 6.5 m long ladder is placed against a wall such that its foot is at a distance of 2.5 m from the wall. Find the height of the wall where the top of the ladder touches it.

Solution : Let AC be the ladder and AB be the wall

$$\therefore \quad AC = 6.5 \text{ cm} = \frac{13}{2} \text{m}$$

$$\text{and} \quad BC = 2.5 = \frac{5}{2} \text{m}$$

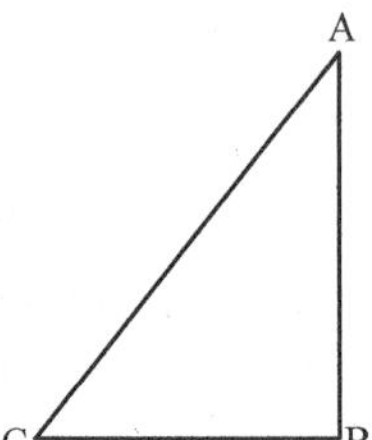

★ are board exam questions from previous years

In right ΔABC,

$$AB^2 + BC^2 = AC^2 \quad \text{[Pythagoras Theorem]}$$

$$AB^2 + \left(\frac{5}{2}\right)^2 = \left(\frac{13}{2}\right)^2$$

$$AB^2 = \frac{169}{4} - \frac{25}{4}$$

$$= \frac{169-25}{4}$$

$$= \frac{144}{4}$$

$$= 36$$

$\therefore$ Required height, $AB = 6$ m **Ans.**

Q. 5. In ΔDEW, $AB \parallel EW$. If $AD = 4$ cm, $DE = 12$ cm and $DW = 24$ cm, then find the value of DB.

Solution : Given, $AD = 4$ cm, $DE = 12$ cm and $DW = 24$ cm.

Let $BD = x$ cm

Since, $BW = DW - BD$

Then, $BW = (24 - x)$ cm.

Similarly, $AE = 12 - 4 = 8$ cm

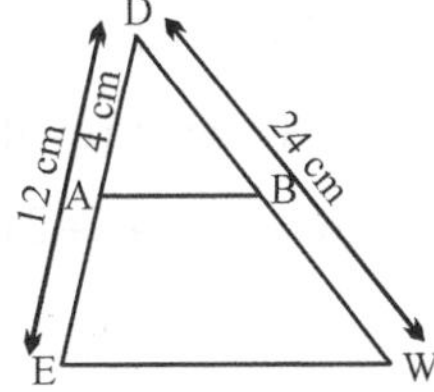

In ΔDEW, $AB \parallel EW$

$$\therefore \quad \frac{AD}{AE} = \frac{BD}{BW} \quad \text{[Thales' theorem]}$$

$$\Rightarrow \quad \frac{4}{8} = \frac{x}{24-x}$$

$$\Rightarrow \quad 8x = 96 - 4x$$

$$\Rightarrow \quad 12x = 96$$

$$\Rightarrow \quad x = \frac{96}{12} = 8$$

$\therefore \quad DB = 8$ cm **Ans.**

Q. 6. In a right triangle ABC, right-angled at B, $BC = 12$ cm and $AB = 5$ cm. Find the radius of the circle inscribed in the triangle.

Solution : Given, ΔABC is right-angled at B.

Applying Pythagoras theorem,

$$AC^2 = BC^2 + AB^2$$

$$\Rightarrow \quad AC^2 = 12^2 + 5^2$$

$$= 144 + 25$$

$$= 169 = 13^2$$

$$\Rightarrow \quad AC = 13 \text{ cm}$$

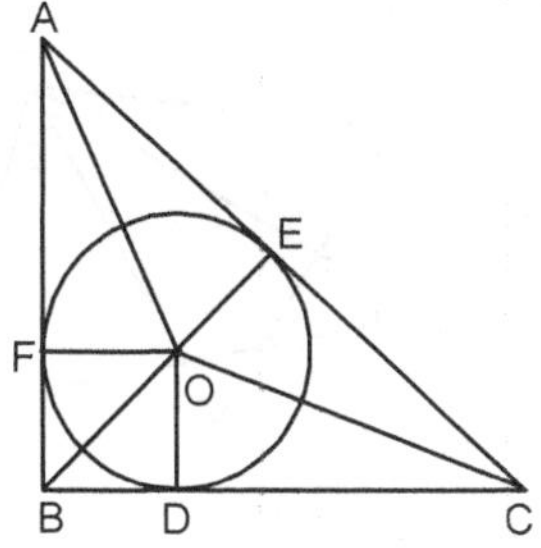

Let O be the centre of the circle inscribed within ΔABC such that OD, OE and OF are the altitude of ΔBOC, ΔCOA and ΔAOB, respectively and $OD = OE = OF = r$.

Thus, Area of ΔABC = Area of ΔBOC + Area of ΔCOA + Area of ΔAOB

$$\Rightarrow \quad \frac{1}{2} AB \times BC = \frac{1}{2} OD \times BC + \frac{1}{2} OE \times AC + \frac{1}{2} OF \times AB$$

$$\Rightarrow \quad \frac{1}{2}(5 \times 12) = \frac{1}{2} r \times 12 + \frac{1}{2} r \times 13 + \frac{1}{2} r \times 5$$

$$\Rightarrow \quad 60 = 12r + 13r + 5r$$

$$\Rightarrow \quad 60 = 30r$$

$$\Rightarrow \quad r = 2.$$ **Ans.**

Q. 7. In ΔABC, D and E are mid-points of AC and BC respectively such that $DE \parallel AB$. If $AD = 2x$, $BE = 2x - 1$, $CD = x + 1$ and $CE = x - 1$, then find the value of x.

Solution : Given, $DE \parallel AB$

$$\text{So,} \quad \frac{AD}{CD} = \frac{BE}{EC} \quad \text{[B.P.T.]}$$

$$\Rightarrow \quad \frac{2x}{x+1} = \frac{2x-1}{x-1}$$

$$\Rightarrow \quad 2x(x-1) = (x+1)(2x-1)$$

$$\Rightarrow \quad 2x^2 - 2x = 2x^2 + 2x - x - 1$$

$$\Rightarrow \quad -2x = x - 1$$

$$\Rightarrow \quad 1 = 3x$$

$$\text{or} \quad x = \frac{1}{3}$$ **Ans.**

Q. 8. On occasion of independence day, Mira made a Rangoli in a design as shown. If $\angle 1 = \angle 2$, $\angle 3 = \angle 4$, show that $PT. QR = PR. ST$.

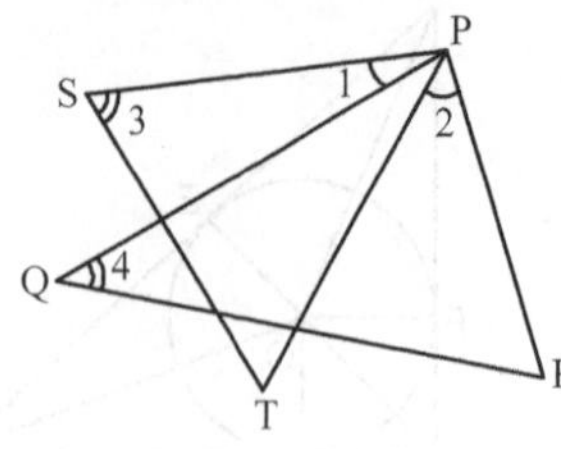

Solution : $\angle 1 = \angle 2$ [Given]

$\Rightarrow \angle 1 + \angle QPT = \angle 2 + \angle QPT$

$\Rightarrow \angle SPT = \angle QPR$

Also $\angle 3 = \angle 4$ [Given]

$\Rightarrow \Delta TSP \sim \Delta RQP$ [By AA similarity]

$\Rightarrow \dfrac{ST}{QR} = \dfrac{PT}{PR}$ [cpct]

$\Rightarrow ST. PR = PT. QR.$ **Hence Proved.**

Q. 9. A man steadily goes 10 m due east and then 24 m due north. Find the distance from the starting point.

Solution : Let the initial position of the man be at O and final position be B.

Since the man goes to 10 m due east and then 24 m due north

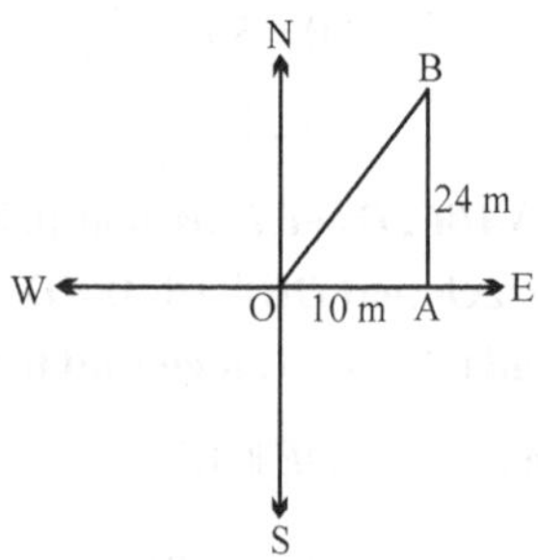

Therefore, ΔAOB is a right angled triangle at angle A.

In ΔAOB

$$OB^2 = OA^2 + AB^2$$
$$= (10)^2 + (24)^2$$
$$= 100 + 576 = 676$$

$\Rightarrow OB = \sqrt{676}$

$\Rightarrow OB = 26$ m

Hence, the man is at a distance of 26 m from the starting point. **Ans.**

Q. 10. A boy walks 12 m due east and 5 m due south. How far is he from the starting point ?

Solution : Applying Pythagoras' theorem

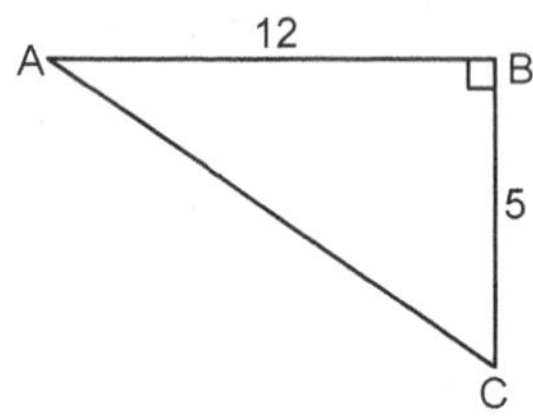

$AC^2 = AB^2 + BC^2$

$\Rightarrow AC^2 = 12^2 + 5^2$

$= 144 + 25$

$= 169 = 13^2$

$\Rightarrow AC = 13$ m. **Ans.**

Q. 11. In ΔPQR, if PS is the internal bisector of $\angle P$ meeting QR at S and $PQ = 15$ cm, $QS = (3 + x)$ cm, $SR = (x - 3)$ cm and $PR = 7$ cm, then find the value of x.

Solution : Since PS is the internal bisector of $\angle P$ and it meets QR at S, therefore

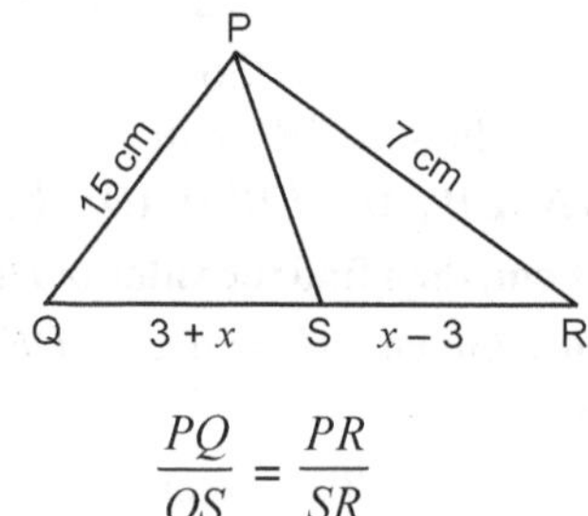

$$\frac{PQ}{QS} = \frac{PR}{SR}$$

[Angle bisector theorem]

$\Rightarrow \dfrac{15}{3+x} = \dfrac{7}{x-3}$

$\Rightarrow 7(3 + x) = 15(x - 3)$

$\Rightarrow 21 + 7x = 15x - 45$

$\Rightarrow 15x - 7x = 45 + 21$

$\Rightarrow 8x = 66$

$\Rightarrow 4x = 33$

$\Rightarrow x = 8.25$ cm. **Ans.**

Q. 12. In ΔABC, $DE \parallel BC$, find the value of x.

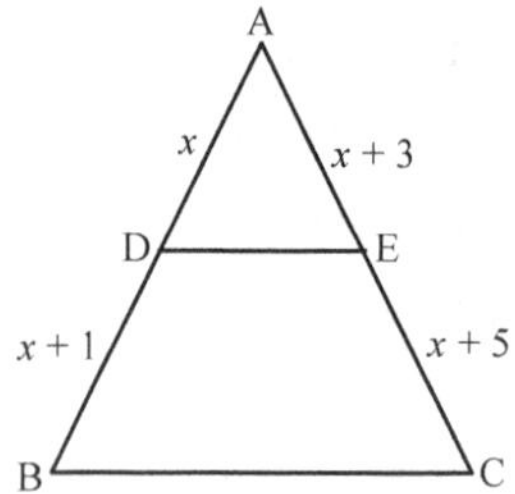

Solution : In ΔABC, $DE \parallel BC$ [Given]

$\therefore \dfrac{AD}{BD} = \dfrac{AE}{EC}$ [Thales theorem]

$$\frac{x}{x+1} = \frac{x+3}{x+5}$$

$$x(x + 5) = (x + 3)(x + 1)$$

$$x^2 + 5x = x^2 + 3x + x + 3$$

$$x^2 + 5x - x^2 - 3x - x = 3$$

$\therefore x = 3$ **Ans.**

Q. 13. In Figure, DE || BC. Find the length of side AD, given that AE = 1·8 cm, BD = 7·2 cm and CE = 5·4 cm.★

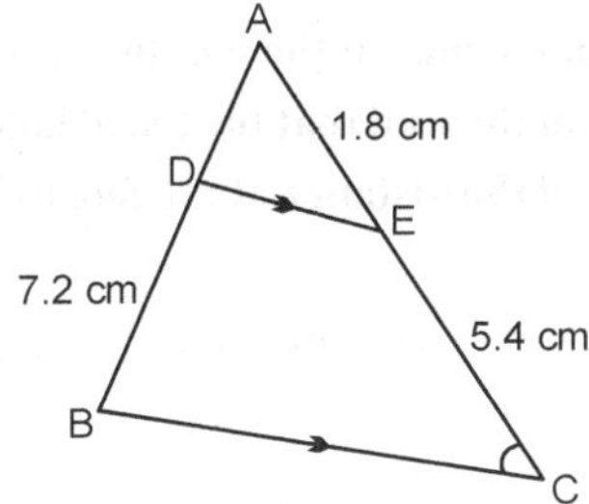

Solution : Given, DE || BC

On applying, Thales theorem, we have

$$\frac{AD}{AB} = \frac{AE}{AC}$$

$$\frac{AD}{AD+7{\cdot}2} = \frac{18}{1{\cdot}8+5{\cdot}4}$$

$$\frac{AD}{AD+7{\cdot}2} = \frac{1{\cdot}8}{7{\cdot}2}$$

$$\frac{AD}{AD+7{\cdot}2} = \frac{1}{4}$$

$$4AD = AD + 7{\cdot}2$$
$$3AD = 7{\cdot}2$$

$$AD = 2{\cdot}4 \text{ cm}$$ **Ans.**

Q. 14. If the bisector of an angle bisects the opposite sides, prove that the triangle is an isosceles.

Solution : Let ABC be a triangle where AD is the bisector of $\angle BAC$ such that

$$BD = DC$$

Since AD is the bisector of $\angle A$, therefore by Angle-Bisector theorem,

$$\frac{AB}{AC} = \frac{BD}{DC} = 1 \qquad [\because BD = DC]$$

or $$\frac{AB}{AC} = 1$$

or $$AB = AC$$

Hence, ΔABC is isosceles triangle. **Hence Proved.**

Q. 15. A vertical row of trees 12 m long casts a shadow 8 m long on the ground. At the same time, a tower casts the shadow 40 m long on the ground. Determine the height of the tower.

Solution : Let AB be the vertical row of trees and AC be its shadow.

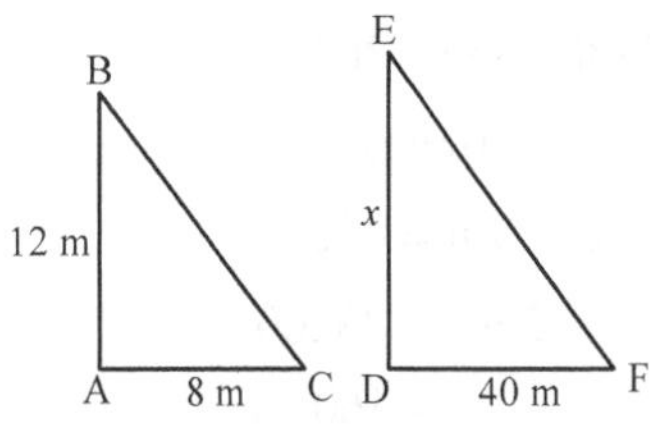

Also, let DE be the vertical tower and DF be its shadow. Join BC and EF.

Let $DE = x$

We have,

$$AB = 12 \text{ m}, AC = 8 \text{ m}$$

and $$DF = 40 \text{ m}, DE = x \text{ m}$$

In ΔABC and ΔDEF, we have

$$\angle A = \angle D = 90°$$

$$\angle C = \angle F$$

$\therefore$ $\Delta ABC \sim \Delta DEF$ [AA criterion]

$\Rightarrow$ $$\frac{AB}{DE} = \frac{AC}{DF}$$ [cpct]

$\Rightarrow$ $$\frac{12}{x} = \frac{8}{40}$$

$\Rightarrow$ $$x = \frac{12 \times 40}{8}$$

$$= 12 \times 5 = 60$$

Thus, height of the tower is 60 m. **Ans.**

Q. 16. A girl of height 90 cm is walking away from the base of a lamp-post at a speed of 1.2 m/s. If the lamp is 3.6 m above the ground, find the length of her shadow after 4 seconds.

Solution : Let AB denote the lamp-post and CD be the height of the girl after walking for 4 seconds away from the lamp-post.

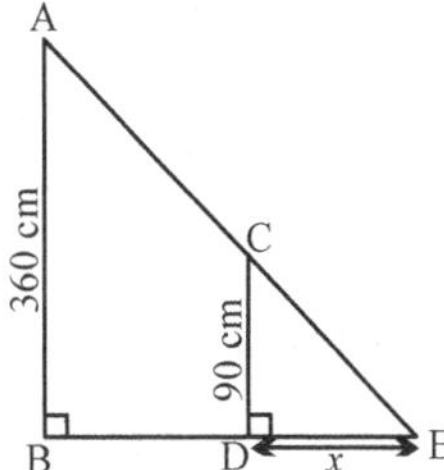

From Fig. $BD = 1.2 \times 4$

$$= 4.8 \text{ m}$$

$$= 480 \text{ cm}.$$

In ΔABE and ΔCDE,

$$\angle B = \angle D$$ [Each 90°]

$$\angle E = \angle E$$ [Common]

$\therefore$ $\Delta ABE \sim \Delta CDE$ [By AA criterion]

$\Rightarrow$ $$\frac{AB}{BE} = \frac{CD}{DE} \Rightarrow \frac{360}{480+x} = \frac{90}{x}$$

★ are board exam questions from previous years

$$90\,(480 + x) = 360x$$
$$43200 + 90x = 360x$$
$$360x - 90x = 43200$$
$$270x = 43200$$
$$x = 160 \text{ cm}$$ **Ans.**

Q. 17. Kitchen garden of Ms. Sanjana is in the form of a triangle as shown. She wants to divide it in two parts; one triangle and one trapezium.

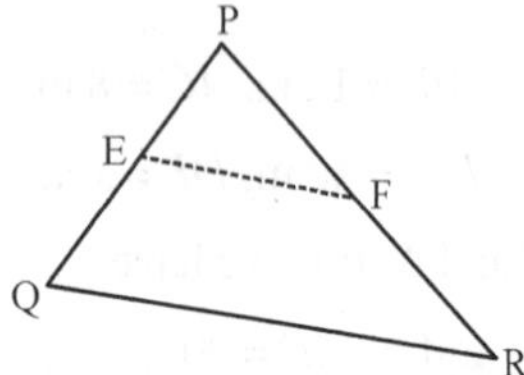

She takes $PE = 4$ m, $QE = 4.5$ m, $PF = 8$ m and $RF = 9$ m. Is $EF \,||\, QR$? Justify your answer.

Solution : $\dfrac{PE}{QE} = \dfrac{4}{4.5} = \dfrac{8}{9}$ ...(i)

$\dfrac{PF}{RF} = \dfrac{8}{9}$ ...(ii)

Form equations (i) and (ii)

$$\frac{PE}{QE} = \frac{PF}{RF}$$

$\therefore$ $EF \,||\, QR$ (yes) [By converse of B.P.T.]

Ans.

Q. 18. Students of a school decided to participate in 'Save girl child' campaign. They decided to decorate a triangular path as shown. If $AB = AC$, $BC^2 = AC \times CD$ prove that $BD = BC$.

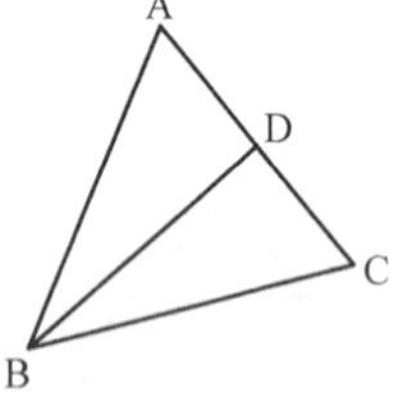

Solution : In ΔABC and ΔBCD

$BC^2 = AC \times CD$ [Given]

$\Rightarrow$ $\dfrac{BC}{CD} = \dfrac{AC}{BC}$

$\angle C = \angle C$ [Common]

$\therefore$ $\Delta ABC \sim \Delta BDC$ [By SAS criterion]

$\Rightarrow$ $\dfrac{AB}{BD} = \dfrac{AC}{BC}$ [cpct]

$\because$ $AB = AC$ [Given]

$\therefore$ $BD = BC$ **Hence Proved.**

Q. 19. An old man was trying to place a ladder 13 m long in such a way that it reaches the window of a building 12 m above the ground. Anoop decided to help the old man. He placed the foot to the ladder at such as distance from the wall that the top of ladder reached the window. Find the distance of the foot of ladder from the wall.

Solution : Using Pythagoras theorem

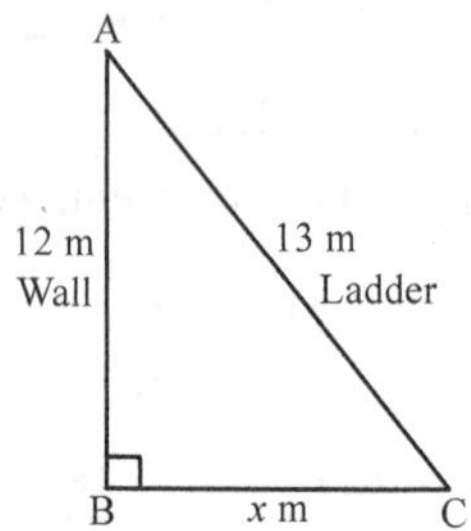

$$(13)^2 = (12)^2 + x^2$$

$\Rightarrow$ $x^2 = 25$

$\therefore$ $x = 5$ m.

So, distance of the foot of ladder from the wall = 5 m.

Ans.

Q. 20. In ΔABC, AD is the bisector of $\angle A$. If $AB = 5.6$ cm, $BD = 3.2$ cm and $BC = 6$ cm, find AC.

Soltion : Given, AD is the bisector of $\angle A$.

Thus, $\angle BAD = \angle CAD$

Now $BC = BD + DC$

or $6 = 3.2 + DC$

or $DC = 2.8$ cm

Since AD is the angle bisector of $\angle A$,

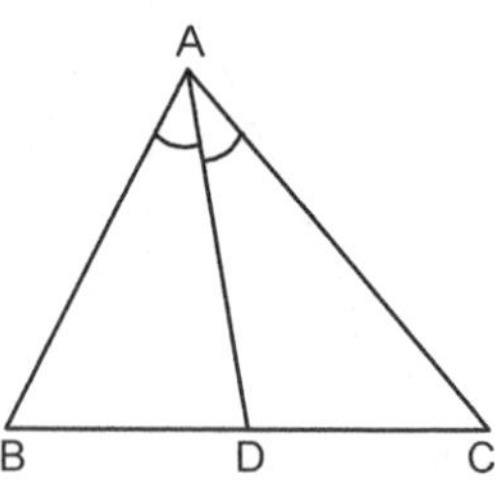

Hence $\dfrac{DB}{DC} = \dfrac{AB}{AC}$ [Angle bisector theorem]

$\Rightarrow$ $\dfrac{3.2}{2.8} = \dfrac{5.6}{AC}$

$\Rightarrow$ $AC = \dfrac{5.6 \times 2.8}{3.2}$

$= 4.9$ cm. **Ans.**

Q. 21. The perimeters of two similar triangles are 25 cm and 15 cm respectively. If one side of the first triangle is 9 cm, find the corresponding side of the second triangle.

Solution : Let ABC and DEF be the two triangles and their corresponding sides be AB and DE.

Thus, $\dfrac{\text{Perimeter of } \Delta ABC}{\text{Perimeter of } \Delta DEF} = \dfrac{\text{Length of side } AB}{\text{Length of side } DE}$

or $\dfrac{25}{15} = \dfrac{9}{\text{Length of side } DE}$

Thus, length of side $DE = \dfrac{15 \times 9}{25}$

$= 5.4$ cm

Thus, the length of the corresponding side is 5.4 cm. **Ans.**

Q. 22. In ΔABC, AD is the bisector of $\angle A$. If $AB = 5.6$ cm, $AC = 4$ cm and $DC = 3$ cm, find BC.

Solution : Given, AD is the bisector of $\angle A$.

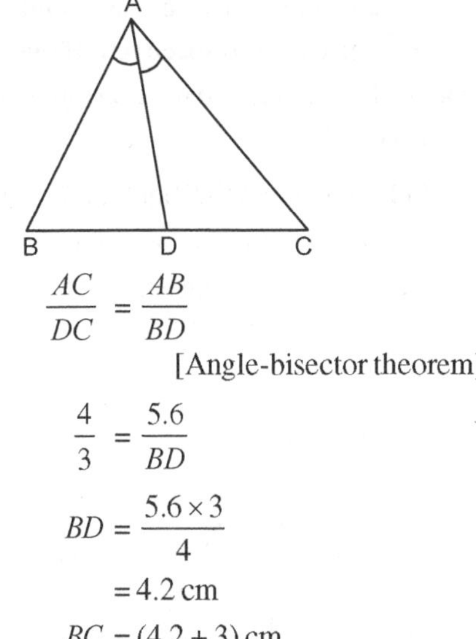

Thus $\dfrac{AC}{DC} = \dfrac{AB}{BD}$ [Angle-bisector theorem]

or $\dfrac{4}{3} = \dfrac{5.6}{BD}$

or $BD = \dfrac{5.6 \times 3}{4}$

$= 4.2$ cm

Hence $BC = (4.2 + 3)$ cm

$= 7.2$ cm. **Ans.**

Q. 23. The perimeters of two similar triangles ABC and PQR are 32 cm and 24 cm respectively. If $PQ = 12$ cm, find AB.

Solution : We know, $\dfrac{AB}{PQ} = \dfrac{\text{Perimeter of } \Delta ABC}{\text{Perimeter of } \Delta PQR}$

$\Rightarrow \dfrac{AB}{12} = \dfrac{32}{24}$

$\Rightarrow AB = \dfrac{12 \times 32}{24}$

$= 16$ cm. **Ans.**

Q. 24. The points $A(4, 7)$, $B(p, 3)$ and $C(7, 3)$ are the vertices of a right triangle, right-angled at B. Find the value of p.

Solution : The given points are $A(4, 7)$, $B(p, 3)$ and $C(7, 3)$.

Since A, B and C are the vertices of a right angled triangle.

$\therefore (AB)^2 + (BC)^2 = (AC)^2$ [By Pythagoras theorem]

$[(p-4)^2 + (3-7)^2] + [(7-p)^2 + (3-3)^2]$
$= [(7-4)^2 + (3-7)^2]$

$(p-4)^2 + (-4)^2 + (7-p)^2 = (3)^2 + (-4)^2$

$p^2 + 16 - 8p + 16 + 49 + p^2 - 14p = 9 + 16$

$2p^2 - 22p + 56 = 0$

$p^2 - 11p + 28 = 0$

$p^2 - 7p - 4p + 28 = 0$

$p(p-7) - 4(p-7) = 0$

$p = 4$ or 7

$p \neq 7$ (as B and C will coincide)

So, $p = 4$. **Ans.**

Q. 25. If the areas of two similar triangles are in the ratio of 25 : 64, find the ratio of their corresponding sides.

Solution : Let ΔABC and ΔDEF be the two similar triangles. We know that the ratio of the areas of two similar triangles is equal to the ratio of the square of their corresponding sides.

$\therefore \dfrac{\text{Area of } \Delta ABC}{\text{Area of } \Delta DEF} = \left(\dfrac{AC}{DF}\right)^2 = \left(\dfrac{AB}{DE}\right)^2 = \left(\dfrac{BC}{EF}\right)^2$

Since $\dfrac{25}{64} = \left(\dfrac{5}{8}\right)^2$

So, $\dfrac{AC}{DF} = \dfrac{AB}{DE} = \dfrac{BC}{EF} = \dfrac{5}{8}$

Thus, the ratio of the corresponding sides is 5 : 8.

Ans.

Q. 26. The area of two similar triangles ΔABC and ΔPQR are 25 cm² and 49 cm² respectively. If $QR = 9.8$ cm, find BC.

Solution : According to the question,

$\Delta ABC \sim \Delta PQR$

We know that the ratio of the areas of two similar triangles is equal to the ratio of the squares of their corresponding sides.

$\therefore \dfrac{\text{Area of } \Delta ABC}{\text{Area of } \Delta PQR} = \left(\dfrac{BC}{QR}\right)^2$

$\Rightarrow \dfrac{25}{49} = \left(\dfrac{BC}{9.8}\right)^2$

$\Rightarrow \left(\dfrac{5}{7}\right)^2 = \left(\dfrac{BC}{9.8}\right)^2$

$\Rightarrow \dfrac{5}{7} = \dfrac{BC}{9.8}$

$\Rightarrow BC = \dfrac{5 \times 9.8}{7}$

$= 5 \times 1.4$

$= 7$ cm. **Ans.**

Q. 27. X and Y are points on the sides AB and AC respectively of a triangle ABC such that $\frac{AX}{AB} = \frac{1}{4}$, $AY = 2$ cm and $YC = 6$ cm. Find whether $XY \parallel BC$ or not.

Solution : Given, $\frac{AX}{AB} = \frac{1}{4}$

i.e., $AX = 1K, AB = 4K$ (K is constant)

$\therefore$ $BX = AB - AX$

$= 4K - 1K = 3K$

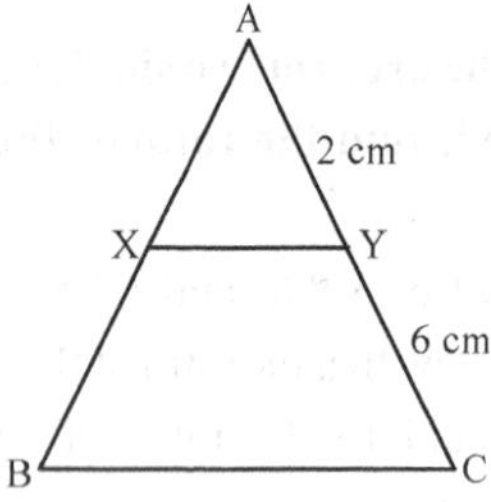

Now, $\frac{AX}{XB} = \frac{1K}{3K} = \frac{1}{3}$

and, $\frac{AY}{YC} = \frac{2}{6} = \frac{1}{3}$

$\because$ $\frac{AX}{XB} = \frac{AY}{YC}$

$\therefore$ $XY \parallel BC$

[By converse of Thales' theorem]

Ans.

Q. 28. The areas of two similar triangles are 81 cm² and 49 cm² respectively. If the altitude of the bigger triangle is 4.5 cm, find the corresponding altitude of the smaller triangle.

Solution : Let ΔABC and ΔDEF be two triangles with AP and DQ being the altitudes of the two and $AP > DQ$.

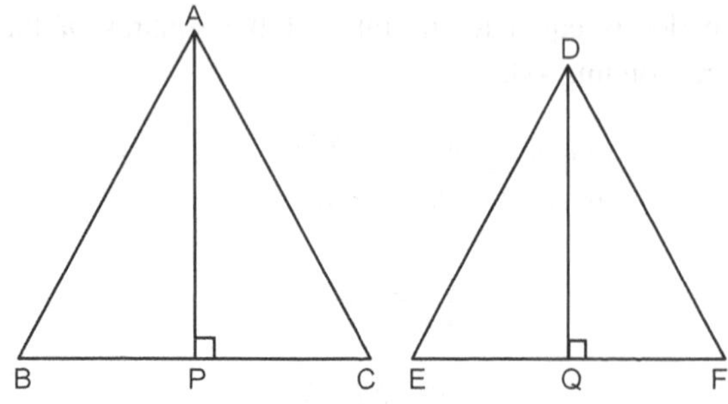

According to the question,

$\Delta ABC \sim \Delta DEF$

Since the ratio of the areas of two similar triangles is equal to the ratio of the squares of their corresponding altitudes.

$$\therefore \quad \frac{\text{Area of } \Delta ABC}{\text{Area of } \Delta DEF} = \left(\frac{AP}{DQ}\right)^2$$

$$\Rightarrow \quad \frac{81}{49} = \left(\frac{4.5}{DQ}\right)^2$$

$$\Rightarrow \quad \left(\frac{9}{7}\right)^2 = \left(\frac{4.5}{DQ}\right)^2$$

$$\Rightarrow \quad \frac{9}{7} = \frac{4.5}{DQ}$$

$$\Rightarrow \quad DQ = \frac{4.5 \times 7}{9}$$

$= 0.5 \times 7$

$= 3.5$ cm. **Ans.**

Q. 29. The areas of two similar triangles are 121 cm² and 64 cm² respectively. If the median of the first triangle is 12.1 cm, find the corresponding median of the other triangle.

Solution : Let ΔLMN and ΔSTU be two triangles with LO and SV being the medians of the two.

According to the question,

$\Delta LMN \sim \Delta STU$

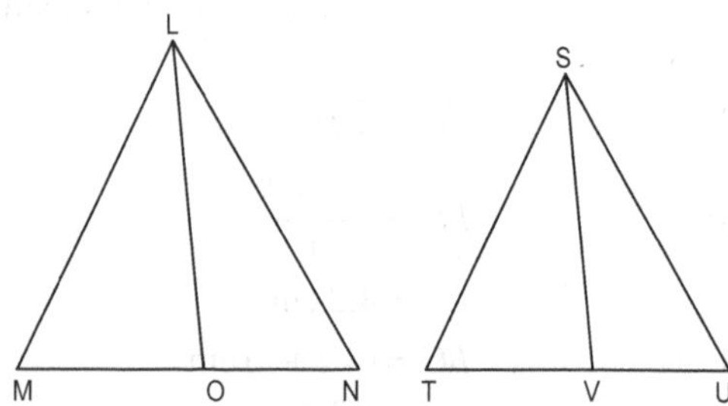

As we know, the ratio of the areas of two similar triangles is equal to the ratio of the squares of their corresponding medians.

$$\therefore \quad \frac{\text{Area of } \Delta LMN}{\text{Area of } \Delta STU} = \left(\frac{LO}{SV}\right)^2$$

$$\Rightarrow \quad \frac{121}{64} = \left(\frac{12.1}{SV}\right)^2$$

$$\Rightarrow \quad \left(\frac{11}{8}\right)^2 = \left(\frac{12.1}{SV}\right)^2$$

$$\Rightarrow \quad \frac{11}{8} = \frac{12.1}{SV}$$

$$\Rightarrow \quad SV = \frac{8 \times 12.1}{11}$$

$= 8 \times 1.1$

$= 8.8$ cm

Hence, the median of the other triangle is 8.8 cm.

Ans.

Q. 30. In a trapezium *ABCD*, *O* is the point of intersection of *AC* and *BD*, *AB* || *CD* and *AB* = 2 × *CD*. If the area of ΔAOB = 84 cm², find the area of ΔCOD.

Solution : In ΔAOB and ΔCOD,

$\angle OAB = \angle OCD$ [alternate angles]

$\angle OBA = \angle ODC$ [alternate angles]

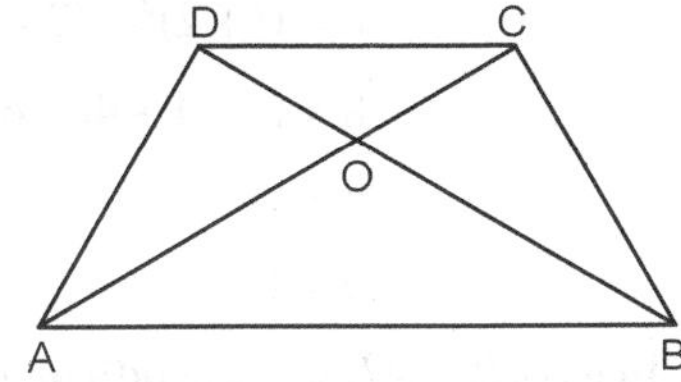

Thus $\Delta AOB \sim \Delta COD$ [AA-similarity]

Hence $\frac{\text{Area of } \Delta AOB}{\text{Area of } \Delta COD} = \frac{AB^2}{CD^2} = \frac{(2CD)^2}{CD^2} = 4$

or Area of $\Delta COD = \frac{1}{4} \times$ Area of ΔAOB

$= \frac{1}{4} \times 84 \text{ cm}^2$

$= 21 \text{ cm}^2$. **Ans.**

Q. 31. In the given figure, *DE* || *BC* and *AD* : *DB* = 5 : 4. Find the ratio of the area of $\Delta DEF : \Delta CFB$.

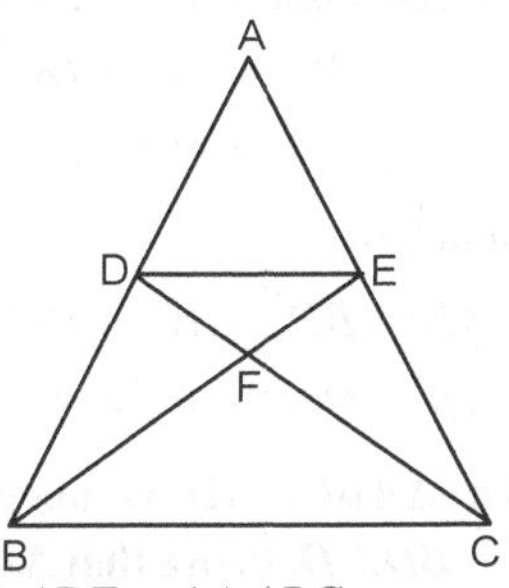

Solution : In ΔADE and ΔABC,

$\angle A = \angle A$ [common]

$\angle ADE = \angle ABC$ [corresponding angles between two parallel lines]

Thus $\Delta ADE \sim \Delta ABC$ [AA similarity]

So $\frac{DE}{BC} = \frac{AD}{AB}$ [cpct]

But $\frac{AD}{DB} = \frac{5}{4}$

$\Rightarrow \frac{DB}{AD} = \frac{4}{5}$

$\Rightarrow \frac{DB}{AD} + 1 = \frac{4}{5} + 1$

$\Rightarrow \frac{DB + AD}{AD} = \frac{9}{5}$

$\frac{AB}{AD} = \frac{9}{5}$

$\Rightarrow \frac{AD}{AB} = \frac{5}{9}$

Similarly in ΔDEF and ΔBFC,

$\angle DFE = \angle BFC$ [vertically opposite angles]

$\angle EDF = \angle FCB$ [alternate angles]

Thus $\Delta DEF \sim \Delta CBF$ [AA similarity]

$\therefore \frac{\text{Area of } \Delta DEF}{\text{Area of } \Delta CBF} = \frac{DE^2}{BC^2}$

$= \frac{5^2}{9^2} = \frac{25}{81}$

Hence, Area of ΔDEF : Area of ΔCBF = 25 : 81. **Ans.**

Q. 32. The areas of two similar triangles *ABC* and *PQR* is in the ratio of 9 : 16. If *BC* = 4.5 cm, find the length of *QR*.

Solution : Given that,

$\Delta ABC \sim \Delta PQR$

$\therefore \frac{\text{Area of } \Delta ABC}{\text{Area of } \Delta PQR} = \frac{BC^2}{QR^2} = \frac{(4.5)^2}{QR^2}$

$\Rightarrow \frac{9}{16} = \frac{(4.5)^2}{QR^2}$

$\Rightarrow \frac{3^2}{4^2} = \frac{(4.5)^2}{QR^2}$

$\Rightarrow QR = \frac{4.5 \times 4}{3}$

$= 6$ cm. **Ans.**

Q. 33. $\Delta ABC \sim \Delta DEF$ and their areas are respectively 100 cm² and 49 cm². If the altitude of ΔABC is 5 cm, find the corresponding altitude of ΔDEF.

Solution : Given that,

$\Delta ABC \sim \Delta DEF$

Let *BG* be the altitude of ΔABC and *EH* the altitutde of ΔDEF.

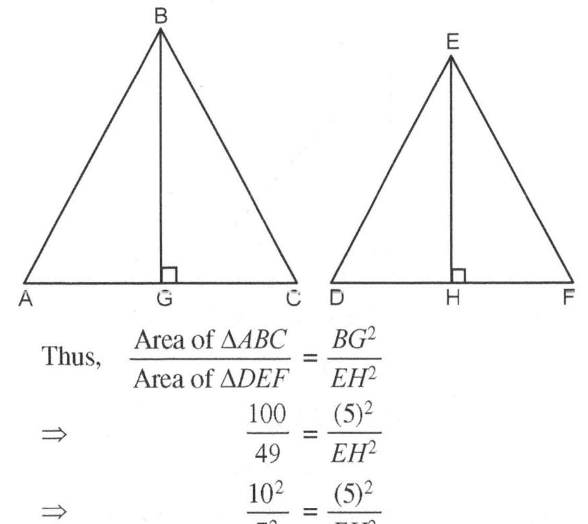

Thus, $\frac{\text{Area of } \Delta ABC}{\text{Area of } \Delta DEF} = \frac{BG^2}{EH^2}$

$\Rightarrow \frac{100}{49} = \frac{(5)^2}{EH^2}$

$\Rightarrow \frac{10^2}{7^2} = \frac{(5)^2}{EH^2}$

$\Rightarrow \qquad EH = \dfrac{7 \times 5}{10}$

$= 3.5$ cm. **Ans.**

Q. 34. The areas of two similar triangles are 81 cm² and 49 cm² respectively. If the altitude of the first triangle is 6.3 cm, find the corresponding altitude of the other.

Solution : Let $\Delta PQR \sim \Delta STU$ be the concerned triangles.

Also, let $QU = 6.3$ be the altitude of ΔPQR and SV the altitude of ΔSTU.

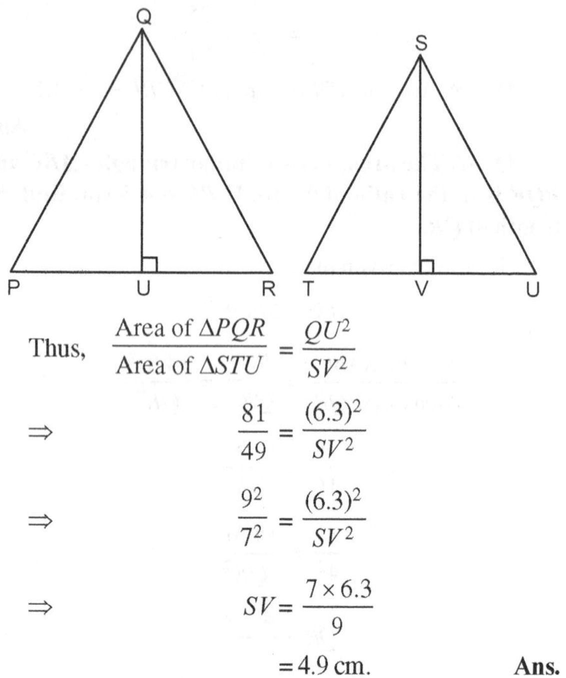

Thus, $\dfrac{\text{Area of } \Delta PQR}{\text{Area of } \Delta STU} = \dfrac{QU^2}{SV^2}$

$\Rightarrow \qquad \dfrac{81}{49} = \dfrac{(6.3)^2}{SV^2}$

$\Rightarrow \qquad \dfrac{9^2}{7^2} = \dfrac{(6.3)^2}{SV^2}$

$\Rightarrow \qquad SV = \dfrac{7 \times 6.3}{9}$

$= 4.9$ cm. **Ans.**

Q. 35. If A (5, 2), B(2, –2) and C(– 2, t) are the vertices of a right angled triangle with $\angle B = 90°$, then find the value of t.

Solution : Given, $A(5, 2)$, $B(2, -2)$ and $C(-2, t)$ are the vertices of a right angled triangle, then,

By Pythagoras theorem,

$$(AC)^2 = (BC)^2 + (AB)^2 \qquad ...(i)$$

Now, $(AC)^2 = (5+2)^2 + (2-t)^2$

$= 49 + (2-t)^2$

$(BC)^2 = (2+2)^2 + (-2-t)^2$

$= 16 + (t+2)^2$

and $(AB)^2 = (5-2)^2 + (2+2)^2$

$= 9 + 16$

$= 25$

Putting these values in (i)

$49 + (2-t)^2 = 16 + (t+2)^2 + 25$

$49 + (2-t)^2 = 41 + (t+2)^2$

$8 = (t+2)^2 - (2-t)^2$

$8 = t^2 + 4 + 4t - 4 - t^2 + 4t$

$8 = 8t$

$t = 1$ **Ans.**

Q. 36. In a ΔABC, AD is perpendicular to BC. Prove that $(AB^2 + CD^2) = (AC^2 + BD^2)$.

Solution : From the right-angled ΔADB, we have

$AB^2 = AD^2 + BD^2$

or $\quad AD^2 = AB^2 - BD^2 \qquad ...(i)$

From the right-angled ΔADC, we have

$AC^2 = AD^2 + DC^2$

or $\quad AD^2 = AC^2 - DC^2 \qquad ...(ii)$

From (i) and (ii),

$AB^2 - BD^2 = AC^2 - DC^2$

or $\quad AB^2 + CD^2 = AC^2 + BD^2$ **Hence Proved.**

Q. 37. In ΔABC, AD is perpendicular to BC such that $AD^2 = BD.CD$. Prove that ΔABC is right-angled at A.

Solution : Since AD is perpendicular to BC,

$\therefore \quad \angle ADB = \angle ADC = 90°$

Also $\quad AB^2 = AD^2 + BD^2 \qquad ...(i)$

and $\quad AC^2 = AD^2 + DC^2 \qquad ...(ii)$

On adding (i) and (ii),

$AB^2 + AC^2 = BD^2 + CD^2 + 2AD^2$

$\Rightarrow \quad AB^2 + AC^2 = BD^2 + CD^2 + 2BD.CD$

[Given $AD^2 = BD.CD$]

$\Rightarrow \quad AB^2 + AC^2 = (BD + CD)^2$

$\Rightarrow \quad AB^2 + AC^2 = BC^2$

Therefore by converse of Pythagoras' theorem, ΔABC is right-angled at A. **Hence Proved.**

Q. 38. ΔABC is right-angled at B and D is the mid-point of BC. Prove that $AC^2 = (4AD^2 - 3AB^2)$.

Solution : In ΔABC,

$\angle B = 90°$

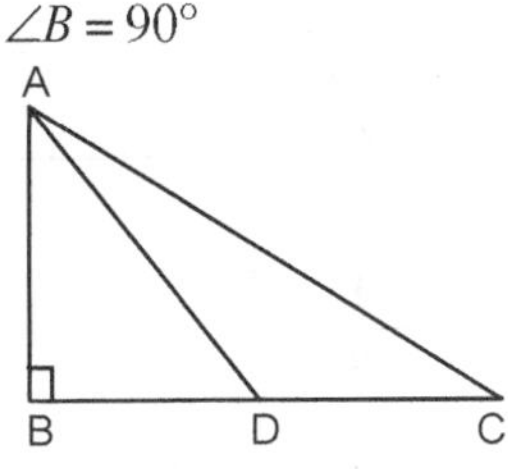

$\therefore$ $AC^2 = AB^2 + BC^2$

$\Rightarrow$ $AC^2 = AB^2 + (2BD)^2$

$[\because BC = BD + DC = 2BD = 2DC]$

$\Rightarrow$ $AC^2 = AB^2 + 4BD^2$

$\Rightarrow$ $AC^2 = AB^2 + 4(AD^2 - AB^2)$

$\Rightarrow$ $AC^2 = AB^2 + 4AD^2 - 4AB^2$

$\Rightarrow$ $AC^2 = 4AD^2 - 3AB^2$. **Hence Proved.**

Q. 39. ΔABC is an isosceles triangles with $AC = BC$. If $AB^2 = 2AC^2$, prove that ΔABC is a right-triangle.

Solution : Given,

$AB^2 = 2AC^2$

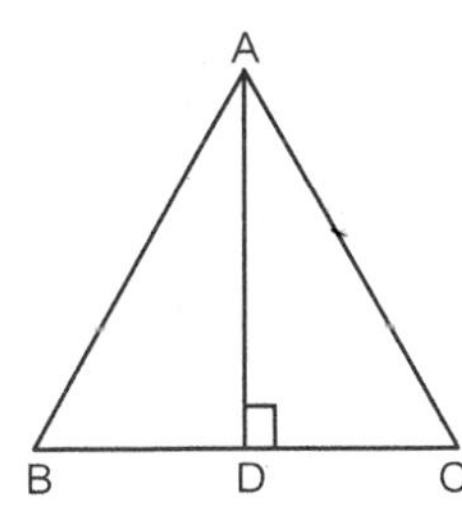

$\Rightarrow$ $AB^2 = AC^2 + AC^2$

$\Rightarrow$ $AB^2 = AC^2 + BC^2$ [Given, $AC = BC$]

Therefore, by converse of Pythagoras theorem, ΔABC is right-angled triangle. **Hence Proved.**

Q. 40. ΔABC is an isosceles triangle with $AB = AC$ = 13 cm. The length of the altitude from A on BC is 5 cm. Find BC.

Solution : Given,

$AB = AC = 13$ cm

Let AD be the altitude on BC.

$\therefore$ $\angle ADB = \angle ADC = 90°$

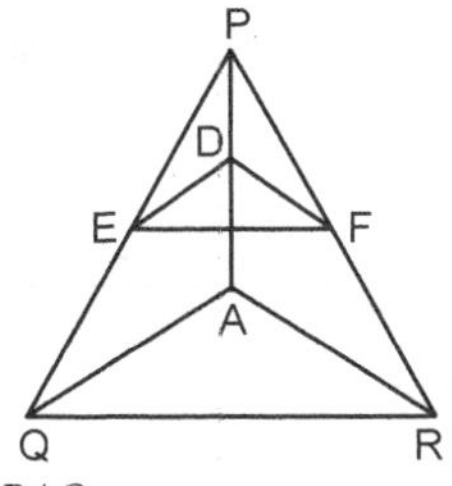

So $AB^2 = AD^2 + BD^2$

$\Rightarrow$ $BD^2 = AB^2 - AD^2$

$\Rightarrow$ $BD^2 = 13^2 - 5^2$

$= 169 - 25$

$= 144 = 12^2$

$\Rightarrow$ $BD = 12$ cm

Similarly, $AC^2 = AD^2 + DC^2$

$\Rightarrow$ $DC^2 = AC^2 - AD^2$

$\Rightarrow$ $DC^2 = 13^2 - 5^2$

$= 169 - 25$

$= 144 = 12^2$

$\Rightarrow$ $DC = 12$ cm

So $BC = BD + DC$

$= (12 + 12)$ cm

$= 24$ cm. **Ans.**

Q. 41. In the given figure, if $DE \parallel AQ$ and $DF \parallel AR$. Prove that $EF \parallel QR$.

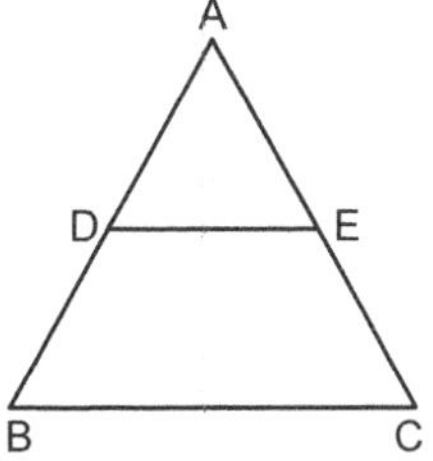

Solution : In ΔPAQ,

$DE \parallel QA$ [Given]

$\therefore$ $\frac{PE}{EQ} = \frac{PD}{DA}$ ...(i)

In ΔPAR, $DE \parallel AR$ [Given)]

$\therefore$ $\frac{PD}{DA} = \frac{PF}{FR}$ ...(ii)

From (i) and (ii),

$\frac{PE}{EQ} = \frac{PF}{FR}$

Thus in ΔPQR,

$EF \parallel QR$.

[By converse of Thales' theorem]]

Hence Proved.

Q. 42. In ΔABC, D and E are points on the sides of AB and AC such that $DE \parallel BC$. If $AD = 2.5$ cm, $BD = 3$ cm, $AE = 3.75$ cm, find the length of AC.

Solution : In ΔABC,

$DE \parallel BC$ [Given]

$\therefore$ $\frac{AD}{BD} = \frac{AE}{EC}$

[Proportionality theorem]

A
D E
B C

$$\Rightarrow \quad \frac{2.5}{3} = \frac{3.75}{EC}$$

$$\Rightarrow \quad EC = \frac{3.75 \times 3}{2.5} = 4.5 \text{ cm}$$

Hence $AC = AE + EC$

$= (3.75 + 4.5)$ cm

$= 8.25$ cm. **Ans.**

Q. 43. In Fig., $DE \parallel BC$, $AD = 1$ cm and $BD = 2$ cm. what is the ratio of the ar (ΔABC) to the ar (ΔADE) ?★

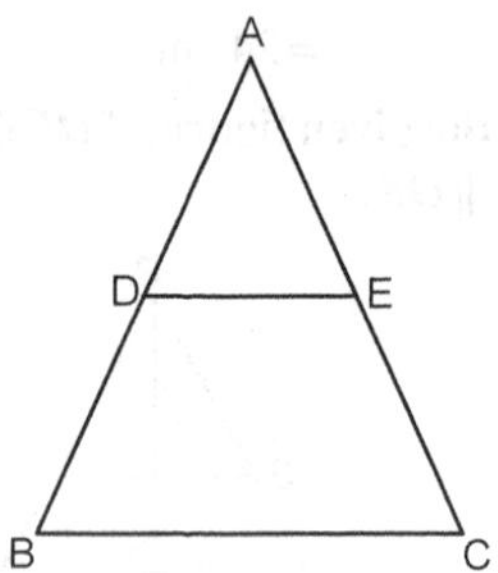

Solution : Given,

$AD = 1$ cm, $BD = 2$ cm

$\therefore \quad AB = 1 + 2 = 3$ cm

Also, $\quad DE \parallel BC$ (Given)

$\therefore \quad \angle ADE = \angle ABC$ ...(i)

(corresponding angles)

In ΔABC and ΔADE

$\angle A = \angle A$ (common)

$\angle ABC = \angle ADE$ [by equation (i)]

$\therefore \quad \Delta ABC \sim \Delta ADE$ (by AA rule)

Now,

$$\frac{\text{ar}(\Delta ABC)}{\text{ar}(\Delta ADE)} = \left(\frac{AB}{AD}\right)^2 \quad [\because \Delta ABC \sim \Delta ADE]$$

or $$\frac{\text{ar}(\Delta ABC)}{\text{ar}(\Delta ADE)} = \left(\frac{3}{1}\right)^2 = \frac{9}{1}$$

$\therefore$ ar (ΔABC) : ar (ΔADE) = 9 : 1 **Ans.**

Q. 44. In ΔABC, D and E are points on AB and AC respectively such that $DE \parallel BC$. If $AD = 2.4$ cm, $AE = 3.2$ cm, $DE = 2$ cm and $BC = 5$ cm, find BD and CE.

Solution : In ΔABC,

$DE \parallel BC$ [Given]

$$\therefore \quad \frac{AD}{AB} = \frac{DE}{BC} \quad \text{[Thales' theorem]}$$

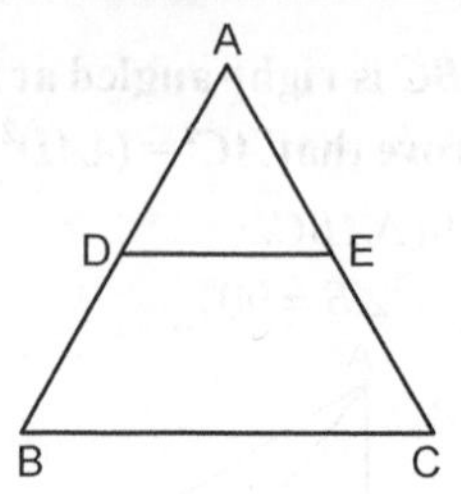

$$\Rightarrow \quad \frac{2.4}{AB} = \frac{2}{5}$$

$\Rightarrow \quad AB = (5 \times 1.2)$ cm

$= 6$ cm

So $\quad BD = AB - AD$

$= (6 - 2.4)$ cm

$= 3.6$ cm

Also, $$\frac{AE}{AC} = \frac{DE}{BC} \quad \text{[Thales' theorem]}$$

$$\Rightarrow \quad \frac{3.2}{AC} = \frac{2}{5}$$

$\Rightarrow \quad AC = (5 \times 1.6)$ cm $= 8$ cm

So $\quad EC = AC - AE$

$= (8 - 3.2)$ cm

$= 4.8$ cm. **Ans.**

Q. 45. In ΔABC, D and E are points on AB and AC respectively such that $DE \parallel BC$. If $BD = CE$. Prove that ΔABC is isosceles.

Solution : In ΔABC, it is given that

$BD = CE$

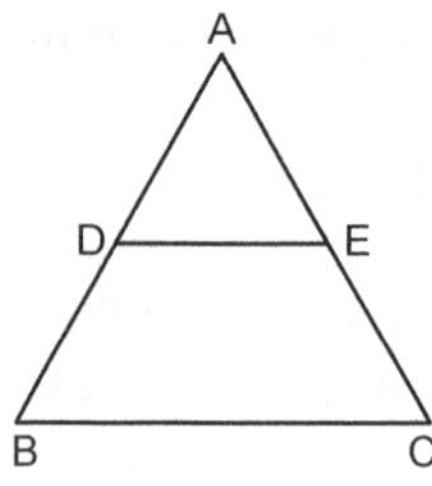

Since, $\quad DE \parallel BC$ [Given]

Therefore, according to Thales' theorem,

$$\frac{AD}{BD} = \frac{AE}{EC}$$

or $$\frac{AD}{EC} = \frac{AE}{EC}$$

or $\quad AD = AE$

Thus $\quad AB = AD + BD$ and $AC = AE + EC$

★ are board exam questions from previous years

or $AB = AC$

[$\because BD = CE$ (given)

and $AD = AE$ (proved above)]

So, ΔABC is isosceles. **Hence Proved.**

Q. 46. In ΔABC, AD is the bisector of $\angle A$ meeting BC at D. Find BC if $AB = 5.6$ cm, $AC = 6$ cm and $DC = 3$ cm.

Solution : In ΔABC,

AD is the bisector of $\angle A$

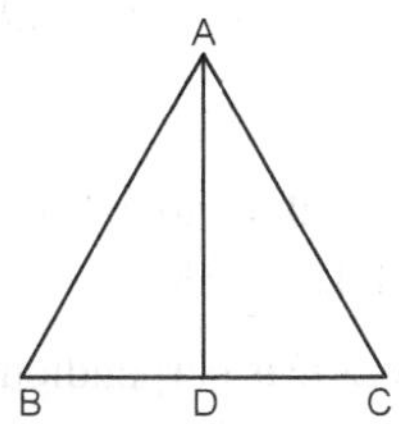

$\therefore \quad \dfrac{AB}{AC} = \dfrac{BD}{DC}$ [Angle bisector theorem]

or $\quad \dfrac{5.6}{6} = \dfrac{BD}{3}$

Thus $\quad BD = 2.8$ cm

So $\quad BC = BD + DC$

$= (3 + 2.8)$ cm

$= 5.8$ cm. **Ans.**

Q. 47. In ΔABC, AD is the bisector of $\angle A$ meeting BC at D. Find BD and DC if $AB = 10$ cm, $AC = 6$ cm and $BC = 12$ cm.

Solution : In ΔABC,

AD is the bisector of $\angle A$

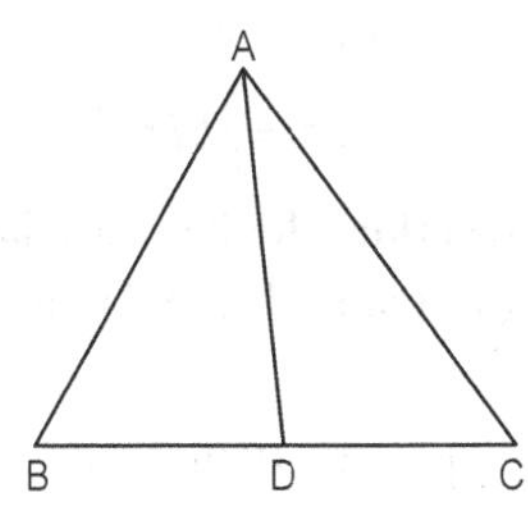

$\therefore \quad \dfrac{AB}{AC} = \dfrac{BD}{DC}$ [Angle bisector theorem]

$\Rightarrow \quad \dfrac{BD}{DC} = \dfrac{10}{6}$

$\Rightarrow \quad \dfrac{BD}{DC} = \dfrac{5}{3}$ ...(i)

But $\quad BC = BD + DC$

$\Rightarrow \quad BD + DC = 12$

So, $\quad BD = 12 - DC$

Substituting in (i),

$$\frac{12 - DC}{DC} = \frac{5}{3}$$

$\Rightarrow \quad 3(12 - DC) = 5DC$

$\Rightarrow \quad 5DC + 3DC = 36$

$\Rightarrow \quad 8DC = 36$

$\Rightarrow \quad DC = \dfrac{36}{8}$

$= \dfrac{9}{2} = 4.5$ cm

Hence $\quad BD = (12 - 4.5)$ cm

$= 7.5$ cm. **Ans.**

Q. 48. In the given figure, $AB \parallel CD$. If $OA = 3x - 19$, $OB = x - 4$, $OC = x - 3$ and $OD = 4$, find x.

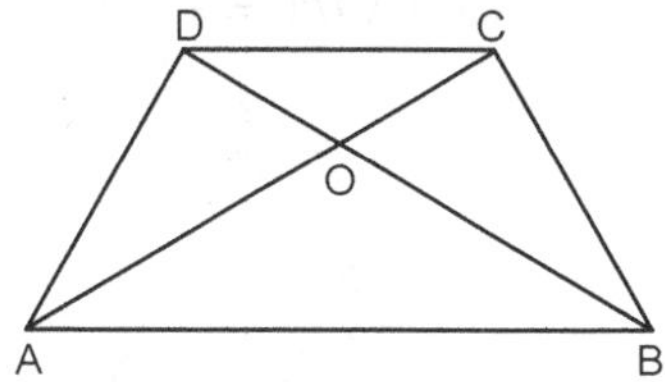

Solution : Since, $AB \parallel DC$ [Given]

So, according to the Basic Proportionality Theorem,

$$\frac{OD}{OB} = \frac{OC}{OA}$$

$\Rightarrow \quad \dfrac{4}{x-4} = \dfrac{x-3}{3x-19}$

$\Rightarrow \quad 4(3x - 19) = (x - 3)(x - 4)$

$\Rightarrow \quad 12x - 76 = x^2 - 3x - 4x + 12$

$\Rightarrow \quad x^2 - 7x + 12 - 12x + 76 = 0$

$\Rightarrow \quad x^2 - 19x + 88 = 0$

$\Rightarrow \quad x^2 - 11x - 8x + 88 = 0$

$\Rightarrow \quad x(x - 11) - 8(x - 11) = 0$

$\Rightarrow \quad (x - 11)(x - 8) = 0$

$\Rightarrow \quad x = 8, 11$

Thus, the value of x is either 8 or 11. **Ans.**

Q. 49. In the given figure, if AB is perpendicular to BC and DE is perpendicular to AC, prove that $\Delta ABC \sim \Delta AED$.

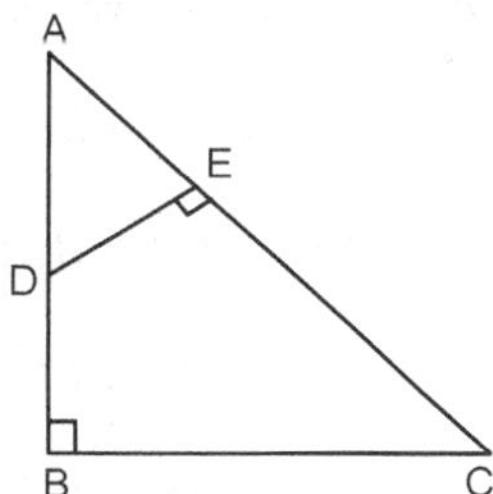

Solution : Given

$$\angle ABC = \angle AED = 90^\circ$$

[AB is perpendicular to BC and DE is perpendicular to AC]

In ΔABC and ΔAED,

$$\angle A = \angle A \quad \text{[common]}$$

$$\angle ABC = \angle AED = 90^\circ$$

Thus, $\Delta ABC \sim \Delta AED$ [AA-Criterion]

Hence Proved.

Q. 50. If ΔABC and ΔBDE are equilateral triangles where D is the mid-point of BC, find the ratio of the area of ΔABC and ΔBDE.

Solution : Let each side of ΔABC be x.

So Area of $\Delta ABC = \frac{\sqrt{3}}{4}x^2$

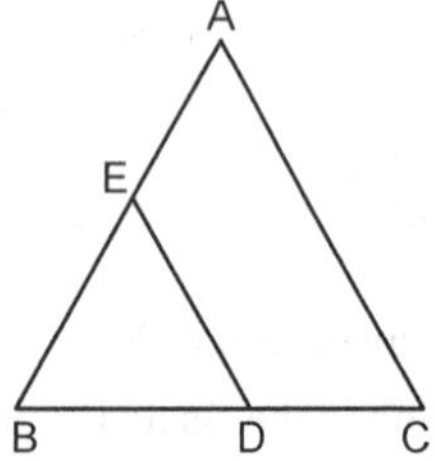

As D is the mid-point of BC,

So, $BD = \frac{x}{2}$

So, Area of $\Delta BDE = \frac{\sqrt{3}}{4}\left(\frac{x^2}{4}\right)$

Hence $\frac{\text{Area of } \Delta ABC}{\text{Area of } \Delta BDE} = \frac{x^2}{\frac{x^2}{4}} = \frac{4}{1}$

Area of ΔABC : Area of $\Delta BDE = 4 : 1$. **Ans.**

Q. 51. ΔABC is a right-angled at B. If D and E be any points on AB and BC, prove that

$$AE^2 + CD^2 = AC^2 + DE^2. \quad \text{[NCERT]}$$

Solution : Applying Pythagoras theorem in ΔABE, in ΔBCD, in ΔBDE and in ΔABC, we get

$$AE^2 = AB^2 + BE^2 \quad ...(i)$$

$$CD^2 = DB^2 + BC^2 \quad ...(ii)$$

$$DE^2 = DB^2 + BE^2 \quad ...(iii)$$

and $$AC^2 = AB^2 + BC^2 \quad ...(iv)$$

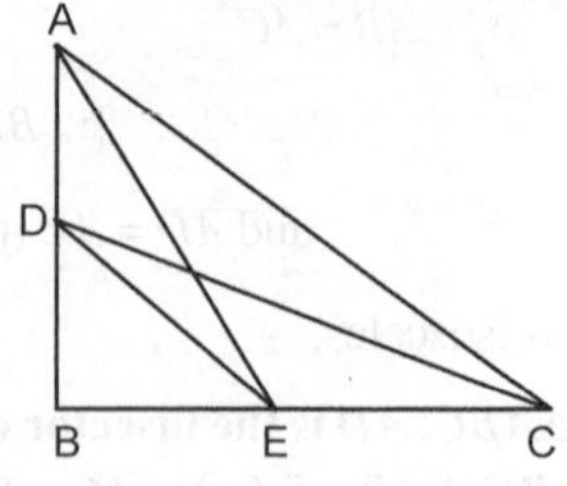

On adding (i) and (ii), we get

$$AE^2 + CD^2 = AB^2 + BE^2 + DB^2 + BC^2$$

$$\Rightarrow AE^2 + CD^2 = (AB^2 + BC^2) + (BE^2 + DB^2)$$

$$\Rightarrow AE^2 + CD^2 = AC^2 + DE^2. \text{[from (iii) and (iv)]}$$

Hence Proved.

Q. 52. In an equilateral ΔABC, AD is perpendicular to BC. Prove that $AD^2 = 3BD^2$.

Solution : As AD is perpendicular to BC,

$\therefore$ $BD = DC$

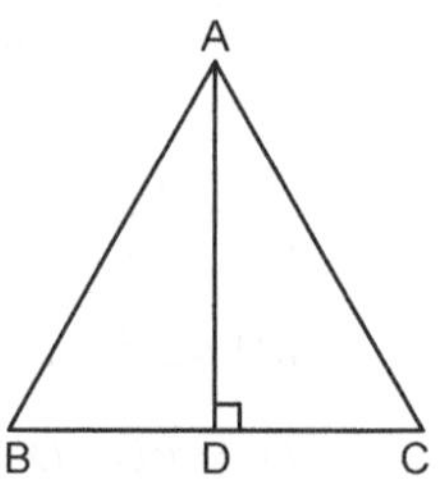

Thus $BC = 2BD = 2DC$

In ΔABD,

$$AB^2 = AD^2 + BD^2 \quad \text{[Pythagoras' theorem]}$$

$$\Rightarrow BC^2 = AD^2 + BD^2 \quad [\because AB = BC]$$

$$\Rightarrow (2BD)^2 = AD^2 + BD^2$$

$$\Rightarrow 4BD^2 = AD^2 + BD^2$$

$$\Rightarrow 3BD^2 = AD^2$$

$$\Rightarrow AD^2 = 3BD^2.$$ **Hence Proved.**

Q. 53. Determine whether the triangle having sides $(a - 1)$ cm, $2\sqrt{a}$ cm and $(a + 1)$ cm is a right-angled triangle.

Solution : Applying Pythagoras' theorem

$$(a-1)^2 + (2\sqrt{a})^2 = a^2 + 1 - 2a + 4a$$

$$= a^2 + 1 + 2a$$

$$= (a + 1)^2$$

Thus, the triangle is right-angled. **Ans.**

Q. 54. The lengths of the diagonals of a rhombus are 30 cm and 40 cm. Find the length of one side of the rhombus.

Solution : Let O be the point of intersection of the diagonals of a rhombus $ABCD$.

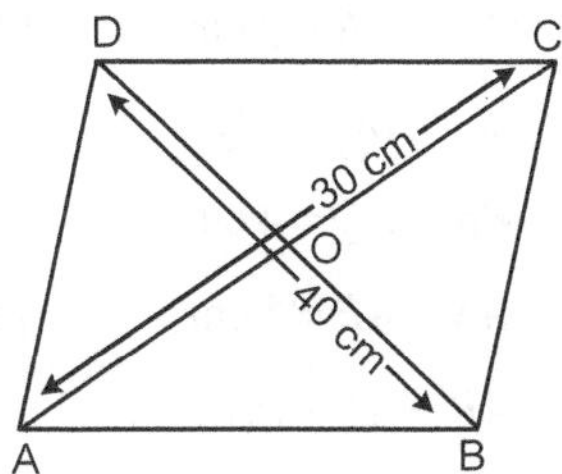

As the diagonals of a rhombus bisect each other.

$\therefore\ OA = OC = \frac{AC}{2}$ 15 cm and $OB = OD = \frac{BO}{2} = 20$ cm

Now, in ΔAOB

$$AB^2 = OA^2 + OB^2$$

[Pythagoras' theorem]

$\Rightarrow \quad AB^2 = 15^2 + 20^2$

$\Rightarrow \quad AB^2 = 225 + 400$

$\Rightarrow \quad AB^2 = 625$

$\Rightarrow \quad AB = \sqrt{625}$

$= 25$ cm. **Ans.**

Q. 55. In the figure given below, $PQ \parallel BC$ and $AP : PB = 1 : 2$. Find $\frac{\text{Area of } \Delta APQ}{\text{Area of } \Delta ABC}$.

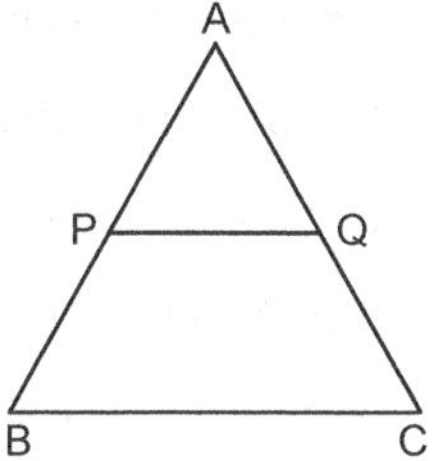

Solution : Given,

$AP : PB = 1 : 2$

or $\quad 2AP = PB$

But $\quad AB = AP + PB$

or $\quad AB = AP + 2AP = 3AP$

$\because \quad PQ \parallel BC$

$\therefore \quad \frac{AP}{AB} = \frac{AQ}{AC}$ [Thales' theorem]

Now, in ΔAPQ and ΔABC

$$\frac{AP}{AB} = \frac{AQ}{AC}$$

and $\quad \angle A = \angle A$ [common]

$\therefore \quad \Delta APQ \sim \Delta ABC$

$\therefore \quad \frac{\text{Area of } \Delta APQ}{\text{Area of } \Delta ABC} = \frac{AP^2}{AB^2}$

or $\quad \frac{\text{Area of } \Delta APQ}{\text{Area of } \Delta ABC} = \frac{AP^2}{(3AP)^2}$

$= \frac{AP^2}{9AP^2}$

$= \frac{1}{9}$. **Ans.**

Q. 56. In the given figure, S and T are points on the sides PQ and PR respectively of ΔPQR such that $PT = 2$ cm, $TR = 4$ cm and ST is parallel to QR. Find the ratio of the areas of ΔPST and ΔPQR.

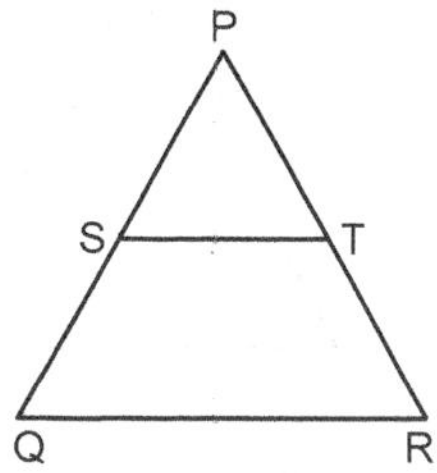

Solution : In ΔPQR,

$$PR = PT + TR$$

or $\quad PR = (2 + 4)$ cm $= 6$ cm

Now, $\quad ST \parallel QR$ [Given]

$\therefore \quad \frac{PS}{PQ} = \frac{PT}{PR}$ [Thales' theorem]

In ΔPST and ΔPQR

$$\frac{PS}{PQ} = \frac{PT}{PR}$$

and $\quad \angle P = \angle P$ [common]

$\therefore \quad \Delta PST \sim \Delta PQR$ [SAS similarity]

Now $\quad \frac{\text{Area of } \Delta PST}{\text{Area of } \Delta PQR} = \frac{PT^2}{PR^2}$

or $\quad \frac{\text{Area of } \Delta PST}{\text{Area of } \Delta PQR} = \frac{2^2}{6^2}$

$= \frac{4}{36} = \frac{1}{9}$. **Ans.**

Q. 57. In the given figure, ΔAHK is similar to ΔABC. If $AK = 10$ cm, $BC = 3.5$ cm and $HK = 7$ cm, find AC.

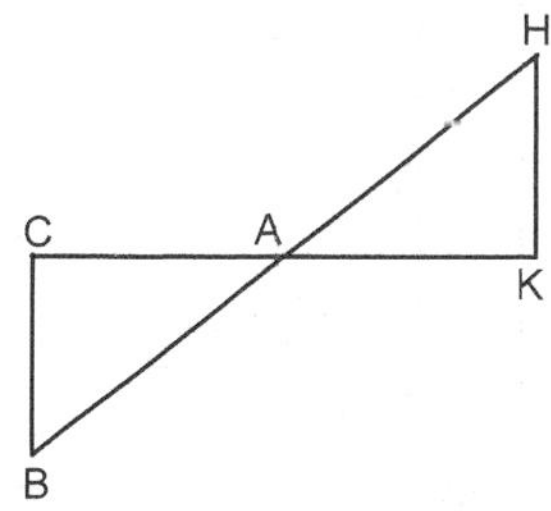

Solution : As $\Delta AHK \sim \Delta ABC$ [Given]

$\therefore \quad \frac{AK}{HK} = \frac{CA}{BC}$ [cpct]

$\Rightarrow \quad \frac{10}{7} = \frac{CA}{3.5}$

$\Rightarrow \quad CA = 5$ cm. **Ans.**

Q. 58. In the figure given below, $DE \parallel BC$ and $AD = \frac{1}{2} BD$. If $BC = 4.5$ cm, find DE.

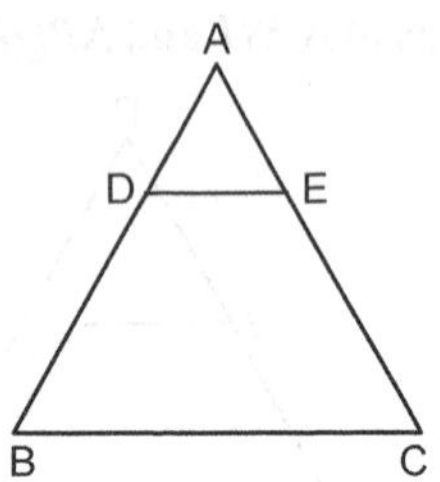

Solution : In ΔABC,

$AB = AD + BD$

But $\quad AD = \frac{1}{2} BD$

or $\quad BD = 2AD$

Thus $\quad AB = AD + 2AD = 3AD$

Now, Since $\quad DE \parallel BC$ [Given]

$\therefore \quad \frac{DE}{BC} = \frac{AD}{AB}$ [Thales' theorem]

or $\quad \frac{DE}{4.5} = \frac{AD}{3AD}$

or $\quad DE = 1.5$ cm. **Ans.**

Chapter 8. Circles

Q. 1. In the given figure, a quadrilateral $ABCD$ is drawn to circumscribe a circle such that its sides AB, BC, CD and AD touch the circle at P, Q, R and S respectively. If $AB = x$ cm, $BC = 7$ cm, $CR = 3$ cm and $AS = 5$ cm, find x.

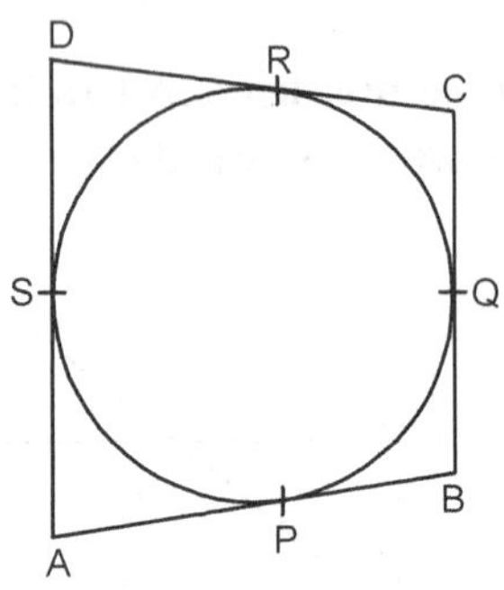

Solution : Given, $AB = x$ cm, $BC = 7$ cm, $CR = 3$ cm and $AS = 5$ cm.

Now, $\quad CQ = CR = 3$ cm [Tangents from C]

Thus $\quad BQ = BC - CQ = (7 - 3)$ cm $= 4$ cm

So $\quad PB = BQ = 4$ cm [Tangents from B]

Also $\quad AS = AP = 5$ cm [Tangents from A]

$\therefore \quad AB = AP + PB$

$= (5 + 4)$ cm

$= 9$ cm.

$\Rightarrow \quad x = 9$ cm **Ans.**

Q. 2. Two concentric circles have radii 5 cm and 3 cm. Find the length of the chord of the larger circle if it touches the smaller circle (tangent).

Solution : Given, $OA = OB = 5$ cm; $OD = 3$ cm

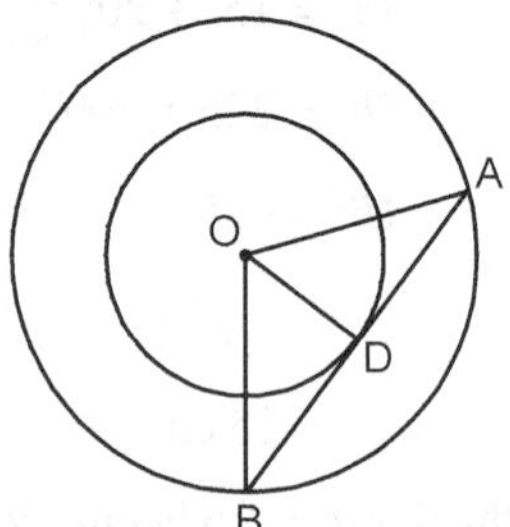

AB is a tangent at D

So $\quad \angle ODB = \angle ODA = 90°$

[The tangent at a point is perpendicular to the radius of a circle through the point of contact]

Applying Pythagoras theorem

$OB^2 = BD^2 + OD^2$

$\Rightarrow \quad BD^2 = OB^2 - OD^2$

$\Rightarrow \quad BD^2 = (5)^2 - (3)^2$

$\Rightarrow \quad BD^2 = 25 - 9 = 16$

$\Rightarrow \quad BD = 4$ cm ...(i)

Similarly, $\quad OA^2 = AD^2 + OD^2$

$\Rightarrow \quad AD^2 = OA^2 - OD^2$

$\Rightarrow \quad AD^2 = (5)^2 - (3)^2$

$\Rightarrow \quad AD^2 = 25 - 9 = 16$

$\Rightarrow \quad AD = 4$ cm ...(ii)

Hence $\quad AB = AD + BD$

$= (4 + 4)$ cm

$= 8$ cm **Ans.**

Q. 3. A chord of a circle of radius 10 cm subtends a right angle at its centre. Find the length of the chord (in cm).

Solution :

Given, $AO = OB = 10$ cm

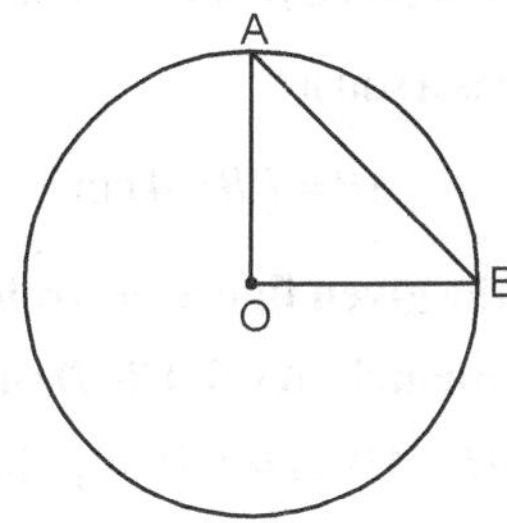

In ΔAOB,

$$AB^2 = AO^2 + OB^2$$

[Pythagoras' theorem]

$\Rightarrow$ $AB^2 = (10)^2 + (10)^2$

$\Rightarrow$ $AB^2 = 100 + 100$

$\Rightarrow$ $AB^2 = 200$

Thus $AB = 10\sqrt{2}$ cm. **Ans.**

Q. 4. Two circles touch each other externally at *P*. *AB* is a common tangent to the circles, touching them at *A* and *B*. Find the value of ∠*APB*.

Solution : Let *PO* meet *AB* at *O* such that *PO* is a tangent.

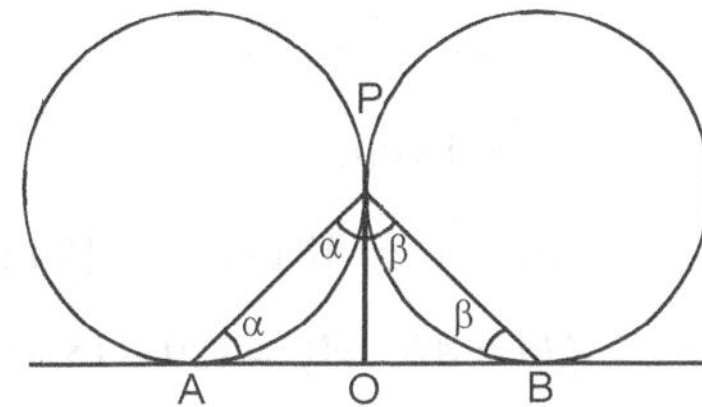

Let $\angle PAO = \alpha$ and $\angle PBO = \beta$

Now,

$OA = OP$

[Tangents from an exterior point]

$\therefore$ In ΔOAP

$\angle PAO = \angle OPA = \alpha$

[Angles opposite to equal sides]

Similarly, $OB = OP$

$\therefore$ $\angle OPB = \angle PBO = \beta$

In ΔAPB

$\angle PAO + \angle APB + \angle PBO = 180^\circ$

[Angle sum property of Δ]

$\Rightarrow$ $\alpha + (\alpha + \beta) + \beta = 180^\circ$

$\Rightarrow$ $2(\alpha + \beta) = 180^\circ$

$\Rightarrow$ $\alpha + \beta = 90^\circ$

$\therefore$ $\angle APB = 90^\circ$ **Ans.**

Q. 5. If the angle between two tangents drawn from an external point *P* to a circle of radius *a* and centre *O*, is 60°, then find the length of *OP*.

Solution : Given, $\angle APB = 60^\circ$

$\Rightarrow$ $\angle APO = 30^\circ$

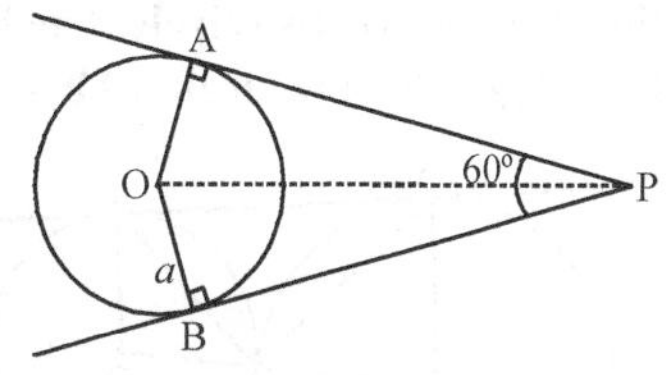

Also, $\angle OAP = 90^\circ$ [Tangent $\perp$ Radius]

$\therefore$ In right angle ΔOAP,

$$\frac{OP}{OA} = \text{cosec } 30^\circ$$

$$\frac{OP}{a} = 2 \Rightarrow OP = 2a.$$ **Ans.**

Q. 6. In the given figure, *QR* is a common tangent to the given circles, touching externally at point *T*. The tangent at *T* meets *QR* at *P*. If *PT* = 3.8 cm, then find the length of *QR*.

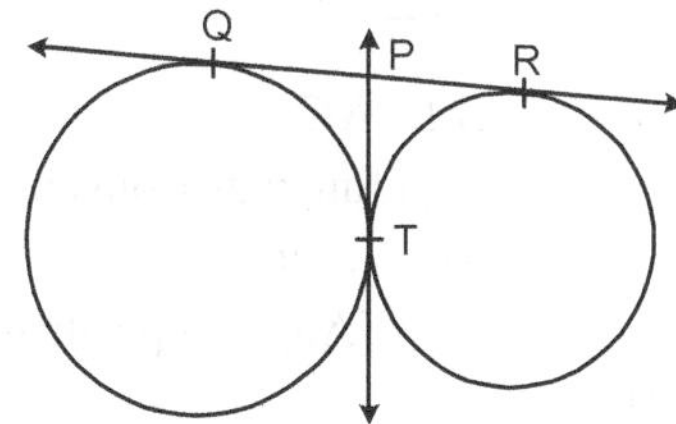

Solution : Since, tangents from an external point to a circle are equal,

$\therefore$ $PT = PQ$ and $PT = PR$

So $PT = PQ = PR = 3.8$ cm

Thus $QR = (3.8 + 3.8)$ cm

$= 7.6$ cm **Ans.**

Q. 7. In the given figure, *PQ* and *PR* are two tangents to a circle with centre *O*. If ∠*QPR* = 46°, then find ∠*QOR*.

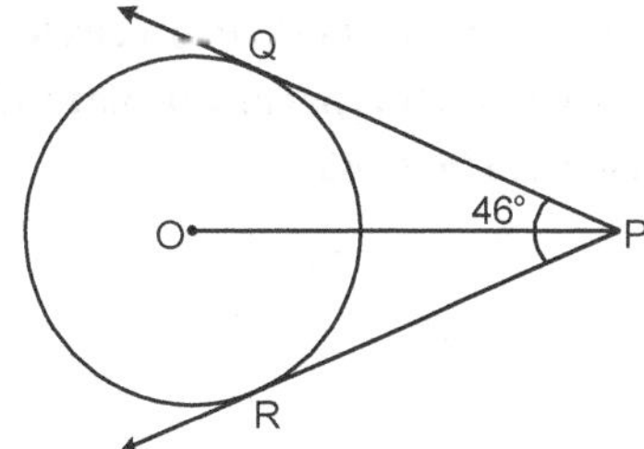

Solution : Join *OP* such that it bisects ∠*OPR*.

Thus $\angle OPQ = \angle OPR = 23^\circ$

Also, $\angle OQP = \angle ORP = 90^\circ$ [Tangent $\perp$ radius]

So $\angle QOP = \angle ROP$

$= 180^\circ - (90^\circ + 23^\circ)$

$= 180^\circ - 113^\circ$

$= 67^\circ$

Hence $\angle QOR = \angle QOP + \angle ROP$

$= 67^\circ + 67^\circ$

$= 134^\circ$ **Ans.**

Q. 8. In the given figure, two circles touch each other at C and AB is a tangent to both circles. Then find $\angle ACB$.

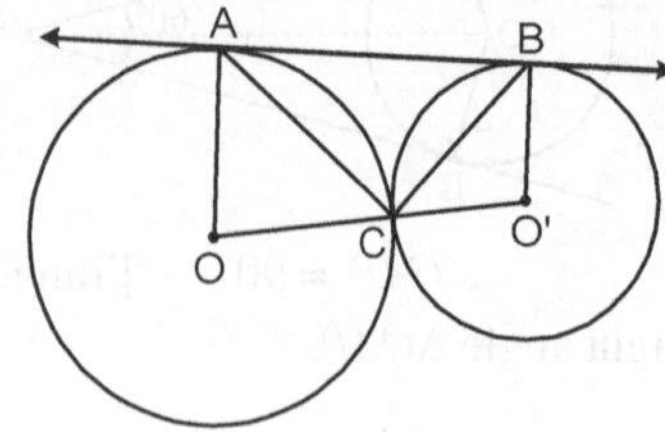

Solution : Let AO and BO' be the two radii that are perpendicular to AB.

Draw a line from point C intersecting AB at N.

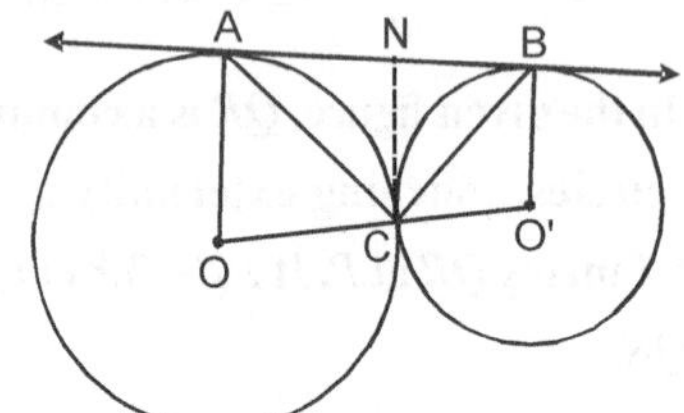

Now,

$$NA = NC$$

[Tangents from external point N]

$$\therefore \quad \angle CAN = \angle ACN$$

[Angle opposite to equal sides]

Similarly,

$$NB = NC$$

$$\therefore \quad \angle CBN = \angle BCN$$

In ΔACB

$$\angle CAB + \angle ACB + \angle ABC = 180°$$

$$\angle CAB + \angle ACN + \angle BCN + \angle ABC = 180°$$

$$2\angle ACN + 2\angle BCN = 180°$$

$$2(\angle ACN + \angle BCN) = 180°$$

$$\angle ACB = 90°$$ **Ans.**

Q. 9. In the given figure, PA and PB are two tangents drawn from an external point P to a circle with centre C and radius 4 cm. If PA is perpendicular to PB, then find the length of each tangent.

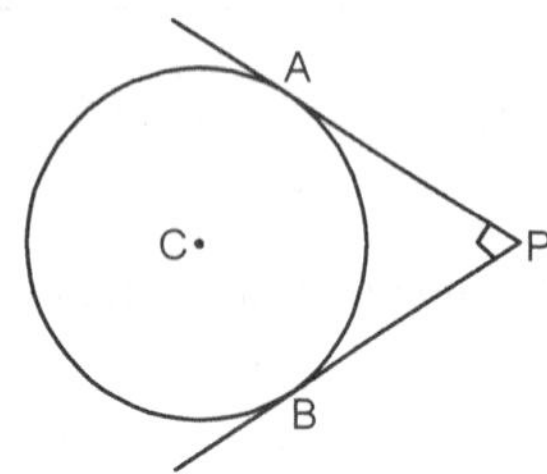

Solution : Given, $AP = BP$

[tangents from external point P]

Join CA and CB such that they are perpendicular to AP and AB respectively.

Given, $\quad CA = CB = 4$ cm [Radii]

$$\therefore \quad \angle ACB = 360° - [90° + 90° + 90°]$$

or $\quad \angle ACB = 90°$

As $APBC$ is a right-angled quadrilateral with its adjacent sides CA and CB being 4 cm.

$\therefore \quad ABCD$ is a square.

$$\therefore \quad AP = PB = 4 \text{ cm}$$ **Ans.**

Q. 10. In the given figure, a circle with centre O is inscribed in a quadrilateral $ABCD$ such that it touches the side BC, AB, AD and CD at points P, Q, R and S respectively. If $AB = 29$ cm, $AD = 23$ cm, $\angle B = 90°$ and $DS = 5$ cm, then find the radius of the circle (in cm).

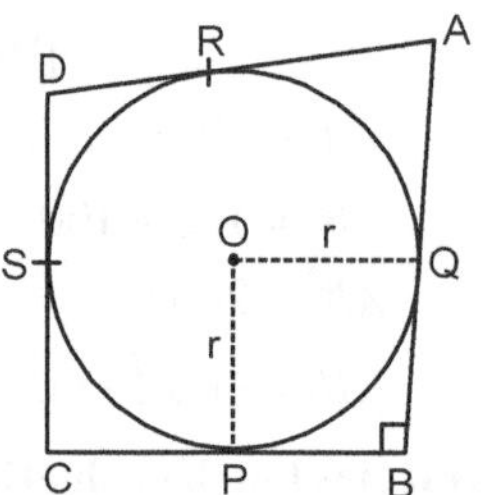

Solution : Given, $AB = 29$ cm, $AD = 23$ cm, $DS = 5$ cm and $\angle B = 90°$

Now, $\quad DS = DR = 5$ cm [tangents from D]

So $\quad AR = AD - DR$

$= (23 - 5)$ cm

$= 18$ cm

Again $\quad AR = AQ = 18$ cm [tangents from A]

So $\quad QB = AB - AQ = (29 - 18)$ cm $= 11$ cm

$$\because \quad \angle OPB = \angle OQB = 90°$$

[OQ and OP are perpendicular to AB and BC respectively]

Thus in the quadrilateral $OQBP$,

$$\angle POQ = 360° - (90° + 90° + 90°) = 90°$$

Hence, it is a right-angled quadrilateral with adjacent sides equal.

So $\quad r = OQ = PB = 11$ cm. **Ans.**

Q. 11. If the difference between the circumference and the radius of a circle is 37 cm, then the radius of the circle. $\left(\text{Use } \pi = \dfrac{22}{7}\right)$

Solution : Let the radius of circle be r cm.

Thus, $\quad$ circumference $= 2\pi r$ cm

Now, $\quad 2\pi r - r = 37$ [Given]

$$\Rightarrow \quad r\left(\frac{44}{7} - 1\right) = 37$$

$$\Rightarrow \quad r\left(\frac{44-7}{7}\right) = 37$$

$$\Rightarrow \quad r = \frac{7 \times 37}{37} = 7 \text{ cm.}$$ **Ans.**

Q. 12. From a point Q, 13 cm away from the centre of a circle, the length of tangent PQ to the circle is 12 cm. Find the radius of the circle (in cm).

Solution : Given, $PQ = 13$ cm and $OQ = 12$ cm

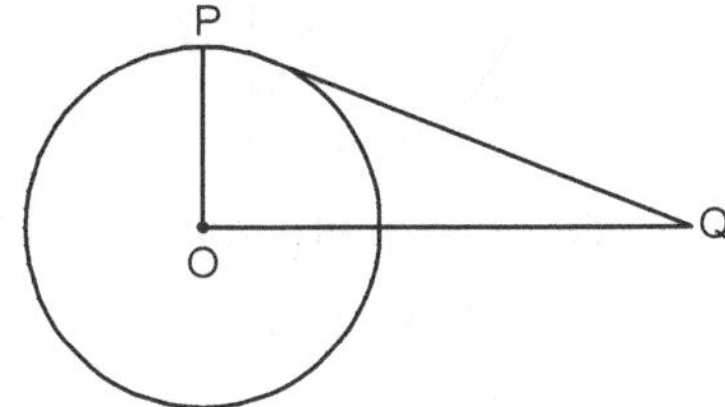

Thus, $r = PO = \sqrt{PQ^2 - OQ^2}$ [Pythagoras theorem]

$= \sqrt{(13)^2 - (12)^2}$

$= \sqrt{169 - 144}$

$= \sqrt{25} = \pm 5$ cm

As length cannot be negative, so the radius of circle is 5 cm. **Ans.**

Q. 13. In the given figure, AP, AQ and BC are tangents to the circle. If $AB = 5$ cm, $AC = 6$ cm and $BC = 4$ cm, then find the length of AP.

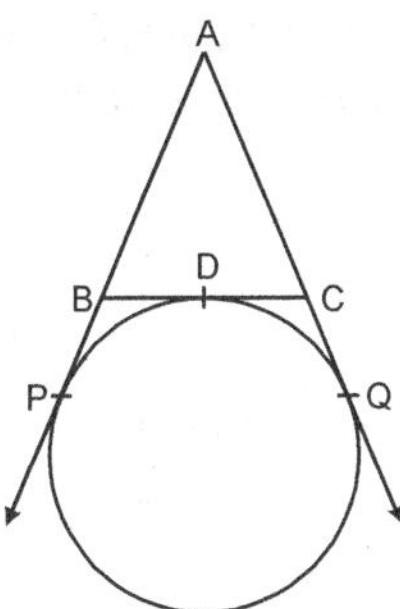

Solution : Given, $AB = 5$ cm, $AC = 6$ cm and $BC = 4$ cm

Let $PB = x$ cm, $CQ = y$ cm and D be the point where the tangent BC touches the circle.

So $AP = (5 + x)$ cm and $AQ = (6 + y)$ cm

Also $PB = BD = x$ cm [tangents from B]

and $CQ = CD = y$ cm [tangents from C]

Now $5 + x = 6 + y$ [$\because AP = AQ$]

or $x - y = 1$ ...(i)

Also $x + y = 4$ [$\because BC = 4$] ...(ii)

Adding (i) and (ii),

$2x = 5$

or $x = 2.5$

Thus $AP = (5 + x)$ cm

$= 7.5$ cm. **Ans.**

Q. 14. In the given figure, the sides AB, BC and CA of a triangle ABC touch a circle at P, Q and R respectively. If $PA = 4$ cm, $BP = 3$ cm and $AC = 11$ cm, then find the length of BC (in cm).

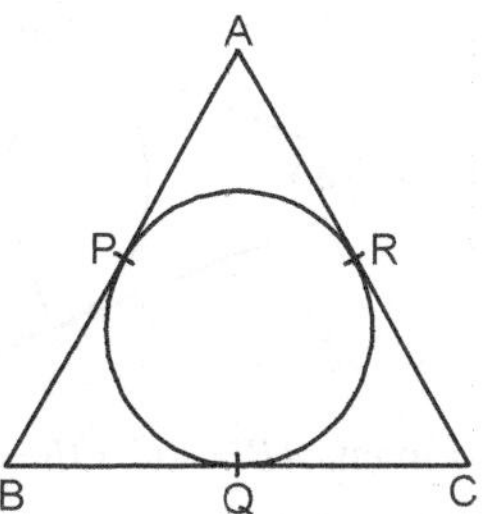

Solution : Given, $PA = 4$ cm, $PB = 3$ cm and $AC = 11$ cm

Now, $BQ = BP = 3$ cm [BQ and BP are tangents from B]

and $AP = AR = 4$ cm [AP and AR are tangents from A]

Thus $RC = AC - AR$

$= (11 - 4)$ cm $= 7$ cm

Also $RC = CQ$ [RC and CQ are tangents from C]

So $CQ = 7$ cm

Hence $BC = BQ + CQ$

$= (3 + 7)$ cm $= 10$ cm **Ans.**

Q. 15. In the given figure, O is the centre of the circle, PQ is a chord and PT is the tangent at P. If $\angle POQ = 70°$, then find $\angle TPQ$.

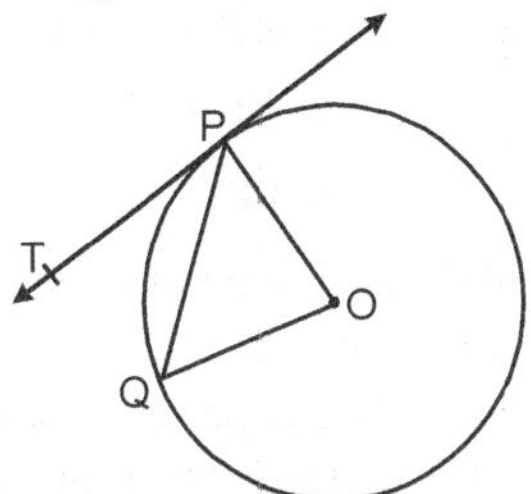

Solution : Given,

$\angle POQ = 70°$

As OP and OQ are the radii of the circle so ΔPOQ is isosceles.

Thus $\angle OPQ = \angle OQP$

Now $180° = \angle POQ + \angle PQO + \angle OPQ$

Hence $180° = \angle OPQ + \angle OQP + 70°$

$\Rightarrow 2\angle OPQ = 110°$

$\Rightarrow \angle OPQ = 55°$

Also as OP is perpendicular to PT,

So, $\angle OPT = 90°$

Thus $\angle TPQ = \angle OPT - \angle OPQ$

or $\angle TPQ = 90° - 55°$

$= 35°$. **Ans.**

Q. 16. In the given figure, AB and AC are tangents to the circle with centre O such that $\angle BAC = 40°$. Find $\angle BOC$.

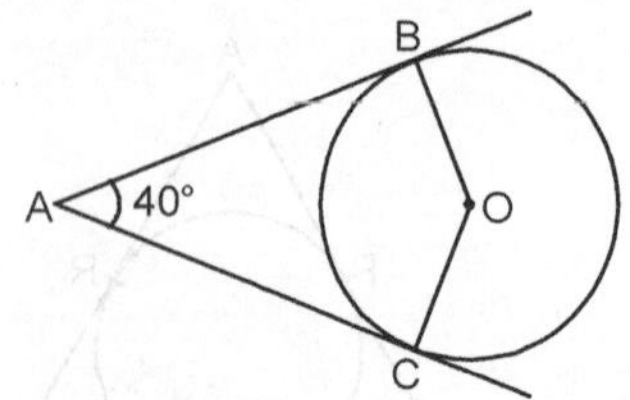

Solution : Given, $AB = AC$, $OB = OC$ and $\angle A = 40°$

Join AO such that it bisects $\angle A$.

Thus, $\angle BAO = \angle OAC = 20°$

Also, as OB and OC are perpendicular to AB and AC respectively.

$\therefore \quad \angle ABO = \angle ACO = 90°$

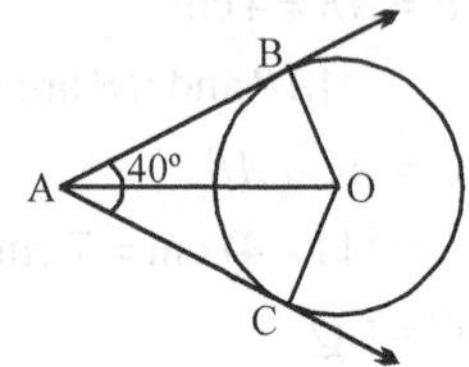

So, in ΔABO

$$\angle BOA = 180° - (90° + 20°)$$

or $\angle BOA = 180° - 110° = 70°$

Similarly, in ΔACO

$$\angle COA = 180° - (90° + 20°)$$

or $\angle COA = 180° - 110° = 70°$

Hence $\angle BOC = \angle BOA + \angle COA$

$= 70° + 70°$

$= 140°$. **Ans.**

Q. 17. Find the perimeter of a square circumscribing a circle of radius *a* cm.

Solution : Let $ABCD$ be the square that inscribes a circle with centre O, touching it at the points E, F, G and H.

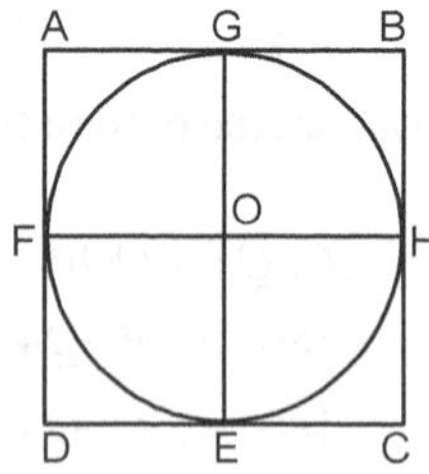

Given, the radius of the circle = a cm

Thus, $OE = OF = OG = OH = a$ cm

Hence, $OG + OE = AD = BC$ and $OF + OH = AB = DC$

So each side of the square = $2a$ cm

Hence, perimeter of the square

$= 4(2a) = 8a$ cm. **Ans.**

Q. 18. In the given figure, *O* is the centre of the circle, *AB* is the chord and *AT* is the tangent at *A*. If $\angle AOB = 100°$, then find $\angle BAT$.

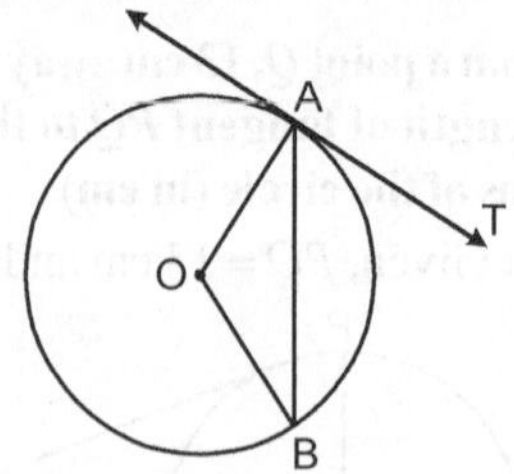

Solution : Given, $\angle AOB = 100°$ and $OA = OB$

[both are the radii of the circle with centre O]

Thus, ΔAOB is an isosceles triangle and

$$\angle OAB = \angle OBA$$

[base angles of an isosceles triangle]

Thus $\angle OAB + \angle OBA = 180° - \angle AOB$

or $2\angle OBA = 180° - 100°$

or $\angle OBA = 40° = \angle OAB$

As OA is perpendicular at AT,

$\therefore \quad \angle OAT = 90°$

Hence $\angle BAT = 90° - 40°$

$= 50°$ **Ans.**

Q. 19. In the given figure, *PA* and *PB* are tangents to the circle with centre *O*. If $\angle APB = 60°$, then find $\angle OAB$.

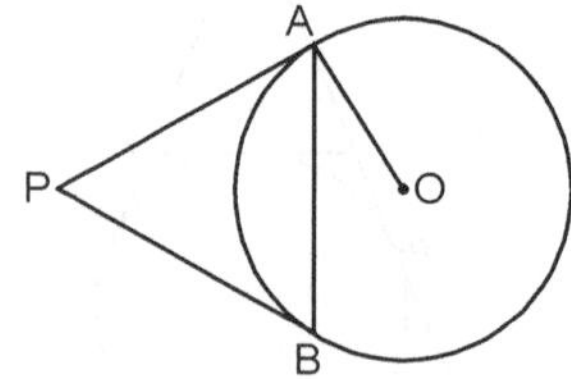

Solution : Given

$\angle APB = 60°$, $AP = PB$

[tangents from P]

In ΔPAB, $PA = PB$

So, ΔPAB is isosceles and $\angle PAB = \angle PBA$

[base angles of isosceles triangles]

In ΔAPB,

$$\angle APB + \angle PAB + \angle PBA = 180°$$

Thus,

$$\angle PAB + \angle PBA = 180° - \angle APB$$

$\Rightarrow \quad 2\angle PAB = 180° - 60° = 120°$

$\Rightarrow \quad \angle PAB = 60° = \angle PBA$

Now, $\angle PAO = 90°$

[OA is perpendicular to AP]

Thus, $\angle OAB = \angle PAO - \angle PAB$

$= 90° - 60°$

$= 30°$. **Ans.**

Q. 20. In the given figure, point *P* is 26 cm away from the centre *O* of a circle and the length *PT* of the tangent drawn from *P* to the circle is 24 cm. Then find the radius of the circle (in cm).

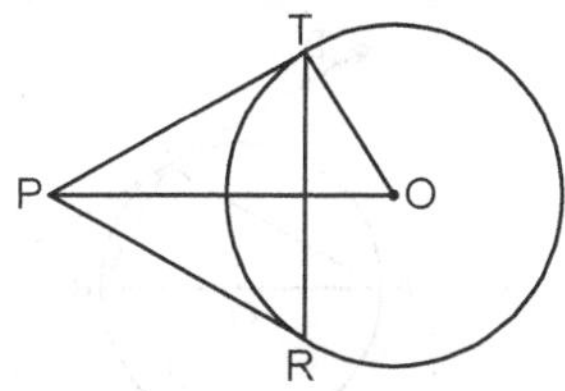

Solution : Given, $OP = 26$ cm and $PT = 24$ cm
Let the radius of circle be r such that

$$OT = r$$

Now, In ΔOTP

$$\angle T = 90^\circ \quad [\because \text{Tangent} \perp \text{radius}]$$

$\therefore$ Applying Pythagoras theorem

$$OP^2 = PT^2 + OT^2$$
$$\Rightarrow \quad OT^2 = OP^2 - PT^2$$
$$\Rightarrow \quad r^2 = (26)^2 - (24)^2$$
$$\Rightarrow \quad r^2 = 676 - 576$$
$$\Rightarrow \quad r^2 = 100$$
$$\Rightarrow \quad r = 10 \text{ cm}$$

[as radius cannot be negative]

Ans.

Q. 21. In the given figure, *TP* and *QT* are two tangents to a circle with centre *O* such that $\angle POQ = 110^\circ$. Then find $\angle PTQ$.

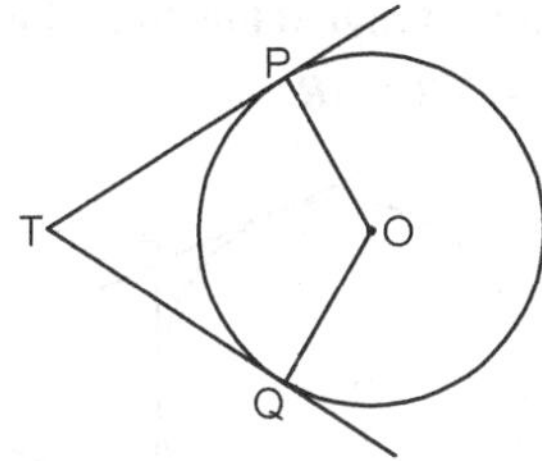

Solution : Given,

$$\angle POQ = 110^\circ$$

Now $\quad TQ = TP$ [both are tangents from T]
Join TO such that it bisects both $\angle PTQ$ and $\angle POQ$

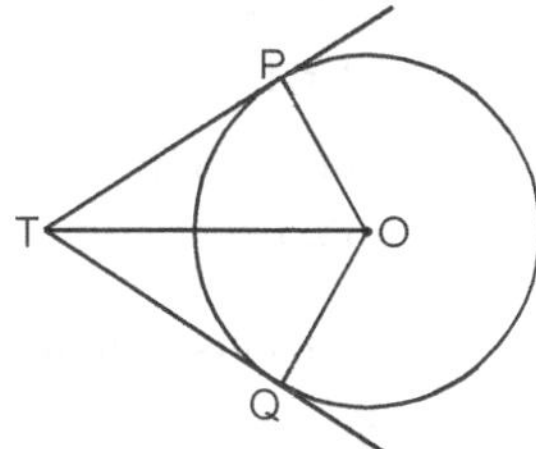

Thus $\quad \angle TOP = 55^\circ$ and $\angle OPT = 90^\circ$

[OP is perpendicular to PT]

Hence $\quad \angle PTO = 180^\circ - (90^\circ + 55^\circ)$
$= 180^\circ - 145^\circ$
$= 35^\circ$

Similarly $\quad \angle QTO = 35^\circ$

Hence $\quad \angle PTQ = (35^\circ + 35^\circ)$
$= 70^\circ$. **Ans.**

Q. 22. From an external point *P*, tangents *PA* and *PB* are drawn to a circle with centre *O*. If $\angle PAB = 50^\circ$, then find $\angle AOB$.

Solution : Since, tangents from an external point are equal

$$\therefore \quad AP = BP$$

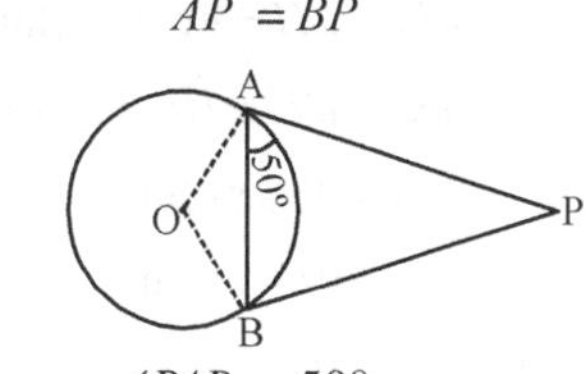

Given, $\quad \angle PAB = 50^\circ$

$\therefore \quad \angle PBA = 50^\circ$

[Angles opposite to equal sides]

In ΔAPB

$$\angle APB = 180^\circ - (50^\circ + 50^\circ) = 80^\circ$$

$\therefore \quad \angle AOB = 180^\circ - 80^\circ$
$= 100$ **Ans.**

Q. 23. In the given figure, *CP* and *CQ* are tangents from an external point *C* to a circle with centre *O*. *AB* is another tangent that touches the circle at *R*. If $CP = 11$ cm and $BR = 4$ cm then find *BC*.

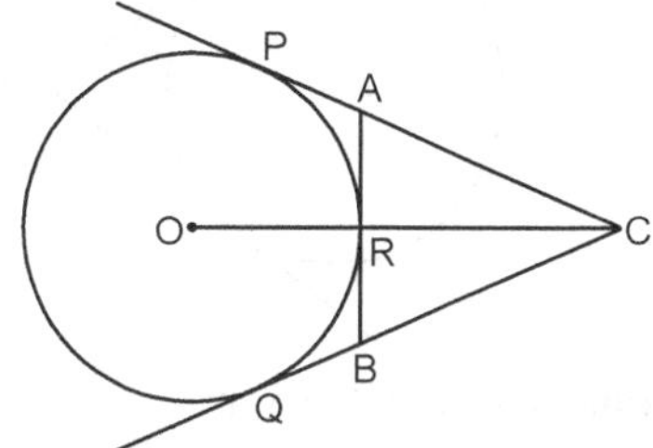

Solution : Given, $CP = 11$ cm and $BR = 4$ cm

Now	$CP = CQ$	[tangents from C]
Thus	$CQ = 11$ cm	
Again	$BR = BQ$	[tangents from B]
Thus	$BQ = 4$ cm	
Hence	$CQ = BQ + BC$	
$\Rightarrow$	$BC = CQ - BQ$	
$\Rightarrow$	$BC = (11 - 4)$ cm	
	$= 7$ cm.	**Ans.**

Q. 24. In the given figure, a circle touches the *DF* of ΔEDF at *H* and touches *ED* and *EF* produced at *K* and *M* respectively. If $EK = 9$ cm, then find the perimeter of ΔEDF.

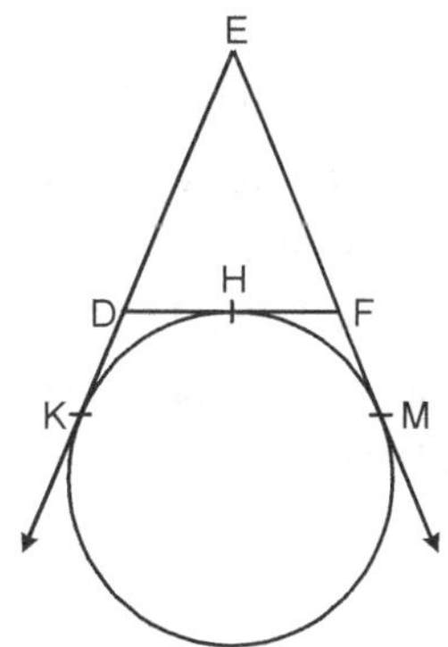

Solution : Given, $EK = 9$ cm

Now $EK = EM = 9$ cm [tangents from E]

Let $DK = x$ cm and $FM = y$ cm

Now $DH = DK = x$ cm [tangents from D]

and $FM = FH = y$ cm [tangents from F]

Thus $DF = DH + HF = (x + y)$ cm

Also $ED = (9 - x)$ cm

and $EF = (9 - y)$ cm

So, perimeter of ΔEDF

$= ED + DF + EF$

$= (9 - x) + (x + y) + (9 - y)$ cm

$= 18$ cm. **Ans.**

Q. 25. In the given figure, *DE* and *DF* are tangents from the external point *D* to a circle with centre *A*. If *DE* = 5 cm and *DE* is perpendicular to *DF*, then find the radius of the circle.

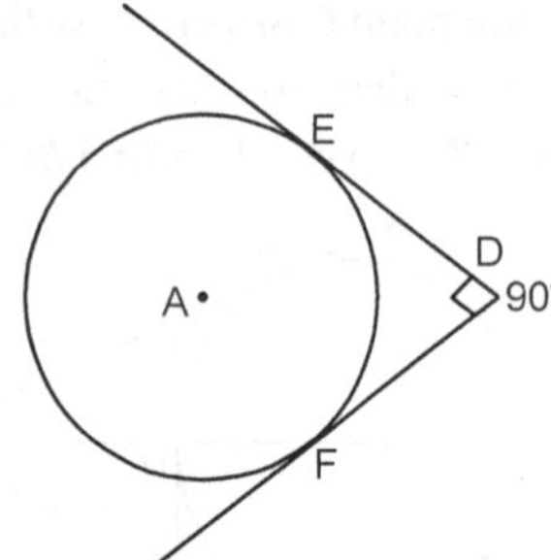

Solution : Given, $DE = 5$ cm

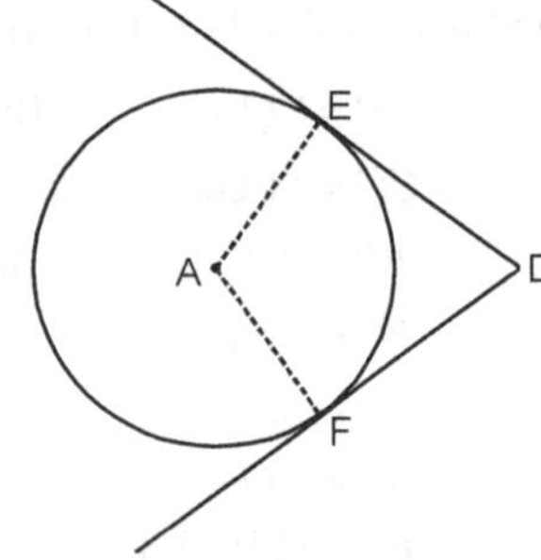

Now $\angle AED = 90°$ [AE is perpendicular to ED]

$\angle AFD = 90°$ [AF is perpendicular to FD]

Thus $\angle EAF = 360° - (\angle AED + \angle AFD + \angle EDF)$

$\Rightarrow \angle EAF = 360° - (90° + 90° + 90°)$

$\Rightarrow \angle EAF = 360° - 270°$

$\Rightarrow \angle EAF = 90°$

Hence *EDFA* is a right-angled quadrilateral with $FD = AE$ and $AF = ED$ *i.e.*, adjacent sides equal.

Hence *EDFA* is a square.

So the radius of the circle is 5 cm **Ans.**

Q. 26. In fig. *PQ* is a tangent at a point *C* to a circle with centre *O*. If *AB* is a diameter and $\angle CAB = 30°$, find $\angle PCA$.

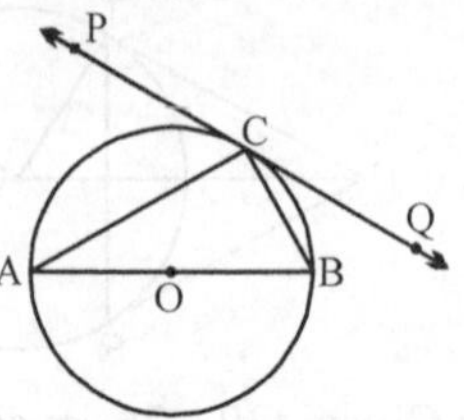

Solution : Given, $\angle CAB = 30°$ and PQ is a tangent at a point C to a circle with centre O.

Since, AB is a diameter

$\therefore \angle ACB = 90°$

Join OC.

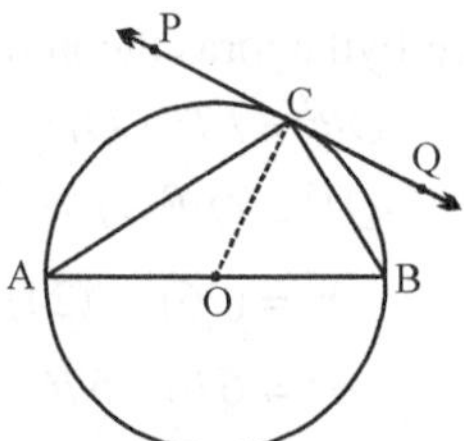

Now, $\angle CAO = \angle ACO = 30°$ [$\because OA = OC$]

and $\angle PCO = 90°$ (tangent is perpendicular to the radius through the point of contact)

$\therefore \angle PCA = \angle PCO - \angle ACO$

$= 90° - 30° = 60°$ **Ans.**

Q. 27. In the given figure, *XP* and *XQ* are two tangents to the circle with centre *O* drawn from an external point *X*. *ARB* is another tangent touching the circle at *R*. Prove that $XA + AR = XB + BR$.

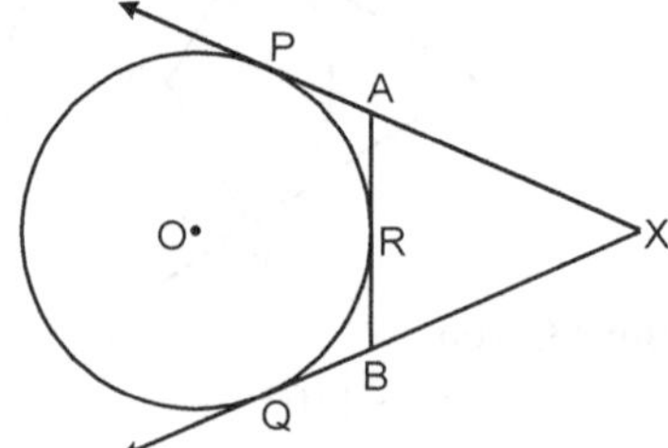

Solution : As XP and XQ are two tangents from X, so

$XP = XQ$

Also, $AP = AR$

[Tangents from the external point A]

and $BR = BQ$

[Tangents from the external point B]

Now, $XP = XQ$ [from above]

$\Rightarrow XA + AP = XB + BQ$

$\Rightarrow XA + AR = XB + BR$

[from above, $AP = AR$ and $BR = BQ$]

Hence Proved.

Q. 28. Prove that the tangents drawn at the end of any diameter are parallel.

Solution : As AC and DF are tangents to B and E, the opposite points of the diameter BE.

$\therefore$ $\angle ABO = \angle OBC = \angle DEO = \angle OEF = 90°$

[the tangent at a point is always perpendicular to the radius at that point]

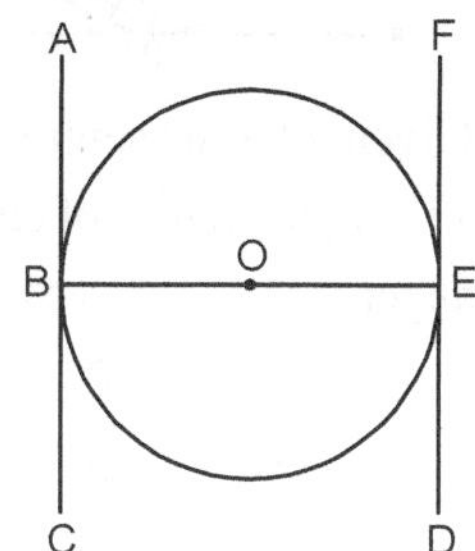

Hence, $\angle ABO = \angle OED$ are alternate interior angles.

$AC \parallel DF$. **Hence Proved.**

Q. 29. In the given figure, *AB* and *CD* are common tangents to the two circles and intersect at *E*. Prove that *AB* = *CD*.

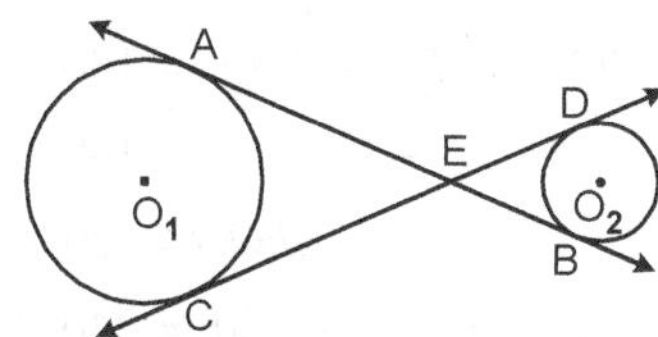

Solution : Given, AB and CD are tangents to circles with centres O_1 and O_2, respectively

$$AB = AE + EB \text{ and } CD = CE + ED$$

Now, as E is the external point from which AE and CE are tangents to the circle with centre O_1, so

$$AE = CE$$

Also, as E is the external point from which ED and EB are tangents to the circle with centre O_2, so

$$EB = ED$$

Thus, $AB = AE + EB = CE + ED = CD.$

Hence Proved.

Q. 30. Prove that the line segment joining the points of contact of two parallel tangents of a circle passes through its centre.

Solution : Given, $AC \parallel DF$

Let O be the centre of the circle. Join OB, OP, OE such that

$$OB = OP = OE = r$$

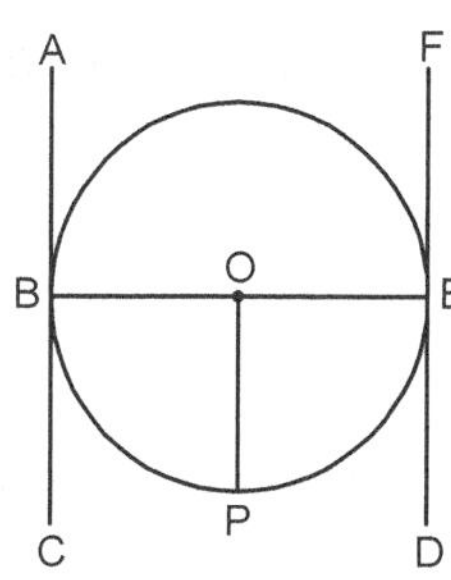

Also, let $OP \parallel BC \parallel ED$

Thus, $\angle CBO + \angle BOP = 180°$ [co-interior angles]

or $90° + \angle BOP = 180°$

[OB is perpendicular to AC]

or $\angle BOP = 90°$

Similarly $\angle POE = 90°$

Thus, $\angle BOE = \angle POE + \angle POB$

$= 90° + 90° = 180°$

Hence, BOE is a striaght line passing through the centre O. **Hence Proved.**

Q. 31. A circle touches all the four sides of a quadrilateral *ABCD*. Prove that

$$\mathbf{AB + CD = BC + DA}$$

Solution : Given, a quadrilateral $ABCD$ and a circle touches its all the four sides at P, Q, R, and S respectively.

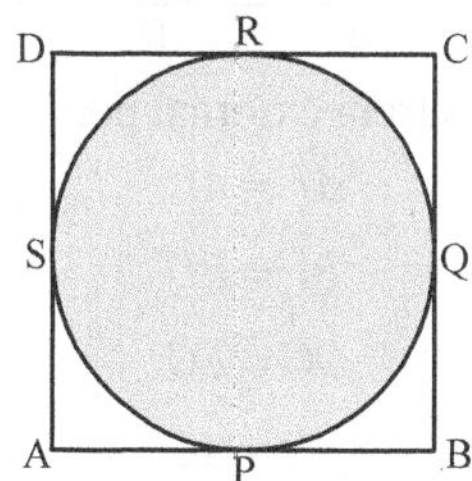

To prove $AB + CD = BC + DA$

L.H.S. $= AB + CD$

$= AP + PB + CR + RD$

$= AS + BQ + CQ + DS$

[Tangents from same external points are always equal.]

$= (AS + SD) + (BQ + QC)$

$= AD + BC$

= R.H.S. **Hence Proved.**

Q. 32. If from an external point *P* of a circle with centre *O*, two tangents *PQ* and *PR* are drawn such that $\angle QPR = 120°$, prove that 2*PQ* = *PO*.

Solution : Construct QS such that QS is perpendicular to OP. Let OQ be the radius such that OQ is perpendicular to QP. Hence

$$\angle OQP = 90°$$

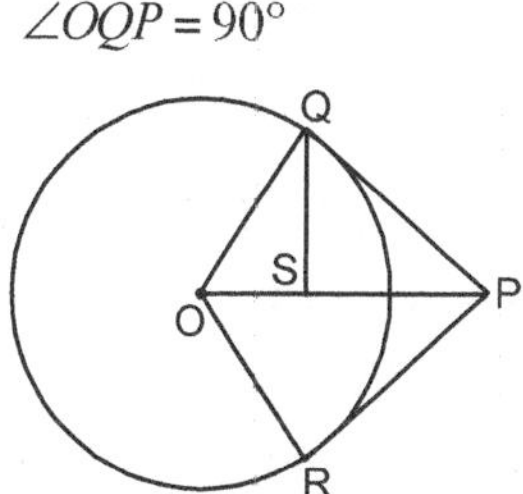

Given, $\angle QPR = 120°$

OP bisects $\angle QPR$ such that

$\angle QPO = 60°$

Thus, $\angle QOP = 180° - (90° + 60°) = 30°$

So, $\sin 30° = \dfrac{QP}{OP}$

$$\Rightarrow \quad \frac{1}{2} = \frac{QP}{OP}$$

$$\Rightarrow \quad OP = 2QP$$

$$\Rightarrow \quad 2PQ = PO.$$

Hence Proved.

Q. 33. The incircle of an isosceles triangle *ABC*, in which *AB* = *AC* touches the sides *BC*, *CA* and *AB* at *D*, *E* and *F* respectively. Prove that *BD* = *DC*.

Solution : Given,

$$AB = AC$$

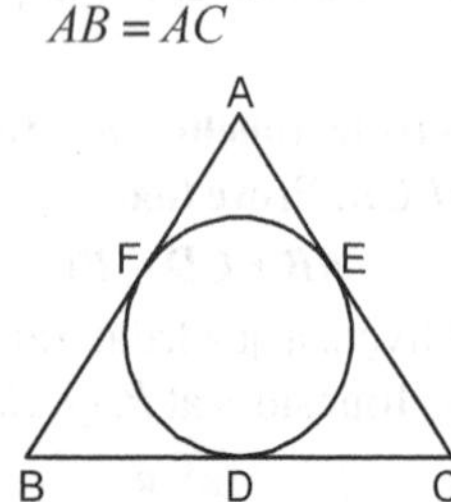

As tangents from external points are equal in length

$\therefore \quad AF = AE$ [tangents from A]

$BF = BD$ [tangents from B]

and $DC = EC$ [tangents from C]

As $AB = AC$

$\Rightarrow \quad AF + BF = AE + EC$

$\Rightarrow \quad AF + BF = AF + EC$ [$\because AF = AE$]

$\Rightarrow \quad BF = EC$

$\Rightarrow \quad BD = DC$

[$\because BF = BD$ and $DC = EC$]

Hence Proved.

Q. 34. In the given figure, *PA* and *PB* are two tangents from an external point *P* to a circle with centre *O*. If $\angle PBA$ = 65°, find $\angle OAB$.

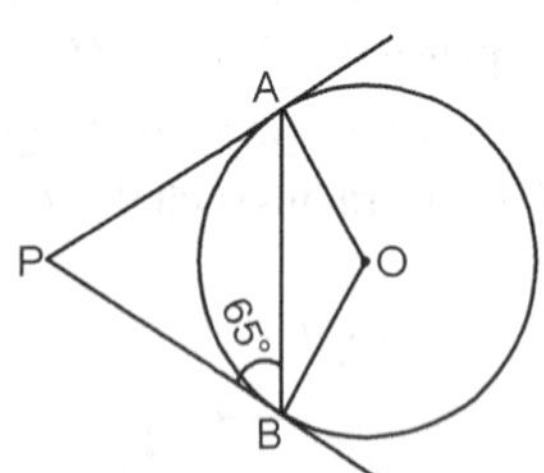

Solution : In ΔPAB, $PA = PB$ [tangents from P]

So, ΔPAB is isosceles.

Thus, $\angle PAB = 65°$

[angles opposite to equal sides in the isosceles triangle are equal]

Now, $\angle OAP = 90°$

[OA is the radius that is perpendicular to tangent PA]

Thus, $\angle OAB = 90° - 65° = 25°$. **Ans.**

Q. 35. In figure, *PQ* is a chord of a circle with centre *O* and *PT* is a tangent. If $\angle QPT$ = 60°, find $\angle PRQ$.

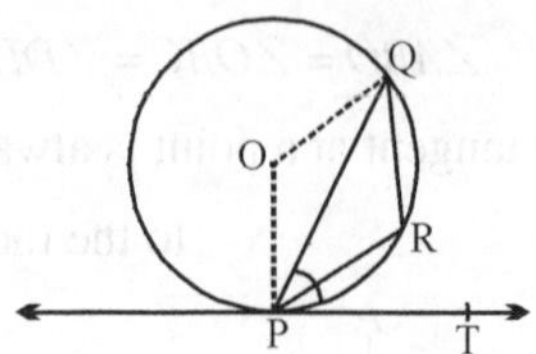

Solution : Given, O is the centre of the given circle.

$\therefore$ OQ and OP are the radius of circle.

$\because$ PT is a tangent

$\therefore \quad OP \perp PT$

So, $\angle OPT = 90°$

$\therefore \quad \angle OPQ = 90° - \angle QPT$

$\angle OPQ = 90° - 60°$

[Given, $\angle QPT = 60°$]

$\Rightarrow \quad \angle OPQ = 30°$

$\therefore \quad \angle OQP = 30°$ [$\because \Delta OPQ$ is isosceles triangle]

Now, in ΔOPQ

$$\angle POQ + \angle OPQ + \angle OQP = 180°$$

$$\angle POQ + 30° + 30° = 180°$$

$$\angle POQ = 120°$$

$$\text{reflex } \angle POQ = 360° - 120° = 240°$$

$$\therefore \quad \angle PRQ = \frac{1}{2}\angle POQ$$

[$\because$ The angle subtended by an arc of a circle at the centre is double the angle subtended by it at any point on the remaining part of the circle]

$$\Rightarrow \quad \angle PRQ = \frac{1}{2} \times 240°$$

Hence, $\angle PRQ = 120°$ **Ans.**

Q. 36. In the given figure, a circle inscribed in ΔABC touches its sides *AB*, *BC* and *AC* at points *D*, *E* and *F* respectively. If *AB* = 12 cm, *BC* = 8 cm and *AC* = 10 cm, find the lengths of *AD*, *BE* and *CF*.

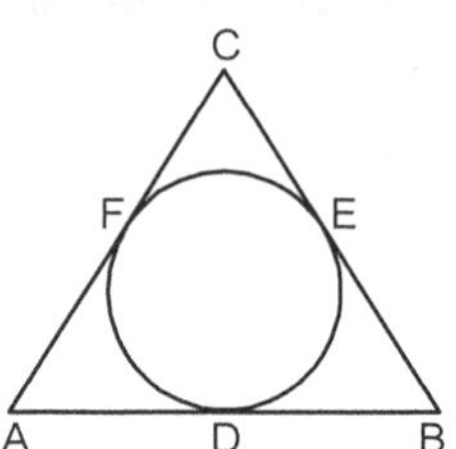

Solution : Given, AB = 12 cm, BC = 8 cm and AC = 10 cm

Let the length of $AD = x$ cm

So $BD = (12 - x)$ cm

Hence $BE = (12 - x)$ cm

[BD and BE are tangents from B and so are equal to each other]

Thus, $CE = BC - BE = [8 - (12 - x)]$ cm

$= (x - 4)$ cm

Also $CF = (x - 4)$ cm

[CE and CF are tangents from C and so are equal to each other]

and $FA = AC - FC$

$= [10 - (x - 4)]$

$= (14 - x)$ cm

Also $FA = AD$

[FA and AD are tangents from A and so are equal to each other]

Hence $14 - x = x$

$\Rightarrow \quad 2x = 14$

$\Rightarrow \quad x = 7$ cm

So $AD = 7$ cm

$BE = (12 - 7)$ cm

$= 5$ cm

and $CF = (7 - 4)$ cm

$= 3$ cm. **Ans.**

Q. 37. Tangents PA and PB are drawn from an external point P to two concentric circles with centre O and radii 8 cm and 5 cm respectively as shown in the given figure. If $AP = 15$ cm, find the length of BP.

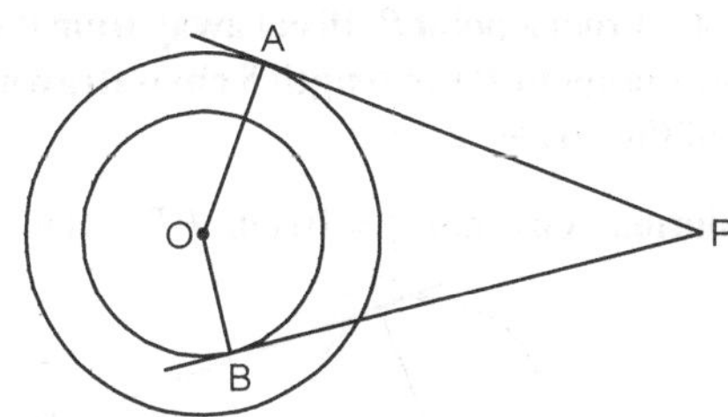

Solution : Given, $AP = 15$ cm, $OA = 8$ cm and $OB = 5$ cm

Join OP.

In ΔAOP and ΔBOP,

Applying Pythagoras' theorem, we have

$$AP^2 + OA^2 = OP^2 = OB^2 + BP^2$$

$\Rightarrow \quad (15)^2 + (8)^2 = (5)^2 + (BP)^2$

$\Rightarrow \quad (BP)^2 = 225 + 64 - 25$

$\Rightarrow \quad (BP)^2 = 225 + 39$

$\Rightarrow \quad (BP)^2 = 264$

$\Rightarrow \quad BP = 16.25$ cm **Ans.**

(as length cannot be negative)

Q. 38. In the given figure, the chord AB of the larger circle is tangent upon the smaller circle at C. If O is the centre of the circles, prove that $AC = CB$.

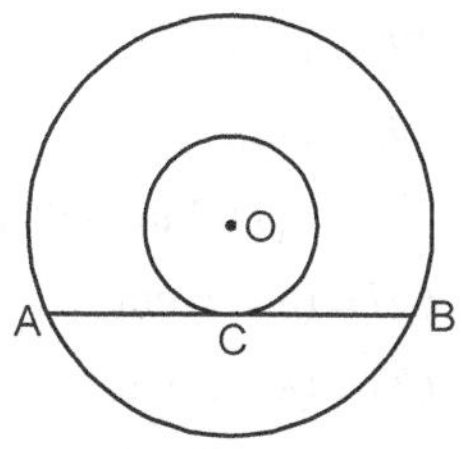

Solution : Let R be the radius of the larger circle and r be the radius of the smaller circle.

Now join OA, OB and OC.

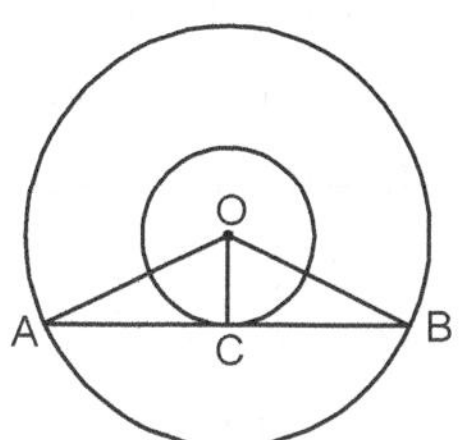

Hence, $OA = OB = R$ and $OC = r$

Since, OC is perpendicular to the tangent AB

$\therefore \quad \angle OCA = \angle OCB = 90°$

In ΔOCA and ΔOCB,

$OC = OC$ [common]

$OA = OB = R$ [proved above]

$\angle OCA = \angle OCB = 90°$ [proved above]

$\Rightarrow \quad \Delta OCA \cong \Delta OCB$ [RHS congruence]

$\Rightarrow \quad AC = CB$

[corresponding sides of congruent triangles]

Hence Proved.

Q. 39. In the given figure, a circle touches all the four sides of a quadrilateral $ABCD$ whose sides are $AB = 6$ cm, $BC = 9$ cm and $CD = 8$ cm. Find the length of side AD.

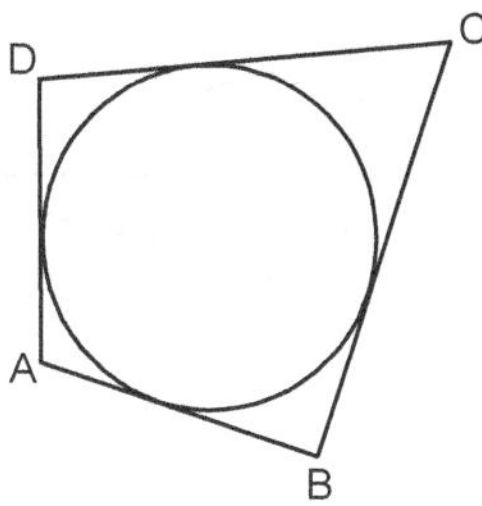

Solution : Given, $AB = 6$ cm, $BC = 9$ cm and $CD = 8$ cm.

Let the circle touch AD at E, AB at F, BC at G and CD at H.

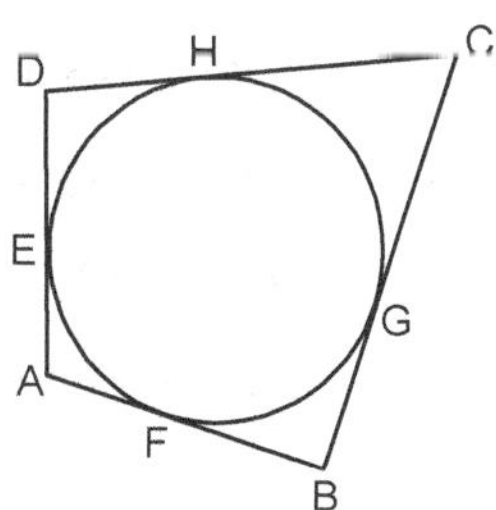

Also let $ED = x$ cm

$\Rightarrow$ $DH = x$ cm

[DH and DE are tangents from D]

$\Rightarrow$ $CH = (8 - x)$ cm

$\Rightarrow$ $CG = (8 - x)$ cm

[CH and CG are tangents from C]

$BG = [9 - (8 - x)]$ cm

$= (1 + x)$ cm

$FB = (1 + x)$ cm

[BG and FB are tangents from B]

$AF = [6 - (1 + x)]$ cm

$= (5 - x)$ cm

$AE = (5 - x)$ cm

[AF and AE are tangents from A]

Now, $AD = [x + (5 - x)]$ cm

$= 5$ cm. **Ans.**

Q. 40. In figure, from an external point *P*, two tangents *PT* and *PS* are drawn to a circle with centre *O* and radius *r*. If *OP* = 2*r*, show that ∠*OTS* = ∠*OST* = 30°.

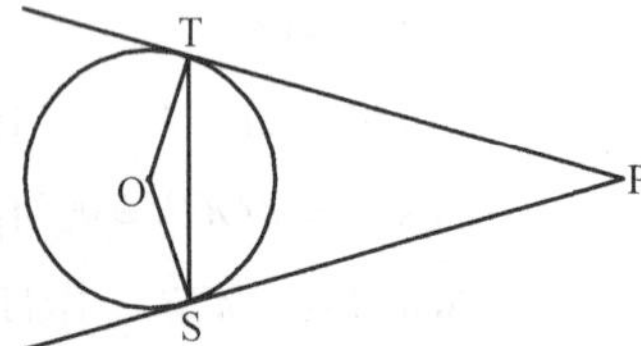

Solution : We have,

$$OP = 2r.$$

Let $\angle TOP = \theta$

In ΔOTP, $\cos\theta = \frac{OT}{OP} = \frac{r}{2r} = \frac{1}{2}$

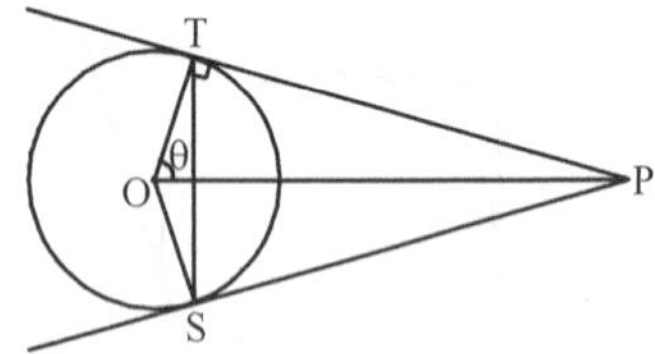

$\Rightarrow$ $\theta = 60°$

Hence, $\angle TOS = 2\theta = 2 \times 60° = 120°$

In ΔTOS

$\angle TOS + \angle OTS + \angle OST = 180°$

$120° + 2\angle OTS = 180°$ [$\because \angle OTS = \angle OST$]

$2\angle OTS = 180° - 120°$

$\angle OTS = 30°$

Hence, $\angle OTS = \angle OST = 30°$

Hence Proved.

Q. 41. Two concentric circles are of radii 7 cm and *r* cm respectively where *r* > 7. A chord of the larger circle of length 48 cm touches the smaller circle. Find the value of *r*.

Solution : Given, $OA = OB = r$ cm, $OC = 7$ cm and $AB = 48$ cm

Now AB, the chord of the larger circle is the tangent of the smaller circle.

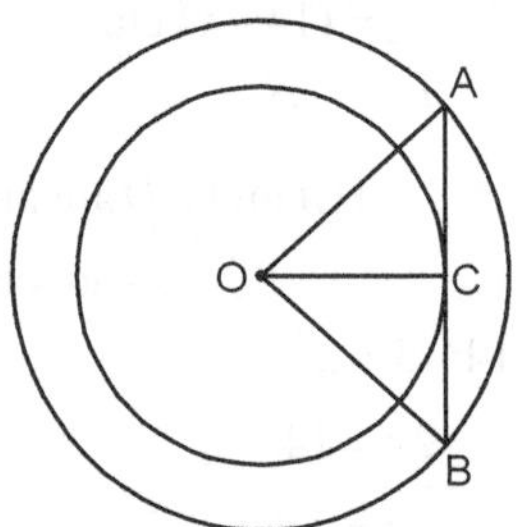

So, OC is perpendicular to AB and bisects AB at C.

Hence, $AC = CB = \frac{48}{2} = 24$ cm

Thus, $AO^2 = AC^2 + OC^2$

$\Rightarrow$ $(AO)^2 = 24^2 + 7^2$

$= 576 + 49$

$= 625$

$\Rightarrow$ $r = AO = 25$ cm **Ans.**

Q. 42. From a point *P*, 10 cm away from the centre of a circle, a tangent *PT* of length 8 cm is drawn. Find the radius of the circle.

Solution : Given, $PO = 10$ cm, $PT = 8$ cm and $TO = r$

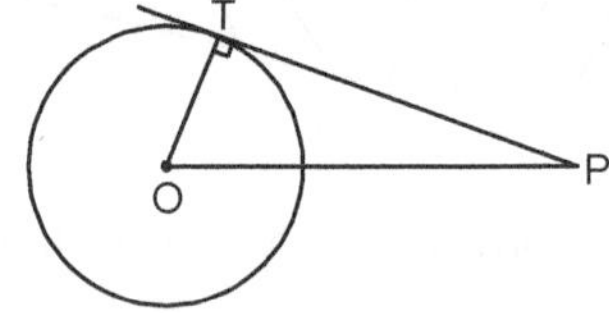

Applying Pythagoras theorem,

$TP^2 + TO^2 = PO^2$

$\Rightarrow$ $(8)^2 + r^2 = (10)^2$

$\Rightarrow$ $r^2 = 100 - 64$

$\Rightarrow$ $r^2 = 36$

$\Rightarrow$ $r = \pm 6$

$\Rightarrow$ $r = 6$ cm [as radius cannot be negative]

Ans.

Q. 43. Prove that the angle between two tangents drawn from an external point to a circle is supplementary to the angle subtended by the line segments joining the points of contact to the centre.

Solution : Let C be the external point from which the tangents AC and BC meet the circle with centre O. Thus,

$AC = BC$

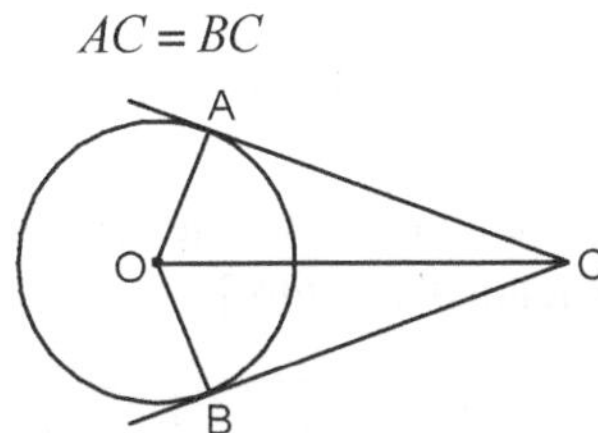

Hence OA and OB, the radii of the circle with centre O are perpendicular to AB and AC.

So, in the quadrilateral $AOBC$,

$$\angle CAO + \angle AOB + \angle OBC + \angle BCA = 360^\circ$$

$$\Rightarrow \quad 90^\circ + \angle AOB + 90^\circ + \angle BCA = 360^\circ$$

$$\Rightarrow \quad 180^\circ + \angle AOB + \angle BCA = 360^\circ$$

$$\Rightarrow \quad \angle AOB + \angle BCA = 180^\circ$$

Hence Proved.

Q. 44. A circle is inscribed in a ΔABC, touching BC, CA and AB at P, Q and R respectively. If AB = 10 cm, AQ = 7 cm and CQ = 5 cm, then find the length of BC.

Solution : Given, AB = 10 cm, AQ = 7 cm and CQ = 5 cm

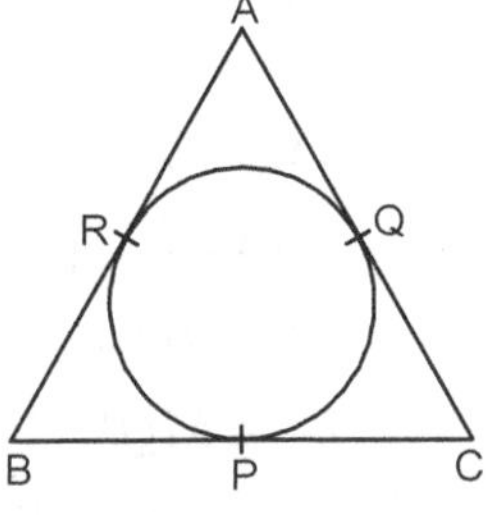

Thus, $PC = CQ = 5$ cm

[as PC and CQ are tangents from the same point C]

Also $AQ = AR = 7$ cm

[as AR and AQ are tangents from the same point A]

Then $BR = (10 - 7)$ cm $= 3$ cm

Also $BP = BR = 3$ cm

[as BP and BR are tangents from the same point B]

Hence $BC = BP + PC = (3 + 5) = 8$ cm. **Ans.**

Q. 45. A circle is touching the side BC of ΔABC at P and touching AB and AC produced at Q and R respectively. Prove that $AQ = \frac{1}{2}$ (perimeter of ΔABC).

Solution : We have,

$AQ = AR$ [tangents from the point A]

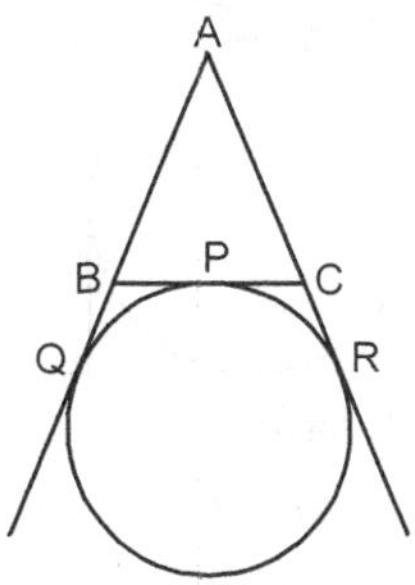

$BQ = BP$ [tangents from the point B]

and $PC = CR$ [tangents from the point C]

Now, perimeter of ΔABC

$$= AB + BC + AC$$
$$= AB + BP + PC + AC$$
$$= AB + BQ + CR + AC$$
$$= AQ + AR$$
$$= 2AQ$$

or $AQ = \frac{1}{2}$(perimeter of ΔABC).

Hence Proved.

Q. 46. In the given figure, PA is a tangent from an external point P to a circle with centre O. If $\angle POB$ = 115°, find $\angle APO$.

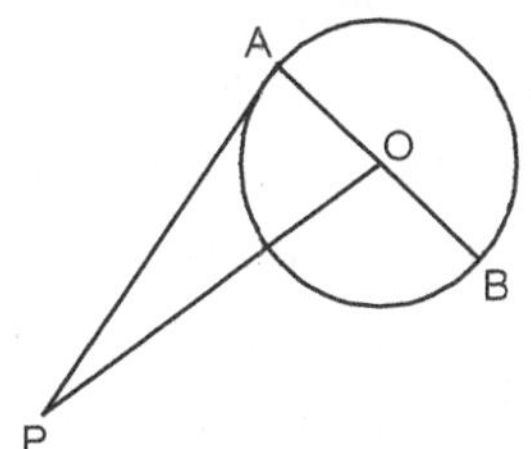

Solution : Given, PA is a tangent to the circle with centre O.

Thus, $\angle PAO = 90^\circ$

[OA, the radius is perpendicular to PA]

Now, $\angle AOP = 180^\circ - 115^\circ$

$= 65^\circ$

[$\angle AOP$ and $\angle BOP$ are supplementary angles]

Thus, $\angle APO = 180^\circ - (90^\circ + 65^\circ)$

[sum of the angles of a triangle is 180°]

$= 180^\circ - 155^\circ$

$= 25^\circ$ **Ans.**

Q. 47. In the given figure, two circles touch each other at the point C. Prove that the common tangent to the circles at C bisects the common tangent PQ.

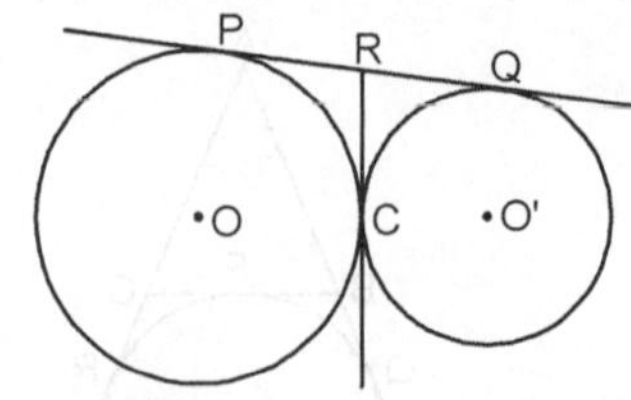

Solution : We have,

$RP = RC$ [tangents from point R]

and $RQ = RC$ [tangents from point R]

Thus $RP = RQ$ [both equal to RC]

Hence, R is the mid-point of the tangent PQ.

Hence Proved.

Q. 48. Tangents PA and PB are drawn from an external point P to two concentric circles with centre O and radii 5 cm and 3 cm respectively, as shown in the given figure. If $AP = 12$ cm, find the length of BP.

Solution : Given, $AP = 12$ cm, $OA = 5$ cm and $OB = 3$ cm.

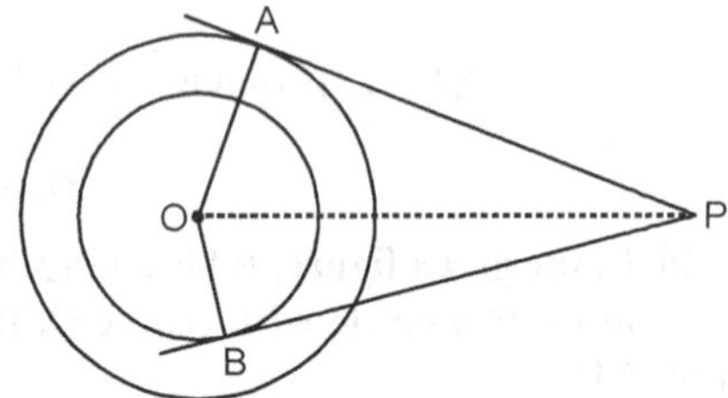

In ΔOAP and ΔOBP,

Applying Pythagoras theorem, we have

$$AP^2 + OA^2 = OP^2 = OB^2 + BP^2$$

$$\Rightarrow \quad (12)^2 + (5)^2 = (3)^2 + (BP)^2$$

$$\Rightarrow \quad (BP)^2 = 144 + 25 - 9$$

$$\Rightarrow \quad (BP)^2 = 169 - 9$$

$$\Rightarrow \quad (BP)^2 = 160$$

$$\Rightarrow \quad BP = 4\sqrt{10} \text{ cm}$$

(as length cannot be negative)

$$\Rightarrow \quad BP = 4(3.16) \text{ cm}$$

$$= 12.64 \text{ cm.}$$ **Ans.**

Q. 49. In given figure, AB is the diameter of a circle with centre O and AT is a tangent. If $\angle AOQ = 58°$, find $\angle ATQ$.

Solution : Given, AB is a diameter of a circle with centre O and AT is a tangent, then

$$\angle BAT = 90°$$

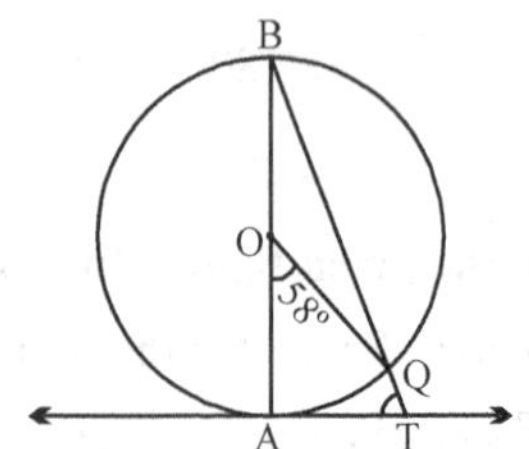

Now, $\angle ABQ = \frac{1}{2}\angle AOQ$

[∵ Angle subtended by an arc on the circle is half of the angle subtended at centre]

$$\angle ABQ = \frac{1}{2} \times 58° = 29°$$

Now, $\angle ATQ = 180° - (\angle ABQ + \angle BAT)$

$$= 180° - (29° + 90°)$$

$\therefore \quad \angle ATQ = 61°$ **Ans.**

Q. 50. From a point T outside a circle of centre O, tangents TP and TQ are drawn to the circle. Prove that OT is the right bisector of line segment PQ.

Solution : Given, TP and TQ are the tangents drawn on a circle with centre O.

To prove : OT is the right bisector of PQ.

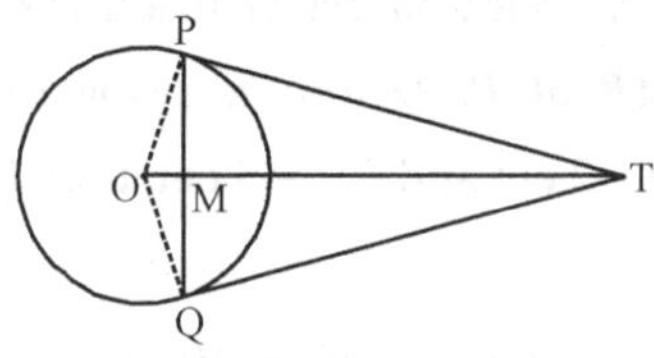

Proof : In ΔTPM and ΔTQM

$TP = TQ$ [tangents drawn from external points are equal]

$TM = TM$ [common]

[Angles opposite to equal sides]

$\angle TPM = TQM$

$\therefore \quad \Delta TPM \cong \Delta TQM$

[By SAS congruence]

$\therefore \quad PM = MQ$

and $\angle PMT = \angle QMT$ [By cpct]

Since, PMQ is a straight line, then

$$\angle PMT + \angle QMT = 180°$$

$\therefore \quad \angle PMT = \angle QMT = 90°$

$\therefore$ OT is the right bisector of PQ. **Hence Proved.**

Q. 51. In fig. AP and BP are tangents to a circle with centre O, such that $AP = 5$ cm and $\angle APB = 60°$. Find the length of chord AB.

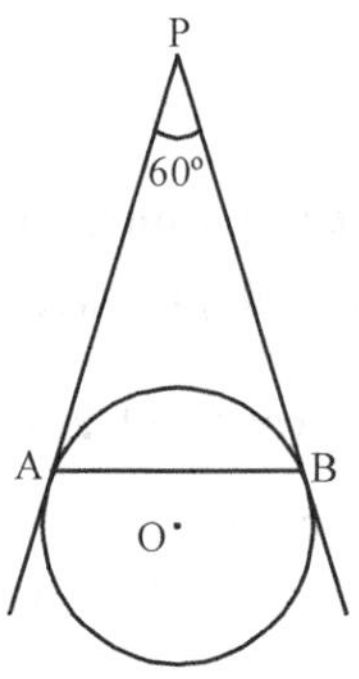

Solution : Given, AP and BP are tangents to a circle with centre O.

$\therefore \quad AP = BP$

Now, $\angle APB = 60°$ [given]

$\therefore \quad \angle PAB = \angle PBA = 60°$ [$\because AP = BP$]

Thus, ΔAPB is an equilateral triangle.

Hence, the length of chord AB is equal to the length of AP *i.e.* 5 cm. **Ans.**

Q. 52. In the given figure, AB is a common tangent to the two circles that intersect at C and D. Also, AD is a tangent to the smaller circle and BD is a tangent to the larger circle. Find the measure of $\angle ADB$.

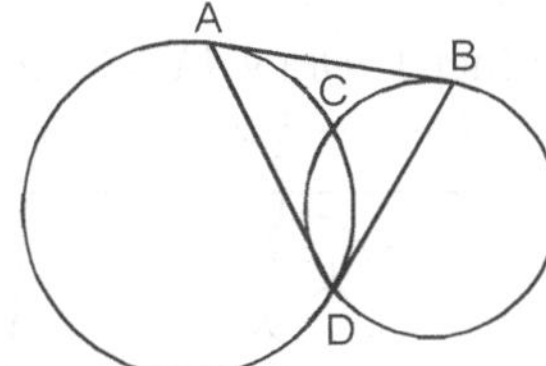

Solution : As AB and BD are tangents to the larger circle from B, so

$AB = BD$

Also, AB and AD are tangents to the smaller circle from A, so

$AD = AB$

Thus $\quad AB = BD = AD$

So ΔABD is equilateral and $\angle A = \angle B = \angle D = 60°$. **Ans.**

Q. 53. In the given figure, PA and PB are tangents to the circle from an external point P. CD is another tangent touching the circle at Q. If $PA = 12$ cm, $QC = QD = 3$cm, then find $PC + PD$.

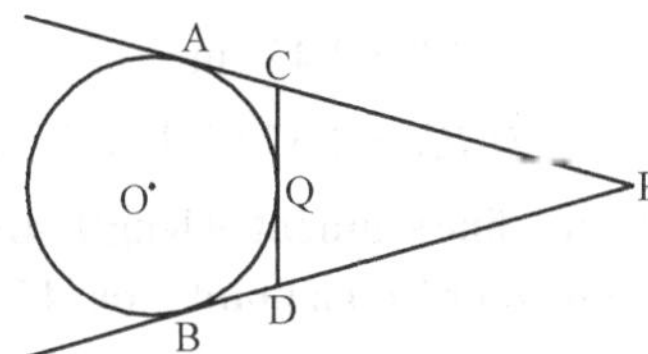

Solution : Given, $\quad PA = PB = 12$ cm

[tangents from P]

$AC = CQ = 3$ cm

[tangents from C]

$BD = QD = 3$ cm

[tangents from D]

So, $\quad PC + PD = (PA - AC) + (PB - BD)$

$= (12 - 3) + (12 - 3)$

$= 9 + 9$

$= 18$ cm. **Ans.**

Q. 54. In figure, two tangents RQ and RP are drawn from an external point R to the circle with centre O. If $\angle PRQ = 120°$, then prove that $OR = PR + RQ$.

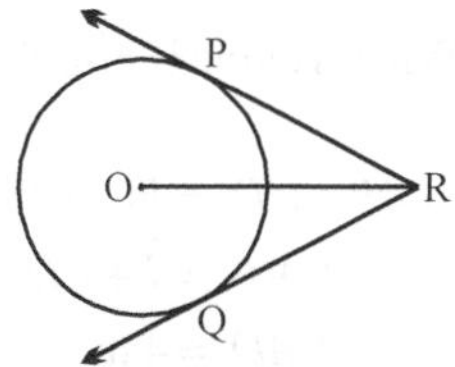

Solution : Given, O is the centre of the circle and $\angle PRQ = 120°$

Construction : Join OP, OQ

To prove : $OP = PR + RQ$.

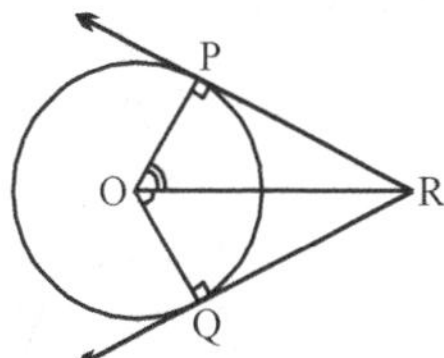

Proof : We know that,

Tangent to a circle is perpendicular to the radius at the point of tangent *i.e.*, $OP \perp RP$ and $OQ \perp RQ$.

$\therefore \quad \angle OPR = \angle OQR = 90°$

Now, in ΔOPR and ΔOQR,

$OP = OQ$ [Radius of circle]

$OR = OR$ [Common]

$\angle OPR = \angle OQR$ [Each 90°]

$\therefore \quad \Delta OPR \cong \Delta OQR$

[By RHS congruence]

So, $\quad PR = QR$ [By C.P.C.T.]

and $\quad \angle ORP = \angle ORQ$

$$= \frac{120°}{2} = 60°$$

Now, in ΔOPR

$$\cos 60° = \frac{PR}{OR} \left[\because \cos\theta = \frac{\text{Base}}{\text{Hypotenuse}}\right]$$

$$\frac{1}{2} = \frac{PR}{OR}$$

$OR = 2PR$

$OR = PR + PR$

$OR = PR + RQ$ [$\because PR = RQ$]

Hence, $\quad OR = PR + RQ$. **Hence Proved.**

Q. 55. A village Panchayt constructed a circular tank to serve as a bird bath. A fencing was made in the shape of a quadrilateral. Sides of the quadrilatieral touched the circle as shown in the figure.

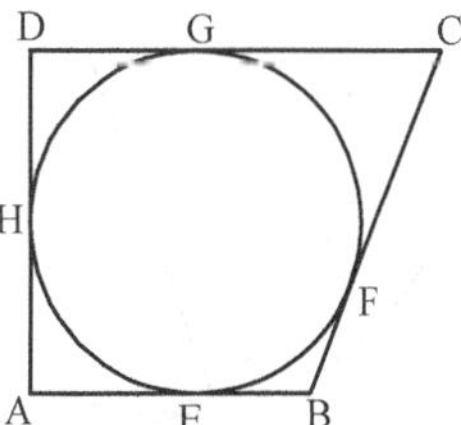

If $AB = 5$ m, $CD = 6$ m, $BC = 7$ m, then find AD.

Solution : When a circle is inscribed in a quadrilateral, then

$$AB + CD = BC + AD$$

$\Rightarrow$ $5 + 6 = 7 + AD$

$\Rightarrow$ $AD = 4$ m **Ans.**

Q. 56. Shipra prepared a project for rain water harvesting. Diagrammatic representation of the project is given in the figure. *PQ* and *PR* are the pipes touching the circular pit. Length of these pipes is 5 m each. What is the perimeter of ΔPMN?

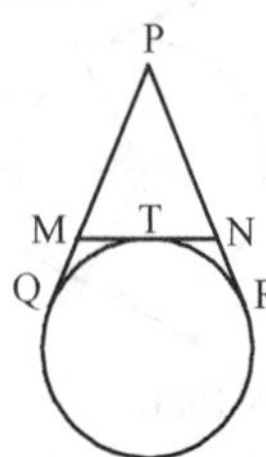

Solution : Here, $MT = MQ$ [Tangents form *M*] ...(i)

and $NT = NR$ [Tangents from *N*] ...(ii)

Now, $PQ + PR = PM + MQ + PN + NR$

$= PM + MT + PN + NT$ [using eq. (i) and (ii)]

$= PM + PN + (MT + NT)$

$= PM + PN + MN$

$=$ Perimeter of Δ *PMN*

$\therefore$ Perimeter of $\Delta PMN = 5 + 5 = 10$ cm. **Ans.**

Q. 57. If figure *PT* is a tangent to a circle. If ($\angle BTA$) = 45° and m($\angle PTB$) = 70°, find m$\angle$(*ABT*).★

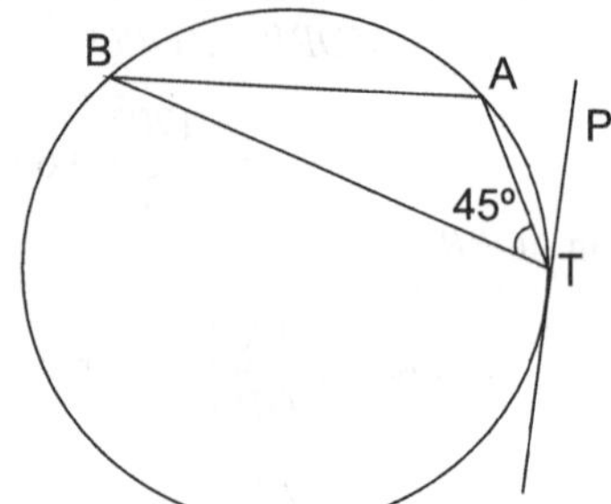

Solution. $\angle ATP = \angle PTB - \angle BTA$

$= 70° - 45° = 25°$

$\therefore$ $\angle ABT = \angle ATP$

(Angles are in alternate segments)

$\Rightarrow$ $\angle ABT = 25°$. **Ans.**

Q. 58. In figure *AT* is a tangent to the circle. If m$\angle ABC = 50°$, $AC = BC$, find $\angle BAT$.

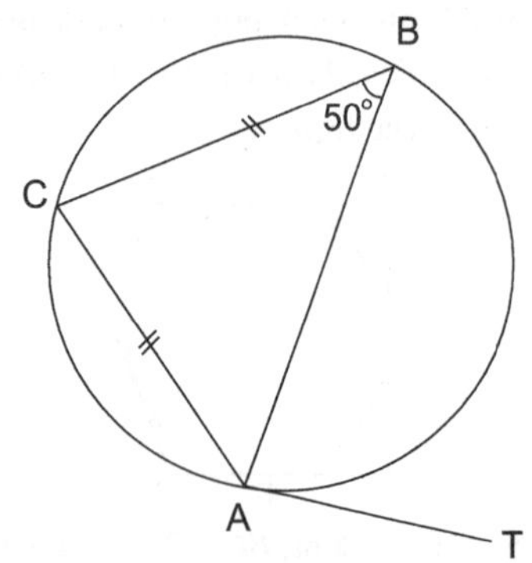

★ are board exam questions from previous years

Solution. Given,

$$AC = BC$$

$\Rightarrow$ $\angle CBA = \angle CAB$

$\Rightarrow$ $\angle CAB = 50°$

$\therefore$ $\angle ACB = 180° - (50° + 50°) = 80°$

Now, $\angle BAT = \angle BCA$

[Angles are in alternate segments]

$\Rightarrow$ $\angle BAT = 80°$. **Ans.**

Chapter 9. Constructions

Q. 1. Draw a line segment of length 6 cm using compass and ruler. Find a point *P* on it which divides it in the ratio of 3 : 4.

Solution : Step I : Draw $AX = 6$ cm.

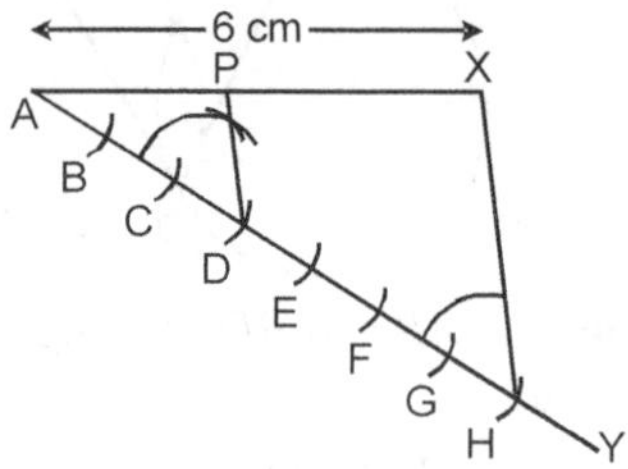

Step II : Draw a line segment *AY* by constructing an acute angle at *A*.

Step III : As the sum of the ratios is $3 + 4 = 7$, draw 7 points on *AY*(*B, C, D, E, F, G, H*) such that

$$AB = BC = CD = DE = EF = GH$$

Step IV : Join *HX* and measure $\angle AHX$.

Step V : Construct a line parallel to *HX* from *D* to *AX* meeting it at *P*.

Step VI : Measure *AP* and *PX* to verify that $AP : PX = 3 : 4$.

Thus, $AP = 2.57$ cm

and $PX = 3.43$ cm

So $AP : PX = 2.57 : 3.43 = 3 : 4$.

Q. 2. Draw a line segment of length *AB* = 7 cm. Using compasses and ruler find a point *P* on *AB* such that

$$AP : AB = 3 : 5.$$

Solution : Step I : Draw $AB = 7$ cm.

Step II : Draw a line segment *AX* by constructing an acute angle at *A*.

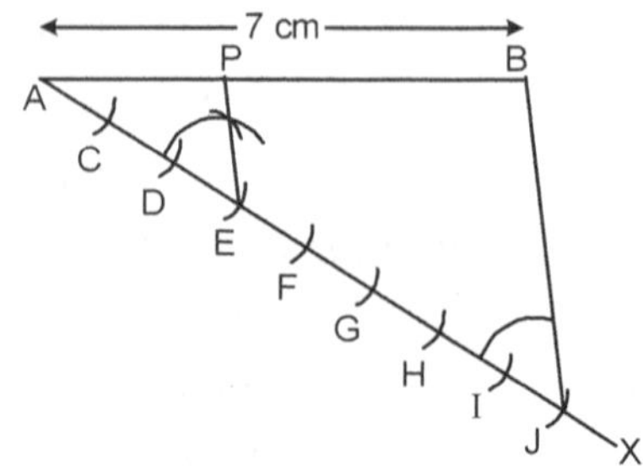

Step III : As the sum of the ratios is $3 + 5 = 8$, draw 8 points on *AX*(*C, D, E, F, G, H, I, J*) such that

$$AC = CD = DE = EF = FG = GH = HI = IJ$$

Step IV : Join JB and measure $\angle AJB$.

Step V : Construct a line parallel to JB from E to AB meeting it at P.

Step VI : Measure AP and BP to verify that $AP : BP = 3 : 5$.

Thus $PB = 4.37$ and $AP = 2.63$ cm

So $AP : PB = 2.63 : 4.37 = 3 : 5$.

Q. 3. Draw a line segment of length 8 cm and divide it internally in the ratio 4 : 5.

Solution : Step I : Draw $AB = 8$ cm

Step II : Draw any ray AX making an acute angle with AB

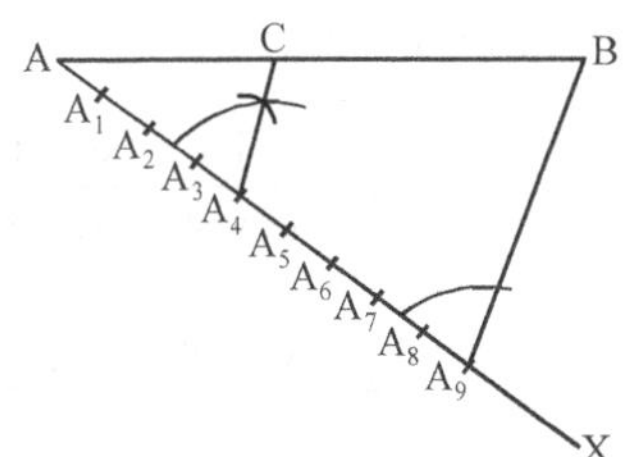

Step III : Draw 9(4 + 5) points on ray AX namely A_1, A_2, A_3, A_4, A_5, A_6, A_7, A_8, A_9 at equal distance.

Step IV : Join BA_9

Step V : Through point A_4, draw a line parallel to A_9B intersecting AB at the point C then $AC : CB = 4 : 5$

Q. 4. Draw a line segment of length 7 cm and divide it internally in the ratio 2 : 3.

Solution : Step I : Draw $AB = 7$ cm

Step II : At A, draw an acute angle with 5 equidistant marks A_1, A_2, A_3, A_4, A_5.

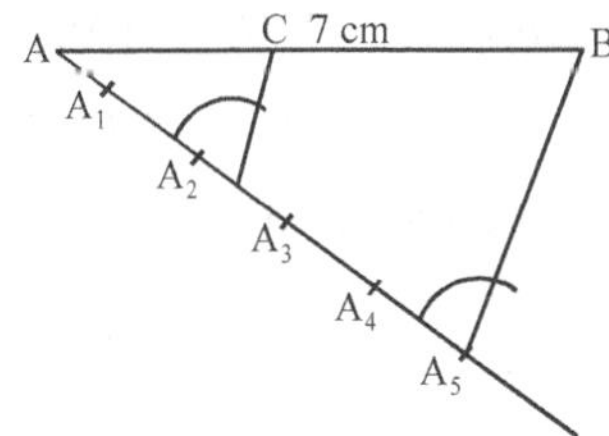

Step III : Join A_5B

Step IV : Draw $A_2C \parallel A_5B$ to get point C on AB.

Step V : Thus, $AC : CB = 2 : 3$.

Chapter 10. Trigonometry

Q. 1. Find the value of $\dfrac{2\tan 30°}{1+\tan^2 30°}$. [NCERT]

Solution : We have,

$$\frac{2\tan 30°}{1+\tan^2 30°} = \frac{2\frac{1}{\sqrt{3}}}{1+\frac{1}{3}}$$

$$= \frac{2\times 3}{\sqrt{3}\times 4}$$

$$= \frac{\sqrt{3}}{2}$$ **Ans.**

Q. 2. Evaluate : $\sin^2 19° + \sin^2 71°$.

Solution : $\sin^2 19° + \sin^2 71°$

$= \sin^2 19° + \sin^2 (90° - 19°)$

$= \sin^2 19° + \cos^2 19°$

$[\because \sin(90° - \theta) = \cos\theta]$

$= 1$ $[\because \sin^2\theta + \cos^2\theta = 1]$

Ans.

Q. 3. Find the value of $(\sin^2 33° + \sin^2 57°)$.★

Solution : $\sin^2 33° + \sin^2 57°$

$= \sin^2 33° + \cos^2 (90° - 57°)$

$= \sin^2 33° + \cos^2 33°$

$= 1$ $[\because \sin^2\theta + \cos^2\theta = 1]$

Ans.

Q. 4. What happens to value of $\cos\theta$ when θ increases from 0° to 90° ?

Solution : $\cos 0° = 1$, $\cos 90° = 0$

When θ increases from 0° to 90°, the value of $\cos\theta$ decreases from 1 to 0. **Ans.**

Q. 5. If ΔABC is right angled at B, what is the value of $\sin(A + C)$?

Solution : $\angle B = 90°$ [Given]

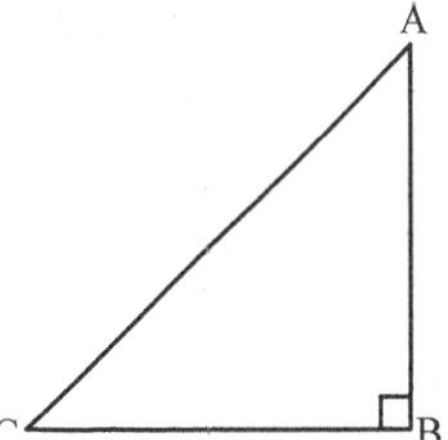

We know that in ΔABC,

$\angle A + \angle B + \angle C = 180°$

[Angle sum property of a Δ]

$\Rightarrow$ $\angle A + 90° + \angle C = 180°$

$\Rightarrow$ $\angle A + \angle C = 180° - 90°$

$\Rightarrow$ $\angle A + \angle C = 90°$

$\Rightarrow$ $\sin(A + C) = \sin 90° = 1$ **Ans.**

Q. 6. If $\sqrt{3}\sin\theta = \cos\theta$, find the value of

$$\frac{3\cos^2\theta + 2\cos\theta}{3\cos\theta + 2}.$$

★ are board exam questions from previous years

Solution : $\sqrt{3}\sin\theta = \cos\theta$ [Given]

$$\Rightarrow \quad \frac{\sin\theta}{\cos\theta} = \frac{1}{\sqrt{3}} \text{ or } \tan\theta = \frac{1}{\sqrt{3}}$$

$$\Rightarrow \quad \tan\theta = \tan 30° \Rightarrow \theta = 30°$$

Now,

$$\frac{3\cos^2\theta + 2\cos\theta}{3\cos\theta + 2} = \frac{\cos\theta(3\cos\theta+2)}{(3\cos\theta+2)} = \cos\theta$$

Put $\theta = 30°$

$$\Rightarrow \quad \cos 30° = \frac{\sqrt{3}}{2}$$ **Ans.**

Q. 7. In the given figure, if $AD = 4$ cm, $\angle D = 90°$, $BD = 3$ cm and $CB = 12$ cm, find cot θ.

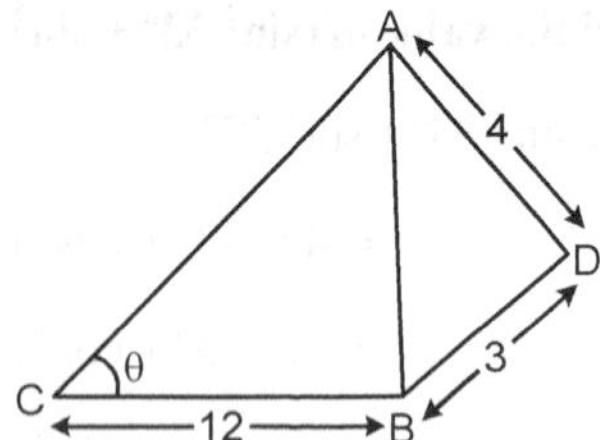

Solution : Given, $AD = 4$ cm, $BD = 3$ cm and $CB = 12$ cm and $\angle D = 90°$

Thus $AB^2 = AD^2 + BD^2$ [Pythagoras theorem]

$$\Rightarrow \quad AB^2 = (4)^2 + (3)^2$$
$$= 16 + 9$$
$$= 25$$
$$= 5^2$$

$$\Rightarrow \quad AB = 5 \text{ cm}$$

Thus $$\cot\theta = \frac{\text{Base}}{\text{Perpendicular}} = \frac{BC}{AB}$$

$$= \frac{12}{5}.$$ **Ans.**

Q. 8. Find the value of $9\sec^2 A - 9\tan^2 A$. **[NCERT]**

Solution : Given,

$9\sec^2 A - 9\tan^2 A = 9(\sec^2 A - \tan^2 A)$

$= 9(1)$ $[\because \sec^2\theta - \tan^2\theta = 1]$

$= 9.$ **Ans.**

Q. 9. Find the value of

(1 + tan θ + sec θ) (1 + cot θ – cosec θ) **[NCERT]**

Solution : Given,

$(1 + \tan\theta + \sec\theta)(1 + \cot\theta - \operatorname{cosec}\theta)$

$= [1 + \tan\theta + \sec\theta + \cot\theta + \tan\theta\cot\theta + \sec\theta\cot\theta - \operatorname{cosec}\theta - \tan\theta\operatorname{cosec}\theta - \sec\theta\operatorname{cosec}\theta]$

$= [1 + \tan\theta + \sec\theta + \cot\theta + 1 + \operatorname{cosec}\theta - \operatorname{cosec}\theta - \sec\theta - \sec\theta\operatorname{cosec}\theta]$

$= [2 + \tan\theta + \cot\theta - \sec\theta\operatorname{cosec}\theta]$

$$= 2 + \frac{\sin\theta}{\cos\theta} + \frac{\cos\theta}{\sin\theta} - \frac{1}{\sin\theta\cos\theta}$$

$$= 2 + \frac{\sin^2\theta + \cos^2\theta - 1}{\sin\theta\cos\theta}$$

$$= 2 + \frac{1-1}{\sin\theta\cos\theta} \quad [\because \sin^2\theta + \cos^2\theta = 1]$$

$= 2.$ **Ans.**

Q. 10. Find (sec A + tan A) (1 – sin A). **[NCERT]**

Solution : Given,

$(\sec A + \tan A)(1 - \sin A)$

$$= \left(\frac{1}{\cos A} + \frac{\sin A}{\cos A}\right)(1 - \sin A)$$

$$= \left(\frac{1 + \sin A}{\cos A}\right)(1 - \sin A)$$

$$= \left(\frac{1 - \sin^2 A}{\cos A}\right)$$

$$= \frac{\cos^2 A}{\cos A} \quad [\because 1 - \sin^2 A = \cos^2 A]$$

$= \cos A.$ **Ans.**

Q. 11. If A, B and C are interior angles of ΔABC, then prove that : $\sin\frac{(A+C)}{2} = \cos\frac{B}{2}$.

Solution : In ΔABC

$\angle A + \angle B + \angle C = 180°$ [Angle sum property]

$\Rightarrow \quad \angle A + \angle C = 180° - \angle B$

Divide by 2 on both sides

$$\frac{\angle A + \angle C}{2} = \frac{180° - \angle B}{2}$$

$$\frac{\angle A + \angle C}{2} = 90° - \frac{\angle B}{2}$$

Taking sin both sides,

$$\sin\left(\frac{\angle A + \angle C}{2}\right) = \sin\left(90° - \frac{\angle B}{2}\right)$$

$$\sin\left(\frac{\angle A + \angle C}{2}\right) = \cos\frac{\angle B}{2} \quad [\because \sin(90° - \theta) = \cos\theta]$$

$$\sin\frac{(A+C)}{2} = \cos\frac{B}{2}$$ **Hence Proved.**

Q. 12. If $x = 3\sin\theta$ and $y = 4\cos\theta$, find the value of $\sqrt{16x^2+9y^2}$.

Solution : Given, $x = 3\sin\theta$

$\Rightarrow \quad x^2 = 9\sin^2\theta$

$\Rightarrow \quad \sin^2\theta = \dfrac{x^2}{9}$ ...(i)

and $\quad y = 4\cos\theta$

$\Rightarrow \quad y^2 = 16\cos^2\theta$

$\Rightarrow \quad \cos^2\theta = \dfrac{y^2}{16}$ ...(ii)

On adding equations (i) and (ii)

$$\sin^2\theta + \cos^2\theta = \frac{x^2}{9} + \frac{y^2}{16}$$

$$1 = \frac{x^2}{9} + \frac{y^2}{16}$$

$[\because \sin^2\theta + \cos^2\theta = 1]$

$$1 = \frac{16x^2+9y^2}{144}$$

$$16x^2 + 9y^2 = 144$$

Taking square root of both sides

$$\sqrt{16x^2+9y^2} = \sqrt{144}$$

Thus, $\sqrt{16x^2+9y^2} = 12$ **Ans.**

Q. 13. If $\cot\theta = \dfrac{15}{8}$, then evaluate $\dfrac{(2+2\sin\theta)(1-\sin\theta)}{(1+\cos\theta)(2-2\cos\theta)}$.

Solution : Given, $\cot\theta = \dfrac{15}{8}$

Now, $\dfrac{(2+2\sin\theta)(1-\sin\theta)}{(1+\cos\theta)(2-2\cos\theta)}$

$$= \frac{2(1+\sin\theta)(1-\sin\theta)}{2(1+\cos\theta)(1-\cos\theta)}$$

$$= \frac{1-\sin^2\theta}{1-\cos^2\theta}$$

$$= \frac{\cos^2\theta}{\sin^2\theta}$$

$$= \cot^2\theta$$

$$= \left(\frac{15}{8}\right)^2 = \frac{225}{64}.$$ **Ans.**

Q. 14. If $\sin A = \dfrac{3}{4}$, calculate sec A.★

Solution: Given,

$$\sin A = \frac{3}{4}$$

We know,

$$\sin A = \frac{P}{H} = \frac{3}{4}$$

$$P = 3k \text{ and } H = 4k$$

$$P^2 + B^2 = H^2$$

[Applying Pythagoras theorem]

$\Rightarrow \quad 9k^2 + B^2 = 16k^2$

$\Rightarrow \quad B^2 = 7k^2$

$\Rightarrow \quad B = \sqrt{7}\,k$

$\therefore \quad \sec A = \dfrac{H}{B} = \dfrac{4k}{\sqrt{7}k} = \dfrac{4}{\sqrt{7}}$ **Ans.**

Q. 15. If $\tan(A+B) = \sqrt{3}$ and $\tan(A-B) = \dfrac{1}{\sqrt{3}}$, find the value of A and B if $0° < A+B < 90°$ and $A > B$.

Solution : Given,

$$\tan(A+B) = \sqrt{3} = \tan 60°$$

or $\quad A + B = 60°$ ...(i)

and $\quad \tan(A-B) = \dfrac{1}{\sqrt{3}} = \tan 30°$

or $\quad (A-B) = 30°$ ...(ii)

Adding (i) and (ii),

$$2A = 90°$$

or $\quad A = 45°$

Thus in (i), $\quad A + B = 60°$

or $\quad B = 60° - 45° = 15°$

Hence $\quad A = 45°$ and $B = 15°$. **Ans.**

Q. 16. If $\tan(A+B) = \sqrt{3}$ and $\tan(A-B) = 1$, find the value of A and B if $0° < A+B < 90°$ and $A > B$.

Solution : Given, $\tan(A+B) = \sqrt{3} = \tan 60°$

or $\quad A + B = 60°$ ...(i)

and $\quad \tan(A-B) = 1 = \tan 45°$

or $\quad (A-B) = 45°$ ...(ii)

Adding (i) and (ii),

$$2A = 105°$$

or $\quad A = 52.5°$

Thus in (i), $\quad A + B = 60°$

or $\quad B = 60° - 52.5° = 7.5°$

Hence $\quad A = 52.5°$ and $B = 7.5°$ **Ans.**

★ are board exam questions from previous years

Q. 17. Prove that : $2\cos^2\theta + \dfrac{2}{1+\cot^2\theta} = 2.$

Solution : Consider,

$$\text{L.H.S.} = 2\cos^2\theta + \frac{2}{1+\cot^2\theta}$$

$$= 2\cos^2\theta + \frac{2\sin^2\theta}{\sin^2\theta+\cos^2\theta} \quad \left[\because \cot\theta = \frac{\cos\theta}{\sin\theta}\right]$$

$$= 2(\sin^2\theta + \cos^2\theta) = 2 = \text{R.H.S.}$$

Hence Proved.

Q. 18. Find A if tan 2A = cot (A – 24°).★

Solution : Given,

$$\tan 2A = \cot(A - 24°)$$

or $\cot(90° - 2A) = \cot(A - 24°)$ $\quad [\because \tan\theta = \cot(90° - \theta)]$

or $90° - 2A = A - 24°$

or $3A = 114°$

$A = 38°$ **Ans.**

Q. 19. Prove that :

$$\frac{\sin^3\theta + \cos^3\theta}{\sin\theta + \cos\theta} + \sin\theta\cos\theta = 1.$$

Solution : Consider,

$$\text{L.H.S.} = \frac{\sin^3\theta+\cos^3\theta}{\sin\theta+\cos\theta} + \sin\theta\cos\theta$$

$$= (\sin^2\theta + \cos^2\theta - \sin\theta\cos\theta) + \sin\theta\cos\theta \quad \left[\because \frac{a^3+b^3}{a+b} = a^2+b^2-ab\right]$$

$= 1 = \text{R.H.S.}$

Hence Proved.

Q. 20. Prove that :

$$(\sin\theta - \text{cosec}\,\theta)(\cos\theta - \sec\theta) = \frac{1}{\tan\theta + \cot\theta}.$$

Solution : Consider,

$$\text{L.H.S.} = (\sin\theta - \text{cosec}\,\theta)(\cos\theta - \sec\theta)$$

$$= \left(\sin\theta - \frac{1}{\sin\theta}\right)\left(\cos\theta - \frac{1}{\cos\theta}\right)$$

$$= \frac{(\sin^2\theta - 1)(\cos^2\theta - 1)}{\sin\theta\cos\theta}$$

$$= \frac{(-\cos^2\theta)(-\sin^2\theta)}{\sin\theta\cos\theta}$$

$$= \sin\theta\cos\theta$$

$$\text{R.H.S.} = \frac{1}{\tan\theta + \cot\theta}$$

$$= \frac{1}{\frac{\sin\theta}{\cos\theta} + \frac{\cos\theta}{\sin\theta}}$$

$$= \frac{\sin\theta\cos\theta}{\sin^2\theta + \cos^2\theta}$$

$$= \sin\theta\cos\theta \quad [\because \sin^2\theta + \cos^2\theta = 1]$$

Thus, $\text{L.H.S.} = \text{R.H.S.} = \sin\theta\cos\theta.$

Hence Proved.

Q. 21. Prove the following identity :

$$\frac{\sin^3\theta + \cos^3\theta}{\sin\theta + \cos\theta} = 1 - \sin\theta.\cos\theta.$$

Solution : $\text{L.H.S.} = \dfrac{\sin^3\theta + \cos^3\theta}{\sin\theta + \cos\theta}$

$$= \frac{(\sin\theta + \cos\theta)(\sin^2\theta + \cos^2\theta - \sin\theta.\cos\theta)}{(\sin\theta + \cos\theta)}$$

$$[\because a^3 + b^3 = (a+b)(a^2+b^2-ab)]$$

$= 1 - \sin\theta.\cos\theta = \text{R.H.S.}$ $\quad [\because \sin^2\theta + \cos^2\theta = 1]$

Hence Proved.

Q. 22. Prove that :

$(\sin\theta + \text{cosec}\,\theta)^2 + (\cos\theta + \sec\theta)^2 = 7 + \tan^2\theta + \cot^2\theta$

Solution : Consider,

$$\text{L.H.S.} = (\sin\theta + \text{cosec}\,\theta)^2 + (\cos\theta + \sec\theta)^2$$

$$= \sin^2\theta + \text{cosec}^2\theta + 2\sin\theta\,\text{cosec}\,\theta + \cos^2\theta + \sec^2\theta + 2\cos\theta\sec\theta$$

$$= (\sin^2\theta + \cos^2\theta) + (1 + \cot^2\theta) + (1 + \tan^2\theta) + 2(\sin\theta\,\text{cosec}\,\theta + \cos\theta\sec\theta)$$

$$= 1 + 1\cot^2\theta + 1 + \tan^2\theta + 4$$

$$= 7 + \cot^2\theta + \tan^2\theta$$

$= \text{R.H.S.}$ **Hence Proved.**

Q. 23. Prove that :

$$\frac{\sin\theta + \cos\theta}{\sin\theta - \cos\theta} + \frac{\sin\theta - \cos\theta}{\sin\theta + \cos\theta} = \frac{2\sec^2\theta}{\tan^2\theta - 1}.$$

Solution : Consider,

$$\text{L.H.S.} = \frac{\sin\theta + \cos\theta}{\sin\theta - \cos\theta} + \frac{\sin\theta - \cos\theta}{\sin\theta + \cos\theta}$$

$$= \frac{(\sin\theta + \cos\theta)^2 + (\sin\theta - \cos\theta)^2}{\sin^2\theta - \cos^2\theta}$$

$$= \frac{2}{\sin^2\theta - \cos^2\theta}$$

$$= \frac{2\sec^2\theta}{\tan^2\theta - 1}$$ [dividing both the numerator and denominator by $\cos^2\theta$]

$= \text{R.H.S}$ **Hence Proved.**

Q. 24. Prove that :

$$\left(1 + \frac{1}{\tan^2 A}\right)\left(1 + \frac{1}{\cot^2 A}\right) = \frac{1}{\sin^2 A - \sin^4 A}.$$

★ are board exam questions from previous years

Solution : Consider,

$$\text{L.H.S.} = \left(1+\frac{1}{\tan^2 A}\right)\left(1+\frac{1}{\cot^2 A}\right)$$
$$= (1+\cot^2 A)(1+\tan^2 A)$$
$$= \operatorname{cosec}^2 A \sec^2 A$$
$$= \frac{1}{\sin^2 A \cos^2 A}$$
$$= \frac{1}{\sin^2 A(1-\sin^2 A)}$$
$$= \frac{1}{\sin^2 A - \sin^4 A} = \text{R.H.S.}$$

Hence Proved.

Q. 25. Prove that :

$$\frac{\cos\theta}{1-\tan\theta}+\frac{\sin\theta}{1-\cot\theta} = \cos\theta+\sin\theta.$$

Solution : Consider,

$$\text{L.H.S.} = \frac{\cos\theta}{1-\tan\theta}+\frac{\sin\theta}{1-\cot\theta}$$
$$= \frac{\cos^2\theta}{\cos\theta-\sin\theta}+\frac{\sin^2\theta}{\sin\theta-\cos\theta}$$
$$= \frac{\cos^2\theta}{\cos\theta-\sin\theta}-\frac{\sin^2\theta}{\cos\theta-\sin\theta}$$
$$= \frac{\cos^2\theta-\sin^2\theta}{\cos\theta-\sin\theta}$$
$$= \frac{(\cos\theta-\sin\theta)(\cos\theta+\sin\theta)}{\cos\theta-\sin\theta}$$
$$= \cos\theta+\sin\theta = \text{R.H.S.}$$

Hence Proved.

Q. 26. Prove that :

$$\frac{\operatorname{cosec}\theta}{\operatorname{cosec}\theta-1}+\frac{\operatorname{cosec}\theta}{\operatorname{cosec}\theta+1} = 2\sec^2\theta.$$

Solution : Consider,

$$\text{L.H.S.} = \frac{\operatorname{cosec}\theta}{\operatorname{cosec}\theta-1}+\frac{\operatorname{cosec}\theta}{\operatorname{cosec}\theta+1}$$
$$= \frac{\operatorname{cosec}\theta(\operatorname{cosec}\theta+1)+\operatorname{cosec}\theta(\operatorname{cosec}\theta-1)}{\operatorname{cosec}^2\theta-1}$$
$$= \frac{2\operatorname{cosec}^2\theta}{\operatorname{cosec}^2\theta-1} = \frac{\frac{2}{\sin^2\theta}}{\frac{1}{\sin^2\theta}-1}$$
$$= \frac{2}{1-\sin^2\theta} = \frac{2}{\cos^2\theta}$$
$$= 2\sec^2\theta = \text{R.H.S.}$$

Hence Proved.

Q. 27. Prove that :

$$\frac{\sin\theta-2\sin^3\theta}{2\cos^3\theta-\cos\theta} = \tan\theta.$$

Solution : Consider,

$$\text{L.H.S.} = \frac{\sin\theta-2\sin^3\theta}{2\cos^3\theta-\cos\theta}$$
$$= \frac{\sin\theta(1-2\sin^2\theta)}{\cos\theta(2\cos^2\theta-1)}$$
$$= \frac{\sin\theta(\sin^2\theta+\cos^2\theta-2\sin^2\theta)}{\cos\theta(2\cos^2\theta-(\cos^2\theta+\sin^2\theta))}$$
$$= \frac{\sin\theta}{\cos\theta}\,\frac{(\cos^2\theta-\sin^2\theta)}{(\cos^2\theta-\sin^2\theta)}$$
$$= \tan\theta = \text{R.H.S.}$$

Hence Proved.

Q. 28. Prove that :

$$\left(\frac{1+\tan^2 A}{1+\cot^2 A}\right) = \left(\frac{1-\tan A}{1-\cot A}\right)^2 = \tan^2 A.$$

Solution : Consider,

$$\left(\frac{1+\tan^2 A}{1+\cot^2 A}\right) = \frac{1+\frac{\sin^2 A}{\cos^2 A}}{1+\frac{\cos^2 A}{\sin^2 A}}$$
$$= \left(\frac{\sin^2 A+\cos^2 A}{\sin^2 A+\cos^2 A}\right)\frac{\sin^2 A}{\cos^2 A}$$
$$= \tan^2 A \qquad \ldots\text{(i)}$$

$$\left(\frac{1-\tan A}{1-\cot A}\right)^2 = \left(\frac{1-\frac{\sin A}{\cos A}}{1-\frac{\cos A}{\sin A}}\right)^2$$
$$= \left[\left(\frac{\cos A-\sin A}{\sin A-\cos A}\right)\left(\frac{\sin A}{\cos A}\right)\right]^2$$
$$= \left[\left(-\frac{\sin A-\cos A}{\sin A-\cos A}\right)(\tan A)\right]^2$$
$$= \tan^2 A \qquad \ldots\text{(ii)}$$

From (i) and (ii),

$$\left(\frac{1+\tan^2 A}{1+\cot^2 A}\right) = \left(\frac{1-\tan A}{1-\cot A}\right)^2 = \tan^2 A.$$

Hence Proved.

Q. 29. Prove that :

$$\left(\frac{\cot A-\cos A}{\cot A+\cos A}\right) = \left(\frac{\operatorname{cosec} A-1}{\operatorname{cosec} A+1}\right).$$

Solution : Consider,

$$\text{L.H.S.} = \left(\frac{\cot A-\cos A}{\cot A+\cos A}\right)$$
$$= \frac{\frac{\cos A}{\sin A}-\cos A}{\frac{\cos A}{\sin A}+\cos A}$$

$$= \left(\frac{\cos A - \sin A \cos A}{\cos A + \sin A \cos A}\right)$$

$$= \left(\frac{1 - \sin A}{1 + \sin A}\right)$$

$$= \frac{\frac{1}{\sin A} - 1}{\frac{1}{\sin A} + 1}$$

[Dividing both numerator and denominator by $\sin A$]

$$= \frac{\text{cosec } A - 1}{\text{cosec } A + 1} = \text{R.H.S.}$$

Hence Proved.

Q. 30. Prove that :

$$(\textbf{cosec } \theta - \textbf{cot } \theta)^2 = \frac{1 - \cos \theta}{1 + \cos \theta}.$$

Solution : Consider,

$$\text{L.H.S.} = (\text{cosec } \theta - \cot \theta)^2$$

$$= \left(\frac{1}{\sin \theta} - \frac{\cos \theta}{\sin \theta}\right)^2$$

$$= \frac{(1 - \cos \theta)^2}{\sin^2 \theta}$$

$$= \frac{(1 - \cos \theta)^2}{1 - \cos^2 \theta}$$

$[\because \sin^2 \theta = 1 - \cos^2 \theta]$

$$= \frac{(1 - \cos \theta)^2}{(1 - \cos \theta)\,(1 + \cos \theta)}$$

$$= \frac{(1 - \cos \theta)}{(1 + \cos \theta)} = \text{R.H.S.}$$

Hence Proved.

Q. 31. Prove that :

$$\frac{1 - \sin \theta}{1 + \sin \theta} = (\sec \theta - \tan \theta)^2.$$

Solution : Consider,

$$\text{R.H.S.} = (\sec \theta - \tan \theta)^2$$

$$= \left(\frac{1}{\cos \theta} - \frac{\sin \theta}{\cos \theta}\right)^2$$

$$= \frac{(1 - \sin \theta)^2}{\cos^2 \theta}$$

$$= \frac{(1 - \sin \theta)^2}{1 - \sin^2 \theta}$$

$[\because \cos^2 \theta = 1 - \sin^2 \theta]$

$$= \frac{(1 - \sin \theta)^2}{(1 - \sin \theta)\,(1 + \sin \theta)}$$

$$= \frac{1 - \sin \theta}{1 + \sin \theta} = \text{L.H.S.}$$

Hence Proved.

Q. 32. Prove that :

$$\sqrt{\frac{1 + \sin \theta}{1 - \sin \theta}} + \sqrt{\frac{1 - \sin \theta}{1 + \sin \theta}} = 2 \sec \theta.$$

Solution : Consider,

$$\text{L.H.S.} = \sqrt{\frac{1 + \sin \theta}{1 - \sin \theta}} + \sqrt{\frac{1 - \sin \theta}{1 + \sin \theta}}$$

$$= \sqrt{\frac{(1 + \sin \theta)(1 + \sin \theta)}{(1 - \sin \theta)(1 + \sin \theta)}} + \sqrt{\frac{(1 - \sin \theta)(1 - \sin \theta)}{(1 + \sin \theta)(1 - \sin \theta)}}$$

$$= \sqrt{\frac{(1 + \sin \theta)(1 + \sin \theta)}{1 - \sin^2 \theta}} + \sqrt{\frac{(1 - \sin \theta)(1 - \sin \theta)}{1 - \sin^2 \theta}}$$

$$= \sqrt{\frac{(1 + \sin \theta)(1 + \sin \theta)}{\cos^2 \theta}} + \sqrt{\frac{(1 - \sin \theta)(1 - \sin \theta)}{\cos^2 \theta}}$$

$$= \frac{(1 + \sin \theta)}{\cos \theta} + \frac{(1 - \sin \theta)}{\cos \theta}$$

$= 2 \sec \theta = \text{R.H.S.}$ **Hence Proved.**

Q. 33. Prove that :

$$\sec \theta\,(1 - \sin \theta)\,(\sec \theta + \tan \theta) = 1.$$

Solution : We have, $\sec \theta\,(1 - \sin \theta)\,(\sec \theta + \tan \theta)$

$$= (\sec \theta - \sin \theta \sec \theta)\,(\sec \theta + \tan \theta)$$

$$= (\sec \theta - \tan \theta)\,(\sec \theta + \tan \theta)$$

$$= \sec^2 \theta - \tan^2 \theta$$

$= 1 = \text{R.H.S.}$ $[\because \sec^2 \theta - \tan^2 \theta = 1]$

Hence Proved.

Q. 34. Prove that :

$$\sin \theta\,(1 + \tan \theta) + \cos \theta\,(1 + \cot \theta) = \sec \theta + \text{cosec } \theta.$$

Solution : Consider,

$$\text{L.H.S.} = \sin \theta\,(1 + \tan \theta) + \cos \theta\,(1 + \cot \theta)$$

$$= \sin \theta + \sin \theta \tan \theta + \cos \theta + \cos \theta \cot \theta$$

$$= \sin \theta + \cos \theta + \frac{\sin^2 \theta}{\cos \theta} + \frac{\cos^2 \theta}{\sin \theta}$$

$$= \sin \theta + \cos \theta + \frac{1 - \cos^2 \theta}{\cos \theta} + \frac{1 - \sin^2 \theta}{\sin \theta}$$

$$= \sin \theta + \cos \theta + \frac{1}{\cos \theta} + \frac{1}{\sin \theta} - \sin \theta - \cos \theta$$

$= \sec \theta + \text{cosec } \theta = \text{R.H.S.}$ **Hence Proved.**

Q. 35. Prove that :

$$\frac{\sin \theta}{1 + \cos \theta} + \frac{1 + \cos \theta}{\sin \theta} = 2 \text{ cosec } \theta.$$

Solution : Consider,

$$\text{L.H.S.} = \frac{\sin \theta}{1 + \cos \theta} + \frac{1 + \cos \theta}{\sin \theta}$$

$$= \frac{\sin^2 \theta + (1 + \cos \theta)^2}{\sin \theta(1 + \cos \theta)}$$

$$= \frac{\sin^2 \theta + 1 + 2 \cos \theta + \cos^2 \theta}{\sin \theta(1 + \cos \theta)}$$

$$= \frac{2+2\cos\theta}{\sin\theta(1+\cos\theta)}$$

$$= \frac{2(1+\cos\theta)}{\sin\theta(1+\cos\theta)}$$

$$= \frac{2}{\sin\theta}$$

$= 2 \operatorname{cosec}\theta = $ R.H.S. **Hence Proved.**

Q. 36. Prove that :

$$\frac{\cos\theta}{1-\tan\theta} + \frac{\sin^2\theta}{\sin\theta-\cos\theta} = \cos\theta + \sin\theta.$$

Solution : Consider,

$$\text{L.H.S.} = \frac{\cos\theta}{1-\tan\theta} + \frac{\sin^2\theta}{\sin\theta-\cos\theta}$$

$$= \frac{\cos^2\theta}{(\cos\theta-\sin\theta)} + \frac{\sin^2\theta}{\sin\theta-\cos\theta}$$

$$= \frac{\sin^2\theta-\cos^2\theta}{\sin\theta-\cos\theta}$$

$$= \frac{(\sin\theta-\cos\theta)(\sin\theta+\cos\theta)}{\sin\theta-\cos\theta}$$

$= \cos\theta + \sin\theta = $ R.H.S. **Hence Proved.**

Q. 37. Prove that :★

$$\sin^6\theta + \cos^6\theta = 1 - 3\sin^2\theta\cos^2\theta.$$

Solution : Consider,

$$\text{L.H.S.} = \sin^6\theta + \cos^6\theta$$

$= (\sin^2\theta)^3 + (\cos^2\theta)^3$

$= (\sin^2\theta + \cos^2\theta)^3 - 3\sin^2\theta\cos^2\theta(\sin^2\theta + \cos^2\theta)$

$= 1^3 - 3\sin^2\theta\cos^2\theta(1)$

$= 1 - 3\sin^2\theta\cos^2\theta = $ R.H.S. **Hence Proved.**

Q. 38. Prove that :

$$\frac{\sin\theta}{\cot\theta+\operatorname{cosec}\theta} - \frac{\sin\theta}{\cot\theta-\operatorname{cosec}\theta} = 2.$$

Solution : Consider,

$$\text{L.H.S.} = \frac{\sin\theta}{\cot\theta+\operatorname{cosec}\theta} - \frac{\sin\theta}{\cot\theta-\operatorname{cosec}\theta}$$

$$= \sin\theta\left[\frac{(\cot\theta-\operatorname{cosec}\theta)-(\cot\theta+\operatorname{cosec}\theta)}{(\cot\theta+\operatorname{cosec}\theta)(\cot\theta-\operatorname{cosec}\theta)}\right]$$

$$= \sin\theta\left[\frac{(-2\operatorname{cosec}\theta)}{\cot^2\theta-\operatorname{cosec}^2\theta}\right]$$

$$= \sin\theta\left[\frac{(-2\operatorname{cosec}\theta)}{-1}\right]$$

$= 2 = $ R.H.S. **Hence Proved.**

★ are board exam questions from previous years

Q. 39. Prove that :

$$(1 - \sin\theta + \cos\theta)^2 = 2(1 + \cos\theta)(1 - \sin\theta).$$

Solution : Consider,

$\text{L.H.S.} = (1 - \sin\theta + \cos\theta)^2$

$= [1 - (\sin\theta - \cos\theta)]^2$

$= 1 + (\sin\theta - \cos\theta)^2 - 2(\sin\theta - \cos\theta)$

$= 1 + \sin^2\theta + \cos^2\theta - 2\sin\theta\cos\theta - 2(\sin\theta - \cos\theta)$

$= 2 - 2\sin\theta\cos\theta - 2(\sin\theta - \cos\theta)$

$= 2[1 - \sin\theta\cos\theta - (\sin\theta - \cos\theta)]$

$\text{R.H.S.} = 2(1 + \cos\theta)(1 - \sin\theta)$

$= 2[1 + \cos\theta - \sin\theta - \sin\theta\cos\theta]$

$= 2[1 - \sin\theta\cos\theta - (\sin\theta - \cos\theta)]$

Hence L.H.S. = R.H.S. **Hence Proved.**

Q. 40. Prove that :

$$\frac{1+\cos\theta-\sin^2\theta}{\sin\theta(1+\cos\theta)} = \cot\theta.$$

Solution : Consider,

$$\text{L.H.S.} = \frac{1+\cos\theta-\sin^2\theta}{\sin\theta(1+\cos\theta)}$$

$$= \frac{\cos\theta+\cos^2\theta}{\sin\theta(1+\cos\theta)}$$

$$= \frac{\cos\theta(1+\cos\theta)}{\sin\theta(1+\cos\theta)}$$

$= \cot\theta = $ R.H.S. **Hence Proved.**

Q. 41. If $x = a\sin\theta + b\cos\theta$ and $y = a\cos\theta - b\sin\theta$, prove that $x^2 + y^2 = a^2 + b^2$.

Solution : Given,

$x = a\sin\theta + b\cos\theta$

and $y = a\cos\theta - b\sin\theta$

Now $x^2 + y^2 = (a\sin\theta + b\cos\theta)^2 + (a\cos\theta - b\sin\theta)^2$

$= a^2\sin^2\theta + b^2\cos^2\theta + 2ab\sin\theta\cos\theta + a^2\cos^2\theta + b^2\sin^2\theta - 2ab\sin\theta\cos\theta$

$= a^2(\sin^2\theta + \cos^2\theta) + b^2(\sin^2\theta + \cos^2\theta)$

$= a^2 + b^2$. **Hence Proved.**

Q. 42. If $\sec\theta + \tan\theta = m$, show that :

$$\frac{m^2-1}{m^2+1} = \sin\theta.$$

Solution : Given, $\sec\theta + \tan\theta = m$

Thus, $\frac{m^2-1}{m^2+1}$

$$= \frac{(\sec\theta+\tan\theta)^2-1}{(\sec\theta+\tan\theta)^2+1}$$

$$= \frac{(1+\sin\theta)^2-\cos^2\theta}{(1+\sin\theta)^2+\cos^2\theta}$$

$$= \frac{1+\sin^2\theta+2\sin\theta-\cos^2\theta}{1+\sin^2\theta+2\sin\theta+\cos^2\theta}$$

$$= \frac{\sin^2\theta+\cos^2\theta+\sin^2\theta+2\sin\theta-\cos^2\theta}{2+2\sin\theta}$$

$[\because \sin^2\theta+\cos^2\theta=1]$

$$= \frac{2\sin^2\theta+2\sin\theta}{2(1+\sin\theta)}$$

$$= \frac{2\sin\theta(1+\sin\theta)}{2(1+\sin\theta)}$$

$= \sin\theta.$ **Hence Proved.**

Q. 43. If $\cos\theta + \sin\theta = \sqrt{2}\cos\theta$, show that

$$\cos\theta - \sin\theta = \sqrt{2}\sin\theta.$$

Solution : Given, $\cos\theta+\sin\theta = \sqrt{2}\cos\theta$

$\Rightarrow (\cos\theta+\sin\theta)^2 = 2\cos^2\theta$

$\Rightarrow \cos^2\theta+\sin^2\theta+2\sin\theta\cos\theta = 2\cos^2\theta$

$\Rightarrow 2\cos^2\theta-\cos^2\theta-\sin^2\theta = 2\sin\theta\cos\theta$

$\Rightarrow \cos^2\theta-\sin^2\theta = 2\sin\theta\cos\theta$

$\Rightarrow (\cos\theta+\sin\theta)(\cos\theta-\sin\theta) = 2\sin\theta\cos\theta$

$\Rightarrow \sqrt{2}\cos\theta(\cos\theta-\sin\theta) = 2\sin\theta\cos\theta$

$\Rightarrow \cos\theta-\sin\theta = \sqrt{2}\sin\theta.$

Hence Proved.

Q. 44. If $\sin\theta + \sin^2\theta = 1$, prove that :

$$\cos^2\theta + \cos^4\theta = 1.$$

Solution : Given,

$\sin\theta+\sin^2\theta = 1$

$\Rightarrow \sin\theta+(1-\cos^2\theta) = 1$

$\Rightarrow \sin\theta = \cos^2\theta$

$\Rightarrow \sin^2\theta = \cos^4\theta$

$\Rightarrow 1-\cos^2\theta = \cos^4\theta$

$\Rightarrow \cos^2\theta+\cos^4\theta = 1.$ **Hence Proved.**

Q. 45. If $a\cos\theta + b\sin\theta = m$ and $a\sin\theta - b\cos\theta = n$, prove that $(m^2+n^2) = (a^2+b^2)$.

Solution : Given, $a\cos\theta+b\sin\theta = m$

and $a\sin\theta-b\cos\theta = n$

So $m^2+n^2 = (a\cos\theta+b\sin\theta)^2+(a\sin\theta-b\cos\theta)^2$

$\Rightarrow m^2+n^2 = a^2\cos^2\theta+b^2\sin^2\theta+2ab\sin\theta\cos\theta+a^2\sin^2\theta+b^2\cos^2\theta-2ab\sin\theta\cos\theta$

$\Rightarrow m^2+n^2 = a^2\cos^2\theta+b^2\sin^2\theta+a^2\sin^2\theta+b^2\cos^2\theta$

$\Rightarrow m^2+n^2 = a^2(\cos^2\theta+\sin^2\theta)+b^2(\cos^2\theta+\sin^2\theta)$

$\Rightarrow m^2+n^2 = a^2+b^2.$ **Hence Proved.**

Q. 46. If $\sin\theta + \cos\theta = \sqrt{2}\sin(90°-\theta)$, show that

$$\cot\theta = (\sqrt{2}+1).$$

Solution : Given,

$\sin\theta+\cos\theta = \sqrt{2}\sin(90°-\theta)$

$\Rightarrow \sin\theta+\cos\theta = \sqrt{2}\cos\theta$

$\Rightarrow \sin\theta = (\sqrt{2}-1)\cos\theta$

Dividing both sides by $\cos\theta$

$\Rightarrow \tan\theta = (\sqrt{2}-1)$

Thus, $\cot\theta = \dfrac{1}{\sqrt{2}-1} = \dfrac{(\sqrt{2}+1)}{(\sqrt{2}-1)(\sqrt{2}+1)}$

$= \dfrac{(\sqrt{2}+1)}{2-1} = (\sqrt{2}+1).$

Hence Proved.

Q. 47. If $\sin\theta + \cos\theta = \sqrt{2}\sin\theta$, show that

$$\sin\theta - \cos\theta = \sqrt{2}\cos\theta.$$

Solution : Given, $\sin\theta+\cos\theta = \sqrt{2}\sin\theta$

$\Rightarrow (\sin\theta+\cos\theta)^2 = 2\sin^2\theta$

$\Rightarrow \sin^2\theta+\cos^2\theta+2\sin\theta\cos\theta = 2\sin^2\theta$

$\Rightarrow \sin^2\theta-\cos^2\theta = 2\sin\theta\cos\theta$

$\Rightarrow (\sin\theta+\cos\theta)(\sin\theta-\cos\theta) = 2\sin\theta\cos\theta$

$\Rightarrow \sqrt{2}\sin\theta(\sin\theta-\cos\theta) = 2\sin\theta\cos\theta$

$\Rightarrow \sin\theta-\cos\theta = \sqrt{2}\cos\theta.$

Hence Proved.

Q. 48. Prove that :

$$\tan 7°\tan 23°\tan 60°\tan 67°\tan 83° = \sqrt{3}.$$

Solution : Consider,

$\tan 7°\tan 23°\tan 60°\tan 67°\tan 83°$

$= \tan(90°-83°)\tan(90°-67°)\tan 60°\tan 67°\tan 83°$

$= \cot 83°\cot 67°\tan 60°\tan 67°\tan 83°$

$= \tan 60°$

$= \sqrt{3}.$ **Hence Proved.**

Q. 49. Prove that :

$$\cot 12°\cot 38°\cot 52°\cot 60°\cot 78° = \frac{1}{\sqrt{3}}.$$

Solution : Consider,

$\cot 12°\cot 38°\cot 52°\cot 60°\cot 78°$

$= \cot(90-78)°\cot(90-52)°\cot 52°\cot 60°\cot 78°$

$= \tan 78°\tan 52°\cot 52°\cot 60°\cot 78°$

$= \cot 60°$

$= \dfrac{1}{\sqrt{3}}.$ **Hence Proved.**

Q. 50. Evaluate :

$$\frac{\cos 58°}{\sin 32°}+\frac{2}{\sqrt{3}}\tan 17°\tan 38°\tan 60°\tan 52°\tan 73° - 3(\sin^2 31°+\sin^2 59°)$$

Solution : Consider,

$$\frac{\cos 58°}{\sin 32°}+\frac{2}{\sqrt{3}}\tan 17°\tan 38°\tan 60°\tan 52°\tan 73° - 3(\sin^2 31°+\sin^2 59°)$$

$= \frac{\cos(90-32)^\circ}{\sin 32^\circ} + \frac{2}{\sqrt{3}} \tan(90-73)^\circ \tan(90-52)^\circ$

$\tan 60^\circ \tan 52^\circ \tan 73^\circ - 3[\sin^2(90-59)^\circ + \sin^2 59^\circ]$

$= \frac{\sin 32^\circ}{\sin 32^\circ} + \frac{2}{\sqrt{3}} \cot 73^\circ \cot 52^\circ \tan 60^\circ \tan 52^\circ \tan 73^\circ$

$- 3[\cos^2 59^\circ + \sin^2 59^\circ]$

$= 1 + \frac{2}{\sqrt{3}} \times \sqrt{3} - 3 = 0.$ **Ans.**

Q. 51. Evaluate :

$$\frac{\tan 50^\circ + \sec 50^\circ}{\cot 40^\circ + \text{cosec } 40^\circ} + \cos 40^\circ \text{ cosec } 50^\circ.$$

Solution : Consider,

$\frac{\tan 50^\circ + \sec 50^\circ}{\cot 40^\circ + \text{cosec } 40^\circ} + \cos 40^\circ \text{ cosec } 50^\circ$

$= \frac{\tan 50^\circ + \sec 50^\circ}{\cot(90-50)^\circ + \text{cosec}(90-50)^\circ}$

$+ \cos(90-50)^\circ \text{ cosec } 50^\circ$

$= \frac{\tan 50^\circ + \sec 50^\circ}{\tan 50^\circ + \sec 50^\circ} + \sin 50^\circ \text{ cosec } 50^\circ = 1 + 1 = 2.$

Ans.

Q. 52. Evaluate :

$$\frac{2 \sin^2 63^\circ + 1 + 2 \sin^2 27^\circ}{3 \cos^2 17^\circ - 2 + 3 \cos^2 73^\circ}.$$

Solution : Consider,

$\frac{2 \sin^2 63^\circ + 1 + 2 \sin^2 27^\circ}{3 \cos^2 17^\circ - 2 + 3 \cos^2 73^\circ}$

$= \frac{2 \sin^2 63^\circ + 1 + 2 \sin^2 (90-63)^\circ}{3 \cos^2 17^\circ - 2 + 3 \cos^2 (90-17)^\circ}$

$= \frac{2 \sin^2 63^\circ + 1 + 2 \cos^2 63^\circ}{3 \cos^2 17^\circ - 2 + 3 \sin^2 17^\circ}$

$= \frac{2+1}{3-2} = 3.$ **Ans.**

Q. 53. Evaluate : $\sin^2 60^\circ + 2 \tan 45^\circ - \cos^2 30^\circ$.★

Solution : We know,

$\sin^2 60^\circ = \frac{\sqrt{3}}{2}$, $\tan 45^\circ = 1$ and $\cos 30^\circ = \frac{\sqrt{3}}{2}$

$\therefore \sin^2 60^\circ + 2 \tan 45^\circ - \cos^2 30^\circ$

$= \left(\frac{\sqrt{3}}{2}\right)^2 + 2(1) - \left(\frac{\sqrt{3}}{2}\right)^2$

$= \frac{3}{4} + 2 - \frac{3}{4}$

$= 2$ **Ans.**

Q. 54. If $\sin 3A = \cos(A - 10^\circ)$, where $3A$ is an acute angle, then find the value of A.

★ are board exam questions from previous years

Solution : Given, $\sin 3A = \cos(A - 10^\circ)$

$\Rightarrow \cos(90^\circ - 3A) = \cos(A - 10^\circ)$

$\Rightarrow (90^\circ - 3A) = (A - 10^\circ)$

$\Rightarrow 4A = 100^\circ$

$\Rightarrow A = 25^\circ.$ **Ans.**

Q. 55. If A, B and C are the angles of the ΔABC, then show that $\sin\left(\frac{B+C}{2}\right) = \cos\left(\frac{A}{2}\right)$.

Solution : As the sum of the angles of a triangle equal to 180°.

So $A + B + C = 180^\circ$

$\Rightarrow B + C = 180^\circ - A$

$\Rightarrow \frac{B+C}{2} = 90^\circ - \frac{A}{2}$

Thus, $\sin\left(\frac{B+C}{2}\right) = \sin\left(90^\circ - \frac{A}{2}\right) = \cos\left(\frac{A}{2}\right)$.

Hence Proved.

Q. 56. Prove that :

$$\tan 35^\circ \tan 40^\circ \tan 45^\circ \tan 50^\circ \tan 55^\circ = 1.$$

Solution : Consider,

L.H.S. $= \tan 35^\circ \tan 40^\circ \tan 45^\circ \tan 50^\circ \tan 55^\circ$

$= \tan(90-55)^\circ \tan(90-50)^\circ \tan 45^\circ \tan 50^\circ \tan 55^\circ$

$= \cot 55^\circ \cot 50^\circ \tan 45^\circ \tan 50^\circ \tan 55^\circ$

$= \tan 45^\circ = 1.$ **Hence Proved.**

Q. 57. Prove that :

$$\tan 5^\circ \tan 25^\circ \tan 30^\circ \tan 65^\circ \tan 85^\circ = \frac{1}{\sqrt{3}}.$$

Solution : Consider,

$= \tan 5^\circ \tan 25^\circ \tan 30^\circ \tan 65^\circ \tan 85^\circ$

$= \tan(90-85)^\circ \tan(90-65)^\circ \tan 30^\circ \tan 65^\circ \tan 85^\circ$

$= \cot 85^\circ \cot 65^\circ \tan 30^\circ \tan 65^\circ \tan 85^\circ$

$= \tan 30^\circ = \frac{1}{\sqrt{3}}.$ **Hence Proved.**

Q. 58. Solve :

$$\frac{\sec^2\theta - \cot^2(90^\circ - \theta)}{\text{cosec}^2 67^\circ - \tan^2 23^\circ} + \sin^2 40^\circ + \sin^2 50^\circ.$$

Solution : Consider,

$\frac{\sec^2\theta - \cot^2(90^\circ - \theta)}{\text{cosec}^2 67^\circ - \tan^2 23^\circ} + \sin^2 40^\circ + \sin^2 50^\circ$

$= \frac{\sec^2\theta - \cot^2(90^\circ - \theta)}{\text{cosec}^2 67^\circ - \tan^2(90-67)^\circ}$

$+ \sin^2 40^\circ + \sin^2(90-40)^\circ$

$= \frac{\sec^2\theta - \tan^2\theta}{\text{cosec}^2 67^\circ - \cot^2 67^\circ} + \sin^2 40^\circ + \cos^2 40^\circ$

$= 1 + 1 = 2.$ **Ans.**

Q. 59. Solve :

$$\frac{2}{3}(\cos^4 30° - \sin^4 45°) - 3(\sin^2 60° - \sec^2 45°) + \frac{1}{4}(\cot^2 30°).$$

Solution : Consider,

$$\frac{2}{3}(\cos^4 30° - \sin^4 45°) - 3(\sin^2 60° - \sec^2 45°) + \frac{1}{4}(\cot^2 30°)$$

$$= \frac{2}{3}\left(\left(\frac{\sqrt{3}}{2}\right)^4 - \left(\frac{1}{\sqrt{2}}\right)^4\right) - 3\left(\left(\frac{\sqrt{3}}{2}\right)^2 - \left(\frac{\sqrt{2}}{1}\right)^2\right) + \frac{1}{4}(\sqrt{3})^2$$

$$= \frac{2}{3}\left(\frac{9}{16} - \frac{1}{4}\right) - 3\left(\frac{3}{4} - 2\right) + \frac{3}{4}$$

$$= \frac{5}{24} + \frac{15}{4} + \frac{3}{4} = \frac{5}{24} + \frac{9}{2} = \frac{5+108}{24} = \frac{113}{24}.$$ **Ans.**

Q. 60. Solve :

$$\frac{\sin 50°}{\cos 40°} + \frac{\text{cosec } 40°}{\sec 50°} - 4 \cos 50° \text{ cosec } 40°.$$

Solution : Consider,

$$\frac{\sin 50°}{\cos 40°} + \frac{\text{cosec } 40°}{\sec 50°} - 4 \cos 50° \text{ cosec } 40°$$

$$= \frac{\sin (90° - 40°)}{\cos 40°} + \frac{\text{cosec } (90° - 50°)}{\sec 50°} - 4 \cos 50° \text{ cosec } (90° - 50°)$$

$$= \frac{\cos 40°}{\cos 40°} + \frac{\sec 50°}{\sec 50°} - 4 \cos 50° \sec 50°$$

$= 1 + 1 - 4 = -2.$ **Ans.**

Q. 61. Prove that :

$$2 \sec^2 \theta - \sec^4 \theta - 2 \text{ cosec}^2 \theta + \text{cosec}^4 \theta = \cot^4 \theta - \tan^4 \theta$$

Solution : Consider,

$2 \sec^2 \theta - \sec^4 \theta - 2 \text{ cosec}^2 \theta + \text{cosec}^4 \theta$

$= \text{cosec}^4 \theta - 2 \text{ cosec}^2 \theta - (\sec^4 \theta - 2 \sec^2 \theta)$

$= \text{cosec}^4 \theta - 2 \text{ cosec}^2 \theta + 1 - (\sec^4 \theta - 2 \sec^2 \theta + 1)$

$= (\text{cosec}^2 \theta - 1)^2 - (\sec^2 \theta - 1)^2$

$= \cot^4 \theta - \tan^4 \theta.$ **Hence Proved.**

Q. 62. Prove that :

$$\frac{\tan \theta + \sec \theta - 1}{\tan \theta - \sec \theta + 1} = \frac{1 + \sin \theta}{\cos \theta}.$$

Solution : Consider, $\frac{\tan \theta + \sec \theta - 1}{\tan \theta - \sec \theta + 1}$

$$= \frac{(\tan \theta + \sec \theta) + (\tan^2 \theta - \sec^2 \theta)}{\tan \theta - \sec \theta + 1}$$

$$= \frac{(\tan \theta + \sec \theta) + (\tan \theta + \sec \theta)(\tan \theta - \sec \theta)}{\tan \theta - \sec \theta + 1}$$

$$= \frac{(\tan \theta + \sec \theta)[1 + (\tan \theta - \sec \theta)]}{\tan \theta - \sec \theta + 1}$$

$$= (\tan \theta + \sec \theta) = \frac{1 + \sin \theta}{\cos \theta}.$$ **Hence Proved.**

Q. 63. Prove that :

$$\frac{\sin^3 \theta + \cos^3 \theta}{\sin \theta + \cos \theta} + \sin \theta \cos \theta = 1.$$

Solution : Consider,

$$\text{L.H.S.} = \frac{\sin^3 \theta + \cos^3 \theta}{\sin \theta + \cos \theta} + \sin \theta \cos \theta$$

$$= \frac{(\sin \theta + \cos \theta)(\sin^2 \theta + \cos^2 \theta - \sin \theta \cos \theta)}{\sin \theta + \cos \theta} + \sin \theta \cos \theta$$

$= 1 - \sin \theta \cos \theta + \sin \theta \cos \theta$

$= 1.$ **Hence Proved.**

Q. 64. Prove that :

$$(1 - \sin \theta + \cos \theta)^2 = 2(1 + \cos \theta)(1 - \sin \theta).$$

Solution : Consider,

$\text{L.H.S.} = (1 - \sin \theta + \cos \theta)^2$

$= [(1 - \sin \theta) + \cos \theta]^2$

$= (1 - \sin \theta)^2 + \cos^2 \theta + 2 \cos \theta (1 - \sin \theta)$

$= 1 + \sin^2 \theta + \cos^2 \theta - 2 \sin \theta + 2 \cos \theta - 2 \sin \theta \cos \theta$

$= 2 - 2 \sin \theta + 2 \cos \theta - 2 \sin \theta \cos \theta$

$= 2(1 - \sin \theta) + 2 \cos \theta (1 - \sin \theta)$

$= 2(1 + \cos \theta)(1 - \sin \theta).$ **Hence Proved.**

Q. 65. If $\sec^2 \theta (1 + \sin \theta)(1 - \sin \theta) = k$, find k.

Solution : Given, $\sec^2 \theta (1 + \sin \theta)(1 - \sin \theta) = k$

$\Rightarrow \quad \sec^2 \theta (1 - \sin^2 \theta) = k$

$\Rightarrow \quad \sec^2 \theta \cos^2 \theta = k$

$\Rightarrow \quad k = 1.$ **Ans.**

Q. 66. Show that

$$\cos^2 \theta - \sin^2 \theta = \frac{2 \tan \theta}{(1 - \tan^2 \theta)}$$ **is not an identity.**

Solution : Taking $\theta = 30°$

$\text{L.H.S.} = \cos^2 \theta - \sin^2 \theta = \cos^2 30° - \sin^2 30°$

$$= \left(\frac{\sqrt{3}}{2}\right)^2 - \left(\frac{1}{2}\right)^2$$

$$= \frac{3}{4} - \frac{1}{4} = \frac{3-1}{4} = \frac{1}{2}$$

$$\text{R.H.S.} = \frac{2 \tan \theta}{1 - \tan^2 \theta}$$

$$= \frac{2 \tan 30}{1 - \tan^2 30}$$

$$= \frac{2\frac{1}{\sqrt{3}}}{1 - \frac{1}{3}} = \frac{\frac{2}{\sqrt{3}}}{\frac{2}{3}}$$

$$= \frac{3}{\sqrt{3}} = \sqrt{3}$$

$\because$ L.H.S. $\neq$ R.H.S.

Hence given expression is not an identity.

Hence Proved.

Chapter 11. Applications of Trigonometry

Q. 1. If the height of a vertical pole is $\sqrt{3}$ times the length of its shadow on the ground, then find the angle of elevation of the Sun at that time.

Solution : Let the length of the shadow be x ft.

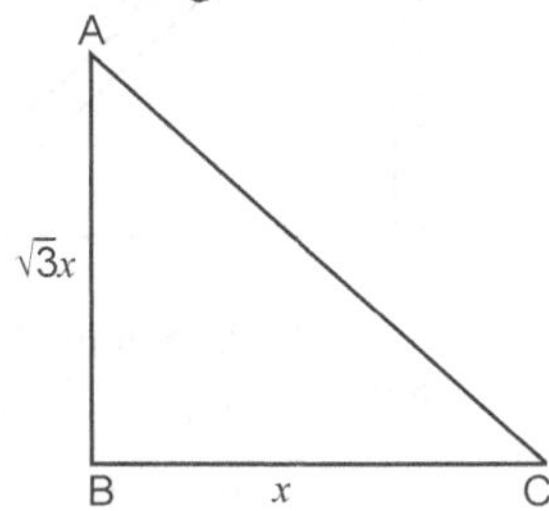

Thus, the height of the vertical pole = $\sqrt{3}x$ ft.

$$\tan\theta = \frac{\text{Perpendicular}}{\text{Base}}$$

$$\tan\theta = \frac{\sqrt{3}x}{x} = \sqrt{3} = \tan 60^\circ$$

or $\theta = 60^\circ$

Thus, the angle of elevation is 60°. **Ans.**

Q. 2. From the top of a cliff 20 m high, the angle of elevation of the top of a tower is found to be equal to the angle of depression of the foot of the tower. Then find height of the tower.

Solution : Let the angle of elevation = angle of depression = x°

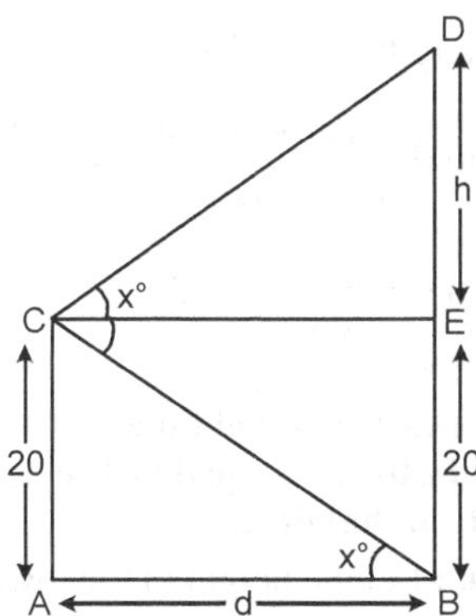

Also, let the distance between the tower (BD) and the cliff (AC) be d m.

Thus, in ΔABC, $\tan x = \dfrac{20}{d}$ ...(i)

Also, in ΔCDE, $\tan x = \dfrac{h}{d}$ ...(ii)

From equations (i) and (ii)

$$\frac{h}{d} = \frac{20}{d}$$

$\Rightarrow$ $h = 20$ m

$\Rightarrow$ Height of the tower is $20 + 20 = 40$ m. **Ans.**

Q. 3. A ladder makes an angle of 60° with the ground when placed against a wall. If the foot of the ladder is 2 m away from the wall, then find the length of the ladder, in metres.

Solution : Given, the distance of the ladder AC from the base of the wall AB is 2 m.

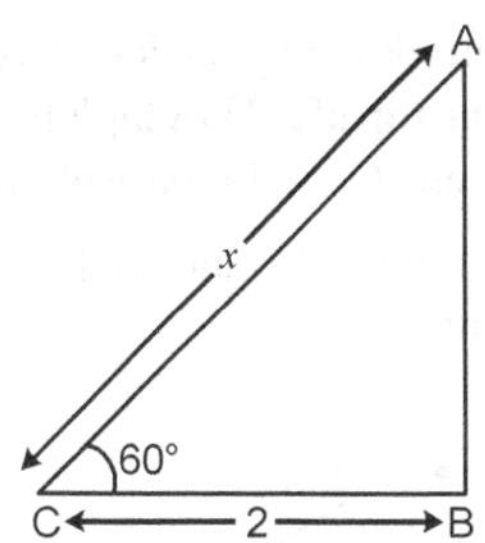

So $\cos 60^\circ = \dfrac{BC}{AC}$

$$\frac{1}{2} = \frac{2}{x}$$

$\Rightarrow$ $x = 2 \times 2$

So, the height of the ladder = 4 m. **Ans.**

Q. 4. A ladder, leaning against a wall, makes an angle of 60° with the horizontal. If the foot of the ladder is 2.5 m away from the wall, find the length of the ladder.

Solution : Let AB be the ladder leaning against a wall AC.

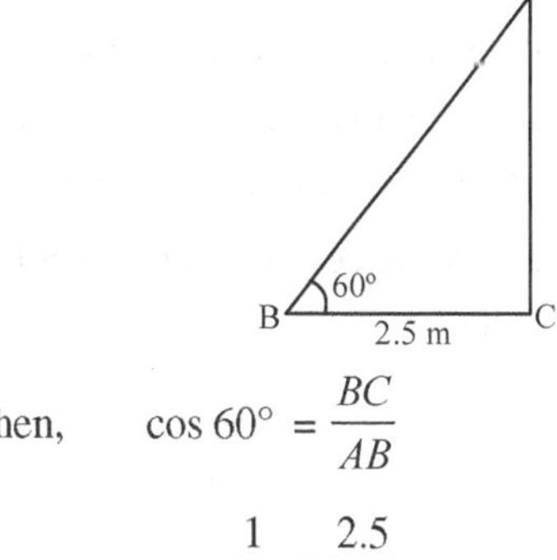

Then, $\cos 60^\circ = \dfrac{BC}{AB}$

$$\frac{1}{2} = \frac{2.5}{AB}$$

$$AB = 2.5 \times 2 = 5 \text{ m}$$

$\therefore$ Length of ladder is 5 m. **Ans.**

Q. 5. If a tower 30 m high, casts a shadow $10\sqrt{3}$ m long on the ground, then what is the angle of elevation of the Sun ?

Solution : In ΔABC,

$$\tan\theta = \frac{AB}{BC}$$

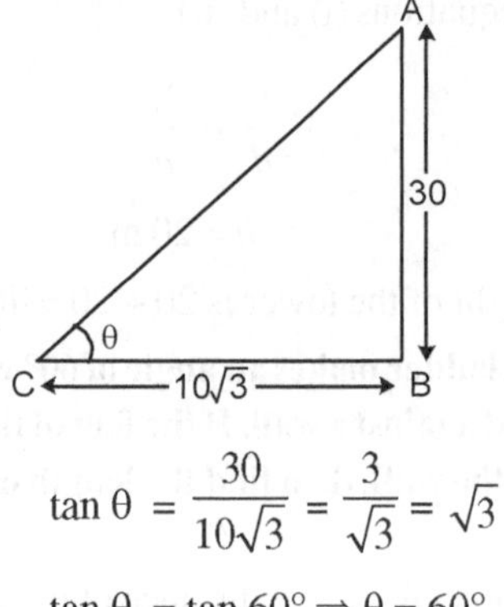

$$\tan\theta = \frac{30}{10\sqrt{3}} = \frac{3}{\sqrt{3}} = \sqrt{3}$$

$$\tan\theta = \tan 60° \Rightarrow \theta = 60°$$

Hence, angle of elevation is 60°. **Ans.**

Q. 6. The angle of depression of a car standing on the ground from the top of a 75 m high tower is 30°. Find the distance of the car from the base of the tower (in m).

Solution : Let the distance of the car from the base of the tower be x m.

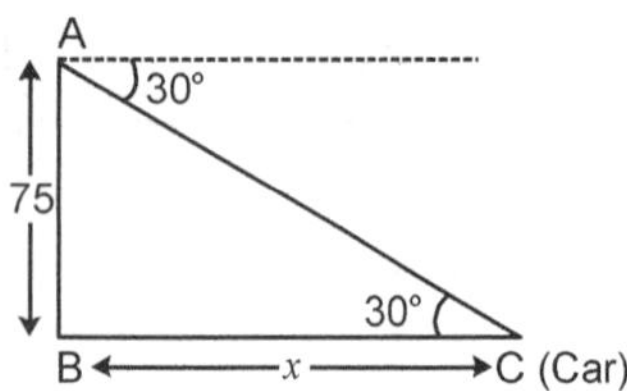

Height of tower = 75 m

$$\tan 30° = \frac{\text{Perpendicular}}{\text{Base}}$$

$$\Rightarrow \quad \frac{1}{\sqrt{3}} = \frac{75}{x}$$

$$\Rightarrow \quad x = 75\sqrt{3}$$

Thus, distance of the car from the base of the tower is $75\sqrt{3}$ m. **Ans.**

Q. 7. The angle of depression of a car parked on the road from the top of a 150 m high tower is 30°. Find the distance of the car from the tower (in metres).

Solution : Let the distance of the parked car from the base of the tower be x m.

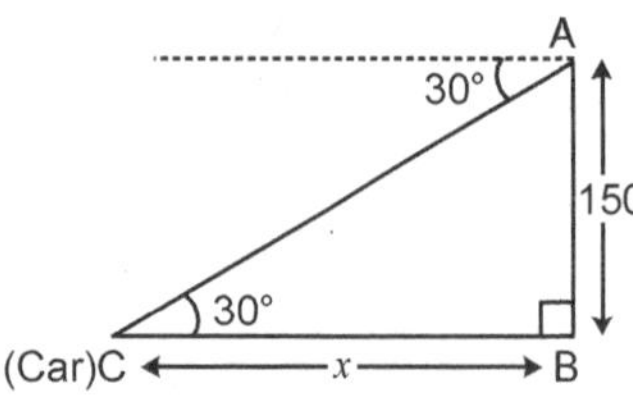

Height of tower = 150 m

$$\tan 30° = \frac{\text{Perpendicular}}{\text{Base}}$$

$$\Rightarrow \quad \frac{1}{\sqrt{3}} = \frac{150}{x}$$

$$\Rightarrow \quad x = 150\sqrt{3}$$

Thus, distance of the parked car from the base of the tower is $150\sqrt{3}$ m. **Ans.**

Q. 8. A kite is flying at a height of 30 m from the ground. The length of the string from the kite to the ground is 60 m. Assuming that there is no slack in the string, then find the angle of elevation of the kite at the ground.

Solution : Given, perpendicular height = 30 m and hypotenuse = 60 m.

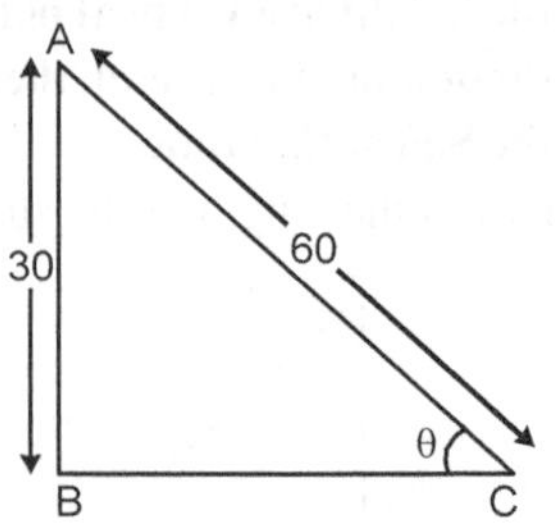

We know, $\sin\theta = \dfrac{\text{Perpendicular}}{\text{Hypotenuse}}$

$$= \frac{30}{60} = \frac{1}{2} = \sin 30°$$

Thus, $\theta = 30°$. **Ans.**

Q. 9. The length of shadow of a tower on the plane ground is $\sqrt{3}$ times the height of the tower. Find the angle of elevation of Sun.

Solution : Let the height of the tower be h m.

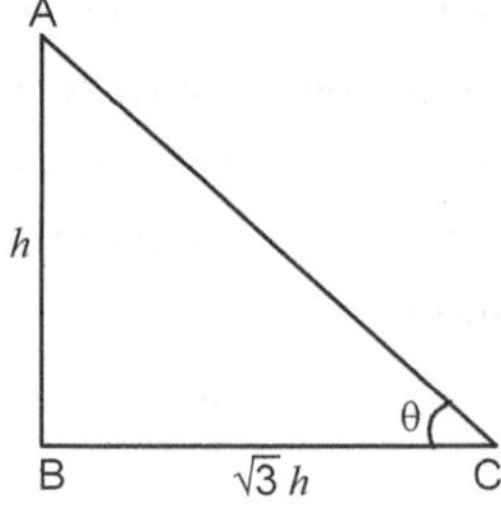

So the length of the shadow will be $\sqrt{3}h$ m.

Thus, $\tan\theta = \dfrac{\text{Perpendicular}}{\text{Base}} = \dfrac{h}{\sqrt{3}h}$

$$\Rightarrow \quad \tan\theta = \frac{1}{\sqrt{3}} = \tan 30°$$

Thus, $\theta = 30°$. **Ans.**

Q. 10. At some point of time in the day, the length of the shadow of a tower is equal to its height. Then find the sun's altitude at that time.

Solution : Let the height of the tower be h m and the length of the shadow be l m.

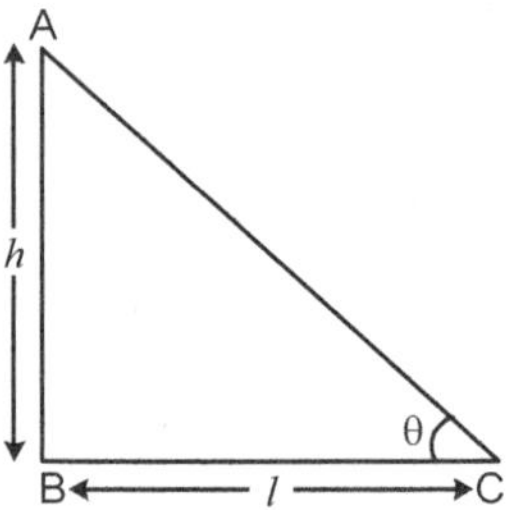

Now $h = l$ (Given)

Thus, $\tan\theta = \dfrac{\text{Perpendicular}}{\text{Base}}$

$\Rightarrow$ $\tan\theta = \dfrac{h}{l} = 1 = \tan 45°$

Hence $\theta = 45°$. **Ans.**

Q. 11. A tower stands vertically on the ground. From a point on the ground which is 25 m away from the foot of the tower, the angle of elevation of the top of the tower is found to be 45°. Then find the height (in metres) of the tower.

Solution : Let the height of the tower be H m.

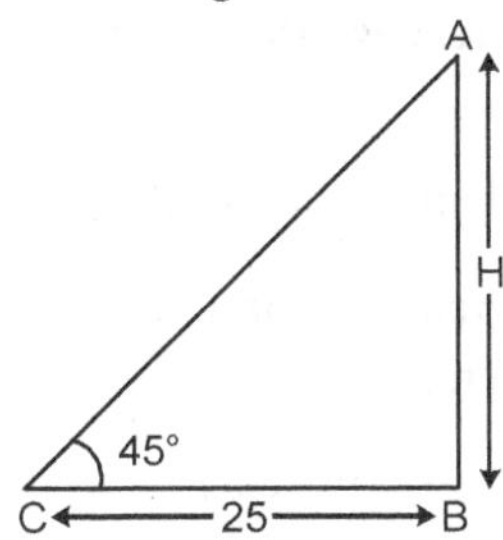

Thus $\tan 45° = \dfrac{\text{Perpendicular}}{\text{Base}}$

or $\dfrac{H}{25} = 1$

Hence $H = 25$ m. **Ans.**

Q. 12. The angle of elevation of the top of a tower from a point on the ground which is 30 m away from the foot of the tower is 45°. Then find height of the tower (in metres).

Solution : Let the height of the tower be H m.

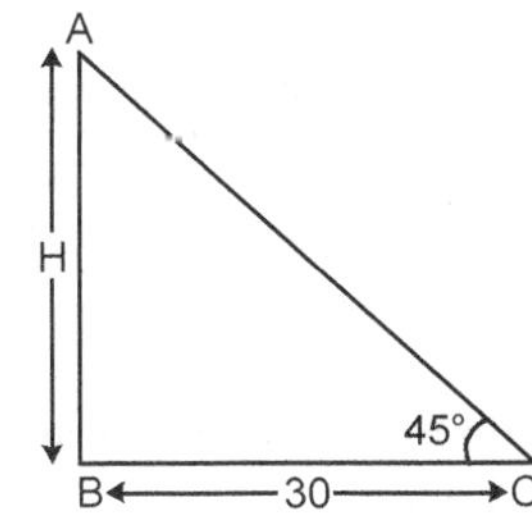

Thus $\tan 45° = \dfrac{\text{Perpendicular}}{\text{Base}}$

or $\dfrac{H}{30} = 1$

Hence $H = 30$ m. **Ans.**

Q. 13. In Fig. AB is a 6 m high pole and CD is a ladder inclined at an angle of 60° to the horizontal and reaches up to a point D of pole. If $AD = 2{\cdot}54$ m, find the length of the ladder. (Use $\sqrt{3} = 1{\cdot}73$)

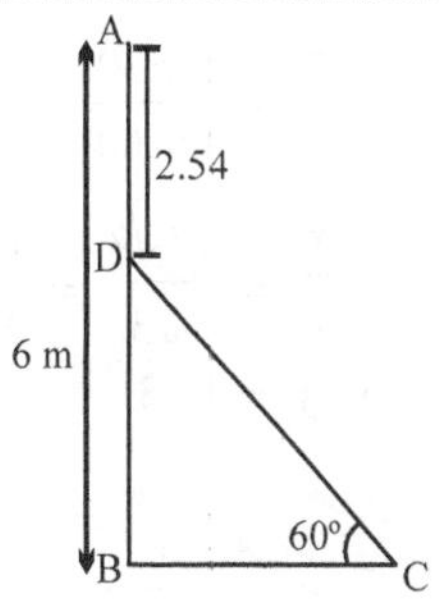

Solution : Given, $AB = 6$ m and $AD = 2.54$ m.

$\therefore DB = (6 - 2.54)\text{ m} = 3{\cdot}46\text{ m}$

In ΔDBC,

$$\sin 60° = \frac{DB}{DC}$$

$$\frac{\sqrt{3}}{2} = \frac{3.46}{DC}$$

$$\Rightarrow \quad DC = \frac{3.46 \times 2}{1.732}$$

$$= 3.995\text{ m} \simeq 4\text{ m}.$$

$\therefore$ The length of the ladder is 4 m **Ans.**

Q. 14. The tops of two towers of height x and y, standing on level ground, subtend angles of 30° and 60° respectively at the centre of the line joining their feet, then find $x : y$.

Solution : Let AB and CD be two towers of height x and y respectively.

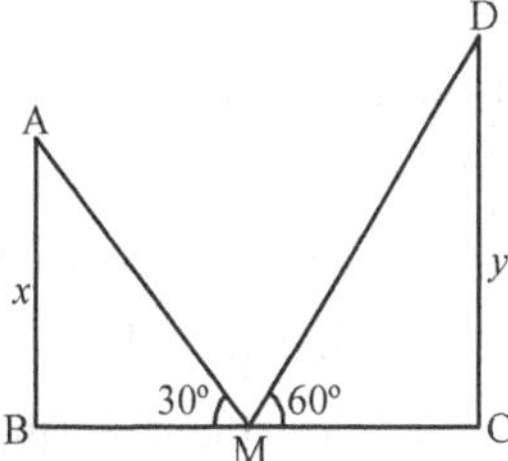

M is the mid-point of BC *i.e.*, $BM = MC$

In ΔABM, we have

$$\frac{AB}{BM} = \tan 30°$$

$$BM = \frac{x}{\tan 30°} \quad \text{...(i)}$$

In ΔCDM, we have

$$\frac{DC}{MC} = \tan 60°$$

$$\frac{y}{MC} = \tan 60°$$

$$MC = \frac{y}{\tan 60°} \quad \text{...(ii)}$$

From equaton (i) and (ii), we get

$$\frac{x}{\tan 30°} = \frac{y}{\tan 60°}$$

$$\frac{x}{y} = \frac{\tan 30°}{\tan 60°}$$

$$\frac{x}{y} = \frac{1/\sqrt{3}}{\sqrt{3}} = \frac{1}{3}$$

$$x : y = 1 : 3.$$ **Ans.**

Chapter 12. Areas related to Circles

Q. 1. The circumference of a circle is 22 cm. Find the area of its quadrant (in cm^2).

Solution : Given, circumference of circle

$$= 2\pi r = 22 \text{ cm}$$

or $$r = \frac{11}{\pi} \text{ cm}$$

$\therefore$ Area of the quadrant

$$= \frac{\pi r^2}{4} = \frac{\pi}{4} \times \frac{11}{\pi} \times \frac{11}{\pi}$$

$$= \frac{11 \times 11 \times 7}{4 \times 22}$$

$$= \frac{77}{8} \text{ cm}^2.$$ **Ans.**

Q. 2. If the area of a circle is equal to sum of the areas of two circles of diameters 10 cm and 24 cm, then find the diameter of the larger circle (in cm).

Solution : Given, the diameters of the two smaller circles = 10 cm and 24 cm.

Thus, radii of the two smaller circles = 5 cm and 12 cm, respectively.

Let the radius of the larger circle be R'.

Now, Area of the larger circle

= Sum of the area of the two smaller circles

$$= \pi(R^2 + r^2)$$

$$= \pi[(12)^2 + (5)^2]$$

$$= \pi[144 + 25]$$

$$= \pi[169]$$

$$= \pi(13)^2$$

$$\Rightarrow \quad \pi(R')^2 = \pi(13)^2$$

$$\Rightarrow \quad (R')^2 = (13)^2$$

$$\Rightarrow \quad R' = 13 \text{ cm}$$

Hence, diameter of the larger circle = 26 cm. **Ans.**

Q. 3. The radii of two circles are 4 cm and 3 cm respectively. Then, find the diameter of the circle having an area equal to the sum of the two circles (in cm).

Solution : Let radii of two circles be r_1 and r_2 respectively.

Then, $r_1 = 4$ cm and $r_2 = 3$ cm [Given]

Let the radius of the larger circle = R

Thus, $$\pi R^2 = \pi[(r_1)^2 + (r_2)^2]$$

$$\Rightarrow \quad R^2 = [(4)^2 + (3)^2]$$

$$\Rightarrow \quad R^2 = [16 + 9]$$

$$\Rightarrow \quad R^2 = 25$$

$$\Rightarrow \quad R = 5 \text{ cm}$$

[as radius cannot be negative]

Thus, the diameter of the larger circle = 10 cm. **Ans.**

Q. 4. If the area of a circle is numerically equal to twice its circumference then find the diameter of the circle.

Solution : Let the radius of the circle be r.

Given, the area of a circle is numerically equal to twice its circumference.

Thus $$\pi r^2 = 2(2\pi r)$$

or $$r = 4 \text{ units}$$

Thus, the diameter of the circle = 8 units. **Ans.**

Q. 5. Find the area of a quadrant of a circle whose circumference is 88 cm.

Solution : Given,

Circumference of circle = 88 cm

$$\Rightarrow \quad 2\pi r = 88 \text{ cm}$$

$$r = \frac{88}{2 \times \frac{22}{7}} = \frac{88 \times 7}{2 \times 22}$$

$$r = 14 \text{ cm}$$

$$\text{Area of quadrant} = \frac{1}{4}\pi r^2$$

$$= \frac{1}{4} \times \frac{22}{7} \times 14 \times 14$$

$$= 154 \text{ cm}^2.$$ **Ans.**

Q. 6. Find the area of annulus whose inner and outer radii are 6 cm and 8 cm.

Solution : Given,

Inner radius of annulus = 6 cm

Outer radius of annulus = 8 cm

$$\text{Area of annulus} = \pi(R^2 - r^2)$$

$$= \pi(8^2 - 6^2)$$

$$= \pi(64 - 36)$$

$$= 28\pi$$

$$= 28 \times \frac{22}{7}$$

$$= 88 \text{ cm}^2$$ **Ans.**

Q. 7. In the given figure, the top of a table in a restaurant is that of a segment of a circle with centre *O* and $\angle BOD = 90°$, $BO = OD = 60$, then find the area of the top of the table. (Use $\pi = 3.14$)

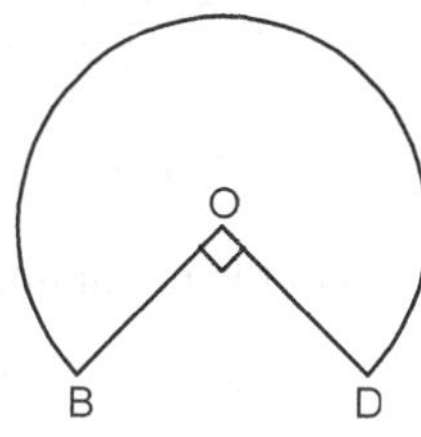

Solution : Area of the top of table

$$= \text{Area of 3 quadrants of circle}$$
$$= \frac{3}{4}\pi r^2$$
$$= \frac{3}{4}\pi(60)^2$$
$$= \frac{3}{4} \times 3.14 \times 3600$$
$$= 8478 \text{ cm}^2.$$ **Ans.**

Q. 8. The perimeter of a sheet of tin in the shape of quadrant of a circle is 12.5 cm then find its area.

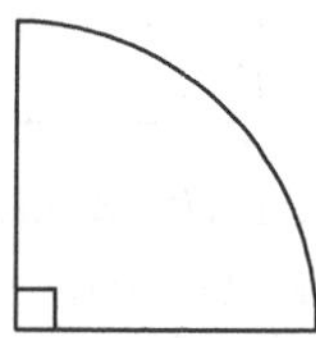

Solution : If r be the radius of the circle.

Then, the perimeter of a quadrant of the circle

$$= \frac{1}{4} \times 2\pi r + 2r$$
$$= \left(\frac{\pi}{2} + 2\right) r = 12.5 \qquad \text{(Given)}$$
$$\Rightarrow \left(\frac{1}{2} \times \frac{22}{7} + 2\right) r = \frac{25}{2}$$
$$\Rightarrow \frac{25r}{7} = \frac{25}{2}$$
$$\Rightarrow r = \frac{7}{2}$$

Area of quadrant $= \frac{1}{4}\pi r^2$

$$= \frac{1}{4} \times \frac{22}{7} \times \left(\frac{7}{2}\right)^2 \text{ cm}^2$$
$$= 9.625 \text{ cm}^2$$ **Ans.**

Q. 9. In the given figure, *OABC* is a quadrant of a circle of radius 7 cm. If *OD* = 4 cm, find the area of the shaded region. $\left(\text{Use } \pi = \frac{22}{7}\right)$

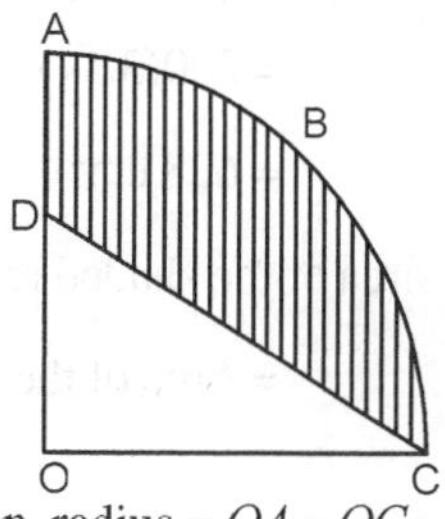

Solution : Given, radius = $OA = OC = 7$ cm and $DO = 4$ cm

Thus, Area of the quadrant *ABCO*

$$= \frac{\pi r^2}{4} = \frac{(7)^2}{4} \times \frac{22}{7}$$
$$= \frac{77}{2} \text{ cm}^2$$

Also, Area of $\Delta COD = \frac{1}{2} \times 7 \times 4 = 14 \text{ cm}^2$

Thus, Area of the shaded region *ABCD*

= Area of the quadrant *ABCO* – Area of ΔCOD

or Area of $ABCD = \frac{77}{2} - 14$

$$= \frac{77 - 28}{2} = \frac{49}{2}$$
$$= 24.5 \text{ cm}^2.$$ **Ans.**

Q. 10. In the given figure, a square *OABC* is inscribed within the quadrant *OPBQ* of a circle with centre *O*. If *OA* = 20 cm, find the area of the shaded region. (Use $\pi = 3.14$)

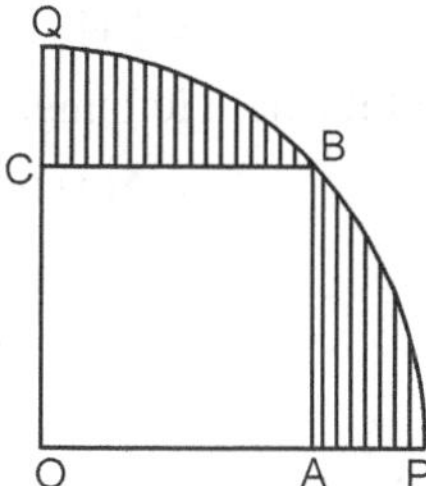

Solution : Given, $OA = 20$ cm and *OABC* is a square.

Thus, *OB* is a diagonal of the square as well as the radius of the quadrant.

So, Area of the square $= 20 \times 20 \text{ cm}^2 = 400 \text{ cm}^2$

Now applying Pythagoras theorem,

$$OB^2 = OA^2 + BA^2$$
$$\Rightarrow OB^2 = (20)^2 + (20)^2$$
$$\Rightarrow OB^2 = 400 + 400$$
$$\Rightarrow OB^2 = 800$$
$$\Rightarrow r^2 = 800$$

Thus, Area of the quadrant

$$= \frac{\pi r^2}{4} = \frac{\pi 800}{4} = 200\pi$$

$= 200(3.14)$

$= 628 \text{ cm}^2$

Hence, the area of the shaded region

= Area of the quadrant

– Area of the square

$= (628 - 400) \text{ cm}^2$

$= 228 \text{ cm}^2$. **Ans.**

Q. 11. A cow is tied with a rope of length 14 m at one corner of a rectangular field of dimensions 20 m × 15 m. Find the area of the field in which the cow cannot graze.

$\left(\textbf{Use } \pi = \frac{22}{7}\right)$

Solution : Area of the field = 20 m × 15 m = 300 m^2

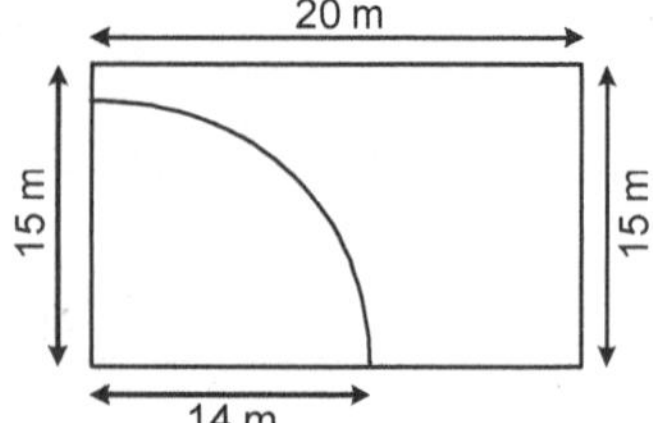

$$\text{Area of the quadrant} = \frac{\pi r^2}{4} = \frac{1}{4} \times \frac{22}{7} \times (14)^2$$

$$= \frac{22 \times 14 \times 14}{7 \times 4}$$

$$= 154 \text{ cm}^2$$

Thus, area of the field left ungrazed

= Area of the rectangle

– Area of the quadrant

$= (300 - 154) \text{ cm}^2 = 146 \text{ cm}^2$.

Ans.

Q. 12. Two circular pieces of equal radii and maximum area touching each other are cut out from a rectangular cardboard of dimensions 14 cm × 7 cm. Find the area of the remaining cardboard. $\left(\textbf{Use } \pi = \frac{22}{7}\right)$

Solution : Given, the dimensions of the cardboard are 14 cm and 7 cm.

Thus, the area of the cardboard = 14 cm × 7 cm

$= 98 \text{ cm}^2$

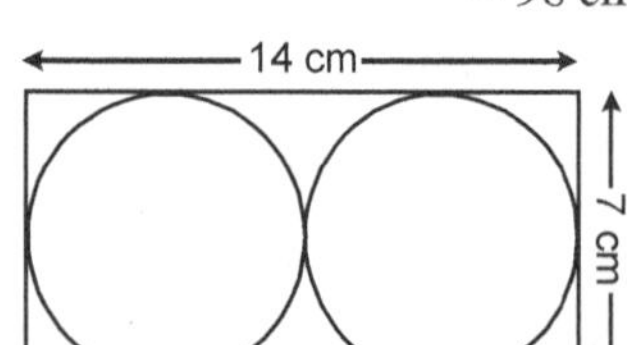

As the maximum area is cut out, so the diameters of the circles are 7 cm each.

Thus, total area of the circles

$= 2\pi r^2$

$$= 2\left(\frac{22}{7}\right)\left(\frac{7}{2}\right)^2$$

$= 77 \text{ cm}^2$

Hence, the area of the remaining board

$= (98 - 77) \text{ cm}^2$

$= 21 \text{ cm}^2$. **Ans.**

Q. 13. In the given figure, a right-angled triangle *ABC* circumscribes a circle of radius *r*. If *AB* and *BC* are of lengths 8 cm and 6 cm respectively, find the value of *r*.

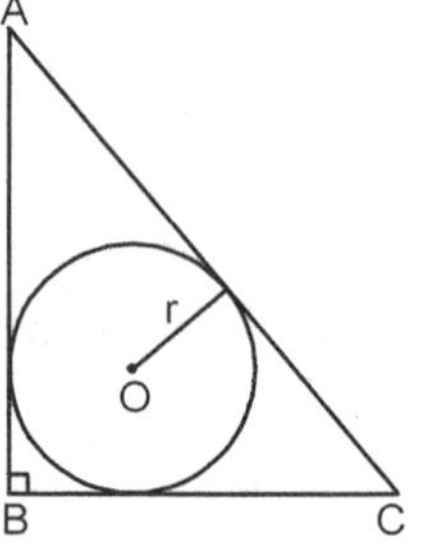

Solution : Given, $AB = 8$ cm and $BC = 6$ cm

Join OC, OB and OA to form ΔOAB, ΔOBC and ΔOCA all having r as the altitudes.

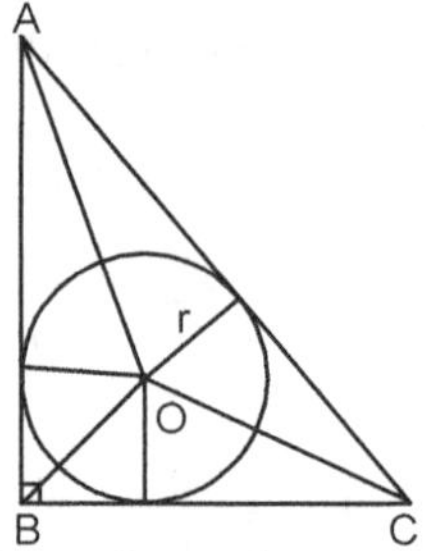

So $\quad AC^2 = AB^2 + BC^2$ [Pythagoras theorem]

$\Rightarrow \quad AC^2 = (8)^2 + (6)^2$

$\Rightarrow \quad AC^2 = 64 + 36 = 100$

$\Rightarrow \quad AC = 10$ cm

[because length cannot be negative]

Now, Area of ΔABC

$= \text{Area of } (\Delta OAB + \Delta OCA + \Delta OBC)$

$$\Rightarrow \quad \frac{1}{2} \times 8 \times 6 = \frac{r}{2}(8 + 10 + 6)$$

$\Rightarrow \quad 48 = r(24)$

$\Rightarrow \quad r = 2$ cm. **Ans.**

Q. 14. In the given figure, *ABCD* is a square of side 4 cm. A quadrant of a circle of radius 1 cm is drawn at each corner of the square and a circle of diameter 2 cm is drawn at the centre. Find the area of the shaded region. (Use $\pi = 3.14$)

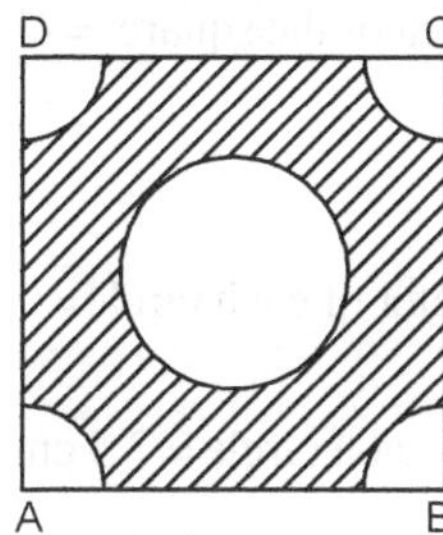

Solution : Area of the square

$= 4 \times 4$
$= 16 \text{ cm}^2$

Area of each quadrant $= \dfrac{\pi \times 1^2}{4}$

$= \dfrac{3.14 \times 1}{4}$

$= 0.785 \text{ cm}^2$

Given, diameter of the circle = 2 cm

Thus, radius = 1 cm

Hence area of the circle $= \pi(1)^2 = 3.14 \text{ cm}^2$

$\therefore$ Area of the shaded region

= Area of the square
– [4 × Area of a quadrant
+ Area of the circle]

$\Rightarrow$ Area of the shaded region

$= 16 - [4 \times (0.785) + 3.14] \text{ cm}^2$
$= 16 - (3.14 + 3.14) \text{ cm}^2$
$= (16 - 6.28) \text{ cm}^2$
$= 9.72 \text{ cm}^2$. **Ans.**

Q. 15. From a rectangular sheet of paper *ABCD* with *AB* = 40 cm and *AD* = 28 cm, a semicircular portion with *BC* as diameter is cut off. Find the area of the remaining paper. $\left(\textbf{Use } \pi = \dfrac{22}{7}\right)$

Solution : Given, $AB = 40$ cm and $AD = 28$ cm

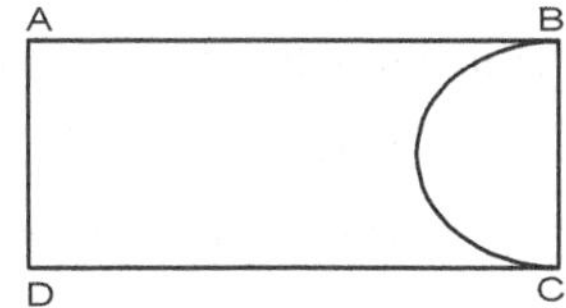

As *BC* is opposite to *AD*, so

$BC = 28$ cm

Now, *BC* is the diameter of the semi-circle.

So the radius of the semi-circle = 14 cm

Hence, Area of the remaining paper

= Area of the rectangle
– Area of the semi-circle

$= \left[(40 \times 28) - \dfrac{1}{2} \times \dfrac{22}{7} \times (14)^2\right] \text{ cm}^2$

$= [1120 - 308] \text{ cm}^2$

$= 812 \text{ cm}^2$. **Ans.**

Q. 16. In the given figure, *OABC* is a square of side 7 cm. If *OAPC* is a quadrant of a circle with centre *O*, then find the area of the shaded region. $\left(\textbf{Use } \pi = \dfrac{22}{7}\right)$

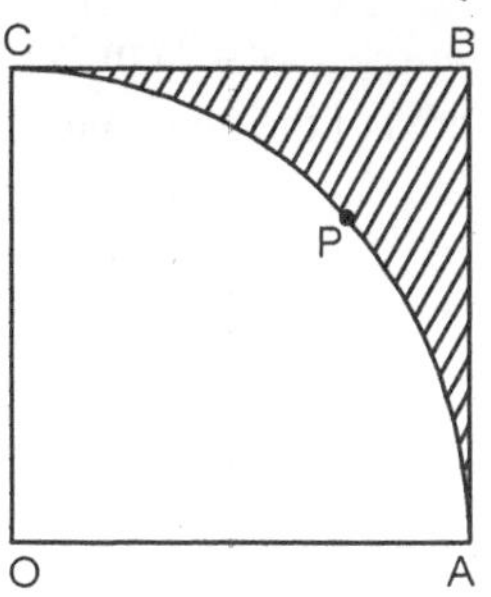

Solution : Given, $OA = AB = BC = CO = 7$ cm

Also $CO = OA$ = Radius of the quadrant

Thus, Area of the shaded region

= Area of square – Area of quadrant

$= (7)^2 - \dfrac{1}{4}\pi(7)^2$

$= 49 - \dfrac{1}{4} \times \dfrac{22}{7} \times (7)^2$

$= 49 - \dfrac{77}{2}$

$= \dfrac{98 - 77}{2} = \dfrac{21}{2}$

$= 10.5 \text{ cm}^2$. **Ans.**

Q. 17. In the given figure, *APB* and *CQD* are semi-circles of diameter 7 cm each, while *ARC* and *BSD* are semi-circles of diameter 14 cm each. Find the perimeter of the figure. $\left(\textbf{Use } \pi = \dfrac{22}{7}\right)$

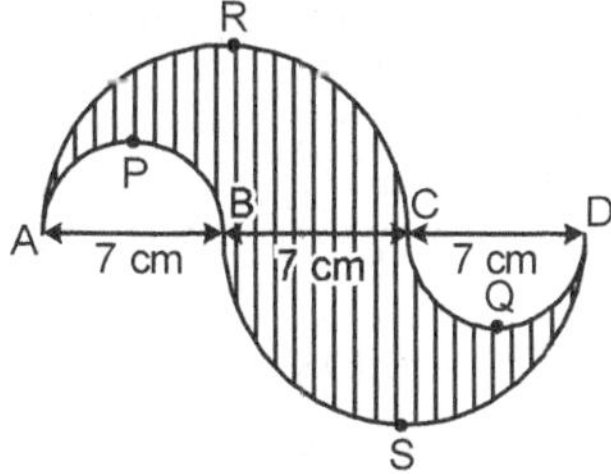

Solution : Given, $d = 7$ cm and $D = 14$ cm

Thus, $r = 3.5$ cm and $R = 7$ cm

Hence, the perimeter of the figure

= Perimeter of $(APB + BSD + DQC + CRA)$

$= \pi(3.5 + 7 + 3.5 + 7)$

$= \dfrac{22}{7}(21)$

$= 66$ cm. **Ans.**

Q. 18. Find the area of a quadrant of a circle where the circumference of the circle is 44 cm. $\left(\textbf{Use } \pi = \dfrac{22}{7}\right)$

Solution : Given,

Circumference = 44 cm = $2\pi r$

Thus, $2\times\frac{22}{7}\times r = 44$

or $r = 7$ cm

Thus, the area of the quadrant

$$= \frac{1}{4}\pi r^2$$

$$= \frac{1}{4}\times\frac{22}{7}(7)^2$$

$$= \frac{77}{2}$$

$= 38.5\text{ cm}^2$. **Ans.**

Q. 19. In the given figure, *OABC* is a quadrant of a circle with centre *O* and radius 3.5 cm. If *OD* = 2 cm, find the area of the shaded region. $\left(\text{Use } \pi = \frac{22}{7}\right)$

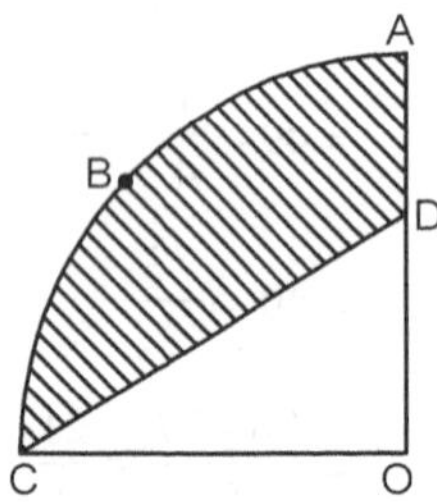

Solution : Given, radius of the circle = $OC = OA = 3.5$ cm and $OD = 2$ cm

Thus, the area of the shaded region

= Area of the quadrant – Area of the ΔODC

$$= \frac{1}{4}\pi r^2 - \frac{1}{2}(b\times h)$$

$$= \frac{1}{4}\times\frac{22}{7}\times(3.5)^2 - \frac{1}{2}(3.5\times 2)\text{ cm}^2$$

$= 6.125\text{ cm}^2$. **Ans.**

Q. 20. In the given figure, *ABCD* is a square of side 14 cm. Find the area of the region not covered by the circles.

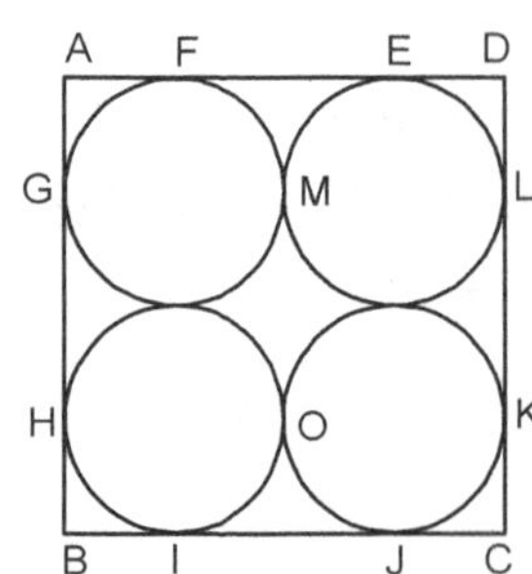

Solution : Given,

$AB = BC = CD = DA = 14$ cm

$\therefore$ Area of the square $= (14\times 14)\text{ cm}^2$

$= 196\text{ cm}^2$

Now, diameter of each circle $= \frac{14}{2} = 7$ cm

$\therefore$ radius of each circle = 3.5 cm

$\therefore$ Total area of all the 4 circles

$$= 4\pi r^2$$

$$= 4\times\frac{22}{7}\times(3.5)^2$$

$= 154\text{ cm}^2$

So, the area of the remaining portion

$= (196 - 154)\text{ cm}^2$

$= 42\text{ cm}^2$. **Ans.**

Q. 21. Find the perimeter of the shaded region, where *ADC*, *AEB* and *BFC* are semi-circles on diameters *AC*, *AB* and *BC* respectively.

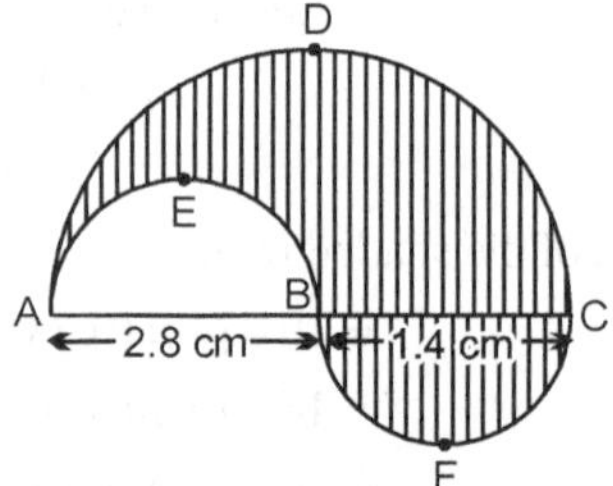

Solution : Given, diameter of *ADC*

$= (2.8 + 1.4)$ cm $= 4.2$ cm

Thus, Radius of *ADC* = 2.1 cm

Diameter of *AEB* = 2.8 cm

Thus, Radius of *AEB* = 1.4 cm

Diameter of *BFC* = 1.4 cm

Thus, Radius of *BFC* = 0.7 cm

Hence, Perimeter of the shaded region

= Perimeter of $(ADC + BFC + AEB)$

$= \pi(2.1 + 0.7 + 1.4)$ cm

$$= \frac{22}{7}(4.2)\text{ cm}$$

= 13.2 cm. **Ans.**

Q. 22. The length of the minute hand of a clock is 14 cm. Find the area swept by the minute hand in 5 minutes.

Solution : Given, length of the minute hand

$= r = 14$ cm

Total number of divisions in the clock = 12

Hence the angle subtended at the centre between two digits covering five minutes $= \frac{360°}{12} = 30°$

Hence, area swept by the minute hand in covering 5 minutes

$$= \frac{\theta}{360°} \pi r^2$$

$$= \frac{30°}{360°} \times \frac{22}{7} \times (14)^2$$

$$= \frac{154}{3}$$

$= 51.33 \text{ cm}^2$. **Ans.**

Q. 23. In a circle of radius 21 cm, an arc subtends an angle of 60° at the centre. Find (i) the length of the arc, and (ii) area of the sector formed by the arc. $\left(\text{Use } \pi = \frac{22}{7}\right)$

Solution : Given, $r = 21$ cm and $\theta = 60°$

(i) Length of an arc $= \frac{\theta}{360°} 2\pi r$

$$= \frac{60°}{360°} \times 2 \times \frac{22}{7} \times 21$$

$= 22$ cm

(ii) Area of the sector $= \frac{\theta}{360°} \times \pi r^2$

$$= \frac{60°}{360°} \times \frac{22}{7} (21)^2$$

$= 231 \text{ cm}^2$. **Ans.**

Q. 24. From each of the two adjacent corners of a square of side 8 cm, a quadrant of a circle of radius 1.4 cm is cut. Another circle of radius 4.2 cm is also cut from the centre as shown in the given figure. Find the area of the remaining portion of the square. $\left(\text{Use } \pi = \frac{22}{7}\right)$

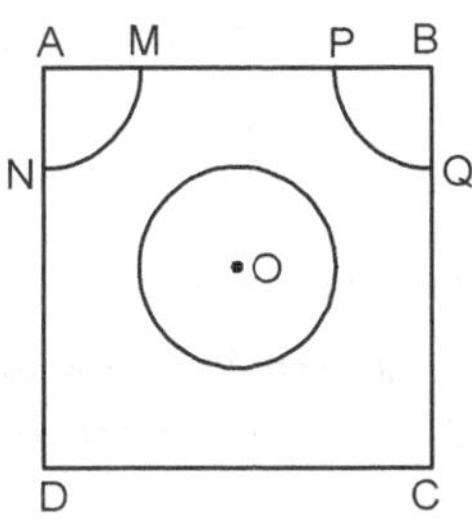

Solution : Given, $AB = BC = CD = DA = 8$ cm, $AM = AN = PB = QB = 1.4$ cm and radius of circle with centre $O = 4.2$ cm

Thus, Area of the remaining portion

= Area of $ABCD$

– [Area of circle + 2 Area of quadrants]

$$= 8 \times 8 - \left\{\frac{22}{7}(4.2)^2 + 2 \times \frac{1}{4} \times \frac{22}{7}(1.4)^2\right\}$$

$= 64 - (55.44 + 3.08) \text{ cm}^2$

$= 5.48 \text{ cm}^2$. **Ans.**

Chapter 13. Surface Areas and Volumes

Q. 1. A rectangular sheet of paper 40 cm × 22 cm is rolled to form a hollow cylinder of height 40 cm. Then find the radius of the cylinder (in cm).

Solution : Given, Area of the sheet of paper

$= (40 \times 22) \text{ cm}^2$

$= 880 \text{ cm}^2$

Height of the cylinder = 40 cm

Let r be the radius of the cylinder

Now, Curved Surface Area of the cylinder $= 2\pi rh$

Thus, $2\left(\frac{22}{7}\right)40r = 40 \times 22$

$\Rightarrow \quad 2\left(\frac{1}{7}\right)r = 1 \times 1$

$\Rightarrow \quad r = \frac{7}{2}$ cm

$= 3.5$ cm. **Ans.**

Q. 2. Find the number of solid spheres each of diameter 6 cm that can be made by melting a solid metal cylinder of height 45 cm and diameter 4 cm.

Solution : Given, height of cylinder = 45 cm

Diameter of cylinder = 4 cm

Diameter of each sphere = 6 cm

Thus, radius of cylinder = 2 cm

And, radius of each sphere= 3 cm

$\therefore$ Number of spheres $= \frac{\text{Volume of cylinder}}{\text{Volume of each sphere}}$

$$= \frac{\pi(2)^2 45}{\frac{4}{3}\pi(3)^3}$$

$= \frac{45}{(3)^2} = 5$. **Ans.**

Q. 3. The surface area of two spheres are in the ratio 16 : 9. Then find ratios of their volumes.

Solution : Let the radius of the larger sphere be R and the smaller be r.

Given, $4\pi R^2 : 4\pi r^2 = 16 : 9$

$\Rightarrow \quad R^2 : r^2 = 4^2 : 3^2$

$\Rightarrow \quad R : r = 4 : 3$

Thus $\frac{4\pi}{3}R^3 : \frac{4\pi}{3}r^3 = 4^3 : 3^3$

or $\quad \frac{4\pi}{3}R^3 : \frac{4\pi}{3}r^3 = 64 : 27$

Then the ratio of their volume is 64 : 27. **Ans.**

Q. 4. A solid right-circular cone is cut into two parts at the middle of its height by a plane parallel to its base. Find the ratio of the volume of the smaller cone to the whole cone.

Solution : Let the height of the bigger cone be $2x$ cm.

Thus, height of the smaller cone = x cm

Also, let the radius of the larger cone be $2r$ cm.

Thus, radius of the smaller cone = r cm [mid-point theorem]

Now, volume of the large cone

$$= \frac{1}{3}\pi(2r)^2(2x)$$

$$= \frac{8\pi}{3}r^2x$$

and volume of the smaller cone

$$= \frac{1}{3}\pi(r)^2(x)$$

$$= \frac{\pi}{3}r^2x$$

Thus, $\frac{\text{Volume of the smaller cone}}{\text{Volume of the larger cone}}$

$$= \frac{\frac{\pi}{3}r^2x}{\frac{8\pi}{3}r^2x} = \frac{1}{8}$$

Hence, the required ratio is 1 : 8. **Ans.**

Q. 5. If the radius of the base of a right-circular cylinder is halved keeping the height same then find the ratio of the volume of the cylinder thus obtained to the volume of the original cylinder.

Solution : Let the original radius of the base be $2r$ cm and the height be h.

Thus, the new radius of the base = r cm

Now $\quad$ Original volume = $\pi(2r)^2h$

New volume = $\pi(r)^2h$

Thus, $\quad \frac{\text{New volume}}{\text{Original volume}} = \frac{\pi r^2 h}{\pi 4 r^2 h}$

$$= \frac{1}{4}$$

Hence, the required ratio is 1 : 4. **Ans.**

Q. 6. Find the radius of the largest circular cone that can be cut out from a cube of edge 4.2 cm.

Solution : Given, the edge of cube = 4.2 cm

Now the largest circular cone will have diameter equal to the measure of the edge.

Thus $\quad$ diameter = 4.2 cm

Hence $\quad$ radius = 2.1 cm. **Ans.**

Q. 7. A sphere of diameter 18 cm is dropped into a cylindrical vessel of diameter 36 cm partly filled with water. If the sphere is completely submerged then by what amount the water level rises (in cm) ?

Solution : Given,

Diameter of sphere = 18 cm

Diameter of cylinder = 36 cm

Thus, radius of sphere= 9 cm

and radius of cylinder = 18 cm

Let the increase in height of water in the cylinder = h cm

Thus, volume of sphere = $\frac{4}{3}\pi(9)^3$

$$= \frac{4}{3}\pi \times 9 \times 9 \times 9 \text{ cm}^3$$

$$= (81)(12)\pi \text{ cm}^3$$

Since, increase in volume of water level in the cylinder = Volume of sphere.

Hence $\quad \pi(18)^2h = (81)(12)\pi$

$\Rightarrow \quad 4h = 12$

$\Rightarrow \quad h = 3$ cm. **Ans.**

Q. 8. A solid is hemispherical at the bottom and conical (of same radius) above it. If the surface area of the two are equal then find the ratio of the radius and the slant height of the conical part.

Solution : Given, Radius of hemisphere

= Radius of cone = r

Also, curved surface area of hemisphere

= Curved surface area of cone

$\Rightarrow \quad 2\pi r^2 = \pi r l$ [l = slant height of the cone]

$\Rightarrow \quad 2r = l$

$\Rightarrow \quad r : l = 1 : 2$

So, the required ratio is 1 : 2. **Ans.**

Q. 9. Volume and surface area of a solid hemisphere are numerically equal. What is the diameter of hemisphere ?

Solution : Let radius of hemisphere be r units.

Volume of hemisphere = S.A. of hemisphere (Given)

$$\frac{2}{3}\pi r^3 = 3\pi r^2$$

$\Rightarrow \quad r = \frac{9}{2}$ or diameter = 9 units

Ans.

Q. 10. If the total surface area of a solid hemisphere is 462 cm², find its volume. $\left(\text{Use } \pi = \frac{22}{7}\right)$

Solution : Given, total surface area of a solid hemisphere = 462 cm²

Let its radius be r.

Thus $3\pi r^2 = 462$

$\Rightarrow \left(\frac{22}{7}\right) r^2 = 154$

$\Rightarrow \left(\frac{1}{7}\right) r^2 = 7$

$\Rightarrow r^2 = (7)^2$

$\Rightarrow r = 7$ cm

So, volume of the hemisphere

$$= \frac{2}{3}\pi r^3$$

$$= \frac{2}{3}\left(\frac{22}{7}\right)(7)^3 \text{ cm}^3$$

$$= \frac{2}{3}\left(\frac{22}{1}\right)(7)^2 \text{ cm}^3$$

$$= \frac{44}{3}(49) \text{ cm}^3$$

$= 718.67$ cm³. **Ans.**

Q. 11. Two circular pieces of equal radii and maximum area touching each other are cut from a rectangular cardboard of dimensions 28 cm × 14 cm. Find the area of the remaining cardboard.

Solution : Given, dimensions of cardboard

= 28 cm × 14 cm

= 392 cm²

As the radii of the circles are equal and they have maximum area, their diameters are equal to the width of the cardboard.

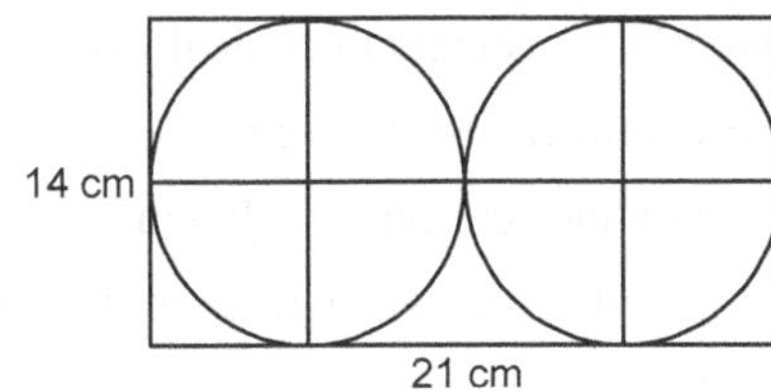

Thus $d = 14$ cm

or $r = 7$ cm

∴ Area of the two circles

$$= 2\pi r^2$$

$$= 2\left(\frac{22}{7}\right)(7)^2 \text{ cm}^2$$

$= 2 \times 22 \times 7$ cm²

= 308 cm²

Thus, area of the remaining portion of the cardboard

= Area of the cardboard – Area of the two circles

= 392 – 308 cm²

= 84 cm². **Ans.**

Q. 12. A solid sphere of radius 10.5 cm is melted and recast into smaller solid cones, each of radius 3.5 cm and height 3 cm. Find the number of cones so formed. $\left(\text{Use } \pi = \frac{22}{7}\right)$

Solution : Given, radius of sphere = 10.5 cm

Radius of small 1 cone = 3.5 cm

Height of 1 small cone = 3 cm

Now, volume of sphere =

$$= \frac{4}{3}\pi(10.5)^3 \text{ cm}^3$$

$$= \frac{4}{3}\left(\frac{22}{7}\right)(10.5)^3 \text{ cm}^3$$

$= 4851$ cm³

Volume of 1 cone $= \frac{1}{3}\pi(3.5)^2 \times 3 \text{ cm}^3$

$$= \left(\frac{22}{7}\right)(3.5)^2 \text{ cm}^3$$

$$= \frac{77}{2} \text{ cm}^3$$

Thus, number of cones $= \dfrac{\text{Volume of sphere}}{\text{Volume of 1 cone}}$

$$= \frac{4851}{\frac{77}{2}}$$

$$= \frac{4851 \times 2}{77}$$

$$= \frac{63 \times 2}{1}$$

= 126. **Ans.**

Q. 13. The volume of a hemisphere is 2425.5 cm³. Find its curved surface area. $\left(\text{Use } \pi = \frac{22}{7}\right)$

Solution : Given, volume = 2425.5 cm³

Let the radius of hemisphere be r.

Thus, $\frac{2}{3}\pi r^3 = 2425.5$

$$\frac{2}{3} \times \frac{22}{7} \times r^3 = \frac{24255}{10}$$

$\Rightarrow \frac{44}{21} r^3 = \frac{4851}{2}$

$\Rightarrow \frac{4}{21} r^3 = \frac{441}{2}$

$\Rightarrow 8r^3 = 441 \times 21$

$$\Rightarrow \quad 8r^3 = 21^3$$

$$\Rightarrow \quad (2r)^3 = (21)^3$$

$$\Rightarrow \quad r = \frac{21}{2} \text{ cm}$$

Thus, the curved surface area

$$= 2\pi\left(\frac{21}{2}\right)^2 \text{ cm}^2$$

$$= 2\left(\frac{22}{7}\right)\left(\frac{21}{2}\right)\left(\frac{21}{2}\right) \text{ cm}^2$$

$$= (33)(21) \text{ cm}^2$$

$$= 693 \text{ cm}^2.$$ **Ans.**

Q. 14. Two cubes, each of volume 27 cm^3 are joined end to end to form a solid. Find the surface area of the resulting cuboid.

Solution : Given, volume of the each cube

$$= 27 \text{ cm}^3$$

$$\Rightarrow \quad a^3 = 27$$

$$\Rightarrow \quad a^3 = (3)^3$$

$$\Rightarrow \quad a = 3 \text{ cm}$$

Thus, length of the larger solid

$$= (3 + 3) \text{ cm} = 6 \text{ cm}$$

Width of the larger solid = 3 cm

Height of the larger solid = 3 cm

Thus, surface area of the resultant cuboid

$$= 2(6 \times 3 + 3 \times 3 + 6 \times 3) \text{ cm}^2$$

$$= 2(18 + 9 + 18) \text{ cm}^2$$

$$= 2(45) \text{ cm}^2$$

$$= 90 \text{ cm}^2.$$ **Ans.**

Q. 15. A cone of height 20 cm and radius of base 5 cm is made up of modelling clay. A child reshapes it in the form of a sphere. Find the diameter of the sphere.

Solution : Given, height of cone = 20 cm

Radius of base of cone = 5 cm

Thus, Volume of cone $= \frac{1}{3}\pi(5)^2\, 20 \text{ cm}^3$

Let the radius of the sphere be r.

Now, the volume of the cone

= Volume of the sphere

Thus $\quad \frac{1}{3}\pi(5)^2\, 20 = \frac{4}{3}\pi(r)^3$

$$\Rightarrow \quad (5)^2\, 20 = 4(r)^3$$

$$\Rightarrow \quad (r)^3 = (5)^2\, 5$$

$$\Rightarrow \quad r = 5 \text{ cm}$$

Thus, diameter of the sphere

$$= 2 \times 5 \text{ cm} = 10 \text{ cm}.$$ **Ans.**

Q. 16. Two cubes each of side 4 cm are joined end to end to form a solid. Find the surface area of the resulting cuboid.

Solution : Given, side of each cube (a) = 4 cm

Thus, length of the larger solid

$$= (4 + 4) \text{ cm} = 8 \text{ cm}$$

Width of the larger solid = 4 cm

Height of the larger solid = 4 cm

Thus, surface area of the resultant cuboid

$$= 2(4 \times 8 + 4 \times 4 + 4 \times 8) \text{ cm}^2$$

$$= 2(32 + 16 + 32) \text{ cm}^2$$

$$= 2 \times 80 \text{ cm}^2$$

$$= 160 \text{ cm}^2.$$ **Ans.**

Q. 17. The dimensions of a metallic cuboid are 100 cm × 80 cm × 64 cm. It is melted and recast into a cube. Find the surface area of the cube.

Solution : Given, the dimensions of a cuboid

$$= 100 \text{ cm} \times 80 \text{ cm} \times 64 \text{ cm}$$

Let the side of a cube be a cm.

Thus $\quad a^3 = 100 \times 80 \times 64 \text{ cm}^3$

$$\Rightarrow \quad a^3 = 1000 \times 8 \times 64 \text{ cm}^3$$

$$\Rightarrow \quad a^3 = 10^3 \times 8^3 \text{ cm}^3$$

$$\Rightarrow \quad a = 10 \times 8 \text{ cm}$$

$$= 80 \text{ cm}$$

$\therefore$ Total surface area of cube $= 6a^2$

$$= 6 \times (80)^2 \text{ cm}^2$$

$$= 38400 \text{ cm}^2.$$ **Ans.**

Q. 18. A conical vessel, whose internal radius is 5 cm and height 24 cm is full of water. The water is emptied into a cylindrical vessel with internal radius 10 cm. Find the height to which the water rises in the cylindrical vessel.

Solution : Given, height of conical vessel = 24 cm

Radius of conical vessel = 5 cm

Radius of cylindrical vessel = 10 cm

Let the height of the water-level in the cylindrical vessel be h cm.

Thus, Volume of water in the cylindrical vessel

= Volume of water in the conical vessel

$$\Rightarrow \quad \pi(10)^2 h = \frac{1}{3}\pi(5)^2\, 24$$

$$\Rightarrow \quad (10)^2 h = (5)^2 8$$

$$\Rightarrow \quad 100h = 25(8)$$

$$\Rightarrow \quad 4h = 8$$

$$\Rightarrow \quad h = 2 \text{ cm}$$

Thus, the water will rise to a height of 2 cm in the cylindrical vessel. **Ans.**

Q. 19. A hemispherical bowl of internal radius 9 cm is full of liquid. The liquid is filled into small cylindrical bottles, each of diameter 3 cm and height 4 cm. How many bottles are needed to empty the bowl ?

Solution : Given, internal radius of hemispherical bowl = 9 cm

Diameter of cylindrical bottles = 3 cm

Thus, Radius of cylindrical bottles = 1.5 cm

Height of cylindrical bottles = 4 cm

Now, number of bottles

$$= \frac{\text{Volume of hemispherical bowl}}{\text{Volume of cylindrical bottle}}$$

$$= \frac{\frac{2}{3}\pi(9)^3}{\pi(1.5)^2 4}$$

$$= \frac{729}{6 \times 2.25}$$

$$= \frac{243}{2 \times 2.25}$$

$= 54.$ **Ans.**

Q. 20. A solid metal cone with radius of base 12 cm and height 24 cm is melted to form solid spherical balls of diameter 6 cm each. Find the number of balls so formed.

Solution : Given, Diameter of spherical balls = 6 cm

Thus, Radius of spherical balls = 3 cm

$$\therefore \text{ Number of balls} = \frac{\text{Volume of cone}}{\text{Volume of one ball}}$$

$$= \frac{\frac{1}{3}\pi(12)^2 24}{\frac{4}{3}\pi(3)^3}$$

$$= \frac{(12)^2 6}{(3)^3} = 32.$$ **Ans.**

Q. 21. The internal and external diameters of a hollow hemispherical shell are 6 cm and 10 cm respectively. It is melted and recast into a solid cone of base diameter 14 cm. Find the height of the cone so formed.

Solution : Given, internal diameter of hemisphere = 6 cm

External diameter of hemisphere = 10 cm

Base diameter of cone = 14 cm

Thus, internal radius of hemisphere = 3 cm

External radius of hemisphere = 5 cm

Base radius of cone = 7 cm

Let the height of the cone be h cm

Thus, Volume of cone = Volume of hemispherical shell

$$\Rightarrow \quad \frac{1}{3}\pi(7)^2 h = \frac{2}{3}\pi[(5)^3 - (3)^3]$$

$$\Rightarrow \quad (7)^2 h = 2[(5)^3 - (3)^3]$$

$$\Rightarrow \quad 49h = 2[125 - 27]$$

$$\Rightarrow \quad 49h = 2 \times 98$$

$$\Rightarrow \quad h = 4 \text{ cm.}$$ **Ans.**

Q. 22. A hemispherical bowl of internal diameter 30 cm is full of liquid. The liquid is filled into small cylindrical bottles each of diameter 5 cm and height 6 cm. How many bottles are needed to empty the bowl ?

Solution : Given, internal diameter of hemispherical bowl = 30 cm

Thus, Radius of hemispherical bowl = 15 cm

Diameter of cylindrical bottles = 5 cm

Thus, Radius of cylindrical bottles = 2.5 cm

Height of cylindrical bottles = 6 cm

Now, number of bottles

$$= \frac{\text{Volume of hemispherical bowl}}{\text{Volume of cylindrical bottle}}$$

$$= \frac{\frac{2}{3}\pi(15)^3}{\pi(2.5)^2 6}$$

$$= \frac{225 \times 15}{9 \times 6.25}$$

$$= \frac{25 \times 15}{6.25} = 60.$$ **Ans.**

Q. 23. A solid metallic sphere of diameter 21 cm is melted and recast into a number of smaller cones each of diameter 3.5 cm and height 3 cm. Find the number of cones so formed.

Solution : Given,

Diameter of sphere = 21 cm

Thus, Radius of sphere = 10.5 cm

Diameter of cone = 3.5 cm

Thus, Radius of cone = 1.75 cm

Height of cone = 3 cm

$$\text{Now, number of cones} = \frac{\text{Volume of sphere}}{\text{Volume of cone}}$$

$$= \frac{\frac{4}{3}\pi(10.5)^3}{\frac{1}{3}\pi 3(1.75)^2}$$

$$= \frac{4 \times (10.5)^3}{3 \times (1.75)^2}$$

$$= \frac{4 \times 6 \times 6 \times 10.5}{3 \times 1}$$

$= 4 \times 36 \times 3.5$

$= 14 \times 36$

$= 504.$ **Ans.**

Q. 24. A spherical cannonball of diameter 28 cm is melted and recast into a right circular conical mould with base diameter of 35 cm. Find the height of the cone.

Solution : Given, diameter of spherical ball = 28 cm

Thus, radius of the spherical ball = 14 cm

Diameter of cone = 35 cm

Thus, radius of cone = 17.5 cm

Let height of cone = h cm

$\therefore$ Volume of cone = Volume of sphere

$$\Rightarrow \frac{1}{3}\pi h(17.5)^2 = \frac{4}{3}\pi(14)^3$$

$$\Rightarrow h(17.5)^2 = 4(14)^3$$

$$\Rightarrow \frac{h \times 175 \times 175}{100} = 4 \times 14 \times 14 \times 14$$

$$\Rightarrow \frac{h \times 1 \times 25}{4} = 4 \times 2 \times 2 \times 14$$

$$\Rightarrow h = 35.84 \text{ cm.}$$ **Ans.**

Q. 25. Three solid metallic spheres of radii 3 cm, 4 cm and 5 cm respectively are melted to form a single solid sphere. Find the diameter of the resulting sphere.

Solution : Given, radii of the smaller spheres = 3 cm, 4 cm and 5 cm

Let the radius of the larger sphere be r cm.

Thus, volume of the larger sphere = Volume of the three smaller spheres taken together

$$\Rightarrow \frac{4}{3}\pi(r)^3 = \frac{4}{3}\pi(3)^3 + \frac{4}{3}\pi(4)^3 + \frac{4}{3}\pi(5)^3$$

$$\Rightarrow \frac{4}{3}\pi(r)^3 = \frac{4}{3}\pi[(3)^3 + (4)^3 + (5)^3]$$

$$\Rightarrow (r)^3 = [27 + 64 + 125]$$

$$\Rightarrow (r)^3 = (216)$$

$$\Rightarrow (r)^3 = (6)^3$$

$$\Rightarrow r = 6 \text{ cm}$$

Thus diameter of the largest sphere = 12 cm. **Ans.**

Q. 26. If the radii of the circular ends of a conical bucket of height 45 cm be 28 cm and 7 cm respectively, find the capacity of the bucket. $\left(\textbf{Use } \pi = \frac{22}{7}\right)$

Solution : Given, height of the bucket, $h = 45$ cm

Radius of the smaller end, $r = 7$ cm

Radius of the larger end, $R = 28$ cm

Thus, volume of the bucket (frustum)

$$= \frac{\pi}{3}(R^2 + r^2 + Rr)h$$

$$= \frac{\pi}{3}[(28)^2 + (7)^2 + (28)(7)]45$$

$$= \frac{1}{3} \times \frac{22}{7}[784 + 49 + 196]45$$

$$= \frac{22}{21}[1029]45$$

$$= 22 \times 49 \times 45$$

$$= 48510 \text{ cm}^3.$$ **Ans.**

Q. 27. If the diameters of the circular ends of a conical bucket of height 15 cm be 56 cm and 42 cm respectively, find the capacity of the bucket. $\left(\textbf{Use } \pi = \frac{22}{7}\right)$

Solution : Given, height of the bucket, $h = 15$ cm

Diameter of the smaller end = 42 cm

Diameter of the larger end = 56 cm

Thus, radius of the smaller end, $r = 21$ cm

and radius of the larger end, $R = 28$ cm

Thus, volume of the bucket (frustum)

$$= \frac{\pi}{3}(R^2 + r^2 + Rr)h$$

$$= \frac{\pi}{3}[(28)^2 + (21)^2 + (28)(21)]15$$

$$= \frac{1}{3} \times \frac{22}{7}[784 + 441 + 588]15$$

$$= \frac{22}{7}[1813]5$$

$$= 28490 \text{ cm}^3$$

Now, $1\, l = 1000 \text{ cm}^3$

Thus, Volume of bucket $= \frac{28490}{1000}\, l$

$= 28.49\, l.$ **Ans.**

Q. 28. A cylindrical pipe has inner diameter of 7 cm and water flows through it at 192.51 litres per minute. Find the rate of flow in the pipe in km/hr.

Solution : Given, inner diameter of cylindrical pipe = 7 cm = 0.07 m.

Thus, inner radius of pipe = 3.5 cm = 0.035 m

Let the length of the column of the water flowing through the pipe be h m.

Since, rate of flow of water = 192.51 litre per minute

So, in 1 minute, volume of water that flows = 192.5 l

Thus in 1 hour (60 minutes), the volume of water that flows

$= 192.5(60) = 11550\ l$

$= 11.55\ m^3$

Now, $\pi(0.035)^2 h = 11.55$

or $h = 3000$ m

Thus, in 1 hour, 3000 m of water flows through the pipe.

Hence, the rate of flow of water = 3000 m/hr

= 3 km/hr. **Ans.**

Q. 29. The internal and external diameters of a hollow spherical shell are 6 cm and 10 cm respectively. It is melted and recast into a solid cylinder of base diameter 14 cm. Find the height of the cylinder so formed.

Solution : Given, internal diameter of sphere = 6 cm

External diameter of sphere = 10 cm

Base diameter of cylinder = 14 cm

Thus, internal radius of sphere, $r_1 = 3$ cm

External radius of sphere, $r_2 = 5$ cm

Base radius of cylinder, $r = 7$ cm

Let the height of the cylinder be h cm.

Thus, volume of cylinder = Volume of spherical shell

$$\pi r^2 h = \frac{4}{3}\pi(r_2{}^3 - r_1{}^3)$$

$$\Rightarrow \quad \pi(7)^2 h = \frac{4}{3}\pi[(5)^3 - (3)^3]$$

$$\Rightarrow \quad 3(7)^2 h = 4[(5)^3 - (3)^3]$$

$$\Rightarrow \quad 3(49)h = 4[125 - 27]$$

$$\Rightarrow \quad 3(49)h = 4[98]$$

$$\Rightarrow \quad 3h = 4[2]$$

$$\Rightarrow \quad h = \frac{8}{3} \text{ cm.}$$ **Ans.**

Q. 30. The surface area of a sphere is 616 cm^2. Find its radius.

Solution : Given, surface area of a sphere = 616 cm^2

Let the radius of the sphere be r cm.

Thus, $4\pi r^2 = 616$

$\Rightarrow \quad \pi r^2 = 154$

$\Rightarrow \quad \frac{1}{7} r^2 = 7$

$\Rightarrow \quad (r)^2 = (7)^2$

$\Rightarrow \quad r = 7$ cm. **Ans.**

Q. 31. A cylinder and a cone are of the same base radius and height. Calculate the ratio of the volume of the cylinder and the cone.

Solution : Given, Radius of cylinder

= Radius of cone = r cm

and Height of cylinder = Height of cone = h cm

Thus,
$$\frac{\text{Volume of cylinder}}{\text{Volume of cone}} = \frac{\pi r^2 h}{\frac{1}{3}\pi r^2 h} = \frac{3}{1}$$

Hence, Volume of cylinder : Volume of cone = 3 : 1.

Ans.

Q. 32. The slant height of the frustum of a cone is 5 cm. If the difference between the radii of its two circular ends is 4 cm, find the height of the frustum.

Solution : Given, slant height of frustum, l= 5 cm

Let radii of two circular ends of frustum be R, r cm

Then, $R - r = 4$ cm (Given)

Let the height of the cone be h cm.

We know, $l = \sqrt{h^2 + (R-r)^2}$

$\Rightarrow \quad 5 = \sqrt{h^2 + (4)^2}$ cm

$\Rightarrow \quad 25 = h^2 + (4)^2$ cm

$\Rightarrow \quad 25 = h^2 + 16$ cm

$\Rightarrow \quad h^2 = 25 - 16$

$\Rightarrow \quad h^2 = 9$

$\Rightarrow \quad h = 3$ cm. **Ans.**

Q. 33. Two types of water tankers are available in a shop. One is in a cubic form of dimensions 1 m × 1 m × 1 m and another is in the cylindrical form of height 1 m and diameter 1 m. Calculate the volume of both the containers. (Use $\pi = 3.14$)

Solution : Dimensions of cubic tank,

$l = 1$ m, $b = 1$ m, $h = 1$ m

$\therefore$ Volume $= lbh = 1 \times 1 \times 1 = 1\ m^3$

Height of cylindrical container = 1 m.

Radius $= \frac{1}{2}$ m

$\therefore$ Volume $= \pi r^2 h$

$= 3.14 \times \frac{1}{2} \times \frac{1}{2} \times 1$

$= 0.785\ m^3$ **Ans.**

Chapter 14. Statistics

Q. 1. Find the class marks of class 10 – 25 and 35 – 55.

Solution : We derive the class mark x_i by the following formula :

$$x_i = \frac{1}{2}(\text{lower limit} + \text{upper limit})$$

Thus $$x_i = \frac{1}{2}(10 + 25)$$

$$= \frac{1}{2}(35) = 17.5$$

and $$x_{ii} = \frac{1}{2}(\text{lower limit} + \text{upper limit})$$

Thus $$x_{ii} = \frac{1}{2}(35 + 55)$$

$$= \frac{1}{2}(90) = 45.$$ **Ans.**

Q. 2. Show that the mode of the series obtained by combining the two series S_1 and S_2 given below is different from that of S_1 and S_2 taken separately :

S_1 : 3, 5, 8, 8, 9, 12, 13, 9, 9

S_2 : 7, 4, 7, 8, 7, 8, 13

Solution : Mode of S_1 series = 9

Mode of S_2 series = 7

After combining S_1 and S_2, the new series will be = 3, 5, 8, 8, 9, 12, 13, 9, 9, 7, 4, 7, 8, 7, 8, 13.

Mode of combined series = 8 (maximum times)

Mode of (S_1, S_2) is different from mode of S_1 and mode of S_2 separately. **Hence Proved.**

Q. 3. Write down the Median Class of the following Frequency Distribution :

Class Interval	Frequency
0 – 10	4
10 – 20	4
20 – 30	8
30 – 40	10
40 – 50	12
50 – 60	8
60 – 70	4

Solution :

Class Interval	Frequency (f_i)	Cumulative Frequency
0–10	4	4
10–20	4	8
20–30	8	16
30–40	10	26
40–50	12	38
50–60	8	46
60–70	4	50
	$N = \Sigma f_i = 50$	

Thus, $$\frac{N}{2} = 25$$

The cumulative frequency just above 25 is 26.

Hence, the Median class is 30 – 40. **Ans.**

Q. 4. From the following probability distribution, find the median class.

Cost of living index	Number of weeks
1400-1550	8
1550-1700	15
1700-1850	21
1850-2000	8

Solution :

Cost of living index	No. of weeks (f)	*c.f.*
1400 – 1550	8	8
1550 – 1700	15	23
1700 – 1850	21	44
1850 – 2000	8	52
	$\Sigma f = 52$	

Here, $N = 52$

$$\Rightarrow \quad \frac{N}{2} = \frac{52}{2} = 26$$

26 will lie in the class interval 1700 – 1850.

∴ Median class 1700 – 1850. **Ans.**

Q. 5. Calculate the value of *p* from the following data :

Class	Frequency
0 – 20	8
20 – 40	15
40 – 60	*p*
60 – 80	12
80 – 100	5
	$N = \Sigma f_i = 60$

Solution : Given, $N = \Sigma f_i = 60$

$\Rightarrow \quad 8 + 15 + p + 12 + 5 = 60$

$\Rightarrow \quad 40 + p = 60$

$\Rightarrow \quad p = 20.$ **Ans.**

Q. 6. If empirical relationship between mean, median and mode is expressed as Mean = *k*(3 Median – Mode), then find the value of *k*.

Solution : Given, Mean = k(3 Median – Mode)

As we know, Mode = 3 Median – 2 Mean

∴ Mean = k[3 Median – (3 Median – 2 Mean)]

Mean = k[3 Median – 3 Median + 2 Mean]

Mean = 2 k Mean

$2k$ Mean – Mean = 0

Mean [$2k$ – 1] = 0

$2k - 1 = 0$

$2k = 1$

$k = 1/2$ **Ans.**

Q. 7. The following table shows the weights (in gm) of a sample of 100 potatoes taken from a large consignment.

Weight	Frequency
50 – 60	8
60 – 70	10
70 – 80	12
80 – 90	16
90 – 100	18
100 – 110	14
110 – 120	12
120 – 130	10

Calculate the cumulative frequency and determine the median class.

Solution : The cumulative frequency is as shown below :

Weight	Frequency	Cumulative Frequency
50 – 60	8	8
60 – 70	10	18
70 – 80	12	30
80 – 90	16	46
90 – 100	18	64
100 – 110	14	78
110 – 120	12	90
120 – 130	10	100
	$N = \Sigma f_i = 100$	

Now $\quad N = \Sigma f_i = 100$

$$\Rightarrow \quad \frac{N}{2} = 50$$

The cumulative frequency just above 50 is 64.

Hence, the median class is 90 – 100. **Ans.**

Q. 8. The contents of 100 matchboxes were checked to determine the number of matchsticks they contained.

Matchboxes	Matchsticks
35	6
36	10
37	18
38	25
39	21
40	12
41	8

Calculate the mean of the number of matchsticks per box and determine how many extra matchsticks would have to be added to the total contents of the 100 boxes to bring up the mean to 39?

Solution :

Class Interval (x_i)	Frequency (f_i)	($f_i \times x_i$)
35	6	210
36	10	360
37	18	666
38	25	950
39	21	819
40	12	480
41	8	328
	$\Sigma f_i = 100$	$\Sigma(f_i \times x_i) = 3813$

Thus, $\quad \text{Mean} = \frac{\Sigma(f_i \times x_i)}{\Sigma f_i}$

$$= \frac{3813}{100}$$

$$= 38.13$$

Now, to make the Mean = 39, $\Sigma(f_i \times x_i)$ should be 3900.

So the number of matchsticks to be added = 3900 – 3813 = 87. **Ans.**

Q. 9. The table below shows the distribution of marks obtained by students in an examination. Calculate the Median marks.

Marks less than	Number of Students
10	5
20	10
30	30
40	60
50	105
60	180
70	270
80	355
90	390
100	400

Solution :

Class Interval	Cumulative Frequency	Frequency (f_i)
0 – 10	5	5
10 – 20	10	10 – 5 = 5
20 – 30	30	30 – 10 = 20
30 – 40	60	60 – 30 = 30
40 – 50	105	105 – 60 = 45
50 – 60	180	180 – 105 = 75
60 – 70	270	270 – 180 = 90
70 – 80	355	355 – 270 = 85
80 – 90	390	390 – 355 = 35
90 – 100	400	400 – 390 = 10
		$N = \Sigma f_i = 400$

Now, $N = \Sigma f_i = 400$

$\Rightarrow \frac{N}{2} = 200$

The cumulative frequency just above 200 is 270.
Hence, the median class is 60 – 70.

So, $l = 60, h = 10, f = 90, F = 180$ and $\frac{N}{2} = 200$

$$\text{Median} = l + \frac{\frac{N}{2} - F}{f} \times h$$

$$= 60 + \frac{200 - 180}{90} \times 10$$

$$= 60 + 2.22$$

$$= 62.22$$ **Ans.**

Q. 10. The table below shows the daily profits (in ₹) of 100 shops. Calculate the Mode.

Profit	Number of Shops
0 – 100	12
100 – 200	18
200 – 300	27
300 – 400	20
400 – 500	17
500 – 600	6

Solution :

Profit	Number of Shops
0 – 100	12
100 – 200	18
200 – 300	27
300 – 400	20
400 – 500	17
500 – 600	6
	$N = \Sigma f_i = 100$

Since 200 – 300 has the highest frequency, it is the Modal Class.

So, $l = 200, h = 100, f = 27, f_1 = 18, f_2 = 20$

Thus, $\text{Mode} = l + \frac{f - f_1}{2f - f_1 - f_2} \times h$

$$= 200 + \frac{27 - 18}{54 - 18 - 20} \times 100$$

$$= 200 + \frac{9}{16} \times 100$$

$$= 200 + 56.25$$

= ₹ 256.25 **Ans.**

Q. 11. The table below shows the distribution of the daily wages (in ₹) of 160 workers. Calculate the Median wage.

Wages	Number of Workers
0 – 10	12
10 – 20	20
20 – 30	30
30 – 40	38
40 – 50	24
50 – 60	16
60 – 70	12
70 – 80	8

Solution :

Class Interval	Frequency (f_i)	Cumulative Frequency
0 – 10	12	12
10 – 20	20	32
20 – 30	30	62
30 – 40	38	100
40 – 50	24	124
50 – 60	16	140
60 – 70	12	152
70 – 80	8	160
	$N = \Sigma f_i = 160$	

Now, $N = \Sigma f_i = 160$

So $\frac{N}{2} = 80$

The cumulative frequency just above 80 is 100.

Hence, the median class is 30 – 40.

So, $l = 30, h = 10, f = 38, F = 62$ and $\frac{N}{2} = 80$

$$\text{Median} = l + \frac{\frac{N}{2} - F}{f} \times h$$

$$= 30 + \frac{80 - 62}{38} \times 10$$

$$= 30 + 4.7$$

= ₹ 34.70. **Ans.**

Q. 12. The table below shows the distribution of marks obtained by students in an examination. Calculate the value of *x* if the mean mark is 18.

Marks	No. of students
5	6
10	4
15	6
20	12
25	x
30	4

Solution : Given, Mean = 18

Marks (x_i)	No. of students (f_i)	Mid - value $f_i \times x_i$
5	6	30
10	4	40
15	6	90
20	12	240
25	x	$25x$
30	4	120
Total	$32+x$	$520+25x$

$$\text{Mean} = \frac{\Sigma f_i x_i}{\Sigma f_i}$$

or $\Sigma f_i \times \text{mean} = \Sigma f_i x_i$

$\Rightarrow$ $(32+x)18 = 520 + 25x$

or $520 + 25x = 576 + 18x$

$\Rightarrow$ $25x - 18x = 576 - 520$

$\Rightarrow$ $x = \frac{56}{7} = 8$ **Ans.**

Q. 13. Using the table given below, calculate the cumulative frequencies of the workers.

Wages	Number of Workers
6500 – 7000	10
7000 – 7500	18
7500 – 8000	22
8000 – 8500	25
8500 – 9000	17
9000 – 9500	10
9500 – 10000	8

Solution :

Wages	Number of Workers	Cumulative Frequency
6500 – 7000	10	10
7000 – 7500	18	28
7500 – 8000	22	50
8000 – 8500	25	75
8500 – 9000	17	92
9000 – 9500	10	102
9500 – 10000	8	110

Ans.

Q. 14. The table gives the frequency distribution of the heights (in cm) of a group of people. Determine the median height.

Height	People
150 – 155	6
155 – 160	12
160 – 165	18
165 – 170	20
170 – 175	13
175 – 180	8
180 – 185	6

Solution :

Class Interval	Frequency (f_i)	Cumulative Frequency
150 – 155	6	6
155 – 160	12	18
160 – 165	18	36
165 – 170	20	56
170 – 175	13	69
175 – 180	8	77
180 – 185	6	83
	$N = \Sigma f_i = 83$	

Now $N = \Sigma f_i = 83$

$\Rightarrow$ $\frac{N}{2} = 41.5$

The cumulative frequency just above 41.5 is 56. Hence, the median class is 165 – 170

So, $l = 165, h = 5, f = 20, F = 36$ and $\frac{N}{2} = 41.5$

$$\text{Median} = l + \frac{\frac{N}{2} - F}{f} \times h$$

$$= 165 + \frac{41.5 - 36}{20} \times 5$$

$$= 165 + 1.375$$

$= 166.375$ cm **Ans.**

Q. 15. Using the data given below, calculate the Median.

Marks	Students
0 – 10	3
10 – 20	8
20 – 30	12
30 – 40	14
40 – 50	10
50 – 60	6
60 – 70	5
70 – 80	2

Solution :

Class Interval	Frequency (f_i)	Cumulative Frequency
0 10	3	3
10 – 20	8	11
20 – 30	12	23
30 – 40	14	37
40 – 50	10	47
50 – 60	6	53
60 – 70	5	58
70 – 80	2	60
	$N = \Sigma f_i = 60$	

Now $N = \Sigma f_i = 60$

$\Rightarrow \frac{N}{2} = 30$

The cumulative frequency just above 30 is 37.
Hence, the median class is 30 – 40.

So, $l = 30, h = 10, f = 14, F = 23$ and $\frac{N}{2} = 30$

$$\text{Median} = l + \frac{\frac{N}{2} - F}{f} \times h$$

$$= 30 + \frac{30-23}{14} \times 10$$

$= 30 + 5 = 35.$ **Ans.**

Q. 16. The following table provides data about the weekly wages (in ₹) of workers in a factory. Calculate the Mean and the Modal Class.

Wages	Number of Workers
50 – 55	5
55 – 60	20
60 – 65	10
65 – 70	10
70 – 75	9
75 – 80	6
80 – 85	12
85 – 90	8

Solution :

Class Interval	Frequency (f_i)	Class Mark (x_i)	($f_i \times x_i$)
50 – 55	5	52.5	262.5
55 – 60	20	57.5	1150
60 – 65	10	62.5	625
65 – 70	10	67.5	675
70 – 75	9	72.5	652.5
75 – 80	6	77.5	465
80 – 85	12	82.5	990
85 – 90	8	87.5	700
	$N = \Sigma f_i = 80$		$\Sigma(f_i \times x_i) = 5520$

Thus $\text{Mean} = \frac{\Sigma(f_i \times x_i)}{\Sigma f_i} = \frac{5520}{80} = ₹\,69$

Since the 55 – 60 has the highest frequency 20, so it is the Modal Class. **Ans.**

Q. 17. The marks obtained by 120 students in a Mathematics test are given below.

Marks	Number of Students
0 – 10	5
10 – 20	9
20 – 30	16
30 – 40	22
40 – 50	26
50 – 60	18
60 – 70	11
70 – 80	6
80 – 90	4
90 – 100	3

Calculate the Median.

Solution :

Class Interval	Frequency (f_i)	Cumulative Frequency
0 – 10	5	5
10 – 20	9	14
20 – 30	16	30
30 – 40	22	52
40 – 50	26	78
50 – 60	18	96
60 – 70	11	107
70 – 80	6	113
80 – 90	4	117
90 – 100	3	120
	$N = \Sigma f_i = 120$	

Now $N = \Sigma f_i = 120$

So $\frac{N}{2} = 60$

The cumulative frequency just above 60 is 78.
Hence, the median class is 40 – 50.

So, $l = 40, h = 10, f = 26, F = 52$ and $\frac{N}{2} = 60$

$$\text{Median} = l + \frac{\frac{N}{2} - F}{f} \times h$$

$$= 40 + \frac{60-52}{26} \times 10$$

$= 40 + 3.08$

$= 43.08$ **Ans.**

Q. 18. Find the Mean of the following frequency distribution :

Class Interval	Frequency
0 – 50	4
50 – 100	8
100 – 150	16
150 – 200	13
200 – 250	6
250 – 300	3

Solution :

Class Interval	Frequency (f_i)	Class Mark (x_i)	$(f_i \times x_i)$
0 – 50	4	25	100
50 – 100	8	75	600
100 – 150	16	125	2000
150 – 200	13	175	2275
200 – 250	6	225	1350
250 – 300	3	275	825
	$N = \Sigma f_i = 50$		$\Sigma(f_i \times x_i) = 7150$

Thus Mean $= \frac{\Sigma(f_i \times x_i)}{\Sigma f_i}$

$= \frac{7150}{50}$

$= 143.$ **Ans.**

Q. 19. Find the Mode of the following frequency distribution :

Class Interval	Frequency
0 – 10	**2**
10 – 20	**8**
20 – 30	**10**
30 – 40	**5**
40 – 50	**4**
50 – 60	**3**

Solution :

Class Interval	Frequency (f_i)
0 – 10	2
10 – 20	8
20 – 30	10
30 – 40	5
40 – 50	4
50 – 60	3
	$N = \Sigma f_i = 32$

Since, 20 – 30 has the highest frequency, it is the modal class.

So, $l = 20, f = 10, f_1 = 8, f_2 = 5, h = 10$

Thus Mode $= l + \frac{f - f_1}{2f - f_1 - f_2} \times h$

$= 20 + \frac{10 - 8}{20 - 8 - 5} \times 10$

$= 20 + \frac{2}{7} \times 10$

$= 20 + 2.86$

$= 22.86$ **Ans.**

Q. 20. Calculate the Mode of the following frequency distribution.

Class Interval	Frequency
0 – 5	**2**
5 – 10	**7**
10 – 15	**18**
15 – 20	**10**
20 – 25	**8**
25 – 30	**5**

Solution :

Class Interval	Frequency (f_i)
0 – 5	2
5 – 10	7
10 – 15	18
15 – 20	10
20 – 25	8
25 – 30	5
	$N = \Sigma f_i = 50$

Since, 10 – 15 has the highest frequency, it is the modal class.

So, $l = 10, h = 5, f = 18, f_1 = 7, f_2 = 10$

Thus Mode $= l + \frac{f - f_1}{2f - f_1 - f_2} \times h$

$= 10 + \frac{18 - 7}{36 - 7 - 10} \times 5$

$= 10 + \frac{11}{19} \times 5$

$= 10 + 2.9$

$= 12.9$ **Ans.**

Q. 21. Find the Mean of the following distribution :

Class Interval	Frequency
20 – 30	**10**
30 – 40	**6**
40 – 50	**8**
50 – 60	**12**
60 – 70	**5**
70 – 80	**9**

Solution :

Class Interval	Frequency (f_i)	Class Mark (x_i)	$(f_i \times x_i)$
20 – 30	10	25	250
30 – 40	6	35	210
40 – 50	8	45	360
50 – 60	12	55	660
60 – 70	5	65	325
70 – 80	9	75	675
	$N = \Sigma f_i = 50$		$\Sigma(f_i \times x_i) = 2480$

Thus, $\text{Mean} = \frac{\Sigma(f_i \times x_i)}{\Sigma f_i}$

$= \frac{2480}{50}$

$= 49.6$ **Ans.**

Q. 22. Find the mean of the following distribution :

Class Interval	Frequency
0 – 10	10
10 – 20	6
20 – 30	8
30 – 40	12
40 – 50	5

Solution :

Class Interval	Frequency	Class Mark (x_i)	$(f_i \times x_i)$
0 – 10	10	5	50
10 – 20	6	15	90
20 – 30	8	25	200
30 – 40	12	35	420
40 – 50	5	45	225
	$\Sigma f_i = 41$		$\Sigma(f_i \times x_i) = 985$

Thus $\text{Mean} = \frac{\Sigma(f_i \times x_i)}{\Sigma f_i}$

$= \frac{985}{41}$

$= 24{\cdot}02.$ **Ans.**

Q. 23. Given below is a cumulative frequency distribution table. Corresponding to it, make an ordinary frequency distribution table.

x	$c.f.$
More than or equal to 0	45
More than or equal to 10	38
More than or equal to 20	29
More than or equal to 30	17
More than or equal to 40	11
More than or equal to 50	6

Solution :

C.I.	Frequency
0 – 10	07(45 – 38)
10 – 20	09(38 – 29)
20 – 30	12(29 – 17)
30 – 40	6(17 – 11)
40 – 50	5(11 – 6)
50 – 60	6(6 – 0)

Q. 24. Atul donates ₹ 1000 per month to a cow shelter, ₹ 2000 per month to blind school, ₹ 3000 per month to a charitable hospital and ₹ 4000 per month to a welfare society and remains for his own purpose. Find the average of his donation.

Solution: Donation for Cow Shelter= ₹ 1000

Donation for Blind School = ₹ 2000

Donation for Charitable Hospital = ₹ 3000

Donation for Welfare Society = ₹ 4000

Total = ₹ 10,000

Thus, Average of his donation $= \frac{10{,}000}{5} =$ ₹ 2500

Ans.

Chapter 15. Probability

Q. 1. In a family of three children, what is the probability of having at least one boy?

Solution : Total number of children = 3

All possible outcomes = *GGG, GGB, GBB, GBG, BBB, BGG, BGB, BBG*

Let the event of getting at least one boy be *E*.

Then the favourable outcomes are

GGB, GBB, GBG, BBB, BBG, BGG, BGB

Hence, number of favourable outcomes = 7

$\therefore \quad P(E) = \frac{7}{8}$ **Ans.**

Q. 2. What is the probability that a number selected at random from the numbers 1, 2, 3, ..., 15 is a multiple of 4?

Solution : Total numbers = 15

Let the event of getting a number multiple of 4 be *E*.

Then, the favourable outcomes are 4, 8, 12

Hence, number of favourable outcomes = 3

$\therefore \quad P(E) = \frac{3}{15} = \frac{1}{5}.$ **Ans.**

Q. 3. If two dice are rolled together, find the probability of getting an even number on both the dice.

Solution : Total number of possible outcomes = (1, 1); (1, 2); (1, 3); (1, 4); (1, 5); (1, 6); (2, 1); (2, 2); (2, 3); (2, 4); (2, 5); (2, 6); (3, 1); (3, 2); (3, 3); (3, 4); (3, 5); (3, 6); (4, 1); (4, 2); (4, 3); (4, 4); (4, 5); (4, 6); (5, 1); (5, 2); (5, 3); (5, 4); (5, 5); (5, 6); (6, 1); (6, 2); (6, 3); (6, 4); (6, 5), (6, 6) = 36

Let *E* be the event of getting an even number on both the dice.

Then, the favourable outcomes are (2, 2); (2, 4); (2, 6); (4, 2); (4, 4); (4, 6); (6, 2); (6, 4); (6, 6).

Hence, the total number of favourable outcomes = 9

$\therefore \quad P(E) = \frac{9}{36} = \frac{1}{4}$ **Ans.**

Q. 4. When a number is selected at random from numbers 1 to 30, what is the probability that it is a prime number ?

Solution : Total numbers = 30

Let the event of getting prime number be E.

Then the favourable outcomes are 2, 3, 5, 7, 11, 13, 17, 19, 23, 29.

Hence number of favourable outcomes = 10

$$\therefore \quad P(E) = \frac{10}{30} = \frac{1}{3} \qquad \textbf{Ans.}$$

Q. 5. Two different coins are tossed simultaneously. Find the probability of getting at least one head.

Solution : Total number of possible outcomes

= HH, HT, TH, TT = 4

Let E be the event of getting at least one head.

Then, the favourable outcomes are HH, HT, TH

Hence, the total number of favourable outcomes = 3

$$\therefore \quad P(E) = \frac{3}{4}. \qquad \textbf{Ans.}$$

Q. 6. A bag contains cards numbered 1 to 25. A card is drawn at random from the bag. Find the probability that the number on the card is divisible by both 2 and 3.

Solution : Total numbers = 25.

Let the event of getting a number divisible by 2 and 3 be E.

Then the favourable outcomes are 6, 12, 18, 24.

Hence, number of favourable outcomes = 4

$$\therefore \quad P(E) = \frac{4}{25} \qquad \textbf{Ans.}$$

Q. 7. What is the probability of getting an even number when a die is thrown once ?

Solution : Total number of possible outcomes

= 1, 2, 3, 4, 5, 6

Let E be the event of getting an even number.

Then the favourable outcomes are 2, 4, 6.

Hence, the total number of favourable outcomes = 3

$$\therefore \quad P(E) = \frac{3}{6} = \frac{1}{2}. \qquad \textbf{Ans.}$$

Q. 8. A box contains 90 discs numbered from 1 to 90. If one disc is drawn at random from the box, then find the probability that it bears a prime number less than 23.

Solution : Total number of possible outcomes = 90

Let E be the event of getting a prime number less than 23.

Then, the favourable outcomes are 2, 3, 5, 7, 11, 13, 17, 19

Hence, the total number of favourable outcomes = 8.

$$\therefore \quad P(E) = \frac{8}{90} = \frac{4}{45}. \qquad \textbf{Ans.}$$

Q. 9. One ticket is drawn at random from a bag containing tickets numbered 1 to 40. Find the probability that the selected ticket has a number that is a multiple of 7.

Solution : The total number of possible outcomes = 40

Let E be the event of getting a ticket number multiple of 7.

Thus, the favourable outcomes are 7, 14, 21, 28, 35.
Hence, the number of favourable outcomes = 5.

$$\therefore \quad P(E) = \frac{5}{40} = \frac{1}{8} \qquad \textbf{Ans.}$$

Q. 10. Two dice are thrown together. What is the probability of getting the same number on both the dice?

Solution : Total possible outcomes = (1, 1); (1, 2); (1, 3); (1, 4); (1, 5); (1, 6); (2, 1); (2, 2); (2, 3); (2, 4); (2, 5); (2, 6); (3, 1); (3, 2); (3, 3); (3, 4); (3, 5); (3, 6); (4, 1); (4, 2); (4, 3); (4, 4); (4, 5); (4, 6); (5, 1); (5, 2); (5, 3); (5, 4); (5, 5); (5, 6); (6, 1); (6, 2); (6, 3); (6, 4); (6, 5), (6, 6) = 36

Let E be the event of getting the same number on both the dice.

Then, the favourable outcomes are (1, 1); (2, 2); (3, 3); (4, 4); (5, 5); (6, 6).

Hence, the total number of favourable outcomes = 6

$$\therefore \quad P(E) = \frac{6}{36} = \frac{1}{6}. \qquad \textbf{Ans.}$$

Q. 11. Cards bearing numbers 2, 3, 4, ..., 11 are kept in a bag. A card is drawn at random from it. Find the probability of getting a card with a prime number.

Solution : Total number of possible outcomes = 10

Let E be the event of getting a card with prime number.

Thus, the favourable outcomes are 2, 3, 5, 7, 11.

Hence, the total number of favourable outcomes = 5

$$\therefore \quad P(E) = \frac{5}{10} = \frac{1}{2} \qquad \textbf{Ans.}$$

Q. 12. A card is drawn at random from a well shuffled pack of 52 playing cards. Find the probability of getting neither a red card nor a queen.

Solution : Total number of possible outcomes = 52

Let E be the event of getting neither a red card nor a queen.

$\therefore$ Number of favourable outcomes

= 52 – (26 + 2) = 24

$$\therefore \quad P(E) = \frac{24}{52} = \frac{6}{13} \qquad \textbf{Ans.}$$

Q. 13. Find the probability of throwing a number greater than 2 with a single dice.

Solution : Total number of possible outcomes = 6

Let E be the event of getting a number greater than 2 on the dice.

Thus, the favourable outcomes are 3, 4, 5, 6.

Hence, the total number of favourable outcomes = 4

$\therefore \quad P(E) = \frac{4}{6} = \frac{2}{3}$. **Ans.**

Q. 14. A letter of English alphabet is chosen at random. Determine the probability that the chosen letter is a consonant.

Solution : Total number of all possible outcomes = 26

Number of consonants = 21

Let E be the event of getting a consonant

$\therefore P$ (getting a consonant) $= P(E) = \frac{21}{26}$ **Ans.**

Q. 15. A card is drawn from a well-shuffled deck of 52 playing cards. Then what is the probability that the card will not be an ace ?

Solution : Total number of possible outcomes = 52

Number of aces in the pack = 4

Thus, the probability of not drawing an ace

$= \frac{52-4}{52} = \frac{12}{13}$ **Ans.**

Q. 16. The probability of selecting a rotten apple randomly from a heap of 900 apples is 0.18. What is the number of rotten apples in the heap ?

Solution : Total apples = 900

$P(E) = 0.18$

$\Rightarrow \quad \frac{\text{No. of rotten apples}}{\text{Total no. of apples}} = 0.18$

$\Rightarrow \quad \frac{\text{No. of rotten apples}}{900} = 0.18$

$\Rightarrow \quad$ No. of rotten apples $= 900 \times 0.18$

$= 162$ **Ans.**

Q. 17. A dice is thrown once. What is the probability of getting a number less than 3 ?

Solution : The possible outcomes are 1, 2, 3, 4, 5, 6.

The sum of the possible outcomes = 6

Let E be the event of getting a number less than 3.

Then, favourable outcomes are 1, 2

Hence, total number of favourable outcomes = 2

$\therefore \quad P(E) = \frac{2}{6} = \frac{1}{3}$ **Ans.**

Q. 18. Cards bearing numbers 3 to 20 are placed in a bag and mixed thoroughly. A card is taken out of the bag at random. What is the probability that the number on the card taken out is an even number ?

Solution : Total number of possible outcomes = 18

Let E be the event of getting an even number of the card.

The favourable outcomes are 4, 6, 8, 10, 12, 14, 16, 18, 20.

Sum of favourable outcomes = 9

$\therefore \quad P(E) = \frac{9}{18} = \frac{1}{2}$ **Ans.**

Q. 19. Cards marked with numbers 3, 4, 5,....., 50 are placed in a box and mixed thoroughly. A card is drawn at random from the box. Find the probability that the selected card bears a perfect square number.

Solution : Total outcomes = 3, 4, 5, ..., 50

Total no. of outcomes = 48

Let E be the event of getting a perfect square number

Favourable outcomes = 4, 9, 16, 25, 36, 49.

No. of favourable outcomes = 6

$\therefore \quad P(E) = \frac{6}{48} = \frac{1}{8}$ **Ans.**

Q. 20. A bag contains 4 red and 6 black balls. A ball is taken out of the bag at random. What is the probability of getting a black ball ?

Solution : Total number of possible outcomes = 10

Let E be the event of getting a black ball

The total number of favourable outcomes = 6

$\therefore \quad P(E) = \frac{6}{10} = \frac{3}{5}$. **Ans.**

Q. 21. A card is drawn out from a well-shuffled deck of 52 cards. What is the probability of getting a black king ?

Solution : Total number of possible outcomes = 52

Let E be the event of getting a black king.

The total number of favourable outcomes = 2

$\therefore \quad P(E) = \frac{2}{52} = \frac{1}{26}$ **Ans.**

Q. 22. Two friends were born in the year 2000. What is the probability that they have the same birthday ?

Solution : Since year 2000 is a leap year.

Total possible cases = 366

Let E be the event of getting the birthdays of two friends on same day.

Favourable case = 1

$\therefore \quad P(E) = \frac{1}{366}$ **Ans.**

Q. 23. What is the probability that two friends have different birthdays ?

Solution : Total number of possible outcomes

$= 365 \times 365$

Let E be the event of getting two friends have different birthdays.

The total number of favourable outcomes

$$= 365 \times (365 - 1) = 365 \times 364$$

$$\therefore \quad P(E) = \frac{365 \times 364}{365 \times 365}$$

$$= \frac{364}{365}$$ **Ans.**

Q. 24. A dice is thrown once. Find the probability of getting a prime number.

Solution : The maximum number of possible outcomes = 6

Let E be the event of getting a prime number on the dice.

Favourable outcomes = 2, 3, 5

Total number of the favourable outcomes = 3

$$\therefore \quad P(E) = \frac{3}{6} = \frac{1}{2}$$ **Ans.**

Q. 25. Two different dice are tossed together. Find the probability that (i) the number of each dice is even, and (ii) the sum of the numbers, appearing on the two dice, is 5.

Solution : Total number of possible outcomes = (1, 1); (1, 2); (1, 3); (1, 4); (1, 5); (1, 6); (2, 1); (2, 2); (2, 3); (2, 4); (2, 5); (2, 6); (3, 1); (3, 2); (3, 3); (3, 4); (3, 5); (3, 6); (4, 1); (4, 2); (4, 3); (4, 4); (4, 5); (4, 6); (5, 1); (5, 2); (5, 3); (5, 4); (5, 5); (5, 6); (6, 1); (6, 2); (6, 3); (6, 4); (6, 5); (6, 6) = 36

(i) Favourable outcomes are (2, 2); (2, 4); (2, 6); (4, 2); (4, 4); (4, 6); (6, 2); (6, 4); (6, 6).

Hence, total number of favourable outcomes = 9

$$\therefore \quad P(E) = \frac{9}{36} = \frac{1}{4}$$

(ii) Favourable outcomes are (1, 4); (2, 3); (3, 2); (4, 1)

Hence, total number of the favourable outcomes = 4

$$\therefore \quad P(E) = \frac{4}{36} = \frac{1}{9}.$$ **Ans.**

Q. 26. Rahim tosses two different coins simultaneously. Find the probability of getting at least one tail.

Solution : Total number of possible outcomes = *HH*; *HT*; *TH*; *TT*

Favourable outcomes are *HT*; *TH*; *TT*

Hence, the total number of favourable outcomes = 3

$$\therefore \quad P(E) = \frac{3}{4}.$$ **Ans.**

Q. 27. Two different dice are rolled simultaneously. Find the probability that the sum of the numbers appearing on the two dice is 10.

Solution : Total number of possible outcomes = (1, 1); (1, 2); (1, 3); (1, 4); (1, 5); (1, 6); (2, 1); (2, 2); (2, 3); (2, 4); (2, 5); (2, 6); (3, 1); (3, 2); (3, 3); (3, 4); (3, 5); (3, 6); (4, 1); (4, 2); (4, 3); (4, 4); (4, 5); (4, 6); (5, 1); (5, 2); (5, 3); (5, 4); (5, 5); (5, 6); (6, 1); (6, 2); (6, 3); (6, 4); (6, 5); (6, 6) = 36

Favourable outcomes are (4, 6); (5, 5); (6, 4)

Hence, the total number of favourable outcomes = 3

$$\therefore \quad P(E) = \frac{3}{36} = \frac{1}{12}.$$ **Ans.**

Q. 28. A card is drawn from a well shuffled pack of 52 playing cards. Find the probability that the drawn card is neither a king nor a queen.

Solution : Total number of cards = 52

Number of kings and queens = 4 + 4 = 8

Thus, number of favourable outcomes

= 52 – 8 = 44

$$\therefore \quad P(E) = \frac{44}{52} = \frac{11}{13}.$$ **Ans.**

Q. 29. A card is drawn from a well shuffled pack of 52 playing cards. Find the probability of getting (i) a red-faced card and (ii) a black-king.

Solution : Total number of cards = 52

(i) Number of red-faced cards = 3 + 3 = 6

$$\therefore \quad P(E) = \frac{6}{52} = \frac{3}{26}$$

(ii) Number of black queens = 2

$$\therefore \quad P(E) = \frac{2}{52} = \frac{1}{26}.$$ **Ans.**

Q. 30. A number is selected at random from the first 50 natural numbers. Find the probability that it is a multiple of three and four.

Solution : Total numbers = 1, 2, 3, ..., 50 = 50

The favourable outcomes are 12, 24, 36, 48

Thus, number of favourable outcomes = 4

$$\therefore \quad P(E) = \frac{4}{50} = \frac{2}{25}.$$ **Ans.**

Q. 31. A card is drawn at random from a well-shuffled pack of 52 cards. Find the probability of getting (i) a red king, and (ii) a queen or a jack.

Solution : Total number of cards = 52

(i) The number of favourable outcomes of drawing a red king = 2

$$\therefore \quad P(E) = \frac{2}{52} = \frac{1}{26}$$

(ii) The number of favourable outcomes of drawing a queen or a jack = 8

$$\therefore \quad P(E) = \frac{8}{52} = \frac{2}{13}.$$ **Ans.**

Q. 32. Two different dice are thrown simultane-ously. Find the probability that the sum of the number appearing on the two dice is 7.

Solution : Total number of possible outcomes = (1, 1); (1, 2); (1, 3); (1, 4); (1, 5); (1, 6); (2, 1); (2, 2); (2, 3); (2, 4); (2, 5); (2, 6); (3, 1); (3, 2); (3, 3); (3, 4); (3, 5); (3, 6); (4, 1); (4, 2); (4, 3); (4, 4); (4, 5); (4, 6); (5, 1); (5, 2); (5, 3); (5, 4); (5, 5); (5, 6); (6, 1); (6, 2); (6, 3); (6, 4); (6, 5); (6, 6) = 36

Favourable outcomes are (1, 6); (2, 5); (3, 4); (4, 3); (5, 2); (6, 1)

Hence, the total number of favourable outcomes = 6

$\therefore \quad P(E) = \frac{6}{36} = \frac{1}{6}.$ **Ans.**

Q. 33. A ticket is drawn at random from a bag containing tickets numbered from 1 to 40. Find the probability that the selected ticket has a number which is a multiple of 5.

Solution : Total number of tickets = 40

Favourable outcomes are 5, 10, 15, 20, 25, 30, 35, 40

Total number of favourable outcomes = 8

$\therefore \quad P(E) = \frac{8}{40} = \frac{1}{5}.$ **Ans.**

Q. 34. Find the probability that a number selected at random from the number 3, 4, 5, ... 25 is a prime.

Solution : The possible outcomes are 3, 4, 5, ... 25.

Hence, total number of possible outcomes = 23.

The favourable outcomes are 3, 5, 7, 11, 13, 17, 19, 23.

Thus, the total number of favourable outcomes = 8

$\therefore \quad P(E) = \frac{8}{23}.$ **Ans.**

Q. 35. Cards each marked with one of the numbers between 6, 7, 8, ..., 15 are placed in a box and mixed thoroughly. One card is drawn at random from the box. What is the probability of getting a card with a number less than 10 ?

Solution : The possible outcomes are 6, 7, 8, ..., 15

Thus, sum of the number of possible outcomes = 10

The favourable outcomes are 6, 7, 8, 9.

Hence, sum of the number of favourable outcomes = 4

$\therefore \quad P(E) = \frac{4}{10} = \frac{2}{5}.$ **Ans.**

Q. 36. A box contains 3 blue, 2 white and 4 red marbles. If a marble is drawn out at random from the box, what is the probability that it will not be a white marble ?

Solution : The sum of the possible outcomes

$= 3 + 2 + 4 = 9$

If the marble drawn out is not to be white then the favourable outcomes are 3 blue marbles and 4 red marbles.

Hence, the sum of the favourable conditions

$= 3 + 4 = 7$

$\therefore \quad P(E) = \frac{7}{9}.$ **Ans.**

★ are board exam questions from previous years

Q. 37. Two dice are thrown simultaneously. Find the probability that the sum of the numbers appearing on the two dice is more than 9.★

Solution : Total number of possible outcomes = (1, 1); (1, 2); (1, 3); (1, 4); (1, 5); (1, 6); (2, 1); (2, 2); (2, 3); (2, 4); (2, 5); (2, 6); (3, 1); (3, 2); (3, 3); (3, 4); (3, 5); (3, 6); (4, 1); (4, 2); (4, 3); (4, 4); (4, 5); (4, 6); (5, 1); (5, 2); (5, 3); (5, 4); (5, 5); (5, 6); (6, 1); (6, 2); (6, 3); (6, 4); (6, 5); (6, 6) = 36

Favourable outcomes are (4, 6); (5, 5); (5, 6); (6, 4); (6, 5); (6, 6)

Hence, the total number of favourable outcomes = 6

$\therefore \quad P(E) = \frac{6}{36} = \frac{1}{6}.$ **Ans.**

Q. 38. Find the probability that a number selected at random from the numbers 1 to 25 is not a prime number.

Solution : The total number of possible outcomes = 25

The favourable conditions are 1, 4, 6, 8, 9, 10, 12, 14, 15, 16, 18, 20, 21, 22, 24, 25

Hence, the sum of the numbers of favourable conditions = 16

$\therefore \quad P(E) = \frac{16}{25}.$ **Ans.**

Q. 39. A card is drawn at random from a well-shuffled pack of 52 cards. Find the probability that the card is neither a red card nor a jack.

Solution : Total number of cards = 52

The number of favourable outcomes of drawing neither a red nor a jack = 52 – (26 + 2) = 24

$\therefore \quad P(E) = \frac{24}{52} = \frac{6}{13}.$ **Ans.**

Q. 40. A card is drawn at random from a well-shuffled pack of 52 cards. Find the probability that the card is neither a queen nor a jack.

Solution : Total number of cards = 52

The number of favourable outcomes of drawing neither a queen nor a jack = 52 – (4 + 4) = 44

$\therefore \quad P(E) = \frac{44}{52} = \frac{11}{13}.$ **Ans.**

Q. 41. A card is drawn at random from a shuffled pack of 52 cards. Find the probability of drawing a (i) face card, and (ii) a card which is neither a king nor a red card.

Solution : Total numbers of cards = 52

(i) The number of favourable outcomes of drawing a face card = 12

$$\therefore \quad P(E) = \frac{12}{52} = \frac{3}{13}$$

(ii) The number of favourable outcomes of drawing neither a king nor a red card = 52 – (2 + 26) = 24

$$\therefore \quad P(E) = \frac{24}{52} = \frac{6}{13}. \qquad \textbf{Ans.}$$

Q. 42. Two different dice are tossed together. Find the probability of getting a doublet.

Solution : Total number of possible outcomes = (1, 1); (1, 2); (1, 3); (1, 4); (1, 5); (1, 6); (2, 1); (2, 2); (2, 3); (2, 4); (2, 5); (2, 6); (3, 1); (3, 2); (3, 3); (3, 4); (3, 5); (3, 6); (4, 1); (4, 2); (4, 3); (4, 4); (4, 5); (4, 6); (5, 1); (5, 2); (5, 3); (5, 4); (5, 5); (5, 6); (6, 1); (6, 2); (6, 3); (6, 4); (6, 5); (6, 6) = 36

The favourable outcomes are (1, 1); (2, 2); (3, 3); (4, 4); (5, 5); (6, 6)

Hence, the sum of the favourable outcomes = 6

$$\therefore \quad P(E) = \frac{6}{36} = \frac{1}{6}. \qquad \textbf{Ans.}$$

Q. 43. From a well-shuffled pack of 52 cards, one is drawn out at random. Find the probability of getting a black queen.

Solution : The sum of the maximum possible outcomes = 52

The sum of the favourable outcomes = 2

$$\therefore \quad P(E) = \frac{2}{52} = \frac{1}{26}. \qquad \textbf{Ans.}$$

Q. 44. A dice is thrown once. Find the probability of getting a number (i) less than three, and (ii) more than four.

Solution : The maximum possible outcomes = 6

(i) The favourable outcomes for a number less than three are 1, 2

Thus, the sum of the favourable outcomes = 2

$$\therefore \quad P(E) = \frac{2}{6} = \frac{1}{3}$$

(ii) The favourable outcomes for a number more than four are 5, 6.

Thus, the sum of the most favourable outcomes = 2

$$\therefore \quad P(E) = \frac{2}{6} = \frac{1}{3}. \qquad \textbf{Ans.}$$

Q. 45. What is the probability that a number selected at random from the number 3, 4, 5, 6, 7, 8, 9 is a multiple of 4 ?

Solution : The maximum possible outcomes = 7

Favourable outcomes are 4, 8.

Thus, the sum of the favourable outcomes = 2

$$\therefore \quad P(E) = \frac{2}{7}. \qquad \textbf{Ans.}$$

Q. 46. Samsung Electronics has launched two new mobile hand sets : Set-I and Set -II. Set-I is cheaper as compared to Set-II. But Set-II has an built-in device to recharge the battery with auto-cut power supply when it is fully charged. In a lot there are 250 pieces Set-I and 100 Set-II. If a mobile is picked at random :

(a) Find the probability of getting Set-I

(b) Find the probability of getting Set-II.

Solution : Total no. of mobile sets = 250 + 100 = 350

(a) Probability of getting set-I = $\frac{250}{350} = \frac{5}{7}$

(b) Probability of getting set-II = $\frac{100}{350} = \frac{2}{7}$ **Ans.**

Q. 47. In answering a question of MCQ test with 4 choices per question, one of them being correct, a student knows the answer, guesses or copies the answer. Suppose a student guesses the answer. What is the probability that his answer is correct ?

Solution : Number of choices = 4

Number of correct choice = 1

$\therefore$ Probability that answer is correct = $\frac{1}{4}$ **Ans.**

Q. 48. Honey goes to school either by a car driven by his driver or uses his bicycle. Probability that he will use the car is $\frac{3}{7}$. What is the probability that he will use his bicycle for going to the school ?

Solution : Probability that he will use his bicycle for going to the school = 1 – Probability that he will use the car = $1 - \frac{3}{7} = \frac{4}{7}$ **Ans.**

●●

Short Questions

| Set 3 |

Chapter 1. Real Numbers

Q. 1. Show that every positive even integer is of the form 2*q* and every positive odd integer is of the form (2*q* + 1), where *q* is some integer. **[NCERT]**

Solution : Let a be the positive integer and $b = 2$.

Therefore, $a = bq + r$ where $0 \le r < b$ is according to Euclid's Division Lemma.

$\Rightarrow \quad a = 2q + r$

$\Rightarrow \quad a = 2q$ when $r = 0$

and $\quad a = 2q + 1$ when $r = 1$

Now $a = 2q$ is even and can be both a positive integer and a negative integer. If we take it to be a positive integer then $a = 2q + 1$ is a positive odd integer.

Hence Proved.

Q. 2. Find the HCF of 1260 and 7344 using Euclid's algorithm.★

Solution : Two numbers are 1260 and 7344

Since 7344 > 1260, we apply the Euclid division lemma to 7344 and 1260, we get

$$7344 = 1260 \times 5 + 1044$$

Also, $$1260 = 1044 \times 1 + 216$$

$$1044 = 216 \times 4 + 180$$

$$216 = 180 \times 1 + 36$$

$$180 = 36 \times 5 + 0$$

Now, remainder is 0, hence our procedure stops here.

$\therefore$ H.C.F. of 7344 and 1260 is 36. **Ans.**

Q. 3. The length, breadth and height of a room are 8m 50cm, 6m 25cm and 4m 75cm respectively. Find the length of the longest rod that can measure the dimensions of the room exactly.★

Solution : To find the length of the longest rod that can measure the dimensions of the room exactly, we have to find HCF.

Length, $L = 8$ m 50 cm = 850 cm = $2^1 \times 5^2 \times 17$

Breadth, $B = 6$ m 25 cm = 625 cm = 5^4

Height, $H = 4$ m 75 cm = 475 cm = $5^2 \times 19$

$\therefore$ HCF of L, B and $H = 5^2 = 25$

$\therefore$ Length of the longest rod = 25 cm **Ans.**

Q. 4. A class of 20 boys and 15 girls is divided into *n* groups so that each group has *x* boys and *y* girls. Find *x*, *y* and *n*.

★ are board exam questions from previous years

Solution : H.C.F. of 20 and 15 = 5

Therefore, the 5 students are in each group.

So, $$n = \frac{20+15}{5} = \frac{35}{5} = 7$$

$$x = \frac{20}{5} = 4$$

and $$y = \frac{15}{5} = 3$$

Hence, $x = 4$, $y = 3$ and $n = 7$. **Ans.**

Q. 5. The L.C.M. of two numbers is 14 times their H.C.F. The sum of L.C.M. and H.C.F. is 750. If one number is 250, then find the other number.

Solution : Let H.C.F. be 'H'

then L.C.M. = 14 H

Sum of L.C.M. and H.C.F. is 750

$\therefore \quad 14\text{H} + \text{H} = 750$

$\Rightarrow \quad 15\text{H} = 750$

$\Rightarrow \quad \text{H} = \frac{750}{15}$

$\Rightarrow \quad \text{H} = 50$

$\therefore \quad \text{L.C.M.} = 14\text{H}$

$= 14 \times 50$

$= 700$

We know,

Product of L.C.M. and H.C.F. = Product of two numbers.

Let other number be y

$$700 \times 50 = 250 \times y$$

$$y = \frac{700 \times 50}{250}$$

$$y = 140.$$

Hence, the other number is 140. **Ans.**

Q. 6. Write the smallest number which is divisible by both 306 and 657.★

Solution : Smallest number which is divisible by 306 and 657 is,

LCM (657, 306)

$657 = 3 \times 3 \times 73$

$306 = 3 \times 3 \times 2 \times 17$

$\text{LCM} = 3 \times 3 \times 73 \times 2 \times 17$

$= 22338$ **Ans.**

Q. 7. In a seminar on the topic 'liberty and equality' the no. of participants in Hindi, Social Science and English are 60, 84 and 108 respectively. Find the minimum

number of rooms required if in each room the same number of participants are to be seated and all of them being in the same subject.

Solution : The number of rooms will be minimum if each room accommodates maximum number of participants. Since in each room the same number of participants are to be seated and all of them must be of the same subject,

Thus, no. of participants in each room

= H.C.F. of 60, 84 and 108

$\because \quad 60 = 2^2 \times 3 \times 5$

$84 = 2^2 \times 3 \times 7$

and $\quad 108 = 2^2 \times 3^3$

$\therefore \quad \text{H.C.F.} = 2^2 \times 3$

$= 4 \times 3$

$= 12$

Thus, in each room 12 participants can be seated

$$\therefore \quad \text{No. of rooms required} = \frac{60 + 84 + 108}{12}$$

$$= \frac{252}{12} = 21 \qquad \textbf{Ans.}$$

Q. 8. Three sets of English, Hindi and Sociology books dealing with cleanliness have to stacked in such a way that all the books are stored topic-wise and height of each stack is the same. The number of English books is 96, number of Hindi books is 240 and the number of Sociology books is 336. Assuming that the books are of same thickness, determine the number of stocks of English, Hindi and Sociology books.

Solution : In order to arrange the books as required, we have to find the largest number that divides 96, 240 and 336 exactly.

Now, largest Number = H.C.F. of 96, 240 and 336

$\because \quad 96 = 2^5 \times 3$

$240 = 2^4 \times 3 \times 5$

and $\quad 336 = 2^4 \times 3 \times 7$

$\therefore \quad \text{H.C.F.} = 2^4 \times 3$

$= 16 \times 3$

$= 48$

So, there must be 48 books in each stack.

$$\therefore \text{ No. of stacks of English books} = \frac{96}{48} = 2$$

$$\text{No. of stacks of Hindi books} = \frac{240}{48} = 5$$

$$\text{No. of stacks of Sociology books} = \frac{336}{48} = 7 \qquad \textbf{Ans.}$$

★ are board exam questions from previous years

Q. 9. Three alarm clocks ring at intervals of 4, 12 and 20 minutes respectively. If they start ringing together, after how much time will they next ring together?

Solution : To find the time when the clocks will next ring together, we have to find LCM of 4, 12 and 20 minutes.

$4 = 2^2$

$12 = 2^2 \times 3$

$20 = 2^2 \times 5$

2	12
2	6
	3

2	20
2	10
	5

$\therefore$ LCM of 4, 12 and 20

$= 2^2 \times 3 \times 5$

$= 60$ minutes.

So, the clocks will ring together again after 60 minutes or one hour. **Ans.**

Q. 10. Find L.C.M. and H.C.F. of 3930 and 1800 by prime factorization method.★

Solution : By prime factorization method,

2	3930
3	1965
5	655
131	131
	1

2	1800
2	900
2	450
3	225
3	75
5	25
5	5
	1

$3930 = 2 \times 3 \times 5 \times 131$

$1800 = 2 \times 2 \times 2 \times 3 \times 3 \times 5 \times 5$

$\therefore \quad \text{HCF} = 2 \times 3 \times 5$

$= 30$

and $\quad \text{LCM} = 2 \times 3 \times 5 \times 131 \times 2 \times 2 \times 3 \times 5$

$= 235800$ **Ans.**

Chapter 2. Polynomials

Q. 1. If $(x + k)$ is the G.C.D. (or H.C.F.) of $x^2 - 2x - 15$ and $x^3 + 27$, find the value of k.★

Solution : Given,

$f(x) = x^2 - 2x - 15 = 0$

and $\quad g(x) = x^3 + 27 = 0$

$\Rightarrow \quad f(x) = x^2 - 5x + 3x - 15 = 0$

and $\quad g(x) = x^3 + 3^3 = 0$

$\Rightarrow \quad f(x) = x(x - 5) + 3(x - 5) = 0$

and $\quad g(x) = (x + 3)^3 - 9x(x + 3) = 0$

$\Rightarrow \quad f(x) = (x + 3)(x - 5) = 0$

and $\quad g(x) = (x + 3)[(x + 3)^2 - 9x] = 0$

Thus, it is evident from the above that $x + 3$ is the highest common factor between the above expressions.

So, $\quad x + 3 = x + k$

$\Rightarrow \quad k = 3.$ **Ans.**

Q. 2. Find the zeros of given polynomial,

$$4\sqrt{3}x^2 + 5x - 2\sqrt{3} = 0.★$$

Solution : Given,

$$f(x) = 4\sqrt{3}x^2 + 5x - 2\sqrt{3} = 0$$
$$\Rightarrow \quad 4\sqrt{3}x^2 + 8x - 3x - 2\sqrt{3} = 0$$
$$\Rightarrow 4x(\sqrt{3}x + 2) - \sqrt{3}(\sqrt{3}x + 2) = 0$$
$$\Rightarrow \quad (\sqrt{3}x + 2)(4x - \sqrt{3}) = 0$$
$$\Rightarrow \quad x = -\frac{2}{\sqrt{3}} \text{ and } \frac{\sqrt{3}}{4}$$

$\therefore$ The zeros of given polynomial are $-\frac{2}{\sqrt{3}}$ and $\frac{\sqrt{3}}{4}$. **Ans.**

Q. 3. If α and β are zeros of the quadratic polynomial $4x^2 + 4x + 1$, then find the quadratic polynomial whose zeros are 2α and 2β.★

Solution : Let $p(x) = 4x^2 + 4x + 1$

$\because$ α, β are zeros of $p(x)$

$$\therefore \quad \alpha + \beta = \text{Sum of zeros} = \frac{-b}{a}$$
$$\Rightarrow \quad \alpha + \beta = \frac{-4}{4} = -1 \quad \text{...(i)}$$
$$\text{Also} \quad \alpha.\beta = \text{Product of zeros} = \frac{c}{a}$$
$$\Rightarrow \quad \alpha.\beta = \frac{1}{4} \quad \text{...(ii)}$$

Now a quadratic polynomial whose zeros are 2α and 2β will be

$$x^2 - (\text{Sum of zeros})x + \text{Product of zeros}$$
$$= x^2 - (2\alpha + 2\beta)x + 2\alpha \times 2\beta$$
$$= x^2 - 2(\alpha + \beta)x + 4(\alpha\beta)$$
$$= x^2 - 2 \times (-1)x + 4 \times \frac{1}{4}$$

[Using eq. (i) and (ii)]

$$= x^2 + 2x + 1.$$ **Ans.**

Q. 4. Find the zeros of the quadratic polynomial $6x^2 - 3 - 7x$ and verify the relationship between the zeros and the coefficients of the polynomial.★

Solution : We have, $6x^2 - 3 - 7x$

$$= 6x^2 - 7x - 3$$
$$= 6x^2 - 9x + 2x - 3$$
$$= 3x(2x - 3) + 1(2x - 3)$$
$$= (2x - 3)(3x + 1)$$

Zeros are :

$$2x - 3 = 0$$

or $$3x + 1 = 0$$

$$\therefore \quad x = \frac{3}{2} \text{ and } x = \frac{-1}{3}$$ **Ans.**

Verification :

$$\text{Sum of zeros } (\alpha + \beta) = \frac{3}{2} + \left(\frac{-1}{3}\right)$$
$$= \frac{9-2}{6} = \frac{7}{6}$$
$$= \frac{-(\text{Coefficient of } x)}{\text{Coefficient of } x^2}$$
$$= \frac{-(-7)}{6} = \frac{7}{6}$$
$$\text{Product of zeros } (\alpha\beta) = \frac{3}{2} \times \frac{-1}{3} = \frac{-3}{6}$$
$$= \frac{\text{Constant term}}{\text{Coefficient of } x^2}$$
$$= \frac{-3}{6}$$

$\therefore$ Relationship holds. **Hence Verified.**

Q. 5. Solve the quadratic polynomial $2x^2 + ax - a^2$ for x.

Solution : Let $f(x) = 2x^2 + ax - a^2 = 0$

$$\Rightarrow \quad 2x^2 + 2ax - ax - a^2 = 0$$
$$\Rightarrow \quad 2x(x + a) - a(x + a) = 0$$
$$\Rightarrow \quad (x + a)(2x - a) = 0$$
$$\therefore \quad x = -a \text{ and } \frac{a}{2}$$

Thus the values of x for the polynomial $2x^2 + ax - a^2$ will be $-a$ and $\frac{a}{2}$. **Ans.**

Q. 6. Obtain the zeros of the quadratic polynomial $\sqrt{3}x^2 - 8x + 4\sqrt{3}$ and verify the relation between its zero and coefficients.

Solution : Let

$$f(x) = \sqrt{3}x^2 - 8x + 4\sqrt{3} = 0$$
$$\Rightarrow \quad \sqrt{3}x^2 - 6x - 2x + 4\sqrt{3} = 0$$
$$\Rightarrow \quad \sqrt{3}x(x - 2\sqrt{3}) - 2(x - 2\sqrt{3}) = 0$$
$$\Rightarrow \quad (\sqrt{3}x - 2)(x - 2\sqrt{3}) = 0$$
$$\Rightarrow \quad x = \frac{2}{\sqrt{3}} \text{ and } 2\sqrt{3}$$

$\therefore$ The zeros of x are $\frac{2}{\sqrt{3}}$ and $2\sqrt{3}$ **Ans**

Verification :

$$\text{Sum of zeros} = \frac{2}{\sqrt{3}} + 2\sqrt{3}$$

★ are board exam questions from previous years

$$= \frac{2+6}{\sqrt{3}} = \frac{8}{\sqrt{3}}$$

$$= \frac{-(\text{Coefficient of } x)}{\text{Coefficient of } x^2}$$

$$= \frac{-(-8)}{\sqrt{3}} = \frac{8}{\sqrt{3}}$$

$$\text{Product of zeros} = \frac{2}{\sqrt{3}} \times 2\sqrt{3}$$

$$= 4$$

$$= \frac{\text{Constant term}}{\text{Coefficient of } x^2}$$

$$= \frac{4\sqrt{3}}{\sqrt{3}} = 4$$ **Hence Verified.**

Q. 7. If one zero of a polynomial $(a^2+9)x^2+13x+6a$ is the reciprocal of the other, then find the value of a.

Solution : Let the zeros of the equation be α and β.

According to the question,

$$\alpha = \frac{1}{\beta} \text{ or } \beta = \frac{1}{\alpha}$$

Thus $$\alpha\beta = \alpha\frac{1}{\alpha} = \beta\frac{1}{\beta} = 1$$

Now $$\alpha+\beta = -\frac{13}{a^2+9} \text{ and } \alpha\beta = \frac{6a}{a^2+9}$$

$\therefore$ $$\frac{6a}{a^2+9} = 1$$

$\Rightarrow$ $$a^2+9 = 6a$$

$\Rightarrow$ $$a^2-6a+9 = 0$$

$\Rightarrow$ $$a^2-3a-3a+9 = 0$$

$\Rightarrow$ $$a(a-3)-3(3-a) = 0$$

$\Rightarrow$ $$(a-3)^2 = 0$$

$\Rightarrow$ $$a = 3.$$ **Ans.**

Q. 8. Find the zeros of the polynomial $f(u) = 4u^2+8u$ and verify the relationship between the zeros and the coefficients.

Solution : We have

$$f(u) = 4u^2+8u = 0$$

or $$4u(u+2) = 0$$

$\therefore$ $$u = 0 \text{ and } -2$$ **Ans.**

Let $$\alpha = 0 \text{ and } \beta = -2$$

Then $$\alpha+\beta = -2 \text{ and } \alpha\beta = 0$$

Verification :

$$\text{Sum of zeros} = -\frac{8}{4} = -2$$

$$= \frac{-(\text{Coeffcient of } x)}{\text{Coefficient of } x^2}$$

$$\text{Product of zeros} = \frac{0}{4} = 0$$

$$= \frac{\text{Constant term}}{\text{Coefficient of } x^2}$$

Hence Verified

Q. 9. Find a quadratic polynomial each with the given numbers as the sum and product of its zeros respectively : [NCERT]

(i) $\frac{1}{4}, -1$ **(ii) $\sqrt{2}, \frac{1}{3}$**

Solution : (i) Let the zeros of the polynomial $f(x)$ be α and β.

$\therefore$ $$\alpha+\beta = \frac{1}{4} \text{ and } \alpha\beta = -1$$

Hence, the required polynomial is

$$f(x) = x^2 - (\text{Sum of the zeros})\,x + \text{Product of the zeros}$$

$\Rightarrow$ $$f(x) = x^2 - \frac{1}{4}x - 1$$

$$= (4x^2-x-4)$$ **Ans.**

(ii) Let the zeros of the polynomial $f(x)$ be α and β.

$\therefore$ $$\alpha+\beta = \sqrt{2} \text{ and } \alpha\beta = \frac{1}{3}$$

Hence, the required polynomial is

$$f(x) = x^2 - (\text{Sum of the zeros})\,x + \text{Product of the zeros}$$

$\Rightarrow$ $$f(x) = x^2 - \sqrt{2}x + \frac{1}{3}$$

$$= (3x^2 - 3\sqrt{2}x + 1).$$ **Ans.**

Q. 10. Find a quadratic polynomial, the sum and product of whose zeros are 0 and $-\frac{3}{5}$ respectively. Hence find the zeros.

Solution : Quadratic polynomial is given as,

$$= x^2 - (\text{Sum of zeros})\,x + \text{Product of zeros}$$

$$= x^2 - (0)\,x + \left(\frac{-3}{5}\right)$$

$$= x^2 - \frac{3}{5}$$

$$= (x)^2 - \left(\sqrt{\frac{3}{5}}\right)^2$$

$$= \left(x - \sqrt{\frac{3}{5}}\right)\left(x + \sqrt{\frac{3}{5}}\right) \quad \left[\begin{array}{l}\text{By applying} \\ (a^2-b^2) = (a+b)(a-b)\end{array}\right]$$

Zeros are,

$$x - \sqrt{\frac{3}{5}} = 0 \quad \text{or} \quad x + \sqrt{\frac{3}{5}} = 0$$

$\Rightarrow$ $$x = \sqrt{\frac{3}{5}} \quad \text{or} \quad x = -\sqrt{\frac{3}{5}}$$ **Ans.**

Q. 11. Find the zeros of the quadratic polynomial $4x^2 - 4x - 3$ and verify the relation between the zeros and the coefficients.

Solution : Given,

$$f(x) = 4x^2 - 4x - 3$$

$\Rightarrow \quad f(x) = 4x^2 - 6x + 2x - 3$

$\Rightarrow \quad f(x) = 2x(2x - 3) + 1(2x - 3)$

$\Rightarrow \quad f(x) = (2x - 3)(2x + 1)$

$\therefore \quad x = \frac{3}{2}, -\frac{1}{2}$

Thus $\quad \alpha = \frac{3}{2}$ and $\beta = -\frac{1}{2}$ **Ans.**

Verification :

Sum of zeros $= \alpha + \beta$

$$= \frac{3}{2} + \frac{(-1)}{2}$$

$$= 1 = \frac{-(\text{Coefficient of } x)}{\text{Coefficient of } x^2}$$

Product of zeros $= \alpha\beta$

$$= \left(\frac{3}{2}\right)\left(\frac{-1}{2}\right)$$

$$= \frac{-3}{4} = \frac{\text{Constant term}}{\text{Coefficient of } x^2}$$

$\therefore$ Relationship holds. **Hence Verified.**

Q. 12. Find the zeros of the quadratic polynomial $5x^2 - 4 - 8x$ and verify the relationship between the zeros and the coefficients of the given polynomial.

Solution : Given,

$$f(x) = 5x^2 - 4 - 8x$$
$$= 5x^2 - 8x - 4$$
$$= 5x^2 - 10x + 2x - 4$$
$$= 5x(x - 2) + 2(x - 2)$$
$$= (x - 2)(5x + 2)$$

$\Rightarrow \quad x = 2, -\frac{2}{5}$

Thus $\quad \alpha = 2$ and $\beta = -\frac{2}{5}$

Verification :

$$\alpha + \beta = \frac{8}{5} \text{ and } \alpha\beta = -\frac{4}{5}$$

$$\text{Sum of zeros} = -\frac{b}{a} = -\left(\frac{-8}{5}\right) = \frac{8}{5}$$

$$= \frac{-(\text{Coefficient of } x)}{\text{Coefficient of } x^2}$$

$$\text{Product of zeros} = \frac{c}{a} = -\frac{4}{5}$$

$$= \frac{\text{Constant term}}{\text{Coefficeint of } x^2}$$

$\therefore$ Relationship holds. **Hence Verified.**

Q. 13. If two zeros of the polynomial $f(x) = x^3 - 4x^2 - 3x + 12$ are $\sqrt{3}$ and $-\sqrt{3}$, then find its third zero.

Solution : Let $\alpha = \sqrt{3}$, $\beta = -\sqrt{3}$ be the given zeros and γ be the third zero. Then,

$$\alpha + \beta + \gamma = -\left(\frac{-4}{1}\right)$$

$$\left[\text{Using } \alpha + \beta + \gamma = -\frac{\text{Coeff. of } x^2}{\text{Coeff. of } x^3}\right]$$

$\Rightarrow \quad \sqrt{3} - \sqrt{3} + \gamma = 4$

$\Rightarrow \quad \gamma = 4$

Hence, third zero is 4. **Ans.**

Q. 14. What must be subtracted or added to $p(x) = 8x^4 + 14x^3 - 2x^2 + 8x - 12$ so that $4x^2 + 3x - 2$ is a factor of $p(x)$?

Solution :

$$4x^2 + 3x - 2 \overline{)\, 8x^4 + 14x^3 - 2x^2 + 8x - 12 \,(} \, 2x^2 + 2x - 1$$

$$\begin{array}{r} 8x^4 + 6x^3 - 4x^2 \\ - \quad - \quad + \\ \hline 8x^3 + 2x^2 + 8x \\ 8x^3 + 6x^2 - 4x \\ - \quad + \\ \hline -4x^2 + 12x - 12 \\ -4x^2 - 3x + 12 \\ + \quad + \quad - \\ \hline 15x - 24 \end{array}$$

Since, on dividing polynomial $p(x) = 8x^4 + 14x^3 - 2x^2 + 8x - 12$ with $4x^2 + 3x - 2$, we get a remainder of $15x - 24$

Therefore, on subtracting $15x - 24$ from $p(x)$, $4x^2 + 3x - 2$ is a factor of $p(x)$. **Ans.**

Q. 15. Show that 3 is a zero of the polynomial $2x^3 - x^2 - 13x - 6$. Hence find all the zeroes of this polynomial.

Solution : Here $\quad p(x) = 2x^3 - x^2 - 13x - 6$

$$p(3) = 2 \times 27 - 4 - 39 - 6$$
$$= 54 - 54 = 0$$

Thus 3 is zero of $p(x)$.

$\Rightarrow (x-3)$ is a factor of $p(x)$.

$$x-3\overline{)2x^3-x^2-13x-6}(2x^2+5x+2$$

$$2x^3-6x^2$$
$$-\quad+$$
$$5x^2-13x$$
$$5x^2-15x$$
$$-\quad+$$
$$2x-6$$
$$2x-6$$
$$-\quad+$$
$$\times$$

$\therefore \quad p(x) = (x-3)(2x^2+5x+2)$

$= (x-3)(2x^2+4x+x+2)$

$(x-3)[2x(x+2)+1(x+2)]$

$(x-3)(2x+1)(x+2)$

Thus, other zeros are $\dfrac{-1}{2}$ and -2. **Ans.**

Q. 16. What real number should be subtracted from the polynomial ($3x^3 + 10x^2 - 14x + 9$) so that ($3x - 2$) divides it exactly ?

Solution : We know that, By division algorithm,

Dividend = Quotient × Divisor + Remainder

or Dividend – Remainder = Quotient × Divisor

Clearly R.H.S. of above result is divisible by the divisor.

$\Rightarrow$ L.H.S. is also divisible by the divisor.

So if we subtract remainder from the dividend, then it will be exactly divisible by the divisor.

So divide $3x^3 + 10x^2 - 14x + 9$ by $3x - 2$

$$x^2+4x-2$$
$$3x-2\overline{)3x^3+10x^2-14x+9}$$
$$3x^2-2x^2$$
$$-\quad+$$
$$12x^2-14x$$
$$12x^2-8x$$
$$-\quad+$$
$$-6x+9$$
$$-6x+4$$
$$+\quad-$$
$$5$$

Quotient $= x^2 + 4x - 2$ and Remainder = 5.

Thus, 5 should be subtracted from the polynomial $3x^3 + 10x^2 - 14x + 9$ so that it is divisible by $3x - 2$. **Ans.**

Q. 17. Find a cubic polynomial when the sum, sum of the products of its zeros taken two at a time and product of its zero are 2, – 7, – 14 respectively. [NCERT]

Solution : Let the zeros of the cubic polynomial be α, β and γ.

$$\therefore \quad \alpha+\beta+\gamma = 2$$
$$\alpha\beta+\beta\gamma+\alpha\gamma = -7$$
$$\alpha\beta\gamma = -14$$

We know, a cubic polynomial is given as,

$$f(x) = x^3-(\alpha+\beta+\gamma)x^2 + (\alpha\beta+\beta\gamma+\alpha\gamma)x - \alpha\beta\gamma$$

$$\therefore \quad f(x) = x^3-2x^2-7x+14$$

Thus, the required cubic polynomial is

$x^3 - 2x^2 - 7x + 14.$ **Ans.**

Q. 18. If the polynomial

$$\boldsymbol{f(x) = (6x^4+8x^3+17x^2+21x+7)}$$

is divided by another polynomial $g(x) = 3x^2 + 4x + 1$, the remainder is $(ax + b)$. Find a and b.

Solution : According to Division Algorithm,

$$f(x) = g(x) \times q(x) + r(x)$$

where $q(x)$ is quotient and $r(x)$ is the remainder.

Dividing $f(x) = 6x^4 + 8x^3 + 17x^2 + 21x + 7$ by

$$g(x) = 3x^2+4x+1$$

$$2x^2+5$$
$$3x^2+4x+1\overline{)6x^4+8x^3+17x^2+21x+7}$$
$$6x^4+8x^3+2x^2$$
$$-\quad-\quad-$$
$$15x^2+21x+7$$
$$15x^2+20x+5$$
$$-\quad-\quad-$$
$$x+2$$

Thus, remainder $= x + 2$

On comparing it with $ax + b$, we get

$a = 1, b = 2.$ **Ans.**

Q. 19. Solve for x :

$$\boldsymbol{3x^2-2\sqrt{6}x+2=0.}$$

Solution : Given, $\quad 3x^2-2\sqrt{6}x+2 = 0$

$$\Rightarrow \quad 3x^2-\sqrt{6}x-\sqrt{6}x+2 = 0$$

$$\Rightarrow \quad \sqrt{3}x(\sqrt{3}x-\sqrt{2})-\sqrt{2}(\sqrt{3}x-\sqrt{2}) = 0$$

$$\Rightarrow \quad (\sqrt{3x}-2)\,(\sqrt{3}x-\sqrt{2}) = 0$$

$$\Rightarrow \quad x = \sqrt{\frac{2}{3}}, \sqrt{\frac{2}{3}}$$ **Ans.**

Q. 20. Find the roots of the polynomial :

$$2\sqrt{3}x^2 - 5x + \sqrt{3} = 0.$$

Solution : We have,

$$2\sqrt{3}x^2 - 5x + \sqrt{3} = 0$$

$$2\sqrt{3}x^2 - 2x - 3x + \sqrt{3} = 0$$

$$2x(\sqrt{3}x - 1) - \sqrt{3}(\sqrt{3}x - 1) = 0$$

$$(\sqrt{3}x - 1)(2x - \sqrt{3}) = 0$$

If $(\sqrt{3}x - 1) = 0$

$\Rightarrow$ $x = \frac{1}{\sqrt{3}}$

If $2x - \sqrt{3} = 0$

$\Rightarrow$ $x = \frac{\sqrt{3}}{2}$

$\therefore$ Roots of given polynomial are $\frac{1}{\sqrt{3}}$ and $\frac{\sqrt{3}}{2}$ **Ans.**

Q. 21. For what value of k are the roots of the quadratic polynomial $kx(x-2)+6=0$ are equal ? **[NCERT]**

Solution : Given,

$$kx(x-2)+6=0$$

$\Rightarrow$ $kx^2 - 2kx + 6 = 0$

Let α and β be the roots of the polynomial.

Hence $\alpha + \beta = \frac{-(-2k)}{k} = 2; \alpha\beta = \frac{6}{k}$

Also $\alpha = \beta$ [Given]

$\Rightarrow$ $\alpha + \alpha = 2$

$\Rightarrow$ $2\alpha = 2$

$\Rightarrow$ $\alpha = \beta = 1$

Hence $\alpha\beta = \frac{6}{k}$

$\Rightarrow$ $1 = \frac{6}{k}$ [$\because \alpha = \beta = 1$]

$\Rightarrow$ $k = 6$

Thus, the roots of the polynomial $kx(x-2)+6=0$ will be equal if $k = 6$. **Ans.**

Q. 22. Using division algorithm, find the quotient and remainder on dividing $f(x)$ by $g(x)$ where $f(x) = 6x^3 + 13x^2 + x - 2$ and $g(x) = 2x + 1$.★

Solution : Given, $f(x) = 6x^3 + 13x^2 + x - 2$ and $g(x) = 2x + 1$

$f(x) \div g(x) \Rightarrow$

$$
\begin{array}{r|l}
2x+1 & 6x^3 + 13x^2 + x - 2 \;(3x^2 + 5x - 2 \\
 & 6x^3 + 3x^2 \\
 & - \quad - \\ \hline
 & 10x^2 + x \\
 & 10x^2 + 5x \\
 & - \quad - \\ \hline
 & -4x - 2 \\
 & -4x - 2 \\
 & + \quad + \\ \hline
 & 0
\end{array}
$$

Quotient = $3x^2 + 5x - 2$, Remainder = 0. **Ans.**

Q. 23. If three zeros of a polynomial $x^4 - x^3 - 3x^2 + 3x$ are $0, \sqrt{3}$ and $-\sqrt{3}$, then find the fourth zero.★

Solution : Let $P(x) = x^4 - x^3 - 3x^2 + 3x$

Given, 0, $\sqrt{3}$ and $-\sqrt{3}$ are three zeros, so

$$x = 0, x = \sqrt{3} \text{ and } x = -\sqrt{3}$$

$\Rightarrow (x - \sqrt{3}) = 0$ and $(x + \sqrt{3}) = 0$

Here, $x(x + \sqrt{3})(x - \sqrt{3})$ will also be the factor of $P(x)$.

or $x(x^2 - 3)$ will be the factor of $P(x)$.

Then

$$
\begin{array}{r|l}
x^3 - 3x & x^4 - x^3 - 3x^2 + 3x \;(x - 1 \\
 & x^4 - 3x^2 \\
 & - \quad + \\ \hline
 & -x^3 + 3x \\
 & -x^3 + 3x \\
 & + \quad - \\ \hline
 & 0
\end{array}
$$

$\therefore$ Quotient = $(x - 1)$

So, fourth zero = $(x - 1) = 0 \Rightarrow x = 1$

Hence, four zeros will be 1, 0, $\sqrt{3}$, $-\sqrt{3}$ **Ans.**

Chapter 3. Pair of Linear Equations in Two Variables

Q. 1. Solve for x and y : $ax + by - a + b = 0$ and $bx - ay - a - b = 0$.

Solution : The equations $ax + by - a + b = 0$ and $bx - ay - a - b = 0$ can be written as

$$ax + by = a - b \quad \text{...(i)}$$

and $$bx - ay = a + b \quad \text{...(ii)}$$

Multiplying equation (i) with a and equation (ii) with b, we have

$$a^2x + aby = a^2 - ab \quad \text{...(iii)}$$

$$b^2x - aby = ab + b^2 \quad \text{...(iv)}$$

★ are board exam questions from previous years

On adding equation (iii) and (iv), we get

$$(a^2 + b^2)x = a^2 + b^2$$

$$\Rightarrow \quad x = 1$$

Substituting $x = 1$ in equation (i), we get

$$a + by = a - b$$

$$\Rightarrow \quad by = -b$$

$$\Rightarrow \quad y = -1$$

Thus, $x = 1$ and $y = -1$. **Ans.**

Q. 2. Solve for x and y :

$$\frac{ax}{b} - \frac{by}{a} = a + b \text{ and } ax - by = 2ab.★$$

Solution : Given,

$$\frac{ax}{b} - \frac{by}{a} = a + b$$

$$\Rightarrow \quad \frac{a^2x - b^2y}{ab} = a + b$$

$$\Rightarrow \quad a^2x - b^2y = a^2b + ab^2 \quad \text{...(i)}$$

and

$$ax - by = 2ab$$

$$\Rightarrow \quad ax = 2ab + by \quad \text{...(ii)}$$

Substituting equation (ii) in (i), we get

$$a(2ab + by) - b^2y = a^2b + ab^2$$

$$\Rightarrow \quad 2a^2b + aby - b^2y = a^2b + ab^2$$

$$\Rightarrow \quad by(a - b) = a^2b + ab^2 - 2a^2b$$

$$\Rightarrow \quad by(a - b) = ab^2 - a^2b$$

$$\Rightarrow \quad by(a - b) = -ab(a - b)$$

$$\Rightarrow \quad y = -a$$

Substituting $y = -a$ in equation (ii), we get

$$ax = 2ab + b(-a)$$

$$ax = 2ab - ab$$

$$\Rightarrow \quad ax = ab$$

$$\Rightarrow \quad x = b$$

Thus $x = b$ and $y = -a$. **Ans.**

Q. 3. Find the value(s) of k so that the pair of equations $x + 2y = 5$ and $3x + ky + 15 = 0$ has a unique solution.★

Solution : Given,

$$x + 2y = 5$$

$$3x + ky + 15 = 0$$

Comparing above equations with

$$a_1x + b_1y + c_1 = 0 \text{ and } a_2x + b_2y + c_2 = 0,$$

We get,

$$a_1 = 1, b_1 = 2, c_1 = -5$$

$$a_2 = 3, b_2 = k, c_3 = 15$$

Condition for the pair of equations to have unique solution is

$$\frac{a_1}{a_2} \neq \frac{b_1}{b_2}$$

$$\frac{1}{3} \neq \frac{2}{k}$$

$$k \neq 6$$

k can have any value except 6. **Ans.**

Q. 4. Solve the following pair of linear equations by the cross-multiplication method :★

$$x + 2y = 2$$

$$x - 3y = 7$$

Solution :

$$x + 2y - 2 = 0$$

$$x - 3y - 7 = 0$$

$$\begin{matrix} 2 & -2 & 1 & 2 \\ -3 & -7 & 1 & -3 \end{matrix}$$

$$\frac{x}{2(-7)-(-3)(-2)} = \frac{y}{-2(1)-(-7)(1)} = \frac{1}{1(-3)-(1)(2)}$$

$$\Rightarrow \quad \frac{x}{-14-6} = \frac{y}{-2+7} = \frac{1}{-3-2}$$

$$\Rightarrow \quad \frac{x}{-20} = \frac{y}{5} = \frac{1}{-5}$$

$$\Rightarrow \quad \frac{x}{-20} = \frac{1}{-5}, \frac{y}{5} = \frac{1}{-5}$$

$$\Rightarrow \quad x = 4, \quad y = -1$$

Thus $x = 4$ and $y = -1$ **Ans.**

Q. 5. Solve for x and y : $0.4x - 1.5y = 6.5$ and $0.3x + 0.2y = 0.9$.

Solution : Given,

$$0.4x - 1.5y = 6.5$$

and

$$0.3x + 0.2y = 0.9$$

Multiplying given equations by 10, we get

$$4x - 15y = 65 \quad \text{...(i)}$$

and

$$3x + 2y = 9 \quad \text{...(ii)}$$

Now, multiplying equation (i) with 2 and equation (ii) with 15, we get

$$8x - 30y = 130 \quad \text{...(iii)}$$

and

$$45x + 30y = 135 \quad \text{...(iv)}$$

Adding equations (iii) and (iv), we get

$$53x = 265$$

$$\Rightarrow \quad x = 5$$

Substituting $x = 5$ in equation (ii), we get

$$3(5) + 2y = 9$$

$$\Rightarrow \quad 15 + 2y = 9$$

$$\Rightarrow \quad 2y = -15 + 9 = -6$$

$$\Rightarrow \quad y = -3$$

Thus $x = 5$ and $y = -3$. **Ans.**

★ are board exam questions from previous years

Q. 6. Solve for x and y:

$$\sqrt{2}x-\sqrt{3}y=0 \text{ and } \sqrt{5}x+\sqrt{2}y=0.$$

Solution : Given, $\sqrt{2}x-\sqrt{3}y=0$ ...(i)

and $\sqrt{5}x+\sqrt{2}y=0$ ...(ii)

Multiplying equation (i) with $\sqrt{2}$ and equation (ii) with $\sqrt{3}$, we get

$$2x-\sqrt{6}y=0 \quad ...(iii)$$

and $\sqrt{15}x+\sqrt{6}y=0$ ...(iv)

Adding equations (iii) and (iv), we get

$$(2+\sqrt{15})x=0$$

$\Rightarrow$ $x=0$

Substituting $x = 0$ in equation (i), we get

$\Rightarrow$ $0\times\sqrt{2}-\sqrt{3}y=0$

$\Rightarrow$ $y=0$

Thus $x=y=0$. **Ans.**

Q. 7. Solve for x and y:

$$\frac{4}{x}+3y=14 \text{ and } \frac{3}{x}-4y=23 \text{ where } x\neq 0.$$

Solution : Given, $\frac{4}{x}+3y=14$ ...(i)

and $\frac{3}{x}-4y=23$ ...(ii)

Let $\frac{1}{x}=p$

Thus equation (i) and (ii), respectively can be written as

$$4p+3y=14 \quad ...(iii)$$

and $3p-4y=23$ ...(iv)

Multiplying equation (iii) with 4 and equation (iv) with 3, we get

$$16p+12y=56 \quad ...(v)$$

$$9p-12y=69 \quad ...(vi)$$

Adding equations (v) and (vi), we get

$$25p=125$$

$\Rightarrow$ $p=5$

$\Rightarrow$ $\frac{1}{x}=5$

$\Rightarrow$ $x=\frac{1}{5}$

Substituting $p = 5$ in equation (iii), we get

$$4(5)+3y=14$$

$\Rightarrow$ $20+3y=14$

$\Rightarrow$ $3y=14-20=-6$

$\Rightarrow$ $y=-2$

Thus, $x=\frac{1}{5}$ and $y=-2$. **Ans.**

Q. 8. Solve for x and y:

$$\frac{2}{x}+\frac{3}{y}=13 \text{ and } \frac{5}{x}-\frac{4}{y}=-2, x\neq 0, y\neq 0.$$

Solution : Given, $\frac{2}{x}+\frac{3}{y}=13$

and $\frac{5}{x}-\frac{4}{y}=-2$

Let $\frac{1}{x}=p$ and $\frac{1}{y}=q$

So $2p+3q=13$ ...(i)

and $5p-4q=-2$ ...(ii)

Multiplying equation (i) with 4 and equation (ii) with 3, we get

$$8p+12q=52 \quad ...(iii)$$

$$15p-12q=-6 \quad ...(iv)$$

Adding equations (iii) and (iv), we get

$$23p=46$$

$\Rightarrow$ $p=2$

Substituting $p = 2$ in equation (i), we get

$$2(2)+3q=13$$

$\Rightarrow$ $4+3q=13$

$\Rightarrow$ $3q=13-4$

$\Rightarrow$ $3q=9$

$\Rightarrow$ $q=3$

Thus, $\frac{1}{x}=2$ and $\frac{1}{y}=3$

$\Rightarrow$ $x=\frac{1}{2}$ and $y=\frac{1}{3}$. **Ans.**

Q. 9. Solve by elimination :

$$3x-y=7$$

$$2x+5y+1=0$$

Solution : Given,

$$3x-y=7 \quad ...(i)$$

$$2x+5y=-1 \quad ...(ii)$$

Multiplying equation (i) by 5 and (ii) by 1, we get

$$2x+5y=-1 \quad ...(iii)$$

$$15x-5y=35 \quad ....(iv)$$

$$17x=34$$

[by adding eqn. (iii) and (iv)]

$$\Rightarrow \quad x = \frac{34}{17} = 2$$

Putting the value of x in (i), we get

$$3(2) - y = 7$$

$$\Rightarrow \quad 6 - y = 7 \Rightarrow -y = 7 - 6$$

$$\therefore \quad y = -1$$

Thus $\quad x = 2$ and $y = -1$ **Ans.**

Q. 10. Solve for x and y :

$$\frac{10}{x+y} + \frac{2}{x-y} = 4 \text{ and } \frac{15}{x+y} - \frac{9}{x-y} = -2.$$

[NCERT]

Solution : Given,

$$\frac{10}{x+y} + \frac{2}{x-y} = 4$$

and
$$\frac{15}{x+y} - \frac{9}{x-y} = -2$$

Let
$$\frac{1}{x+y} = u \text{ and } \frac{1}{x-y} = v$$

Thus
$$10u + 2v = 4$$

$$\Rightarrow \quad 5u + v = 2 \quad \ldots\text{(i)}$$

And
$$15u - 9v = -2 \quad \ldots\text{(ii)}$$

Multiplying equation (i) by 9, we get

$$45u + 9v = 18 \quad \ldots\text{(iii)}$$

Adding equations (ii) and (iii), we get

$$60u = 16$$

$$\Rightarrow \quad 15u = 4$$

$$\Rightarrow \quad u = \frac{4}{15}$$

Thus
$$\frac{1}{x+y} = \frac{4}{15}$$

$$\Rightarrow \quad 4x + 4y = 15 \quad \ldots\text{(iv)}$$

Substituting $u = \frac{4}{15}$ in equation (iii), we get

$$45\left(\frac{4}{15}\right) + 9v = 18$$

$$\Rightarrow \quad 12 + 9v = 18$$

$$\Rightarrow \quad 9v = 6$$

$$\Rightarrow \quad v = \frac{2}{3}$$

Thus
$$\frac{1}{x-y} = \frac{2}{3}$$

$$\Rightarrow \quad 2x - 2y = 3$$

$$\Rightarrow \quad 4x - 4y = 6 \quad \ldots\text{(v)}$$

Adding equations (iv) and (v), we get

$$8x = 21$$

$$\Rightarrow \quad x = \frac{21}{8} = 2\frac{5}{8}$$

Substituting $x = \frac{21}{8}$ in equation (iv), we get

$$4\left(\frac{21}{8}\right) + 4y = 15$$

$$\Rightarrow \quad 4y = 15 - \frac{21}{2} = \frac{9}{2}$$

$$\Rightarrow \quad y = \frac{9}{8} = 1\frac{1}{8}$$

Thus $\quad x = 2\frac{5}{8}$ and $y = 1\frac{1}{8}$. **Ans.**

Q. 11. Solve the following system of equations for x and y :★

$$\frac{5}{x-1} + \frac{1}{y-2} = 2 \text{ and } \frac{6}{x-1} - \frac{3}{y-2} = 1 \; (x \neq 0, y \neq 0).$$

Solution : Given,
$$\frac{5}{x-1} + \frac{1}{y-2} = 2 \quad \ldots\text{(i)}$$

and
$$\frac{6}{x-1} - \frac{3}{y-2} = 1 \quad \ldots\text{(ii)}$$

Multiplying equation (i) with 3, we get

$$\frac{15}{x-1} + \frac{3}{y-2} = 6 \quad \ldots\text{(iii)}$$

Adding equations (ii) and (iii), we get

$$\frac{21}{x-1} = 7$$

$$\Rightarrow \quad x - 1 - 3$$

$$\Rightarrow \quad x = 4$$

Substituting $x = 4$ in equation (i), we have

$$\frac{5}{4-1} + \frac{1}{y-2} = 2$$

$$\Rightarrow \quad \frac{5}{3} + \frac{1}{y-2} = 2$$

$$\Rightarrow \quad \frac{5(y-2)+3}{3(y-2)} = 2$$

$$\Rightarrow \quad 5(y-2) + 3 = 6(y-2)$$

$$\Rightarrow \quad y - 2 = 3$$

$$\Rightarrow \quad y = 5$$

Thus $\quad x = 4$ and $y = 5$. **Ans.**

★ are board exam questions from previous years

Q. 12. Solve for x and y :

$$\frac{x+1}{2}+\frac{y-1}{3}=8 \text{ and } \frac{x-1}{3}+\frac{y+1}{2}=9.$$

Solution : Given, $\frac{x+1}{2}+\frac{y-1}{3}=8$

$\Rightarrow \quad \frac{3(x+1)+2(y-1)}{6}=8$

$\Rightarrow \quad 3(x+1)+2(y-1)=48$

$\Rightarrow \quad 3x+3+2y-2=48$

$\Rightarrow \quad 3x+2y+1=48$

$\Rightarrow \quad 3x+2y=47$ ...(i)

and $\quad \frac{x-1}{3}+\frac{y+1}{2}=9$

$\Rightarrow \quad \frac{2(x-1)+3(y+1)}{6}=9$

$\Rightarrow \quad 2(x-1)+3(y+1)=54$

$\Rightarrow \quad 2x-2+3y+3=54$

$\Rightarrow \quad 2x+3y+1=54$

$\Rightarrow \quad 2x+3y=53$ ...(ii)

Adding equations (i) and (ii), we get

$5x+5y=100$

$\Rightarrow \quad 5(x+y)=100$

$\Rightarrow \quad x+y=20$ ...(iii)

Also subtracting equation (ii) from equation (i), we get

$x-y=-6$ ...(iv)

Adding equations (iii) and (iv), we get

$2x=14$

$\Rightarrow \quad x=7$

Substituting $x = 7$ in equation (iii), we get

$7+y=20$

$\Rightarrow \quad y=20-7=13$

Thus $\quad x=7$ and $y=13$. **Ans.**

Q. 13. Solve for x and y :

$$(a-b)x+(a+b)y=a^2-2ab-b^2$$

and $\quad (a+b)(x+y)=a^2+b^2$.★

Solution : Given,

$(a-b)x+(a+b)y=a^2-2ab-b^2$ ...(i)

and $\quad (a+b)(x+y)=a^2+b^2$

or $\quad (a+b)x+(a+b)y=a^2+b^2$ ...(ii)

Subtracting equation (ii) from equation (i), we get

$(a-b)x-(a+b)x=a^2-2ab-b^2-a^2-b^2$

$\Rightarrow \quad (a-b-a-b)x=-2ab-2b^2$

$\Rightarrow \quad -2bx=-2b(a-b)$

$\Rightarrow \quad x=a-b$

Substituting $x = a - b$ in equation (i), we get

$(a-b)(a-b)+(a+b)y=a^2-2ab-b^2$

$\Rightarrow \quad a^2+b^2-2ab+(a+b)y=a^2-2ab-b^2$

$\Rightarrow \quad (a+b)y=-2b^2$

$\Rightarrow \quad y=-\frac{2b^2}{a+b}$

Thus, $\quad x=a-b$ and $y=-\frac{2b^2}{a+b}$. **Ans.**

Q. 14. Solve for x and y :★

$$\frac{4}{x}+3y=8 \text{ and } \frac{6}{x}-4y=-5.$$

Solution : Given, $\frac{4}{x}+3y=8$ ...(i)

and $\quad \frac{6}{x}-4y=-5$ ...(ii)

Multiplying equation (i) with 4 and equation (ii) with 3, we get

$\frac{16}{x}+12y=32$ ...(iii)

$\frac{18}{x}-12y=-15$ ...(iv)

Adding equations (iii) and (iv), we get

$\frac{34}{x}=17$

$\Rightarrow \quad x=2$

Substituting $x = 2$ in equation (i), we get

$\frac{4}{2}+3y=8$

$\Rightarrow \quad 2+3y=8$

$\Rightarrow \quad 3y=6$

$\Rightarrow \quad y=2$

Thus $\quad x=y=2$. **Ans.**

Q. 15. Solve by cross-multiplication :

$$\frac{57}{x+y}+\frac{6}{x-y}=5 \text{ and } \frac{38}{x+y}+\frac{21}{x-y}=9.$$

Solution : Given,

$\frac{57}{x+y}+\frac{6}{x-y}=5$

and $\quad \frac{38}{x+y}+\frac{21}{x-y}=9$

Let $\quad \frac{1}{x+y}=a$ and $\frac{1}{x-y}=b$

Thus $\quad 57a+6b-5=0$ ...(i)

and $\quad 38a+21b-9=0$ ...(ii)

By cross multiplication,

★ are board exam questions from previous years

$$\frac{a}{6(-9)-21(-5)} = \frac{b}{38(-5)-57(-9)}$$

$$= \frac{1}{(57)(21)-(38)(6)}$$

$$\Rightarrow \quad \frac{a}{-54+105} = \frac{b}{-190+513} = \frac{1}{1197-228}$$

$$\Rightarrow \quad \frac{a}{51} = \frac{b}{323} = \frac{1}{969}$$

$$\Rightarrow \quad a = \frac{51}{969} = \frac{1}{19} = \frac{1}{x+y}$$

$$\Rightarrow \quad x+y = 19$$

$$\Rightarrow \quad x+y-19 = 0 \quad \text{...(iii)}$$

and $$b = \frac{323}{969} = \frac{1}{3} = \frac{1}{x-y}$$

$$\Rightarrow \quad x-y = 3$$

$$\Rightarrow \quad x-y-3 = 0 \quad \text{...(iv)}$$

Again solving equations (ii) and (iv) by cross multipliation, we get

$$\Rightarrow \quad \frac{x}{-3-19} = \frac{y}{-19+3} = \frac{1}{-1-1}$$

$$\Rightarrow \quad x = 11 \text{ and } y = 8.$$ **Ans.**

Q. 16. Determine the value of a and b for which the given system of equations has infinitely many solutions :

$(2a-1)x + 3y - 5 = 0$ and $3x + (b-1)y - 2 = 0$.

Solution : Given,

$$(2a-1)x + 3y - 5 = 0 \quad \text{...(i)}$$

and $$3x + (b-1)y - 2 = 0 \quad \text{...(ii)}$$

For this pair of equations to have infinite number of solutions, it must satisfy the condition of

$$\frac{2a-1}{3} = \frac{3}{b-1} = \frac{5}{2}$$

$$\Rightarrow \quad 2(2a-1) = 15$$

$$\Rightarrow \quad 4a-2 = 15$$

$$\Rightarrow \quad 4a = 17$$

$$\Rightarrow \quad a = 4\frac{1}{4}$$

and $$5(b-1) = 6$$

$$\Rightarrow \quad 5b-5 = 6$$

$$\Rightarrow \quad 5b = 11$$

$$\Rightarrow \quad b = 2\frac{1}{5}$$

Thus, the pair of equations will have infinite number of solutions for $a = 4\frac{1}{4}$ and $b = 2\frac{1}{5}$. **Ans.**

★ are board exam questions from previous years

Q. 17. Solve the following pair of equations by reducing them to a pair of linear equations :★

$$\frac{1}{x} - \frac{4}{y} = 2$$

$$\frac{1}{x} + \frac{3}{y} = 9$$

Solution : Given, $$\frac{1}{x} - \frac{4}{y} = 2$$

$$\frac{1}{x} + \frac{3}{y} = 9$$

Let $$\frac{1}{x} = u, \ \frac{1}{y} = v$$

So, $$u - 4v = 2 \quad \text{...(i)}$$

$$u + 3v = 9 \quad \text{...(ii)}$$

On subtracting equations (i) and (ii), we get

$$u - 4v = 2$$
$$u + 3v = 9$$
$$- \quad - \quad -$$
$$-7v = -7$$
$$v = 1$$

Substituting $v = 1$ in equation (i), we have

$$u - 4 \times 1 = 2$$
$$u - 4 = 2$$
$$u = 2 + 4$$
$$u = 6$$

So $$v = 1 \Rightarrow \frac{1}{y} = 1, \Rightarrow y = 1$$

$$u = 6 \Rightarrow \frac{1}{x} = 6, \Rightarrow x = \frac{1}{6}$$

Thus, $x = \frac{1}{6}$ and $y = 1$. **Ans.**

Q. 18. Determine the values of a and b for which the given system of equations have infinitely many solutions : $2x - 3y = 7$ and $(a+b)x - (a+b-3)y = 4a + b$.

Solution : Given,

$$2x - 3y = 7 \quad \text{...(i)}$$

$$\text{and} (a+b)x - (a+b-3)y = 4a + b \quad \text{...(ii)}$$

For this pair of equations to have infinite number of solutions, it must satisfy the condition of

$$\frac{2}{a+b} = \frac{3}{a+b-3} = \frac{7}{4a+b}$$

$$\Rightarrow \quad 2(4a+b) = 7(a+b)$$

$$\Rightarrow \quad 8a + 2b = 7a + 7b$$

$$\Rightarrow \quad a = 5b$$

$\Rightarrow \quad a - 5b = 0 \quad \ldots(iii)$

and $\quad 2(a + b - 3) = 3(a + b)$

$\Rightarrow \quad 2a + 2b - 6 = 3a + 3b$

$\Rightarrow \quad a + b = -6 \quad \ldots(iv)$

Subtracting equation (iv) from equation (iii), we get

$$-5b - b = 0 - (-6)$$

$\Rightarrow \quad -6b = 6$

$\Rightarrow \quad b = -1$

Substituting $b = -1$ in equation (iv), we have

$$a - 1 = -6$$

$\Rightarrow \quad a = -5$

Thus, the pair of equations will have infinite number of solutions for $a = -5$ and $b = -1$. **Ans.**

Q. 19. Find the value of *k* for which the following pair of linear equations have infinitely many solutions.★

$$\mathbf{2x + 3y = 7, (k + 1)x + (2k - 1)y = 4k + 1}$$

Solution. Given,

$$2x + 3y = 7 \text{ and } (k + 1)x + (2k - 1)y = 4k + 1$$

On comparing above equations with $a_1x + b_1y + c_1 = 0$ and $a_2x + b_2y + c_2 = 0$, we get

$$a_1 = 2, \quad b_1 = 3 \quad c_1 = -7$$
$$a_2 = k + 1 \quad b_2 = 2k - 1 \quad c_2 = -(4k + 1)$$

For infinitely many solutions

$$\frac{a_1}{a_2} = \frac{b_1}{b_2} = \frac{c_1}{c_2}$$

$\therefore \quad \dfrac{2}{k+1} = \dfrac{3}{2k-1} = \dfrac{-7}{-(4k+1)}$

$\Rightarrow \quad 2(2k - 1) = 3(k + 1)$

$$4k - 2 = 3k + 3$$

$$k = 5$$

or $\quad 3(4k + 1) = 7(2k - 1)$

$\Rightarrow \quad k = 5$

Hence, $\quad k = 5$. **Ans.**

Q. 20. Determine the values of *a* and *b* for which the given system of equations has infinitely many solutions : 3x + 4y = 12 and (a + b)x + 2(a – b)y = 5a – 1.

Solution : Given, $3x + 4y = 12 \quad \ldots(i)$

and $\quad (a + b)x + 2(a - b)y = 5a - 1 \quad \ldots(ii)$

For this pair of equations to have infinite number of solutions, it must satisfy the condition of

$$\frac{3}{a+b} = \frac{2}{a-b} = \frac{12}{5a-1}$$

$\Rightarrow \quad (a + b)12 = 3(5a - 1)$

$\Rightarrow \quad 12a + 12b = 15a - 3$

$\Rightarrow \quad 3a - 12b = 3$

$\Rightarrow \quad a - 4b = 1 \quad \ldots(iii)$

and $\quad 2(5a - 1) = 12(a - b)$

$\Rightarrow \quad 5a - 1 = 6a - 6b$

$\Rightarrow \quad a - 6b = -1 \quad \ldots(iv)$

Subtracting equation (iv) from equation (iii), we get

$$2b = 2$$

$\Rightarrow \quad b = 1$

Substituting $b = 1$ in equation (iii), we get

$$a - 4(1) = 1$$

$\Rightarrow \quad a - 4 = 1$

$\Rightarrow \quad a = 5$

Thus, the pair of equations will have infinite number of solutions for $a = 5$ and $b = 1$. **Ans.**

Q. 21. Determine the values of *a* and *b* for which the given system of equations has infinitely many solutions: 2x + 3y = 7 and (a – b)x + (a + b)y = 3a + b – 1. [NCERT]

Solution : Given, $2x + 3y = 7 \quad \ldots(i)$

and $\quad (a - b)x + (a + b)y = 3a + b - 1 \quad \ldots(ii)$

For this pair of equations to have infinite number of solutions, it must satisfy the condition of

$$\frac{2}{a-b} = \frac{3}{a+b} = \frac{7}{3a+b-1}$$

$\Rightarrow \quad 2(a + b) = 3(a - b)$

$\Rightarrow \quad 2a + 2b = 3a - 3b$

$\Rightarrow \quad a = 5b \quad \ldots(iii)$

and $\quad 2(3a + b - 1) = 7(a - b)$

$\Rightarrow \quad 6a + 2b - 2 = 7a - 7b$

$\Rightarrow \quad a - 9b = -2 \quad \ldots(iv)$

Substituting $a = 5b$ in equation (iv), we get

$$5b - 9b = -2$$

$\Rightarrow \quad -4b = -2$

$\Rightarrow \quad 2b = 1$

$\Rightarrow \quad b = \dfrac{1}{2}$

Thus $\quad a = 5\left(\dfrac{1}{2}\right) = \dfrac{5}{2}$

Thus, the pair of equations will have infinite number of solutions for $a = \dfrac{5}{2}$ and $b = \dfrac{1}{2}$. **Ans.**

Q. 22. Determine the value of *a* for which the given system of equations has infinitely many solutions : 2x + 3y = 7 and (a – 1)x + (a + 2)y = 3a.★

Solution : Given, $2x + 3y = 7 \quad \ldots(i)$

and $\quad (a - 1)x + (a + 2)y = 3a \quad \ldots(ii)$

★ are board exam questions from previous years

For this pair of equations to have infinite number of solutions, it must satisfy the condition of

$$\frac{2}{a-1} = \frac{3}{a+2} = \frac{7}{3a}$$

$\Rightarrow$ $2(a+2) = 3(a-1)$

$\Rightarrow$ $2a+4 = 3a-3$

$\Rightarrow$ $a = 7$

Also $\frac{3}{a+2} = \frac{7}{3a}$

$9a = 7a + 14$

$2a = 14$

$a = 7$

Thus, the pair of equations will have infinite number of solutions for $a = 7$. **Ans.**

Q. 23. $7x - 5y - 4 = 0$ is given. Write another linear equation, so that the lines represented by the pair are :★

(i) Intersecting (ii) Coincident (iii) Parallel.

Solution : Given equation is,

$7x - 5y - 4 = 0$

(i) $7x + 3y + 2 = 0$ $\left[\because \text{ Here, } \frac{a_1}{a_2} \neq \frac{b_1}{b_2}\right]$

(ii) $14x - 10y - 8 = 0$ $\left[\because \text{ Here, } \frac{a_1}{a_2} = \frac{b_1}{b_2} = \frac{c_1}{c_2}\right]$

(iii) $7x - 5y + 3 = 0$ $\left[\because \text{ Here, } \frac{a_1}{a_2} = \frac{b_1}{b_2} \neq \frac{c_1}{c_2}\right]$

Q. 24. Determine the value of *a* for which the given system of equations has infinitely many solutions : $2x + 3y = 7$ and $(a + b + 1)x + (a + 2b + 2)y = 4(a + b) + 1$.

Solution : Given, $2x + 3y = 7$ ...(i)

and $(a + b + 1)x + (a + 2b + 2)y = 4(a + b) + 1$ ...(ii)

For this pair of equations to have infinite number of solutions, it must satisfy the condition of

$$\frac{2}{a+b+1} = \frac{3}{a+2b+2} = \frac{7}{4(a+b)+1}$$

$\Rightarrow$ $2(a + 2b + 2) = 3(a + b + 1)$

$\Rightarrow$ $2a + 4b + 4 = 3a + 3b + 3$

$\Rightarrow$ $a - b = 1$ ...(iii)

and $2(4a + 4b + 1) = 7(a + b + 1)$

$\Rightarrow$ $8a + 8b + 2 = 7a + 7b + 7$

$\Rightarrow$ $a + b = 5$ ...(iv)

Adding equations (iii) and (iv), we get

$2a = 6$

$\Rightarrow$ $a = 3$

Substituting $a = 3$ in equation (iv), we get

$3 + b = 5$

$\Rightarrow$ $b = 2$

Thus, the pair of equations will have infinite number of solutions for $a = 3$ and $b = 2$. **Ans.**

Q. 25. For what value of *k*, will the following pair of equations have infinitely many solutions :★

$2x + 3y = 7$ and $(k + 2)x - 3(1 - k)y = 5k + 1$

Solution : Given, The system of equations is

$2x + 3y = 7$ and $(k + 2)x - 3(1 - k)y = 5k + 1$

These equations are of the form $a_1x + b_1y + c_1 = 0$

and $a_2x + b_2y + c_2 = 0$

where, $a_1 = 2, b_1 = 3, c_1 = -7$

$a_2 = (k + 2), b_2 = -3(1 - k), c_2 = -(5k + 1)$

Since, the given system of equations have infinitely many solutions.

$$\therefore \quad \frac{a_1}{a_2} = \frac{b_1}{b_2} = \frac{c_1}{c_2}$$

$$\frac{2}{k+2} = \frac{3}{-3(1-k)} = \frac{-7}{-(5k+1)}$$

$$\frac{2}{k+2} = \frac{3}{-3(1-k)} \text{ and } \frac{3}{-3(1-k)} = \frac{7}{(5k+1)}$$

$-6(1 - k) = 3k + 6$ and $3(5k + 1) = -21(1 - k)$

$-6 + 6k = 3k + 6$ $15k + 3 = -21 + 21k$

$3k = 12$ $6k = 24$

$k = 4$ $k = 4$

Hence, the given system of equations has infinitely many solutions when $k = 4$. **Ans.**

Q. 26. Determine the value of *k* for which the given system of equations has infinitely many solutions :

$5x + 2y = 2k$ and $2(k + 1)x + ky = (3k + 4)$.

Solution : Given, $5x + 2y = 2k$ ...(i)

and $2(k + 1)x + ky = (3k + 4)$ ...(ii)

For the pair of equations to have infinite number of solutions, it must satisfy the condition of

$$\frac{5}{2(k+1)} = \frac{2}{k} = \frac{2k}{3k+4}$$

$\Rightarrow$ $\frac{2}{k} = \frac{2k}{3k+4}$

$\Rightarrow$ $2(3k + 4) = 2k^2$

$\Rightarrow$ $6k + 8 = 2k^2$

$\Rightarrow$ $2k^2 - 6k - 8 = 0$

$\Rightarrow$ $k^2 - 3k - 4 = 0$

★ are board exam questions from previous years

$$\Rightarrow \quad k^2 - 4k + k - 4 = 0$$

$$k(k-4) + 1(K-4) = 0$$

$$(k-4)(k+1) = 0$$

$$\Rightarrow \quad k = 4, -1$$

and $\quad \dfrac{5}{2(k+1)} = \dfrac{2}{k}$

$$\Rightarrow \quad 4(k+1) = 5k$$

$$\Rightarrow \quad 4k + 4 = 5k$$

$$\Rightarrow \quad k = 4$$

Thus, the pair of equations will have infinite number of solutions for $k = 4$. **Ans.**

Q. 27. Find c if the system of equations $cx + 3y + (3 - c) = 0$, $12x + cy - c = 0$ has infinitely many solutions ?★

Solution : The given equations are

$$cx + 3y + (3 - c) = 0$$

and $\quad 12x + cy - c = 0$

On comparing with equation $a_1x + b_1y + c_1 = 0$ and equation $a_2x + b_2y + c_2 = 0$, we get

$$a_1 = c, b_1 = 3, c_1 = 3 - c$$

and $\quad a_2 = 12, b_2 = c, c_2 = -c$

For infinitely many solutions

$$\frac{a_1}{a_2} = \frac{b_1}{b_2} = \frac{c_1}{c_2}$$

or $\quad \dfrac{c}{12} = \dfrac{3}{c} = \dfrac{3-c}{-c}$

$$\frac{c}{12} = \frac{3}{c} \quad \text{or} \quad \frac{3}{c} = \frac{3-c}{-c}$$

$$\Rightarrow \quad c^2 = 36 \qquad \Rightarrow \quad -3c = 3c - c^2$$

$$\Rightarrow \quad c = \pm 6 \qquad \Rightarrow \quad -6c = -c^2$$

$$\Rightarrow \quad c^2 - 6c = 0$$

$$\Rightarrow \quad c(c-6) = 0$$

$$\Rightarrow \quad c = 0 \text{ or } c = 6$$

So, from both the above cases

$$c = 6$$ **Ans.**

Q. 28. Determine the value of k for which the given system of equations has infinitely many solutions :

$x + (k+1)y = 5$ and $(k+1)x + 9y = (8k - 1)$.

Solution : Given, $\quad x + (k+1)y = 5 \quad$ ...(i)

and $\quad (k+1)x + 9y = (8k - 1) \quad$ ...(ii)

For this pair of equations to have infinite number of solutions, it must satisfy the condition of

$$\frac{1}{k+1} = \frac{k+1}{9} = \frac{5}{8k-1}$$

$$\Rightarrow \quad \frac{1}{k+1} = \frac{5}{8k-1}$$

$$\Rightarrow \quad 5(k+1) = 8k - 1$$

$$\Rightarrow \quad 5k + 5 = 8k - 1$$

$$\Rightarrow \quad 3k = 6$$

$$\Rightarrow \quad k = 2$$

and $\quad \dfrac{1}{k+1} = \dfrac{k+1}{9}$

$$(k+1)^2 = 9$$

$$k + 1 = \pm 3$$

$$\Rightarrow \quad k = 2$$

or $\quad k = -4$

(Rejected as it doesn't satisfy the condition)

Thus, the pair of equations will have infinite number of solutions for $k = 2$. **Ans.**

Q. 29. The sum of the digits of a two digit number is 8 and the difference between the number and that formed by reversing the digits is 18. Find the number.★

Solution : Let $\quad$ Unit's digit $= x$

Ten's digit $= y$

So, original number = unit's digit + 10 × ten's digit

$$= x + 10y$$

According to question,

Sum of digits = 8

$$\Rightarrow \quad x + y = 8$$

$$\Rightarrow \quad y = 8 - x \quad \text{...(i)}$$

On reversing the digits, unit's digit $= y$

Ten's digit $= x$

So, $\quad$ New number $= 10x + y$

According to question,

Difference = 18

$$\Rightarrow \quad x + 10y - (10x + y) = 18$$

$$\Rightarrow \quad x + 10y - 10x - y = 18$$

$$\Rightarrow \quad 9y - 9x = 18$$

$$\Rightarrow \quad y - x = 2 \quad \text{...(ii)}$$

By adding equations (i) and (ii), we get

$$2y = 10$$

$$y = \frac{10}{2} \Rightarrow y = 5$$

Substituting $y = 5$ in equation (i), we get

$$x + 5 = 8$$

$$\Rightarrow \quad x = 8 - 5$$

$$\Rightarrow \quad x = 3$$

$\therefore$ Original number $= 10y + x$

$$= 10 \times 5 + 3$$

★ are board exam questions from previous years

$= 50 + 3$

$= 53$ **Ans.**

Q. 30. The well monitored total expenditure per month of a household consists of a fixed rent of the house and mess charges depending upon the number of people sharing the house. The total monthly expenditure is ₹ 3900 for 2 people and ₹ 7500 for 5 people. Find the rent of the house.

Solution : Let the monthly rent of the house be ₹ x and the mess charges per head per month be ₹ y.

According to the question,

$x + 2y = 3900$ ...(i)

and $x + 5y = 7500$ ...(ii)

Subtracting equation (ii) from equation (i), we get

$-3y = -3600$

$\Rightarrow y = 1200$

substituting $y = 1200$ in equation (i), we get

$x + 2(1200) = 3900$

$\Rightarrow x + 2400 = 3900$

$\Rightarrow x = 3900 - 2400$

$\Rightarrow x = 1500$

Hence, monthly rent = ₹ 1500 **Ans.**

Q. 31. The ratio of income of two persons is 11 : 7 and the ratio of their expenditures is 9 : 5. If each of them manages to save ₹ 400 per month, find their monthly incomes.

Solution : Let the income of two persons be $11x$ and $7x$.

Also, the expenditures of two persons be $9y$ and $5y$, respectively.

According to the question,

$11x - 9y = 400$ ...(i)

and $7x - 5y = 400$ ...(ii)

From equations (i) and (ii)

$11x - 9y = 7x - 5y$

$\Rightarrow 4x = 4y$

$\Rightarrow x = y$ ...(iii)

Substituting $x = y$ in equation (ii), we get

$7y - 5y = 400$

$\Rightarrow 2y = 400$

$\Rightarrow y = 200$

$\therefore$ From equation (iii),

$x = y = 200$

Thus, their monthly income is

$11 \times 200 = 2200$

and $7 \times 200 = 1400.$ **Ans.**

Q. 32. The taxi charges in a city comprise of a fixed charge together with the charges for the distance covered. For a journey of 10 km the charge paid is ₹ 75 and for a journey of 15 km the charge paid is ₹ 110. What will a person have to pay for travelling a distance of 25 km ?

Solution : Let the fixed charges of taxi be ₹ x and the running charges be ₹ y per km.

According to the question,

$x + 10y = 75$ ...(i)

$x + 15y = 110$ ...(ii)

On subtracting equation (ii) from equation (i), we get

$-5y = -35$

$\Rightarrow y = 7$

Substituting $y = 7$ in equation (i), we get

$x + 10(7) = 75$

$x + 70 = 75$

$x = 75 - 70$

$x = 5$

Thus, fixed charge is ₹ 5 and the running charges is ₹ 7 per km.

$\therefore$ Total charges for travelling a distance of 25 km

$= x + 25y$

$= 5 + 25 \times 7$

$= 5 + 175$

$=$ ₹ 180 **Ans.**

Q. 33. An honest person invested some amount at the rate of 12% simple interest and some other amount at the rate of 10% simple interest. He received yearly interest of 130, but if he had interchanged amounts invested, he would have received 4 more as interest. How much amount did he invest at different rates ?

Solution : Suppose the person invested ₹ x at the rate of 12% simple interest and ₹ y at the rate of 10% simple interest, then

$$\text{Yearly interest} = \frac{12x}{100} + \frac{10y}{100}$$

Given that yearly interest = ₹ 130

$$\therefore \quad \frac{12x + 10y}{100} = 130$$

$12x + 10y = 13000$

or $6x + 5y = 6500$ ...(i)

Also if the investment is interchanged, then yearly interest is ₹ 134 *i.e.*,

$$\frac{10x}{100} + \frac{12y}{100} = 134$$

$$10x + 12y = 13400$$

$$5x + 6y = 6700 \quad ...(ii)$$

On adding equations (i) and (ii), we get

$$11x + 11y = 13200$$

$$\Rightarrow \quad x + y = 1200 \quad ...(iii)$$

Also subtracting equations (i) and (ii), we get

$$x - y = -200 \quad ...(iv)$$

On adding equations (iii) and (iv), we get

$$2x = 1000$$

$$\Rightarrow \quad x = 500$$

Substituting $x = 500$ in equation (iii), we get

$$500 + y = 1200$$

$$\Rightarrow \quad y = 1200 - 500$$

$$\Rightarrow \quad y = 700$$

Thus, the person invested ₹ 500 at the rate of 12% per year and ₹ 700 at the rate of 10% per year. **Ans.**

Chapter 4. Quadratic Equations

Q. 1. Solve : $\frac{3}{x+1} - \frac{1}{2} = \frac{2}{3x-1}$, where $x \neq -1, \frac{1}{3}$.★

Solution : Given, $\frac{3}{x+1} - \frac{1}{2} = \frac{2}{3x-1}$

$$\Rightarrow \quad \frac{6-(x+1)}{2(x+1)} = \frac{2}{3x-1}$$

$$\Rightarrow \quad [6-(x+1)](3x-1) = 4(x+1)$$

$$\Rightarrow \quad 6(3x-1) - (3x-1)(x+1) = 4(x+1)$$

$$\Rightarrow \quad 18x - 6 - (3x^2 - x + 3x - 1) = 4x + 4$$

$$\Rightarrow \quad 14x - 10 - 3x^2 - 2x + 1 = 0$$

$$\Rightarrow \quad -3x^2 + 12x - 9 = 0$$

$$\Rightarrow \quad 3x^2 - 12x + 9 = 0$$

$$\Rightarrow \quad x^2 - 4x + 3 = 0$$

$$\Rightarrow \quad x^2 - 3x - x + 3 = 0$$

$$\Rightarrow \quad x(x-3) - 1(x-3) = 0$$

$$\Rightarrow \quad (x-3)(x-1) = 0$$

$$\Rightarrow \quad x = 1, 3.$$

Ans.

Q. 2. Solve : $\frac{4}{x} - 3 = \frac{5}{2x+3}$, where $x \neq 0, -\frac{3}{2}$.★★

Solution : Given, $\frac{4}{x} - 3 = \frac{5}{2x+3}$

$$\Rightarrow \quad \frac{4-3x}{x} = \frac{5}{2x+3}$$

$$\Rightarrow \quad 4(2x+3) - 3x(2x+3) = 5x$$

$$\Rightarrow \quad 8x + 12 - 6x^2 - 9x = 5x$$

$$\Rightarrow \quad -6x^2 - 6x + 12 = 0$$

$$\Rightarrow \quad 6x^2 + 6x - 12 = 0$$

$$\Rightarrow \quad x^2 + x - 2 = 0$$

$$\Rightarrow \quad x^2 + 2x - x - 2 = 0$$

$$\Rightarrow \quad x(x+2) - 1(x+2) = 0$$

$$\Rightarrow \quad (x+2)(x-1) = 0$$

$$\Rightarrow \quad x = 1, -2.$$

Ans.

Q. 3. Solve for x :★

$$\frac{2x}{x-3} + \frac{1}{2x+3} + \frac{3x+9}{(x-3)(2x+3)} = 0, x \neq 3, -\frac{3}{2}$$

Solution : We have,

$$\frac{2x}{x-3} + \frac{1}{2x+3} + \frac{3x+9}{(x-3)(2x+3)} = 0$$

$$2x(2x+3) + (x-3) + (3x+9) = 0$$

$$4x^2 + 6x + x - 3 + 3x + 9 = 0$$

$$4x^2 + 10x + 6 = 0$$

$$2x^2 + 5x + 3 = 0$$

$$2x^2 + 2x + 3x + 3 = 0$$

$$2x(x+1) + 3(x+1) = 0$$

$$(2x+3)(x+1) = 0$$

$$\Rightarrow \quad x = -1, \frac{-3}{2}$$

$$\Rightarrow \quad x = -1$$

[∵ Given $x \neq -3/2$] **Ans.**

Q. 4. Solve : $\frac{14}{x+3} - 1 = \frac{5}{x+1}$, where $x \neq -3, -1$.

Solution : Given, $\frac{14}{x+3} - 1 = \frac{5}{x+1}$

$$\Rightarrow \quad \frac{14-(x+3)}{x+3} = \frac{5}{x+1}$$

$$\Rightarrow \quad 14(x+1) - (x+3)(x+1) = 5(x+3)$$

$$\Rightarrow \quad 14x + 14 - (x^2 + x + 3x + 3) = 5x + 15$$

$$\Rightarrow \quad 9x - 1 - x^2 - 4x - 3 = 0$$

$$\Rightarrow \quad -x^2 + 5x - 4 = 0$$

$$\Rightarrow \quad x^2 - 5x + 4 = 0$$

$$\Rightarrow \quad x^2 - 4x - x + 4 = 0$$

$$\Rightarrow \quad x(x-4) - 1(x-4) = 0$$

$$\Rightarrow \quad (x-4)(x-1) = 0$$

$$\Rightarrow \quad x = 1, 4.$$

Ans.

Q. 5. Solve : $x^2 - 4ax - b^2 + 4a^2 = 0$.★

Solution : Given, $x^2 - 4ax - b^2 + 4a^2 = 0$

$$\Rightarrow \quad x^2 - 4ax - (b^2 - 4a^2) = 0$$

$$\Rightarrow \quad x^2 - (b+2a)x + (b-2a)x - (b-2a)(b+2a) = 0$$

$$\Rightarrow \quad x[x-(b+2a)] + (b-2a)[x-(b+2a)] = 0$$

$$\Rightarrow \quad x[x-b-2a] + (b-2a)[x-b-2a] = 0$$

$$\Rightarrow \quad [x-b-2a][x+b-2a] = 0$$

$$\Rightarrow \quad x = 2a + b \text{ and } 2a - b.$$

Ans.

★ are board exam questions from previous years
★★ are frequently asked board exam questions

Q. 6. The sum of two natural numbers is 8 and their product is 15. Find the numbers.★

Solution : Let one of the numbers be x.

So, the other number is $(8 - x)$.

Now, $x(8 - x) = 15$

$\Rightarrow 8x - x^2 = 15$

$\Rightarrow x^2 - 8x + 15 = 0$

$\Rightarrow x^2 - 5x - 3x + 15 = 0$

$\Rightarrow x(x - 5) - 3(x - 5) = 0$

$\Rightarrow (x - 5)(x - 3) = 0$

$\Rightarrow x = 3 \text{ or } 5$

$\therefore 8 - x = 5 \text{ or } 3$

Thus, the two numbers are 3 and 5. **Ans.**

Q. 7. Solve : $4x^2 - 4ax + (a^2 - b^2) = 0$.★

Solution : Given,

$4x^2 - 4ax + (a^2 - b^2) = 0$

$\Rightarrow 4x^2 - 4ax + (a - b)(a + b) = 0$

$\Rightarrow 4x^2 - 2(a - b)x - 2(a + b)x + (a - b)(a + b) = 0$

$\Rightarrow 2x[2x - (a - b)] - (a + b)[2x - (a - b)] = 0$

$\Rightarrow [2x - (a - b)][2x - (a + b)] = 0$

$\Rightarrow 2x = a - b \text{ and } a + b$

$\Rightarrow x = \frac{a-b}{2} \text{ and } \frac{a+b}{2}$. **Ans.**

Q. 8. Solve : $3x^2 - 2\sqrt{6}x + 2 = 0$.★★ **[NCERT]**

Solution : Given, $3x^2 - 2\sqrt{6}x + 2 = 0$

$\Rightarrow 3x^2 - \sqrt{6}x - \sqrt{6}x + 2 = 0$

$\Rightarrow \sqrt{3}x(\sqrt{3}x - \sqrt{2}) - \sqrt{2}(\sqrt{3}x - \sqrt{2}) = 0$

$\Rightarrow (\sqrt{3}x - \sqrt{2})^2 = 0$

$\Rightarrow \sqrt{3}x - \sqrt{2} = 0$

$\Rightarrow \sqrt{3}x = \sqrt{2}$

$\Rightarrow x = \sqrt{\frac{2}{3}}$

$\therefore x = \sqrt{\frac{2}{3}} \text{ and } \sqrt{\frac{2}{3}}$ **Ans.**

Q. 9. Solve : $\sqrt{7}y^2 - 6y - 13\sqrt{7} = 0$.

Solution : Given,

$\sqrt{7}y^2 - 6y - 13\sqrt{7} = 0$

$\Rightarrow \sqrt{7}y^2 - 13y + 7y - 13\sqrt{7} = 0$

$\Rightarrow y(\sqrt{7}y - 13) + \sqrt{7}(\sqrt{7}y - 13) = 0$

$\Rightarrow (\sqrt{7}y - 13)(y + \sqrt{7}) = 0$

$\Rightarrow y = \frac{13}{\sqrt{7}} \text{ and } -\sqrt{7}$. **Ans.**

★ are board exam questions from previous years

★★ are frequently asked board exam questions

Q. 10. The sum of a number and its reciprocal is $\frac{10}{3}$. Find the number.

Solution : Let the number be x. So, its reciprocal is $\frac{1}{x}$.

Now, $x + \frac{1}{x} = \frac{10}{3}$

$\Rightarrow \frac{x^2+1}{x} = \frac{10}{3}$

$\Rightarrow 3x^2 + 3 = 10x$

$\Rightarrow 3x^2 - 10x + 3 = 0$

$\Rightarrow 3x^2 - 9x - x + 3 = 0$

$\Rightarrow 3x(x - 3) - 1(x - 3) = 0$

$\Rightarrow (x - 3)(3x - 1) = 0$

$\Rightarrow x = 3 \text{ and } \frac{1}{3}$. **Ans.**

Q. 11. Divide 12 into two parts such that the sum of their squares is 74.

Solution : Let the two parts be x and $(12 - x)$.

So, $x^2 + (12 - x)^2 = 74$

$\Rightarrow x^2 + 144 + x^2 - 24x - 74 = 0$

$\Rightarrow 2x^2 - 24x + 70 = 0$

$\Rightarrow x^2 - 12x + 35 = 0$

$\Rightarrow x^2 - 7x - 5x + 35 = 0$

$\Rightarrow x(x - 7) - 5(x - 7) = 0$

$\Rightarrow (x - 7)(x - 5) = 0$

$\Rightarrow x = 5 \text{ and } 7$. **Ans.**

Q. 12. The sum of two numbers is 15 and their reciprocals is $\frac{3}{10}$. Find the numbers.

Solution : Let the numbers be x and $(15 - x)$. So, their reciprocals are $\frac{1}{x}$ and $\frac{1}{15-x}$ respectivley.

Now, $\frac{1}{x} + \frac{1}{15-x} = \frac{3}{10}$

$\Rightarrow \frac{15-x+x}{x(15-x)} = \frac{3}{10}$

$\frac{5}{x(15-x)} = \frac{1}{10}$

$\Rightarrow x(15 - x) = 50$

$\Rightarrow 15x - x^2 = 50$

$\Rightarrow x^2 - 15x + 50 = 0$

$\Rightarrow x^2 - 10x - 5x + 50 = 0$

$\Rightarrow x(x - 10) - 5(x - 10) = 0$

$\Rightarrow (x - 5)(x - 10) = 0$

$\Rightarrow x = 5 \text{ and } 10$. **Ans.**

Q. 13. The sum of the squares of two consecutive natural numbers is 421. Find the numbers.

Solution : Let the two consecutive natural numbers be x and $(x + 1)$.

Thus, $x^2 + (x + 1)^2 = 421$

$\Rightarrow x^2 + x^2 + 2x + 1 = 421$

$\Rightarrow 2x^2 + 2x - 420 = 0$

$\Rightarrow x^2 + x - 210 = 0$

$\Rightarrow x^2 + 15x - 14x - 210 = 0$

$\Rightarrow x(x + 15) - 14(x + 15) = 0$

$\Rightarrow (x + 15)(x - 14) = 0$

$\Rightarrow x = 14$ and -15

As -15 is not a natural number so, the two numbers will be 14 and 15. **Ans.**

Q. 14. The sum of the squares of two consecutive odd numbers is 394. Find the numbers.

Solution : Let the two consecutive odd numbers be x and $(x + 2)$.

Thus $x^2 + (x + 2)^2 = 394$

$\Rightarrow x^2 + x^2 + 4x + 4 = 394$

$\Rightarrow 2x^2 + 4x - 390 = 0$

$\Rightarrow x^2 + 2x - 195 = 0$

$\Rightarrow x^2 + 15x - 13x - 195 = 0$

$\Rightarrow x(x + 15) - 13(x + 15) = 0$

$\Rightarrow (x + 15)(x - 13) = 0$

$\Rightarrow x = 13$ and -15

So, the numbers are either 13 and 15 or -15 and -13. **Ans.**

Q. 15. In a class test, the sum of Kamal's marks in Mathematics and English is 40. Had he secured 3 more marks in Mathematics and 4 less in English then, the product of the marks in both the tests would have been 360. Find the marks obtained by him in the two subjects separately.

Solution : Let the marks obtained in Mathematics be x.

So the marks obtained in English is $40 - x$.

Now, $(x + 3)(40 - x - 4) = 360$

$\Rightarrow (x + 3)(36 - x) = 360$

$\Rightarrow 36x + 108 - x^2 - 3x = 360$

$\Rightarrow x^2 - 33x + 252 = 0$

$\Rightarrow x^2 - 21x - 12x + 252 = 0$

$\Rightarrow x(x - 21) - 12(x - 21) = 0$

$\Rightarrow (x - 12)(x - 21) = 0$

$\Rightarrow x = 12$ and 21

Thus, the marks obtained in Mathematics are either 12 and 21 and in Science are 28 or 19. **Ans.**

Q. 16. A dealer sells a toy for ₹ 24 and gains as much percent as the cost price of the toy. Find the cost price of the toy.

Solution : Suppose the cost price of the toy be ₹x, then gain is $x\%$

We know $\text{Gain} = \left(x \times \frac{x}{100}\right)$

$= \left(\frac{x^2}{100}\right)$

Now, S.P. = C.P. + Gain

$= x + \frac{x^2}{100}$

As, S.P. = ₹ 24 (Given)

$\therefore x + \frac{x^2}{100} = 24$

$\Rightarrow x^2 + 100x - 2400 = 0$

$x^2 + 120x - 20x - 2400 = 0$

$x(x + 120) - 20(x + 120) = 0$

$(x - 20)(x + 120) = 0$

$x = 20 \quad (\because x > 0)$

Hence, the cost price of the toy will be ₹ 20. **Ans.**

Q. 17. The difference of the squares of two natural numbers is 45. The square of the smaller number is 4 times the larger number. Find the numbers.

Solution : Let the smaller number be x and the larger be y.

Now, $x^2 = 4y$

Also, $y^2 - x^2 = 45$

$\Rightarrow y^2 - 4y = 45$

$\Rightarrow y^2 - 4y - 45 = 0$

$\Rightarrow y^2 - 9y + 5y - 45 = 0$

$\Rightarrow y(y - 9) + 5(y - 9) = 0$

$\Rightarrow (y - 9)(y + 5) = 0$

$\Rightarrow y = 9$ or -5

As natural numbers are not negative, so the larger number is 9.

Thus, $x^2 = 4(9) = 36$

$\Rightarrow x = 6$ or -6

Again as natural numbers are not negative, so the smaller number is 6.

So, the two natural numbers are 9 and 6. **Ans.**

Q. 18. Three-eighth of the students of a class opted for visiting an old age home. Sixteen students opted for having a nature walk. Square root of total number of students in the class opted for tree plantation in the school. The number of students who visited the old age home is same as the number of students who went for a nature walk and did tree plantation. Find the total number of students.

Solution : Suppose the total number of students is x

Number of students who opted for visiting old age home $= \frac{3}{8}x$

Number of students who opted for having a nature walk = 16

Number of students who opted for tree plantation $= \sqrt{x}$

So, $\frac{3}{8}x = 16 + \sqrt{x}$

$\therefore \quad 3x - 8\sqrt{x} - 128 = 0$

Suppose, $\sqrt{x} = y$

$\therefore \quad 3y^2 - 8y - 128 = 0$

$\therefore \quad (y-8)(3y+16) = 0$

$y = 8 \qquad [\because y > 0]$

Hence, $x = 64$

Number of students = 64 **Ans.**

Q. 19. A charity trust decides to build the player hall having a carpet area of 300 sq m with its length 1 m more than twice its breadth. Find the length and breadth of the hall.

Solution : Let breadth of the hall be x m

$\therefore$ Length of the hall = $(2x + 1)$ m

Since, Area $= l \times b$

$\Rightarrow \quad (2x+1) \times x = 300$

$\Rightarrow \quad 2x^2 + x - 300 = 0$

$\Rightarrow \quad 2x^2 + 25x - 24x - 300 = 0$

$\Rightarrow \quad x(2x+25) - 12(2x+25) = 0$

$\Rightarrow \quad (x-12)(2x+25) = 0$

$\Rightarrow \quad x = 12, x = -\frac{25}{2}$ (Rejected)

$\therefore \quad x = 12$ m $\Rightarrow$ breadth = 12 m,

Length $= (2 \times 12 + 1) = 25$ m **Ans.**

Q. 20. Students of class X collected ₹ 18000. They wanted to divide it equally among a certain number of students residing in slums area. When they started distributing the amount, 20 more students from nearby slums also joined. Now each student got ₹ 240 less. Find the number of students living in the slum.

Solution : Let number of students living in slums area be x

$\therefore$ Share per student = ₹ $\frac{18000}{x}$

When number of students $= x + 20$

Then, share per student = ₹ $\frac{18000}{x+20}$

According to question,

$$\frac{18000}{x} - \frac{18000}{x+20} = 240$$

$$\Rightarrow \frac{18000x + 360000 - 18000x}{(x+20)(x)} = 240$$

$\Rightarrow \quad 360000 = 240(x^2 + 20x)$

$\Rightarrow \quad 1500 = x^2 + 20x$

$\Rightarrow \quad x^2 + 20x - 1500 = 0$

$\Rightarrow \quad x^2 + 50x - 30x - 1500 = 0$

$\Rightarrow \quad x(x+50) - 30(x+50) = 0$

$\Rightarrow \quad (x+50)(x-30) = 0$

$\Rightarrow x = 30$ or $x = -50$ (Rejected)

$\Rightarrow x = 30$

$\therefore$ Number of students living in the slum area = 30. **Ans.**

Q. 21. A passenger train takes 2 hours less for a journey of 300 km if its speed is increased by 5 km/hr from its usual speed. Find its usual speed.★

Solution : Let the usual speed be x km/hr.

Thus Time taken $= \frac{300}{x}$ hr.

Increased speed is $(x + 5)$ km/hr.

Thus New time $= \frac{300}{x+5}$ hr.

According to question,

$$\frac{300}{x} - \frac{300}{x+5} = 2$$

$$\Rightarrow 300\left(\frac{x+5-x}{x(x+5)}\right) = 2$$

$\Rightarrow \quad 1500 = 2x(x+5)$

$\Rightarrow \quad x(x+5) = 750$

$\Rightarrow \quad x^2 + 5x - 750 = 0$

$\Rightarrow \quad x^2 + 30x - 25x - 750 = 0$

$\Rightarrow \quad x(x+30) - 25(x+30) = 0$

$\Rightarrow \quad (x+30)(x-25) = 0$

$\Rightarrow \quad x = 25$ or -30

As speed cannot be negative, so the correct answer is 25 km/hr. **Ans.**

Q. 22. A train travels a distance of 360 km at uniform speed. If its speed is increased by 5 km/hr, the journey will take 1 hour less. Find the original speed of the train.★

[NCERT]

Solution : Let the original speed be x km/hr.

Thus Time taken $= \frac{360}{x}$ hr.

★ are board exam questions from previous years

Increased speed is $(x + 5)$ km/hr.

Thus New time $= \frac{360}{x+5}$ hr.

According to question,

$$\Rightarrow \frac{360}{x} - \frac{360}{x+5} = 1$$

$$\Rightarrow 360\left(\frac{x+5-x}{x(x+5)}\right) = 1$$

$$\Rightarrow 1800 = x(x+5)$$

$$\Rightarrow x(x+5) = 1800$$

$$\Rightarrow x^2 + 5x - 1800 = 0$$

$$\Rightarrow x^2 + 45x - 40x - 1800 = 0$$

$$\Rightarrow x(x+45) - 40(x+45) = 0$$

$$\Rightarrow (x+45)(x-40) = 0$$

$$\Rightarrow x = 40 \text{ or } -45$$

As speed cannot be negative, so the correct answer is 40 km/hr. **Ans.**

Q. 23. The distance between Mumbai and Pune is 192 km. Travelling by the Deccan Queen, it takes 48 minutes less than another train. Calculate the speed of the Deccan Queen if the speeds of the two trains differ by 20 km/hr.★

Solution : Let the speed of the other train be x km/hr.

Thus, the speed of the Deccan Queen is $(x + 20)$ km/hr.

Time taken by the other train $= \frac{192}{x}$ hr.

Time taken by the Deccan Queen $= \frac{192}{x+20}$ hr.

According to question,

$$\Rightarrow \frac{192}{x} - \frac{192}{x+20} = \frac{48}{60}$$

$$\Rightarrow 192\left(\frac{x+20-x}{x(x+20)}\right) = \frac{48}{60}$$

$$\Rightarrow 3840\left(\frac{60}{48}\right) = x(x+20)$$

$$\Rightarrow 4800 = x^2 + 20x$$

$$\Rightarrow x^2 + 20x - 4800 = 0$$

$$\Rightarrow x^2 + 80x - 60x - 4800 = 0$$

$$\Rightarrow x(x+80) - 60(x+80) = 0$$

$$\Rightarrow (x+80)(x-60) = 0$$

$$\Rightarrow x = 60 \text{ or } -80$$

As speed cannot be negative, so the correct answer is 60 km/hr. **Ans.**

Q. 24. The difference of the squares of two numbers is 88. If the larger number is 5 more than twice the smaller number then find the two numbers.★

★ are board exam questions from previous years

Solution : Let the larger number be x and the smaller number be y.

Given, $x = 2y + 5$

and $x^2 - y^2 = 88$

$$\Rightarrow (2y+5)^2 - y^2 = 88$$

$$\Rightarrow 4y^2 + 25 + 20y - y^2 = 88$$

$$\Rightarrow 3y^2 + 20y - 63 = 0$$

$$\Rightarrow 3y^2 + 27y - 7y - 63 = 0$$

$$\Rightarrow 3y(y+9) - 7(y+9) = 0$$

$$\Rightarrow (y+9)(3y-7) = 0$$

$$\Rightarrow y = -9 \text{ or } \frac{7}{3}$$

Thus, $x = -18 + 5 = -13$

or $x = 2\left(\frac{7}{3}\right) + 5 = \frac{29}{3}$

Hence, the two numbers are -13 and -9 or $\frac{7}{3}$ and $\frac{29}{3}$. **Ans.**

Q. 25. Three consecutive natural numbers are such that the square of the middle number exceeds the difference of the squares of the other two by 60. Find the numbers.★

Solution : Let the three consecutive natural numbers be x, $x + 1$ and $x + 2$.

According to the question,

$$(x+1)^2 - [(x+2)^2 - x^2] = 60$$

$$x^2 + 1 + 2x - [(x+2-x)(x+2+x)] = 60$$

$$x^2 + 2x + 1 - [2(2+2x)] = 60$$

$$x^2 + 2x + 1 - 4 - 4x = 60$$

$$x^2 - 2x - 63 = 0$$

$$x^2 - 9x + 7x - 63 = 0$$

$$x(x-9) + 7(x-9) = 0$$

$$(x+7)(x-9) = 0$$

$$\therefore \quad x = 9 \text{ or } -7$$

$$\Rightarrow \quad x = 9$$

($\because$ numbers are natural)

$\therefore$ Numbers are 9, 10, 11 **Ans.**

Q. 26. The sum of two numbers is 16 and sum of their reciprocals is $\frac{1}{3}$ · Find the numbers.

Solution : Let the numbers be x and $(16-x)$. So, their reciprocals are $\frac{1}{x}$ and $\frac{1}{16-x}$ respectively.

Now, $$\frac{1}{x}+\frac{1}{16-x} = \frac{1}{3}$$

$$\Rightarrow \frac{16-x+x}{x(16-x)} = \frac{1}{3}$$

$$\Rightarrow \frac{16}{x(16-x)} = \frac{1}{3}$$

$$\Rightarrow x(16-x) = 48$$

$$\Rightarrow 16x - x^2 = 48$$

$$\Rightarrow x^2 - 16x + 48 = 0$$

$$\Rightarrow x^2 - 12x - 4x + 48 = 0$$

$$\Rightarrow x(x-12) - 4(x-12) = 0$$

$$\Rightarrow (x-4)(x-12) = 0$$

$$\Rightarrow x = 4 \text{ and } 12.$$ **Ans.**

Q. 27. If the roots of the quadratic equation★

$$(a-b)x^2 + (b-c)x + (c-a) = 0$$

are equal, prove that $2a = b + c$.

Solution : Given equation is,

$(a-b)x^2 + (b-c)x + (c-a) = 0$

By comparing the given equation with $ax^2 + bx + c = 0$, we get

$A = a - b, B = b - c, C = c - a$

Since the roots of the given quadratic equation are equal,

$$\therefore (b-c)^2 - 4(c-a)(a-b) = 0$$

$$[\because B^2 - 4AC = 0]$$

$$b^2 + c^2 - 2bc - 4(ac - a^2 - bc + ab) = 0$$

$$b^2 + c^2 - 2bc - 4ac + 4a^2 + 4bc - 4ab = 0$$

$$(b^2 + c^2 + 2bc) - 4a(b+c) + 4a^2 = 0$$

$$(b+c)^2 - 4a(b+c) + (2a)^2 = 0$$

$$[(b+c) - 2a]^2 = 0$$

$$b + c - 2a = 0$$

i.e. $2a = b + c$ **Hence Proved.**

Q. 28. A girl is twice as old as her sister. Four years hence the product of their ages (in years) will be 160. Find their present ages.★

Solution : Let the sister's age be x. Thus, the girl's age is $2x$.

Four years hence,

Sister's age = $(x+4)$

and Girl's age = $(2x+4)$

Now $(x+4)(2x+4) = 160$

$$\Rightarrow 2x^2 + 8x + 4x + 16 = 160$$

$$\Rightarrow 2x^2 + 12x + 16 = 160$$

$$\Rightarrow 2x^2 + 12x - 144 = 0$$

$$\Rightarrow x^2 + 6x - 72 = 0$$

$$\Rightarrow x^2 + 12x - 6x - 72 = 0$$

$$\Rightarrow x(x+12) - 6(x+12) = 0$$

$$\Rightarrow (x+12)(x-6) = 0$$

$$\Rightarrow x = 6 \text{ or } -12$$

$$\Rightarrow x = 6$$

(as age is always positive)

Thus, the sister's age is 6 years and the girl's age is $2 \times 6 = 12$ year. **Ans.**

Q. 29 Sumit is 3 times as old as his son. Five years later, he shall be two and a half times as old as his son. How old is Sumit at present?★

Solution : Let age of Sumit be x years and age of his son be y years. Then, according to question we have,

$x = 3y$...(i)

Five years later,

$$x + 5 = 2\frac{1}{2}(y+5) \quad \text{...(ii)}$$

On putting $x = 3y$ in equation (ii)

$$3y + 5 = \frac{5}{2}(y+5)$$

$$3y + 5 = \frac{5y}{2} + \frac{25}{2}$$

$$3y - \frac{5y}{2} = \frac{25}{2} - 5$$

$$\frac{6y-5y}{2} = \frac{25-10}{2} = \frac{15}{2}$$

$$\frac{y}{2} = \frac{15}{2}$$

$$y = 15 \text{ years}$$

Then, age of sumit is

$$3 \times y = 3 \times 15$$

$$= 45 \text{ years}$$ **Ans.**

Q. 30. If the price of a toy is reduced by ₹ 2, a person can buy two more toys for ₹ 360. Find the original price of the toy.

Solution : Let the original price of the toy be ₹ x.

Thus, number of toys that can be bought for ₹ 360 = $\frac{360}{x}$

Now, reduced price of the toy = ₹ $(x-2)$

Hence, number of toys = $\frac{360}{x-2}$

★ are board exam questions from previous years

Thus $\frac{360}{x} + 2 = \frac{360}{x-2}$

$\Rightarrow \frac{360}{x-2} - \frac{360}{x} = 2$

$\Rightarrow 360\left(\frac{x-x+2}{x(x-2)}\right) = 2$

$\Rightarrow 360\left(\frac{2}{x(x-2)}\right) = 2$

$\Rightarrow 360\left(\frac{1}{x(x-2)}\right) = 1$

$\Rightarrow x(x-2) = 360$

$\Rightarrow x^2 - 2x - 360 = 0$

$\Rightarrow x^2 - 20x + 18x - 360 = 0$

$\Rightarrow x(x-20) + 18(x-20) = 0$

$\Rightarrow (x-20)(x+18) = 0$

$\Rightarrow x = 20 \text{ or } -18$

$\Rightarrow x = 20$

(as price is always positive)

Hence, the original price of the toy is ₹ 20. **Ans.**

Q. 31. If $ad \neq bc$, then prove that the equation★

$$(a^2 + b^2)x^2 + 2(ac + bd)x + (c^2 + d^2) = 0$$

has no real roots.

Solution : Given,

$(a^2 + b^2)x^2 + 2(ac + bd)x + (c^2 + d^2) = 0, ad \neq bc$

$D = b^2 - 4ac$

$= [2(ac + bd)]^2 - 4(a^2 + b^2)(c^2 + d^2)$

$= 4[a^2c^2 + b^2d^2 + 2abcd] - 4[a^2c^2 + a^2d^2 + b^2c^2 + b^2d^2]$

$= 4[a^2c^2 + b^2d^2 + 2abcd - a^2c^2 - a^2d^2 - b^2c^2 - b^2d^2]$

$= 4[-a^2d^2 - b^2c^2 + 2abcd]$

$= -4(a^2d^2 + b^2c^2 - 2abcd)$

$= -4(ad - bc)^{20}$

$\because D$ is negative

Hence, given equation has no real roots.

Hence Proved.

Chapter 5. Arithmetic Progressions

Q. 1. If the 8th term of an A.P. is 31 and the 15th term is 16 more than the 11th term, find the A.P.★

Solution : Let the first term be a and common difference be d.

Thus, $t_8 = a + (8-1)d = 31$

$\Rightarrow a + 7d = 31$ ...(i)

and $t_{15} = 16 + t_{11}$

$\Rightarrow a + (15-1)d = 16 + a + (11-1)d$

$\Rightarrow a + 14d = 16 + a + 10d$

$\Rightarrow 4d = 16$

$\Rightarrow d = 4$

Substituting $d = 4$ in (i), we get

$a + 7(4) = 31$

$\Rightarrow a + 28 = 31$

$\Rightarrow a = 3$

Thus, the A.P. is 3, 7, 11, 15 ... **Ans.**

Q. 2. Which term of the A.P. 3, 10, 17, ... will be 84 more than the 13th term ?★

Solution : Given A.P. is 3, 10, 17, ...

So, $a = 3$ and $d = 7$

Thus, $t_{13} = 3 + (13-1)7$

$\Rightarrow t_{13} = 3 + (12)7$

$\Rightarrow t_{13} = 3 + 84 = 87$

Now, $87 + 84 = 171$

Let 171 be the nth term.

So, $t_n = 3 + (n-1)7 = 171$

$\Rightarrow 3 + 7n - 7 = 171$

$\Rightarrow 7n = 171 - 3 + 7$

$\Rightarrow 7n = 175$

$\Rightarrow n = 25$

Thus, the 25th term is 84 more than the 13th term.

Ans.

Q. 3. Which term of the A.P. 9, 12, 15, ... will be 39 more than the 36th term ?★

Solution : Given, A.P. is 9, 12, 15.....

$a = 9, d = 3$

Thus, $t_{36} = \{9 + (36-1)3\}$

$\Rightarrow t_{36} = 9 + 105$

$\Rightarrow t_{36} = 114$

Now, $114 + 39 = 153$

Let the nth term be 153.

So, $t_n = 9 + (n-1)3 = 153$

$\Rightarrow 9 + 3n - 3 = 153$

$\Rightarrow 3n = 153 - 9 + 3$

$\Rightarrow 3n = 147$

$\Rightarrow n = 49$

Thus, the required term is the 49th term. **Ans.**

Q. 4. Which term of the A.P. 8, 14, 20, ... will be 72 more than the 41st term ?★

Solution : Given, $a = 8, d = 6$

Thus, $t_{41} = 8 + (41-1)6$

$\Rightarrow t_{41} = 8 + 240$

★ are board exam questions from previous years

$\Rightarrow \quad t_{41} = 248$

Now, $\quad 72 + 248 = 320$

Let the nth term be 320.

So, $\quad t_n = 8 + (n-1)6 = 320$

$\Rightarrow \quad 2 + 6n = 320$

$\Rightarrow \quad 6n = 318$

$\Rightarrow \quad n = 53.$

Thus, the required term is 53 term **Ans.**

Q. 5. The sum of the 5th and the 9th term of an A.P. is 30. If its 25th term is three times the 8th term, find the A.P.★

Solution : Let a be the first term and d be the common difference.

$\therefore \quad T_5 = a + (5-1)d$

$\Rightarrow \quad T_5 = a + 4d$

and $\quad T_9 = a + (9-1)d$

$\Rightarrow \quad T_9 = a + 8d$

Now, $\quad T_5 + T_9 = 30$ [Given]

$\Rightarrow \quad a + 4d + a + 8d = 30$

$\Rightarrow \quad 2a + 12d = 30$

$\Rightarrow \quad a + 6d = 15 \quad \ldots(i)$

Also, $\quad T_{25} = a + (25-1)d$

$\Rightarrow \quad T_{25} = a + 24d$

and $\quad T_8 = a + (8-1)d$

$\Rightarrow \quad T_8 = a + 7d$

Now, $\quad T_{25} = 3T_8$ [Given]

$\Rightarrow \quad a + 24d = 3(a + 7d)$

$\Rightarrow \quad a + 24d = 3a + 21d$

$\Rightarrow \quad 2a - 3d = 0$

On multiplying the above equation by 2, we get

$4a - 6d = 0 \quad \ldots(ii)$

Adding equations (i) and (ii), we get

$5a = 15$

$\Rightarrow \quad a = 3$

and $\quad d = \frac{2(3)}{3} = 2$

So the required A.P. is 3, 5, 7, 9, ... **Ans.**

Q. 6. The sum of the first three terms of an A.P. is 48. If the product of the first and the second terms exceeds four times the third term by 12, find the A.P.★

Solution : Given, $S_3 = 48$

$$S_3 = \frac{3}{2}\{2a + (3-1)d\}$$

$$\Rightarrow \quad 48 = \frac{3}{2}\{2a + 2d\}$$

$\Rightarrow \quad 48 = 3(a + d)$

$\Rightarrow \quad a + d = 16$

$\Rightarrow \quad d = 16 - a$

Now, $\quad T_2 = a + (2-1)\,d = a + d$

$T_3 = a + (3-1)\,d = a + 2d$

Now, $\quad a(a + d) = 4(a + 2d) + 12$ [Given]

$\Rightarrow \quad 16a = 4a + 8d + 12 \quad [\because a + d = 16]$

$\Rightarrow \quad 12a = 8d + 12$

$\Rightarrow \quad 3a = 2d + 3$

$\Rightarrow \quad 3a = 2(16 - a) + 3 \quad [\because d = 16 - a]$

$\Rightarrow \quad 3a = 32 - 2a + 3$

$\Rightarrow \quad 5a = 35$

$\Rightarrow \quad a = 7$

Hence, $\quad d = 16 - 7 = 9$

Hence, the required A.P. is 7, 16, 25,..... **Ans.**

Q. 7. Which term of the A.P. 3, 15, 27, 39, ... will be 120 more than its 21st term ?★

Solution : The given A.P. is 3, 15, 27, 39, ...

Here $a = 3$, $d = 12$

$\therefore \quad a_{21} = a + 20d$

$= 3 + 20 \times 12$

$= 3 + 240 = 243$

Now, $\quad a_n = a_{21} + 120$

$= 243 + 120 = 363$

$a_n = a + (n-1)\,12$

or $\quad 360 = (n-1)\,12$

or $\quad n - 1 = 30$

$n = 31$

Hence, the term which is 120 more than its 21st term will be its 31st term. **Ans.**

Q. 8. Find the sum of all multiples of 7 lying between 500 and 900.

Solution : Multiples of 7 lying between 500 and 900 are

504, 511, 518, 525, 532, ..., 896

We observe that the above series is an A.P. with

$a = 504$, $d = 7$ and $T_n = 896$

Thus, $\quad T_n = 504 + (n-1)7 = 896$

$\Rightarrow \quad 7(n-1) = 392$

$\Rightarrow \quad n - 1 = 56$

$\Rightarrow \quad n = 57$

Thus, $\quad S_{57} = \frac{57}{2}\{2(504) + (57-1)7\}$

★ are board exam questions from previous years

$$= \frac{57}{2}\{1008 + 392\}$$

$$= \frac{57}{2}(1400)$$

$= 39900.$ **Ans.**

Q. 9. Find the sum of the first *n* terms of an A.P. whose n^{th} term is $5n - 1$. Find the sum of the first 20 terms.★

Solution : Given, $T_n = 5n - 1$

We know that, $T_n = a + (n - 1)d$

Thus, $a + (n - 1)d = 5n - 1$

$\Rightarrow \quad a + nd - d = 5n - 1$

$\Rightarrow \quad nd + (a - d) = 5n - 1$

On comparing both sides, we get

$\Rightarrow \quad d = 5$ and $a - d = -1$

$\Rightarrow \quad a = 5 - 1 = 4$

Thus, $S_{20} = \frac{20}{2}\{2(4) + (20 - 1)5\}$

$= 10(8 + 95)$

$= 1030.$ **Ans.**

Q. 10. Find the value of the middle term of the following A.P. : – 6, – 2, 2, ..., 58.★

Solution : From the given A.P. we have,

$a = -6, d = 4$ and $T_n = 58$

Thus $T_n = -6 + (n - 1)4 = 58$

$-6 + 4n - 4 = 58$

$\Rightarrow \quad 4n = 68$

$\Rightarrow \quad n = 17$

Thus, the middle term would be $\left(\frac{n+1}{2}\right)^{th}$ *i.e*, 9th term.

Hence, $T_9 = \{-6 + (9 - 1)4\}$

$= -6 + 32 = 26.$ **Ans.**

Q. 11. The 6th term of an A.P. is – 10 and its 10th term is – 26. Determine the 15th term of the A.P.

Solution : Let the first term be a and the common difference be d.

Thus, $T_6 = a + (6 - 1)d = -10$ [Given]

$\Rightarrow \quad a + 5d = -10$ ...(i)

and $T_{10} = a + (10 - 1)d = -26$ [Given]

$\Rightarrow \quad a + 9d = -26$ ...(ii)

Subtracting equation (i) from equation (ii), we get

$4d = -16$

$\Rightarrow \quad d = -4$

From equation (i),

$a + 5d = -10$

$\Rightarrow \quad a + 5(-4) = -10$

$\Rightarrow \quad a - 20 = -10$

$\Rightarrow \quad a = -10 + 20 = 10$

Now, $T_{15} = 10 + (15 - 1)(-4)$

$= 10 - 4(14)$

$= 10 - 56$

$= -46.$ **Ans.**

Q. 12. In a given A.P., if the pth term is q and the qth term is p, then show that the nth term is $(p + q - n)$.★★

Solution : Let the first term be a and the common difference be d.

Thus, $T_p = a + (p - 1)d = q$

$\Rightarrow \quad a - d = q - pd$

and $T_q = a + (q - 1)d = p$

$\Rightarrow \quad a - d = p - qd$

From equation (i) and (ii),

$q - pd = p - qd$

$\Rightarrow \quad d(q - p) = p - q$

$\Rightarrow \quad d = -1$

and $a = p + q - 1 = q + p - 1$

So, $T_n = \{a + (n - 1)d\}$

$= p + q - 1 + (n - 1)(-1)$

$= p + q - 1 - n + 1$

$= p + q - n.$ **Hence Proved.**

Q. 13. If the 10th term of an A.P. is 52 and the 17th term is 20 more than the 13th term, find the A.P.

Solution : Let a be the first term and d be the common difference.

So, $T_{10} = a + (10 - 1)d = 52$ [Given]

$\Rightarrow \quad a + 9d = 52$ ...(i)

Now, $T_{17} = T_{13} + 20$ [Given]

$\Rightarrow \quad a + (17 - 1)d = a + (13 - 1)d + 20$

$a + 16d = a + 12d + 20$

$\Rightarrow \quad 16d = 12d + 20$

$\Rightarrow \quad 4d = 3d + 5$

$\Rightarrow \quad d = 5$

From equation (i),

$\Rightarrow \quad a = 52 - 9(5)$

$= 52 - 45 = 7$

Hence, the required A.P. is 7, 12, 17, 22, ... **Ans.**

Q. 14. The first term of an A.P. is 5, the last term is 45 and the sum of all its terms is 400. Find the number of terms and the common difference of the A.P.

★ are board exam questions from previous years

★★ are frequently asked board exam questions

Solution : Let a be the first term and d be the common difference.

Then, $a = 5$, $l = 45$ and $S_n = 400$ [Given]

$$\therefore \quad \frac{n}{2}(a+l) = 400$$

$$\Rightarrow \quad \frac{n}{2}(5+45) = 400$$

$$\Rightarrow \quad 400 = \frac{n}{2}[50]$$

$$\Rightarrow \quad 25n = 400 \Rightarrow n = \frac{400}{25}$$

$$\Rightarrow \quad n = 16$$

Now, $l = a + (n-1)d$

$$\Rightarrow \quad 45 = 5 + (16-1)d$$

$$\Rightarrow \quad 45 - 5 = 15d$$

$$\Rightarrow \quad 15d = 40$$

$$\Rightarrow \quad d = \frac{8}{3}$$

So $n = 16$ and $d = \frac{8}{3}$ **Ans.**

Q. 15. The sum of the three numbers in A.P. is 21 and their product is 231. Find the numbers.

Solution : Let the numbers be $(a-d)$, a, $(a+d)$.

So, the first term $= a - d$

and the common difference $= d$

We know, $S_n = \frac{n}{2}\{2a + (n-1)d\}$

$$\Rightarrow \quad S_3 = \frac{3}{2}\{2(a-d) + (3-1)d\}$$

$$\Rightarrow \quad 21 = \frac{3}{2}\{2a - 2d + 2d\} = 3a \text{ [Given]}$$

$$\Rightarrow \quad a = 7$$

Now, $(a-d)\,a(a+d) = 231$ [Given]

$$\Rightarrow \quad (7-d)\,7(7+d) = 231$$

$$\Rightarrow \quad 49 - d^2 = 33$$

$$\Rightarrow \quad d^2 = 16$$

$$\Rightarrow \quad d = \pm 4$$

Thus, the numbers are either 3, 7, 11 or 11, 7, 3. **Ans.**

Q. 16. Find the sum of all two-digit odd positive numbers.

Solution : Two-digit odd positive integers are, 11, 13, 15, ... 99.

The above series is an A.P. with

$$a = 11,\ d = 2 \text{ and } T_n = 99$$

Now, $T_n = 11 + (n-1)2 = 99$

$$\Rightarrow \quad (n-1)2 = 88$$

$$\Rightarrow \quad n - 1 = 44$$

$$\Rightarrow \quad n = 45$$

$$\therefore \quad S_{45} = \frac{45}{2}\{2(11) + (45-1)(2)\}$$

$$= \frac{45}{2}\{2(11) + (44)(2)\}$$

$$= 45\{11 + 44\}$$

$$= 45 \times 55$$

$$= 2475.$$ **Ans.**

Q. 17. The 16th term of an A.P. is five times its third term. If its 10th term is 41, then find the sum of its first fifteen terms.★

Solution : Given that 16th term of an A.P. is five time its third term.

$$\therefore \quad a + (16-1)d = 5[a + (3-1)d]$$

$$\Rightarrow \quad a + 15d = 5[a + 2d]$$

$$\Rightarrow \quad a + 15d = 5a + 10d$$

$$\Rightarrow \quad 4a - 5d = 0 \quad \text{...(i)}$$

Also given that,

$$T_{10} = 41$$

$$\Rightarrow \quad a + (10-1)d = 41$$

$$\Rightarrow \quad a + 9d = 41 \quad \text{...(ii)}$$

On multiplying equation (ii) by 4, we get

$$4a + 36d = 164 \quad \text{...(iii)}$$

Subtracting equation (iii) from (i), we get

$$\begin{array}{r} 4a - 5d = 0 \\ 4a + 36d = 164 \\ - \quad - \quad - \\ \hline -41d = -164 \end{array}$$

$$\Rightarrow \quad d = 4$$

On substituting the value of d in (i), we get

$$4a - 5 \times 4 = 0$$

$$4a = 20$$

$$a = 5$$

Now, $S_{15} = \frac{15}{2}[2a + (15-1)d]$

$$S_{15} = \frac{15}{2}(2 \times 5 + 14 \times 4)$$

$$= \frac{15}{2}2(5 + 14 \times 2)$$

$$= 15(5 + 28) = 15 \times 33$$

★ are board exam questions from previous years

$S_{15} = 495$

Hence, sum of first fifteen terms is 495. **Ans.**

Q. 18. How many terms of the A.P. 3, 5, 7, 9, ... must be added to get the sum of 120 ?

Solution : Given A.P. is, 3, 5, 7, 9, ...

Thus, $a = 3$ and $d = 2$

Now, $S_n = \frac{n}{2}\{2(3) + (n-1)(2)\}$

$\Rightarrow n\{3 + (n-1)\} = 120$

$\Rightarrow n(n+2) = 120$

$\Rightarrow n^2 + 2n - 120 = 0$

$\Rightarrow n^2 + 12n - 10n - 10 = 0$

$\Rightarrow n(n+12) - 10(n+12) = 0$

$\Rightarrow (n+12)(n-10) = 0$

$\Rightarrow n = 10 \text{ or } -12$

Here, the given A.P. series is increasing, so n cannot be negative.

Hence, so the required answer is 10. **Ans.**

Q. 19. Find, how many two digit natural numbers are divisible by 7.★

Solution. Two digit numbers which are divisible by 7 are

14, 21, 28, ... 98

It is an A.P., such that

$a = 14,\ a_n = 98;\ d = 21 - 14 = 7$

$\because a_n = a + (n-1)d$

$\therefore 98 = 14 + (n-1) \times 7$

$84 = (n-1)\,7$

or $n - 1 = 12$

$n = 13$

Hence, there are 13 two digit numbers, divisible by 7. **Ans.**

Q. 20. If the sum of the first n terms of an A.P. is given by $S_n = \frac{n}{2}(3n+5)$, then find its 25th term.

Solution : Let the first term be a and the common difference be d.

It is given that, $S_n = \frac{n}{2}\{2a + (n-1)d\} = \frac{n}{2}(3n+5)$

$\Rightarrow 2a + (n-1)d = 3n + 5$

$\Rightarrow (2a - d) + dn = 5 + 3n$

On comparing, we get

$d = 3 \text{ and } 2a - d = 5$

$\Rightarrow 2a = 8$

$\Rightarrow a = 4$

$\therefore T_{25} = 4 + (25-1)3$

$= 4 + 24 \times 3$

$= 76.$ **Ans.**

Q. 21. The 14th term of an A.P. is twice its 8th term. If its 6th term is – 8, then find the sum of its first 20 terms.★

Solution : In the given AP, let first term = a and common difference = d

Then, $T_n = a + (n-1)d$

$\Rightarrow T_{14} = a + (14-1)d = a + 13d$

and $T_8 = a + (8-1)d = a + 7d$

Now, it is given that,

$T_{14} = 2T_8$

$\Rightarrow a + 13d = 2(a + 7d)$

$\Rightarrow a + 13d = 2a + 14d$

$a = -d$ ...(i)

Also, $T_6 = a + (6-1)d$

$\Rightarrow a + 5d = -8$ ...(ii)

Substituting the value of a from equation (i) and equation (ii), we get

$-d + 5d = -8$

$4d = -8$

$d = -2$

From equation (i),

$a = -d = -(-2) = 2$

$\therefore$ Sum of first 20 terms is

$S_{20} = \frac{n}{2}\{2a + (n-1)d\}$

$= \frac{20}{2}\{2 \times 2 + (20-1)(-2)\}$

$= 10\{4 - 38\}$

$= -340$ **Ans.**

Q. 22. The sum of the first 7 terms of an A.P. is 49 and the sum of the first 17 terms is 289. Find the sum of first n terms.★★ [NCERT]

Solution : Let the first terms be a and the common difference be d.

Now, it is given that,

$S_7 = \frac{7}{2}\{2a + (7-1)d\} = 49$

$\Rightarrow 49 = \frac{7}{2}\{2a + 6d\}$

$\Rightarrow a + 3d = 7$ ...(i)

★ are board exam questions from previous years

★★ are frequently asked board exam questions

and $$S_{17} = \frac{17}{2}\{2a + (17-1)d\} = 289$$

$\Rightarrow$ $$289 = \frac{17}{2}\{2a + (17-1)d\}$$

$\Rightarrow$ $$a + 8d = 17 \qquad \ldots\text{(ii)}$$

Subtracting equation (i) from equation (ii), we have

$$5d = 10$$

$\Rightarrow$ $$d = 2$$

From equation (i),

$$a = 7 - 3d$$
$$= 7 - 3(2)$$
$$= 7 - 6 = 1$$

$\therefore$ $$S_n = \frac{n}{2}\{2(1) + (n-1)2\}$$

$\Rightarrow$ $$S_n = \frac{n}{2}\{2 + 2n - 2\} = n^2$$

Hence, the sum of n terms will be n^2. **Ans.**

Q. 23. If the sum of first n terms of an A.P. is n^2, then find its 10th term.★

Solution : Given, $S_n = n^2$, $S_{n-1} = (n-1)^2$

$$a_n = S_n - S_{n-1}$$
$$= n^2 - (n-1)^2$$
$$= n^2 - [n^2 - 2n + 1]$$
$$= n^2 - n^2 + 2n - 1$$
$$a_n = 2n - 1$$

Put $n = 10$,

$\therefore$ $$a_{10} = 2 \times 10 - 1 = 19$$

Hence 10th term = 19 **Ans.**

Q. 24. The sum of the first 9 terms of an A.P. is 81 and the sum of the first 20 terms is 400. Find the first term and the common difference.

Solution : Let the first term be a and the common difference be d.

Now, it is given that,

$$S_9 = \frac{9}{2}\{2a + (9-1)d\} = 81$$

$\Rightarrow$ $$81 = \frac{9}{2}\{2a + 8d\}$$

$\Rightarrow$ $$a + 4d = 9 \qquad \ldots\text{(i)}$$

and $$S_{20} = \frac{20}{2}\{2a + (20-1)d\} = 400$$

$\Rightarrow$ $$400 = 10\{2a + 19d\}$$

$\Rightarrow$ $$2a + 19d = 40 \qquad \ldots\text{(ii)}$$

Multiplying equation (i) with 2 and then subtracting from equation (ii), we get

$$11d = 22$$

$\Rightarrow$ $$d = 2$$

From equation (i),

$$a = 9 - 4(2)$$
$$= 9 - 8 = 1.$$

Ans.

Q. 25. The 9th term of an A.P. is equal to six times its 2nd term. If its 5th term is 22, find the A.P.

Solution : Let the first term be a and the common difference be d.

Given, $$T_9 = 6T_2$$

$\Rightarrow$ $$\{a + (9-1)d\} = 6\{a + (2-1)d\}$$

$\Rightarrow$ $$a + 8d = 6a + 6d$$

$\Rightarrow$ $$5a = 2d \qquad \ldots\text{(i)}$$

and $$T_5 = \{a + (5-1)d\} = 22$$

$\Rightarrow$ $$a + 4d = 22$$

$\Rightarrow$ $$a + 2(2d) = 22$$

$\Rightarrow$ $$a + 10a = 22 \qquad [\because 5a = 2d]$$

$\Rightarrow$ $$11a = 22$$

$\Rightarrow$ $$a = 2$$

Substituting $a = 2$ in equation (i), we get

$$10 = 2d$$

$\Rightarrow$ $$d = 5$$

Hence, the required A.P. is 2, 7, 12, 17, … **Ans.**

Q. 26. The 24th term of an A.P. is twice its 10th term. Show that its 72nd term is 4 times its 15th term.

Solution : Let the first term be a and the common difference be d.

Given, $$T_{24} = 2T_{10}$$

$\Rightarrow$ $$\{a + (24-1)d\} = 2\{a + (10-1)d\}$$

$\Rightarrow$ $$a + 23d = 2a + 18d$$

$\Rightarrow$ $$a = 5d \qquad \ldots\text{(i)}$$

Now, $$T_{15} = \{a + (15-1)d\}$$

$\Rightarrow$ $$T_{15} = a + 14d$$

$\Rightarrow$ $$T_{15} = 5d + 14d \qquad [\because a = 5d]$$

$\Rightarrow$ $$T_{15} = 19d$$

and $$T_{72} = a + (72-1)d$$

$\Rightarrow$ $$T_{72} = a + 71d$$

$\Rightarrow$ $$T_{72} = 5d + 71d$$

$\Rightarrow$ $$T_{72} = 76d = 4(19d)$$

$\Rightarrow$ $$T_{72} = 4T_{15}.$$

Hence Proved.

★ are board exam questions from previous years

Q. 27. The digits of a positive number of three digits are in A.P. and their sum is 15. The number obtained by reversing the digits is 594 less than the original number. Find the number.★

Solution : Let the three digits of a positive number be $a-d, a, a+d$

$\therefore\ a-d+a+a+d = 3a = 15$ [Given]

$\Rightarrow\ a = 5$

Original number $= 100(a-d)+10(a)+a+d$

$= 100a-100d+10a+a+d$

$= 111a-99d$

and number obtained by reversing the digits

$= 100(a+d)+10(a)+a-d$

$= 100a+100d+10a+a-d$

$= 111a+99d$

According to the question,

$(111a-99d)-(111a+99d) = 594$

$-198d = 594$

$d = -3$

$\therefore$ Original number is $111(5)-99(-3)$

i.e, 852 **Ans.**

Q. 28. If the sum of the first *n* terms of an A.P. is given by $S_n = \frac{n}{2}(3n+13)$, then find its 25th term.

Solution : Let the first term be *a* and the common difference be *d*.

Given, $S_n = \frac{n}{2}\{2a+(n-1)d\} = \frac{n}{2}(3n+13)$

$\Rightarrow\ 2a+(n-1)d = 3n+13$

$\Rightarrow\ (2a-d)+dn = 13+3n$

On comparing, we get

$d = 3$ and $2a-d = 13$

$\Rightarrow\ 2a = 16$

$\Rightarrow\ a = 8$

$\therefore\ T_{25} = 8+(25-1)3$

$T_{25} = 8+24\times 3 = 80.$ **Ans.**

Q. 29. If S_n, denotes the sum of first *n* terms of an A.P., prove that $S_{12} = 3(S_8 - S_4)$.★

Solution : Let be the first term be *a* and the common difference be *d*,

We know, $S_n = \frac{n}{2}[2a+(n-1)d]$

$\therefore\ S_{12} = \frac{12}{2}[2a+(12-1)d]$

$= 6(2a+11d)$

$= 12a+66d$

$S_8 = \frac{8}{2}[2a+(8-1)d]$

$= 4(2a+7d)$

$= 8a+28d$

and $S_4 = \frac{4}{2}[2a+(4-1)d]$

$= 2(2a+3d) = 4a+6d$

Given, $3(S_8 - S_4)$

$= 3(8a+28d-4a-6d)$

$= 3(4a+22d)$

$= 12a+66d$

$= S_{12}$ **Hence Proved.**

Q. 30. The nth term of an A.P. is given by $(-4n+15)$. Find the sum of the first 20 terms of this A.P.

Solution : Let the first term be *a* and the common difference be *d*.

Given, $T_n = (-4n+15)$

$\{a+(n-1)d\} = (-4n+15)$

$\Rightarrow\ a+nd-d = -4n+15$

$\Rightarrow\ (a-d)+nd = 15-4n$

On comparing, we get

$d = -4$

and $a-d = 15$

$\Rightarrow\ a = 15+d$

$\Rightarrow\ a = 15-4 = 11$

$\therefore\ S_{20} = \frac{20}{2}\{2(11)+(20-1)(-4)\}$

$= 10\{22-76\}$

$= -540.$ **Ans.**

Q. 31. Pradeep repays the total loan of ₹ 1, 18000 by paying every month starting with the first instalment of ₹ 1000. He increases the instalment by ₹ 100 every month. What amount will he pay as the last instalment of loan ?

Solution : Ist instalment = ₹ 1000

IInd instalment = ₹ (1000 + 100) = ₹ 1100

IIIrd instalment = ₹ (1100 + 100) = ₹ 1200

Let the number of instalments be *n*.

$\therefore$ Sum $= 1000+1100+1200+\ldots\ldots+n^{th}$ term

★ are board exam questions from previous years

These terms are in AP with $a = 1000$, $d = 100$.

$$\therefore \text{ Sum} = \frac{n}{2}[2 \times 1000 + (n-1)100]$$

$$\Rightarrow \frac{n}{2}(2000 + 100n - 100) = 118000$$

$$\Rightarrow n(1900 + 100n) = 236000$$

$$\Rightarrow 100n^2 + 1900n - 236000 = 0$$

$$\Rightarrow n^2 + 19n - 2360 = 0$$

$$\Rightarrow n^2 + 59n - 40n - 2360 = 0$$

$$\Rightarrow (n + 59)(n - 40) = 0$$

$\Rightarrow n = 40$ or $n = -59$ (Rejected)

$\therefore$ Last instalment, $a_{40} = a + 39d$

$= 1000 + 39 \times 100 = ₹ 4900$ **Ans.**

Q. 32. Saurav gets pocket money from his father every day. Out of the pocket money, he saves ₹ 2.75 on first day and on each succeeding day he increases his saving by 25 paise. Find

(i) the amount saved by Saurav on 14th day,

(ii) the amount saved by Saurav on 25th day,

(iii) the total amount saved by Saurav in 30 days.

Solution : Money saved on Ist day = ₹ 2.75

Money saved on IInd day

= ₹ (2.75 + 0.25) = ₹ 3.00

Money saved on IIIrd day

= ₹ (3.00 + 0.25) = ₹ 3.25.

Money saved by Saurav form an AP with $a = 2.75$ and $d = 0.25$

(i) Money saved on 14th day

$= a + 13d$

$= 2.75 + 13 \times 0.25$

$= 2.75 + 3.25$

$= ₹ 6.00$ **Ans.**

(ii) Money saved on 25th day

$= a + 24d$

$= 2.75 + 24 \times 0.25$

$= ₹ 8.75$ **Ans.**

(iii) Total amount of money saved in 30 days

$$= \frac{30}{2}(2 \times 2.75 + 29 \times 0.25)$$

$= ₹ 191.25$ **Ans.**

Q. 33. In a school, students thought of planting trees in and around the school to reduce air pollution. It was decided that the number of trees, that each section of each class will plant, will be the same as the class, in which they are studing, *e.g.*, a section of class I will plant 1 tree, a section of class II will plant 2 trees and so on till class XII. There are three sections of each class. How many trees will be planted by the students ?

Solution : Number of trees planted by students of class I

$= 3 \times 1 = 3$

Number of trees planted by students of class II

$= 3 \times 2 = 6$

Number of trees planted by students of class III

$= 3 \times 3 = 9$

Number of trees planted by students of class XII

$= 3 \times 12 = 36$

Total number of trees planted

$= 3 + 6 + 9 + ... + 36$

These are 12 terms of an A.P. with $a = 3$ and last term $= 36$

$$\Rightarrow l = 36$$

$$\therefore \text{ Sum} = \frac{n}{2}(a + l)$$

$$= \frac{12}{2}(3 + 36)$$

$= 234$ **Ans.**

Q. 34. In a potato race, a bucket is placed at the starting point, which is 5 m from the first potato and the other potatoes are placed 3 m apart in a straight line. There are ten potatoes in the line.

Each competitor starts from the bucket, picks up the nearest potato, runs back with it, drops it in the bucket, runs back to pick up the next potato, runs to the bucket to drop it in, and continues in the same way until all the potatoes are in the bucket. What is the total distance the competitor has to run ?

Solution : We have,

d_1 = Distance run by the competitor to pick up first potato = 2×5 m

d_2 = Distance run by the competitor to pick up second potato = $2(5 + 3)$ m

d_3 = Distance run by the competitor to pick up third potato = $2(5 + 2 \times 3)$ m

d_4 = Distance run by the competitor to pick up fourth potato = $2(5 + 3 \times 3)$ m

d_n = Distance run by the competitor to pick up nth potato = $2\{5 + (n - 1) \times 3\}$ m

Total distance run by the competitor to pick up potatoes

$= d_1 + d_2 + d_3 + ... + d_n$

$= 2 \times 5 + 2(5 + 3) + 2(5 + 2 \times 3) + 2(5 + 3 \times 3) + ...$

$+ 2\{5 + (n - 1) \times 3\}$ metres

$= 2[5 + \{5 + 3\} + \{5 + (2 \times 3)\} + \{5 + (3 \times 3)\} + ...$
$+\{5 + (n-1) \times 3\}]$

$= 2[(5 + 5 + ... + 5) + \{3 + (2 \times 3) + (3 \times 3) + ...$
$+ (n-1) \times 3]$

$= 2[(5 + 5 + ... + 5)\} 3 + 1 + 2 + 3 + ... (n-1)\}]$

$= 2\left[5n + 3\left(\frac{n-1}{2}\right)\{1 + (n-1)\}\right]$

$= 2\left[5n + \frac{3n(n-1)}{2}\right]$

$= [10n + 3n(n-1)]$

$= 3n^2 + 7n$

$= n(3n + 7)$ metres

Since, $n = 10$

$\therefore$ Total distance run by the competitor

$= 10(3(10) + 7)$

$= 370$ m. **Ans.**

Q. 35. Sham was late by 5 minutes in joining his duty on the first working day. On the second day, he was late by 10 minutes, on third day by 15 minutes and so on. After 25 working days he was shunted out of the job. Find the total working time avoided by Sham.

Solution : On first day, Sham was late by 5 minutes

On second day, he was late by 10 minutes

On third day, he was late by 15 minutes

and so on

The time by which sham was late for his office forms an A.P.

$\therefore\ a = 5, d = 10 - 5 = 5, n = 25$

So, total working time avoided by Sham,

$$S_{25} = \frac{25}{2}\{2a + (25-1)d\}$$

$$= \frac{25}{2}\{2 \times 5 + (25-1)5\}$$

$$= \frac{25}{2}\{2 \times 5 + 24 \times 5\}$$

$$= 25\{5 + 12 \times 5\}$$

$$= 25\{5 + 60\}$$

$$= 25 \times 65$$

$= 1625$ minutes **Ans.**

Chapter 6. Coordinate Geometry

Q. 1. Find the ratio in which the line segment joining the points $A(3, -3)$ and $B(-2, 7)$ is divided by the X-axis. Also, find the coordinate of the point of division.★

Solution : Let AB be divided by the X-axis in the ratio $p : 1$ at the point K.

Thus, the coordinates of the point

$$K = \left(\frac{mx_2 + nx_1}{m+n}, \frac{my_2 + ny_1}{m+n}\right)$$

$$= \left(\frac{p(-2)+3}{p+1}, \frac{7p+(-3)}{p+1}\right)$$

$$= \left(\frac{3-2p}{p+1}, \frac{7p-3}{p+1}\right)$$

However as K lies in the X-axis, its coordinate is $(x, 0)$.

$\Rightarrow \quad \frac{7p-3}{p+1} = 0$

$\Rightarrow \quad 7p - 3 = 0$

$\Rightarrow \quad 7p = 3$

$\Rightarrow \quad p = \frac{3}{7}$

Thus, the required ratio $= \frac{3}{7} : 1 = 3 : 7$

Now, $\quad x = \frac{3-2p}{p+1}$

$\Rightarrow \quad x = \frac{3 - 2\left(\frac{3}{7}\right)}{\left(\frac{3}{7}\right) + 1}$

$= \frac{21-6}{3+7}$

$= \frac{15}{10} = \frac{3}{2}$

Hence, $\quad P = \left(\frac{3}{2}, 0\right)$. **Ans.**

Q. 2. Find the ratio in which the segment joining the points (1, – 3) and (4, 5) is divided by x-axis ? Also find the coordinates of this point on x-axis.★

Solution : Let the given points be A (1, – 3) and B (4, – 5) and the line-segment joining by these points is divided by x-axis, so the co-ordinate of the point of intersection will be $P(x, 0)$.

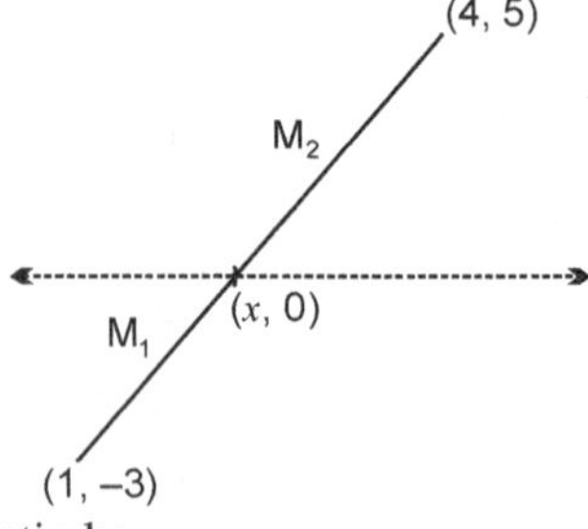

Let the ratio be $m_1 : m_2$

So, By section formula

$$0 = \frac{m_1y_2 + m_2y_1}{m_1 + m_2}$$

$$0 = \frac{5m_1 - 3m_2}{m_1 + m_2}$$

or $\quad 5m_1 - 3m_2 = 0$

or $\quad \frac{m_1}{m_2} = \frac{3}{5}$

$\therefore$ Required ratio is $3 : 5$ **Ans.**

★ are board exam questions from previous years

Now to find the co-ordinates of this point on x-axis

$$\therefore \quad x = \frac{3\times4+5\times1}{3+5}$$

$$x = \frac{12+5}{8}$$

$$x = \frac{17}{8}$$

$\therefore$ The required point is $\left(\frac{17}{8}, 0\right)$. **Ans.**

Q. 3. If the point $P(k-1, 2)$ is equidistant from the points $A(3, k)$ and $B(k, 5)$, find the value of k.★

Solution : As P is equidistant from A and B,

$$\therefore \quad AP = BP$$

$$\Rightarrow \sqrt{[3-(k-1)]^2+(k-2)^2} = \sqrt{(k-1-k)^2+(2-5)^2}$$

$$\Rightarrow (4-k)^2+(k-2)^2 = 1+9$$

$$\Rightarrow k^2+16-8k+4+k^2-4k = 10$$

$$\Rightarrow 2k^2-12k+10 = 0$$

$$\Rightarrow k^2-6k+5 = 0$$

$$\Rightarrow k^2-5k-k+5 = 0$$

$$\Rightarrow k(k-5)-1(k-5) = 0$$

$$\Rightarrow (k-5)(k-1) = 0$$

$$\Rightarrow k = 1, 5$$

Thus, the value of k are 1 or 5. **Ans.**

Q. 4. Find the value of p for which the points $(3p+1, p)$, $(p+2, p-5)$ and $(p+1, -p)$ are collinear.

Solution : Since the points $(3p+1, p)$, $(p+2, p-5)$ and $(p+1, -p)$ are collinear, therefore, we have

$$x_1(y_2-y_3)+x_2(y_3-y_1)+x_3(y_1-y_2) = 0$$

$$\Rightarrow (3p+1)(p-5+p)+(p+2)(-p-p) + (p+1)(p-p+5) = 0$$

$$\Rightarrow (3p+1)(2p-5)+(p+2)(-2p)+(p+1)(5) = 0$$

$$\Rightarrow 6p^2+2p-15p-5-2p^2-4p+5p+5 = 0$$

$$\Rightarrow 4p^2-12p = 0$$

$$\Rightarrow 4p^2 = 12p$$

$$\Rightarrow 4p(p-3) = 0$$

Hence, the value of p is 3 or 0. **Ans.**

Q. 5. Find a relation between x and y if the points A(x, y), B$(-4, 6)$ and C$(-2, 3)$ are collinear.★

Solution : Given, A(x, y), B$(-4, 6)$, C$(-2, 3)$

$$x_1 = x,\ y_1 = y,\ x_2 = -4,\ y_2 = 6, x_3 = -2,\ y_3 = 3$$

If these points are collinear, then area of triangle made by these points is 0.

$$\frac{1}{2}[(x_1(y_2-y_3)+x_2(y_3-y_1)+x_3(y_1-y_2)] = 0$$

$$\frac{1}{2}[x(6-3)+(-4)(3-y)+(-2)(y-6)] = 0$$

$$3x-12+4y-2y+12 = 0$$

$$3x+4y-2y = 0$$

$$3x+2y = 0$$

$$3x = -2y$$

Ans.

Q. 6. Points $A(-1, y)$ and $B(5, 7)$ lie on a circle with centre $O(2, -3y)$. Find the values of y. Hence find the radius of the circle.★

Solution : As A and B lie on a circle so AO and BO form the radii.

$$\text{Thus,} \quad AO = BO$$

$$\Rightarrow \sqrt{[2-(-1)]^2+(-3y-y)^2} = \sqrt{(2-5)^2+(-3y-7)^2}$$

$$\Rightarrow (2+1)^2+(-4y)^2 = (-3)^2+(-3y-7)^2$$

$$\Rightarrow 9+16y^2 = 9+9y^2+49+42y$$

$$\Rightarrow 7y^2-42y-49 = 0$$

$$\Rightarrow y^2-6y-7 = 0$$

$$\Rightarrow y^2-7y+y-7 = 0$$

$$\Rightarrow y(y-7)+1(y-7) = 0$$

$$\Rightarrow (y-7)(y+1) = 0$$

$$\Rightarrow y = -1, 7$$

Thus, the value of y are -1 or 7.

$$\text{So,} \quad AO = BO = \sqrt{[2-(-1)]^2+(-3y-y)^2} = \sqrt{(2-5)^2+(-3y-7)^2}$$

When $y = -1$,

$$AO = BO = \sqrt{(2+1)^2+(3+1)^2} = \sqrt{(2-5)^2+(3-7)^2}$$

$$\Rightarrow AO = BO = \sqrt{(-3)^2+(-4)^2} = \sqrt{9+16} = 5 \text{ units}$$

When $y = 7$,

$$AO = BO = \sqrt{[2-(-1)]^2+(-21-7)^2}$$

$$= \sqrt{(2-5)^2+(-21-7)^2}$$

$$= \sqrt{(-3)^2+(-28)^2}$$

$$= \sqrt{9+784}$$

★ are board exam questions from previous years

$= \sqrt{793}$

$= 28.16$ units

Hence, the radius of the circle is either 5 units or 28.16 units. **Ans.**

Q. 7. Given figure shows the arrangement of desks in a classroom. Ashima, Bharti and Camella are seated at *A*(3, 1), *B*(6, 4) and *C*(8, 6) respectively. Do you think they are seated in a line ? Give reason for your answer.★

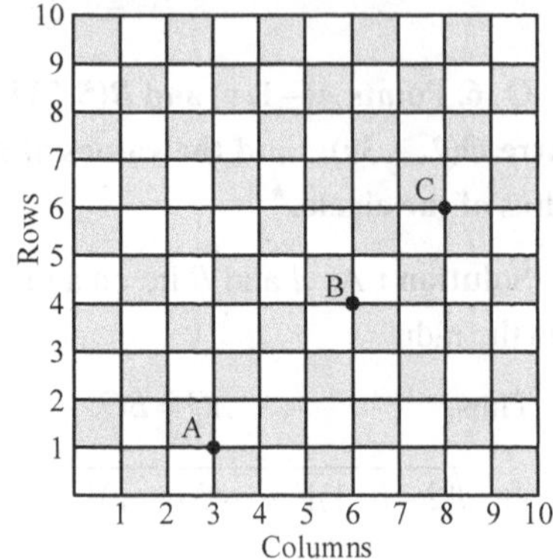

Solution : By distance formula,

$$AB = \sqrt{(6-3)^2 + (4-1)^2}$$

$$= \sqrt{9+9} = 3\sqrt{2} \text{ units}$$

and $$BC = \sqrt{(8-6)^2 + (6-4)^2}$$

$$= \sqrt{4+4} = 2\sqrt{2} \text{ units}$$

Now, $$AC = \sqrt{(8-3)^2 + (6-1)^2}$$

$$= \sqrt{25+25}$$

$$= 5\sqrt{2} \text{ units}$$

Now, $AB + BC = 3\sqrt{2} + 2\sqrt{2} = 5\sqrt{2} = AC$

Therefore, *A*, *B* and *C* are collinear.

Hence, Ashima, Bharti and Camella are seated a line. **Ans.**

Q. 8. Three villages having nearly the same population are situated at the positions (1, 2) , (5, 4) and (3, 8). The Block Development Officer wants to set-up a community centre in such a way that it is equally convenient to reach it for the villagers of all the three villages. Find the position where the community centre should be set-up.

Solution : Suppose the position coordinate of the community centre is (x, y).

Since, three villages are situated at the positions (1, 2), (5, 4) and (3, 8).

So, distance between community centre and first village.

= Distance between community centre and second village

$$\sqrt{(x-1)^2 + (y-2)^2} = \sqrt{(x-5)^2 + (y-4)^2}$$

After squaring on both sides, we get

$\Rightarrow (x-1)^2 + (y-2)^2 = (x-5)^2 + (y-4)^2$

$\Rightarrow x^2 + 1 - 2x + y^2 + 4 - 4y$

$= x^2 + 25 - 10x + y^2 + 16 - 8y$

$\Rightarrow \quad 8x + 4y = 36$

$\Rightarrow \quad 2x + y = 9$ ...(i)

Also, distance between community centre and second village

= Distance between community centre and third village

$$\sqrt{(x-5)^2 + (y-4)^2} = \sqrt{(x-3)^2 + (y-8)^2}$$

$\Rightarrow x^2 + 25 - 10x + y^2 - 8y + 16$

$= x^2 + 9 - 6x + y^2 - 16y + 64$

$\Rightarrow \quad 4x - 8y = -32$

$\Rightarrow \quad x - 2y = -8$ ...(ii)

After solving (i) and (ii), we get

$$x = 2, y = 5$$

Hence, the position coordinate of the community centre will be (2, 5). **Ans.**

Q. 9. If the points *P*(– 3, 9), *Q*(*a*, *b*) and *R*(4, – 5) are collinear and *a* + *b* = 1, find the values of *a* and *b*.★

Solution : Since *P*(– 3, 9), *Q*(*a*, *b*) and *R*(4, – 5) are collinear

$\therefore \quad x_1(y_2 - y_3) + x_2(y_3 - y_1) + x_3(y_1 - y_2) = 0$

$\Rightarrow (-3)[b - (-5)] + a[(-5) - 9] + 4[9 - b] = 0$

$\Rightarrow \quad (-3)[b + 5] - 14a + 36 - 4b = 0$

$\Rightarrow \quad -3b - 15 - 14a + 36 - 4b = 0$

$\Rightarrow \quad -14a - 7b = -21$

$\Rightarrow \quad 2a + b = 3$ ...(i)

$\Rightarrow \quad 2a + b = 3(a + b)$ $[\because a + b = 1]$

$\Rightarrow \quad 2a + b = 3a + 3b$

$\Rightarrow \quad a + 2b = 0$

$\Rightarrow \quad a = -2b$ ...(ii)

Substituting $a = -2b$ in (i), we have

$$-4b + b = 3$$

$\Rightarrow \quad -3b = 3$

$\Rightarrow \quad b = -1$

and $\quad a = 2$ [from (ii)] **Ans.**

Q. 10. Points *P*, *Q*, *R* and *S* divide the line segment joining the points *A*(1, 2) and *B*(6, 7) in 5 equal parts. Find the coordinates of the points *P*, *Q* or *R*.

Solution : Let the points *P*, *Q*, *R* and *S* divide the join of *A*(1, 2) and *B*(6, 7) into five equal parts. Then *P*(*x*, *y*) divides *AB* in the ratio 1 : 4.

A(1, 2) P Q R S B(6, 7)

★ are board exam questions from previous years

$\therefore \quad P = \left(\frac{1\times6+4\times1}{1+4}, \frac{1\times7+4\times2}{1+4}\right)$

$P = \left(\frac{10}{5}, \frac{15}{5}\right)$

$\therefore \quad P = (2, 3)$

Also, $Q(p, q)$ divides AB in the ratio 2 : 3.

$\therefore \quad Q = \left(\frac{2\times6+3\times1}{2+3}, \frac{2\times7+3\times2}{2+3}\right)$

$Q = \left(\frac{15}{5}, \frac{20}{5}\right)$

$\therefore \quad Q = (3, 4)$

and $R(r, s)$ divides AB in the ratio 3 : 2.

$\therefore \quad R = \left(\frac{3\times6+2\times1}{3+2}, \frac{3\times7+2\times2}{3+2}\right)$

$R = \left(\frac{20}{5}, \frac{25}{5}\right)$

$\therefore \quad R = (4, 5).$ **Ans.**

Q. 11. If the point $P(2, 2)$ is equidistant from the points $A(-2, k)$ and $B(-2k, -3)$, find k. Also, find the length of AP.★

Solution : As P is equidistant from both A and B,

$$AP = PB$$

$\Rightarrow \sqrt{(2+2)^2+(2-k)^2} = \sqrt{(2+2k)^2+(2+3)^2}$

$\Rightarrow \quad 16 + k^2 + 4 - 4k = 4k^2 + 4 + 8k + 25$

$\Rightarrow \quad 3k^2 + 12k + 9 = 0$

$\Rightarrow \quad k^2 + 4k + 3 = 0$

$\Rightarrow \quad k^2 + 3k + k + 3 = 0$

$\Rightarrow \quad k(k + 3) + 1(k + 3) = 0$

$\Rightarrow \quad (k + 3)(k + 1) = 0$

$\Rightarrow \quad k = -3, -1$

When $k = -1$, $AP = \sqrt{(2+2)^2+(2+1)^2}$

$= \sqrt{(4)^2+(3)^2}$

$= \sqrt{16+9}$

$= \sqrt{25}$

$= 5$ units

When $k = -3$, $AP = \sqrt{(2+2)^2+(2+3)^2}$

$= \sqrt{(4)^2+(5)^2}$

$= \sqrt{16+25}$

$= \sqrt{41}$ units

Thus, $\quad AP = 5$ or $\sqrt{41}$ units. **Ans.**

★ are board exam questions from previous years

Q. 12. Find a point P on the Y-axis which is equidistant from $A(4, 8)$ and $B(-6, 6)$. Also find the distance AP.

Solution : As the point P lies on – Y-axis, so let the coordinates be $(0, y)$.

Thus, $\quad AP = PB$

$\Rightarrow \sqrt{(4-0)^2+(8-y)^2} = \sqrt{(0+6)^2+(y-6)^2}$

$\Rightarrow \quad 16 + (8-y)^2 = 36 + (y-6)^2$

$\Rightarrow \quad (8-y)^2 - (y-6)^2 = 20$

$\Rightarrow (8 - y - y + 6)(8 - y + y - 6) = 20$

$\Rightarrow \quad (14 - 2y)(2) = 20$

$\Rightarrow \quad 28 - 4y = 20$

$\Rightarrow \quad 4y = 8$

$\Rightarrow \quad y = \frac{8}{4} = 2.$

Thus, the point P is (0, 2)

Now, $\quad AP. = \sqrt{(4-0)^2+(8-2)^2}$

$= \sqrt{16+36}$

$= 2\sqrt{13}$ units **Ans.**

Q. 13. Ayush starts walking from his house to office. Instead of going to the office directly, he goes to a bank first, from there to his daughter's school and then reaches the office. What is the extra distance travelled by Ayush in reaching his office ? (Assume that all distances covered are in straight lines)

If the house is situated at (2, 4), bank at (5, 8), school at (13, 14) and office at (13, 26) and coordinates are in km.

Solution : Let his house is at $A(2, 4)$, bank at $B(5, 8)$, school at $C(13, 14)$ and office at $D(13, 26)$.

$AB = \sqrt{(5-2)^2+(8-4)^2}$

$= \sqrt{9+16} = 5$ km

$BC = \sqrt{(13-5)^2+(14-8)^2}$

$= \sqrt{100} = 10$ km

$CD = \sqrt{(13-13)^2+(26-14)^2}$

$= \sqrt{12^2} = 12$ km

$AD - \sqrt{(13-2)^2+(26-4)^2}$

$= \sqrt{(11)^2+(22)^2}$

$= 11\sqrt{5}$ km

$= 24.59$ km

Distance covered by Ayush

$$= 5 + 10 + 12$$
$$= 27 \text{ km}$$

Extra distance $= 27 - 24.59$

$= 2.41$ km. **Ans.**

Q. 14. Prove that the diagonals of a rectangle *ABCD* with vertices *A*(2, – 1), *B*(5, – 1), *C*(5, 6) and *D*(2, 6) are equal and bisect each other.★

Solution : Given, $A(2, -1), B(5, -1), C(5, 6)$ and $D(2, 6)$

$$AC = \sqrt{(5-2)^2 + (6+1)^2}$$
$$= \sqrt{(3)^2 + (7)^2}$$
$$= \sqrt{9+49} = \sqrt{58}$$
$$BD = \sqrt{(2-5)^2 + (6+1)^2}$$
$$= \sqrt{(3)^2 + (7)^2}$$
$$= \sqrt{9+49}$$
$$= \sqrt{58}$$

Thus, $AC = BD$.

Let E be the mid-point of AC and F the mid-point of BD.

Thus, $$E = \left(\frac{2+5}{2}, \frac{6-1}{2}\right) = \left(\frac{7}{2}, \frac{5}{2}\right)$$

Also, $$F = \left(\frac{5+2}{2}, \frac{6-1}{2}\right) = \left(\frac{7}{2}, \frac{5}{2}\right)$$

Hence, AC and BD bisect each other at the point $\left(\frac{7}{2}, \frac{5}{2}\right)$. **Hence Proved,**

Q. 15. Find the coordinates of a point *P*, which lies on the line segment joining the points *A*(–2, –2) and *B*(2, – 4) such that $AP = \frac{3}{7}AB$.★★

Solution : Given, $A = (-2, -2)$, $B = (2, -4)$ and $AP = \frac{3}{7}AB$

A(–2, –2) P B(2, –4)

$$\Rightarrow \quad 7AP = 3AB = 3AP + 3BP$$
$$4AP = 3BP$$
$$\Rightarrow \quad \frac{AP}{BP} = \frac{3}{4}$$

Thus, $AP : PB = 3 : 4$

Let the coordinates of P be (x, y).

So, $$x = \frac{3(2)+4(-2)}{3+4} \text{ and } y = \frac{3(-4)+4(-2)}{3+4}$$

$$\Rightarrow \quad x = \frac{6-8}{7} \text{ and } y = \frac{-12-8}{7}$$

$$\Rightarrow \quad x = \frac{-2}{7} \text{ and } y = \frac{-20}{7}$$

Hence, the coordinates of P are $\left(\frac{-2}{7}, \frac{-20}{7}\right)$. **Ans.**

Q. 16. Find the area of the quadrilateral *ABCD* whose vertices are *A*(– 3, – 1), *B*(– 2, – 4), *C*(4, – 1) and *D*(3, 4).★★

Solution : Given, $A(-3, -1), B(-2, -4), C(4, -1)$ and $D(3, 4)$

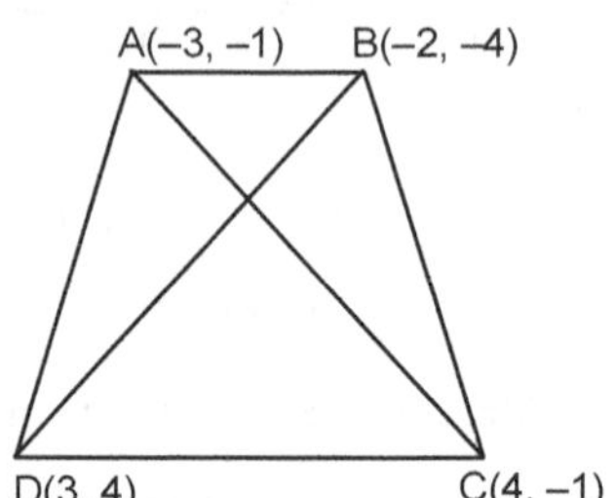

Area of $ABCD$ = Area of ΔABC + Area of ΔACD

$$= \frac{1}{2}|-3(-4+1) - 2(-1+1) + 4(-1+4)|$$
$$+ \frac{1}{2}|-3(-1-4) + 4(4+1) + 3(-1+1)|$$
$$= \frac{1}{2}|9 + 12 + 15 + 20|$$
$$= \frac{56}{2} = 28 \text{ sq. units.}$$ **Ans.**

Q. 17. A point *P* divides the line segment joining the points *A*(3, – 5) and *B*(– 4, 8) such that $\frac{AP}{PB} = \frac{k}{1}$. If *P* lies on the line $x + y = 0$, then find the value of *k*.

Solution : Given,

$$A = (3, -5);\ B = (-4, 8);\ \frac{AP}{PB} = \frac{k}{1} \text{ and } x + y = 0$$

As, $$\frac{AP}{PB} = \frac{k}{1}$$

So, $$AP : BP = k : 1$$

Let the coordinates of point P be (x, y).

A(3, –5) P(x, y) B(–4, 8)
k 1

Thus, $$x = \frac{1(3)+k(-4)}{k+1} \text{ and } y = \frac{1(-5)+k(8)}{k+1}$$

or $$x = \frac{3-4k}{k+1} \text{ and } y = \frac{8k-5}{k+1}$$

Now, $$x + y = 0$$

$$\Rightarrow \quad \frac{3-4k}{k+1} + \frac{8k-5}{k+1} = 0$$

$$\Rightarrow \quad \frac{4k-2}{k+1} = 0$$

★ are board exam questions from previous years
★★ are frequently asked board exam questions

$\Rightarrow \quad 4k - 2 = 0$

$\Rightarrow \quad 4k = 2$

$\Rightarrow \quad k = \frac{1}{2}$. **Ans.**

Q. 18. If the vertices of a triangle are (1, – 3), (4 *p*) and (– 9, 7) and its area is 15 sq. units, find the value of *p*.

Solution : Given, the vertices of triangle are (1, – 3), (4, *p*) and (– 9, 7) and area is 15 sq. units.

Thus, $15 = \frac{1}{2} |1(p-7) + 4(7+3) - 9(-3-p)|$

$\Rightarrow \quad 30 = |p - 7 + 40 + 27 + 9p|$

$\Rightarrow \quad 30 = |10p + 60|$

Either $10p = -30$

$\Rightarrow \quad p = -3$

or $30 = -10p - 60$

$\Rightarrow \quad p = -9.$

Thus, $p = -3$ or -9 **Ans.**

Q. 19. Find the area of a triangle whose vertices are given as (1, – 1) (– 4, 6) and (– 3, – 5).★

Solution :

$x_1 = 1, y_1 = -1, x_2 = -4, y_2 = 6, x_3 = -3, y_3 = -5.$

Area of triangle

$= \frac{1}{2}[(x_1(y_2 - y_3) + x_2(y_3 - y_1) + x_3(y_1 - y_2)]$

$= \frac{1}{2}[1(6+5) + (-4)(-5+1) + (-3)(-1-6)]$

$= \frac{1}{2}[11 + 16 + 21]$

$= \frac{1}{2} \times 48$

= 24 square unit. **Ans.**

Q. 20. Point *P*(*x*, 4) lies on the line segment joining the points *A*(– 5, 8) and *B*(4, – 10). Find the ratio in which point *P* divides the line segment *AB*. Also, find the value of *x*.

Solution : Given, $A = (-5, 8)$, $B = (4, -10)$ and $P = (x, 4)$

Let $AP : PB = m : n$

Now, $P_y = \frac{m(-10) + 8n}{m + n}$

$\Rightarrow \quad 4 = \frac{8n - 10m}{m + n}$

$\Rightarrow \quad 4(m + n) = 8n - 10m$

$\Rightarrow \quad 4m + 4n = 8n - 10m$

$\Rightarrow \quad 14m = 4n$

$\Rightarrow \quad 7m = 2n$

$\Rightarrow \quad \frac{m}{n} = \frac{2}{7}$

$\Rightarrow \quad m : n = 2 : 7$

Thus, $P_x = \frac{2(4) + 7(-5)}{2 + 7}$

$\Rightarrow \quad x = \frac{8 - 35}{9}$

$\Rightarrow \quad x = -\frac{27}{9} = -3$ **Ans.**

Q. 21. Find the area of the triangle formed by joining the mid-points of the sides of the triangle whose vertices are *A*(2, 1), *B*(4, 3) and *C*(2, 5).

Solution : Let *A*(2, 1), *B*(4, 3) and *C*(2, 5) be the vertices of ΔABC. Let *D, E, F* be the mid-point of sides *BC, CA, AB* respectively.

By mid-point formula,

Coordinates of $D = \left(\frac{4+2}{2}, \frac{3+5}{2}\right)$

$= \left(\frac{6}{2}, \frac{8}{2}\right) = (3, 4)$

Coordinates of $E = \left(\frac{2+2}{2}, \frac{1+5}{2}\right)$

$= \left(\frac{4}{2}, \frac{6}{2}\right) = (2, 3)$

Coordinates of $F = \left(\frac{2+4}{2}, \frac{1+3}{2}\right)$

$= \left(\frac{6}{2}, \frac{4}{2}\right) = (3, 2)$

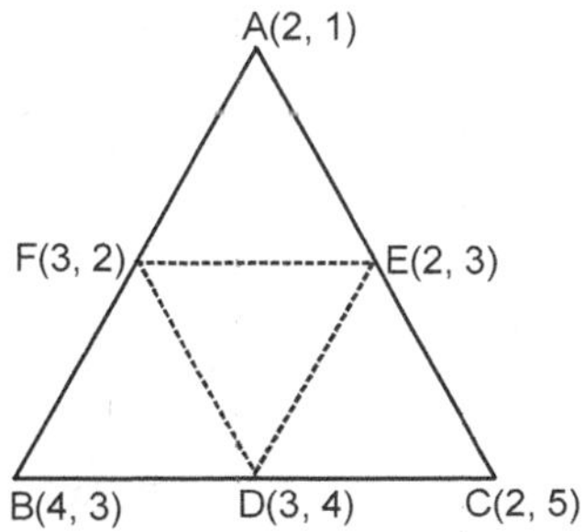

★ are board exam questions from previous years

Now, Area of ΔDEF

$$= \frac{1}{2}|3(3-2)+2(2-4)+3(4-3)|$$

$$= \frac{1}{2}|3-4+3|$$

$$= \frac{1}{2}\times 2 = 1 \text{ sq. unit.}$$ **Ans.**

Q. 22. To raise social awareness about the hazards of smoking, a school decided to start "No smoking campaign". A student is asked to prepare a campaign banner in the shape of a triangle shown in the figure. If cost of 1 cm² of banner is ₹ 2, find the cost of the banner.

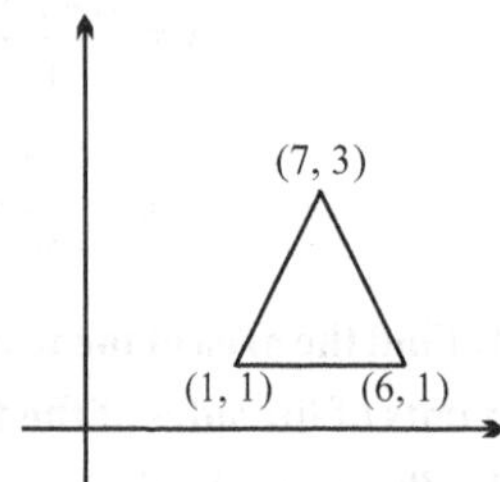

Solution :

$$\text{Area of } \Delta = \frac{1}{2}|x_1(y_2-y_3)+x_2(y_3-y_1)+x_3(y_1-y_2)|$$

$$= \frac{1}{2}|7(1-1)+1(1-3)+6(3-1)|$$

$$= \frac{1}{2}|10| = 5 \text{ cm}^2$$

∴ Cost of banner = 2 × 5 = ₹ 10 **Ans.**

Q. 23. If two vertices of an equilateral triangle are (3, 0) and (6, 0), find the third vertex.

Solution : Given two vertices of an equilateral triangle are (3, 0) and (6, 0).

Let the third vertex be (x, y).

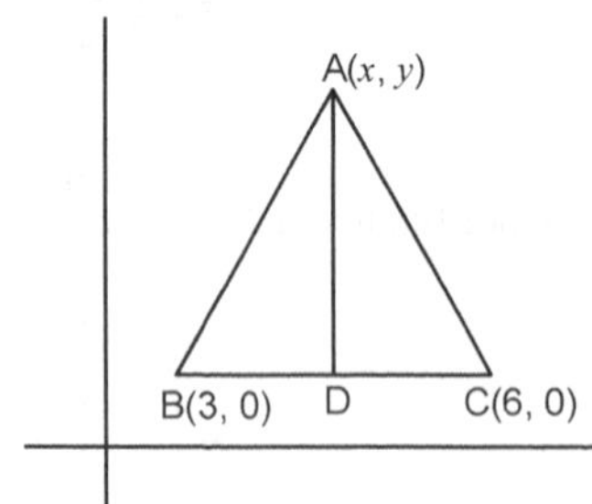

Thus,

$$BC = \sqrt{(6-3)^2+(0-0)^2} = 3$$

$$AB = \sqrt{(x-3)^2+(y-0)^2} = \sqrt{x^2-6x+9+y^2}$$

$$AC = \sqrt{(x-6)^2+(y-0)^2} = \sqrt{x^2-12x+36+y^2}$$

Now, $AB = AC$

So, $x^2-6x+9+y^2 = x^2-12x+36+y^2$

$\Rightarrow -2x+3 = -4x+12$

$\Rightarrow 2x = 9$

$\Rightarrow x = \frac{9}{2}$

Also, $AB = BC$

So, $x^2-6x+9+y^2 = 3$

$\Rightarrow x^2-6x+y^2 = 3-9$

$\Rightarrow \left(\frac{9}{2}\right)^2 - 6\left(\frac{9}{2}\right) + y^2 = -6$

$\Rightarrow 81-108+4y^2 = -24$

$-27+4y^2 = -24$

$\Rightarrow 4y^2 = 3$

$\Rightarrow y^2 = \frac{3}{4}$

$\Rightarrow y = \pm\frac{\sqrt{3}}{2}$

Thus, $P = \left(\frac{9}{2}, \pm\frac{\sqrt{3}}{2}\right)$. **Ans.**

Q. 24. Show that the points $A(3, 0)$, $B(6, 4)$ and $C(-1, 3)$ are the vertices of a right-angled triangle. Also, prove that these are the vertices of an isosceles triangle.

Solution : Given, $A = (3, 0)$, $B = (6, 4)$ and $C = (-1, 3)$

Now,

$$AB = \sqrt{(6-3)^2+(4-0)^2} = \sqrt{(3)^2+(4)^2} = 5 \text{ units}$$

$$BC = \sqrt{(6+1)^2+(4-3)^2} = \sqrt{(7)^2+(1)^2} = \sqrt{50} \text{ units}$$

and

$$AC = \sqrt{(3+1)^2+(0-3)^2} = \sqrt{(4)^2+(3)^2} = 5 \text{ units}$$

Thus, $AC = AB$

Hence, it is proved that ΔABC is an isosceles triangle.

Also, $AC^2 + AB^2 = 25 + 25 = 50$

and $BC^2 = (\sqrt{50})^2 = 50$

Thus, $BC^2 = AC^2 + AB^2$ follows the Pythagoras' theorem.

Hence, it is proved that ΔABC is an isosceles right-angled triangle. **Hence Proved.**

★ are board exam questions from previous years

Q. 25. The x-coordinate of a point P is twice the y-coordinate. If P is equidistant from $Q(2, -5)$ and $R(-3, 6)$ then find the coordinates of P.

Solution : Let the coordinates of P be (x, y).

Now, $x = 2y$ [Given]

Thus, $P = (2y, y)$

Given, $Q = (2, -5)$ and $R = (-3, 6)$

Now, $PQ = QR$ [Given]

$$\Rightarrow \sqrt{(2-2y)^2 + (-5-y)^2} = \sqrt{(2y+3)^2 + (y-6)^2}$$

$$\Rightarrow 4 - 8y + 4y^2 + 25 + 10y + y^2 = 4y^2 + 12y + 9 + y^2 - 12y + 36$$

$\Rightarrow$ $29 + 2y = 45$

$\Rightarrow$ $2y = 16$

$\Rightarrow$ $y = 8$

So, $x = 2 \times 8 = 16$

Thus, $P = (16, 8)$ **Ans.**

Q. 26. Find the value of x such that $PQ = QR$ where the coordinates of P, Q and R are $(6, -1)$, $(1, 3)$ and $(x, 8)$ respectively.

Solution : Given, $P = (6, -1)$, $Q = (1, 3)$, $R = (x, 8)$ and $PQ = QR$

Now, $PQ = QR$

$$\therefore \sqrt{(6-1)^2 + (-1-3)^2} = \sqrt{(1-x)^2 + (3-8)^2}$$

$\Rightarrow$ $(5)^2 + (-4)^2 = (1-x)^2 + (-5)^2$

$\Rightarrow$ $25 + 16 = (1-x)^2 + 25$

$\Rightarrow$ $(1-x)^2 = 4^2$

$\Rightarrow$ $1 - x = \pm 4$

$\Rightarrow$ $x = -3$ or 5

Thus, the value of x is -3 or 5. **Ans.**

Q. 27. In what ratio does the point $\left(\frac{24}{11}, y\right)$ divide the line segment joining the points $P(2, -2)$ and $Q(3, 7)$? Also find the value of y.★

Solution : Let point $R\left(\frac{24}{11}, y\right)$ divides PQ in the ratio $k : 1$

P (2, –2) —k— R$\left(\frac{24}{11}, y\right)$ —1— Q (3, 7)

Thus,
$$\left(\frac{24}{11}, y\right) = \left(\frac{k(3)+1(2)}{k+1}, \frac{k(7)+1(-2)}{k+1}\right)$$

$$= \left(\frac{3k+2}{k+1}, \frac{7k-2}{k+1}\right)$$

$\Rightarrow$ $\frac{3k+2}{k+1} = \frac{24}{11}$

$\Rightarrow$ $11(3k+2) = 24(k+1)$

$\Rightarrow$ $33k + 22 = 24k + 24$

$\Rightarrow$ $33k - 24k = 24 - 22$

$\Rightarrow$ $9k = 2$

$\Rightarrow$ $k = 2/9$

Thus, $k : 1 = \frac{2}{9} : 1$

$= 2 : 9$

Now,
$$y = \frac{7k-2}{k+1} = \frac{7\left(\frac{2}{9}\right) - 2}{\frac{2}{9} + 1}$$

$$= \frac{\frac{14}{9} - 2}{\frac{2}{9} + 1} = \frac{\frac{14-18}{9}}{\frac{2+9}{9}}$$

$$= \frac{-4}{11}$$

Line PQ divides in the ratio $2 : 9$ and value of

$y = \frac{-4}{11}$. **Ans.**

Q. 28. Prove that the points $(a, b + c)$, $(b, c + a)$ and $(c, a + b)$ are collinear.★

Solution : Given,

$A = (a, b + c)$, $B = (b, c + a)$ and $C = (c, a + b)$

For A, B and C to be collinear, they must satisfy the condition of collinearity.

Thus, Area of $\Delta ABC = 0$

$$\frac{1}{2}|a(c + a - a - b) + b(a + b - b - c) + c(b + c - c - a)| = 0$$

$\Rightarrow$ $a(c - b) + b(a - c) + c(b - a) = 0$

$\Rightarrow$ $ac - ab + ab - bc + bc - ac = 0$

$\Rightarrow$ $0 = 0$

As both L.H.S. = R.H.S. = 0, the three points are collinear. **Hence Proved.**

Q. 29. For what value of x will the points $(x, -1)$, $(2, 1)$ and $(4, 5)$ lie on a line ?

Solution : Given, $A = (x, -1)$, $B = (2, 1)$ and $C = (4, 5)$

For A, B and C to be collinear, they must satisfy the condition of collinearity.

Thus, Area of $\Delta ABC = 0$

$$\Rightarrow \frac{1}{2}|x(1-5) + 2(5+1) + 4(-1-1)| = 0$$

$\Rightarrow$ $-4x + 12 - 8 = 0$

$\Rightarrow$ $-4x + 4 = 0$

★ are board exam questions from previous years

$\Rightarrow \quad -4x = -4$

$\Rightarrow \quad x = 1$

Thus, the points will be collinear if $x = 1$. **Ans.**

Q. 30. If $P(x, y)$ is any point on the line joining the points $A(a, 0)$ and $B(0, b)$, then show that $\frac{x}{a}+\frac{y}{b}=1$.

Solution : Given, $A = (a, 0)$ and $B = (0, b)$. Also $P = (x, y)$ is collinear with A and B.

Thus, Area of $\Delta PAB = 0$

$$\Rightarrow \frac{1}{2}|x(0-b)+a(b-y)+0(y-0)| = 0$$

$$\Rightarrow -bx+ab-ay = 0$$

$$\Rightarrow ab = ay+bx$$

$$\Rightarrow \frac{ab}{ab} = \frac{ay}{ab}+\frac{bx}{ab}$$

[dividing both sides by ab]

$$\Rightarrow 1 = \frac{y}{b}+\frac{x}{a}$$

$$\Rightarrow \frac{x}{a}+\frac{y}{b} = 1.$$ **Hence Proved.**

Q. 31. If the points (p, q), (m, n) and $(p-m, q-n)$ are collinear, show that $pn = qm$.★

Solution : Given, (p, q), (m, n) and $(p-m, q-n)$ are collinear.

Thus,

$$\frac{1}{2}|p[n-(q-n)]+m[(q-n)-q]+(p-m)(q-n)| = 0$$

$$\Rightarrow p(2n-q)+m(-n)+pq-mq-pn+mn = 0$$

$$\Rightarrow 2pn-pq-mn+pq-mq-pn+mn = 0$$

$$\Rightarrow pn-mq = 0$$

$$\therefore pn = mq.$$ **Hence Proved.**

Q. 32. Find the area of a parallelogram $ABCD$ if three of its vertices are $A(2, 4)$, $B(2+\sqrt{3}, 5)$ and $C(2, 6)$.★

Solution : Given, $A = (2, 4)$, $B = (2+\sqrt{3}, 5)$ and $C = (2, 6)$

As AC is the diagonal of the parallelogram, so AC divides $ABCD$ into two triangles of equal areas.

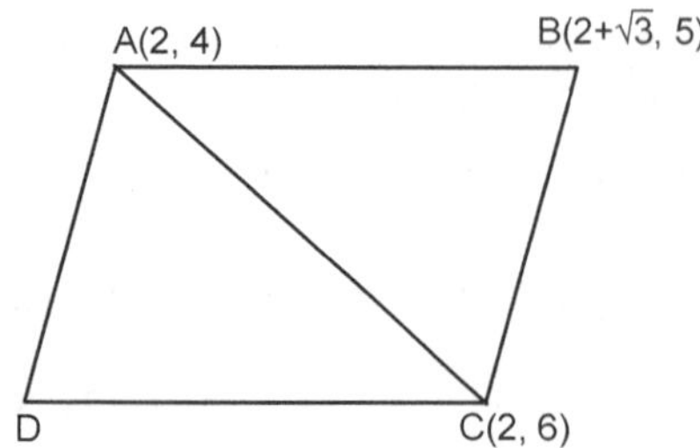

Thus, Area of $ABCD = 2$ Area of ΔABC

$$= 2\times\frac{1}{2}[2(5-6)+(2+\sqrt{3})(6-4)+2(4-5)]$$

$$= [2(-1)+2(2+\sqrt{3})+2(-1)]$$

$$= [-2+4+2\sqrt{3}-2]$$

$$= 2\sqrt{3} \text{ sq. units.}$$ **Ans.**

Q. 33. In fig, ABC is a triangle coordinates of whose vertex A are $(0, -1)$. D and E respectively are the mid-points of the sides AB and AC and their coordinates are $(1, 0)$ and $(0, 1)$ respectively. If F is the mid-point of BC, find the areas of ΔABC and ΔDEF.★

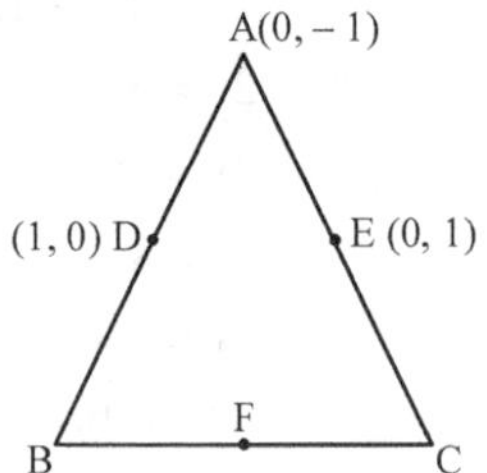

Solution : Given, the coordinates of vertex A $(0, -1)$ and mid-points D $(1, 0)$ and E $(0, 1)$ respectively.

Let coordinates of B be (x, y)

Since, D is the mid-point of AB

then, $\left(\frac{x+0}{2}, \frac{y-1}{2}\right) = (1, 0)$

which gives B $(2, 1)$.

Similarly, E is the mid-point of AC

Let coordinates of C are (x', y')

then, $\left(\frac{x'+0}{2} \frac{y'-1}{2}\right) = (0, 1)$

which gives C $(0, 3)$

Now, Area of $\Delta ABC = \frac{1}{2}|0(1-3)+2(3+1)+0(-1-1)|$

$= 4$ sq. units.

Now, F is the mid-point of BC.

$\Rightarrow$ Coordinates of $F = \left(\frac{2+0}{2}, \frac{1+3}{2}\right) = (1, 2)$

$\therefore$ Area of $\Delta DEF = \frac{1}{2}|1(1-2)+0(2-0)+1(0-1)|$

$= \frac{|-2|}{2} = 1$ sq. unit. **Ans.**

★ are board exam questions from previous years

Chapter 7. Triangles

Q. 1. In triangle *ABC*, if $AP \perp BC$ and $AC^2 = BC^2 - AB^2$, then prove that $PA^2 = PB \times CP$.★

Solution : $AC^2 = BC^2 - AB^2$ (Given)

$\Rightarrow$ $AC^2 + AB^2 = BC^2$

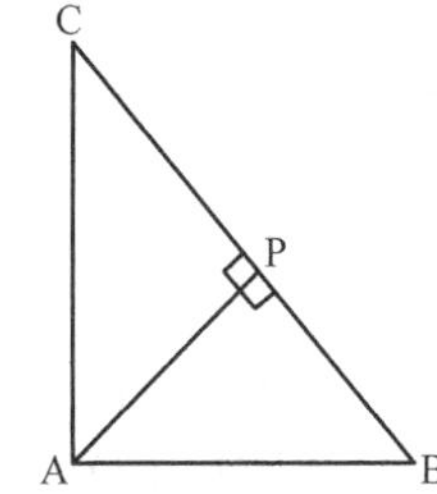

$\therefore$ $\angle BAC = 90°$

[By converse of Pythagoras' theorem]

Now, $\Delta APB \sim \Delta CPA$

[If a perpendicular is drawn from the vertex of the right angle of a triangle to the hypotenuse then triangles on both sides of the perpendicular are similar to the whole triangle and to each other]

$\Rightarrow$ $\frac{AP}{CP} = \frac{PB}{PA}$

[In similar triangle, corresponding sides are proportional]

$\Rightarrow$ $PA^2 = PB \, . \, CP$ **Hence Proved.**

Q. 2. In the figure of ΔABC, *D* divides *CA* in the ratio 4 : 3. If $DE \parallel BC$, then find ar (*BCDE*) : ar (ΔABC).★

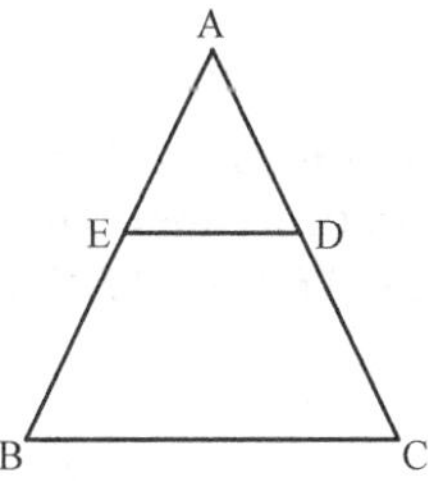

Solution : Given, *D* divides *CA* in 4 : 3

Thus, let $CD = 4K$

and $DA = 3K$

Also, $DE \parallel BC$ [Given]

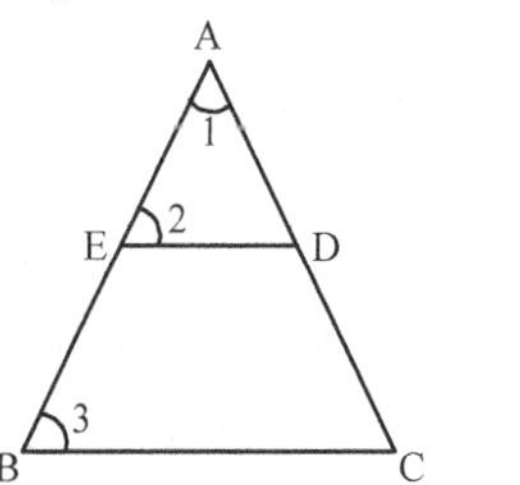

Now,

In ΔAED and ΔABC

$\angle 1 = \angle 1$ [Common]

$\angle 2 = \angle 3$ [Corresponding angles]

$\therefore$ $\Delta AED \sim \Delta ABC$ [A A similarity]

$\Rightarrow$ $\frac{\text{ar}(\Delta AED)}{\text{ar}(\Delta ABC)} = \frac{AD^2}{AC^2}$

[The ratio of the areas of two similar triangles is equal to the square of the ratio of their corresponding sides.]

$= \frac{(3K)^2}{(7K)^2} = \frac{9K^2}{49K^2}$

$\Rightarrow$ $\frac{\text{ar}(\Delta AED)}{\text{ar}(\Delta ABC)} = \frac{9}{49}$

Let ar (ΔAED) = $9p$ and ar (ΔABC) = $49p$

$\therefore$ ar (*BCDE*) = $49p - 9p = 40p$

$\Rightarrow$ $\frac{\text{ar}(BCDE)}{\text{ar}(\Delta ABC)} = \frac{40p}{49p}$

$\therefore$ ar (*BCDE*) : ar (ΔABC) = 40 : 49 **Ans.**

Q. 3. In the given figure, $DE \parallel BC$ and $\frac{AD}{DB} = \frac{3}{5}$ · If $AC = 4.8$ cm, find the length of *AE*.★

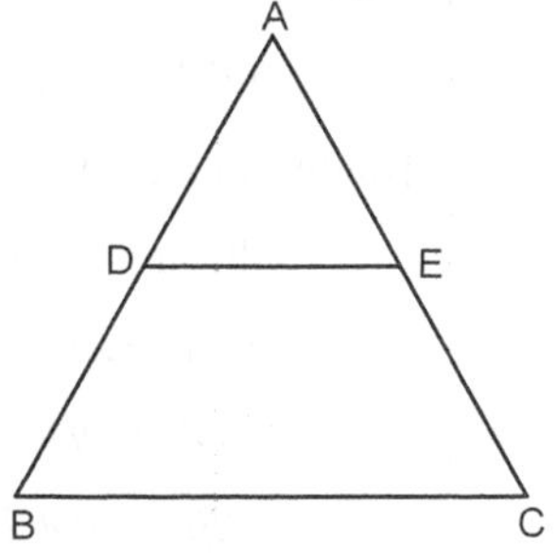

Solution : Let $AE = x$ cm

Then $EC = AC - AE = (4.8 - x)$ cm

In ΔABC, $DE \parallel BC$

So, $\frac{AD}{DB} = \frac{AE}{EC}$ [Thales' theorem]

$\Rightarrow$ $\frac{3}{5} = \frac{x}{4.8 - x}$

$\Rightarrow$ $3(4.8 - x) = 5x$

$\Rightarrow$ $5x = 14.4 - 3x$

$\Rightarrow$ $8x = 14.4$

$\Rightarrow$ $x = 1.8$ cm

Thus, the length of $AE = 1.8$ cm. **Ans.**

Q. 4. The incircle of an isosceles triangle *ABC*, in which *AB* = *AC*, touches the sides *BC*, *CA* and *AB* at *D*, *E* and *F* respectively. Prove that *BD* = *DC*.★

★ are board exam questions from previous years

Solution : Given that, ΔABC is isosceles with $AB = AC$. Also, the incircle touches the sides BC, CA and AB at D, E and F respectively. Let O be the centre of the circle. So, $OD = OE = OF = r$.

Now O is the point of intersection of angle bisectors of ΔABC. Thus, OB bisects $\angle B$ and OC bisects $\angle C$.

In ΔABC, $AB = AC$ [Given]

$\angle C = \angle B$

[Angles opposite to equal sides are equal]

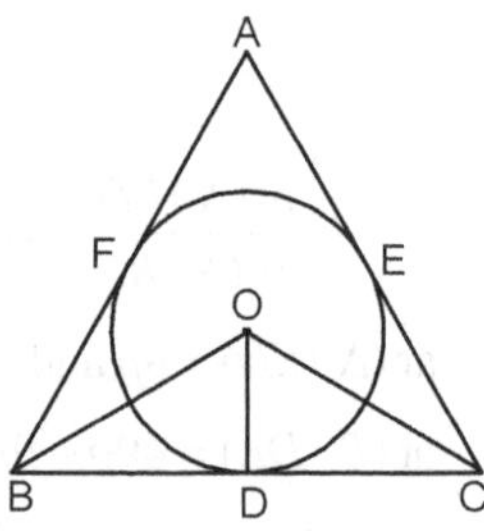

So $\frac{1}{2}\angle C = \frac{1}{2}\angle B$

$\angle OCD = \angle OBD$

[OB bisects $\angle B$ and OC bisects $\angle C$]

In ΔOBD and ΔOCD,

$\angle ODB = \angle ODC$

[Radius of a circle is always perpendicular to the tangent at the point of contact]

Thus $\angle OBD = \angle OCD$ [Proved above]

$OD = OD$ [Common]

$\Delta OBD \cong \Delta OCD$

So $BD = DC$. **[cpct] Hence Proved.**

Q. 5. State whether the given pairs of triangles are similar or not. In case of similarity, mention the criterion.★

(i)

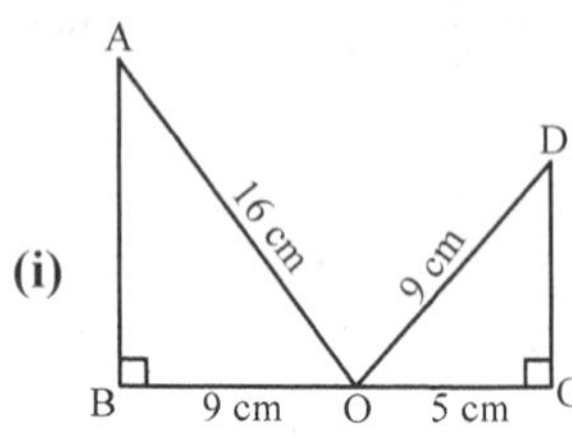

(ii)

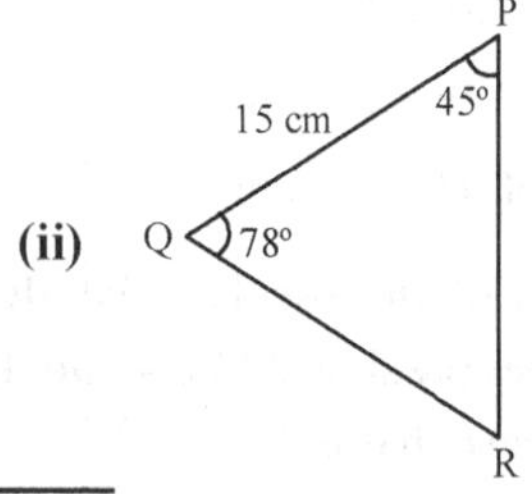

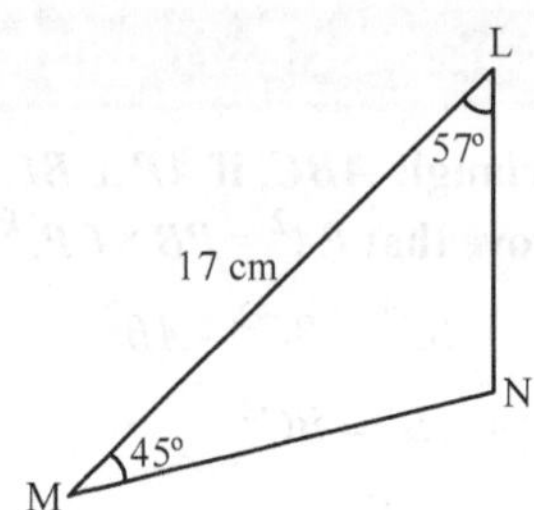

Solution :

(i) $\frac{AO}{DO} = \frac{16}{9}$ and $\frac{BO}{CO} = \frac{9}{5}$

$\because$ $\frac{AO}{DO} \neq \frac{BO}{CO}$

$\therefore$ Given triangles are not similar. **Ans.**

(ii) In ΔPQR, $\angle P + \angle Q + \angle R = 180°$

[Angle-sum property of a Δ]

$\Rightarrow$ $45° + 78° + \angle R = 180°$

$\angle R = 180° - 45° - 78° = 78°$

In ΔLMN,

$\angle L + \angle M + \angle N = 180°$

[Angle-sum property of a Δ]

$\Rightarrow$ $57° + 45° + \angle N = 180°$

$\angle N = 180° - 57° - 45° = 78°$

Now, in ΔPQR and ΔLMN

$\angle P = \angle M$ [Each 45°]

$\angle Q = \angle N$ [Each 78°]

$\angle R = \angle L$ [Each 57°]

By AAA similarity criterion

$\Delta PQR \sim \Delta MNL$ **Ans.**

Q. 6. In the given figure, *DEFG* is a square and $\angle BAC = 90°$. Prove that : (i) $\Delta AGF \sim \Delta DBG$, (ii) $\Delta AGF \sim \Delta EFC$, (iii) $\Delta DBG \sim \Delta EFC$, (iv) $DE^2 = BD \times EC$.★

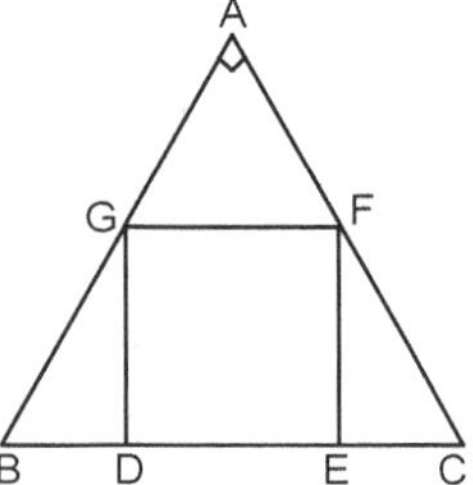

Solution : (i) In ΔAGF and ΔDBG, we have

$\angle GAF = \angle BDG = 90°$

and $\angle AGF = \angle GBD$

[corresponding angles between parallel lines GF and BC with AB being the transversal]

Thus, by AA similarity criterion,

★ are board exam questions from previous years

$\Delta AGF \sim \Delta DBG.$

(ii) In ΔAGF and ΔEFC, we have

$$\angle GAF = \angle FEC = 90°$$

and $\angle AFG = \angle FCE$

[corresponding angles between parallel lines GF and BC with AC being the transversal]

Thus, $\Delta AGF \sim \Delta EFC$

(iii) From (i) and (ii), we have

$$\Delta AGF \sim \Delta DBG$$

and $\Delta AGF \sim \Delta EFC$

Thus $\Delta DBG \sim \Delta EFC$

(iv) From (iii), we have

$$\Delta DBG \sim \Delta EFC$$

Thus $\dfrac{BD}{FE} = \dfrac{DG}{EC}$ [cpct]

$\Rightarrow \dfrac{BD}{DE} = \dfrac{DE}{EC}$

[$DE = DG = FE = GF$ as they are the sides of a square]

$\Rightarrow DE^2 = BD \times EC.$ **Hence Proved.**

Q. 7. Two right-angled triangles, *ABC* and *DBC*, right angled at *A* and *D* respectively are drawn on the same hypotenuse *BC* and on the same side of *BC*. If *AC* and *BD* intersect at *P*, prove that $AP \times PC = PD \times BP$.

Solution : In ΔBAP and ΔCDP, we have

$\angle BAP = \angle CDP = 90°$ [given]

$\angle BPA = \angle CPD$

[vertically opposite angles]

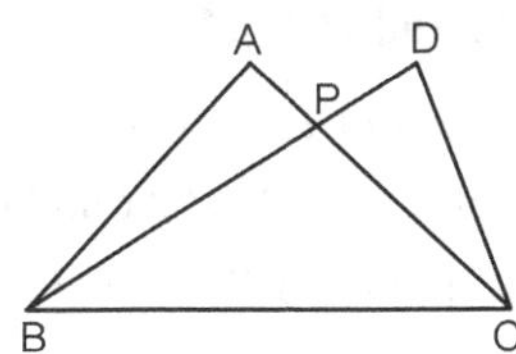

by AA similarity criterion,

So, $\Delta BAP \sim \Delta CDP$

$\therefore \dfrac{AP}{DP} = \dfrac{BP}{CP}$ [cpct]

$\Rightarrow AP \times CP = BP \times DP$

$\Rightarrow AP \times PC = BP \times PD.$ **Hence Proved.**

Q. 8. In the given figure, *DB* is perpendicular to *BC*, *DE* is perpendicular to *AB* and *AC* is perpendicular to *BC*. Prove that $\dfrac{BE}{DE} = \dfrac{AC}{BC}$.

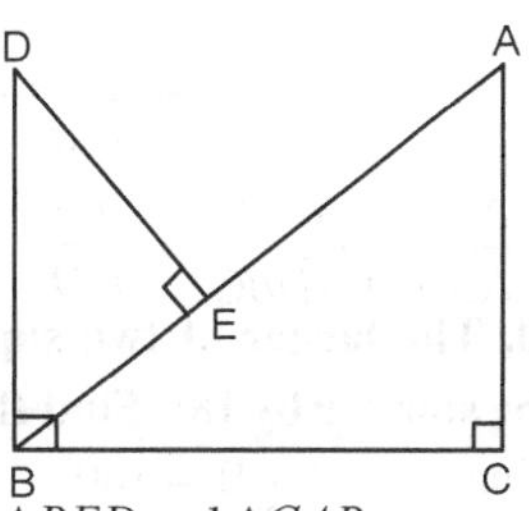

Solution : In ΔBED and ΔCAB,

$\angle BED = \angle BCA = 90°$ [Given]

$\angle DBE = \angle BAC$ [Alternative angles]

by AA similarity axiom,

So, $\Delta BED \sim \Delta ACB$

$\dfrac{BE}{AC} = \dfrac{DE}{BC}$ [cpct]

$\Rightarrow \dfrac{BE}{DE} = \dfrac{AC}{BC}.$ **Hence Proved.**

Q. 9. In the given figure, ΔABC and ΔDBC are on the *BC*. If *AD* intersects *BC* at *O*, prove that :★★

$$\frac{\textbf{Area of } \Delta ABC}{\textbf{Area of } \Delta DBC} = \frac{AO}{DO}.$$

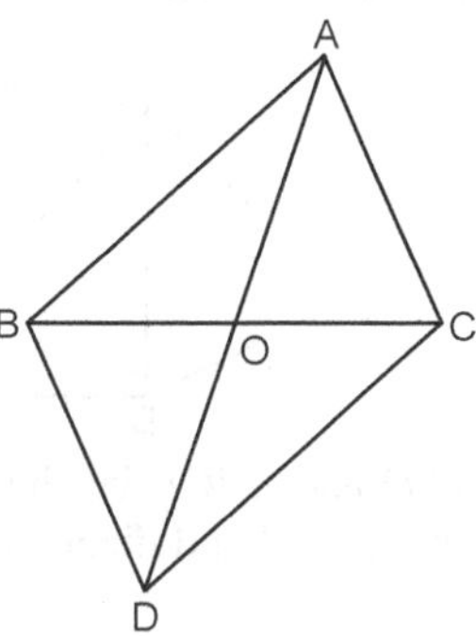

Solution : Let AL and DM be perpendicular to BC in ΔABC and ΔDBC respectively

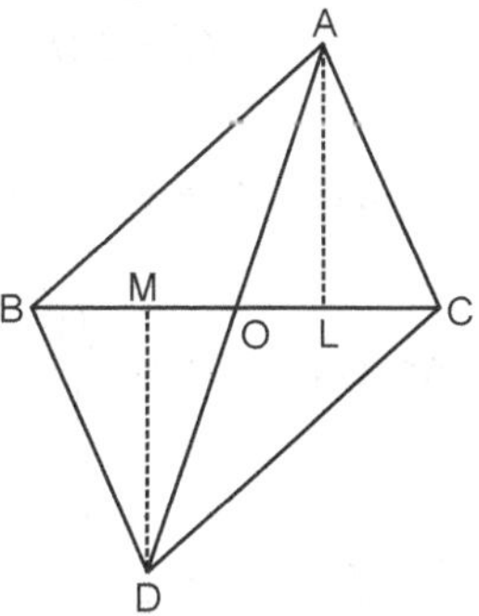

Now, $\angle AOL = \angle DOM$

and $\angle OMD = \angle OLA = 90°$

$\therefore \Delta ALO \sim \Delta DMO$ [AA similarity]

Hence $\dfrac{AL}{DM} = \dfrac{AO}{DO}$ [cpct]

$$\Rightarrow \frac{\text{Area of } \Delta ABC}{\text{Area of } \Delta DBC} = \frac{\frac{1}{2} \times BC \times AL}{\frac{1}{2} \times BC \times DM}$$

★★ are frequently asked board exam questions

$$= \frac{AL}{DM} = \frac{AO}{DO}$$

$$\Rightarrow \quad \frac{\text{Area of } \Delta ABC}{\text{Area of } \Delta DBC} = \frac{AO}{DO}.$$ **Hence Proved.**

Q. 10. The larger of two supplementary angles exceeds the smaller by 18°. Find the angles.★

Solution : $A + B = 180°$ ...(i)

Given, $A = B + 18°$

On putting $A = B + 18°$ in equation (i), we get

$$B + 18° + B = 180°$$
$$2B + 18° = 180°$$
$$2B = 162°$$
$$B = 81°$$
$$A = B + 18°$$
$$\Rightarrow \quad A = 99°$$ **Ans.**

Q. 11. Prove that the area of an equilateral triangle described on a side of a right-angled isosceles triangle is half the area of the equilateral triangle described on the hypotenuse.★

Solution : Let ABC be a right-angled isosceles triangle with $AB = BC = x$.

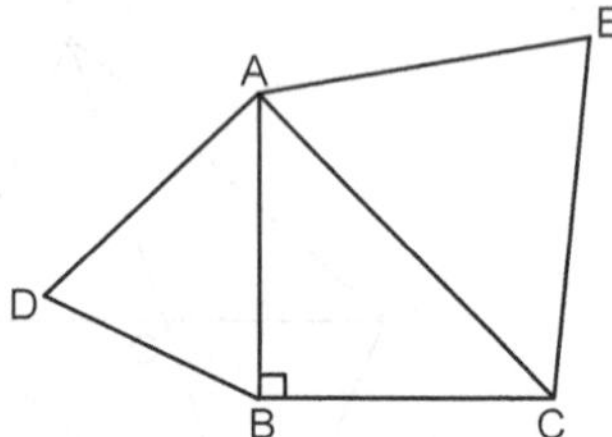

Let ΔABD and ΔACE be the equilateral triangles described on side AB and diagonal AC, respectively of ΔABC.

Now $AB^2 + BC^2 = AC^2$ [Pythagoras' theorem]

$$\Rightarrow \quad AC^2 = 2AB^2$$

Also Area of $\Delta ACE = \frac{\sqrt{3}}{4} AC^2$

and Area of $\Delta ABD = \frac{\sqrt{3}}{4} AB^2$

$$\Rightarrow \quad \frac{\text{Area of } \Delta ACE}{\text{Area of } \Delta ABD} = \frac{\frac{\sqrt{3}}{4} AC^2}{\frac{\sqrt{3}}{4} AB^2} = \frac{AC^2}{AB^2}$$

$$= \frac{AC^2}{AB^2} = \frac{2AB^2}{AB^2} = 2$$

$$\Rightarrow \quad \text{Area of } \Delta ABD = \frac{1}{2} \times \text{Area of } \Delta ACE$$

Hence Proved.

Q. 12. In the given figure, *ABC* is a triangle and *PQ* is a straight line meeting *AB* at *P* and *AC* at *Q* and parallel to *BC*. If *AP* = 1 cm, *PB* = 3 cm, *AQ* = 1.5 cm, *QC* = 4.5 cm, prove that the area of Δ*APQ* is $\frac{1}{16}$ of the area of Δ*ABC*.

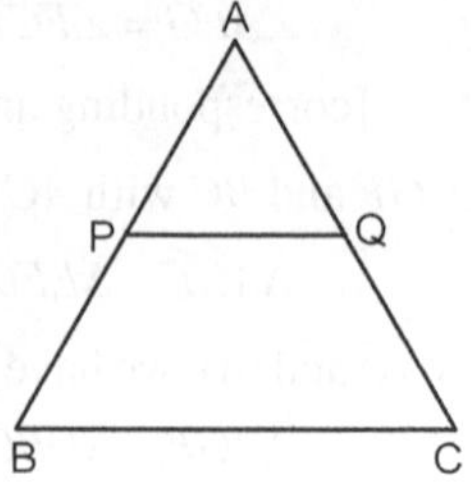

Solution : Given, $AP = 1$ cm, $PB = 3$ cm, $AQ = 1.5$ cm, $QC = 4.5$ cm

Also, $PQ \parallel BC$

$\therefore \quad \angle APQ = \angle ABC$ [Corresponding angles]

$AB = AP + BP = 1 + 3 = 4$ cm

Now, $AC = AQ + QC = 1.5 + 4.5 = 6$ cm

In ΔABC and ΔAPQ,

$\angle A = \angle A$ [Common]

$\angle APQ = \angle ABC$ [Prove above]

$\Rightarrow \quad \Delta ABC \sim \Delta APQ$ [AA similarity]

$$\Rightarrow \quad \frac{\text{Area of } \Delta ABC}{\text{Area of } \Delta APQ} = \frac{AB^2}{AP^2}$$

$$\Rightarrow \quad \frac{\text{Area of } \Delta ABC}{\text{Area of } \Delta APQ} = \frac{4^2}{1^2} = \frac{16}{1}$$

$$\Rightarrow \quad \text{Area of } \Delta ABC = \frac{16}{1} \text{ Area of } \Delta APQ$$

$$\Rightarrow \quad \text{Area of } \Delta APQ = \frac{1}{16} \text{ Area of } \Delta ABC$$

Hence Proved.

Q. 13. Find the area of the triangle *PQR* with *Q* (3, 2) and the mid-points of the sides through *Q* being (2, –1) and (1, 2).★

Solution : Let $P(x_1, y_1)$, $Q(3, 2)$ and $R(x_2, y_2)$ be the vertices of a triangle PQR and let $A(2, -1)$ and $B(1, 2)$ be the mid-points of PQ and QR respectively.

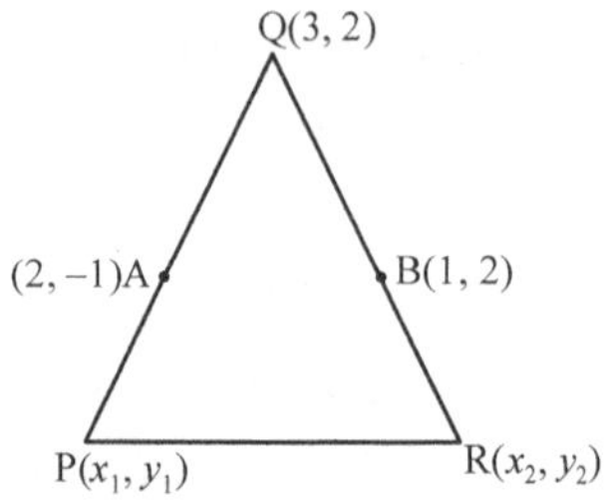

$\because$ A is the mid-point of PQ

$$\therefore \quad \frac{3 + x_1}{2} = 2, \frac{2 + y_1}{2} = -1.$$

$$\Rightarrow \quad x_1 = 1, y_1 = -4$$

★ are board exam questions from previous years

So, $P(1, -4)$

$\because$ B is the mid-point of QR

$\therefore \quad \frac{3+x_2}{2} = 1, \frac{2+y_2}{2} = 2$

$\Rightarrow \quad x_2 = -1, y_2 = 2$

So, $R(-1, 2)$

Thus, Area of ΔPQR

$$= \frac{1}{2}|1(2-2) - 1(2+4) + 3(-4-2)|$$

$$= \frac{1}{2}|1(0) - 1(6) + 3(-6)|$$

$$= \frac{1}{2}|-6-18|$$

$$= \frac{24}{2} = 12 \text{ sq. units}$$ **Ans.**

Q. 14. Two trees of height *a* and *b* are *p* metre apart. Prove that the height of the point of intersection of the lines joining the top of each tree to the foot of the opposite trees is given by $\frac{ab}{a+b}$ m.

Solution : Let AB and CD be the two trees of height a and b metre respectively that are p metre apart. Let the lines AD and BC meet at O such that $OL = h$ metre

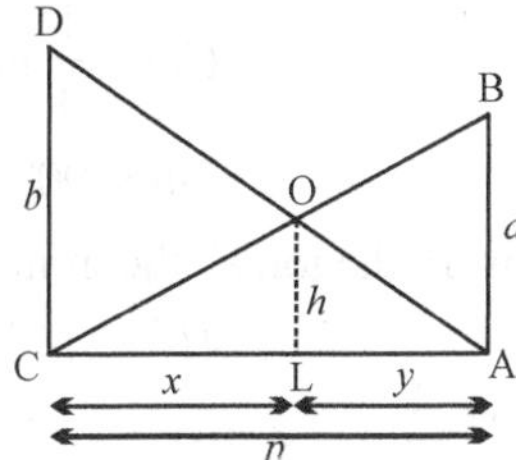

Let $CL = x$ and $LA = y$

then $x + y = p$

In ΔABC and ΔLOC, we have

$\angle CAB = \angle CLO$ [Each 90°]

$\angle C = \angle C$ [Common]

$\Rightarrow \quad \Delta CAB \sim \Delta CLO$ [By AA axiom]

$\therefore \quad \frac{CA}{CL} = \frac{AB}{LO}$ [cpct]

$\Rightarrow \quad \frac{p}{x} = \frac{a}{h}$

$\Rightarrow \quad x = \frac{ph}{a}$ ...(i)

Also, in ΔALO and ΔACD

$\angle ALO = \angle ACD$ [Each 90°]

$\angle A = \angle A$ [Common]

$\Delta ALO \sim \Delta ACD$ [By AA axiom]

$\therefore \quad \frac{AL}{AC} = \frac{OL}{DC}$ [cpct]

$\Rightarrow \quad \frac{y}{p} = \frac{h}{b}$

$\Rightarrow \quad y = \frac{ph}{b}$ ...(ii)

Adding equations (i) and (ii), we get

$$x + y = \frac{ph}{a} + \frac{ph}{b}$$

$$x + y = ph\left(\frac{1}{a} + \frac{1}{b}\right)$$

$$p = ph\left(\frac{1}{a} + \frac{1}{b}\right) \quad [\because x + y = p]$$

$\Rightarrow \quad \frac{1}{h} = \frac{1}{a} + \frac{1}{b}$

$$= \frac{a+b}{ab}$$

$\Rightarrow \quad h = \frac{ab}{a+b}$ m **Hence Proved.**

Q. 15. Two poles of height 6 m and 11 m stand vertically on a plane ground. If the distance between them is 12 m, find the distance between their tops. [NCERT]

Solution : Let the two poles be represented by EF and ST such that

$$EF = 11 \text{ m and } ST = 6 \text{ m}$$

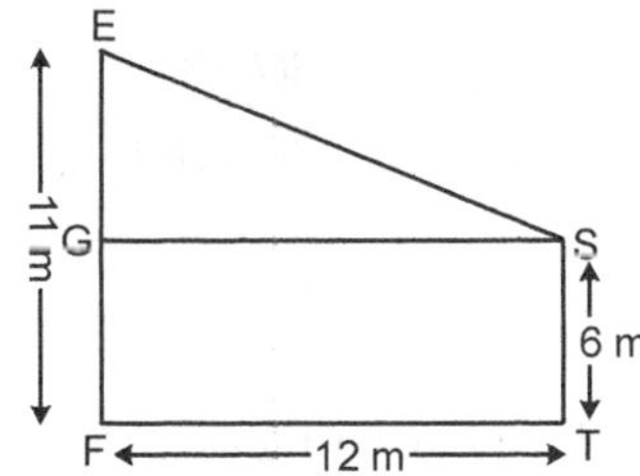

Now, $\quad FT = 12$ m

As $\quad GS \parallel FT$

$\Rightarrow \quad GS = 12$ m

$\Rightarrow \quad GF = 6$ m $\quad [GF = ST]$

$\Rightarrow \quad EG = EF - GF$

$= (11 - 6)$ m

$= 5$ m

The distance between the tips of the poles is represented by ES.

Now, $\quad ES^2 = EG^2 + GS^2$

[Pythagoras' theorem]

$\Rightarrow \quad ES^2 = 5^2 + 12^2$

$$= 25 + 144$$
$$= 169 = 13^2$$
$$\Rightarrow \quad ES = 13 \text{ m}$$

So the distance between the tips of the poles is 13 m.

Ans.

Q. 16. In a rhombus of side 10 cm, one of the diagonals is 12 cm long. Find the length of the second diagonal.

Solution : Let $ABCD$ be the rhombus with $AB = BC = CD = AD = 10$ cm.

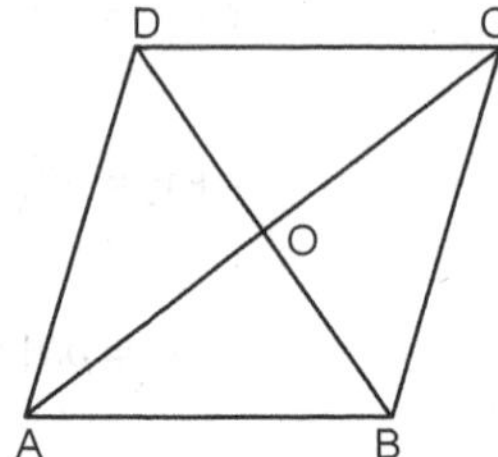

Let O be the point of their intersection such that
$$\angle AOB = \angle BOC = \angle COD = \angle DOA = 90°$$
[Diagonal of a rhombus bisect each other]

Let AC and BD be the two diagonals.

Also let $\quad BD = 12$ cm

So $\quad OD = 6$ cm

[O is the mid-point of BD]

Hence, by Pythagoras' theorem,
$$AO^2 = AD^2 - OD^2$$
$$\Rightarrow \quad AO^2 = 10^2 - 6^2$$
$$= 100 - 36$$
$$= 64 = 8^2$$
$$\Rightarrow \quad AO = 8$$
$$\Rightarrow \quad AC = AO + OC$$
$$= 2AO = (2 \times 8) \text{ cm}$$
$$= 16 \text{ cm.}$$

Ans.

Q. 17. In given figure, $EB \perp AC$, $BG \perp AE$ and $CF \perp AE$ Prove that :

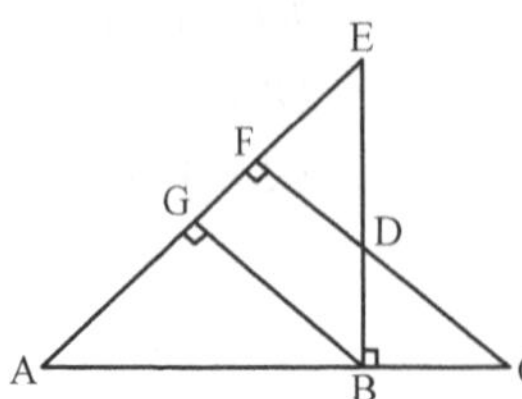

(i) $\Delta ABG \sim \Delta DCB$

(ii) $\dfrac{BC}{BD} = \dfrac{BE}{BA}$

Solution : Given, $EB \perp AC$, $BG \perp AE$ and $CF \perp AE$.

(i) In ΔABG and ΔDCB,

$\angle 2 = \angle 5$ [Each 90°]

$\angle 6 = \angle 4$ [Corresponding angles]

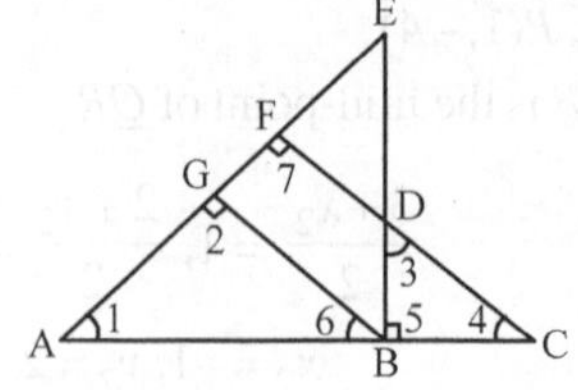

$\therefore \quad \Delta ABG \sim \Delta DCB$ [By AA similarity]

Hence Proved.

$\angle 1 = \angle 3$ [cpct]

(ii) In ΔABE and ΔDBC

$\angle 1 = \angle 3$ [Proved above]

$\angle ABE = \angle 5$ [Each 90°]

$\therefore \quad \Delta ABE \sim \Delta DBC$ [By AA similarity]

In similar Δs, corresponding sides are proportional

$$\therefore \quad \frac{BC}{BD} = \frac{BE}{BA}$$

Hence Proved

Q. 18. In ΔABC, AD is perpendicular to BC and $BD = 3CD$. Prove that $2AB^2 = 2AC^2 + BC^2$.★

Solution : Given that
$$BD = 3CD$$

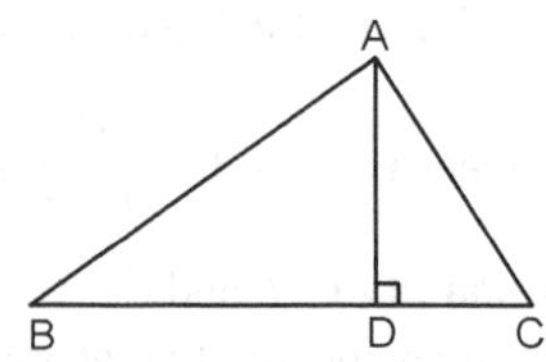

So $\quad BC = BD + CD = 3CD + CD = 4CD$

$$\Rightarrow \quad CD = \frac{1}{4}BC \quad \text{...(i)}$$

In ΔADB, $\quad \angle ADB = 90°$

So by Pythagoras' theorem,
$$AB^2 = AD^2 + BD^2 \quad \text{...(ii)}$$

In ΔADC, $\quad \angle ADC = 90°$

So by Pythagoras' theorem,
$$AC^2 = AD^2 + DC^2 \quad \text{...(iii)}$$

Subtracting equation (iii) from equation (ii), we get

Thus $\quad AB^2 - AC^2 = (AD^2 + BD^2) - (AD^2 + DC^2)$
$$\Rightarrow \quad AB^2 - AC^2 = BD^2 - CD^2$$
$$\Rightarrow \quad AB^2 - AC^2 = (3CD)^2 - CD^2$$
$$\Rightarrow \quad AB^2 - AC^2 = 9CD^2 - CD^2$$
$$\Rightarrow \quad AB^2 - AC^2 = 8CD^2$$
$$\Rightarrow \quad AB^2 - AC^2 = 8\left(\frac{1}{4}BC\right)^2$$
[From equation (i)]
$$\Rightarrow \quad AB^2 - AC^2 = 8 \times \frac{1}{16}BC^2$$
$$\Rightarrow \quad AB^2 - AC^2 = \frac{1}{2}BC^2$$

★ are board exam questions from previous years

$\Rightarrow \quad 2(AB^2 - AC^2) = BC^2$

$\Rightarrow \quad 2AB^2 - 2AC^2 = BC^2$

$\Rightarrow \quad 2AB^2 = 2AC^2 + BC^2.$

Hence Proved.

Q. 19. ΔABC is a right-angled triangle in which $\angle C = 90°$ and CD is perpendicular to AB. If $BC = a$, $CA = b$, $AB = c$ and $CD = p$, then prove that :

(i) $cp = ab$ and (ii) $\frac{1}{p^2} = \frac{1}{a^2} + \frac{1}{b^2}$.

Solution : (i) Area of $\Delta ABC = \frac{1}{2} \times AB \times CD$

$$= \frac{1}{2} \times BC \times AC$$

$\Rightarrow \quad \frac{1}{2} \times c \times p = \frac{1}{2} \times a \times b$

$\Rightarrow \quad cp = ab.$ **Hence Proved.**

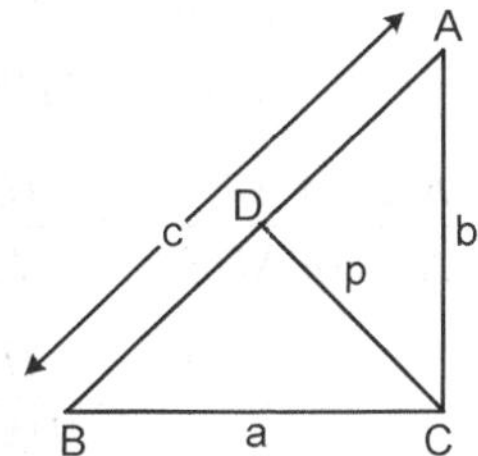

(ii) Now dividing $cp = ab$ by abp, we get

$$\frac{c}{ab} = \frac{1}{p}$$

$\Rightarrow \quad \frac{1}{p^2} = \frac{c^2}{a^2b^2} = \frac{a^2+b^2}{a^2b^2}$ $[\because AB^2 = BC^2 + AC^2]$

$\Rightarrow \quad \frac{1}{p^2} = \frac{a^2}{a^2b^2} + \frac{b^2}{a^2b^2}$

$\Rightarrow \quad \frac{1}{p^2} = \frac{1}{b^2} + \frac{1}{a^2}.$ **Hence Proved.**

Q. 20. In an equilateral triangle, prove that three times the square of one side is equal to four times the square of one of its altitudes.★★

Solution : Let ABC be an equilateral triangle such that $AB = BC = AC$ and AD is perpendicular to BC.

Then in ΔABD and ΔACD,

$\angle B = \angle C = 60°$

$AD = AD$ [Common]

$\angle ADB = \angle ADC$ [Each 90°]

$\Rightarrow \Delta ABD \cong \Delta ACD$

$\Rightarrow \quad BD = DC = \frac{1}{2}BC$

In ΔABD, $\quad AB^2 = AD^2 + BD^2$

[Pythagoras' theorem]

$\Rightarrow \quad AB^2 = AD^2 + \left(\frac{1}{2}BC\right)^2$

$\Rightarrow \quad AB^2 = AD^2 + \frac{1}{4}BC^2$

$\Rightarrow \quad 4AB^2 = 4AD^2 + BC^2$

$\Rightarrow \quad 4AB^2 = 4AD^2 + AB^2$ $[AB = BC]$

$\Rightarrow \quad 3AB^2 = 4AD^2.$ **Hence Proved.**

Q. 21. In the given figure, $PQ \parallel BC$ and $PR \parallel CD$. Prove that : (i) $\frac{AR}{AD} = \frac{AQ}{AB}$, (ii) $\frac{QB}{AQ} = \frac{DR}{AR}$.

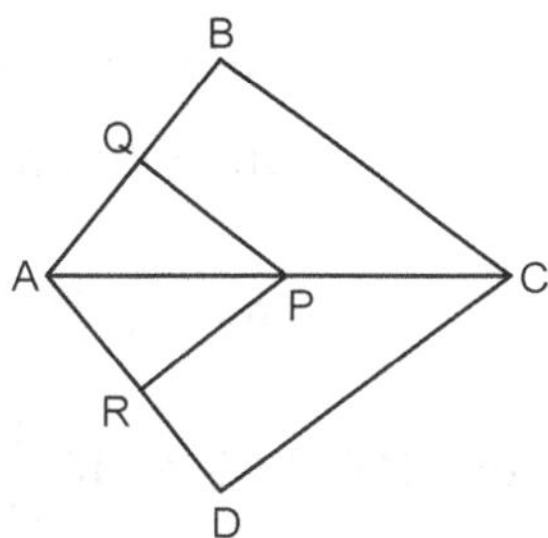

Solution : (i) In ΔADC,

$PR \parallel CD$

$\therefore \quad \frac{AR}{AD} = \frac{AP}{AC}$ [Thales' theorem]...(i)

and in ΔABC,

$PQ \parallel BC$

$\therefore \quad \frac{AP}{AC} = \frac{AQ}{AB}$ [Thales' theorem]...(ii)

From equations (i) and (ii),

$$\frac{AR}{AD} = \frac{AQ}{AB}.$$ **Hence Proved.**

(ii) In ΔABC,

$PQ \parallel DC$

$\therefore \quad \frac{QB}{AQ} = \frac{PC}{AP}$ [Thales' thereom]...(i)

and in ΔADC, $\quad PR \parallel DC$

$\therefore \quad \frac{PC}{AP} = \frac{DR}{AR}$ [Thales' theorem]...(ii)

From equation (i) and (ii),

$$\frac{QB}{AQ} = \frac{DR}{AR}.$$ **Hence Proved.**

Q. 22. In the given figure, P is the mid-point of BC and Q is the mid-point of AP. If BQ meets AC at R, prove that $RA = \frac{1}{3}CA$.

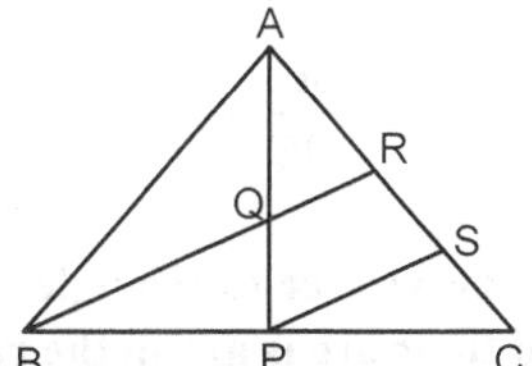

★★ are frequently asked board exam questions

Solution : Let $PS \parallel BR$ meet AC at S.

In ΔBCR, P is the mid-point of BC and $PS \parallel BR$.

$\therefore$ S is the mid-point of CR.

$\Rightarrow \quad CS = SR \quad$ ...(i)

Similarly, in ΔAPS, Q is the mid-point of AP and $QR \parallel PS$.

$\therefore$ R is the mid-point of AS.

$\Rightarrow \quad AR = RS \quad$ ...(ii)

From equations (i) and (ii),

$$AR = RS = CS$$

Thus $\quad AC = AR + RS + SC = 3AR$

$\Rightarrow \quad RA = \frac{1}{3}CA.$ **Hence Proved.**

Q. 23. In ΔABC, $DE \parallel BC$. If $\frac{AD}{DB} = \frac{2}{3}$, find $\frac{BC}{DE}$.

Solution : In ΔABC and ΔADE,

$\angle A = \angle A$ [common]

$\angle ADE = \angle B$ [Corresponding angles]

$\angle AED = \angle C$ [Corresponding angles]

Thus, $\quad \Delta ABC \sim \Delta ADE$ [AAA-criterion]

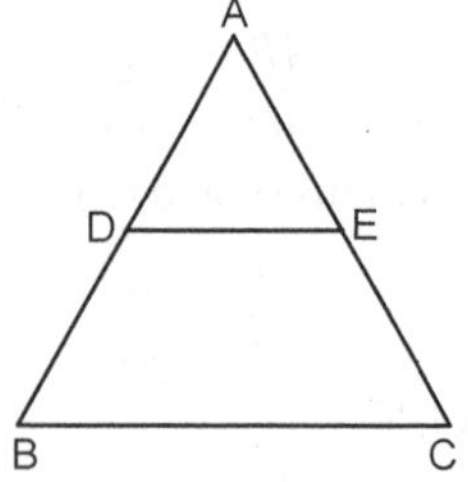

Now $\quad \frac{AD}{DB} = \frac{2}{3}$

$\Rightarrow \quad 3AD = 2DB$

Also, $\quad AB = AD + DB$

$\Rightarrow \quad 2AB = 2AD + 2DB$

$\Rightarrow \quad 2AB = 2AD + 3AD \quad [3AD = 2DB]$

$\Rightarrow \quad 2AB = 5AD$

$\Rightarrow \quad \frac{AB}{AD} = \frac{5}{2}$

Since $\quad \Delta ABC \sim \Delta ADE,$

Then $\quad \frac{AB}{AD} = \frac{BC}{DE}$ [cpct]

$\Rightarrow \quad \frac{BC}{DE} = \frac{5}{2} \quad \left[\because \frac{AB}{AD} = \frac{5}{2}\right]$

Ans.

Q. 24. Two isosceles triangles have equal vertical angles and their areas are in the ratio of 16 : 25. Find the ratio of their corresponding heights.

Solution : Let the ratio of the area of ΔABC and ΔDEF be 16 : 25. Also

$$\angle A = \angle D$$

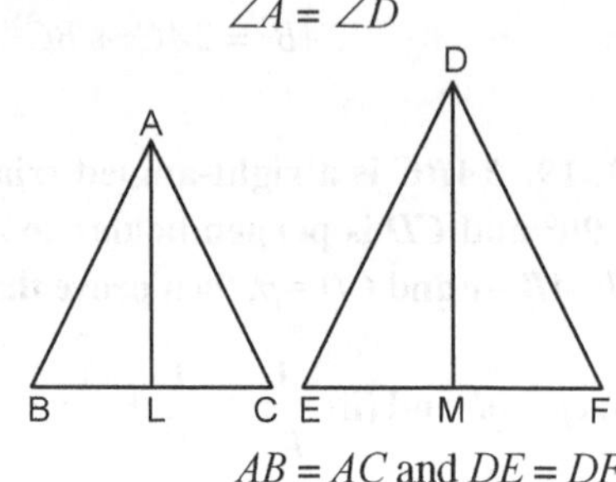

Now $\quad AB = AC$ and $DE = DF$

[$\because$ Δs are isosceles]

$\Rightarrow \quad \frac{AB}{AC} = \frac{DE}{DF} = 1$

So in ΔABC and ΔDEF,

$\angle A = \angle D$ [Given]

$\frac{AB}{AC} = \frac{DE}{DF}$ [Proved above]

$\therefore \quad \Delta ABC \sim \Delta DEF$

$\Rightarrow \quad \frac{\text{Area of } \Delta ABC}{\text{Area of } \Delta DEF} = \frac{AL^2}{DM^2}$

$\Rightarrow \quad \frac{16}{25} = \frac{AL^2}{DM^2}$ [Given]

$\Rightarrow \quad \frac{4^2}{5^2} = \frac{AL^2}{DM^2}$

$\Rightarrow \quad \frac{4}{5} = \frac{AL}{DM}$

$\Rightarrow \quad AL : DM = 4 : 5.$ **Ans.**

Q. 25. In the given figure, $DE \parallel BC$ and $AD : DB = 5 : 4$. Find $\frac{\text{Area of } \Delta DEF}{\text{Area of } \Delta CFB}$.

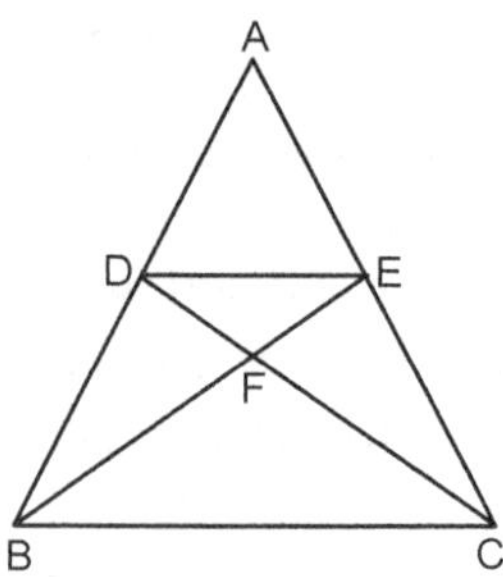

Solution : Given, $DE \parallel BC$

So $\angle ADE = \angle ABC$ and $\angle AED = \angle ACB$

[Corresponding angles]

Thus, in ΔABC and ΔADE,

$\angle A = \angle A$ [Common]

$\angle ADE = \angle ABC$

[Corresponding angles]

$\angle AED = \angle ACB$

[Corresponding angles]

$\Rightarrow \quad \Delta ABC \sim \Delta ADE$

$\therefore \quad \dfrac{AD}{AB} = \dfrac{DE}{BC}$ [cpct]

We have $\dfrac{AD}{DB} = \dfrac{5}{4}$ [Given]

$\Rightarrow \quad \dfrac{DB}{AD} = \dfrac{4}{5}$

$\Rightarrow \quad \dfrac{DB}{AD} + 1 = \dfrac{4}{5} + 1$

$\Rightarrow \quad \dfrac{DB + AD}{AD} = \dfrac{4+5}{5}$

$\Rightarrow \quad \dfrac{AB}{AD} = \dfrac{9}{5}$

$\Rightarrow \quad \dfrac{AD}{AB} = \dfrac{5}{9} = \dfrac{DE}{BC}$

In ΔDFE and ΔCFB,

$\angle DEB = \angle EBC$ [Alternate angles]

and $\angle DFE = \angle BFC$

[Vertically opposite angles]

$\Rightarrow \quad \Delta DFE \sim \Delta CFB$ [AA – Criteria]

$\therefore \quad \dfrac{\text{Area of } \Delta DEF}{\text{Area of } \Delta CFB} = \dfrac{DE^2}{BC^2}$

or $\dfrac{\text{Area of } \Delta DEF}{\text{Area of } \Delta CFB} = \left(\dfrac{5}{9}\right)^2 = \dfrac{25}{81}$. **Ans.**

Q. 26. In ΔABC, $\angle C = 90°$. If D is the mid-point of BC, prove that $AB^2 = 4AD^2 - 3AC^2$.★

Solution : Given,

Δ is the mid-point of BC

$\therefore \quad CD = DB$

In ΔABC,

$AB^2 = BC^2 + AC^2$

[Pythagoras theorem]

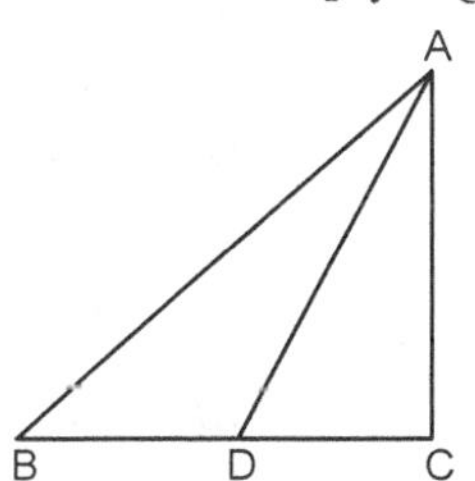

$\Rightarrow \quad AB^2 = (CD + DB)^2 + AC^2$

$[\because BC = CD + DB]$

$\Rightarrow \quad AB^2 = (2CD)^2 + AC^2 \quad [\because CD = DB]$

$\Rightarrow \quad AB^2 = 4CD^2 + AC^2 \quad \ldots(i)$

★ are board exam questions from previous years

Also, in ΔADC

$AD^2 = AC^2 + CD^2$

$\Rightarrow \quad CD^2 = AD^2 - AC^2 \quad \ldots(ii)$

Multiplying equation (ii) by 4 and then adding it to in equation (i), we get

$\Rightarrow \quad AB^2 + 4CD^2 = 4CD^2 + AC^2 + 4AD^2 - 4AC^2$

$\Rightarrow \quad AB^2 = 4AD^2 - 3AC^2$. **Hence Proved.**

Chapter 8. Circles

Q. 1. Prove that the tangents drawn at the end points of a chord of a circle make equal angles with the chord.

Solution : Given, a circle of radius OA and centred at O with chord AB and tangents PQ and RS are drawn from point A and B respectively.

Draw $OM \perp AB$, and join OA and OB.

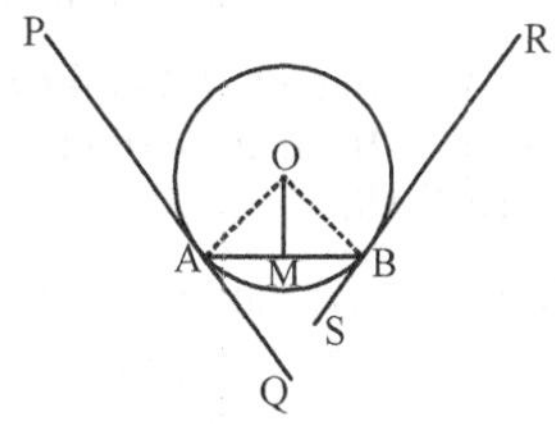

Now, in ΔOAM and ΔOMB,

$OA = OB$ [Radii]

$OM = OM$ [Common]

$\angle OMA = \angle OMB$ [Each 90°]

$\therefore \quad \Delta OAM \cong \Delta OMB$ [R.H.S cong.]

$\therefore \quad \angle OAM = \angle OBM$ ... (i) [cpct]

Also, $\angle OAP = \angle OBR = 90°$ ...(ii)

[line joining point of contact of tangent to centre is perpendicular on it].

On adding (i) and (ii),

$\angle OAM + \angle OAP = \angle OBM + \angle OBR$

$\therefore \quad \angle PAB = \angle RBA$ **Hence Proved.**

Q. 2. In the given figure, a circle is inscribed in ΔPQR with $PQ = 10$ cm, $QR = 8$ cm and $PR = 12$ cm. Find the length QM, RN and PL.

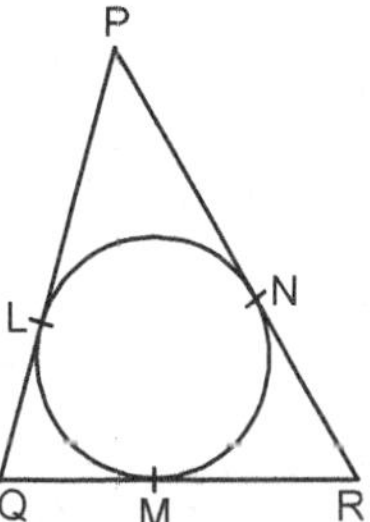

Solution : Given, $PQ = 10$ cm, $PR = 12$ cm and $QR = 8$ cm.

Let the length of $QM = x$ cm

Thus, $RM = (8 - x)$ cm

and $QL = x$ cm

[$\because$ LQ and QM are tangents from Q]

$RN = (8 - x)$ cm

[$\because$ RN and RM are tangents from R]

Now, $PN = PR - RN = [12 - (8 - x)]$ cm

$= (4 + x)$ cm

and $PN = PL$

[$\because$ PL and PN are tangents from P]

Again, $QL = PQ - PL = [10 - (4 + x)]$ cm

$= (6 - x)$ cm

But $QL = x$ cm [From above]

$\therefore$ $x = 6 - x$

$\Rightarrow$ $2x = 6$

$\Rightarrow$ $x = 3$ cm

$\therefore$ $QM = 3$ cm,

$RN = (8 - 3)$ cm $= 5$ cm

and $PL = (4 + 3)$ cm $= 7$ cm. **Ans.**

Q. 3. Two tangents *PA* and *PB* are drawn to a circle with centre *O* from an external point *P*. Prove that $\angle APB = 2\angle OAB$.★★

Solution : As PA and PB are two tangents drawn from an external point P on a circle with centre O, so

$$PA = PB$$

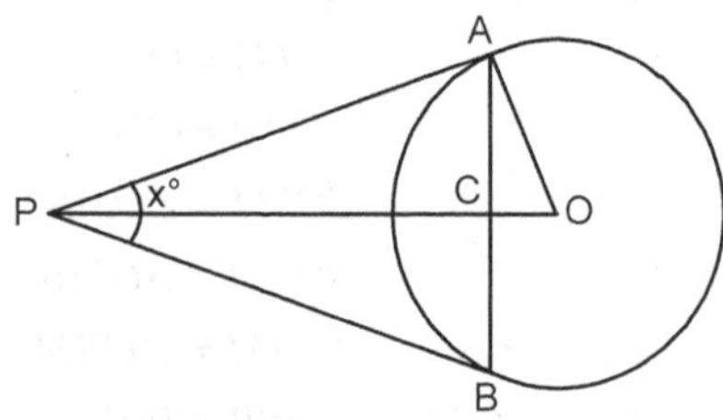

Construct PO such that PO is perpendicular to AB at C and bisects $\angle APB$.

Let $\angle APB = x°$

Thus, $\angle OPA = \dfrac{x°}{2}$

Also, as AO is perpendicular to PA, so

$\angle OAP = 90°$

Hence, $\angle AOP = 180° - \left(90° + \dfrac{x°}{2}\right) = 90° - \dfrac{x°}{2}$

But $\angle ACO = 90°$ [$\because$ CO perpendicular to AB]

Thus, $\angle OAC = 180° - \left[90° + 90° - \dfrac{x°}{2}\right]$

$= \dfrac{x°}{2} = \angle OAB$

Hence, $\dfrac{\angle APB}{2} = \angle OAB$

or $\angle APB = 2\angle OAB.$ **Hence Proved.**

Q. 4. In the given figure, *O* is the centre of two concentric circles. *AB* is a chord of the larger circle touching the smaller circle at *C*. Prove that *AC* = *BC*.

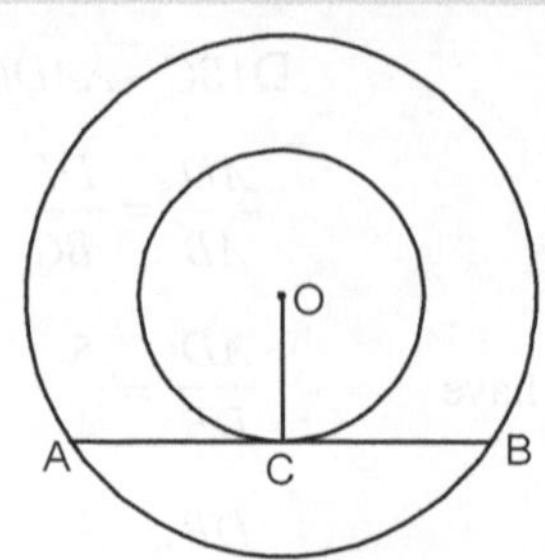

Solution : Given, OC is the radius of the smaller circle and is perpendicular to AB because AB is a tangent to it.

Also, AO and OB are the radii of the larger circle.

Thus, construct OA and OB.

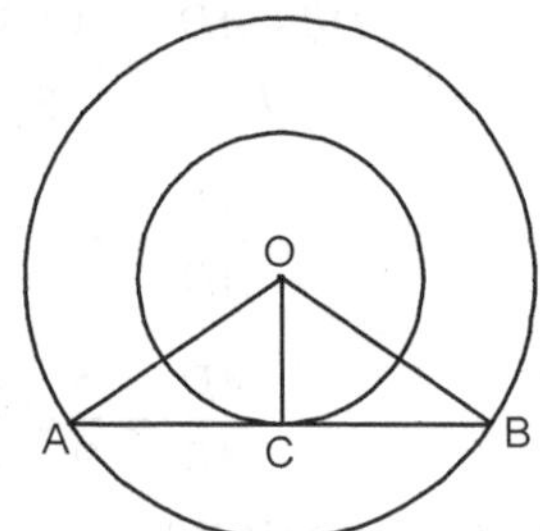

Hence, in ΔOAC and ΔOBC,

$OC = OC$ [Common]

$OA = OB$ [Radii]

$\angle OCA = \angle OCB = 90°$ [Proved above]

Hence $\Delta OAC \cong \Delta OBC$ [R.H.S.]

So, $AC = BC$

[corresponding sides of congruent triangles]

Hence Proved.

Q. 5. In the given figure, a circle is inscribed in a quadrilateral *ABCD* in which $\angle B = 90°$. If *AD* = 23 cm, *AB* = 29 cm and *DS* = 5 cm, find the value of radius (*r*).

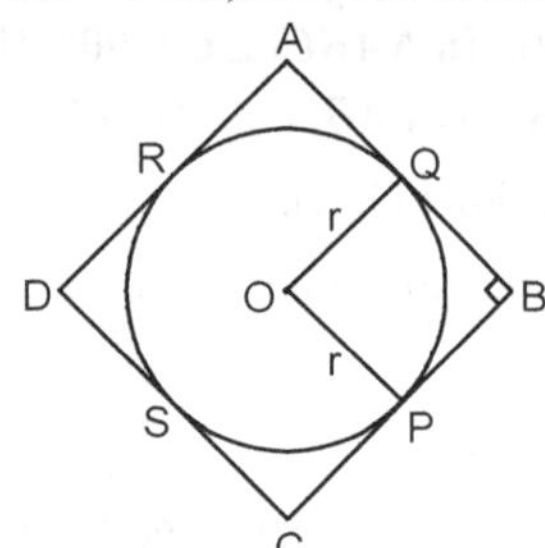

Solution : Given, $\angle B = 90°$, $AD = 23$ cm, $AB = 29$ cm, $DS = 5$ cm and $OQ = OP = r$.

Now, $DS = DR = 5$ cm [tangents from D]

Thus, $AR = AD - DR = (23 - 5)$ cm $= 18$ cm

Hence, $AQ = 18$ cm [tangents from A]

or $BQ = AB - AQ = (29 - 18)$ cm $= 11$ cm

Also, $\angle OQB = \angle OPB = 90°$

[both the radii are perpendicular to tangents AB and BC respectively]

Hence, in the quadrilateral $POQB$,

$\angle QOP = 360° - (90° + 90° + 90°)$

★ are board exam questions from previous years

★★ are frequently asked board exam questions

$= 360° - 270° = 90°$

In quadrilateral *POQB*,

$\angle QOP = \angle OPQ = \angle PBQ = \angle BQO = 90°$

and $BQ = PB$ [tangents from *B*]

i.e. all angles are 90° and adjacent sides are equal.

Hence, *POQB* is a square

$\therefore$ $OQ = BQ = 11$ cm

But $OQ = OP$ [radii of the same circle]

So, $OQ = OP = PB = QB = 11$ cm

Thus, the radius of the circle is 11 cm. **Ans.**

Q. 6. From an external point *P*, tangents *PA* and *PB* are drawn to a circle with centre *O*. If *CD* is the tangent to the circle at a point *E* and *PA* = 14 cm, find the perimeter of Δ*PCD*.

Solution : We have,

$AP = PB$ [Tangents from point *P*]

$AC = CE$ [Tangents from point *C*]

$BD = DE$ [Tangents from point *D*]

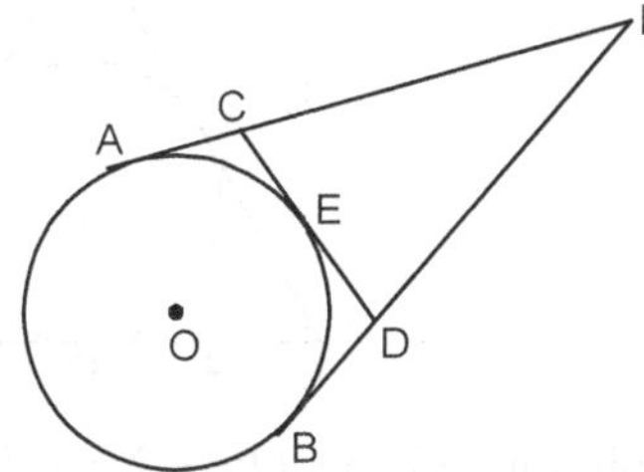

Thus, $CD = CE + ED = AC + BD$

Now, $CP = AP - AC = 14 - AC$

and $PD = BP - BD = 14 - BD$

[$AP = BP$; from above]

Now, perimeter of Δ*PCD*

$= CP + CD + DP$

$= 14 - AC + AC + BD + 14 - BD$

$= 28$ cm. **Ans.**

Q. 7. A circle is inscribed in a Δ*ABC* touching *AB*, *BC* and *AC* at *P*, *Q* and *R* respectively. If *AB* = 10 cm, *AR* = 7 cm and *CR* = 5 cm, find the length of *BC*.★

Solution : Given, $AB = 10$ cm, $AR = 7$ cm and $CR = 5$ cm

Now, $AP = AR = 7$ cm [Tangents from *A*]

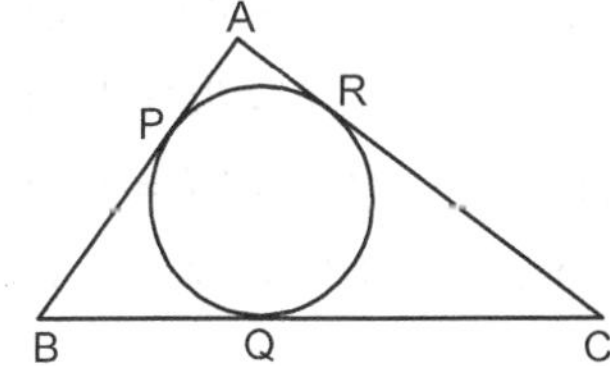

Thus, $PB = AB - AP = (10 - 7)$ cm $= 3$ cm

Also, $PB = BQ = 3$ cm [Tangents from *B*]

★ are board exam questions from previous years

★★ are frequently asked board exam questions

and $QC = RC = 5$ cm [Tangents from *C*]

Hence, $BC = BQ + QC = (3 + 5)$ cm

$= 8$ cm. **Ans.**

Q. 8. For the inauguration of the Eco-friendly week in a school, badges were given to volunteers. Nelu made these badges in the shape of a triangle with a circle inscribed in it (as shown in figure). A message supporting tree plantation was written in the circle. If *BD* = 6 cm, *CD* = 8 cm and *AE* = 3 cm. Find the length of *AB* and *AC*.★★

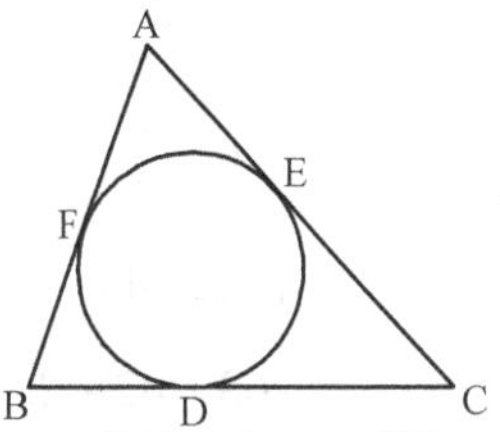

Solution : Given, $BD = 6$ cm, $CD = 8$ cm and $AE = 3$ cm.

Now, $BD = BF$ [Tangents from *B*]

$\Rightarrow$ $BF = 6$ cm

Also, $AF = AE$ [Tangents from *A*]

$\Rightarrow$ $AF = 3$ cm

Now, $AB = AF + BF$

$= 3 + 6 = 9$ cm.

and $CE = CD$ [Tangents from *C*]

$\Rightarrow$ $CE = 8$ cm

Now, $AC = AE + CE$

$= 3 + 8 = 11$ cm **Ans.**

Q. 9. In the given figure, a Δ*PQR* is drawn to circumscribe a circle of radius 6 cm such that the segments *QT* and *TR* into which *QR* is divided by the point of contact are of lengths 12 cm and 9 cm respectively. If the area of Δ*PQR* = 189 cm², then find the lengths of sides *PQ* and *PR*.★★

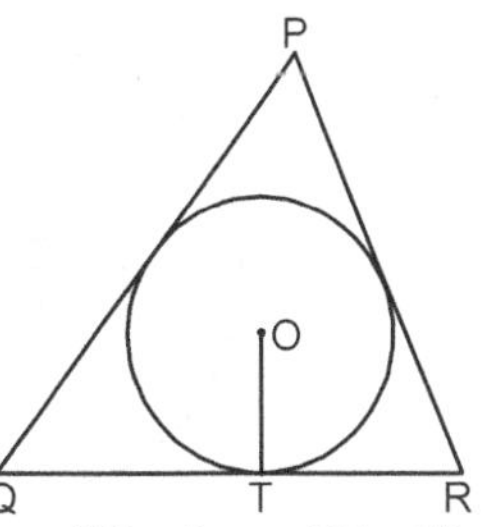

Solution : Given, $OT = 6$ cm, $QT = 12$ cm, $TR = 9$ cm and area of $\Delta PQR = 189$ cm^2.

Let *PQ* and *PR* be tangents to the circle at *S* and *U* respectively.

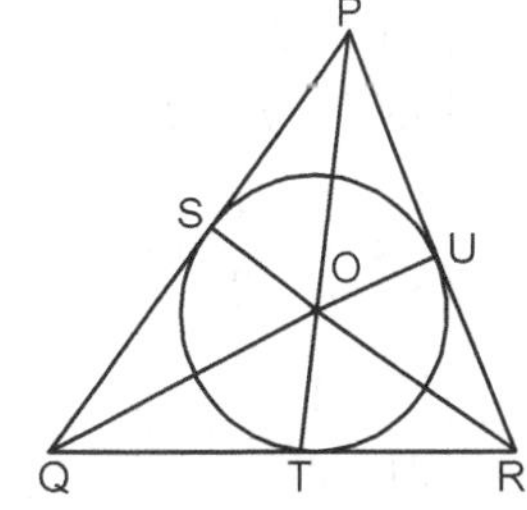

Join PO, QO and RO.

So, $SO = UO = TO = 6$ cm

Now, $PS = PU = x$ (say) [Tangents from P]

Also, $QS = QT = 12$ cm [Tangents from Q]

and $UR = RT = 9$ cm [Tangents from R]

Area of ΔPQR = Area of ΔOQR + Area of ΔPOQ + Area of ΔPOR

$$\Rightarrow \quad 189 = \frac{1}{2}[(12+9)6 + (12+x)6 + (9+x)6]$$

$$\Rightarrow \quad 189 = \frac{6}{2}[(12+9) + (12+x) + (9+x)]$$

$$\Rightarrow \quad 189 = 3[21 + 21 + 2x]$$

$$\Rightarrow \quad 63 = 42 + 2x$$

$$\Rightarrow \quad 2x = 21$$

$$\Rightarrow \quad x = 10.5 \text{ cm}$$

Thus, the length of $PQ = PS + QS$

$= (10.5 + 12)$ cm

$= 22.5$ cm

and the length of $PR = PU + UR$

$= (10.5 + 9)$ cm

$= 19.5$ cm. **Ans.**

Q. 10. A chord of a circle of radius 14 cm, subtends an angle of 120° at the centre. Find the area of the corresponding minor segment of the circle.

[Use $\pi = \frac{22}{7}$ and $\sqrt{3} = 1.73$].

Solution : Given, $r = 14$ and $\angle COB = 120°$

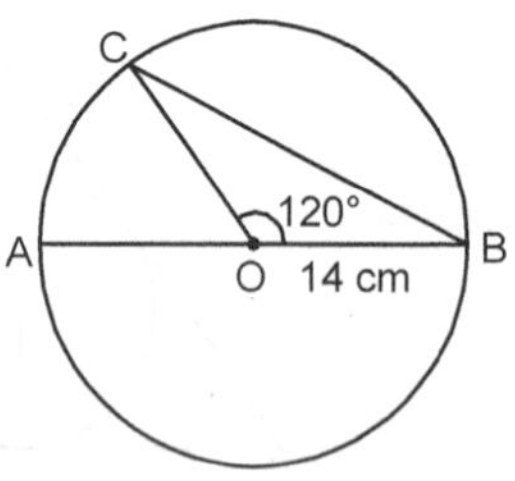

Thus, area of the corresponding minor segment

$$= \left\{\frac{\theta}{360°} \times \pi - \frac{1}{2}\sin\theta\right\} r^2$$

$$= \left\{\frac{120°}{360°} \times \frac{22}{7} - \frac{1}{2}\sin 120°\right\} (14)^2 \text{ cm}^2$$

$$= \left\{\frac{1}{3} \times \frac{22}{7} - \frac{1}{2}\sin(90° + 30°)\right\} 196 \text{ cm}^2$$

$$= \left\{\frac{22}{21} - \frac{1}{2}\sin(90° - (-30°))\right\} 196 \text{ cm}^2$$

$$= \left\{\frac{22}{21} - \frac{1}{2}\cos(-30°)\right\} 196 \text{ cm}^2$$

$$= \left\{\frac{22}{21} - \frac{\sqrt{3}}{4}\right\} 196 \text{ cm}^2$$

$$= \left\{\frac{88 - 21\sqrt{3}}{84}\right\} 196 \text{ cm}^2$$

$$= \left\{\frac{88 - 21(1.73)}{21}\right\} 49 \text{ cm}^2$$

$$= \left\{\frac{88 - 36.33}{3}\right\} 7 \text{ cm}^2$$

$= 120{\cdot}56$ cm^2. **Ans.**

Q. 11. In the given figure, a triangle ΔABC is drawn to circumscribe a circle of radius 2 cm such that the segments BD and DC into which BC is divided by the point of contact are of lengths 4 cm and 3 cm respectively. If the area of ΔABC = 21 cm², then find the lengths of sides AB and AC.

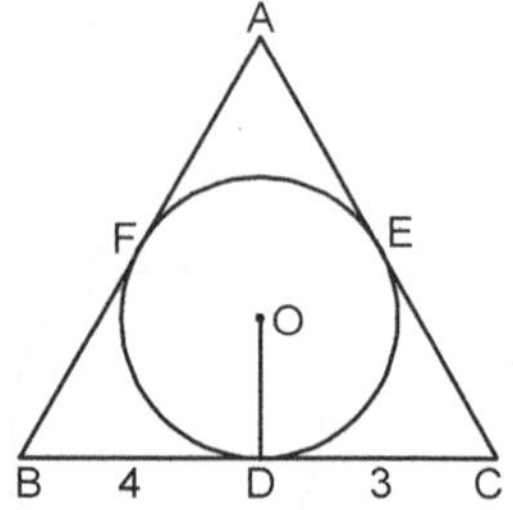

Solution : Given, $OD = 2$ cm, $BD = 4$ cm, $DC = 3$ cm and area of $\Delta ABC = 21$ cm^2

Let AB and AC be tangents to the circle at F and E respectively.

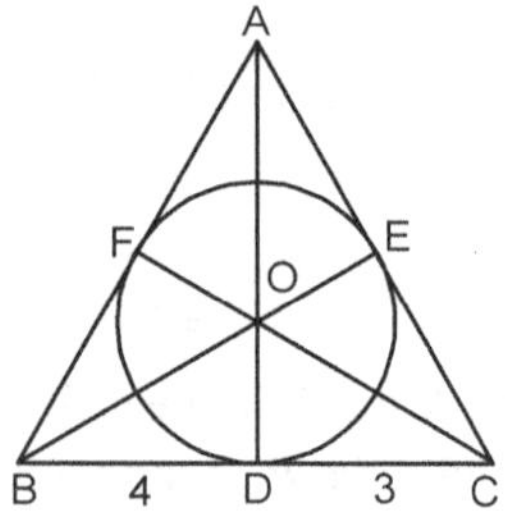

Join AO, OE and OF.

So, $OD = OF = OE = 2$ cm

Now, $AF = AE = x$ (say) [Tangents from A]

Also, $BF = BD = 4$ cm [Tangents from Q]

and $CD = CE = 3$ cm [Tangents from R]

Area of ΔABC = Area of ΔBOC + Area of ΔAOB + Area of ΔCOA

$$\Rightarrow \quad 21 = \frac{1}{2}[(4+3)2 + (4+x)2 + (3+x)2]$$

$\Rightarrow \quad 21 = \frac{2}{2}[(4+3)+(4+x)+(3+x)]$

$\Rightarrow \quad 21 = [7+7+2x]$

$\Rightarrow \quad 21 = 14+2x$

$\Rightarrow \quad 2x = 7$

$\Rightarrow \quad x = 3.5$ cm

Thus, the length of $AB = AF + BF$

$= (3.5+4)$ cm

$= 7.5$ cm

and the length of $AC = (3.5+3)$ cm

$= 6.5$ cm. **Ans.**

Q. 12. Find the area of the major segment *APB* in the figure of a circle of radius 35 cm and $\angle AOB = 90°$.

$\left(\text{Use } \pi = \frac{22}{7}\right)$.

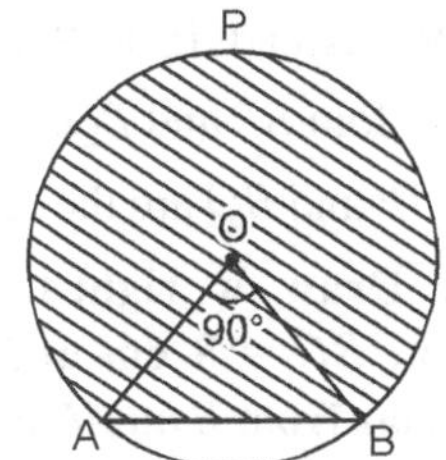

Solution : Given, $r = 35$ cm and $\angle AOB = 90°$

Thus, the area of the major segment

= Area of the circle

– Area of the minor segment

$= \pi r^2 - \left\{\frac{\theta}{360°}\times\pi - \frac{1}{2}\sin\theta\right\} r^2 \text{ cm}^2$

$= r^2\left[\pi - \left\{\frac{\theta}{360°}\times\pi - \frac{1}{2}\sin\theta\right\}\right] \text{ cm}^2$

$= (35)^2\left[\frac{22}{7} - \left\{\frac{90°}{360°}\times\frac{22}{7} - \frac{1}{2}\sin 90°\right\}\right] \text{ cm}^2$

$= 1225\left[\frac{22}{7} - \left\{\frac{1}{4}\times\frac{22}{7} - \frac{1}{2}\right\}\right] \text{ cm}^2$

$= 1225\left[\frac{22}{7} - \frac{22}{28} + \frac{1}{2}\right] \text{ cm}^2$

$= 1225\left[\frac{88-22+16}{28}\right] \text{ cm}^2$

$= 1225\left[\frac{80}{28}\right] \text{ cm}^2$

$= 1225\left[\frac{40}{14}\right] \text{ cm}^2$

$= 3500 \text{ cm}^2$. **Ans.**

Q. 13. In the given figure, a triangle *ABC* is drawn to circumscribe a circle of radius 10 cm such that the segments *BP* and *PC* into which *BC* is divided by the point of contact *P* are of lengths 15 cm and 20 cm respectively. If the area of $\Delta ABC = 525$ cm^2, then find the lengths of the sides *AB* and *AC*.

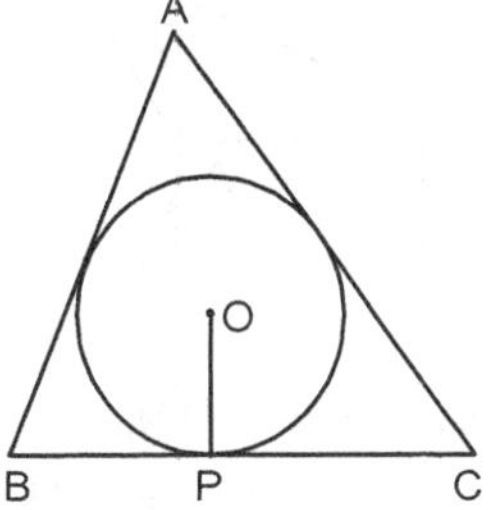

Solution : Given, $r = OP = 10$ cm, $BP = 15$ cm, $PC = 20$ cm and area of $\Delta ABC = 525$ cm^2.

Let the circle touch AB at Q and AC at R such that

$AQ = AR$ [Tangents from A]

$BQ = BP$ [Tangents from B]

and $CR = CP$ [Tangents from C]

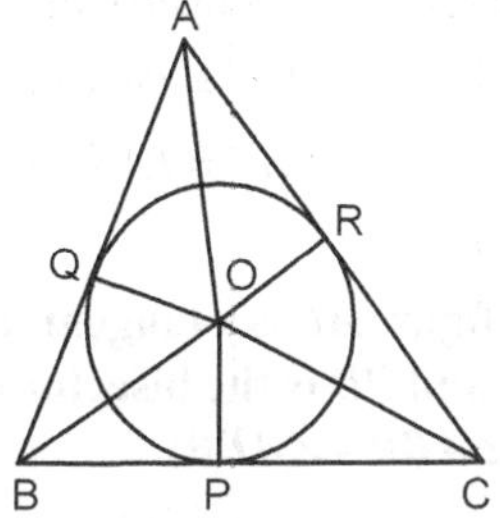

Join QO and RO such that,

$QO = RO = OP = 10$ cm

Also, join AO, OB and OC.

Let $AQ = AR$ be x cm

Thus, $AB = (x+15)$ cm

and $AC = (x+20)$ cm

and $BC = (20+15)$ cm $= 35$ cm

Now, Area of ΔABC = Area of ΔBOC + Area of ΔAOB + Area of ΔAOC

$\Rightarrow \quad 525 = \frac{1}{2}[(15+20)10 + (15+x)10 + (20+x)10]$

$\Rightarrow \quad 525 = \frac{10}{2}[(15+20)+(15+x)+(20+x)]$

$\Rightarrow \quad 525 = 5[70+2x]$

$\Rightarrow \quad 105 = 70+2x$

$\Rightarrow \quad 2x = 35$

$\Rightarrow \quad x = 17.5$ cm

Hence, $AB = (17.5+15)$ cm

and $AC = (17.5+20)$ cm

or $AB = 32.5$ cm and $AC = 37.5$ cm. **Ans.**

14. In the figure, PM is a tangent to the circle and $PA = AM$. Prove that:

(i) ΔPMB is isosceles.

(ii) $PA \times PB = MB^2$.

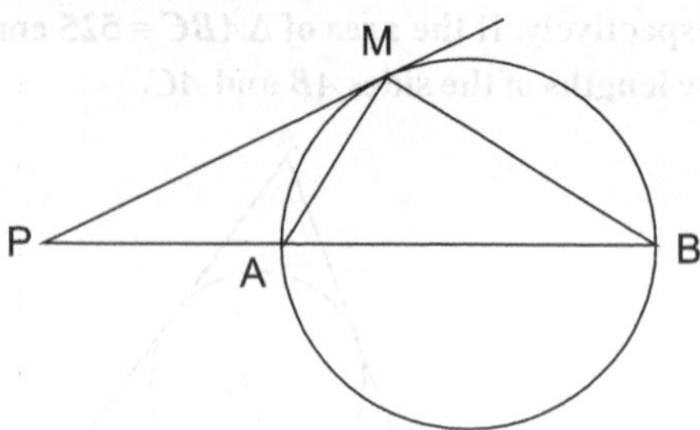

Solution. (i) In ΔPAM,

$$\angle APM = \angle AMP \qquad \ldots\text{(i)}$$

$$PA = AM \qquad \text{(Given)}$$

by alternate segment property of tangent

$$\angle ABM = \angle AMP$$

$$\therefore \quad \angle APM = \angle ABM \qquad \text{[from (i) and (ii)]}$$

$$\therefore \quad PM = MB$$

i.e., ΔPMB is an isosceles. **Hence Proved.**

(ii) By rectangle property of tangent and chord

$$PM^2 = PA \times PB$$

$$\therefore \quad MB^2 = PA \times PB \qquad [\because PM = MB]$$

Hence Proved.

15. In figure AP is a tangent to the circle at P, ABC is a secant and PD is the bisector of $\angle BPC$. Prove that $\angle BPD = (\angle ABP - \angle APB)$.

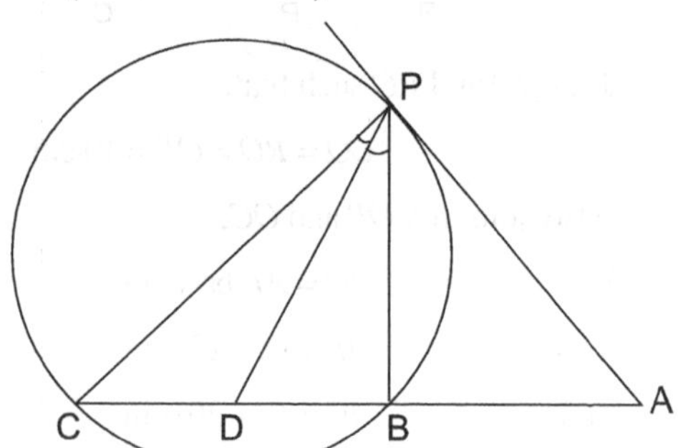

Solution. Since, $\angle APB$ and $\angle BCP$ are angles in alternate segments of chord PB.

$$\therefore \quad \angle APB = \angle BCP \qquad \ldots\text{(i)}$$

Since, PD is bisector of $\angle BPC$.

$$\therefore \quad \angle CPB = 2\angle BPD \qquad \ldots\text{(ii)}$$

In DPCB, side CB has been produced to A, forming exterior angle $\angle APB$.

$$\therefore \quad \angle ABP = \angle BCP + \angle CPB$$

$$\Rightarrow \quad \angle ABP = \angle APB + 2\angle BPD$$

[Using (i) and (ii)]

$$\Rightarrow \quad 2\angle BPD = \angle ABP - \angle APB$$

$$\Rightarrow \quad \angle BPD = (\angle ABP - \angle APB).$$

Hence Proved.

★ are board exam questions from previous years

Chapter 9. Constructions

Q. 1. Draw a pair of tangents to a circle of radius 6 cm which are inclined to each other at an angle of 60°.★

Solution : Step I : Draw a line $OA = 6$ cm. Construct a circle with O as the centre and OA as the radius.

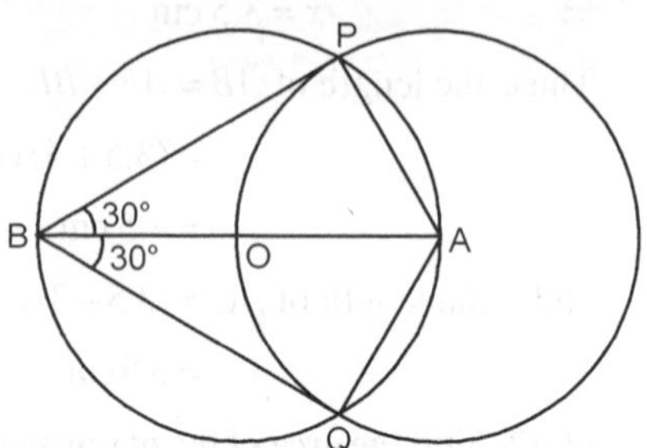

Step II : From point A, construct another circle with AO as the radius.

Step III : Let the two circles intersect at points P and Q.

Step IV : Extend AO to B, where $OA = OB$.

Step V : Join BP and BQ.

Thus, BP and BQ are the required tangents

Q. 2. Draw a right-triangle ABC in which $AB = 6$ cm, $BC = 8$ cm and $\angle B = 90°$. Draw BD perpendicular from B on AC and draw a circle passing through the points B, C and D. Construct tangents from A to this circle.

Solution : Step I : Draw $AB = 6$ cm.

Step II : From B, construct $\angle ABP = 90°$.

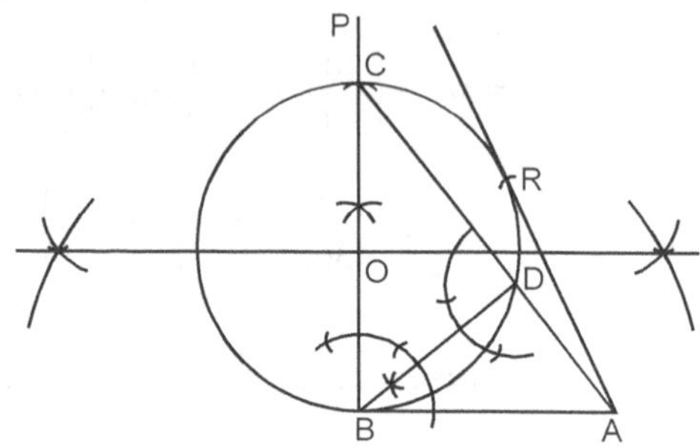

Step III : With B as centre and radius 8 cm, draw an arc, intersecting BP at C.

Step IV : Join AC.

Step V : Draw perpendicular bisector of BC. Let O be the mid-point of BC.

Step VI : Taking O as centre and OB as radius draw a circle. The point of intersection of the circle and side AC is D.

Step VII : Draw perpendicular from D on AC and join BD.

Step VIII : Construct $AR = AB$, where R is the point of intersection between the circle and the tangent AR.

Q. 3. Draw a line-segment AB of length 8 cm. Taking A as centre draw a circle of radius 4 cm and taking B as centre, draw another circle B of radius 3 cm. Construct tangents to each circle from the centre of the other circle.

Solution : Step I : Draw $AB = 8$ cm.

Step II : Construct a circle of radius 4 cm from point A.

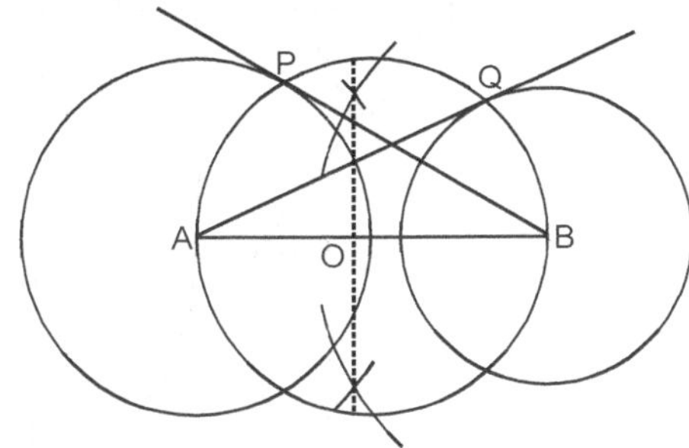

Step III : Construct another circle of radius 3 cm from point B.

Step IV : Now draw perpendicular bisector of AB. Let O be the mid-point of AB.

Step V : Keeping O as the centre, draw circle with $AO = OB$ as the radius.

Step VI : Mark point of intersection of this circle with the larger circle as P.

Step VII : Draw PB as tangent to the larger circle.

Step VIII : Similarly, mark the point of intersection of the circle with centre O with the smaller circle as Q.

Step IX : Draw AQ as the tangent to the smaller circle.

Q. 4. Construct a triangle with sides 5 cm, 5.5 cm and 6.5 cm. Now, construct another triangle whose sides are $\frac{3}{5}$ times the corresponding sides of the given triangle.★

Solution : Step I : Draw $AB = 6.5$ cm with the help of ruler and pencil.

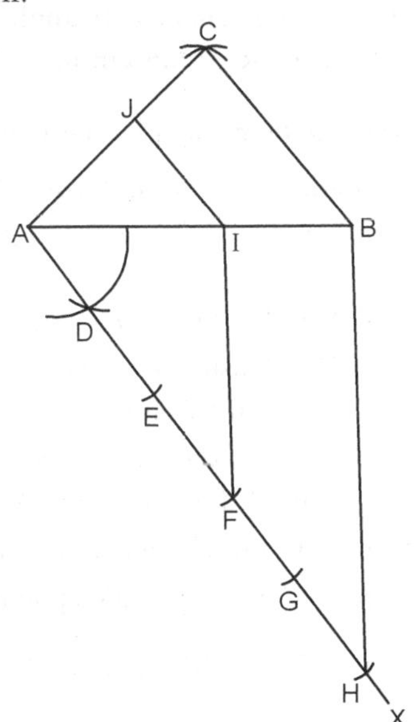

Step II : Measure out 5.5 cm and keeping the compass point at B, draw an arc on one side of the line-segment AB.

Step III : Measure 5 cm and keeping the compass point at A, draw an arc on the same side of the line-segment AB as the other arc.

Step IV : The point of intersection of the two arcs is point C. Join AC and BC. Thus, ΔABC is obtained.

Step V : Below AB, make an acute angle $\angle BAX$.

Step VI : Along AX, marks 5 points (greater of 3 and 5 in $\frac{3}{5}$) D, E, F, G, H such that $AD = DE = EF = FG = GH$.

Step VII : Join HB.

Step VIII : From F, draw $FI \parallel HB$, meeting AB at I.

Step IX : From I, draw $IJ \parallel BC$.

Thus, ΔAIJ is the required triangle.

Q. 5. Draw a line-segment AB of length 7 cm. Taking A as centre, draw a circle of radius 3 cm and taking B as centre, draw another circle B of radius 2 cm. Construct tangents to each circle from the centre of the other circle.★★

Solution : Step I : Draw $AB = 7$ cm.

Step II : Construct circle of radius 3 cm from point A.

Step III : Construct circle of radius 2 cm from point B.

Step IV : Now draw perpendicular bisector of AB. Let O be the mid-point of AB.

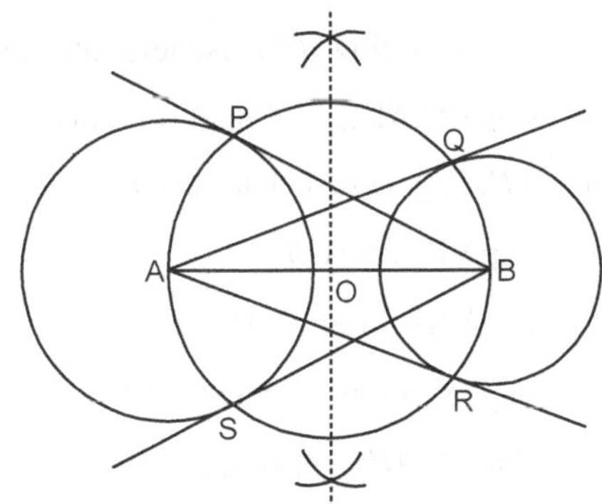

Step V : Keeping O as the centre, draw a circle with $AO = OB$ as the radius.

Step VI : Mark points of intersection of this circle with the larger circle as P and S.

Step VII : Draw PB and SB as tangents to the larger circle.

Step VIII : Similarly, mark the points of intersection of the circle with centre O with the smaller circle as Q and R.

★ are board exam questions from previous years
★★ are frequently asked board exam questions

Step IX : Draw AQ and AR as tangents to the smaller circle.

Q. 6. Construct an isosceles triangle *ABC* whose base is 8 cm and altitude is 4 cm. Then construct another triangle whose sides are $\frac{3}{4}$ times the corresponding sides of the ΔABC.

Solution : Step I : Draw $AB = 8$ cm.

Step II : Draw perpendicualr bisector of AB as XY, bisecting AB at M.

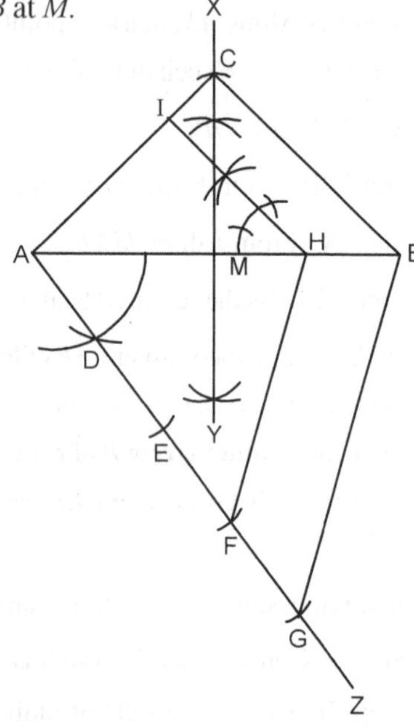

Step III : With M as centre and radius 4 cm, draw an arc, intersecting XY at C.

Step IV : Join AC and BC. Thus, ΔABC is obtained.

Step V : Below AB make an acute angle $\angle BAZ$.

Step VI : Along AZ, mark 4 points (greater of 3 and 4 in $\frac{3}{4}$) D, E, F, G such that $AD = DE = EF = FG$.

Step VII : Join GB.

Step VIII : From F, draw $FH \parallel GB$, meeting AB at H.

Step IX : From H, draw $HI \parallel BC$.

Thus, ΔAHI is the required triangle.

Q. 7. Construct a right triangle *ABC* in which the sides other than the hypotenuse are of lengths 3 cm and 4 cm. Then construct another triangle whose sides are $\frac{3}{5}$ times the corresponding sides of the ΔABC.★★

Solution : Step I : Draw $AB = 4$ cm with the help of ruler and pencil.

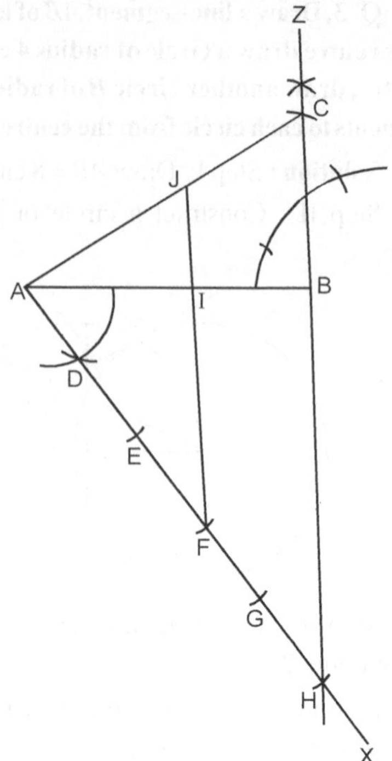

Step II : Construct $\angle ABZ = 90°$.

Step III : Measure 3 cm and keeping the compass point at B, draw an arc on BZ.

Step IV : The point of intersection of the arc and the line BZ is point C. Join AC. Thus, ΔABC is obtained.

Step V : Below AB make an acute angle $\angle BAX$.

Step VI : Along AX mark 5 points (greater of 3 and 5 in $\frac{3}{5}$) D, E, F, G, H such that $AD = DE = EF = FG = GH$.

Step VII : Join HB.

Step VIII : From F draw $FI \parallel HB$ meeting AB at I.

Step IX : From I, draw $IJ \parallel BC$.

Thus, ΔAIJ is the required triangle.

Q. 8. Construct a triangle *ABC* with sides *AB* = 4.5 cm, *BC* = 6.5 cm and ∠*ABC* = 60°. Then construct another triangle whose sides are $\frac{3}{4}$ times the corresponding sides of the ΔABC.★★

Solution : Step I : Draw $BC = 6.5$ cm with the help of ruler and pencil.

Step II : Construct $\angle CBZ = 60°$.

Step III : Measure 4.5 cm and keeping the compass point at B, draw an arc on BZ.

Step IV : The point of intersection of the arc and the line BZ is point A. Join AC. Thus, ΔABC is obtained.

Step V : Below BC, make an acute angle $\angle CBX$.

Step VI : Along BX mark 4 points (greater of 3 and 4 in $\frac{3}{4}$) D, E, F, G such that $BD = DE = EF = FG$.

★ are board exam questions from previous years
★★ are frequently asked board exam questions

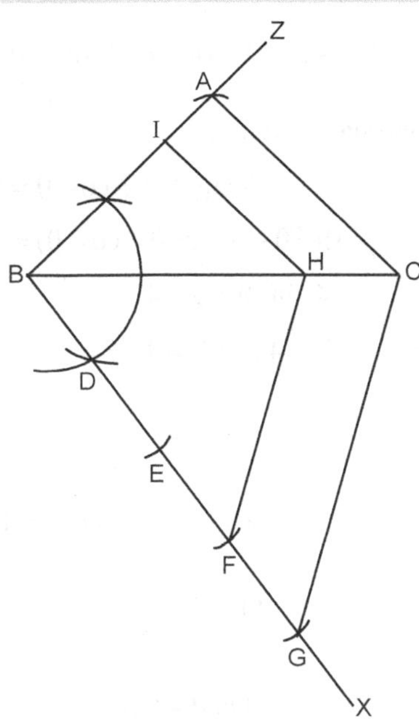

Step VII : Join GC.

Step VIII : From F, draw $FH \parallel GC$ meeting BC at H.

Step IX : From H, draw $HI \parallel AC$.

Thus, ΔBHI is the required triangle.

Q. 9. Construct an isosceles triangle ABC whose base is 9 cm and altitude is 5 cm. Then construct another triangle whose sides are $\frac{3}{4}$ times the corresponding sides of the ΔABC.

Solution : Step I : Draw $AB = 9$ cm.

Step II : Draw perpendicular bisector of AB as YZ, bisecting AB at O.

Step III : With O as centre and radius 5 cm, draw an arc, intersecting YZ at C.

Step IV : Join AC and BC. Thus, ΔABC is obtained.

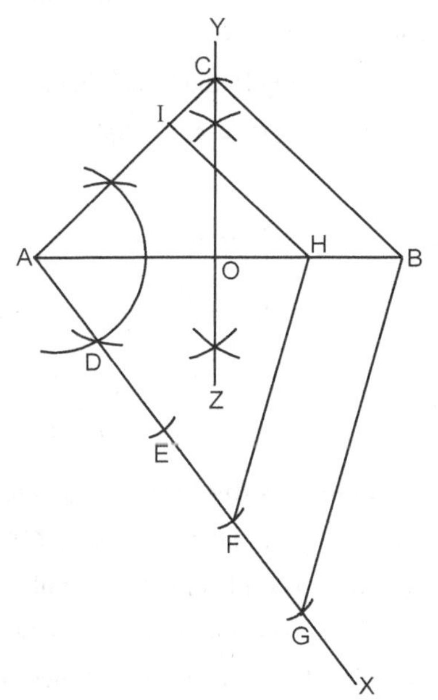

Step V : Below AB, make an acute angle $\angle BAX$.

Step VI : Along AX mark 4 points (greater of 3 and 4 in $\frac{3}{4}$) D, E, F, G such that $AD = DE = EF = FG$.

Step VII : Join GB.

Step VIII : From F, draw $FH \parallel GB$ meeting AB at H.

Step IX : From H, draw $HI \parallel BC$.

Thus, ΔAHI is the required triangle.

Q. 10. Draw a circle of radius 5 cm. Take a point P on it. Without using the centre of the circle, construct a tangent at the point P.

Solution : Step I : Draw a circle with O as the centre and radius $OP = 5$ cm.

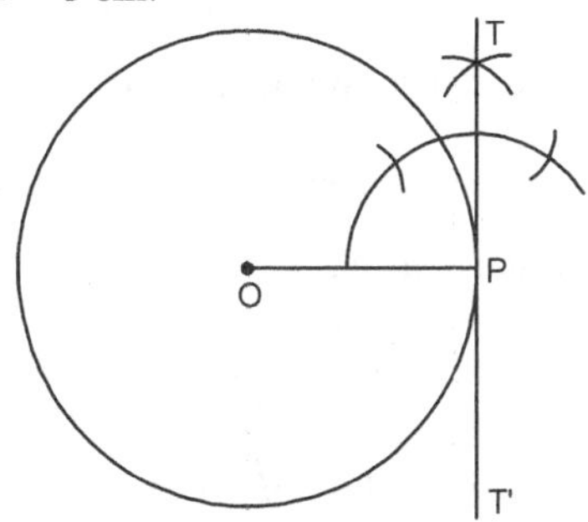

Step II : From P, construct a perpendicular and produce it to T.

Thus, PT is the required tangent.

Q. 11. Construct a right triangle ABC in which the sides other than the hypotenuse are of lengths 8 cm and 6 cm. Then construct another triangle whose sides are $\frac{5}{3}$ times the corresponding sides of the ΔABC.★★

Solution : Step I : Draw $AB = 8$ cm with the help of ruler and pencil.

Step II : Construct $\angle ABZ = 90°$.

Step III : Measure 6 cm and keeping the compass point at B, draw an arc on BZ.

Step IV : The point of intersection of the arc and the line BZ is point C. Join AC. Thus, ΔABC is obtained.

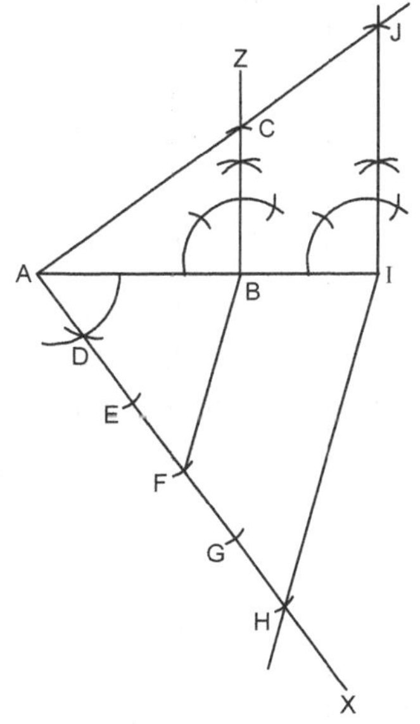

★★ are frequently asked board exam questions

Step V : Below AB, make an acute angle $\angle BAX$.

Step VI : Along AX, mark 5 points (greater of 3 and 5 in $\frac{5}{3}$) D, E, F, G, H such that $AD = DE = EF = FG = GH$.

Step VII : Join FB.

Step VIII : From H, draw $HI \parallel FB$, meeting AB extended at I.

Step IX : From I, drawn $IJ \parallel BC$, meeting AC extended at J.

Thus, ΔAIJ is the required triangle.

Chapter 10. Trigonometry

Q. 1. In ΔABC, it is given that $\angle C = 90°$ and $\tan A = \frac{1}{\sqrt{3}}$. Find the value of ($\sin A \cos B + \cos A \sin B$).

Solution : Given,

$$\tan A = \frac{1}{\sqrt{3}} = \frac{\text{Perpendicular}}{\text{Base}} = \frac{BC}{AC}$$

Thus $AC = \sqrt{3}BC$

Applying Pythagoras' theorem

$$AB^2 = BC^2 + AC^2$$

$$\Rightarrow \quad AB^2 = (1k)^2 + (\sqrt{3}k)^2 = 1k^2 + 3k^2 = 4k^2$$

$$\Rightarrow \quad AB = 2k$$

So $$\sin A = \frac{BC}{AB} = \frac{1k}{2k} = \frac{1}{2};$$

$$\cos A = \frac{AC}{AB} = \frac{\sqrt{3}k}{2k} = \frac{\sqrt{3}}{2};$$

$$\sin B = \frac{AC}{AB} = \frac{\sqrt{3}k}{2k} = \frac{\sqrt{3}}{2}$$

and $$\cos B = \frac{BC}{AB} = \frac{1k}{2k} = \frac{1}{2}$$

Hence, ($\sin A \cos B + \cos A \sin B$)

$$= \left[\frac{1}{2} \times \frac{1}{2} + \frac{\sqrt{3}}{2} \times \frac{\sqrt{3}}{2}\right]$$

$$= \frac{1}{4} + \frac{3}{4} = 1.$$ **Ans.**

Q. 2. If $7\sin^2\theta + 3\cos^2\theta = 4$, show that $\tan\theta = \frac{1}{\sqrt{3}}$.

Solution : Given

$$7\sin^2\theta + 3\cos^2\theta = 4$$

$$\Rightarrow \quad 4\sin^2\theta + 3(\sin^2\theta + \cos^2\theta) = 4$$

$$\Rightarrow \quad 4\sin^2\theta + 3 = 4$$

$$\Rightarrow \quad 4\sin^2\theta = 4 - 3 = 1$$

$$\Rightarrow \quad \sin^2\theta = \frac{1}{4}$$

$$\therefore \quad \cos^2\theta = 1 - \sin^2\theta = 1 - \frac{1}{4} = \frac{3}{4}$$

Hence $$\tan^2\theta = \frac{1}{3}$$

$$\Rightarrow \quad \tan\theta = \frac{1}{\sqrt{3}}.$$ **Hence Proved.**

Q. 3. Find the value of sin 30° geometrically.★★

Solution : To find the value of sin 30° geometrically, let us consider an equilateral ΔABC with each side having a length of $2a$. So each angle equals 60°. Let AD be a perpendicular from A on BC. Since the triangle is equilateral, AD is the bisector of $\angle A$ and D is the mid-point of BC.

Thus $BD = DC = a$ and $\angle BAD = 30°$

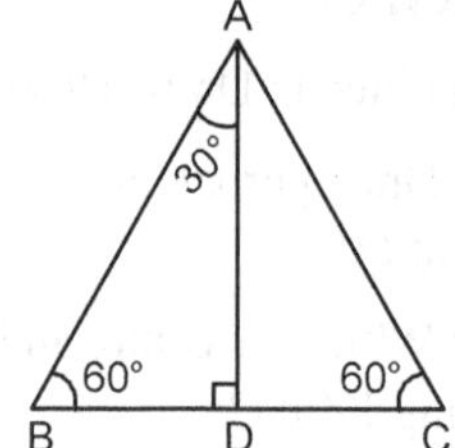

Thus in ΔABD, $\angle D$ is a right angle, $AB = 2a$ is the hypotenuse and $BD = a$ is the base.

Hence by using Pythagoras' theorem,

$$AB^2 = AD^2 + BD^2$$

$$\Rightarrow \quad (2a^2) = AD^2 + a^2$$

$$\Rightarrow \quad AD^2 = 3a^2$$

$$\Rightarrow \quad AD = \sqrt{3}a$$

Thus $$\sin 30° = \frac{BD}{AB} = \frac{a}{2a} = \frac{1}{2}.$$ **Ans.**

Q. 4. Find the value of sin 60° geometrically.★★

Solution : To find the value of sin 60° geometrically, let us consider an equilateral ΔABC with each side having a length of $2a$. So each angle equals 60°. Let AD be a perpendicular from A on BC. Since the triangle is

★★ are frequently asked board exam questions

equilateral, AD is the bisector of $\angle A$ and D is the mid-point of BC.

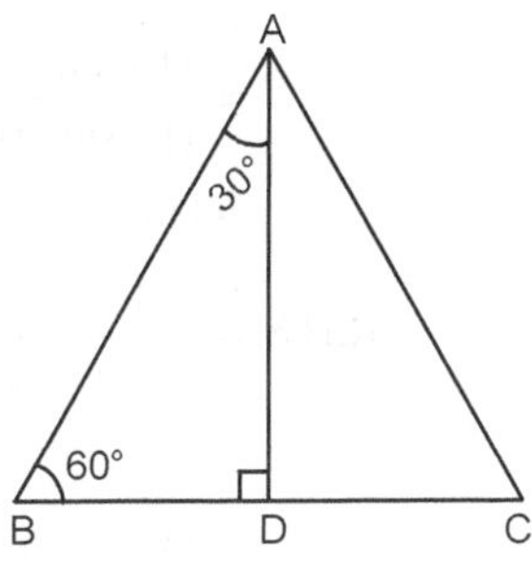

Thus $\quad BD = DC = a$ and $\angle BAD = 30°$

Thus in ΔABD, $\angle D$ is a right angle, $AB = 2a$ is the hypotenuse and $BD = a$ is the base.

Hence by using Pythagoras' theorem,

$$AB^2 = AD^2 + BD^2$$

$$\Rightarrow \quad (2a^2) = AD^2 + a^2$$

$$\Rightarrow \quad AD^2 = 3a^2$$

$$\Rightarrow \quad AD = \sqrt{3}a$$

Thus, $\quad \sin 60° = \dfrac{AD}{AB} = \dfrac{\sqrt{3}a}{2a} = \dfrac{\sqrt{3}}{2}$. **Ans.**

Q. 5. If $\sin\theta = \dfrac{12}{13}$, $0° < \theta < 90°$, find the value of :★

$$\frac{\sin^2\theta - \cos^2\theta}{2\sin\theta.\cos\ \theta} \times \frac{1}{\tan^2\theta}.$$

Solution : Given, $\sin\theta = \dfrac{12}{13}$

$$\Rightarrow \quad \frac{P}{H} = \frac{12}{13}$$

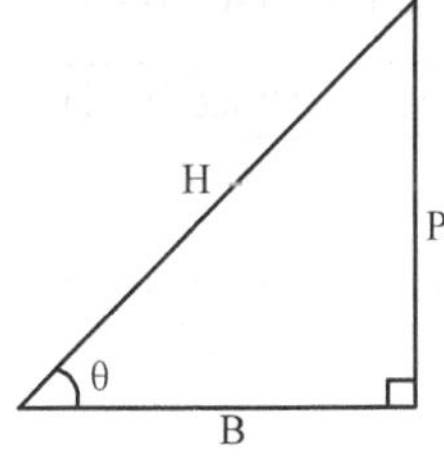

Let, $\quad P = 12K,\ H = 13K$

$P^2 + B^2 = H^2$ [Pythagoras' theorem]

$$(12K)^2 + B^2 = (13K)^2$$

$$144K^2 + B^2 = 169K^2$$

$$B^2 = 169\ K^2 - 144K^2$$

$$= 25\ K^2$$

$$B = 5K$$

$$\therefore \quad \cos\theta = \frac{B}{H} = \frac{5K}{13K} = \frac{5}{13}$$

and $\quad \tan\theta = \dfrac{P}{B} = \dfrac{12K}{5K} = \dfrac{12}{5}$

★ are board exam questions from previous years

★★ are frequently asked board exam questions

Now, $\dfrac{\sin^2\theta - \cos^2\theta}{2\sin\theta.\cos\theta} \times \dfrac{1}{\tan^2\theta}$

$$= \frac{\left(\frac{12}{13}\right)^2 - \left(\frac{5}{13}\right)^2}{2\left(\frac{12}{13}\right)\left(\frac{5}{13}\right)} \times \frac{1}{\left(\frac{12}{5}\right)^2}$$

$$= \frac{\frac{144-25}{169}}{\frac{120}{169}} \times \frac{25}{144}$$

$$= \frac{119}{120} \times \frac{25}{144}$$

$$= \frac{595}{3456}$$ **Ans.**

Q. 6. Prove that :★★

$(1 + \cot\theta - \text{cosec}\ \theta)\ (1 + \tan\theta + \sec\theta) = 2.$

Solution : Consider,

L.H.S. $= (1 + \cot\theta - \text{cosec}\ \theta)\ (1 + \tan\theta + \sec\theta)$

$$= \left(1 + \frac{\cos\theta}{\sin\theta} - \frac{1}{\sin\theta}\right)\left(1 + \frac{\sin\theta}{\cos\theta} + \frac{1}{\cos\theta}\right)$$

$$= \left(\frac{\sin\theta + \cos\theta - 1}{\sin\theta}\right)\left(\frac{\cos\theta + \sin\theta + 1}{\cos\theta}\right)$$

$$= \frac{(\sin\theta + \cos\theta)^2 - 1}{\sin\theta\cos\theta}$$

$$= \frac{\sin^2\theta + \cos^2\theta + 2\sin\theta\cos\theta - 1}{\sin\theta\cos\theta}$$

$$= \frac{1 + 2\sin\theta\cos\theta - 1}{\sin\theta\cos\theta}$$

$$= \frac{2\sin\theta\cos\theta}{\sin\theta\cos\theta} = 2$$

= R.H.S. **Hence Proved.**

Q. 7. A school organised maths quiz to reward students participating in co-curricular activities group and a group with 100% attendence.

One of the question asked is as follows :

If $\sin\theta + \sin^2\theta = 1$, then find the value of

$\cos^2\theta + \cos^4\theta$.

Solution : Given,

$$\sin\theta + \sin^2\theta = 1 \quad ...(i)$$

$$\Rightarrow \quad \sin\theta = 1 - \sin^2\theta \quad ...(ii)$$

Now, $\cos^2\theta + \cos^4\theta = \cos^2\theta(1 + \cos^2\theta)$

$\Rightarrow (1 - \sin^2\theta)\ (1 + 1 - \sin^2\theta) = \sin\theta(1 + \sin\theta)$

[Using (ii)]

$= \sin\theta + \sin^2\theta = 1$ [From (i)]

Ans.

Q. 8. Prove that :

$$\frac{1}{\text{cosec}\,\theta - \cot\theta} - \frac{1}{\sin\theta} = \frac{1}{\sin\theta} - \frac{1}{\text{cosec}\,\theta + \cot\theta}.$$

Solution : Consider,

$$\text{L.H.S.} = \frac{1}{\text{cosec}\,\theta - \cot\theta} - \frac{1}{\sin\theta}$$

$$= \frac{\sin\theta}{1-\cos\theta} - \frac{1}{\sin\theta}$$

$$= \frac{\sin^2\theta - 1 + \cos\theta}{(1-\cos\theta)\sin\theta}$$

$$= \frac{\cos\theta - (1-\sin^2\theta)}{(1-\cos\theta)\sin\theta}$$

$$= \frac{\cos\theta - \cos^2\theta}{(1-\cos\theta)\sin\theta}$$

$$= \frac{(1-\cos\theta)\cos\theta}{(1-\cos\theta)\sin\theta}$$

$$= \cot\theta$$

$$\text{R.H.S.} = \frac{1}{\sin\theta} - \frac{1}{\text{cosec}\,\theta + \cot\theta}$$

$$= \frac{1}{\sin\theta} - \frac{\sin\theta}{1+\cos\theta}$$

$$= \frac{1+\cos\theta - \sin^2\theta}{(1+\cos\theta)\sin\theta}$$

$$= \frac{\cos\theta + (1-\sin^2\theta)}{(1+\cos\theta)\sin\theta}$$

$$= \frac{\cos\theta + \cos^2\theta}{(1+\cos\theta)\sin\theta}$$

$$= \frac{(1+\cos\theta)\cos\theta}{(1+\cos\theta)\sin\theta} = \cot\theta$$

Thus, L.H.S. = R.H.S. **Hence Proved.**

Q. 9. Prove that :

$$\frac{1}{\sec x - \tan x} - \frac{1}{\cos x} = \frac{1}{\cos x} - \frac{1}{\sec x + \tan x}.$$

Solution : Consider,

$$\text{L.H.S.} = \frac{1}{\sec x - \tan x} - \frac{1}{\cos x}$$

$$= \frac{\cos x}{1-\sin x} - \frac{1}{\cos x}$$

$$= \frac{\cos^2 x - 1 + \sin x}{(1-\sin x)\cos x}$$

$$= \frac{\sin x - (1-\cos^2 x)}{(1-\sin x)\cos x}$$

$$= \frac{\sin x - \sin^2 x}{(1-\sin x)\cos x}$$

$$= \frac{(1-\sin x)\sin x}{(1-\sin x)\cos x}$$

$$= \tan x$$

$$\text{R.H.S.} = \frac{1}{\cos x} - \frac{1}{\sec x + \tan x}$$

$$= \frac{1}{\cos x} - \frac{\cos x}{1+\sin x}$$

$$= \frac{1+\sin x - \cos^2 x}{(1+\sin x)\cos x}$$

$$= \frac{\sin x + (1-\cos^2 x)}{(1+\sin x)\cos x}$$

$$= \frac{\sin x + \sin^2 x}{(1+\sin x)\cos x}$$

$$= \frac{(1+\sin x)\sin x}{(1+\sin x)\cos x}$$

$$= \tan x$$

Thus, L.H.S. = R.H.S. **Hence Proved.**

Q. 10. Prove that :

$$\sqrt{\frac{\sec\theta - 1}{\sec\theta + 1}} + \sqrt{\frac{\sec\theta + 1}{\sec\theta - 1}} = 2\,\text{cosec}\,\theta.$$

Solution : Consider,

$$\text{L.H.S.} = \sqrt{\frac{\sec\theta - 1}{\sec\theta + 1}} + \sqrt{\frac{\sec\theta + 1}{\sec\theta - 1}}$$

$$= \sqrt{\frac{(\sec\theta-1)(\sec\theta-1)}{(\sec\theta+1)(\sec\theta-1)}} + \sqrt{\frac{(\sec\theta+1)(\sec\theta+1)}{(\sec\theta-1)(\sec\theta+1)}}$$

$$= \sqrt{\frac{(\sec\theta-1)(\sec\theta-1)}{\sec^2\theta - 1}} + \sqrt{\frac{(\sec\theta+1)(\sec\theta+1)}{\sec^2\theta - 1}}$$

$$= \sqrt{\frac{(\sec\theta-1)(\sec\theta-1)}{\tan^2\theta}} + \sqrt{\frac{(\sec\theta+1)(\sec\theta+1)}{\tan^2\theta}}$$

$[\because \tan^2\theta = \sec^2\theta - 1]$

$$= \frac{(\sec\theta - 1)}{\tan\theta} + \frac{(\sec\theta + 1)}{\tan\theta}$$

$$= \frac{2\sec\theta}{\tan\theta} = 2\,\text{cosec}\,\theta$$

= R.H.S. **Hence Proved.**

Q. 11. Prove that :

$$(\tan^2 A - \tan^2 B) = \frac{\sin^2 A - \sin^2 B}{\cos^2 A \cos^2 B} = \frac{\cos^2 B - \cos^2 A}{\cos^2 A \cos^2 B}.$$

Solution : Consider,

$(\tan^2 A - \tan^2 B)$

$$= \frac{\sin^2 A}{\cos^2 A} - \frac{\sin^2 B}{\cos^2 B}$$

$$= \frac{\sin^2 A \cos^2 B - \sin^2 B \cos^2 A}{\cos^2 A \cos^2 B}$$

$$= \frac{\sin^2 A(1-\sin^2 B) - \sin^2 B(1-\sin^2 A)}{\cos^2 A \cos^2 B}$$

$$= \frac{\sin^2 A - \sin^2 A \sin^2 B - \sin^2 B + \sin^2 A \sin^2 B}{\cos^2 A \cos^2 B}$$

$$= \frac{\sin^2 A - \sin^2 B}{\cos^2 A \cos^2 B} \quad \ldots(i)$$

Also, $(\tan^2 A - \tan^2 B)$

$$= \frac{\sin^2 A}{\cos^2 A} - \frac{\sin^2 B}{\cos^2 B}$$

$$= \frac{\sin^2 A \cos^2 B - \sin^2 B \cos^2 A}{\cos^2 A \cos^2 B}$$

$$= \frac{\cos^2 B(1-\cos^2 A) - \cos^2 A(1-\cos^2 B)}{\cos^2 A \cos^2 B}$$

$$= \frac{\cos^2 B - \cos^2 A \cos^2 B - \cos^2 A + \cos^2 A \cos^2 B}{\cos^2 A \cos^2 B}$$

$$= \frac{\cos^2 B - \cos^2 A}{\cos^2 A \cos^2 B} \quad \ldots(ii)$$

Thus, from equations (i) and (ii),

$$(\tan^2 A - \tan^2 B) = \frac{\sin^2 A - \sin^2 B}{\cos^2 A \cos^2 B}$$

$$= \frac{\cos^2 B - \cos^2 A}{\cos^2 A \cos^2 B}.$$

Hence Proved.

Q. 12. Prove that:

$$\frac{\tan\theta}{1-\cot\theta} + \frac{\cot\theta}{1-\tan\theta} = 1 + \sec\theta\,\mathrm{cosec}\,\theta.$$

Solution : Consider,

$$\text{L.H.S.} = \frac{\tan\theta}{1-\cot\theta} + \frac{\cot\theta}{1-\tan\theta}$$

$$\frac{\frac{\sin\theta}{\cos\theta}}{1-\frac{\cos\theta}{\sin\theta}} + \frac{\frac{\cos\theta}{\sin\theta}}{1-\frac{\sin\theta}{\cos\theta}}$$

$$= \frac{\sin^2\theta}{\cot\theta(\sin\theta-\cos\theta)} + \frac{\cos^2\theta}{\sin\theta(\cos\theta-\sin\theta)}$$

$$= \frac{1-\cos^2\theta}{\cos\theta(\sin\theta-\cos\theta)} + \frac{1-\sin^2\theta}{\sin\theta(\cos\theta-\sin\theta)}$$

$$= \frac{1}{\cos\theta(\sin\theta-\cos\theta)} - \frac{\cos\theta}{\sin\theta-\cos\theta} + \frac{1}{\sin\theta(\cos\theta-\sin\theta)} - \frac{\sin\theta}{\cos\theta-\sin\theta}$$

$$= \frac{1}{(\sin\theta-\cos\theta)}\left[\frac{1}{\cos\theta} - \frac{1}{\sin\theta}\right] + \frac{\sin\theta-\cos\theta}{(\sin\theta-\cos\theta)}$$

$$= \frac{1}{(\sin\theta-\cos\theta)}\left[\frac{\sin\theta-\cos\theta}{\sin\theta\cos\theta}\right] + 1$$

$$= \left[\frac{1}{\sin\theta\cos\theta}\right] + 1$$

$= 1 + \sec\theta\,\mathrm{cosec}\,\theta$

= R.H.S. **Hence Proved.**

Q. 13. Prove that:

$$\frac{1-\sin\theta}{1+\sin\theta} = (\sec\theta - \tan\theta)^2.$$

Solution : Consider,

$$\text{L.H.S.} = \frac{1-\sin\theta}{1+\sin\theta}$$

$$= \frac{(1-\sin\theta)(1-\sin\theta)}{(1+\sin\theta)(1-\sin\theta)}$$

$$= \frac{(1-\sin\theta)(1-\sin\theta)}{1-\sin^2\theta}$$

$$= \frac{(1-\sin\theta)^2}{\cos^2\theta}$$

$$\text{R.H.S.} = (\sec\theta - \tan\theta)^2$$

$$= \left(\frac{1}{\cos\theta} - \frac{\sin\theta}{\cos\theta}\right)^2$$

$$= \frac{(1-\sin\theta)^2}{\cos^2\theta}$$

Thus, L.H.S. = R.H.S. **Hence Proved.**

Q. 14. Prove that:

$$\frac{\mathrm{cosec}\,\theta + \cot\theta}{\mathrm{cosec}\,\theta - \cot\theta} = (\mathrm{cosec}\,\theta + \cot\theta)^2$$

$$= 1 + 2\cot^2\theta + 2\,\mathrm{cosec}\,\theta\cot\theta.$$

Solution : Consider,

$$\frac{\mathrm{cosec}\,\theta + \cot\theta}{\mathrm{cosec}\,\theta - \cot\theta}$$

$$= \frac{(\mathrm{cosec}\,\theta + \cot\theta)(\mathrm{cosec}\,\theta + \cot\theta)}{(\mathrm{cosec}\,\theta - \cot\theta)(\mathrm{cosec}\,\theta + \cot\theta)}$$

$$= \frac{(\mathrm{cosec}\,\theta + \cot\theta)^2}{\mathrm{cosec}^2\theta - \cot^2\theta}$$

$$= (\mathrm{cosec}\,\theta + \cot\theta)^2 \quad \ldots(i)$$

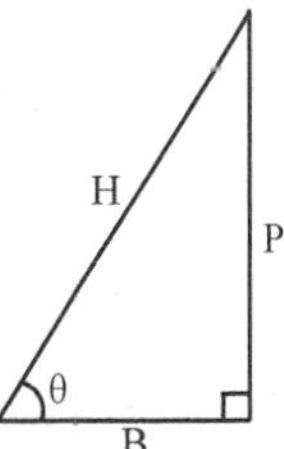

$1 + 2\cot^2\theta + 2\,\mathrm{cosec}\,\theta\cot\theta$

$= \text{cosec}^2\theta + \cot^2\theta + 2\,\text{cosec}\,\theta\cot\theta$

$= (\text{cosec}\,\theta + \cot\theta)^2$ ...(ii)

From equations (i) and (ii)

$$\frac{\text{cosec}\,\theta + \cot\theta}{\text{cosec}\,\theta - \cot\theta} = (\text{cosec}\,\theta + \cot\theta)^2$$

$$= 1 + 2\cot^2\theta + 2\,\text{cosec}\,\theta\cot\theta.$$

Hence Proved.

Q. 15. Prove that :

$$\frac{\sin\theta + 1 - \cos\theta}{\cos\theta - 1 + \sin\theta} = \frac{1+\sin\theta}{\cos\theta}.$$

Solution : Consider,

$$\text{L.H.S.} = \frac{\sin\theta + 1 - \cos\theta}{\cos\theta - 1 + \sin\theta}$$

$$= \frac{(\sin\theta - \cos\theta) + 1}{(\cos\theta + \sin\theta) - 1}$$

$$= \frac{[(\sin\theta - \cos\theta) + 1]\,[(\cos\theta + \sin\theta) + 1]}{[(\cos\theta + \sin\theta) - 1]\,[(\cos\theta + \sin\theta) + 1]}$$

$$= \frac{(\sin\theta - \cos\theta)(\sin\theta + \cos\theta) + (\sin\theta + \cos\theta) + (\sin\theta - \cos\theta) + 1}{(\cos\theta + \sin\theta)^2 - 1}$$

$$= \frac{\sin^2\theta - \cos^2\theta + \sin^2\theta + \cos^2\theta + 2\sin\theta}{\sin^2\theta + \cos^2\theta + 2\sin\theta\cos\theta - \sin^2\theta - \cos^2\theta}$$

$$= \frac{2\sin^2\theta + 2\sin\theta}{2\sin\theta\cos\theta}$$

$$= \frac{2\sin\theta(1+\sin\theta)}{2\sin\theta\cos\theta}$$

$$= \frac{1+\sin\theta}{\cos\theta}.$$

Hence Proved.

Q. 16. If $\tan\theta + \sin\theta = m$ and $\tan\theta - \sin\theta = n$, show that $(m^2 - n^2) = \sqrt{mn}$.

Solution : Given,

$\tan\theta + \sin\theta = m$ and $\tan\theta - \sin\theta = n$

Now, $(m^2 - n^2) = (\tan\theta + \sin\theta)^2 - (\tan\theta - \sin\theta)^2$

$= \tan^2\theta + \sin^2\theta + 2\tan\theta\sin\theta + 2\tan\theta\sin\theta - \tan^2\theta - \sin^2\theta$

$= 4\tan\theta\sin\theta$

and $4\sqrt{mn} = 4\sqrt{(\tan\theta + \sin\theta)(\tan\theta - \sin\theta)}$

$= 4\sqrt{\tan^2\theta - \sin^2\theta}$

$= 4\sqrt{\dfrac{\sin^2\theta}{\cos^2\theta} - \sin^2\theta}$

$= 4\tan\theta\sqrt{1 - \cos^2\theta}$

$= 4\tan\theta\sqrt{\sin^2\theta}$

$= 4\tan\theta\sin\theta$

Thus, $(m^2 - n^2) = 4\sqrt{mn}$ **Hence Proved.**

Q. 17. If $a = \sin\theta + \cos\theta$; $b = \sin^3\theta + \cos^3\theta$, then show that $(3a - 2b) = a^3$.

Solution : Given,

$a = \sin\theta + \cos\theta$ ...(i)

$b = \sin^3\theta + \cos^3\theta$ ...(ii)

Here $(\sin\theta + \cos\theta)^2 = a^2$

$\Rightarrow \sin^2\theta + \cos^2\theta + 2\sin\theta\cos\theta = a^2$

$\Rightarrow \sin\theta\cos\theta = \dfrac{a^2 - 1}{2}$ ...(iii)

$[\because \sin^2\theta + \cos^2\theta = 1]$

Also, $(\sin\theta + \cos\theta)^3 = a^3$

$\Rightarrow \sin^3\theta + \cos^3\theta + 3\sin\theta\cos\theta(\sin\theta + \cos\theta) = a^3$

$\Rightarrow \sin^3\theta + \cos^3\theta + 3\left(\dfrac{a^2-1}{2}\right)a = a^3$

[From (i) and (iii)]

$\Rightarrow b + \dfrac{3a}{2}(a^2 - 1) = a^3$

$\Rightarrow 2b + 3a(a^2 - 1) = 2a^3$ [From (ii)]

$\Rightarrow 2b + 3a^3 - 3a = 2a^3$

$\Rightarrow (3a - 2b) = a^3$ **Hence Proved.**

Q. 18. If $a\cos\theta - b\sin\theta = c$, prove that

$$(a\sin\theta + b\cos\theta) = \sqrt{a^2 + b^2 - c^2}.$$

Solution : Given, $a\cos\theta - b\sin\theta = c$ ...(i)

Now, $a^2 + b^2 - c^2 = a^2 + b^2 - (a\cos\theta - b\sin\theta)^2$

[From (i)]

$= a^2 + b^2 - (a^2\cos^2\theta + b^2\sin^2\theta - 2ab\sin\theta\cos\theta)$

$= a^2 + b^2 - [a^2(1 - \sin^2\theta) + b^2(1 - \cos^2\theta) - 2ab\sin\theta\cos\theta]$

$= a^2 + b^2 - [a^2 - a^2\sin^2\theta + b^2 - b^2\cos^2\theta - 2ab\sin\theta\cos\theta]$

$= a^2 + b^2 - a^2 + a^2\sin^2\theta - b^2 + b^2\cos^2\theta + 2ab\sin\theta\cos\theta]$

$= a^2\sin^2\theta + b^2\cos^2\theta + 2ab\sin\theta\cos\theta$

$= (a\sin\theta + b\cos\theta)^2$

Thus, $a^2 + b^2 - c^2 = (a\sin\theta + b\cos\theta)^2$

or $\sqrt{a^2 + b^2 - c^2} = (a\sin\theta + b\cos\theta)$

$\Rightarrow (a\sin\theta + b\cos\theta) = \sqrt{a^2 + b^2 - c^2}.$

Hence Proved.

Q. 19. In a ΔABC, right-angled at A, if $\tan C = \sqrt{3}$, find the value of $\sin B\cos C + \cos B\sin C$.

Solution : Given,

$$\tan C = \sqrt{3} = \frac{AB}{AC}$$

or $$AB = \sqrt{3}AC$$

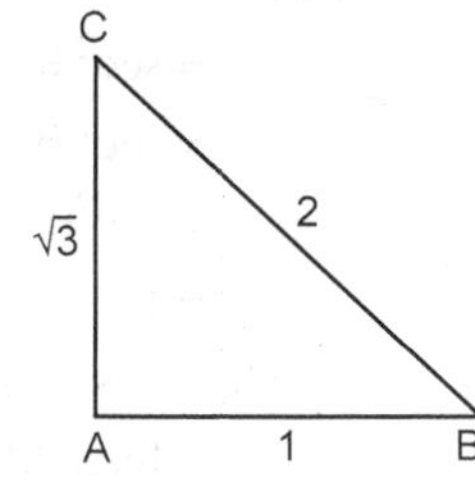

Thus, $$CB^2 = AB^2 + AC^2$$ [Pythagoras' theorem]

$$\Rightarrow \quad CB^2 = (\sqrt{3}AC)^2 + AC^2 = 4AC^2$$

$$\Rightarrow \quad CB = 2AC$$

Hence $$\sin B = \frac{AC}{BC} = \frac{\sqrt{3}}{2},\ \cos C = \frac{AC}{BC} = \frac{\sqrt{3}}{2},$$

$$\cos B = \frac{AB}{BC} = \frac{1}{2} \text{ and } \sin C = \frac{AB}{BC} = \frac{1}{2}$$

So, $\sin B \cos C + \cos B \sin C$

$$= \frac{\sqrt{3}}{2} \times \frac{\sqrt{3}}{2} + \frac{1}{2} \times \frac{1}{2}$$

$$= \frac{3}{2} + \frac{1}{2} = \frac{4}{2} = 2.$$ **Ans.**

Q. 20. If $\sec\theta + \tan\theta = p$, prove that $\sin\theta = \frac{p^2 - 1}{p^2 + 1}$.★

Solution : Given,

$\sec\theta + \tan\theta = p$

Now, $$\text{R.H.S.} = \frac{p^2 - 1}{p^2 + 1}$$

$$= \frac{(\sec\theta + \tan\theta)^2 - 1}{(\sec\theta + \tan\theta)^2 + 1}$$

$$= \frac{\sec^2\theta + \tan^2\theta + 2\sec\theta\tan\theta - 1}{\sec^2\theta + \tan^2\theta + 2\sec\theta\tan\theta + 1}$$

$[\because (a + b)^2 = a^2 + b^2 + 2ab]$

$$= \frac{(\sec^2\theta - 1) + \tan^2\theta + 2\sec\theta\tan\theta}{\sec^2\theta + (1 + \tan^2\theta) + 2\sec\theta\tan\theta}$$

$$= \frac{\tan^2\theta + \tan^2\theta + 2\sec\theta\tan\theta}{\sec^2\theta + \sec^2\theta + 2\sec\theta\tan\theta} \quad \begin{bmatrix} \sec^2\theta - 1 = \tan^2\theta \\ \sec^2\theta = 1 + \tan^2\theta \end{bmatrix}$$

$$= \frac{2\tan^2\theta + 2\sec\theta\tan\theta}{2\sec^2\theta + 2\sec\theta\tan\theta}$$

$$= \frac{2\tan\theta(\tan\theta + \sec\theta)}{2\sec\theta(\sec\theta + \tan\theta)} = \frac{\tan\theta}{\sec\theta}$$

$$= \frac{\frac{\sin\theta}{\cos\theta}}{\frac{1}{\cos\theta}}$$

$= \sin\theta = \text{L.H.S.}$ **Hence Proved.**

Q. 21. Without using trigonometric tables, evaluate :

$$\frac{\cos 58^\circ}{\sin 32^\circ} + \frac{\sin 22^\circ}{\cos 68^\circ} - \frac{\cos 38^\circ \text{ cosec } 52^\circ}{\tan 18^\circ \tan 35^\circ \tan 60^\circ \tan 72^\circ \tan 55^\circ}$$

Solution : We have,

$$\frac{\cos 58^\circ}{\sin 32^\circ} + \frac{\sin 22^\circ}{\cos 68^\circ} - \frac{\cos 38^\circ \text{ cosec } 52^\circ}{\tan 18^\circ \tan 35^\circ \tan 60^\circ \tan 72^\circ \tan 55^\circ}$$

$$= \frac{\cos(90 - 32)^\circ}{\sin 32^\circ} + \frac{\sin(90 - 68)^\circ}{\cos 68^\circ} - \frac{\cos(90 - 52)^\circ \text{ cosec } 52^\circ}{\tan(90 - 72)^\circ \tan(90 - 55)^\circ \tan 60^\circ \tan 72^\circ \tan 55^\circ}$$

$$= \frac{\sin 32^\circ}{\sin 32^\circ} + \frac{\cos 68^\circ}{\cos 68^\circ} - \frac{\sin 52^\circ \text{ cosec } 52^\circ}{\cot 72^\circ \cot 55^\circ \tan 60^\circ \tan 72^\circ \tan 35^\circ}$$

$\because \cos(90 - \theta) = \sin\theta$
$\sin(90 - \theta) = \sin\theta$
$\tan(90 - \theta) = \cot\theta$

$$= 1 + 1 - \frac{1}{\sqrt{3}} = 2 - \frac{1}{\sqrt{3}}.$$ **Ans.**

Q. 22. Solve :

$$\frac{\sin 15^\circ \cos 75^\circ + \cos 15^\circ \sin 75^\circ}{\tan 5^\circ \tan 30^\circ \tan 35^\circ \tan 55^\circ \tan 85^\circ}.$$

Solution : We have,

$$\frac{\sin 15^\circ \cos 75^\circ + \cos 15^\circ \sin 75^\circ}{\tan 5^\circ \tan 30^\circ \tan 35^\circ \tan 55^\circ \tan 85^\circ}$$

$$= \frac{\sin 15^\circ \cos(90 - 15)^\circ + \cos 15^\circ \sin(90 - 15)^\circ}{\tan 5^\circ \tan 30^\circ \tan 35^\circ \tan(90 - 35)^\circ \tan(90 - 5)^\circ}$$

$$= \frac{\sin 15^\circ \sin 15^\circ + \cos 15^\circ \cos 15^\circ}{\tan 5^\circ \tan 30^\circ \tan 35^\circ \cot 35^\circ \cot 5^\circ}$$

$$= \frac{\sin^2 15^\circ + \cos^2 15^\circ}{\tan 30^\circ}$$

$$= \frac{1}{\tan 30^\circ} \quad [\because \sin^2\theta + \cos^2\theta = 1]$$

$$= \frac{1}{1/\sqrt{3}} = \sqrt{3}$$ **Ans.**

Q. 23. Solve :

$$-\tan\theta \cot(90 - \theta) + \frac{\sec\theta \text{ cosec}(90 - \theta) + \sin^2 35^\circ + \sin^2 55^\circ}{\tan 10^\circ \tan 20^\circ \tan 30^\circ \tan 70^\circ \tan 80^\circ}$$

★ are board exam questions from previous years

Solution : We have, $\dfrac{-\tan\theta\cot(90-\theta)+\sec\theta\,\text{cosec}\,(90-\theta)+\sin^2 35°+\sin^2 55°}{\tan 10°\tan 20°\tan 30°\tan 70°\tan 80°}$

$$= \frac{-\tan\theta\tan\theta+\sec\theta\sec\theta+\sin^2 35°+\sin^2(90-35)°}{\tan 10°\tan 20°\tan 30°\tan(90-20)°\tan(90-10)°}$$

$$= \frac{1+\sin^2 35°+\cos^2 35°}{\tan 10°\tan 20°\tan 30°\cot 20°\cot 10°}$$

$$= \frac{1+1}{\tan 30°} = 2\sqrt{3}.$$ **Ans.**

Q. 24. Solve :

$$\mathbf{\frac{2}{3}\,cosec^2 58° - \frac{2}{3}\cot 58°\tan 32° - \frac{5}{3}\tan 13°\tan 37°\tan 45°\tan 53°\tan 77°.}$$ ★

Solution : Consider,

$$\frac{2}{3}\text{cosec}^2 58° - \frac{2}{3}\cot 58°\tan 32° - \frac{5}{3}\tan 13°\tan 37°\tan 45°\tan 53°\tan 77°$$

$$= \frac{2}{3}\text{cosec}^2 58° - \frac{2}{3}\cot 58°\tan(90-58)° - \frac{5}{3}\tan 13°\tan 37°\tan 45°\tan(90-37)°\tan(90-13)°$$

$$= \frac{2}{3}\text{cosec}^2 58° - \frac{2}{3}\cot 58°\cot 58° - \frac{5}{3}\tan 13°\tan 37°\tan 45°\cot 37°\cot 13°$$

$$= \frac{2}{3}\text{cosec}^2 58° - \frac{2}{3}\cot^2 58° - \frac{5}{3}\tan 45°$$

$$= \frac{2}{3}(\text{cosec}^2 58° - \cot^2 58°) - \frac{5}{3}\tan 45°$$

$$= \frac{2}{3} - \frac{5}{3} = -1. \quad [\because \text{cosec}^2\theta - \cot^2\theta = 1]$$

Ans.

Q. 25. If $x = a\sec\theta + b\tan\theta$ and $y = a\tan\theta + b\sec\theta$, prove that $x^2 - y^2 = a^2 - b^2$.★★

Solution : L.H.S. $= x^2 - y^2$

$= (a\sec\theta + b\tan\theta)^2 - (a\tan\theta + b\sec\theta)^2$

$= a^2\sec^2\theta + b^2\tan^2\theta + 2ab\sec\theta\tan\theta - a^2\tan^2\theta - b^2\sec^2\theta - 2ab\sec\theta\tan\theta$

$= a^2\sec^2\theta - a^2\tan^2\theta + b^2\tan^2\theta - b^2\sec^2\theta$

$= a^2(\sec^2\theta - \tan^2\theta) + b^2(\tan^2\theta - \sec^2\theta)$

$= a^2 - b^2 =$ R.H.S. **Hence Proved.**

★ are board exam questions from previous years

★★ are frequently asked board exam questions

Q. 26. Prove that :

$\sec^2\theta - \cot^2(90° - \theta) = \cos^2(90° - \theta) + \cos^2\theta$.

Solution : Consider,

L.H.S. $= \sec^2\theta - \cot^2(90° - \theta)$

$= \sec^2\theta - [\cot(90-\theta)]^2$

$= \sec^2\theta - (\tan\theta)^2$

$= \sec^2\theta - \tan^2\theta$

$= 1$

R.H.S. $= \cos^2(90-\theta) + \cos^2\theta$

$= [\cos(90-\theta)]^2 + \cos^2\theta$

$= (\sin\theta)^2 + \cos^2\theta$

$= \sin^2\theta + \cos^2\theta$

$= 1$

Hence, L.H.S. = R.H.S. **Hence Proved.**

Q. 27. If $\cos A = \dfrac{7}{25}$, find the value of $\tan A + \cot A$.

Solution : Given,

$$\cos A = \frac{7}{25}$$

$$\Rightarrow \quad \cos^2 A = \frac{49}{625}$$

We know, $\sin^2 A = 1 - \cos^2 A$

$$\Rightarrow \quad \sin^2 A = 1 - \frac{49}{625} = \frac{625-49}{625} = \frac{576}{625}$$

$$\Rightarrow \quad \sin A = \frac{24}{25}$$

Also, $\tan A = \dfrac{\sin A}{\cos A} = \dfrac{\frac{24}{25}}{\frac{7}{25}} = \dfrac{24}{7}$

and $\cot A = \dfrac{1}{\tan A} = \dfrac{7}{24}$

Hence, $\tan A + \cot A = \dfrac{24}{7} + \dfrac{7}{24} = \dfrac{625}{168}$. **Ans.**

Chapter 11. Applications of Trigonometry

Q. 1. The angle of elevation of an aeroplane from a point on the ground is 60°. After a flight of 30 seconds, the angle of elevation become 30°. If the aeroplane is flying at a constant height of $3000\sqrt{3}$ m, find the speed of the aeroplane.★

Solution : Let the ground distance between the aeroplane A and the point E be x m.

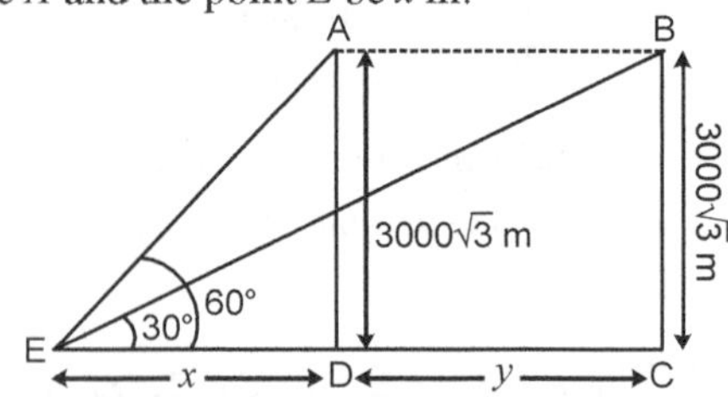

Given, height AD is $3000\sqrt{3}$ m and the angle of elevation is 60°.

So, in ΔAED

$$\tan 60° = \frac{3000\sqrt{3}}{x}$$

$$\sqrt{3} = \frac{3000\sqrt{3}}{x}$$

$$\Rightarrow \quad x = 3000 \text{ m}$$

Let the new distance (DC) covered by aeroplane in 30 seconds be y m.

So $$\tan 30° = \frac{3000\sqrt{3}}{3000+y}$$

$$\frac{1}{\sqrt{3}} = \frac{3000\sqrt{3}}{3000+y}$$

$$\Rightarrow \quad 3000 + y = (3000\sqrt{3})\sqrt{3}$$

$$\Rightarrow \quad 3000 + y = 9000$$

$$\Rightarrow \quad y = 9000 - 3000 = 6000 \text{ m}$$

Thus, distance covered in 30 seconds = 6000 m

Hence, $$\text{Speed} = \frac{6000}{30} = 200 \text{ m/sec} = 200 \times \frac{18}{5} \text{ km/hr}$$

Hence the speed of the aeroplane is 720 km/hr. **Ans.**

Q. 2. Two ships are there in the sea on either side of a lighthouse in such a way that the ships and the light-house are in the same straight line. The angles of depression of two ships as observed from the top of the lighthouse are 60° and 45°. If the height of the lighthouse is 200 m, find the distance between the two ships.★

[Use $\sqrt{3} = 1.73$]

Solution : Given, the height of the lighthouse AC is 200 m. Let the distance of the first ship B from the light- house be x m and the distance of the second ship D from the lighthouse be y m.

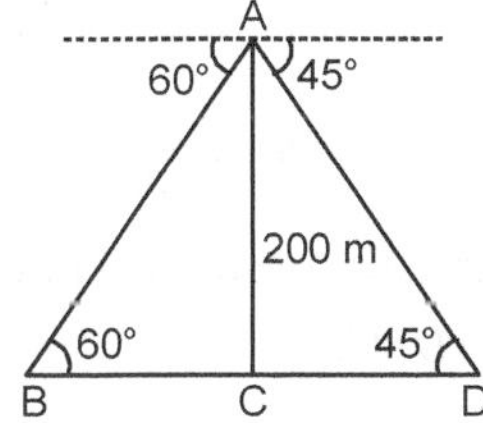

Since the angle of depression is 60° with the first ship, so

In ΔABC, we have

$$\tan 60° = \frac{200}{x}$$

★ are board exam questions from previous years

$$\Rightarrow \quad \sqrt{3} = \frac{200}{x}$$

$$\Rightarrow \quad x = \frac{200}{\sqrt{3}} \quad \ldots\text{(i)}$$

Also, as the angle of depression is 45° with the second ship, so

In ΔACD, we have

$$\tan 45° = \frac{200}{y}$$

$$\Rightarrow \quad 1 = \frac{200}{y}$$

$$\Rightarrow \quad y = 200 \text{ m} \quad \ldots\text{(ii)}$$

Thus, from equations (i) and (ii), the distance between the two ships is

$$\frac{200}{\sqrt{3}} + 200 = 200\left(1 + \frac{1}{\sqrt{3}}\right)$$

$$= 200\left(\frac{\sqrt{3}+1}{\sqrt{3}}\right)$$

$$= 200\left(\frac{1.73+1}{1.73}\right)$$

$$= 200\left(\frac{2.73}{1.73}\right)$$

$$= 200(1.578)$$

$$= 315.6 \text{ m.}$$ **Ans.**

Q. 3. A highway leads to the foot of 300 m high tower. An obeservatory is set at the top of the tower. It sees a car moving towards it at an angle of depression of 30°. After 15 seconds, angle of depression becomes 60°. Find the distance travelled by the car during this time.

Solution : Let $AB = 300$ m is the tower. Initially, car is at C and after 15 seconds, it reaches at D.

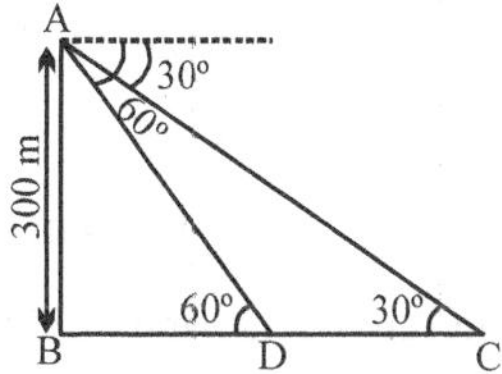

In right ΔABC,

$$\frac{AB}{BC} = \tan 30°$$

$$\Rightarrow \quad \frac{300}{BC} = \frac{1}{\sqrt{3}}$$

$$\Rightarrow \quad BC = 300\sqrt{3} \text{ m}$$

In right ΔABD,

$$\frac{AB}{BD} = \tan 60^\circ$$

$$\frac{300}{BD} = \sqrt{3}$$

$$\Rightarrow \quad BD = \frac{300}{\sqrt{3}} \text{ m}$$

$\therefore$ Distance covered by car

$$= DC = BC - BD$$

$$= 300\sqrt{3} - \frac{300}{\sqrt{3}} = \frac{600}{\sqrt{3}}$$

$$= 200\sqrt{3} \text{ m}$$ **Ans.**

Q. 4. The horizontal distance between two poles is 15 m. The angle of depression of the top of the first pole as seen from the top of the second pole is 30°. If the height of the second pole is 24 m, find the height of the first pole.

[Use $\sqrt{3} = 1.732$]

Solution : Let the height of the first pole be H m.

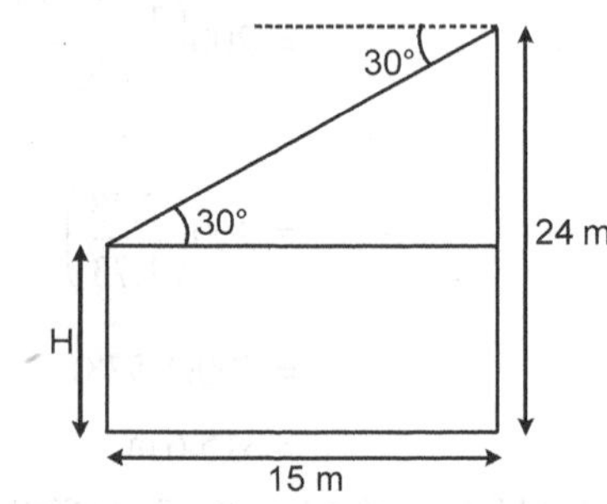

Thus $\quad \tan 30^\circ = \dfrac{24 - H}{15} = \dfrac{1}{\sqrt{3}}$

$$\Rightarrow \quad \sqrt{3}(24 - H) = 15$$

$$\Rightarrow \quad 24 - H = 5\sqrt{3}$$

$$\Rightarrow \quad H = 24 - 5\sqrt{3}$$

$$= [24 - 5(1.732)] \text{ m}$$

$$= (24 - 8.66 \text{ m})$$

$$= 15.34 \text{ m}$$

So, the height of the first pole is 15.34 m. **Ans.**

Q. 5. The angle of elevation of the top of a tower at a point on the ground is 30°. If the height of the tower is tripled, find the angle of elevation of the top of the same point.★

Solution : Let the height of the tower AB be h m and the distance between the tower and the point of observation C on the ground be y m.

★ are board exam questions from previous years

So $\quad \tan 30^\circ = \dfrac{h}{y}$

$$\frac{1}{\sqrt{3}} = \frac{h}{y}$$

$$\Rightarrow \quad y = h\sqrt{3} \quad \text{...(i)}$$

Let the new height be DB and the new angle of elevation be x°.

So $\quad \tan x^\circ = \dfrac{3h}{y}$

$$\Rightarrow \quad \tan x^\circ = \frac{3h}{h\sqrt{3}}$$

$$= \sqrt{3} = \tan 60^\circ$$

$$\Rightarrow \quad x^\circ = 60^\circ$$

Thus, the angle of elevation is 60° from the same point when the height of the tower is tripled. **Ans.**

Q. 6. A contract was awarded to constuct a vertical pillar at a horizontal distance of 100 m from a fixed point. It was decided that angle of elevation of the top of the complete pillar from this point to be 60°. Contractor finished the job by making a pillar such that angle of elevation of its top was 45°. Find the height of the pillar to be increased as per the terms of contract.

Solution : Let AB is the required pillar and $BC = 100$ m such that

$$\angle ACB = 60^\circ$$

Let contractor constructed pillar BD.

In right ΔDBC,

$$\frac{DB}{BC} = \tan 45^\circ$$

$$\Rightarrow \quad \frac{DB}{100} = 1 \Rightarrow DB = 100 \text{ m}$$

In right ΔABC,

$$\frac{AB}{BC} = \tan 60^\circ$$

$$\Rightarrow \quad \frac{AB}{100} = \sqrt{3}$$

$$\Rightarrow \quad AB = 100\sqrt{3} \text{ m}$$

Height of pillar to be increase

$= 100\sqrt{3} - 100 = 100(\sqrt{3} - 1)$ m. **Ans.**

Q. 7. A fire in a building AB is reported on telephone to fire stations P and Q, 20 km apart from each other on a straight road. P observes that the fire is at angle of 60° to the road and Q observes that it is at an angle of 45° to the road. Which station should send its team and how much will this team have to travel ?

Solution : Let $BA \perp PQ$. A is a point on the road PQ (P, A and Q are points on the ground and AB is the height of the building).

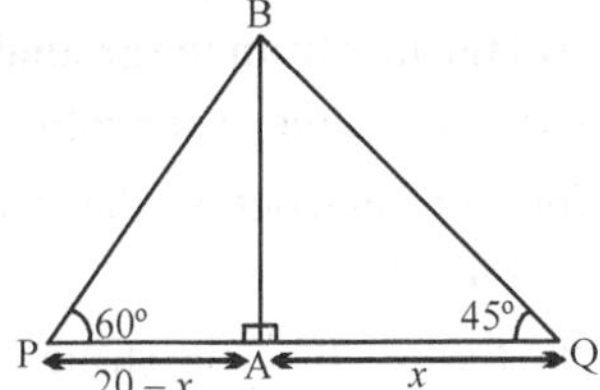

Let distance between Q and A be x

Then, distance between P and $A = 20 - x$

In right ΔBAQ,

$$\frac{BA}{AQ} = \tan 45^\circ$$

$$BA = AQ = x \text{ km}$$

In right ΔBAP,

$$\frac{BA}{AP} = \tan 60^\circ$$

$$\Rightarrow \quad \frac{x}{20-x} = \sqrt{3}$$

$$\Rightarrow \quad x = \sqrt{3}(20-x)$$

$$\Rightarrow \quad x + \sqrt{3}x = 20\sqrt{3}$$

$$\Rightarrow \quad x = \frac{20\sqrt{3}}{\sqrt{3}+1}$$

$$\Rightarrow \quad x = \frac{20(\sqrt{3})(\sqrt{3}-1)}{3-1}$$

$$= 30 - 10\sqrt{3}$$

$$= 30 - 10 \times 1.732$$

$$= 12.68 \text{ km}$$

Thus, $AQ = 12.68$ km

and $AP = 20 - x$

$= 20 - 12.68$

$= 7.32$ km

Hence, station P should send its team and, distance covered $= 7.32$ km **Ans.**

Q. 8. A ladder of length 6 m makes an angle of 45° with the floor while leaning against one wall of a room. If the foot of the ladder is kept fixed on the floor and it is made to lean against the opposite wall of the room, it makes an angle of 60° with the floor. Find the distance between these two walls of the room.

Solution : Let the distance between the first wall and the ladder be L_1 and the distance between the ladder and the second wall be L_2.

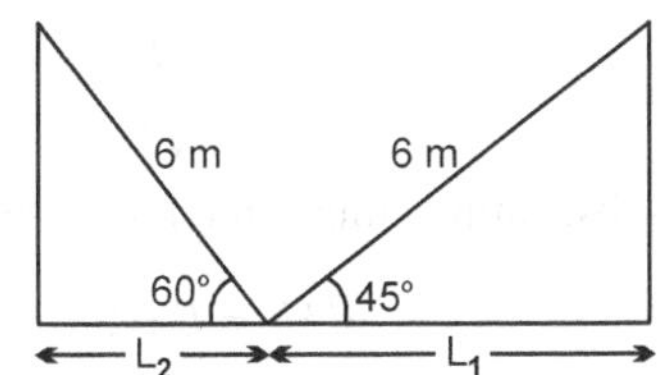

Thus, $\cos 45^\circ = \dfrac{\text{base}}{\text{hypotenuse}} = \dfrac{L_1}{6}$

$$\Rightarrow \quad \frac{1}{\sqrt{2}} = \frac{L_1}{6}$$

$$\Rightarrow \quad L_1 = 3\sqrt{2} \text{ m}$$

and $\cos 60^\circ = \dfrac{\text{base}}{\text{hypotenuse}} = \dfrac{L_2}{6}$

$$\frac{1}{2} = \frac{L_2}{6}$$

or $L_2 = 3$ m

Thus, the total distance between the walls

$= (3 + 3\sqrt{2})$ m

$= 3(1 + \sqrt{2})$ m

$= 3(2.414)$ m $\quad [\because \sqrt{2} = 1.414]$

$= 7.242$ m

Thus, the distance between the walls is 7.242 m. **Ans.**

Q. 9. From the top of a vertical tower, the angles of depression of two cars in the same straight line with the base of the tower at an instant are found to be 45° and 60°. If the cars are 100 m apart and are on the same side of the tower, find the height of the tower. [Use $\sqrt{3} = 1.73$]

Solution :

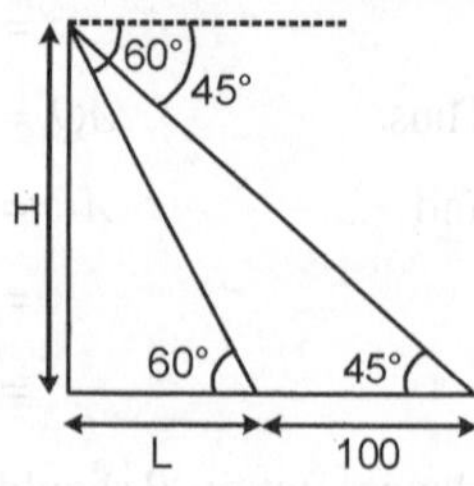

We have, $\tan 45° = \dfrac{H}{L+100}$

$$1 = \frac{H}{L+100}$$

$\Rightarrow$ $L + 100 = H$ ...(i)

and $\tan 60° = \dfrac{H}{L}$

$$\sqrt{3} = \frac{H}{L}$$

$\Rightarrow$ $\sqrt{3}L = H$

$\Rightarrow$ $L = \dfrac{H}{\sqrt{3}}$ ...(ii)

Substituting equation (i) in equation (ii), we have

$$\sqrt{3}L = L + 100$$

$\Rightarrow$ $\sqrt{3}L - L = 100$

$\Rightarrow$ $(\sqrt{3} - 1)L = 100$

$\Rightarrow$ $L = \dfrac{100}{0.732}$

$= 136.61$ m

Thus $H = 236{\cdot}61$ m [From (i)]

Hence, the height of the tower is 236.61 m. **Ans.**

Q. 10. From the top of a tower 100 m high, a man observes two cars on the opposite sides of the tower with angles of depression 30° and 45° respectively. Find the distance between the cars. [Use $\sqrt{3}$ =1.73]

Solution : Let the distance of one car from the base of the tower be L_1 and the distance of the other car be L_2.

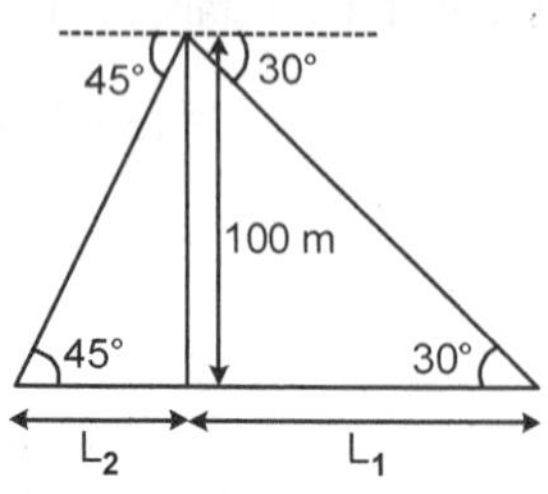

Now, $\tan 30° = \dfrac{100}{L_1}$

$$\frac{1}{\sqrt{3}} = \frac{100}{L_1}$$

$\Rightarrow$ $L_1 = 100\sqrt{3}$ ...(i)

and $\tan 45° = \dfrac{100}{L_2}$

$$1 = \frac{100}{L_2}$$

$\Rightarrow$ $L_2 = 100$ ...(ii)

Adding equations (i) and (ii),

$$L_1 + L_2 = 100\sqrt{3} + 100$$
$$= 100(\sqrt{3}+1)$$
$$= 100(1{\cdot}732 + 1)$$
$$= 100\,(2{\cdot}732)$$
$$= 273{\cdot}2$$

Thus, the distance between the two cars is 273·2 m. **Ans.**

Q. 11. A kite is flying with a string of length 200 m. If the thread makes an angle 30° with the ground, find the distance of the kite from the ground level. (Here, assume that the string is along a straight line)

Solution : Let h denote the distance of the kite from the ground level.

In the figure, AC is the string.

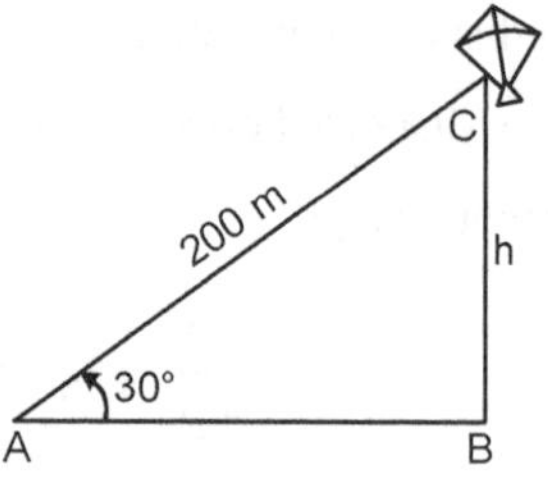

Given, $\angle CAB = 30°$ and $AC = 200$ m

In the ΔCAB,

$$\sin 30° = \frac{h}{200}$$

$\Rightarrow$ $h = 200 \sin 30°$

$\therefore$ $h = 200 \times \dfrac{1}{2} = 100$ m

Hence, the distance of the kite from the ground level is 100 m. **Ans.**

Q. 12. From a point on the ground 40 m away from the foot of a tower, the angle of elevation of the top of the tower is 30°. The angle of elevation of the top of a water tank (on the top of the tower) is 45°. Find the (i) height of the tower, (ii) the depth of the tank.

Solution : Let BC be the tower of height h m and CD be the water tank of height h_1 m. Let A be a point on the ground at a distance of 40 m away from the foot B of the tower.

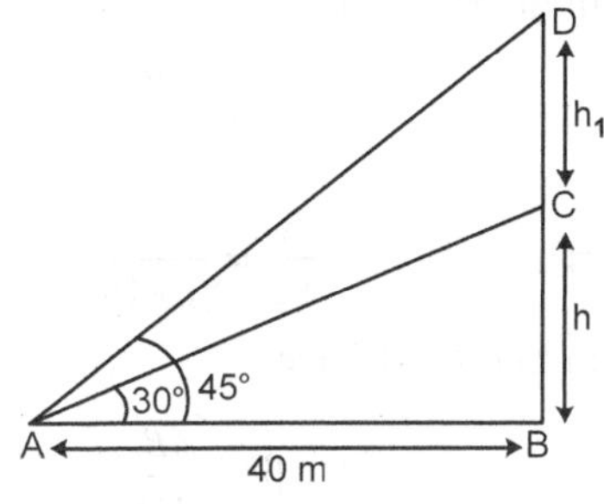

In ΔABD, $\tan 45° = \frac{BD}{AB}$

$$1 = \frac{h+h_1}{40}$$

$$h + h_1 = 40 \text{ m} \quad \ldots\text{(i)}$$

In ΔABC, we have

$$\tan 30° = \frac{BC}{AB}$$

$$\frac{1}{\sqrt{3}} = \frac{h}{40}$$

$$h = \frac{40}{\sqrt{3}} \text{ m}$$

$$= \frac{40\sqrt{3}}{3} \text{ m}$$

$$= 23.1 \text{ m}$$

On putting the value of h in equation (i), we have

$$23.1 + h_1 = 40$$

$$h_1 = (40 - 23.1) \text{ m}$$

$$= 16.9 \text{ m}$$

Thus, the height of the tower is $h = 23.1$ m and the depth of the tank is $h_1 = 16.9$ m. **Ans.**

Q. 13. A man standing on the deck of a ship, which is 10 m above water level. He observes the angle of elevation of the top of a hill as 60° and the angle of depression of the base of the hill as 30°. Calculate the distance of the hill from the ship and the height of the hill.★

Solution : Let the man is standing on the deck of a ship at point A and let CD be the hill.

Given, $\angle EAD = 60°, \angle BCA = 30°$

$AB = 10$ m

In ΔAED, we have

$$\tan 60° = \frac{DE}{EA}$$

$$\sqrt{3} = \frac{h}{x}$$

$$\Rightarrow \quad h = \sqrt{3}x \quad \ldots\text{(i)}$$

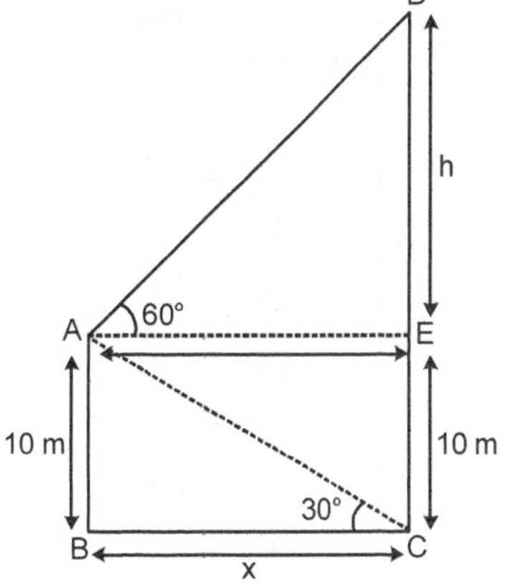

In ΔABC, we have

$$\tan 30° = \frac{AB}{BC}$$

$$\frac{1}{\sqrt{3}} = \frac{10}{x}$$

$$\Rightarrow \quad x = 10\sqrt{3} \quad \ldots\text{(ii)}$$

On putting $x = 10\sqrt{3}$ in equation (i), we get

$$h = \sqrt{3} \times 10\sqrt{3} = 30$$

$$DE = 30 \text{ m}$$

Thus, $CD = CE + ED$

$$= 10 + 30$$

$$= 40 \text{ m}$$

The distance of the hill from the ship is $10\sqrt{3}$ m and the height of the hill is 40 m. **Ans.**

Q. 14. A man on a cliff observes a boat at an angle of depression of 30° which is approaching the shore to the point immediately beneath the observer with a uniform speed. Six minutes later, the angle of depression of the boat is found to be 60°. Find the time taken by the boat to reach the shore.

Solution : Suppose OA be the cliff and P be the initial position of the boat where the angle of depression is 30°. After 6 minutes, the boat reaches to Q where the angle of depression at Q is 60°.

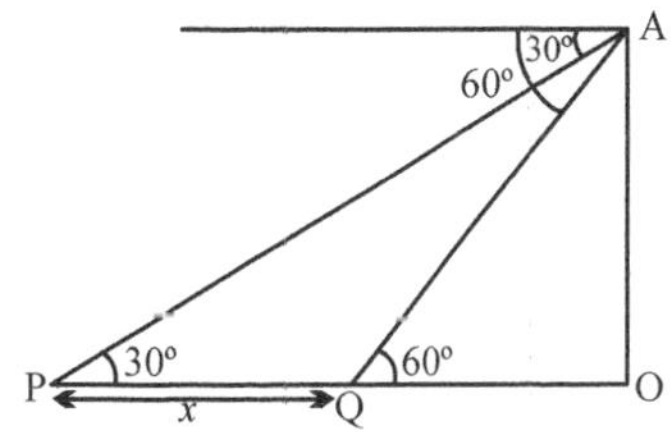

Suppose, $PQ = x$ m.

In ΔPOA and ΔQOA,

$$\tan 30° = \frac{OA}{OP}$$

★ are board exam questions from previous years

and $\tan 60° = \frac{OA}{OQ}$

$$\therefore \quad \frac{1}{\sqrt{3}} = \frac{OA}{OP} \text{ and } \sqrt{3} = \frac{OA}{OQ}$$

$$OA = \frac{OP}{\sqrt{3}} \text{ and } OA = \sqrt{3}OQ$$

$$\frac{OP}{\sqrt{3}} = \sqrt{3}OQ$$

$$OP = 3OQ$$

$$PQ = OP - OQ$$

$$= OP - \frac{OP}{3} \qquad \left[OQ = \frac{1}{3}OP\right]$$

$$= \frac{2}{3}OP$$

Consider the speed of the boat of v m/minute, so, PQ = Distance covered by the boat in 6 minutes.

$$PQ = 6v$$

$$\frac{2}{3}(OP) = 6v$$

$$OP = 9v$$

So, time elapsed by the boat to reach at the shore is given by,

$$T = \frac{OP}{v}$$

$$\Rightarrow \quad T = \frac{9v}{v} \text{ minutes}$$

$$= 9 \text{ minutes}$$ **Ans.**

Q. 15. At a point on level ground, the angle of elevation of a vertical tower is found to be such that its tangent is 5/12. On walking 192 metres towards the tower, the tangent of the angle of elevation is 3/4. Find the height of the tower.

Solution : Let AB be the tower and let the angle of elevation of its top at C be α. Let D be a point at a distance of 192 metres from C such that the angle of elevation of the top of the tower at D be β.

Let h be the height of the tower and $AD = x$.

Given that $\tan\alpha = \frac{5}{12}$ and $\tan\beta = \frac{3}{4}$

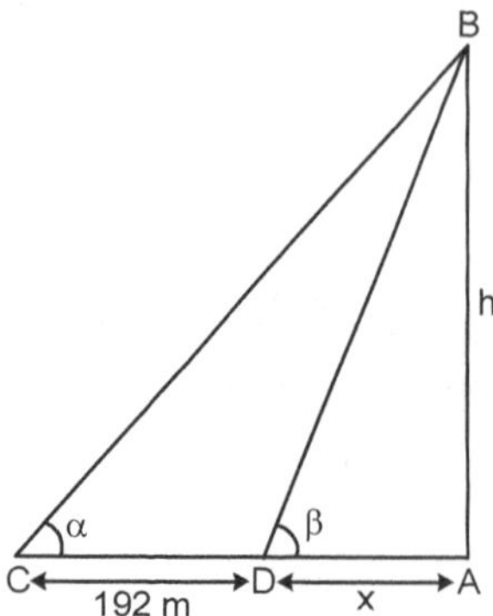

In ΔCAB, we have

$$\tan\alpha = \frac{AB}{AC}$$

$$\Rightarrow \quad \frac{5}{12} = \frac{h}{x+192} \quad \text{...(i)}$$

In ΔDAB, we have

$$\tan\beta = \frac{AB}{AD}$$

$$\Rightarrow \quad \frac{3}{4} = \frac{h}{x}$$

$$3x = 4h$$

$$\Rightarrow \quad x = \frac{4h}{3}$$

On putting the value of x in equation (i), we get

$$\frac{5}{12} = \frac{h}{\frac{4h}{3}+192}$$

$$\Rightarrow \quad 5\left(\frac{4h}{3}+192\right) = 12h$$

$$\Rightarrow \quad \frac{5(4h+576)}{3} = 12h$$

$$\Rightarrow \quad 20h + 2880 = 36h$$

$$16h = 2880$$

$$\Rightarrow \quad h = 180$$

Thus, the height of the tower is 180 m. **Ans.**

Q. 16. The angles of depression of the top and bottom of a 50 m high building from the top of a tower are 45° and 60° respectively. Find the height of the tower and the horizontal distance between the tower and the building.

[Use $\sqrt{3} = 1.73$].

Solution : Let AB and CD be the tower and high building respectively.

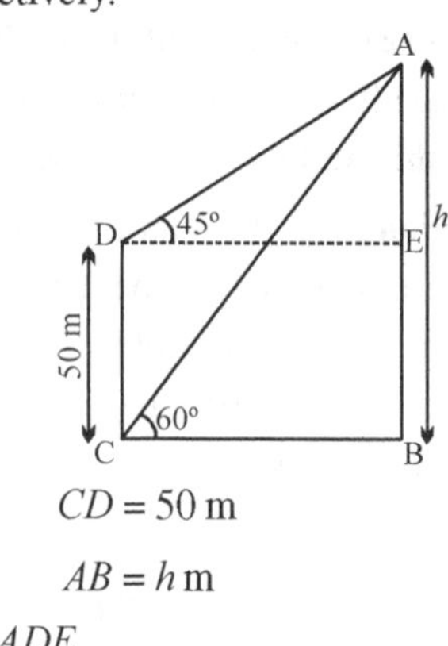

Given, $CD = 50$ m

Let, $AB = h$ m

Then, in ΔADE

$$\tan 45° = \frac{AE}{DE}$$

$$1 = \frac{h-50}{DE}$$

$$DE = h - 50 \quad ...(i)$$

and, in ΔACB

$$\tan 60^\circ = \frac{AB}{CB}$$

$$\sqrt{3} = \frac{h}{CB}$$

$$CB = \frac{h}{\sqrt{3}} \quad ...(iii)$$

From equations (i) and (ii), we get

$$h - 50 = \frac{h}{\sqrt{3}} \quad [\because CB = DE]$$

$$h - \frac{h}{\sqrt{3}} = 50$$

$$\left(\frac{\sqrt{3}-1}{\sqrt{3}}\right) h = 50$$

$$h = \frac{50\sqrt{3}}{\sqrt{3}-1}$$

$$= \frac{50\sqrt{3}}{\sqrt{3}-1} \times \frac{\sqrt{3}+1}{\sqrt{3}+1}$$

$$= 25 \times 3 + 25\sqrt{3}$$

$$\Rightarrow \quad h = 75 + 25\,(1.73)$$

$$= 118.25 \text{ m}$$

From (i), $DE = 118.25 - 50$

$$= 68.25\text{m}$$

Hence, the height of the tower is 118.25 m and the horizontal distance between the tower and the building is 68.25 m. **Ans.**

Q. 17. A 1.2 m tall girl spots a balloon moving with the wind in a horizontal line at a height of 88.2 m from the ground. The angle of elevation of the balloon from the eyes of the girl at any instant is 60°. After sometime, the angle of elevation reduces to 30°. Find the distance travelled by the balloon during the interval.

Solution : Suppose P be the position of the balloon if its angle of elevation from the eyes of the girl is 60° and Q be the position if angle of elevation is 30°.

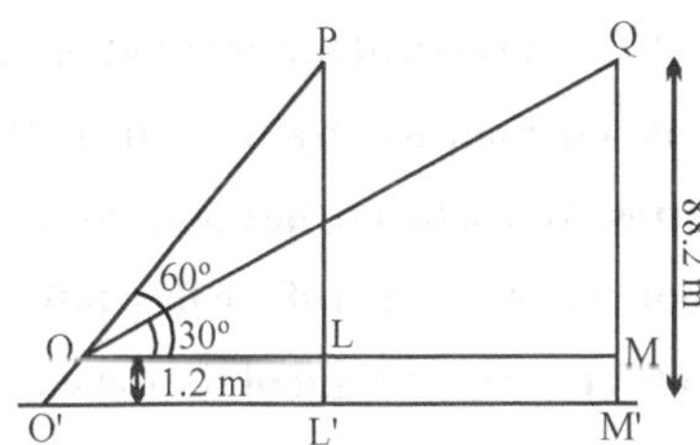

In ΔOLP, $\tan 60^\circ = \frac{PL}{OL}$

$$\therefore \quad \sqrt{3} = \frac{PL' - LL'}{OL}$$

$$= \frac{88.2 - 1.2}{OL} = \frac{87}{OL}$$

So, $OL = \frac{87}{\sqrt{3}}$

In ΔOMQ, $\tan 30^\circ = \frac{QM}{OM}$

$$= \frac{QM' - MM'}{OM}$$

$$\therefore \quad \frac{1}{\sqrt{3}} = \frac{88.2 - 1.2}{OM}$$

So, $OM = 87\sqrt{3}$

Therefore, distance covered by the balloon,

$$PQ = OM - OL$$

$$= \left(87\sqrt{3} - \frac{87}{\sqrt{3}}\right) \text{m}$$

$$= 87\left(\sqrt{3} - \frac{1}{\sqrt{3}}\right) \text{m}$$

$$= \frac{174}{\sqrt{3}} \text{ m} = 58\sqrt{3} \text{ m}$$ **Ans.**

Q. 18. A straight highway leads to the foot of a tower. A watchman standing at the top of the tower observes a car at an angle of depression of 30°, which is approaching the foot of the tower with a uniform speed. Two minutes later, the angle of depression was found to be 60°. The watchman suspects that some terrorists are approaching the tower. It needs half a minute for the watchman to inform the security staff to be on the alert. How much time the car will take to reach the foot of the tower?

Solution : Let AB be the tower. Suppose C and D are the two positions of the car.

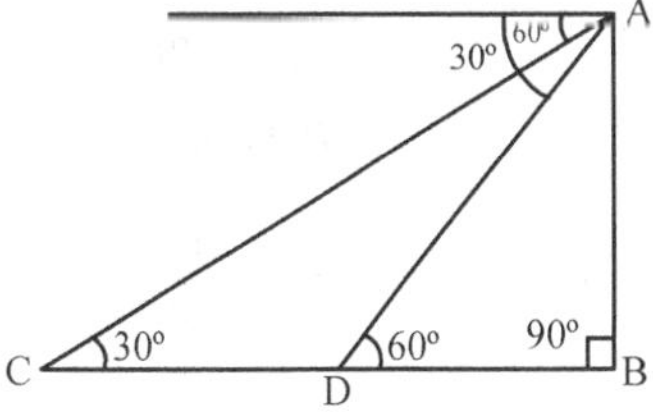

$\therefore$ In ΔCBA

$$\tan 30^\circ = \frac{AB}{BC}$$

$$\frac{1}{\sqrt{3}} = \frac{AB}{BC} \quad ...(i)$$

and in ΔDBA,

$$\tan 60^\circ = \frac{AB}{BD}$$

$$\sqrt{3} = \frac{AB}{BD} \quad ...(ii)$$

After dividing equation (ii) by equation (i), we get

$$3 = \frac{BC}{BD}$$

$\Rightarrow$ $BC = 3\,BD$

$\Rightarrow$ $BD + CD = 3BD$ $[\because BC = BD + CD]$

$\Rightarrow$ $BD = \frac{1}{2}CD$

As time elasped by car in travelling from C to D is 2 minutes. So the time taken by car in travelling from D to B will be 1 minute because $BD = \frac{1}{2}CD$.

Therefore, the car will take one minute to reach the foot of the tower. **Ans.**

Q. 19. From a window A, 10 m above ground, the angle of elevation of the top C to a tower is x°, where $\tan x^\circ = 5/2$ and the angle of depression of the foot D of the tower is y°, where $\tan y^\circ = 1/4$. Calculate the height CD of the tower in meters.

Solution : From figure,

$\angle BDA = \angle DAE = y^\circ$ [alternate angles]

and $AB = DE$

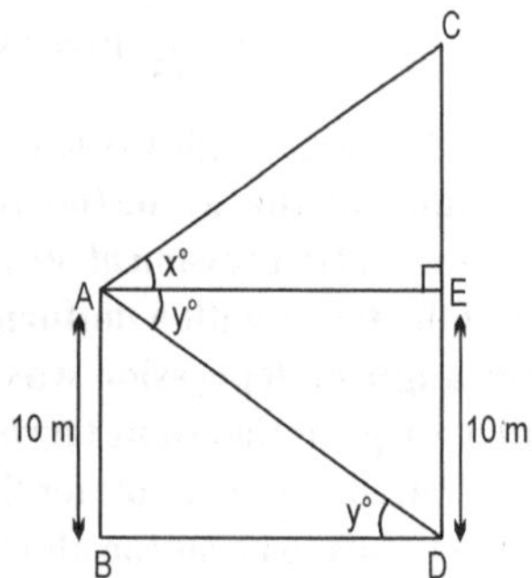

In ΔABD, $\tan y^\circ = \frac{AB}{BD}$

$\Rightarrow$ $\frac{1}{4} = \frac{10}{BD}$

or $BD = 40\text{ m} = AE$

In ΔAEC, $\tan x^\circ = \frac{CE}{AE}$

$$\frac{5}{2} = \frac{CE}{40}$$

$\Rightarrow$ $CE = 5 \times 20 = 100\text{ m}$

Thus, $CD = CE + DE$

$= 100\text{ m} + 10\text{ m}$

$= 110\text{ m}.$

Thus the height of tower is 110 m. **Ans.**

Q. 20. A pole being broken by the wind, the top struck the ground making an angle of 30° and at a distance of 8 m from the foot of the pole. Find the whole height of the pole.

Solution : Let ABC be the pole. When broken at B by the wind, let its top A strike the ground.

Given, $\angle CAB = 30^\circ$ and $AC = 8$ m

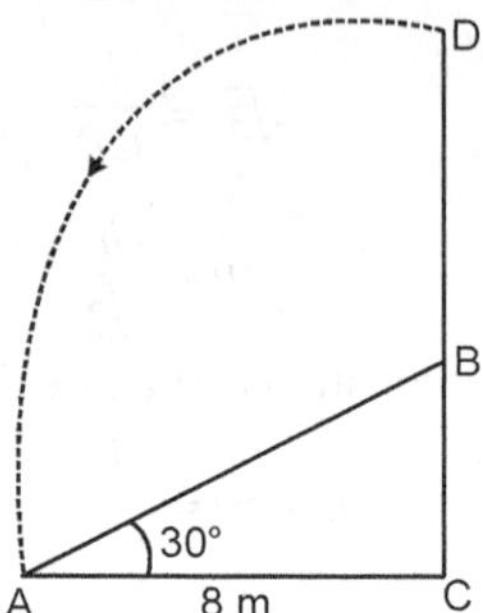

In ΔACB, we have

$$\tan 30^\circ = \frac{BC}{AC}$$

$\Rightarrow$ $\frac{1}{\sqrt{3}} = \frac{BC}{8}$

$\Rightarrow$ $BC = \frac{8}{\sqrt{3}}$ m

Also $\cos 30^\circ = \frac{AC}{AB}$

$\Rightarrow$ $\frac{\sqrt{3}}{2} = \frac{8}{AB}$

$\Rightarrow$ $AB = \frac{16}{\sqrt{3}}$ m

$\therefore$ Height of the pole $= AB + BC$

$= \left(\frac{16}{\sqrt{3}} + \frac{8}{\sqrt{3}}\right)$ m

$= \frac{24}{\sqrt{3}}$ m

$= 8\sqrt{3}$ m

$= 8 \times 1.732$

$= 13.856$ m. **Ans.**

Q. 21. A vertical tower stands on a horizontal plane and is surmounted by a vertical flagstaff of height 5 metres. At a point on the plane, the angles of elevation of the bottom and the top of the flagstaff are respectively 30° and 60°. Find the height of the tower.

Solution : Let height of the tower $= BC = h$ m

Height of flagstaff, $DC = 5$ m

$\therefore$ $\angle BAC = 30^\circ$

and $\angle BAD = 60^\circ$

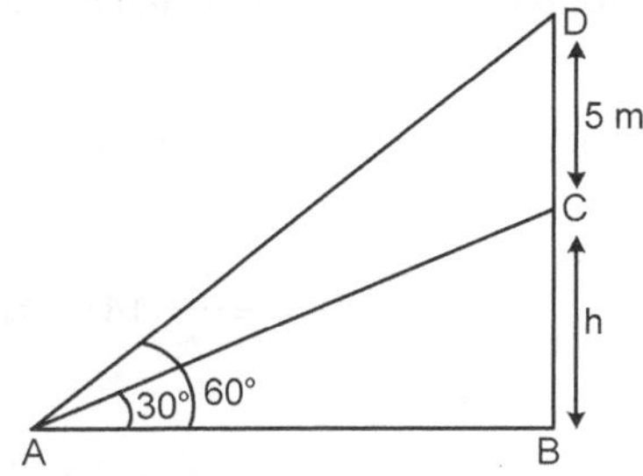

In ΔABC, we have

$$\Rightarrow \quad \tan 30^\circ = \frac{BC}{AB}$$

$$\therefore \quad \frac{1}{\sqrt{3}} = \frac{h}{AB}$$

$$\Rightarrow \quad AB = \sqrt{3}h \quad \text{...(i)}$$

In ΔABD, we have

$$\tan 60^\circ = \frac{BD}{AB}$$

$$\sqrt{3} = \frac{h+5}{AB}$$

$$\Rightarrow \quad AB = \frac{h+5}{\sqrt{3}} \quad \text{...(ii)}$$

From equations (i) and (ii),

$$\sqrt{3}h = \frac{h+5}{\sqrt{3}}$$

$$\Rightarrow \quad 3h = h+5$$

$$\Rightarrow \quad 2h = 5$$

$$\Rightarrow \quad h = 2.5 \text{ m.}$$

Ans.

Q. 22. On a straight line passing through the foot of a tower, two points *C* and *D* are at distances of 4 m and 16 m from the foot respectively. If the angles of elevation from *C* and *D* of the top of the tower are complementary, then find the height of the tower.

Solution : Let AB be the tower.

Since, the angles made by points C and D are complementary,

Therefore, $\angle ACB = \theta$

and $\quad \angle ADB = 90^\circ - \theta$

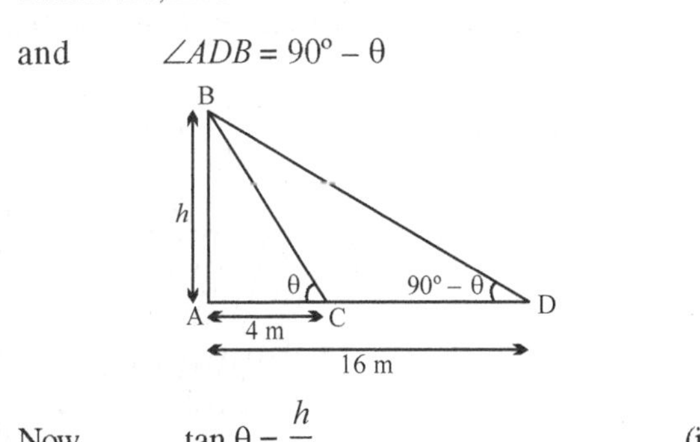

Now, $\quad \tan\theta = \frac{h}{4} \quad$ ...(i)

$$\tan(90 - \theta) = \frac{h}{16}$$

$$\Rightarrow \quad \cot\theta = \frac{h}{16} \quad \text{...(ii)}$$

Multiplying equations (i) and (ii), we get

$$\frac{h}{4} \times \frac{h}{16} = \cot\theta \times \tan\theta$$

$$\frac{h^2}{64} = 1$$

$$\left[\because \cot\theta \times \tan\theta = \frac{1}{\tan\theta} \times \tan\theta = 1\right]$$

$$\Rightarrow \quad h^2 = 64 \Rightarrow h = 8 \text{ m}$$

Thus, height of tower = 8 m. **Ans.**

Chapter 12. Areas Related to Circles

Q. 1. *PS* is the diameter of a circle of radius 6 cm. *Q* and *R* are points on the diameter such that *PQ*, *QR* and *RS* are equal. Semicircles are drawn with *PQ* and *QS* as diameters as shown in the figure. Find the perimeter of the shaded region. [Use $\pi = 3.14$]

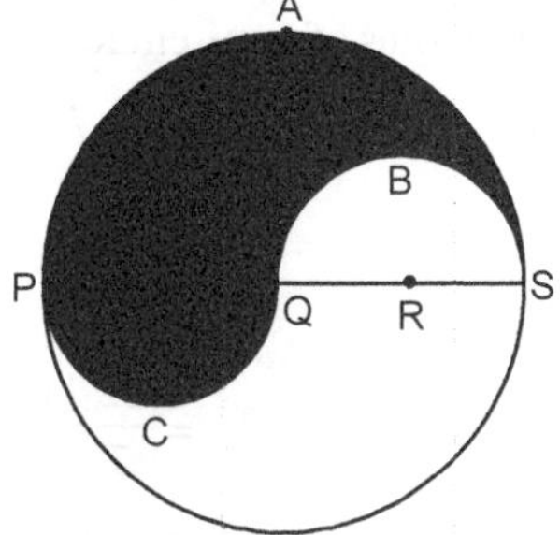

Solution : Given,

$$r = 6 \text{ cm}, PQ = QR = RS$$

and $\quad PS = PQ + QR + RS$

Hence, $\quad PS = 2r = 12$ cm

Now, $\quad PS = PQ + QR + RS$

$$\Rightarrow \quad 12 = 3PQ = 3QR = 3RS$$

$$\Rightarrow \quad PQ = QR = RS = 4 \text{ cm}$$

Thus, PQ is the diameter of the smaller semicircle and $(QR + RS)$ is the diameter of the larger semicircle.

So, the perimeter of the shaded region

$= \text{Perimeter of } (PAS + PCQ + QBS)$

$= [\pi(6) + \pi(2) + \pi(4)]$ cm

$= 12\pi$ cm

$= 12(3.14)$ cm

$= 37.68$ cm. **Ans.**

Q. 2. In the given figure, *ABDC* is a quadrant of a circle of radius 28 cm and a semi-circle *BEC* is drawn with *BC* as diameter. Find the area of the shaded region.

$$\left[\text{Use } \pi = \frac{22}{7}\right]$$

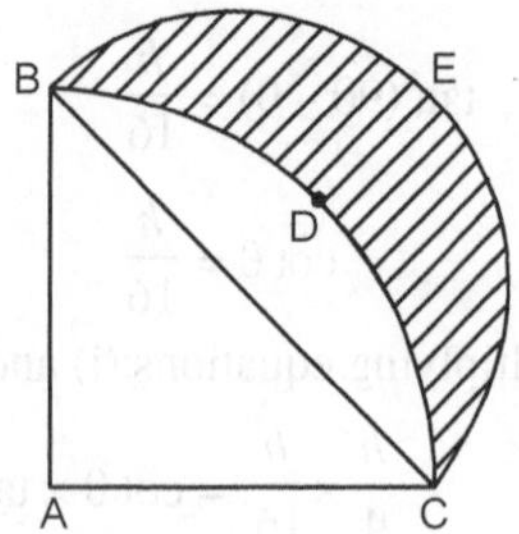

Solution : Given, radius = 28 cm

Thus, $AB = AC = 28$ cm

Now, area of quadrant $ABDC$

$$= \frac{\pi r^2}{4} = \frac{22}{7} \times \frac{(28)^2}{4}$$

$$= 616 \text{ cm}^2$$

Also, applying Pythagoras' theorem, we get

$$BC^2 = AB^2 + AC^2$$

$$\Rightarrow \quad BC^2 = (28)^2 + (28)^2$$

$$\Rightarrow \quad BC^2 = 2(28)^2$$

$$\Rightarrow \quad BC = 28\sqrt{2} \text{ cm}$$

Thus, radius of semi-circle $BEC = 14\sqrt{2}$ cm.

Hence, area of semi-circle BEC

$$= \frac{\pi r^2}{2}$$

$$= \frac{22}{7} \times \frac{(14\sqrt{2})^2}{2}$$

$$= 616 \text{ cm}^2$$

and Area of $\Delta ABC = \frac{1}{2} \times 28 \times 28 = 392 \text{ cm}^2$

Thus, the area of the shaded region

= Area of semi-circle BEC

+ Area of ΔABC

– Area of quadrant $ABCD$

$= 616 \text{ cm}^2 + 392 \text{ cm}^2 - 616 \text{ cm}^2$

$= 392 \text{ cm}^2$. **Ans.**

Q. 3. In fig. find the area of the shaded region, enclosed between two concentric circles of radii 7 cm and 14 cm where $\angle AOC = 40°$. $\left[\text{Use } \pi = \frac{22}{7}\right]$

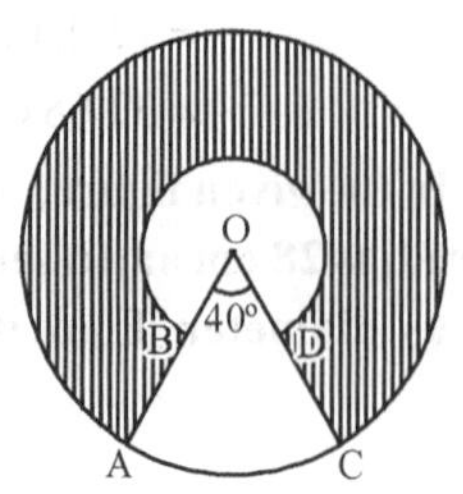

Solution : Given, $r = 7$cm and $R = 14$ cm.

$$\text{Area of shaded region} = \pi (R^2 - r^2) \times \frac{\theta}{360°}$$

$$= \frac{22}{7}(14^2 - 7^2) \times \frac{(360° - 40°)}{360°}$$

$$= \frac{22}{7} \times 7 \times 21 \times \frac{320°}{360°}$$

$$= 410.67 \text{ cm}^2$$ **Ans.**

Q. 4. In Fig, *ABCD* is a square of side 14 cm. Semi-circles are drawn with each side of square as diameter. Find the area of the shaded region. $\left[\text{Use } \pi = \frac{22}{7}\right]$

Solution : Given, a square $ABCD$ of side 14 cm

Then, Area of square $= (\text{Side})^2$

$= (14)^2 = 196 \text{ cm}^2$

$$2[\text{Area of semi-circle}] = 2\left(\frac{1}{2}\pi r^2\right)$$

$$= \pi r^2$$

$$= \frac{22}{7} \times \frac{14}{2} \times \frac{14}{2}$$

$$= 154 \text{ cm}^2$$

Now, area of shaded region

= 2[Area of square – 2(Area of semi-circle)]

= 2[196 – 154]

$= 2 \times 42 = 84 \text{ cm}^2$ **Ans.**

Q. 5. In the given figure, *ABCD* is a trapezium of area 24.5 sq. cm. $AD \parallel BC$, $\angle DAB = 90°$, $AD = 10$ cm and $BC = 4$ cm. If *ABE* is a quadrant of a circle, find the area of the shaded region. $\left[\text{Use } \pi = \frac{22}{7}\right]$.

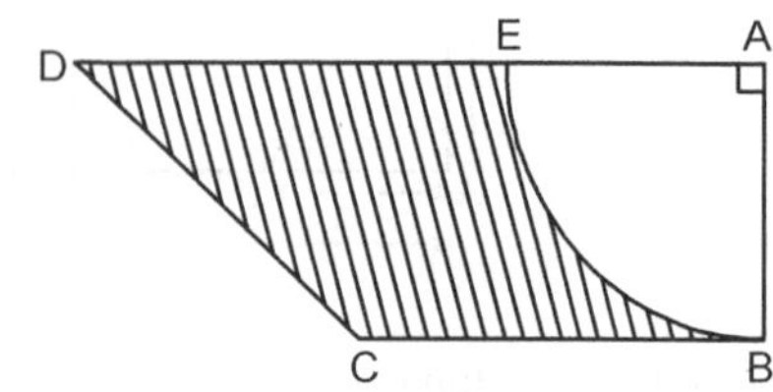

Solution : Given,

Area of $ABCD = 24.5$ sq. cm, $AD \parallel BC$,

$\angle DAB = 90°$, $AD = 10$ cm and $BC = 4$ cm

Now, area of trapezium $ABCD$

$$= \frac{1}{2} \times \text{(Sum of parallel sides)} \times \text{(Distance between parallel sides)}$$

$$24.5 = \frac{1}{2} \times (10 + 4) \times (AB)$$

$$24.5 = \frac{1}{2} \times (14) \times (AB)$$

$$AB = \frac{24.5}{7}$$

$$= 3.5 \text{ cm}$$

Again, AB is the radius of the quadrant of the circle.

So, area of quadrant $ABE = \frac{\pi r^2}{4}$

$$= \frac{22}{7} \times \frac{(3\cdot5)^2}{4} = 9\cdot625 \text{ cm}^2$$

Thus, area of the shaded region

= Area of trapezium $ABCD$

– Area of quadrant ABE

$= (24.5 - 9\cdot625)$ cm^2

$= 14.875$ cm^2. **Ans.**

Q. 6. In the given figure, two concentric circles with centre *O* have radii of 21 cm and 42 cm. If $\angle AOB = 60°$, find the area of the shaded region.★★ $\left[\text{Use } \pi = \frac{22}{7}\right]$

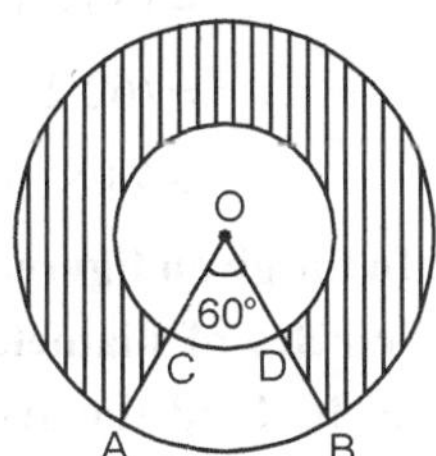

Solution : Given, $AO = OB = 42$ cm, $CO = OD = 21$ cm and $\angle AOB = 60°$.

Now, area of the larger circle

$$= \frac{22}{7}(42)^2 = \frac{22 \times 42 \times 42}{7}$$

$$= 22 \times 42 \times 6$$

$$- 5544 \text{ cm}^2$$

Area of the smaller circle

$$= \frac{22}{7}(21)^2 = \frac{22 \times 21 \times 21}{7}$$

$$= 22 \times 21 \times 3$$

$$= 1386 \text{ cm}^2$$

$$\text{Area of } ACDB = \frac{\theta}{360°} \times \pi(R^2 - r^2)$$

$$= \frac{60°}{360°} \times \frac{22}{7} \{(42)^2 - (21)^2\}$$

$$= \frac{11}{21} \{(63)(21)\}$$

$$= 693 \text{ cm}^2$$

Thus, the area of the shaded region

= Area of the larger circle

– [Area of the smaller circle

+ Area of $ACDB$]

$= [5544 - (1386 + 693)]$ cm^2

$= (5544 - 2079)$ cm^2

$= 3465$ cm^2. **Ans.**

Q. 7. A farmer has a field in the form of an equilateral triangle. He makes pits for rainwater harvesting in three corners of his field as shown in shaded part. Find the area of shaded part. $\left[\text{Use } \pi = \frac{22}{7}\right]$

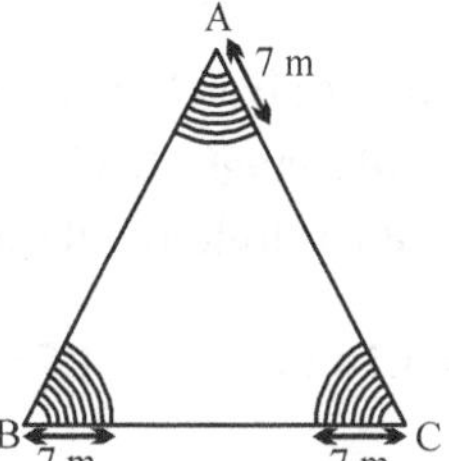

Solution : Area of one sector $= \frac{\theta}{360} \times \pi r^2$

Here $\theta = 60°$, $r = 7$ m

$$\therefore \quad \text{Area of the sector} = \frac{60°}{360°} \times \frac{22}{7} \times 7 \times 7$$

$$= \frac{1}{6} \times 227 = \frac{77}{3} \text{ m}^2$$

$$\therefore \quad \text{Area of 3 sectors} = 3 \times \frac{77}{3} = 77 \text{ m}^2$$

Hence, total area of pits (shaded part) = 77 m^2. **Ans.**

Q. 8. For the inauguration of an Eco-club to the school, badges were given to teachers. Seema made these badges in the shape of an equilateral triangle with a circle of radius $\sqrt{3}$ cm inscribed in it as shown in the figure. Find the area of the shaded portion.

$[\text{Use } \pi = 3.14]$

★★ are frequently asked board exam questions

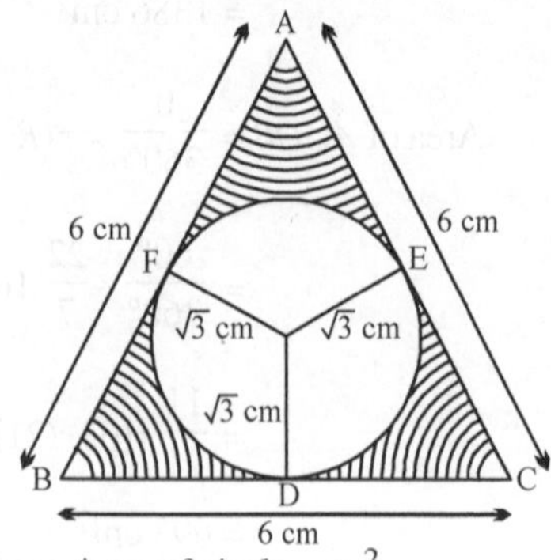

Solution : Area of circle $= \pi r^2$

$$= 3.14 \times (\sqrt{3})^2 = 9.42 \text{ cm}^2$$

Area of equilateral triangle $= \dfrac{\sqrt{3}}{4} \times \text{side}^2$

$$= \frac{\sqrt{3}}{4} \times (6)^2$$

$$= 9\sqrt{3} \text{ cm}^2$$

$$= 9 \times 1.73 = 15.57 \text{ cm}^2$$

Area of shaded part

= Area of triangle – Area of circle

= 15.57 – 9.42

$= 6.15 \text{ cm}^2$ **Ans.**

Q. 9. In the given figure, a circle is inscribed within an equilateral triangle *ABC* of side 12 cm. Find the radius of the inscribed circle and the area of the shaded region.

$\left[\textbf{Use } \pi = 3.14 \textbf{ and } \sqrt{3} = 1.73\right]$

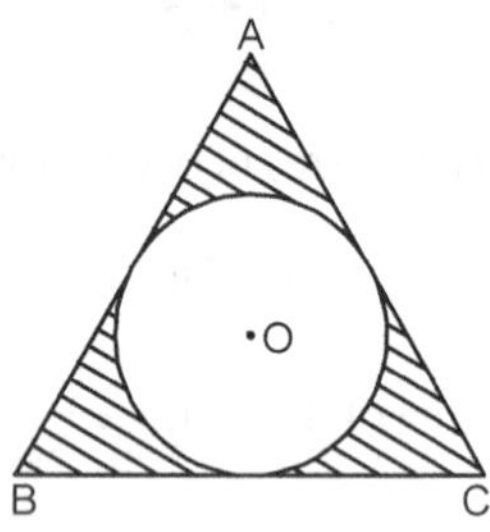

Solution : Let AD be the perpendicular to BC.

Thus, D is the mid-point of BC.

$\therefore$ $BD = DC = 6$ cm

Also, $AB = 12$ cm (Given)

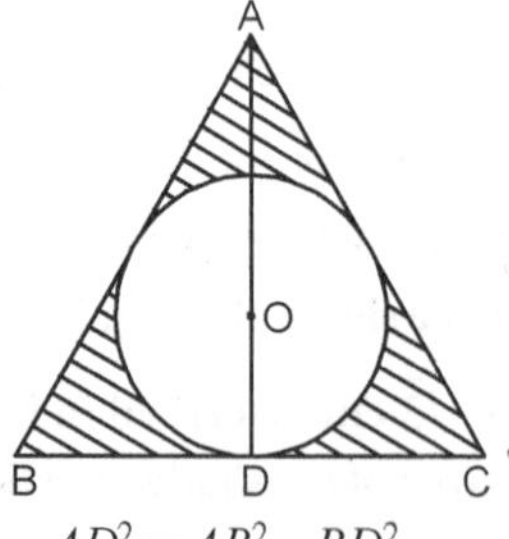

Hence, $AD^2 = AB^2 - BD^2$

[By Pythagoras' Theorem]

$\Rightarrow$ $AD^2 = (12)^2 - 6^2$

$\Rightarrow$ $AD^2 = 144 - 36$

$\Rightarrow$ $AD^2 = 108$ cm

$\Rightarrow$ $AD = \sqrt{108} \text{ cm} = 6\sqrt{3} \text{ cm}$

As O is the centre of the circle, it is also the centroid of the equilateral triangle.

Thus, $AO : OD = 2 : 1$

Hence, the sum of the ratios,

$$AD = AO + OD = 2 + 1 = 3$$

and $OD = \dfrac{AD}{3}$

$$= \frac{6\sqrt{3}}{3} = 2\sqrt{3} \text{ cm} = r$$

Now, area of $\Delta ABC = \dfrac{\sqrt{3}a^2}{4}$

$$= \frac{\sqrt{3}(12)^2}{4}$$

$$= \frac{144\sqrt{3}}{4}$$

$$= 36\sqrt{3} \text{ cm}^2$$

and Area of the circle $= \pi r^2$

$$= \pi(2\sqrt{3})^2$$

$$= 12\pi \text{ cm}^2$$

Thus, the area of the shaded region

= Area of ΔABC – Area of the circle

$= (36 \times 1.73) - (3.14 \times 12) \text{ cm}^2$

$= (62.28 - 37.68) \text{ cm}^2$

$= 24.6 \text{ cm}^2$. **Ans.**

Q. 10. In the given figure, *PSR*, *RTQ* and *PAQ* are three semi-circles of diameters 10 cm, 3 cm and 7 cm respectively. Find the area of the shaded region and the perimeter of the boundary of the shape. [Use π = 3.14]

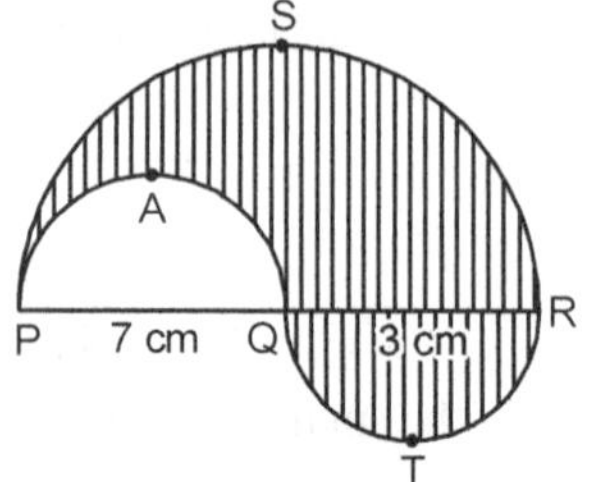

Solution : Given, Diameter of $PSR = 10$ cm

$\therefore$ Radius of $PSR = 5$ cm

Diameter of $PAQ = 7$ cm

$\therefore$ Radius of $PAQ = 3.5$ cm

Diameter of $RTQ = 3$ cm

$\therefore$ Radius of $RTQ = 1.5$ cm

Hence, Area of the shaded region

$$= \text{Area of } PSR + \text{Area of } RTQ - \text{Area of } PAQ$$

$$= \frac{3.14}{2}[(5)^2 + (1.5)^2 - (3.5)^2]\text{ cm}^2$$

$$= \frac{3.14}{2}[25 + 2.25 - 12.25]\text{ cm}^2$$

$$= 1.57[15]\text{ cm}^2$$

$$= 23.55\text{ cm}^2$$

and Perimeter of the figure

$$= \text{Perimeter of } PSR + \text{Perimeter of } RTQ + \text{Perimeter of } PAQ$$

$$= 3.14\,(5 + 1.5 + 3.5)\text{ cm}$$

$$= 3.14\,(10)\text{ cm}$$

$$= 31.4\text{ cm.}$$ **Ans.**

Q. 11. In the given figure from a rectangular region *ABCD* with *AB* = 20 cm, a right-angled triangle, *AED* with *AE* = 9 cm and *DE* = 12 cm is cut off. On the other end taking *BC* as diameter, a semi-circle is added on the outside of the rectangle. Find the area of the shaded region.

[Use $\pi = 3.14$]

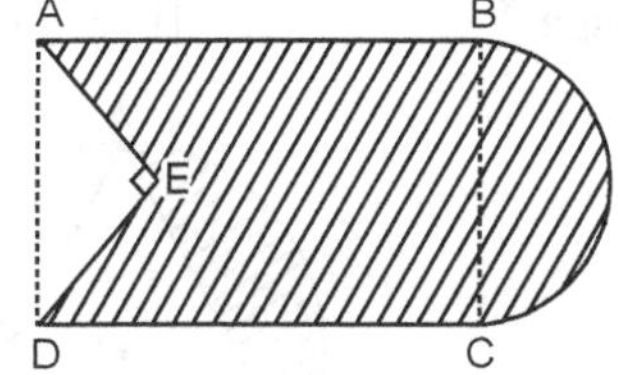

Solution : Given, $AB = 20$ cm, $AE = 9$ cm and $DE = 12$ cm. Also, BC is the diameter of the semi-circle.

In ΔAED,

Applying Pythagoras' theorem,

$$AD^2 = AE^2 + DE^2$$

$\Rightarrow$ $AD^2 = (9)^2 + (12)^2$

$\Rightarrow$ $AD^2 = 81 + 144$

$\Rightarrow$ $AD^2 = 225$

$\Rightarrow$ $AD = 15$ cm

[as length cannot be negative]

So Area of rectangle $= 20 \times 15 = 300\text{ cm}^2$

and area of triangle $= \frac{1}{2} \times 9 \times 12 = 54\text{ cm}^2$

Also, $BC = AD = 15$ cm

So, $r = 7.5$ cm

$\therefore$ Area of the semi-circle

$$= \frac{\pi r^2}{2} = \frac{(3.14)(7.5)^2}{2}$$

$$= 1.57(56.25)$$

$$= 88.31\text{ cm}^2$$

Hence, area of the shaded region

$$= \text{Area of rectangle} + \text{Area of semi-circle} - \text{Area of triangle}$$

$$= (300 + 88.31 - 54)\text{ cm}^2$$

$$= 334.31\text{ cm}^2.$$ **Ans.**

Q. 12. All the vertices of a rhombus lie on a circle. Find the area of the rhombus, if the area of the circle is 1256 cm². **[Use $\pi = 3.14$]**

Solution : Given that the area of the circle is 1256 cm².

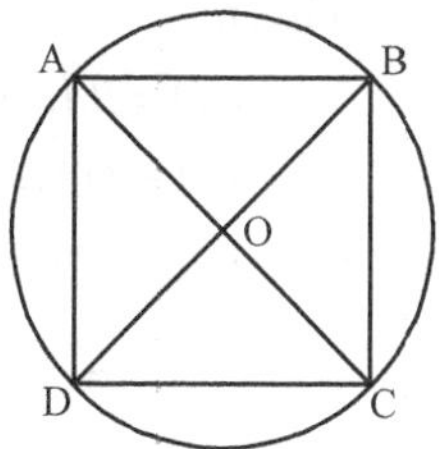

$\because$ Area of the circle $= \pi r^2$

$\Rightarrow$ $1256 = 3.14 \times r^2$

$$1256 = \frac{314}{100} \times r^2$$

$$r^2 = \frac{1256 \times 100}{314}$$

$$r = \sqrt{400}$$

$$r - 20\text{ cm}$$

Now, $ABCD$ are the vertices of a rhombus.

$\therefore$ $\angle A = \angle C$ ...(i)

[$\because$ opposite angles of rhombus are equal]

But $ABCD$ lie on the circle.

So, $ABCD$ is called cyclic quadrilateral

$\therefore$ $\angle A + \angle C = 180°$ ...(ii)

$\Rightarrow$ $\angle A + \angle A = 180°$ [using (i)]

$$2\angle A = 180°$$

$$\angle A = 90°$$

and $\angle C = 90°$

So, BD is a diameter of circle.

[$\because$ The angle in a semi-circle is a right triangle]

Now, Area of rhombus $= \frac{1}{2} \times$ Product of diagonals

$$= \frac{1}{2} \times 40 \times 40$$

$= 800 \text{ cm}^2$

Hence, Area of rhombus is 800 cm^2. **Ans.**

Q. 13. In the given figure, *AB* and *CD* are the two diameters of the circle with centre *O* which are perpendicular to each other. *OB* is the diameter of the smaller circle. If *OA* = 7 cm, find the area of the shaded region. $\left[\text{Use } \pi = \frac{22}{7}\right]$

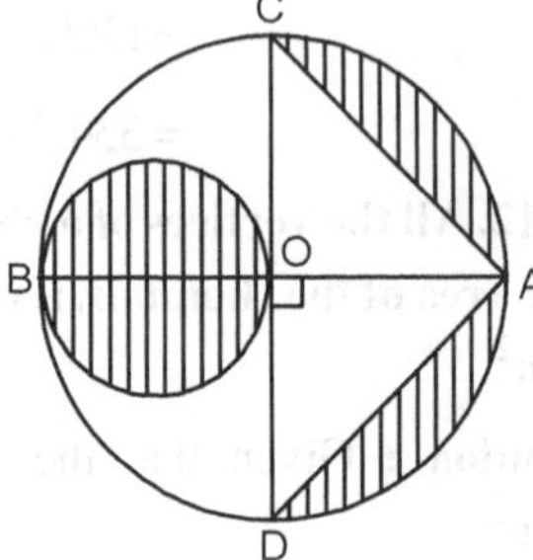

Solution : Given,

$CD = BA, OA = 7 \text{ cm}$

Now, $OA = OB = CO = DO = 7 \text{ cm}$

Area of semi-circle *CAD*

$$= \frac{\pi r^2}{2} = \frac{1}{2} \times \frac{22}{7} \times (7)^2$$

$$= 77 \text{ cm}^2$$

$$\text{Area of } \Delta CAD = \frac{1}{2} \times b \times h$$

$$= \frac{1}{2} \times 14 \times 7 = 49 \text{ cm}^2$$

$$\text{Radius of the smaller circle} = \frac{7}{2} \text{ cm} = 3.5 \text{ cm}$$

Thus, area of the smaller circle

$$= \pi r^2 = \frac{22}{7} \times (3.5)^2$$

$$= 38.5 \text{ cm}^2$$

Hence, area of the shaded region

= Area of smaller circle + (Area of semi-circle – Area of triangle)

$$= [38.5 + (77 - 49)] \text{ cm}^2$$

$$= 66.5 \text{ cm}^2.$$ **Ans.**

Q. 14. In a circle of radius 21 cm, an arc subtends an angle of 60° at the centre. Find : (i) the length of the arc, (ii) area of the sector formed by the arc. $\left[\text{Use } \pi = \frac{22}{7}\right]$

★ are board exam questions from previous years

Solution : Given, $r = 21$ cm and angle of the arc = 60°

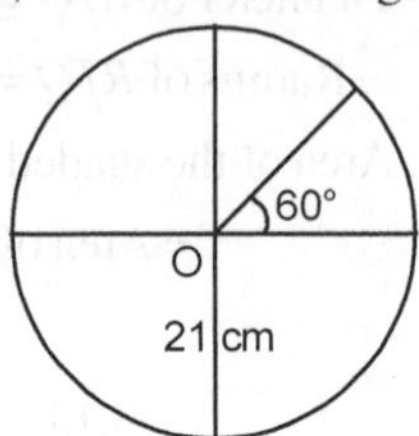

(i) Length of the arc

$$= \frac{\theta}{360°} \times 2\pi r$$

$$= \frac{60°}{360°} \times 2\left(\frac{22}{7}\right)(21)$$

$$= 22 \text{ cm}$$ **Ans.**

(ii) Area of the sector formed by the arc

$$= \frac{\theta}{360°} \times \pi r^2$$

$$= \frac{60°}{360°} \times \left(\frac{22}{7}\right)(21)^2$$

$$= 231 \text{ cm}^2.$$ **Ans.**

Q. 15. On a circular table cover of radius 42 cm, a design is made by a girl leaving an equilateral triangle *ABC* in the middle as shown in the figure. It was decided that the payment be made to the girl be proportional to the covered area of the design. Find the covered area of the design.★ **[Use $\sqrt{3} = 1.73$ and $\pi = \frac{22}{7}$]**

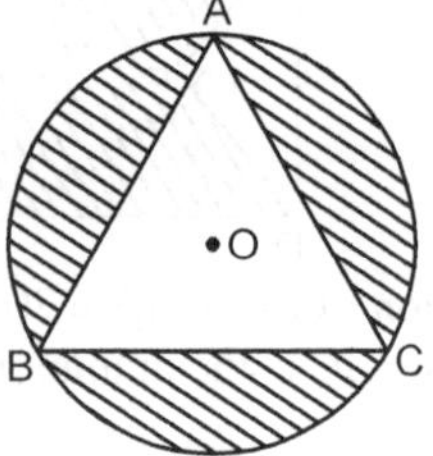

Solution : Given,

Radius = 42 cm

So, area of the circle $= \pi r^2 = \frac{22}{7} \times 42 \times 42 \text{ cm}^2$

$= 5544 \text{ cm}^2$

Now, join *AD* such that it passes through *O* and is perpendicular upon *BC* at *D*.

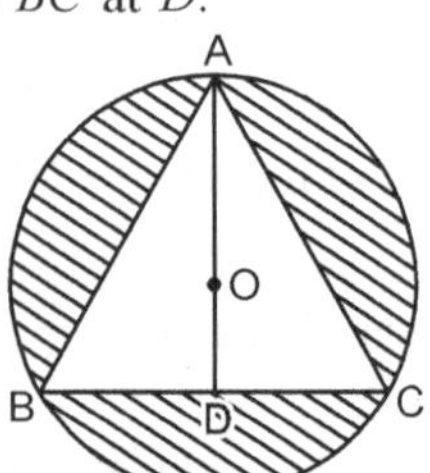

As *O* is the centre of the circle, it is also the centroid of the equilateral triangle.

Thus, $AO : OD = 2 : 1$

$$\Rightarrow \quad OD = \frac{AO}{2} = \frac{42}{2} = 21 \text{ cm}$$

$$\Rightarrow \quad AD = OA + OD$$

$$= 42 + 21 = 63 \text{ cm}$$

Thus, as ABC is an equilateral triangle,

$$\angle A = \angle B = \angle C = 60°$$

Thus, $\sin 60° = \frac{AD}{AB}$

$$\Rightarrow \quad \frac{\sqrt{3}}{2} = \frac{63}{AB}$$

$$\Rightarrow \quad AB = \frac{126}{\sqrt{3}} \text{ cm} = a$$

Now, area of the equilateral triangle

$$= \frac{\sqrt{3}a^2}{4}$$

$$= \frac{\sqrt{3}}{4} \times \frac{126}{\sqrt{3}} \times \frac{126}{\sqrt{3}}$$

$$= \frac{3969}{\sqrt{3}}$$

$$= 2294.22 \text{ cm}^2$$

So the area covered by the design

= Area of circle

– Area of equilateral triangle

$$= (5544 - 2294.22) \text{ cm}^2$$

$$= 3249.78 \text{ cm}^2.$$ **Ans.**

Q. 16. *ABCD* is a field in the shape of a trapezium. *AD* || *BC*, ∠*ABC* = 90° and ∠*ADC* = 60°. Four sectors are formed with centres *A*, *B*, *C* and *D* as shown in figure. The radius of each sector is 14 m.

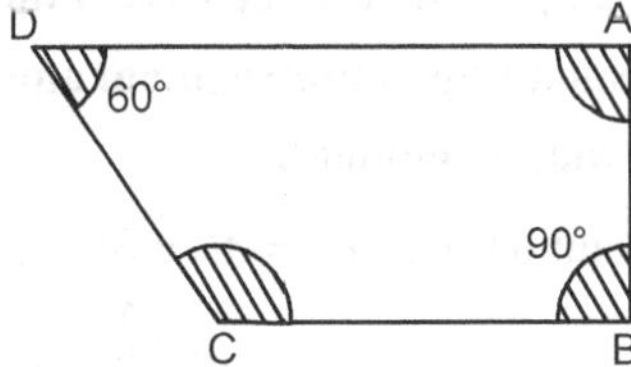

Find the following :

(i) Total area of the four sectors.

(ii) Area of the remaining portion given that *AD* = 55 m, *BC* = 45 m and *AB* = 30 m.

Solution : Area of quadrant A

= Area of quadrant B

$$= \frac{\pi r^2}{4} = \frac{1}{4} \times \frac{22}{7} \times (14)^2$$

$$= 154 \text{ cm}^2$$

$$\text{Area of sector } D = \frac{\theta}{360°} \times \pi r^2$$

$$= \frac{60°}{360°} \times \frac{22}{7} \times (14)^2$$

$$= \frac{1}{6} \times \frac{22}{7} \times (14)^2$$

$$= 102.67 \text{ cm}^2$$

$$\text{Angle of sector } C = 360° - (90° + 90° + 60°)$$

$$= 120°$$

$$\therefore \quad \text{Area of sector } C = \frac{\theta}{360°} \times \pi r^2$$

$$= \frac{120°}{360°} \times \frac{22}{7} \times (14)^2$$

$$= \frac{1}{3} \times \frac{22}{7} \times (14)^2$$

$$= 205.33 \text{ cm}^2$$

(i) Total area of the four sectors

$$= (154 + 154 + 102.67 + 205.33) \text{ cm}^2$$

$$= 616 \text{ cm}^2$$

(ii) Area of the remaining portion

= Area of the trapezium

– Total area of the four sectors

$$= \left[\frac{1}{2} \times (55 + 45) \times (30) - 616\right] \text{ cm}^2$$

$$= \left[\frac{1}{2} \times (100) \times (30) - 616\right] \text{ cm}^2$$

$$= [1500 - 616] \text{ cm}^2$$

$$= 884 \text{ cm}^2.$$ **Ans.**

Q. 17. In the given figure, *O* is the centre of the circle, *AC* = 24 cm, *AB* = 7 cm and ∠*BOD* = 90°. Find the area of the shaded region. [Use π = 3.14].

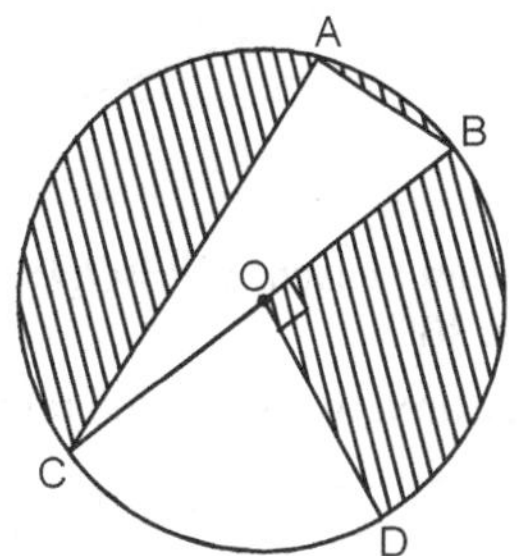

Solution : Given, AC = 24 cm, AB = 7 cm and $\angle BOD = 90°$

Also, the angle in a semi-circle is always 90°. So

$$\angle CAB = 90°$$

Thus, ΔABC is a right-angled triangle.

$$\text{Hence, Area of } \Delta ABC = \frac{1}{2} \times 24 \times 7 = 84 \text{ cm}^2$$

$$\text{Now,} \quad CB^2 = AB^2 + AC^2$$

[By Pythagoras' Theorem]

$\Rightarrow \qquad CB^2 = (7)^2 + (24)^2$

$\Rightarrow \qquad CB^2 = 49 + 576 = 625$

$\Rightarrow \qquad CB = 25$ cm

[$\because$ length cannot be negative]

As CB passes through the centre and touches opposite ends of the circle, it is the diameter.

So, radius = 12.5 cm = OD

Hence, area of the circle

$= \pi(12.5)^2$

$= 3.14 \times 12.5 \times 12.5$

$= 490.63$ cm^2

and area of quadrant COD

$$= \frac{\pi r^2}{4} = \frac{1}{4} \times 3.14 \times (12.5)^2$$

$= 122.66$ cm^2

Hence, area of the shaded region

= Area of the circle – [Area of the ΔABC + Area of quadrant COD]

$= 490.63 - [84 + 122.66]$ cm^2

$= 283.97$ cm$^2 \approx 284$ cm^2. **Ans.**

Q. 18. In the given figure, *PQ* and *AB* respectively are the arcs of two concentric circles having radii 7 cm and 3.5 cm and centre *O*. If $\angle POQ = 30°$, then find the area of the shaded region. $\left[\text{Use } \pi = \frac{22}{7}\right]$

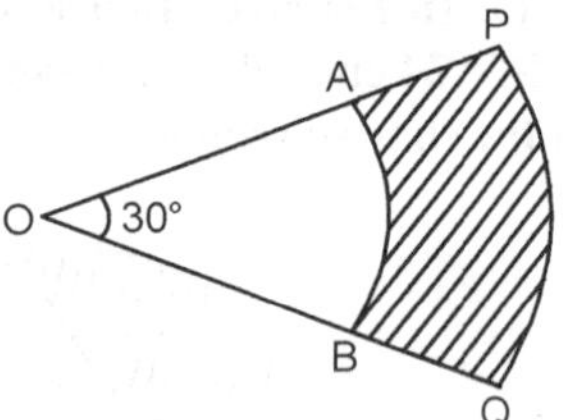

Solution : Given, $OP = OQ = 7$ cm, $OA = OB = 3.5$ cm and $\angle O = 30°$

Thus, area of the shaded region

$$= \frac{\theta}{360°} \times \pi(R^2 - r^2)$$

$$= \frac{30°}{360°} \times \frac{22}{7} (7^2 - 3.5^2)$$

$$= \frac{1}{12} \times \frac{22}{7} (36{\cdot}75)$$

$= 9{\cdot}625$ cm^2. **Ans.**

Q. 19. In fig. find the area of the shaded region. **[Use $\pi = 3.14$]**

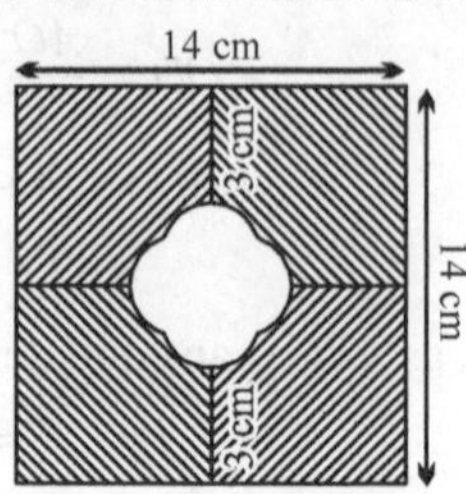

Solution : Area of square ($ABCD$)

$= 14 \times 14$

$= 196$ cm^2

Area of small square (EFGH)

$= (7 - 3)^2$

$= 4 \times 4 = 16$ cm^2

$$4(\text{Area of semicircle}) = 4 \times \frac{\pi r^2}{2}$$

$= 2 \times 3.14 \times 2 \times 2$

$= 25.12$ cm^2

$\therefore$ Required area (shaded) = Area of square – Area of small square – 4 × Area of semicircle.

$= (196 - 16 - 25.12)$ cm^2

$= 154.88$ cm^2 **Ans.**

Q. 20. A chord of a circle of radius 10 cm subtends a right angle at the centre. Find the area of the corresponding minor segment and hence find the area of the major segment. **[Use $\pi = 3.14$]**

Solution : Given, $r = AO = BO = 10$ cm and $\angle AOB = 90°$

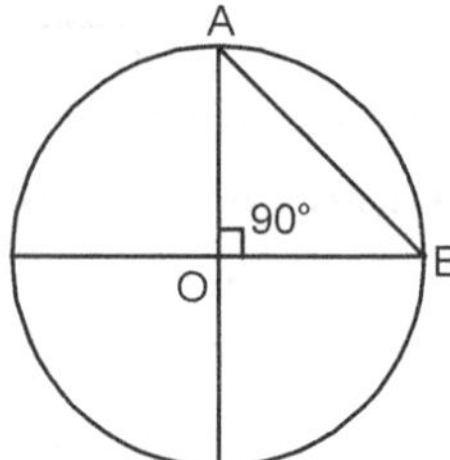

Thus, the area of the minor segment

$$= \left\{\frac{90°}{360°} \times \pi - \frac{1}{2} \sin 90°\right\} (10)^2$$

$$= \left\{\frac{3.14}{4} - \frac{1}{2}\right\} (10)^2$$

$$= \left\{\frac{1.57}{2} - \frac{1}{2}\right\} (10)^2$$

$$= \left\{\frac{0.57}{2}\right\} 100$$

$$= \frac{57}{2} = 28.5 \text{ cm}^2$$

Thus, the area of the major segment

= Area of the circle

– Area of the minor segment

$= (\pi r^2 - 28.5) \text{ cm}^2$

$= [3.14(10)^2 - 28.5] \text{ cm}^2$

$= [314 - 28.5] \text{ cm}^2$

$= 285.5 \text{ cm}^2$. **Ans.**

Q. 21. Three semi-circles each of diameter 3 cm and a circle of diameter 4.5 cm and a semi-circle of radius 4.5 cm are drawn in the given figure. Find the area of the shaded region.

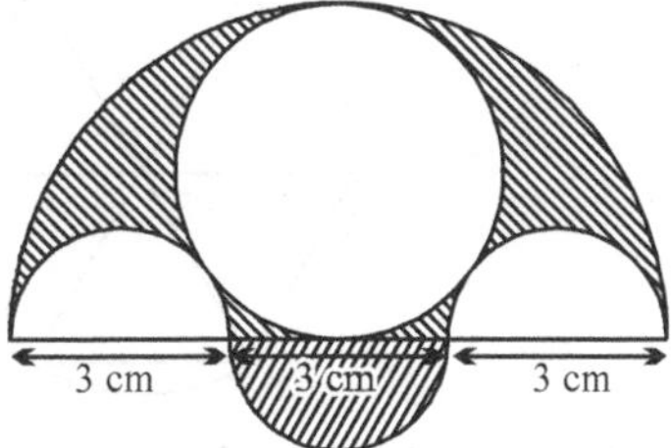

Solution : Given, radius of large semi-circle = 4.5 cm

$\therefore$ Area of large semi-circle $= \frac{1}{2}\pi R^2$

$$= \frac{1}{2} \times \frac{22}{7} \times 4.5 \times 4.5$$

Diameter of inner circle = 4.5 cm

$\Rightarrow$ $r = \frac{4.5}{2}$ cm

$\therefore$ Area of inner circle $= \pi r^2$

$$= \frac{22}{7} \times \frac{4.5}{2} \times \frac{4.5}{2}$$

Diameter of small semi-circle= 3 cm.

$r = \frac{3}{2}$ cm

$\therefore$ Area of small semi-circle $= \frac{1}{2}\pi r^2$

$$= \frac{1}{2} \times \frac{22}{7} \times \frac{3}{2} \times \frac{3}{2}$$

Area of shaded region = Area of large semi-circle + Area of 1small semi-circle – Area of inner circle – Area of 2 small semi-circle.

$$= \frac{1}{2} \times \frac{22}{7} \times 4.5 \times 4.5 + \frac{1}{2} \times \frac{22}{7} \times \frac{3}{2} \times \frac{3}{2}$$

$$- \frac{22}{7} \times \frac{4.5}{2} \times \frac{4.5}{2} - 2 \times \frac{1}{2} \times \frac{22}{7} \times \frac{3}{2} \times \frac{3}{2}$$

$$= \frac{1}{2} \times \frac{22}{7} \times \left[20.25 + \frac{9}{4}\right] - \frac{22}{7}\left[\frac{20.25}{4} + \frac{9}{4}\right]$$

$$= \frac{11}{7} \times \frac{90}{4} - \frac{22}{7} \times \frac{29.25}{4}$$

$$= \frac{990 - 643.5}{28} = \frac{346.5}{28}$$

$= 12.375 \text{ cm}^2$ (approx) **Ans.**

Q. 22. Find the area of *RSP* and *PTQ* in the given figure, if *PR* = 24 cm, *PQ* = 7 cm and *O* is the centre of the circle.

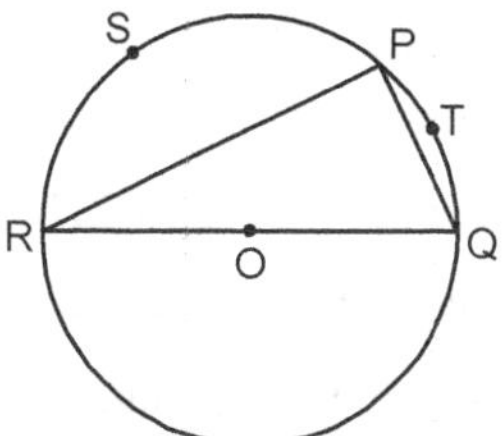

Solution : Given, $PR = 24$ cm, $PQ = 7$ cm and O is the centre of the circle.

Thus, area of $\Delta RPQ = \frac{1}{2} \times RP \times PQ$

$$= \frac{1}{2} \times 24 \times 7 = 84 \text{ cm}^2$$

Now, applying Pythagoras' theorem,

$RQ^2 = RP^2 + PQ^2$

$= (24)^2 + (7)^2$

$= 576 + 49$

$= 625$

$\Rightarrow$ $RQ = 25$ cm

[as length cannot be negative]

Thus, $OQ = 12.5$ cm

Thus, area of semi-circle RPQ

$$= \frac{\pi}{2}r^2 = \frac{1}{2} \times \frac{22}{7} \times (12.5)^2 \ (12.5)^2$$

$= 245.54 \text{ cm}^2$

Hence, area of RSP and PTQ

= Area of semi-circle RPQ

– Area of triangle RPQ

$= (245.54 - 84) \text{ cm}^2$

$= 161.54 \text{ cm}^2$. **Ans.**

Q. 23. Four equal circles are described about the four corners of a square such that each touches two of the others as shown in the figure. Find the area of the region at the centre if each side of the square measures 14 cm. [NCERT]

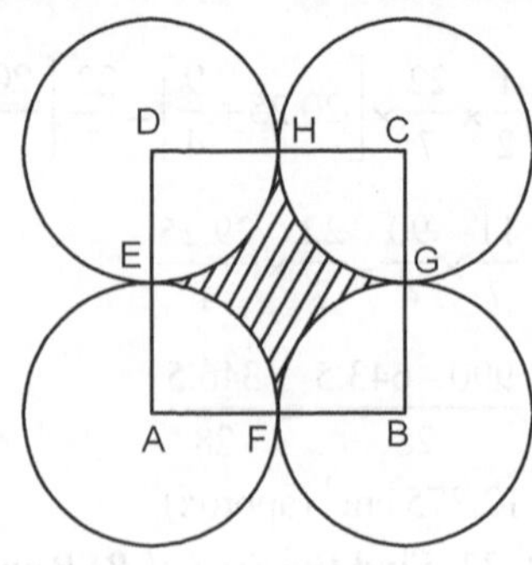

Solution : Given, $AB = BC = CD = DA = 14$ cm

Also, radius of circle with centre A = radius of circle with centre B = radius of circle with centre C = radius of circle with centre D

Hence, $DH = HC = CG = GB = FB = AF = AE = ED = 7$ cm

Now, area of square $ABCD = (14 \times 14)$ cm^2 = 196 cm^2

and area of the 4 quadrants

$$= 4\left(\frac{\pi}{4}r^2\right)$$

$$= 4 \times \frac{1}{4} \times \frac{22}{7} \times (7)^2$$

$$= 154 \text{ cm}^2$$

Hence, the area of the central region

$$= (196 - 154) \text{ cm}^2$$

$$= 42 \text{ cm}^2.$$ **Ans.**

Q. 24. Find the area of the minor segment of a circle of radius 14 cm, when its central angle is 60°. Also find the area of the corresponding major segment.

$$\left[\textbf{Use } \pi = \frac{22}{7}\right]$$

Solution : Let ACB be the given arc subtending an angle of 60° at the centre.

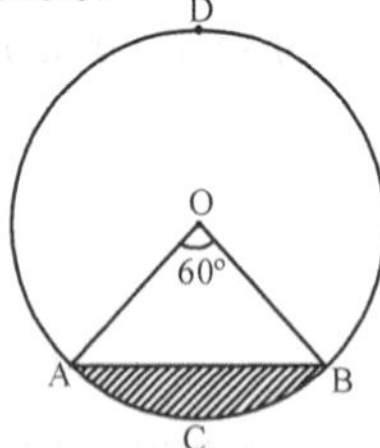

Here, $r = 14$ cm and $\theta = 60°$.

Area of the minor segment $ACBA$

= (Area of the sector $OACBO$) – (area of ΔOAB)

$$= \frac{\pi r^2 \theta}{360°} - \frac{1}{2}r^2 \sin\theta$$

$$= \frac{22}{7} \times 14 \times 14 \times \frac{60°}{360°} - \frac{1}{2} \times 14 \times 14 \times \sin 60°$$

$$= \frac{308}{3} - 7 \times 14 \times \frac{\sqrt{3}}{2}$$

$$= \frac{308}{3} - 49\sqrt{3}$$

$$= 17.89 \text{ cm}^2$$

Area of the major segment $BDAB$

= Area of circle – Area of minor segment $ACBA$

$$= \pi r^2 - 17.89$$

$$= \frac{22}{7} \times 14 \times 14 - 17.89$$

$$= 616 - 17.89$$

$$= 598.11 \approx 598 \text{ cm}^2.$$ **Ans.**

Q. 25. A round table cover had six equal designs as shown in the given figure. If the radius of the cover is 35 cm, then find the total area of the remaining portion.

[Use $\sqrt{3} = 1.732$ and $\pi = 3.14$]

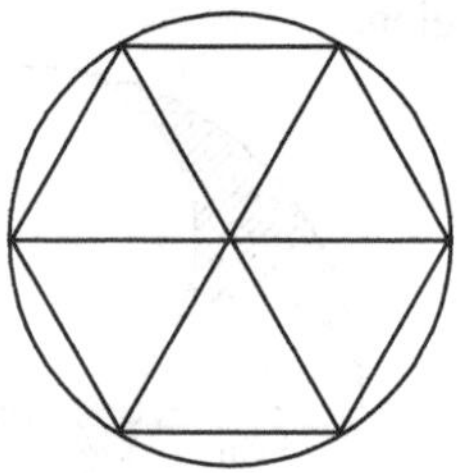

Solution : Given, $r = 35$ cm.

As the inscribed polygon is a hexagon of equal sides, the angle subtended at the centre by each side is 60°.

Thus, all the triangles are equilateral and the length of each side is 35 cm.

Hence, the total surface area of the remaining portion

= Area of the circle

– Area of the hexagon

$$= \pi r^2 - 6 \times \frac{\sqrt{3}}{4}r^2$$

$$= \pi r^2 - \frac{3\sqrt{3}}{2}r^2$$

$$= r^2\left(3.14 - \frac{3(1.732)}{2}\right)$$

$$= (35)^2(0.542)$$

$$= 663.95 \text{ cm}^2.$$ **Ans.**

Q. 26. Calculate the area other than the area common between two quadrants of circles of radius 16 cm each which is shown as the shaded region in the given figure.

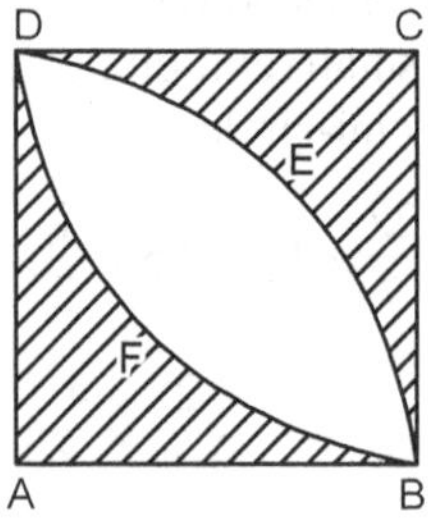

Solution : Area of $ABCD = (16 \times 16)\ \text{cm}^2 = 256\ \text{cm}^2$

Area of quadrant $DFBC$

$$= \text{Area of quadrant } DEBA$$

$$= \frac{\pi}{4}r^2 = \frac{1}{4} \times \frac{22}{7}(16)^2$$

$$= 201.14\ \text{cm}^2$$

Now, Area of the shaded region

= [Area of square $ABCD$ – Area of quadrant $DFBC$]

+ [Area of square $ABCD$ – Area of quadrant $DEBA$]

$= [(256 - 201.14) + (256 - 201.14)]\ \text{cm}^2$

$= 109.72\ \text{cm}^2$. **Ans.**

Q. 27. In the given figure, OE = 20 cm. In sector *OSFT*, square *GOEF* is inscribed. Find the area of the shaded region. [Use $\pi = 3.14$].

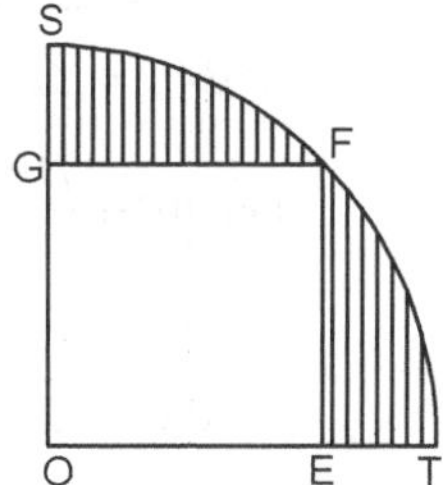

Solution : Given, $OE = 20$ cm and $GOEF$ is a square.

Thus, OF is a diagonal of the square as well as the radius of the quadrant.

So, Area of the square $= 20 \times 20\ \text{cm}^2 = 400\ \text{cm}^2$

Applying Pythagoras' theorem,

$$OF^2 = OE^2 + EF^2$$

$\Rightarrow$ $OF^2 = (20)^2 + (20)^2$

$\Rightarrow$ $OF^2 - 400 + 400$

$\Rightarrow$ $OF^2 = 800$

$\Rightarrow$ $r^2 = 800$

Thus, area of the quadrant

$$= \frac{\pi r^2}{4} = \frac{\pi 800}{4} = 200\pi$$

$$= 200(3.14)$$

$$= 628\ \text{cm}^2$$

Hence, the area of the shaded region

= Area of the quadrant

– Area of the square

$= (628 - 400)\ \text{cm}^2$

$= 228\ \text{cm}^2$. **Ans.**

Q. 28. Calculate the area of the shaded portion. Each of the quadrants have radius = 7 cm. $\left[\text{Use } \pi = \frac{22}{7}\right]$

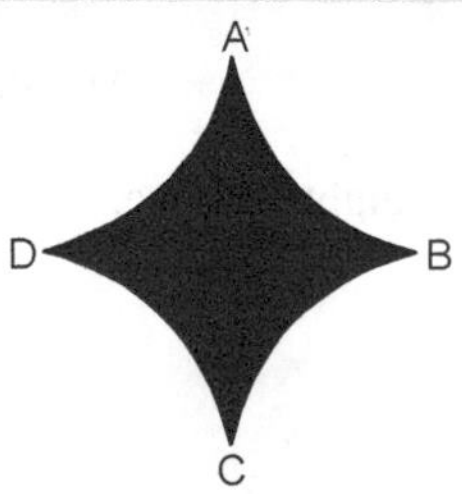

Solution : Join $AM, AN, NB, BO, OC, CP, PD$ and DM such that $AM = AN = NB = BO = OC = CP = PD = DM = 7$ cm

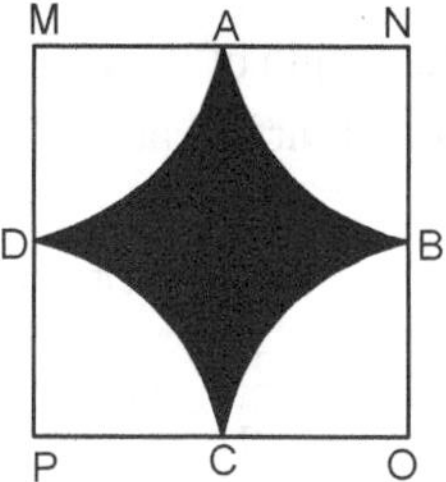

Thus, $MN = MA + AN = 14$ cm

and $MP = MD + DP = 14$ cm

Similarly, $PO = PC + CO = 14$ cm

and $ON = OB + BN = 14$ cm

Hence, $MNOP$ is a square.

Thus, area of $MNOP = 14 \times 14\ \text{cm}^2 = 196\ \text{cm}^2$.

Now, $r = 7$ cm

Hence,

Area of quadrant ($MDA = DPC = COB = BNA$)

$$= \frac{1}{4} \times \frac{22}{7} \times 7 \times 7\ \text{cm}^2$$

$$= \frac{77}{2}\ \text{cm}^2$$

Hence,

Area of quadrants ($MDA + DPC + COB + BNA$)

$$= 4 \times \frac{77}{2}\ \text{cm}^2$$

$$= 154\ \text{cm}^2$$

Thus, Area of the shaded region

= Area of $MNOP$ – Area of quadrants

($MDA + DPC + COB + BNA$)

$= (196 - 154)\ \text{cm}^2 = 42\ \text{cm}^2$. **Ans.**

Chapter 13. Surface Areas and Volumes

Q. 1. A 5 m wide cloth is used to make a conical tent of base diameter 14 m and height 24 m. Find the cost of cloth used at the rate of ₹ 25 per m. $\left[\text{Use } \pi = \frac{22}{7}\right]$

Solution : Given, width of cloth = 5 m

Base diameter of tent = 14 m

$\therefore$ Radius $= \frac{14}{2} = 7$ cm

Height = 24 m

Slant height of cone, $l' = \sqrt{h^2 + r^2}$

$= \sqrt{(24)^2 + (7)^2}$

$= \sqrt{576 + 49}$

$= \sqrt{625}$

$= 25$ cm

Let the length of the cloth be l m.

$\therefore$ Curved surface area of cone

= Area of cloth

$\Rightarrow \quad \pi r l' = l \times b$

$\Rightarrow \quad \frac{22}{7} \times 7 \times 25 = l \times 5$

$\Rightarrow \quad 22 \times 25 = l \times 5$

$\Rightarrow \quad l = \frac{22 \times 25}{5}$ m

$\Rightarrow \quad l = 110$ m

Thus, total cost of the cloth = ₹ (25 × 110) = ₹ 2750.

Ans.

Q. 2. A girl empties a cylindrical bucket, full of sand, of base radius 18 cm and height 32 cm on the floor, to form a conical heap of sand. If the height of this heap is 24 cm, find the slant height correct to two decimal places.★★

Solution : Given, base radius of cylindrical bucket = 18 cm

Height of cylindrical bucket = 32 cm

Height of conical heap = 24 cm

Let the radius of the cone be r cm.

Now, Volume of cylinder = Volume of cone

$\Rightarrow \quad \pi(18)^2 \times 32 = \frac{1}{3}\pi r^2 \times 24$

$\Rightarrow \quad r^2 = \frac{18 \times 18 \times 32 \times 3}{24}$

$= 18 \times 18 \times 4$

$\Rightarrow \quad r = \sqrt{18 \times 18 \times 2 \times 2}$

$= 18 \times 2 = 36$ cm

Thus, slant height of the cone

$= \sqrt{(24)^2 + (36)^2}$

$= \sqrt{576 + 1296}$

$= \sqrt{1872}$

$= 43.27$ cm. **Ans.**

★ are board exam questions from previous years

★★ are frequently asked board exam questions

Q. 3. A solid metallic right-circular cone, 20 cm high, whose vertical angle is 60° is cut into two parts at the middle of its heights by a plane parallel to its base. If the frustum so obtained be drawn into a wire of diameter $\frac{1}{12}$ cm, find the length of the wire.★

Solution : Given $AP = 20$ cm, $AO = OP = 10$ cm and $\angle BAC = \angle DAE = 60°$

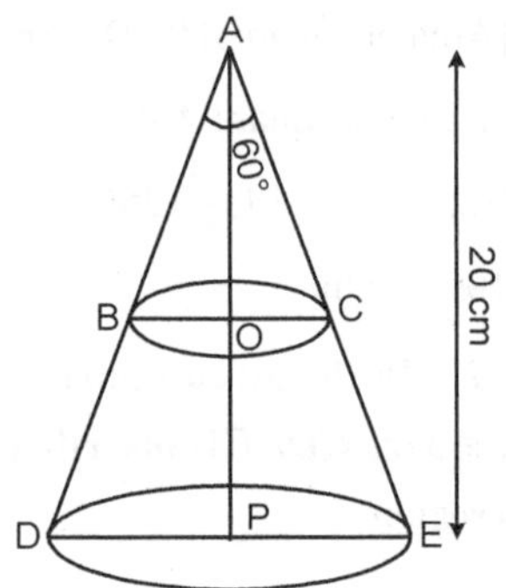

Thus, $\quad \angle BAO = \angle DAP = 30°$

Now, $\quad \tan 30° = \frac{DP}{AP}$

$\Rightarrow \quad \frac{1}{\sqrt{3}} = \frac{DP}{20}$

$\Rightarrow \quad DP = \frac{20}{\sqrt{3}}$ cm

Thus, $\quad r_1 = \frac{20}{\sqrt{3}}$ cm

Also, $\quad \tan 30° = \frac{OB}{OA}$

$\Rightarrow \quad \frac{1}{\sqrt{3}} = \frac{OB}{10}$

$\Rightarrow \quad OB = \frac{10}{\sqrt{3}}$ cm

Thus, $\quad r_2 = \frac{10}{\sqrt{3}}$ cm

Hence, volume of the frustum $BCDE$

$\frac{\pi}{3}h(r_1^2 + r_2^2 + r_2 r_2)$

$= \frac{10\pi}{3}\left(\frac{400}{3} + \frac{100}{3} + \frac{20}{\sqrt{3}} \times \frac{10}{\sqrt{3}}\right)$ cm^3

$= \frac{10\pi}{3}\left(\frac{500}{3} + \frac{200}{3}\right)$ cm^3

$= \frac{10\pi}{3}\left(\frac{700}{3}\right)$ cm^3

Diameter of the wire = $\frac{1}{12}$ cm

Thus, radius of the wire $= \frac{1}{24}$ cm

Let the length of the wire $= h$ cm

Thus, volume of the wire $= \pi\left(\frac{1}{24}\right)\left(\frac{1}{24}\right) h \text{ cm}^3$

Now, $\pi\left(\frac{1}{24}\right)\left(\frac{1}{24}\right) h = \frac{10\pi}{3}\left(\frac{700}{3}\right)$

or $\left(\frac{1}{8}\right)\left(\frac{1}{8}\right) h = \frac{10}{1}\left(\frac{700}{1}\right)$

or $h = 7000 \times 64$ cm

$= 448000$ cm

$= 4480$ m

$= 4$ km 480 m. **Ans.**

Q. 4. The largest possible sphere is carved out of a solid wooden cube of side 7 cm. Find the volume of the wood left. $\left[\textbf{Use } \pi = \frac{22}{7}\right]$

Solution : Length of each side of the cube = 7 cm

∴ Largest possible diameter of the sphere = 7 cm

∴ Radius of the sphere = 3.5 cm

Thus, volume of cube $= (7)^3 \text{ cm}^3$

$= 343 \text{ cm}^3$

and volume of sphere $= \frac{4}{3} \times \frac{22}{7} (3.5)^3 \text{ cm}^3$

$= \frac{539}{3} \text{ cm}^3$

∴ Remaining volume of wood

$= \left(343 - \frac{539}{3}\right) \text{ cm}^3$

$= \frac{1029 - 539}{3} \text{ cm}^3$

$= 163.33 \text{ cm}^3$. **Ans.**

Q. 5. Due to heavy floods in a state, thousands were rendered homeless. 50 schools collectively offered to the state government to provide place and the canvas for 1500 tents to be fixed by the government and decided to share the whole expenditure equally. The lower part of each tent is cylindrical of base radius 2.8 m and height 3.5 m, with conical upper part of same base radius but of height 2.1 m. If the canvas used to make the tents costs ₹ 120 per sq. m, find the amount shared by each school to set-up the tents. $\left[\textbf{Use } \pi = \frac{22}{7}\right]$

Solution : Radius of the base of cylinder $(r) = 2.8$ m

Radius of the base of the cone $(r) = 2.8$ m

Height of the cylinder $(h) = 3.5$ m

Height of the cone $(H) = 2.1$ m

Slant height of conical part (l)

$= \sqrt{r^2 + H^2}$

$= \sqrt{(2.8)^2 + (2.1)^2}$

$= \sqrt{7.84 + 4.41}$

$= \sqrt{12.25}$

$= 3.5$ m

Area of canvas used to make tent

$= CSA$ of cylinder + CSA of cone

$= 2\pi rh + \pi rl$

$= 2 \times \frac{22}{7} \times 2.8 \times 3.5 + \frac{22}{7} \times 2.8 \times 3.5$

$= 61.6 + 30.8$

$= 92.4 \text{ m}^2$

Cost of 1500 tents at ₹ 120 per sq. m

$= 1500 \times 120 \times 92.4$

= ₹ 16,632,000

Share of each school to set-up the tents

$= \frac{16632000}{50}$

= ₹ 332,640 **Ans.**

Q. 6. A housing society collects rain water from the roof of its building of area 22 m × 20 m in a cylindrical vessel of diameter 2 m and height 3.5 m and then pumps this water into the main water tank so that everyone can use it. On a particular day, the rain water collected from the roof just fills the cylindrical vessel to the brim. Calculate the rainfall in cm.★★

Solution : Given,

Area of the roof = 22 m × 20 m

Height of cylindrical vessel = 3.5 m

Diameter of the vessel = 2 m

∴ Radius of the vessel = 1 m

Let the total rainfall be h m.

Thus, Volume of rain water collected on the roof

= Volume of the cylindrical vessel

$\Rightarrow \quad 22 \times 20 \times h = \pi(1)3.5$

$\Rightarrow \quad 22 \times 20 \times h = \frac{22}{7}(3.5)$

$\Rightarrow \quad 20h = 0.5$

$\Rightarrow \quad h = \frac{1}{40}$ m

$= 2.5$ cm. **Ans.**

Q. 7. A hemispherical bowl of internal radius 9 cm is full of water. Its contents are emptied in a cylindrical vessel of internal radius 6 cm. Find the height of the water in the cylindrical vessel.

★★ are frequently asked board exam questions

Solution : Given, internal radius of hemispherical bowl = 9 cm

Internal radius of cylindrical vessel = 6 cm

Let the height of the water in the cylindrical vessel be h cm.

Now, Volume of the hemispherical vessel

= Volume of water in the cylindrical vessel

$$\Rightarrow \quad \frac{2}{3}\pi R^3 = \pi r^2 h$$

$$\Rightarrow \quad \frac{2}{3}\pi(9)^3 = \pi(6)^2 h$$

$$\Rightarrow \quad \frac{2}{3}(9)^3 = (6)^2 h$$

$$\Rightarrow \quad h = \frac{2\times9\times9\times9}{3\times6\times6} \text{ cm}$$

$$= \frac{27}{2} \text{ cm}$$

$$\Rightarrow \quad h = 13.5 \text{ cm}$$

Hence, the height of the water level in the cylindrical vessel = 13.5 cm. **Ans.**

Q. 8. From a solid cylinder of height 7 cm and base diameter of 12 cm, a conical cavity of same height and same base diameter is hollowed out. Find the total surface area of the remaining solid.

$$\left[\textbf{Use } \pi = \frac{22}{7}\right]$$

Solution : Given,

Height of cylinder = Height of cone = 7 cm

Base diameter of cylinder

= Base diameter of cone

= 12 cm

Thus, base radius = 6 cm

Thus, total surface area of the remaining solid = Total surface area of cylinder + Curved surface area of cone – Area of the base of the cylinder and cone

$$= 2\pi r(h + r) + \pi rl - \pi r^2$$

$$= 2\pi\, 6(7 + 6) + \pi 6(\sqrt{6^2 + 7^2}) - \pi(6)^2 \text{ cm}^2$$

$$= \pi[12(13) + 6(\sqrt{36 + 49}) - 36] \text{ cm}^2$$

$$= \pi[156 + 6(\sqrt{85}) - 36] \text{ cm}^2$$

$$= \pi[156 + 55.317 - 36] \text{ cm}^2$$

$$= \pi[175.317] \text{ cm}^2$$

$$= 550.997 \text{ cm}^2.$$ **Ans.**

Q. 9. An open metal bucket is in the shape of a frustum of height 21 cm with radii of its lower and upper ends as 10 cm and 20 cm respectively. Find the cost of milk that can completely fill the bucket at ₹ 30 per litre.

$$\left[\textbf{Use } \pi = \frac{22}{7}\right]$$

Solution : Given, height of the frustum = 21 cm

Radius of lower end = 10 cm

Radius of upper end = 20 cm

Cost of milk per litre = ₹ 30

Thus, volume of the frustum

$$= \frac{\pi}{3}[R_1^2 + R_2^2 + R_1R_2]H$$

$$= \frac{1}{3}\times\frac{22}{7}[(10)^2 + (20)^2 + (10)(20)]21 \text{ cm}^3$$

$$= \frac{22}{21}[100 + 400 + 200]21 \text{ cm}^3$$

$$= \frac{22}{21}[700]21 \text{ cm}^3$$

$$= 22(700) \text{ cm}^3$$

$$= 15400 \text{ cm}^3$$

$$= 15.4l \qquad [\because 1l = 1000 \text{ cm}^3]$$

∴ The cost of the milk in completely filling up the frustum = ₹ 30 × 15.4 = ₹ 462. **Ans.**

Q. 10. The dimensions of a solid iron cuboid are 4.4 m × 2.6 m × 1.0 m. It is melted and recast into a hollow cylindrical pipe of 30 cm inner radius and thickness 5 cm. Find the length of the pipe.

Solution : Volume of cuboid = $4.4 \times 2.6 \times 1\text{m}^3$

Inner radius of cylindrical pipe 'r' = 30 cm

Thickness of pipe = 5 cm

Outer radius of cylindrical 'R' = 30 + 5 = 35 cm

$$= \frac{35}{100} \text{ m}$$

∴ Volume of material used

$$= \pi(R^2 - r^2)h$$

$$= \frac{22}{7}\left[\left(\frac{35}{100}\right)^2 - \left(\frac{30}{100}\right)^2\right] h \text{ m}^3$$

$$= \frac{22}{7}\left(\frac{65}{100}\right)\left(\frac{5}{100}\right) h \text{ m}^3$$

Now, $\frac{22}{7}\left(\frac{65}{100}\right)\left(\frac{5}{100}\right) h = 4.4 \times 2.6$

$$\Rightarrow \quad h = \frac{7\times4.4\times2.6\times100\times100}{22\times65\times5}$$

$$\Rightarrow \quad h = 112 \text{ m}$$ **Ans.**

Q. 11 The radii of two circular ends of a bucket of height 15 cm are 14 cm and r cm (r < 14 cm). If the volume of the bucket is 5390 cm³, find the value of r.

$$\left[\textbf{Use } \pi = \frac{22}{7}\right]$$

Solution : Given, height of bucket = 15 cm

Larger radius (R) = 14 cm

Smaller radius = r cm

and Volume of bucket = 5390 cm^3

Since, (r < 14 cm) the bucket is in the shape of frustum.

$$\therefore \quad \frac{\pi}{3}[R^2 + r^2 + R \times r]\, h = 5390$$

$$\Rightarrow \frac{\pi}{3}[(14)^2 + (r)^2 + (14)(r)]15 = 5390$$

$$\Rightarrow \frac{22}{21}[196 + r^2 + 14r]15 = 5390$$

$$\Rightarrow \frac{2}{21}[196 + r^2 + 14r]15 = 490$$

$$\Rightarrow \frac{1}{7}[196 + r^2 + 14r]5 = 245$$

$$\Rightarrow \frac{1}{7}[196 + r^2 + 14r] = 49$$

$$\Rightarrow [196 + r^2 + 14r] = 343$$

$$\Rightarrow r^2 + 14r - 147 = 0$$

$$\Rightarrow r^2 + 21r - 7r - 147 = 0$$

$$\Rightarrow r(r + 21) - 7(r + 21) = 0$$

$$\Rightarrow (r + 21)(r - 7) = 0$$

$$\Rightarrow r = 7, -21$$

Since, radius cannot be negative, so r = 7 cm. **Ans.**

Q. 12. A solid toy is in the form of a right circular cylinder with a hemispherical shape at one end and a cone at the other. Their common diameter is 4.2 cm and the heights of the cylindrical and conical portions are 12 cm and 7 cm respectively. Find the volume of the given toy.

$\left[\text{Use } \pi = \frac{22}{7}\right]$ **[NCERT]**

Solution : Given,

Diameter of hemisphere = Diameter of cone

= 4.2 cm

Thus, Radius of hemisphere = Radius of cylinder = Radius of cone = 2.1 cm

Height of the cylinder = 12 cm

Height of the cone = 7 cm

Hence, volume of the given toy

= Volume of (hemisphere + cylinder + cone)

$$= \frac{2}{3}\pi r^2 + \pi r^2 h + \frac{1}{3}\pi r^2 H$$

$$= \left[\frac{2}{3}\pi(2.1)^3 + \pi(2.1)^2(12) + \frac{1}{3}\pi(2.1)^2(7)\right] \text{ cm}^3$$

$$= \pi(2.1)^2\left[\frac{2}{3}(2.1) + (12) + \frac{1}{3}(7)\right] \text{ cm}^3$$

$$= \left(\frac{22}{7}\right)\left(\frac{21}{10}\right)\left(\frac{21}{10}\right)\left[\frac{2}{3}\left(\frac{21}{10}\right) + (12) + \frac{1}{3}(7)\right] \text{ cm}^3$$

$$= 22\left(\frac{3}{10}\right)\left(\frac{21}{10}\right)\left[\frac{7}{5} + 12 + \frac{7}{3}\right] \text{ cm}^3$$

$$= 22\left(\frac{3}{10}\right)\left(\frac{21}{10}\right)\left[\frac{21 + 180 + 35}{15}\right] \text{ cm}^3$$

$$= 22\left(\frac{3}{10}\right)\left(\frac{21}{10}\right)\left[\frac{236}{15}\right] \text{ cm}^3$$

$$= 22\left(\frac{1}{10}\right)\left(\frac{21}{10}\right)\left[\frac{236}{5}\right] \text{ cm}^3$$

= 218.064 cm^3. **Ans.**

Q. 13. In fig. a tent is in the shape of a cylinder surmounted by a conical top of same diameter. If the height and diameter of cylindrical part are 2.1 m and 3 m respectively and the slant height of conical part is 2.8 m, find the cost of canvas needed to make the tent if the canvas is available at the rate of ₹ 500 per sq. metre.

$\left[\textbf{Use } \pi = \frac{22}{7}\right]$

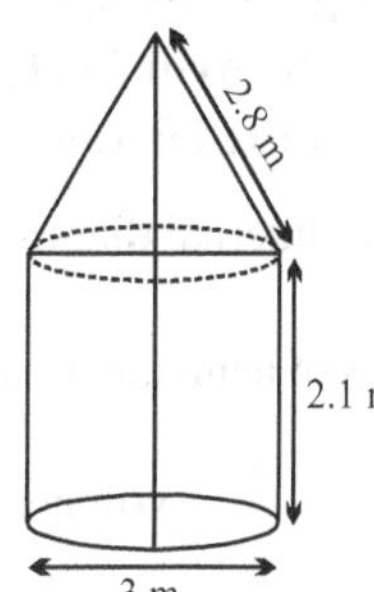

Solution : We have, height (h) and diameter (d) of cylinder as 2.1 m and 3 m respectively.

$\therefore$ Radius = $\frac{3}{2}$ m

and, slant height of concial part is 2.8 m.

Area of canvas needed = C.S.A. of (cylinder + cone)

$$= 2\pi rh + \pi rl$$

$$= 2 \times \frac{22}{7} \times \frac{3}{2} \times 2.1 + \frac{22}{7} \times \frac{3}{2} \times 2.8$$

$$= \frac{22}{7}(6.3 + 4.2)$$

$$= \frac{22}{7} \times 10.5 = 33 \text{ m}^2$$

$\therefore$ Cost of canvas needed at the rate of ₹ 500 per m^2

= ₹ (33 × 500)

= ₹16500 **Ans.**

Q. 14. A solid is made up of a cube surmounted by a hemisphere. If the edges of the cube equal to 5 cm each and the diameter of the hemisphere is 4.2 cm, find the total surface area. $\left[\textbf{Use } \pi = \frac{22}{7}\right]$ **[NCERT]**

Solution : Given, side of the cube = 5 cm

Diameter of the hemisphere = 4.2 cm

Hence, radius of the hemisphere = 2.1 cm

Thus, total surface area of the solid

= Total surface area of cube

+ Curved surface area of hemisphere

– Area of the base of the hemisphere

$$= [6(5)^2 + 2\pi(2.1)^2 - \pi(2.1)^2]\ \text{cm}^2$$

$$= \left[6(25) + \left(\frac{22}{7}\right)(2.1)^2\right] \text{cm}^2$$

$$= \left[150 + \left(\frac{22}{7}\right)\left(\frac{21}{10}\right)\left(\frac{21}{10}\right)\right] \text{cm}^2$$

$$= \left[150 + 22\left(\frac{3}{10}\right)\left(\frac{21}{10}\right)\right] \text{cm}^2$$

$= [150 + 13.86]\ \text{cm}^2$

$= 163.86\ \text{cm}^2$. **Ans.**

Q. 15. A hemispherical bowl of internal diameter 36 cm contains liquid. This liquid is filled into 72 cylindrical bottles of diameter 6 cm. Find the height of each bottle, if 10% liquid is wasted in this transfer.

Solution : Internal diameter of hemispherical bowl

= 36 cm

∴ Radius of hemispherical bowl (r) = 18 cm

$$\text{Volume of liquid} = \frac{2}{3}\pi r^3$$

$$= \frac{2}{3} \times \pi \times (18)^3$$

Now, Diameter of bottle = 6 cm

∴ Radius of bottle = 3 cm

Now, Volume of cylindrical bottle $= \pi R^2 h$

$= \pi 3^2 h$

$= 9\pi h$

Volume of liquid transferred

= Volume of liquid – 10% volume of liquid

$$= \frac{2}{3}\pi(18)^3 - \frac{10}{100}\left(\frac{2}{3}\pi(18)^3\right)$$

$$= \frac{2}{3}\pi(18)^3\left(1 - \frac{10}{100}\right)$$

$$= \frac{2}{3}\pi(18)^3 \times \frac{9}{10}$$

Volume of liquid transferred $= \pi \times (18)^3 \times \frac{3}{5}$

Number of cylindrical bottles

$$= \frac{\text{Volume of liquid to be transferred}}{\text{Volume of a bottle}}$$

$$72 = \frac{\pi \times 18 \times 18 \times 18 \times \frac{3}{5}}{9\pi h}$$

$$h = \frac{27}{5} = 5.4\ \text{cm}$$

Hence, height of each bottle will be 5.4 cm. **Ans.**

Q. 16. The cost of fencing a circular field at the rate of ₹ 20 per metre is ₹ 4400. Government authorities acquired the field for some public project and allotted to the owner, a square plot of each side 62 m 5 cm. Find the area of the circular field and the square field.

$\left[\text{Use } \pi = \frac{22}{7}\right]$

Solution : Since the cost of fencing a circular field at the rate of ₹ 20 per metre is ₹ 4400.

So, the circumference of circular field

$$= \frac{4400}{20}\ \text{m}$$

$= 220$ m

As circumference of a circle having radius $r = 2\pi r$

∴ $2\pi r = 220$

⇒ $2 \times \frac{22}{7} \times r = 220$

⇒ $r = 35$ m

Therefore, the area of the circular field $= \pi r^2$

$$= \frac{22}{7} \times 35 \times 35\ \text{m}^2$$

$= 22 \times 5 \times 35\ \text{m}^2$

$= 3850\ \text{m}^2$

Hence the area of the circular field is 3850 m^2

∵ The area of square plot

= 62 m 5 cm × 62 m 5 cm

$= 62.05 \times 62{\cdot}05\ \text{m}^2$

$= 3850.2025\ \text{m}^2$

Q. 17. A cubical block of side 10 cm is surmounted by a hemisphere. What is the largest diameter that the hemisphere can have ? Find the cost of painting the total surface area of the solid so formed, at the rate of ₹ 5 per 100 sq. cm. [Use $\pi = 3.14$]

Solution : Side of the cubical block (a) = 10 cm

Since the cube is surmounted by a hemisphere, therefore, the greatest diameter of the hemisphere should be equal to the side of the cube.

∴ Diameter of the sphere = 10 cm

⇒ Radius of the sphere (r) = 5 cm

Total surface area of solid = T.S.A. of the cube + C.S.A. of hemisphere – Inner cross-section area of hemisphere

$$= 6a^2 + 2\pi r^2 - \pi r^2$$

$$= 6a^2 + \pi r^2$$

$$= 6(10)^2 + 3.14(5)^2$$

$$= 600 + 25 \times 3.14$$

$$= 600 + 78.5$$

$$= 678.5 \text{ cm}^2$$

Cost of painting per square metre is ₹ 5

$$\text{Total cost for painting} = \frac{₹\ 678.5}{100} \times 5$$

$$= ₹\ 33.92$$

Hence, total cost for painting will be ₹ 33.92. **Ans.**

Q. 18. 504 cones each of diameter 3.5 cm and height 3 cm are melted and recast into a metallic sphere. Find the diameter of the sphere and hence find its surface area. $\left[\textbf{Use } \pi = \frac{22}{7}\right]$

Solution : Diameter of each cone (d) = 3.5 cm

$\therefore$ Radius of each cone (r) = $\frac{3.5}{2} = \frac{7}{4}$ cm

Height of each cone (h) = 3 cm

Volume of 504 cones = 504 × Volume of one cone

$$= 504 \times \frac{1}{3}\pi r^2 h$$

$$= 504 \times \frac{1}{3} \times \frac{22}{7} \times \frac{7}{4} \times \frac{7}{4} \times 3$$

Let radius of sphere be R cm.

$\therefore$ Volume of sphere = Volume of 504 cones

$$\frac{4}{3} \times \frac{22}{7} \times R^3 = 504 \times \frac{1}{3} \times \frac{22}{7} \times \frac{7}{4} \times \frac{7}{4} \times 3$$

$$R = 3\sqrt{\frac{504 \times 7 \times 7 \times 3}{4 \times 4 \times 4}}$$

$$R = 3\sqrt{\frac{3 \times 3 \times 7 \times 7 \times 7 \times 3}{2 \times 2 \times 2}}$$

$$R = \frac{21}{2} \text{ cm}$$

Now, surface area of sphere = $4\pi R^2$

$$= 4 \times \frac{22}{7} \times \frac{21}{2} \times \frac{21}{2}$$

$$= 63 \times 22$$

$$= 1386 \text{ cm}^2$$

Hence, surface area of sphere is 1386 cm^2. **Ans.**

Q. 19. The sum of the radius of base and height of a solid right circular cylinder is 37 cm. If the total surface area of the solid cylinder is 1628 sq. cm, find the volume of the cylinder. $\left[\textbf{Use } \pi = \frac{22}{7}\right]$

Solution : Let the radius of base and height of a solid cylinder be r and h respectively.

Now, we have, $r + h = 37$ cm ...(i)

and, T.S.A. of solid cylinder= $2\pi r(r + h) = 1628$ cm^2

$\Rightarrow$ $2\pi r(37) = 1628$ [Using (i)]

$$\Rightarrow \quad r = \frac{1628}{37 \times 2 \times \frac{22}{7}}$$

$$r = 7 \text{ cm}$$

$\therefore$ Volume of the cylinder $= \pi r^2 h$

$$= \frac{22}{7} \times 7 \times 7 \times 30$$

[using eq. (i), $h = 30$]

$$= 4620 \text{ cm}^3$$ **Ans.**

Q. 20. A well of diameter 4 m is dug 21 m deep. The earth taken out of it has been spread evenly all around it in the shape of a circular ring of width 3 m to form an embankment. Find the height of the embankment.

Solution : Given, diameter and height of cylindrical well are 4 m and 21 m respectively.

Now, the earth has been taken out to spread evenly all around.

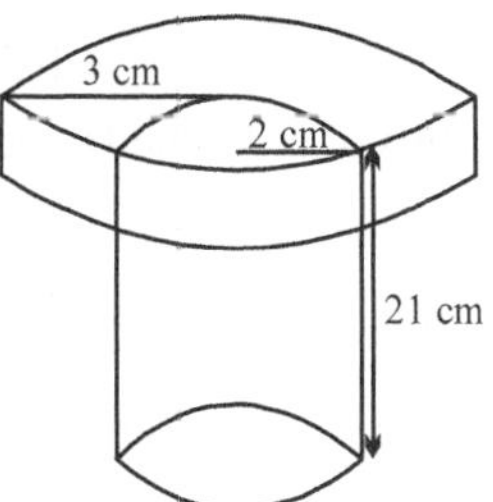

Then, volume of earth dug out $= \frac{22}{7} \times \frac{4}{2} \times \frac{4}{2} \times 21$

$$= 264 \text{ m}^3$$

and the volume of embankment of width 3 m which forms a shape of circular ring = $\pi\,[(5)^2 - (2)^2] \times h$

[$\because$ Outer radius = 2 + 3 = 5 cm]

$$= \frac{22}{7}(25 - 4) \times h = 66\,h \text{ m}$$

$\because$ Volume of earth dug out = Volume of embankment

$\Rightarrow$ $264 = 66h$

$\Rightarrow \qquad h = \frac{264}{66} = 4\text{ m}$

Hence, the height of the embankment is 4 m. **Ans.**

Q. 21. The adjoining figure depicts an archery target marked with its five scoring areas from the centre outwards as Gold, Red, Blue, Black and white. The diameter of the region representing Gold score is 21 cm and each of the other bands is 10.5 cm wide. Find the area of each of the five scoring regions.

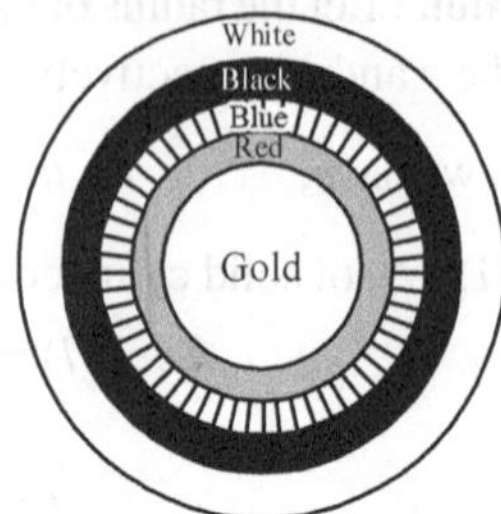

Solution : Since diameter of the Gold score = 21 cm

$\therefore \qquad \text{Radius} = \frac{21}{2} = 10.5\text{ cm}$

So, area of the Gold scoring region

$= \pi r^2$

$= \frac{22}{7} \times 10.5 \times 10.5$

$= 346.5\text{ cm}^2$

Now, radius of combined Gold and Red region

$= 10.5 + 10.5$

$= 21\text{ cm}$

Therefore, area of Red scoring region = Area of combined Gold and Red regions – Area of Gold region

$= \left(\frac{22}{7} \times 21 \times 21\right) - 346.5$

$= 1386 - 346.5$

$= 1039.5\text{ cm}^2$

$\because$ Radius of combined Gold, Red and Blue regions

$= 21 + 10.5$

$= 31.5\text{ cm}$

Thus, area of Blue scoring region = Area of combined Gold, Red and Blue region – Area of combined Gold and Red region

$= \left(\frac{22}{7} \times 31.5 \times 31.5\right) - 1386$

$= 3118.5 - 1386$

$= 1732.5\text{ cm}^2$

$\because$ Radius of combined Gold, Red, Blue and Black reigon

$= 31.5 + 10.5$

$= 42\text{ cm}$

Thus, area of Black scoring regions = Area of combined Gold, Red, Blue and Black regions – Area of combined Gold, Red and Blue regions

$= \left(\frac{22}{7} \times 42 \times 42\right) - 3118.5$

$= 5544 - 3118.5$

$= 2425.5\text{ cm}^2$

$\because$ Radius of combined Gold, Red, Blue, Black and White regions

$= 42 + 10.5$

$= 52.5\text{ cm}$

Hence, area of white scoring region = Area of combined Gold, Red, Blue, Black and White region – Area of combined Gold, Red, Blue and Black regions

$= \left(\frac{22}{7} \times 52.5 \times 52.5\right) - 5544$

$= 8662.5 - 5544$

$= 3118.5\text{ cm}^2$ **Ans.**

Q. 22. The internal and external radii of a hollow sphere are 3 cm and 5 cm respectively. The sphere is melted to form a solid cylinder of height $\frac{8}{3}$ cm. Find the diameter and the curved surface area of the cylinder.

$\left[\text{Use } \pi = \frac{22}{7}\right]$

Solution : Given, external radius of sphere = 5 cm

Internal radius of sphere = 3 cm

Height of the solid cylinder = $\frac{8}{3}$ cm

Let the radius of the cylinder be r cm.

Thus, volume of hollow sphere = Volume of cylinder

$\Rightarrow \qquad \frac{4}{3}\pi[(5)^3 - (3)^3] = \frac{1}{3}\pi r^2(8)$

$\Rightarrow \qquad [125 - 27] = 2r^2$

$\Rightarrow \qquad 2r^2 = 98$

$\Rightarrow \qquad r^2 = 49$

$\Rightarrow \qquad r^2 = 7^2$

$\Rightarrow \qquad r = 7\text{ cm}$

$\therefore$ Diameter of the cylinder = 14 cm

Now, curved surface area of the cylinder

$= 2\pi(7)\left(\frac{8}{3}\right)\text{ cm}^2$

$= 2\left(\frac{22}{7}\right)(7)\left(\frac{8}{3}\right)\text{ cm}^2$

$= 2\left(\frac{22}{1}\right)\left(\frac{8}{3}\right)\text{ cm}^2$

$= \frac{352}{3}$ cm^2

$= 117.33$ cm^2. **Ans.**

Q. 23. In fig. is a decorative block, made up of two solids – a cube and a hemisphere. The base of the block is a cube of side 6 cm and the hemisphere fixed on the top has a diameter of 3.5 cm. Find the total surface area of the block. $\left[\text{Use } \pi = \frac{22}{7}\right]$

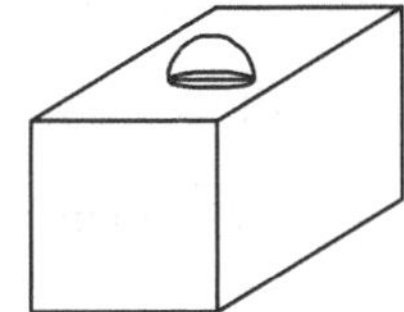

Solution : Given, side of a cube = 6 cm

and the diameter of hemisphere = 3.5 cm

$\therefore$ Radius $= \frac{3.5}{2}$ cm

Now, total surface area of decorative block

= Total surface area of cube

– Surface area of base of hemisphere

+ C.S.A of hemisphere

$$= (6)^3 - \frac{22}{7} \times \frac{3.5}{2} \times \frac{3.5}{2} + 2 \times \frac{22}{7} \times \frac{3.5}{2} \times \frac{3.5}{2}$$

$$= 216 - \frac{22 \times 7}{16} + \frac{22 \times 7}{8}$$

$$= 216 + \frac{154}{16}$$

$= 225.625$ cm^2 **Ans.**

Q. 24. A container in the shape of a frustum of a cone having diameters of its two circular faces as 35 cm and 30 cm and vertical height 14 cm is completely filled with oil. If each cm^3 of oil has mass 1.2 g, then find the cost of oil in the container at the rate of ₹ 40 per kg.

Solution : Given, height of the frustum = 14 cm

Diameter of smaller side = 30 cm

$\therefore$ Radius of smaller side = 15 cm

Diameter of larger side = 35 cm

$\therefore$ Radius of larger side = 17.5 cm

1 cm^3 of oil = 1.2 g

Rate of oil = ₹ 40/kg

Thus, volume of oil

$$= \frac{\pi}{3}(R + r^2 + Rr)\,h$$

$$= \frac{\pi}{3}[(17.5)^2 + (15)^2 + (17.5)(15)]14 \text{ cm}^3$$

$$= \frac{22}{21}[306.25 + 225 + 262.5]14 \text{ cm}^3$$

★ are board exam questions from previous years

$$= \frac{44}{3}[793.75] \text{ cm}^3$$

Thus, weight of oil $= \left[\frac{44}{3}(793.75)\right](1.2)$ g

= 13970 g

= 13 kg 970 g

Thus, price of the oil = ₹ 40(13.97)

= ₹ 558.80 **Ans.**

Q. 25. The diameter of a metallic sphere is 6 cm. The sphere is melted and drawn into a wire of uniform cross-section. If the length of the wire is 36 m, find its radius.★

Solution : Given, diameter of the sphere = 6 cm

Thus, Radius of the sphere = 3 cm

and Length of the wire = 36 m = 3600 cm

Let the radius of the wire be r.

Hence, Volume of the wire = Volume of the sphere

$$\Rightarrow \quad \pi r^2(3600) = \frac{4}{3}\pi(3)^3$$

$$\Rightarrow \quad r^2(3600) = \frac{4}{3}(3)^3$$

$$\Rightarrow \quad r^2(3600) = 4(3)^2$$

$$\Rightarrow \quad r^2(3600) = 36$$

$$\Rightarrow \quad r^2(100) = 1$$

$$\Rightarrow \quad r^2 = \left(\frac{1}{10}\right)^2$$

$$\Rightarrow \quad r = \frac{1}{10} \text{ cm}$$

$\Rightarrow \quad r = 0.1$ cm = 1 mm. **Ans.**

Q. 26. The difference between the outer and inner curved surface area of a hollow right-circular cylinder 14 cm long is 88 cm^2. If the volume of metal used in making the cylinder is 176 cm^3, find the outer and inner diameters of the cylinder. $\left[\text{Use } \pi = \frac{22}{7}\right]$

Solution : Given, volume of the cylinder = 176 cm^3

Height of the cylinder = 14 cm

and difference between outer and inner curved surface area = 88 cm^2

Now, let the outer radius be R cm and the inner radius be r cm

$$\text{Thus,} \quad \frac{22}{7}(R^2 - r^2)14 = 176$$

$$\Rightarrow \quad (R^2 - r^2) = 4$$

$$\Rightarrow \quad (R - r)(R + r) = 4 \quad \ldots\text{(i)}$$

Now, Outer curved surface area – Inner curved surface area = 88 cm^2

$\Rightarrow \quad 2\pi R(14) - 2\pi r(14) = 88$

$\Rightarrow \quad 2\frac{22}{7}(14)(R-r) = 88$

$\Rightarrow \quad (R-r) = 1 \quad$ ...(ii)

Substituting equation (ii) in (i), we get

$(R+r) = 4 \quad$ ...(iii)

Adding equations (ii) and (iii), we get

$2R = 5$

or $\quad R = 2.5$ cm

and $\quad r = 1.5$ cm

Thus, the outer diameter = 5 cm

and the inner diameter = 3 cm. **Ans.**

Q. 27. A sphere of diameter 12 cm, is dropped in a right circular cylindrical vessel, partly filled with water. If the sphere is completely submerged in water, the water level in the cylindrical vessel rises by $3\frac{5}{9}$ cm. Find the diameter of the cylindrical vessel.★

Solution : Given, diameter of sphere = 12 cm

Then, radius of sphere $(r) = \frac{12}{2} = 6$ cm

$\therefore$ Volume of sphere $= \frac{4}{3}\pi r^3$

$= \frac{4}{3} \times \pi \times (6)^3$ cm^3

Now, sphere is completely submerged in water and rise in water in cylindrical vessel is $3\frac{5}{9}$ cm.

Volume of sphere = Volume of change in water level in cylindrical vessel

$$\frac{4}{3}\pi \times (6)^3 = \pi r^2 \times \frac{32}{9}$$

$$r^2 = \frac{4\times6\times6\times6\times9}{3\times32}$$

$$r = \sqrt{81}$$

$$r = 9 \text{ cm.}$$

$\therefore$ Diameter of the cylindrical vessel is 18 cm. **Ans.**

Q. 28. From a solid cylinder of height 10 cm and base radius of 6 cm, a conical cavity of same height and same base diameter is hollowed out. Find the volume and the total surface area of the remaining solid. [Use $\pi = 3.14$]

Solution : Given,

Height of the cylinder = Height of the cone

= 10 cm

Base radius of the cylinder = Base radius of the cone

= 6 cm

Thus, total surface area of the remaining solid

= Total surface area of the cylinder

+ Curved surface area of the cone

– Area of the base of the cylinder and cone

$= 2\pi 6(10+6) + \pi 6(\sqrt{6^2+10^2}) - \pi(6)^2$ cm^2

$= \pi[12(16) + 6(\sqrt{36+100}) - 36]$ cm^2

$= \pi[192 + 6(\sqrt{136}) - 36]$ cm^2

$= \pi[156 + 6(2\sqrt{34})]$ cm^2

$= 12\pi[13 + \sqrt{34}]$ cm^2

$= 12(3.14)\,[13 + \sqrt{34}]$ cm^2

= 709.55 cm^2

and volume of the remaining solid

= Volume of the cylinder – Volume of the cone

$= \left[\pi(6)^2 10 - \frac{1}{3}\pi(6)^2 10\right]$ cm^3

$= \left[\frac{2}{3}\pi(6)^2 10\right]$ cm^3

$= \left[\frac{2}{1}\pi(12)10\right]$ cm^3

= 753.6 cm^3. **Ans.**

Q. 29. In fig. from the top of a solid cone of height 12 cm base radius 6 cm, a cone of height 4 cm is removed by a plane parallel to the base. Find the total surface area of the remaining solid.

$\left[\text{Use } \pi = \frac{22}{7} \text{ and } \sqrt{5} = 2.236\right]$

Solution : Height of the given cone = 12 cm

and the radius of the base = 6 cm.

Let the radius of the base of the smaller cone be x cm and height is 4 cm.

Now, $\quad \Delta ACD \sim \Delta ARQ$

$\therefore \quad \frac{DC}{QR} = \frac{AD}{AQ}$

$$\frac{6}{x} = \frac{12}{4} \Rightarrow x = 2 \text{ cm}$$

★ are board exam questions from previous years

$\therefore \quad l = RC = \sqrt{h^2 + (R-r^2)}$

$= \sqrt{8^2 + (6-2)^2}$

$= \sqrt{64+16}$

$= 4\sqrt{5}$

Total surface area of frustum $PRCB$

$= [\pi l(R+r) + \pi r^2 + \pi R^2]$

$= \frac{22}{7} \times 4\sqrt{5}\,(6+2) + \frac{22}{7} \times (2)^2 + \frac{22}{7} + (6)^2$

$= \frac{22}{7}[32 \times 2.236 + 4 + 36]$

$= \frac{22}{7}(111.552)$

$= 350.592 \text{ cm}^2$ **Ans.**

Q. 30. A cylindrical vessel with internal diameter 10 cm and height 10.5 cm is full of water. A solid cone of base diameter 7 cm and height 6 cm is completely immersed in the vessel. Find the volume of water displaced and the volume remaining. $\left[\textbf{Use } \pi = \frac{22}{7}\right]$

Solution : Given, diameter of cylindrical vessel = 10 cm

Height of cylindrical vessel = 10.5 cm

Diameter of cone = 7 cm

$\therefore$ Radius of cone = 3.5 cm

Height of cone = 6 cm

Total volume of water in the cylinder

= Volume of cylinder

$= \frac{22}{7}(5)^2(10.5) \text{ cm}^3$

$= \frac{22}{1}(25)(1.5) \text{ cm}^3$

$= 33(25) \text{ cm}^3$

$= 825 \text{ cm}^3$

Volume of water displaced

= Volume of the cone

$= \frac{1}{3}\left(\frac{22}{7}\right)(3.5)^2 6 \text{ cm}^3$

$= 77 \text{ cm}^3$

Thus, volume of water left behind

$= (825 - 77) \text{ cm}^3$

$= 748 \text{ cm}^3$. **Ans.**

★ are board exam questions from previous years

Q. 31. Water in a canal, 5.4 m wide and 1.8 m deep, is flowing with a speed of 25 km/hour. How much area can it irrigate in 40 minutes, if 10 cm of standing water is required for irrigation ?★

Solution : Volume of water flowing in 40 min

$= 5.4 \times 1.8 \times 25000 \times \frac{40}{60} \text{ m}^3$

$= 162000 \text{ m}^3$

Height of standing water = 10 cm = 0.10 m

$\therefore$ Area of be irrigated $= \frac{162000}{0.10}$

$= 1620000 \text{ m}^2$

Given, Width of canal = 5.4 m

Depth of canal = 1.8 m

Speed of flowing water = 25 km/h

$\therefore$ Length of water in canal in 1 hr

= 25 km

= 25000 m

Volume of water flown out from canal in 1 hr

$= l \times b \times h$

$= 5.4 \times 1.8 \times 25000$

$= 243000 \text{ m}^3$

Volume of water in 40 min $= 243000 \times \frac{40}{60}$

$= 162000 \text{ m}^3$

Thus, area irrigated with 10 cm standing water in field

$= \frac{\text{Volume}}{\text{Height}} = \frac{162000 \times 100}{10} \text{ m}^2$

$= 1620000 \text{ m}^2$

= 162 hectare **Ans.**

Chapter 14. Statistics

Q. 1. The marks attained by 40 students in a short assessment is given below where *a* and *b* are missing. If the mean of the distribution is 7.2, find *a* and *b*.

Marks	5	6	7	8	9
No. of students	6	*a*	16	13	*b*

Solution :

Class Interval (x_i)	Frequency (f_i)	$(f_i \times x_i)$
5	6	30
6	a	$6a$
7	16	112
8	13	104
9	b	$9b$
	$\Sigma f_i = 35 + a + b$ $= 40$	$\Sigma(f_i \times x_i)$ $= 246 + 6a + 9b$

We know that, Mean $= \dfrac{\Sigma(f_i \times x_i)}{\Sigma f_i}$

$$7.2 = \frac{246+6a+9b}{40}$$

$\Rightarrow \quad 246 + 6a + 9b = 40(7.2)$

$\Rightarrow \quad 246 + 6a + 9b = 288$

$\Rightarrow \quad 6a + 9b = 42$

$\Rightarrow \quad 2a + 3b = 14 \quad \ldots\text{(i)}$

Also $\quad 35 + a + b = 40$

$\Rightarrow \quad a + b = 5$

$\Rightarrow \quad 2a + 2b = 10 \quad \ldots\text{(ii)}$

Subtracting equation (ii) from equation (i),

$$b = 4$$

and $\quad a = 5 - 4 = 1.$ **Ans.**

Q. 2. Find the mean of children per family from the data given below:

No. of children	0	1	2	3	4	5
No. of family	5	11	25	12	5	2

Solution :

No. of children (x_i)	No. of families (f_i)	f_ix_i
0	5	0
1	11	11
2	25	50
3	12	36
4	5	20
5	2	10
Total	$\Sigma f_i = 60$	$\Sigma f_ix_i = 127$

Mean $= \dfrac{\Sigma f_ix_i}{\Sigma f_i} = \dfrac{127}{60} = 2.12$ (approx.) **Ans.**

Q. 3. Calculate the mean of the following data :

Class Interval	Frequency
0 – 10	12
10 – 20	16
20 – 30	6
30 – 40	7
40 – 50	9

Solution :

Class Interval	Frequency (f_i)	Class Marks (x_i)	$(f_i \times x_i)$
0 – 10	12	5	60
10 – 20	16	15	240
20 – 30	6	25	150
30 – 40	7	35	245
40 – 50	9	45	405
	$\Sigma f_i = 50$		$\Sigma(f_i \times x_i) = 1100$

We know that,

$$\text{Mean} = \frac{\Sigma(f_i \times x_i)}{\Sigma f_i} = \frac{1100}{50} = 22.$$ **Ans.**

Q. 4. If the arithmetic mean of the following frequency distribution is 54, determine the value of *p*.

Class	Frequency
0 – 20	7
20 – 40	*p*
40 – 60	10
60 – 80	9
80 – 100	13

Solution :

Class Interval	Frequency (f_i)	Class Marks (x_i)	$(f_i \times x_i)$
0 – 20	7	10	70
20 – 40	p	30	$30p$
40 – 60	10	50	500
60 – 80	9	70	630
80 – 100	13	90	1170
	$\Sigma f_i = 39 + p$		$\Sigma(f_i \times x_i) = 2370 + 30p$

We know that, Mean $= \dfrac{\Sigma(f_i \times x_i)}{\Sigma f_i}$

$$= \frac{2370+30p}{39+p} = 54$$

$\Rightarrow \quad 2370 + 30p = 54(39 + p)$

$\Rightarrow \quad 2370 + 30p = 2106 + 54p$

$\Rightarrow \quad 24p = 264$

$\Rightarrow \quad p = 11.$ **Ans.**

Q. 5. If the arithmetic mean of the following frequency distribution is 62.8 and the sum of all frequencies is 50, determine the value of f_1 and f_2.

Class	Frequency
0 – 20	5
20 – 40	f_1
40 – 60	10
60 – 80	f_2
80 – 100	7
100 – 120	8
Total	50

Solution : We have,

$$5 + f_1 + f_2 + 10 + 7 + 8 = 50$$

$\Rightarrow \quad 30 + f_1 + f_2 = 50$

$\Rightarrow \quad f_1 + f_2 = 20$

$\Rightarrow \quad f_2 = 20 - f_1$

Class Interval	Frequency (f_i)	Class Marks (x_i)	($f_i \times x_i$)
0 – 20	5	10	50
20 – 40	f_1	30	$30f_1$
40 – 60	10	50	500
60 – 80	$f_2 = 20 - f_1$	70	$1400 - 70f_1$
80 – 100	7	90	630
100 – 120	8	110	880
	$\Sigma f_i = 30 + f_1 + f_2 = 50$		$\Sigma(f_i \times x_i) = 3460 - 40f_1$

We know that,

$$\text{Mean} = \frac{\Sigma(f_i \times x_i)}{\Sigma f_i}$$

$$62.8 = \frac{3460 - 40f_1}{50}$$

$\Rightarrow$ $3460 - 40f_1 = 50(62.8)$

$\Rightarrow$ $3460 - 40f_1 = 3140$

$\Rightarrow$ $40f_1 = 320$

$\Rightarrow$ $f_1 = 8$

and $f_2 = 20 - 8$

$= 12.$ **Ans.**

Q. 6. Calculate the arithmetic mean of the following frequency distribution by the step-deviation method :

Class	0 – 50	50 – 100	100 – 150	150 – 200	200 – 250	250 – 300
Frequency	17	35	43	40	21	24

Solution :

Class Interval	Frequency (f_i)	Mid - Value (x_i)	$u_i = \frac{x_i - A}{h}$; h = [upper limit – lower limit]	$f_i \times u_i$
0 – 50	17	25	– 2	– 34
50 – 100	35	75	– 1	– 35
100 – 150	43	$125 = A$	0	0
150 – 200	40	175	1	40
200 – 250	21	225	2	42
250 – 300	24	275	3	72
	$\Sigma f_i = 180$			$\Sigma(f_i \times u_i) = 85$

Thus $A = 125$, $h = 50$, $\Sigma f_i = 180$, $\Sigma(f_i \times u_i) = 85$

We know that,

$$\text{Mean} = A + \left\{h \times \frac{\Sigma(f_i \times u_i)}{\Sigma f_i}\right\}$$

$$= 125 + \left\{50 \times \frac{85}{180}\right\}$$

$= 125 + 23.61$

$= 148.61.$ **Ans.**

Q. 7. Calculate the mean by using the direct method :

Class Interval	Frequency
0 – 10	3
10 – 20	5
20 – 30	9
30 – 40	5
40 – 50	3

Solution :

Class Interval	Frequency (f_i)	Class Marks (x_i)	($f_i \times x_i$)
0 – 10	3	5	15
10 – 20	5	15	75
20 – 30	9	25	225
30 – 40	5	35	175
40 – 50	3	45	135
	$\Sigma f_i = 25$		$\Sigma(f_i \times x_i) = 625$

Thus $\text{Mean} = \frac{\Sigma(f_i \times x_i)}{\Sigma f_i}$

$= \frac{625}{25} = 25.$ **Ans.**

Q. 8. If the arithmetic mean of the following frequency distribution is 50, determine the value of f_1 and f_2.

Class	Frequency
0 – 20	17
20 – 40	f_1
40 – 60	32
60 – 80	f_2
80 – 100	19
Total	120

Solution : We have

$$17 + f_1 + f_2 + 32 + 19 = 120$$
$$\Rightarrow \quad 68 + f_1 + f_2 = 120$$
$$\Rightarrow \quad f_1 + f_2 = 52$$
$$\Rightarrow \quad f_2 = 52 - f_1$$

Class Interval	Frequency (f_i)	Class Marks (x_i)	$(f_i \times x_i)$
0 – 20	17	10	170
20 – 40	f_1	30	$30f_1$
40 – 60	32	50	1600
60 – 80	$f_2 = 52 - f_1$	70	$3640 - 70f_1$
80 – 100	19	90	1710
	$\Sigma f_i = 68 + f_1 + f_2 = 120$		$\Sigma(f_i \times x_i) = 7120 - 40f_1$

Thus $\quad \text{Mean} = \dfrac{\Sigma(f_i \times x_i)}{\Sigma f_i}$

$$50 = \frac{7120 - 40f_1}{120}$$
$$\Rightarrow \quad 7120 - 40f_1 = 50(120)$$
$$\Rightarrow \quad 7120 - 40f_1 = 6000$$
$$\Rightarrow \quad 40f_1 = 1120$$
$$\Rightarrow \quad f_1 = 28$$
and $\quad f_2 = 52 - 28$

$f_2 = 24.$ **Ans.**

Q. 9. Find the mean and median for the following data :★

Class	0 – 4	4 – 8	8 – 12	12 – 16	16 – 20
Frequency	3	5	9	5	3

Solution :

Class	Frequency (f_i)	X_i	f_iX_i	$c.f_i$
0 – 4	3	2	6	3
4 – 8	5	6	30	$cf = 8$
8 – 12	$f = 9$	10	90	17
12 – 16	5	14	70	22
20 – 20	3	18	54	25
	$\Sigma f_i = 25$		$\Sigma f_iX_i = 250$	

$$\text{Mean } (\overline{X}) = \frac{\Sigma f_i X_i}{\Sigma f_i} = \frac{250}{25} = 10$$

Now, $\quad n = \Sigma f_i = 25$

$$\frac{n}{2} = \frac{25}{2} = 12.5$$

Cumulative frequency just above 12.5 is 17 which lies in class interval 8 – 12.

∴ Median class is 8 – 12

$$\text{Median} = l + \frac{\frac{n}{2} - c.f}{f} \times h$$

Here, $l = 8, h = 4, cf = 8, f = 9$

$$= 8 + \frac{12.5 - 8}{9} \times 4$$
$$= 8 + \frac{4.5}{9} \times 4$$
$$= 8 + 2 = 10$$ **Ans.**

Q. 10. The lengths of leaves of a plant are measured correct to the nearest mm and the data obtained is represented as the following frequency distribution :★

Length (in mm)	No. of leaves
110 – 115	2
115 – 120	6
120 – 125	10
125 – 130	13
130 – 135	6
135 – 140	3
140 – 145	2

Draw a 'more than type' ogive for the above data.

Solution :

Length (in mm)	No. of leaves
More than 110	42
More than 115	40
More than 120	34
More than 125	24
More than 130	11
More than 135	5
More than 140	2

★ are board exam questions from previous years

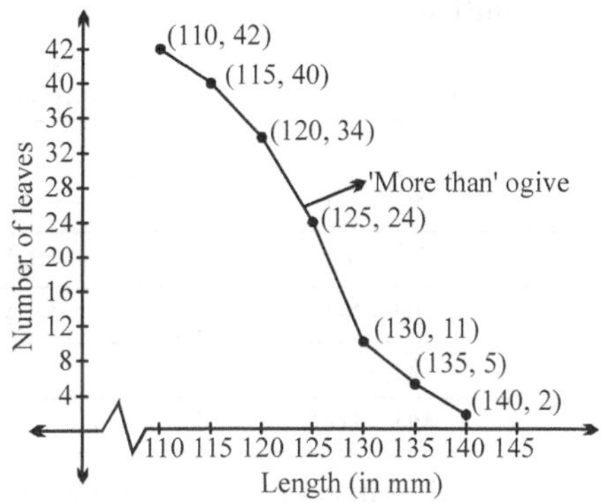

Q. 11. Find the mean of the following distribution by Assumed Mean Method :

Class	Frequency
10 – 20	8
20 – 30	7
30 – 40	12
40 – 50	23
50 – 60	11
60 – 70	13
70 – 80	8
80 – 90	6
90 – 100	12

Solution :

Class interval	Frequency (f_i)	X_i	$d_i = X_i - 55$	$f_i d_i$
10 – 20	8	15	–40	–320
20 – 30	7	25	–30	–210
30 – 40	12	35	–20	–240
40 – 50	23	45	–10	–230
50 – 60	11	55	0	0
60 – 70	13	65	10	130
70 – 80	8	75	20	160
80 – 90	6	85	30	180
90 – 100	12	95	40	480
	$\Sigma f_i = 100$			$\Sigma f_i d_i = -50$

Let $A = 55$

$$\text{Mean} = A + \frac{\Sigma f_i d_i}{\Sigma f_i} = 55 + \left(\frac{-50}{100}\right)$$

$$= 55 - \frac{5}{10}$$

$$= 55 - 0.5$$

$$= 54.5$$ **Ans.**

Q. 12. The average score of boys in the examination of a school is 71 and that of the girls is 73. The average score of the school in the examination is 71.8. Find the ratio of number of boys fo the number of girls who appeared in the examination.

Solution : Let the number of boys = n_1

and number of girls = n_2

Average boys' score $= 71 = \overline{X_1}$ (Let)

Average girls' score $= 73 = \overline{X_2}$ (Let)

$$\text{Combined mean} = \frac{n_1\overline{X_1} + n_2\overline{X_2}}{n_1 + n_2}$$

$$71.8 = \frac{n_1(71) + n_2(73)}{n_1 + n_2}$$

$$71n_1 + 73n_2 = 71.8n_1 + 71.8n_2$$
$$71n_1 - 71.8n_1 = 71.8n_2 - 73n_2$$
$$-0.8n_1 = -1.2n_2$$

$$\frac{n_1}{n_2} = \frac{1.2}{0.8} \Rightarrow \frac{n_1}{n_2} = \frac{3}{2}$$

$\Rightarrow \quad n_1 : n_2 = 3 : 2$

$\therefore$ No. of boys : No. of girls = 3 : 2 **Ans.**

Q. 13. Following table gives the ages in years of militants operating in a certain area of a country.

Age (in years)	40 – 43	43 – 46	46 – 49	49 – 52	52 – 54
Number of militants	31	58	60	k	27

If mean of the above distribution is 47.2, find how many militants in the age groups 49-52 are active in the area ?

Solution :

Class Interval	Frequency (f_i)	x_i	$f_i x_i$
40 – 42	31	41.5	1286.5
43 – 46	58	44.5	2581
46 – 49	60	47.5	2850
49 – 52	k	50.5	$50.5k$
52 – 55	27	53.5	1444.5
	$\Sigma f_i = 176 + k$		$\Sigma f_i x_i = 8162 + 50.5k$

Mean ($\bar{x}$) = 47.2 [Given]

We know that $$\bar{x} = \frac{\Sigma f_i x_i}{\Sigma f_i}$$

$$\Rightarrow \quad 47.2 = \frac{8162 + 50.5k}{176 + k}$$

$$\Rightarrow \quad 47.2\,(176 + k) = 8162 + 50.5k$$

$$\Rightarrow \quad 8307.2 + 47.2k = 8162 + 50.5k$$

$$\Rightarrow \quad 8307.2 - 8162 = 50.5k - 47.2k$$

$$\Rightarrow \quad 145.2 = 3.3k$$

$$\Rightarrow \quad k = \frac{145.2}{3.3} = 44$$

Thus, there are 44 militants operating in the age group 49 – 52. **Ans.**

Q. 14. One hundred students took an examination. Compute values of *x* and *y* from the following data if mean is 42.

Marks obtained	No. of students
0 – 10	7
10 – 20	10
20 – 30	x
30 – 40	13
40 – 50	y
50 – 60	10
60 – 70	14
70 – 80	9

Solution :

Class Interval	x_i	f_i	$u_i = \frac{x_i - A}{h}$	$f_i u_i$
0 – 10	5	7	–2	–14
10 – 20	15	10	–1	–10
20 – 30	25 = A	x	0	0
30 – 40	35	13	1	13
40 – 50	45	y	2	$2y$
50 – 60	55	10	3	30
60 – 70	65	14	4	56
70 – 80	75	9	5	45
		$\Sigma f_i = 63 + x + y$		$\Sigma f_i u_i = 120 + 2y$

Here $A = 25$, $h = 10$ and $\overline{x} = 42$

We know

$$\Sigma f_i = 100$$

i.e. $$63 + x + y = 100$$

i.e. $$x + y = 37 \quad \text{...(i)}$$

$$\text{Mean } (\overline{x}) = A + \frac{\Sigma f_i u_i}{\Sigma f_i} \times h$$

$$42 = 25 + \frac{120 + 2y}{100} \times 10$$

$$\Rightarrow \quad 17 = \frac{120 + 2y}{10}$$

$$\Rightarrow \quad 170 = 120 + 2y$$

$$\Rightarrow \quad 50 = 2y$$

$$\Rightarrow \quad y = 25$$

From (i) $$x = 12$$

Hence, $x = 12$ and $y = 25$. **Ans.**

Q. 15. The weekly pocket money (in Rupees) of 50 students is given below. Calculate the Mode.

Pocket money	Frequency
40 – 50	2
50 – 60	8
60 – 70	12
70 – 80	14
80 – 90	8
90 – 100	6

★ are board exam questions from previous years

Solution :

Class Interval	Frequency (f_i)
40 – 50	2
50 – 60	8
60 – 70	12
70 – 80	14
80 – 90	8
90 – 100	6

Since class interval 70 – 80 has the highest frequency, therefore, it is the Modal Class.

We know that,

$$\text{Mode} = l + \frac{f_1 - f_0}{2f_1 - f_0 - f_2} \times h$$

$$= 70 + \frac{14 - 12}{28 - 12 - 8} \times 10$$

$$= 70 + \frac{2}{8} \times 10$$

$$= 70 + 2.5$$

$$= ₹\, 72.5$$ **Ans.**

Q. 16. Find the mode of the following frequency distribution :★

Class interval	Frequency
25-30	25
30-35	34
35-40	50
40-45	42
45-50	38
50-55	14

Solution :

Class interval	Frequency
25 – 30	25
30 – 35	34
35 – 40	50
40 – 45	42
45 – 50	38
50 – 55	14

Here, maximum frequency is 50.

So, 35 – 40 will be the modal class.

$l = 35, f_0 = 34, f_1 = 50, f_2 = 42$ and $h = 5$

$$\text{Mode} = l + \left(\frac{f_1 - f_0}{2f_1 - f_0 - f_2}\right) \times h$$

$$= 35 + \left(\frac{50 - 34}{2 \times 50 - 34 - 42}\right) \times 5$$

$$= 35 + \left(\frac{16}{100 - 76}\right) \times 5$$

$$= 35+\frac{16}{24}\times 5$$

$$= 35+\frac{80}{24}$$

$$= 35 + 3{\cdot}3$$

$$= 38{\cdot}33$$ **Ans.**

Q. 17. The following table shows the age distribution of cases of a certain disease admitted during a year in a particular hospital :

Class	5 – 14	15 – 24	25 – 34	35 – 44	45 – 54	55 – 64
Frequency	6	11	21	23	14	5

Find the average age for which maximum cases occurred.

Solution :

Age (in years)	No. of cases	
4.5 – 14.5	6	
14.5 – 24.5	11	
24.5 – 34.5	21	→ Modal Class
34.5 – 44.5	23	
44.5 – 54.5	14	
54.5 – 64.5	5	

Here, Highest frequency group = 34.5 – 44.5

∴ Modal class = 34.5 – 44.5

Thus, $l = 34.5, h = 10, f_1 = 23, f_0 = 21, f_2 = 14$

$$\text{Mode } = l + \left(\frac{f_1 - f_0}{2f_1 - f_0 - f_2}\right)\times h$$

$$= 34.5 + \left(\frac{23-21}{46-21-14}\right)\times 10$$

$$= 34.5 + \frac{2}{11}\times 10$$

$$= 34{\cdot}5 + 1.81$$

$$= 36.31$$ **Ans.**

Q.18. The table below gives the distribution of villages under different heights from sea level in a certain region

Heights	200	600	1000	1400	1800	2200
No.of village	142	265	560	271	89	16

Compute the mean height of the region.

Solution :

Height (x_i)	No.of villages (f_i)	$u_i = \frac{x_i - A}{h}$	$f_i u_i$
200	142	– 3	– 426
600	265	– 2	– 530
1000	560	– 1	– 560
1400	271	0	0
1800	89	1	89
2200	16	2	32
	$\Sigma f_i = 1343$		$\Sigma f_i u_i = -1395$

We know that,

$$\text{Mean } = A + \left\{\frac{\Sigma f_i u_i}{\Sigma f_i}\right\}\times h$$

$$= 1400 + \frac{-1395}{1343}\times 400$$

$$= 1400 - 415.49$$

$$= 984.51$$ **Ans.**

Q. 19. For the month of February, a class teacher of Class IX has the following absentee record for 45 students. Find the mean number of days, a student was absent.

Number of days of absent	Number of students
0 — 4	18
4 — 8	3
8 — 12	6
12 — 16	2
16 — 20	0
20 — 24	1

Solution :

C.I.	f_i	x_i (mid-value)	$d = x_i - A$	$f_i \times d_i$
0 – 4	18	2	– 12	– 216
4 – 8	3	6	– 8	– 24
8 – 12	6	10	– 4	– 24
12 – 16	2	$A = 14$	0	00
16 – 20	0	18	4	00
20 – 24	1	22	8	08
	$\Sigma f_i = 30$			$\Sigma f_i d_i = -256$

$$\text{Mean } = A + \frac{\Sigma f_i d_i}{\Sigma f_i}$$

$$= 14 + \left(\frac{-256}{30}\right)$$

$$= 14 - 8.53$$

$$= 5.47$$ **Ans.**

Q. 20. Find the missing frequency (x) of the following distribution, if mode is 34.5.

Marks obtained	0 – 10	10 – 20	20 – 30	30 – 40	40 – 50
Name of students	4	8	10	x	8

Solution :

C.I.	Frequency
0 – 10	4
10 – 20	8
20 – 30	$10 = f_0$
$l = 30 – 40$	$x = f_1$
40 – 50	$8 = f_2$

We have, mode = 34.5

$\therefore$ Modal class = 30 – 40

We know,

$$\text{Mode} = l + \left(\frac{f_1 - f_0}{2f_1 - f_0 - f_2}\right)h$$

$$34.5 = 30 + \left(\frac{x-10}{2x-10-8}\right)10$$

$$4.5 = \left(\frac{x-10}{2x-18}\right)10$$

$$4.5 = \left(\frac{x-10}{x-9}\right)5$$

$$0.9 = \frac{x-10}{x-9}$$

$$0.9(x-9) = x - 10$$

$$0.9x - 8.1 = x - 10$$

$$x - 0.9x = 10 - 8.1$$

$$0.1x = 1.9$$

$\Rightarrow$ $x = 19$ **Ans.**

Chapter 15. Probability

Q. 1. A box contains 100 red cards, 200 yellow cards and 50 blue cards. If a card is drawn at random from the box, then find the probability that it will be :

(i) a blue card,

(ii) not a yellow card, and

(iii) neither yellow nor a blue card.

Solution : Total number of cards

= 100 + 200 + 50 = 350

(i) Possible outcomes of drawing a blue card = 50

Hence, $P(E) = \frac{50}{350} = \frac{1}{7}$.

(ii) Possible outcomes of not drawing a yellow card = 350 – 200 = 150.

Hence, $P(E) = \frac{150}{350} = \frac{3}{7}$.

(iii) Possible outcomes of drawing neither yellow nor a blue card = 350 – (200 + 50) = 350 – 250 = 100

Hence, $P(E) = \frac{100}{350} = \frac{2}{7}$. **Ans.**

Q. 2. All kings, queens and aces are removed from a pack of 52 cards. The remaining cards are well-shuffled and then a card is drawn from it. Find the probability that the drawn card is (i) a black face card, and (ii) a red card.

Solution : Total number of cards = 52

Number of cards removed = 4 + 4 + 4 = 12

Hence, total number of remaining cards

= 52 – 12 = 40.

(i) Now, number of black face cards (Jack card)

= 1 + 1 = 2

Hence, $P(E) = \frac{2}{40} = \frac{1}{20}$.

(ii) And number of red cards = 13 of hearts + 13 of diamond – (1 king of hearts + 1 king of diamonds + 1 queen of hearts + 1 queen of diamonds + 1 ace of hearts + 1 ace of diamonds) = 20

Hence, $P(E) = \frac{20}{40} = \frac{1}{2}$. **Ans.**

Q. 3. Cards marked with numbers 5, 6, 7, ..., 74 are placed in a bag and mixed thoroughly. One card is drawn at random from the bag. Find the probability that the number on the card is a perfect square.

Solution : Total number of possible outcomes

= 5, 6, 7, 8, 9, ..., 74 = 70

Now, the favourable outcomes = 9, 16, 25, 36, 49, 64

Hence, the total number of favourable outcomes = 6

$\therefore$ $P(E) = \frac{6}{70} = \frac{3}{35}$. **Ans.**

Q. 4. Two dice are rolled together. Find the probability of getting such numbers on the two dice whose product is 12.

Solution : Total number of possible outcomes = (1, 1); (1, 2); (1, 3); (1, 4); (1, 5); (1, 6); (2, 1); (2, 2); (2, 3); (2, 4); (2, 5); (2, 6); (3, 1); (3, 2); (3, 3); (3, 4); (3, 5); (3, 6); (4, 1); (4, 2); (4, 3); (4, 4); (4, 5); (4, 6); (5, 1); (5, 2); (5, 3); (5, 4); (5, 5); (5, 6); (6, 1); (6, 2); (6, 3); (6, 4); (6, 5); (6, 6) = 36

Now, the favourable outcomes = (2, 6); (3, 4); (4, 3); (6, 2).

Thus, the total number of favourable outcomes = 4

Hence, $P(E) = \frac{4}{36} = \frac{1}{9}$. **Ans.**

Q. 5. A box contains 80 discs numbered from 1 to 80. If one disc is drawn at random from the box, find the probability that the number it bears is a perfect square.

Solution : Total number of possible outcomes = 1, 2, 3, 4, 5, 6, 7, 8, 9, ..., 80 = 80

Now, the favourable outcomes = 1, 4, 9, 16, 25, 36, 49, 64

Hence, the total number of favourable outcomes = 8

$\therefore$ $P(E) = \frac{8}{80} = \frac{1}{10}$. **Ans.**

Q. 6. Two dice are rolled together. Find the probability of getting such number on the two dice whose product is a perfect square.

Solution : Total number of possible outcomes = (1, 1); (1, 2); (1, 3); (1, 4); (1, 5); (1, 6); (2, 1); (2, 2); (2, 3); (2, 4); (2, 5); (2, 6); (3, 1); (3, 2); (3, 3); (3, 4); (3, 5); (3, 6); (4, 1); (4, 2); (4, 3); (4, 4); (4, 5); (4, 6); (5, 1); (5, 2); (5, 3); (5, 4); (5, 5); (5, 6); (6, 1); (6, 2); (6, 3); (6, 4); (6, 5); (6, 6) = 36

Now, the favourable outcomes = (1, 1); (2, 2); (3, 3); (4, 4); (5, 5); (6, 6); (1, 4); (4, 1)

Thus, the total number of favourable outcomes = 8

$$\therefore \quad P(E) = \frac{8}{36} = \frac{2}{9}. \qquad \textbf{Ans.}$$

Q. 7. A game consists of tossing a coin three times and noting the outcome each time. Hanif wins if he gets three heads or three tails and loses otherwise. Calculate the probability that Hanif will lose the game.★★

[NCERT]

Solution : The possibilities are *HHH, HHT, HTT, THT, TTH, TTT, HTH, THH*

Thus, the total number of possibilities = 8

The favourable outcomes for Hanif's loss are *HHT, HTT, THT, TTH, HTH, THH*

Thus, the total number of favourable outcomes = 6

$$\therefore \quad P(E) = \frac{6}{8} = \frac{3}{4}. \qquad \textbf{Ans.}$$

Q. 8. There are 100 cards in a bag on which numbers from 1 to 100 are written. A card is taken out from the bag at random. Find the probability that the number on the selected card (i) is divisible by 9 and is a perfect square (ii) is a prime number greater than 80.★

Solution : Number of possible outcomes = 100

(i) Let E_1 be the event of getting a number divisible by 9 and is a perfect square.

∴ Favourable outcomes = {9, 36, 81}

Number of favourable outcomes = 3

$$\therefore \quad P(E_1) = \frac{3}{100} \qquad \textbf{Ans.}$$

(ii) Let E_2 be the event of getting a prime number greater than 80.

∴ Favourable outcomes = {83, 89, 97}

Number of favourable outcomes = 3

$$\therefore \quad P(E_2) = \frac{3}{100} \qquad \textbf{Ans.}$$

Q. 9. A bag contains 5 red balls and some blue balls. If the probability of drawing a blue ball from the bag is thrice that of a red ball, find the number of blue balls in the bag.

Solution : Let the number of blue balls be x.

Thus, sum of the possible outcomes = $5 + x$

Now, the sum of the favourable outcomes if the red ball is drawn = 5.

$$\therefore \quad P(R) = \frac{5}{5+x}$$

Again, the sum of the favourable outcomes if the blue ball is drawn = x.

$$\therefore \quad P(B) = \frac{x}{5+x}$$

Now, $\quad P(B) = 3P(R) \quad$ [Given]

$$\Rightarrow \quad \frac{x}{5+x} = 3\left(\frac{5}{5+x}\right)$$

$$\Rightarrow \quad x = 15$$

Hence, the number of blue balls = 15. **Ans.**

Q. 10. One card is drawn at random from a well-shuffled deck of 52 cards. Find the probability of getting (i) a red king and (ii) a black queen, (iii) a jack of hearts, and (iv) a red face card.

Solution : Total number of cards = 52

(i) The number of probable outcomes of drawing a red king = 2.

$$\therefore \quad P(E) = \frac{2}{52} = \frac{1}{26}. \qquad \textbf{Ans.}$$

(ii) The number of probable outcomes of drawing a black queen = 2.

$$\therefore \quad P(E) = \frac{2}{52} = \frac{1}{26}. \qquad \textbf{Ans.}$$

(iii) The number of probable outcomes of drawing a jack of hearts = 1.

$$\therefore \quad P(E) = \frac{1}{52}. \qquad \textbf{Ans.}$$

(iv) The number of probable outcomes of drawing a red face card = 6.

$$\therefore \quad P(E) = \frac{6}{52} = \frac{3}{26}. \qquad \textbf{Ans.}$$

Q. 11. From a pack of 52 playing cards, Jacks, Queens and kings of red colour are removed. From the remaining, a card is drawn at random. Find the probability that drawn card is :

(i) a black king (ii) a card of red colour (iii) a card of black colour

Solution : Since, Jacks, Queens and Kings of red colour are removed. Then,

Total number of possible outcomes = 52 – 6 = 46

(i) Let E_1 be the event of getting a black king.

∴ Favourable outcomes = king of spade and king of club.

★ are board exam questions from previous years

★★ are frequently asked board exam questions

No. of favourable outcomes = 2

$$P(E_1) = \frac{2}{46} = \frac{1}{23} \qquad \textbf{Ans.}$$

(ii) Let E_2 be the event of getting a card of red colour

$\therefore$ Favourable outcomes = 10 cards of heart and 10 cards of diamonur.

No. of favouable outcomes = 20

$$P(E_2) = \frac{20}{46} = \frac{10}{23} \qquad \textbf{Ans.}$$

(iii) Let E_3 be the event of getting a card of black colour

$\therefore$ Favourable outcomes = 13 cards of spade and 13 cards of club.

No. of favourable outcomes = 26

$$P(E_3) = \frac{26}{46} = \frac{13}{23} \qquad \textbf{Ans.}$$

Q. 12. It is known that a box of 200 electric bulbs contains 16 defective bulbs. One bulb is taken out at random from the box. What is the probability that the bulb drawn is (i) defective, and (ii) non-defective ?

Solution : The total number of possible outcomes = 200

(i) Total number of defective bulbs = 16

$$\therefore \qquad P(E) = \frac{16}{200} = \frac{2}{25}. \qquad \textbf{Ans.}$$

(ii) Total number of non-defective bulbs = 200 – 16 = 184

$$\therefore \qquad P(E) = \frac{184}{200} = \frac{23}{25}. \qquad \textbf{Ans.}$$

Q. 13. What is the probability that two friends will have their birthdays : (i) on the same day, (ii) on different days (ignoring the leap year) ?

Solution : The total number of possible outcomes = 365 × 365 days.

(i) The total number of favourable outcomes when they have birthdays on the same day = 365 days

$$\therefore \qquad P(E) = \frac{365}{365 \times 365} = \frac{1}{365}. \qquad \textbf{Ans.}$$

(ii) Out of 365 days, A can have birthday on any day, hence B can have birthday on any other day of (365 – 1) *i.e.*, 364 days.

Thus, the total number of favourable outcomes when they have birthdays on different days

$$= 365(365 - 1)$$
$$= 365 \times 364 \text{ days}$$

$$\therefore \qquad P(E) = \frac{365 \times 364}{365 \times 365} = \frac{364}{365}. \qquad \textbf{Ans.}$$

Q. 14. A bag contains cards numbered from 1 to 30. If a card is drawn at random from the bag after mixing them thoroughly, find the probability that the number on the card is not divisible by 3.★

★ are board exam questions from previous years

Solution : All possible outcomes = 1, 2, 3, ..., 30

Sum of all possible outcomes = 30

The multiples of 3 are 3, 6, 9, 12, 15, 18, 21, 24, 27, 30

The total number of favourable outcomes

$$= (30 - 10) = 20$$

$$\therefore \qquad P(E) = \frac{20}{30} = \frac{2}{3}. \qquad \textbf{Ans.}$$

Q. 15. Find the probability of getting 53 Fridays in a leap year.

Solution : Total number of days in a leap year = 366

Total number of days in a week = 7 days

Thus, the number of weeks and days in a leap year

$$= \frac{366}{7} = 52 \text{ weeks } 2 \text{ days}$$

Thus, the two days can be Sunday and Monday; Monday and Tuesday; Tuesday and Wednesday; Wednesday and Thursday; Thursday and Friday; Friday and Saturday; Saturday and Sunday

So, the sum of the possible outcomes = 7

and the favourable outcomes are Thursday and Friday; Friday and Saturday

Therefore, the sum of the favourable outcomes = 2

$$\therefore \qquad P(E) = \frac{2}{7}. \qquad \textbf{Ans.}$$

Q. 16. A card is drawn at random from a well-shuffled pack of 52 cards. Find the probability of getting (i) a card of spades or an ace, (ii) a red king, (iii) either a king or a queen, and (iv) neither a king nor a queen.

Solution : Total number of cards = 52

(i) The number of probable outcomes of drawing a card of spades or an ace = 13 + 3 = 16

$$\therefore \qquad P(E) = \frac{16}{52} = \frac{4}{13}. \qquad \textbf{Ans.}$$

(ii) The number of probable outcomes of drawing a red king = 2

$$\therefore \qquad P(E) = \frac{2}{52} = \frac{1}{26}. \qquad \textbf{Ans.}$$

(iii) The number of probable outcomes of drawing a king or a queen = 4 + 4 = 8

$$\therefore \qquad P(E) = \frac{8}{52} = \frac{2}{13}. \qquad \textbf{Ans.}$$

(iv) The number of probable outcomes of drawing neither a king nor a queen = 52 – (4 + 4) = 44

$$\therefore \qquad P(E) = \frac{44}{52} = \frac{11}{13}. \qquad \textbf{Ans.}$$

Q. 17. Cards marked with numbers 13, 14, 15, ..., 60 are placed in a box and mixed thoroughly. One card is drawn at random from the box. Find the probability that the number on the card is (i) divisible by 5, and (ii) a number that is a perfect square.

Solution : The number of possible outcomes = 48

(i) The favourable outcomes for the number to be divisible by 5 are 15, 20, 25, 30, 35, 40, 45, 50, 55, 60

Hence, the sum of the number of outcomes of favourable condition = 10

$$\therefore \quad P(E) = \frac{10}{48} = \frac{5}{24}. \quad \textbf{Ans.}$$

(ii) The favourable outcomes for the number to be a perfect square are 16, 25, 36, 49.

Hence, the sum of the number of outcomes of favourable conditions = 4

$$\therefore \quad P(E) = \frac{4}{48} = \frac{1}{12}. \quad \textbf{Ans.}$$

Q. 18. Cards numbered 1, 3, 5, ..., 35 are kept in a bag. If a card is drawn at random from the bag, find the probability that the number of the card is : (i) a prime number less than 15, and (ii) a number divisible by 3 and 5.

Solution : All possible outcomes = 1, 3, ..., 35

Sum of all possible outcomes = 18

(i) The favourable outcomes of a prime number less than 15 = 3, 5, 7, 11, 13.

Thus, total number of favourable outcomes = 5

$$\therefore \quad P(E) = \frac{5}{18}. \quad \textbf{Ans.}$$

(ii) The favourable outcomes of a number divisible by 3 and 5 = 15, 30.

Thus, total number of favourable outcomes = 2.

$$\therefore \quad P(E) = \frac{2}{18} = \frac{1}{9}. \quad \textbf{Ans.}$$

Q. 19. A bag contains tickets numbered 11, 12, 13, ..., 30. A ticket is drawn at random from the bag. Find the probability that the number on the ticket is a (i) multiple of 7, and (ii) is greater than 15 and a multiple of 5.

Solution : The number of possible outcomes = 20

(i) The favourable outcomes for the number to be a multiple of 7 are 14, 21, 28.

Hence, the sum of the number of outcomes of favourable conditions = 3.

$$\therefore \quad P(E) = \frac{3}{20}. \quad \textbf{Ans.}$$

(ii) The favourable outcomes for the number to be greater than 15 and a multiple of 5 are 20, 25, 30.

Hence, the sum of the number of outcomes of favourable conditions = 3.

$$\therefore \quad P(E) = \frac{3}{20}. \quad \textbf{Ans.}$$

Q. 20. Cards marked with numbers 3, 5, 7, ..., 35, 37 are placed in a box and mixed thoroughly. One card is drawn at random from the bag. Find the probability that the number on the card is a prime number.

Solution : Total number of possible outcomes = 3, 5, 7, ..., 35, 37 = 18

Thus, the favourable outcomes are 3, 5, 7, 11, 13, 17, 19, 23, 29, 31, 37.

Hence, the total number of favourable outcomes = 11

$$\therefore \quad P(E) = \frac{11}{18}. \quad \textbf{Ans.}$$

Q. 21. Cards marked with numbers 1, 2, 3, 4, ..., 20 are well-shuffled and a card is drawn from it at random. What is the probability that the number on the card is (i) a prime number, (ii) divisible by 3, and (iii) a perfect square.

Solution : The sum of the maximum possible outcomes = 20.

(i) The most favourable outcomes for a prime number are 2, 3, 5, 7, 11, 13, 17, 19.

Thus, the sum of the most favourable outcomes for a prime number = 8.

$$\therefore \quad P(E) = \frac{8}{20} = \frac{2}{5}. \quad \textbf{Ans.}$$

(ii) The most favourable outcomes for a number divisible by 3 are 3, 6, 9, 12, 15, 18.

Thus, the sum of the most favourable outcomes for a number divisible by 3 = 6.

$$\therefore \quad P(E) = \frac{6}{20} = \frac{3}{10}. \quad \textbf{Ans.}$$

(iii) The most favourable outcomes for a perfect square are 1, 4, 9, 16.

Thus, the sum of the most favourable outcomes for a perfect square = 4.

$$\therefore \quad P(E) = \frac{4}{20} = \frac{1}{5}. \quad \textbf{Ans.}$$

Q. 22. A die is thrown twice. Find the probability that★

(i) 5 will come up at least once.

(ii) 5 will not come up either time.

Solution : When two dice are thrown simultaneously, all possible outcomes are

(1, 1), (1, 2), (1, 3), (1, 4), (1, 5), (1, 6)

(2, 1), (2, 2), (2, 3), (2, 4), (2, 5), (2, 6)

(3, 1), (3, 2), (3, 3), (3, 4), (3, 5), (3, 6)

(4, 1), (4, 2), (4, 3), (4, 4), (4, 5), (4, 6)

(5, 1), (5, 2), (5, 3), (5, 4), (5, 5), (5, 6)

★ are board exam questions from previous years

(6, 1), (6, 2), (6, 3), (6, 4), (6, 5), (6, 6)

Total number of outcomes = 36

Total outcomes where 5 comes up at least once = 11

(i) Probability that 5 will come up at least once

$$= \frac{\text{Total outcomes where 5 will come up}}{\text{Total number of outcomes}}$$

$$= \frac{11}{36}$$

Total outcomes where 5 will not come up

$= 36 - 11 = 25$

(ii) Probability that 5 will not come up either time

$$= \frac{\text{Total outcomes where 5 will not come up}}{\text{Total number of outcomes}}$$

$$= \frac{25}{36}$$ **Ans.**

Q. 23. From a pack of 52 playing cards, all cards that are multiple of 3 are removed. Then a card is drawn at random from it. What is the probability that the card drawn is a (i) face card and (ii) an even-numbered red card ?

Solution : The maximum number of possible conditions = 52.

Number of cards which are multiples of 3 = 3, 6, 9 each of clubs, spades, hearts and diamonds = 12.

Thus, the new maximum number of possible conditions = 52 – 12 = 40.

(i) The sum of the most favourable outcomes for face cards = 12.

$\therefore \quad P(E) = \frac{12}{40} = \frac{3}{10}.$ **Ans.**

(ii) The sum of the most favourable outcomes for even numbered red cards = 2, 4, 8 and 10 of hearts and diamonds each = 8

$\therefore \quad P(E) = \frac{8}{40} = \frac{1}{5}.$ **Ans.**

Q. 24. A bag contains 15 white and some black balls. If the probability of drawing a black ball from the bag is thrice that of drawing a white ball, find the number of black balls in the bag.

Solution : Let the number of black balls in the bag be n.

$\therefore$ Total number of balls are $15 + n$

$\because P(\text{Black ball}) = 3 \times P(\text{White ball})$ [Given]

$$\Rightarrow \quad \frac{n}{15+n} = 3 \times \frac{15}{15+n}$$

$\Rightarrow \quad n = 45.$ **Ans.**

Q. 25. The probability of selecting a blue marble at random from a jar that contains only blue, black and green marbles is $\frac{1}{5}$. The probability of selecting a black marble at random from the same jar is $\frac{1}{4}$. If the jar contains 11 green marbles, find the total number of marbles in the jar. ★

Solution : Let probability of selecting a blue marble, black marble and green marble are P(x), P(y), P(z) respecitvely.

$$(x) = \frac{1}{5},\ P(y) = \frac{1}{4} \qquad \text{(Given)}$$

We know,

$$P(x) + P(y) + P(z) = 1$$

$$\frac{1}{5} + \frac{1}{4} + P(z) = 1$$

$$\frac{4+5}{20} + P(z) = 1$$

$$\frac{9}{20} + P(z) = 1$$

$$\Rightarrow \quad P(z) = 1 - \frac{9}{20}$$

$$= \frac{20-9}{20}$$

$$P(z) = \frac{11}{20}$$

$$\frac{\text{No. of green marbles}}{\text{Total no. of marbles}} = \frac{11}{20}$$

$$\frac{11}{\text{Total no. of marbles}} = \frac{11}{20}$$

($\because$ No. of green marbles = 11)

Total no. of marbles = 20

$\therefore$ There are 20 marbles in the jar. **Ans.**

Q. 26. A group consists of 12 persons, of which 3 are extremely patient, other 6 are extremely honest and rest are extremely kind. A person from the group is selected at random. Assuming that each person is equally likely to be selected, find the probability of selecting a person who is (i) extremely patient (ii) extremely kind or honest.

Solution : Number of extremely patient persons = 3

Number of extremely honest persons = 6

Number of extremely kind persons = 12 – (6 + 3) = 3

Total number of ways to select a person = 12

(i) Number of ways to select one extremely patient person = 3

★ **are board exam questions from previous years**

$\therefore$ Probability of selecting one extremely patient person $= \frac{3}{12} = \frac{1}{4}$ **Ans.**

(ii) Number of ways of selecting one extremely kind or honest person = 6 + 3 = 9

$\therefore$ Probability of selecting a person who is extremely kind or honest $= \frac{9}{12} = \frac{3}{4}$ **Ans.**

Q. 27. A game of chance consists of spinning an arrow which comes to rest pointing at one of the number 1, 2, 3, 4, 5, 6, 7, 8 (see figure) and these are equally likely outcomes. What is the probability that it will point at :

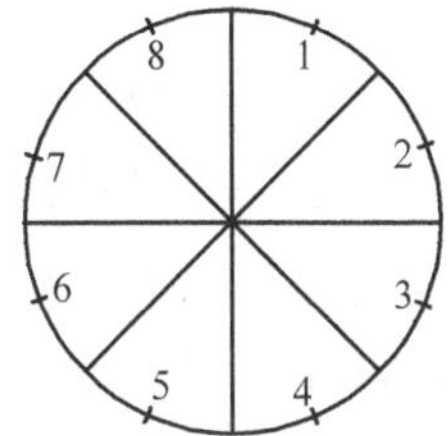

(i) 8 ?

(ii) an odd number ?

(iii) a number greater than 2 ?

(iv) a number less than 9 ?

Solution : (i) Here, total number of points = 8

As, number of favourable outcomes when an arrow will point at 8 = 1

So, P (arrow will point at 8)

$$= \frac{\text{Number of favourable outcomes}}{\text{Total number of possible outcomes}}$$

$$= \frac{1}{8}$$ **Ans.**

(ii) Here, number of odd number points = 4

So, P(arrow will point at odd number)

$$= \frac{\text{Number of favourable outcomes}}{\text{Total number of possible outcomes}}$$

$$= \frac{4}{8} = \frac{1}{2}$$ **Ans.**

(iii) Here, number of points greater than 2 = 6

So, P(arrows will point at a number greater than 2)

$$= \frac{\text{Number of favourable outcomes}}{\text{Total number of possible outcomes}}$$

$$= \frac{6}{8} = \frac{3}{4}$$ **Ans.**

(iv) Here, number of point less than 9 = 8.

So, P(arrow will point at a number less than 9)

$$= \frac{\text{Number of favourable outcomes}}{\text{Total number of possible outcomes}}$$

$$= \frac{8}{8} = 1$$ **Ans.**

Q. 28. A die is thrown once. Find the probability of getting a number which (i) is a prime number (ii) lies between 2 and 6.★

Solution : In throwing a die

Total possible outcomes = 6

i.e., $S = \{1, 2, 3, 4, 5, 6\}$

Prime number 2, 3, 5

$$\therefore \quad P\text{ (Prime No.)} = \frac{\text{favourable outcomes}}{\text{Total possible outcomes}}$$

$$= \frac{3}{6} = \frac{1}{2}$$

Numbers between 2 and 6 are 3, 4,5

P (Numbers between 2 and 6) $= \frac{3}{6} = \frac{1}{2}$ **Ans.**

★ are board exam questions from previous years

Long Questions

| Set 4 |

Chapter 1. Real Numbers

Q. 1. Use Euclid's Division lemma to show that the square of any positive integer is either of the form 3*m* or (3*m* + 1) for some integer *m*. [NCERT]

Solution : Let a be any positive integer.

Then by the principle of Euclid's Division Algorithm corresponding to the positive integers a and 3, there exist non-negative integers such that

$$a = 3q + r \text{ where } 0 \le r < 3$$

Case 1 : If $r = 0$, then

$$a = 3q$$

or $a^2 = 9q^2 = 3(3q^2)$

$= 3m$

where $m = 3q^2$ and is an integer.

Case 2 : If $r = 1$, then

$$a = 3q + 1$$

or $a^2 = (3q + 1)^2$

$= 9q^2 + 6q + 1$

$= 3(3q^2 + 2q) + 1$

$= 3m + 1$

where $m = 3q^2 + 2q$ and is also an integer.

Case 3 : If $r = 2$, then

$$a = 3q + 2$$

or $a^2 = (3q + 2)^2$

$= 9q^2 + 12q + 4$

$= 3(3q^2 + 4q + 1) + 1$

$= 3m + 1$

where $m = 3q^2 + 4q + 1$ and is also an integer

Hence the square of any positive integer is either of the form $3m$ or $(3m + 1)$ for some integer m.

Hence Proved.

Q. 2. Use Euclid's Division Lemma to show that the cube of any positive integer is either of the form 9*m*, 9*m* + 1 or 9*m* + 8 for some integer *m*. [NCERT]

OR

Using Euclid's Division Lemma, show that the cube of any positive integer is of the form 9*q*, (9*q* + 1) or (9*q* + 8) for some integer *q*.

Solution : Let a any positive number and $b = 3$ such that

$$a = 3q + r \text{ where } r = 0, 1, 2$$

Case I : When $r = 0$

$\therefore$ $a = 3q$

$a^3 = 27q^3 = 9(3q^3) = 9m$ where $m = 3q^3$

Case II : When $r = 1$

$a = 3q + 1$

$\Rightarrow$ $a^3 = (3q + 1)^3$

$= 27q^3 + 27q^2 + 9q + 1$

$= 9(3q^3 + 3q^2 + q) + 1$

$\Rightarrow$ $a^3 = 9m + 1$ where $m = 3q^3 + 3q^2 + q$

Case III : When $r = 2$

$a = 3q + 2$

$\Rightarrow$ $a^3 = (3q + 2)^3$

$\Rightarrow$ $a^3 = 27q^3 + 54q^2 + 36q + 8$

$\Rightarrow$ $a^3 = 9(3q^3 + 6q^2 + 4q) + 8$

$\Rightarrow$ $a^3 = 9m + 8$ where $m = 3q^3 + 6q^2 + 4q$

Hence, cube of any integer is either $9m$, $9m + 1$ or $9m + 8$ form where m is some integer. **Hence Proved.**

Q. 3. Show that one and only one out of *n*, (*n* + 1) and (*n* + 2) is divisible by 3, where *n* is any positive integer.★

Solution : Let n, $n + 1$, $n + 2$ be three consecutive positive integers. We know that n is of the form $3q$, $3q + 1$ or $3q + 2$.

Case I : When $n = 3q$

In this case, n is divisible by 3,

but $n + 1$ and $n + 2$ are not divisible by 3.

Case II : When $n = 3q + 1$

In this case

$n + 2 = (3q + 1) + 2 = 3q + 3.$

$= 3(q + 1)$

$(n + 2)$ is divisible by 3,

but n and $n + 1$ are not divisible by 3.

Case III : When $n = 3q + 2$

In the case,

$n + 1 = (3q + 2) + 1$

$= 3q + 3$

$= 3(q + 1)$

$(n + 1)$ is divisible by 3,

but n and $n + 2$ are not divisible by 3.

Hence, only one out of n, $n + 1$ and $n + 2$ is divisible by 3. **Hence Proved.**

★ are board exam questions from previous years

Q. 4. Prove that $\sqrt{2}$ is an irrational number.★★

[NCERT]

Solution : To prove that $\sqrt{2}$ is an irrational number, let us assume its opposite *i.e.*, $\sqrt{2}$ is a rational number. Also let it be $\frac{a}{b}$ in its simplest form where a and b are co-prime numbers having an H.C.F. of 1. Also let $b \neq 0$.

$$\therefore \quad \sqrt{2} = \frac{a}{b}$$

$$\Rightarrow \quad (\sqrt{2})^2 = \left(\frac{a}{b}\right)^2$$

$$\Rightarrow \quad 2 = \frac{a^2}{b^2}$$

$$\Rightarrow \quad 2b^2 = a^2$$

Thus $\quad 2 \mid a^2 \quad [\because 2 \mid 2b^2 \text{ and } 2b^2 = a^2]$

$\Rightarrow \quad 2 \mid a \quad$ ...(i)

[From theorem, if p is a prime member and 'a' is a positive integer then, if p divides a^2 then p divides a]

Now let $a = 2c$ for some integer c.

Replacing the value of a in the equation $2b^2 = a^2$ we have

$$2b^2 = (2c)^2$$

$$\Rightarrow \quad 2b^2 = 4c^2$$

$$\Rightarrow \quad b^2 = 2c^2$$

$\Rightarrow \quad 2 \mid b^2 \quad [\because 2 \mid 2c^2 \text{ and } 2c^2 = b^2]$

$\Rightarrow \quad 2 \mid b \quad$ ...(ii)

[From theorem, if p is a prime member and 'a' is a positive integer then, if p divides a^2 then p divides a]

Hence, from (i) and (ii), we find that 2 is a common factor of both a and b. However, this contradicts the fact that a and b have only 1 as their common factor. Such a contradiction arises by considering $\sqrt{2}$ as a rational number. Hence, it is proved that $\sqrt{2}$ is an irrational number. **Hence Proved.**

Q. 5. Prove that $\frac{2+\sqrt{3}}{5}$ is an irrational number, given that $\sqrt{3}$ is an irrational number.★

Solution. Let $\frac{2+\sqrt{3}}{5}$ is a rational number

$\therefore \quad \frac{2+\sqrt{3}}{5} = \frac{a}{b}$, where a and b are co-prime numbers.

or $\quad 2+\sqrt{3} = \frac{5a}{b}$

$$\sqrt{3} = \frac{5a}{b} - \frac{2}{1}$$

$$\sqrt{3} = \frac{5a-2b}{b}$$

In R.H.S., a, b, 2 and 5 are integers.

$\therefore$ R.H.S. is a rational number but L.H.S. $= \sqrt{3}$, which is given that $\sqrt{3}$ is an irrational

So, it is a contradiction.

Hence, $\frac{2+\sqrt{3}}{5}$ is an irrational number. **Ans.**

Q. 6. Prove that $\sqrt{5}$ is an irrational number. Hence show that $3+2\sqrt{5}$ is also an irrational number.★

Solution : Let $\sqrt{5}$ be a rational number.

So, $\quad \sqrt{5} = \frac{p}{q}$

On squaring both sides

$$5 = \frac{p^2}{q^2}$$

$$q^2 = \frac{p^2}{5}$$

$\Rightarrow$ 5 is a factor of p^2

$\Rightarrow$ 5 is a factor of p.

Now, again let $p = 5c$.

So, $\quad \sqrt{5} = \frac{5c}{q}$

On squaring both sides

$$5 = \frac{25c^2}{q^2}$$

$$q^2 = 5c^2$$

$$c^2 = \frac{q^2}{5}$$

$\Rightarrow$ 5 is factor of q^2

$\Rightarrow$ 5 is a factor of q.

Here 5 is a common factor of p, q which contradicts the fact that p, q are co-prime.

Hence our assumption is wrong, $\sqrt{5}$ is an irrational number.

Now, we have to show that $3 + 2\sqrt{5}$ is an irrational number. So let us assume that $3 + 2\sqrt{5}$ is a rational number.

$$\Rightarrow \quad 3 + 2\sqrt{5} = \frac{p}{q}$$

★ are board exam questions from previous years

★★ are frequently asked board exam questions

$$\Rightarrow \quad 2\sqrt{5} = \frac{p}{q} - 3$$

$$\Rightarrow \quad 2\sqrt{5} = \frac{p-3q}{q}$$

$$\Rightarrow \quad \sqrt{5} = \frac{p-3q}{2q}$$

$\frac{p-3q}{2q}$ is in the rational form of $\frac{p}{q}$, so $\sqrt{5}$ is a rational number but we have already proved that $\sqrt{5}$ is an irrational number so contradiction arises because we supposed wrong that $3 + 2\sqrt{5}$ is a rational number. So we can say that $3 + 2\sqrt{5}$ is an irrational number.

Hence Proved.

Q. 7. Prove that $2 + 5\sqrt{3}$ is an irrational number, given that $\sqrt{3}$ is an irrational number.★

Solution : Let $2+5\sqrt{3} = r$, where, r is rational.

$$(2+5\sqrt{3})^2 = r^2$$

$$40 + 75 + 20\sqrt{3} = r^2$$

$$79 + 20\sqrt{3} = r^2$$

$$20\sqrt{3} = r^2 - 79$$

$$\sqrt{3} = \frac{r^2-79}{20}$$

Now, $\frac{r^2-79}{20}$ is a rational number. So, $\sqrt{3}$ must also be a rational number. But $\sqrt{3}$ is an irrational number (Given).

So, our assumption is wrong.

$\therefore$ $2+5\sqrt{3}$ is an irrational number.

Hence Proved.

Q. 8. Find HCF of numbers 134791, 6341 and 6339 by Euclid's division algorithm.★

Solution : First we find HCF of 6339 and 6341 by Euclid's division method

```
6339)6341(1
     6339
        2)6339( 3169
          6
          --
           3
           2
          --
           13
           12
          ---
            19
            18
           ----
           1)2(2
             2
            ---
             0
```

★ are board exam questions from previous years

$$6341 > 6339$$

$$\Rightarrow \quad 6341 = 6339 \times 1 + 2$$

Also, $\quad 6339 = 2 \times 3169 + 1$

$$2 = 1 \times 2 + 0$$

$\therefore$ HCF of 6341 and 6339 is 1.

Now, we find the HCF of 134791 and 1

$$134791 = 1 \times 134791 + 0$$

$\therefore$ HCF of 134791 and 1 is 1.

Hence, HCF of given three numbers is 1. **Ans.**

Chapter 2. Polynomials

Q. 1. On dividing $(x^3 - 3x^2 + x + 2)$ by a polynomial $g(x)$, the quotient and remainder are $(x-2)$ and $(-2x+4)$ respectively. Find the value of $g(x)$.

Solution : Let

$$f(x) = x^3 - 3x^2 + x + 2$$

$$q(x) = x - 2; \; r(x) = -2x + 4$$

By Euclid's division lemma,

$$f(x) = g(x) \times q(x) + r(x)$$

$$\Rightarrow \quad g(x) \times q(x) = f(x) - r(x)$$

$$\Rightarrow \quad g(x) = \frac{f(x)-r(x)}{q(x)} \quad \ldots(i)$$

Now, $f(x) - r(x) = x^3 - 3x^2 + x + 2 + 2x - 4$

$$= x^3 - 3x^2 + 3x - 2$$

$$= x^3 - 2x^2 - x^2 + 2x + x - 2$$

$$= x^2(x-2) - x(x-2) + 1(x-2)$$

$$= (x-2)(x^2 - x + 1)$$

$$\therefore \quad \frac{f(x)-r(x)}{q(x)} = \frac{(x-2)(x^2-x+1)}{(x-2)}$$

$$= (x^2 - x + 1) = g(x)$$

Thus, the value of $g(x)$ is $(x^2 - x + 1)$. **Ans.**

Q. 2. For what value of k, is the polynomial

$$f(x) = 3x^4 - 9x^3 + x^2 + 15x + k$$

completely divisible by $3x^2 - 5$?★

Solution : Given,

$$f(x) = 3x^4 - 9x^3 + x^2 + 15x + k$$

It is completely divisible by $3x^2 - 5$

Let $\quad g(x) = 3x^2 - 5$

$$3x^2 - 5) \overline{3x^4 - 9x^3 + x^2 + 15x + k} (x^2 - 3x + 2$$

$$\begin{array}{r} 3x^4 + 0 \quad - 5x^2 \\ - \qquad + \\ \hline - 9x^3 + 6x^2 + 15x \\ - 9x^3 \qquad + 15x \\ + \qquad - \\ \hline 6x^2 + k \\ 6x^2 - 10 \\ - \quad + \\ \hline k + 10 \end{array}$$

$\because f(x)$ is completely divisible by $g(x)$ then

$$k + 10 = 0$$

$$k = - 10$$ **Ans.**

Q. 3. Find all zeros of the polynomial $3x^4 + 6x^3 - 2x^2 - 10x - 5$, if two of its zeros are $\sqrt{\frac{5}{3}}$ and $-\sqrt{\frac{5}{3}}$.

Solution : Since $\sqrt{\frac{5}{3}}$ and $-\sqrt{\frac{5}{3}}$ are two zeros of $f(x)$

$$\therefore \quad \left(x+\sqrt{\frac{5}{3}}\right)\left(x-\sqrt{\frac{5}{3}}\right) = \left(x^2 - \frac{5}{3}\right)$$

$$= \frac{1}{3}(3x^2 - 5)$$

is a factor of $f(x)$. So, $(3x^2 - 5)$ is factor of $f(x)$.

Divide $f(x)$ by $(3x^2 - 5)$

$$\begin{array}{r} x^2 + 2x + 1 \\ 3x^2 - 5 \overline{)3x^4 + 6x^3 - 2x^2 - 10x - 5} \\ 3x^4 + 0x^3 - 5x^2 \\ - \quad - \quad + \\ \hline 6x^3 + 3x^2 - 10x - 5 \\ 6x^3 + 0x^2 - 10x \\ + \quad - \quad + \\ \hline 3x^2 - 5 \\ 3x^2 - 5 \\ - \quad + \\ \hline 0 \end{array}$$

By division algorithm

$$3x^4 + 6x^3 - 2x^2 - 10x - 5$$
$$= (x^2 + 2x + 1)(3x^2 - 5)$$
$$= (x + 1)^2 (3x^2 - 5).$$

All zeros of $f(x)$ are $x = -1, -1, \sqrt{\frac{5}{3}}, -\sqrt{\frac{5}{3}}$. **Ans.**

Q. 4. Obtain all other zeros of the polynomial $x^4 + 6x^3 + x^2 - 24x - 20$, if two of its zeros are $+2$ and -5.★

★ are board exam questions from previous years

Solution : Given, 2, – 5 are the zeros of polynomial

$p(x) = x^4 + 6x^3 + x^2 - 24x - 20$

So, $(x - 2)$ and $(x + 5)$ are factors of $p(x)$

$\Rightarrow (x - 2)(x + 5)$ is also a factor of $p(x)$

$(x - 2)(x + 5) = x^2 + 3x - 10$

$$x^2 + 3x - 10) \overline{x^4 + 6x^3 + x^2 - 24x - 20} (x^2 + 3x + 2$$

$$\begin{array}{r} x^4 + 3x^3 - 10x^2 \\ - \quad - \quad + \\ \hline 3x^3 + 11x^2 - 24x - 20 \\ 3x^3 + 9x^2 - 30x \\ - \quad - \quad + \\ \hline 2x^2 + 6x - 20 \\ 2x^2 + 6x - 20 \\ - \quad - \quad + \\ \hline 0 \end{array}$$

So, by remainder theorem,

Dividend = Divisor × Quotient + Remainder

$x^4 + 6x^3 + x^2 - 24x - 20$

$= (x^2 + 3x - 10) \times (x^2 + 3x + 2) + 0$

$= x^2 + 5x - 2x - 10 (x^2 + 2x + x + 2)$

$= [x(x + 5) - 2(x + 5)][x(x + 2) + 1(x + 2)]$

$= (x - 2)(x + 5)(x + 2)(x + 1)$

So, other zeros are – 2 and – 1. **Ans.**

Q. 5. Find all zeroes of the polynomial $3x^3 + 10x^2 - 9x - 4$ if one of its zero is 1.★

Solution. Given,

$$P(x) = 3x^3 + 10x^2 - 9x - 4$$

$x = 1$ is a zero of $P(x)$

$\therefore (x - 1)$ is a factor of $P(x)$

To find other zeroes, we divide $P(x)$ by $(x - 1)$

$$x - 1) \overline{3x^3 + 10x^2 - 9x - 4} (3x^2 + 13x + 4$$

$$\begin{array}{r} 3x^3 - 3x^2 \\ - \quad + \\ \hline 13x^2 - 9x \\ 13x^2 - 13x \\ - \quad + \\ \hline 4x - 4 \\ 4x - 4 \\ - \quad + \\ \hline 0 \end{array}$$

$$P(x) = (x - 1)(3x^2 + 13x + 4)$$
$$= (x - 1)(3x^2 + 12x + x + 4)$$

$= (x-1)\{3x(x+4)+1(x+4)\}$

$= (x-1)(x+4)(3x+1)$

other zeroes are $x + 4 = 0$

$x = -4,$

and $3x + 1 = 0$

$x = -\frac{1}{3}$

$\therefore$ other zeroes are $x = -4$ and $x = -\frac{1}{3}$. **Ans.**

Q. 6. Verify that 3, – 1 and $-\frac{1}{3}$ are the zeros of the cubic polynomial $p(x) = 3x^3 - 5x^2 - 11x - 3$ and also verify the relationship between the zeros and their coefficients. [NCERT]

Solution : In order to verify that 3, – 1 and $-\frac{1}{3}$ are the zeros of the cubic polynomial $p(x) = 3x^3 - 5x^2 - 11x - 3$, we have to substitute $x = 3, -1, -\frac{1}{3}$ in $p(x)$ and ensure that the result is 0.

Thus $p(x) = 3x^3 - 5x^2 - 11x - 3$

$$p(3) = 3(3)^3 - 5(3)^2 - 11 \times 3 - 3$$
$$= 3 \times 27 - 5 \times 9 - 33 - 3$$
$$= 81 - 45 - 36$$
$$= 81 - 81 = 0$$
$$p(-1) = 3(-1)^3 - 5(-1)^2 - 11(-1) - 3$$
$$= -3 - 5 + 11 - 3$$
$$= -11 + 11 = 0$$
$$p\left(-\frac{1}{3}\right) = 3\left(-\frac{1}{3}\right)^3 - 5\left(-\frac{1}{3}\right)^2 - 11\left(-\frac{1}{3}\right) - 3$$
$$= -\frac{1}{9} - \frac{5}{9} + \frac{11}{3} - 3$$
$$= \frac{-1-5+33-27}{9}$$
$$= \frac{33-33}{9} = 0$$

Thus, it is verified that 3, – 1 and $-\frac{1}{3}$ are the zeros of the cubic polynomial.

$$p(x) = 3x^3 - 5x^2 - 11x - 3$$

To verify the relationship between the zeros and their coefficients, let the zeros be α, β and γ such that,

$$\alpha = 3, \beta = -1 \text{ and } \gamma = -\frac{1}{3}$$

$$\therefore \quad \alpha + \beta + \gamma = 3 - 1 - \frac{1}{3}$$
$$= 2 - \frac{1}{3} = \frac{5}{3}$$
$$= -\left\{-\frac{5}{3}\right\} = -\frac{\text{Coefficient of } x^2}{\text{Coefficient of } x^3}$$

$$\alpha\beta + \beta\gamma + \alpha\gamma = 3(-1) + (-1)\left(-\frac{1}{3}\right) + 3\left(-\frac{1}{3}\right)$$
$$= -3 + \frac{1}{3} - 1$$
$$= -4 + \frac{1}{3}$$
$$= \frac{-12+1}{3} = -\frac{11}{3}$$
$$= \frac{\text{Coefficient of } x}{\text{Coefficient of } x^3}$$

$$\alpha\beta\gamma = 3(-1)\left(-\frac{1}{3}\right) = 1$$
$$= -\frac{\text{Constant term}}{\text{Coefficient of } x^3}$$

Hence the relationship between the zeros and the their coefficients has also been verified.

Hence Verified.

Q. 7. Verify that the numbers $\frac{1}{2}$, 1 and – 2 are the zeros of the cubic polynomial $f(x) = 2x^3 + x^2 - 5x + 2$. Also verify the relationship between the zeros and the coefficients. [NCERT]

Solution : To verify $\frac{1}{2}$, 1, – 2 are the zeros of given polynomial, we must substitute $x = \frac{1}{2}$, 1, – 2 in $f(x)$ and see whether the result is 0 or not.

$$\therefore \quad f\left(\frac{1}{2}\right) = 2\left(\frac{1}{2}\right)^3 + \left(\frac{1}{2}\right)^2 - 5\left(\frac{1}{2}\right) + 2$$
$$= \frac{1}{4} + \frac{1}{4} - \frac{5}{2} + 2$$
$$= \frac{1}{2} - \frac{5}{2} + 2$$
$$= -2 + 2 = 0$$
$$f(1) = 2(1)^3 + (1)^2 - 5(1) + 2$$
$$= 2 + 1 - 5 + 2$$

$$= 0$$

$$f(-2) = 2(-2)^3 + (-2)^2 - 5(-2) + 2$$

$$= -16 + 4 + 10 + 2$$

$$= 0$$

Hence, as all the outcomes are 0, it is verified that $\frac{1}{2}$, 1 and – 2 are the zeros of the cubic polynomial

$$f(x) = 2x^3 + x^2 - 5x + 2$$

Now, $f(x) = 2x^3 + x^2 - 5x + 2$

Let α, β and γ be the zeros of $f(x)$.

So, $\alpha + \beta + \gamma = \frac{1}{2} + 1 + (-2)$

$$= -1 + \frac{1}{2} = -\frac{1}{2}$$

$$= -\frac{\text{Coefficient of } x^2}{\text{Coefficient of } x^3}$$

$$\alpha\beta + \beta\gamma + \alpha\gamma = \frac{1}{2} \times 1 + 1 \times (-2) + \frac{1}{2} \times (-2)$$

$$= \frac{1}{2} - 2 - 1 = -\frac{5}{2}$$

$$= \frac{\text{Coefficient of } x}{\text{Coefficient of } x^3}$$

$$\alpha\beta\gamma = \frac{1}{2} \times 1 \times (-2) = -1$$

$$= -\frac{\text{Constant term}}{\text{Coefficient of } x^3}$$

Hence, the relationship between the zeros and the cocfficicnts is also verified. **Hence Verified.**

Q. 8. Find the zeroes of the quadratic polynomia $7y^2 - \frac{11}{3}y - \frac{2}{3}$ and verfiy the relationship between the zeroes and the coefficients.★

Solution : The given polynomial is

$$p(y) = 7y^2 - \frac{11}{3}y - \frac{2}{3}$$

$\because$ $$P(y) = 0$$

$$7y^2 - \frac{11}{3}y - \frac{2}{3} = 0$$

$$21y^2 - 11y - 2 = 0$$

$$21y^2 - 14y + 3y - 2 = 0$$

$$7y(3y - 2) + 1(3y - 2) = 0$$

★ are board exam questions from previous years

$$(3y - 2)(7y + 1) = 0$$

$$y = \frac{2}{3}, -\frac{1}{7}$$

So zeroes of P(y) are $\frac{2}{3}, -\frac{1}{7}$ **Ans.**

Verification : On comparing $7y^2 - \frac{11}{3}y - \frac{2}{3}$ with $ax^2 + bx + c$, we get

$$a = 7, b = \frac{11}{3}, c = \frac{2}{3}$$

$$\text{Sum of zeroes} = \frac{-b}{a}$$

$$\frac{2}{3} + \left(-\frac{1}{7}\right) = \frac{-\left(\frac{-11}{3}\right)}{7}$$

$$\frac{14-3}{21} = \frac{11}{3 \times 7}$$

$$\frac{11}{21} = \frac{11}{21}$$

$$\text{Product of zeroes} = \frac{c}{a}$$

$$\frac{2}{3} \times \left(-\frac{1}{7}\right) = \frac{\left(-\frac{2}{3}\right)}{7}$$

$$\frac{-2}{21} = \frac{-2}{3 \times 7}$$

$$\frac{-2}{21} = \frac{-2}{21}$$

Hence Verified.

Q. 9. Find the quotient and remainder on dividing $f(x)$ by $g(x)$ in the following :

(i) $f(x) = x^3 - 6x^2 + 11x - 6$; $g(x) = x + 2$ [NCERT]

(ii) $f(x) = x^3 - 3x^2 + 5x - 3$; $g(x) = x^2 - 2$ [NCERT]

Solution : (i) Given, $f(x) = x^3 - 6x^2 + 11x - 6$ and $g(x) = x + 2$

To find quotient and remainder,

Divide $f(x)$ by $g(x)$

$$\begin{array}{r|l} & x^2 - 8x + 27 \\ x+2 & x^3 - 6x^2 + 11x - 6 \\ & x^3 + 2x^2 \\ & - \quad - \\ \hline & -8x^2 + 11x \\ & -8x^2 - 16x \\ & + \quad + \\ \hline & 27x - 6 \\ & 27x + 54 \\ & - \quad - \\ \hline & -60 \end{array}$$

Hence, by division algorithm,

$$f(x) = g(x)\,q(x) + r(x)$$

$$f(x) = (x^2 - 8x + 27)\,(x + 2) - 60$$

$\Rightarrow$ Quotient $= x^2 - 8x + 27$, remainder $= -60$. **Ans.**

(ii) Given, $f(x) = x^3 - 3x^2 + 5x - 3$ and $g(x) = x^2 - 2$

To find quotient and remainder,

Divide $f(x)$ by $g(x)$

$$\begin{array}{r|l} & x - 3 \\ x^2 - 2 & x^3 - 3x^2 + 5x - 3 \\ & x^3 \quad\quad - 2x \\ & - \quad\quad\quad + \\ \hline & -3x^2 + 7x - 3 \\ & -3x^2 \quad\quad + 6 \\ & + \quad\quad\quad - \\ \hline & \quad\quad 7x - 9 \end{array}$$

Hence quotient $= x - 3$, remainder $= 7x - 9$. **Ans.**

Q. 10. If two zeros of the polynomial $f(x) = (x^4 - 6x^3 - 26x^2 + 138x - 35)$ are $(2 \pm \sqrt{3})$, find other zeros.

Solution : It is given that $(2+\sqrt{3})$ and $(2-\sqrt{3})$ are two zeros of $f(x)$.

$$\therefore \{x-(2+\sqrt{3})\}\{x-(2-\sqrt{3})\}$$

$$= (x-2-\sqrt{3})\,(x-2+\sqrt{3})$$

$$= (x-2)^2 - (\sqrt{3})^2$$

$$= x^2 - 4x + 1$$

$\therefore x^2 - 4x + 1$ is a factor of $f(x)$ Hence, $f(x)$ will be divisible by $x - 4x + 1$.

Let us now divide $f(x)$ by $x^2 - 4x + 1$.

$$\begin{array}{r|l} & x^2 - 2x - 35 \\ x^2 - 4x + 1 & x^4 - 6x^3 - 26x^2 + 138x - 35 \\ & x^4 - 4x^3 + x^2 \\ & - \quad + \quad - \\ \hline & -2x^3 - 27x^2 + 138x \\ & -2x^3 + 8x^2 - 2x \\ & + \quad - \quad + \\ \hline & -35x^2 + 140x - 35 \\ & -35x^2 + 140x - 35 \\ & + \quad - \quad + \\ \hline & 0 \end{array}$$

$\therefore \quad f(x) = (x^2 - 4x + 1)\,(x^2 - 2x - 35)$

Hence, other two zeros of $f(x)$ are the zeros of the polynomial $x^2 - 2x - 35 = 0$.

$\Rightarrow \quad x^2 - 2x - 35 = x^2 - 7x + 5x - 35 = (x - 7)\,(x + 5) = 0$

$\Rightarrow \quad x = 7, -5$

Hence, other two zeros of $f(x)$ are 7 and -5. **Ans.**

Q. 11. When divided by $(x-3)$, the polynomials $(x^3 - px^2 + x + 6)$ and $(2x^3 - x^2 - (p+3)x - 6)$ leaves the same remainder. Find the value of p.

Solution : Let,

$$f(x) = x^3 - px^2 + x + 6$$

and $\quad g(x) = 2x^3 - x^2 - (p + 3)x - 6$

Now, $\quad x - 3 = 0$

$\Rightarrow \quad x = 3$

Replacing x with 3 in $f(x)$, we have

$$f(3) = (3)^3 - p(3)^2 + 3 + 6$$
$$= 27 - 9p + 9$$
$$= 36 - 9p = 9(4 - p)$$

and in $g(x)$, we have

$$g(3) = 2(3)^3 - (3)^2 - (p + 3)\,(3) - 6$$
$$= 54 - 9 - 3p - 9 - 6$$
$$= 30 - 3p = 3(10 - p)$$

As per the question,

$$9(4 - p) = 3(10 - p)$$

$\Rightarrow \quad 3(4 - p) = 10 - p$

$\Rightarrow \quad 12 - 3p = 10 - p$

$\Rightarrow \quad 3p - p = 12 - 10$

$\Rightarrow \quad 2p = 2$

$\Rightarrow \quad p = 1$

Thus, the value of p is 1. **Ans.**

Q. 12. Find the values of a and b so that $x^4 + x^3 + 8x^2 + ax - b$ is divisible by $x^2 + 1$.★

Solution :

$$\begin{array}{r|l} & x^2 + x + 7 \\ x^2 + 1 & x^4 + x^3 + 8x^2 + ax - b \\ & +x^4 \quad\quad + x^2 \\ & - \quad\quad\quad - \\ \hline & x^3 + 7x^2 + ax - b \\ & +x^3 \quad\quad + x \\ & - \quad\quad\quad - \\ \hline & 7x^2 + (a - 1)\,x - b \\ & +7x^2 \quad\quad\quad + 7 \\ & - \quad\quad\quad\quad - \\ \hline & (a - 1)\,x - b - 7 \end{array}$$

Since $x^4 + x^3 + 8x^2 + ax - b$ is divisible by $x^2 + 1$

$\therefore \quad$ Remainder $= 0$

$\Rightarrow \quad (a - 1)\,x - b - 7 = 0$

or $(a - 1)\,x + (-b - 7) = 0.\,x + 0$

Comparing the coefficients on both sides, we get

$a - 1 = 0, \quad -b - 7 = 0$

$\Rightarrow a = 1 \quad \Rightarrow \quad -7 = b$

$\therefore a = 1, b = -7$ **Ans.**

Q. 13. If $(x-2)$ is a factor of the expression $2x^3 + ax^2 + bx - 14$ and when divided by $(x-3)$, the remainder is 52, find the values of a and b.

Solution : Let $f(x) = 2x^3 + ax^2 + bx - 14$

Since $(x - 2)$ is a factor of $f(x)$,

$\therefore \quad f(2) = 0$

Replacing x by 2 in $f(x)$, we have

★ are board exam questions from previous years

$\Rightarrow \quad f(2) = 2(2)^3 + a(2)^2 + b(2) - 14$

$= 16 + 4a + 2b - 14$

$= 2(1 + 2a + b)$

$\Rightarrow \quad 2(1 + 2a + b) = 0$

$\Rightarrow \quad 2a + b = -1 \quad \text{...(i)}$

Also, when $f(x)$ is divided by $(x - 3)$, the remainder is 52.

$\Rightarrow \quad f(3) = 52$

$f(3) = 2(3)^3 + a(3)^2 + b(3) - 14$

$= 54 + 9a + 3b - 14$

$= 9a + 3b + 40$

$\Rightarrow \quad 9a + 3b + 40 = 52$

$\Rightarrow \quad 9a + 3b = 12$

$\Rightarrow \quad 3a + b = 4$

$\Rightarrow \quad a + (2a + b) = 4$

$\Rightarrow \quad a - 1 = 4 \quad$ [From (i)]

$\Rightarrow \quad a = 5$

and $\quad b = -1 - 10 = -11$

$\therefore \quad a = 5$ and $b = -11$. **Ans.**

Q. 14. Find the values of k for which the polynomial $(k + 4)x^2 + (k + 1)x + 1$ has equal roots. Also find these roots.★

Solution : Let $f(x) = (k + 4)x^2 + (k + 1)x + 1$

Let the roots of $f(x)$ be α and β.

Thus, $\quad \alpha + \beta = -\frac{k+1}{k+4}$

and $\quad \alpha\beta = \frac{1}{k+4}$

According to question,

$\alpha = \beta$

Hence, $\quad 2\alpha = -\frac{k+1}{k+4}$

$\Rightarrow \quad \alpha = -\frac{k+1}{2(k+4)} \quad \text{...(i)}$

and $\quad \alpha\beta = \frac{1}{k+4}$

$\Rightarrow \quad -\frac{k+1}{2(k+4)}\beta = \frac{1}{k+4} \quad$ [From (i)]

$\Rightarrow \quad \beta = -\frac{2}{k+1} \quad \text{...(ii)}$

From (i) and (ii),

$$-\frac{k+1}{2(k+4)} = -\frac{2}{k+1}$$

$\Rightarrow \quad (k + 1)^2 = 4(k + 4)$

$\Rightarrow \quad k^2 + 2k + 1 = 4k + 16$

$\Rightarrow \quad k^2 + 2k - 4k + 1 - 16 = 0$

★ are board exam questions from previous years

$\Rightarrow \quad k^2 - 2k - 15 = 0$

$\Rightarrow \quad k^2 - 5k + 3k - 15 = 0$

$\Rightarrow \quad k(k - 5) + 3(k - 5) = 0$

$\Rightarrow \quad (k - 5)(k + 3) = 0$

$\Rightarrow \quad k = -3, 5$

Thus, the polynomial $(k + 4)x^2 + (k + 1)x + 1$ will have equal roots when k equals -3 and 5. **Ans.**

Q. 15. Find the value of k such that the polynomial $x^2 - (k + 6)x + 2(2k - 1)$ has sum of its zeros equal to half to their product.★

Solution : The given quadratic polynomial is

$x^2 - (k + 6)x + 2(2k - 1)$

Comparing with $ax^2 + bx + c$, we get

$a = 1, b = -(k + 6)$ and $c = 2(2k + 1)$

Let the zeroes of the polynomial be α and β we know that

$$\alpha + \beta = -\frac{b}{a}$$

$$= \frac{k+6}{1}$$

or $\quad \alpha + \beta = k + 6 \quad \text{...(i)}$

Also, $\quad \alpha\beta = \frac{c}{a}$

$$= \frac{2(2k-1)}{1}$$

or $\quad \alpha\beta = 2(2k - 1) \quad \text{...(ii)}$

According to question

$$\text{Sum of zeroes} = \frac{1}{2} \text{ of their product}$$

$\therefore \quad \alpha + \beta = \frac{1}{2}\alpha\beta$

or $\quad k + 6 = \frac{1}{2} \times 2(2k - 1)$

[using equations (i) & (ii)]

or $\quad k + 6 = 2k - 1$

$k = 7$ **Ans.**

Q. 16. What must be subtracted from $8x^4 + 14x^3 - 2x^2 + 7x - 8$ so that the resulting polynomial is exactly divisible by $4x^2 + 3x - 2$?

Solution : We know that

Dividend = Quotient × Divisor + Remainder

$\Rightarrow$ Dividend – Remainder = Quotient × Divisor

Clearly, R.H.S. of the above result is divisible by the

divisor.

Therefore, L.H.S. is also divisible by the divisor. Thus, if we subtract remainder from the dividend then it will be exactly divisible by the divisor.

Dividing $8x^4 + 14x^3 - 2x^2 + 7x - 8$ by $4x^2 + 3x - 2$, we get

$$\begin{array}{r|l} & 2x^2+2x-1 \\ \hline 4x^2+3x-2 & 8x^4+14x^3-2x^2+7x-8 \\ & 8x^4+6x^3-4x^2 \\ & -\quad -\quad + \\ \hline & 8x^3+2x^2+7x-8 \\ & 8x^3+6x^2-4x \\ & -\quad -\quad + \\ \hline & -4x^2+11x-8 \\ & -4x^2-3x+2 \\ & +\quad +\quad - \\ \hline & 14x-10 \end{array}$$

Quotient = $2x^2 + 2x - 1$ and Remainder = $14x - 10$

Thus, if we subtract the remainder $14x - 10$ from $8x^4 + 14x^3 - 2x^2 + 7x - 8$, it will be divisible by $4x^2 + 3x - 2$. **Ans.**

Chapter 3. Pair of Linear Equations in Two Variables

Q. 1. Solve the following systems of equations graphically : $2x + 3y = 2$ and $x - 2y = 8$.

Solution : Let XOX' and YOY' be the X-axis and Y-axis respectively.

Now, $2x + 3y = 2$

$\Rightarrow$ $2x = 2 - 3y$

$\Rightarrow$ $x = \dfrac{2-3y}{2}$

If $y = 0$, $x = 1$

If $y = 2$, $x = -2$

If $y = -2$, $x = 4$

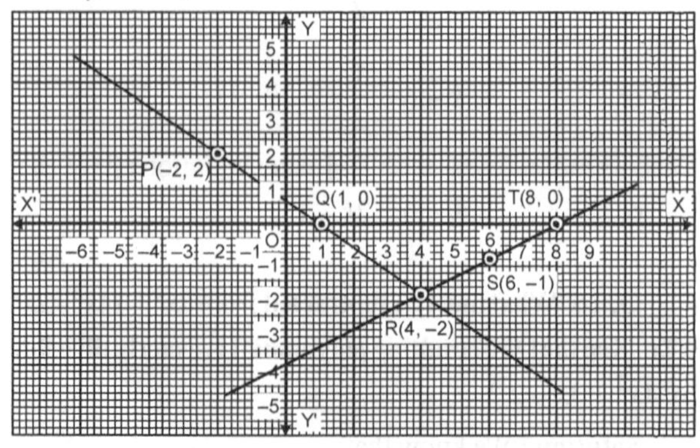

Therefore,

x	– 2	1	4
y	2	0	– 2

Thus, plotting the points $P(-2, 2)$, $Q(1, 0)$ and $R(4, -2)$ on the graph paper, we get the graph of $2x + 3y = 2$, which is represented by PR.

Now, $x - 2y = 8$

$\Rightarrow$ $2y = x - 8$

$\Rightarrow$ $y = \dfrac{x-8}{2}$

If $x = 4$, $y = -2$

If $x = 6$, $y = -1$

If $x = 8$, $y = 0$

Therefore,

x	4	6	8
y	– 2	– 1	0

Thus, plotting the points $R(4, -2)$, $S(6, -1)$ and $T(8, 0)$ on the graph paper, we get the graph of $x - 2y = 8$, which is represented by RT.

Hence, the solution is $(4, -2)$ as this is the point of intersection. **Ans.**

Q. 2. Check graphically whether the pair of equations $3x - 2y + 2 = 0$ and $\frac{3}{2}x - y + 3 = 0$ is consistent. Also find the coordinates of the points where the graphs of the equations meet the Y-axis.★

Solution : Let XOX' and YOY' be the X-axis and Y-axis respectively.

Now, $3x - 2y + 2 = 0$

$\Rightarrow$ $3x = 2y - 2$

$\Rightarrow$ $x = \dfrac{2(y-1)}{3}$

If $y = -2$, $x = -2$

If $y = 1$, $x = 0$

If $y = 4$, $x = 2$

Therefore,

x	– 2	0	2
y	– 2	1	4

Thus, plotting the points $P(-2, -2)$, $Q(0, 1)$ and $R(2, 4)$ on the graph paper, we get the graph of $3x - 2y + 2 = 0$, which is represented by PR.

★ are board exam questions from previous years

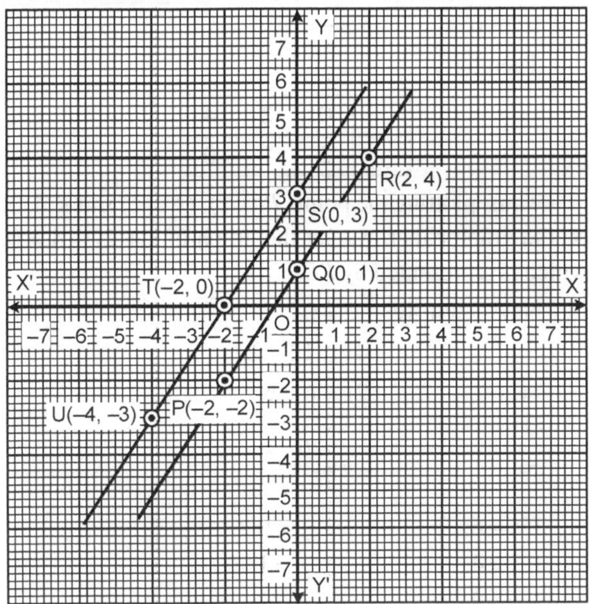

Now, $\frac{3}{2}x - y + 3 = 0$

$\Rightarrow \quad 3x - 2y + 6 = 0$

$\Rightarrow \quad y = \frac{3(x+2)}{2}$

If $x = 0$, $\quad y = 3$

If $x = -2$, $\quad y = 0$

If $x = -4$, $\quad y = -3$

Therefore,

x	0	– 2	– 4
y	3	0	– 3

Thus, plotting the points $S(0, 3)$, $T(-2, 0)$ and $U(-4, -3)$ on the graph paper, we get the graph of $\frac{3}{2}x - y + 3 = 0$, which is represented by SU.

Since, the lines do not intersect each other at any point, so the pair of equations is inconsistent.

Also the lines representing the graphs of the equations, meet the Y-axis at points (0, 1) and (0, 3). **Ans.**

Q. 3. Solve the following pair of linear equations graphically : $2x - y = 1$ and $x - y = -1$. Shade the region between the lines and the Y-axis.

Solution : Let XOX' and YOY' be the X-axis and Y-axis respectively.

Now, $\quad 2x - y = 1$

$\Rightarrow \quad 2x = y + 1$

$\Rightarrow \quad x = \frac{y+1}{2}$

If $y = -1$, $\quad x = 0$

If $y = 1$, $\quad x = 1$

If $y = 3$, $\quad x = 2$

★ are board exam questions from previous years

Therefore,

x	0	1	2
y	– 1	1	3

Thus, plotting the points $P(0, -1)$, $Q(1, 1)$ and $R(2, 3)$ on the graph paper, we get the graph of $2x - y = 1$, which is represented by PR.

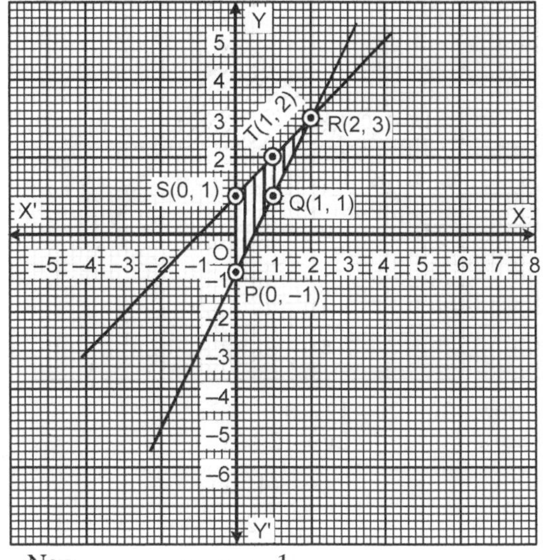

Now, $\quad x - y = -1$

$\Rightarrow \quad y = x + 1$

If $\quad x = 0$, $\quad y = 1$

If $\quad x = 1$, $\quad y = 2$

If $\quad x = 2$, $\quad y = 3$

Therefore,

x	0	1	2
y	1	2	3

Thus, plotting the points $S(0, 1)$, $T(1, 2)$ and $R(2, 3)$ on the graph paper, we get the graph of $x - y = -1$, which is represented by ST.

The solution of the pair of equations by (2, 3).

Shaded region is ΔPRS. **Ans.**

Q. 4. Solve the following system of linear equations graphically : $x - y + 1 = 0$ and $3x + 2y - 12 = 0$. Calculate the area of the region bounded by these lines and the X-axis.★

Solution : Let XOX' and YOY' be the X-axis and Y-axis respectively.

Now, $\quad x - y + 1 = 0$

$\Rightarrow \quad x = y - 1$

If $y = 1$, $\quad x = 0$

If $y = 2$, $\quad x = 1$

If $y = 3$, $\quad x - 2$

Therefore,

x	0	1	2
y	1	2	3

Thus, plotting the points $P(0, 1)$, $Q(1, 2)$ and $R(2, 3)$ on the graph paper, we get the graph of $x - y + 1 = 0$, which is represented by PR.

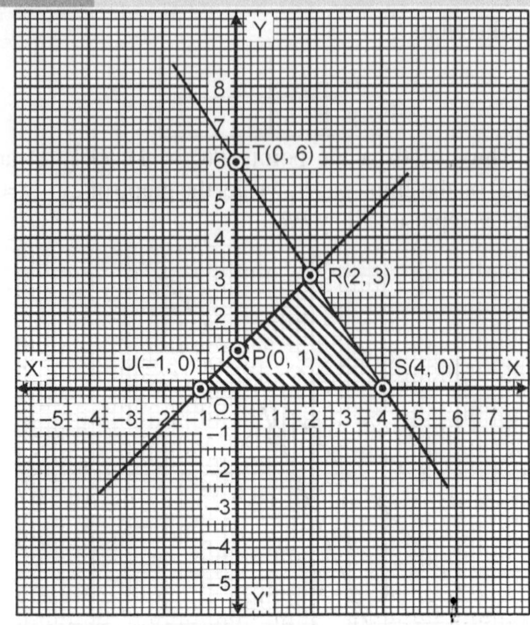

Now, $3x + 2y - 12 = 0$

$\Rightarrow \quad 2y = 12 - 3x$

$\Rightarrow \quad y = \dfrac{3(4-x)}{2}$

If $x = 4$, $\quad y = 0$

If $x = 2$, $\quad y = 3$

If $x = 0$, $\quad y = 6$

Therefore,

x	4	2	0
y	0	3	6

Thus, plotting the points $S(4, 0)$, $R(2, 3)$ and $T(0, 6)$ on the graph paper, we get the graph of $3x + 2y - 12 = 0$, which is represented by ST.

The solution of the pair of equations is given by (2, 3).

Hence, the coordinates of the triangle formed by these two lines and the X-axis are $(-1, 0)$, $(2, 3)$ and $S(4, 0)$.

$$\text{Area} = \frac{1}{2} \times \text{Base} \times \text{Height}$$

Now, $\quad$ Base $= 4 - (-1) = 5$

Height $= 3$

Thus, $\quad \text{Area} = \dfrac{1}{2} \times 5 \times 3$

$= 7.5$ sq. units. **Ans.**

Q. 5. The path of a train A is given by the equation $x + 2y - 4 = 0$ and of B by the equation $2x + 4y - 12 = 0$. Represent this situation graphically. [NCERT]

Solution : Let XOX' and YOY' be the X-axis and Y-axis respectively.

Now, $\quad x + 2y - 4 = 0$

$\Rightarrow \quad x = 4 - 2y$

If $y = 2$, $\quad x = 0$

If $y = 1$, $\quad x = 2$

If $y = 0$, $\quad x = 4$

Therefore,

x	0	2	4
y	2	1	0

Thus, plotting the points $P(0, 2)$, $Q(2, 1)$ and $R(4, 0)$ on the graph paper, we get the graph of $x + 2y - 4 = 0$, which is represented by PR.

Now, $\quad 2x + 4y - 12 = 0$

$\Rightarrow \quad 4y = 12 - 2x$

$\Rightarrow \quad 2y = 6 - x$

$\Rightarrow \quad y = \dfrac{6-x}{2}$

If $x = 6$, $\quad y = 0$

If $x = 4$, $\quad y = 1$

If $x = 2$, $\quad y = 2$

Therefore,

x	6	4	2
y	0	1	2

Thus, plotting the points $S(6, 0)$, $T(4, 1)$ and $U(2, 2)$ on the graph paper, we get the graph of $2x + 4y = 12$, which is represented by SU.

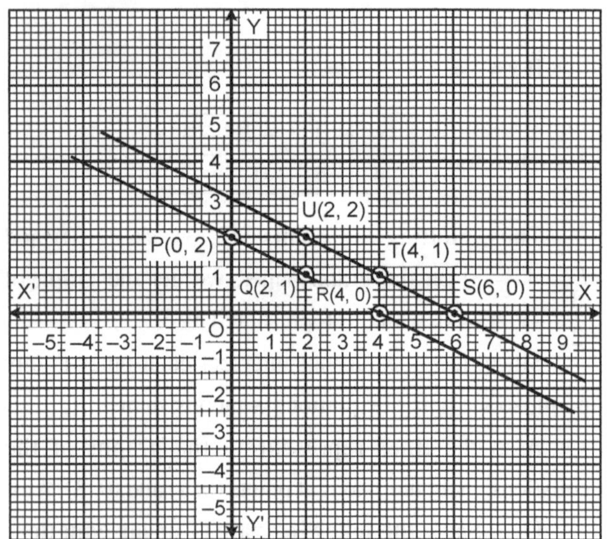

Since the lines do not meet, it is evident that they are parallel and so do not have any solution. **Ans.**

Q. 6. Draw the graph of the following pair of linear equations :

$$x + 3y = 6 \text{ and } 2x - 3y = 2$$

Find the ratio of the areas of the two triangles formed by first line, $x = 0$, $y = 0$ and second line, $x = 0$, $y = 0$.★

Solution :

First line	**Second line**
$x + 3y = 6$	$2x - 3y = 12$
$\Rightarrow x = 6 - 3y$	$\Rightarrow 2x = 12 + 3y$
	$\Rightarrow x = \dfrac{12+3y}{2}$

★ are board exam questions from previous years

x	6	3	0
y	0	1	2

x	6	3	0
y	0	–2	–4

(6, 0), (3, 1), (0, 2)   (6, 0), (3, – 2), (0, – 4),

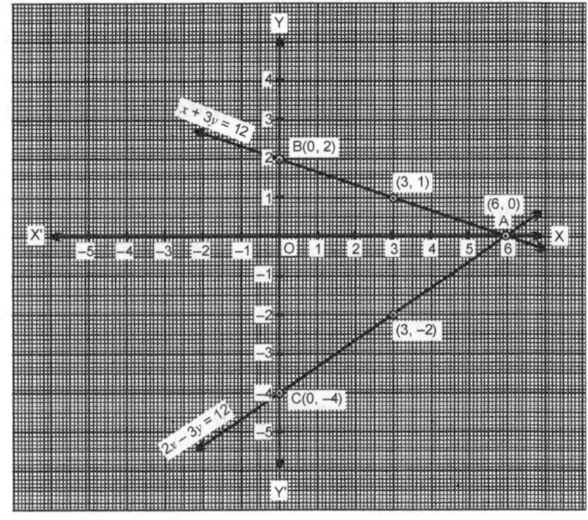

Area of triangle

$$= \frac{1}{2} \times \text{Base} \times \text{Corresponding altitude}$$

$$\therefore \quad \frac{\text{Area of } \Delta AOB}{\text{Area of } \Delta AOC} = \frac{1/2 \times OA \times OB}{1/2 \times OA \times OC}$$

$$\Rightarrow \quad \frac{OB}{OC} = \frac{2}{4} = \frac{1}{2}$$

$\therefore$ Required ratio = 1 : 2 **Ans.**

Q. 7. Show graphically that the following system of equations has infinitely many solutions : $2x + 3y = 6$ and $4x + 6y = 12$.★

Solution : Let XOX' and YOY' be the X-axis and Y-axis respectively.

Now, $2x + 3y = 6$

$\Rightarrow$ $2x = 6 - 3y$

$\Rightarrow$ $x = \frac{6-3y}{2}$

If $y = 4$, $x = -3$

If $y = 2$, $x = 0$

If $y = 0$, $x = 3$

Therefore,

x	– 3	0	3
y	4	2	0

★ are board exam questions from previous years

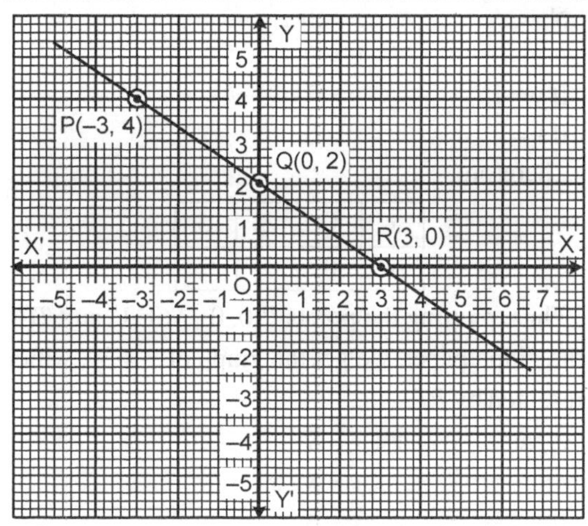

Thus, plotting the points $P(-3, 4)$, $Q(0, 2)$ and $R(3, 0)$ on the graph paper, we get the graph of $2x + 3y = 6$, which is represented by PR.

Now, $4x + 6y = 12$

$\Rightarrow$ $2x + 3y = 6$

Thus, this equation coincide with the graph of the first equation, so the number of solutions is infinite. **Ans.**

Q. 8. Determine graphically the coordinates of the vertices of a triangle, the equations of whose sides are $y = x$, $3y = x$ and $y + x = 8$.

Solution : Let XOX' and YOY' be the X-axis and Y-axis respectively.

Now, $y = x$

If $x = 0$, $y = 0$

If $x = 1$, $y = 1$

If $x = 2$, $y = 2$

Therefore,

x	0	1	2
y	0	1	2

Thus, plotting the points $P(0, 0)$, $Q(1, 1)$ and $R(2, 2)$ on the graph paper, we get the graph of $y = x$, which is represented by PR.

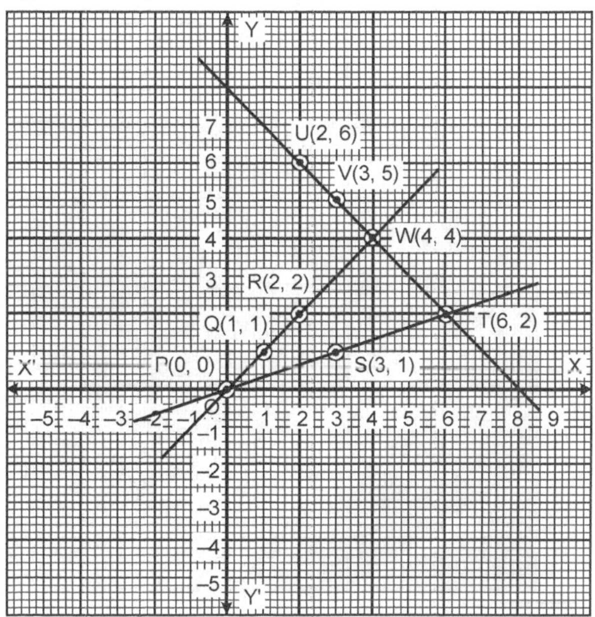

Now, $x = 3y$

If $y = 0$, $x = 0$

If $y = 1$, $x = 3$

If $y = 2$, $x = 6$

Therefore

x	0	3	6
y	0	1	2

Thus, plotting the points $P(0, 0)$, $S(3, 1)$ and $T(6, 2)$ on the graph paper, we get the graph of $x = 3y$, which is represented by PT.

Again, $y + x = 8$

$\Rightarrow$ $y = 8 - x$

If $x = 2$, $y = 6$

If $x = 3$, $y = 5$

If $x = 4$, $y = 4$

Therefore,

x	2	3	4
y	6	5	4

Thus, plotting the points $U(2, 6)$, $V(3, 5)$ and $W(4, 4)$ on the graph paper, we get the graph of $y + x = 8$, which is represented by UT.

The vertices of the triangle thus formed are (0, 0), (4, 4) and (6, 2). **Ans.**

Q. 9. Draw graph of the following pair of linear equations :

$$y = 2(x - 1)$$
$$4x + y = 4$$

Also write the coordinate of the points where these lines meets X-axis and Y-axis.★

Solution : $y = 2(x - 1)$

So,

x	1	2	3	4
y	0	2	4	6

and $4x + y = 4$

or $y = 4 - 4x$

x	1	2	1/2
y	0	–4	2

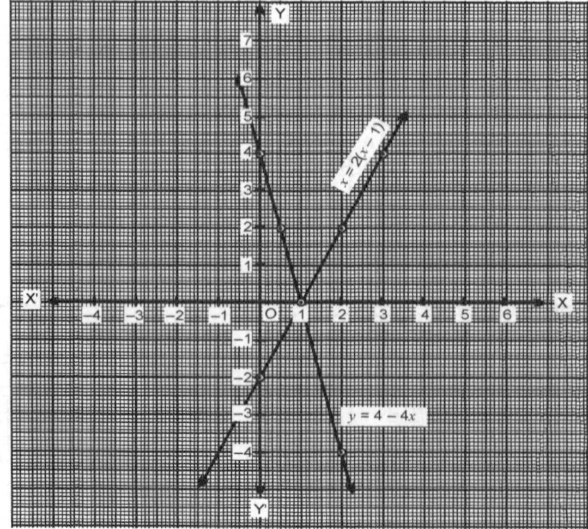

★ are board exam questions from previous years

Coordinates of point where lines meets

Line 1: X-axis = (1, 0)

and Y-axis = (0, – 2)

Line 2 : X-axis = (1, 0)

Y-axis = (0, 4) **Ans.**

Q. 10. The sum of the numerator and denominator of a fraction is 3 less than twice the denominator. If the numerator and denominator are decreased by 1, the numerator becomes half the denominator. Determine the fraction.★

Solution : Let the fraction be $\frac{n}{d}$.

Now, $n + d = 2d - 3$

$\Rightarrow$ $n - d = -3$ ...(i)

Also, $\frac{n-1}{d-1} = \frac{1}{2}$

$\Rightarrow$ $2n - 2 = d - 1$

$\Rightarrow$ $2n - d = 1$ ...(ii)

Subtracting equation (i) from equation (ii), we get

$$2n - d - n + d = 1 + 3$$
$$n = 4$$

Substituting $n = 4$ in equation (i), we get

$$4 - d = -3$$

$\Rightarrow$ $d = 7$

Thus, the fraction is $\frac{4}{7}$. **Ans.**

Q. 11. The sum of the digits of a two-digit number is 12. The number obtained by interchanging its digits exceeds the given number by 18. Find the number.★

Solution : Let the ten's digit be x and the unit's digit be y.

Thus, the given number is $(10x + y)$.

Now, $x + y = 12$ ...(i)

After interchanging the places of the digits, the number is $(10y + x)$.

Now, $10y + x = 10x + y + 18$

$\Rightarrow$ $9x - 9y = -18$

$\Rightarrow$ $x - y = -2$ ...(ii)

Adding equations (i) and (ii), we get

$$2x = 10$$

$\Rightarrow$ $x = 5$

Substituting $x = 5$ in equation (i), we have

$$5 + y = 12$$

$\Rightarrow$ $y = 12 - 5 = 7$

Thus, the required number is 57. **Ans.**

Q. 12. Five years ago, A was thrice as old as B and ten years later, A shall be twice as old as B. What are the present ages of A and B ?

Solution : Let the present ages of A and B be x and y years respectively.

Five years ago,

$$x - 5 = 3(y - 5)$$

$$\Rightarrow \quad x - 5 = 3y - 15$$

$$\Rightarrow \quad x - 3y = -10 \quad \text{...(i)}$$

Ten years later,

$$x + 10 = 2(y + 10)$$

$$\Rightarrow \quad x + 10 = 2y + 20$$

$$\Rightarrow \quad x - 2y = 10 \quad \text{...(ii)}$$

Subtracting equation (ii) from equation (i), we get

$$-3y - (-2y) = -10 - 10$$

$$\Rightarrow \quad -y = -20$$

$$\Rightarrow \quad y = 20$$

Substituting $y = 20$ in equation (i), we get

$$x - 3(20) = -10$$

$$\Rightarrow \quad x - 60 = -10$$

$$\Rightarrow \quad x = 60 - 10$$

$$\Rightarrow \quad x = 50$$

Thus, the ages of A and B are 50 years and 20 years respectively. **Ans.**

Q. 13. A father's age is three times the sum of the ages of his two children. After 5 years his age will be two times the sum of their ages. Find the present age of the father.★

Solution : Let the present age of father be x years and sum of ages of his two children be y years

According to question

$$x = 3y \quad \text{...(i)}$$

After 5 years

Father's age = $(x + 5)$ years

Sum of ages of two children = $(y + 5 + 5)$ years

= $(y + 10)$ years

In 2nd case

According to question

$$x + 5 = 2(y + 10)$$

or $$x + 5 = 2y + 20$$

or $$x - 2y = 15$$

or $$3y - 2y = 15 \quad \{\text{Using equations (i)}\}$$

$$y = 15$$

Now from equation (i)

$$x = 3y \quad (\text{Put } y = 15)$$

or $$x = 3 \times 15$$

$$x = 45$$

So, Present age of father = 45 years. **Ans.**

Q. 14. Places A and B are 160 km apart on the highway. One car starts from A and another from B at the same time. If they travel in the same direction, they meet in 8 hours. If they travel in opposite directions, they meet in 2 hours. Find the speed of each car.

Solution : Let the speed of car A be x km/hr and of B be y km/hr.

★ are board exam questions from previous years

Applying the formula for speed, time and distance :

$$\text{Speed } (s) = \frac{\text{Distance } (d)}{\text{Time } (t)}$$

$$d = s \times t$$

In 8 hours time, A travels $8x$ km and B travels $8y$ km.

Thus, when they travel in the same direction,

$$8x - 8y = 160$$

$$\Rightarrow \quad x - y = 20 \quad \text{...(i)}$$

In 2 hours, A travels $2x$ km and B travels $2y$ km.

Thus, when they travel in opposite directions,

$$2x + 2y = 160$$

$$\Rightarrow \quad x + y = 80 \quad \text{...(ii)}$$

Adding equations (i) and (ii), we get

$$2x = 100$$

$$\Rightarrow \quad x = 50$$

Substituting $x = 50$ in equation (ii), we get

$$50 + y = 80$$

$$\Rightarrow \quad y = 30$$

Thus, the speeds of car A and B are 50 km/hr and 30 km/hr respectively. **Ans.**

Q. 15. Abdul travelled 300 km by train and 200 km by taxi, taking 5 hours and 30 minutes. If he had travelled 260 km by train and 240 km by taxi, he would have taken 6 minutes longer. Find the speed of the train and the taxi.

Solution : Let the speed of the train be x km/hr and that of the taxi be y km/hr.

Time taken by the train to travel 300 km = $\frac{300}{x}$

Time taken by the train to travel 260 km = $\frac{260}{x}$

Time taken by the taxi to travel 200 km = $\frac{200}{y}$

Time taken by the taxi to travel 240 km = $\frac{240}{y}$

According to the question,

$$\frac{300}{x} + \frac{200}{y} = 5\frac{30}{60} = 5\frac{1}{2} = \frac{11}{2}$$

$$\Rightarrow \quad 600y + 400x = 11xy \quad \text{...(i)}$$

and $$\frac{260}{x} + \frac{240}{y} = 5\frac{36}{60} = 5\frac{6}{10} = \frac{56}{10}$$

$$\Rightarrow \quad 2600y + 2400x = 56xy$$

$$\Rightarrow \quad 650y + 600x = 14xy$$

$$\Rightarrow \quad 325y + 300x = 7xy \quad \text{...(ii)}$$

Multiplying equation (i) with 3 and equation (ii) with 4, we get

$$1200x + 1800y = 33xy \quad \text{...(iii)}$$

$$1200x + 1300y = 28xy \quad \text{...(iv)}$$

Subtracting equation (iv) from equation (iii),

$$500y = 5xy$$

$$\Rightarrow \quad x = 100$$

Substituting $x = 100$ in $\frac{300}{x}+\frac{200}{y}=\frac{11}{2}$, we have

$$\frac{300}{100}+\frac{200}{y} = \frac{11}{2}$$

$$\Rightarrow \quad 3+\frac{200}{y} = \frac{11}{2}$$

$$\Rightarrow \quad \frac{200}{y} = \frac{11}{2}-3$$

$$\Rightarrow \quad \frac{200}{y} = \frac{5}{2}$$

$$\Rightarrow \quad y = 80$$

Thus, the speed of the train and taxi are 100 km/hr and 80 km/hr respectively. **Ans.**

Q. 16. A number consists of two-digits. When it is divided by the sum of its digits, the quotient is 6 with no remainder. When the number is diminished by 9, the digits are reversed. Find the number.

Solution : Let the ten's digit be x and the unit's digit be y.

Thus, the given number is $(10x + y)$ and its reverse is $(10y + x)$.

According to the question,

$$\frac{10x+y}{x+y} = 6$$

$$\Rightarrow \quad 10x + y = 6x + 6y$$

$$\Rightarrow \quad 4x - 5y = 0 \quad \text{...(i)}$$

Also, $$10x + y - 9 = 10y + x$$

$$\Rightarrow \quad 9x - 9y = 9$$

$$\Rightarrow \quad x - y = 1 \quad \text{...(ii)}$$

Substituting $x - y = 1$ in equation (i), we get

$$4 - y = 0$$

$$[\because 4x - 5y = 4(x-y) - y]$$

$$\Rightarrow \quad y = 4$$

Substituting $y = 4$ in equation (ii), we get

$$x - 4 = 1$$

$$\Rightarrow \quad x = 5$$

Thus, the number is 54. **Ans.**

Q. 17. A fraction becomes $\frac{1}{3}$, if 2 is added to both the numerator and the denominator and $\frac{2}{5}$, if 3 is added to both the numerator and the denominator. Find the fraction.★

Solution : Let the fraction be $\frac{n}{d}$.

Now, $$\frac{n+2}{d+2} = \frac{1}{3}$$

$$\Rightarrow \quad 3(n+2) = (d+2)$$

$$\Rightarrow \quad 3n - d = -4 \quad \text{...(i)}$$

Also, $$\frac{n+3}{d+3} = \frac{2}{5}$$

$$\Rightarrow \quad 5(n+3) = 2(d+3)$$

$$\Rightarrow \quad 5n - 2d = -9 \quad \text{...(ii)}$$

Multiplying equation (i) by 2, we get

$$6n - 2d = -8 \quad \text{...(iii)}$$

Subtracting equation (ii) from equation (iii), we get

$$6n - 2d - 5n + 2d = -8 + 9$$

$$\Rightarrow \quad n = 1$$

Substituting $n = 1$ in equation (i), we get

$$3 - d = -4$$

$$\Rightarrow \quad d = 7$$

Thus, the fraction is $\frac{1}{7}$. **Ans.**

Q. 18. A fraction becomes $\frac{1}{3}$ when 2 is subtracted from the numerator and it becomes $\frac{1}{2}$ when 1 is subtracted from the denominator. Find the fraction.★

Solution : Let the fraction be $\frac{x}{y}$

According to question

$$\frac{x-2}{y} = \frac{1}{3}$$

or $$3(x-2) = y$$

or $$3x - y = 6 \quad \text{...(i)}$$

again, According to question

$$\frac{x}{y-1} = \frac{1}{2}$$

or $$2x = y - 1$$

or $$2x - y = -1 \quad \text{...(ii)}$$

On solving equation (i) and (ii), we get

$$x = 7, y = 15$$

$\therefore$ The required fraction is $\frac{7}{15}$ **Ans.**

Q. 19. The car hire charges in a city comprise of a fixed charge together with the charge for the distance covered. For a journey of 12 km, the charge paid is ₹ 89 and for a journey of 20 km, the charge paid is ₹ 145. What will a person pay for travelling a distance of 30 km ?

★ are board exam questions from previous years

Solution : Let the fixed charge be ₹ x and the variable charge be ₹ y.

For 12 km, $x + 12y = 89$ ...(i)

For 20 km, $x + 20y = 145$ ...(ii)

Subtracting equation (i) from equation (ii), we get

$$8y = 56$$

$\Rightarrow$ $y = 7$

Substituting $y = 7$ in equation (i), we get

$$x + 12 \times 7 = 89$$

$\Rightarrow$ $x = 89 - 84 = 5$

Thus, for a 30 km journey, the fare would be

$x + 30y = 5 + 30 \times 7 = 5 + 210 =$ ₹ 215. **Ans.**

Q. 20. A boat goes 30 km upstream and 44 km downstream in 10 hours. The same boat goes 40 km upstream and 55 km downstream in 13 hours. On this information, some students guessed the speed of the boat in still water as 8·5 km/h and speed of the stream as 3·8 km/h. Do you agree with their guess ?★

Solution : Let the speed of boat in still water = x km/hr.

Let the speed of stream = y km/hr.

$\therefore$ Speed of boat in upstream = $(x - y)$ km/hr

Speed of boat in downstream = $(x + y)$ km/hr

Time taken to cover 30 km upstream = $\frac{30}{x-y}$ hrs.

Time taken to cover 40 km downstream = $\frac{44}{x+y}$ hrs.

Total time taken = 10 hrs.

$\Rightarrow$ $\frac{30}{x-y} + \frac{44}{x+y} = 10$ ...(i)

Now, time taken to cover 55 km downstream

$= \frac{55}{x+y}$ hrs.

Time taken to cover 40 km upstream = $\frac{40}{x-y}$ hrs.

Total time taken = 13 hrs.

$\Rightarrow$ $\frac{40}{x-y} + \frac{55}{x+y} = 13$ ...(ii)

Let $\frac{1}{x-y} = u$, $\frac{1}{x+y} = v$.

$\therefore$ $30u + 44v = 10$

and $40u + 55v = 13$

or $15u + 22v = 5$ ...(iii)

and $8u + 11v = \frac{13}{5}$ ...(iv)

★ are board exam questions from previous years

Multiplying equation (iii) by 8 and equation (iv) by 15, we get

$$120u + 176v = 40$$
$$120u + 165v = 39$$
$$- \quad - \quad -$$ [On subtracting]
$$11v = 1$$

$$v = \frac{1}{11}$$

Putting the value of v in equation (iii), we get

$$15u + 22 \times \frac{1}{11} = 5$$

$\Rightarrow$ $15u + 2 = 5$

$\Rightarrow$ $15u = 3$

$\Rightarrow$ $u = \frac{3}{15}$

or $u = \frac{1}{5}$

Now, $v = \frac{1}{11}$

$\Rightarrow$ $\frac{1}{x+y} = \frac{1}{11}$

$\Rightarrow$ $x + y = 11$ ...(v)

and $u = \frac{1}{5}$

$\Rightarrow$ $\frac{1}{x-y} = \frac{1}{5}$

$\Rightarrow$ $x - y = 5$ ...(vi)

On adding equations (v) and (vi), we get

$$2x = 16$$

or $x = 8.$

Putting the value of x in equation (v), we get

$$8 + y = 11$$
$$y = 11 - 8$$
$$y = 3$$

Thus, the speed of boat in still water = 8 km/hr.

The speed of stream = 3 km/hr. **Ans.**

Q. 21. The owner of a taxi company decides to run all the taxis on CNG fuel instead of petrol/ diesel. The taxi charges in city comprises of fixed charges together with the charge for the distance covered. For a journey of 13 km, the charge paid is ₹129 and for a journey of 22 km, the charge paid is ₹ 210. What will a person have to pay for travelling a distance of 32 km?★

Solution : Let the fixed charge be ₹ x

Let the charge for the distance = ₹ y per km

According to the question,

For a journey of 13 km,

$$x + 13y = 129$$

$$\Rightarrow \quad x = 129 - 13y \quad ...(i)$$

For a journey of 22 km,

$$x + 22y = 210$$

$$129 - 13y + 22y = 210 \quad \text{[From (i)]}$$

$$9y = 210 - 129 = 81$$

$$9y = 81$$

$$\Rightarrow \quad y = 9$$

∴ The charge for the distance, y = ₹ 9

From equation (i),

$$x = 129 - 13(9)$$

$$= 129 - 117 = 12$$

∴ Fixed charge, x = ₹ 12

To pay for travelling a distance of 32 km

$$= x + 32y$$

$$= 12 + 32(9)$$

$$= 12 + 288 = ₹\ 300 \quad \textbf{Ans.}$$

Q. 22. Ten years hence, a man's age will be twice the age of his son. Ten years ago, the man was four times as old as his son. Find their present ages.★

Solution : Let the man's present age be x years and his son's be y years respectively.

Ten years ago :

$$x - 10 = 4(y - 10)$$

$$\Rightarrow \quad x - 4y = -30 \quad ...(i)$$

Ten years hence :

$$x + 10 = 2(y + 10)$$

$$x - 2y = 10 \quad ...(ii)$$

Subtracting equation (i) from equation (ii), we get

$$x - 2y - x + 4y = 10 + 30$$

$$\Rightarrow \quad 2y = 40$$

$$\Rightarrow \quad y = 20$$

Substituting $y = 20$ in equation (ii), we get

$$x - 40 = 10$$

$$\Rightarrow \quad x = 50$$

Thus, the present age of the father is 50 years and that of son is 20 years. **Ans.**

Q. 23. The monthly incomes of A and B are in the ratio of 5 : 4 and their monthly expenditures are in the ratio of 7 : 5. If each saves ₹ 3,000 per month, find the monthly income of each.

★ are board exam questions from previous years

Solution : Let total income be represented as x and total expenditure be represented as y.

So, incomes of $A : B = 5x : 4x$

and expenditures of $A : B = 7y : 5y$

$$\text{Now,} \quad 5x - 7y = 3000 \quad ...(i)$$

$$\text{and} \quad 4x - 5y = 3000 \quad ...(ii)$$

Multiplying equation (i) with 5 and equation (ii) with 7, we get

$$25x - 35y = 15000 \quad ...(iii)$$

$$28x - 35y = 21000 \quad ...(iv)$$

Subtracting equation (iii) from equation (iv), we get

$$3x = 6000$$

$$\Rightarrow \quad x = 2000$$

Thus, the monthly incomes of $A : B = 5(2000) : 4(2000) = 10000 : 8000$

So, A's monthly income is ₹ 10000 and B's monthly income is ₹ 8000. **Ans.**

Q. 24. A man sold a chair and a table together for ₹ 760, thereby making a profit of 25% on the chair and 10% on the table. Had they been sold together for ₹ 767.50, he would have made a profit of 10% on the chair and 25% on the table. Find the cost price of each.

Solution : Let the cost of each chair be x and the cost of each table be y.

Thus, S.P. of chair with 10% profit

$$= x + \frac{10x}{100} = x + \frac{1x}{10} = \frac{11x}{10}$$

and S.P. of chair with 25% profit $= x + \frac{25x}{100} = x + \frac{x}{4} = \frac{5x}{4}$

= S.P. of chair

Also, S.P. of table with 10% profit

$$= y + \frac{10y}{100} = y + \frac{1y}{10} = \frac{11y}{10}$$

and S.P. of table with 25% profit

$$= y + \frac{25y}{100} = y + \frac{y}{4} = \frac{5y}{4}$$

$$\text{Now,} \quad \frac{5x}{4} + \frac{11y}{10} = 760$$

$$\Rightarrow \quad \frac{50x + 44y}{40} = 760$$

$$\Rightarrow \quad 25x + 22y = 760 \times 20$$

$$\Rightarrow \quad 25x + 22y = 15200 \quad ...(i)$$

$$\text{and} \quad \frac{11x}{10} + \frac{5y}{4} = 767.50$$

$$\Rightarrow \quad \frac{44x + 50y}{40} = 767.50$$

$$\Rightarrow \quad 22x + 25y = 7675 \times 2$$

$$\Rightarrow \quad 22x + 25y = 15350 \quad ...(ii)$$

Adding equations (i) and (ii), we get

$$47x + 47y = 30550$$

$$\Rightarrow \quad x + y = 650 \quad \text{...(iii)}$$

Subtracting equation (ii) from equation (i), we get

$$3x - 3y = -150$$

$$\text{or} \quad x - y = -50 \quad \text{...(iv)}$$

Adding equations (iii) and (iv), we get

$$2x = 600$$

$$\Rightarrow \quad x = 300$$

Substituting $x = 300$ in equation (iii), we get

$$300 + y = 650$$

$$\Rightarrow \quad y = 350$$

Thus, the cost of each chair is ₹ 300 and that of each table is ₹ 350. **Ans.**

Q. 25. Two water taps together can fill a tank in $1\frac{7}{8}$ hours. The tap with longer diameter takes 2 hours less than the tap with smaller one to fill the tank separately. Find the time in which each tap can fill the tank separately.★

Solution : Let the tap A with longer diameter take x hours and the tap B with smaller diameter take $(x + 2)$ hours to fill the tank.

∴ Portion of tank filled by the tap A in 1 hr.

$$= \frac{1}{x}$$

and Portion of tank filled by the tan B in 1 hr.

$$= \frac{1}{x+2}$$

Portion of the tank filled by both taps in 1 hr.

$$= \frac{1}{x} + \frac{1}{x+2}$$

$$= \frac{x+2+x}{x(x+2)}$$

Time taken by both taps to fill the tank

$$= 1\frac{7}{8} \text{ hrs} = \frac{15}{8} \text{hrs}$$

∴ Portion of the tank filled by both in 1 hr.

$$= \frac{8}{15}$$

According to question,

$$\frac{2x+2}{x(x+2)} = \frac{8}{15}$$

$$\frac{2(x+1)}{x(x+2)} = \frac{8}{15}$$

$$15x + 15 = 4x^2 + 8x$$

$$\text{or} \quad 4x^2 - 7x - 15 = 0$$

$$4x^2 - 12x + 5x - 15 = 0$$

$$4x(x-3) + 5(x-3) = 0$$

$$(4x+5)(x-3) = 0$$

$$\Rightarrow \quad 4x + 5 = 0 \text{ or } x - 3 = 0$$

$\Rightarrow x = \frac{-5}{4}$ Since, time can not be negative hence, neglected this value is ; $x = 3$

Hence, the time taken with longer diameter tap = 3 hours

and the time taken with smaller diameter tap = 5 hours. **Ans.**

Q. 26. Places A and B are 70 km apart on the highway. One car starts from A and another from B at the same time. If they travel in the same direction, they meet in 7 hours. If they travel in opposite directions, they meet in 1 hour. Find the speed of each car.

Solution : Let the speed of car A be x km/hr and of B be y km/hr respectively.

Applying the formula for speed, time and distance :

$$\text{Speed } (s) = \frac{\text{Distance } (d)}{\text{Time } (t)}$$

$$d = s \times t$$

In 7 hours time, A travels $7x$ km and B travels $7y$ km.

Thus, when they travel in the same direction,

$$7x - 7y = 70$$

$$\Rightarrow \quad x - y = 10 \quad \text{...(i)}$$

In 1 hour, A travels x km and B travels y km.

Thus, when they travel in opposite directions,

$$\Rightarrow \quad x + y = 70 \quad \text{...(ii)}$$

Adding equations (i) and (ii), we get

$$2x = 80$$

$$\Rightarrow \quad x = 40$$

Substituting $x = 40$ in equation (i), we get

$$40 + y = 70$$

$$\Rightarrow \quad y = 30.$$

Thus, the speeds of A and B are 40 km/hr and 30 km/hr respectively. **Ans.**

Q. 27. A man travels 370 km, partly by train and partly by car. If he covers 250 km by train and 120 cm by car, it takes him 4 hours. But if he travels 130 km by train and 240 km by car, he takes 18 minutes longer. Find the speed of the train and the car.

★ are board exam questions from previous years

Solution : Let the speed of the train be x km/hr and that of the car be y km/hr.

Time taken by the train to travel 250 km = $\frac{250}{x}$

Time taken by the train to travel 130 km = $\frac{130}{x}$

Time taken by the car to travel 120 km = $\frac{120}{y}$

Time taken by the car to travel 240 km = $\frac{240}{y}$

According to the question,

$$\frac{250}{x}+\frac{120}{y}=4$$

$$\Rightarrow \quad 250y + 120x = 4xy$$

$$\Rightarrow \quad 120x + 250y = 4xy \quad \text{...(i)}$$

and $\frac{130}{x}+\frac{240}{y}=4\frac{18}{60}=4\frac{3}{10}=\frac{43}{10}$

[4 hours and 18 minutes

$$=\left(4+\frac{18}{60}\right)h=\left(4+\frac{3}{10}\right)h=4.3h]$$

$$\Rightarrow \quad 1300y + 2400x = 43xy$$

$$\Rightarrow \quad 2400x + 1300y = 43xy \quad \text{...(ii)}$$

Multiplying equation (i) with 20, we get

$$2400x + 5000y = 80xy \quad \text{...(iii)}$$

Subtracting equation (ii) from equation (iii),

$$3700y = 37xy$$

$$\Rightarrow \quad x = 100$$

Substituting $x = 100$ in $\frac{250}{x}+\frac{120}{y}=4$, we have

$$\frac{250}{100}+\frac{120}{y}=4$$

$$\Rightarrow \quad \frac{5}{2}+\frac{120}{y}=4$$

$$\Rightarrow \quad \frac{120}{y}=4-\frac{5}{2}$$

$$\Rightarrow \quad \frac{120}{y}=\frac{3}{2}$$

$$\Rightarrow \quad y=80$$

Thus, the speeds of the train and taxi are 100 km/hr and 80 km/hr respectively. **Ans.**

Q. 28. Find the four angles of a cyclic quadrilateral *ABCD* in which $\angle A = (x + y + 10)°$, $\angle B = (y + 20)°$, $\angle C = (x + y - 30)°$ and $\angle D = (x + y)°$.

Solution : In cyclic quadrilateral *ABCD*,

$$\angle A + \angle C = 180°$$

[Sum of opposite angles is equal to 180°]

$$\Rightarrow \quad (x + y + 10)° + (x + y - 30)° = 180°$$

$$\Rightarrow \quad 2x + 2y = 200°$$

$$\Rightarrow \quad x + y = 100° \quad \text{...(i)}$$

Similarly, $\angle B + \angle D = 180°$

$$(y + 20)° + (x + y)° = 180°$$

$$\Rightarrow \quad x + 2y = 160° \quad \text{...(ii)}$$

Subtracting equation (i) from equation (ii), we get

$$y = 60°$$

Substituting $y = 60°$ in equation (i), we get

$$x + 60° = 100°$$

$$\Rightarrow \quad x = 100° - 60°$$

$$\Rightarrow \quad x = 40°$$

Thus, $\angle A = (40 + 60 + 10)°$, $\angle B = (60 + 20)°$, $\angle C = (40 + 60 - 30)°$ and $\angle D = (40 + 60)°$

or $\angle A = 110°$, $\angle B = 80°$, $\angle C = 70°$ and $\angle D = 100°$. **Ans.**

Q. 29. The sum of the numerator and denominator of a fraction is 4 more than twice the numerator. If the numerator and denominator are increased by 3, then they are in the ratio 2 : 3. Determine the fraction.★★

Solution : Let the fraction be $\frac{n}{d}$.

Now, according to the question,

$$n + d = 2n + 4$$

$$\Rightarrow \quad -n + d = 4 \quad \text{...(i)}$$

Also, $\frac{n+3}{d+3}=\frac{2}{3}$

$$\Rightarrow \quad 3n + 9 = 2d + 6$$

$$\Rightarrow \quad 3n - 2d = -3 \quad \text{...(ii)}$$

Multiplying equation (i) by 2, we get

$$-2n + 2d = 8 \quad \text{...(iii)}$$

Adding equations (ii) and (iii), we get

$$n = 5$$

Substituting $n = 5$ in equation (i), we get

$$-5 + d = 4$$

$$\Rightarrow \quad d = 9$$

Thus, the fraction is $\frac{5}{9}$. **Ans.**

Chapter 4. Quadratic Equations

Q. 1. Solve for x :★

$$\frac{1}{x+1}+\frac{2}{x+2}=\frac{4}{x+4}, x \neq -1, -2, -4$$

★ are board exam questions from previous years

★★ are frequently asked board exam questions

Solution : We have,

$$\frac{1}{x+1}+\frac{2}{x+2}=\frac{4}{x+4}, x\neq -1, -2, -4$$

$$(x+2)(x+4)+2(x+1)(x+4)=4(x+1)(x+2)$$

$$x^2+2x+4x+8+2(x^2+x+4x+4) = 4(x^2+x+2x+2)$$

$$x^2+6x+8+2x^2+10x+8=4x^2+12x+8$$

$$3x^2+16x+16=4x^2+12x+8$$

$$x^2-4x-8=0$$

$$\therefore \quad x=\frac{4\pm\sqrt{16+32}}{2}$$

$$x=\frac{4\pm\sqrt{48}}{2}=2\pm 2\sqrt{3}$$

$$\therefore \quad x=2\pm 2\sqrt{3}.$$ **Ans.**

Q. 2. Solve for x : $3\left(\frac{7x+1}{5x-3}\right)-4\left(\frac{5x-3}{7x+1}\right)=11$, where $x\neq \frac{3}{5}, -\frac{1}{7}$.★

Solution : Given, $3\left(\frac{7x+1}{5x-3}\right)-4\left(\frac{5x-3}{7x+1}\right)=11$

$$\Rightarrow \frac{3(7x+1)^2-4(5x-3)^2}{(7x+1)(5x-3)}=11$$

$$\Rightarrow 3(7x+1)^2-4(5x-3)^2=11(7x+1)(5x-3)$$

$$\Rightarrow 3[49x^2+1+14x]-4[25x^2+9-30x] = 11[35x^2-16x-3]$$

$$\Rightarrow 147x^2+3+42x-100x^2-36+120x = 385x^2-176x-33$$

$$\Rightarrow 47x^2-33+162x=385x^2-176x-33$$

$$\Rightarrow 338x^2=338x$$

$$\Rightarrow 338x(x-1)=0$$

$$\Rightarrow x=1, 0.$$ **Ans.**

Q. 3. Solve for x : $2\left(\frac{2x-1}{x+3}\right)-3\left(\frac{x+3}{2x-1}\right)=5$, where $x\neq \frac{1}{2}, -3$.★★

Solution : Given,

$$2\left(\frac{2x-1}{x+3}\right)-3\left(\frac{x+3}{2x-1}\right)=5$$

$$\Rightarrow \frac{2(2x-1)^2-3(x+3)^2}{(x+3)(2x-1)}=5$$

$$\Rightarrow 2(2x-1)^2-3(x+3)^2=5(x+3)(2x-1)$$

$$\Rightarrow 2[4x^2+1-4x]-3[x^2+9+6x] = 5[2x^2+6x-x-3]$$

$$\Rightarrow 8x^2+2-8x-3x^2-27-18x = 10x^2+25x-15$$

$$\Rightarrow 5x^2-25-26x=10x^2+25x-15$$

$$\Rightarrow 5x^2+51x+10=0$$

$$\Rightarrow 5x^2+50x+x+10=0$$

$$\Rightarrow 5x(x+10)+1(x+10)=0$$

$$\Rightarrow (5x+1)(x+10)=0$$

$$\Rightarrow x=-10, -\frac{1}{5}.$$ **Ans.**

Q. 4. Solve for x :

$$\frac{1}{2a+b+2x}=\frac{1}{2a}+\frac{1}{b}+\frac{1}{2x}.$$

Solution : Given,

$$\frac{1}{2a+b+2x}=\frac{1}{2a}+\frac{1}{b}+\frac{1}{2x}$$

$$\Rightarrow \frac{1}{2a+b+2x}=\frac{bx+2ax+ab}{2abx}$$

$$2abx=(2a+b+2x)(bx+2ax+ab)$$

$$\Rightarrow 2abx=2abx+b^2x+2bx^2+4a^2x + 2bax+4ax^2+2a^2b+ab^2+2abx$$

$$\Rightarrow 0=2abx+b^2x+2bx^2+4a^2x + 2abx+4ax^2+2a^2b+ab^2$$

$$\Rightarrow (4ax^2+2bx^2)+(4a^2x+4abx+b^2x) + (2a^2b+ab^2)=0$$

$$\Rightarrow 2x^2(2a+b)+x[4a^2+4ab+b^2]+ab(2a+b)=0$$

$$\Rightarrow 2x^2(2a+b)+x[4a^2+2ab+2ab+b^2] + ab(2a+b)=0$$

$$\Rightarrow 2x^2(2a+b)+x[2a(2a+b)+b(2a+b)] + ab(2a+b)=0$$

$$\Rightarrow 2x^2(2a+b)+x(2a+b)^2+ab(2a+b)=0$$

$$\Rightarrow 2x^2+x(2a+b)+ab=0$$

$$\Rightarrow x=\frac{-(2a+b)\pm\sqrt{(2a+b)^2-4(2ab)}}{4}$$

$$\left[\text{Applying } x=\frac{-b\pm\sqrt{b^2-4ac}}{2a}\right]$$

$$\Rightarrow x=\frac{-(2a+b)\pm\sqrt{4a^2+4ab+b^2-8ab}}{4}$$

$$\Rightarrow x=\frac{-(2a+b)\pm\sqrt{4a^2+b^2-4ab}}{4}$$

★ are board exam questions from previous years

★★ are frequently asked board exam questions

$$\Rightarrow \quad x = \frac{-(2a+b) \pm (2a-b)}{4}$$

$$= \frac{-2a-b+2a-b}{4} \text{ and } \frac{-2a-b-2a+b}{4}$$

$$\Rightarrow \quad x = \frac{-2b}{4} \text{ and } \frac{-4a}{4}$$

$$\Rightarrow \quad x = -\frac{b}{2} \text{ and } -a.$$ **Ans.**

Q. 5. The numerator of a fraction is 3 less than the denominator. If 1 is added to the denominator, the fraction is decreased by $\frac{1}{15}$. Find the fraction.

Solution : Let the denominator of the fraction be x.

Then, the numerator is $x - 3$.

Hence, the fraction is $\frac{x-3}{x}$.

Now,
$$\frac{x-3}{x+1} = \frac{x-3}{x} - \frac{1}{15}$$

$$\Rightarrow \quad (x-3)\left(\frac{1}{x+1} - \frac{1}{x}\right) = -\frac{1}{15}$$

$$\Rightarrow \quad \frac{(x-3)(x-x-1)}{x(x+1)} = -\frac{1}{15}$$

$$\Rightarrow \quad -\frac{(x-3)}{(x+1)x} = -\frac{1}{15}$$

$$\Rightarrow \quad \frac{(x-3)}{(x+1)x} = \frac{1}{15}$$

$$\Rightarrow \quad 15(x-3) = x(x+1)$$

$$\Rightarrow \quad 15x - 45 = x^2 + x$$

$$\Rightarrow \quad x^2 - 15x + x + 45 = 0$$

$$\Rightarrow \quad x^2 - 14x + 45 = 0$$

$$\Rightarrow \quad x^2 - 9x - 5x + 45 = 0$$

$$\Rightarrow \quad x(x-9) - 5(x-9) = 0$$

$$\Rightarrow \quad (x-9)(x-5) = 0$$

$$\Rightarrow \quad x = 5 \text{ or } 9$$

Thus, the fraction is $\frac{2}{5}$ or $\frac{2}{3}$. **Ans.**

Q. 6. A shopkeeper buys some books for ₹ 80. If he had bought 4 more books for the same amount then the price of each book would have been reduced by ₹ 1. Find the number of books he bought.

Solution : Let the number of books be x.

Thus, the price of 1 book = ₹ $\frac{80}{x}$.

If the number of books were $x + 4$, then the price per book = ₹ $\frac{80}{x+4}$

According to the question,

$$\frac{80}{x+4} = \frac{80}{x} - 1$$

$$\Rightarrow \quad \frac{80}{x} - \frac{80}{x+4} = 1$$

$$\Rightarrow \quad 80\left(\frac{1}{x} - \frac{1}{x+4}\right) = 1$$

$$\Rightarrow \quad 80\left(\frac{x+4-x}{x(x+4)}\right) = 1$$

$$\Rightarrow \quad x(x+4) = 320$$

$$\Rightarrow \quad x^2 + 4x - 320 = 0$$

$$\Rightarrow \quad x^2 + 20x - 16x - 320 = 0$$

$$\Rightarrow \quad x(x+20) - 16(x+20) = 0$$

$$\Rightarrow \quad (x+20)(x-16) = 0$$

$$\Rightarrow \quad x = 16, -20$$

As the number of books bought cannot be negative, so the correct answer is 16. **Ans.**

Q. 7. *A* takes 6 days less than *B* to do a work. If both *A* and *B* working together can do it in 4 days, how many days will *B* take to finish it?

Solution : Let *B* can finish a work in x days

So, *A* can finish work in $(x - 6)$ days.

Together they finish the work in 4 days

According to the question,

$$\frac{1}{x} + \frac{1}{x-6} = \frac{1}{4}$$

$$\frac{x-6+x}{(x)(x-6)} = \frac{1}{4}$$

$$4(2x-6) = x^2 - 6x$$

$$8x - 24 = x^2 - 6x$$

$$x^2 - 14x + 24 = 0$$

$$x^2 - 12x - 2x + 24 = 0$$

$$x(x-12) - 2(x-12) = 0$$

$$(x-12)(x-2) = 0$$

Either $x - 12 = 0$ or $x - 2 = 0$

$x = 12$ or $x = 2$; Rejected

Hence, *B* can finish work in 12 days.

A can finish work in 6 days. **Ans.**

Q. 8. Write all the values of p for which the quadratic equation $x^2 + px + 16 = 0$ has equal roots. Find the roots of the equation so obtained. ★

Solution : Given, equation is $x^2 + px + 16 = 0$

This is of the form $ax^2 + bx + c = 0$

where, $a = 1$, $b = p$ and $c = 16$

★ are board exam questions from previous years

$\therefore$ $D = b^2 - 4ac$
$= p^2 - 4 \times 1 \times 16$
$= p^2 - 64$

for equal roots, we have

$D = 0$
$p^2 - 64 = 0$
$p^2 = 64$
$p = \pm 8$

Putting $p = 8$ in given equation we have,

$x^2 + 8x + 16 = 0$
$(x + 4)^2 = 0$
$x + 4 = 0$
$x = -4$

Now, putting $p = -8$ in the given equation, we get

$x^2 - 8x + 16 = 0$
$(x - 4)^2 = 0$
$x = 4$

$\therefore$ Required roots are -4 and -4 or 4 and 4. **Ans.**

Q. 9. Two water taps together take 6 hours to fill a tank. If the tap with the larger diameter takes 9 hours lesser than the tap with the smaller diameter, then find the time in which each tap can separately fill the tap.

Solution : Let the smaller tap fill the tank in x hr and the total capacity of the tank be 1 unit.

Then in 1 hr, it can fill $\frac{1}{x}$ unit.

So the larger tank takes $(x - 9)$ hours to fill the tank.

Then in 1 hr, it can fill $\frac{1}{x-9}$ unit.

Together they can fill the tank in 6 hours.

Then in 1 hour, they can fill $\frac{1}{6}$ unit of the tank.

Thus, $\frac{1}{x} + \frac{1}{x-9} = \frac{1}{6}$

$\Rightarrow \frac{x-9+x}{x(x-9)} = \frac{1}{6}$

$\Rightarrow \frac{2x-9}{x(x-9)} = \frac{1}{6}$

$\Rightarrow 6(2x - 9) = x(x - 9)$

$\Rightarrow 12x - 54 = x^2 - 9x$

$\Rightarrow x^2 - 9x - 12x + 54 = 0$

$\Rightarrow x^2 - 21x + 54 = 0$

$\Rightarrow x^2 - 18x - 3x + 54 = 0$

$\Rightarrow x(x - 18) - 3(x - 18) = 0$

$\Rightarrow (x - 18)(x - 3) = 0$

$\Rightarrow x = 3, 18$

As 9 hours less than 3 hours is -6 which is not possible, so the smaller tap takes 18 hours to fill the tank and the larger takes 9 hours to do so. **Ans.**

Q. 10. Solve for x :★

$$\frac{1}{x+1} + \frac{3}{5x+1} = \frac{5}{x+4}, x \neq -1, -\frac{1}{5}, -4$$

Solution : Given,

$$\frac{1}{x+1} + \frac{3}{5x+1} = \frac{5}{x+4}$$

$\Rightarrow \frac{1}{x+1} - \frac{5}{x+4} = \frac{-3}{5x+1}$

$\Rightarrow \frac{(x+4)-5(x+1)}{(x+1)(x+4)} = \frac{-3}{5x+1}$

$\Rightarrow \frac{x+4-5x-5}{x^2+5x+4} = \frac{-3}{5x+1}$

$\Rightarrow \frac{-(4x+1)}{x^2+5x+4} = \frac{-3}{5x+1}$

$\Rightarrow (4x + 1)(5x + 1) = 3(x^2 + 5x + 4)$

$\Rightarrow 20x^2 + 4x + 5x + 1 = 3x^2 + 15x + 12$

$\Rightarrow 17x^2 - 6x - 11 = 0$

$\Rightarrow 17x^2 - 17x + 11x - 11 = 0$

$\Rightarrow 17x(x - 1) + 11(x - 1) = 0$

$\Rightarrow (x - 1)(17x + 11) = 0$

Either $x = 1$ or $x = \frac{-11}{17}$

Given $x \neq 1$, so $x = \frac{-11}{17}$ **Ans.**

Q. 11. Solve $\frac{x-1}{x-2} + \frac{x-3}{x-4} = \frac{10}{3}$, where $x \neq 2, 4$.★

Solution : Given, $\frac{x-1}{x-2} + \frac{x-3}{x-4} = \frac{10}{3}$

$\Rightarrow \frac{(x-1)(x-4)+(x-3)(x-2)}{(x-2)(x-4)} = \frac{10}{3}$

$\Rightarrow 3[(x - 1)(x - 4) + (x - 3)(x - 2)]$
$= 10[(x - 2)(x - 4)]$

$\Rightarrow 3[x^2 - x - 4x + 4 + x^2 - 3x - 2x + 6]$
$= 10[x^2 - 2x - 4x + 8]$

$\Rightarrow 3[2x^2 - 10x + 10] = 10[x^2 - 6x + 8]$

$\Rightarrow 6x^2 - 30x + 30 = 10x^2 - 60x + 80$

$\Rightarrow 4x^2 - 30x + 50 = 0$

$\Rightarrow 2x^2 - 15x + 25 = 0$

$\Rightarrow 2x^2 - 10x - 5x + 25 = 0$

$\Rightarrow 2x(x - 5) - 5(x - 5) = 0$

$\Rightarrow (x - 5)(2x - 5) = 0$

★ are board exam questions from previous years

$\Rightarrow \quad x = 5, \frac{5}{2}.$ **Ans.**

Q. 12. Solve for x :

$$\left(\frac{4x-3}{2x+1}\right) - 10\left(\frac{2x+1}{4x-3}\right) = 3, \text{ where } x \neq -\frac{1}{2}, \frac{3}{4}.$$

Solution : Given,

$$\left(\frac{4x-3}{2x+1}\right) - 10\left(\frac{2x+1}{4x-3}\right) = 3$$

$$\Rightarrow \quad \frac{(4x-3)^2 - 10(2x+1)^2}{(2x+1)(4x-3)} = 3$$

$$\Rightarrow \quad (4x-3)^2 - 10(2x+1)^2 = 3(2x+1)(4x-3)$$

$$\Rightarrow [16x^2 + 9 - 24x] - 10[4x^2 + 1 + 4x] = 3[8x^2 + 4x - 6x - 3]$$

$$\Rightarrow 16x^2 + 9 - 24x - 40x^2 - 10 - 40x = 24x^2 - 6x - 9$$

$$\Rightarrow \quad 48x^2 + 58x - 8 = 0$$

$$\Rightarrow \quad 24x^2 + 29x - 4 = 0$$

$$\Rightarrow \quad 24x^2 + 32x - 3x - 4 = 0$$

$$\Rightarrow \quad 8x(3x+4) - 1(3x+4) = 0$$

$$\Rightarrow \quad (8x-1)(3x+4) = 0$$

$$\Rightarrow \quad x = -\frac{4}{3}, \frac{1}{8}.$$ **Ans.**

Q. 13. Solve :★★

$$9x^2 - 9(a+b)x + (2a^2 + 5ab + 2b^2) = 0.$$

Solution : Given,

$$9x^2 - 9(a+b)x + (2a^2 + 5ab + 2b^2) = 0$$

Comparing with $Ax^2 + Bx + C = 0$, we get

$A = 9, B = -9(a+b)$ and $C = (2a^2 + 5ab + 2b^2)$

$\therefore \quad D = B^2 - 4AC$

$\Rightarrow \quad D = [-9(a+b)]^2 - 4(9)(2a^2 + 5ab + 2b^2)$

$\Rightarrow \quad D = 81(a+b)^2 - 36(2a^2 + 5ab + 2b^2)$

$\Rightarrow \quad D = 81[a^2 + b^2 + 2ab] - 72a^2 - 180ab - 72b^2$

$\Rightarrow \quad D = 81a^2 + 81b^2 + 162ab - 72a^2 - 180ab - 72b^2$

$\Rightarrow \quad D = 9a^2 + 9b^2 - 18ab$

$\Rightarrow \quad D = 9(a^2 + b^2 - 2ab)$

$\Rightarrow \quad D = 9(a-b)^2 \geq 0$

Thus $x = \frac{-B+\sqrt{D}}{2A}$ and $x = \frac{-B-\sqrt{D}}{2A}$

$$x = \frac{9(a+b)+3(a-b)}{18} \text{ and } \frac{9(a+b)-3(a-b)}{18}$$

$$\Rightarrow \quad x = \frac{6(2a+b)}{18} \text{ and } \frac{6(a+2b)}{18}$$

$$\Rightarrow \quad x = \frac{(2a+b)}{3} \text{ and } \frac{(a+2b)}{3}.$$ **Ans.**

★ are board exam questions from previous years
★★ are frequently asked board exam questions

Q. 14. Solve for x :★

$$x^2 + 5x - (a^2 + a - 6) = 0.$$

Solution : Taking $(a^2 + a - 6)$

$$= a^2 + 3a - 2a - 6$$
$$= a(a+3) - 2(a+3)$$
$$= (a+3)(a-2)$$

$$\therefore \quad x^2 + 5x - (a+3)(a-2) = 0$$

$$x^2 + (a+3)x - (a-2)x - (a+3)(a-2) = 0$$

$$x[x + (a+3)] - (a-2)[x + (a+3)] = 0$$

$$(x - a + 2)(x + a + 3) = 0$$

Hence, $\quad x - a + 2 = 0$ and $x + a + 3 = 0$

$$x = a - 2 \quad \text{and } x = -(a+3)$$

Required values of x are $(a-2), -(a+3)$ **Ans.**

Q. 15. In a class test, the sum of Arun's marks in Hindi and English is 30. Had he got 2 marks more in Hindi and 3 marks less in English, the product of the marks would have been 210. Find his marks in the two subjects.★

Solution : Let Arun marks in hindi be x and marks in english be y.

Then, according to question, we have

$$x + y = 30 \quad ...(i)$$

$$(x+2)(y-3) = 210 \quad ...(ii)$$

From equation (i) put $x = 30 - y$ in equation (ii),

$$(30 - y + 2)(y - 3) = 210$$

$$(32 - y)(y - 3) = 210$$

$$32y - 96 - y^2 + 3y = 210$$

$$y^2 - 35y + 306 = 0$$

$$y^2 - 18y - 17y + 306 = 0$$

$$y(y - 18) - 17(y - 18) = 0$$

$$(y - 18)(y - 17) = 0$$

$$y = 18, 17$$

Put $y = 18$ and 17 in equation (i), we get

$$x = 12, 13$$

Hence, his marks in hindi can be 12 and 13 and in english his marks can be 18 and 17. **Ans.**

Q. 16. A two-digit number is such that the product of its digits is 18. When 63 is subtracted from the number, the digits interchange their places. Find the number.

Solution : Let the ten's digit be x and the unit's digit be y. Thus, the number is $10x + y$.

Now $\quad xy = 18$

$$\Rightarrow \quad x = \frac{18}{y} \quad ...(i)$$

Also $\quad (10x + y) - 63 = (10y + x)$

$$\Rightarrow \quad 10x + y - 63 = 10y + x$$

$$\Rightarrow \quad 9x - 9y = 63$$

$$\Rightarrow \quad x - y = 7$$

$$\Rightarrow \quad \frac{18}{y} - y = 7 \qquad \text{[From (i)]}$$

$$\Rightarrow \quad 18 - y^2 = 7y$$

$$\Rightarrow \quad y^2 + 7y - 18 = 0$$

$$\Rightarrow \quad y^2 + 9y - 2y - 18 = 0$$

$$\Rightarrow \quad y(y+9) - 2(y+9) = 0$$

$$\Rightarrow \quad (y-2)(y+9) = 0$$

$$\Rightarrow \quad y = 2 \text{ or } -9$$

Thus, $x = 9 \text{ or } -2$

As digit cannot be negative,

$\therefore x = 9, y = 2$

So, the required number is $10x + y = 10.9 + 2 = 92$ **Ans.**

Q. 17. The denominator of a fraction is one more than twice the numerator. If the sum of the fraction and its reciprocal is $2\frac{16}{21}$, find the fraction.

Solution : Let the numerator be x. So the denominator is $2x + 1$.

Now $$\frac{x}{2x+1} + \frac{2x+1}{x} = 2\frac{16}{21}$$

$$\Rightarrow \quad \frac{x}{2x+1} + \frac{2x+1}{x} = \frac{58}{21}$$

$$\Rightarrow \quad \frac{x^2 + (2x+1)^2}{x(2x+1)} = \frac{58}{21}$$

$$\Rightarrow \quad \frac{x^2 + 4x^2 + 4x + 1}{x(2x+1)} = \frac{58}{21}$$

$$\Rightarrow \quad \frac{5x^2 + 4x + 1}{x(2x+1)} = \frac{58}{21}$$

$$\Rightarrow \quad 21(5x^2 + 4x + 1) = 58x(2x + 1)$$

$$\Rightarrow \quad 105x^2 + 84x + 21 = 116x^2 + 58x$$

$$\Rightarrow \quad 11x^2 - 26x - 21 = 0$$

$$\Rightarrow \quad 11x^2 - 33x + 7x - 21 = 0$$

$$\Rightarrow \quad 11x(x - 3) + 7(x - 3) = 0$$

$$\Rightarrow \quad (x - 3)(11x + 7) = 0$$

$$\Rightarrow \quad x = 3 \text{ and } -\frac{7}{11}$$

So the numerator is 3. [As a fraction cannot be negative]

Hence, the fraction is $\frac{3}{7}$. **Ans.**

Q. 18. 7 years ago, Varun's age was five times the square of Swati's age. 3 years hence, Swati's age will be two-fifths of Varun's age. Find their present ages.

Solution : Let Swati's present age be x years and Varun's be y years.

7 years ago, Swati's age was $(x - 7)$ years and Varun's was $(y - 7)$.

According to the question,

$$\Rightarrow \quad y - 7 = 5(x - 7)^2 \qquad ...(i)$$

3 years later, Swati's age will be $(x + 3)$ and Varun's will be $(y + 3)$.

According to the question,

$$x + 3 = \frac{2}{5}(y + 3)$$

$$\Rightarrow \quad 5x + 15 = 2y + 6$$

$$\Rightarrow \quad 2y = 5x + 9$$

$$\Rightarrow \quad y = \frac{5x+9}{2} \qquad ...(ii)$$

Substituting the value of y in equation (i),

Hence, $$5(x - 7)^2 = \frac{5x+9}{2} - 7$$

$$\Rightarrow \quad 10(x - 7)^2 = 5x + 9 - 14$$

$$\Rightarrow \quad 10[x^2 - 14x + 49] = 5x - 5$$

$$\Rightarrow \quad 10x^2 - 140x + 490 = 5x - 5$$

$$\Rightarrow \quad 10x^2 - 145x + 495 = 0$$

$$\Rightarrow \quad 2x^2 - 29x + 99 = 0$$

$$\Rightarrow \quad 2x^2 - 18x - 11x + 99 = 0$$

$$\Rightarrow \quad 2x(x - 9) - 11(x - 9) = 0$$

$$\Rightarrow \quad (x - 9)(2x - 11) = 0$$

$$\Rightarrow \quad x = 9 \text{ and } \frac{11}{2}$$

If Swati's present age is $5\frac{1}{2}$ years, then 7 years ago it would be $-1\frac{1}{2}$ years, which is not possible. Thus, her present age is 9 years.

Thus, Varun's present age $\frac{5(9)+9}{2} = \frac{45+9}{2} = \frac{54}{2} =$ 27 years. **Ans.**

Q. 19. A train travels 288 km at a uniform speed. If the speed has been 4 km/hr more, it would have taken 1 hour less for the same journey. Find the speed of the train.

Solution : Let the speed of the train be x km/hr.

The distance travelled by the train is 288 km.

Thus, $$\text{time} = \frac{288}{x} \text{ hr.}$$

and when speed is $(x+4)$ km/hr, time taken = $\frac{288}{x+4}$

Accroding to the question,

$$\frac{288}{x} = \frac{288}{x+4} + 1$$

$$\Rightarrow \quad 288\left(\frac{1}{x} - \frac{1}{x+4}\right) = 1$$

$$\Rightarrow \quad 288\left(\frac{x+4-x}{x(x+4)}\right) = 1$$

$$\Rightarrow \quad 288(4) = x^2 + 4x$$

$$\Rightarrow \quad x^2 + 4x - 1152 = 0$$

$$\Rightarrow x^2 + 36x - 32x - 1152 = 0$$

$$\Rightarrow x(x+36) - 32(x+36) = 0$$

$$\Rightarrow \quad (x+36)(x-32) = 0$$

$$\Rightarrow \quad x = 32 \text{ and } -36$$

As speed cannot be negative, so the speed of the train is 32 km/hr. **Ans.**

Q. 20. An aeroplane left 30 minutes later than its scheduled time and in order to reach its destination 1500 km away on time, it had to increase its speed by 250 km/hr. Determine its usual speed.

Solution : Let the speed of the flight be x km/hr.

Distance to be covered is 1500 km.

Thus, time taken = $\frac{1500}{x}$ hr.

If flight is delayed by 30 minutes, *i.e.*, $\frac{1}{2}$ hour,

Then, speed of flight = $x + 250$

$\because$ Time taken = $\frac{1500}{x+250}$

Now, according to the question,

$$\frac{1500}{x} - \frac{1500}{x+250} = \frac{1}{2}$$

$$\Rightarrow 1500\left(\frac{1}{x} - \frac{1}{x+250}\right) = \frac{1}{2}$$

$$\Rightarrow 1500\left(\frac{x+250-x}{x(x+250)}\right) = \frac{1}{2}$$

$$\Rightarrow \quad 1500 \times 250 \times 2 = x(x+250)$$

$$\Rightarrow \quad 750000 = x^2 + 250x$$

$$\Rightarrow x^2 + 250x - 750000 = 0$$

$$\Rightarrow x^2 + 1000x - 750x - 750000 = 0$$

$$\Rightarrow x(x+1000) - 750(x+1000) = 0$$

$$\Rightarrow \quad (x+1000)(x-750) = 0$$

$$\Rightarrow \quad x = 750, -1000$$

Since speed cannot be negative, so the speed of the aeroplane is 750 km/hr. **Ans.**

Q. 21. The sum of the areas of two squares is 260 m². If the difference of their perimeters is 24 m, then find the sides of the two squares.

Solution : Let the sides of the two squares be x m and y m respectively.

Thus area of first square = x^2 m^2

and perimeter of first square = $4x$ m

Also, area of second square = y^2 m^2

and perimeter of second square = $4y$ m

Now, $x^2 + y^2 = 260$ ...(i)

and $4(x - y) = 24$

$\Rightarrow$ $y = x - 6$ ...(ii)

Now, $x^2 + y^2 = 260$ [From (i)]

$\Rightarrow$ $x^2 + (x-6)^2 = 260$ [Using (ii)]

$$\Rightarrow \quad x^2 + x^2 - 12x + 36 = 260$$

$$\Rightarrow \quad 2x^2 - 12x - 224 = 0$$

$$\Rightarrow \quad x^2 - 6x - 112 = 0$$

$$\Rightarrow x^2 - 14x + 8x - 112 = 0$$

$$\Rightarrow x(x-14) + 8(x-14) = 0$$

$$\Rightarrow \quad (x+8)(x-14) = 0$$

$$\Rightarrow \quad x = -8 \text{ and } 14$$

As length cannot be negative, so we take $x = 14$ m

and $y = 14 - 6 = 8$ m

Thus, the sides of the two squares are 14 m and 8 m long. **Ans.**

Q. 22. The diagonal of a rectangular field is 60 m more than the shorter side. If the longer side is 30 m more than the shorter side, find the sides of the field.

Solution : As the field is rectangular, so all the angles are of 90°.

Hence the triangle formed by the diagonal and the two sides is a right-angled triangle.

Let the length of the shorter side be x m.

Thus the length of the longer side is $(x + 30)$ m and the diagonal is $(x + 60)$ m.

Applying Pythagoras' theorem,

$$(x+60)^2 = (x+30)^2 + x^2$$

$$\Rightarrow \quad x^2 = (x+60)^2 - (x+30)^2$$

$\Rightarrow \quad x^2 = (x+60-x-30)(x+60+x+30)$

$\Rightarrow \quad x^2 = 30(2x+90)$

$\Rightarrow \quad x^2 = 60x + 2700$

$\Rightarrow \quad x^2 - 60x - 2700 = 0$

$\Rightarrow \quad x^2 - 90x + 30x - 2700 = 0$

$\Rightarrow \quad x(x-90) + 30(x-90) = 0$

$\Rightarrow \quad (x-90(x+30) = 0$

$\Rightarrow \quad x = 90 \text{ and } -30$

As length cannot be negative, so the length of the smaller side is 90 m and the length of the larger side is 120 m. **Ans.**

Q. 23. The hypotenuse of a right-angled triangle is 6 cm more than twice the shortest side. If the third side is 2 cm less than the hypotenuse, find the sides of the triangle.

Solution : Let the shortest side be x cm.

So the length of the hypotenuse is $(2x + 6)$ cm and the third side is $(2x + 4)$ cm.

Now, $x^2 + (2x+4)^2 = (2x+6)^2$

$\Rightarrow \quad x^2 = (2x+6)^2 - (2x+4)^2$

$\Rightarrow \quad x^2 = (2x+6-2x-4)(2x+6+2x+4)$

$\Rightarrow \quad x^2 = 2(4x+10)$

$\Rightarrow \quad x^2 - 8x - 20 = 0$

$\Rightarrow \quad x^2 - 10x + 2x - 20 = 0$

$\Rightarrow \quad x(x-10) + 2(x-10) = 0$

$\Rightarrow \quad (x-10)(x+2) = 0$

$\Rightarrow \quad x = 10 \text{ and } -2$

As length cannot be negative, so we take $x = 10$ cm.

Hence length of shortest side is 10 cm, length of hypotenuse is 26 cm and length of the third side is 24 cm. **Ans.**

Q. 24. A person on tour has ₹ 4,200 for his expenses. If he extends his tour by 3 days, he has to cut down his daily expenses by ₹ 70. Find the original duration of the tour.

Solution : Let the original duration of the tour be x days.

Thus, daily expenses = ₹ $\frac{4200}{x}$

Now, extended duration = $x + 3$ days

Thus, new daily expenses = ₹ $\frac{4200}{x+3}$

Accroding to the question,

$$\frac{4200}{x+3} = \frac{4200}{x} - 70$$

$$\Rightarrow \quad \frac{4200}{x} - \frac{4200}{x+3} = 70$$

$$\Rightarrow \quad 4200\left(\frac{1}{x} - \frac{1}{x+3}\right) = 70$$

$$\Rightarrow \quad 60\left(\frac{x+3-x}{x(x+3)}\right) = 1$$

$$\Rightarrow \quad 180 = x^2 + 3x$$

$$\Rightarrow \quad x^2 + 3x - 180 = 0$$

$$\Rightarrow \quad x^2 + 15x - 12x - 180 = 0$$

$$\Rightarrow \quad x(x+15) - 12(x+15) = 0$$

$$\Rightarrow \quad (x+15)(x-12) = 0$$

$$\Rightarrow \quad x = 12 \text{ and } -15$$

As the number of days cannot be negative, so the original duration of the tour is 12 days. **Ans.**

Q. 25. Speed of a boat in still water is 15 km/h. It goes 30 km upstream and returns back at the same point in 4 hours 30 minutes. Find the speed of the stream.★

Solution : Let the speed of stream be x.

According to the question,

$$\frac{30}{x+15} + \frac{30}{15-x} = 4 + \frac{30}{60}$$

$$\left[\because \text{ 4 hours 30 mins} = 4\frac{30}{60} \text{ hours} = \frac{41}{2}\frac{9}{2} \text{ hours}\right.$$

$$\left. = 4\,\frac{30}{60} = 4\,\frac{1}{2} = \frac{9}{2} \text{ hours}\right]$$

$$\Rightarrow \quad 30\,\frac{(15-x+x+15)}{(15)^2-(x)^2} = \frac{9}{2}$$

$$\Rightarrow \quad \frac{900}{225-x^2} = \frac{9}{2}$$

$$\Rightarrow \quad \frac{2}{9} \times 900 = 225 - x^2$$

$$200 = 225 - x^2$$

or $\quad -x^2 = 200 - 225 = -25$

or $\quad x = 5$ km/hr

Thus, speed of stream is 5 km/hr. **Ans.**

Q. 26. The difference of two numbers is 4 and the difference of their reciprocals is $\frac{4}{21}$ · Find the numbers.

Solution : Let one number be x.

So the other number is $(4 + x)$.

According to question,

★ are board exam questions from previous years

$$\frac{1}{x}-\frac{1}{4+x}=\frac{4}{21}$$

$$\Rightarrow \quad \frac{4+x-x}{x(4+x)}=\frac{4}{21}$$

$$\Rightarrow \quad \frac{4}{x(4+x)}=\frac{4}{21}$$

$$\Rightarrow \quad \frac{1}{x(4+x)}=\frac{1}{21}$$

$$\Rightarrow \quad 21=x(4+x)$$

$$\Rightarrow \quad x^2+4x-21=0$$

$$\Rightarrow \quad x^2+7x-3x-21=0$$

$$\Rightarrow \quad x(x+7)-3(x+7)=0$$

$$\Rightarrow \quad (x+7)(x-3)=0$$

$$\Rightarrow \quad x=3 \text{ or } -7$$

Thus, the other number is 7 or – 3. But – 3 and – 7 do not satisfy the second condition. So, the two numbers are 7 and 3. **Ans.**

Q. 27. A rectangular park is to be designed whose breadth is 3 m less than its length. Its area is to be 4 square metres more than the area of a park that has already been made in the shape of an isosceles triangle with its base as the breadth of the rectangular park and of altitude 12 m. Find the length and breadth of the rectangular park.★

Solution : Let the length of the rectangular park be x m. Then, breadth = $(x – 3)$ m

Area of rectangular park = $x(x-3)\text{ m}^2$

and area of isosceles triangular park

$$=\frac{1}{2}(x-3)\times 12\text{ m}^2$$

$$=6(x-3)\text{ m}^2$$

Acoording to the question,

$$x(x-3)-6(x-3)=4$$

$$x^2-3x-6x+18=4$$

$$x^2-9x+14=0$$

$$x^2-7x-2x+14=0$$

$$x(x-7)-2(x-7)=0$$

$$(x-2)(x-7)=0$$

$$x=2 \text{ or } 7$$

$$\therefore \quad x=7\text{ m}$$

(as breadth can't be negative)

and $x-3=(7-3)\text{ m}=4\text{ m}$

Hence, length and breadth of the rectangular park is 7 m and 4 m respectively. **Ans.**

Q. 28. 300 apples were distributed equally among a certain number of students. Had there been 10 more students, each would have received one apple less. Find the number of students.

Solution : Let the number of students be x.

So, the number of apples per student is $\frac{300}{x}$.

If the number of students were $x + 10$ then the number of apples per student would have been $\frac{300}{x+10}$.

According to the question,

$$\frac{300}{x+10}+1=\frac{300}{x}$$

$$\Rightarrow \quad \frac{300}{x}-\frac{300}{x+10}=1$$

$$\Rightarrow \quad 300\left(\frac{x+10-x}{x(x+10)}\right)=1$$

$$\Rightarrow \quad 300\left(\frac{10}{x(x+10)}\right)=1$$

$$\Rightarrow \quad x(x+10)=3000$$

$$\Rightarrow \quad x^2+10x=3000$$

$$\Rightarrow \quad x^2+10x-3000=0$$

$$\Rightarrow \quad x^2+60x-50x-3000=0$$

$$\Rightarrow \quad x(x+60)-50(x+60)=0$$

$$\Rightarrow \quad (x+60)(x-50)=0$$

$$\Rightarrow \quad x=50 \text{ and } -60$$

Since number of students cannot be negative, so the correct answer is 50. **Ans.**

Q. 29. The total cost of a certain length of a piece of cloth is ₹ 200. If the piece was 5 m longer and each metre of cloth costs ₹ 2 less, the cost of the piece would have remained unchanged. How long is the piece and what is its original rate per metre?★

Solution : Let the original length of piece of cloth is x m and rate of cloth is ₹ y per metre.

Then according to question, we have

$$x\times y=200 \quad \text{...(i)}$$

and if length be 5 m longer and each meter of cloth be ₹ 2 less then

$$(x+5)(y-2)=200$$

$$(x+5)(y-2)=200$$

$$xy-2x+5y-10=200 \quad \text{...(ii)}$$

On equating equation (i) and (ii), we have

$$xy=xy-2x+5y-10$$

$$\Rightarrow \quad 2x-5y=-10 \quad \text{...(iii)}$$

$$\left(y=\frac{200}{x}\right) \text{ from equation (i)}$$

$$\Rightarrow \quad 2x-5\times\frac{200}{x}=-10$$

$$\Rightarrow \quad 2x-\frac{1000}{x}=-10$$

★ are board exam questions from previous years

$$\Rightarrow \quad 2x^2 - 1000 = -10x$$
$$\Rightarrow \quad 2x^2 + 10x - 1000 = 0$$
$$\Rightarrow \quad x^2 + 5x - 500 = 0$$
$$\Rightarrow \quad x^2 + 25x - 20x - 500 = 0$$
$$\Rightarrow \quad x(x+25) - 20(x+25) = 0$$
$$(x+25)(x-20) = 0$$
$$x = 20$$

($x \neq -25$ length of cloth can never be negative)

$$\therefore \quad x \times y = 200$$
$$20 \times y = 200$$
$$y = 10$$

Thus, length of the piece of cloth is 20 m and original price per metre is ₹ 10. **Ans.**

Q. 30. Two taps running together can fill a tank in $3\frac{1}{13}$ hours. If one tap takes 3 hours more than the other to fill the tank, then how much time will each tap take to fill the tank?★

Solution : Let one tap fill the tank in x hrs.

So, other tap will take $(x + 3)$ hrs

Together they fill the tank in $3\frac{1}{13} = \frac{40}{13}$ hours

According to the question,

$$\frac{1}{x} + \frac{1}{x+3} = \frac{13}{40}$$
$$\frac{x+3+x}{(x)(x+3)} = \frac{13}{40}$$
$$\frac{2x+3}{x^2+3x} = \frac{13}{40}$$
$$13x^2 + 39x - 80x + 120$$
$$13x^2 - 41x - 120 = 0$$
$$13x^2 - 65x + 24x - 120 = 0$$
$$13x(x-5) + 24(x-5) = 0$$
$$(x-5)(13x+24) = 0$$

Either $x - 5 = 0$ or $13x + 24 = 0$

$x = 5, x = -24/13$ (Rejected)

Thus, one tap fill the tank in 5 hrs.

So, other tap fill the tank in $5 + 3 = 8$ hrs. **Ans.**

Q. 31. The sum of the areas of 2 squares is 640 m^2. If the difference of their perimeters is 64 m, then find the sides of the two squares. [NCERT]

Solution : Let the sides of the two squares be x m and y m respectively.

Thus, area of first square $= x^2$ m^2

and perimeter of first square $= 4x$ m

Also area of second square $= y^2$ m^2

and perimeter of second square $= 4y$ m

Now, $x^2 + y^2 = 640$ ...(i)

and $4(x - y) = 64$

$\Rightarrow \quad y = x - 16$ ...(ii)

From equation (i),

$$x^2 + y^2 = 640$$
$$\Rightarrow \quad x^2 + (x-16)^2 = 640$$
$$\Rightarrow \quad x^2 + x^2 - 32x + 256 = 640 \quad \text{[Using (i)]}$$
$$\Rightarrow \quad 2x^2 - 32x - 384 = 0$$
$$\Rightarrow \quad x^2 - 16x - 192 = 0$$
$$\Rightarrow \quad x^2 - 24x + 8x - 192 = 0$$
$$\Rightarrow \quad x(x-24) + 8(x-24) = 0$$
$$\Rightarrow \quad (x+8)(x-24) = 0$$
$$\Rightarrow \quad x = -8 \text{ and } 24$$

As length cannot be negative so we take $x = 24$ m

and $y = 24 - 16 = 8$ m

Thus, the sides of the two squares are 24 m and 8 m long. **Ans.**

Q. 32. Solve : $\frac{1}{x} - \frac{1}{x-2} = 3$, where $x \neq 0, 2$.★ [NCERT]

Solution : Given,

$$\frac{1}{x} - \frac{1}{x-2} = 3$$
$$\Rightarrow \quad \frac{1}{x} = 3 + \frac{1}{x-2}$$
$$\Rightarrow \quad \frac{1}{x} = \frac{3(x-2)+1}{x-2}$$
$$\Rightarrow \quad \frac{1}{x} = \frac{3x-6+1}{x-2}$$
$$\Rightarrow \quad \frac{1}{x} = \frac{3x-5}{x-2}$$
$$\Rightarrow \quad x - 2 = 3x^2 - 5x$$
$$\Rightarrow \quad 3x^2 - 5x - x + 2 = 0$$
$$\Rightarrow \quad 3x^2 - 6x + 2 = 0$$
$$\Rightarrow \quad x = \frac{-(-6) \pm \sqrt{(-6)^2 - 4(3)(2)}}{2(3)}$$
$$\Rightarrow \quad x = \frac{6 \pm \sqrt{36-24}}{6}$$
$$\Rightarrow \quad x = \frac{6 \pm \sqrt{12}}{6}$$
$$\Rightarrow \quad x = \frac{6 \pm 2\sqrt{3}}{6}$$

★ are board exam questions from previous years

$$\Rightarrow \quad x = \frac{3 \pm \sqrt{3}}{3}$$

$$\Rightarrow \quad x = \frac{\sqrt{3} \pm 1}{\sqrt{3}}.$$ **Ans.**

Q. 33. A passenger while boarding a plane hurt herself and the captain called for immediate medical attention. Thus the plane left 30 minutes behind schedule. In order to reach its destination 1500 km away on time, the speed was increased by 100 km/hr from its usual speed. Find the usual speed.★

Solution : Let the usual speed be x km/hr.

We know, $\text{Time} = \frac{\text{Distance}}{\text{Speed}}$

So, Usual time taken $= \frac{1500}{x}$ hr.

Increased speed $= (x + 100)$ km/hr

Thus, new time taken $= \frac{1500}{x+100}$

Now, according to the question

$$\frac{1500}{x+100} + \frac{1}{2} = \frac{1500}{x}$$

$$\Rightarrow \quad \frac{1500}{x} - \frac{1500}{x+100} = \frac{1}{2}$$

$$\Rightarrow \quad 1500\left(\frac{1}{x} - \frac{1}{x+100}\right) = \frac{1}{2}$$

$$\Rightarrow \quad 1500\left(\frac{x+100-x}{x(x+100)}\right) = \frac{1}{2}$$

$$\Rightarrow \quad 1500\left(\frac{100}{x(x+100)}\right) = \frac{1}{2}$$

$$\Rightarrow \quad x(x + 100) = 300000$$

$$\Rightarrow \quad x^2 + 100x - 300000 = 0$$

$$\Rightarrow x^2 + 600x - 500x - 300000 = 0$$

$$\Rightarrow \quad x(x + 600) - 500(x + 600) = 0$$

$$\Rightarrow \quad (x + 600)(x - 500) = 0$$

$$\Rightarrow \quad x = 500 \text{ and } -600$$

As speed cannot be negative, so the original speed of the plane is 500 km/hr. **Ans.**

Q. 34. An aeroplane left 50 minutes later than its scheduled time and in order to reach the destination, 1250 km away, on time, its speed was increased by 250 km/hr. Find its usual speed.★

Solution : Let the usual speed be x km/hr.

We know, $\text{Time} = \frac{\text{Distance}}{\text{Speed}}$

So, Usual time taken $= \frac{1250}{x}$ hr.

Increased speed $= (x + 250)$ km/hr

Thus, New time taken $= \frac{1250}{x+250}$

Now, according to the question,

$$\frac{1250}{x+250} + \frac{50}{60} = \frac{1250}{x}$$

$$\Rightarrow \quad \frac{1250}{x} - \frac{1250}{x+250} = \frac{5}{6}$$

$$\Rightarrow \quad 1250\left(\frac{1}{x} - \frac{1}{x+250}\right) = \frac{5}{6}$$

$$\Rightarrow \quad 1250\left(\frac{x+250-x}{x(x+250)}\right) = \frac{5}{6}$$

$$\Rightarrow \quad 1250\left(\frac{250}{x(x+250)}\right) = \frac{5}{6}$$

$$\Rightarrow \quad x(x + 250) = 375000$$

$$\Rightarrow \quad x^2 + 250x - 375000 = 0$$

$$\Rightarrow \quad x^2 + 750x - 500x - 375000 = 0$$

$$\Rightarrow \quad x(x + 750) - 500(x + 750) = 0$$

$$\Rightarrow \quad (x + 750)(x - 500) = 0$$

$$\Rightarrow \quad x = 500 \text{ and } -750$$

As speed cannot be negative, so the original speed of the plane is 500 km/hr. **Ans.**

Q. 35. A bus travels at a certain average speed for a distance of 75 km and then travels a distance of 90 km at an average speed of 10 km/h more than the first speed. If it takes 3 hours to complete the total journey, find its first speed. ★

Solution : Let x be the initial speed of the bus.

We know that

$$\text{Speed} = \frac{\text{Distance}}{\text{Time}}$$

or

$$\text{Time} = \frac{\text{Distance}}{\text{Speed}}$$

So, according to the question,

$$3 = \frac{75}{x} + \frac{90}{x+10}$$

$$\Rightarrow \quad 3 = \frac{75(x+10)+90x}{x(x+10)}$$

$$\Rightarrow \quad 3(x)(x + 10) = 75x + 750 + 90x$$

$$\Rightarrow \quad 3x^2 + 30x = 165x + 750$$

$$\Rightarrow \quad 3x^2 - 135x - 750 = 0$$

$$\Rightarrow \quad x^2 - 45x - 250 = 0$$

$$\Rightarrow \quad x^2 - 50x + 5x - 250 = 0$$

★ are board exam questions from previous years

$\Rightarrow \quad x(x-50)+5(x-50)=0$

$\Rightarrow \quad (x+5)(x-50)=0$

$\Rightarrow \quad x=-5 \text{ or } x=50$

Since, speed cannot be negative

So, $\quad x=50$

Hence, the initial speed of bus is 50 km/hr. **Ans.**

Q. 36. The sum of the areas of 2 squares is 400 m². If the difference of their perimeters is 16 m, then find the sides of the two squares.

Solution : Let the sides of the two squares be x m and y m respectively.

Thus, area of first square = x^2 m²

and perimeter of first square = $4x$ m

Also area of second square = y^2 m²

and perimeter of second square = $4y$ m

Now, $\quad x^2+y^2=400 \quad$ ...(i)

and $\quad 4(x-y)=16$

$\Rightarrow \quad y=x-4 \quad$ ...(ii)

From equation (i),

$x^2+y^2=400$

$\Rightarrow \quad x^2+(x-4)^2=400 \quad$ [From (ii)]

$\Rightarrow \quad x^2+x^2-8x+16=400$

$\Rightarrow \quad 2x^2-8x-384=0$

$\Rightarrow \quad x^2-4x-192=0$

$\Rightarrow \quad x^2-16x+12x-192=0$

$\Rightarrow \quad x(x-16)+12(x-16)=0$

$\Rightarrow \quad (x+12)(x-16)=0$

$\Rightarrow \quad x=-12 \text{ and } 16$

As length cannot be negative so we take $x = 16$ m

and $\quad y = 16-4 = 12$ m

Thus, the sides of the two squares are 16 m and 12 m long. **Ans.**

Q. 37. Two pipes together can fill a tank in $11\frac{1}{9}$ minutes. If one pipe takes 5 minutes more than the other to fill tank separately, then find the time in which each pipe can fill the tank separately.★

Solution : Let the smaller tap fill the tank in x minutes and the total capacity of the tank be 1 sq. unit.

Then in 1 minute, smaller tap can fill $\frac{1}{x}$ sq. unit.

So the other pipe takes $(x-5)$ minutes to fill the tank.

Then in 1 minute, it can fill $\frac{1}{x-5}$ sq. unit.

Together they can fill the tank is $11\frac{1}{9}$ minutes

$\left(=\frac{100}{9} \text{ minutes}\right)$

Then in 1 minute, they can fill $\frac{9}{100}$ of the tank.

Thus, $\quad \frac{1}{x}+\frac{1}{x-5}=\frac{9}{100}$

$\Rightarrow \quad \frac{x-5+x}{x(x-5)}=\frac{9}{100}$

$\frac{2x-5}{x(x-5)}=\frac{9}{100}$

$200x-500=9x^2-45x$

$\Rightarrow \quad 9x^2-245x+500=0$

$\Rightarrow \quad 9x^2-225x-20x+500=0$

$\Rightarrow \quad 9x(x-25)-20(x-25)=0$

$\Rightarrow \quad (x-25)(9x-20)=0$

$\Rightarrow \quad x=25 \text{ or } \frac{20}{9}.$

As 5 minutes less than $\frac{20}{9}$ minutes is less than 0 which is not possible, so the first pipe takes 25 minutes to fill the tank and the second takes 20 minutes to so do. **Ans.**

Q. 38. The difference between two natural numbers is 5 and the difference between their reciprocals is $\frac{5}{14}$. Find the numbers.★

Solution : Let the natural numbers be x and y respectively.

Now, $\quad x-y=5 \quad$ ...(i)

Also, $\quad \frac{1}{y}-\frac{1}{x}=\frac{5}{14}$

$\Rightarrow \quad 14x-14y=5xy$

$\Rightarrow \quad 14(x-y)=5xy$

$\Rightarrow \quad 14\times 5=5xy \quad$ [From (i)]

$\Rightarrow \quad xy=14$

$\Rightarrow \quad x=\frac{14}{y}$

Substituting $x=\frac{14}{y}$ in equation (i), we get

$\frac{14}{y}-y=5$

$\Rightarrow \quad y-\frac{14}{y}+5=0$

$\Rightarrow \quad y^2+5y-14=0$

$\Rightarrow \quad y^2+7y-2y-14=0$

$\Rightarrow \quad y(y+7)-2(y+7)=0$

$\Rightarrow \quad (y+7)(y-2)=0$

★ are board exam questions from previous years

$\Rightarrow \quad y = 2, -7$

Since natural numbers cannot be –ve, so $y = 2$.

Thus, from equation (i),

$$x - y = 5$$

$$\Rightarrow \quad x - 2 = 5$$

$$\Rightarrow \quad x = 7$$

So, the two numbers are 7 and 2. **Ans.**

Q. 39. The difference between two natural numbers is 3 and the difference between their reciprocals is $\frac{3}{28}$. Find the numbers.

Solution : Let the natural numbers be x and y respectively.

Thus, $\quad x - y = 3 \quad$ ...(i)

Also, $\quad \frac{1}{y} - \frac{1}{x} = \frac{3}{28}$

$$\Rightarrow \quad 28x - 28y = 3xy$$

$$\Rightarrow \quad 28(x - y) = 3xy$$

$$\Rightarrow \quad 28 \times 3 = 3xy \quad \text{[From (i)]}$$

$$\Rightarrow \quad xy = 28$$

$$\Rightarrow \quad x = \frac{28}{y}$$

Substituting $x = \frac{28}{y}$ in equation (i), we get

$$\frac{28}{y} - y = 3$$

$$\Rightarrow \quad y - \frac{28}{y} + 3 = 0$$

$$\Rightarrow \quad y^2 + 3y - 28 = 0$$

$$\Rightarrow \quad y^2 + 7y - 4y - 28 = 0$$

$$\Rightarrow \quad y(y + 7) - 4(y + 7) = 0$$

$$\Rightarrow \quad (y + 7)(y - 4) = 0$$

$$\Rightarrow \quad y = 4, -7$$

Since natural numbers cannot be –ve, so $y = 4$

Thus, from equation (i),

$$x - y = 3$$

$$\Rightarrow \quad x - 4 = 3$$

$$\Rightarrow \quad x = 7$$

So, the two numbers are 7 and 4. **Ans.**

Q. 40. A two-digit number is such that the product of its digits is 20. If 9 is added to the number, the digit interchange their places. Find the number.★

Solution : Let the ten's digit be x and the unit's digit be y.

Thus, the given number is $(10x + y)$ and its reverse is $(10y + x)$.

So, $\quad xy = 20$

$$\Rightarrow \quad x = \frac{20}{y} \quad \text{...(i)}$$

and $\quad 10x + y + 9 = 10y + x$

$$\Rightarrow \quad 9x - 9y = -9$$

$$\Rightarrow \quad x - y = -1$$

$$\Rightarrow \quad \frac{20}{y} - y = -1 \quad \text{[From (i)]}$$

$$\Rightarrow \quad 20 - y^2 = -y$$

$$\Rightarrow \quad y^2 - y - 20 = 0$$

$$\Rightarrow \quad y^2 - 5y + 4y - 20 = 0$$

$$\Rightarrow \quad y(y - 5) + 4(y - 5) = 0$$

$$\Rightarrow \quad (y - 5)(y + 4) = 0$$

$$\Rightarrow \quad y = 5, -4$$

So, $\quad x = \frac{20}{5}, -\frac{20}{4}$

$$\Rightarrow \quad x = 4, -5$$

Thus, the number is 45 or – 54. **Ans.**

Chapter 5. Arithmetic Progressions

Q. 1. In an A.P. of 50 terms, the sum of the first 10 terms is 210 and the sum of the last 15 terms is 2565. Find the A.P.★

Solution : Let a be the first term and d be the common difference.

Now, $\quad S_{10} = \frac{10}{2}[2a + (10 - 1)d]$

$$\Rightarrow \quad 210 = 5(2a + 9d)$$

$$\Rightarrow \quad 2a + 9d = 42 \quad \text{...(i)}$$

and 15th term from the end is $(50 - 15 + 1) = 36^{th}$ term from the beginning.

So, $\quad t_{36} = a + (36 - 1)d$

$$t_{36} = a + 35d$$

Hence, the sum of the last 15 terms

$$= \frac{15}{2}\{2t_{36} + (15 - 1)d\}$$

$$\Rightarrow \quad \frac{15}{2}\{2(a + 35d) + 14d\} = 2565$$

$$\Rightarrow \quad 15\{a + 35d + 7d\} = 2565$$

$$a + 42d = 171$$

$$\Rightarrow \quad 2a + 84d = 342 \quad \text{...(ii)}$$

Subtracting (i) from (ii), we have

$$75d = 300$$

$$\Rightarrow \quad d = 4$$

From (i), $\quad 2a + 9d = 42$

★ are board exam questions from previous years

$\Rightarrow \quad 2a = 42 - 9d$

$\Rightarrow \quad 2a = 42 - 9 \times 4$

$\Rightarrow \quad 2a = 42 - 36 = 6$

$\Rightarrow \quad a = 3$

Thus, the A.P. is 3, 7, 11, ... 199. **Ans.**

Q. 2. If S_n denotes the sum of the first n terms of an A.P., prove that $S_{30} = 3(S_{20} - S_{10})$.★

Solution : Let a be the first term of the series and d be the common difference.

$\therefore \quad S_n = \frac{n}{2}\{2a + (n-1)d\}$

So, $\quad S_{30} = \frac{30}{2}\{2a + (30-1)d\}$

$\Rightarrow \quad S_{30} = 15\{2a + 29d]$ ...(i)

$S_{20} = \frac{20}{2}\{2a + (20-1)d\}$

$\Rightarrow \quad S_{20} = 10\{2a + 19d\}$ ...(ii)

and $\quad S_{10} = \frac{10}{2}\{2a + (10-1)d\}$

$\Rightarrow \quad S_{10} = 5\{2a + 9d\}$ ...(iii)

Now, $\quad S_{20} - S_{10} = 10\{2a + 19d\} - 5\{2a + 9d\}$

$= 20a + 190d - 10a - 45d$

$= 10a + 145d$

$= 5(2a + 29d)$ [From (iii)]

$3(S_{20} - S_{10}) = 3[5\ (2a + 29d]$

$S30 = 15\ (2a + 9b)$ [From (i)]

Thus, $3(S_{20} - S_{10}) = S_{30}$ **Hence Proved.**

Q. 3. Find the sum of the middlemost terms of the A.P. $-\frac{4}{3}, -1, -\frac{2}{3}, \ldots, 4\frac{1}{3}$.★

Solution : The given A.P. is $-\frac{4}{3}, -1, -\frac{2}{3}, \ldots, 4\frac{1}{3}$

Thus, $\quad a = -\frac{4}{3},\ d = -1 - \left(-\frac{4}{3}\right)$

$= -1 + \frac{4}{3} = \frac{1}{3}$ and $t_n = 4\frac{1}{3}$

$t_n = a + (n-1)d$

$\Rightarrow \quad 4\frac{1}{3} = -\frac{4}{3} + (n-1)\frac{1}{3}$

$\Rightarrow \quad \frac{13}{3} = -\frac{4}{3} + \frac{1}{3}n - \frac{1}{3}$

$\Rightarrow \quad \frac{13}{3} = -\frac{5}{3} + \frac{1}{3}n$

$\Rightarrow \quad \frac{13}{3} + \frac{5}{3} = \frac{1}{3}n$

$\Rightarrow \quad n = 18$

Thus, the middlemost terms of the A.P. are the 9th and 10th terms.

So $\quad t_9 = a + (9-1)d = a + 8d$

$= -\frac{4}{3} + \frac{8}{3} = \frac{4}{3}$

and $\quad t_{10} = a + (10-1)d = a + 9d$

$= -\frac{4}{3} + \frac{9}{3} = \frac{5}{3}$

Thus, $\quad t_9 + t_{10} = \frac{4}{3} + \frac{5}{3} = \frac{9}{3} = 3$ **Ans.**

Q. 4. Find the common difference of an A.P. whose first term is 5 and the sum of the first 4 terms is half of the sum of the next four terms.★

Solution : Given, $a = 5$

Thus, $\quad S_4 = \frac{4}{2}\{2a + (4-1)d\}$

$= 2(2a + 3d)$

$t_5 = \{a + (5-1)d\}$

$= (a + 4d)$

For the next set of 4 numbers,

$a = t_5$ and d remains the same

Thus, $\quad S_4 = \frac{4}{2}\{2(a + 4d) + (4-1)d\}$

$= 2(2a + 8d + 3d)$

$= 2(2a + 11d)$

According to the question,

$\frac{1}{2} \times 2(2a + 11d) = 2(2a + 3d)$

$\Rightarrow \quad 2a + 11d = 2(2a + 3d)$

$\Rightarrow \quad 2a + 11d = 4a + 6d$

$\Rightarrow \quad 5d = 2a = 2(5)$

$\Rightarrow \quad d = 2$

Thus the common difference is 2. **Ans.**

Q. 5. The sum of the 4th and 8th terms of an A.P. is 24 and the sum of the 6th and 10th terms is 44. Find the first three terms of the A.P.★

Solution : Let the first term be a, the common difference be d and the number of terms be n.

Thus, $\quad t_4 = a + (4-1)d - a + 3d$

$t_8 = a + (8-1)d = a + 7d$

$t_6 = a + (6-1)\mathrm{d} = a + 5d$

$t_{10} = a + (10-1)d = a + 9d$

Now, $\quad t_4 + t_8 = 24$

$\Rightarrow a + 3d + a + 7d = 24$

★ are board exam questions from previous years

$\Rightarrow \quad 2a + 10d = 24$

$\Rightarrow \quad a + 5d = 12 \quad \ldots(i)$

and $\quad t_6 + t_{10} = 44$

$\Rightarrow a + 5d + a + 9d = 44$

$\Rightarrow \quad 2a + 14d = 44$

$\Rightarrow \quad a + 7d = 22 \quad \ldots(ii)$

Subtracting (i) from (ii), we have

$2d = 10$

$\Rightarrow \quad d = 5$

From (i) $\quad a = 12 - 5(5) = 12 - 25 = -13$

Hence, first three terms of the A.P. are – 13, – 8, – 3.

Ans.

Q. 6. The sum of the first 6 terms of an A.P. is 42. If its 10th and 30th terms are in the ratio 1 : 3, find the 1st and 13th terms.★

Solution : Let a be the first term and d be the common difference of the A.P.

So, $\quad S_6 = \frac{6}{2}\{2a + (6-1)d\}$

$\Rightarrow \quad 42 = \frac{6}{2}\{2a + 5d\}$

$\Rightarrow \quad 42 = 3(2a + 5d)$

$\Rightarrow \quad 2a + 5d = 14 \quad \ldots(i)$

$t_{10} = a + (10-1)d$

$\Rightarrow \quad t_{10} = a + 9d$

$t_{30} = a + (30-1)d$

$\Rightarrow \quad t_{30} = a + 29d$

Now $\quad t_{10} : t_{30} = 1 : 3$

$\Rightarrow a + 9d : a + 29d = 1 : 3$

$\Rightarrow \quad \frac{a+9d}{a+29d} = \frac{1}{3}$

$\Rightarrow \quad 3a + 27d = a + 29d$

$\Rightarrow \quad 2a = 2d$

$\Rightarrow \quad a = d \quad \ldots(ii)$

Substituting (ii) in (i),

$2a + 5a = 14$

$\Rightarrow \quad 7a = 14$

$\Rightarrow \quad a = 2$

Thus, $\quad d = 2$

Now $\quad t_{13} = 2 + (13-1)2$

$\Rightarrow \quad t_{13} = 2 + 24 = 26$

So the first term is 2 and the 13th term is 26. **Ans.**

Q. 7. Find the sum of all three-digit natural numbers, which are divisible by 13.★

Solution : Three-digit natural numbers which are divisible by 13 are 104, 117, 130, 143, ..., 988.

The above series is an A.P.

So, $\quad a = 104, d = 13$ and $t_n = 988$

We know that,

$t_n = a + (n-1)d$

$\Rightarrow \quad 988 = 104 + (n-1)13$

$\Rightarrow \quad 76 = 8 + n - 1$

$\Rightarrow \quad n = 76 - 8 + 1$

$\Rightarrow \quad n = 69$

$\therefore \quad S_{69} = \frac{69}{2}\{2(104) + (69-1)(13)\}$

$= 69\{104 + (34)(13)\}$

$= 69\{104 + 442\}$

$= 69 \times 546$

$= 37674.$ **Ans.**

Q. 8. If the ratio of the sum of the first n terms of two A.Ps. is $(7n+1) : (4n+27)$, then find the ratio of their 9th terms.★

Solution : Let the first terms be a and a' and d and d' be their respective common differences.

Then, $\quad \frac{S_n}{S'_n} = \frac{\frac{n}{2}\{2a+(n-1)d\}}{\frac{n}{2}\{2a'+(n-1)d'\}}$

$= \frac{7n+1}{4n+27}$

$\Rightarrow \quad \frac{a+\left(\frac{n-1}{2}\right)d}{a'+\left(\frac{n-1}{2}\right)d'} = \frac{7n+1}{4n+27}$

To get ratio of 9th terms, replacing $\frac{n-1}{2} = 8$

$\Rightarrow \quad n = 17$

Hence $\frac{t_9}{t'_9} = \frac{a+8d}{a'+8d'} = \frac{7\times17+1}{4\times17+27} = \frac{120}{95} = \frac{24}{19}$

Ans.

Q. 9. Find the number of terms of the A.P. 64, 60, 56, ... so that sum is 544. Explain the double answer.★

Solution : Given A.P. is 64, 60, 56, ...

So, $\quad a = 64, d = -4$ and $S_n = 544$

Let the number of terms be n.

Thus, $\quad S_n = \frac{n}{2}\{2(64) + (n-1)(-4)\} = 544$

$\Rightarrow \frac{n}{2}\{128 - 4n + 4\} = 544$

$\Rightarrow \quad \frac{n}{2}\, 132 - 4n = 544$

★ are board exam questions from previous years

$$\Rightarrow \quad \frac{4n}{2}\{33-n\} = 544$$

$$\Rightarrow \quad \frac{n}{2}\{33-n\} = 136$$

$$\Rightarrow \quad 33n - n^2 = 272$$

$$\Rightarrow n^2 - 33n + 272 = 0$$

$$\Rightarrow \quad n^2 - 17n - 16n + 272 = 0$$

$$\Rightarrow \quad n(n-17) - 16(n-17) = 0$$

$$\Rightarrow \quad (n-17)(n-16) = 0$$

$$\Rightarrow \quad n = 16 \text{ or } 17$$

The double answer is because of the zero. It does not add to the total sum but is definitely a number in the series (17th term).

$$t_n = 64 + (n-1)(-4) = 0$$

$$\Rightarrow \quad 64 - (4n-4) = 0$$

$$\Rightarrow \quad 68 - 4n = 0$$

$$\Rightarrow \quad n = 17.$$

i.e., 17th term is zero. So sum of all 16 and 17 terms is zero as 0 will not contribute to the sum. **Ans.**

Q. 10. In an A.P., the first term is 22, n^{th} term is – 11 and sum of first n terms is 66. Find n and the common difference.★

Solution : Given, $a = 22$, $t_n = -11$ and $S_n = 66$

Now, $t_n = 22 + (n-1)d = -11$

$$\Rightarrow \quad (n-1)d = -33 \quad \text{...(i)}$$

and $S_n = \frac{n}{2}\{2(22) + (n-1)d\} = 66$

$$\Rightarrow n\{2(22) + (n-1)d\} = 132$$

$$\Rightarrow \quad n\{44 - 33\} = 132 \quad \text{[From (i)]}$$

$$\Rightarrow \quad 11n = 132$$

$$\Rightarrow \quad n = 12$$

Substituting $n = 12$ in (i), we get

$$(12-1)d = -33$$

$$\Rightarrow \quad 11d = -33$$

$$\Rightarrow \quad d = -3.$$ **Ans.**

Q. 11. Find the sum of the following :

$$\left(1-\frac{1}{n}\right)+\left(1-\frac{2}{n}\right)+\left(1-\frac{3}{n}\right) + \dots \textbf{ upto } n \textbf{ terms.}★$$

Solution : Given,

$$\left(1-\frac{1}{n}\right)+\left(1-\frac{2}{n}\right)+\left(1-\frac{3}{n}\right)+\dots\left(1-\frac{n}{n}\right)$$

$$= (1+1+1+\dots+1) - \left(\frac{1}{n}+\frac{2}{n}+\frac{3}{n}+\frac{4}{n}+\dots+\frac{n}{n}\right)$$

★ are board exam questions from previous years

$$= n - \left[\frac{n}{2}\left\{2\left(\frac{1}{n}\right)+(n-1)\frac{1}{n}\right\}\right]$$

$$= n - \left[\frac{n}{2}\left\{\frac{2}{n}-\frac{1}{n}+1\right\}\right]$$

$$= n - \left[\frac{n}{2}\left\{\frac{1}{n}+1\right\}\right]$$

$$= n - \frac{1+n}{2}$$

$$= \frac{2n-n-1}{2n}$$

$$= \frac{n-1}{2}$$

Thus, the sum of $\left(1-\frac{1}{n}\right)+\left(1-\frac{2}{n}\right)+\left(1-\frac{3}{n}\right) + \dots\left(1-\frac{n}{n}\right)$ is $\frac{n-1}{2}$. **Ans.**

Q. 12. If mth term of an A.P. is $\frac{1}{n}$ and nth term is $\frac{1}{m}$, then find the sum of its first mn terms.★

Solution : Let a and d be the first term and common difference respectively of the given A.P.

Then, $\frac{1}{n} = m^{th}$ term $\Rightarrow \frac{1}{n} = a + (m-1)d$ ...(i)

$\frac{1}{m} = n^{th}$ term $\Rightarrow \frac{1}{m} = a + (n-1)d$ ...(ii)

Subtracting equation (ii) from equation (i),

$$\frac{1}{n} - \frac{1}{m} = (m-n)d$$

$$\Rightarrow \quad \frac{m-n}{nm} = (m-n)d$$

$$\Rightarrow \quad d = \frac{1}{mn}$$

Putting $d = \frac{1}{mn}$ in equation (i), we get

$$\frac{1}{n} = a + (m-1)\frac{1}{mn}$$

$$\Rightarrow \quad \frac{1}{n} = a + \frac{1}{n} - \frac{1}{mn}$$

$$\Rightarrow \quad a = \frac{1}{mn}$$

∴ Sum of first mn terms

$$= \frac{mn}{2}[2a + (mn-1)d]$$

$$= \frac{mn}{2}\left[\frac{2}{mn}+(mn-1)\frac{1}{mn}\right]$$

$$= \frac{mn}{2}\left[\frac{2}{mn}+1-\frac{1}{mn}\right]$$

$$= \frac{mn}{2}\left(\frac{1}{mn}+1\right)$$

$$= \frac{1+mn}{2}$$ **Ans.**

Q. 13. How many terms of the Arithmetic Progession 45, 39, 33, ... must be taken so that theris um is 180 ? Explain the double answer.★

Solution : Given A.P. is

45, 39, 33, ...

Here $a = 45$, $d = 39 - 45 = -6$

Let $S_n = 180$

$$\frac{n}{2}[2a+(n-1)d] = 180$$

$$\frac{n}{2}[2 \times 45 + (n-1)(-6)] = 180$$

$$\frac{n}{2}[90 - 6n + 6] = 180$$

$$\frac{n}{2}[96 - 6n] = 180$$

$$n(96 - 6n) = 360$$

$$96n - 6n^2 = 360$$

$$6n^2 - 96n + 360 = 0$$

On dividing the above equation by 6

$$n^2 - 16n + 60 = 0$$

$$n^2 - 10n - 6n + 60 = 0$$

$$n(n-10) - 6(n-10) = 0$$

$$(n-10)(n-6) = 0$$

$$n = 10, 6$$

∴ Sum of first 10 terms = Sum of first 6 terms

= 180

This means that the sum of all terms from 7th to 10th is zero. **Ans.**

Q. 14. Which term of the Arithmetic Progression – 7, – 12, – 17, – 22, ... will be – 82 ? Is – 100 any term of the A.P. ? Give reason for your answer.★

Solution : Given A.P. is

– 7, – 12, – 17, – 22, ...

Here $a = -7$, $d = -12 - (-7)$

$= -12 + 7$

$= -5$

Let $T_n = -82$

∴ $T_n = a + (n-1)d$

$$-82 = -7 + (n-1)(-5)$$

$$-82 = -7 - 5n + 5$$

$$-82 = -2 - 5n$$

$$-82 + 2 = -5n$$

$$-80 = -5n$$

$$n = 16$$

Therefore, 16th term will be – 82.

Let $T_n = -100$

Again, $T_n = a + (n-1)d$

$$-100 = -7 + (n-1)(-5)$$

$$-100 = -7 - 5n + 5$$

$$-100 = -2 - 5n$$

$$-100 + 2 = -5n$$

$$-98 = -5n$$

$$n = \frac{98}{5}$$

But the number of terms can not be in fraction.

So, – 100 can not be the term of this A.P. **Ans.**

Q. 15. If the *p*th term of an A.P. is $\frac{1}{q}$ and *q*th term is $\frac{1}{p}$, prove that the sum of first *pq* terms of the A.P. is $\left(\frac{pq+1}{2}\right)$.★

Solution : Let a be the first term and d is common difference.

Then, $a_p = \frac{1}{q} \Rightarrow a + (p-1)d = \frac{1}{q}$ ...(i)

and $a_q = \frac{1}{p} \Rightarrow a + (q-1)d = \frac{1}{p}$ ...(ii)

Subtracting equation (ii) from equation (i)

$$pd - qd + = \frac{1}{q} - \frac{1}{p} = \frac{p-q}{pq}$$

$$(p-q)d = \frac{p-q}{pq}$$

$$\Rightarrow \quad d = \frac{1}{pq}$$

Putting the value of d in equation (i), we get

$$a + (p-1)\frac{1}{pq} = \frac{1}{q}$$

$$\Rightarrow \quad a = \frac{1}{q} - \frac{p}{pq} + \frac{1}{pq}$$

★ are board exam questions from previous years

$$\Rightarrow \quad a = \frac{1}{pq}$$

$\therefore$ Sum of first pq terms

$$S_{pq} = \frac{pq}{2}[2a+(pq-1)d]$$

$$= \frac{pq}{2}\left(\frac{2}{pq}+(pq-1)\frac{1}{pq}\right)$$

$$= \frac{pq}{2}\left(\frac{2}{pq}+\frac{pq}{pq}-\frac{1}{pq}\right)$$

$$= \frac{pq}{2}\left(\frac{1}{pq}+1\right)$$

$$S_{pq} = \frac{pq}{2}\left(\frac{1+pq}{pq}\right)$$

$$= \left(\frac{pq+1}{2}\right) \quad \textbf{Hence Proved}$$

Q. 16. In an A.P., the sum of the first 10 terms is – 150 and the sum of the next ten terms is – 550. Find the A.P.★

Solution : Let the first term be a and the common difference be d.

Thus, $S_{10} = \frac{10}{2}\{2a+(10-1)d\} = -150$

$\Rightarrow \quad 2a+9d = -30 \quad$ ...(i)

and $S_{20} = \frac{20}{2}\{2a+(20-1)d\}$

$= 10\{2a+19d\}$

Sum of next ten terms

$S_{10}' = S_{20} - S_{10}$

$-550 = S_{20} + 150$

$S_{20} = -550 - 150 = -700$

$\Rightarrow \quad 10\{2a+19d\} = -700$

$\Rightarrow \quad 2a+19d = -70 \quad$ ...(ii)

Subtracting (i) from (ii), we have

$10d = -40$

$\Rightarrow \quad d = -4$

Substituting $d = -4$ in (i), we have

$2a+9(-4) = -30$

$\Rightarrow \quad 2a - 36 = -30$

$\Rightarrow \quad 2a = 6$

$\Rightarrow \quad a = 3$

Thus, the A.P. is 3, – 1, – 5, – 9, ... **Ans.**

Q. 17. The sum of first n terms of an A.P. is $5n^2 + 3n$. If m^{th} term is 168, find the value of m and also the 20th term.★

Solution : Given, $S_n = 5n^2 + 3n$ and $t_m = 168$

Thus, $S_n = \frac{n}{2}\{2a+(n-1)d\} = 5n^2 + 3n$

$\Rightarrow n\{2a+(n-1)d\} = 2n(5n+3)$

$\Rightarrow \quad 2a + dn - d = 10n + 6$

$\Rightarrow \quad (2a-d) + dn = 6 + 10n$

On comparing both sides, we get

$d = 10$

and $2a - d = 6$

$\Rightarrow \quad 2a = d + 6$

$\Rightarrow \quad 2a = 10 + 6$

$\Rightarrow \quad a = 8$

Now, $t_m = a + (m-1)d = 168$

$\Rightarrow \quad 8 + (m-1)10 = 168$

$\Rightarrow \quad 10(m-1) = 160$

$\Rightarrow \quad m - 1 = 16$

$\Rightarrow \quad m = 17$

Also, $t_{20} = 8 + (20-1)10 = 198.$ **Ans.**

Q. 18. The sum of first q terms of an A.P. is $63q - 3q^2$. If its pth term is – 60, find the value of p. Also find the 11th term of this A.P.★

Solution : Let the first term be a and the common difference be d.

Given, $S_q = 63q - 3q^2 \quad$ ...(i)

and $t_p = -60 \quad$ ...(ii)

From (i), $S_q = \frac{q}{2}\{2a+(q-1)d\}$

$= 63q - 3q^2$

$\Rightarrow \quad q\{2a+(q-1)d\} = 6q(21-q)$

$\Rightarrow \quad 2a + qd - d = 126 - 6q$

$\Rightarrow \quad (2a-d) + qd = 126 - 6q$

On comparing both sides, we get

$d = -6$

and $2a - d = 126$

$\Rightarrow \quad 2a = 126 + d$

$\Rightarrow \quad 2a = 126 + (-6)$

$\Rightarrow \quad 2a = 120$

$\Rightarrow \quad a = 60$

From (ii), $t_p = a + (p-1)d$

$= -60$

$\Rightarrow \quad 60 + (p-1)(-6) = -60$

★ are board exam questions from previous years

$\Rightarrow \quad -6(p-1) = -120$

$\Rightarrow \quad p-1 = 20$

$\Rightarrow \quad p = 21$

Also, $\quad t_{11} = 60 + (11-1)(-6)$

$= 60 - 60 = 0.$ **Ans.**

Q. 19. If the sum of first four terms of an A.P. is 40 and that of first 14 terms is 280. Find the sum of first *n* terms.★

Solution : Given, $S_4 = 40$ and $S_{14} = 280$.

If a be the first term and d be the common difference of an A.P.

Then, Sum of $n$term $(S_n) = \frac{n}{2}[2a+(n-1)d]$

or $\quad S_n = \frac{n}{2}[2a+(n-1)d]$

or $\quad S_4 = \frac{4}{2}[2a+3d]$

or $\quad 40 = 2(2a+3d)$

or $\quad 2a + 3d = 20 \quad$ ...(i)

Also, Sum of first 14 terms = 280

$$S_{14} = \frac{14}{2}[2a+(13)d]$$

$$280 = 7(2a+13d)$$

or $\quad 2a + 13d = 40 \quad$ ...(ii)

On solving equation (i) and (ii), we get

$a = 7, d = 2$

Now, sum of n terms

$$S_n = \frac{n}{2}[2a+(n-1)d]$$

On putting $a = 7, d = 2$

$$S_n = \frac{n}{2}[14+(n-1)2]$$

$$= n[7+n-1]$$

$$= n(n+6)$$

$$= n^2 + 6n$$

Hence, the sum of first n terms is $n^2 + 6n$ **Ans.**

Q. 20. If the sum of first *m* terms of an A.P. is the same as the sum of its first *n* terms, show that the sum of its first (*m* + *n*) terms is zero.★

Solution : Let a be first term and d be common difference of given A.P.

Then, $\quad S_m = S_n \quad$ (Given)

$$\frac{m}{2}\{2a+(m-1)d\} = \frac{n}{2}\{2a+(n-1)d\}$$

$$\Rightarrow \quad \frac{2am}{2} + \frac{m}{2}(m-1)d - \frac{2an}{2} - \frac{n}{2}(n-1)d = 0$$

$$2am - 2an + \{m(m-1) - n(n-1)\}d = 0$$

$$2a(m-n) + (m^2 - m - n^2 + n)d = 0$$

$$2a(m-n) + \{m^2 - n^2 - (m-n\}d = 0$$

$$2a(m-n) + (m-n)(m+n-1)d = 0$$

$$(m-n)\{2a+(m+n-1)d\} = 0$$

$$2a + (m+n-1)d = 0$$

Now, $\quad S_{m+n} = \frac{m+n}{2}\{2a+(m+n-1)d\}$

$$= \frac{m+n}{2} \times 0 = 0.$$

Hence Proved.

Q. 21. The houses in a row are numbered consecutively from 1 to 49. Show that there exists a value of *X* such that sum of numbers of houses proceeding the house numbered *X* is equal to sum of the numbers of houses following *X*.★

Solution : Given, the houses in a row numbered consecutively from 1 to 49.

Now, sum of numbers preceding the number X

$$= \frac{X(X-1)}{2}$$

and sum of numbers following the number X

$$= \frac{49(50)}{2} - \frac{X(X-1)}{2} - X$$

$$= \frac{2450 - X^2 + X - 2X}{2}$$

$$= \frac{2450 - X^2 - X}{2}$$

According to the question,

Sum of no's preceding X = Sum of no's following X

$$\frac{X(X-1)}{2} = \frac{2450 - X^2 - X}{2}$$

$$X^2 - X = 2450 - X^2 - X$$

$$2X^2 = 2450$$

$$X^2 = 1225$$

$$X = 35$$

Hence, at $X = 35$, sum of number of houses preceeding the house no. X is equal to sum of the number of houses following X. **Ans.**

Q. 22. A thief runs with a uniform speed of 100 m/ minute. After one minute, a policeman runs after the thief to catch him. He goes with a speed of 100 m/minute

★ are board exam questions from previous years

in the first minute and increases his speed by 10 m/minute every succeeding minute. After how many minutes the policeman will catch the thief ?★

Solution : Let total time be n minutes.

Since policeman runs after 1 minute so he will catch the thief in $(n-1)$ minutes.

Total distance covered by thief = 100 m/minute × n minute

$$= (100\, n)\text{ m}$$

Now, total distance covered by the policeman = (100)m + (100 + 10)m + (100 + 10 + 10)m + ...+ $(n-1)$ terms

i.e., 100 + 110 + 120 + ... + $(n-1)$ terms

$$\therefore S_{n-1} = \frac{n-1}{2}\,[2 \times 100 + (n-2)\,10]$$

Now, $\frac{n-1}{2}[200 + (n-2)\,10] = 100\,n.$

$\Rightarrow$ $(n-1)(200 + 10n - 20) = 200\,n.$

$\Rightarrow 200\,n - 200 + 10n^2 - 10n + 20 - 20n = 200n$

$\Rightarrow$ $10n^2 - 30n - 180 = 0$

$\Rightarrow$ $n^2 - 3n - 18 = 0$

$\Rightarrow$ $n^2 - (6-3)\,n - 18 = 0$

$\Rightarrow$ $n^2 - 6n + 3n - 18 = 0$

$\Rightarrow$ $n(n-6) + 3(n-6) = 0$

$\Rightarrow$ $(n+3)(n-6) = 0$

$\therefore$ $n = 6$ or $n = -3$ (neglect)

Hence, policeman will catch the thief in (6 – 1) *i.e.*, 5 minutes. **Ans.**

Chapter 6. Coordinate Geometry

Q. 1. Find the ratio in which the point $P(x, 2)$ divides the line segment joining the points $A(12, 5)$ and $B(4, -3)$. Also the value of x.★

Solution : Let the required ratio be $m : n$.

Now using section formula, we have

$$\text{Coordinates of } P = \left(\frac{12n+4m}{m+n}, \frac{5n-3m}{m+n}\right)$$

But it is given that

$$P = (x, 2)$$

$$\Rightarrow \quad \frac{12n+4m}{m+n} = x$$

and $\frac{5n-3m}{m+n} = 2$

$\Rightarrow$ $5n - 3m = 2m + 2n$

$\Rightarrow$ $5m = 3n$

$\Rightarrow$ $\frac{m}{n} = \frac{3}{5}$

Required ratio is 3 : 5

Now, $\frac{12n+4m}{m+n} = x$

$$\Rightarrow \quad \frac{n\left(12+\frac{4m}{n}\right)}{n\left(\frac{m}{n}+1\right)} = x$$

$$\Rightarrow \quad \frac{12+\frac{4\times 3}{5}}{\frac{3}{5}+1} = x$$

$$\Rightarrow \quad \frac{60+12}{3+5} = x$$

$$\Rightarrow \quad \frac{72}{8} = x$$

$\Rightarrow$ $x = 9$ **Ans.**

Q. 2. The mid-point P of the line segment joining the points $A(-10, 4)$ and $B(-2, 0)$ lies on the line segment joining the points $C(-9, -4)$ and $D(-4, y)$. Find the ratio in which P divides CD. Also, find the value of y.

Solution : Since, P is the mid-point of AB

Thus, $P = \left(\frac{-10-2}{2}, \frac{4+0}{2}\right)$

$$= \left(\frac{-12}{2}, \frac{4}{2}\right) = (-6, 2)$$

Also let CD be divided in the ratio $s : t$.

Now, by section formula,

$$-6 = \frac{t(-9)+s(-4)}{s+t}$$

$\Rightarrow$ $-6(s+t) = -9t - 4s$

$\Rightarrow$ $6t + 6s = 9t + 4s$

$\Rightarrow$ $3t - 2s = 0$

$\Rightarrow$ $2s = 3t$

$\Rightarrow$ $\frac{s}{t} = \frac{3}{2}$

$\Rightarrow$ $s : t = 3 : 2$

Hence, the required ratio is 3 : 2.

and $2 = \frac{t(-4)+\Delta \times y}{S+t}$

$$\Rightarrow \quad 2 = \frac{2(-4)+3y}{2+3}$$

$\Rightarrow$ $10 = -8 + 3y$

★ are board exam questions from previous years

$\Rightarrow \quad 3y = 18$

$\Rightarrow \quad y = 6.$ **Ans.**

Q. 3. If $A(4, 2)$, $B(7, 6)$ and $C(1, 4)$ are the vertices of ΔABC and AD is its median, prove that the median AD divides ΔABC into two triangles of equal areas.★

Solution : Given, $A(4, 2)$, $B(7, 6)$ and $C(1, 4)$ are the vertices of ΔABC and AD is its median.

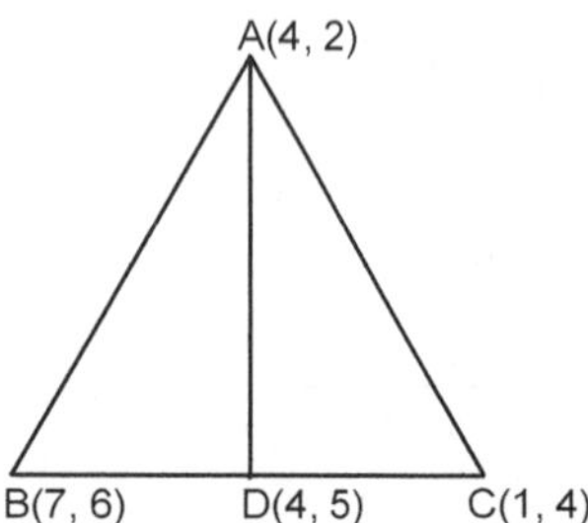

Thus, AD bisects BC at D such that D is its mid-point.
So, the coordinates of D

$$= \left(\frac{7+1}{2}, \frac{6+4}{2}\right) = (4, 5)$$

Now, Area of ΔABC

$$= \frac{1}{2}|4(6-4) + 7(4-2) + 1(2-6)|$$

$$= \frac{1}{2}|4(2) + 7(2) + 1(-4)|$$

$$= \frac{1}{2}|8 + 14 - 4| = 9 \text{ sq. units}$$

Also, Area of ΔADB

$$= \frac{1}{2}|4(6-5) + 7(5-2) + 4(2-6)|$$

$$= \frac{1}{2}|4(1) + 7(3) + 4(-4)|$$

$$= \frac{1}{2}|4 + 21 - 16|$$

$= 4.5$ sq. units

Also, Area of ΔADC

$$= \frac{1}{2}|4(4-5) + 1(5-2) + 4(2-4)|$$

$$= \frac{1}{2}|4(-1) + 1(3) + 4(-2)|$$

$$= \frac{1}{2}|-4 + 3 - 8|$$

$= 4.5$ sq. units

$\Rightarrow$ ar (ΔADB) = ar (ΔADC) = 4.5 sq. units

Hence Proved.

★ **are board exam questions from previous years**

Q. 4. If $A(4, -6)$, $B(3, -2)$ and $C(5, 2)$ are the vertices of ΔABC and AD is its median, prove that the median AD divides ΔABC into two triangles of equal areas.★

Solution : Given $A(4, -6)$, $B(3, -2)$ and $C(5, 2)$ are the vertices of ΔABC and AD is its median.

Thus, AD bisects BC at D such that D is its mid-point.

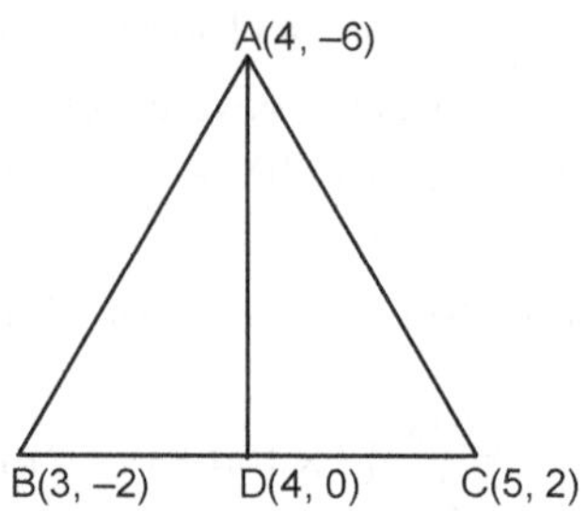

So the coordinates of D

$$= \left(\frac{3+5}{2}, \frac{-2+2}{2}\right)$$

$= (4, 0)$

Now, Area of ΔABC

$$= \frac{1}{2}|4(-2-2) + 3(2+6) + 5(-6+2)|$$

$$= \frac{1}{2}|4(-4) + 3(8) + 5(-4)|$$

$$= \frac{1}{2}|-16 + 24 - 20|$$

$= 6$ sq. units

[as area cannot be negative]

Also, Area of ΔADB

$$= \frac{1}{2}|4(0+2) + 4(-2+6) + 3(-6-0)|$$

$$= \frac{1}{2}|4(2) + 4(4) + 3(-6)|$$

$$= \frac{1}{2}|8 + 16 - 18|$$

$= 3$ sq. units.

Also, Area of ΔADC

$$= \frac{1}{2}|4(0-2) + 4(2+6) + 5(-6+0)|$$

$$= \frac{1}{2}|4(-2) + 4(8) + 5(-6)|$$

$$= \frac{1}{2}|-8 + 32 - 30|$$

$= 3$ sq. units

$\Rightarrow$ ar (ΔADB) = ar (ΔADC) = 3 sq. units

Hence Proved.

Q. 5. Three vertices of a parallelogram *ABCD* are *A*(6, 1), *B*(8, 2) and *C*(9, 4). If *E* is the mid-point of *DC*, find the area of ΔADE.★

Solution : As *ABCD* is a parallelogram, so

$$AB = CD, AD = BC$$

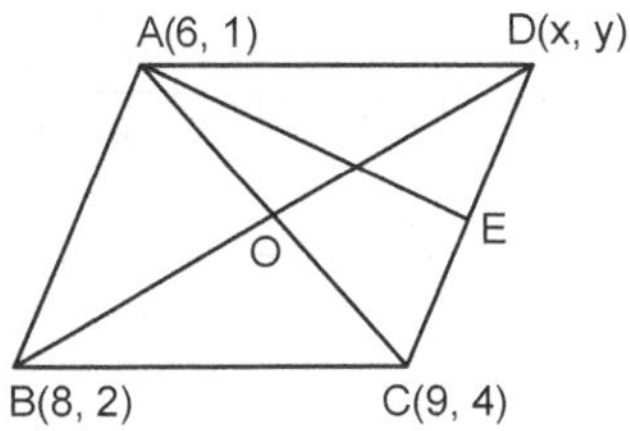

Area of ΔABC = Area of ΔADC

Let $\quad D = (x, y)$

Thus, Area of $\Delta ABC = \frac{1}{2}[6(2-4)+8(4-1)+9(1-2)]$

$$= \frac{1}{2}[6(-2)+8(3)+9(-1)]$$

$$= \frac{1}{2}[-12+24-9]$$

$$= \frac{3}{2} \text{ sq. units}$$

Hence, Area of ΔABC

$$= \text{Area of } \Delta ADC$$

$$= \frac{3}{2} \text{ sq. units}$$

As *E* is the mid-point of *DC*, so

$$DE = EC$$

Thus, Area of ΔAEC = Area of ΔAED

and Area of ΔADC = Area of ΔAEC + Area of ΔAED

$$\Rightarrow \quad \frac{3}{2} \text{ sq. units} = 2 \times \text{Area of } \Delta AED$$

$$\Rightarrow \quad \text{Area of } \Delta AED = \frac{3}{4} \text{ sq. units.}$$ **Ans.**

Q. 6. Show that the points (1, 1), (– 1, 5), (7, 9) and (9, 5) taken in that order are the vertices of rectangle.★

Solution : Given, $A = (1, 1)$, $B = (-1, 5)$, $C = (7, 9)$ $D = (9, 5)$

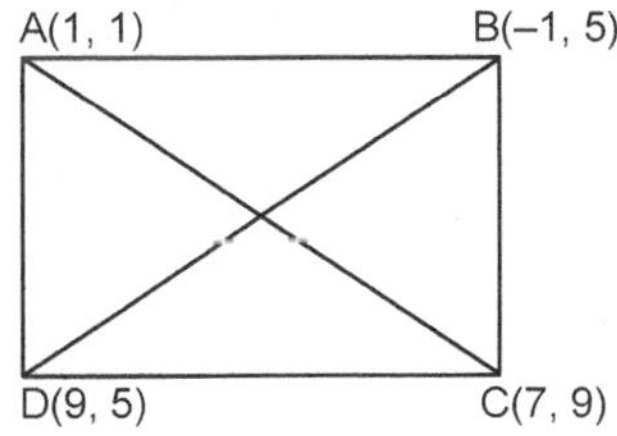

Thus, $AB = \sqrt{(1+1)^2+(1-5)^2}$

$$= \sqrt{(2)^2+(-4)^2}$$

$$= \sqrt{20} = 2\sqrt{5} \text{ units}$$

$$BC = \sqrt{(-1-7)^2+(5-9)^2}$$

$$= \sqrt{(-8)^2+(-4)^2}$$

$$= \sqrt{80} = 4\sqrt{5} \text{ units}$$

$$CD = \sqrt{(7-9)^2+(9-5)^2}$$

$$= \sqrt{(-2)^2+(4)^2}$$

$$= \sqrt{20} = 2\sqrt{5} \text{ units}$$

$$DA = \sqrt{(9-1)^2+(5-1)^2}$$

$$= \sqrt{(8)^2+(4)^2}$$

$$= \sqrt{80} = 4\sqrt{5} \text{ units}$$

$$AC = \sqrt{(7-1)^2+(9-1)^2}$$

$$= \sqrt{(6)^2+(8)^2}$$

$$= \sqrt{100} = 10 \text{ units}$$

$$BD = \sqrt{(9+1)^2+(5-5)^2} = \sqrt{(10)^2}$$

$$= \sqrt{100} = 10 \text{ units}$$

Hence $AB = CD$; $AC = BD$ and $BC = DA$

Thus, as two pair of opposite sides are equal and the diagonals are also equal, so the figure is a rectangle.

Hence Proved.

Q. 7. Show that the points *A*(6, 2), *B*(2, 1), *C*(1, 5) and *D*(5, 6) are the vertices of a square.★★

Solution : Given,

$$A = (6, 2), B = (2, 1), C = (1, 5) \text{ and } D = (5, 6)$$

A(6, 2)
B(2, 1)
D(5, 6)
C(1, 5)

Thus $\quad AB = \sqrt{(6-2)^2+(2-1)^2} = \sqrt{(4)^2+(1)^2}$

$$= \sqrt{17} \text{ units}$$

$$BC = \sqrt{(2-1)^2+(1-5)^2} = \sqrt{(1)^2+(-4)^2}$$

$$= \sqrt{17} \text{ units}$$

$$CD = \sqrt{(1-5)^2+(5-6)^2} = \sqrt{(-4)^2+(-1)^2}$$

$$= \sqrt{17} \text{ units}$$

$$DA = \sqrt{(6-5)^2+(2-6)^2} = \sqrt{(1)^2+(-4)^2}$$

★ are board exam questions from previous years

★★ are frequently asked board exam questions

$= \sqrt{17}$ units

$AC = \sqrt{(6-1)^2 + (2-5)^2} = \sqrt{(5)^2 + (-3)^2}$

$= \sqrt{34}$ units

$BD = \sqrt{(2-5)^2 + (1-6)^2} = \sqrt{(-3)^2 + (-5)^2}$

$= \sqrt{34}$ units

Hence $AB = CD = BC = DA$ and $AC = BD$

Thus, as all the sides are equal and the diagonals are also equal to each other, so the figure is a square.

Hence Proved.

Q. 8. If the points $A(k + 1, 2k)$, $B(3k, 2k + 3)$ and $C(5k - 1, 5k)$ are collinear, then find the value of k.★

Solution : Since $A(k + 1, 2k)$, $B(3k, 2k + 3)$ and $C(5k - 1, 5k)$ are collinear points, so area of triangle formed by them is 0.

$$\Delta = \frac{1}{2}|x_1(y_2 - y_3) + x_2(y_3 - y_1) + x_3(y_1 - y_2)|$$

$$\Delta = \frac{1}{2}|(k + 1)(2k + 3 - 5k) + 3k(5k - 2k) + (5k - 1)(2k - 2k - 3)|$$

$$\Delta = \frac{1}{2}|(k + 1)(3 - 3k) + 3k(3k) + (5k - 1)(-3)|$$

$$\Delta = \frac{1}{2}|3k + 3 - 3k^2 - 3k + 9k^2 - 15k + 3|$$

$$0 = \frac{1}{2}|6k^2 - 15k + 6|$$

$6k^2 - 15k + 6 = 0$

$\Rightarrow 6k^2 - 12k - 3k + 6 = 0$

$\Rightarrow 6k(k - 2) - 3(k - 2) = 0$

$\Rightarrow (k - 2)(6k - 3) = 0$

$k = 2$ or $k = \frac{1}{2}$ **Ans.**

Q. 9. Point A lies on the line segment XY joining X(6, – 6) and Y (– 4, – 1) in such a way that $\frac{XA}{XY} = \frac{2}{5}$. If point A also lies on the line $3x + k(y + 1) = 0$, find the value of k.★

Solution : Given,

$$\frac{XA}{XY} = \frac{2}{5}$$

$$\frac{XA}{XA + AY} = \frac{2}{5}$$

$5XA = 2XA + 2AY$

$3XA = 2AY$

$$\frac{XA}{AY} = \frac{2}{3}$$

$XA : AY = 2 : 3$

So A divides XY in ratio 2 : 3

Here, $m = 2$, $n = 3$, $x_1 = 6$, $y_1 = -6$, $x_2 = -4$ and $y_2 = -1$

Coordinates of Point A are $\left(\frac{mx_2 + nx_1}{m+n}, \frac{my_2 + ny_1}{m+n}\right)$

$$\Rightarrow \left(\frac{2 \times (-4) + 3(6)}{2+3}, \frac{2(-1) + 3(-6)}{2+3}\right)$$

$$\Rightarrow \left(\frac{-8+18}{5}, \frac{-2-18}{5}\right) = (2, -4)$$

Since, point A(2, – 4) lies on line $3x + k(y + 1) = 0$.

Therefore it will satisfy the equation.

On putting $x = 2$ and $y = -4$ in the equation, we get

$3 \times 2 + k(-4 + 1) = 0$

$6 - 3k = 0$

$3k = 6$

$k = 2$ **Ans.**

Q. 10. Find the coordinates of the points of trisection of the line segment joining the points $A(-5, 6)$ and $B(4, -3)$.★

Solution : Given, $A = (-5, 6)$ and $B = (4, -3)$

Let the required co-ordinates be $P(x, y)$ and $Q(j, k)$.

A(–5, 6) P(x, y) Q (j, k) B(4, –3)

Now, $AP = PQ = QB$

Thus, $AQ = 2QB$ and $2AP = PB$

$$\Rightarrow \frac{AQ}{QB} = \frac{2}{1}$$

and $$\frac{AP}{PB} = \frac{1}{2}$$

$$\Rightarrow x = \frac{2(-5) + 1(4)}{2+1}$$

and $$y = \frac{2(6) + 1(-3)}{2+1}$$

$$\Rightarrow x = \frac{-10+4}{3} \text{ and } y = \frac{12-3}{2+1}$$

$$\Rightarrow x = \frac{-6}{3} \text{ and } y = \frac{9}{3}$$

$\Rightarrow P(x, y) = (-2, 3)$

Similarly, $$j = \frac{1(-5) + 2(4)}{2+1}$$

and $$k = \frac{1(6) + 2(-3)}{2+1}$$

★ are board exam questions from previous years

$$\Rightarrow \qquad j = \frac{-5+8}{3} \text{ and } k = \frac{6-6}{2+1}$$

$$\Rightarrow \qquad j = 1 \text{ and } k = 0$$

$$\Rightarrow \qquad Q(j, k) = (1, 0).$$ **Ans.**

Q. 11. Find the ratio in which the line $x - 3y = 0$ divides the line segment joining the points $(-2, -5)$ and $(6, 3)$. Find the coordinates of the point of intersection.★

Solution : Let required ratio be $k : 1$.

By section formula, we have

$$x = \frac{mx_2 + nx_1}{m+n}, \; y = \frac{my_2 + ny_1}{m+n}$$

Here, $x_1 = -2,\ x_2 = 6,\ y_1 = -5,\ y_2 = 3$

$m = k,\ n = 1$

$$\Rightarrow \qquad x = \frac{k(6)+(-2)}{k+1} = \frac{6k-2}{k-1}$$

$$\Rightarrow \qquad y = \frac{k(3)+(-5)}{k+1} = \frac{3k-5}{k+1}$$

$\left(\frac{6k-2}{k+1}, \frac{3k-5}{k+1}\right)$ points lie on the line $x - 3y = 0$

$$\therefore \qquad \left(\frac{6k-2}{k+1}\right) - 3\left(\frac{3k-5}{k+1}\right) = 0$$

$$\frac{6k-2}{k+1} - \frac{(9k-15)}{k+1} = 0$$

$$6k - 2 - 9k + 15 = 0$$

$$-3k + 13 = 0$$

$$k = \frac{13}{3}$$

Hence required ratio is $\left(\frac{13}{3}, 1\right)$ *i.e.*, (13, 3)

Here, intersection point are,

$$x = \frac{6k-2}{k+1} = \frac{\frac{6\times 13}{3} - 2}{\frac{13}{3}+1}$$

$$= \frac{(26-2)\times 3}{16}$$

$$= \frac{72}{16} = \frac{9}{2}$$

$$y = \frac{3k-5}{k+1} = \frac{3\times\frac{13}{3} - 5}{\frac{13}{3}+1}$$

$$= \frac{(13-5)\times 3}{13+3} = \frac{24}{16}$$

$$= \frac{3}{2}$$

$\therefore$ Intersection point is $\left(\frac{9}{2}, \frac{3}{2}\right)$. **Ans.**

Q. 12. If $(-2, -1)$, $(a, 0)$, $(4, b)$ and $(1, 2)$ are the vertices of parallelogram taken in order, find the value of a and b.★

Solution : Given, $A = (-2, -1)$, $B = (a, 0)$, $C = (4, b)$ and $D = (1, 2)$.

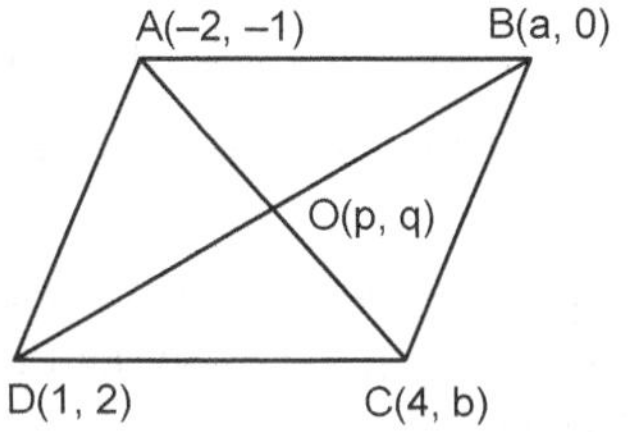

As $ABCD$ is a parallelogram then its diagonals, AC and BD bisect each other.

Let the point of bisection of diagonals be $O(p, q)$.

$$\text{Thus,} \qquad p = \frac{1+a}{2} = \frac{4-2}{2}$$

$$\Rightarrow \qquad p = 1 + a = 2$$

$$\Rightarrow \qquad a = 1$$

$$\text{and} \qquad q = \frac{2+0}{2} = \frac{b-1}{2}$$

$$\Rightarrow \qquad q = 2 = b - 1$$

$$\Rightarrow \qquad q = b = 3$$

Thus, $a = 1$ and $b = 3$. **Ans.**

Q. 13. Prove that the points $A(3, 0)$, $B(4, 5)$, $C(-1, 4)$ and $D(-2, -1)$ taken in order, are the vertices of a rhombus. Also find its area.★ [NCERT]

Solution : Given, $A = (3, 0)$, $B = (4, 5)$, $C = (-1, 4)$ and $D = (-2, -1)$

Now, $AB = \sqrt{(4-3)^2 + (5-0)^2} = \sqrt{(1)^2 + (5)^2}$

$= \sqrt{26}$ units

$BC = \sqrt{(-1-4)^2 + (4-5)^2} = \sqrt{(-5)^2 + (-1)^2}$

$= \sqrt{26}$ units

$CD = \sqrt{(-2+1)^2 + (-1-4)^2} = \sqrt{(-1)^2 + (-5)^2}$

$= \sqrt{26}$ units

and $DA = \sqrt{(-2-3)^2 + (-1-0)^2} = \sqrt{(-5)^2 + (-1)^2}$

$= \sqrt{26}$ units

Also, $AC = \sqrt{(3+1)^2 + (0-4)^2} = \sqrt{(4)^2 + (-4)^2}$

$= \sqrt{32}$ units

★ are board exam questions from previous years

and $\quad BD = \sqrt{(4+2)^2+(5+1)^2} = \sqrt{(6)^2+(6)^2}$

$= \sqrt{72}$ units

Thus, the sides of the quadrilateral are of equal length but the diagonals are unequal.

Hence, $ABCD$ is proved to be a rhombus.

Now, Area of a rhombus $= \frac{1}{2}[\sqrt{32}\times\sqrt{72}]$

$= \frac{1}{2}\sqrt{2304}$ sq. units

$= 24$ sq. units **Ans.**

Q. 14. Find the coordinates of the centroid of a triangle.★ [NCERT]

Solution : Let the vertices of ΔABC be $A = (x_1, y_1)$, $B = (x_2, y_2)$ and $C = (x_3, y_3)$. Also, let the vertices of the centroid be $G = (x, y)$.

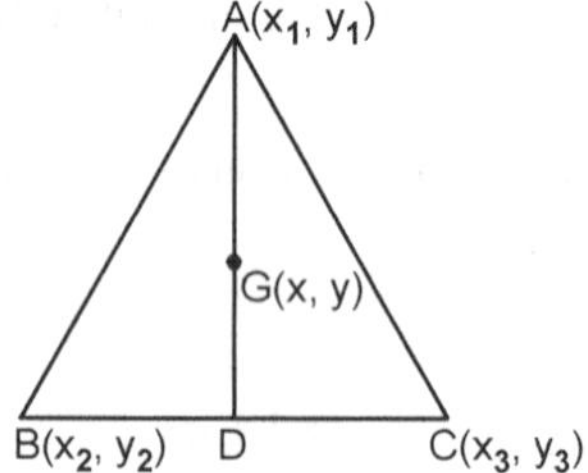

Let AD be the median that bisects BC.

Thus $\quad BD = DC$

Hence $\quad D = \left(\frac{x_2+x_3}{2}, \frac{y_2+y_3}{2}\right)$

If G is the centroid of ΔABC, then

$AG : GD = 2 : 1$

Thus,

$$= \left(\frac{2\left(\frac{x_2+x_3}{2}\right)+1(x_1)}{2+1}, \frac{2\left(\frac{y_2+y_3}{2}\right)+1(y_1)}{2+1}\right)$$

$$= \left(\frac{x_1+x_2+x_3}{3}, \frac{y_1+y_2+y_3}{3}\right)$$ **Ans.**

Q. 15. If $P(-5,-3)$, $Q(-4,-6)$, $R(2,-3)$ and $S(1,2)$ are the vertices of a quadrilateral $PQRS$, find its area.★

Solution : We have $P(-5,-3)$, $Q(-4,-6)$, $R(2,-3)$ and $S(1, 2)$ are the vertices of a quadrilateral Join P and R. Then,

Area of quad. $PQRS$ = (Area of ΔPQR) + (Area of ΔPRS)

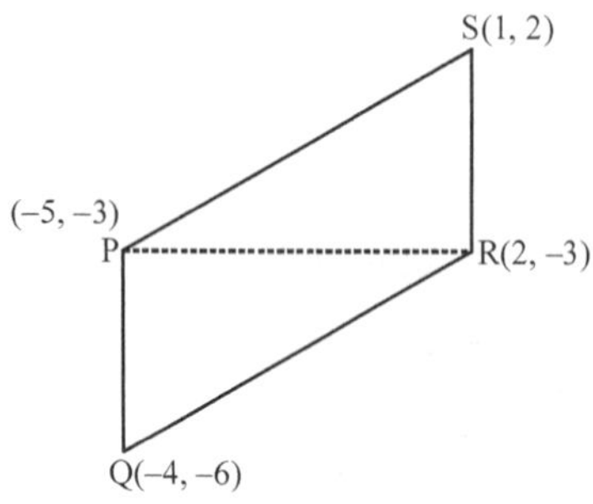

★ are board exam questions from previous years

Area of ΔPQR

$= \frac{1}{2}|-5(-6+3)-4(-3+3)+2(-3+6)|$

$= \frac{1}{2}|-5(-3)-4(0)+2(3)|$

$= \frac{1}{2}|15+6|$

$= \frac{21}{2}$ sq. units

and, area of ΔPRS

$= \frac{1}{2}|-5(-3-2)+2(2+3)+1(-3+3)|$

$= \frac{1}{2}|-5(-5)+2(5)+1(0)|$

$= \frac{1}{2}|25+10| = \frac{35}{2}$ sq. units

Hence, area of quad. $PQRS$ = Area of ΔPQR + Area of ΔPRS)

$= \left(\frac{21}{2}+\frac{35}{2}\right)$ sq. units

$= \frac{56}{2}$ sq. units

$= 28$ sq. units. **Ans.**

Q. 16. The line segment joining the points $A(4,-5)$ and $B(4, 5)$ is divided by the point P such that $\frac{AP}{AB} = \frac{2}{5}$. Find the coordinates of P.★

Solution : Given, $A = (4,-5)$, $B = (4, 5)$ and $\frac{AP}{AB} = \frac{2}{5}$.

Now, $\quad AB = AP + PB$

A(4, –5) 2 P(x, y) 3 B(4, 5)

Thus $\quad \frac{AP}{AP+PB} = \frac{2}{5}$

$\Rightarrow \quad \frac{AP+PB}{AP} = \frac{5}{2}$

$\Rightarrow \quad 1+\frac{PB}{AP} = \frac{5}{2}$

$\Rightarrow \quad \frac{PB}{AP} = \frac{5}{2}-1$

$\Rightarrow \quad \frac{PB}{AP} = \frac{3}{2}$

Thus $\quad PB : AP = 3 : 2$

Let $\quad P = (x, y)$

$\Rightarrow \quad x = \frac{3\times4+2\times4}{3+2}$

and $y = \frac{3(-5)+2(5)}{3+2}$

$\Rightarrow$ $x = \frac{12+8}{5}$ and $y = \frac{-15+10}{5}$

$\Rightarrow$ $x = \frac{20}{5}$ and $y = \frac{-5}{5}$

$\Rightarrow$ $x = 4$ and $y = -1$

Thus $P = (4, -1)$. **Ans.**

Q. 17. Points P, Q, R and S divide the line segment joining the points $A(1, 6)$ and $B(5, -2)$ in 4 equal parts. Find the co-ordinates of the points P, Q and R.★

Solution : Given, $A(1, 6)$ and $B(5, -2)$

A(1, 6) P(x, y) Q(a, b) R(m, n) B(5, –2)

Let $P = (x, y)$, $Q = (a, b)$ and $R = (m, n)$

Now, $AP = PQ = QR = RB$

Thus $AP : PB = 1 : 3$

So $x = \frac{3\times1+1\times5}{1+3}$

and $y = \frac{3(6)+1(-2)}{1+3}$

$\Rightarrow$ $x = \frac{8}{4}$ and $y = \frac{16}{4}$

$\Rightarrow$ $x = 2$ and $y = 4$

$\Rightarrow$ $P(x, y) = (2, 4)$

Now, $PQ : QB = 1 : 1$

So, $a = \frac{1\times1+1\times5}{1+1}$

and $b = \frac{1(6)+1(-2)}{1+1}$

$\Rightarrow$ $a = \frac{6}{2}$ and $b = \frac{4}{2}$

$\Rightarrow$ $a = 3$ and $b = 2$

$\Rightarrow$ $Q(a, b) = (3, 2)$

and $QR : RB = 1 : 1$

So, $m = \frac{1(3)+1(5)}{1+1}$ and $n = \frac{1(2)+1(-2)}{1+1}$

$\Rightarrow$ $m = \frac{8}{2}$ and $n = \frac{0}{2}$

$\Rightarrow$ $m = 4$ and $n = 0$

$\Rightarrow$ $R(m, n) = (4, 0)$. **Ans.**

Q. 18. Find the ratio in which the y-axis divides the line segment joining the points $(-1, -4)$ and $(5, -6)$. Also find the coordinates of the point of intersection.★

★ are board exam questions from previous years

Solution : Let the y-axis cuts the line joining point $A(-1, -4)$ and point $B(5, -6)$ in the ratio $k : 1$ at the point $P(0, y)$.

Then, by section fromula, we have

$$x = \frac{mx_2+nx_1}{m+n}$$

$\Rightarrow$ $0 = \frac{k(5)+(-1)}{k+1}$

$\Rightarrow$ $0 = \frac{5k-1}{k+1}$

$\Rightarrow$ $5k - 1 = 0$

$\Rightarrow$ $k = \frac{1}{5}$

Then the required ratio is $\left(\frac{1}{5}:1\right)$ *i.e.*, $(1 : 5)$

Again, by section formula, we have

$$y = \frac{my_2+ny_1}{m+n}$$

$$= \frac{1(-6)+5(-4)}{1+5}$$

$$= \frac{-6-20}{6}$$

$$= \frac{-26}{6} = \frac{-13}{3}$$

Hence, the intersection co-ordinates is $\left(0, -\frac{13}{3}\right)$

Ans.

Q. 19. Determine the ratio in which the line $x - y - 2 = 0$ divides the line segment joining the points $A(3, -1)$ and $B(8, 9)$.★

Solution : Given, $A = (3, -1)$, $B = (8, 9)$ and $x - y - 2 = 0$

Let the line $x - y - 2 = 0$ intersect the line segment AB at R in the ratio $p : 1$.

Now, $R_x = \frac{1(3)+p(8)}{p+1}$ and $R_y = \frac{1(-1)+p(9)}{p+1}$

$\Rightarrow$ $R_x = \frac{3+8p}{p+1}$ and $R_y = \frac{-1+9p}{p+1}$

Thus, $R = \left(\frac{3+8p}{p+1}, \frac{9p-1}{p+1}\right)$

Now substituting the value of R in the given equation, we get

$$\frac{3+8p}{p+1} - \frac{9p-1}{p+1} - 2 = 0$$

$\Rightarrow 3 + 8p - 9p + 1 - 2(p + 1) = 0$

$\Rightarrow$ $4 - p - 2p - 2 = 0$

$\Rightarrow \quad 3p = 2$

$\Rightarrow \quad p = \frac{2}{3}$

Thus, the ratio is $\frac{2}{3} : 1 = 2 : 3$. **Ans.**

Q. 20. Find the area of the triangle formed by joining the mid-points of the sides of the triangle whose vertices are (2, 2) and (4, 4) and (2, 6).★

Solution : Given, $A = (2, 2)$, $B = (4, 4)$ and $C = (2, 6)$.

Let D, E and F be the mid-points of AB, BC and CD respectively.

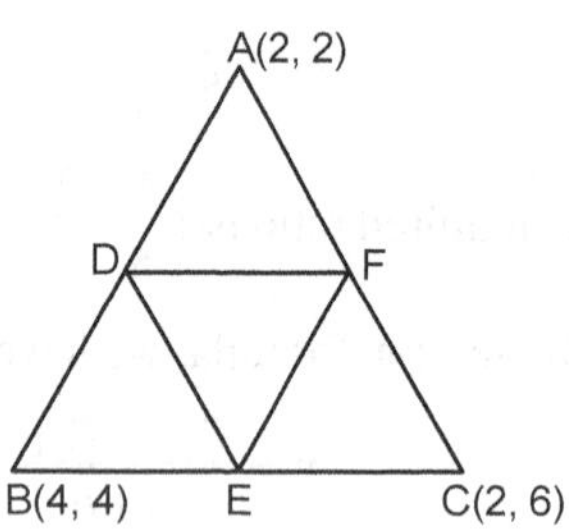

Thus, $\quad D = \left(\frac{2+4}{2}, \frac{2+4}{2}\right) = (3, 3)$

$E = \left(\frac{2+4}{2}, \frac{6+4}{2}\right) = (3, 5)$

and $\quad F = \left(\frac{2+2}{2}, \frac{2+6}{2}\right) = (2, 4)$

Hence, Area of ΔDEF

$= \frac{1}{2}|3(5-4) + (3(4-3) + 2(3-5)|$

$= \frac{1}{2}|3 + 3 - 4|$

$= 1$ sq. units **Ans.**

Q. 21. If $a \neq b \neq 0$, prove that the points (a, a^2), (b, b^2) (c, c) will not be collinear.★

Solution : Given points are $A(a, a^2)$, $B(b, b^2)$ and $C(c, c^2)$.

Now, Area of $\Delta ABC =$

$\frac{1}{2}| a(b^2 - c^2) + b(c^2 - a^2) + c(a^2 - b^2) |$

$= \frac{1}{2}|a(b-c)(b+c) - a^2(b-c) - bc(b-c)|$

$= \frac{1}{2}|(b-c)(a(b+c) - a^2 - bc)|$

$= \frac{1}{2}|(b-c)(ab + ac - a^2 - bc)|$

★ are board exam questions from previous years

$= \frac{1}{2}|(b-c)(a-b)(c-a)|$

This can never be zero, as $a \neq b \neq c$.

Hence, these points can never be collinear.

Hence Proved.

Q. 22. Show that the points (0, – 1), (6, 7),(– 2, 3) and (8, 3) taken in that order, are the vertices of a rectangle.★

Solution : Given, $A = (0, -1)$, $B = (6, 7)$, $C = (-2, 3)$ and $D = (8, 3)$

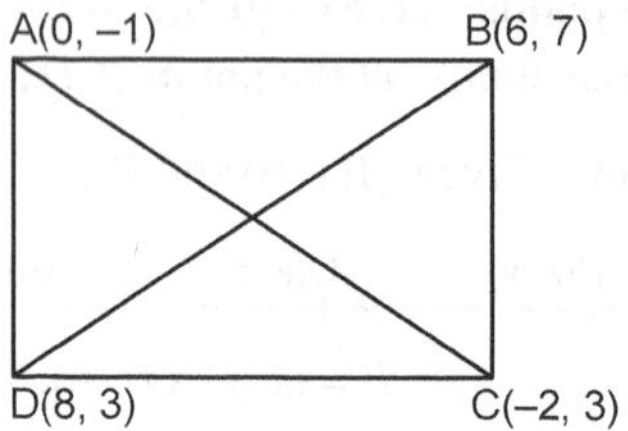

Thus, $\quad AB = \sqrt{(6-0)^2 + (7+1)^2} = \sqrt{(6)^2 + (8)^2}$

$= \sqrt{100} = 10$ units

$BC = \sqrt{(-2-6)^2 + (3-7)^2} = \sqrt{(8)^2 + (4)^2}$

$= \sqrt{80} = 4\sqrt{5}$ units

$CD = \sqrt{(8+2)^2 + (3-3)^2} = \sqrt{(10)^2}$

$= \sqrt{100} = 10$ units

$DA = \sqrt{(8-0)^2 + (3+1)^2} = \sqrt{(8)^2 + (4)^2}$

$= \sqrt{80} = 4\sqrt{5}$ units

$AC = \sqrt{(-2-0)^2 + (3+1)^2} = \sqrt{(2)^2 + (4)^2}$

$= \sqrt{20} = 2\sqrt{5}$ units

$BD = \sqrt{(8-6)^2 + (3-7)^2} = \sqrt{(2)^2 + (-4)^2}$

$= \sqrt{20} = 2\sqrt{5}$ units

Hence $\quad AB = CD$; $AC = BD$ and $BC = DA$

Thus, as two pairs of opposite sides are equal and the diagonals are also equal to each other, so the figure is a rectangle. **Hence Proved.**

Q. 23. If $A(5, -1)$, $B(-3, -2)$ and $C(-1, 8)$ are the vertices of a triangle ABC, find the length of the median through A and the coordinates of the centroid.★

Solution : Given, $A = (5, -1)$, $B = (-3, -2)$ and $C = (-1, 8)$.

Let D be the point through which the median from A meets BC.

Thus, D is the mid-point of BC.

Hence $\quad D = \left(\frac{-3-1}{2}, \frac{-2+8}{2}\right)$

$= (-2, 3)$

So, $AD = \sqrt{(-2-5)^2 + (3+1)^2}$

$= \sqrt{(-7)^2 + (4)^2}$

$= \sqrt{65}$ units

Let O be the centroid of the triangle such that

$O = (x, y)$

Thus, $$(x, y) = \left(\frac{5-3-1}{3}, \frac{-1-2+8}{3}\right)$$

$$= \left(\frac{1}{3}, \frac{5}{3}\right)$$

Hence, the centroid of the triangle is $\left(\frac{1}{3}, \frac{5}{3}\right)$. **Ans.**

Q. 24. Find the ratio in which the line segment joining the points $A(-2, -3)$ and $B(5, 6)$ is divided by (i) the X-axis and (ii) the Y-axis. Also, find the coordinates of the point of division in each case.

Solution : (i) Let AB be divided by the X-axis in the ratio $p : 1$ at the point K.

Thus, the coordinates of the point K

$$= \left(\frac{mx_2 + nx_1}{m+n}, \frac{my_2 + ny_1}{m+n}\right)$$

$$= \left(\frac{p(5) - 2}{p+1}, \frac{p(6) - 3}{p+1}\right)$$

$$= \left(\frac{5p-2}{p+1}, \frac{6p-3}{p+1}\right)$$

But P lies in the X-axis, so its coordinates are $(x, 0)$.

Thus, $$\frac{6p-3}{p+1} = 0$$

$\Rightarrow$ $6p - 3 = 0$

$\Rightarrow$ $6p = 3$

$\Rightarrow$ $p = \frac{3}{6}$

$\Rightarrow$ $p = \frac{1}{2}$

Thus, the required ratio is $\frac{1}{2} : 1 = 1 : 2$

Now, $$x = \frac{5p-2}{p+1}$$

$\Rightarrow$ $$x = \frac{5\left(\frac{1}{2}\right) - 2}{\left(\frac{1}{2}\right) + 1}$$

$\Rightarrow$ $x = \frac{1}{3}$

Hence, $$(x, y) = \left(\frac{1}{3}, 0\right)$$

(ii) Let AB be divided by the Y-axis in the ratio $1 : p$ at the point K.

Thus, the coordinates of the point K

$$= \left(\frac{mx_2 + nx_1}{m+n}, \frac{my_2 + ny_1}{m+n}\right)$$

$$= \left(\frac{1(5) - 2(p)}{1+p}, \frac{1(6) - 3p}{1+p}\right)$$

$$= \left(\frac{5-2p}{1+p}, \frac{6-3p}{1+p}\right)$$

But P lies in the Y-axis, so its coordinate are $(0, y)$.

Thus, $$\frac{5-2p}{1+p} = 0$$

$\Rightarrow$ $5 - 2p = 0$

$\Rightarrow$ $2p = 5$

$\Rightarrow$ $p = \frac{5}{2}$

Thus, the required ratio is $\frac{5}{2} : 1 = 5 : 2$

Now, $$y = \frac{6-3p}{1+p}$$

$\Rightarrow$ $$y = \frac{6 - 3\left(\frac{5}{2}\right)}{1 + \left(\frac{5}{2}\right)}$$

$\Rightarrow$ $y = -\frac{3}{7}$

Thus, $$(x, y) = \left(0, -\frac{3}{7}\right).$$ **Ans.**

Q. 25. If the coordinates of the mid-points of the sides of a triangle are (3, 4), (4, 6) and (5, 7). Find its vertices.★

Solution : Given, $D = (3, 4)$, $E = (4, 6)$ and $F = (5, 7)$

Let $A = (x, y)$, $B = (a, b)$ and $C = (m, n)$

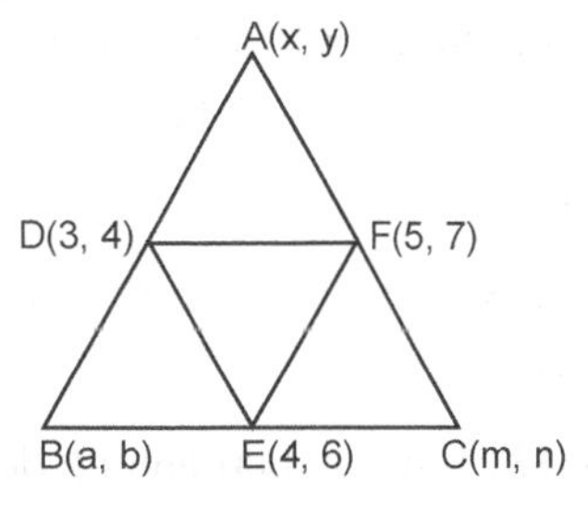

Now, $$\frac{x+a}{2} = 3$$

$\Rightarrow$ $x + a = 6$ ...(i)

$$\frac{y+b}{2} = 4$$

$\Rightarrow$ $y + b = 8$ ...(ii)

★ are board exam questions from previous years

$$\frac{a+m}{2}=4$$

$\Rightarrow$ $a+m=8$ ...(iii)

$$\frac{b+n}{2}=6$$

$\Rightarrow$ $b+n=12$ ...(iv)

$$\frac{x+m}{2}=5$$

$\Rightarrow$ $x+m=10$ ...(v)

$$\frac{y+n}{2}=7$$

$\Rightarrow$ $y+n=14$ ...(vi)

Adding equations (i) and (v), we get

$$x+a+x+m=6+10$$

$\Rightarrow$ $2x+(a+m)=16$

$\Rightarrow$ $2x+8=16$ [Using (iii)]

$\Rightarrow$ $2x=8$

$\Rightarrow$ $x=4$

Thus, substituting $x=4$ in (i), we get

$$4+a=6$$

$\Rightarrow$ $a=2$

and substituting $x=4$ in (v), we get

$$4+m=10$$

$\Rightarrow$ $m=6$

Similarly, adding (ii) and (vi), we get

$$y+b+y+n=8+14$$

$\Rightarrow$ $2y+(b+n)=22$

$\Rightarrow$ $2y+12=22$ [Using (iv)]

$\Rightarrow$ $2y=10$

$\Rightarrow$ $y=5$

Thus, substituting $y=5$ in (ii), we get

$$5+b=8$$

$\Rightarrow$ $b=3$

and substituting $y=5$ in (vi), we get

$$5+n=14$$

$\Rightarrow$ $n=9$

Hence, $A=(x, y)=(4, 5)$; $B=(a, b)=(2, 3)$ and $C=(m, n)=(6, 9)$. **Ans.**

Q. 26. The line segment joining the points $P(3, 3)$ and $Q(6, -6)$ is trisected at the points A and B such that A is nearer to P. Find the coordinates of A and B. If A also lies on the line $2x+y+k=0$, find the value of k.★

Solution : Given, $P=(3, 3)$, $Q=(6, -6)$

Let $A=(x, y)$ and $B=(m, n)$

P(3, 3) A(x, y) B(m, n) Q(6, –6)

Now, $PA=AB=BQ$

Thus, $PA:AQ=1:2$ and $PB:BQ=2:1$

$\therefore$ $x=\frac{2(3)+1(6)}{2+1}$ and $y=\frac{2(3)+1(-6)}{2+1}$

$\Rightarrow$ $x=\frac{12}{3}$ and $y=\frac{0}{3}$

$\Rightarrow$ $x=4$ and $y=0$

Thus, $A=(4, 0)$

Also, $PB:BQ=2:1$

$\therefore$ $m=\frac{1(3)+2(6)}{1+2}$ and $n=\frac{1(3)+2(-6)}{1+2}$

$\Rightarrow$ $m=\frac{18}{3}$ and $n=\frac{9}{3}$

$\Rightarrow$ $m=6$ and $n=3$

Thus $B=(6, 3)$

Since point A lies on given line, therefore, substituting the values of A (4, 0) in equation $2x+y+k=0$, we get

$$2(4)+(0)+k=0$$

$\Rightarrow$ $8+k=0$

$\Rightarrow$ $k=-8.$ **Ans.**

Q. 27. The line segment joining the points $P(2, 1)$ and $Q(5, -8)$ is trisected at the points A and B such that A is nearer to P. Find the coordinates of A and B. If A also lies on the line $2x-y+k=0$, find the value of k.★

Solution : Given, $P=(2, 1)$, $Q=(5, -8)$

Let $A=(x, y)$ and $B=(m, n)$

P(2, 1) A(x, y) B(m, n) Q(5, –8)

Now, $PA=AB=BQ$

Thus, $PA:AQ=1:2$ and $PB:BQ=2:1$

$\therefore$ $x=\frac{2(2)+1(5)}{2+1}$ and $y=\frac{2(1)+1(-8)}{2+1}$

$\Rightarrow$ $x=\frac{9}{3}$ and $y=\frac{-6}{3}$

$\Rightarrow$ $x=3$ and $y=-2$

Thus, $A=(3, -2)$

Also, $PB:BQ=2:1$

$\therefore$ $m=\frac{1(2)+2(5)}{1+2}$ and $n=\frac{1(1)+2(-8)}{1+2}$

★ are board exam questions from previous years

$\Rightarrow \quad m = \frac{12}{3}$ and $n = \frac{-15}{3}$

$\Rightarrow \quad m = 4$ and $n = 5$

Thus, $\quad B = (4, -5)$

Since the point A lies on given line, substituting the values of $A(3, -2)$ in equation

$2x - y + k = 0$, we get

$$2(3) - (-2) + k = 0$$

$\Rightarrow \quad 8 + k = 0$

$\Rightarrow \quad k = -8.$ **Ans.**

Q. 28. Find the ratio in which the point $P(-1, y)$ divides the line segment joining the points $A(-3, 10)$ and $B(6, -8)$. Also, find the value of y.★

Solution : Given, $A(-3, 10)$, $B(6-8)$, $P(-1, y)$.

Let the required ratio be $m : n$.

Now, using section formula, we have

$$P = \left(\frac{6m-3n}{m+n}, \frac{-8m+10n}{m+n}\right)$$

But it is given that

$$P = (-1, y)$$

So, $\quad \frac{6m-3n}{m+n} = -1$

$\Rightarrow \quad 6m - 3n = -1(m+n)$

$\Rightarrow \quad 6m - 3n = -m - n$

$\Rightarrow \quad 7m = 2n$

$\Rightarrow \quad \frac{m}{n} = \frac{2}{7}$

$\Rightarrow \quad m : n = 2 : 7$

So, the required ratio = 2 : 7

Now, $\quad y = \frac{-8m+10n}{m+n}$

$\Rightarrow \quad y = \frac{-8(2)+10(7)}{2+7}$

$\Rightarrow \quad y = \frac{-16+70}{9}$

$\Rightarrow \quad y = \frac{54}{9}$

$\Rightarrow \quad y = 6$

Thus, the value of y is 6. **Ans.**

Q. 29. If fig, the vertices of ΔABC are $A(4, 6)$, $B(1, 5)$ and $C(7, 2)$. A line-segment DE is drawn to intersect the sides AB and AC at D and E respectively such that $\frac{AD}{AB} = \frac{AE}{AC} = \frac{1}{3}$. Calculate the area of ΔADE and compare it with area of ΔABC.★

★ are board exam questions from previous years

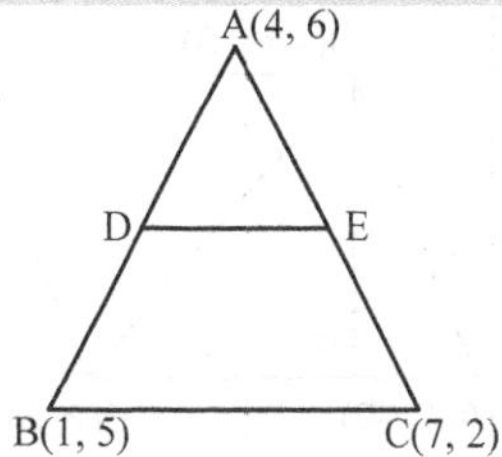

Solution : We have, the vertices of ΔABC are $A(4, 6)$, $B(1, 5)$ and $C(7, 2)$

and $\quad \frac{AD}{AB} = \frac{AE}{AC} = \frac{1}{3}$

Then, coordinates of D

$$= \left(\frac{1(1)+2(4)}{1+2}, \frac{1(5)+2(6)}{1+2}\right)$$

$$= \left(\frac{1+8}{3}, \frac{5+12}{3}\right) = \left(3, \frac{17}{3}\right)$$

and coordinates of E

$$= \left(\frac{1(7)+2(4)}{1+2}, \frac{1(2)+2(6)}{1+2}\right)$$

$$= \left(\frac{7+8}{3}, \frac{2+12}{3}\right) = \left(5, \frac{14}{3}\right)$$

Now, Area of ΔADE

$$= \frac{1}{2}\left|4\left(\frac{17}{3} - \frac{14}{3}\right) + 3\left(\frac{14}{3} - 6\right) + 5\left(6 - \frac{17}{3}\right)\right|$$

$$= \frac{1}{2}\left|4(1) + 3\left(-\frac{4}{3}\right) + 5\left(\frac{1}{3}\right)\right|$$

$$= \frac{5}{6} \text{ sq. units}$$

and Area of ΔABC

$$= \frac{1}{2}|4(5-2) + 1(2-6) + 7(6-5)|$$

$$= \frac{1}{2}|4(3) + 1(-4) + 7(1)|$$

$$= \frac{15}{2} \text{ sq. units}$$

$$\therefore \quad \frac{\text{ar}(\Delta ADE)}{\text{ar}(\Delta ABC)} = \frac{5/6}{15/2} = \frac{1}{9}$$

i.e., ar (ΔADE) : ar $(\Delta ABC) = 1 : 9$ **Ans.**

Q. 30. If the points $A(1, -2)$, $B(2, 3)$, $C(-3, 2)$ and $D(-4, -3)$ are the vertices of the parallelogram $ABCD$, then taking AB as the base, find the height of the parallelogram.★

Solution : Given, $A = (1, -2)$, $B = (2, 3)$, $C = (-3, 2)$ and $D = (-4, -3)$.

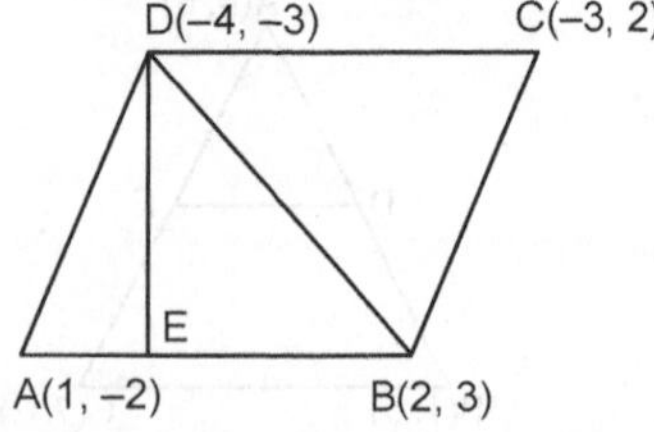

Let DE be the height of the parallelogram.

Now, Area of $\Delta ABD = \frac{1}{2}[AB \times DE]$

$$\Rightarrow \quad DE = \frac{2 \times \text{Area of } \Delta ABD}{AB} \quad \ldots\text{(i)}$$

Also, Area of $\Delta ABD = \frac{1}{2}|1(3+3) + 2(-3+2) - 4(-2-3)|$

$= \frac{1}{2}|6 - 2 + 20|$

$= \frac{1}{2}|24|$

$= 12$ sq. units

and $AB = \sqrt{(2-1)^2 + (3+2)^2}$

$= \sqrt{(1)^2 + (5)^2}$

$= \sqrt{1+25}$

$= \sqrt{26}$ units

Thus, from (i),

$DE = \frac{2(12)}{\sqrt{26}}$ units

$= \frac{24}{\sqrt{26}}$ units. **Ans.**

Q. 31. Three vertices of a parallelogram *ABCD* are *A*(3, – 4), *B*(– 1, – 3) and *C*(– 6, 2). Find the coordinates of vertex *D* and also the area of *ABCD*.★

Solution : $ABCD$ is a parallelogram.

So, $AB = CD, AD = BC$ ...(i)

and Area of ΔABC = Area of ΔADC ...(ii)

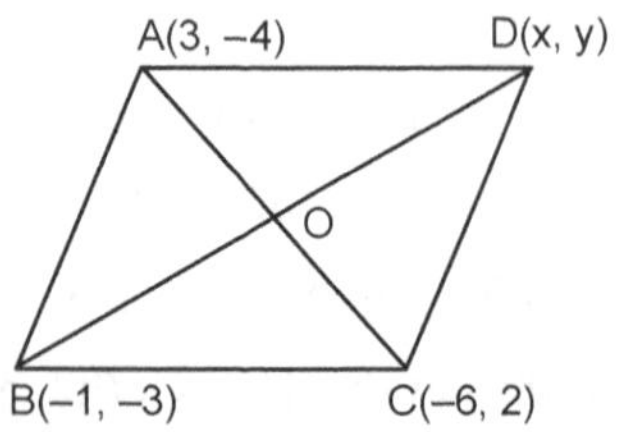

Let $D = (x, y)$ and O be the point of intersection of the two diagonals of the parallelogram $ABCD$.

Hence, $O = \left(\frac{x-1}{2}, \frac{y-3}{2}\right) = \left(\frac{3-6}{2}, \frac{-4+2}{2}\right)$

$\Rightarrow \quad \frac{x-1}{2} = \frac{3-6}{2}$

$\Rightarrow \quad x - 1 = -3$

$\Rightarrow \quad x = -2$

and $\quad \frac{y-3}{2} = \frac{-4+2}{2}$

$\Rightarrow \quad y - 3 = -2$

$\Rightarrow \quad y = 1$

Thus, $\quad D = (x, y) = (-2, 1)$

Now, Area of $ABCD$ = Area of ΔABC + Area of ΔADC

Area of $ABCD$ = 2(Area of ΔABC) [Using (ii)]

$= 2 \times \frac{1}{2}|3(-3-2) - 1(2+4) - 6(-4+3)$

$= |3(-5) - 1(6) - 6(-1)|$

$= |-15 - 6 + 6$

$= |-15|$

$= 15$ sq. units. **Ans.**

Chapter 7. Triangles

Q. 1. If a line is drawn parallel to one side of a triangle to intersect the other two sides at distinct points then prove that the other two sides are divided in the same ratio.★

Solution : In ΔABC, let $DE \parallel BC$ intersect AB and AC at D and E respectively.

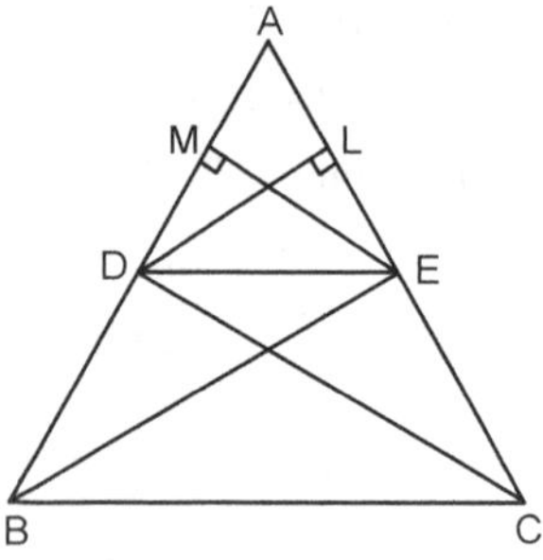

To prove : $\frac{AD}{DB} = \frac{AE}{EC}$

Join BE and CD. Also draw EM perpendicular to AB and DL perpendicular to AC.

Now, Area of $\Delta ADE = \frac{1}{2} \times AD \times EM$

and Area of $\Delta DBE = \frac{1}{2} \times DB \times EM$

Thus, $\frac{\text{Area of } \Delta ADE}{\text{Area of } \Delta DBE} = \frac{\frac{1}{2} \times AD \times EM}{\frac{1}{2} \times DB \times EM} = \frac{AD}{DB}$ ...(i)

★ are board exam questions from previous years

Again, Area of $\Delta AED = \frac{1}{2} \times AE \times DL$

and Area of $\Delta ECD = \frac{1}{2} \times EC \times DL$

Thus, $\frac{\text{Area of } \Delta AED}{\text{Area of } \Delta ECD} = \frac{\frac{1}{2} \times AE \times DL}{\frac{1}{2} \times EC \times DL} = \frac{AE}{EC}$ ...(ii)

As ΔDEB and ΔDEC are on the same base DE and between the same parallel lines DE and BC, their areas are the same.

Thus, area of ΔDEB = area of ΔDEC ...(iii)

So, from equation (i), (ii) and (iii), we have

$$\frac{AD}{DB} = \frac{AE}{EC}.$$ **Hence Proved.**

Q. 2. In the given figure, *ABC* is drawn to circumscribe a circle of radius 2 cm such that the segments *BD* and *DC* into which *BC* is divided by the point of contact *D* are of lengths 4 cm and 3 cm respectively. If the area of ΔABC = 21 cm² then find the lengths of the side *AB* and *AC*.

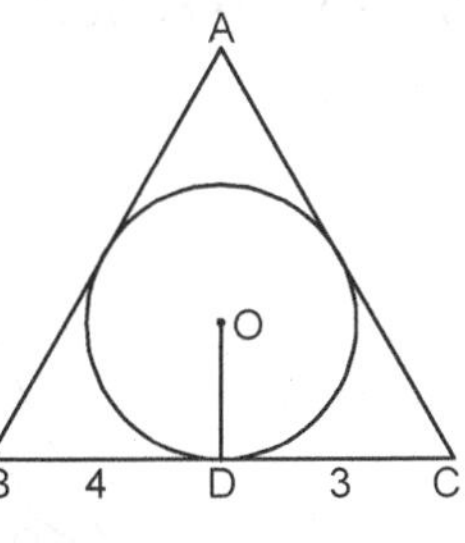

Solution : Given, $OF = OD = OE = 2$ cm

[radii of the circumscribed circle]

OD is perpendicular to BC

[radius of a circle is always perpendicular to the tangent at the point of contact]

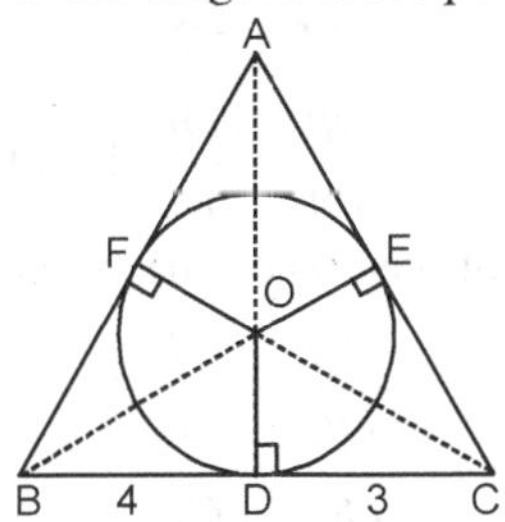

Therefore, by Pythagoras' theorem,

In ΔBOD

$$OB^2 = BD^2 + OD^2$$
$$\Rightarrow \quad OB^2 = 4^2 + 2^2$$
$$= 16 + 4 = 20$$

Also, in ΔBOF

$$\Rightarrow \quad BF^2 = OB^2 - OF^2$$
$$\Rightarrow \quad BF^2 = 20 - 4 = 16$$
$$\Rightarrow \quad BF = 4 \text{ cm}$$

Again, in ΔDOC

$$\Rightarrow \quad CO^2 = DC^2 + OD^2$$
$$\Rightarrow \quad CO^2 = 4 + 9 = 13$$

Also, in ΔCOE

$$\Rightarrow \quad EC^2 = CO^2 - OE^2$$
$$\Rightarrow \quad EC^2 = 13 - 4 = 9$$
$$\Rightarrow \quad EC = 3 \text{ cm}$$

Now, Area of $\Delta BOC = \frac{1}{2} \times 7 \times 2 = 7$ cm²

Area of $\Delta FOB = \frac{1}{2} \times 4 \times 2 = 4$ cm²

and Area of $\Delta EOC = \frac{1}{2} \times 3 \times 2 = 3$ cm²

Thus, Area of quadrilateral $AFOE$

= Area of ΔABC – [Areas of Δs $BOC + FOB + EOC$]

or Area of $AFOE = 21 - [7 + 4 + 3]$

$= 21 - 14 = 7$ cm²

Again, area of quadrilateral $AFOE$

= Area of ($\Delta OFA + \Delta OEA$)

= 7 cm² ...(i)

In ΔOFA and ΔOEA,

$AO = AO$ [Common]

$OF = OE$ [Radii of the circle]

$\angle AFO = \angle AEO$

[Perpendicular from the centre of the circle to the tangent AB and AC respectively]

$\Rightarrow \quad \Delta OFA \cong \Delta OEA$

$\Rightarrow$ Area of ΔOFA = Area of ΔOEA

From (i),

Area of ($\Delta OFA + \Delta OEA$) = 7 cm²

$\Rightarrow$ 2 (Area of ΔOFA) = $2\Delta OEA$ = 7 cm²

$\Rightarrow$ Area of $\Delta OFA = \Delta OEA$ = 3.5 cm²

Hence $3.5 = \frac{1}{2} \times 2 \times AF$

$= \frac{1}{2} \times 2 \times AE$

$\Rightarrow \quad AF = AE = 3.5$ cm

So $AB = AF + FB = 3.5 + 4 = 7.5$ cm

and $AC = AE + EC = 3.5 + 3 = 6.5$ cm. **Ans.**

Q. 3. In ΔABC, $DE \parallel BC$ such that $AD = (4x - 3)$ cm, $AE = (8x - 7)$ cm, $BD = (3x - 1)$ cm and $CE = (5x - 3)$ cm. Find the value of x.★

Solution : In ΔABC, we have

$$DE \parallel BC$$

So, $\frac{AD}{DB} = \frac{AE}{EC}$ [Thales' theorem]

★ are board exam questions from previous years

$$\Rightarrow \quad \frac{4x-3}{3x-1} = \frac{8x-7}{5x-3}$$

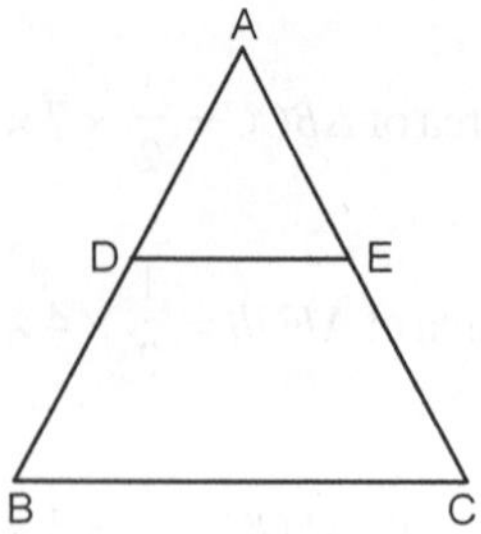

$$\Rightarrow \quad (4x-3)(5x-3) = (8x-7)(3x-1)$$
$$\Rightarrow 20x^2 - 15x - 12x + 9 = 24x^2 - 21x - 8x + 7$$
$$\Rightarrow \quad 4x^2 - 2x - 2 = 0$$
$$\Rightarrow \quad 2x^2 - x - 1 = 0$$
$$\Rightarrow \quad 2x^2 - 2x + x - 1 = 0$$
$$\Rightarrow 2x(x-1) + 1(x-1) = 0$$
$$\Rightarrow \quad (x-1)(2x+1) = 0$$
$$\Rightarrow \quad x = 1$$

[as length cannot be negative]

So, the value of $x = 1$. **Ans.**

Q. 4. In the given figure, $AD = 3$ cm, $AE = 5$ cm, $BD = 4$ cm, $CE = 4$ cm, $CF = 2$ cm, $BF = 2.5$ cm, then find the pair of parallel lines and hence their lengths.★

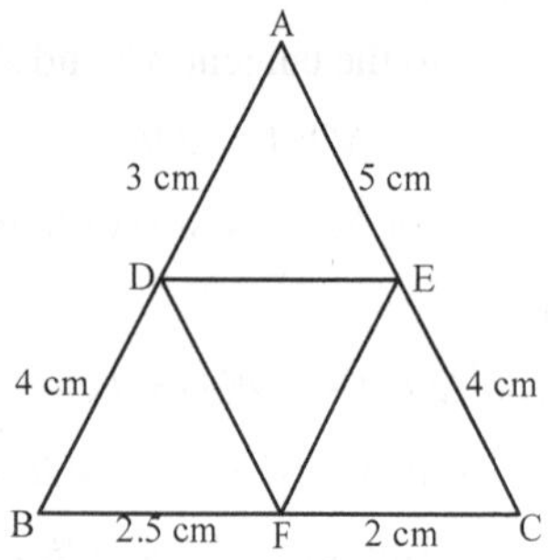

Solution : We have,

$AD = 3$ cm, $AE = 5$ cm, $BD = 4$ cm, $CE = 4$ cm, $CF = 2$ cm and $BF = 2.5$ cm

So, $\quad \frac{EC}{EA} = \frac{4}{5}$ and $\frac{CF}{FB} = \frac{2}{2.5} = \frac{4}{5}$

$$\Rightarrow \quad \frac{EC}{EA} = \frac{CF}{FB}$$

Now in ΔABC

$$\because \quad \frac{EC}{EA} = \frac{CF}{FB}$$

$\therefore \quad EF \parallel AB \quad$ [Converse of Thales' theorem]

Also, $\frac{CE}{CA} = \frac{4}{4+5} = \frac{4}{9} \quad$ ...(i)

$$\frac{CF}{CB} = \frac{2}{2+2.5} = \frac{2}{4.5} = \frac{4}{9}$$

In ΔECF and ΔACB

$$\frac{CE}{CA} = \frac{CF}{CB}$$

$\angle ECF = \angle ACB \quad$ [Common]

$\therefore \quad \Delta CFE \sim \Delta CBA \quad$ [SAS similarity]

$$\Rightarrow \quad \frac{EF}{AB} = \frac{CE}{CA}$$

[In similar Δ's, corresponding sides are proportional]

$\Rightarrow \quad \frac{EF}{7} = \frac{4}{9} \quad$ [$\because AB = 3 + 4 = 7$ cm]

$\therefore \quad EF = \frac{28}{9}$ cm and $AB = 7$ cm **Ans.**

Q. 5. In the given figure, $PQ \parallel AB$ and $PR \parallel AC$. Prove that $QR \parallel BC$.★

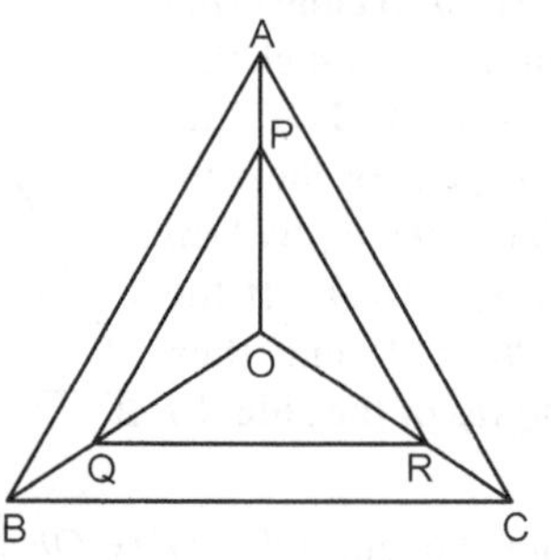

Solution : In ΔOAB,

$$PQ \parallel AB$$

$\therefore \quad \frac{OP}{PA} = \frac{OQ}{QB} \quad$ [Thales' theorem] ...(i)

In ΔOAC, $\quad PR \parallel AC$

$\therefore \quad \frac{OP}{PA} = \frac{OR}{RC} \quad$ [Thales' theorem] ...(ii)

From equations (i) and (ii), we get

$$\frac{OQ}{QB} = \frac{OR}{RC}$$

In ΔOBC, $\quad \frac{OQ}{QB} = \frac{OR}{RC}$

$\Rightarrow \quad QR \parallel BC.$

[Converse of Thales' theorem]

Hence Proved.

Q. 6. In the figure, $\angle BED = \angle BDE$ and E divides BC in the ratio 2 : 1. Prove that $AF \times BE = 2\,AD \times CF$.★

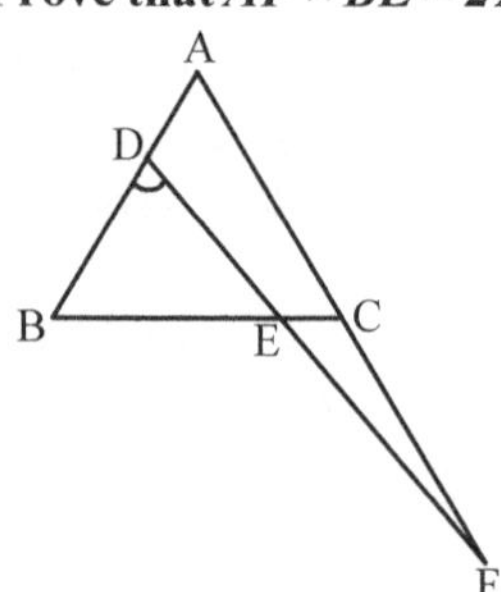

★ are board exam questions from previous years

Solution :

Since, E divides BC in the ratio 2 : 1

$\therefore \quad \frac{BE}{EC} = \frac{2}{1}$ ...(i)

$\angle BED = \angle BDE$ [Given]

$\Rightarrow \quad BD = BE$ ...(ii)

[Sides opposite to equal angles]

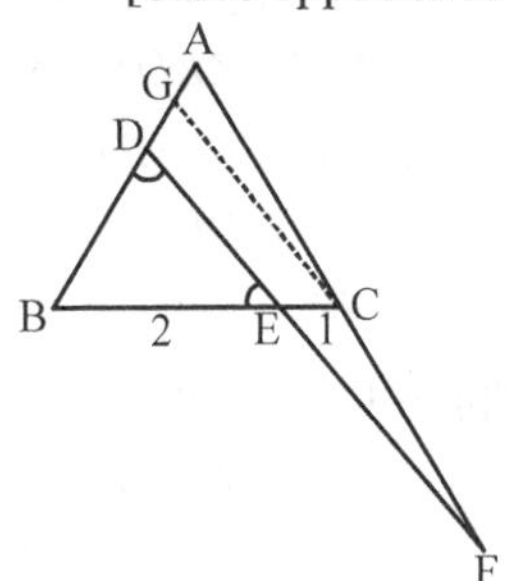

In ΔCBG,

$DE \parallel CG$ [By construction]

$\Rightarrow \quad \frac{BD}{DG} = \frac{BE}{EC}$ [Thale's theorem]

$\Rightarrow \quad \frac{BD}{DG} = \frac{BE}{EC} = \frac{2}{1}$ [From eq. (i)]

$\Rightarrow \quad \frac{BD}{DG} = \frac{2}{1}$

$2DG = BD$

$2DG = BE$ [From eq. (ii)]

$DG = \frac{1}{2} BE$ ...(iii)

In ΔADF,

$CG \parallel DF$ [By construction]

$\Rightarrow \quad \frac{AG}{GD} = \frac{AC}{CF}$ [Thales' theorem]

$\Rightarrow \quad \frac{AG}{GD} + 1 = \frac{AC}{CF} + 1$ [Adding 1 on both sides]

$\Rightarrow \quad \frac{AG + GD}{GD} = \frac{AC + CF}{CF}$

$\Rightarrow \quad \frac{AD}{GD} = \frac{AF}{CF}$

$\Rightarrow \quad AF \times GD = AD \times CF$

$\Rightarrow \quad AF \times \frac{BE}{2} = AD \times CF$ [From (iii)]

$\Rightarrow \quad AF \times BE = 2\, AD \times CF.$ **Hence Proved.**

Q. 7. In the given figure, $AB \parallel DE$ and $BD \parallel EF$. Prove that $DC^2 = CF \times AC$.★

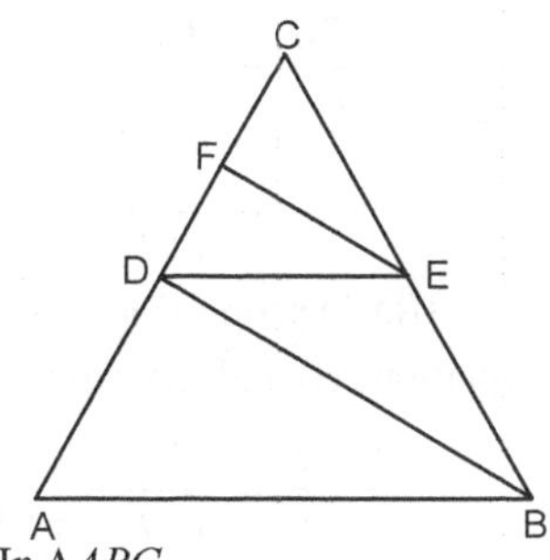

Solution : In ΔABC,

$AB \parallel DE$

$\therefore \quad \frac{CD}{DA} = \frac{CE}{EB}$ [Thales' theorem] ...(i)

In ΔCDB, $BD \parallel FE$

$\therefore \quad \frac{CF}{FD} = \frac{CE}{EB}$ [Thales' theorem] ...(ii)

From equations (i) and (ii),

$\frac{CD}{DA} = \frac{CF}{FD}$

$\Rightarrow \quad \frac{DA}{CD} = \frac{FD}{CF}$

$\Rightarrow \quad \frac{DA}{CD} + 1 = \frac{FD}{CF} + 1$

$\Rightarrow \quad \frac{DA + CD}{CD} = \frac{FD + CF}{CF}$

$\Rightarrow \quad \frac{AC}{CD} = \frac{CD}{CF}$

$\Rightarrow \quad DC^2 = CF \times AC.$ **Hence Proved.**

Q. 8. In the given figure, PA, QB and RC are perpendicular to AC such that $PA = x$, $RC = y$, $QB = z$, $AB = a$ and $BC = b$. Prove that :★★

$$\frac{1}{x} + \frac{1}{y} = \frac{1}{z}.$$

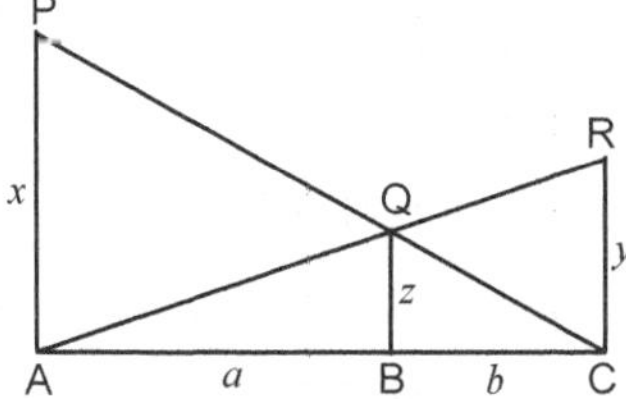

Solution : As PA, QB and RC are perpendicular to AC

$\therefore \quad PA \parallel QB \parallel RC$

Thus, in ΔPAC,

$QB \parallel PA$

$\therefore \quad \frac{QB}{PA} = \frac{BC}{AC}$

[Basic proportionality theorem]

or $\quad \frac{z}{x} = \frac{b}{a + b}$ ...(i)

Similarly, in ΔRAC,

$QB \parallel RC$

★ are board exam questions from previous years

★★ are frequently asked board exam questions

$\therefore \quad \frac{QB}{RC} = \frac{AB}{AC}$

or $\quad \frac{z}{y} = \frac{a}{a+b}$

[Basic proportionality theorem] ...(ii)

Adding equations (i) and (ii), we get

$$\frac{z}{x} + \frac{z}{y} = \frac{b}{a+b} + \frac{a}{a+b}$$

$$\Rightarrow \quad z\left(\frac{1}{x} + \frac{1}{y}\right) = \frac{a+b}{a+b}$$

$$\Rightarrow \quad \frac{1}{x} + \frac{1}{y} = \frac{1}{z}.$$

Hence Proved.

Q. 9. The diagonals of a quadrilateral *ABCD* intersect each other at the point *O* such that $\frac{AO}{OC} = \frac{BO}{OD}$. Show that *ABCD* is a trapezium.★★

Solution : Let $EO \parallel AB \parallel DC$ meet AD at E.

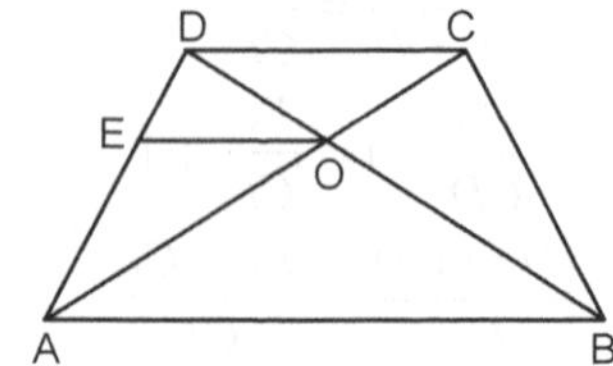

So, in ΔADC, $\quad EO \parallel DC$

So, $\quad \frac{AO}{OC} = \frac{AE}{ED}$ [Thales' theorem]...(i)

But $\quad \frac{AO}{OC} = \frac{BO}{OD}$ [Given]

So, $\quad \frac{BO}{OD} = \frac{AE}{ED}$

Hence, $\frac{BO}{OD} = \frac{AE}{ED}$ and BO, OD, AE and ED are segments of ΔDAB.

So, $\quad EO \parallel AB$

[Converse of Thales' theorem]

But $\quad EO \parallel DC$

Thus, $\quad AB \parallel DC$

As there are only two parallel sides in this quadrilateral, it is a trapezium. **Hence Proved.**

Q. 10. Through the mid-point *M* of the side *CD* of a parallelogram *ABCD*, the line *BM* is drawn intersecting *AC* at *L* and *AD* produced at *E* and *AD = DE*. Prove that *EL* = 2*BL*.★★

Solution : In ΔBMC and ΔEMD, we have

$\angle 1 = \angle 2$

[Vertically opposite angles]

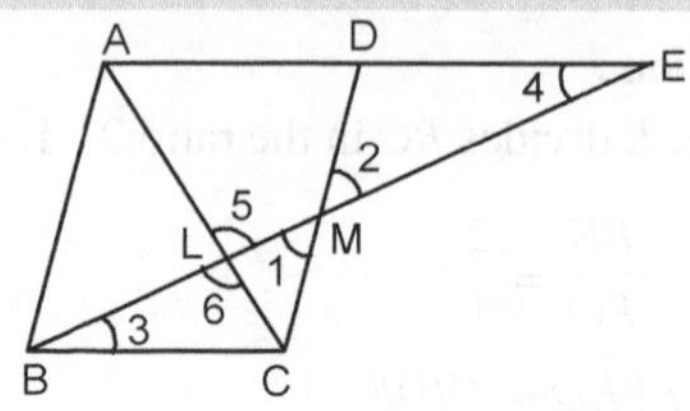

$MC = MD$

[M being the mid-point of CD]

$\angle BCM = \angle EDM$ [Alternative angles]

Thus, $\quad \Delta BMC \cong \Delta EMD$ [ASA]

$\Rightarrow \quad BC = DE$

Again, $\quad BC = AD$

[Opposite sides of the parallelogram $ABCD$]

$\therefore \quad BC = AD = DE$

So, $\quad AE = AD + DE = 2BC$ ...(i)

Again, in ΔAEL and ΔCBL,

$\angle 5 = \angle 6$

[Vertically opposite angles]

$\angle 3 = \angle 4$ [Alternative angles]

So, $\quad \Delta AEL \sim \Delta CBL$

$\therefore \quad \frac{EL}{BL} = \frac{AE}{BC} = \frac{2BC}{BC} = 2$ [From (i)]

Thus, $\quad EL = 2BL$. **Hence Proved.**

Q. 11. Prove that the ratio of the areas of two similar triangles equals the ratio of the squares of their corresponding sides.★★ [NCERT]

Solution : Given, two triangles ABC and DEF such that $\Delta ABC \sim \Delta DEF$.

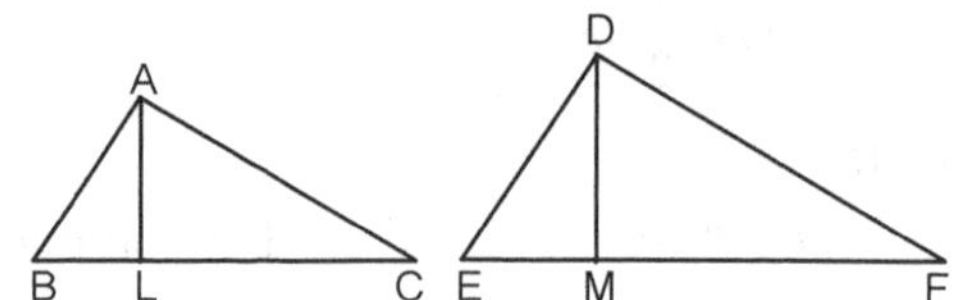

To prove : $\frac{\text{Area }(\Delta ABC)}{\text{Area }(\Delta DEF)} = \frac{AB^2}{DE^2}$

$$= \frac{BC^2}{EF^2} = \frac{AC^2}{DF^2}$$

Construction : Draw $AL \perp BC$ and $DM \perp EF$.

Proof : Since $\Delta ABC \sim \Delta DEF$, it implies that they are equiangular and their sides are proportional.

Hence, $\angle A = \angle D$, $\angle B = \angle E$, $\angle C = \angle F$, $\angle BAL = \angle EDM$, $\angle CAL = \angle MDF$

and $\quad \frac{AB}{DE} = \frac{BC}{EF} = \frac{AC}{DF}$

Now, $\quad$ Area of $\Delta ABC = \frac{1}{2} \times BC \times AL$

and $\quad$ Area of $\Delta FED = \frac{1}{2} \times EF \times DM$

$$\therefore \quad \frac{\text{Area of } \Delta ABC}{\text{Area of } \Delta FED} = \frac{BC \times AL}{EF \times DM}$$

$$= \frac{BC}{EF} \times \frac{AL}{DM} \quad ...(i)$$

In ΔALC and ΔDMF,

$\angle ALC = \angle DMF = 90°$ [By construction]

$\angle C = \angle F$ [Proved above]

So, $\Delta ALC \sim \Delta DMF$

$\therefore \quad \frac{AL}{DM} = \frac{AC}{DF}$

But $\quad \frac{AC}{DF} = \frac{AB}{DE} = \frac{BC}{EF}$

$\therefore \quad \frac{AL}{DM} = \frac{BC}{EF}$ ....(ii)

Substituting (ii) in (i), we have

$$\therefore \quad \frac{\text{Area of } \Delta ABC}{\text{Area of } \Delta FED} = \frac{BC}{EF} \times \frac{BC}{EF} = \left(\frac{BC}{EF}\right)^2$$

Similarly,

$$\frac{\text{Area of } \Delta ABC}{\text{Area of } \Delta FED} = \left(\frac{AC}{DF}\right)^2 = \left(\frac{AB}{DE}\right)^2 = \left(\frac{BC}{EF}\right)^2$$

Hence Proved.

Q. 12. Prove that in a right-angled triangle, the square of the hypotenuse is equal to the sum of the squares of the other two sides. ★★

Solution : Let ΔABC be a right-angled triangle such that AB is perpendicular to BC.

So, $\angle B = 90°$

Also, let BD be a perpendicular on the hypotenuse AC.

Now, in ΔABC and ΔADB,

$\angle A = \angle A$ [Common]

$\angle ABC = \angle ADB$ [Each 90°]

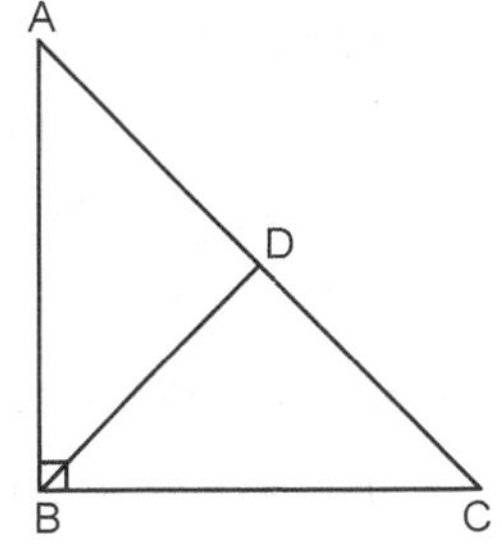

So, by AA axiom

$\Delta ABC \sim \Delta ADB$

Thus, $\quad \frac{AD}{AB} = \frac{AB}{AC}$ [cpct]

$\Rightarrow \quad AB^2 = AD \times AC$ ...(i)

Again, in ΔABC and ΔBDC,

$\angle C = \angle C$ [Common]

$\angle ABC = \angle BDC = 90°$ [Each 90°]

★ are board exam questions from previous years

★★ are frequently asked board exam questions

So, by AA axiom

$\Delta ABC \sim \Delta BDC$

Thus, $\quad \frac{DC}{BC} = \frac{BC}{AC}$ [cpct]

$\Rightarrow \quad BC^2 = DC \times AC$ ...(ii)

Adding equations (i) and (ii), we get

$AB^2 + BC^2 = (AD \times AC) + (DC \times AC)$

$\Rightarrow \quad AB^2 + BC^2 = (AD + DC)AC = AC^2$

Thus, $\quad AB^2 + BC^2 = AC^2$. **Hence Proved.**

Q. 13. Prove that in a triangle, if the square of one side is equal to the sum of the squares of the other two sides then the angle opposite to the first side is a right angle.★★

Solution : Given that, $AC^2 = AB^2 + BC^2$

Let DEF be a triangle such that

$DE = AB$, $EF = BC$ and $\angle E = 90°$

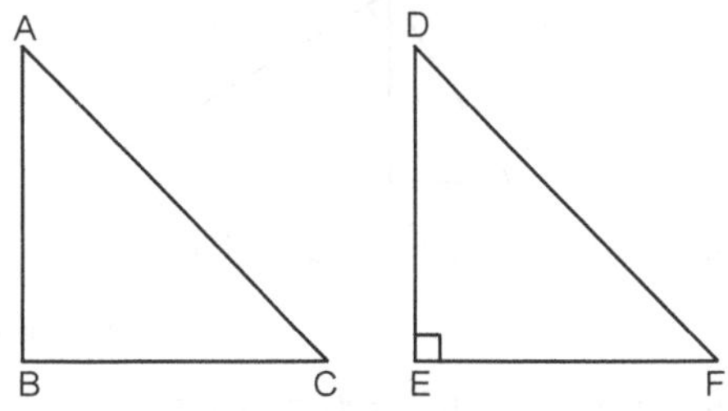

So, by Pythagoras theorem,

$DF^2 = DE^2 + EF^2$

$\Rightarrow \quad DF^2 = AB^2 + BC^2$

[$\because$ $DE = AB$ and $EF = BC$] ...(i)

But $\quad AC^2 = AB^2 + BC^2$ ...(ii)

So, $\quad DF^2 = AC^2$

or $\quad DF = AC$

Now, in ΔABC and ΔDEF,

$DF = AC$ [Proved above]

$DE = AB$ [By construction]

$EF = BC$ [By construction]

$\therefore \quad \Delta ABC \cong \Delta DEF$

$\Rightarrow \quad \angle B = \angle E = 90°$. **Hence Proved.**

Q. 14. *P* and *Q* are points on the sides *CA* and *CB* respectively of a Δ*ABC*, right-angled triangle at *C*. Prove that★★

$$(AQ^2 + BP^2) = (AB^2 + PQ^2). \quad \text{[NCERT]}$$

Solution : By Pythagoras theorem, we have

$AB^2 = AC^2 + BC^2$ ...(i)

$PQ^2 = PC^2 + QC^2$ ...(ii)

$AQ^2 = AC^2 + QC^2$ ...(iii)

$BP^2 = PC^2 + BC^2$ ...(iv)

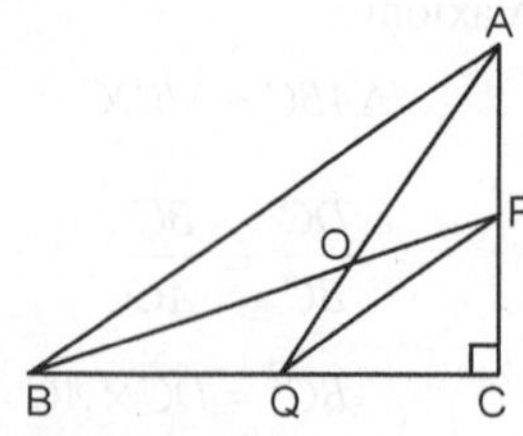

Adding equations (iii) and (iv),

$$AQ^2 + BP^2 = AC^2 + QC^2 + PC^2 + BC^2$$

$$\Rightarrow \quad AQ^2 + BP^2 = AC^2 + BC^2 + QC^2 + PC^2$$

$$\Rightarrow \quad AQ^2 + BP^2 = AB^2 + PQ^2 \quad \text{[From (i) and (ii)]}$$

Hence Proved.

Q. 15. Given a ΔABC in which $\angle A = 90°$ and AD is perpendicular to BC. Prove that $AD^2 = BD.CD$.★★

Solution : In ΔBAC,

$$\angle A = 90°$$

$$\therefore \quad BC^2 = AB^2 + AC^2 \quad ...(i)$$

[Pythagoras theorem]

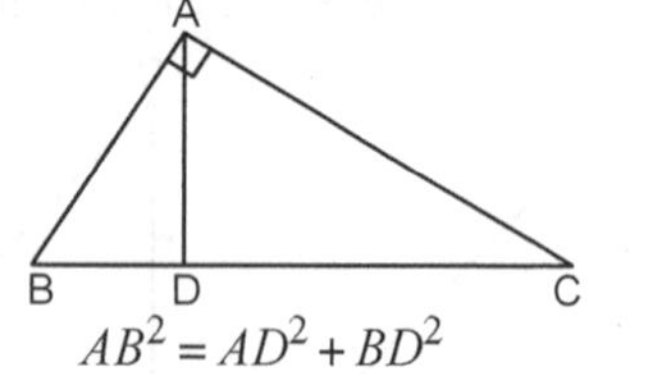

In ΔADB, $\quad AB^2 = AD^2 + BD^2 \quad$ ...(ii)

[Pythagoras theorem]

In ΔADC, $\quad AC^2 = AD^2 + CD^2 \quad$ ...(iii)

Adding equations (ii) and (iii), we get

$$AB^2 + AC^2 = AD^2 + BD^2 + AD^2 + CD^2$$

$$\Rightarrow \quad BC^2 = 2AD^2 + BD^2 + CD^2 \quad \text{[From (i)]}$$

$$\Rightarrow \quad (BD + CD)^2 = 2AD^2 + BD^2 + CD^2$$

$$\Rightarrow BD^2 + CD^2 + 2BD.CD = 2AD^2 + BD^2 + CD^2$$

$$\Rightarrow \quad AD^2 = BD.CD.$$ **Hence Proved.**

Q. 16. In Fig. $\angle ACB = 90°$ and $CD \perp AB$, prove that $CD^2 = BD \times AD$.★

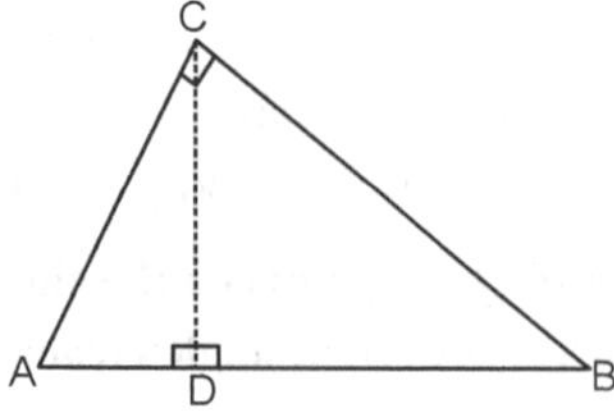

Solution : Given, A ΔACB in which $\angle ACB = 90°$ and $CD \perp AB$

To prove : $CD^2 = BD \times AD$

Proof : In ΔADC and ΔACB

$\angle A = \angle A$ (common)

$\angle ADC = \angle ACB$ (90° each)

$\therefore$ $\Delta ADC \sim \Delta ACB$ (By AA rule) ...(i)

Similarly,

$\Delta CDB \sim \Delta ACB$ (By AA rule) ...(ii)

From equation (i) and (ii)

$$\Delta ADC \sim \Delta CDB$$

$$\therefore \quad \frac{AD}{CD} = \frac{CD}{DB}$$

(by the definition of similarity of traingles)

or $\quad CD^2 = AD \cdot BD$

or $\quad CD^2 = BD \times AD$ **Hence Proved.**

Q. 17. In the given figure, ΔPQR is right-angled at Q and the points S and T trisect the side QR. Prove that

$$8PT^2 = 3PR^2 + 5PS^2.★$$

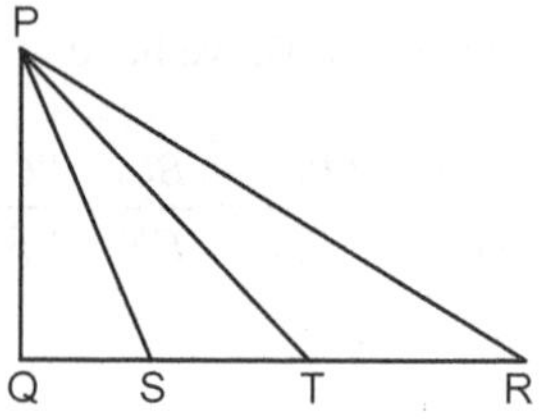

Solution : Since, S and T trisect the line QR

$\therefore$ $\quad QS = ST = TR$

Let $\quad QS = ST = TR = x$

Then, $\quad QT = x + x = 2x$

and $\quad QR = 2x + x = 3x$

Now, in ΔPQS

$$PS^2 = PQ^2 + QS^2$$

[Pythagoras theorem]

$$\Rightarrow \quad PS^2 = PQ^2 + x^2 \quad ...(i)$$

In ΔPQT, $\quad PT^2 = PQ^2 + QT^2$

[Pythagoras theorem]

$$\Rightarrow \quad PT^2 = PQ^2 + (2x)^2$$

$$\Rightarrow \quad PT^2 = PQ^2 + 4x^2 \quad ...(ii)$$

In ΔPQR $\quad PR^2 = PQ^2 + QR^2$

[Pythagoras theorem]

$$\Rightarrow \quad PR^2 = PQ^2 + (3x)^2$$

$$\Rightarrow \quad PR^2 = PQ^2 + 9x^2 \quad ...(iii)$$

Now, $3PR^2 + 5PS^2 = 3(PQ^2 + 9x^2) + 5(PQ^2 + x^2)$

$$\Rightarrow \quad 3PR^2 + 5PS^2 = 3PQ^2 + 27x^2 + 5PQ^2 + 5x^2$$

$$\Rightarrow \quad 3PR^2 + 5PS^2 = 8PQ^2 + 32x^2$$

$$\Rightarrow \quad 3PR^2 + 5PS^2 = 8(PQ^2 + 4x^2)$$

$$\Rightarrow \quad 3PR^2 + 5PS^2 = 8PT^2. \quad \text{[From (ii)]}$$

Hence Proved.

Q. 18. In an equilateral triangle with side a, prove that (i) altitude = $\frac{\sqrt{3}}{2}a$, and (ii) area = $\frac{\sqrt{3}}{4}a^2$.★

Solution : Let ΔABC be an equilateral triangle with side a and AD perpendicular to BC.

★ are board exam questions from previous years

★★ are frequently asked board exam questions

Then, $AB = AC = BC = a$

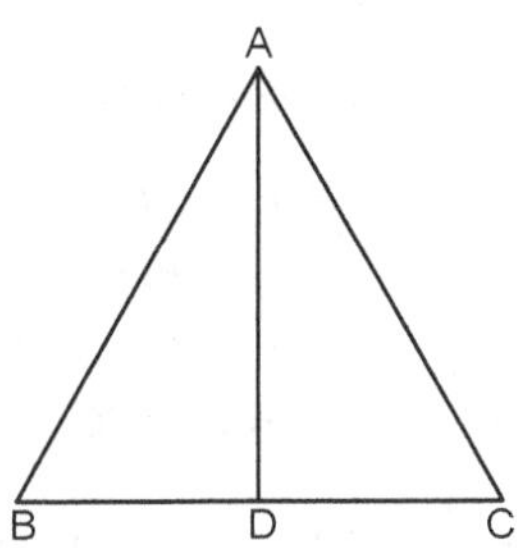

In ΔADB and ΔADC,

$$AB = AC \quad \text{[Given]}$$
$$\angle B = \angle C = 60^\circ$$
$$\angle ADB = \angle ADC = 90^\circ$$
$$\therefore \quad \Delta ADB \cong \Delta ADC$$

So, $BD = DC = \dfrac{a}{2}$

(i) In ΔADB, we have

$$AB^2 = AD^2 + BD^2 \quad \text{[Pythagoras theorem]}$$
$$\Rightarrow \quad AD^2 = AB^2 - BD^2$$
$$\Rightarrow \quad AD^2 = a^2 - \left(\frac{a}{2}\right)^2$$
$$\Rightarrow \quad AD^2 = a^2 - \frac{a^2}{4}$$
$$\Rightarrow \quad AD^2 = \frac{4a^2 - a^2}{4} = \frac{3a^2}{4}$$
$$\Rightarrow \quad AD = \frac{\sqrt{3}a}{2}$$

Thus, altitude $AD = \dfrac{\sqrt{3}a}{2}$

(ii) Area of $\Delta ABC = \dfrac{1}{2}$ Base × Height

$$= \frac{1}{2} \times BC \times AD$$
$$= \frac{1}{2} \times a \times \frac{\sqrt{3}a}{2} = \frac{\sqrt{3}a^2}{4} \text{ sq. units.}$$

Ans.

Q. 19. (i) *O* is any point inside a rectangle *ABCD*. Prove that : $OB^2 + OD^2 = OA^2 + OC^2$.★

(ii) If *OB* = 6 cm, *OD* = 8 cm, *OA* = 5 cm, find the length of *OC*.★

Solution : Let $QP \parallel AD \parallel BC$ be perpendicular to AB and CD and pass through O.

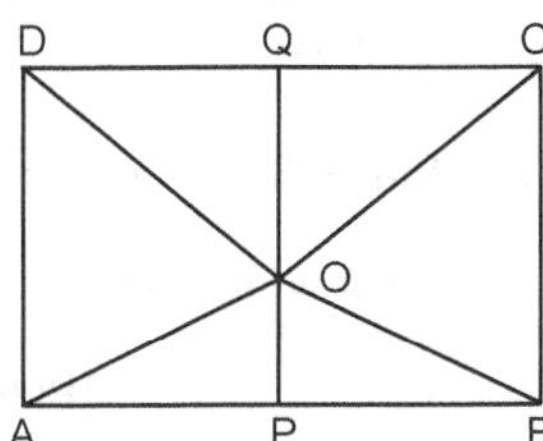

★ are board exam questions from previous years

So, QO is the altitude of ΔCOD and OP is the altitude of ΔBOA.

(i) Now,

$$OB^2 = OP^2 + PB^2 \quad \text{[Pythagoras theorem] ...(i)}$$
$$OA^2 = OP^2 + AP^2 \quad \text{[Pythagoras theorem] ...(ii)}$$
$$OC^2 = OQ^2 + QC^2 \quad \text{[Pythagoras theorem] ...(iii)}$$
$$OD^2 = OQ^2 + DQ^2 \quad \text{[Pythagoras theorem] ...(iv)}$$

Thus, $OB^2 + OD^2 = OQ^2 + DQ^2 + OP^2 + PB^2$

$$\Rightarrow \quad OB^2 + OD^2 = OQ^2 + PB^2 + DQ^2 + OP^2$$
$$\Rightarrow \quad OB^2 + OD^2 = OQ^2 + QC^2 + AP^2 + OP^2$$

[$\because$ $AP = DQ$ and $PB = QC$; opp sides of rectangle]

$$\Rightarrow \quad OB^2 + OD^2 = OC^2 + OA^2 \quad \text{[From (ii) and (iii)]}$$

Hence Proved.

(ii) $\because$ $OB^2 + OD^2 = OC^2 + OA^2$ [Proved above]

Let $OC = x$ cm

$$\Rightarrow \quad (6)^2 + (8)^2 = x^2 + (5)^2$$
$$\Rightarrow \quad 36 + 64 = x^2 + 25$$
$$\Rightarrow \quad x^2 + 25 = 100$$
$$\Rightarrow \quad x^2 = 75$$
$$\Rightarrow \quad x = 5\sqrt{3} \text{ cm.}$$

Ans.

Q. 20. In the given figure, *O* is a point inside Δ*PQR* such that ∠*POR* = 90°, *OP* = 6 cm and *OR* = 8 cm. If *PQ* = 24 cm and *QR* = 26 cm, then prove that Δ*PQR* is right-angled.★

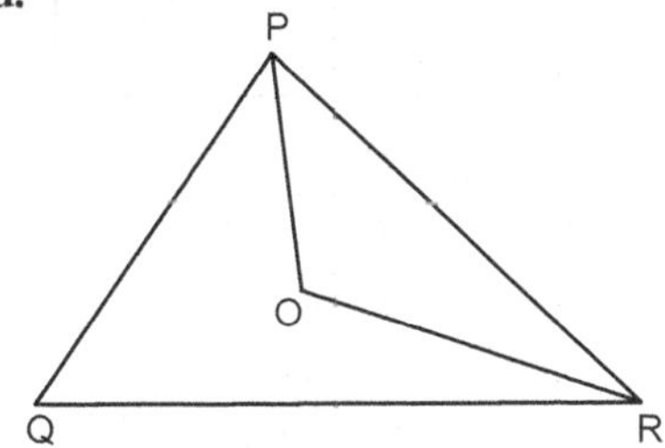

Solution : In ΔPOR,

$$\angle O = 90^\circ$$
$$\therefore \quad PR^2 = OP^2 + OR^2 \quad \text{[Pythagoras theorem]}$$
$$\Rightarrow \quad PR^2 = 8^2 + 6^2$$
$$= 64 + 36$$
$$= 100 = 10^2$$
$$\Rightarrow \quad PR = 10 \text{ cm}$$

If ΔPQR is right-angled then it should satisfy one of the three conditions below :

$$PR^2 = QP^2 + QR^2$$
$$PQ^2 = PR^2 + QR^2$$
$$QR^2 = PQ^2 + PR^2$$

On substituting the values of PR, PQ and QR in the above equations we get,

(i) $PR^2 = QP^2 + QR^2$

$\Rightarrow 10^2 = 24^2 + 26^2$

This is invalid.

(ii) $PQ^2 = PR^2 + OR^2$

$\Rightarrow 24^2 = 10^2 + 26^2$

This is also invalid.

(iii) $QR^2 = PQ^2 + PR^2$

$\Rightarrow 26^2 = 24^2 + 10^2$

$\Rightarrow 26^2 = 576 + 100$

$\Rightarrow 26^2 = 676 = 26^2$

This is true.

Therefore, by Pythagoras theorem, ΔPQR is a right-angled triangle. **Hence Proved.**

Q. 21. If a perpendicular is drawn from the vertex containing the right angle of a right triangle to the hypotenuse then prove that the triangle on each side of the perpendicular are similar to each other and to the original triangle. Also prove that the square of the perpendicular is equal to the product of the lengths of the two parts of the hypotenuse.

Solution : Given, $\angle B = \angle BDC = \angle BDA = 90°$

[$\because$ AB is perpendicular to BC and BD is perpendicular to AC]

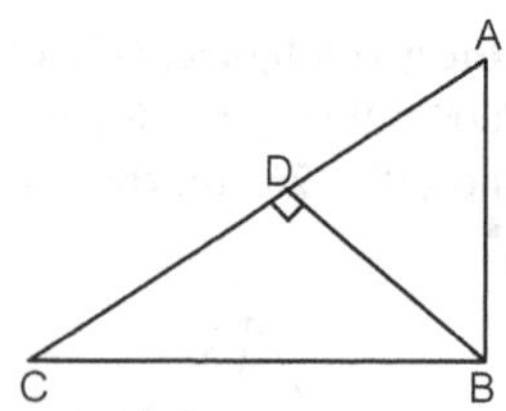

(i) In ΔADB and ΔBDC,

$\angle ABD + \angle DBC = \angle B = 90°$ ...(i)

and $\angle C + \angle CDB + \angle CBD = 180°$

$\Rightarrow \angle C + 90° + \angle CBD = 180°$

$\Rightarrow \angle C + \angle CBD = 90°$ ...(ii)

From (i) and (ii),

$\angle ABD + \angle DBC = \angle C + \angle CBD$

or $\angle ABD = \angle C$

Thus, in the two triangles,

$\angle BDC = \angle BDA = 90°$

and $\angle ABD = \angle C$

So, $\Delta ADB \sim \Delta BDC$ [AA-Criteria]

(ii) In ΔBDC and ΔABC,

$\angle C = \angle C$ [Common]

$\angle B = \angle BDC$ [Each 90°]

So, $\Delta BDC \sim \Delta ABC$ [AA-Criteria]

(iii) In ΔADB and ΔABC,

$\angle A = \angle A$ [Common]

$\angle B = \angle ADB$ [Each 90°]

So, $\Delta ADB \sim \Delta ABC$ [AA-Criteria]

(iv) As $\Delta ADB \sim \Delta BDC$

$\therefore \dfrac{BD}{AD} = \dfrac{CD}{BD}$

$\Rightarrow BD^2 = AD \times CD.$ **Hence Proved.**

Q. 22. In an equilateral ΔABC, E is any point on BC such that $BE = \frac{1}{4}BC$. Prove that $16\,AE^2 = 13\,AB^2$.★

Solution : Here $BE = \frac{1}{4}BC$ [Given]

Draw $AD \perp BC$

In ΔAED, by Pythagoras theorem,

$AE^2 = AD^2 + DE^2$ ...(i)

In ΔADB

$AB^2 = AD^2 + BD^2$

$AB^2 = AE^2 - DE^2 + BD^2$ [Using (i)]

$AB^2 = AE^2 - DE^2 + (BE + DE)^2$

$AB^2 = AE^2 - DE^2 + BE^2 + DE^2 + 2BE.DE$

$AB^2 = AE^2 + BE^2 + 2BE.DE$

$AB^2 = AE^2 + \left(\dfrac{BC}{4}\right)^2 + 2\dfrac{BC}{4}.(BD - BE)$

$AB^2 = AE^2 + \dfrac{BC^2}{16} + \dfrac{BC}{2}\left(\dfrac{BC}{2} - \dfrac{BC}{4}\right)$

$AB^2 = AE^2 + \dfrac{AB^2}{16} + \dfrac{AB}{2}\left[\dfrac{2AB - AB}{4}\right]$

$AB^2 = AE^2 + \dfrac{AB^2}{16} + \dfrac{AB}{2} \times \dfrac{AB}{4}$

$AB^2 - \dfrac{AB^2}{16} - \dfrac{AB^2}{8} = AE^2$

$\dfrac{16AB^2 - AB^2 - 2AB^2}{16} = AE^2$

$16AB^2 - 3AB^2 = 16AE^2$

$13AB^2 = 16AE^2$ **Hence Proved.**

Q. 23. *P* and *Q* are the mid-points of the sides *CA* and *CB* respectively of a ΔABC, right-angled at *C*. Prove that : (i) $4AQ^2 = 4AC^2 + BC^2$, (ii) $4BP^2 = 4BC^2 + AC^2$ and (iii) $4(AQ^2 + BP^2) = 5AB^2$.★★ [NCERT]

Solution : (i) Since AQC is a right-angled triangle.

$\therefore AQ^2 = AC^2 + QC^2$

[Pythagoras theorem]

★ are board exam questions from previous years

★★ are frequently asked board exam questions

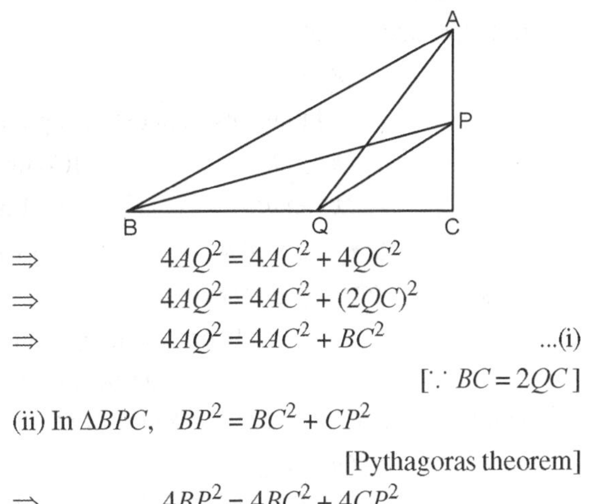

$\Rightarrow \quad 4AQ^2 = 4AC^2 + 4QC^2$

$\Rightarrow \quad 4AQ^2 = 4AC^2 + (2QC)^2$

$\Rightarrow \quad 4AQ^2 = 4AC^2 + BC^2 \quad$ ...(i)

$[\because BC = 2QC]$

(ii) In ΔBPC, $BP^2 = BC^2 + CP^2$

[Pythagoras theorem]

$\Rightarrow \quad 4BP^2 = 4BC^2 + 4CP^2$

$\Rightarrow \quad 4BP^2 = 4BC^2 + (2CP)^2$

$\Rightarrow \quad 4BP^2 = 4BC^2 + AC^2 \quad$ ...(ii)

$[\because AC = 2CP]$

(iii) Adding equations (i) and (ii),

$4AQ^2 + 4BP^2 = 4AC^2 + BC^2 + 4BC^2 + AC^2$

$\Rightarrow \quad 4(AQ^2 + BP^2) = 5(AC^2 + BC^2)$

$\Rightarrow \quad 4(AQ^2 + BP^2) = 5AB^2 \quad [\because AB^2 = AC^2 + BC^2]$

Hence Proved.

Q. 24. *ABC* is an isosceles triangle, right-angled at *B*. Similar triangles *ACD* and *ABE* are constructed on sides *AC* and *AB* respectively. Find the ratio between the areas of Δ*ABE* and Δ*ACD*.★★

Solution : Let $AB = BC = x$

So, $AC^2 = AB^2 + BC^2$

[Pythagoras theorem]

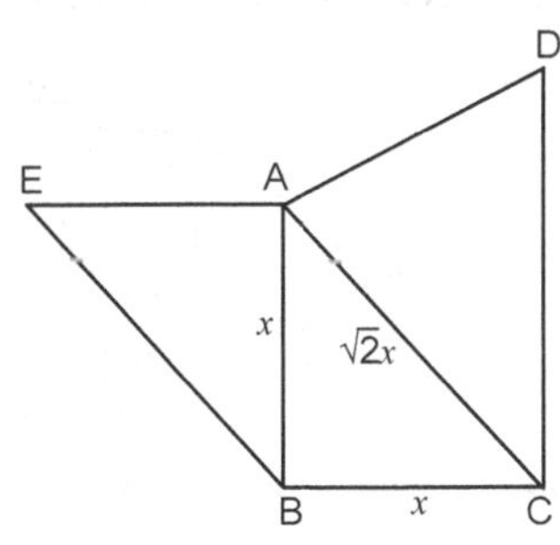

$\Rightarrow \quad AC^2 = x^2 + x^2$

$\Rightarrow \quad AC^2 = 2x^2$

$\Rightarrow \quad AC = \sqrt{2}x$

Given that $\Delta ABE \sim \Delta ACD$

Thus, $\dfrac{\text{Area of } \Delta ABE}{\text{Area of } \Delta ACD} = \dfrac{AB^2}{AC^2}$

$\Rightarrow \quad \dfrac{\text{Area of } \Delta ABE}{\text{Area of } \Delta ACD} = \dfrac{x^2}{(\sqrt{2}x)^2}$

$\Rightarrow \quad \dfrac{\text{Area of } \Delta ABE}{\text{Area of } \Delta ACD} = \dfrac{1}{2}$

Thus, Area of ΔABE : Area of $\Delta ACD = 1:2$. **Ans.**

Q. 25. Diagonals of a trapezium *PQRS* intersect each other at the point O, PQ || RS and PQ = 3RS. Find the ratio of the areas of traingles POQ and ROS.★

Solution : Given, PQRS is a trapezium where PQ || RS and diagonals intersect at O and PQ = 3RS

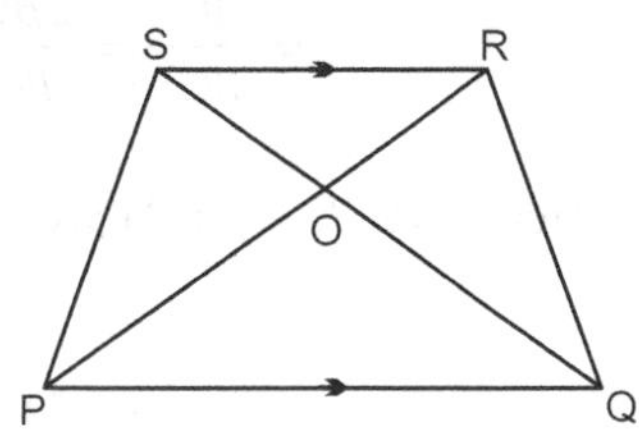

In ΔPOQ and ΔROS, we have

$\angle ROS = \angle POQ$

(vertically opposite angles)

$\angle OQP = \angle OSR \quad$ (alternate angles)

Hence, ΔPOQ ~ ΔROS by AA similarity then, If two triangles are similar, then ratio of areas is equal to the ratio of square of its corresponding sides.

Then,

$\dfrac{\text{area of } \Delta PQQ}{\text{area of } \Delta ROS} = \dfrac{(PQ)^2}{(RS)^2}$

$= \dfrac{(3RS)^2}{(RS)^2} = \dfrac{9}{1}$

$= 9:1$ **Ans.**

Q. 26. Prove that the area of an equilateral triangle described on one side of a square is equal to half the area of an equilateral triangle described on one of its diagonals.★

Solution : Let *ABCD* be the square and *ACF* and *CBE* be the equilateral triangles that have been described on the diagonal and one side of the square respectively.

$\Rightarrow \quad \Delta ACF \sim \Delta BCE \quad$ [Equilateral triangles]

Now, *AC* is the diagonal of square *ABCD*, so

$AC^2 = AB^2 + BC^2$

$\Rightarrow \quad AC^2 = 2BC^2 \quad [\because AB = BC] \quad$ ...(i)

Area of $\Delta ACF = \dfrac{\sqrt{3}}{4} AC^2$

and Area of $\Delta BCE = \dfrac{\sqrt{3}}{4} BC^2$

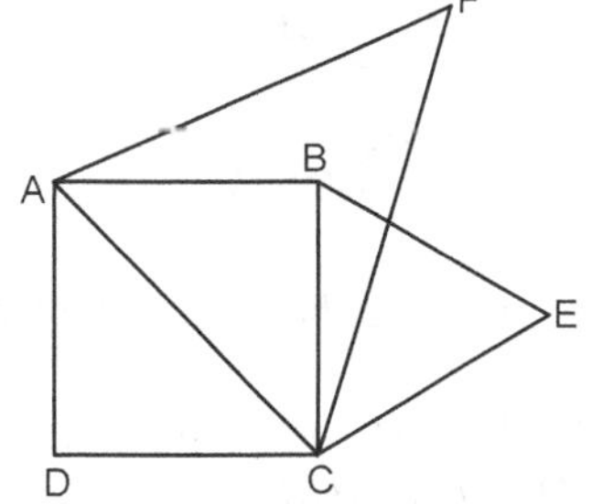

★ are board exam questions from previous years

★★ are frequently asked board exam questions

$$\Rightarrow \quad \frac{\text{Area of } \Delta ACF}{\text{Area of } \Delta BCE} = \frac{\frac{\sqrt{3}}{4}AC^2}{\frac{\sqrt{3}}{4}BC^2}$$

$$= \frac{AC^2}{BC^2} = \frac{2BC^2}{BC^2} = 2$$

$$\Rightarrow \quad \text{Area of } \Delta BCE = \frac{1}{2} \times \text{Area of } \Delta ACF.$$

Hence Proved.

Chapter 8. Circles

Q. 1. Prove that the tangent at any point to a circle is perpendicular to the radius through the point of contact.★

Solution : Let AB be the tangent to the circle with centre O and let point of contact of the tangent and the circle be P.

Let Q be a point on AB, outside the circle and let R be the point of intersection of OQ with the circumference of the circle.

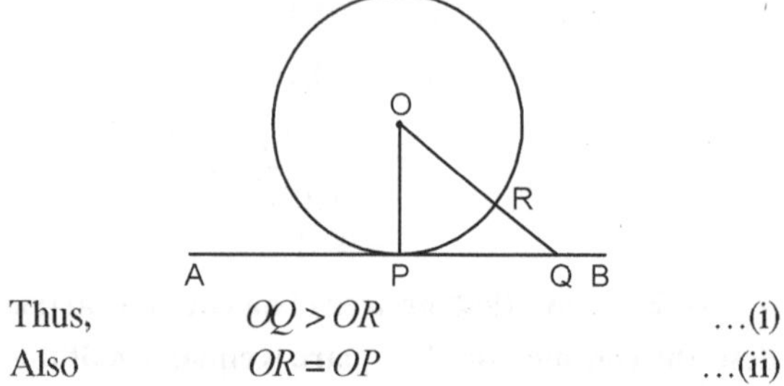

Thus, $OQ > OR$ ...(i)

Also $OR = OP$ ...(ii)

[Both are radii of the circle with with the same centre]

Now, $OQ > OP$ [From (i) and (ii)]

So, OP is perpendicular to AB as it is the shortest distance between the point O and the line AB.

Hence Proved.

Q. 2. Prove that the opposite sides of a quadrilateral circumscribing a circle subtend supplementary angles at the centre of the circle.★★ **[NCERT]**

Solution : Given, a circle with centre O touches the opposite sides of a quadrilateral at points P, Q, R and S.

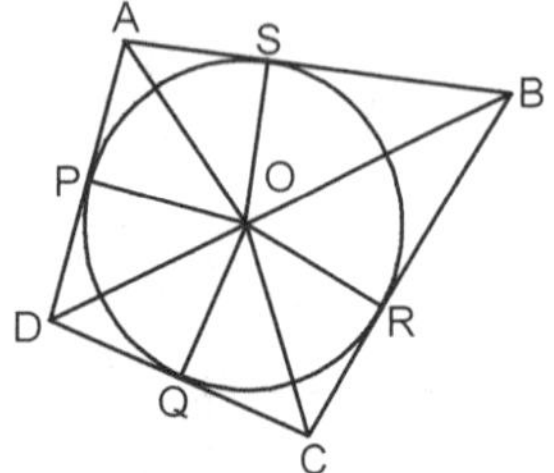

In order to prove the above, the following need to be satisfied :

$\angle AOB + \angle COD = 180°$ and $\angle AOD + \angle BOC = 180°$

Join OS, OP, OQ and OR.

In ΔOQC and ΔOCR,

$QC = CR$ [Tangents from external point C]

$OC = OC$ [Common]

$OQ = OR$ [Radii]

Thus $\Delta OQC \cong \Delta OCR$ [S.S.S.]

Hence $\angle QOC = \angle COR$

[Corresponding angles of congruent triangles]

Similarly, $\angle ROB = \angle SOB$, $\angle SOA = \angle POA$

and $\angle POD = \angle QOD$

Now $\angle QOC + \angle COR + \angle ROB + \angle SOB + \angle SOA + \angle POA + \angle POD + \angle QOD = 360°$

$\Rightarrow \quad 2(\angle QOC + \angle ROB + \angle SOA + \angle POD) = 360°$

and $\quad 2(\angle COR + \angle SOB + \angle POA + \angle QOD) = 360°$

$\Rightarrow \quad \angle QOC + \angle ROB + \angle SOA + \angle POD = 180°$

and $\quad \angle COR + \angle SOB + \angle POA + \angle QOD = 180°$

$\Rightarrow \quad \angle QOC + \angle SOB + \angle SOA + \angle QOD = 180°$

and $\quad \angle COR + \angle ROB + \angle POA + \angle POD = 180°$

$\Rightarrow \quad \angle QOC + \angle QOD + \angle SOA + \angle SOB = 180°$

and $\quad \angle COR + \angle ROB + \angle POA + \angle POD = 180°$

$\Rightarrow \quad \angle AOB + \angle COD = 180°$

and $\quad \angle AOD + \angle BOC = 180°$

Hence Proved.

Q. 3. In the given figure, PQ is a chord of length 16 cm and the radius of the circle is 10 cm. The tangents at P and Q intersect at a point T. Find the length of TP.★

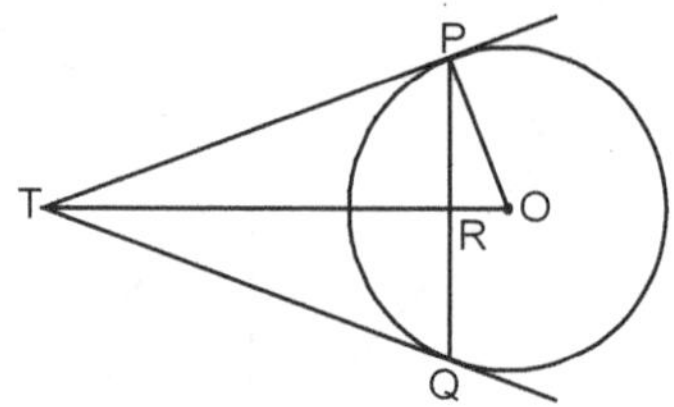

Solution : Given, r = 10 cm and length of chord PQ = 16 cm

As TO bisects PQ at R, so $PR = RQ = 8$ cm

Let $TP = x$ cm and $TR = y$ cm

Now $OR^2 = PO^2 - PR^2$

[Pythagoras theorem ; $\angle PRO = 90°$]

$\Rightarrow \quad OR^2 = (10)^2 - (8)^2$

$\Rightarrow \quad OR^2 = 100 - 64$

$\Rightarrow \quad OR^2 = 36$

$\Rightarrow \quad OR = 6$ cm

Thus, $TO = TR + RO = (y + 6)$ cm

Also $TP^2 = TR^2 + PR^2$

$\Rightarrow \quad x^2 = y^2 + 8^2$

★ are board exam questions from previous years

★★ are frequently asked board exam questions

$\Rightarrow \quad x^2 = y^2 + 64 \quad \ldots(i)$

Again $\quad TO^2 = OP^2 + TP^2$

$\Rightarrow \quad (y+6)^2 = 10^2 + x^2$

$\Rightarrow \quad (y+6)^2 = 10^2 + y^2 + 64 \quad$ [From (i)]

$\Rightarrow \quad y^2 + 36 + 12y = y^2 + 164$

$\Rightarrow \quad 36 + 12y = 164$

$\Rightarrow \quad 12y = 128$

$\Rightarrow \quad 3y = 32$

$\Rightarrow \quad y = \frac{32}{3}$ cm

Hence $\quad TP^2 = x^2 = y^2 + 64$

$\Rightarrow \quad TP^2 = x^2 = \left(\frac{32}{3}\right)^2 + 64$

$\Rightarrow \quad x^2 = \frac{1024}{9} + 64$

$\Rightarrow \quad x^2 = \frac{1024+576}{9}$

$= \frac{1600}{9}$

$\Rightarrow \quad x = \frac{40}{3}$

Hence $\quad TP = 13.33$ cm. **Ans.**

Q. 4. In Fig. PQ is a chord of length 8 cm of a circle of radius 5 cm and centre O. The tangents at P and Q intersect at point T. Find the length of TP.★

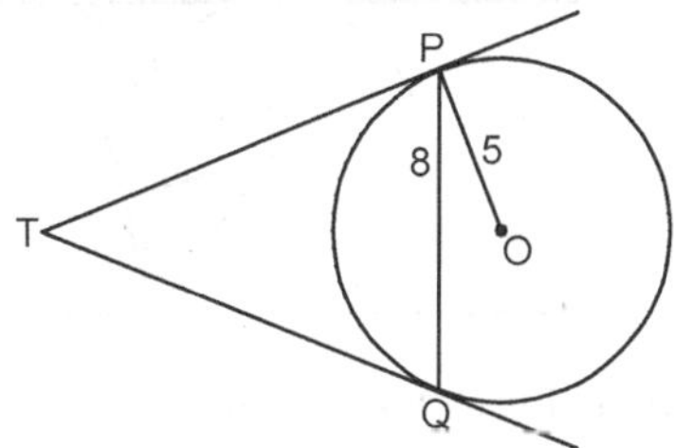

Solution : Join OT, let it intersect PQ at the point R

Now, ΔTPQ is an isosceles triangle and TO is the angle bisector of ∠PTQ. So, OT ⊥ PQ and therefore, OT bisects PQ

$\therefore \quad PR = RQ = 4$ cm

Also, $\quad OR = \sqrt{OP^2 - PR^2} = \sqrt{5^2 - 4^2}$

$= \sqrt{25-16}$

$= \sqrt{9} = 3$ cm

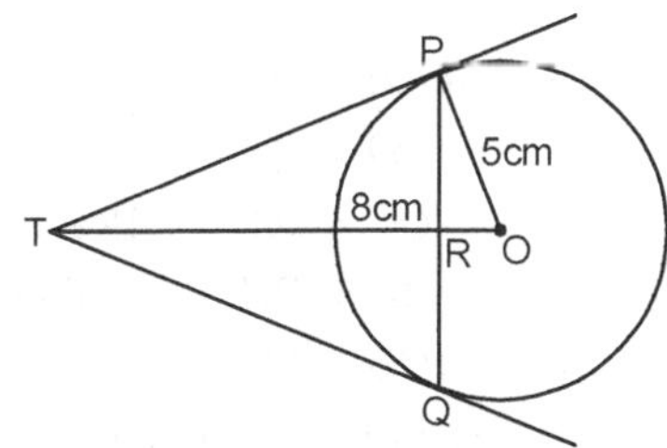

Now, $\angle TPR + \angle RPO = 90° = \angle TPR + \angle PTR$

[$\because$ In ΔTRP, ∠TRP = 90°]

$\Rightarrow \quad \angle RPO = \angle PTR$

So, $\quad \Delta TRP \sim \Delta PRO \quad$ (By AA rule)

$\therefore \quad \frac{TP}{PO} = \frac{RP}{RO}$

or $\quad \frac{TP}{5} = \frac{4}{3}$, or $TP = \frac{20}{3}$ cm

Hence, the length of $TP = \frac{20}{3}$ cm **Ans.**

Q. 5. Prove that the lengths of the tangents drawn from an external point to a circle are equal.★★

Solution : Let P be the external point from which tangents PA and PB are drawn on the circle with centre O.

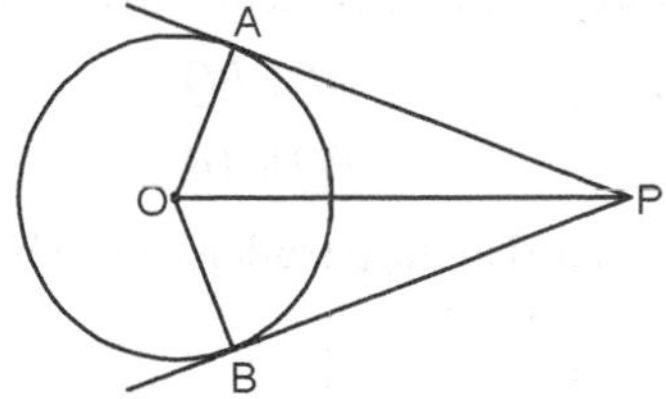

Thus, OB is perpendicular to PB and OA is perpendicular to PA.

Hence, $\quad \angle OBP = \angle OAP = 90°$

In ΔPOB and ΔPOA,

$OA = OB$ [Radii of the circle]

$OP = OP$ [Common side]

$\angle OBP = \angle OAP$ [Each 90°]

$\therefore \quad \Delta POB \cong \Delta POA$ [S.A.S.]

So, $\quad PA = PB$ [By cpct]

Hence Proved.

Q. 6. Prove that the tangent drawn at the mid-point of an arc of a circle is parallel to the chord joining the end points of the arc.★

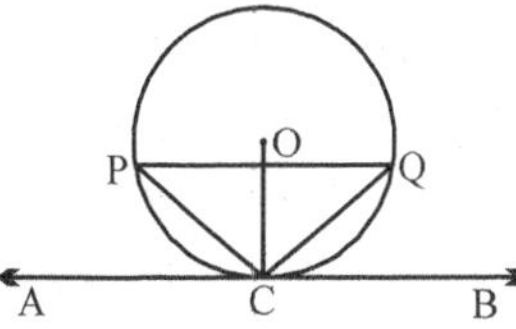

Solution: Given, C is the mid-point of the minor arc PQ and O is the centre of the circle and AB is tangent to the circle through point C.

Construction : Draw PC and QC.

To prove : $\quad PQ \parallel AB$

Proof : It is given that C is the mid-point of the arc PQ.

So, $\quad$ Minor arc PC = Minor arc QC

★ are board exam questions from previous years

★★ are frequently asked board exam questions

$\Rightarrow \qquad PC = QC$

Hence, ΔPQC is an isosceles triangle.

Thus, the perpendicular bisector of the side PQ of ΔPQC passes through vertex C.

But we know that the perpendicular bisector of a chrod passes through centre of the circle.

So, the perpendicular bisector of PQ passes through the centre O of the centre.

Thus, the perpendicular bisector of PQ passes through the points O and C.

$\Rightarrow \qquad PQ \perp OC \qquad$ ...(i)

AB is perpendicular to the circle through the point C on the circle

$\Rightarrow \qquad AB \perp OC \qquad$ ...(ii)

From equations (i) and (ii), the chord PQ and tangent AB of the circle are perpendicular to the same line OC.

Hence, $\qquad AB \parallel PQ$

or $\qquad PQ \parallel AB \qquad$ **Hence Proved.**

Q. 7. Prove that the parallelogram circumscribing a circle is a rhombus.★ [NCERT]

Solution : Given, $ABCD$ is a parallelogram where $AB = DC$ and $AD = BC$.

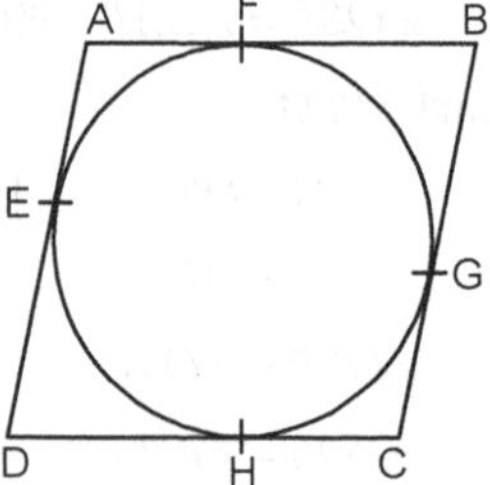

Now, $\qquad AF = AE \qquad$ [Tangents from A] ...(i)

$BF = BG \qquad$ [Tangents from B] ...(ii)

$CH = CG \qquad$ [Tangents from C] ...(iii)

$DH = DE \qquad$ [Tangents from D] ...(iv)

Adding (i), (ii), (iii) and (iv), we get

$$AF + BF + CH + DH = AE + BG + CG + DE$$

$\Rightarrow \qquad AB + CD = AD + BC$

$\Rightarrow \qquad 2AB = 2BC$

$[\because AB = DC \text{ and } AD = BC]$

$\Rightarrow \qquad AB = BC = CD = AD$

So, $ABCD$ is a rhombus. $\qquad$ **Hence Proved.**

Q. 8. In figure, two equal circles, with centres O and O', touch each other at X. OO' produced meets the circle with centre O' at A. AC is tangent to the circle with centre O at the point C. $O'D$ is perpendicular to AC. Find the value of $\frac{DO'}{CO}$.

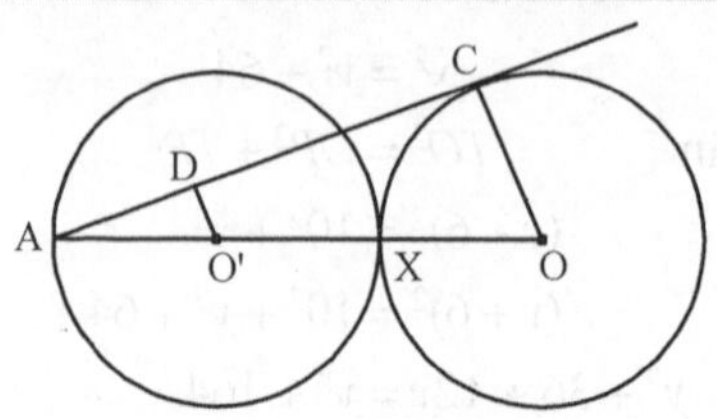

Solution : Given, AC is tangent to the circle with centre O. and $O'D$ is perpendicular to AC.

then, $\qquad \angle ACO = 90°$

Also, $\qquad \angle ADO' = 90°$

In $\Delta AO'D$ and ΔAOC

$\angle ACO = \angle ADO' \qquad$ [Each 90°]

$\angle CAO = \angle DAO' \qquad$ [Common angle]

$\therefore \qquad \Delta AO'D \sim \Delta AOC \qquad$ [AA criterion]

$\Rightarrow \qquad \frac{AO'}{AO} = \frac{DO'}{CO} \qquad$ [cpct]

$\Rightarrow \qquad \frac{AO'}{3 \cdot AO'} = \frac{DO'}{CO} \qquad \begin{bmatrix} AX = 2AO' \\ OX = AO' \end{bmatrix}$

$\frac{DO'}{CO} = \frac{1}{3} \qquad$ **Ans.**

Q. 9. In the given figure, l and m are two parallel tangents to a circle with centre O, touching the circle at A and B respectively. Another tangent at C intersects the line l at D and m at E. Prove that $\angle DOE = 90°$.★★ [NCERT]

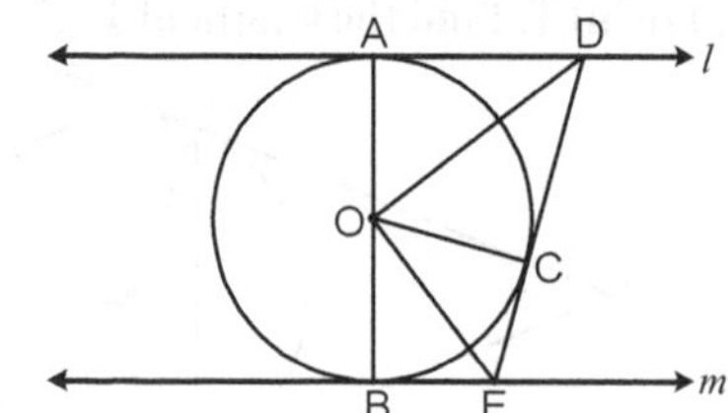

Solution : Let $\angle OEB = \theta$ and $\angle ADO = \phi$

Join OC such that OC is perpendicular to DE. Hence

$$\angle ECO = \angle OCD = 90°$$

Now, in ΔOBE and ΔECO,

$OE = OE \qquad$ [Common]

$OC = OB \qquad$ [Radii of the circle]

$BE = EC \qquad$ [Tangents from E]

Thus, $\qquad \Delta OBE \cong \Delta ECO \qquad$ [SSS]

So $\qquad \angle OEB = \angle OEC = \theta$

Hence $\qquad \angle EOC = 180° - (90° + \theta) = 90° - \theta \qquad$ ...(i)

Similarly, in ΔAOD and ΔOCD,

$OD = OD \qquad$ [Common]

$OC = OA \qquad$ [Radii of the circle]

$AD = DC \qquad$ [Tangents from D]

Thus, $\qquad \Delta AOD \cong \Delta OCD \qquad$ [SSS]

So $\qquad \angle ADO = \angle CDO = \phi$

★ are board exam questions from previous years

★★ are frequently asked board exam questions

Hence $\angle COD = 180° - (90° + \phi) = 90° - \phi$ ...(ii)

As AB is perpendicular to both l and m and passes through the centre

$\therefore \quad \angle BOE + \angle EOC + \angle COD + \angle DOA = 180°$

$\Rightarrow [180° - (90° + \theta)] + 90° - \theta + 90° - \phi + [180° - (90 + \phi)] = 180°$

$\Rightarrow 90° - \theta + 90° - \theta + 90° - \phi + 90° - \phi = 180°$

$\Rightarrow \quad 360° - 2(\theta + \phi) = 180°$

$\Rightarrow \quad 2(\theta + \phi) = 180°$

$\Rightarrow \quad \theta + \phi = 90°$ ...(iii)

In ΔOED, $\angle DOE = 180° - (\theta + \phi)$

$\Rightarrow \quad \angle DOE = 180° - 90°$ [From (iii)]

$\Rightarrow \quad \angle DOE = 90°$. **Hence Proved.**

Q. 10. In Figure, PQ and RS are two parallel tangents to a circle with centre O and another tangent AB with point of contact C intersecting PQ at A and RS at B. Prove that ∠AOB = 90°.★

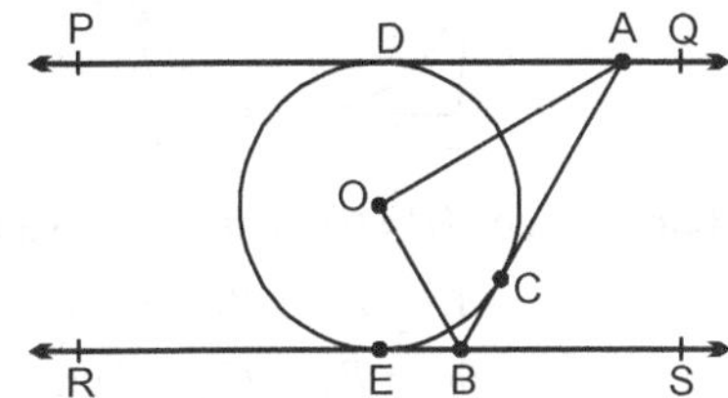

Solution : Given, PQ || RS

To prove : ∠AOB = 90°

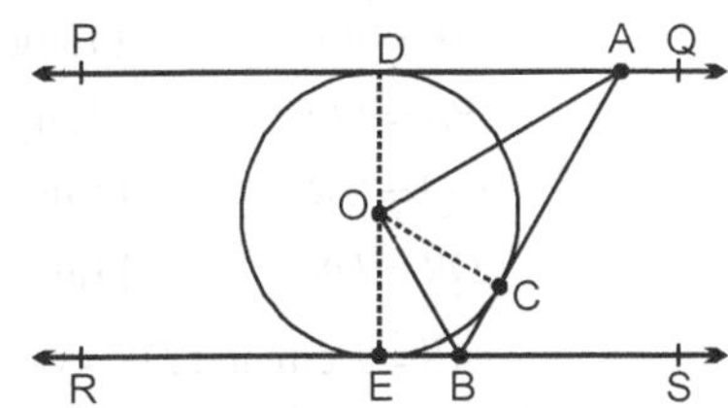

Construction : Join O and C, D and E

In ΔODA and ΔOCA

OD = OC (radii of circle)
OA = OA (common)
AD = AC

(tangent drawn from same point)

By SSS congruency

ΔODA ≅ ΔOCA

Then, ∠DOA = ∠AOC ...(i)

Similiarly, in ΔEOB and ΔBOC, we have

ΔEOB ≅ ΔBOC

∠EOB = ∠BOC ...(ii)

EOD is a diameter of circle, therefore it is a straight line.

Hence,

$\angle DOA + \angle AOC + \angle EOB + \angle BOC = 180°$

$2(\angle AOC) + 2(\angle BOC) = 180°$

$\angle AOC + \angle BOC = 90°$

$\angle AOB = 90°$. **Hence Proved.**

Q. 11. In the given figure, PQ is a chord of length 4.8 cm and the radius of the circle is 3 cm. The tangents at P and Q intersect at a point T. Find the length of TP.★

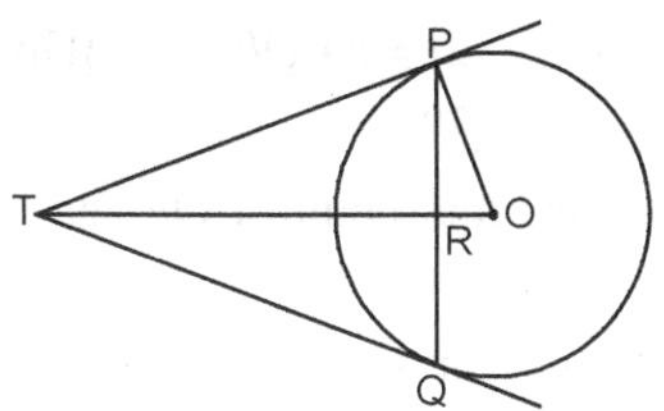

Solution : Given, r = 3 cm and length of chord PQ = 4.8 cm

As TO bisects PQ at R, so

$PR = RQ = 2.4$ cm

Let $TP = x$ cm and $TR = y$ cm

Now, $OR^2 = PO^2 - PR^2$

[Pythagoras theorem]

$\Rightarrow \quad OR^2 = (3)^2 - (2.4)^2$

$\Rightarrow \quad OR^2 = 9 - 5.76$

$\Rightarrow \quad OR^2 = 3.24$

$\Rightarrow \quad OR = 1.8$ cm

Thus, $TO = (y + 1.8)$ cm

Also $TP^2 = TR^2 + PR^2$

$\Rightarrow \quad x^2 = y^2 + (2.4)^2$

$\Rightarrow \quad x^2 = y^2 + 5.76$ ...(i)

Again $TO^2 = OP^2 + TP^2$

$\Rightarrow \quad (y + 1.8)^2 = 3^2 + x^2$

$\Rightarrow \quad (y + 1.8)^2 = 9 + y^2 + 5.76$ [Using (i)]

$\Rightarrow y^2 + 3.24 + 3.6y = y^2 + 14.76$

$\Rightarrow \quad 3.6y = 11.52$

$\Rightarrow \quad y = 3.2$

Hence $TP^2 = x^2 = y^2 + 5.76$

$\Rightarrow \quad TP^2 = x^2 = (3.2)^2 + 5.76$

$\Rightarrow \quad x^2 = 10.24 + 5.76 = 16$

$\Rightarrow \quad x = 4$ cm

$\Rightarrow \quad TP = 4$ cm. **Ans.**

Q. 12. From a point P, two tangents PA and PB are drawn to a circle with centre O and radius r. If $OP = 2r$, show that ΔAPB is equilateral.★

Solution : Given, $OP = OQ + QP$,

$OQ = OA = OB$

[Radius of the circle]

and $AP = BP$ [Tangents from P]

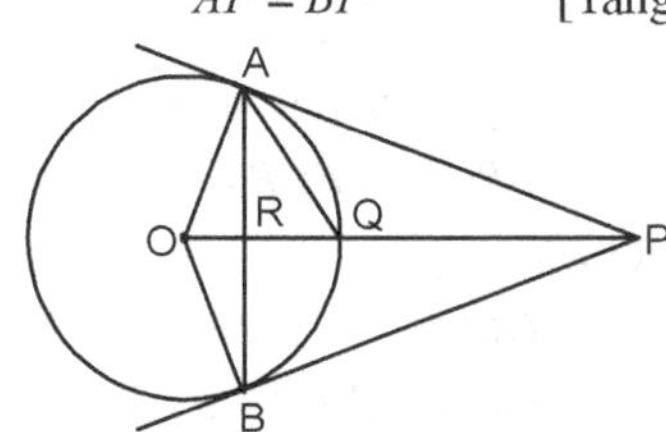

★ are board exam questions from previous years

Let OP meet the circle at Q and join OA and OQ.

Thus, OA is perpendicular to AP.

$\Rightarrow \quad \angle OAP = 90°$

Now, $\quad OP = OQ + QP \quad$ ...(i)

But, $\quad OQ = OA = OB \quad$ [Radii of the circle]

$\therefore$ From (i) $\quad 2r = r + QP \quad$ [Given; $OP = 2r$]

$\Rightarrow \quad QP = r$

Thus Q is the mid-point of OP and is equidistant from O, A and P.

As ΔAOP is right-angled triangle in which OP is hypotenuse and mid-point of hypotenuse of a right-angled triangle is equidistant from its all vertices.

$\therefore \quad OA = OQ = AQ = r$

$\Rightarrow \Delta AOQ$ is equilateral.

Thus, $\quad \angle AOQ = \angle AOP = 60°$

[Angles of an equilateral triangle]

Hence $\quad \angle OPA = 180° - (90° + 60°) = 30°$

Similarly, it can be proved that $\angle OPB = 30°$

So, $\quad \angle P = \angle OPA + \angle OPB = 60°$

Also as $\quad AP = PB$

So, $\quad \angle PAB = \angle PBA$

[base angles of an isosceles triangle]

Now, $\quad \angle PAB = \angle PBA = \dfrac{180° - 60°}{2} = 60°$

Hence, ΔAPB is equilateral as all its angles are equal to 60°. **Hence Proved.**

Q. 13. In figure, tangents *PQ* and *PR* are drawn from an external point *P* to a circle with centre *O*, such that $\angle RPQ = 30°$. A chord *RS* is drawn parallel to the tangent *PQ*. Find $\angle RQS$.★

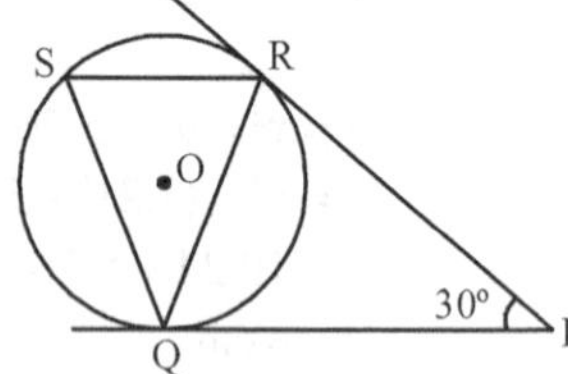

Solution : We have, $\quad PR = PQ$

and $\quad \angle PRQ = \angle PQR$

In ΔPQR,

$\angle PRQ + \angle PQR + \angle RPQ = 180°$

$2\angle PRQ + 30° = 180°$

$\angle PRQ = \dfrac{180° - 30°}{2} = 75°$

$\because SR \parallel QP$ and QR is a transversal

$\therefore \quad \angle SRQ = \angle PQR = 75°$

Join OR, OQ.

$\therefore \quad \angle ORQ = \angle RQO = 90° - 75° = 15°$

$\therefore \quad \angle QOR = (180° - 2 \times 15°)$

$= 180° - 30° = 150°$

$\angle QSR = \dfrac{1}{2}\angle QOR$

$= 75°$

[Angle subtended on arc is half the angle subtended on centre]

$\therefore$ In ΔSQR

$\angle RQS = 180° - (\angle SRQ + \angle RSQ)$

$= 180° - (75° + 75°)$

$\therefore \quad \angle RQS = 30°$ **Ans.**

Q. 14. In the given figure, a circle touches all four sides of a quadrilateral *ABCD* with *AB* = 6 cm, *BC* = 7 cm and *CD* = 4 cm. Find *AD*.★

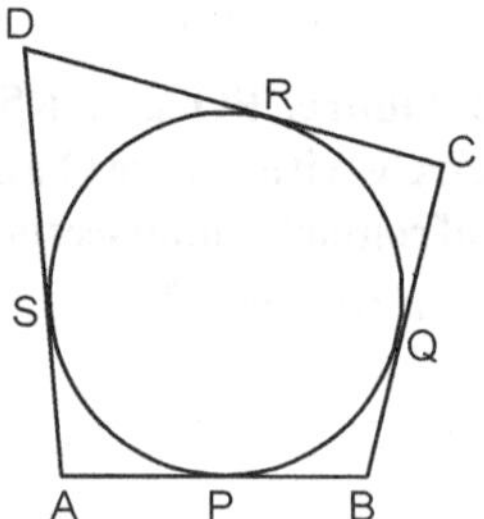

Solution : Given, $AB = 6$ cm, $BC = 7$ cm and $CD = 4$ cm

Now $\quad AB = AP + PB = 6$ cm

$BC = BQ + CQ = 7$ cm

$CD = CR + DR = 4$ cm

$AD = AS + DS$

Now, $\quad AP = AS \quad$ [Tangents from A]

$PB = BQ \quad$ [Tangents from B]

$CQ = CR \quad$ [Tangents from C]

$DR = DS \quad$ [Tangents from D]

Let $\quad AS = x$ cm and $DS = y$ cm

$\Rightarrow \quad AP = x$ cm

$\Rightarrow \quad PB = (6 - x)$ cm

So $\quad BQ = (6 - x)$ cm

and $\quad QC = [7 - (6 - x)]$ cm $= (1 + x)$ cm

Thus, $\quad CR = (1 + x)$ cm

$\Rightarrow \quad DR = [4 - (1 + x)]$ cm $= (3 - x)$ cm

Finally, $\quad DS = (3 - x)$ cm

But $\quad DS = y$ cm

$\Rightarrow \quad y = (3 - x)$ cm

Hence $\quad AD = (x + y)$ cm

$= [x + (3 - x)]$ cm

$= 3$ cm. **Ans.**

Q. 15. In the given figure, *PT* touches a circle with centre *O* at *R*. The diameter *SQ* when produced meets *PT* at *P*. If $\angle SPR = x°$ and $\angle QRP = y°$, show that $x° + 2y° = 90°$.★

★ are board exam questions from previous years

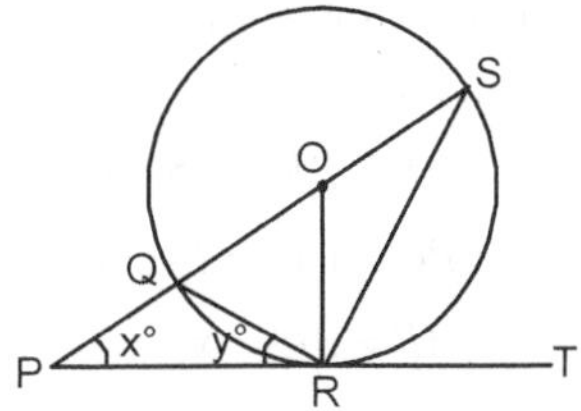

Solution : Given, $\angle SPR = x^\circ$ and $\angle QRP = y^\circ$

Also, OR is the radius that touches tangent PT at R.

Hence, OR is perpendicular to PT.

Similarly, as SQ is the diameter, so $\angle SRQ = 90^\circ$

Now, $\angle PQR = 180^\circ - (x^\circ + y^\circ)$

[Sum of angles of a triangle is equal to 180°]

and $\angle OQR = 180^\circ - \angle PQR$ [Linear pair]

$= 180^\circ [180^\circ - (x^\circ + y^\circ)]$

$= (x^\circ + y^\circ)$

Now, in ΔQOR,

$OQ = OR$ [radii of the circle]

$\Rightarrow \angle ORQ = (x^\circ + y^\circ)$

[Angles opposite to equal sides are always equal]

As OR is perpendicular to PT,

$\therefore \angle ORP = 90^\circ$

Thus, $\angle ORP = \angle ORQ + \angle QRP$

$\Rightarrow (x^\circ + y^\circ) + y^\circ = 90^\circ$

$\Rightarrow x^\circ + 2y^\circ = 90^\circ$. **Hence Proved.**

Q. 16. *ABC* is a triangle where *AB* = 10 cm, *BC* = 8 cm and *AC* = 6 cm. Three circles are drawn touching each other with the vertices being the centres. Find the radii of the three circles.

Solution : Given, $AB = 10$ cm, $BC = 8$ cm and $AC = 6$ cm

Let $BP = x$ cm

$\Rightarrow AP = AB - BP = (10 - x)$ cm

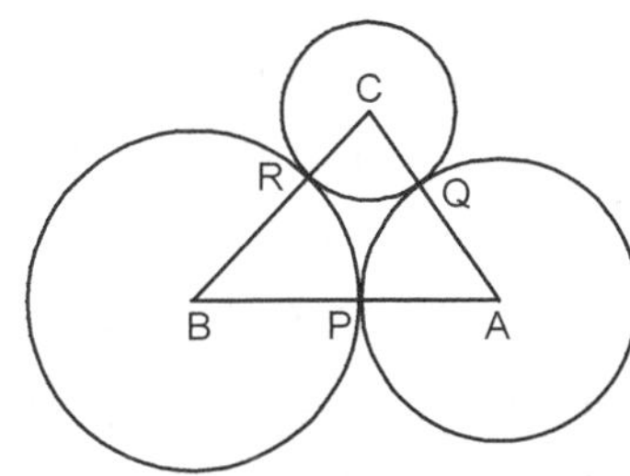

Also $BR = BP = x$ cm

[BR and BP are the radii of the largest circle]

$\Rightarrow RC = BC - BR = (8 - x)$ cm

Now $RC = CQ = (8 - x)$ cm

[RC and CQ are the radii of the smallest circle]

Also $AC = CQ + QA$

$\Rightarrow 6 = [(8 - x) + (10 - x)]$ cm

$\Rightarrow 18 - 2x = 6$

$\Rightarrow 2x = 12$

$\Rightarrow x = 6$ cm

Hence, radii of circle with centre $B = 6$ cm

Radii of circle with centre $A = (10 - 6) = 4$ cm

and radii of circle with centre $C = (8 - 6) = 2$ cm. **Ans.**

Q. 17. In Fig. a circle is inscribed in a ΔABC having sides BC = 8 cm, AB = 10 cm and AC = 12 cm. Find the lengths BL, CM and AN.★

Solution. A circle is inscribed in a DABC

AB = 10 cm, BC = 8 cm and AC = 12 cm

Let AN = AM = z

BN = BL = x

CL = CM = y

(Tangents drawn from an exterior points are equal in length.)

Perimeter of D

$= AB + BC + CA = 10 + 8 + 12$

or $x + z + x + y + y + z = 30$

$2(x + y + z) = 30$

$x + y + z = 15$ ...(i)

Also, AB = 10 cm

or $x + z = 10$ ...(ii)

and AC = 12

or $y + z = 12$ ...(iii)

and BC = 8 cm

$x + y = 8$ ...(iv)

From equation (i) and (ii),

$y = 5$ cm

From equation (i) and (iii)

$x = 3$ cm

From equation (i) and (iv)

$z = 7$ cm

So, BL = 3 cm, CM = 5 cm, AN = 7 cm. **Ans.**

Q. 18. In fig., *O* is the centre of a circle of radius 5 cm. *T* is a point such that *OT* = 13 cm and *OT* intersects circle at *E*. If *AB* is a tangent to the circle at *E*, find the length of *AB*, where *TP* and *TQ* are two tangents to the circle.★

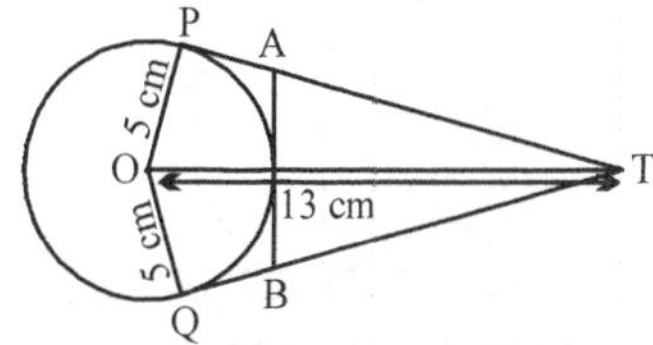

★ are board exam questions from previous years

Solution : Given, a circle with centre of radius 5 cm and $OT = 13$ cm.

Since, PT is a tangent at P and OP is a radius through P.

$\therefore \quad OP \perp PT$

In ΔOPT,

$$(PT)^2 = (OT)^2 - (OP)^2$$

$$\Rightarrow \quad PT = \sqrt{(13)^2 - (5)^2}$$

$$\Rightarrow \quad PT = \sqrt{169 - 25}$$

$$= \sqrt{144}$$

$$\Rightarrow \quad PT = 12 \text{ cm}$$

and, $TE = OT - OE = (13 - 5)$ cm $= 8$ cm

Now, $\quad PA = AE$

Let $\quad PA = AE = x$

Then, in ΔAET,

$$(AT)^2 = (AE)^2 + (ET)^2$$

$$(12 - x)^2 = (x)^2 + (8)^2$$

$$144 + x^2 - 24x = x^2 + 64$$

$$24x = 80$$

$$AE = x = 3.33 \text{ cm}$$

$$\therefore \quad AB = 2AE$$

$$= 2 \times 3.33$$

$$= 6.66 \text{ cm}$$

Ans.

Q. 19. In the given figure, *ABC* is drawn to circumscribe a circle of radius 10 cm such that the segments *BP* and *PC* into which *BC* is divided by the point of contact *P* are of lengths 15 cm and 20 cm respectively. If the area of $\Delta ABC = 525$ cm², then find the lengths of the sides *AB* and *AC*.★

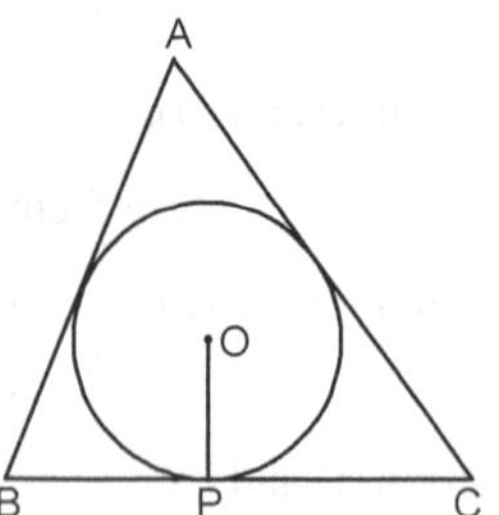

Solution : Given, $\quad OP = OQ = OR = 10$ cm

[radii of the circumscribed circle]

$BP = 15$ cm, $PC = 20$ cm

and $\quad$ ar $(\Delta ABC) = 525$ cm^2

Since, tangents drawn from an external point to a circle are equal in length

$$\therefore \quad BP = BR = 15 \text{ cm}$$

$$PC = CQ = 20 \text{ cm}$$

$$\text{and} \quad AR = AQ = x$$

Now, Area of $\Delta BOC = \frac{1}{2} \times \text{Base} \times \text{Height}$

$$= \frac{1}{2} \times BC \times OP$$

$$= \frac{1}{2} \times (15 + 20) \times 10$$

$$= 5 \times 35$$

$$= 175 \text{ cm}^2$$

Area of $\Delta AOB = \frac{1}{2} \times AB \times OR$

$$= \frac{1}{2} \times (x + 15) \times 10$$

$$= 5(x + 15) \text{ cm}^2$$

and Area of $\Delta AOC = \frac{1}{2} \times AC \times OC$

$$= \frac{1}{2} \times (x + 20) \times 10$$

$$= 5(x + 20) \text{ cm}^2$$

But, Area of ΔABC = Area of $(\Delta BOC + \Delta AOB + \Delta AOC)$

$$\Rightarrow \quad 525 = 175 + 5(x + 15) + 5(x + 20)$$

$$350 = 5(2x + 35)$$

$$\Rightarrow \quad 70 = 2x + 35$$

$$\Rightarrow \quad 2x = 35$$

$$\Rightarrow \quad x = 17.5 \text{ cm}$$

Thus, $\quad AB = x + 15$

$$= 17.5 + 15 = 32.5 \text{ cm}$$

and $\quad AC = x + 20$

$$= 17.5 + 20 = 37.5 \text{ cm}$$

Ans.

Q. 20. In the given figure, a ΔABC is drawn to circumscribe a circle of radius 4 cm such that the segments *BD* and *DC* are of lengths 8 cm and 6 cm, respectively. Find the lengths of the sides *AB* and *AC*

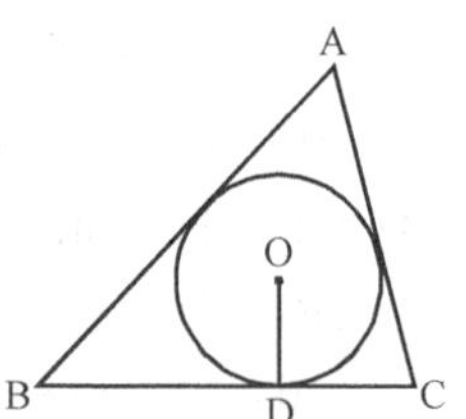

★ are board exam questions from previous years

Solution : Given,

$$OD = 4 \text{ cm}, BD = 8 \text{ cm and } DC = 6 \text{ cm}$$

Let $AF = AE = x$ cm [Tangents from A]

So $BC = (6 + 8)$ cm $= 14$ cm

Thus, $AB = (x + 8)$ cm

[$\because$ $FB = BD$ as both are tangents from B]

and $AC = (x + 6)$ cm

[$\because$ $C = EC$ as both are tangents from C]

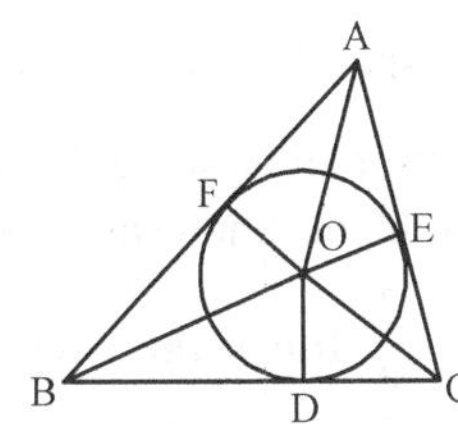

Applying Heron's formula

$$s = \frac{x + 8 + 14 + x + 6}{2}$$

$$= \frac{2x + 28}{2}$$

$$= (x + 14)$$

So, Area of ΔABC

$$= \sqrt{(x+14)(x+14-14)(x+14-x-8)(x+14-x-6)}$$

$\Rightarrow$ Area of $\Delta ABC = \sqrt{(x+14)(x)(14-8)(14-6)}$

$\Rightarrow$ Area of $\Delta ABC = \sqrt{(x+14)(x)6 \times 8}$

$\Rightarrow$ Area of $\Delta ABC = \sqrt{48x(x+14)}$ cm^2

Area of $\Delta ABOC = \frac{1}{2} \times 14 \times 4 = 28$ cm^2

Area of $\Delta BOC = \frac{1}{2} \times (x + 6) \times 4 = 2(x + 6)$ cm^2

Area of $\Delta AOB = \frac{1}{2} \times (x + 8) \times 4 = 2(x + 8)$ x cm^2

Again, Area of ΔABC = Area of ΔBOC + Area of ΔAOC + Area of ΔAOB

$\Rightarrow$ $\sqrt{48x(x+14)} = 28 + 2x + 12 + 2x + 16$

$\Rightarrow$ $\sqrt{48x(x+14)} = 4x + 56 = 4(x + 14)$

$\Rightarrow$ $48x(x + 14) = 16(x + 14)^2$

$\Rightarrow$ $3x(x + 14) = (x + 14)^2$

$\Rightarrow$ $(x + 14)^2 - 3x(x + 14) = 0$

$\Rightarrow$ $(x + 14)[(x + 14) - 3x] = 0$

$\Rightarrow$ $(x + 14)[14 - 2x] = 0$

$\Rightarrow$ $2(x + 14)(7 - x) = 0$

$\Rightarrow$ $x = 7$

So, $AB = 7 + 8 = 15$ cm

and $AC = 7 + 6 = 13$ cm.

★ are board exam questions from previous years

21. In figure, TA is a tangent to a circle from the point T and TBC is a secant to the circle. If AD is the bisector of $\angle BAC$, prove that ΔADT is isosceles.★

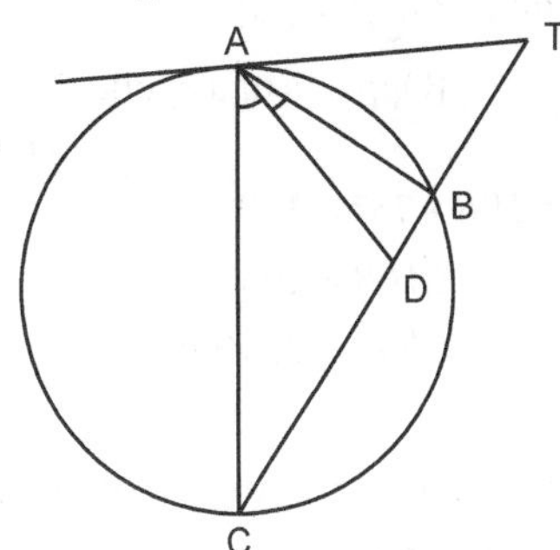

Solution. In order to prove that ΔADT is isosceles *i.e.*, $TA = TD$, it is sufficient to show that

$$\angle TAD = \angle TDA$$

Since $\angle TAB$ and $\angle BCA$ are angles in the alternate segments of chord AB.

$\therefore$ $\angle TAB = \angle BCA$ ...(i)

It is given that AD is the bisector of $\angle BAC$.

$\therefore$ $\angle BAD = \angle CAD$ ...(ii)

Now, $\angle TAD = \angle TAB + \angle BAD$

$\Rightarrow$ $\angle TAD = \angle BCA + \angle CAD$

[Using (i) and (ii)]

$\Rightarrow$ $\angle TAD = \angle DCA + \angle CAD$

[$\because$ $\angle BCA = \angle DCA$]

$\Rightarrow$ $\angle TAD = 180° - \angle CDA$

[In ΔCAD, $\angle CAD + \angle DCA + \angle CDA = 180°$

$\therefore$ $\angle CAD + \angle BCA = 180° - \angle CDA$]

$\Rightarrow$ $\angle TAD = \angle TDA$

[$\because$ $\angle CDA + \angle TDA = 180°$]

$\Rightarrow$ $TD = TA$

Hence, ΔADT is isosceles. **Hence Proved.**

22. In the adjoining diagram, TA and TB are tangents, O is the centre. If $\angle PAT = 35°$ and $\angle PBT = 40°$, calculate:

(i) $\angle AQP$ **(ii) $\angle BQP$**

(iii) $\angle AQB$ **(iv) $\angle APB$**

(v) $\angle AOB$ **(vi) $\angle ATB$**

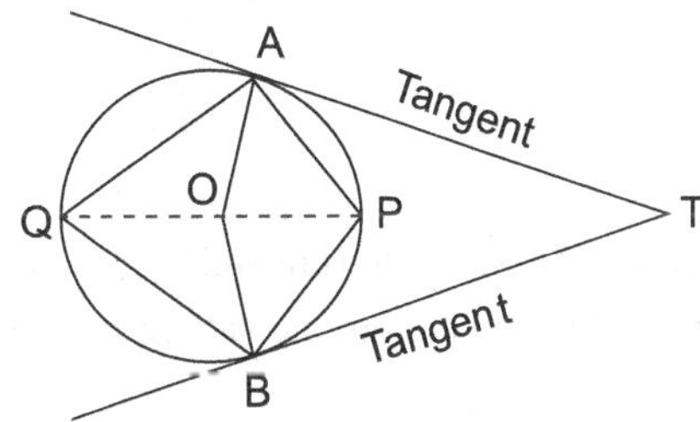

Solution. (i) $\angle AQP = \angle PAT = 35°$

[Angles are in alternate segment] **Ans.**

(ii) $\angle BQP = \angle PBT = 40°$

[Angles are in alternate segment] **Ans.**

(iii) $\angle AQB = \angle AQP + \angle BQP$

$= 35° + 40° = 75°.$ **Ans.**

(iv) $\angle APB + \angle AQB = 180°$

[Opp. ∠s of a cyclic quadrilateral are supplementary]

$\therefore \quad \angle APB + 75° = 180°$

$\therefore \quad \angle APB = 105°.$ **Ans.**

(v) $\angle AOB = 2\angle AQB$

$= 2(75°) = 150°$ **Ans.**

[Angle at the centre = 2 Angle at the circumference]

(vi) In quadrilateral *AOBT*:

$\angle ATB = 360° - (\angle OAT + \angle OBT + \angle AOB)$

$= 360° - (90° + 150° + 90°) = 30°$

[$\angle OAT = \angle OBT = 90°$, radius ⊥ tangent] **Ans.**

Chapter 9. Constructions

Q. 1. Construct a triangle *ABC* with sides *AB* = 5 cm, *BC* = 6 cm and *AC* = 7 cm. Now construct another triangle whose sides are $\frac{3}{5}$ times the corresponding sides of the Δ*ABC*.★★

Solution :

Step I : Draw *AC* = 7 cm with the help of ruler and pencil.

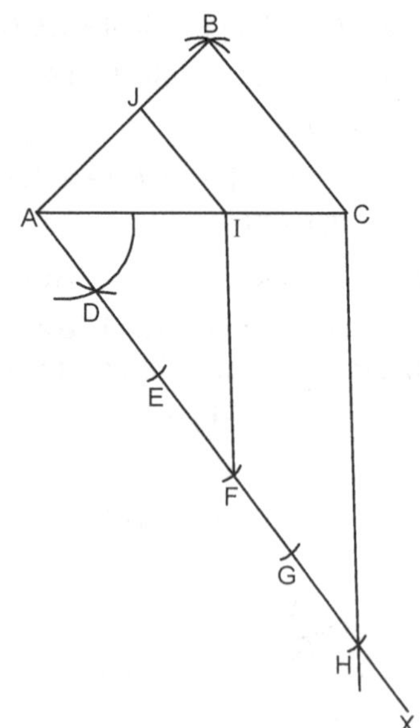

Step II : Measure out 6 cm and keeping the compass point at *C*, draw an arc on one side of the line-segment *AC*.

Step III : Measure 5 cm and keeping the compass point at *A*, draw an arc on the same side of the line-segment *AC* as the other arc.

Step IV : The point of intersection of the two arcs is point *B*. Join *AB* and *CB*.

★ are board exam questions from previous years

★★ are frequently asked board exam questions

Thus, Δ*ABC* is obtained.

Step V : Below *AC*, make an acute angle ∠*CAX*.

Step VI : Along *AX*, mark 5 points (greater of 3 and 5 in $\frac{3}{5}$) *D, E, F, G, H* such that *AD* = *DE* = *EF* = *FG* = *GH*.

Step VII : Join *HC*.

Step VIII : From *F*, draw *FI* || *HC* meeting *AC* at *I*.

Step IX : From *I*, draw *IJ* || *BC* meeting *AB* at *J*.

Thus, Δ*AIJ* is the required triangle whose sides are $\frac{3}{5}$ times the corresponding sides of Δ*ABC*.

Q. 2. Construct an isosceles triangle whose base is 6 cm and altitude 4 cm. Then construct another triangle whose sides are $\frac{3}{4}$ times the corresponding sides of isosceles triangle.★

Solution :

Step I : Draw a line segment *AB* = 6 cm.

Step II : Draw a perpendicular bisector *PQ* of *AB*, intersecting *AB* at *O*.

Step III : Draw an arc at a distance 4 cm (from *O*) intersecting *PQ* at *C*.

Step IV : Join *CA* and *CB*.

Thus, Δ*ABC* is obtained.

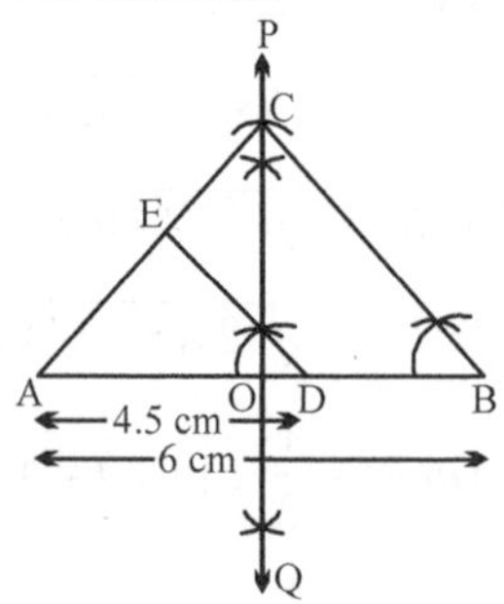

Step V : Mark *D* on *AB*, such that $AD = \frac{3}{4}$

$AB = \left(\frac{3}{4} \times 6\right)$ cm $= \frac{9}{2}$ cm $= 4.5$ cm

Step VI : Draw *DE* || *BC*, cutting *AC* at *E*.

Then, Δ*ADE* is the required triangle similar to Δ*ABC* such that each side of Δ*ADE* is $\frac{3}{4}$ times the corresponding side of Δ*ABC*.

Q. 3. Construct a ΔABC in which CA = 6 cm, AB = 5 cm and ∠BAC = 45°. Then construct a triangle whose sides are $\frac{3}{5}$ of the corresponding sides of ΔABC.★

Solution : Steps of Construction are as follows :

Setp I : Draw AB = 5 cm

Setp II : At the point A draw ∠BAX = 45°

Setp III : From AX cut off AC = 6 cm

Setp IV : Join BC, ΔABC is formed with given data.

Setp V : Draw $\overrightarrow{AY}$ making any acute angle with AB as shown in the figure.

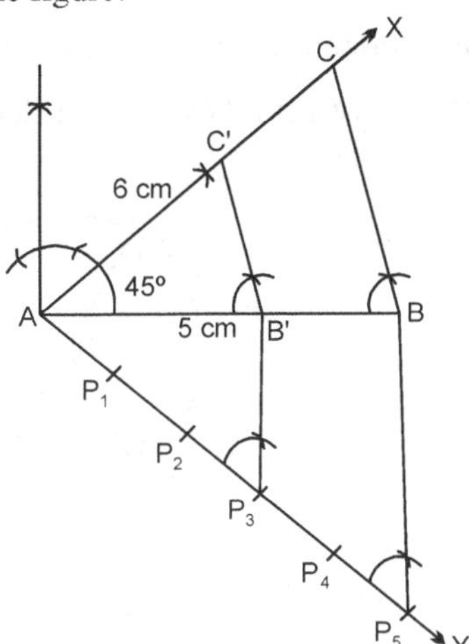

Setp VI : Draw 5 arcs P_1, P_2, P_3, P_4 and P_5 with equal intervals.

Setp VII : Join BP_5.

Setp VIII : Draw $P_3B' \parallel P_5B$ meeting AB at B′.

Setp IX : From B′ draw $B'C' \parallel BC$ meeting AC at C′.

$\therefore \quad \Delta AB'C' \sim \Delta ABC$

Hence ΔAB′C′ is the required triangle. **Ans.**

Q. 4. Construct a ΔABC in which AB = 6 cm, $\angle A$ = 30 and $\angle B$ = 60°. Construct another $\angle A'B'C'$ similar to ΔABC with base AB' = 8 cm.★

Solution :

Step I : Draw a line segment $AB = 6$ cm.

Step II : Construct $\angle ABP = 60°$ and $\angle QAB = 30°$

Step III : Join AC and BC such that C is the intersection point of BP and AQ.

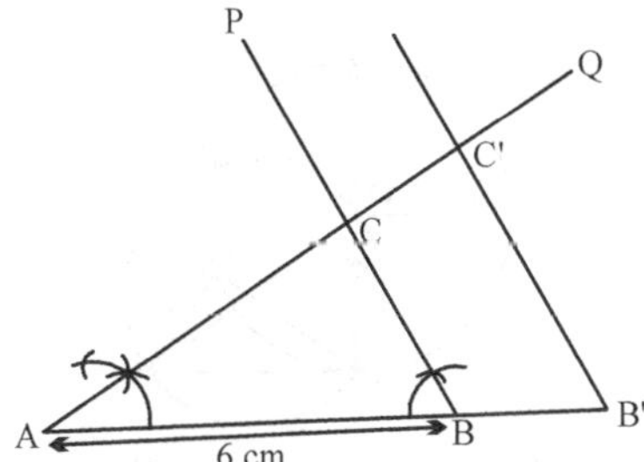

Thus, ΔABC is the required triangle.

Step IV : Extend AB to B', such that $AB' = 8$ cm

Step V : Draw $B'C' \parallel BC$ cutting AC produced at C'.

Then, $\Delta AB'C$ is the required triangle similar to ΔABC.

Q. 5. Construct a triangle ABC with side $BC = 7$ cm, $\angle B = 45°$ and $\angle C = 60°$. Then construct another triangle whose sides are $\frac{3}{5}$ times the corresponding sides of the ΔABC.★

Solution : Step I : Draw a line segment BC = 7 cm with the help of ruler and pencil.

Step II : Construct $\angle B = 45°$ and draw an arc.

Step III : Construct $\angle C = 60°$ and draw an arc.

Step IV : The point of intersection of the two arcs is point A. Join AC and AB.

Thus, ΔABC is obtained.

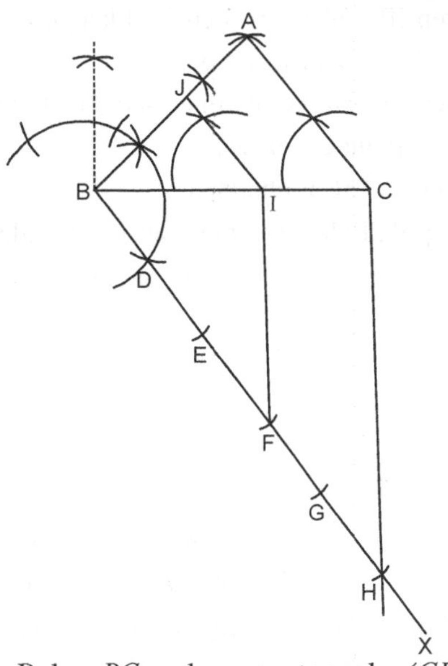

Step V : Below BC, make an acute angle $\angle CBX$.

Step VI : Along BX, mark 5 points (greater of 3 and 5 in $\frac{3}{5}$) D, E, F, G, H such that $BD = DE = EF = FG = GH$.

Step VII : Join HC.

Step VIII : From F, draw $FI \parallel HC$, meeting BC at I.

Step IX : From I, draw $IJ \parallel AC$ meeting AB at J.

Thus, ΔBIJ is the required triangle whose sides are $\frac{3}{5}$ of corresponding sides of ΔABC.

Q. 6. Construct a triangle ABC with sides AB = 5 cm, BC = 6 cm and $\angle ABC = 60°$. Then construct another triangle whose sides are $\frac{5}{7}$ times the corresponding sides of the ΔABC.★

Solution :

Step I : Draw a line segment $BC = 6$ cm with the help of ruler and pencil.

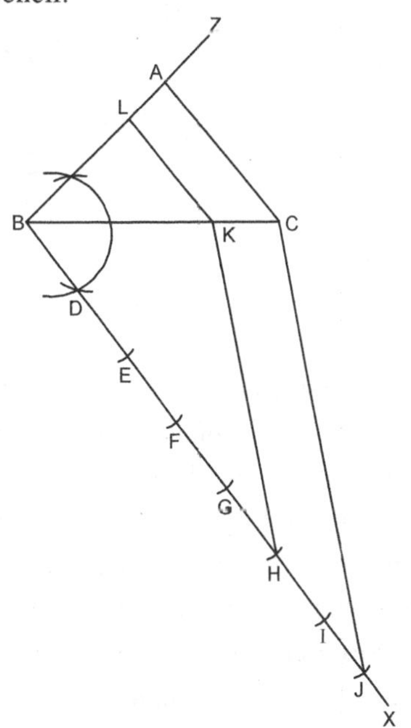

Step II : Construct $\angle ZBC = 60°$.

★ are board exam questions from previous years

Step III : Measure 5 cm and keeping the compass point at B, draw an arc on BZ.

Step IV : The point of intersection of the arc and the line BZ is point C. Join AC.

Thus, ΔABC is obtained.

Step V : Below BC, make an acute angle $\angle CBX$.

Step VI : Along BX, mark 7 points (greater of 5 and 7 in $\frac{5}{7}$) D, E, F, G, H, I, J such that

$$BD = DE = EF = FG = GH = HI = IJ$$

Step VII : Join JC.

Step VIII : From H, draw $HK \parallel JC$ meeting BC at K.

Step IX : From K, draw $KL \parallel AC$ meeting AB at L.

Thus, ΔKLB is the required triangle whose sides are $\frac{5}{7}$ of corresponding sides of ΔABC.

Q. 7. Draw two concentric circles of radii 2 cm and 5 cm. Taking a point on the outer circle, construct a pair of tangents to the other. Also measure the length of one of the tangents.★★

Solution : Step I : Draw a circle with centre O and radius OQ = 5 cm and another circle with centre O and radius OP = 2 cm.

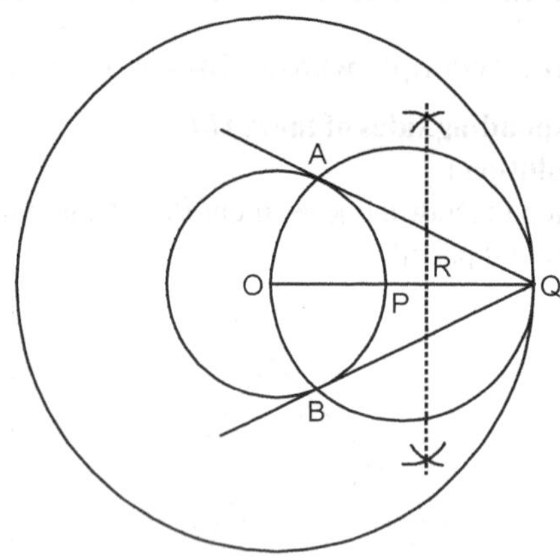

Step II : Draw perpendicular bisector of OQ bisecting it at R.

Step III : Draw a circle from R with radius $OR = RQ$.

Step IV : Mark the points of intersection of the circle with centre R and the smaller circle as A and B.

Step V : Join AQ and BQ and measure them.

AQ and BQ are the tangents to the smaller circle from the point Q on the larger circle and $AQ = BQ$ = 4.5 cm.

Q. 8. Draw a circle of radius 4 cm. Draw two tangents to the circle inclined at an angle of 60° to each other.★

Solution :

Step I : Draw a circle with O as centre and radius 4 cm.

Step II : Draw any diameter AOB of this circle.

Step III : Construct $\angle BOC = 60°$ such that radius OC meets the circle at C.

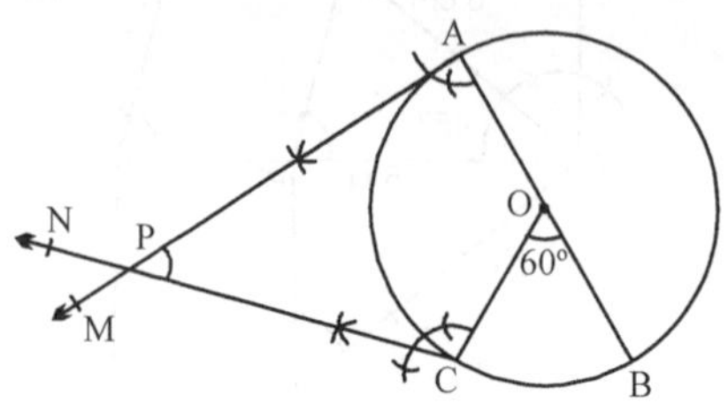

Step IV : Draw $AM \perp AB$ and $CN \perp OC$.

Let AM and CN intersect each other at P.

Then, PA and PC are the required tangents to the given circle and inclined at an angle of 60° each other.

Q. 9. Construct a right triangle ABC with AB = 6cm, BC = 8 cm and $\angle B$ = 90°. Draw BD, the perpendicular from B on AC. Draw the circle through B, C and D and construct the tangents from A to this circle.

Solution : Step I : Draw a line segment AB = 6 cm.

Step II : Construct $\angle XBA = 90°$

Step III : Taking B as centre, draw an arc of radius 8 cm intersecting XB at C.

Step IV : Join BC and AC.

Thus, $\angle ABC$ is obtained.

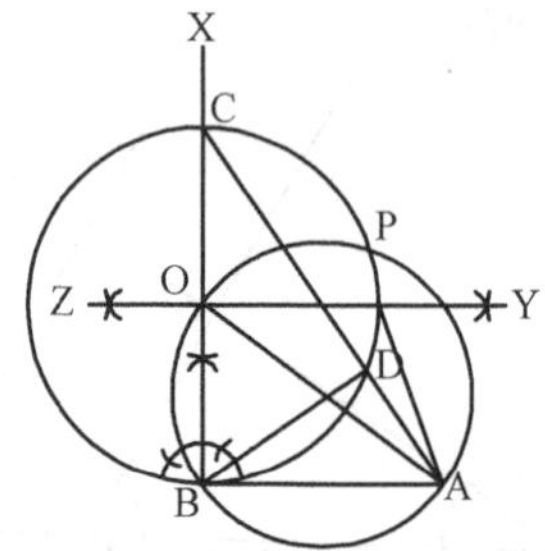

Step V : Draw a perpendicular BD to AC.

Step VI : Draw perpendicular bisector YZ of BC, bisecting it at O.

Step VII : Taking O as centre, $OB = OC$ as radius, draw a circle passing through points B, C and D.

Step VIII : Join AO, and taking OA as diameter, draw another circle.

Let the two circles meets at points B and P.

Step IX : Join AP and AB.

Thus, AP and AB are the required tangents to the circle with centre O.

Q. 10. Draw a triangle ABC with BC = 7 cm, $\angle B = 45°$ and $\angle A = 105°$. Then construct a triangle whose sides are $\frac{4}{5}$ times the corresponding sides of ΔABC.★

★ are board exam questions from previous years

★★ are frequently asked board exam questions

Solution :

$\angle B = 45^\circ$ and $\angle A = 105^\circ$

$\because$ Sum of angles of triangles is 180°

$\therefore \quad \angle A + \angle B + \angle C = 180^\circ$

$\Rightarrow 105^\circ + 45^\circ + \angle C = 180^\circ$

$\therefore \quad \angle C = 180^\circ - (105^\circ + 45^\circ)$

$\Rightarrow \quad \angle C = 30^\circ$

Step I : Draw a line segment $BC = 7$ cm.

Step II : Construct $\angle B = 45^\circ$ and $\angle C = 30^\circ$

Step III : A is the intersecting point of rays through B and C.

Thus, ΔABC is obtained.

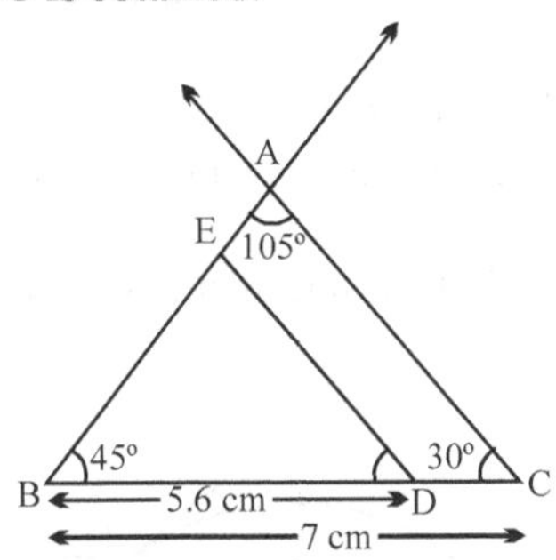

Step IV : Draw D on BC such that $BD = \frac{4}{5}$ $BC = \left(\frac{4}{5} \times 7\right)$ cm $= \frac{28}{5}$ cm $= 5.6$ cm

Step V : Draw $DE \parallel CA$, cutting BA at E.

Then, ΔBDE is the required triangle similar to ΔABC such that each side of ΔBDE is $\frac{4}{5}$ times the corres ponding side of ΔABC.

Q. 11. Draw a line segment AB of length 7 cm. Taking A as centre, draw a circle of radius 3 cm and taking B as centre, draw another circle of radius 2.5 cm. Construct tangents to each circle from the centre of the other circle.★

Solution :

Step I : Draw a line segment $AB = 7$ cm.

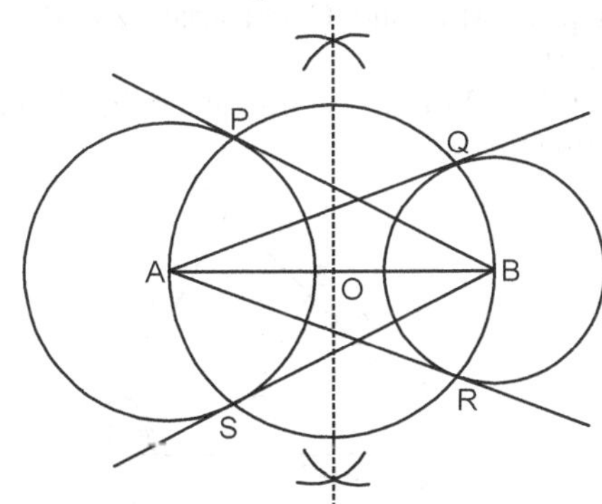

Step II : With A as centre and radius 3 cm , draw a circle.

Step III : With B as centre and radius 2.5 cm, draw another circle.

Step IV : Draw perpendicular bisector of AB, meeting AB at O.

Step V : Keeping O as the centre, draw a circle with $AO = OB$ as the radius.

Step VI : Mark points of intersection of this circle with the larger circle as P and S.

Step VII : Draw PB and SB as tangents to the larger circle.

Step VIII : Similarly, mark the points of intersection of the circle with centre O with the smaller circle as Q and R.

Step IX : Draw AQ and AR as tangents to the smaller circle.

Q. 12. Construct a triangle ABC with sides $AB = 5$ cm, $BC = 6$ cm and $\angle ABC = 60^\circ$. Then construct another triangle whose sides are $\frac{3}{4}$ times the corresponding sides of the ΔABC.★★

Solution :

Step I : Draw a line segment $BC = 6$ cm with the help of ruler and pencil.

Step II : Construct $\angle CBZ = 60^\circ$.

Step III : Measure 5 cm and keeping the compass point at B, draw an arc on BZ.

Step IV : The point of intersection of the arc and the line BZ is point A. Join AC.

Thus, ΔABC is obtained.

Step V : Below BC make an acute angle $\angle CBX$.

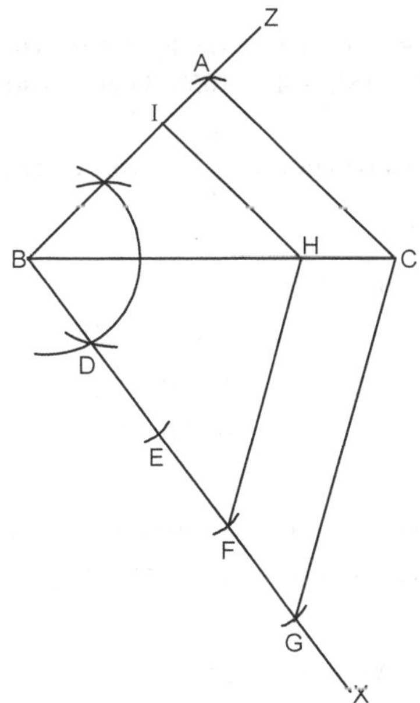

Step VI : Along BX mark 4 points (greater of 3 and 4 in $\frac{3}{4}$) D, E, F, G such that $BD = DE = EF = FG$.

Step VII : Join GC.

Step VIII : From F, draw $FH \parallel GC$ meeting BC at H.

★ are board exam questions from previous years

★★ are frequently asked board exam questions

Step IX : From H, draw $HI \parallel AC$ meeting AB at I.

Thus, ΔBHI is the required triangle whose sides are $\frac{3}{4}$ of corresponding sides of ΔABC.

Q. 13. Draw a pair of tangents to a circle of radius 4.5 cm which are inclined to each other at an angle of 45°.★

Solution : Step I : Draw a circle with radius $OA = 4.5$ cm and centre O.

Step II : Construct $OB = OA$ and $\angle AOB = 360° - (90° + 90° + 45°) = 360° - 225° = 135°$.

Step III : Construct perpendiculars from A and B and let them meet at C.

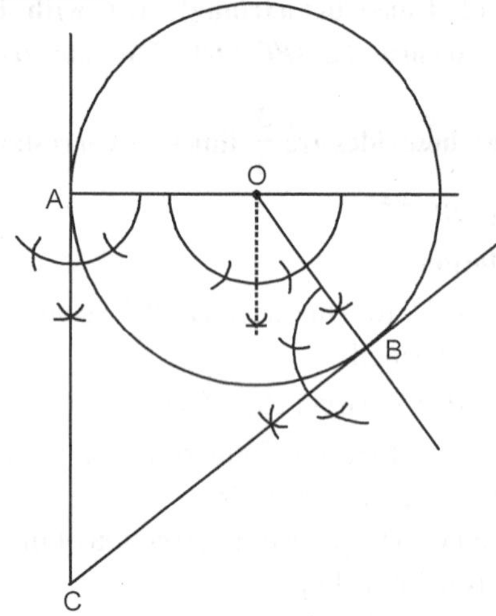

Thus, AC and BC are the required tangents from C to the circle with centre O.

Q. 14. Construct a triangle ABC with side $BC = 7$ cm, $\angle B = 45°$, $\angle A = 105°$. Then construct another triangle whose sides are $\frac{3}{4}$ times the corresponding sides of the ΔABC.★

Solution : $BC = 7$ cm, $\angle B = 45°$, $\angle A = 105°$

$$\angle C = 180° - (\angle B + \angle A)$$
$$= 180° - (45° + 105°)$$
$$= 180 - 150°$$
$$= 30°$$

Step I : Draw a line segment $BC = 7$ cm.

Step II : Draw an angle 45° at B and 30° at C. They intersect of A.

Thus, ΔABC is obtained.

Step III : Draw an acute angle at B.

Step IV : Divide angle ray in 4 equal parts as B_1, B_2, B_3, and B_4 such that $BB_1 = B_1B_2 = B_2B_3 = B_3B_4$.,

Step V : Join B_4 to C.

Step VI : From B_3, draw a line parallel to B_4C intersecting BC at C'.

Step VII : From C', draw a line parallel to CA intersecting AB ray at A'.

Step VIII : Thus, $\Delta A'B'C'$ is required triangle whose sides are $\frac{3}{4}$ of corresponding sides of ΔABC.

Q. 15. Draw a pair of tangents to a circle of radius 4.5 cm from a point P at a distance of 6.2 cm from its centre.

Solution :

Step I : Draw a circle with radius ON = 4.5 cm and centre O.

Step II : Mark point P outside the circle such that $OP = 6.2$ cm.

Step III : Bisect OP at M.

Step IV : Draw a circle with M as its centre and radius $= MP = MO$.

Step V : Mark the intersections of the two circles at Q and R.

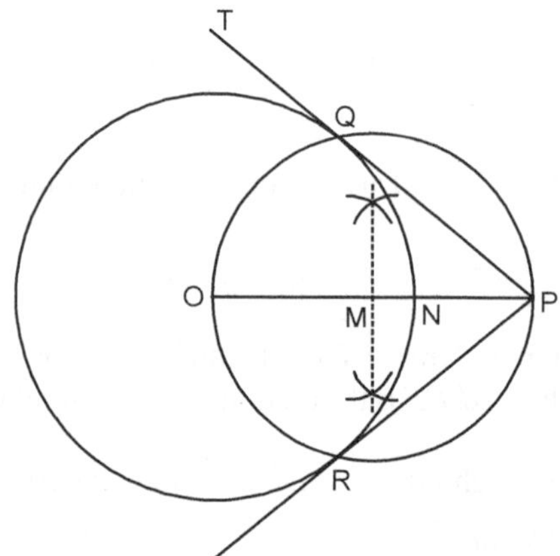

Step VI : Join PQ and PR to form the required tangents.

Q. 16. Construct an equilateral ΔABC with each side 5 cm. Then construct another triangle whose sides are $\frac{2}{3}$ times the corresponding sides of ΔABC.★

Solution : Steps for construction are as follows :

Step I : Draw a line sgement BC = 5 cm

Step II : At B and C construct ∠CBX = 60° and ∠BCX = 60°

Step III : With B as centre and radius 5 cm, draw an arc cutting ray BX at A.

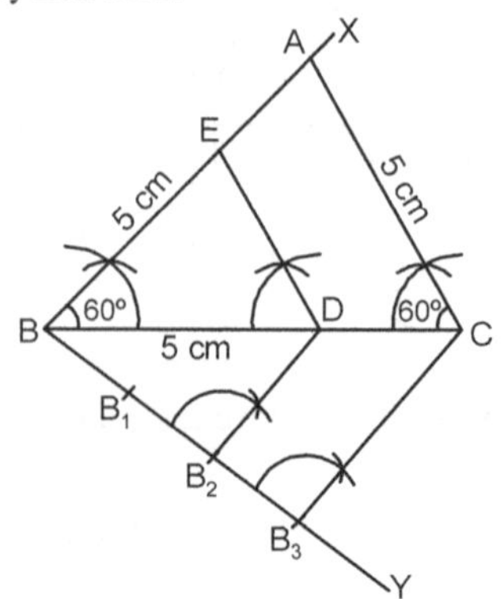

Step IV : Join AC.

Thus an equilateral ΔABC is obtained.

★ are board exam questions from previous years

Step V : Below BC, make an acute angle ∠CBY.

Step VI : Along BY, mark off 3 points B_1, B_2, B_3

Such that BB_1, B_1B_2, B_2B_3 are equal.

Step VII : Join B_3C.

Step VIII :From B_2 draw $B_2D \parallel B_3C$, meeting BC at D.

Step IX : From D, draw DE || CA, metting AB at E.

Then ΔEBD is the required triangle, each of whose sides is $\frac{2}{3}$ of the corresponding side of ΔABC.

Chapter 10. Trigonometry

Q. 1. In ΔOPQ right angled at *P*, *OP* = 7 cm, *OQ*–*PQ* = 1cm. Determine the values of sin *Q* and cos *Q*. [NCERT]

Solution : In ΔOPQ, we have

$$OQ^2 = OP^2 + PQ^2$$

$$(PQ+1)^2 = OP^2 + PQ^2$$

$$[\because OQ - PQ = 1 \Rightarrow OQ = 1 + PQ]$$

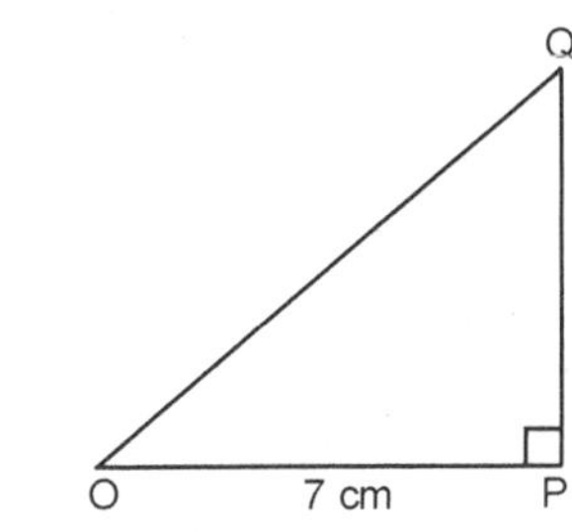

$$PQ^2 + 2PQ + 1 = OP^2 + PQ^2$$

$$2PQ + 1 = 49$$

$$2PQ = 48$$

$$PQ = 24 \text{ cm}$$

$$OQ - PQ = 1 \text{ cm}$$

$$OQ = (PQ+1) \text{ cm} = 25 \text{ cm}$$

Thus, $\sin Q = \frac{OP}{OQ} = \frac{7}{25}$

and $\cos Q = \frac{PQ}{OQ} = \frac{24}{25}$. **Ans.**

Q. 2. In a ΔABC right angled at *C*, if tan *A* = $\frac{1}{\sqrt{3}}$, find the value of sin *A* cos *B* + cos *A* sin *B*.★

Solution : Draw a ΔABC, right angled at *C* in which $\tan A = \frac{1}{\sqrt{3}}$.

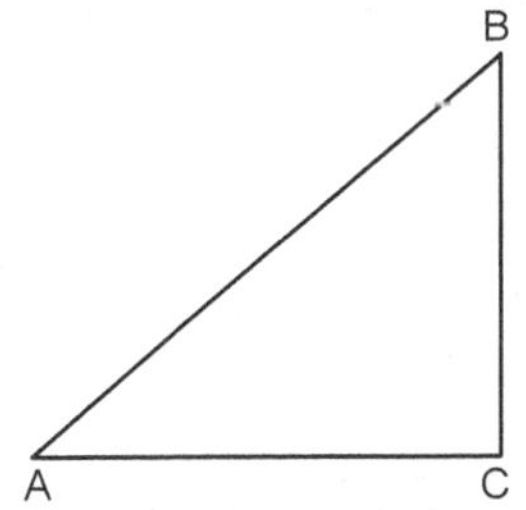

Now, $\tan A = \frac{1}{\sqrt{3}}$

$\Rightarrow \frac{BC}{AC} = \frac{1}{\sqrt{3}}$

$\Rightarrow BC = k \text{ and } AC = \sqrt{3}k$ ...(i)

In ΔABC by Pythagoras theorem, we have

$$AB^2 = AC^2 + BC^2$$

$\Rightarrow AB^2 = (\sqrt{3}k)^2 + k^2$

$\Rightarrow AB^2 = 3k^2 + k^2$

$\Rightarrow AB^2 = 4k^2$

$\Rightarrow AB = 2k$

To find the trigonometric ratios of ∠A, we have

Base = $AC = \sqrt{3}k$,

Perpendicular = $BC = k$

and Hypotenuse = $AB = 2k$

$\therefore \sin A = \frac{BC}{AB} = \frac{k}{2k} = \frac{1}{2}$

and $\cos A = \frac{AC}{AB} = \frac{\sqrt{3}k}{2k} = \frac{\sqrt{3}}{2}$

When we consider the trigonometric ratios of ∠B, we have

Base = $BC = k$,

Perpendicular $AC = \sqrt{3}k$

and Hypotenuse = $AB = 2k$

$\therefore \cos B = \frac{BC}{AB} = \frac{k}{2k} = \frac{1}{2}$

and $\sin B = \frac{AC}{AB} = \frac{\sqrt{3}k}{2k} = \frac{\sqrt{3}}{2}$

$\therefore \sin A \cos B + \cos A \sin B$

$$= \frac{1}{2} \times \frac{1}{2} + \frac{\sqrt{3}}{2} \times \frac{\sqrt{3}}{2}$$

$$= \frac{1}{4} + \frac{3}{4} = 1.$$ **Ans.**

Q. 3. If ∠B and ∠Q are acute angles such that sin *B* = sin *Q*, then prove that ∠B = ∠Q. [NCERT]

Solution : Consider two right angled triangles *ABC* and *PQR* such that sin *B* = sin *Q*.

We have, $\sin B = \frac{AC}{AB}$ and $\sin Q = \frac{PR}{PQ}$

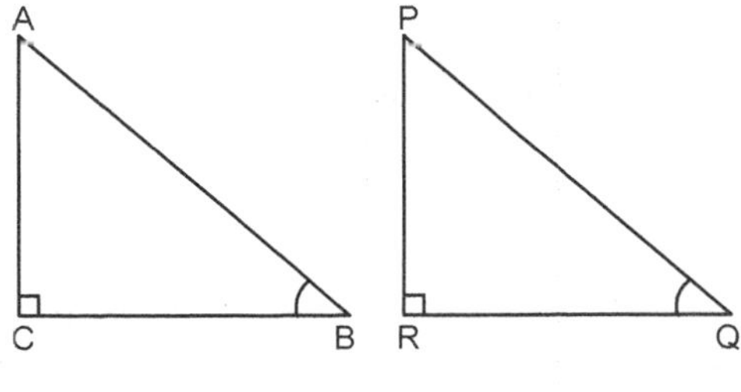

$\because \sin B = \sin Q$

★ are board exam questions from previous years

$$\Rightarrow \quad \frac{AC}{AB} = \frac{PR}{PQ}$$

$$\Rightarrow \quad \frac{AC}{PR} = \frac{AB}{PQ} = k \text{ (say)}, \quad ...(i)$$

$$\Rightarrow \quad AC = kPR \text{ and } AB = kPQ$$

In ΔABC and ΔPQR by Pythagoras theorem, we have

$$AB^2 = AC^2 + BC^2$$

and $$PQ^2 = PR^2 + QR^2$$

$$\Rightarrow \quad BC = \sqrt{AB^2 - AC^2}$$

and $$QR = \sqrt{PQ^2 - PR^2}$$

$$\Rightarrow \quad \frac{BC}{QR} = \frac{\sqrt{AB^2 - AC^2}}{\sqrt{PQ^2 - PR^2}}$$

$$= \frac{\sqrt{k^2 PQ^2 - k^2 PR^2}}{\sqrt{PQ^2 - PR^2}}$$

$$\Rightarrow \quad \frac{BC}{QR} = \frac{k\sqrt{PQ^2 - PR^2}}{\sqrt{PQ^2 - PR^2}} = k \quad ...(ii)$$

From (i) and (ii), we have

$$\frac{AC}{PR} = \frac{AB}{PQ} = \frac{BC}{QR}$$

$$\Rightarrow \quad \Delta ACB \sim \Delta PRQ$$

$$\Rightarrow \quad \angle B = \angle Q.$$ **Hence Proved.**

Q. 4. Find the value of :★

$$\left(\frac{3\tan 41^\circ}{\cot 49^\circ}\right)^2 - \left(\frac{\sin 35^\circ \sec 55^\circ}{\tan 10^\circ \tan 20^\circ \tan 60^\circ \tan 70^\circ \tan 80^\circ}\right).$$

Solution :

$$\left(\frac{3\tan 41^\circ}{\cot 49^\circ}\right)^2 - \left(\frac{\sin 35^\circ . \sec 55^\circ}{\tan 10^\circ . \tan 20^\circ . \tan 60^\circ . \tan 70^\circ . \tan 80^\circ}\right)$$

$$= \left(\frac{3\cot(90^\circ - 41^\circ)}{\cot 49^\circ}\right)^2 - \left(\frac{\sin 35^\circ .\text{cosec}\,(90^\circ - 55^\circ)}{\cot(90^\circ - 10^\circ).\cot(90^\circ - 20^\circ) .\tan 60^\circ \tan 10^\circ . \tan 80^\circ}\right)$$

$$\left(\because \cot(90^\circ - \theta) = \tan\theta, \quad \sec(90^\circ - \theta) = \text{cosec}\theta\right)$$

$$= \left(\frac{3\cot 49^\circ}{\cot 49^\circ}\right)^2 - \left(\frac{\sin 35^\circ .\text{cosec } 35^\circ}{\cot 80^\circ . \cot 70^\circ . \tan 60^\circ . \tan 70^\circ . \tan 80^\circ}\right)^2$$

$$= 9 - \left(\frac{\sin 35^\circ . \tan 80^\circ - \tan 70^\circ}{\cot 35^\circ . \cot 60^\circ . \tan 70^\circ . \tan 80^\circ}\right)^2$$

$$\left[\because \sin\theta = \frac{1}{\cos\theta}, \quad \tan\theta = \frac{1}{\cot\theta}\right]$$

$$= 9 - \left(\frac{1}{\tan 60^\circ}\right)^2 \quad \left(\because \tan 60^\circ = \frac{\sqrt{3}}{2}\right)$$

$$= 9 - \left(\frac{2}{\sqrt{3}}\right)^2$$

$$= 9 - \frac{4}{3}$$

$$= \frac{23}{3}$$ **Ans.**

Q. 5. Evaluate :★

$$\left(\frac{\mathbf{3\sin 43^\circ}}{\mathbf{\cos 47^\circ}}\right)^2 - \frac{\mathbf{\cos 37^\circ \text{ cosec } 53^\circ}}{\mathbf{\tan 5^\circ \tan 25^\circ \tan 45^\circ \tan 85^\circ}}.$$

Solution : Given,

$$\left(\frac{3\sin 43^\circ}{\cos 47^\circ}\right)^2 - \frac{\cos 37^\circ \text{ cosec } 53^\circ}{\tan 5^\circ \tan 25^\circ \tan 45^\circ \tan 65^\circ \tan 85^\circ}$$

$$= \left\{\frac{3\cos(90^\circ - 43^\circ)}{\cos 47^\circ}\right\}^2 - \frac{\sin(90^\circ - 37^\circ)\text{cosec } 53^\circ}{\cot(90^\circ - 5^\circ)\cot(90^\circ - 25^\circ)\tan 45^\circ \tan 65^\circ \tan 85^\circ}$$

$$= \left\{\frac{3\cos 47^\circ}{\cos 47^\circ}\right\}^2 - \frac{\sin 53^\circ \text{ cosec } 53^\circ}{\cot 85^\circ \cot 65^\circ \tan 45^\circ \tan 65^\circ \tan 85^\circ}$$

$$[\because \cos(90^\circ - \theta) = \sin\theta] \sin(90^\circ - \theta) = \cos\theta, \quad \cot(90^\circ - \theta) = \tan\theta]$$

$$= (3)^2 - \frac{\sin 53^\circ \times \frac{1}{\sin 53^\circ}}{\frac{1}{\tan 85^\circ} \times \frac{1}{\tan 65^\circ} \times \tan 45^\circ \times \tan 65^\circ \times \tan 85^\circ}$$

$$\left(\because \sin\theta = \frac{1}{\text{cosec}\,\theta}, \tan\theta = \frac{1}{\cot\theta}\right)$$

$$= 9 - \frac{1}{\tan 45^\circ}$$

$$= 9 - 1 \quad (\because \tan 45^\circ = 1)$$

$$= 8$$ **Ans.**

Q. 6. Prove the following trigonometric identity :

$\sin A\,(1 + \tan A) + \cos A\,(1 + \cot A) = \sec A + \text{cosec} A$.★

Solution : L.H.S. $= \sin A\,(1 + \tan A) + \cos A\,(1 + \cot A)$

$$= \sin A\left(1 + \frac{\sin A}{\cos A}\right) + \cos A\left(1 + \frac{\cos A}{\sin A}\right)$$

$$= \sin A\left(\frac{\cos A + \sin A}{\cos A}\right) + \cos A\left(\frac{\sin A + \cos A}{\sin A}\right)$$

★ are board exam questions from previous years

$$= (\sin A + \cos A)\left(\frac{\sin A}{\cos A} + \frac{\cos A}{\sin A}\right)$$

$$= (\sin A + \cos A)\left(\frac{\sin^2 A + \cos^2 A}{\cos A \sin A}\right)$$

$$= \frac{\sin A + \cos A}{\cos A \sin A} \quad [\because \sin^2 A + \cos^2 A = 1]$$

$$= \frac{\sin A}{\cos A \sin A} + \frac{\cos A}{\sin A \cos A}$$

$$= \frac{1}{\cos A} + \frac{1}{\sin A}$$

$= \sec A + \text{cosec} A$

= R.H.S. **Hence Proved.**

Q. 7. Prove that $(\sin\theta + \text{cosec}\,\theta)^2 + (\cos\theta + \sec\theta)^2 = 7 + \tan^2\theta + \cot^2\theta$.★

Solution : L.H.S.

$= (\sin\theta + \text{cosec}\,\theta)^2 + (\cos\theta + \sec\theta)^2$

$= \sin^2\theta + \text{cosec}^2\theta + 2.\sin\theta.\text{cosec}\,\theta + + \cos^2\theta + \sec^2\theta + 2\cos\theta\sec\theta.$

$[\because (a+b)^2 = a^2 + b^2 + 2ab]$

$= \sin^2\theta + \cos^2\theta + \text{cosec}^2\theta + \sec^2\theta$

$$+2\sin\theta\cdot\frac{1}{\sin\theta} + 2\cdot\cos\theta\cdot\frac{1}{\cos\theta}$$

$$[\because \text{cosec}\,\theta = \frac{1}{\sin\theta}$$

$$\sec\theta = \frac{1}{\cos\theta}]$$

$= 1 + 1 + \cot^2\theta + 1 + \tan^2\theta + 4$

$[\because \text{cosec}^2\theta = 1 + \cot^2\theta$

$\sec^2\theta = 1 + \tan^2\theta]$

$= 7 + \tan^2\theta + \cot^2\theta$ (R.H.S.) **Hence Proved.**

Q. 8. A rhombus of side 20 cm has two angles of 60° each. Find the length of the diagonal.

Solution : Let $ABCD$ be a rhombus of side 20 cm and $\angle BAD = \angle BCD = 60°$.

We know that the diagonals of a rhombus are perpendicular bisector of each other. Also, AC and BD are bisectors of $\angle BAD$ and $\angle ABC$ respectively.

So, ΔAOB is a right triangle such that $\angle BAO = 30°$, $\angle AOB = 90°$ and $AB = 20$ cm.

$$\therefore \quad \cos\angle BAO = \frac{OA}{AB}$$

$$\Rightarrow \quad \cos 30° = \frac{OA}{20}$$

$$\Rightarrow \quad \frac{\sqrt{3}}{2} = \frac{OA}{20}$$

$$\Rightarrow \quad OA = \frac{\sqrt{3}}{2} \times 20 = 10\sqrt{3}$$

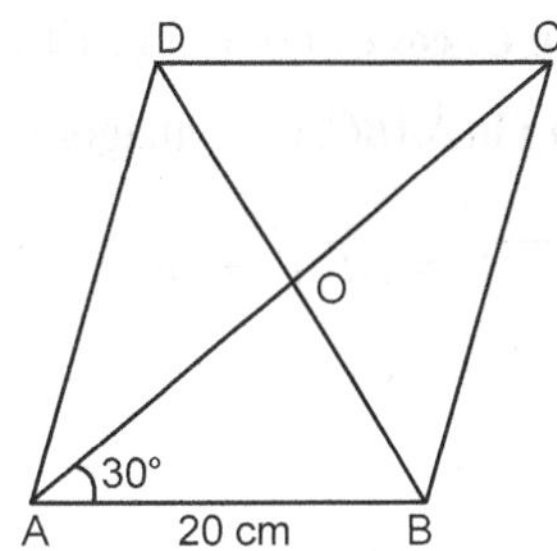

$$\text{Also,} \quad \sin\angle BAO = \frac{OB}{AB}$$

$$\Rightarrow \quad \sin 30° = \frac{OB}{20}$$

$$\Rightarrow \quad \frac{1}{2} = \frac{OB}{20}$$

$$\Rightarrow \quad BO = \frac{20}{2} = 10$$

$\therefore \quad AC = 2.\,AO = 2 \times 10\sqrt{3} = 20\sqrt{3}$ cm

$BD = 2.\,BO = 2 \times 10 = 20$ cm. **Ans.**

Q. 9. Prove that★

$$\frac{\tan^2 A}{\tan^2 A - 1} + \frac{\text{cosec}^2 A}{\sec^2 A - \text{cosec}^2 A} = \frac{1}{1 - 2\cos^2 A}.$$

Solution :

$$\text{L.H.S.} = \frac{\tan^2 A}{\tan^2 A - 1} + \frac{\text{cosec}^2 A}{\sec^2 A - \text{cosec}^2 A}$$

$$= \frac{\frac{\sin^2 A}{\cos^2 A}}{\frac{\sin^2 A \quad \cos^2 A}{\cos^2 A}} + \frac{\frac{1}{\sin^2 A}}{\frac{1}{\cos^2 A} - \frac{1}{\sin^2 A}}$$

$$= \frac{\sin^2 A}{\sin^2 A - \cos^2 A} + \frac{\frac{1}{\sin^2 A}}{\frac{\sin^2 A - \cos^2 A}{\sin^2 A \cdot \cos^2 A}}$$

$$= \frac{\sin^2 A}{\sin^2 A - \cos^2 A} + \frac{1}{\sin^2 A} \times \frac{\sin^2 A \cdot \cos^2 A}{\sin^2 A - \cos^2 A}$$

$$= \frac{\sin^2 A + \cos^2 A}{\sin^2 A - \cos^2 A} = \frac{1}{\sin^2 A - \cos^2 A}$$

$$= \frac{1}{1 - \cos^2 A - \cos^2 A} = \frac{1}{1 - 2\cos^2 A}$$

= R.H. S. $[\because \sin^2 A = 1 - \cos^2 A]$

Hence Proved.

★ are board exam questions from previous years

Q. 10. In the ΔABC (see figure), $\angle A$ = right angle, $AB = \sqrt{x}$ and $BC = \sqrt{x+5}$. Evaluate★

$$\sin C.\cos C.\tan C + \cos^2 C.\sin A$$

Solution : In ΔABC, by Pythagoras theorem,

$$\left(\sqrt{x+5}\right)^2 = \left(\sqrt{x}\right)^2 + AC^2$$

$$x+5 = x + AC^2$$

$$5 = AC^2$$

or $\quad AC = \sqrt{5}$

$$\sin C = \frac{\sqrt{x}}{\sqrt{x+5}};\ \cos C = \frac{\sqrt{5}}{\sqrt{x+5}};$$

$$\tan C = \frac{\sqrt{x}}{\sqrt{5}}$$

and $\quad \sin A = \sin 90° = 1$

Thus, $\sin C \cos C \tan C + \cos^2 C \sin A$

$$= \frac{\sqrt{x}}{\sqrt{x+5}}\frac{\sqrt{5}}{\sqrt{x+5}}\frac{\sqrt{x}}{\sqrt{5}} + \left(\frac{\sqrt{5}}{\sqrt{x+5}}\right)^2.1$$

$$= \frac{x}{x+5} + \frac{5}{x+5}$$

$$= \frac{x+5}{x+5} = 1$$ **Ans.**

Q. 11. If $\frac{\cos B}{\sin A} = n$ and $\frac{\cos B}{\cos A} = m$, then show that $(m^2+n^2)\cos^2 A = n^2$.★

Solution : Given, $\quad n = \frac{\cos B}{\sin A};\ m = \frac{\cos B}{\cos A}$

So, $\quad m^2 = \frac{\cos^2 B}{\cos^2 A};\ n^2 = \frac{\cos^2 B}{\sin^2 A}$

L.H.S. $= (m^2+n^2)\cos^2 A$

$$= \left(\frac{\cos^2 B}{\sin^2 A} + \frac{\cos^2 B}{\sin^2 A}\right)\cos^2 A$$

$$= \frac{\left(\sin^2 A\cos^2 B + \cos^2 A\cos^2 B\right)}{\cos^2 A\sin^2 A} \times \cos^2 A$$

$$= \frac{\cos^2 B\left(\sin^2 A + \cos^2 A\right)}{\sin^2 A}$$

$$= \frac{\cos^2 B}{\sin^2 A} \qquad [\because \sin^2 A + \cos^2 A = 1]$$

$= n^2 =$ R.H.S. **Hence Proved.**

Q. 12. $(\tan A + \mathbf{cosec}\ B)^2 - (\mathbf{cot}\ B - \mathbf{sec}\ A)^2$

$$= 2\tan A\cot B(\mathrm{cosec}\, A + \sec B).$$

★ are board exam questions from previous years

Solution : L.H.S. $= (\tan A + \mathrm{cosec}\, B)^2 - (\cot B - \sec A)^2$

$= \tan^2 A + \mathrm{cosec}^2 B + 2\tan A.\mathrm{cosec}\, B - (\cot^2 B + \sec^2 A - 2\cot B.\sec A)$

$= \tan^2 A + \mathrm{cosec}^2 B + 2\tan A.\mathrm{cosec}\, B - \cot^2 B - \sec^2 A + 2\cot B\sec A$

$= (\mathrm{cosec}^2 B - \cot^2 B) - (\sec^2 A - \tan^2 A) + 2\tan A.\mathrm{cosec}\, B + 2\cot B\sec A$

$= 1 - 1 + 2\tan A.\mathrm{cosec}\, B + 2\cot B.\sec A$

$= 2\tan A.\mathrm{cosec}\, B + 2\cot B.\sec A$

$$= 2\left(\frac{\sin A}{\cos A}\times\frac{1}{\sin B} + \frac{\cos B}{\sin B}\times\frac{1}{\cos A}\right)$$

$$= 2\left(\frac{\sin A + \cos B}{\cos A.\sin B}\right) \qquad ...(i)$$

R.H.S. $= 2\tan A.\cot B.(\mathrm{cosec}\, A + \sec B)$

$= 2\tan A.\cot B.\ \mathrm{cosec}\, A + 2\tan A.\ \cot B.\sec B$

$$= 2\frac{\sin A}{\cos A}\cdot\frac{\cos B}{\sin B}\cdot\frac{1}{\sin A} + \frac{2\sin A}{\cos A}\cdot\frac{\cos B}{\sin B}\cdot\frac{1}{\cos B}$$

$$= 2\left(\frac{\cos B}{\cos A.\sin B} + \frac{\sin A}{\cos A.\sin B}\right)$$

$$= 2\left(\frac{\cos B + \sin A}{\cos A.\sin A}\right) \qquad ...(ii)$$

From (i) and (ii),

L.H.S. = R.H.S. **Hence Proved.**

Q. 13. Prove that :★

$$\frac{\tan\theta}{1-\cot\theta} + \frac{\cot\theta}{1-\tan\theta} = 1 + \sec\ \mathrm{cosc}.$$

Solution : $\frac{\tan\theta}{1-\cot\theta} + \frac{\cot\theta}{1-\tan\theta} = 1 + \sec\theta.\ \mathrm{cosec}\,\theta$

$$\text{L.H.S.} = \frac{\tan\theta}{1-\cot\theta} + \frac{\cot\theta}{1-\tan\theta} = \frac{\frac{\sin\theta}{\cos\theta}}{1-\frac{\cos\theta}{\sin\theta}} + \frac{\frac{\cos\theta}{\sin\theta}}{1-\frac{\sin\theta}{\cos\theta}}$$

$$= \frac{\frac{\sin\theta}{\cos\theta}}{\frac{\sin\theta-\cos\theta}{\sin\theta}} + \frac{\frac{\cos\theta}{\sin\theta}}{\frac{\cos\theta-\sin\theta}{\cos\theta}}$$

$$= \frac{\sin^2\theta}{\cos\theta\,(\sin\theta-\cos\theta)} + \frac{\cos^2\theta}{\sin\theta(\cos\theta-\sin\theta)}$$

$$= \frac{\sin^2\theta}{\cos\theta\,(\sin\theta-\cos\theta)} - \frac{\cos^2\theta}{\sin\theta\,(\sin\theta-\cos\theta)}$$

$$= \frac{1}{\sin\theta - \cos\theta}\left[\frac{\sin^2\theta}{\cos\theta} - \frac{\cos^2\theta}{\sin\theta}\right]$$

$$= \frac{1}{\sin\theta - \cos\theta}\left[\frac{\sin^3\theta - \cos^3\theta}{\cos\theta.\sin\theta}\right]$$

$$= \frac{[\sin\theta - \cos\theta][\sin^2\theta + \cos^2\theta + \sin\theta.\cos\theta]}{(\sin\theta - \cos\theta)\cdot(\cos\theta \times \sin\theta)}$$

$$[\because a^3 - b^3 = (a-b)(a^2 + b^2 + ab)]$$

$$= \frac{\sin^2\theta + \cos^2\theta + \sin\theta\cdot\cos\theta}{(\cos\theta \times \sin\theta)}$$

$$[\because \sin^2\theta + \cos^2\theta = 1]$$

$$= \frac{1 + \sin\theta.\cos\theta}{\cos\theta \times \sin\theta} = \frac{1}{\cos\theta.\sin\theta} + \frac{\sin\theta\cdot\cos\theta}{\sin\theta\cdot\cos\theta}$$

$$= 1 + \sec\theta.\operatorname{cosec}\theta$$

$$\left(\because \frac{1}{\cos\theta} = \sec\theta, \frac{1}{\sin\theta} = \operatorname{cosec}\theta\right)$$

Hence Proved.

Q. 14. If $\tan(A+B) = \sqrt{3}$ and $\tan(A-B) = \frac{1}{\sqrt{3}}$, where $0 < A+B < 90^\circ$, $A > B$, find A and B. Also calculate $\tan A.\sin(A+B) + \cos.A\tan(A-B)$.★

Solution : Given, $\tan(A+B) = \sqrt{3}$, $\tan(A-B) = \frac{1}{\sqrt{3}}$

$\Rightarrow \quad \tan(A+B) = \tan 60^\circ$

$(A+B) = 60^\circ \quad$ ...(i)

and $\quad \tan(A-B) = \frac{1}{\sqrt{3}}$

$\Rightarrow \tan(A-B) = \tan 30^\circ$

$\Rightarrow \quad (A-B) = 30^\circ \quad$ ...(ii)

On adding equation (i) and (ii),

$A + B = 60^\circ$

$A - B = 30^\circ$

$+ \quad 2A = 90^\circ$

$\Rightarrow \quad A = \frac{90^\circ}{2} = 45^\circ$

From equation (i),

$A + B = 60^\circ$

$45^\circ + B = 60^\circ$

$B = 15^\circ$

$\therefore \quad A = 45^\circ, B = 15^\circ$

★ are board exam questions from previous years

Now, $\tan A.\sin(A+B) + \cos A.\tan(A-B)$

$$= \tan 45^\circ.\sin(60^\circ) + \cos 45^\circ.\tan(30^\circ)$$

$$= 1 \times \frac{\sqrt{3}}{2} + \frac{1}{\sqrt{2}} \times \frac{1}{\sqrt{3}}$$

$$= \frac{\sqrt{3}}{2} + \frac{1}{\sqrt{6}} \times \frac{\sqrt{6}}{\sqrt{6}} = \frac{\sqrt{3}}{2} + \frac{\sqrt{6}}{6}$$

$$= \frac{3\sqrt{3} + \sqrt{6}}{6}$$

Ans.

Q. 15. If $\frac{x}{a}\cos\theta + \frac{y}{b}\sin\theta = 1$ and $\frac{x}{a}\sin\theta - \frac{y}{b}\cos\theta = 1$, prove that $\frac{x^2}{a^2} + \frac{y^2}{b^2} = 2$.★

Solution : Given,

$$\frac{x}{a}\cos\theta + \frac{y}{b}\sin\theta = 1$$

Squaring both sides, we get

$$\frac{x^2}{a^2}\cos^2\theta + \frac{y^2}{b^2}\sin^2\theta + \frac{2xy}{ab}\cos\theta.\sin\theta = 1 \quad \text{...(i)}$$

and $\quad \frac{x}{a}\sin\theta - \frac{y}{b}\cos\theta = 1$

Squaring both sides, we get

$$\frac{x^2}{a^2}\sin^2\theta + \frac{y^2}{b^2}\cos^2\theta - \frac{2xy}{ab}\cos\theta\sin\theta = 1 \quad \text{...(ii)}$$

Adding equations (i) and (ii), we get

$$\frac{x^2}{a^2}(\cos^2\theta + \sin^2\theta) + \frac{y^2}{b^2}(\sin^2\theta + \cos^2\theta) = 2$$

$$\frac{x^2}{a^2} + \frac{y^2}{b^2} = 2 \quad [\because \sin^2\theta + \cos^2\theta = 1]$$

Hence Proved.

Q. 16. If $\sec\theta = x + \frac{1}{4x}$, prove that

$\sec\theta + \tan\theta = 2x$ or $\frac{1}{2x}$.

Solution : Given,

$$\sec\theta = x + \frac{1}{4x}$$

On squaring both sides, we get

$$\sec^2\theta = \left(x + \frac{1}{4x}\right)^2$$

We know that

$$\tan^2\theta = \sec^2\theta - 1$$

$$\Rightarrow \quad \tan^2\theta = \left(x + \frac{1}{4x}\right)^2 - 1$$

$$\Rightarrow \quad \tan^2\theta = x^2 + \frac{1}{16x^2} + \frac{1}{2} - 1$$

$$\Rightarrow \quad \tan^2\theta = x^2 + \frac{1}{16x^2} - \frac{1}{2}$$

$$\Rightarrow \quad \tan^2\theta = \left(x - \frac{1}{4x}\right)^2$$

$$\Rightarrow \quad \tan\theta = \pm\left(x - \frac{1}{4x}\right)$$

$$\Rightarrow \quad \tan\theta = x - \frac{1}{4x} \text{ or } \tan\theta = -\left(x - \frac{1}{4x}\right)$$

When $\tan\theta = \left(x - \frac{1}{4x}\right)$

Then, $\sec\theta + \tan\theta = x + \frac{1}{4x} + x - \frac{1}{4x} = 2x$

When $\tan\theta = -\left(x - \frac{1}{4x}\right)$

Then $\sec\theta + \tan\theta = \left(x + \frac{1}{4x}\right) - \left(x - \frac{1}{4x}\right) = \frac{2}{4x}$

$$= \frac{1}{2x}$$

Hence, $\sec\theta + \tan\theta = 2x$ or $\frac{1}{2x}$. **Hence Proved.**

Q. 17. If $\sec\theta + \tan\theta = p$, show that $\frac{p^2-1}{p^2+1} = \sin\theta$.★

Solution : Given,

$$\sec\theta + \tan\theta = p$$

Consider, $\frac{p^2-1}{p^2+1} = \frac{(\sec\theta + \tan\theta)^2 - 1}{(\sec\theta + \tan\theta)^2 + 1}$

$$= \frac{\sec^2\theta + \tan^2\theta + 2\sec\theta.\tan\theta - 1}{\sec^2\theta + \tan^2\theta + 2\sec\theta.\tan\theta + 1}$$

$$= \frac{(\sec^2\theta - 1) + \tan^2\theta + 2\sec\theta.\tan\theta}{\sec^2\theta + 2\sec\theta.\tan\theta + (1 + \tan^2\theta)}$$

$$= \frac{\tan^2\theta + \tan^2\theta + 2\tan\theta\sec\theta}{\sec^2\theta + 2\sec\theta\tan\theta + \sec^2\theta}$$

$$= \frac{2\tan^2\theta + 2\tan\theta.\sec\theta}{2\sec^2\theta + 2\sec\theta.\tan\theta}$$

$$= \frac{2\tan\theta(\tan\theta + \sec\theta)}{2\sec\theta(\sec\theta + \tan\theta)}$$

$$= \frac{\tan\theta}{\sec\theta}$$

$$= \frac{\sin\theta.\cos\theta}{\cos\theta}$$

$= \sin\theta$. **Hence Proved.**

Q. 18. If $\operatorname{cosec}\theta - \sin\theta = m$ and $\sec\theta - \cos\theta = n$, prove that $(m^2n)^{2/3} + (mn^2)^{2/3} = 1$.

Solution : Given,

$$\operatorname{cosec}\theta - \sin\theta = m$$

$$\Rightarrow \quad \frac{1}{\sin\theta} - \sin\theta = m$$

$$\Rightarrow \quad \frac{1 - \sin^2\theta}{\sin\theta} = m$$

$$\Rightarrow \quad \frac{\cos^2\theta}{\sin\theta} = m$$

Also, $\sec\theta - \cos\theta = n$

$$\Rightarrow \quad \frac{1}{\cos\theta} - \cos\theta = n$$

$$\Rightarrow \quad \frac{1 - \cos^2\theta}{\cos\theta} = n$$

$$\Rightarrow \quad \frac{\sin^2\theta}{\cos\theta} = n$$

Now, consider,

$$\text{L.H.S.} = (m^2n)^{2/3} + (mn^2)^{2/3}$$

$$= \left\{\left(\frac{\cos^2\theta}{\sin\theta}\right)^2 \frac{\sin^2\theta}{\cos\theta}\right\}^{2/3} + \left\{\frac{\cos^2\theta}{\sin\theta} \times \left(\frac{\sin^2\theta}{\cos\theta}\right)^2\right\}^{2/3}$$

$$= \left(\frac{\cos^4\theta.\sin^2\theta}{\sin^2\theta.\cos\theta}\right)^{2/3} + \left(\frac{\cos^2\theta}{\sin\theta} \times \frac{\sin^4\theta}{\cos^2\theta}\right)^{2/3}$$

$$= (\cos^3\theta)^{2/3} + (\sin^3\theta)^{2/3}$$

$$= \cos^2\theta + \sin^2\theta$$

$= 1 = \text{R.H.S.}$ **Hence Proved.**

Q. 19. Prove that :★

$$\frac{\sec A - 1}{\sec A + 1} = \left(\frac{\sin A}{1 + \cos A}\right)^2 = (\cot A - \operatorname{cosec} A)^2$$

Solution : L.H.S. $= \frac{\sec A - 1}{\sec A + 1}$

$$= \frac{\frac{1}{\cos A} - 1}{\frac{1}{\cos A} + 1}$$

$$= \frac{\frac{1 - \cos A}{\cos A}}{\frac{1 + \cos A}{\cos A}}$$

$$= \frac{1 - \cos A}{1 + \cos A}$$

★ are board exam questions from previous years

$$= \frac{(1-\cos A)(1+\cos A)}{(1+\cos A)(1+\cos A)}$$

$$= \frac{1-\cos^2 A}{(1+\cos A)^2}$$

$$= \frac{\sin^2 A}{(1+\cos A)^2} \; [\because 1-\cos^2 A = \sin^2 A]$$

$$= \left(\frac{\sin A}{1+\cos A}\right)^2$$

and $$\left(\frac{\sin A}{1+\cos A}\right)^2 = \left[\left(\frac{\sin A}{1+\cos A}\right)\times\frac{(1-\cos A)}{(1-\cos A)}\right]^2$$

$$= \left[\frac{\sin A(1-\cos A)}{1-\cos^2 A}\right]^2$$

$$= \left[\frac{\sin A(1-\cos A)}{\sin^2 A}\right]^2$$

$$= \left[\frac{1-\cos A}{\sin A}\right]^2$$

$= (\text{cosec } A - \cot A)^2$

$= (-1)^2 (\cot A - \text{cosec } A)^2$

$= (\cot A - \text{cosec } A)^2 = \text{R.H.S.}$ **Hence Proved.**

Chapter 11. Applications of Trigonometry

Q. 1. The angle of elevation of the top of a chimney from the foot of a tower is 60° and the angle of depression of the foot of the chimney from the top of the tower is 30°. If the height of the tower is 40 m, find the height of the chimney.★

Solution : Let AB be the tower of height 40 m and CD be the chimney of height h m.

Let the distance between the foot of the tower and the foot of the chimney be x m.

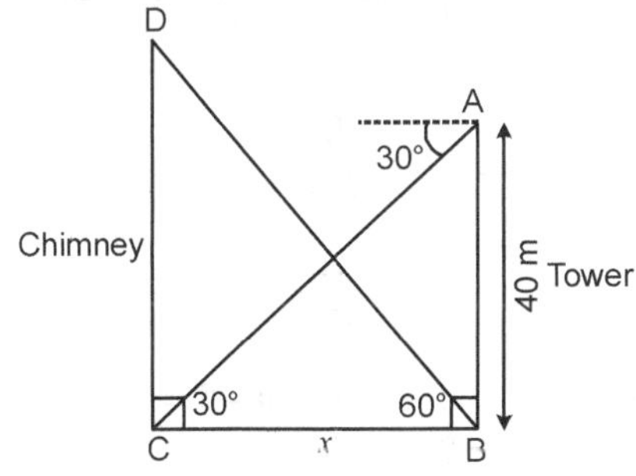

$\therefore$ As per the question,

In ΔABC, $\tan 30° = \frac{AB}{BC}$

$\Rightarrow \quad \frac{1}{\sqrt{3}} = \frac{40}{x}$

$\Rightarrow \quad x = 40\sqrt{3}$

In ΔBCD, $\tan 60° = \frac{CD}{BC}$

$\Rightarrow \quad \sqrt{3} = \frac{h}{x}$

$\Rightarrow \quad h = \sqrt{3}x = \sqrt{3}\times 40\sqrt{3}$

$= 120 \text{ m}$

Thus, the height of the chimney is 120 m. **Ans.**

Q. 2. The angles of elevation and depression of the top and the bottom of a tower from the top of a building, 60 m high are 30° and 60° respectively. Find the difference between the height of the building and the tower and the distance between them.★★

Solution : Let AD be the building of height 60m, BE be the tower and x be the distance between them.

Thus, Base $= x$ m

Let $EC = h$ m

So, height of tower $= (60 + h)$ m

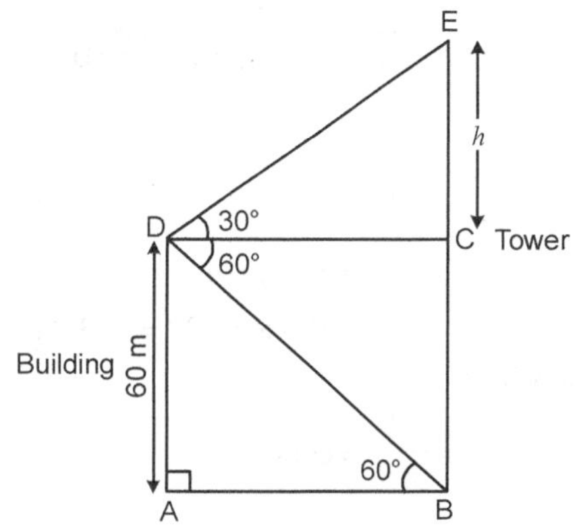

Now, in ΔEDC

$$\tan 30° = \frac{EC}{DC}$$

$$\frac{1}{\sqrt{3}} = \frac{h}{x}$$

$\Rightarrow \quad x = h\sqrt{3}$ ...(i)

In ΔADB

$$\tan 60° = \frac{AD}{AB}$$

$$\sqrt{3} = \frac{60}{x}$$

$\Rightarrow \quad x = \frac{60}{\sqrt{3}} = 20\sqrt{3} \text{ m}$ ...(ii)

Fro equation (a) and (ii)

$$20\sqrt{3} = h\sqrt{3}$$

★ are board exam questions from previous years

★★ are frequently asked board exam questions

$\Rightarrow \qquad h = 20$ m

So, the difference between the height of the tower and the building is 20 m and the distance between the tower and the building is $20\sqrt{3}$ m. **Ans.**

Q. 3. An aeroplane is flying at a height of 300 m above the ground. Flying at this height, the angles of depression from the aeroplane of two points on both banks of a river in opposite directions are 45° and 60° respectively. Find the width of the river.★

[Use $\sqrt{3} = 1.732$]

Solution : Let aeroplane be at A, 300 m high from a river, C and D be the opposite banks of river.

In ΔABC,

$$\frac{BC}{AB} = \cot 60^\circ$$

$$\Rightarrow \quad \frac{x}{300} = \frac{1}{\sqrt{3}} \Rightarrow x = \frac{300}{\sqrt{3}} \times \frac{\sqrt{3}}{\sqrt{3}}$$

$$= 100\sqrt{3} \text{ m}$$

$$= 100 \times 1.732 = 173.2 \text{ m}$$

In ΔABD,

$$\Rightarrow \quad \frac{BD}{AB} = \cot 45^\circ$$

$$\Rightarrow \quad \frac{y}{300} = 1 \Rightarrow y = 300$$

Width of river $= x + y$

$= 173.2 + 300$

$= 473.2$ m **Ans.**

Q. 4. The shadow of a tower standing on a level ground is found to be 30 m longer when the sun's altitude is 30° than when altitude is 60°. Find the height of the tower.★

Solution : Let the height of the tower be H and the length of the shadow at 60° be L.

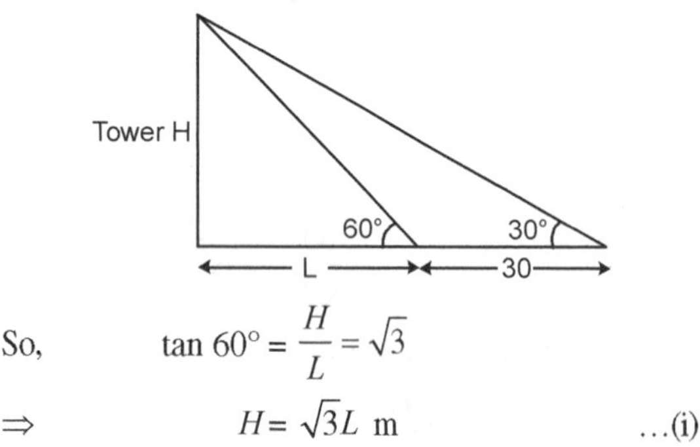

So, $\qquad \tan 60^\circ = \frac{H}{L} = \sqrt{3}$

$\Rightarrow \qquad H = \sqrt{3}L$ m $\qquad$ ...(i)

and $\qquad \tan 30^\circ = \frac{H}{L+30} = \frac{1}{\sqrt{3}}$

$\Rightarrow \qquad H = \frac{L+30}{\sqrt{3}}$ m $\qquad$ ...(ii)

From equations (i) and (ii)

$$\frac{L+30}{\sqrt{3}} = \sqrt{3}L$$

$\Rightarrow \qquad L + 30 = 3L$

$\Rightarrow \qquad 2L = 30$

$\Rightarrow \qquad L = 15$ m

Thus, $\qquad H = \sqrt{3}(15)$ m $= 15(1.732)$ m

$= 25.98 \text{ m} \approx 26$ m. **Ans.**

Q. 5. A moving boat is observed from the top of a 150 m high cliff moving away from the cliff. The angle of depression of the boat changes from 60° to 45° in 2 minutes. Find the speed of the boat in m/h.★

Solution : Let AB be the cliff and C, D be the two position of the boat during a period of 2 minutes.

So, $\qquad AB = 150$ m

In ΔABC,

$$\tan 60^\circ = \frac{AB}{BC}$$

$$\sqrt{3} = \frac{150}{BC}$$

$$\Rightarrow \quad BC = \frac{150}{\sqrt{3}} \text{ m}$$

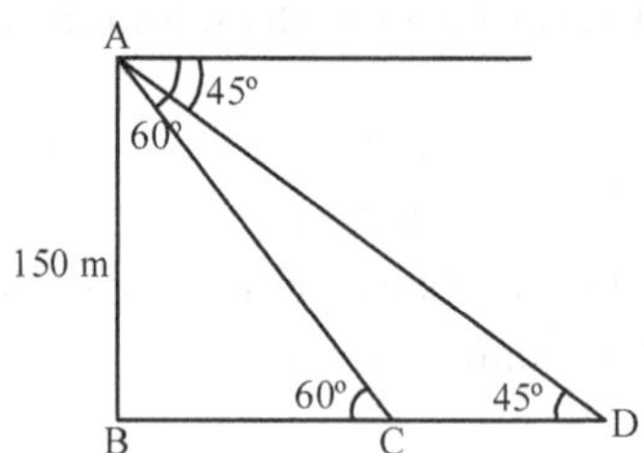

In ΔABD,

$$\tan 45^\circ = \frac{AB}{BD}$$

$$1 = \frac{AB}{BD}$$

$\Rightarrow \qquad AB = BD \; [\tan 45^\circ = 1]$

$\Rightarrow \qquad BD = 150$ m

Distance covered in 2 min $= BD - BC$

$$= 150 - \frac{150}{\sqrt{3}} = \frac{150\sqrt{3} - 150}{\sqrt{3}}$$

Distance covered in 1 hour $= \frac{150(\sqrt{3}-1)}{\sqrt{3} \times 2} \times 60$ m

★ are board exam questions from previous years

$$\text{Speed} = \frac{4500(\sqrt{3}-1)}{\sqrt{3}}$$

$$= 4500 - 1500\sqrt{3}$$
$$= 4500 - 2598$$
$$= 1902 \text{ m/hr}$$

Hence, the speed of boat is 1902 m/hr. **Ans.**

Q. 6. A man in a boat rowing away from a light house 100 m high takes 2 minutes to change the angle of elevation of the top of the light house from 60° to 30°. Find the speed of the boat in metres per minute. ★

[Use $\sqrt{3} = 1{\cdot}732$]

Solution : Let AB be the light house C and D be the two positions of the boat, such that,

CD = x m and BC = y m

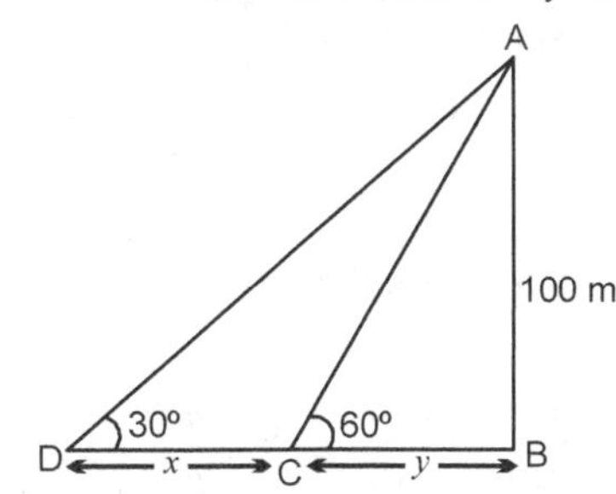

Now, In ΔABC

$$\tan 60° = \frac{AB}{BC}$$

$$\sqrt{3} = \frac{100}{y}$$

$$y = \frac{100}{\sqrt{3}} \quad ...(i)$$

In ΔABD

$$\tan 30° = \frac{AB}{BD}$$

$$\frac{1}{\sqrt{3}} = \frac{100}{x+y}$$

$$\Rightarrow \quad x + y = 100\sqrt{3}$$

or $$y = 100\sqrt{3} - x \quad ...(ii)$$

From equatin (i) and (ii)

$$\frac{100}{\sqrt{3}} = 100\sqrt{3} - x$$

$$\Rightarrow \quad x = 100\sqrt{3} - \frac{100}{\sqrt{3}}$$

$$= 100\left(\sqrt{3} - \frac{1}{\sqrt{3}}\right)$$

$$= 100 \times \frac{2}{\sqrt{3}} = \frac{200}{\sqrt{3}}$$

$$= 115{\cdot}48 \text{ metres}$$

∵ Time taken to cover 115·48 m = 2 min

$$\therefore \quad \text{Speed of boat} = \frac{115{\cdot}48}{2} = 57{\cdot}74 \text{ m/min}$$

Hence, speed of boat = 57·74 m/min **Ans.**

Q. 7. The angle of elevation of the top of a vertical tower from a point on the ground is 60°. From another point 10 m vertically above the first, its angle of elevation is 30°. Find the height of the tower.★

Solution : Let AD be the tower and B, C be the two observation points such that C is 10 m above B.

So, $BC = 10$ m

Let $AD = H$ m and $AB = L$ m.

In ΔABD,

$$\tan 60° = \frac{AD}{AB}$$

$$\sqrt{3} = \frac{H}{L}$$

$$\Rightarrow \quad L = \frac{H}{\sqrt{3}} \quad ...(i)$$

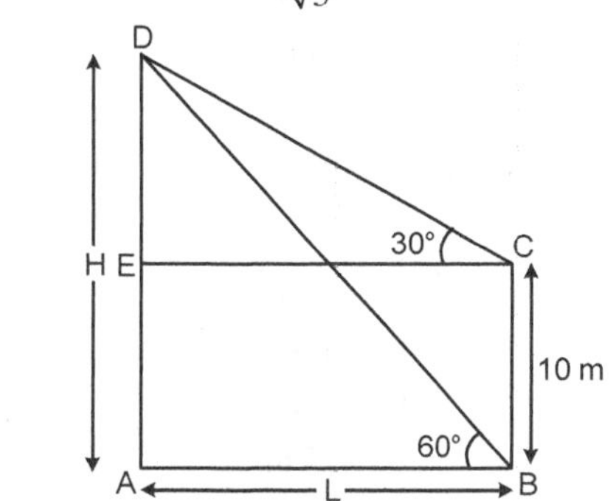

In ΔDEC,

$$\tan 30° = \frac{ED}{EC}$$

and $$\frac{1}{\sqrt{3}} = \frac{H-10}{L}$$

$$\Rightarrow \quad L = \sqrt{3}(H-10) \quad ...(ii)$$

From equations (i) and (ii),

$$\frac{H}{\sqrt{3}} = \sqrt{3}(H-10)$$

$$\Rightarrow \quad H = 3(H-10)$$
$$\Rightarrow \quad H = 3H - 30$$
$$\Rightarrow \quad 2H = 30$$
$$\Rightarrow \quad H = 15 \text{ m}$$

Thus, the height of the tower is 15 m. **Ans.**

Q. 8. Two poles of equal heights are standing opposite to each other on either side of the road, which is 100 m wide. From a point between them on the road, the angles of elevation of the top of the poles are 60° and 30° respectively. Find the height of the poles.

★ are board exam questions from previous years

Solution : Let the height of each of the poles be Hm.

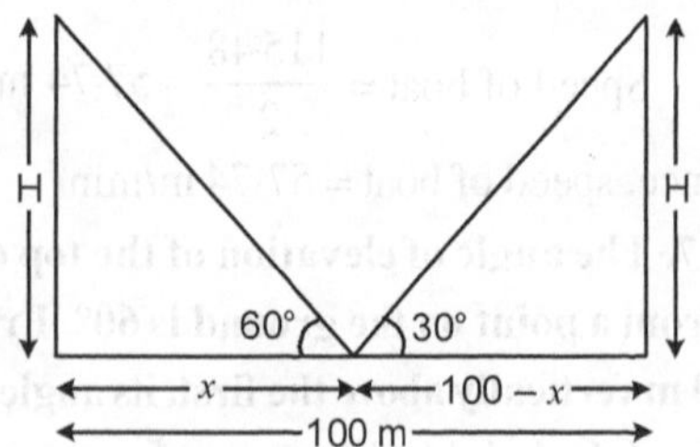

Let the distance of the point from one pole be x m and from the second pole be $(100 - x)$ m.

Now, $\tan 60° = \dfrac{H}{x}$

$$\sqrt{3} = \frac{H}{x}$$

$\Rightarrow$ $H = \sqrt{3}x$ ...(i)

Also, $\tan 30° = \dfrac{H}{100 - x}$

$$\frac{1}{\sqrt{3}} = \frac{H}{100 - x}$$

$\Rightarrow$ $H = \dfrac{100 - x}{\sqrt{3}}$ ...(ii)

From equations (i) and (ii),

$$\sqrt{3}x = \frac{100 - x}{\sqrt{3}}$$

$\Rightarrow$ $3x = 100 - x$

$\Rightarrow$ $4x = 100$

$\Rightarrow$ $x = 25$ m

Thus, $H = \sqrt{3}(25)$

$= 25(1.732) = 43.3$ m. **Ans.**

Q. 9. A ladder rests against a wall at an angle having measure α with the ground. Its foot is pulled away from the wall by *a* m keeping ladder on the ground. By doing this, its upper end on the wall slides down by *b* m. Now the ladder makes an angle of measure β with the ground.

Prove that $\dfrac{a}{b} = \dfrac{\cos\beta - \cos\alpha}{\sin\alpha - \sin\beta}$.

Solution : Let AB be the wall and AC be the ladder.

$\therefore$ $AC = DF = x$ m (say)

and $AD = b$ m, $CF = a$ m $BD = y$ and $BC = z$ m (say)

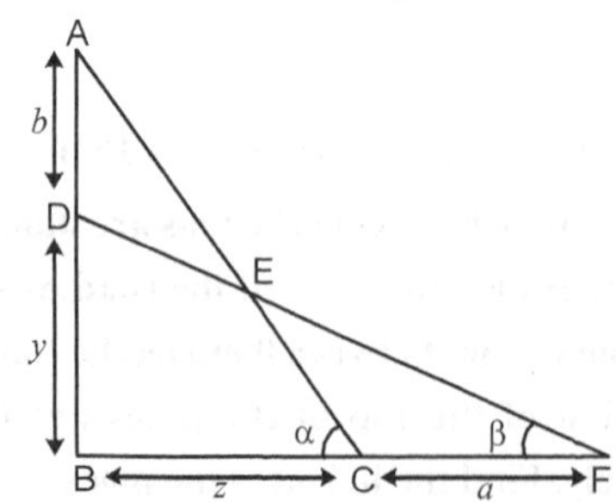

In ΔABC,

$$\cos\alpha = \frac{BC}{AC} = \frac{z}{x}$$

and $\sin\alpha = \dfrac{AB}{AC} = \dfrac{y+b}{x}$

In ΔDBF,

$$\cos\beta = \frac{BF}{DF} = \frac{z+a}{x}$$

and $\sin\beta = \dfrac{BD}{DF} = \dfrac{y}{x}$

So, $\cos\alpha - \cos\beta = \dfrac{z}{x} - \dfrac{z+a}{x}$

$$= \frac{z - z - a}{x} = -\frac{a}{x}$$

$\therefore$ $\cos\beta - \cos\alpha = \dfrac{a}{x}$

and $\sin\alpha - \sin\beta = \dfrac{y+b}{x} - \dfrac{y}{x}$

$$= \frac{y + b - y}{x}$$

$\Rightarrow$ $\sin\alpha - \sin\beta = \dfrac{b}{x}$

$$\frac{\cos\beta - \cos\alpha}{\sin\alpha - \sin\beta} = \frac{\frac{a}{x}}{\frac{b}{x}} = \frac{a}{b}$$

$\therefore$ $\dfrac{\cos\beta - \cos\alpha}{\sin\alpha - \sin\beta} = \dfrac{a}{b}$ **Hence Proved.**

Q. 10. From the top of a tower, the angles of depression of two objects on either sides of the tower are found to be α and β(α > β). If the distance between them is *d* m, show that the height of the tower is given by $h = \dfrac{d\tan\alpha\tan\beta}{\tan\alpha + \tan\beta}$.

Solution : Let AD be the tower and B, C be the tow objects on either sides of AD.

Given, $BC = d = BD + DC$ and $AD = h$

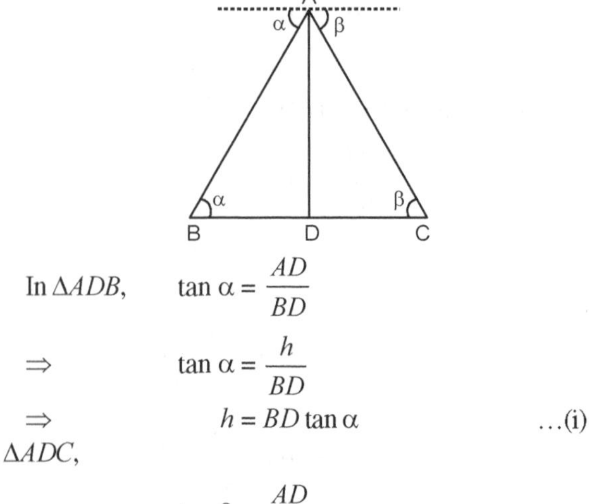

In ΔADB, $\tan\alpha = \dfrac{AD}{BD}$

$\Rightarrow$ $\tan\alpha = \dfrac{h}{BD}$

$\Rightarrow$ $h = BD\tan\alpha$ ...(i)

In ΔADC,

$$\tan\beta = \frac{AD}{DC}$$

$$\Rightarrow \quad \tan\beta = \frac{h}{DC}$$

$$\Rightarrow \quad h = DC\tan\beta \quad \ldots\text{(ii)}$$

Thus, from equations (i) and (ii), we get

$$BD\tan\alpha = DC\tan\beta$$

$$\Rightarrow \quad (d - DC)\tan\alpha = DC\tan\beta$$

$$\Rightarrow \quad d\tan\alpha = DC(\tan\alpha + \tan\beta)$$

$$\Rightarrow \quad DC = \frac{d\tan\alpha}{\tan\alpha + \tan\beta}$$

Substituting the value of DC in (ii), we get

$$h = \frac{d\tan\alpha}{\tan\alpha + \tan\beta}\tan\beta$$

$$\Rightarrow \quad h = \frac{d\tan\alpha\tan\beta}{\tan\alpha + \tan\beta}.$$

Hence Proved.

Q. 11. From a point P on the ground, the angle of elevation of the top of a tower is 30° and that of the top of a flag staff fixed on the top of the tower is 60°. If the length of the flag staff is 5 m, find the height of the tower.★

Solution : Let CB be the tower of x m and AC be the flag staff of 5 m.

Then, in ΔCPB

$$\tan 30^\circ = \frac{x}{PB}$$

$$PB = \frac{x}{\tan 30^\circ} = \sqrt{3}x \quad \ldots\text{(i)}$$

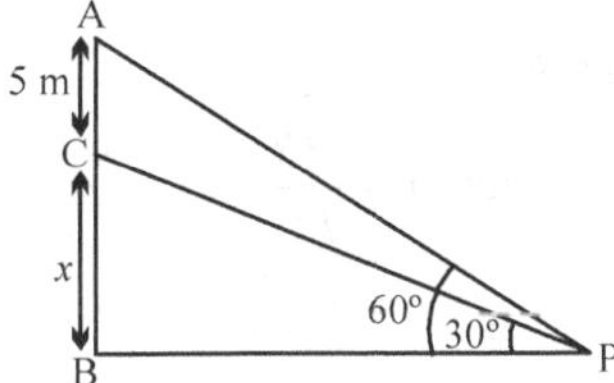

In ΔAPB

$$\tan 60^\circ = \frac{5+x}{PB}$$

$$PB = \frac{5+x}{\sqrt{3}} \quad \ldots\text{(ii)}$$

From equations (i) and (ii)

$$\sqrt{3}x = \frac{x+5}{\sqrt{3}}$$

$$\Rightarrow \quad 3x = x + 5$$

$$\Rightarrow \quad 3x - x = 5$$

$$\Rightarrow \quad 2x = 5$$

$$\Rightarrow \quad x = \frac{5}{2} = 2.5$$

∴ Height of the tower is 2.5 m. **Ans.**

★ are board exam questions from previous years

Q. 12. A flagstaff of height h is surmounted on a pole. From a point at the base of the pole and some distance away from it, the angle of elevation of the top of the pole is α and that of the flagstaff is β. If the height of the pole is l m and the distance of the point from the base is d, prove that $h = d(\tan\beta - \tan\alpha)$.

Solution : Let BD be the pole and AD be the flagstaff.

Let $\quad BD = l$

Now, in ΔBDC,

$$\tan\alpha = \frac{BD}{BC}$$

$$\Rightarrow \quad \tan\alpha = \frac{l}{d}$$

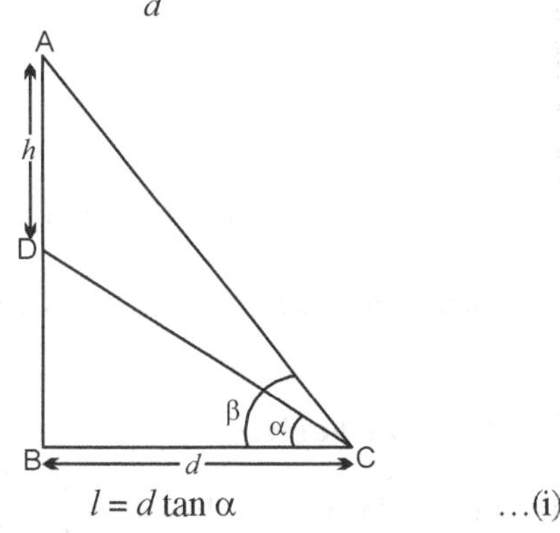

$$\Rightarrow \quad l = d\tan\alpha \quad \ldots\text{(i)}$$

and in ΔABC,

$$\tan\beta = \frac{AB}{BC}$$

$$\tan\beta = \frac{l+h}{d}$$

$$\Rightarrow \quad l + h = d\tan\beta \quad \ldots\text{(ii)}$$

Substituting the value of l from eq. (i) in eq. (ii), we have

$$d\tan\alpha + h = d\tan\beta$$

$$\Rightarrow \quad h = d(\tan\beta - \tan\alpha)$$

Hence Proved.

Q. 13. Amit, standing on a horizontal plane, finds a bird flying at a distance of 200 m from him at an elevation of 30°. Deepak standing on the roof of a 50 m high building, finds the angle of elevation of the same bird to be 45°. Amit and Deepak are on opposite sides of the bird. Find the distance of the bird from Deepak.★

Solution : Let Amit be at C point and bird is at A point. Such that ∠ACB = 30°. AB is the height of bird from point B on ground and deepak is at D point, DE is the building of height 50m.

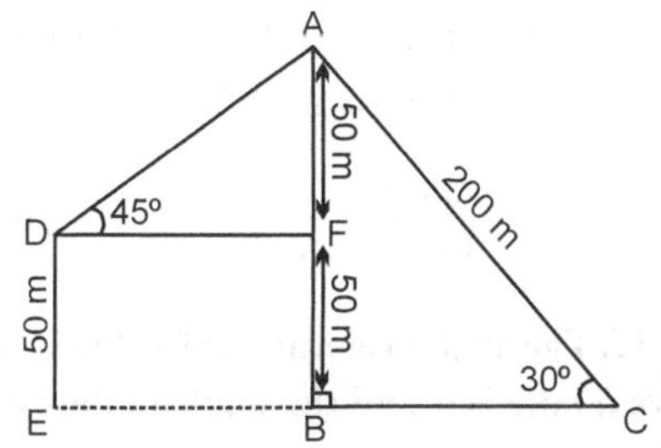

Now, In right triangle ABC, we have

$$\sin 30° = \frac{P}{H} = \frac{AB}{AC}$$

$$AB = 100 \text{ m}$$

In right ΔAFD, we have

$$\frac{1}{2} = \frac{AB}{200}$$

$(\because AB = AF + BF$

$100 = AF + 50$

$AF = 50)$

$$\frac{1}{\sqrt{2}} = \frac{50}{AD}$$

$$AD = 50\sqrt{2} \text{ m}$$

Hence, the distance of bird from Deepak is $50\sqrt{2}$ m.

Ans.

Q. 14. In the given figure, ΔABC is right-angled at D and $BD : DC$ is 2 : 3. Show that $\frac{\tan\theta}{\tan\phi} = \frac{2}{3}$.

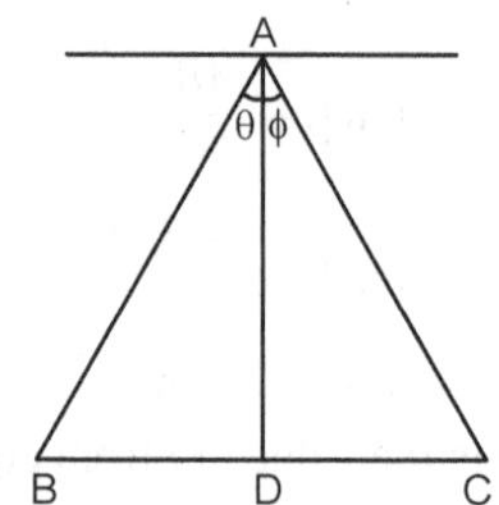

Solution : Given, $BD : DC = 2 : 3$

Thus, in ΔABD,

$$\tan\theta = \frac{BD}{AD}$$

$$\Rightarrow \quad BD = AD\tan\theta \quad \ldots\text{(i)}$$

and in ΔADC,

$$\tan\phi = \frac{DC}{AD}$$

$$\Rightarrow \quad DC = AD\tan\phi \quad \ldots\text{(ii)}$$

Now, $\quad BD : DC = 2 : 3$

$$\Rightarrow \quad \frac{BD}{DC} = \frac{2}{3}$$

$$\Rightarrow \quad 3BD = 2DC$$

$$\Rightarrow \quad 3AD\tan\theta = 2AD\tan\phi \quad \text{[Using (i) and (ii)]}$$

$$\Rightarrow \quad 3\tan\theta = 2\tan\phi$$

$$\Rightarrow \quad \frac{\tan\theta}{\tan\phi} = \frac{2}{3}.$$

Hence Proved.

Q. 15. The angle of elevation of a cloud from a point h m above a lake is α and the angle of depression of the reflection of the cloud in the lake is β. Prove that the distance of the cloud from the point of observation is $\frac{2h\sec\alpha}{\tan\beta - \tan\alpha}$ m.★

Solution : Let DE be the surface of the lake.

Thus, $\quad DE = BC = x$ m (say)

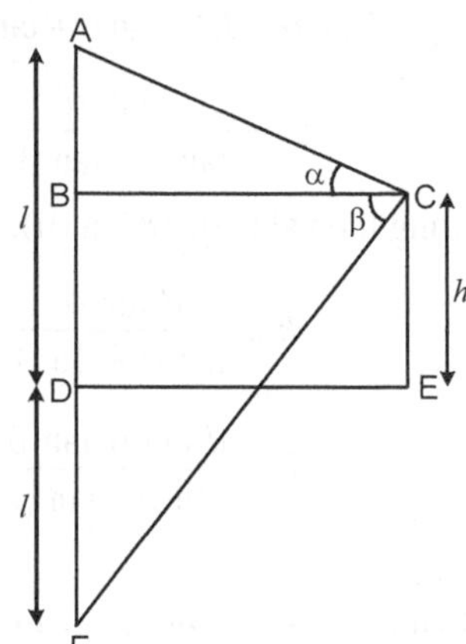

Also let AD be the height (l) of the cloud from the surface of the lake, DF be the length of the image of the cloud below the surface of the lake and AC be the distance of the cloud from the point of observation C.

Thus $\quad AD = DF$

[From the laws of plane mirrors, an image is as far behind a plane mirror as the object is in front of it]

So $\quad AD = DF = l$ m

Given, $\quad \angle BCF = \beta$ and $\angle BCA = \alpha$

Now, in ΔABC,

$$\tan\alpha = \frac{AB}{BC}$$

$$\Rightarrow \quad \tan\alpha = \frac{l-h}{x}$$

$$\Rightarrow \quad x = \frac{l-h}{\tan\alpha} \text{ m} \quad \ldots\text{(i)}$$

Similarly, in ΔBCF,

$$\tan\beta = \frac{FB}{BC}$$

$$\Rightarrow \quad \tan\beta = \frac{l+h}{x}$$

$$\Rightarrow \quad x = \frac{l+h}{\tan\beta} \text{ m} \quad \ldots\text{(ii)}$$

From equations (i) and (ii), we get

$$\frac{l-h}{\tan\alpha} = \frac{l+h}{\tan\beta}$$

$$\Rightarrow \quad (l-h)\tan\beta = (l+h)\tan\alpha$$

$$\Rightarrow l\tan\beta - h\tan\beta = l\tan\alpha + h\tan\alpha$$

$$\Rightarrow l\tan\beta - l\tan\alpha = h\tan\beta + h\tan\alpha$$

★ are board exam questions from previous years

$$\Rightarrow l(\tan\beta - \tan\alpha) = h(\tan\beta + \tan\alpha)$$

$$\Rightarrow \quad l = \frac{h(\tan\beta + \tan\alpha)}{\tan\beta - \tan\alpha}$$

Now,
$$x = \frac{\dfrac{h(\tan\beta + \tan\alpha)}{\tan\beta - \tan\alpha} - h}{\tan\alpha} \quad \text{[From eq. (i)]}$$

$$\Rightarrow \quad x = \frac{h(\tan\beta + \tan\alpha) - h(\tan\beta - \tan\alpha)}{\tan\alpha(\tan\beta - \tan\alpha)}$$

$$\Rightarrow \quad x = h\left[\frac{\tan\beta + \tan\alpha - \tan\beta + \tan\alpha}{\tan\alpha(\tan\beta - \tan\alpha)}\right]$$

$$\Rightarrow \quad x = h\left[\frac{2\tan\alpha}{\tan\alpha(\tan\beta - \tan\alpha)}\right]$$

$$\Rightarrow \quad x = \frac{2h}{\tan\beta - \tan\alpha}$$

Thus, in ΔABC,

$$\sec\alpha = \frac{AC}{BC}$$

$$\Rightarrow \quad \sec\alpha = \frac{AC}{x}$$

$$\Rightarrow \quad AC = x\sec\alpha$$

$$\Rightarrow \quad AC = \frac{2h\sec\alpha}{\tan\beta - \tan\alpha}.$$

Hence Proved.

Q. 16. From an aeroplane, vertically above a straight horizontal road, the angles of depression of two consecutive milestones on opposite sides of the aeroplane are observed to be α and β. Show that the height in miles of the aeroplane above the road is given by $\dfrac{\tan\beta\tan\alpha}{\tan\beta + \tan\alpha}$.★

Solution : Let A be the position of aeroplane and B, D be the two consecutive milestone.

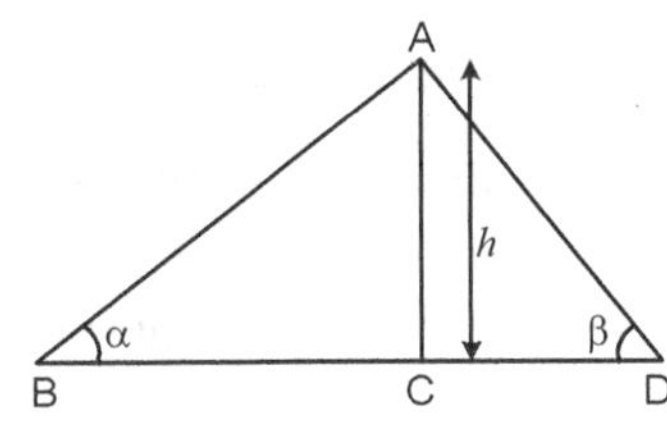

Let $AC = h$ m

Given, $BD = BC + CD = 1$ mile

$\Rightarrow$ $BC = 1 - CD$

Now, in ΔABC

$$\tan\alpha = \frac{AC}{BC}$$

$$\Rightarrow \quad AC = BC\tan\alpha \quad \text{...(i)}$$

and in ΔADC,

$$\tan\beta = \frac{AC}{CD}$$

$$\Rightarrow \quad AC = CD\tan\beta \quad \text{...(ii)}$$

Thus, from equations (i) and (ii), we have

$$BC\tan\alpha = CD\tan\beta$$

$$\Rightarrow \quad 1 - CD = \frac{CD\tan\beta}{\tan\alpha}$$

$$\Rightarrow \quad \frac{CD\tan\beta}{\tan\alpha} + CD = 1$$

$$\Rightarrow CD\left(\frac{\tan\alpha + \tan\beta}{\tan\alpha}\right) = 1$$

$$\Rightarrow \quad CD = \frac{\tan\alpha}{\tan\alpha + \tan\beta}$$

Hence, from equation (ii), we get

$$AC = CD\tan\beta$$

$$\Rightarrow \quad AC = \frac{\tan\alpha}{\tan\alpha + \tan\beta}\tan\beta$$

$$\Rightarrow \quad AC = \frac{\tan\beta\tan\alpha}{\tan\beta + \tan\alpha} \text{ miles.}$$

Hence Proved.

Q. 17. An observer finds the angle of elevation of the top of the tower from a certain point on the ground as 30°. If the observer moves 20 m towards the base of the tower, the angle of elevation of the top increases by 15°, find the height of the tower.★

Solution : Let AB be tower of height h

In ΔABC,

$$\frac{AB}{BC} = \tan 45^\circ$$

$$\Rightarrow \quad \frac{h}{BC} = 1$$

$$\Rightarrow \quad h = BC$$

In ΔABD,

$$\Rightarrow \quad \frac{AB}{BD} = \tan 30^\circ$$

$$\Rightarrow \quad \frac{h}{BD} = \frac{1}{\sqrt{3}}$$

$$\Rightarrow \quad BD = \sqrt{3}h$$

Now, $CD = BD - BC$

$$\Rightarrow \quad 20 = \sqrt{3}h - h$$

$$\Rightarrow \quad 20 = h(\sqrt{3} - 1)$$

$$\Rightarrow h = \frac{20}{\sqrt{3}-1} \times \frac{\sqrt{3}+1}{\sqrt{3}+1} = \frac{20(\sqrt{3}+1)}{(\sqrt{3})^2 - (1)^2}$$

$$= \frac{20(\sqrt{3}+1)}{2} = 10(\sqrt{3}+1)\text{ m}$$

Ans.

★ are board exam questions from previous years

Q. 18. The angles of elevation of the top of a tower from two points on the ground, at distance *a* m and *b* m from the base of the tower and in the same straight line, are complementary. Prove that the height of the tower is $\sqrt{ab}$ m.★

Solution : Let AB be the tower and C, D be the two observation points on the ground.

Since the two angles are complementary, so they sum up to 90°.

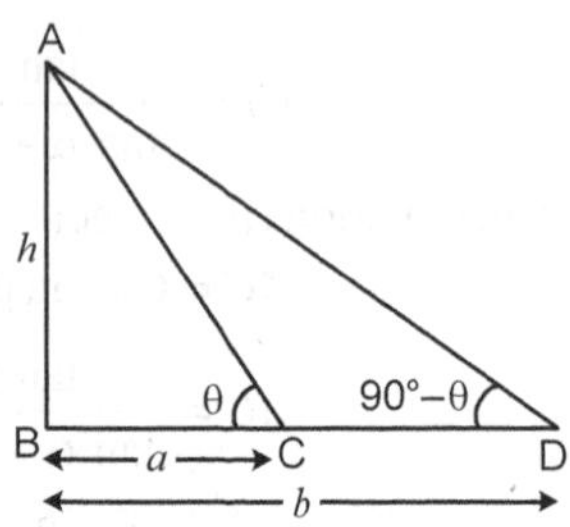

Thus, let one angle be θ. So, the other angle is 90° – θ.

Also, let $AB = h$

So, in ΔABC,

$$\tan\theta = \frac{AB}{BC} = \frac{h}{a}$$

$$\Rightarrow \quad h = a\tan\theta \qquad \ldots\text{(i)}$$

and in ΔABD,

$$\tan(90^\circ - \theta) = \frac{AB}{BD} = \frac{h}{b}$$

$$\Rightarrow \quad h = b\cot\theta \qquad \ldots\text{(ii)}$$

Multiplying equation (i) and (ii), we get

$$h^2 = ab\tan\theta\cot\theta$$

$$\Rightarrow \quad h^2 = ab$$

$$\Rightarrow \quad h = \sqrt{ab}.$$

Hence Proved.

Q. 19. An aeroplane when flying at a height of *H* m from the ground passes vertically above another aeroplane at an instant when the angles of the elevation of the two planes from the point on the ground are θ and ϕ respectively. Prove that the vertical distance between the two planes is $\dfrac{H(\tan\theta - \tan\phi)}{\tan\theta}$.

Solution : Let AB be the distance between the two planes and D be the point of observation on the ground.

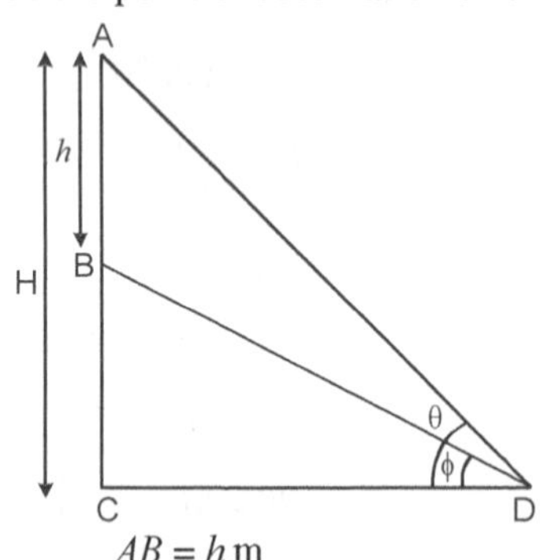

So $\quad AB = h$ m

Let BC be the height of the second plane from the ground and $BC = (H - h)$ m

Now, in ΔBDC,

$$\tan\phi = \frac{BC}{CD} = \frac{H-h}{CD}$$

$$\Rightarrow \quad CD = (H-h)\cot\phi \qquad \ldots\text{(i)}$$

and, in ΔADC

$$\tan\theta = \frac{AC}{CD} = \frac{H}{CD}$$

$$\Rightarrow \quad CD = H\cot\theta \qquad \ldots\text{(ii)}$$

From equations (i) and (ii), we have

$$(H-h)\cot\phi = H\cot\theta$$

$$\Rightarrow H(\cot\phi - \cot\theta) = h\cot\phi$$

$$\Rightarrow \quad h = \frac{H(\cot\phi - \cot\theta)}{\cot\phi}$$

$$h = \frac{H\left(\dfrac{1}{\tan\phi} - \dfrac{1}{\tan\theta}\right)}{\dfrac{1}{\tan\phi}}$$

$$= \frac{H(\tan\theta - \tan\phi)\tan\phi}{\tan\theta.\,\tan\phi}$$

$$\Rightarrow \quad h = \frac{H(\tan\theta - \tan\phi)}{\tan\theta}.$$

Hence Proved.

Q. 20. From a window, *h* m above the ground of a house in a street, the angles of elevation and depression of the top and foot of another house on the opposite side of the street are θ and ϕ. Show that the height of the opposite house is *h*(1 + tan θ cot ϕ) m.★★

Solution : Let the height of the opposite house AC be l m and the distance between both the house $(CD = BE)$ be x m.

As $BCDE$ forms a rectangular shape, so

$$BC = DE = h\text{ m}$$

Also, $\quad \angle BCE = 180^\circ - (90^\circ + \phi) = 90^\circ - \phi$

Thus, $\tan(90^\circ - \phi) = \dfrac{EB}{BC}$ [taking BC as the base]

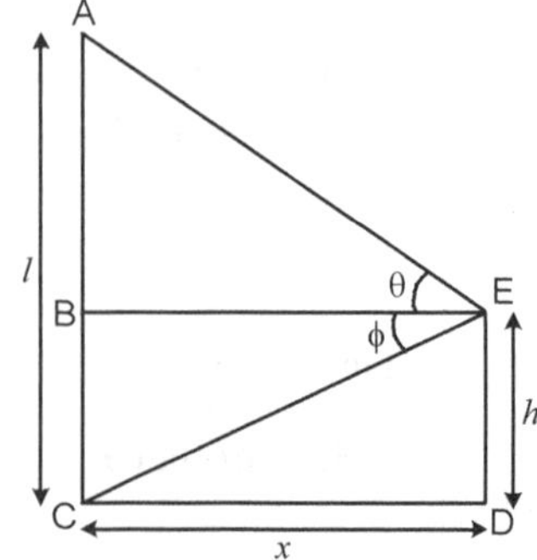

$$\Rightarrow \quad \tan(90^\circ - \phi) = \frac{x}{h}$$

★ are board exam questions from previous years

★★ are frequently asked board exam questions

$\Rightarrow \quad x = h \tan (90° - \phi)$ m

$\Rightarrow \quad x = h \cot \phi \quad$ ...(i)

Now, in ΔABE,

$$\tan \theta = \frac{AB}{BE}$$

$$\Rightarrow \quad \tan \theta = \frac{l-h}{x}$$

$\Rightarrow \quad l - h = x \tan \theta$

$\Rightarrow \quad l = x \tan \theta + h$

$\Rightarrow \quad l = h \cot \phi . \tan \theta + h$

[Using (i)]

$\Rightarrow \quad l = h(1 + \tan \theta . \cot \phi)$ m.

Hence Proved.

Q. 21. The angle of elevation of a cloud from a point *h* m above a lake is α and the angle of depression of the reflection of the cloud in the lake is β. Prove that the height of the cloud is $\frac{h(\tan \beta + \tan \alpha)}{\tan \beta - \tan \alpha}$ m.

Solution : Let *DE* be the surface of the lake.

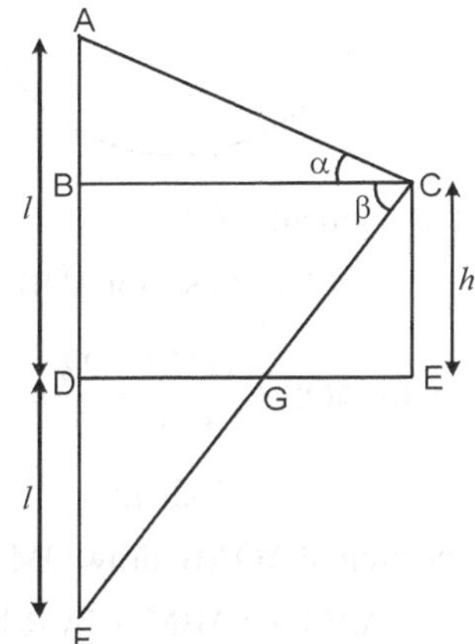

Thus, $\quad DE = BC = x$ m (say)

Also let *AD* be the height (l) of the cloud from the surface of the lake and *DF* be the length of the image of the cloud below the surface of the lake.

Thus, $\quad AD = DF$

[From the laws of plane mirrors, an image is far behind a plane mirror as the object is in front of it.]

So, $\quad AD = DF = l$ m

Given, $\quad \angle BCF = \beta$ and $\angle BCA = \alpha$

Now, in ΔABC,

$$\tan \alpha = \frac{AB}{BC}$$

$$\Rightarrow \quad \tan \alpha = \frac{l-h}{x}$$

$$\Rightarrow \quad x = \frac{l-h}{\tan \alpha} \text{ m} \quad \text{...(i)}$$

Similarly, in ΔBCF,

$$\tan \beta = \frac{FB}{BC}$$

$$\Rightarrow \quad \tan \beta = \frac{l+h}{x}$$

$$\Rightarrow \quad x = \frac{l+h}{\tan \beta} \text{ m} \quad \text{...(ii)}$$

From (i) and (ii), we get

$$\frac{l-h}{\tan \alpha} = \frac{l+h}{\tan \beta}$$

$\Rightarrow \quad (l - h) \tan \beta = (l + h) \tan \alpha$

$\Rightarrow l \tan \beta - h \tan \beta = l \tan \alpha + h \tan \alpha$

$\Rightarrow l \tan \beta - l \tan \alpha = h \tan \beta + h \tan \alpha$

$\Rightarrow l(\tan \beta - \tan \alpha) = h(\tan \beta + \tan \alpha)$

$$\Rightarrow \quad l = \frac{h(\tan \beta + \tan \alpha)}{\tan \beta - \tan \alpha}.$$

Hence Proved.

Q. 22. The angle of elevation of the top of a tower from a point on the same level as the foot of the tower is α. On advacing '*p*' m towards the foot of the tower, the angle of elevation becomes β. Show that the height '*h*' of the tower is given by★

$$h = \left(\frac{p \tan \alpha . \tan \beta}{\tan \beta - \tan \alpha}\right) \text{ m}$$

Also determine the height of the tower when $p = 150$ m, $\alpha = 30°$ and $\beta = 60°$.

Solution : Let *AB* be the tower of height *h* m and *C*, *D* be the observation points.

Also, let $BC = l$ m

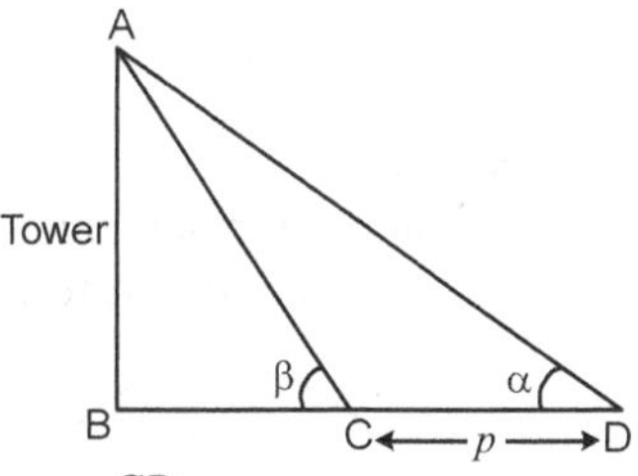

Given, $\quad CD = p$ m

Thus, in ΔABD,

$$\tan \alpha = \frac{AB}{BD} = \frac{AB}{BC + CD} = \frac{h}{BC + p}$$

$\Rightarrow \quad h = (BC + p) \tan \alpha \quad$ ...(i)

and in ΔABC,

$$\tan \beta = \frac{AB}{BC} = \frac{h}{BC}$$

$\Rightarrow \quad h = BC \tan \beta$

$\Rightarrow \quad BC = h \cot \beta \quad$ ...(ii)

Substituting (ii) and (i), we get

$h = (h \cot \beta + p) \tan \alpha$

$\Rightarrow \quad h = h \cot \beta . \tan \alpha + p \tan \alpha$

$\Rightarrow h - h \cot \beta . \tan \alpha = p \tan \alpha$

$\Rightarrow h(1 - \cot \beta . \tan \alpha) = p \tan \alpha$

$$\Rightarrow \quad h = \frac{p \tan \alpha}{(1 - \cot \beta . \tan \alpha)}$$

★ are board exam questions from previous years

$\Rightarrow \quad h = \dfrac{p \tan\alpha . \tan\beta}{\tan\beta - \tan\alpha}$ m.

Hence Proved.

Now substituting $p = 150$ m, $\alpha = 30°$ and $\beta = 60°$ in

$$h = \frac{p \tan\alpha \tan\beta}{(\tan\beta - \tan\alpha)}, \text{ we have}$$

$\Rightarrow \quad h = \dfrac{150 \tan 30° \tan 60°}{(\tan 60° - \tan 30°)}$

$\Rightarrow \quad h = \dfrac{150\left(\frac{1}{\sqrt{3}}\right)(\sqrt{3})}{\left(\sqrt{3} - \frac{1}{\sqrt{3}}\right)}$

$\Rightarrow \quad h = \dfrac{150\sqrt{3}}{(3-1)}$

$\Rightarrow \quad h = 75\sqrt{3}$

$\Rightarrow \quad h = 75(1.732)$

$\Rightarrow \quad h = 129.9$ m

Hence, the height of the tower is 129.9 m. **Ans.**

Chapter 12. Areas related to Circles

Q. 1. Find the area of the segment of a circle formed by a chord of length 5 cm subtending an angle of 90° at the centre.★ [Use $\pi = 3.14$]

Solution : Given, Length of the chord = 5 cm and θ = 90°

Now,

Area of the segment

= Area of the corresponding sector

– Area of the corresponding triangle

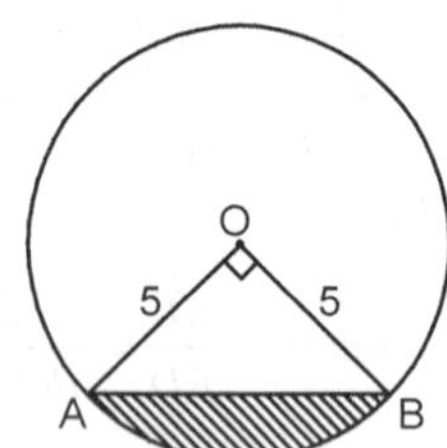

Area of the sector $= \dfrac{\theta}{360°} \times \pi r^2$

$= \dfrac{90°}{360°} \times \pi(5)^2$

$= \dfrac{1}{4}\pi(25) = \dfrac{25\pi}{4}$

Area of the triangle $= \dfrac{1}{2} \times$ Base $\times$ Height

$= \dfrac{1}{2} \times 5 \times 5 = \dfrac{25}{2}$

$\Rightarrow$ Area of the segment $= \dfrac{25\pi}{4} - \dfrac{25}{2}$

$= 25\left(\dfrac{3.14}{4} - \dfrac{1}{2}\right)$

$= 25\left(\dfrac{1.57}{2} - \dfrac{1}{2}\right)$

$= 25\left(\dfrac{0.57}{2}\right)$

$= 7.125$ cm^2. **Ans.**

Q. 2. Find the area of the segment shown in Fig. if radius of the circle is 21 cm and $\angle AOB = 120°$.★

$\left(\text{Use } \pi = \dfrac{22}{7}\right)$

Solution : Given, Radius of the circle = 21 cm and $\angle AOB = 120°$

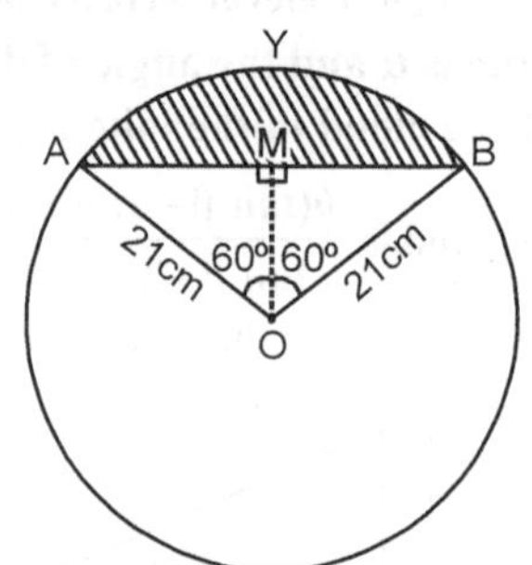

Area of the segment AYB

= Area of sector AOB – Area of ΔAOB

Area of sector AOB $= \dfrac{120°}{360°} \times \dfrac{22}{7} \times 21 \times 21$

$= 462$ cm^2

To find the area of ΔOAB, draw OM $\perp$ AB

$\Delta AMO \cong \Delta BMO$ (by R.H.S.)

$\Delta AOM = \Delta BOM = \dfrac{1}{2} \times 120° = 60°$

From ΔOMA, $\dfrac{OM}{OA} = \cos 60°$

$\dfrac{OM}{21} = \dfrac{1}{2}$

$OM = \dfrac{21}{2}$ cm

Also, $\dfrac{AM}{OA} = \sin 60°$

$AM = 21 \times \dfrac{\sqrt{3}}{2}$

or $AB = 2 \times AM = 21\sqrt{3}$ cm

So, area of $\Delta OAB = \dfrac{1}{2} \times AB \times OM$

$= \dfrac{1}{2} \times 21\sqrt{3} \times \dfrac{21}{2}$

★ are board exam questions from previous years

$$= \frac{441}{4}\sqrt{3}\ \text{cm}^2$$

$$\therefore \text{Area of segment} = \left(462 - \frac{441}{4}\sqrt{3}\right)\ \text{cm}^2$$

$$= \frac{21}{4}(88 - 21\sqrt{3})\ \text{cm}^2$$

$$= 271{\cdot}04\ \text{cm}^2 \qquad \textbf{Ans.}$$

Q. 3. In the given figure, three circles each of radius 3.5 cm are drawn in such a way that each of them just touches the other two. Find the area enclosed between these three circles (shaded region). ★

$$\left[\text{Use } \pi = \frac{22}{7}\right]$$

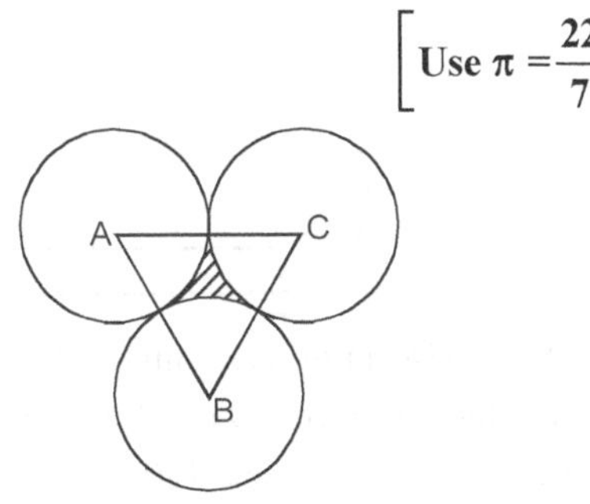

Solution : Given, $r = 3.5$ cm

Join the three centres size of the circles such that

$$AB = BC = CA = (3.5 + 3.5)\ \text{cm} = 7\ \text{cm}$$

Hence, ΔABC is an equilateral triangle where

$$\angle A = \angle B = \angle C = 60°$$

Let the points where the circles touch each other be X, Y and Z.

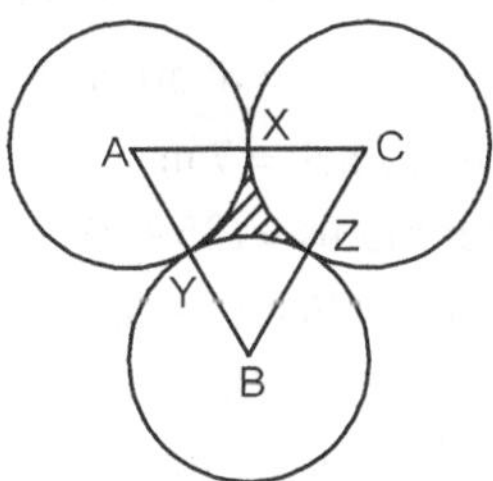

Thus XAY, YBZ and ZCX are the three sectors of the three circles.

Now, $\quad \text{Area of } \Delta ABC = \frac{\sqrt{3}}{4}(7)^2\ \text{cm}^2$

$\Rightarrow \quad \text{Area of } \Delta ABC = \frac{49\sqrt{3}}{4}\ \text{cm}^2 = 21.28\ \text{cm}^2$

Also, Area of sector XAY

= Area of sector YBZ = Area of sector ZCX

$$= \frac{60°}{360°} \times \frac{22}{7}(3.5)^2\ \text{cm}^2$$

$$= \frac{1}{6} \times \frac{22}{7}\left(\frac{7}{2}\right)^2\ \text{cm}^2$$

$$= 6{\cdot}42\ \text{cm}^2$$

Thus, Area of the shaded region

$$= \text{Area of } \Delta ABC - 3 \times \text{Area of sector } XAY$$

$$= [21.28 - 3(6.42)]\ \text{cm}^2$$

$$= [21.28 - 19.26]\ \text{cm}^2$$

$$= 2.02\ \text{cm}^2 \approx 2\ \text{cm}^2. \qquad \textbf{Ans.}$$

Q. 4. From a thin metallic piece in the shape of a trapezium *ABCD* in which *AB* || *CD* and $\angle BCD = 90°$, a quarter circle *BFEC* is removed as shown in the figure. Given, *AB* = *BC* = 3.5 cm and *DE* = 2 cm, calculate the area of the remaining part of the metal sheet. ★

$$\left[\text{Use } \pi = \frac{22}{7}\right]$$

Solution : Given, $AB = BC = 3.5$ cm and $DE = 2$ cm

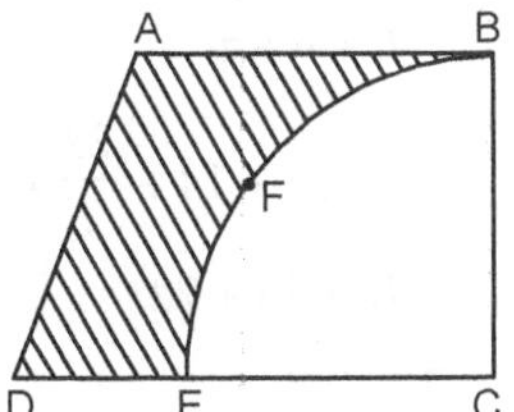

As BC and EC are the radii of the circle, so

$$BC = EC$$

Thus, $\qquad DC = (3.5 + 2)\ \text{cm} = 5.5\ \text{cm}$

Hence, Area of the remaining part of the metal sheet

= Area of the trapezium – Area of the quadrant

$$= \frac{1}{2} \times (\text{sum of parallel sides}) \times (\text{distance between parallel sides}) - \frac{1}{4}\pi r^2$$

$$= \frac{1}{2} \times (3.5 + 5.5) \times 3.5 - \frac{1}{4} \times \frac{22}{7} \times (3.5)^2$$

$$= \frac{1}{2} \times 9 \times 3.5 - \frac{11 \times 0.5 \times 3.5}{2}$$

$$= \frac{31.5}{2} - \frac{19.25}{2}$$

$$= \frac{12.25}{2}$$

$$= 6.125\ \text{cm}^2. \qquad \textbf{Ans.}$$

Q. 5. In the given figure, *ABCD* is a rectangle of dimensions 21 cm × 14 cm. A semicircle is drawn with *BC* as diameter. Find the area and the perimeter of the shaded region in the figure.★

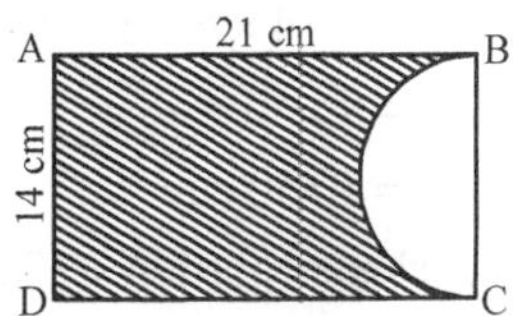

★ are board exam questions from previous years

Solution : Given, $AB = DC = 21$ cm

$AD = BC = 14$ cm

Radius, $r = \frac{BC}{2} = \frac{14}{2} = 7$ cm

Area of shaded region

= Area of rectangle $ABCD$ – Area of semi-circle

$$= l \times b - \frac{1}{2}\pi r^2$$

$$= 21 \times 14 - \frac{1}{2} \times \frac{22}{7} \times 7 \times 7$$

$$= 294 - 77$$

$$= 217 \text{ cm}^2.$$

Perimeter of shaded region

$$= 2l + b + \pi r$$

$$= 2 \times 21 + 14 + \frac{22}{7} \times 7$$

$$= 42 + 14 + 22$$

$$= 78 \text{ cm}$$ **Ans.**

Q. 6. Find the area of the shaded region in Fig. if ABCD is a rectnagle with sides 8 cm and 6 cm and D is the centre of circle.★ [Take π = 3·14]

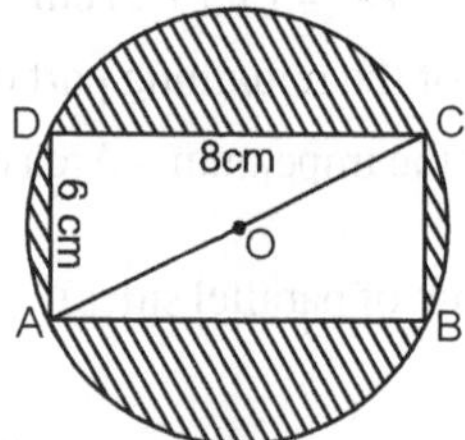

Solution : Given, ABCD is a rectangle with sides AB = 8 cm and BC = 6 cm

In ΔABC,

$$AC^2 = 8^2 + 6^2$$

(By Pythagoras Theorem)

$$= 64 + 36$$

$$\Rightarrow \quad AC^2 = 100$$

$$\Rightarrow \quad AC = 10 \text{ cm}$$

The diagonal of the rectangle will be the diameter of the circle.

$\therefore$ radius of the circle $= \frac{10}{2} = 5$ cm

Area of the shaded portion

= Area of circle – Area of Rectangle

$$= \pi r^2 - l \times b$$

$$= 3{\cdot}14 \times 5 \times 5 - 8 \times 6$$

$$= 78{\cdot}50 - 48$$

$$= 30{\cdot}50 \text{ cm}^2$$

Hence, Area of shaded portion = $30{\cdot}5$ cm^2 **Ans.**

★ **are board exam questions from previous years**

Q. 7. The inner perimeter of a running track as shown in the figure, is 340 m. The length of each straight portion is 60 m and the curved portions are semi-circles. If the track is 7 m wide, find the area of the track. Also, find the perimeter of the outer track.

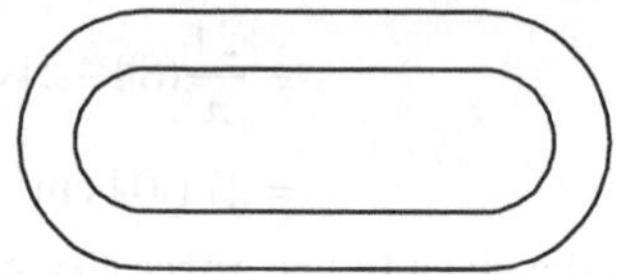

Solution : Given, $AB = DC = 60$ m and perimeter of $ABCD = 340$ m

Thus, length of each inner curved portion

$$= \frac{(340 - 60 - 60)}{2} \text{ cm} = 110 \text{ m}$$

Q P
B A
C D
R S

Now, AD and BC are semi-circles.

Let the radius of AD = radius of $BC = r$ m

$$\Rightarrow \quad \pi r = 110$$

$$\Rightarrow \quad r = 35 \text{ m}$$

Hence, distance between AB and CD

$$= (35 + 35) \text{ m} = 70 \text{ m}$$

Thus, area of inner portion of the track

$$= \left[(60 \times 70) + 2 \times \frac{1}{2} \times \frac{22}{7} \times (35)^2\right]$$

$$= [4200 + 3850]$$

$$= 8050 \text{ m}^2$$

Now, width of track = 7 m

$\therefore$ outer radius = (35 + 7) cm = 42 m

and Length of $QR = (7 + 35 + 35 + 7)$ m

$$= 84 \text{ m}$$

Thus, outer area of the track

$$= \left[(60 \times 84) + \frac{22}{7} \times (42)^2\right]$$

$$= (5040 + 5544) = 10584 \text{ m}^2$$

Hence, area of the race track

$$= (10584 - 8050) \text{ m}^2$$

$$= 2534 \text{ m}^2$$

Now, perimeter of outer track

$$= \left[2(60) + 2 \times \frac{22}{7} \times 42\right]$$

$$= [120 + 264]$$

$$= 384 \text{ m.}$$ **Ans.**

Q. 8. In the given figure, Δ*ABC* is right-angled triangle at *A*. Semi-circles are drawn on *AB*, *AC* and *BC*.

It is given that $AB = 3$ cm and $AC = 4$ cm. Find the area of the shaded region.★★

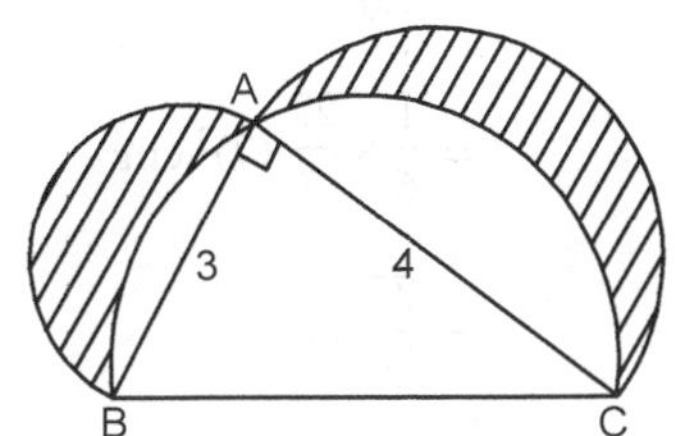

Solution : Given, $AB = 3$ cm, $AC = 4$ cm

So, area of $\Delta ABC = \frac{1}{2} \times 3 \times 4 = 6 \text{ cm}^2$

By Pythagoras' theorem,

$$BC^2 = AB^2 + AC^2$$

$$\Rightarrow BC^2 = (3)^2 + (4)^2$$

$$\Rightarrow BC^2 = 9 + 16 = 25 = 5^2$$

$$\Rightarrow BC = 5 \text{ cm}$$

[$\because$ length cannot be negative]

So, radius of semi-circle with BC as diameter = 2.5 cm

Hence, area of semi-circle with BC as diameter

$$= \frac{1}{2}\pi r^2 = \frac{1}{2} \times \frac{22}{7} \times (2.5)^2 = \frac{275}{28} \text{ cm}^2$$

Similarly, radius of semi-circle with AB as diameter = 1.5 cm

Hence area of semi-circle with AB a diameter

$$= \frac{1}{2}\pi r^2 = \frac{1}{2} \times \frac{22}{2} \times (1.5)^2 = \frac{99}{28} \text{ cm}^2$$

and radius of semi-circle with AC as diameter = 2 cm

Hence, area of semi-circle with AC as diameter

$$= \frac{1}{2}\pi r^2 = \frac{1}{2} \times \frac{22}{7} \times (2)^2 = \frac{44}{7} \text{ cm}^2$$

$\therefore$ The area of the shaded region

= Area of (semi-circle BA + semi-circle AC + ΔABC – semi-circle BAC)

$$= \left(\frac{99}{28} + \frac{44}{7} + 6 - \frac{275}{28}\right)$$

$$= \left(\frac{99 + 176 + 168 - 275}{28}\right)$$

$$= \frac{168}{28} = 6 \text{ cm}^2.$$ **Ans.**

Q. 9. In the given figure, ΔABC is a right-angled at A. Find the area of the shaded region if $AB = 6$ cm, $BC = 10$ cm and O is the centre of the incircle of ΔABC.★

[Use $\pi = 3.14$]

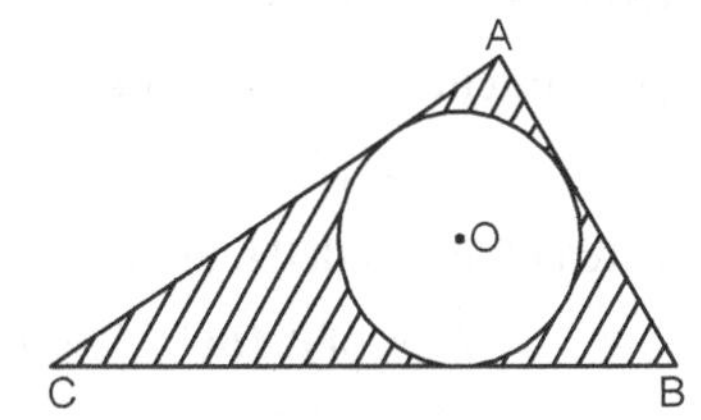

Solution : Given, $AB = 6$ cm and $BC = 10$ cm

By Pythagoras theorem,

$$AC^2 = BC^2 - AB^2$$

$$\Rightarrow AC^2 = (10)^2 - (6)^2$$

$$\Rightarrow AC^2 = 100 - 36$$

$$\Rightarrow AC^2 = 64 = 8^2$$

$$\Rightarrow AC = 8 \text{ cm}$$

[$\because$ length cannot be negative]

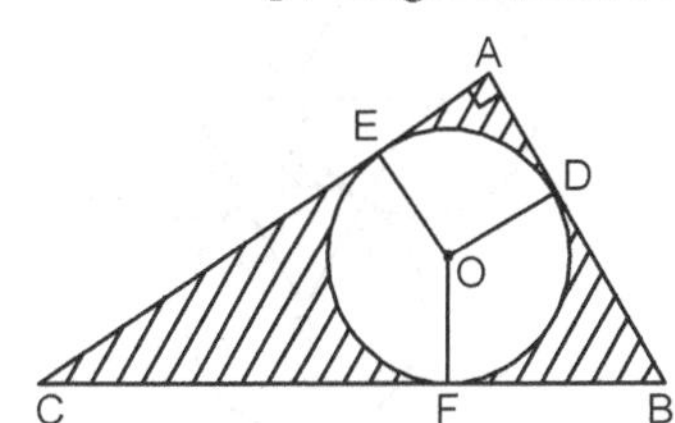

Now, Area of $\Delta ABC = \frac{1}{2} \times 8 \times 6 = 24 \text{ cm}^2$

Also let $AD = x$ cm

Now, $AD = EA$ [Tangents from A]

$\Rightarrow x = EA$

and $EA = AC - EC = (8 - EC)$ cm

$\Rightarrow x = (8 - EC)$ cm

Again $EC = CF$ [Tangents from C]

$\Rightarrow EC = (BC - FB)$

$\Rightarrow EC = (10 - FB)$ cm

Thus, $8 - x = (10 - FB)$ cm

$\Rightarrow x = (FB - 2)$ cm

and $FB = BD$ [Tangents from B]

$\Rightarrow FB = AB - AD$

$\Rightarrow FB = 6 - x$

$\Rightarrow x + 2 = 6 - x$

$\Rightarrow 2x = 4$

$\Rightarrow x = 2$ cm

Now in the quadrilateral $ADOE$,

★ are board exam questions from previous years

$\angle OEA = \angle ODA = 90°$

[OD and OE are perpendicular to AB and AC respectively]

$\angle EAD = 90°$ [Given]

Thus, $\angle EOD = 360° - (90° + 90° + 90°) = 90°$

Hence $ADOE$ is a right angled quadrilateral *i.e.,* is a square and so its opposite sides are equal.

$\therefore \quad AD = EO = 2$ cm

Thus, radius of the circle = 2 cm

So, area of the circle = $\pi r^2 = (3.14 \times 4)$ cm^2

$= 12.56$ cm^2

Hence, area of the shaded region

= Area of ΔABC – Area of the circle

$= (24 - 12.56)$ cm^2

$= 11.44$ cm^2. **Ans.**

Q. 10. The area of an equilateral triangle is $49\sqrt{3}$ cm^2. Taking each angular point as centre, circles are drawn with radius equal to half the length of the side of the triangle. Find the area of the triangle not included in the circles.★[Take $\sqrt{3} = 1.73$]

Solution : Let the length of each side of triangle be a.

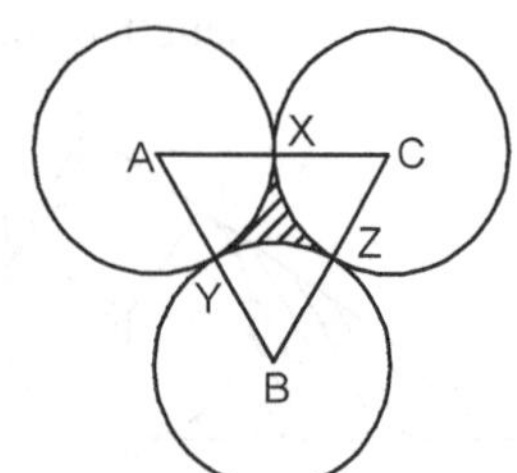

Thus, the area of the equilateral triangle

$$= \frac{\sqrt{3}}{4}a^2 = 49\sqrt{3}$$

$\Rightarrow \quad a^2 = 49 \times 4$

$\Rightarrow \quad a = (7 \times 2)$ cm = 14 cm

Thus, radius of each side = 7 cm

Let ABC be the equilateral triangle.

Then, $\quad AB = BC = CA = 14$ cm

and $\quad \angle A = \angle B = \angle C = 60°$

Let the points where the circles touch each other be X, Y and Z.

Thus, XAY, YBZ and ZCX are the three sectors of the three circles.

Now, Area of sector XAY

= Area of sector YBZ

= Area of sector ZCX

$$= \frac{60°}{360°} \times \frac{22}{7}(7)^2 \text{ cm}^2$$

$$= \frac{1}{6} \times \frac{22}{7}(7)^2 \text{ cm}^2$$

$$= \frac{77}{3} \text{ cm}^2$$

$= 25.67$ cm^2

Thus, Area of the shaded region

= Area of $\Delta ABC - 3 \times$ Area of sector XAY

$= [49\sqrt{3} - 3(25.67)]$

$= [49(1.73) - 77.01]$

$= 7.76$ cm^2. **Ans.**

Q. 11. In the given figure, ΔABC is a right-angled triangle with $\angle B = 90°$, $AB = 48$ cm and $BC = 14$ cm. With AC as diameter, a semi-circle is drawn and with BC as radius, a quarter of a circle is drawn. Find the area of the shaded region.★ $\left[\text{Use } \pi = \frac{22}{7}\right]$

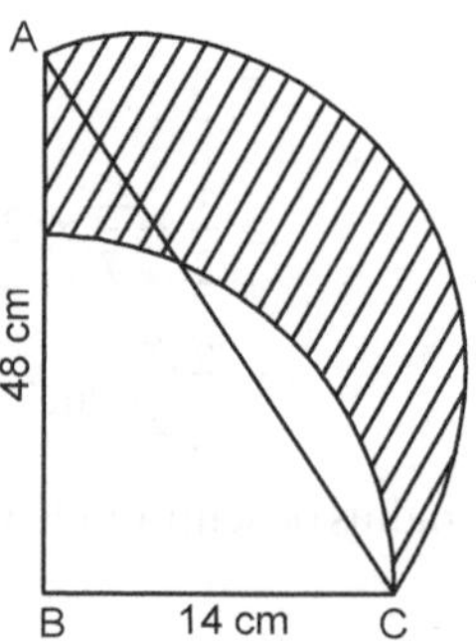

Solution : Given, $AB = 48$ cm and $BC = 14$ cm

Now, $\quad AC^2 = AB^2 + BC^2$

[Pythagoras theorem]

$\Rightarrow \quad AC^2 = (48)^2 + (14)^2$

$\Rightarrow \quad AC^2 = 2304 + 196 = 2500$

$\Rightarrow \quad AC = 50$ cm

Thus, area of $\Delta ABC = \frac{1}{2} \times 48 \times 14 = 336$ cm^2

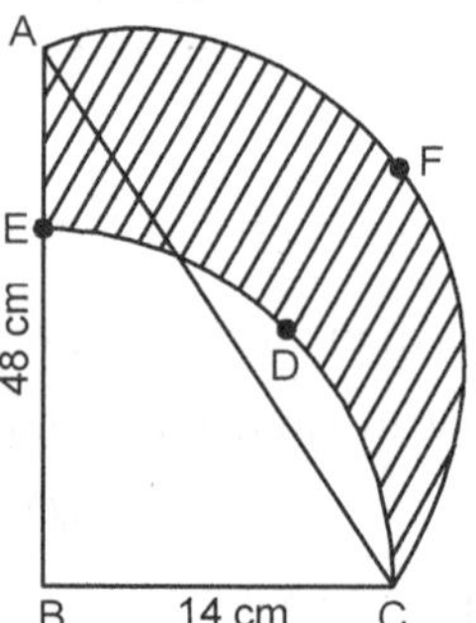

Area of quadrant BCD

★ are board exam questions from previous years

$$= \frac{1}{4} \times \frac{22}{7} \times 14 \times 14$$
$$= 154 \text{ cm}^2$$

Area of semi-circle ACF

$$= \frac{1}{2} \times \frac{22}{7} \times 25 \times 25$$

$$= 982.14 \text{ cm}^2$$

Hence, Area of $AFCDE$

$$= \text{Area of semi-circle } ACF + \text{Area of } \Delta ABC - \text{Area of quadrant } BCD$$
$$= (982.14 + 336 - 154) \text{ cm}^2$$
$$= 1164.14 \text{ cm}^2.$$ **Ans.**

Q. 12. In a circular table cover of radius 70 cm, a design is formed, leaving an equilateral ΔABC in the middle, as shown in the figure. Find the total area of the design.★

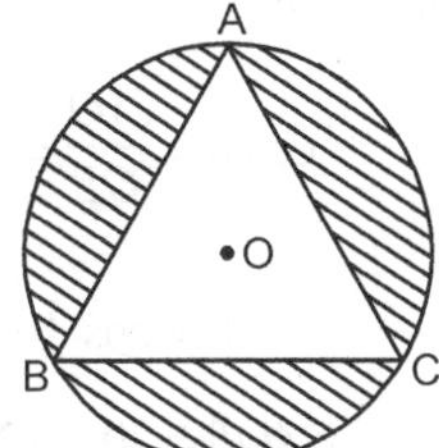

Solution : Given, radius = 70 cm

So, area of the circle $= \pi r^2 = \frac{22}{7} \times 70 \times 70$

$$= 15400 \text{ cm}^2$$

Join OD perpendicular to BC.

Also, as ABC is an equilateral triangle,

$\therefore$ $\angle A = \angle B = \angle C = 60°$

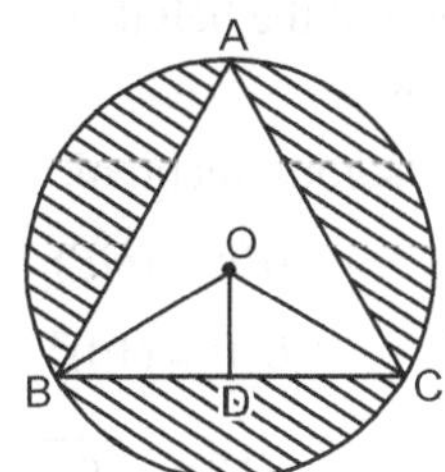

Thus, $\angle OBC = 30°$

$$\cos 30° = \frac{BD}{OB}$$

$\Rightarrow$ $\frac{\sqrt{3}}{2} = \frac{BD}{70}$

$\Rightarrow$ $BD = 35\sqrt{3}$ cm

Thus, $BC = 70\sqrt{3}$ cm

Now, area of the equilateral triangle

$$= \frac{\sqrt{3}a^2}{4} = \frac{\sqrt{3}}{4} \times 70\sqrt{3} \times 70\sqrt{3}$$

$$= 3675\sqrt{3}$$
$$= 6365.1 \text{ cm}^2$$

So, the area covered by the design

$$= (15400 - 6365.1) \text{ cm}^2$$
$$= 9034.9 \text{ cm}^2.$$ **Ans.**

Q. 13. In the adjoining figure, ΔABC is rightangled at A. Find the area of the shaded region if AB = 6 cm, BC = 10 cm and I is the centre of the incircle of ΔABC.★

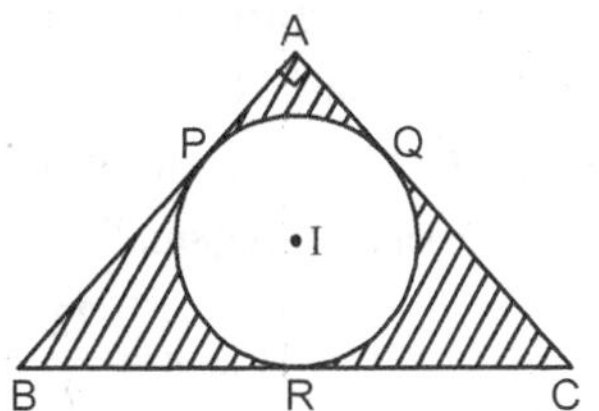

Solution : Given, $AB = 6$ cm, $BC = 10$ cm

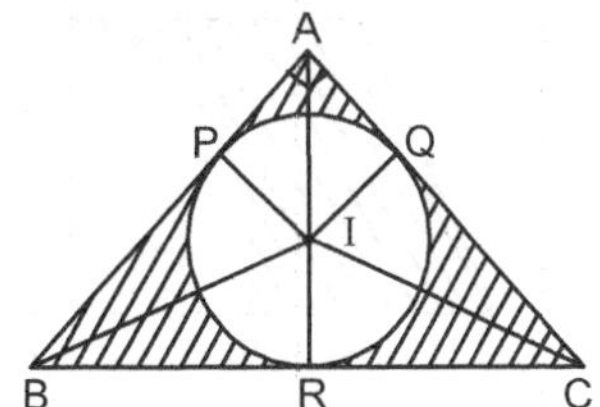

Thus, applying Pythagoras theorem,

$$AC^2 = BC^2 - AB^2$$

$\Rightarrow$ $AC^2 = (10)^2 - (6)^2$
$= 100 - 36 = 64$

$\Rightarrow$ $AC^2 = 8^2$

$\Rightarrow$ $AC = 8$ cm

Thus, area of $\Delta ABC = \frac{1}{2} \times 8 \times 6$
$= 24 \text{ cm}^2$

Now, join PI, RI and QI such that they are the radii of the inscribed circle and are perpendicular to AB, BC and AC respectively.

So, $PI = RI = QI = r$ (say)

Also, join BI, CI and AI to form three different triangles having r as the height.

Thus, Area of ΔABC = Area of ΔAIB + Area of ΔAIC + Area of ΔBIC

$\Rightarrow$ $24 = \frac{r}{2}[6 + 8 + 10]$

$\Rightarrow$ $24 = \frac{r}{2}[24]$

$\Rightarrow$ $r = 2$ cm

Thus, Area of the circle $= \pi(2)^2 = \frac{88}{7}$ cm^2

Hence, the area of the shaded region

★ are board exam questions from previous years

$$= \left(24 - \frac{88}{7}\right) \text{cm}^2$$

$$= \left(\frac{168 - 88}{7}\right) \text{cm}^2$$

$$= \frac{80}{7} \text{cm}^2$$

$= 11.42 \text{ cm}^2$. **Ans.**

Q. 14. In the given figure, *ABCD* is a trapezium with *AB* || *DC* and ∠*BCD* = 60°. If *BFEC* is a sector of a circle with centre *C*, *AB* = *BC* = 7 cm and *DE* = 4 cm, then find the area of the shaded region.★

[Use $\pi = \frac{22}{7}$ and $\sqrt{3} = 1.732$]

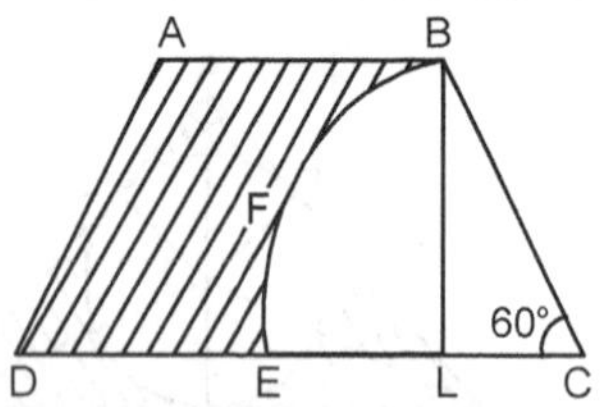

Solution : Given, $AB \parallel DC$, $\angle BCD = 60°$, $AB = BC =$ 7 cm and $DE = 4$ cm

As *BFEC* is a sector of a circle, so

$BC = EC$ [Radii]

Hence, $EC = 7$ cm and $DC = (7 + 4) = 11$ cm

Now, in ΔBLC

$$\frac{BL}{BC} = \sin 60°$$

$$\Rightarrow \quad BL = BC \sin 60°$$

$$\Rightarrow \quad BL = \frac{7\sqrt{3}}{2} \text{ cm}$$

Hence, Area of the shaded region

= Area of the Trapezium – Area of the Sector

$= \frac{1}{2} \times$ (Sum of parallel sides)

$\times$ (Distance between parallel sides) $- \frac{\theta}{360°} \times \pi r^2$

$$= \left[\frac{1}{2} \times (7 + 11) \times \left(\frac{7\sqrt{3}}{2}\right) - \frac{60°}{360°} \times \frac{22}{7}(7)^2\right] \text{cm}^2$$

$$= \left[18 \times \left(\frac{7\sqrt{3}}{4}\right) - \frac{1}{6} \times 22 \times 7\right] \text{cm}^2$$

$$= \left[\frac{63\sqrt{3}}{2} - \frac{77}{3}\right] \text{cm}^2$$

$$= \frac{189\sqrt{3} - 154}{6} \text{ cm}^2$$

$= 28.89 \text{ cm}^2$. **Ans.**

★ are board exam questions from previous years

Q. 15. An elastic belt is placed around the rim of a pulley of radius 5 cm. From one point *C* on the belt, the elastic belt is pulled directly away from the centre *O* of the pulley until it is at *P*, 10 cm from the point *O*. Find the length of the belt that is still in contact with the pulley. Also find the shaded area.★

[Use $\pi = 3.14$ and $\sqrt{3} = 1.73$]

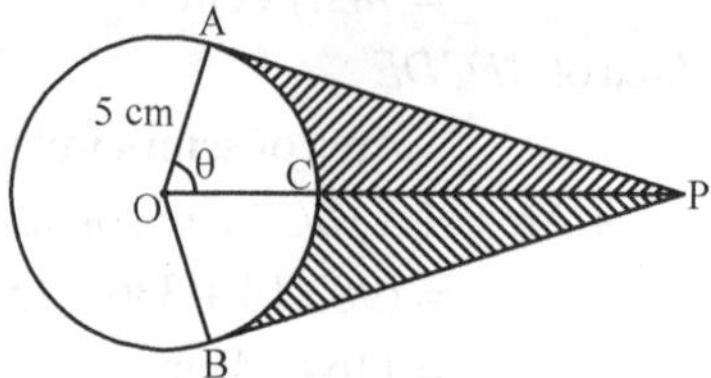

Solution : Given, a circular pulley of radius 5 cm with centre *O*.

$\therefore \quad AO = OB = OC = 5$ cm

and $\quad OP = 10$ cm

Now, in right ΔAOP,

$$\cos\theta = \frac{AO}{OP} = \frac{5}{10} = \frac{1}{2}$$

$$\therefore \quad \theta = \cos^{-1}\left(\frac{1}{2}\right) = 60°$$

$$\therefore \quad \angle AOB = 2\theta = 120°$$

$\Rightarrow$ Reflex $\angle AOB = 360° - 120° = 240°$

$$\text{Length of major arc } \widehat{AB} = \frac{2\pi r}{360°} \text{ reflex } \angle AOB$$

$$= \frac{2 \times 3.14 \times 5 \times 240°}{360°}$$

$$= 20.93 \text{ cm}$$

Hence, length of the belt that is still in contact with pulley = 20.93 cm.

Now, by Pythagoras theorem,

$$(AP)^2 = (OP)^2 - (OA)^2$$

$$(AP)^2 = (10)^2 - (5)^2$$

$$AP = \sqrt{100 - 25}$$

$$= \sqrt{75} = 5\sqrt{3} \text{ cm}$$

$$\therefore \quad \text{Area of } \Delta AOP = \frac{1}{2} \times 5 \times 5\sqrt{3}$$

$$= \frac{25\sqrt{3}}{2} \text{ cm}$$

Also, Area of ΔBOP = Area of ΔAOP

and, area of quad. $AOBP$ = 2(Area of ΔAOP)

$$= 2 \times \frac{25\sqrt{3}}{2} = 25\sqrt{3} \text{ cm}^2$$

$= 43.25 \text{ cm}^2$

$$\text{Area of sector } ACBO = \frac{\angle AOB}{360^\circ}\pi r^2$$

$$= \frac{120^\circ}{360^\circ} \times 3.14 \times 5 \times 5$$

$$= 26.16 \text{ cm}^2$$

$\therefore$ Area of shaded region = Area of quad. $AOBP$ – Area of sector $ACBO$

$$= (43.25 - 26.16) \text{ cm}^2$$

$$= 17.09 \text{ cm}^2 \quad \textbf{Ans.}$$

Q. 16. In Figure, a square OABC is inscribed in a quadrant OPBQ. If OA = 15 cm, find the area of the shaded region.★ **[Use π = 3·14]**

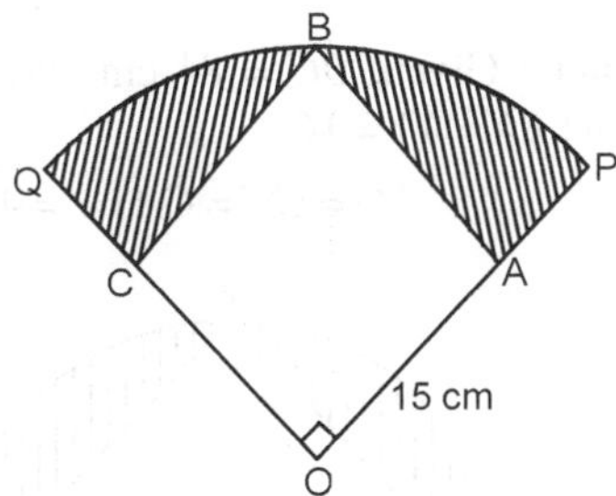

Solution : Given, OABC is a square with OA = 15 cm

OB = radius = r

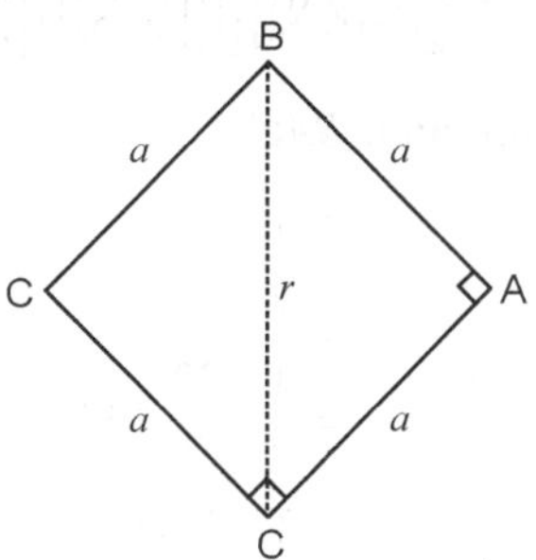

Let side of square be a then,

$$a^2 + a^2 = r^2$$
$$2a^2 = r^2$$
$$r = \sqrt{2}\,a$$
$$r = 15\sqrt{2} \text{ cm} \quad (\because a = 15 \text{ cm})$$

$$\text{Area of square} = \text{Side} \times \text{Side}$$
$$= 15 \times 15$$
$$= 225 \text{ cm}^2$$

$$\text{Area of quadrant OPBQ} - \frac{1}{4} \times \pi r^2$$

$$= \frac{1}{4} \times 3{\cdot}14 \times 15\sqrt{2} \times 15\sqrt{2}$$

$$= \frac{225 \times 2 \times 3{\cdot}14}{4}$$

$$= 225 \times 1{\cdot}57$$
$$= 353{\cdot}25 \text{ cm}^2$$

Area of shaded region

= Area of quadrant OPBQ – Area of square

$= 353{\cdot}25 - 225$

$= 128{\cdot}25 \text{ cm}^2$ **Ans.**

Q. 17. In Figrue, ABCD is a square with side $2\sqrt{2}$ cm and inscribed in a circle. Find the area of the shaded region.★ **(Use π = 3·14)**

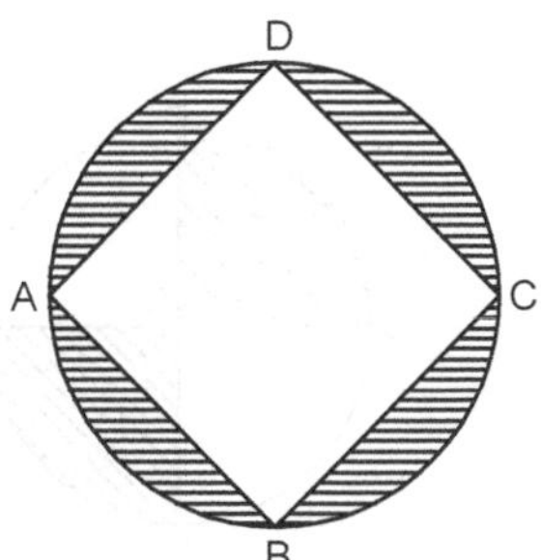

Solution : Given, ABCD is a square with side $2\sqrt{2}$ cm.

$\because$ BD $= 2r$

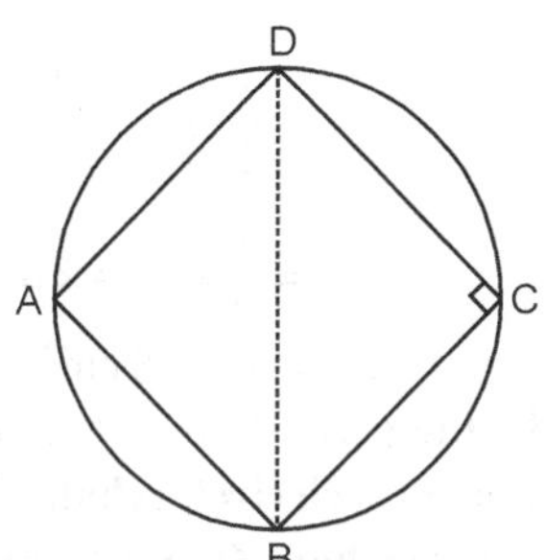

In ΔBDC,

$$BD^2 = DC^2 + BC^2$$
$$\Rightarrow \quad 4r^2 = 2(DC)^2$$
$$(\because DC = CB = \text{Side} = 2\sqrt{2})$$
$$\Rightarrow \quad 4r^2 = 2 \times 2\sqrt{2} \times 2\sqrt{2}$$
$$\Rightarrow \quad 4r^2 = 8 \times 2$$
$$\Rightarrow \quad 4r^2 = 16$$
$$\Rightarrow \quad r^2 = 4$$
$$r = 2 \text{ cm}$$

$$\text{Area of square BCDA} = \text{Side} \times \text{Side}$$
$$= DC \times BC$$
$$= 2\sqrt{2} \times 2\sqrt{2}$$
$$= 8 \text{ cm}^2$$

$$\text{Area of circle} = \pi r^2$$
$$= 3{\cdot\cdot}14 \times 2 \times 2$$
$$= 12{\cdot}56 \text{ cm}^2$$

★ are board exam questions from previous years

Area of shaded region

= Area of circle – Area of square.

= 12·56 – 8

= 4·56 cm^2 **Ans.**

Q. 18. *AC* and *BD* are two perpendicular diameters of a circle with centre *O*. If *AC* = 16 cm, calculate the area and perimeter of the shaded part. [Take π = 3.14]

Solution : Given, $AC = BD = 16$ cm

Thus, $AO = OC = DO = OB = r = 8$ cm

Hence, the area of the shaded part

= Area of quadrant *AOD*

+ Area of quadrant *COB*

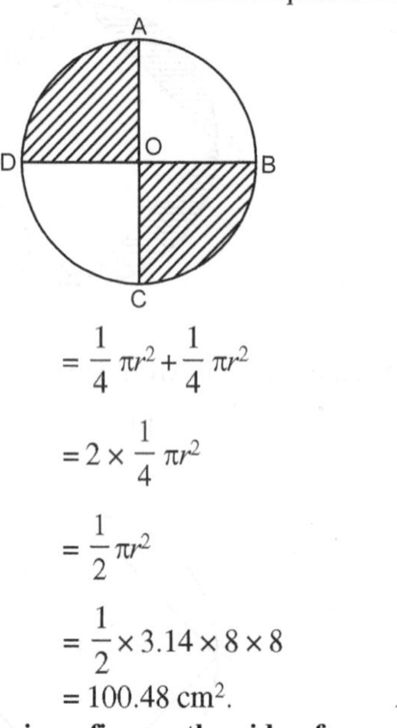

$$= \frac{1}{4}\pi r^2 + \frac{1}{4}\pi r^2$$

$$= 2 \times \frac{1}{4}\pi r^2$$

$$= \frac{1}{2}\pi r^2$$

$$= \frac{1}{2} \times 3.14 \times 8 \times 8$$

= 100.48 cm^2. **Ans.**

Q. 19. In the given figure, the side of square is 28 cm and radius of each circle is half of the length of the side of the square where *O* and *O'* are centres of the circles. Find the area of shaded region.★

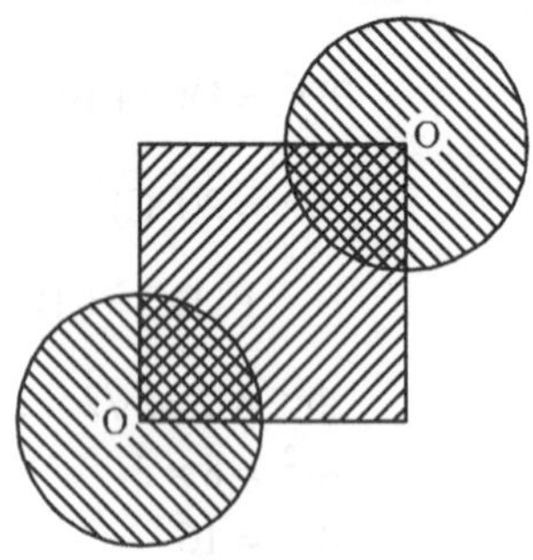

Solution : Area of shaded region

= 2 × (Area of circle) + Area of square

– 2 × (Area of quadrant)

$$= 2 \times \pi r^2 + (\text{Side})^2 - 2\left(\frac{1}{4} \times \pi r^2\right)$$

$$= 2\pi r^2 - \frac{1}{2}\pi r^2 + (\text{Side})^2$$

$$= \frac{3}{2}\pi r^2 + (\text{Side})^2$$

★ are board exam questions from previous years

Now, $r = \frac{1}{2}(\text{Side}) = 14$ cm

[∵ Side = 28 cm]

$$\therefore \text{ Area of shaded region} = \frac{3}{2} \times \frac{22}{7} \times 14 \times 14 + 28 \times 28$$

= 924 + 784

= 1708 cm^2 **Ans**

Q. 20. A doorway is decorated as shown in the figure. There are four semicircles. *BC*, the diameter of the largest semicircle is of length 84 cm. Centres of the three equal semicircles lie on *BC*. *ABC* is an isosceles triangle with *AB* = *AC*. If *BO* = *OC*, find the area of the shaded region.

$$\left[\textbf{Take } \pi = \frac{22}{7}\right]$$

Solution : Given, $BC = 84$ cm, $BO = OC = OA$, $AB = BC$ and $BX = XY = YC$

$\Rightarrow$ $BO = OC = OA = r = 42$ cm

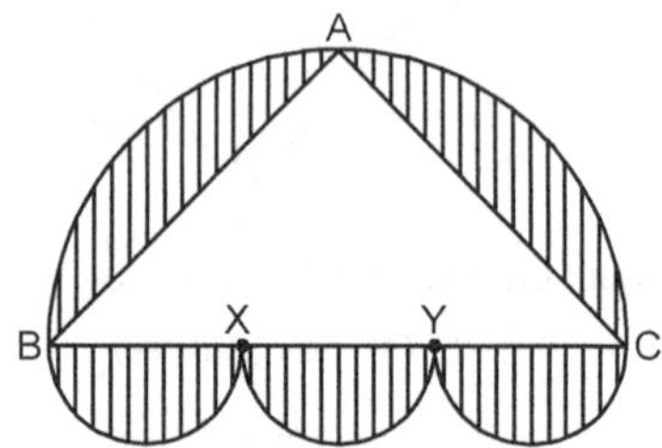

Area of semi-circle *BAC*

$$= \frac{1}{2}\pi r^2$$

$$= \frac{1}{2} \times \frac{22}{7} \times 42 \times 42$$

= 2772 cm^2

Also, area of $\Delta ABC = \frac{1}{2} \times BC \times OA$

$$= \frac{1}{2} \times 84 \times 42$$

= 42 × 42

= 1764 cm^2

Diameter of each of three equal semicircles

$$BX = XY = YC = \frac{84}{3} \text{ cm}$$

= 28 cm

∴ Radii of all three small semi-circles = 14 cm each

Hence, area of each small semi-circle

$$= \frac{1}{2} \times \frac{22}{7} \times 14 \times 14 \text{ cm}^2$$

= 308 cm^2

Thus, Area of the shaded region = Area of large semicircle + Areas of three small semicircles – Area of triangle

$$= [2772 + 3 \times 308 - 1764]$$

$$= 1932 \text{ cm}^2. \qquad \textbf{Ans.}$$

Q. 21. In the given figure, O is the centre of the circle with $AC = 24$ cm, $AB = 7$ cm and $\angle BOD = 90°$. Find the area of the shaded region.★

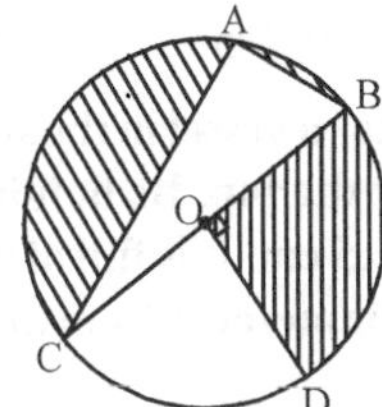

Solution : Given, $C(O, OB)$ with $AC = 24$ cm,

$AB = 7$ cm and $\angle BOD = 90°$

$$\angle CAB = 90° \quad \text{[Angle in semi-circle]}$$

$\therefore$ In ΔCAB

Using Pythagoras theorem,

$$BC^2 = AC^2 + AB^2$$

$$= (24)^2 + (7)^2$$

$$= 576 + 49$$

$$= 625$$

$$\Rightarrow \quad BC = 25 \text{ cm}$$

Radius of circle $= OB = OD = OC = \frac{25}{2}$ cm

Area of shaded region

= Area of semi-circle with diameter BC

– Area of ΔCAB + Area of sector BOD

$$= \frac{1}{2}\pi r^2 - \frac{1}{2} \times BC \times AB + \frac{90°}{360°}\pi r^2$$

$$= \frac{1}{2}\pi r^2 - \frac{1}{2} BC \times AD + \frac{1}{4}\pi r^2$$

$$= \frac{3}{4}\pi r^2 - \frac{1}{2} BC \times AB$$

$$= \frac{3}{4} \times \frac{22}{7} \times \frac{25}{2} \times \frac{25}{2} - \frac{1}{2} \times 24 \times 7$$

$$= \frac{20625}{56} - 84$$

$$= \frac{20625 - 4704}{56}$$

$$= \frac{15921}{56}$$

$$= 248.3 \text{ cm}^2 \text{ (approx)} \qquad \textbf{Ans.}$$

★ are board exam questions from previous years

Q. 22. In fig., a sector OAP of a circle is shown with centre O, containing $\angle\theta$. AB is perpendicular to the radius OA and meets OP produced at B. Prove that the perimeter of shaded region is $r\left[\tan\theta + \sec\theta + \frac{\pi\theta}{180} - 1\right]$ ★

Solution : Given, the radius of circle with centre O is r and $\angle POA = \theta$.

Then, length of the arc $\widehat{PA} = \frac{\theta}{360°} 2\pi r = \frac{\theta}{180°}\pi r$

and $\tan\theta = \frac{AB}{r}$

$AB = r\tan\theta$

and $\sec\theta = \frac{OB}{r}$

$OB = r\sec\theta$

Now, $PB = OB - OP$

$= r\sec\theta - r$

$\therefore$ Perimeter of shaded region

$$= AB + PB + \widehat{PA}$$

$$= r\tan\theta + r\sec\theta - r + \frac{\pi r\theta}{180°}$$

$$= r\left[\tan\theta + \sec\theta + \frac{\pi\theta}{180} - 1\right]$$

Hence Proved.

Chapter 13. Surface Areas and Volumes

Q. 1. A hemispherical depression is cut out from one face of a cubical block of side 7 cm such that the diameter of the hemisphere is equal to the edge of the cube. Find the surface area of the remaining solid.★ $\left[\text{Use } \pi = \frac{22}{7}\right]$

Solution : Given, side of the cube = 7 cm

Diameter of the hemisphere = 7 cm

Thus, radius of the hemisphere = 3.5 cm

Total surface area of the cube $= 6(7)^2$ cm²

Curved surface area of the hemisphere

$= 2\pi(3.5)^2$ cm²

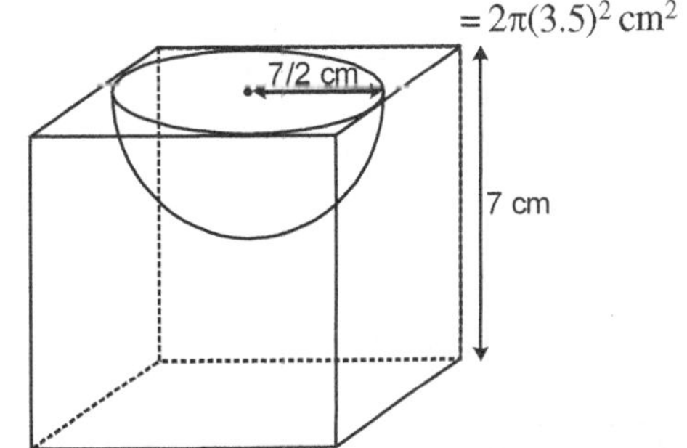

Thus, total surface area area of the remaining solid
= Total surface area of cube + Curved surface area of the hemisphere – Area of the top face of the hemisphere

$= [6(7)^2 + 2\pi(3.5)^2 - \pi(3.5)^2]\ \text{cm}^2$

$= \left[6(7)^2 + \left(\frac{22}{7}\right)(3.5)^2\right] \text{cm}^2$

$= \left[294 + \frac{77}{2}\right] \text{cm}^2$

$= 332.5\ \text{cm}^2.$ **Ans.**

Q. 2. A metallic bucket open at the top of height 24 cm is in the shape of the frustum of a cone, the radii of whose upper and lower circular ends being 14 cm and 7 cm respectively. Find : (i) the volume of water which can completely fill the bucket and (ii) the area of the metal sheet used to make the bucket.★

$\left[\textbf{Use } \pi = \frac{22}{7}\right]$

Solution : Given, height of frustum, $h = 24$ cm
Upper radius of frustum, $R = 14$ cm
Lower radius of the frustum, $r = 7$ cm

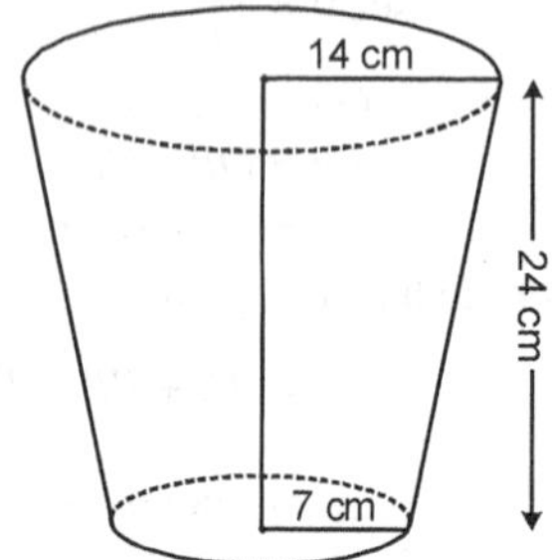

Slant height, $l = \sqrt{h^2 + (R-r)^2}$

$= \sqrt{(24)^2 + (14-7)^2}$

$= \sqrt{576 + 49}$

$= \sqrt{625}$

$= 25$

(i) Thus, volume of frustum

$= \frac{\pi}{3}(R^2 + Rr + r^2)h$

$= \frac{\pi}{3}(14^2 + 14 \times 7 + 7^2)24\ \text{cm}^3$

$= \left(\frac{22}{7}\right)(196 + 98 + 49)8\ \text{cm}^3$

$= \frac{22}{7} \times 343 \times 8\ \text{cm}^3$

$= 8624\ \text{cm}^3$

(ii) Area of the metal sheet required

= Lateral surface area + Area of the base

$= \pi(R + r)l + \pi r^2$

$= \pi(14 + 7)25 + \pi(7)^2\ \text{cm}^2$

$= \pi[(21)25 + 49]\ \text{cm}^2$

$= \left(\frac{22}{7}\right)[525 + 49]\ \text{cm}^2$

$= \frac{22}{7} \times 574\ \text{cm}^2$

$= 1804\ \text{cm}^2.$ **Ans.**

Q. 3. Sushant has a vessel in the shape of an inverted cone that is open at the top. Its height is 11 cm and the radius of the top is 2.5 cm. It is full of water and metallic spherical balls of diameter 0.5 cm are put in the vessel such that $\frac{2}{5}$ of the water flows out. Find the number of balls that were put in the vessel.★

Solution :

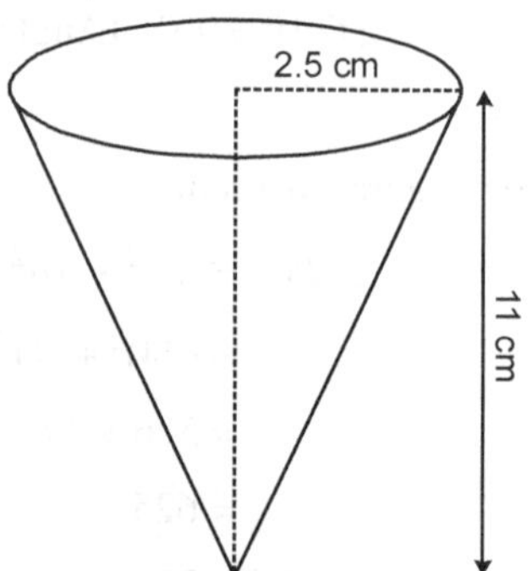

Given, Height of the cone = 11 cm
Radius of the top of the cone = 2.5 cm
Diameter of each ball = 0.5 cm
Thus, radius of each ball = 0.25 cm
Volume of water that flows out

$= \frac{2}{5}$(Volume of the cone)

Now, volume of the water in the cone $= \frac{1}{3}\pi r^2 h$

$= \frac{\pi}{3}(2.5)^2(11)\ \text{cm}^3$

$= \frac{11\pi}{3}(2.5 \times 2.5)\ \text{cm}^3$

Thus, the volume of water that flows out

$= \frac{11}{3}\pi(2.5 \times 2.5)\left(\frac{2}{5}\right)\ \text{cm}^3$

= Total volume of all the spherical balls

Now, volume of 1 spherical ball

★ are board exam questions from previous years

$$= \left(\frac{4}{3}\right)\pi(0.25)^3 \text{ cm}^3$$

$$= \left(\frac{4}{3}\right)\pi(0.25 \times 0.25 \times 0.25) \text{ cm}^3$$

Hence the number of spherical balls

$$= \frac{\left(\frac{1}{3}\right)11\pi(2.5\times2.5)\left(\frac{2}{5}\right)}{\left(\frac{4}{3}\right)\pi(0.25\times0.25\times0.25)}$$

$$= \frac{11(2.5\times2.5)}{10(0.25\times0.25\times0.25)}$$

$$= 440$$

Thus, number of balls = 440. **Ans.**

Q. 4. Two spheres of same metal weigh 1 kg and 7 kg. The radius of the smaller sphere is 3 cm. The two spheres are melted to form a single big sphere. Find the diameter of the new sphere.★

Solution : Given, radius of small sphere be

$$r = 3 \text{ cm}$$

Then,

$$\text{Volume of small sphere} = \frac{4}{3}\pi r^3$$

$$= \frac{4}{3}\pi(3)^3$$

$$= 36\pi \text{ cm}^3$$

$$\text{Density of small sphere} = \frac{\text{Mass of sphere}}{\text{Volume of sphere}}$$

$$= \frac{1}{36\pi} \text{ kg/cm}^3$$

$\because$ Both spheres are made by same metal, then their densities will be same.

Let radius of bigger sphere = r' then,

Density of bigger sphere

$$= \frac{\text{Mass of bigger sphere}}{\text{Volume of bigger sphere}}$$

$$\frac{1}{36\pi} = \frac{7}{4/3\pi r'^3}$$

$$(r')^3 = 189$$

Then according to question, we have,

Volume of bigger sphere + Volume of smaller shpere

= Volume of new sphere.

$$\frac{4}{3}\pi(r')^3 + \frac{4}{3}\pi r^3 = \frac{4}{3}\pi R^3$$

$$\Rightarrow \quad (r')^3 + r^3 = R^3$$

$$\Rightarrow \quad 189 + 27 = R^3$$

$$\Rightarrow \quad 216 = R^3$$

$$\Rightarrow \quad R = 6$$

$$\Rightarrow \quad D = 6 \times 2$$

$$= 12$$

Radius of new sphere is 6 cm.

So, diameter is 12 cm. **Ans.**

Q. 5. A container open at the top is in the form of a frustum of a cone of height 24 cm with radii of its lower and upper circular ends as 8 cm and 20 cm respectively. Find the cost of milk which can completely fill the container at the rate of ₹ 21 per litre.★

$$\left[\textbf{Use } \pi = \frac{22}{7}\right]$$

Solution : Given, height of the frustum = 24 cm

Radius of lower end = 8 cm

Radius of upper end = 20 cm

Rate of milk = ₹ 21 per litre

Now,

$$\text{Volume of frustum} = \frac{\pi}{3}(20^2 + 8^2 + 20 \times 8)24 \text{ cm}^3$$

$$= \frac{22}{7}(20^2 + 8^2 + 20 \times 8)8 \text{ cm}^3$$

$$= \frac{22}{7}(400 + 64 + 160)8 \text{ cm}^3$$

$$= \frac{22}{7 \times 1000} \times 624 \times 8 \text{ litre}$$

$$\text{Thus, the cost of milk} = ₹ \frac{22}{7 \times 1000} \times 624 \times 8 \times 21$$

$$= ₹\ 329.47 \approx ₹\ 330$$ **Ans.**

Q. 6. Three cubes of a metal whose edges are in the ratio of 3 : 4 : 5 are melted and converted into a single cube whose diagonal is 12$\sqrt{3}$ cm. Find the edges of the three cubes.★

Solution : Given, diagonal of the larger cube = $12\sqrt{3}$ cm

Ratio of the sides of the three smaller cube = 3 : 4 : 5

Let the common factor between the edges of the smaller cubes be x cm.

Sum of ratios of the edges of the smaller cubes

$$= 3 + 4 + 5 = 12$$

Thus, $\text{Edge of 1}^{\text{st}} \text{ cube} = \frac{3x}{12}$

$\text{Edge of 2}^{\text{nd}} \text{ cube} = \frac{4x}{12}$

and $\text{Edge of 3}^{\text{rd}} \text{ cube} = \frac{5x}{12}$

★ are board exam questions from previous years

Thus, volume of the 1st cube = $\left(\frac{3x}{12}\right)^3 = \frac{27x^3}{1728}$ cm^3

Volume of the 2nd cube = $\left(\frac{4x}{12}\right)^3 = \frac{64x^3}{1728}$ cm^3

and Volume of the 3rd cube = $\left(\frac{5x}{12}\right)^3 = \frac{125x^3}{1728}$ cm^3

Now, diagonal of the larger cube = $12\sqrt{3} = a\sqrt{3}$

Thus, side of the larger cube (a) = 12

Volume of the larger cube = $(12)^3 = 1728$ cm^3

$$\text{Thus,} \quad \left(\frac{27x^3}{1728}+\frac{64x^3}{1728}+\frac{125x^3}{1728}\right) = 1728$$

$$\Rightarrow \quad \left(\frac{27x^3+64x^3+125x^3}{1728}\right) = 1728$$

$$\Rightarrow \quad \left(\frac{216x^3}{1728}\right) = 1728$$

$$\Rightarrow \quad \left(\frac{x^3}{8}\right) = 1728$$

$$\Rightarrow \quad \left(\frac{x}{2}\right)^3 = (12)^3$$

$$\Rightarrow \quad \frac{x}{2} = 12$$

$$\Rightarrow \quad x = 24 \text{ cm}$$

Thus, Edge of 1st cube = $\frac{3x}{12} = \frac{3\times 24}{12} = 6$ cm

Edge of 2nd cube = $\frac{4x}{12} = \frac{4\times 24}{12} = 8$ cm

and Edge of 3rd cube = $\frac{5x}{12} = \frac{5\times 24}{12} = 10$ cm. **Ans.**

Q. 7. Water is flowing through a cylindrical pipe of internal diameter 2 cm into a cylindrical tank of base radius 40 cm at the rate of 0.4 m/s. Determine the rise in the water level in the tank in half an hour.★

Solution : Given, internal diameter of cylindrical pipe = 2 cm

Thus, radius of the cylindrical pipe = 1 cm

Rate of flow of water through the pipe

= 0.4 m/s = 40 cm/s

Base radius of the tank = 40 cm

Time = 30 min = (30 × 60) sec = 1800 sec

Let the rise in water level = h m

Now, volume of water that flows through the pipe in 1 sec

$= \pi(1)^2 (40)$ cm^3

$= 40\pi$ cm^3

★ are board exam questions from previous years

Thus, volume of water that flows through the pipe in 1800 sec = $(1800)40\pi$ cm^3

Thus, volume of water that collects in the tank in 1800 sec $= (1800)40\pi$ cm^3

$$\therefore \quad \pi(40)^2 h = (1800)40\pi$$

$$\Rightarrow \quad (40)^2 h = (1800)40$$

$$\Rightarrow \quad 40h = 1800$$

$$\Rightarrow \quad h = 45 \text{ cm}$$

Thus, rise in the water level in the tank in half an hour is 45 cm. **Ans.**

Q. 8. A hemispherical tank full of water is emptied by a pipe at the rate of $\frac{25}{7}$ l/s. How much time will it take to empty half the tank if the diameter of the base of the tank is 3 m ?★

Solution : Given, diameter of the tank = 3 m

∴ Radius = 1.5 m

Rate of flow of water = $\frac{25}{7}$ l/s

Thus, Volume = $\frac{2}{3}\pi(1.5)^3$ m^3

$= 2.25\pi$ m^3

$= 2250\pi\, l$ $\quad [\because 1 \text{ m}^3 = 1000\, l]$

Thus, half of the volume of the tank = $1125\pi\, l$

Thus, the time taken to empty half of the tank

$$= \frac{1125\pi}{\frac{25}{7}} \text{ sec}$$

$$= \frac{1125\times\frac{22}{7}}{\frac{25}{7}} \text{ sec}$$

$$= \frac{1125\times 22}{25} \text{ sec}$$

= 990 sec

= 16 minutes 30 seconds. **Ans.**

Q. 9. A military tent of height 8.25 m is in the form of a right circular cylinder of base diameter 30 m and height 5.5 m surmounted by a right circular cone of the same base radius. Find the length of the canvas used in making the tent, if the breadth of the canvas is 1.5 m.★

Solution : Given, total height of the tent = 8.25 m

Height of the cylindrical structure = 5.5 m

Base diameter of cylindrical structure

= Base diameter of conical structure = 30 m

Breadth of the canvas = 1.5 m

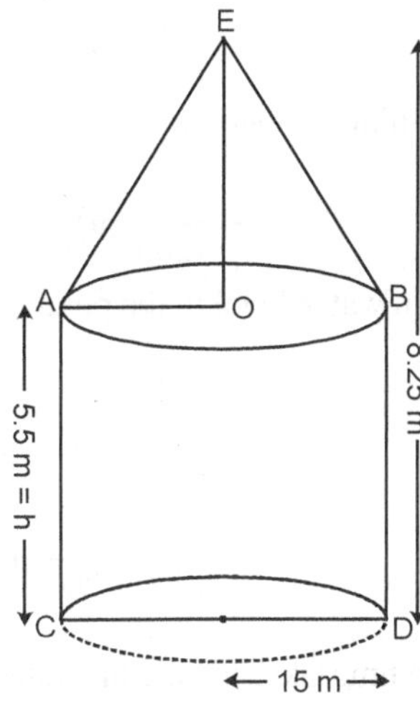

Thus, height of the conical structure

$= (8.25 - 5.50)$ m

$= 2.75$ m

Radius of cylindrical structure

= Radius of conical structure

= 15 m

Thus, surface area of the tent

= Curved surface area of the (cylindrical structure + conical structure)

$= 2\pi rh + \pi rl$

$= 2\pi(15)(5.5) + \pi 15\sqrt{(15)^2 + (2.75)^2}$

$= 15\pi\left[2 \times 5.5 + \sqrt{232.56}\right]$

$= 15\pi\,[11 + 15.25]$

$= 15\pi \times 26.25$

$= 1237.5 \text{ m}^2$

Thus, the length of canvas required

$= \dfrac{1237.5}{1.5}$ m

$= 825$ m. **Ans.**

Q. 10. A solid iron pole consists of a cylinder of height 220 cm and base diameter 24 cm, which is surmounted by another cylinder of height 60 cm and radius 8 cm. Find the mass of the pole, given that 1 cm^3 of iron has approximately 8 gm mass.★ (Use $\pi = 3{\cdot}14$)

Solution : Let AB be the iron pole of height 220 cm with base radius 12 cm and there is an other cylinder CD of height 60 cm whose base radius is 8 cm.

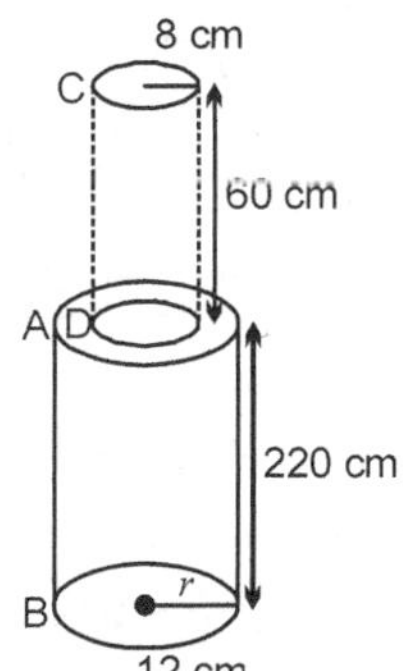

Volume of AB pole $= \pi r_1^2 h_1$

$= 3{\cdot}14 \times 12 \times 12 \times 220$

$= 99475{\cdot}2 \text{ cm}^3$

Volume of CD pole $= \pi r_2^2 h_2$

$= 3{\cdot}14 \times 8 \times 8 \times 60$

$= 12057{\cdot}6 \text{ cm}^3$

Total volume of the poles

$= 99475{\cdot}2 + 12057{\cdot}6$

$= 111532{\cdot}8 \text{ cm}^3$

It is given that,

Mass of 1 cm^3 of iron = 8 gm

Then mass of 111532·8 cm^3 of iron

$= 111532{\cdot}8 \times 8$ gm

Then total mass of the pole is $= 111532{\cdot}8 \times 8$ gm

$= 892262{\cdot}4$ gm

$= 892{\cdot}2624$ kg **Ans.**

Q. 11. A solid is in the shape of a cone surmounted on a hemisphere, the radius of each of them being 3.5 cm and the total height of the solid is 9.5 cm. Find the volume of the solid. ★ $\left[\textbf{Use } \pi = \dfrac{22}{7}\right]$

Solution : Given, radius of hemisphere

= radius of cone = 3.5 cm

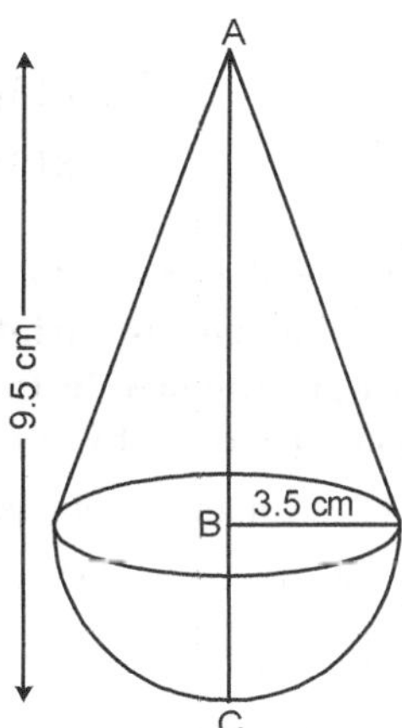

Total height of the solid = 9.5 cm

Height of the hemisphere

= Radius of the hemisphere

= 3.5 cm

∴ Height of the cone $= (9.5 - 3.5)$ cm $= 6$ cm

Hence, volume of the solid

= Volume of the hemisphere

+ Volume of the cone

$= \dfrac{2}{3}\pi r^3 + \dfrac{1}{3}\pi r^2 h$

$= \dfrac{1}{3}\pi r^2(2r + h)$

$= \dfrac{1}{3}\pi(3.5)^2\,[2(3.5) + (6)]$

★ are board exam questions from previous years

$$= \frac{1}{3}\pi(3.5)^2[7+6]$$

$$= \frac{1}{3}\left(\frac{22}{7}\right)(3.5)^2(13)$$

$$= \frac{77\times 13}{6}$$

$= 166.83$ cm^3. **Ans.**

Q. 12. A bucket is in the shape of frustum of a cone and it can hold 28.49 *l* of water. If the radii of its circular ends are 28 cm and 21 cm, find the height of the bucket.★

$$\left[\textbf{Use } \pi = \frac{22}{7}\right]$$

Solution : Given, volume of bucket

$= 28.49\ l = 28490$ cm^3

$[\because 1\ l = 1000 \text{ cm}^3]$

Radius of larger end (R) = 28 cm

Radius of smaller end (r) = 21 cm

Let the height of the frustum be h cm.

Thus,

$$\frac{\pi}{3}[(28)^2 + (21)^2 + (28)(21)]h = 28490$$

$$\Rightarrow \quad \frac{22}{21}[784 + 441 + 588]h = 28490$$

$$\Rightarrow \quad \frac{1}{21}[784 + 441 + 588]h = 1295$$

$$\Rightarrow \quad \frac{1}{21}[1813]h = 1295$$

$$\Rightarrow \quad [1813]h = 27195$$

$$\Rightarrow \quad h = 15 \text{ cm.}$$ **Ans.**

Q. 13. Water is flowing at the rate of 15 km/hr through a pipe of diameter 14 cm into a cuboidal pond which is 50 m long and 44 m wide. In what time will the level of water in the pond rise by 21 cm ?★★

Solution : Given, diameter of pipe = 14 cm = 0.14 m

$\therefore$ Radius of the pipe = 0.07 m

Length of pond = 50 m

Breadth of pond = 44 m

Height of water in the pond = 21 cm = 0.21 m

Rate of flow of water = 15 km/hr

$$= \frac{15\times 5}{18} \text{ m/sec}$$

$$= \frac{25}{6} \text{ m/sec}$$

Thus in 1 second, the volume of water that flows through the pipe

$$= \pi(0.07)^2\,\frac{25}{6}\ \text{m}^3$$

$$= \frac{22}{7}(0.07)^2\,\frac{25}{6}\ \text{m}^3$$

$$= \frac{77}{1200}\ \text{m}^3$$

Thus, volume of water flows into the pond in 1 sec.

$$= \frac{77}{1200}\ \text{m}^3$$

$\therefore$ Rise in water-level in the pond in 1 sec.

$$= \frac{\frac{77}{1200}}{44\times 50}\ \text{m}$$

$$= \frac{7}{1200\times 200}\ \text{m}$$

Thus, the time taken to raise the water level by $\frac{21}{100}$ m

$$= \frac{\frac{21}{100}}{\frac{7}{1200\times 200}}\ \text{sec}$$

$$= \frac{21}{100}\times\frac{1200\times 200}{7}\ \text{sec}$$

$$= \frac{3}{1}\times\frac{1200\times 2}{1}\ \text{sec}$$

$= 7200$ sec

$$= \frac{7200}{60}\ \text{min}$$

$= 120$ min $= 2$ hours. **Ans.**

Q. 14. From a solid right circular cylinder of height 2.4 cm and radius 0.7 cm, a right circular cone of same height and same radius is cut out. Find the total surface area of the remaining solid.★

Solution : Given, height of cylinder = 2.4 cm

Radius of base = 0.7 cm

and slant height, $l = \sqrt{h^2 + r^2}$

$$= \sqrt{(2.4)^2 + (0.7)^2}$$

$$= \sqrt{5.76 + 0.49}$$

$$= \sqrt{6.25}$$

$= 2.5$ cm

2.4 cm

0.7 cm

Total surface area of remaining solid

= CSA of cylinder + CSA of cone + Area of top

$$= 2\pi rh + \pi rl + \pi r^2$$

★ are board exam questions from previous years

★★ are frequently asked board exam questions

$= \pi r [2h + l + r]$

$= \frac{22}{7} \times 0.7 [2 \times 2.4 + 2.5 + 0.7]$

$= 2.2 [4.8 + 2.5 + 0.7]$

$= 2.2 \times 8$

$= 17.6 \text{ cm}^2$ **Ans.**

Q. 15. From a solid cylinder of height 15 cm and diameter 16 cm, a conical cavity of the same height and same diameter is hollowed out. Find the total surface area of the remaining solid.★ [Take $\pi = 3.14$]

Solution : Given, height of solid cylinder and cone $(h) = 15$ cm

Diameter of cylinder and cone = 16 cm

∴ Radius of cylinder and cone $(r) = 8$ cm

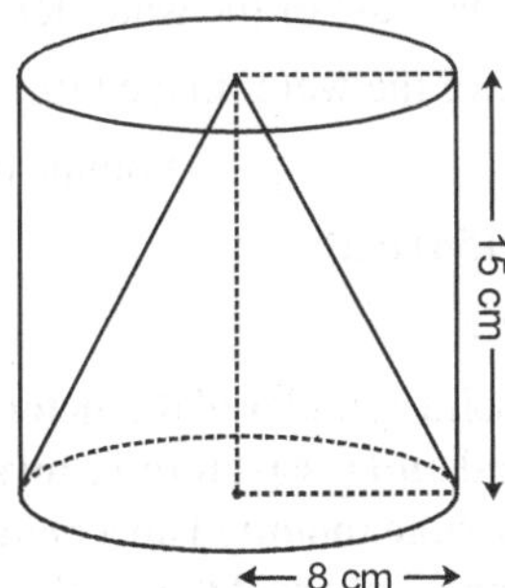

Thus, curved surface area of cylinder

$= 2\pi rh$

$= 2 \times \pi \times 8 \times 15 \text{ cm}^2$

$= 240\,\pi \text{ cm}^2$

and curved surface area of cone

$= \pi(8)(\sqrt{(8)^2 + (15)^2}) \text{ cm}^2$

$= \pi(8)(\sqrt{64 + 225}) \text{ cm}^2$

$= \pi(8)(\sqrt{289}) \text{ cm}^2$

$= \pi(8)(17) \text{ cm}^2$

$= 136\pi \text{ cm}^2$

and the area of the top of cylinder

$= \pi(8)^2 \text{ cm}^2$

$= 64\pi \text{ cm}^2$

Hence, the total surface area of the remaining solid

$= 240\pi + 136\pi + 64\pi$

$= 440\pi \text{ cm}^2$

$= 440(3.14) \text{ cm}^2$

$= 1381.6 \text{ cm}^2.$ **Ans.**

Q. 16. Water is flowing at the rate of 6 km/hr through a pipe of diameter 14 cm into a rectangular tank which is 60 m long and 22 m wide. In what time will the level of water in the pond rise by 7 cm ?★

Solution : Given, diameter of pipe = 14 cm = 0.14 m

∴ Radius of the pipe = 0.07 m

Length of tank = 60 m

Breadth of tank = 22 m

Height of water in the tank

= 7 cm = 0.07 m

Rate of flow of water

= 6 km/hr

= 100 m/min

Volume of water flows through pipe in 1 minute (V_1)

$= \frac{22}{7} \times (0.07)^2 \times 100$

$= 1.54$

Volume of water in the tank with 7cm height (V_2)

$= 60 \times 22 \times \frac{7}{100}$

$= 13.2 \times 7$

Time taken $= \frac{V_2}{V_1}$

$= \frac{13.2 \times 7}{1.54}$

$= 60$ min

$= 1$ hr **Ans.**

Q. 17. A hollow sphere of internal and external diameters 4 cm and 8 cm respectively is melted to form a cone of base diameter 8 cm. Find the height and slant height of the cone.★

Solution : Given, external diameter of sphere = 8 cm

Internal diameter of sphere = 4 cm

Diameter of cone = 8 cm

Thus, external radius of sphere = 4 cm

Internal radius of sphere = 2 cm

Base radius of cone = 4 cm

Let the height of the cone be h cm.

Now, volume of hollow sphere = Volume of cone

$\Rightarrow \frac{4}{3}\pi[(4)^3 - (2)^3] = \frac{1}{3}\pi(4)^2 h$

$\Rightarrow 64 - 8 = 4h$

$\Rightarrow h = 14$ cm

Let the slant height of the cone be l cm.

Thus, $l = \sqrt{(4)^2 + (14)^2}$ cm

$= \sqrt{16 + 196}$ cm

$= \sqrt{212}$ cm

$= 2\sqrt{53}$ cm

Thus, the height of the cone is 14 cm and the slant height is $2\sqrt{53}$ cm. **Ans.**

Q. 18. A toy is in the shape of a cone mounted on a hemisphere, the radius of each of them being 3.5 cm and the total height of the solid is 15.5 cm. Find the volume and total surface area of the toy.★★

$\left[\text{Use } \pi = \frac{22}{7}\right]$

★ are board exam questions from previous years

Solution : Given, radius of hemisphere = Radius of cone = 3.5 cm

Total height of the solid = 15.5 cm

$\because$ Height of the hemisphere

= Radius of the hemisphere = 3.5 cm

$\therefore$ Height of the cone = (15.5 – 3.5) cm = 12 cm

Hence, volume of the solid

= Volume of the hemisphere + Volume of the cone

$$= \left[\frac{2}{3}\pi(3.5)^3 + \frac{1}{3}\pi(3.5)^2(12)\right]$$

$$= \frac{1}{3}\pi(3.5)^2[2(3.5) + (12)]$$

$$= \frac{1}{3}\pi(3.5)^2[7 + 12]$$

$$= \frac{1}{3}\pi(3.5)^2[19]$$

$$= \frac{1}{3}\left(\frac{22}{7}\right)(3.5)^2(19)$$

$$= \frac{77 \times 19}{6}$$

$$= 243.83 \text{ cm}^3$$

Total surface area of the toy

= Curved surface area of the (hemisphere + cone)

$$= [2\pi(3.5)^2 + \pi(3.5)\sqrt{(3.5)^2 + (12)^2}]$$

$$= \pi(3.5)[2.(3.5) + \sqrt{12.25 + 144}]$$

$$= 11[7 + \sqrt{156.25}]$$

$$= 11[7 + 12.5] = 11[19.5]$$

= 214.5 cm². **Ans.**

Q. 19. A cylindrical tub of radius 5 cm and length 9.8 cm is full of water. A solid in the form of a right circular cone mounted on a hemisphere is immersed into the tub. If the radius of the hemisphere is 3.5 cm and the height of the cone is 5 cm, find the volume of water left in the tub.★

$$\left[\textbf{Use } \pi = \frac{22}{7}\right]$$

Solution :

Given, Radius of cylindrical tub = 5 cm

Height of cylindrical tub = 9.8 cm

Radius of hemisphere = 3.5 cm

Height of cone = 5 cm

Thus, volume of water in the tub

= Volume of the cylinder

$$= \pi(5)^2(9.8)$$

$$= \frac{22}{7}(25)(9.8)$$

$$= 22 \times 35$$

$$= 770 \text{ cm}^3$$

Now, volume of water displaced

= Volume of the solid

= Volume of (hemisphere + cone)

$$= \left[\frac{2}{3}\pi(3.5)^3 + \frac{1}{3}\pi(3.5)^2 5\right]$$

$$= \frac{1}{3}\pi(3.5)^2[2(3.5) + 5]$$

$$= \frac{22}{21}(3.5)^2[7 + 5]$$

$$= \frac{22}{21}(3.5)^2[12]$$

$$= 154 \text{ cm}^3$$

Thus, the volume of the water left in the tub

= Volume of the water in the tub – Volume of water displaced

= (770 – 154) cm³

= 616 cm³. **Ans.**

Q. 20. A solid cylinder of diameter 12 cm and height 15 cm is melted and recast into 12 toys in the shape of a right-circular cone mounted on a hemisphere. Find the radius of the hemisphere if the total height of the cone is 3 times the radius.★★

Solution : Given,

Diameter of the cylinder = 12 cm

$\therefore$ Radius = 6 cm

Height of the cylinder = 15 cm

Number of toys = 12

Let the radius of the hemisphere be r cm

Thus, height of the cone = $3r$ cm

Now, volume of cylinder = $\pi(6)^2$ 15 cm³

= (36) (15)π cm³

= 540π cm³

Now, total volume of 12 toys

= 12[Volume of hemisphere + Volume of cone]

Thus, total volume of 12 toys

$$= 12\left[\frac{1}{3}\pi 3r(r)^2 + \frac{2}{3}\pi r^3\right] \text{ cm}^3$$

$$= 4[3\pi r^3 + 2\pi r^3] \text{ cm}^3$$

$$= 20\pi r^3 \text{ cm}^3$$

Now, $20\pi r^3 = 540\pi$

$\Rightarrow r^3 = 27$

$\Rightarrow r^3 = (3)^3$

$\Rightarrow r = 3$ cm

★ are board exam questions from previous years

★★ are frequently asked board exam questions

Thus, radius of the hemisphere = 3 cm

and height of the cone = (3)(3) cm = 9 cm

∴ Total height of the toy

= (9 + 3) cm = 12 cm. **Ans.**

Q. 21. A well of diameter 4 m is dug 14 m deep. The earth taken out is spread evenly all around the well to form a 40 cm high embankment. Find the width of the embankment.★

Solution : We have, diameter of well = 4 m

∴ Radius = 2 m

and height = 14 m.

Volume of earth taken out after digging the well

$$= \frac{22}{7} \times 2 \times 2 \times 14$$

$$= 176 \text{ m}^3$$

Let x be the width of the embankment formed by the earth taken out.

Then, volume of embankment

$$= \frac{22}{7}[(2+x)^2 - (2)^2] \times \frac{40}{100} = 176$$

$$\Rightarrow \frac{22}{7}[4 + x^2 + 4x - 4] \times \frac{2}{5} = 176$$

$$\Rightarrow x^2 + 4x = \frac{176 \times 5 \times 7}{22 \times 2}$$

$$\Rightarrow x^2 + 4x - 140 = 0$$

$$\Rightarrow x^2 + 14x - 10x - 140 = 0$$

$$\Rightarrow x(x + 14) - 10(x + 14) = 0$$

$$\Rightarrow (x + 14)(x - 10) = 0$$

$$\Rightarrow x = -14 \text{ or } 10$$

$$x = -14 \text{(neglect)}$$

$$\therefore x = 10$$

Hence, width of embankment = 10 m. **Ans.**

Q. 22. Water is flowing at the rate of 2.52 km/h through a cylindrical pipe into a cylindrical tank, the radius of whose base in 40 cm. If the increase in the level of water in the tank, in half an hour is 3.15 m, find the internal diameter of the pipe.★

Solution : Let the internal radius of the pipe be x m.

Radius of base of tank = 40 cm = $\frac{2}{5}$ m

Speed of water flowing through the pipe

= 2.52 km/hr

$$= \frac{2.52}{2} \times 1000$$

= 1260 m in half an hour

Volume of water flown through pipe in half an hour

$$= \pi r^2 h$$

$$= \frac{22}{7} \times x \times x \times 1260$$

$$= 3960\, x^2$$

Level of water raised in the tank = 3.15 m = $\frac{315}{100}$ m

Volume of water in the tank with $\frac{315}{100}$ m height

$$= \pi \times \frac{2}{5} \times \frac{2}{5} \times \frac{315}{100}$$

Now, $\pi \times \frac{2}{5} \times \frac{2}{5} \times \frac{315}{100} = 3960\, x^2$

$$\Rightarrow x^2 = \frac{22 \times 2 \times 2 \times 315}{7 \times 5 \times 5 \times 100 \times 3960}$$

$$\Rightarrow x^2 = \frac{4}{10000}$$

$$x = \frac{2}{100}$$

$$= 0.02 \text{ m}$$

Internal diameter of the pipe = 0.04 m = 4 cm **Ans.**

Q. 23. A solid is composed of a cylinder with hemispherical ends. If the whole length of the solid is 104 cm and the radius of each of its hemispherical ends is 7 cm, find the cost of polishing its surface at the rate of ₹ 10 per dm². ★ $\left[\text{Use } \pi = \frac{22}{7}\right]$

Solution : Given, length of solid (hemisphere + cylinder + hemisphere) = 104 cm

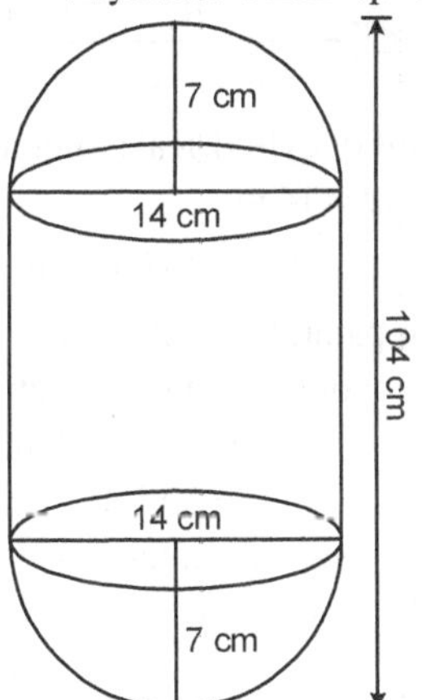

Radius of the hemisphere = Radius of cylinder = 7cm

Rate of polishing = ₹ 10 per dm²

Thus, height of the cylinder = [104 – 2(7)] cm

= 90 cm

Now, total surface area to be polished

= 2[Curved surface area of a hemisphere]

+ Curved surface area of the cylinder

$$= [2\{2\pi(7)^2\} + 2\pi(7)(90)] \text{ cm}^2$$

$$= 2\pi(7)[2(7) + 90] \text{ cm}^2$$

$$= 2\left(\frac{22}{7}\right)(7)[14 + 90] \text{ cm}^2$$

★ are board exam questions from previous years

$= 44 \times 104 \text{ cm}^2$

$= 4576 \text{ cm}^2$

Now, $10 \text{ cm} = 1 \text{ dm}$

Thus, $100 \text{ cm}^2 = 1 \text{ dm}^2$

So, $4576 \text{ cm}^2 = 45.76 \text{ dm}^2$

Thus, total cost of polishing the structure at ₹ 10 per dm^2 = ₹ (10×45.76) = ₹ 457.60. **Ans.**

Q. 24. A tent is in the form of a rightcircular cylinder of base radius 14 m and height 3 m surmounted by a right circular cone of the same base radius. The total height of the tent is 13.5 m. Find the cost of the canvas used in making the tent at ₹ 80 per square metre and the cost of painting it at ₹ 2 per square metre.★★ $\left[\text{Use } \pi = \frac{22}{7}\right]$

Solution : Given, height of the cylindrical structure = 3 m

Total height of the tent = 13.5 m

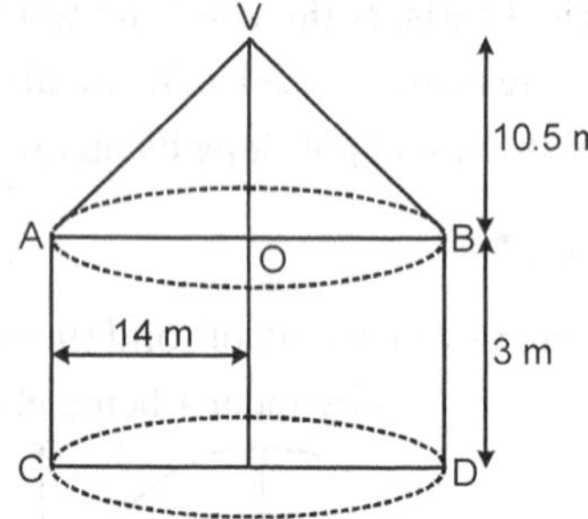

Base radius of cylindrical structure = Base radius of conical structure = 14 m

Rate of canvas = ₹ 80 per m^2

Rate of painting = ₹ 2 per m^2

Now, height of the conical structure

$= (13.5 - 3) \text{ m}$

$= 10.5 \text{ m}$

∴ Total surface area of the tent

= Curved surface area of the (cylindrical structure + conical structure)

$= 2\pi rh + \pi rl$

$= \pi r(2h + l)$

$= \pi(14)[2(3) + (\sqrt{(14)^2 + (10.5)^2}\,] \text{ m}^2$

$= \pi(14)(6 + \sqrt{196 + 110.25}) \text{ m}^2$

$= (14)\pi(6 + \sqrt{306.25}) \text{ m}^2$

$= (14)\frac{22}{7}(6 + \sqrt{(17.5)^2}) \text{ m}^2$

$= 44(6 + 17.5) \text{ m}^2$

$= 44 \times 23.5 \text{ m}^2$

$= 1034 \text{ m}^2$

Thus, the cost of canvas = ₹ (80×1034)

= ₹ 82720

and, the cost of painting = ₹ (2×1034)

= ₹ 2068. **Ans.**

Q. 25. A container shaped like a cylinder having diameter 12 cm and height 15 cm is full of ice-cream. This ice-cream is to be filled into cones of height 12 cm and diameter 6 cm having a hemispherical shape on the top. Find the number of such cones that can be filled with ice-cream.★★ $\left[\text{Use } \pi = \frac{22}{7}\right]$

Solution : Given,

Diameter of cylinder = 12 cm

∴ Radius of cylinder = 6 cm

Height of cylinder = 15 cm

Height of cone = 12 cm

Diameter of cone = 6 cm

∴ Radius of cone = 3 cm

Diameter of hemisphere = 6 cm

∴ Radius of hemisphere = 3 cm

Hence, volume of cylinder $= \pi(6)^2\, 15 \text{ cm}^3$

$= \pi(36)15 \text{ cm}^3$

$= 540\pi \text{ cm}^3$

Volume of cone $= \frac{1}{3}\pi(3)^2 12 \text{ cm}^3$

$= 36\pi \text{ cm}^3$

Volume of hemisphere $= \frac{2}{3}\pi(3)^3 \text{ cm}^3$

$= 18\pi \text{ cm}^3$

∴ Volume of icecream in each cone

$= (36\pi + 18\pi) \text{ cm}^3$

$= 54\pi \text{ cm}^3$

Thus, number of icecream cones

$= \frac{\text{Total volume of icecream in the cylinder}}{\text{Total volume of icecream in each cone}}$

$= \frac{540\pi}{54\pi} = 10$

Hence, the number of icecream cones = 10. **Ans.**

Q. 26. A gulabjamun when ready for eating contains sugar syrup of about 30% of its volume. Find approximately how much syrup would be found in 45 such gulabjamuns if each of them is shaped like a cylinder with two hemispherical ends. The complete length of each of them is 5 cm and the diameter is 2.8 cm.★ $\left[\text{Use } \pi = \frac{22}{7}\right]$ **[NCERT]**

Solution : Given, length of each gulabjamun = 5 cm

Diameter of each gulabjamun = 2.8 cm

Thus, radius of each gulabjamun = 1.4 cm

Total number of gulabjamuns = 45

Percentage of syrup in each gulabjamun = 30%

★ are board exam questions from previous years

★★ are frequently asked board exam questions

and length of the cylindrical part

$$= [5 - (1.4 + 1.4)] \text{ cm}$$
$$= (5 - 2.8) \text{ cm}$$
$$= 2.2 \text{ cm}$$

$\therefore$ Volume of 45 gulabjamun

= 45[Volume of (cylindrical portion + 2 hemispherical ends)]

$$= 45\left[\pi(1.4)^2 2.2 + 2\left\{\frac{2}{3}\pi(1.4)^3\right\}\right] \text{ cm}^3$$

$$= 45\pi(1.4)^2\left[2.2 + \frac{4}{3}(1.4)\right] \text{ cm}^3$$

$$= (45)\frac{22}{7}(1.4)(1.4)\left[2.2 + \frac{5.6}{3}\right] \text{ cm}^3$$

$$= (45)\frac{44}{100}(14)\left[\frac{6.6 + 5.6}{3}\right] \text{ cm}^3$$

$$= (45)\frac{44}{100}(14)\left[\frac{12.2}{3}\right] \text{ cm}^3$$

$$= 1127.28 \text{ cm}^3$$

Hence, volume of syrup in 45 gulabjamuns

$$= 30\% \text{ of } 1127.28 \text{ cm}^3$$

$$= \frac{30}{100} \times 1127.28 \text{ cm}^3$$

$$= 338.18 \text{ cm}^3.$$ **Ans.**

Q. 27. A toy is in the shape of a cone mounted on a hemisphere, the diameter of each of them being 7 cm and the total height of the toy is 14.5 cm. Find the volume and total surface area of the toy.★★

$$\left[\text{Use } \pi = \frac{22}{7}\right]$$

Solution : Given, diameter of hemisphere = Diameter of cone = 7 cm

Thus, radius of hemisphere = Radius of cone = 3.5 cm

Total height of the toy = 14.5 cm

$\because$ Height of the hemisphere = Radius of the hemisphere = 3.5 cm

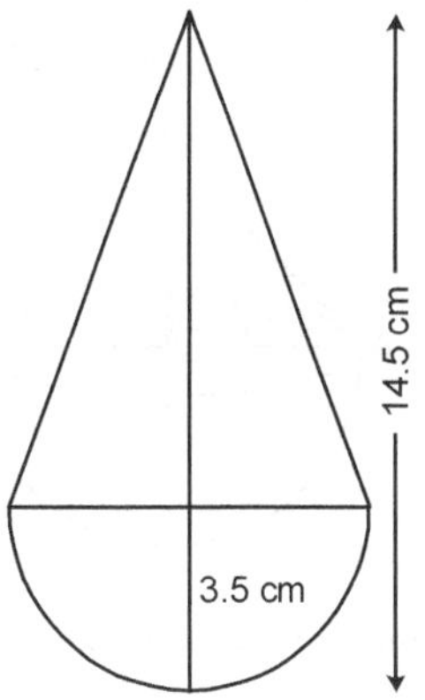

$\therefore$ Height of the cone = (14.5 – 3.5) cm = 11 cm

Hence, volume of the solid

= Volume of the hemisphere + Volume of the cone

$$= \left[\frac{2}{3}\pi(3.5)^3 + \frac{1}{3}\pi(3.5)^2(11)\right] \text{ cm}^3$$

$$= \frac{1}{3}\pi(3.5)^2[2(3.5) + (11)] \text{ cm}^3$$

$$= \frac{1}{3}\pi(3.5)^2[7 + 11] \text{ cm}^3$$

$$= \frac{1}{3}\pi(3.5)^2[18] \text{ cm}^3$$

$$= \frac{1}{3}\left(\frac{22}{7}\right)(3.5)^2(18) \text{ cm}^3$$

$$= 231 \text{ cm}^3$$

Total surface area of the toy

= Curved surface area of the (hemisphere + cone)

$$= [2\pi(3.5)^2 + \pi(3.5)\sqrt{(3.5)^2 + (11)^2}] \text{ cm}^2$$

$$= \pi(3.5)[2(3.5) + \sqrt{12.25 + 121}] \text{ cm}^2$$

$$= \frac{22}{7} \times 3.5[7 + \sqrt{133.25}] \text{ cm}^2$$

$$= 11[7 + 11.54] \text{ cm}^2$$

$$= 11[18.54] \text{ cm}^2$$

$$= 203.94 \text{ cm}^2.$$ **Ans.**

Q. 28. A solid is in the form of a right-circular cone mounted on a hemisphere. The radius of the hemisphere is 2.1 cm and the height of the cone is 4 cm. The solid is placed in a cylindrical tub full of water in such a way that the whole solid is submerged in water. If the radius of the cylinder is 5 cm and its height is 9.8 cm, find the volume of water left in the tub.★★

$$\left[\text{Use } \pi = \frac{22}{7}\right]$$

Solution : Given, radius of the cylinder = 5 cm

Height of the cylinder = 9.8 cm

Radius of the hemisphere and cone = 2.1 cm

★ are board exam questions from previous years

★★ are frequently asked board exam questions

Height of cone = 4 cm

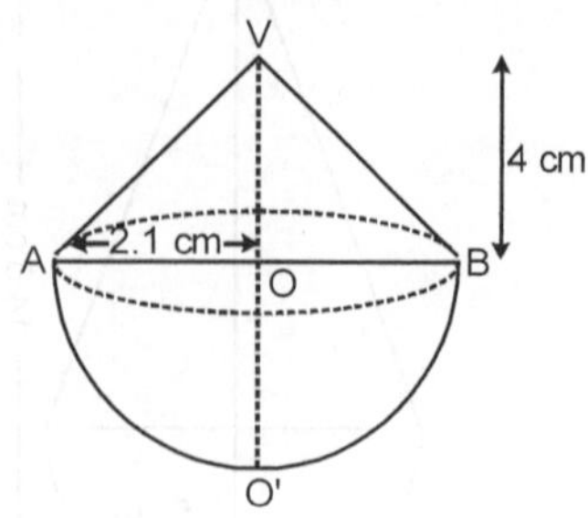

Thus, volume of water in the tub

$= \text{Volume of the cylinder}$

$= \pi(5)^2 (9.8) \text{ cm}^3$

$= \frac{22}{7}(25)(9.8) \text{ cm}^3$

$= 22(35) \text{ cm}^3$

$= 770 \text{ cm}^3$

Now, volume of water displaced

$= \text{Volume of the solid}$

$= \text{Volume of (hemisphere + cone)}$

$= \left[\frac{2}{3}\pi(2.1)^3 + \frac{1}{3}\pi(2.1)^2 4\right] \text{ cm}^3$

$= \frac{1}{3}\pi(2.1)^2 [2(2.1)+4] \text{ cm}^3$

$= \frac{22}{21}(2.1)^2 [4.2+4] \text{ cm}^3$

$= (2.1)(2.2)(8.2) \text{ cm}^3$

$= 37.88 \text{ cm}^3$

Thus the volume of the water left in the tub

$= \text{Volume of water in the tub} - \text{Volume of water displaced}$

$= (770 - 37.88) \text{ cm}^3$

$= 732.12 \text{ cm}^3$. **Ans.**

Q. 29. A solid is in the form of a cylinder with hemispherical ends. The total height of the solid is 20 cm and the diameter of the cylinder is 7 cm. Find the total volume of the solid.★ $\left[\text{Use } \pi = \frac{22}{7}\right]$

Solution : ABCD is a cylinder and BFC and AED are two hemisphere which has radius $(r) = \frac{7}{2}$ cm

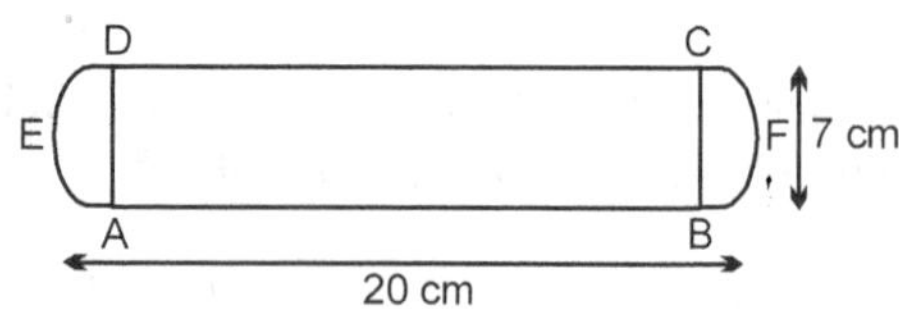

Here, $AB = 20 - 2 \times \frac{7}{2}$

$h = 13 \text{ cm}, r = \frac{7}{2} \qquad \left(r = \frac{d}{2}\right)$

Volume of cylinder $= \pi r^2 h$

$= \frac{22}{7} \times \frac{7}{2} \times \frac{7}{2} \times 13$

$= \frac{11 \times 13 \times 7}{2}$

$= \frac{1001}{2}$

$= 500{\cdot}5 \text{ cm}^3$

Volume of two hemisphere $= 2 \times \frac{2}{3}\pi r^3$

$= 2 \times \frac{2}{3} \times \frac{22}{7} \times \frac{7}{2} \times \frac{7}{2} \times \frac{7}{2}$

$= \frac{49 \times 11}{3}$

$= \frac{539}{3}$

$= 179{\cdot}67 \text{ cm}^3$

Total volume of solid

= Volume of two hemisphere + Volume of cylinder

$= 179{\cdot}67 + 500{\cdot}5$

$= 680{\cdot}{\cdot}17 \text{ cm}^3$ **Ans.**

Q. 30. A farmer connects a pipe of internal diameter 25 cm from a canal into a cylindrical tank which is 12 m in diameter and 2.5 m deep. If the water flows through the pipe at the rate of 3.6 km/hr, in how much time will the tank be completely filled ? Also find the cost of water if the irrigation department charges at the rate of ₹ 0.07per m³.★ $\left[\text{Use } \pi = \frac{22}{7}\right]$

Solution : Given, internal diameter of the pipe =25 cm $= \frac{1}{4}$ m

Diameter of the tank = 12 m

Height of tank = 2.5 m

Rate of flow of water = 3.6 km/hr

$= \frac{3.6 \times 5}{18} \text{ m/s} = 1 \text{ m/s}$

Thus, internal radius of pipe = 12.5 cm $= \frac{1}{8}$ m

and Radius of tank = 6 m

Now, volume of water flowing through the pipe in 1 second

$= \pi\left(\frac{1}{8}\right)^2 1 \text{ m}^3$

$= \frac{\pi}{64} \text{ m}^3$

★ are board exam questions from previous years

Volume of the tank = $\pi(6)^2 (2.5)\ m^3$

$= 90\pi\ m^3$

Thus, time taken to fill the tank

$$= \frac{\text{Volume of tank}}{\text{Volume of water flowing in 1 sec}}$$

$$= \frac{90\pi}{\frac{\pi}{64}} \text{ sec}$$

$= 90(64)$ sec

$= 5860$ sec

$= 96$ min

$= 1$ hour and 36 minutes

Now, cost of water = Volume of water × Rate

$= ₹\ [90\pi \times 0.07]$

$$= ₹\left[90 \times \frac{22}{7} \times \frac{7}{100}\right]$$

$= ₹\ 19.80.$ **Ans.**

Q. 31. A juice-seller serves his customers using a glass whose inner diameter is 5 cm but the bottom of the glass has a raised hemispherical portion that reduces its capacity. If the height of the glass is 10 cm, find the apparent and actual capacities of the glass.★

[Use $\pi = 3.14$] **[NCERT]**

Solution : Given, height of glass = 10 cm

Diameter of glass = Diameter of the hemisphere = 5cm

Thus, radius of glass = Radius of the hemisphere = 2.5 cm

Now, apparent capacity of the glass

$= \pi(2.5)^2\ 10\ cm^3$

$= 196.25\ cm^3$

And, actual capacity of the glass

= Apparent capacity of the glass – Volume of the hemisphere

$$= \left[196.25 - \frac{2}{3}\pi(2.5)^3\right] cm^3$$

$= [196.25 - 32.71]\ cm^3$

$= 163.54\ cm^3$ **Ans.**

Q. 32. A cone of radius 10 cm is divided into two parts by drawing a plane through the mid-point of its axis parallel to its base. Compare the volumes of the two parts.★★

Solution : Given, radius of the cone $(CP) = (PB) = 10$ cm

Let the height of the cone (AP) be h cm

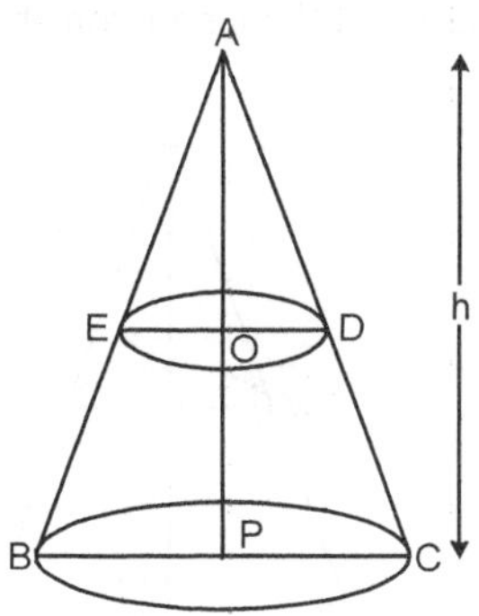

Thus, $AO = OP$

Since $OD \parallel PC$

$\Rightarrow \Delta AOD \sim \Delta APC$

$$\Rightarrow \frac{OD}{PC} = \frac{AO}{AP} = \frac{AD}{AC}$$

$$\Rightarrow \frac{OD}{10} = \frac{AO}{2AO}$$

$\Rightarrow OD = 5$ cm

Thus, Volume of cone $AED = \frac{1}{3}\pi(5)^2\frac{h}{2}\ cm^3$

and Volume of frustum $BEDC$

$$= \frac{1}{3}\pi[(10)^2 + (5)^2 + (10)(5)]\frac{h}{2}\ cm^3$$

$$= \frac{1}{3}\pi[100 + 25 + 50]\frac{h}{2}\ cm^3$$

$$= \frac{1}{3}\pi(175)\frac{h}{2}\ cm^3$$

$$\therefore \frac{\text{Volume of } AED}{\text{Volume of } BEDC} = \frac{\frac{1}{3}\pi(5)^2\frac{h}{2}}{\frac{1}{3}\pi(175)\frac{h}{2}}$$

$$\Rightarrow \frac{\text{Volume of } AED}{\text{Volume of } BEDC} = \frac{5^2}{175}$$

$$\Rightarrow \frac{\text{Volume of } AED}{\text{Volume of } BEDC} = \frac{25}{175}$$

$$\Rightarrow \frac{\text{Volume of } AED}{\text{Volume of } BEDC} = \frac{1}{7}$$

$\Rightarrow$ Volume of AED : Volume of $BEDC = 1 : 7$. **Ans.**

Q. 33. The height of a cone is 30 cm. A small cone is cut off at the top by a plane parallel to the base. If its volume is $\frac{1}{27}$ of the volume of the given cone, at what height above the base is the section made ?★★

Solution : Given, height of the larger cone = 30 cm

Ratio of volumes of the smaller and larger cones = 1 : 27

★ are board exam questions from previous years

★★ are frequently asked board exam questions

Let the radius of the larger cone be R.

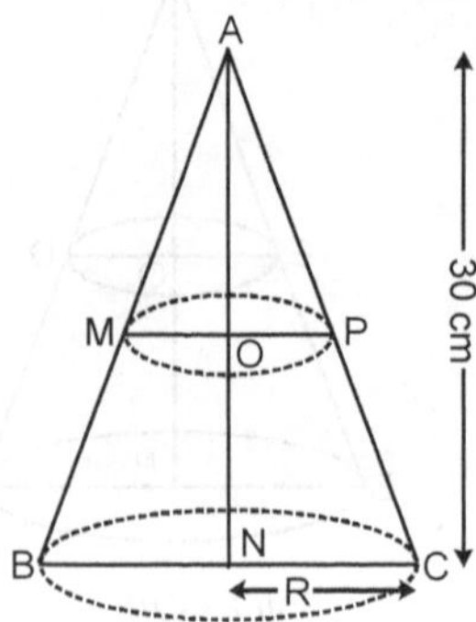

Also let the radius and the height of the smaller cone be r cm and h cm respectively.

Since $OP \parallel NC$

$\Rightarrow \quad \Delta AOP \sim \Delta ANC$

$\Rightarrow \quad \frac{AO}{AN} = \frac{OP}{NC}$

$\Rightarrow \quad \frac{h}{30} = \frac{r}{R}$

$\Rightarrow \quad h = \frac{30r}{R}$ cm

Now, $\frac{\text{Volume of } ABC}{\text{Volume of } AMP} = \frac{\frac{1}{3}\pi(R)^2(30)}{\frac{1}{3}\pi(r)^2 h}$

$= \frac{(R)^2(30)}{(r)^2 h}$

$= \frac{(R)^2(30)}{(r)^2 \frac{30r}{R}}$

$\frac{27}{1} = \frac{(R)^3}{(r)^3}$

$\Rightarrow \quad \frac{(3)^3}{1} = \frac{(R)^3}{(r)^3}$

$\Rightarrow \quad R = 3r$

Thus, $h = \frac{30r}{R}$

$= \frac{30r}{3r}$

$= 10$ cm.

$H - h = 30 - 10 = 20$ cm

Thus, the section of the cone is made at a height of 20 cm above the base. **Ans.**

Q. 34. The radii of the circular ends of a solid frustum of a cone are 33 cm and 27 cm and its slant height is 10 cm. Find the volume and the total surface area. ★

$\left[\textbf{Use } \pi = \frac{22}{7}\right]$

Solution : Given, radius of upper circle = 33 cm

Radius of lower circle = 27 cm

Slant height = 10 cm

Let the height of the cone be h cm.

We know,

Slant height $= \sqrt{h^2 + (33-27)^2}$

$\Rightarrow \quad 10 = \sqrt{h^2 + (33-27)^2}$

$\Rightarrow \quad 100 = h^2 + (6)^2$

$\Rightarrow \quad h^2 = 100 - 36$

$= 64$

$\Rightarrow \quad h^2 = 8^2$

$\Rightarrow \quad h = 8$ cm

Hence, volume of the frustum

$= \frac{\pi}{3}[R + r^2 + Rr]h$

$= \frac{\pi}{3}[(33)^2 + (27)^2 + (33)(27)]8 \text{ cm}^3$

$= \frac{\pi}{3}[1089 + 729 + 891]8 \text{ cm}^3$

$= \frac{22}{21}[2709]8 \text{ cm}^3$

$= 22704 \text{ cm}^3$

Now, total surface area of the frustum

$= \pi[l(R + r) + R^2 + r^2]$

$= \pi\{(33 + 27)10 + (33)^2 + (27)^2\} \text{ cm}^2$

$= \pi\{60 \times 10 + 1089 + 729\} \text{ cm}^2$

$= \frac{22}{7}(600 + 1089 + 729) \text{ cm}^2$

$= \frac{22}{7} \times 2418 \text{ cm}^2$

$= 7599.43 \text{ cm}^2$. **Ans.**

Q. 35. A bucket open at the top and made of a metal sheet is in the shape of a frustum. The depth of the bucket is 16 cm and the radii of its upper and lower circular ends are 20 cm and 8 cm respectively. Find the cost of metal used in it at the rate of ₹ 15 per 100 cm². ★★

[Use $\pi = 3.14$]

Solution : Given, depth of the bucket (h) = 16 cm

Radius of upper circular end (R) = 20 cm

Radius of lower circular end (r) = 8 cm

Rate of metal sheet = ₹ 15 per 100 cm²

= ₹ 0.15 per cm²

Now, slant height of the frustum

$= \sqrt{(16)^2 + (20-8)^2}$

$= \sqrt{(16)^2 + (12)^2}$

$= \sqrt{256 + 144}$

★ are board exam questions from previous years

★★ are frequently asked board exam questions

$= \sqrt{400}$

$= 20$ cm

Thus, the surface area to be covered with metal

= Lateral surface area + Area of the base

$= [\pi(20+8)20 + \pi(8)^2]$ cm^2

$= \pi[(28)20 + 64]$ cm^2

$= \pi[560 + 64]$ cm^2

$= (3.14)(624)$ cm^2

$= 1959.36$ cm^2

Thus, total cost = ₹ (1959.36 × 0.15)

= ₹ 293.90. **Ans.**

Q. 36. A tent consists of a frustum of a cone surmounted by a cone. If the diameters of the upper and lower circular ends of the frustum are 14 m and 26 m respectively, the height of the frustum is 8 m and the slant height of the surmounted conical portion is 12 m, find the area of the canvas required to make the tent assuming that the radii of the upper circular end and the base of the surmounted conical portion being equal.★★ [Use $\pi = 3.14$]

Solution : Given, diameter of the upper circular end = 14 m.

∴ Radius of the upper end = 7 m

Diameter of the lower circular end = 26 m

∴ Radius of the lower end = 13 m

Height of frustum = 8 m

Slant height of the cone = 12 m

Also radius of the cone = 7 m

Hence, slant height of the frustum

$= \sqrt{(8)^2 + (13-7)^2}$

$= \sqrt{(8)^2 + (6)^2}$

$= \sqrt{64+36}$

$= \sqrt{100} = 10$ m

∴ Total area of the canvas required

= Lateral surface area of the frustum + Curved surface area of the cone

$= [\pi(13+7)10 + \pi(7)12]$ cm^2

$= \pi[(20)10 + (7)12]$ cm^2

$= \pi[200 + 84]$ cm^2

$= 284\pi$ cm^2

$= 284(3.14)$ cm^2

$= 891.76$ cm^2. **Ans.**

Q. 37. A milk container is in the form of a frustum of a cone of height 18 cm with radii of its upper and lower ends being 8 cm and 32 cm respectively. Find the amount of milk which can completely fill the container and its cost at the rate of ₹ 20 per litre.★ [Use $\pi = 3.14$]

★ are board exam questions from previous years

★★ are frequently asked board exam questions

Solution : Given, height of frustum = 18 cm

Radius of upper end = 8 cm

Radius of lower end = 32 cm

Rate of milk = ₹ 20 per l

Thus, volume of frustum

$= \frac{\pi}{3}[(32)^2 + (8)^2 + (32)(8)]18$ cm^3

$= \frac{\pi}{3}[1024 + 64 + 256]18$ cm^3

$= \frac{\pi}{1}[1344]6$ cm^3

$= 25320.96$ cm^3

$= 25.32\ l$

Hence, cost of milk = ₹ (25.32)20

= ₹ 506.40. **Ans.**

Q. 38. A hemispherical depression is cut out from one face of a cubical block of side 21 cm such that the diameter of the hemisphere is equal to the edge of the cube. Find the volume and the total surface area of the remaining solid.★ $\left[\text{Use } \pi = \frac{22}{7}\right]$

Solution : Given, side of the cube = 21 cm

Diameter of the hemisphere = 21 cm

Thus, radius of the hemisphere = 10.5 cm

Volume of the cube $= (21)^3$ cm^3

$= 9261$ cm^3

Volume of the hemisphere

$= \frac{2}{3}\pi(10.5)^3$ cm^3

$= 2425.5$ cm^3

Thus, volume of the remaining block

$= (9261 - 2425.5)$ cm^3

$= 6835.5$ cm^3

Total surface area of the cube $= 6(21)^2$ cm^2

Curved surface area of the hemisphere

$= 2\pi(10.5)^2$ cm^2

Thus total surface area of the remaining solid

= Total surface area of cube + Curved surface area of the hemisphere – Area of the top face of the hemisphere

$= [6(21)^2 + 2\pi(10.5)^2 - \pi(10.5)^2]$ cm^2

$= \left[6(21)^2 + \left(\frac{22}{7}\right)(10.5)^2\right]$ cm^2

$= [2646 + 346.5]$ cm^2

$= 2992.5$ cm^2. **Ans.**

Q. 39. A bucket made of aluminium is in the shape of frustum. The height of the bucket is 20 cm and the radii of the two ends are 25 cm and 10 cm. Find the cost of

making the bucket if the aluminium sheet costs ₹ 70 per 100 cm².★ [Use π = 3.14]

Solution : Given, height of the bucket = 20 cm

Radius of the larger end = 25 cm

Radius of the smaller end = 10 cm

Rate of aluminium = ₹ 70 per 100 cm²

Now, slant height = $\sqrt{(20)^2+(25-10)^2}$

$= \sqrt{(20)^2+(15)^2}$

$= \sqrt{400+225}$

$= \sqrt{625}$

= 25 cm

Thus, surface area of the aluminium

= Curved surface area of the frustum + Area of the lower circular end

$= \pi[(25+10)25+\pi(10)^2]$ cm²

$= \pi[(35)25+(10)^2]$ cm²

$= \pi[975]$ cm²

= 3061.5 cm²

Thus, total cost of the aluminium sheet

$= ₹\ \frac{70}{100}(3061.5)$

= ₹ 2143.05. **Ans.**

Q. 40. A bucket made of a metal sheet is in the shape of a frustum. The height of the bucket is 16 cm and the diameters of the two ends are 40 cm and 10 cm. Find the cost of making the bucket if the metal sheet cost ₹ 20 per 100 cm² and the volume of the bucket.★★ [Use π = 3.14]

Solution : Given, height of the bucket = 16 cm

Diameter of the larger end = 40 cm

∴ Radius of the larger end = 20 cm

Diameter of the smaller end = 16 cm

∴ Radius of the smaller end = 8 cm

Rate of metal = ₹ 20 per 100 cm²

Now, slant height = $\sqrt{(16)^2+(20-8)^2}$

$= \sqrt{(16)^2+(12)^2}$

$= \sqrt{256+144}$

$= \sqrt{400}$

= 20 cm

Thus, surface area of the metal sheet

= Curved surface area of the frustum + Area of the lower circular end

$= [\pi(20+8)20+\pi(8)^2]$ cm²

$= \pi[(28)20+(8)^2]$ cm²

$= \pi[624]$ cm²

= 1959.36 cm²

Thus, total cost of the metal sheet

$= ₹\left(\frac{20}{100}\times 1959.36\right)$

= ₹ 391.87

Volume of the bucket

$= \frac{\pi}{3}[(20)^2+(8)^2+(20)(8)]16\text{ cm}^3$

$= \frac{\pi}{3}[400+64+160]16\text{ cm}^3$

$= \frac{\pi}{3}[624]16\text{ cm}^3$

$= 208(16)\pi$ cm³

= 10449.92 cm³. **Ans.**

Q. 41. A solid is in the shape of a frustum of a cone. The diameters of the two circular ends are 60 cm and 36 cm and the length is 9 cm. Find the surface area and the volume.★

Solution : Given, length of the frustum = 9 cm

Diameter of the larger end = 60 cm

∴ Radius of the larger end = 30 cm

Diameter of the smaller end = 36 cm

∴ Radius of the smaller end = 18 cm

slant height = $\sqrt{(9)^2+(30-18)^2}$ cm

$= \sqrt{(9)^2+(12)^2}$ cm

$= \sqrt{81+144}$ cm

$= \sqrt{225}$ cm

= 15 cm

Hence, total surface area

$= \pi[(30+18)15+(30)^2+(18)^2]$ cm²

$= \pi[(48)15+900+324]$ cm²

$= \pi[720+900+324]$ cm²

= 6104.16 cm²

and Volume $= \frac{\pi}{3}[(30)^2+(18)^2+(30)(18)]9\text{ cm}^3$

$= \frac{\pi}{3}[900+324+540]9\text{ cm}^3$

$= \frac{\pi}{3}[1764]9\text{ cm}^3$

$= 5292\pi$ cm³

= 16616.88 cm³. **Ans.**

★ are board exam questions from previous years

★★ are frequently asked board exam questions

Q. 42. A milk container made of metal and in the shape of a frustum has a volume of 10459 $\frac{3}{7}$ cm³. The radii of its lower and upper circular ends are 8 cm and 20 cm respectively. Find the cost of metal used in making the container at the rate of ₹ 1.40 per cm².★

$$\left[\textbf{Use } \pi = \frac{22}{7}\right]$$

Solution : Given,

Volume of the frustum $= 10459\frac{3}{7}$ cm³ $= \frac{73216}{7}$ cm³

Radius of the lower circular end = 8 cm

Radius of the upper circular end = 20 cm

Rate of metal sheet = ₹ 1.40 per cm²

Let the height of the frustum be h.

Thus, $\frac{\pi}{3}[(8)^2+(20)^2+8(20)]h = \frac{73216}{7}$

$\Rightarrow \frac{22}{21}[64+400+160]h = \frac{73216}{7}$

$\Rightarrow \frac{22}{3}[624]h = \frac{73216}{1}$

$\Rightarrow 22[208]h = 73216$

$\Rightarrow h = 16$ cm

Now, let l cm be the slant height.

Thus, $l = \sqrt{(16)^2+(20-8)^2}$

$= \sqrt{(16)^2+(12)^2}$

$= \sqrt{256+144}$

$= \sqrt{400}$

= 20 cm

Thus, the surface area that needs to be covered with metal

= Lateral surface area + Area of the lower end

$= [\pi(20+8)20+\pi(8)^2]$ cm²

$= \pi[(28)(20)+64]$ cm²

$= \pi(624)$ cm²

Thus, the total cost of the metal sheet

$= ₹\,(1.40\times 624\pi)$

$= ₹\left[1.40\times 624\left(\frac{22}{7}\right)\right]$

= ₹ 2745.60 **Ans.**

Q. 43. Water in a canal, 6 m wide and 1·5 m deep, is flowing with a speed of 10 km/hour. How much area will it irrigate in 30 minutes, if 8 cm standing water is needed ?★

Solution : Let b be the width and h be the depth of the canal

$\therefore b = 6$m and $h = 1\cdot5$ m

Water is flowing with a speed = 10 km/h

= 10,000 m/h

⇒ Length of water flowing in 1 hr = 10,000 m

Length (l) of water flowing in $\frac{1}{2}$hr = 5,000 m

Volume of water flowing in 30 min. $= l\times b\times h$

$= 5{,}000\times 6\times 1\cdot5$ m³

Let the area irrigated in 30 min $\left(\frac{1}{2}\text{hr}\right)$ be x m²

Volume of water required for irrigation

= Volume of water flowing in 30 min.

$\therefore \quad x\times\frac{8}{100} = 5{,}000\times 6\times 1\cdot5$

or $\quad x = \frac{5{,}000\times 6\times 1\cdot5\times 100}{8}$ m²

$x = 562500$ m² = 56·25 hectares.

($\because$ 1 hactare = 10^4 m²)

Hence, the canal will irrigate 56·25 hectares in 30 min. **Ans.**

Chapter 14. Statistics

Q. 1. The following table gives the distribution of expenditures of different families on education. Find the mean expenditure on education by a family.★

Class	Frequency
1000 – 1500	24
1500 – 2000	10
2000 – 2500	33
2500 – 3000	28
3000 – 3500	30
3500 – 4000	22
4000 – 4500	16
4500 – 5000	7

Solution :

Class Interval	Frequency (f_i)	Mid-Value (x_i)	$u_i = \frac{x_i - A}{h}$; h = upper limit – lower limit	$f_i \times u_i$
1000–1500	24	1250	– 4	– 96
1500–2000	10	1750	– 3	– 30
2000–2500	33	2250	– 2	– 66
2500–3000	28	2750	– 1	– 28
3000–3500	30	3250 = A	0	0

★ are board exam questions from previous years

3500 – 4000	22	3750	1	22
4000 – 4500	16	4250	2	32
4500 – 5000	7	4750	3	21
	$\Sigma f_i = 170$			$\Sigma(f_i \times u_i) = -145$

Thus, $A = 3250, h = 500,$

$\Sigma f_i = 170, \Sigma(f_i \times u_i) = -145$

Now, $\text{Mean} = A + \left\{h \times \frac{\Sigma(f_i \times u_i)}{\Sigma f_i}\right\}$

$= 3250 + \left\{500 \times \frac{(-145)}{170}\right\}$

$= 3250 - 426.47$

$=$ ₹ 2823.53. **Ans.**

Q. 2. The arithmetic mean of the following frequency distribution is 53. Find the value of *k*.★

Class	Frequency
0 – 20	12
20 – 40	15
40 – 60	32
60 – 80	k
80 – 100	13

Solution. Given, Median = 53

Class	Frequency f_i	Mid-value	$f_i x_i$
0 – 20	12	10	120
20 – 40	15	30	450
40 – 60	32	50	1600
60 – 80	k	70	$70k$
80 – 100	13	90	1170
	$72 + k$		$3340 + 70k$

$$\text{Mean} = \frac{\Sigma f_i x_i}{\Sigma f_i}$$

$$53 = \frac{3340 + 70k}{72 + k}$$

$\Rightarrow \quad 53(72 + k) = 3340 + 70k$

$\Rightarrow \quad 3816 + 53k = 3340 + 70k$

$\Rightarrow \quad k = 28$

Hence $k = 28$. **Ans.**

★ are board exam questions from previous years

Q. 3. Find the Arithmetic mean for the following frequency distribution :★

Class	Frequency
25 – 29	14
30 – 34	22
35 – 39	16
40 – 44	6
45 – 49	5
50 – 54	3
55 – 59	4

Solution : As this series is an inclusive one, we can make it exclusive by adding 0.5 to the upper limit and subtracting 0.5 from the lower limit of each class interval. Thus, we have

Class Interval	Frequency (f_i)	Mid Value (x_i)	$u_i = \frac{x_i - A}{h}$; h = upper limit – lower limit	$f_i \times u_i$
24.5 – 29.5	14	27	– 3	– 42
29.5 – 34.5	22	32	– 2	– 44
34.5 – 39.5	16	37	– 1	– 16
39.5 – 44.5	6	$42 = A$	0	0
44.5 – 49.5	5	47	1	5
49.5 – 54.5	3	52	2	6
54.5 – 59.5	4	57	3	12
	$\Sigma f_i = 70$			$\Sigma(f_i \times u_i) = -79$

Thus, $A = 42, h = 5, \Sigma f_i = 70, \Sigma(f_i \times u_i) = -79$

$\text{Mean} = A + \left\{h \times \frac{\Sigma(f_i \times u_i)}{\Sigma f_i}\right\}$

$= 42 + \left\{5 \times \frac{(-79)}{70}\right\}$

$= 42 - 5.64$

$= 36.36.$ **Ans.**

Q. 4. The marks obtained by 100 students in an examination are given below :★

Marks	Number of Students
30 – 35	14
35 – 40	16
40 – 45	28
45 – 50	23
50 – 55	18
55 – 60	8
60 – 65	3

Find the mean marks of the students.

Solution :

Class Interval (Marks)	No. of Students (f_i)	x_i	$f_i x_i$
30 – 35	14	32·5	455
35 – 40	16	37·5	600
40 – 45	28	42·5	1190
45 – 50	23	47·5	1092·5
50 – 55	18	52·5	945
55 – 60	8	57·5	460
60 – 65	3	62·5	187·5
	$\Sigma f_i = 110$		$\Sigma f_i x_i = 4930$

$$\text{Mean} = \frac{\Sigma f_i x_i}{\Sigma f_i} = \frac{4930}{110}$$

$= 44{\cdot}81$ **Ans.**

Q. 5. Find the Arithmetic mean for the following frequency distribution using Step-Deviation Method :

Class	No. of Students
4 – 8	2
8 – 12	12
12 – 16	15
16 – 20	25
20 – 24	18
24 – 28	12
28 – 32	13
32 – 36	3

Solution :

Class Interval	Frequency (f_i)	Mid Value (x_i)	$u_i = \frac{x_i - A}{h}$; h = upper limit – lower limit	$f_i \times u_i$
4 – 8	2	6	– 3	– 6
8 – 12	12	10	– 2	– 24
12 – 16	15	14	– 1	– 15
16 – 20	25	18 = A	0	0
20 – 24	18	22	1	18
24 – 28	12	26	2	24
28 – 32	13	30	3	39
32 – 36	3	34	4	12
	$\Sigma f_i = 100$			$\Sigma(f_i \times u_i) = 48$

Thus, $A = 18, h = 4, \Sigma f_i = 100, \Sigma(f_i \times u_i) = 48$

$$\text{Mean} = A + \left\{h \times \frac{\Sigma(f_i \times u_i)}{\Sigma f_i}\right\} = 18 + \left\{4 \times \frac{48}{100}\right\}$$

$= 18 + 1.92 = 19.92$ **Ans.**

★ are board exam questions from previous years

Q. 6. The table below shows the daily expenditure on food of 25 households in a locality. Find the mean daily expenditure of food.★

Daily Expenditure (in ‘) :	Number of Households :
100 – 150	4
150 – 200	5
200 – 250	12
250 – 300	2
300 – 350	2

Solution :

Daily Expenditure	No. of Households (fi)	Mid-value (x_i)	$f_i x_i$
100 – 150	4	125	500
150 – 200	5	175	875
200 – 250	12	225	2700
250 – 300	2	275	550
300 – 350	2	325	650
	$\Sigma f_i = 25$		$\Sigma f_i x_i = 5275$

$$\text{Mean } (\bar{x}) = \frac{\Sigma f_i x_i}{\Sigma f_i}$$

$$= \frac{5275}{25} = 211$$

Hence, Mean = 211 **Ans.**

Q. 7. Find the mean, median and mode of the following data :★

Class	Frequency
0 – 10	5
10 – 20	10
20 – 30	18
30 – 40	30
40 – 50	20
50 – 60	12
60 – 70	5

Solution :

Class Interval	Frequency (f_i)	Cumulative Frequency	Class Mark (x_i)	($f_i \times x_i$)
0 – 10	5	5	5	25
10 – 20	10	15	15	150
20 – 30	18	33	25	450
30 – 40	30	63	35	1050
40 – 50	20	83	45	900
50 – 60	12	95	55	660
60 – 70	5	100	65	325
	$N = \Sigma f_i = 100$			$\Sigma(f_i \times x_i) = 3560$

(i) $\text{Mean} = \dfrac{\Sigma(f_i \times x_i)}{\Sigma f_i} = \dfrac{3560}{100}$

$= 35.6$

(ii) Now, $N = \Sigma f_i = 100$

So, $\dfrac{N}{2} = 50$

The cumulative frequency just above 50 is 63.
Hence, the Median Class is 30 – 40.

So, $l = 30, h = 10, f = 30, F = 33$ and $\dfrac{N}{2} = 50$

$$\text{Median} = l + \dfrac{\frac{N}{2} - F}{f} \times h$$

$$= 30 + \dfrac{50 - 33}{30} \times 10$$

$$= 30 + 5.67 = 35.67$$

(iii) We know, Mode = 3 Median – 2 Mean

$\Rightarrow$ Mode = 3(35.67) – 2(35.6)

= 107.01 – 71.2

= 35.81. **Ans.**

Q. 8. Find the mode of following frequency distribution.★

Class	Frequency
0 – 10	8
10 – 20	10
20 – 30	10
30 – 40	16
40 – 50	12
50 – 60	6
60 – 70	7

Solution : The given frequency distribution table is

Class	Frequency
0 – 10	8
10 – 20	10
20 – 30	10
30 – 40	16
40 – 50	12
50 – 60	6
60 – 70	7

Here, the maximum class frequency is 16.

$\therefore$ Modal class = 30 – 40

$\therefore$ lower limit (l) of modal class = 40

Class size (h) = 10

Frequency (f_1) of the modal class = 16

Frequency (f_0) of preceding class = 10

Frequency (f_2) of succeeding class = 12 ,

$$\text{Mode} = l + \left(\dfrac{f_1 - f_0}{2f_1 - f_0 - f_2}\right) \times h$$

$$= 30 + \left(\dfrac{16 - 10}{32 - 10 - 12}\right) \times 10$$

$$= 30 + \dfrac{6}{32 - 22} \times 10$$

$$= 30 + \dfrac{6}{10} \times 10$$

$$= 30 + \dfrac{6}{10} \times 10$$

$$= 30 + 6 = 36$$

Hence, Mode = 36. **Ans.**

Q. 9. 100 surnames were randomly picked from a local telephone directory and the distribution of the number of letters of the English alphabet in the surnames was obtained as follows :

Letters	Surnames
1 – 4	6
4 – 7	30
7 – 10	40
10 – 13	16
13 – 16	4
16 – 19	4

Determine the median and mean number of letters in the surname. Also find the modal size of the surnames.★

Solution :

Class Interval	Frequency (f_i)	Cumulative Frequency	Class Mark (x_i)	($f_i \times x_i$)
1–4	6	6	2.5	15
4–7	30	36	5.5	165
7–10	40	76	8.5	340
10–13	16	92	11.5	184
13–16	4	96	14.5	58
16–19	4	100	17.5	70
	$N = \Sigma f_i = 100$			$\Sigma(f_i \times x_i) = 832$

(i) $\text{Mean} = \dfrac{\Sigma(f_i \times x_i)}{\Sigma f_i}$

$= \dfrac{832}{100} = 8.32$

(ii) Now, $N = \Sigma f_i = 100$

So, $\dfrac{N}{2} = 50$

The cumulative frequency just above 50 is 76.
Hence, the Median Class is 7 – 10.

So, $l = 7, h = 3, f = 40, F = 36$ and $\dfrac{N}{2} = 50$

$$\text{Median} = l + \dfrac{\frac{N}{2} - F}{f} \times h$$

★ are board exam questions from previous years

$$= 7 + \frac{50-36}{40} \times 3$$

$$= 7 + 1.05 = 8.05$$

(iii) Mode = 3 Median – 2 Mean

$\Rightarrow$ Mode = 3(8.05) – 2(8.32)

= 24.15 – 16.64

= 7.51 **Ans.**

Q. 10. The following table provides the daily income (in Rupees) of 50 workers in a factory :

Income	No. of Workers
100 – 120	12
120 – 140	14
140 – 160	8
160 – 180	6
180 – 200	10

Determine the mean, median and mode of the above data.★

Solution :

Class Interval	Frequency (f_i)	Cumulative Frequency	Class Mark (x_i)	($f_i \times x_i$)
100 – 120	12	12	110	1320
120 – 140	14	26	130	1820
140 – 160	8	34	150	1200
160 – 180	6	40	170	1020
180 – 200	10	50	190	1900
	$N = \Sigma f_i$ = 50			$\Sigma(f_i \times x_i)$ = 7260

(i) $$\text{Mean} = \frac{\Sigma(f_i \times x_i)}{\Sigma f_i}$$

$$= \frac{7260}{50} = ₹\ 145.2$$

(ii) Now, $N = \Sigma f_i = 50$

So, $$\frac{N}{2} = 25$$

The cumulative frequency just above 25 is 26.

Hence, the Median Class is 120 – 140.

So, $l = 120, h = 20, f = 14, F = 12$ and $\frac{N}{2} = 25$

$$\text{Median} = l + \frac{\frac{N}{2} - F}{f} \times h$$

$$= 120 + \frac{25-12}{14} \times 20$$

$$= 120 + 18.57$$

$$= ₹\ 138.57$$

(iii) Mode = 3 Median – 2 Mean

$\Rightarrow$ Mode = 3(138.57) – 2(145.2)

= 415.71 – 290.4

= ₹ 125.31 **Ans.**

Q. 11. The following table shows the daily expenditure on food of 30 households in a locality :★

Expenditure	Households
100 – 150	6
150 – 200	7
200 – 250	12
250 – 300	3
300 – 350	2

Determine the mean, median and mode of the above data.

Solution :

Class Interval	Frequency (f_i)	Cumulative Frequency	Class Mark (x_i)	($f_i \times x_i$)
100 – 150	6	6	125	750
150 – 200	7	13	175	1225
200 – 250	12	25	225	2700
250 – 300	3	28	275	825
300 – 350	2	30	325	650
	$N = \Sigma f_i$ = 30			$\Sigma(f_i \times x_i)$ = 6150

(i) $$\text{Mean} = \frac{\Sigma(f_i \times x_i)}{\Sigma f_i}$$

$$= \frac{6150}{30}$$

$$= ₹\ 205$$

(ii) Now, $N = \Sigma f_i = 30$

So, $$\frac{N}{2} = 15$$

The cumulative frequency just above 15 is 25.

Hence, the Median Class is 200 – 250.

So $l = 200, h = 50, f = 12, F = 13$ and $\frac{N}{2} = 15$

$$\text{Median} = l + \frac{\frac{N}{2} - F}{f} \times h$$

$$= 200 + \frac{15-13}{12} \times 50$$

$$= 200 + 8.33$$

$$= ₹\ 208.33$$

★ are board exam questions from previous years

(iii) Mode = 3 Median – 2 Mean

⇒ Mode = 3(208.33) – 2(205)

= 624.99 – 410

= ₹ 214.99 **Ans.**

Q. 12. The following table gives the production yield per hectare of wheat of 100 farms of a village :★

Production yield	No. of Farms
40 – 45	4
45 – 50	6
50 – 55	16
55 – 60	20
60 – 65	30
65 – 70	24

Change the distribution to a 'more than type' distribution and draw its ogive.

Solution : The 'more than type' of series can be prepared as follows :

Production Yield (per hectare)	Cumulative Frequency
More than 65	24
More than 60	54
More than 55	74
More than 50	90
More than 45	96
More than 40	100

On the graph paper, we plot the points $A(40, 100)$, $B(45, 96)$, $C(50, 90)$, $D(55, 74)$, $E(60, 54)$ and $F(65, 24)$.

Joining all the points gives a 'More Than, ogive'.

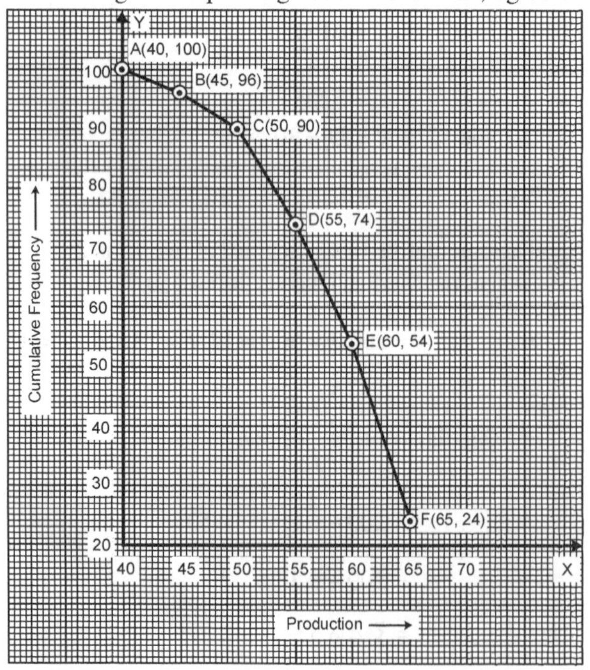

★ are board exam questions from previous years

Q. 13. During a medical check-up of 35 students of a class, their weights were recorded as follows :

Weight	No. of Students
38 – 40	3
40 – 42	2
42 – 44	4
44 – 46	5
46 – 48	14
48 – 50	4
50 – 52	3

Draw a 'less than type' and a 'more than type' ogive from the given data. Hence obtain the median weight from the graph.★

Solution : The 'Less Than' series is as follows :

Weight (in kg)	No. of Students
Less than 40	3
Less than 42	5
Less than 44	9
Less than 46	14
Less than 48	28
Less than 50	32
Less than 52	35

Thus, the points to be plotted are $A(40, 3)$, $B(42, 5)$, $C(44, 9)$, $D(46, 14)$, $E(48, 28)$, $F(50, 32)$ and $G(52, 35)$.

Joining the above points gives the 'Less Than' curve.

Similarly, the 'More Than' series is as follows :

Weight (in kg)	No. of Students
More than 38	35
More than 40	32
More than 42	30
More than 44	26
More than 46	21
More than 48	7
More than 50	3

Thus, the points to be plotted are $P(38, 35)$, $Q(40, 32)$, $R(42, 30)$, $S(44, 26)$, $T(46, 21)$, $U(48, 7)$ and $V(50, 3)$.

Joining the above points gives the 'More Than' curve.

Let the point of intersection of the two curves be L.

Then the Median weight can be found by drawing a perpendicular from L on to the X-axis.

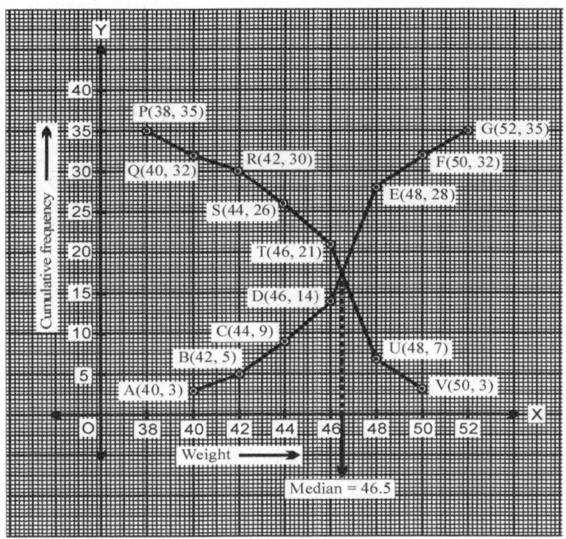

Thus, the Median Weight, $M = 46.5$ kg. **Ans.**

Q. 14. The following distribution gives the daily income of 50 workers of a factory.★

Daily Income (in ₹)	Number of Workers
200 – 220	12
220 – 240	14
240 – 260	8
260 – 280	6
280 – 300	10

Convert the distirbution above to a 'less than type' cumulative frequency distribution and draw its ogive.

Solution :

Daily Income	Cumulative Frequency
less than 220	4
less than 240	5
less than 260	12
less than 280	2
less than 300	2

Less than type frequency distribution table.

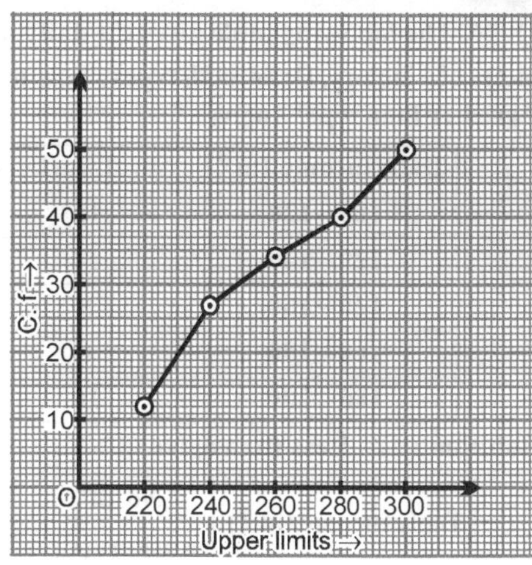

Q. 15. Change the following data into 'less than type' distribution and draw its ogive :★

Class Interval	Frequency
30 – 40	7
40 – 50	5
50 – 60	8
60 – 70	10
70 – 80	6
80 – 90	6
90 – 100	8

Solution :

Class Interval	Frequency
less than 40	7
less than 50	12
less than 60	20
less than 70	30
less than 80	36
less than 90	42
less than 100	50

On graph paper, we take the scale.

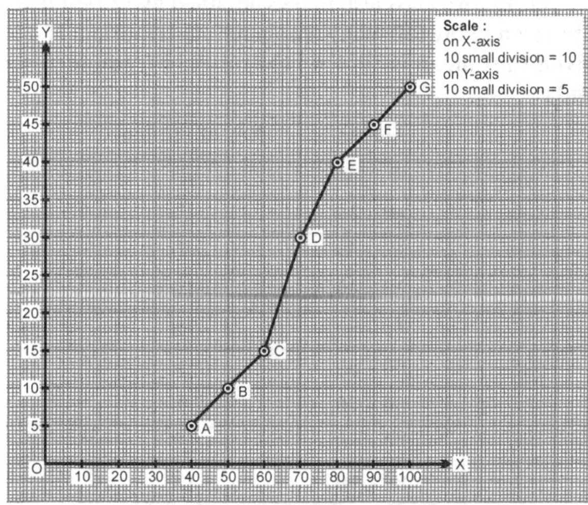

★ are board exam questions from previous years

Q. 16. The following table gives the daily income of 50 wokers of a factory. Draw both types ("less than type" and "greater than type") ogives.★

Daily income (in ₹)	No. of workers
100 – 120	12
120 – 140	14
140 – 160	8
160 – 180	6
180 – 200	10

Solution : Less than ogive

Daily income (in ₹)	No. of workers (*c.f.*)
Less than 120	12
Less than140	26
Less than160	34
Less than180	40
Less than 200	50

More than ogive

Daily income (in ₹)	No. of workers (*c.f.*)
More than 100	50
More than120	38
More than140	24
More than160	16
More than180	10

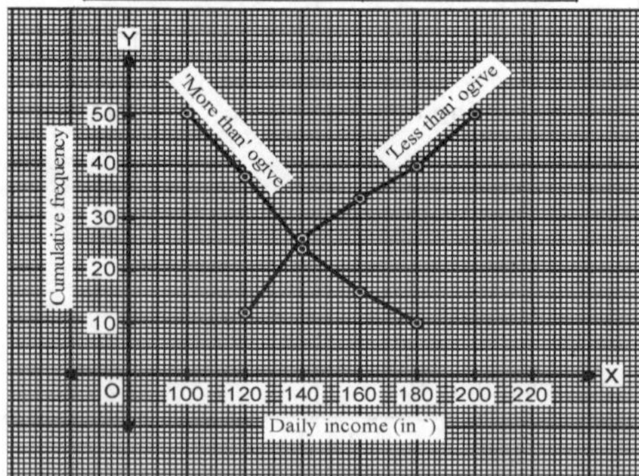

Q. 17. In a class test, marks obtained by 120 students are given in the following frequency distribution. If it is given that mean is 59, find the missing frequencies x and y.

Marks	No. of students
0 – 10	1
10 – 20	3
20 – 30	7
30 – 40	10
40 – 50	15
50 – 60	x
60 – 70	9
70 – 80	27
80 – 90	18
90 – 100	y

Solution :

Marks	No. of students (f_i)	X_i	$d_i = \frac{X_i - 55}{10}$	$f_i d_i$
0–10	1	5	–5	–5
10–20	3	15	–4	–12
20–30	7	25	–3	–21
30–40	10	35	–2	–20
40–50	15	45	–1	–15
50–60	x	$A = 55$	0	0
60–70	9	65	1	9
70–80	27	75	2	54
80–90	18	85	3	54
90–100	y	95	4	$4y$
	$\Sigma f_i = 90 + x + y$			$\Sigma f_i d_i = 44 + 4y$

$$\Sigma f_i = 90 + x + y$$

But $\Sigma f_i = 120$ [Given]

$\therefore \quad 90 + x + y = 120$

$$x = 120 - 90 - y = 30 - y \qquad ...(i)$$

$$\text{Mean} = A + \frac{\Sigma f_i d_i}{\Sigma f_i} \times h$$

$$\Rightarrow \quad 59 = 55 + \left(\frac{44 + 4y}{120} \times 10\right)$$

$$59 - 55 = \frac{4(11 + y)}{12}$$

$$4 \times 3 = 11 + y$$

$$\Rightarrow \quad y = 12 - 11 = 1$$

From eq. (i),

$$x = 30 - 1 = 29$$

$\therefore \quad x = 29, y = 1$ **Ans.**

Q. 18. Find the mean, median and mode of the following data :★

Class Interval	Frequency
0 – 20	6
20 – 40	8
40 – 60	10
60 – 80	12
80 – 100	6
100 – 120	5
120 – 140	3

★ are board exam questions from previous years

Solution :

Class Interval	Frequency (f_i)	Cumulative Frequency	Class Mark (x_i)	$(f_i \times x_i)$
0 – 20	6	6	10	60
20 – 40	8	14	30	240
40 – 60	10	24	50	500
60 – 80	12	36	70	840
80 – 100	6	42	90	540
100 – 120	5	47	110	550
120 – 140	3	50	130	390
	$N = \Sigma f_i = 50$			$\Sigma(f_i \times x_i) = 3120$

(i) $$\text{Mean} = \frac{\Sigma(f_i \times x_i)}{\Sigma f_i} = \frac{3120}{50} = 62.4$$

(ii) Now $N = \Sigma f_i = 50$

So, $\frac{N}{2} = 25$

The cumulative frequency just above 25 is 36.

Hence, the Median Class is 60 – 80.

So $l = 60$, $h = 20$, $f = 12$, $F = 24$ and $\frac{N}{2} = 25$

$$\text{Median} = l + \frac{\frac{N}{2} - F}{f} \times h$$

$$= 60 + \frac{25 - 24}{12} \times 20$$

$$= 60 + 1.67$$

$$= 61.67$$

(iii) Mode = 3 Median – 2 Mean

$\Rightarrow$ Mode = 3(61.67) – 2(62.4)

= 185.01 – 124.8

= 60.21 **Ans.**

Q. 19. Given below are the weights (in grams) of 50 apples. Calculate the mean weight by the Step-Deviation Method.★

Weight	No. of Apples
80 – 85	5
85 – 90	8
90 – 95	10
95 – 100	12
100 – 105	8
105 – 110	4
110 – 115	3

Solution :

Class Interval	Frequency (f_i)	Mid Value (x_i)	$u_i = \frac{x_i - A}{h}$; h = upper limit – lower limit	$f_i \times u_i$
80 – 85	5	82.5	– 3	– 15
85 – 90	8	87.5	– 2	– 16
90 – 95	10	92.5	– 1	– 10
95 – 100	12	97.5 = A	0	0
100 – 105	8	102.5	1	8
105 – 110	4	107.5	2	8
110 – 115	3	112.5	3	9
	$\Sigma f_i = 50$			$\Sigma(f_i \times u_i) = -16$

Thus, $A = 97.5$, $h = 5$, $\Sigma f_i = 50$, $\Sigma(f_i \times u_i) = -16$

$$\text{Mean} = A + \left\{h \times \frac{\Sigma(f_i \times u_i)}{\Sigma f_i}\right\}$$

$$= 97.5 + \left\{5 \times \frac{(-16)}{50}\right\}$$

$$= 97.5 - 1.6$$

$$= 95.9 \text{ gm} \approx 96 \text{ gm.}$$ **Ans.**

Q. 20. Following table shows marks (out of 100) of students in a class test :

Marks	No. of students
More than or equal to 0	80
More than or equal to 10	77
More than or equal to 20	72
More than or equal to 30	65
More than or equal to 40	55
More than or equal to 50	43
More than or equal to 60	28
More than or equal to 70	16
More than or equal to 80	10
More than or equal to 90	8
More than or equal to 100	0

Draw a 'more than type' ogive. From the curve, find the median. Also, check the value of the median by actual calcultaion.★

★ are board exam questions from previous years

Solution :

More than type	C.I.	No. of students	Frequency	*c.f.*
More than or equal to 0	0–10	80	3	3
More than or equal to 10	10–20	77	5	8
More than or equal to 20	20–30	72	7	15
More than or equal to 30	30–40	65	10	25
More than or equal to 40	40–50	55	12	37
More than or equal to 50	50–60	43	15	52
More than or equal to 60	60–70	28	12	64
More than or equal to 70	70–80	16	6	70
More than or equal to 80	80–90	10	2	72
More than or equal to 90	90–100	8	8	80
More than or equal to 100	100–110	0	0	

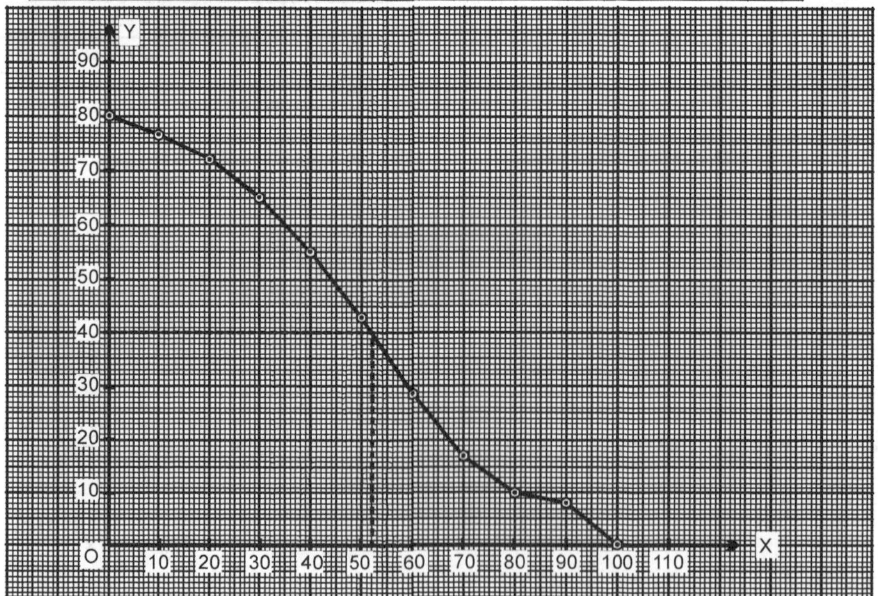

Median will be 52.

Median by actual calculation :

$$N = 80$$

$$\therefore \quad \frac{N}{2} = \frac{80}{2} = 40$$

The cumulative frequency just above 40 is 52.

Hence So, median class will be 50 – 60

$$l = 50, h = 10, f = 15, F. = 37,$$

$$\text{Median} = l + \left[h \times \frac{\left(\frac{N}{2} - F.\right)}{f} \right]$$

$$= 50 + \left[10 \frac{(40-37)}{15} \right]$$

$$= 50 + 10 \times \frac{3}{15}$$

$$= 50 + 2$$

$$= 52 \qquad \textbf{Hence Verified.}$$

Q. 21. From the following data, find the median age of 100 residents of a colony who took part in 'Swachch Bharat Abhiyan :★

Age (in yrs.) More than or equal to	No. of residents
0	50
10	46
20	40
30	20
40	10
50	3

Solution : First convert the given table into C.I. Table.

C.I.	Frequency	*c.f.*
0 – 10	4	4
10 – 20	6	10
20 – 30	20	30
30 – 40	10	40
40 – 50	7	47
50 – 60	3	50

★ are board exam questions from previous years

Here, $\frac{N}{2} = \frac{50}{2} = 25$

Cumulative frequency just above 25 is 30.

Hence, the median class is 20 – 30.

So, $l = 20, h = 10, f = 20, F = 10, \frac{N}{2} = 25$

$$\text{Median} = l + \frac{\frac{N}{2} - F}{f} \times h$$

$$= 20 + \frac{25-10}{20} \times 10$$

$$= 20 + \frac{15}{2}$$

$$= 27.5$$ **Ans.**

Q. 22. The distribution given below shows the marks obtained by 25 students in an aptitude test. Find the Mean, Median and Mode of the distribution.

Marks	No. of Students
50 – 60	**4**
60 – 70	**8**
70 – 80	**14**
80 – 90	**19**
90 – 100	**5**

Solution :

Class Interval	Frequency (f_i)	Cumulative Frequency	Class Mark (x_i)	($f_i \times x_i$)
50 – 60	4	4	55	220
60 – 70	8	12	65	520
70 – 80	14	26	75	1050
80 – 90	19	45	85	1615
90 – 100	5	50	95	475
	$N = \Sigma f_i = 50$			$\Sigma(f_i \times x_i) = 3880$

(i) $\text{Mean} = \frac{\Sigma(f_i \times x_i)}{\Sigma f_i}$

$$= \frac{3880}{50}$$

$= 77.6$ marks $\approx$ 78 marks

(ii) Now, $N = \Sigma f_i = 50$

So, $\frac{N}{2} = 25$

The cumulative frequency just above 25 is 26.

Hence, the Median Class is 70 – 80.

So $l = 70, h = 10, f = 14, F = 12$ and $\frac{N}{2} = 25$

$$\text{Median} = l + \frac{\frac{N}{2} - F}{f} \times h$$

$$= 70 + \frac{25-12}{14} \times 10$$

$$= 70 + 9.29$$

$= 79.29$ marks

$\approx$ 79 marks.

(iii) Since the highest frequency is 19, so the Modal Class is 80 – 90.

So $l = 80, h = 10, f = 19, f_1 = 14, f_2 = 5$

$$\text{Mode} = l + \frac{f - f_1}{2f - f_1 - f_2} \times h$$

$$= 80 + \frac{19-14}{38-14-5} \times 10$$

$$= 80 + \frac{5}{19} \times 10$$

$$= 80 + 2.63$$

$= 82.63$ marks $\approx$ 83 marks. **Ans.**

Q. 23. Mode of the following frequency distribution is 65 and sum of all the frequencies is 70. Find the missing frequencies x and y.★

Class Interval	Frequency
0 — 20	**8**
20 — 40	**11**
40 — 60	x
60 — 80	**12**
80 — 100	y
100 — 120	**9**
120 — 140	**9**
140 — 160	**5**

Solution :

Class Interval	Frequency
0 — 20	8
20 — 40	11
40 — 60	$x(f_0)$
60 — 80	12 (f_1)
80 — 100	$y(f_2)$
100 — 120	9
120 — 140	9
140 — 160	5
	$\Sigma f = 70$

★ are board exam questions from previous years

Here,

$$8+11+x+12+y+9+9+5=70 \quad \text{[Given]}$$

$$54+x+y=70$$

$$x+y=70-54=16 \quad \text{...(i)}$$

$$\text{Mode}=65 \quad \text{[Given]}$$

$\therefore$ Modal Class is 60 – 80

So, $l=60, h=20, f_0=x, f_1=12, f_2=y$

$$\text{Mode}=l+\frac{f_1-f_0}{2f_1-f_0-f_2}\times h$$

$$\therefore \quad 65=60+\frac{12-x}{2(12)-x-y}\times 20$$

$$65-60=\frac{12-x}{24-(x+y)}\times 20$$

$$5=\frac{12-x}{24-16}\times 20$$

[From equation (i)]

$$5=\frac{12-x}{8}\times 20$$

$$\Rightarrow \quad 5\times\frac{2}{5}=12-x$$

$$2=12-x$$

$$\therefore \quad x=12-2=10$$

$$x+y=16 \quad \text{[From equation (i)]}$$

$$10+y=16$$

$$\Rightarrow \quad y=16-10=6$$

$$\therefore \quad x=10, y=6$$

Ans.

Chapter 15. Probability

Q. 1. 2 red queens and 2 black jacks are removed from a pack of 52 playing cards. A card is drawn at random from the remaining cards after reshuffling them. Find the probability that the card drawn is (i) a king, (ii) a red colour, (iii) a face card, and (iv) a queen.★

Solution : Total numbers of cards = 52

Number of cards removed = 2 red queens and 2 black jacks = 4

Number of cards remaining = 52 – 4 = 48

Thus, number of black cards = 26 – 2 = 24

and number of red cards = 26 – 2 = 24

(i) Possible outcomes of drawing a king = king of hearts (red), king of diamonds (red), king of clubs (black) and king of spades (black) = 4

Hence $$P(E)=\frac{4}{48}=\frac{1}{12}$$

★ are board exam questions from previous years

(ii) Possible outcomes of drawing a red colour = 24

Hence $$P(E)=\frac{24}{48}=\frac{1}{2}$$

(iii) Possible outcomes of drawing a face card = 4 kings + 2 queens + 2 jack = 8

Hence $$P(E)=\frac{8}{48}=\frac{1}{6}$$

(iv) Possible outcomes of drawing a queen = 2

Hence $$P(E)=\frac{2}{48}=\frac{1}{24}$$

Ans.

Q. 2. Five cards the ten, jack, queen, king and ace of diamonds are well shuffled with their face downwards. One card is then picked up in random. (i) What is the probability that the drawn card is a queen? (ii) If the queen is drawn and put aside and a second card is drawn, find the probability that the second card is (a) an ace, and (b) a queen.

Solution : Total number of cards = 5

(i) Possible outcomes of drawing a queen = 1

Hence, $$P(E)=\frac{1}{5}$$

(ii) Now the queen is drawn and set aside.

So, the total number of cards = 4

(a) Possible outcomes of drawing an ace = 1

Hence $$P(E)=\frac{1}{4}$$

(b) Possible outcomes of drawing a queen = 0

Hence, $P(E)=0$

This is an impossible event. **Ans.**

Q. 3. Cards numbered from 11 to 60 are kept in a box. If a card is drawn at random from the box, find the probability that the number on the card is : (i) an odd number, (ii) a perfect square, (iii) divisible by 5, and (iv) a prime number less than 20.★

Solution : All possible outcomes

= {11, 12, 13, ..., 60}

Sum of all possible outcomes = 50

(i) The favourable outcomes = {11, 13, 15, 17, 19, 21, 23, 25, 27, 29, 31, 33, 35, 37, 39, 41, 43, 45, 47, 49, 51, 53, 55, 57, 59}

Thus, total number of favourable outcomes = 25

Hence, $$P(E)=\frac{25}{50}=\frac{1}{2}$$

(ii) The favourable outcomes = {16, 25, 36, 49}

Thus, total number of favourable outcomes = 4

Hence, $P(E) = \frac{4}{50} = \frac{2}{25}$

(iii) The favourable outcomes = {15, 20, 25, 30, 35, 40, 45, 50, 55, 60}

Thus, total number of favourable outcomes = 10

Hence, $P(E) = \frac{10}{50} = \frac{1}{5}$

(iv) The favourable outcomes = 11, 13, 17, 19

Thus, total number of favourable outcomes = 4

Hence $P(E) = \frac{4}{50} = \frac{2}{25}$ **Ans.**

Q. 4. A piggy bank contains hundred 50p coins, fifty ₹ 1 coins, twenty ₹ 2 coins and ten ₹ 5 coins. If it is equally likely that one of the coins will fall out when the bank is turned upside down, find the probability that the coin which fell (i) will be a 50p coin, (ii) will be of value more than ₹ 1, (iii) will be value less than ₹ 5, and (iv) will be a ₹ 1 or ₹ 2 coin.★ [NCERT]

Solution : Total number of coins = (100 + 50 + 20 + 10) = 180

(i) Possible number of outcomes of a 50p coin falling out = 100

Hence, $P(E) = \frac{100}{180} = \frac{5}{9}$

(ii) Possible number of outcomes of a coin falling out having value more than ₹ 1 = (20 + 10) = 30

Hence, $P(E) = \frac{30}{180} = \frac{1}{6}$

(iii) Possible number of outcomes of a coin falling out having value less than ₹ 5 = (100 + 50 + 20) = 170

Hence, $P(E) = \frac{170}{180} = \frac{17}{18}$

(iv) Possible number of outcomes of a coin falling out having value either ₹ 1 or ₹ 2 = (50 + 20) = 70

Hence, $P(E) = \frac{70}{180} = \frac{7}{18}$ **Ans.**

Q. 5. All the black face cards are removed from a pack of 52 playing cards. The remaining cards are well-shuffled and then a card is drawn at random from them. Find the probability of getting a (i) face card, (ii) red card, (iii) black card, and (iv) king.★

Solution : Total number of cards = 52

Number of cards removed = 2 black kings + 2 black queens + 2 black jacks = 6

★ are board exam questions from previous years

Number of cards remaining = 52 – 6 = 46

Thus, remaining number of face cards = 2 red kings + 2 red queens + 2 red jacks = 6

Number of red cards = 26

Number of black cards = 26 – 6 = 20

Number of kings = 2 red kings = 2

(i) Possible number of outcomes of drawing a face card = 6

Hence, $P(E) = \frac{6}{46} = \frac{3}{23}$

(ii) Possible number of outcomes of drawing a red card = 26

Hence, $P(E) = \frac{26}{46} = \frac{13}{23}$

(iii) Possible number of outcomes of drawing a black card = 20

Hence, $P(E) = \frac{20}{46} = \frac{10}{23}$

(iv) Possible number of outcomes of drawing a king = 2

Hence, $P(E) = \frac{2}{46} = \frac{1}{23}$. **Ans.**

Q. 6. A bag contains cards numbered from 1 to 49. If a card is drawn at random from the bag after mixing them thoroughly, find the probability that the number on the card is (i) an odd number (ii) multiple of 5 (iii) a perfect square, and (iv) an even prime number.★

Solution : All possible outcomes = {1, 2, 3, ..., 49}

∴ Sum of all possible outcomes = 49

(i) The favourable outcomes = {1, 3, 5, 7, 9, 11, 13, 15, 17, 19, 21, 23, 25, 27, 29, 31, 33, 35, 37, 39, 41, 43, 45, 47, 49}

Thus, total number of favourable outcomes = 25

Hence, $P(E) = \frac{25}{49}$

(ii) The favourable outcomes = {5, 10, 15, 20, 25, 30, 35, 40, 45}

Thus, total number of favourable outcomes = 9

Hence, $P(E) = \frac{9}{49}$

(iii) The favourable outcomes = {1, 4, 9, 16, 25, 36, 49}

Thus, total number of favourable outcomes = 7

Hence $P(E) = \frac{7}{49} = \frac{1}{7}$

(iv) The favourable outcomes = 2

Thus, total number of favourable outcomes = 1

Hence $P(E) = \frac{1}{49}$. **Ans.**

Q. 7. A die is rolled twice. Find the probability that (i) 5 will not come up either time, and (ii) 5 will come up exactly once.★

Solution : Total possible outcomes = (1, 1); (1, 2); (1, 3); (1, 4); (1, 5); (1, 6); (2, 1); (2, 2); (2, 3); (2, 4); (2, 5); (2, 6); (3, 1); (3, 2); (3, 3); (3, 4); (3, 5); (3, 6); (4, 1); (4, 2); (4, 3); (4, 4); (4, 5); (4, 6); (5, 1); (5, 2); (5, 3); (5, 4); (5, 5); (5, 6); (6, 1); (6, 2); (6, 3); (6, 4); (6, 5); (6, 6)

Thus, total number of possible outcomes = 36

(i) Now, the total number of favourable outcomes = 25

Hence, $P(E) = \frac{25}{36}$

(ii) The favourable outcomes are (1, 5); (2, 5); (3, 5); (4, 5); (6, 5); (5, 1); (5, 2); (5, 3); (5, 4); (5, 6)

Thus, the total number of favourable outcomes = 10

Hence, $P(E) = \frac{10}{36} = \frac{5}{18}$. **Ans.**

Q. 8. All the red face cards are removed from a pack of 52 playing cards. The remaining cards are well-shuffled and then a card is drawn at random from them. Find the probability of drawing a (i) red coloured card, (ii) queen, (iii) ace, and (iv) face card.

Solution : Total number of cards = 52

Number of cards removed = 2 red kings + 2 red queens + 2 red jacks = 6

Number of cards remaining = 52 – 6 = 46

Number of red cards = 26 – 6 = 20

Number of queens = 2 black queens = 2

Number of aces = 4

Thus, remaining number of face cards = 2 black kings + 2 black queens + 2 black jacks = 6

(i) Possible number of outcomes of drawing a red coloured card = 20

Hence, $P(E) = \frac{20}{46} = \frac{10}{23}$

(ii) Possible number of outcomes of drawing a queen = 2

Hence, $P(E) = \frac{2}{46} = \frac{1}{23}$

(iii) Possible number of outcomes of drawing an ace = 4

Hence, $P(E) = \frac{4}{46} = \frac{2}{23}$

(iv) Possible number of outcomes of drawing a face card = 6

Hence, $P(E) = \frac{6}{46} = \frac{3}{23}$. **Ans.**

Q. 9. A bag contains cards numbered from 1 to 30. If a card is drawn at random from the bag after mixing them thoroughly, find the probability that the number on the card is : (i) not divisible by 3, (ii) a prime number greater than 7 and (iii) not a perfect square.★

Solution : All possible outcomes = {1, 2, 3, ..., 30}

∴ Sum of all possible outcomes = 30

(i) The multiples of 3 are 3, 6, 9, 12, 15, 18, 21, 24, 27, 30

Thus, total number of favourable outcomes

= (30 – 10) = 20

Hence, $P(E) = \frac{20}{30} = \frac{2}{3}$

(ii) Prime numbers greater than 7 are 11, 13, 17, 19, 23, 29

Thus, total number of favourable outcomes = 6

Hence, $P(E) = \frac{6}{30} = \frac{1}{5}$

(iii) The perfect squares are 1, 4, 9, 16, 25

Thus, total number of favourable outcomes

= (30 – 5) = 25

Hence, $P(E) = \frac{25}{30} = \frac{5}{6}$. **Ans.**

Q. 10. Two different dice are thrown together. Find the probability that the numbers obtained have

(i) even sum, and

(ii) even product.★

Solution :

(i) **For even sum :** Favourable outcomes are

(1, 1), (1, 3), (1, 5), (2, 2), (2, 4),(2, 6), (3, 1),(3, 3), (3, 5), (4, 2), (4, 4), (4, 6), (5, 1), (5, 3), (5, 5), (6, 2), (6, 4), (6, 6)

No. of favourable outcomes = 18

$$P(\text{even sum}) = \frac{\text{Favourable outcomes}}{\text{Total outcomes}}$$

$$= \frac{18}{36} = \frac{1}{2}$$ **Ans.**

★ are board exam questions from previous years

(ii) **For even product :** Favourable outcomes are

(1, 2), (1, 4), (1, 6), (2, 1), (2, 2), (2, 3), (2, 4), (2, 5), (2, 6), (3, 2), (3, 4), (3, 6), (4, 1), (4, 2), (4, 3) ,(4, 4), (4, 5), (4, 6), (5, 2), (5 ,4), (5, 6), (6, 1), (6, 2), (6, 3), (6, 4), (6, 5), (6, 6).

No. of favourable outcomes = 27

$$P(\text{even product}) = \frac{\text{Favourable outcomes}}{\text{Total outcomes}}$$

$$= \frac{27}{36} = \frac{3}{4} \quad \textbf{Ans.}$$

Q. 11. A piggy bank contains hundred 50p coins, seventy ₹ 1 coins, fifty ₹ 2 coins and thirty ₹ 5 coins. If it is equally likely that one of the coins will fall out when the bank is turned upside down, find the probability that the coin which fell (i) will be a ₹ 1 coin, (ii) will not be a ₹ 5 coin, (iii) will be a 50p or ₹ 2 coin.★

Solution : Total number of coins

$$= (100 + 70 + 50 + 30) = 250$$

(i) Possible number of outcomes of a ₹ 1 coin falling out = 70

Hence, $P(E) = \frac{70}{250} = \frac{7}{25}$

(ii) Possible number of outcomes of a coin falling out not having value of ₹ 5 = (100 + 70 + 50) = 220

Hence, $P(E) = \frac{220}{250} = \frac{22}{25}$

(iii) Possible number of outcomes of a coin falling out having value either 50p or ₹ 2 = (100 + 50) = 150

Hence, $P(E) = \frac{150}{250} = \frac{3}{5}$. **Ans.**

Q. 12. Seventeen cards numbered 1, 2, 3, 4, …, 16, 17 are put in a box and mixed thoroughly. One person draws a card from the box. Find the probability that the number on the card is (i) odd, (ii) a prime, (iii) divisible by 3 and (iv) divisible by 2 and 3 both.★

Solution : All possible outcomes = {1, 2, 3, …, 17}

∴ Sum of all possible outcomes = 17

(i) The favourable outcomes = {1, 3, 5, 7, 9, 11, 13, 15, 17}

Thus, total number of favourable outcomes = 9

Hence, $P(E) = \frac{9}{17}$

(ii) The favourable outcomes = {2, 3, 5, 7, 11, 13, 17}

Thus, total number of favourable outcomes = 7

Hence $P(E) = \frac{7}{17}$

(iii) The favourable outcomes = {3, 6, 9, 12, 15}

Thus total number of favourable outcomes = 5

Hence, $P(E) = \frac{5}{17}$

(iv) The favourable outcomes = {6, 12}

Thus, total number of favourable outcomes = 2

Hence $P(E) = \frac{2}{17}$. **Ans.**

Q. 13. Find the probability that a number selected at random from the numbers 1, 2, 3, 4, …, 34, 35 is a (i) prime, (ii) multiple of 7, and (iii) divisible by 3 or 5.★

Solution : All possible outcomes = {1, 2, 3, …, 34, 35}

∴ Sum of all possible outcomes = 35

(i) The favourable outcomes = {2, 3, 5, 7, 11, 13, 17, 19, 23, 29, 31}

Thus, total number of favourable outcomes = 11

Hence, $P(E) = \frac{11}{35}$

(ii) The favourable outcomes = {7, 14, 21, 28, 35}

Thus total number of favourable outcomes = 5

Hence, $P(E) = \frac{5}{35} = \frac{1}{7}$

(iii) The favourable outcomes = {3, 5, 6, 9, 10, 12, 15, 18, 20, 21, 24, 25, 27, 30, 33, 35}

Thus, total number of favourable outcomes = 16

Hence $P(E) = \frac{16}{35}$. **Ans.**

Q. 14. A bag contains 6 red balls, 8 white balls, 5 green balls and 3 black balls. One ball is drawn at random from the bag. Find the probability that the ball drawn out from the bag is (i) white, (ii) red or black, (iii) not green, and (iv) neither white nor black.★

Solution : The sum of the total outcomes

$$= 6 + 8 + 5 + 3 = 22$$

(i) The sum of the favourable outcomes = 8

∴ $P(E) = \frac{8}{22} = \frac{4}{11}$

(ii) The sum of the favourable outcomes

$$= 6 + 3 = 9$$

∴ $P(E) = \frac{9}{22}$

(iii) The sum of the favourable outcomes

$$= 6 + 8 + 3 = 17$$

★ are board exam questions from previous years

$$\therefore \quad P(E) = \frac{17}{22}$$

(iv) The sum of the favourable outcomes

$$= 6 + 5 = 11$$

$$\therefore \quad P(E) = \frac{11}{22} = \frac{1}{2}. \quad \textbf{Ans.}$$

Q. 15. From a pack of 52 playing cards jacks, queen, kings and aces of red colour are removed. From the remaining, a card is drawn at random. Find the probability that the card drawn is (i) a black queen, (ii) a red card, (iii) a ten, and (iv) a face card.★

Solution : Total number of cards = 52

Cards removed = 2 + 2 + 2 + 2 = 8

Thus, number of possible outcomes = 52 – 8 = 44

(i) The total number of favourable outcomes for black queens = 2 [1 of spades and 1 of clubs]

$$\therefore \quad P(E) = \frac{2}{44} = \frac{1}{22}$$

(ii) Total number of favourable outcomes for a red card = 26 – 8 = 18

$$\therefore \quad P(E) = \frac{18}{44} = \frac{9}{22}$$

(iii) Total number of favourable outcomes for ten = 4

$$\therefore \quad P(E) = \frac{4}{44} = \frac{1}{11}$$

(iv) Total number of favourable outcomes for face cards = 6[1 king + 1 queen + 1 jack from Spades and Clubs each]

$$\therefore \quad P(E) = \frac{6}{44} = \frac{3}{22}. \quad \textbf{Ans.}$$

Q. 16. A card is drawn at random from a well-shuffled deck of playing cards. Find the probability that the card drawn is :

(i) A card of spade or an ace

(ii) A black king

(iii) Neither a jack nor a king

(iv) Either a king or a queen★

Solution : Let S be the sample space of drawing a card from a well-shuffled deck

Then, $S = 52$

(i) There are 13 spade cards and 4 ace's in a deck.

As a ace of spade is included in 13 spade cards, so, there are 13 spade cards and 3 ace's

A card of spade or an ace can be drawn in

13 + 4 – 1 = 16 (ways)

Probability of drawing a card of spade or an ace

$$P = \frac{16}{52} = \frac{4}{13}$$

(ii) There are 2 black king cards in a deck.

Probability of drawing a black king

$$P = \frac{2}{52} = \frac{1}{26}$$

(iii) There are 4 jack and 4 king cards in a deck.

So, there are 52 – 8 = 44 cards which are neither jack nor king.

∴ Probability of drawing a card which is neither a jack nor a king

$$P = \frac{44}{52} = \frac{11}{13}$$

(iv) There are 4 queen and 4 king cards in a deck.

So, there are 8 cards which are either king or queen.

Probability of drawing a card which is either king or a queen

$$P = \frac{8}{52} = \frac{2}{13} \quad \textbf{Ans.}$$

Q. 17. A bag contains 7 black, 5 red and 3 white balls. One ball is drawn at random from the bag. Find the probability that the ball drawn out from the bag is (i) red (ii) white or black and (iii) not black.★

Solution : The sum of the total outcomes

$$= 7 + 5 + 3 = 15$$

(i) The sum of the favourable outcomes for red balls = 5

$$\therefore \quad P(E) = \frac{5}{15} = \frac{1}{3}$$

(ii) The sum of the favourable outcomes for white or black balls = 3 + 7 = 10

$$\therefore \quad P(E) = \frac{10}{15} = \frac{2}{3}$$

(iii) The sum of the favourable outcomes for balls not black = 5 + 3 = 8

$$\therefore \quad P(E) = \frac{8}{15}. \quad \textbf{Ans.}$$

Q. 18. A bag contains 7 red, 5 white and 3 black balls. One ball is drawn at random from the bag. Find the probability that the ball drawn out from the bag is (i) red or white, (ii) not black, and (iii) neither white nor black.★

Solution : The sum of the total outcomes

$$= 7 + 5 + 3 = 15$$

(i) The sum of the favourable outcomes for red or white balls = 7 + 5 = 12

$$\therefore \quad P(E) = \frac{12}{15} = \frac{4}{5}$$

(ii) The sum of the favourable outcomes for balls not black = 7 + 5 = 12

★ are board exam questions from previous years

$\therefore \quad P(E) = \frac{12}{15} = \frac{4}{5}$

(iii) The sum of the favourable outcomes for balls neither white nor black = 7

$\therefore \quad P(E) = \frac{7}{15}$. **Ans.**

Q. 19. Two different dice are thrown together. Find the probability that the numbers obtained

(i) have a sum less than 7.

(ii) have a product less than 16.

(iii) is a doublet of odd numbers.★

Solution : Total number of possible outcomes in each case = 6 × 6 = 36

(i) Having a sum less than 7

Possible outcomes are,

(1, 1), (1, 2), (1, 3), (1, 4), (1, 5)

(2 ,1), (2, 2), (2, 3), (2, 4)

(3, 1), (3, 2), (3, 3)

(4, 1), (4, 1)

(5, 1)

$\therefore \quad n(E) = 15$

So, probability $= \frac{25}{36}$

(ii) Having a product less than 16

Possible outcomes are,

(1, 1), (1, 2), (1, 3), (1, 4), (1, 5), (1, 6)

(2, 1), (2, 2), (2, 3), (2, 4), (2, 5), (2, 6)

(3, 1), (3, 2), (3, 3), (3, 4), (3, 5),

(4, 1), (4, 2), (4, 3),

(5, 1), (5, 2), (5, 3),

(6, 1), (6, 2),

$\therefore \quad n(E) = 25$

So, probability $= \frac{25}{36}$

(iii) A doublet of odd numbers

Possible outcomes are (1, 1), (3, 3), (5, 5)

$\therefore \quad n(E) = 3$

So, probability $= \frac{3}{36} = \frac{1}{12}$ **Ans.**

Q. 20. A box contains 20 balls bearing numbers from 1 to 20. A ball is drawn at random from the box. Find the probability that the number on the ball is (i) an odd number, (ii) divisible by 2 or 3, (iii) not divisible by 10, and (iv) a prime number.★★

Solution : The total number of possible outcomes = 20

(i) The favourable outcomes for numbers that are odd are 1, 3, 5, 7, 9, 11, 13, 15, 17, 19

Hence, the sum of the number of outcomes of favourable outcomes = 10

★ are board exam questions from previous years

$\therefore \quad P(E) = \frac{10}{20} = \frac{1}{2}$

(ii) The favourable outcomes for numbers divisible by 2 or 3 are 2, 3, 4, 6, 8, 9, 10, 12, 14, 15, 16, 18, 20.

Hence, the sum of the number of outcomes of favourable outcomes = 13

$\therefore \quad P(E) = \frac{13}{20}$

(iii) The favourable outcomes for numbers that are not divisible by 10 are 1, 2, 3, 4, 5, 6, 7, 8, 9, 11, 12, 13, 14, 15, 16, 17, 18, 19.

Hence, the sum of the number of outcomes of favourable outcomes = 18.

$\therefore \quad P(E) = \frac{18}{20} = \frac{9}{10}$

(iv) The favourable outcomes for prime numbers are 2, 3, 5, 7, 11, 13, 17, 19.

Hence, the sum of the number of outcomes of favourable outcomes = 8

$\therefore \quad P(E) = \frac{8}{20} = \frac{2}{5}$. **Ans.**

Q. 21. Red kings, queens and jacks are removed from a pack of 52 playing cards. A card is drawn at random from the remaining cards after reshuffling them. Find the probability that the card drawn is (i) a king, (ii) red, and (iii) a spade.★

Solution : Total numbers of cards = 52

Number of cards removed = 6

$\therefore$ Number of cards remaining = 52 – 6 = 46

(i) Possible outcomes of drawing a king = 2

Hence, $P(E) = \frac{2}{46} = \frac{1}{23}$

(ii) Possible number of outcomes of drawing a red card = 26 – 6 = 20

Hence, $P(E) = \frac{20}{46} = \frac{10}{23}$

(iii) Possible number of outcomes of drawing a spade = 13

Hence, $P(E) = \frac{13}{46}$. **Ans.**

Q. 22. A bag contains 3 red balls, 5 black balls and 4 white balls. A ball is drawn at random from the bag. What is the probability that the ball drawn is (i) white, (ii) red, (iii) black, and (iv) not red ?★

Solution : The sum of the total possible outcomes = 3 + 5 + 4 = 12

(i) The sum of the favourable outcomes for drawing a white ball = 4

$\therefore \quad P(E) = \frac{4}{12} = \frac{1}{3}$

(ii) The sum of the favourable outcomes for drawing a red ball = 3

$$\therefore \quad P(E) = \frac{3}{12} = \frac{1}{4}$$

(iii) The sum of the favourable outcomes for drawing a black ball = 5

$$\therefore \quad P(E) = \frac{5}{12}$$

(iv) The sum of the favourable outcomes for drawing a ball not red = 9

$$\therefore \quad P(E) = \frac{9}{12} = \frac{3}{4}.$$ **Ans.**

Q. 23. A bag contains 6 red balls and some blue balls. If the probability of drawing a blue ball from the bag is double that of a red ball, find the number of blue balls in the bag.★

Solution : Let the number of blue balls be x.

Thus, the sum of the possible outcomes = $6 + x$

Now, the sum of the favourable outcomes if the red ball is drawn = 6

$$\therefore \quad P(E) = \frac{6}{6+x}$$

Again. the sum of the favourable outcomes if the blue ball is drawn = x

$$\therefore \quad P(E) = \frac{x}{6+x}$$

According to question,

$$\frac{x}{6+x} = 2\left(\frac{6}{6+x}\right)$$

$$\Rightarrow \quad x = 12$$

Hence, the number of blue balls = 12. **Ans.**

●●

★ are board exam questions from previous years

Evaluation and Analysis Based Questions

| Set 5 |

Chapter 1. Real Numbers

Q. 1. Show that $5 - 2\sqrt{3}$ is an irrational number.

Solution : Let $5 - 2\sqrt{3} = \frac{p}{q}$ be a rational number.

$$\therefore 5 - \frac{p}{q} = 2\sqrt{3} \text{ or } \frac{5q-p}{2q} = \sqrt{3},$$

which is contradiction because $\sqrt{3}$ is an irrational number and $\frac{5q-p}{2q}$ is a rational.

$\therefore$ Our supposition is wrong and hence, $5 - 2\sqrt{3}$ is irrational. **Hence Proved.**

Q. 2. Show that there is no positive integer n for which $\sqrt{n-1}+\sqrt{n+1}$ is rational.

Solution : Let there be a positive integer n for which $\sqrt{n-1}+\sqrt{n+1}$ is a rational number.

Then, $$\sqrt{n-1}+\sqrt{n+1} = \frac{p}{q} \quad ...(i)$$

where p, q are integers and $q \neq 0$

$$\Rightarrow \frac{1}{\sqrt{n-1}+\sqrt{n+1}} = \frac{q}{p}$$

$$\Rightarrow \frac{\sqrt{n-1}-\sqrt{n+1}}{(\sqrt{n-1}+\sqrt{n+1})\times(\sqrt{n-1}-\sqrt{n+1})} = \frac{q}{p}$$

$$\Rightarrow \frac{\sqrt{n-1}-\sqrt{n+1}}{(n-1)-(n+1)} = \frac{q}{p}$$

$$\Rightarrow \frac{\sqrt{n-1}-\sqrt{n+1}}{n-1-n-1} = \frac{q}{p}$$

$$\Rightarrow \frac{\sqrt{n+1}-\sqrt{n-1}}{2} = \frac{q}{p}$$

$$\Rightarrow \sqrt{n+1}-\sqrt{n-1} = \frac{2q}{p} \quad ...(ii)$$

Adding equation (i) and (ii), we get

$$\sqrt{n-1}+\sqrt{n+1}+\sqrt{n+1}-\sqrt{n-1} = \frac{p}{q}+\frac{2q}{p}$$

$$\Rightarrow 2\sqrt{n+1} = \frac{p^2+2q^2}{pq}$$

$$\Rightarrow \sqrt{n+1} = \frac{p^2+2q^2}{2pq}$$

$\Rightarrow \sqrt{n+1}$ is rational number as $\frac{p^2+2q^2}{2pq}$ is rational

$\Rightarrow n+1$ is a perfect square of a positive integer ...(A)

Again, subtracting equation (ii) from (i), we get

$$\sqrt{n-1}+\sqrt{n+1}-\sqrt{n+1}+\sqrt{n-1} = \frac{p}{q}-\frac{2q}{q}$$

$$\Rightarrow 2\sqrt{n-1} = \frac{p^2-2p^2}{pq}$$

$\Rightarrow \sqrt{n-1}$ is rational number as $\frac{p^2-2q^2}{2pq}$ is rational.

$\Rightarrow \sqrt{n-1}$ is also perfect square of a positive integer. ...(B)

From equations (A) and (B),

$\sqrt{n+1}$ and $\sqrt{n-1}$ are perfect squares of positive integer. It contradict the fact that two perfect squares differ at least by 3.

Hence, there is no positive integer n for which $\sqrt{n-1}+\sqrt{n+1}$ is rational. **Hence Proved.**

Q. 3. Prove that $15 + 17\sqrt{3}$ is an irrational number.★

Solution : Let $\sqrt{3} = \frac{a}{b}$, where a and b are coprime integers, $b \neq 0$.

Squaring both sides, we get

$$3 = \frac{a^2}{b^2}.$$

Multiplying with b on both sides, we get

$$3b = \frac{a^2}{b}$$

L.H.S. = $3 \times b$ = Integer

$$\text{R.H.S.} = \frac{a^2}{b} = \frac{\text{Integer}}{\text{Integer}} = \text{Rational number}$$

$\because$ LH.S. $\neq$ R.H.S.

$\therefore$ Our supposition is wrong.

$\Rightarrow \sqrt{3}$ is irrational.

Let $15 + 17\sqrt{3}$ is a rational number.

$\therefore 15 + 17\sqrt{3} = \frac{a}{b}$, where a and b are co-prime, $b \neq 0$

$$\Rightarrow 17\sqrt{3} = \frac{a}{b} - 15$$

$$\sqrt{3} = \frac{a-15b}{17b}$$

$\frac{a-15b}{17b}$ is rational number.

But $\sqrt{3}$ is irrational.

i.e., $$\sqrt{3} \neq \frac{a-15b}{17b}$$

$\therefore$ Our supposition is wrong.

$\Rightarrow 15 + 17\sqrt{3}$ is irrational. **Hence Proved.**

★ are board exam questions from previous years

Q. 4. Let a, b, c, k be rational numbers such that k is not a perfect cube.

If $a + bk^{1/3} + ck^{2/3} = 0$, then prove that $a = b = c = 0$.

Solution : Given, $a + bk^{1/3} + ck^{2/3} = 0$ ...(i)

Multiplying both sides by $k^{1/3}$, we have

$$ak^{1/3} + bk^{2/3} + ck = 0 \quad ...(ii)$$

Multiplying equation (i) by b and equation (ii) by c and then subtracting, we have

$$(ab + b^2k^{1/3} + bck^{2/3}) - (ack^{1/3} + bck^{2/3} + c^2k) = 0$$

$\Rightarrow$ $(b^2 - ac)k^{1/3} + ab - c^2k = 0$

$\Rightarrow$ $b^2 - ac = 0$ and $ab - c^2k = 0$

[Since $k^{1/3}$ is irrational]

$\Rightarrow$ $b^2 = ac$ and $ab = c^2k$

$\Rightarrow$ $b^2 = ac$ and $a^2b^2 = c^4k^2$

$\Rightarrow$ $a^2(ac) = c^4k^2$

[By putting $b^2 = ac$ in $a^2b^2 = c^4k^2$]

$\Rightarrow$ $a^3c - k^2c^4 = 0$

$\Rightarrow$ $(a^3 - k^2c^3)\,c = 0$

$\Rightarrow$ $a^3 - k^2c^3 = 0$, or $c = 0$

Now, $a^3 - k^2c^3 = 0 \Rightarrow k^2 = \dfrac{a^3}{c^3}$

$\Rightarrow$ $(k^2)^{1/3} = \left(\dfrac{a^3}{c^3}\right)^{1/3} \Rightarrow k^{2/3} = \dfrac{a}{c}$

This is impossible as $k^{2/3}$ is irrational and $\dfrac{a}{c}$ is rational.

$\therefore$ $a^3 - k^2c^3 \neq 0$

Hence, $c = 0$

Substituting $c = 0$ in $b^2 - ac = 0$, we get $b = 0$.

Substituting $b = 0$ and $c = 0$ in $a + bk^{1/3} + ck^{2/3} = 0$, we get $a = 0$

Hence, $a = b = c = 0$. **Hence Proved.**

Q. 5. Find the largest positive integer that will divide 398, 436 and 542 leaving remainders 7, 11 and 15 respectively.

Solution : It is given that on dividing 398 by the required number, there is a remainder of 7. This means that 398 –7 = 391 is exactly divisible by the required number. In other words, required number is a factor of 391.

Similarly, required positive integer is a factor of 436 – 11 = 425 and 542 – 15 = 527.

Clearly, required number is the HCF of 391, 425 and 527.

Using the factor tree, we get the prime factorisation of 391, 425 and 527 as follows :

$391 = 17 \times 23$, $425 = 5^2 \times 17$ and $527 = 17 \times 31$

$\therefore$ HCF of 391, 425 and 527 is 17.

Hence, required number = 17. **Ans.**

Q. 6. Find the HCF of 81 and 237 and express it as a linear combination of 81 and 237.

Solution : Given integers are 81 and 237 such that $81 < 237$. Applying Euclid's division lemma to 81 and 237, we get

$$237 = 81 \times 2 + 75 \quad ...(i) \qquad \left[\because 81\overline{)237}(2,\ \ 162,\ \ 75\right]$$

Since the remainder $75 \neq 0$. So, consider the divisor 81 and the remainder 75 and apply division lemma to get

$$81 = 75 \times 1 + 6 \quad ...(ii) \qquad \left[\because 75\overline{)81}(1,\ \ 75,\ \ 6\right]$$

We consider the new divisor 75 and the new remainder 6 and apply division lemma to get

$$75 = 6 \times 12 + 3 \quad ...(iii) \qquad \left[\because 6\overline{)75}(12,\ \ 72,\ \ 3\right]$$

We consider the new divisor 6 and the new remainder 3 and apply division lemma to get

$$6 = 3 \times 2 + 0 \quad ...(iv) \qquad \left[\because 3\overline{)6}(2,\ \ 6,\ \ 0\right]$$

The remainder at this stage is zero. So, the last divisor *i.e.*, 3 is the HCF of 81 and 237.

To represent the HCF as a linear combination of the given two numbers, we start from the last but one step and successively eliminate the previous remainders as follows :

From equation (iii), we have

$$3 = 75 - 6 \times 12$$

$\Rightarrow$ $3 = 75 - (81 - 75 \times 1) \times 12$

[Substituting $6 = 81 - 75 \times 1$ obtained from (ii)]

$\Rightarrow$ $3 = 75 - 12 \times 81 + 12 \times 75$

$\Rightarrow$ $3 = 13 \times 75 - 12 \times 81$

$\Rightarrow$ $3 = 13 \times (237 - 81 \times 2) - 12 \times 81$

[Substituting $75 = 237 - 81 \times 2$ obtained from (i)]

$\Rightarrow$ $3 = 13 \times 237 - 26 \times 81 - 12 \times 81$

$\Rightarrow$ $3 = 13 \times 237 - 38 \times 81$

$\Rightarrow$ $3 = 237x + 81y$, where $x = 13$ and $y = -38$. **Ans.**

Chapter 2. Polynomials

Q. 1. If α and β are the zeros of the quadratic polynomial $f(x) = 2x^2 - 5x + 7$, find a polynomial whose zeroes are $2\alpha + 3\beta$ and $3\alpha + 2\beta$.

Solution : Since, α and β are zeros of polynomial

$$f(x) = 2x^2 - 5x + 7$$

So, $\alpha + \beta = -\left(-\frac{5}{2}\right)$ and $\alpha\beta = \frac{7}{2}$

Let S and P denote respectively, the sum and product of zeros of the required polynomial.

Then, polynomial is

$$p(x) = k(x^2 - Sx + P)$$

Now, $S = (2\alpha + 3\beta) + (3\alpha + 2\beta)$

$$= 5(\alpha + \beta) = 5 \times \frac{5}{2} = \frac{25}{2}$$

and $P = (2\alpha + 3\beta)(3\alpha + 2\beta)$

$\Rightarrow$ $P = 6(\alpha^2 + \beta^2) + 13\alpha\beta$

$$= 6[(\alpha + \beta)^2 - 2\alpha\beta] + 13\alpha\beta$$

$$= 6(\alpha + \beta)^2 - 12\alpha\beta + 13\alpha\beta$$

$$= 6(\alpha + \beta)^2 + \alpha\beta$$

$\Rightarrow$ $P = 6 \times \left(\frac{5}{2}\right)^2 + \frac{7}{2}$

$$= \frac{75}{2} + \frac{7}{2} = \frac{82}{2} = 41$$

Hence, the required polynomial is given by

$$p(x) = k(x^2 - Sx + P)$$

or $p(x) = k\left(x^2 - \frac{25}{2}x + 41\right)$

where k is any non-zero real number. **Ans.**

Q. 2. If 1 and – 1 are zeros of polynomial

$$Lx^4 + Mx^3 + Nx^2 + Rx + P = 0,$$

show that $L + N + P = M + R = 0$.

Solution : Let $p(x) = Lx4 - Mx3 + Nx2 + Rx + P$

Now, it is given that 1 and – 1 are the zeros of given polynomial.

$\therefore$ $L(1)^4 - M(1)^3 + N(1)^2 + R(1) + P = 0$

$\Rightarrow$ $L + M + N + R + P = 0$ ...(i)

Also $p(-1) = 0$

$\Rightarrow$ $L(-1)^4 + M(-1)^3 + N(-1)^2 + R(-1) + P = 0$

$\Rightarrow$ $L - M + N - R + P = 0$

$\Rightarrow$ $L + N + P = M + R$ ...(ii)

From equations (i) and (ii)

$$2(M + R) = 0$$

$\Rightarrow$ $M + R = 0$

From equation (ii),

$$L + N + P = 0$$

Hence, $L + N + P = M + R = 0.$ **Hence Proved.**

Q. 3. Form a cubic polynomial having sum, sum of the product of its zeros taken two at a time, and the product of its zeros are 2, –7 and – 14 respectively.

Solution : Let the cubic polynomial be

$$ax^3 + bx^2 + cx + d \quad ...(i)$$

Let α, β, γ be the zeros of the required cubic polynomial, then

$$\alpha + \beta + \gamma = \frac{-b}{a};$$

$$\alpha\beta + \beta\gamma + \gamma\alpha = \frac{c}{a};$$

and $\alpha\beta\gamma = \frac{-d}{a}.$

Now, according to the question,

$$\alpha + \beta + \gamma = 2$$

$\Rightarrow$ $\frac{-b}{a} = 2 \Rightarrow \frac{b}{a} = -2$

$$\alpha\beta + \beta\gamma + \gamma\alpha = -7 \Rightarrow \frac{c}{a} = -7$$

and $\alpha\beta\gamma = -14 \Rightarrow \frac{-d}{a} = -14 \Rightarrow \frac{d}{a} = 14$

If $a = 1$, then $b = -2$, $c = -7$, $d = 14$

Substituting the values of a, b c and d in equation (i), we get the required cubic polynomial

$$x^3 - 2x^2 - 7x + 14$$ **Ans.**

Q. 4. Draw the graph of the quadratic polynomial $f(x) = 3 - 2x - x^2$

Solution : Let $y = f(x)$ or $y = 3 - 2x - x^2$

Let us list a few values of $y = 3 - 2x - x^2$ corresponding to a few values of x as follows :

x	–5	–4	–3	–2	–1	0	1	2	3	4
$y = 3 - 2x - x^2$	–12	–5	0	3	4	3	0	–5	–12	–21

Thus, the following points lie on the graph of polynomial $y = 3 - 2x - x^2$

(– 5, – 12), (– 4, – 5), (– 3, 0), (– 2, 3), (– 1, 4), (0, 3), (1, 0), (2, – 5), (3, –12) and (4, –21).

Let us plot these points on a graph paper and draw a smooth free hand curve passing through these points to obtain the graph of $y = 3 - 2x - x^2$. The curve thus obtained represents a parabola as shown in figure.

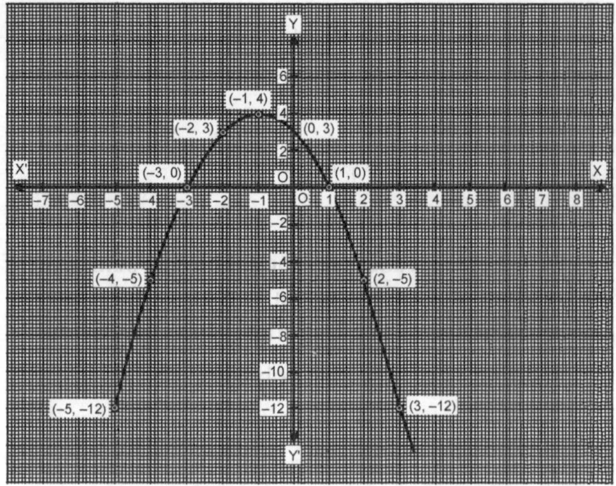

Q. 5. If α, β, γ be zeros of polynomial $6x^3 + 3x^2 - 5x + 1$, then find the value of $\alpha^{-1} + \beta^{-1} + \gamma^{-1}$.

Solution : Let $p(x) = 6x^3 + 3x^2 - 5x + 1$

So $a = 6, b = 3, c = -5, d = 1$

$\because$ α, β and γ are zeros of the polynomial $p(x)$

$$\therefore \quad \alpha+\beta+\gamma = \frac{-b}{a} = \frac{-3}{6} = \frac{-1}{2}$$

$$\alpha\beta+\alpha\gamma+\beta\gamma = \frac{c}{a} = \frac{-5}{6}$$

and
$$\alpha\beta\gamma = \frac{-d}{a} = \frac{-1}{6}$$

Now
$$\alpha^{-1}+\beta^{-1}+\gamma^{-1} = \frac{1}{\alpha}+\frac{1}{\beta}+\frac{1}{\gamma}$$
$$= \frac{\beta\gamma+\alpha\gamma+\alpha\beta}{\alpha\beta\gamma}$$
$$= \frac{-5/6}{-1/6} = 5$$ **Ans.**

Q. 6. Find the zeros of the polynomial $f(x) = x^3 - 12x^2 + 39x - 28$, if it is given that the zeros are in A.P.

Solution : Given $f(x) = x^3 - 12x^2 + 39x - 28$

$\Rightarrow \quad a = 1, b = -12, c = 39, d = -28$

Let α, β, γ be the zeros of the given polynomial

If α, β, γ are in A.P. , then

$$\beta - \alpha = \gamma - \beta$$
$$\Rightarrow \quad 2\beta = \alpha + \gamma \qquad ...(i)$$

Now,
$$\alpha+\beta+\gamma = \frac{-b}{a} = \frac{-(-12)}{1} = 12$$
$$\Rightarrow \quad \alpha+\gamma = 12-\beta \qquad ...(ii)$$

From equation (i) and (ii)

$$2\beta = 12 - \beta \Rightarrow 3\beta = 12 \Rightarrow \beta = 4$$

Putting the value of β in equation (i), we get

$$2\times 4 = \alpha+\gamma$$
$$8 = \alpha+\gamma \qquad ...(iii)$$

Now,
$$\alpha\beta\gamma = -\frac{d}{a} = \frac{-(-28)}{1} = 28$$

$$(\alpha\gamma)4 = 28 \Rightarrow \alpha\gamma = 7 \text{ or } \gamma = \frac{7}{\alpha} \qquad ...(iv)$$

Putting the value of $\gamma = \frac{7}{\alpha}$ in equation (iii), we get

$$8 = \alpha + \frac{7}{\alpha} \Rightarrow 8\alpha = \alpha^2 + 7$$
$$\Rightarrow \quad \alpha^2 - 8\alpha + 7 = 0$$
$$\Rightarrow \quad \alpha^2 - 7\alpha - 1\alpha + 7 = 0$$
$$\Rightarrow \quad \alpha(\alpha-7) - 1(\alpha-7) = 0$$
$$\Rightarrow \quad (\alpha-1)(\alpha-7) = 0$$
$$\Rightarrow \quad \alpha = 1 \text{ or } \alpha = 7$$

Putting α = 1 in equation (iv), we get

$$\gamma = \frac{7}{1}$$

or $\gamma = 7$

$\therefore$ zeros are 1, 7, 4.

Putting α = 7 in equation (iv), we get

$$\gamma = \frac{7}{7}$$

or $\gamma = 1$

$\therefore$ zeros are 7, 4, 1. **Ans.**

Q. 7. If the polynomial $f(x) = x^4 - 6x^3 + 16x^2 - 25x + 10$ is divided by another polynomial $x^2 - 2x + k$, the remainder comes out to be $x + a$. Find k and a. **[NCERT]**

Solution : By division algorithm, we have

Dividend = Divisor × Quotient + Remainder

$\Rightarrow$ Dividend – Remainder = Divisor × Quotient

$\Rightarrow$ (Dividend – Remainder) is always divisible by the divisor.

When $f(x) = x^4 - 6x^3 + 16x^2 - 25x + 10$ is divided by $x^2 - 2x + k$, the remainder comes out to be $x + a$

$$\therefore f(x) - (x+a) = x^4 - 6x^3 + 16x^2 - 25x + 10 - (x+a)$$
$$= x^4 - 6x^3 + 16x^2 - 25x + 10 - x - a$$
$$= x^4 - 6x^3 + 16x^2 - 26x + 10 - a$$

is exactly divisible by $x^2 - 2x + k$.

Let us now divide $x^4 - 6x^3 + 16x^2 - 26x + 10 - a$ by $x^2 - 2x + k$.

$$\begin{array}{r|l} & x^2 - 4x + (8-k) \\ \hline x^2 - 2x + k & x^4 - 6x^3 + 16x^2 - 26x + 10 - a \\ & x^4 - 2x^3 + kx^2 \\ & - \quad + \quad - \\ \hline & -4x^3 + (16-k)x^2 - 26x + 10 - a \\ & -4x^3 + 8x^2 \qquad -4kx \\ & + \quad - \quad + \\ \hline & (8-k)x^2 - (26-4k)x + 10 - a \\ & (8-k)x^2 - (16-2k)x + (8k-k^2) \\ & - \quad + \quad - \\ \hline & (-10+2k)x + (10-a-8k+k^2) \end{array}$$

For $f(x) - (x + a) = x^4 - 6x^3 + 16x^2 - 26x + 10 - a$ to be exactly divisible by $x^2 - 2x + k$, we must have

$(-10 + 2k)x + (10 - a - 8k + k^2) = 0$ for all values of x

$\Rightarrow$ $-10 + 2k = 0$ and $10 - a - 8k + k^2 = 0$

$\Rightarrow$ $k = 5$ and $10 - a - 8k + k^2 = 0$

$\Rightarrow$ $k = 5$ and $10 - a - 40 + 25 = 0$

$\Rightarrow$ $k = 5$ and $a = -5$. **Ans.**

Q. 8. If one zero of the polynomial $(k + 1)x^2 - 5x + 5$ is multiplicative inverse of the other, then find the zeros of $kx^2 - 3kx + 9$, where k is constant.

Solution : Here $f(x) = (k + 1)x^2 - 5x + 5$

$\Rightarrow$ $a = k + 1, b = -5, c = 5$

Let one zero $= \alpha$

$\therefore$ According to question, other zero $= \dfrac{1}{\alpha}$

Now, product of zeroes $= \dfrac{c}{a}$

$\Rightarrow$ $\alpha \times \dfrac{1}{\alpha} = \dfrac{5}{k+1}$

$\Rightarrow$ $1 = \dfrac{5}{k+1}$

$\Rightarrow$ $k + 1 = 5$

$\Rightarrow$ $k = 4$

Now putting $k = 4$ in polynomial

$p(x) = kx^2 - 3kx + 9$

we get $p(x) = 4x^2 - 12x + 9$

For zeros of $p(x)$,

$4x^2 - 12x + 9 = 0$

$\Rightarrow$ $(2x - 3)(2x - 3) = 0$

$\Rightarrow$ $x = \dfrac{3}{2}, x = \dfrac{3}{2}$

Hence, zeros are $\dfrac{3}{2}, \dfrac{3}{2}$. **Ans.**

Q. 9. $p(x)$ is a polynomial of degree more than 2. When $p(x)$ is divided by $x - 2$, it leaves remainder 1. When it is divided by $x - 3$, it leaves a remainder 3. Find the remainder when $p(x)$ is divided by $(x - 2)(x - 3)$.

Solution : Let remainder $r(x) = ax + b$, when $p(x)$ is divided by $(x - 2)(x - 3)$.

According to division algorithm,

$\Rightarrow$ $p(x) = (x - 2)(x - 3) \,.\, q(x) + r(x)$

$\Rightarrow$ $p(x) = (x - 2)(x - 3)\, q(x) + ax + b$

Now, $p(2) = 1$ [Given]

$\Rightarrow$ $(2 - 2)(2 - 3)\, q(2) + a \times 2 + b = 1$

$\Rightarrow$ $2a + b = 1$

Also, $p(3) = 3$ [Given]

$\Rightarrow$ $(3 - 2)(3 - 3)\, q(3) + a \times 3 + b = 3$

$\Rightarrow$ $3a + b = 3$

Solving equations (i) and (ii), we get

$a = 2, b = -3$

$\Rightarrow$ $r(x) = 2x - 3$. **Ans.**

Q. 10. Find the polynomial of least degree which should be subtracted from the polynomial $x^4 + 2x^3 - 4x^2 + 6x - 3$ so that it is exactly divisible by $x^2 - x + 1$.

Solution : Let $p(x) = x^4 + 2x^3 - 4x^2 + 6x - 3$

and $g(x) = x^2 - x + 1$

Now,

$$
\begin{array}{r|l}
x^2 - x + 1 & x^4 + 2x^3 - 4x^2 + 6x - 3 \;(\; x^2 + 3x - 2 \\
 & x^4 - x^3 + x^2 \\
 & - \quad + \quad - \\
\hline
 & 3x^3 - 5x^2 + 6x - 3 \\
 & 3x^3 - 3x^2 + 3x \\
 & - \quad + \quad - \\
\hline
 & -2x^2 + 3x - 2 \\
 & -2x^2 + 2x - 2 \\
 & + \quad - \quad + \\
\hline
 & x - 1
\end{array}
$$

Since, remainder $= x - 1$

Therefore, $x - 1$ should be subtracted, so that $p(x)$ is completely divisible by $g(x)$. **Ans.**

Chapter 3. Pair of Linear Equations in Two Variables

Q. 1. X takes 3 hours more than Y to walk 30 km. But, if X doubles his pace, he is ahead of Y in $1\frac{1}{2}$ hours. Find their speed of walking.

Solution : Let the speed of X and Y be x km/hr and y km/hr respectively. Then,

Time taken by X to cover 30 km $= \dfrac{30}{x}$ hrs

and, Time taken by Y to cover 30 km $= \dfrac{30}{y}$ hrs

According to the given conditions we have

$$\frac{30}{x} - \frac{30}{y} = 3 \Rightarrow \frac{10}{x} - \frac{10}{y} = 1 \quad \text{...(i)}$$

If X doubles his pace, then speed of X is $2x$ km/hr

Times taken by X to cover 30 km $= \dfrac{30}{2x}$ hrs

Time taken by Y to cover 30 km $= \dfrac{30}{y}$ hrs

According to the given conditions, we have

$$\frac{30}{y} - \frac{30}{2x} = 1\frac{1}{2}$$

$$\Rightarrow \quad \frac{30}{y} - \frac{30}{2x} = \frac{3}{2}$$

$$\Rightarrow \quad \frac{10}{y}-\frac{5}{x}=\frac{1}{2}$$

$$\Rightarrow \quad -\frac{10}{x}+\frac{20}{y}=1 \quad \text{...(ii)}$$

Putting $\frac{1}{x}=u$ and $\frac{1}{y}=v$ in equations (i) and (ii), we get

$$10u-10v=1 \Rightarrow 10u-10v-1=0 \quad \text{...(iii)}$$
$$-10u+20v=1 \Rightarrow -10u+20v-1=0 \quad \text{...(iv)}$$

Adding equations (iii) and (iv), we get

$$10v-2=0 \Rightarrow v=\frac{1}{5}$$

Putting $v=\frac{1}{5}$ in equation (iii), we get

$$10u-3=0 \Rightarrow u=\frac{3}{10}$$

$$u=\frac{3}{10} \Rightarrow x=\frac{10}{3} \text{ and } v=\frac{1}{5} \Rightarrow y=5$$

Hence X's speed $=\frac{10}{3}$ km/hr and Y's speed = 5 km/hr

Ans.

Q. 2. A train covered a certain distance at a uniform speed. If the train would have been 6 km/h faster, it would have taken 4 hours less than the scheduled time. And, if the train was slower by 6 km/h, it would have taken 6 hours more than the scheduled time. Find the length of the journey.

Solution : Let the actual speed of the train be x km/hr and the actual time taken be y hours. Then,

Distance covered = (xy) km ...(i)

[$\because$ Distance = Speed × Time]

If the speed is increased by 6 km/hr, then time of journey is reduced by 4 hours *i.e.,* when speed is $(x+6)$ km/hr, time of journey is $(y-4)$ hours.

Distance covered = $(x+6)(y-4)$

$xy=(x+6)(y-4)$ [Using (i)]

$$-4x+6y-24=0$$

$$-2x+3y-12=0 \quad \text{...(ii)}$$

When the speed is reduced by 6 km/hr, then the time of journey is increased by 6 hours *i.e.,* when speed is $(x-6)$ km/hr, time of journey is $(y-6)$ hours.

Distance covered = $(x-6)(y+6)$

$xy=(x-6)(y+6)$ [Using (i)]

$$6x-6y-36=0$$

$$x-y-6=0 \quad \text{...(iii)}$$

Thus, we obtain the following system of equations :

$$-2x+3y-12=0$$
$$x-y-6=0$$

By using cross-multiplication, we have

$$\frac{x}{3\times(-6)-(-1)\times-12}=\frac{-y}{(-2)\times(-6)-1\times(-12)}=\frac{1}{(-2)\times(-1)-1\times 3}$$

$$\Rightarrow \quad \frac{x}{-30}=\frac{-y}{24}=\frac{1}{-1}$$

$$\Rightarrow \quad x=30 \text{ and } y=24.$$

Putting the values of x and y in equation (i), we obtain

Distance = (30×24) km = 720 km.

Hence, the length of the journey is 720 km. **Ans.**

Q. 3. 8 men and 12 boys can finish a piece of work in 10 days while 6 men and 8 boys can finish it in 14 days. Find the time taken by one man alone and that by one boy alone to finish the work.

Solution : Let one man alone can finish the work in x days and one boy alone can finish the work in y days. Then,

One day work of one man $=\frac{1}{x}$,

One day work of one boy $=\frac{1}{y}$

$\therefore$ One day work of 8 men $=\frac{8}{x}$,

One day work of 12 boys $=\frac{12}{y}$

Since, 8 men and 12 boys can finish the work in 10 days

$$\therefore \quad 10\left(\frac{8}{x}+\frac{12}{y}\right)=1 \Rightarrow \frac{80}{x}+\frac{120}{y}=1 \quad \text{...(i)}$$

Again, 6 men and 8 boys can finish the work in 14 days

$$\therefore \quad 14\left(\frac{6}{x}+\frac{8}{y}\right)=1 \Rightarrow \frac{84}{x}+\frac{112}{y}=1 \quad \text{...(ii)}$$

Put $\frac{1}{x}=u$ and $\frac{1}{y}=v$ in equations (i) and (ii), we get

$$80u+120v-1=0 \text{ and } 84u+112v-1=0$$

By using cross-multiplication, we have

$$\frac{u}{120\times(-1)-112\times(-1)}=\frac{-v}{80\times(-1)-84\times(-1)}=\frac{1}{80\times 112-84\times 120}$$

$$\Rightarrow \quad \frac{u}{-120+112}=\frac{-v}{-80+84}=\frac{1}{80\times 112-84\times 120}$$

$$\Rightarrow \quad \frac{u}{-8} = \frac{-v}{4} = \frac{1}{-1120}$$

$$\Rightarrow \quad u = \frac{-8}{-1120} = \frac{1}{140}$$

and $$v = \frac{-4}{-1120} = \frac{1}{280}$$

We have, $u = \frac{1}{140} \Rightarrow \frac{1}{x} = \frac{1}{140} \Rightarrow x = 140$

and $v = \frac{1}{280} \Rightarrow \frac{1}{y} = \frac{1}{280} \Rightarrow y = 280.$

Hence, one man alone can finish the work in 140 days and one boy alone can finish the work in 280 days.

Ans.

Q. 4. A boat covers 25 km upstream and 44 km downstream in 9 hours. Also, it covers 15 km upstream and 22 km downstream in 5 hours. Find the speed of the boat in still water and that of the stream.

Solution : Let the speed of the boat in still water be x km/h and that of the stream be y km/h. Then,

Speed upstream = $(x - y)$ km/h

Speed downstream = $(x + y)$ km/h

Now, time taken to cover 25 km upstream

$$= \frac{25}{x-y} \text{ hours}$$

Time taken to cover 44 km downstream = $\frac{44}{x+y}$ hours

The total time of journey is 9 hours

$$\therefore \quad \frac{25}{x-y} + \frac{44}{x+y} = 9 \quad \text{...(i)}$$

Time taken to cover 15 km upstream = $\frac{15}{x-y}$

Time taken to cover 22 km downstream = $\frac{22}{x+y}$

In this case, total time of journey is 5 hours

$$\therefore \quad \frac{15}{x-y} + \frac{22}{x+y} = 5 \quad \text{...(ii)}$$

Put $\frac{1}{x-y} = u$ and $\frac{1}{x+y} = v$ in equations (i) and (ii), we get

$$25u + 44v = 9$$

$$\Rightarrow \quad 25u + 44v - 9 = 0 \quad \text{...(iii)}$$

and $$15u + 22v = 5$$

$$\Rightarrow \quad 15u + 22v - 5 = 0 \quad \text{...(iv)}$$

By cross-multiplication, we have

$$\frac{u}{44 \times (-5) - 22 \times (-9)} = \frac{-v}{25 \times (-5) - 15 \times (-9)}$$

$$= \frac{1}{25 \times 22 - 15 \times 44}$$

$$\Rightarrow \quad \frac{u}{-220 + 198} = \frac{-v}{-125 + 135} = \frac{1}{550 - 660}$$

$$\Rightarrow \quad \frac{u}{-22} = \frac{-v}{10} = \frac{1}{-110}$$

$$\Rightarrow \quad \frac{u}{22} = \frac{v}{10} = \frac{1}{110}$$

$$\Rightarrow \quad \frac{u}{22} = \frac{1}{110} \quad \text{and} \quad \frac{v}{10} = \frac{1}{110}$$

$$\Rightarrow \quad u = \frac{22}{110} = \frac{1}{5} \quad \text{and} \quad v = \frac{1}{11}$$

We have, $$u = \frac{1}{5}$$

$$\Rightarrow \quad \frac{1}{x-y} = \frac{1}{5}$$

$$\Rightarrow \quad x - y = 5 \quad \text{...(v)}$$

and $v = \frac{1}{11} \Rightarrow \frac{1}{x+y} = \frac{1}{11} \Rightarrow x + y = 11 \quad \text{...(vi)}$

Solving equations (v) and (vi), we get $x = 8$ and $y = 3$.

Hence, speed of the boat in still water is 8 km/h and speed of the stream is 3 km/h. **Ans.**

Q. 5. Students of a class are made to stand in rows. If one student is extra in each row, there would be 2 rows less. If one student is less in each row, there would be 3 rows more. Find the number of students in the class.

Solution : Let total number of rows be y

and total number of students in each row be x.

$\therefore$ Total number of students = xy

Case I : If one student is extra in each row, there would be 2 rows less.

Now, number of rows = $(y - 2)$

and, number of students in each row = $(x + 1)$

Total number of students

= Number of rows × Number of students in each row

$$xy = (y - 2)(x + 1)$$

$$\Rightarrow \quad xy = xy + y - 2x - 2$$

$$\Rightarrow \quad xy - xy - y + 2x = -2$$

$$\Rightarrow \quad 2x - y = -2 \quad \text{...(i)}$$

Case II : If one student is less in each row, there would be 3 rows more.

Now, number of rows = $(y + 3)$

and, number of students in each row = $(x - 1)$

Total number of students

= Number of rows

× Number of students in each row

$$xy = (y + 3)(x - 1)$$

$$\Rightarrow \quad xy = xy - y + 3x - 3$$

$$\Rightarrow \quad xy - xy + y - 3x = -3$$

$$\Rightarrow \quad -3x + y = -3 \quad \text{...(ii)}$$

On adding equations (i) and (ii), we have

$$2x - y = -2$$

$$-3x + y = -3$$

$$-x = -5$$

or $\quad x = 5$

Putting the value of x in equation (i), we get

$$2(5) - y = -2$$

$$\Rightarrow \quad 10 - y = -2$$

$$-y = -2 - 10$$

$$\Rightarrow \quad -y = -12$$

or $\quad y = 12$

∴ Total number of students in the class $5 \times 12 = 60$.

Ans.

Q. 6. Draw the graph of $2x + y = 6$ and $2x - y + 2 = 0$. Shade the region bounded by these lines and X-axis. Find the area of the shaded region.

Solution : We have, $2x + y = 6 \Rightarrow y = 6 - 2x$

When $x = 0$, we have

$$y = 6 - 2 \times 0 = 6$$

When $x = 3$, we have

$$y = 6 - 2 \times 3 = 0$$

When $x = 2$, we have

$$y = 6 - 2 \times 2 = 2$$

Thus, we get the following table :

x	0	3	2
y	6	0	2

Now, we plot the points $A(0, 6)$, $B(3, 0)$ and $C(2, 2)$ on the graph paper. We join A, B and C and extend it on both sides to obtain the graph of the equation $2x + y = 6$.

Also, we have, $2x - y + 2 = 0 \Rightarrow y = 2x + 2$

When $x = 0$, we have

$$y = 2 \times 0 + 2 = 2$$

When $x = -1$, we have

$$y = 2 \times (-1) + 2 = 0$$

When $x = 1$, we have

$$y = 2 \times 1 + 2 = 4$$

Thus, we have the following table :

x	0	–1	1
y	2	0	4

Now, we plot the points $D(0, 2)$, $E(-1, 0)$ and $F(1, 4)$ on the same graph paper. We join D, E and F and extend it on both sides to obtain the graph of the equation $2x - y + 2 = 0$.

It is evident from the graph that the two lines intersect at point $F(1, 4)$. The area enclosed by the given lines and X-axis is shown in Fig.

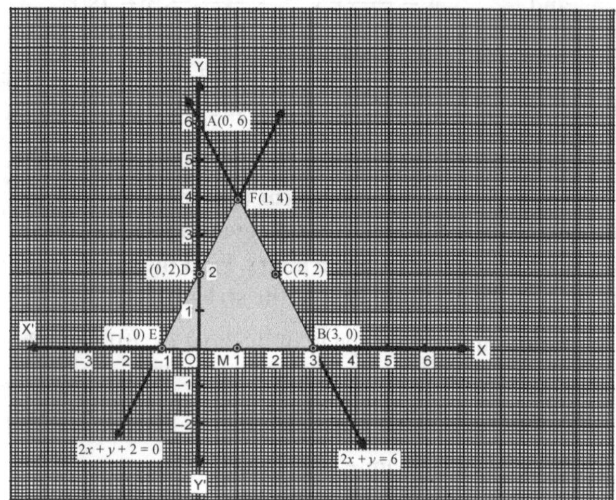

Thus, $x = 1, y = 4$ is the solution of the given system of equations. Draw FM perpendicular from F on X-axis.

Clearly, we have

$FM = y$-coordinate of point $F(1, 4) = 4$

and $\quad BE = 4$

∴ Area of the shaded region = Area of ΔFBE

$$\Rightarrow \text{Area of the shaded region} = \frac{1}{2}(\text{Base} \times \text{Height})$$

$$= \frac{1}{2}(BE \times FM)$$

$$= \left(\frac{1}{2} \times 4 \times 4\right) \text{ sq. units}$$

$$= 8 \text{ sq. units.}$$

Ans.

Q. 7. The ages of two friends Ani and Biju differ by 3 years. Ani's father Dharam is twice as old as Ani and Biju is twice as old as his sister Cathy. The ages of Cathy and Dharam differ by 30 years. Find the ages of Ani and Biju. [NCERT]

Solution : Let the ages of Ani and Biju be x and y years respectively.

Then, $\quad x - y = \pm 3$.

Age of Dharam = $2x$ years

Age of Cathy = $\frac{y}{2}$ years

Clearly, Dharam is older than Cathy.

$\therefore\ 2x - \frac{y}{2} = 30 \Rightarrow \frac{4x-y}{2} = 30 \Rightarrow 4x - y = 60$

Thus, we have following two systems of linear equations

$$x - y = 3 \quad \text{...(i)}$$
$$4x - y = 60 \quad \text{...(ii)}$$

and

$$x - y = -3 \quad \text{...(iii)}$$
$$4x - y = 60 \quad \text{...(iv)}$$

Subtracting equation (i) from (ii), we get

$$\begin{array}{r} 4x - y = 60 \\ x - y = 3 \\ - \quad + \quad - \\ \hline 3x = 57 \end{array} \Rightarrow x = 19$$

Putting $x = 19$ in equation (i), we get

$$19 - y = 3 \Rightarrow y = 16$$

Now, subtracting equation (iii) from (iv)

$$\begin{array}{r} 4x - y = 60 \\ x - y = -3 \\ - \quad + \quad + \\ \hline 3x = 63 \end{array} \Rightarrow x = 21$$

Putting $x = 21$ in equation (iii), we get

$$21 - y = -3 \Rightarrow y = 24$$

Hence, age of Ani = 19 years and age of Biju = 16 years or age of Ani = 21 years and age of Biju = 24 years.

Ans.

Q. 8. A train covered a certain distance at a uniform speed. If the train would have been 10 km/h faster, it would have taken 2 hours less than the scheduled time. And, if the train was slower by 10 km/h, it would have taken 3 hours more than the scheduled time. Find the distance covered by the train. [NCERT]

Solution : Let the actual speed of the train be x km/h and actual time taken be y hours. Then,

Distance covered = Speed × Time = xy km ...(i)

Case I : When speed is $(x + 10)$ km/h, then time taken is $(y - 2)$ hours

$\therefore$ Distance covered $= (x + 10)(y - 2)$

$\Rightarrow xy = (x + 10)(y - 2)$ [Using (i)]

$\Rightarrow xy = xy - 2x + 10y - 20$

$\Rightarrow 2x - 10y = -20$

$\Rightarrow x - 5y = -10$ ...(ii)

Case II : When speed is $(x - 10)$ km/h, then time taken is $(y + 3)$ hours.

$\therefore$ Distance covered $= (x - 10)(y + 3)$

$\Rightarrow xy = (x - 10)(y + 3)$ [Using (i)]

$\Rightarrow xy = xy + 3x - 10y - 30$

$\Rightarrow 3x - 10y = 30$ ...(iii)

Multiplying equation (ii) by 2 and subtracting it from (iii), we get

$$\begin{array}{r} 3x - 10y = 30 \\ 2x - 10y = -20 \\ - \quad + \quad + \\ \hline x = 50 \end{array}$$

Putting $x = 50$ in equation (ii), we get

$50 - 5y = -10$

$\Rightarrow 50 + 10 = 5y$

$\Rightarrow 60 = 5y$

$\Rightarrow y = 12$

$\therefore$ Distance covered by the train = 50 × 12 km = 600 km.

Ans.

Q. 9. On selling a tea-set at 5% loss and a lemon-set at 15% gain, a crockery seller gains ₹ 7. If he sells the tea-set at 5% gain and the lemon-set at 10% gain, he gains ₹ 13. Find the actual rate of the tea-set and the lemon set.

Solution : Let the cost price of the tea-set and the lemon-set be ₹ x and ₹ y respectively.

Case I : When tea set is sold at 5% loss and lemon-set at 15% gain.

$$\text{Loss on tea-set} = ₹\ \frac{5x}{100} = ₹\ \frac{x}{20}$$

$$\text{Gain on lemon-set} = ₹\ \frac{15y}{100} = ₹\ \frac{3y}{20}$$

$$\text{Net gain} = ₹\ \frac{3y}{20} - \frac{x}{20}$$

According to the question,

$$\frac{3y}{20} - \frac{x}{20} = 7$$

$\Rightarrow 3y - x = 140$

$\Rightarrow x - 3y + 140 = 0$ ...(i)

Case II : When tea-set is sold at 5% gain and the lemon-set at 10% gain.

$$\text{Gain on tea-set} = ₹\ \frac{5x}{100} = ₹\ \frac{x}{20}$$

$$\text{Gain on lemon-set} = ₹\ \frac{10y}{100} = ₹\ \frac{y}{10}$$

$$\text{Total gain} = ₹\ \frac{x}{20} + \frac{y}{10}$$

According to the question,

$$\frac{x}{20} + \frac{y}{10} = 13$$

$\Rightarrow x + 2y = 260$

$\Rightarrow x + 2y - 260 = 0$ ...(ii)

Subtracting equation (ii) from equation (i), we get

$$-5y + 400 = 0 \Rightarrow y = 80$$

Putting $y = 80$ in equation (i), we get

$$x - 240 + 140 = 0 \Rightarrow x = 100$$

Hence, cost prices of tea-set and lemon-set are ₹ 100 and ₹ 80 respectively.

Ans.

Q. 10. A two digit number is obtained by either multiplying sum of the digits by 8 and adding 1 or by multiplying the difference of the digits by 13 and adding 2. Find the number.

Solution : Let the two- digit number be $10x + y$.

According to the question,

$$10x + y = 8(x + y) + 1 \quad ...(i)$$

or

$$10x + y = 13(x - y) + 2 \quad ...(ii)$$

From equation (i), we have

$$10x + y = 8x + 8y + 1$$

$$2x - 7y = 1 \quad ...(iii)$$

From equation (ii), we have

$$10x + y = 13x - 13y + 2$$

$$-3x + 14y = 2 \quad ...(iv)$$

Multiplying equation (iii) by 2 and adding it to equation (iv), we get

$$4x - 14y = 2$$

$$-3x + 14y = 2$$

$$x = 4$$

Putting $x = 4$ in equation (iii), we get

$$2.4 - 7y = 1$$

$$-7y = -7$$

$$y = 1$$

$\therefore$ Two- digit number $= 10x + y$

$$= 10(4) + 1$$

$$= 41$$ **Ans.**

Q. 11. Solve the following pairs of equations by reducing them to a pair of linear equations :

$$\frac{1}{3x+y} + \frac{1}{3x-y} = \frac{3}{4};$$

$$\frac{1}{2(3x+y)} - \frac{1}{2(3x-y)} = \frac{-1}{8}$$

Solution : Given pair of linear equations are,

$$\frac{1}{3x+y} + \frac{1}{3x-y} = \frac{3}{4};$$

and

$$\frac{1}{2(3x+y)} - \frac{1}{2(3x-y)} = \frac{-1}{8}$$

Let

$$\frac{1}{3x+y} = a \quad ...(i)$$

and

$$\frac{1}{3x-y} = b \quad ...(ii)$$

$$\therefore \quad a + b = \frac{3}{4}$$

and

$$\frac{1}{2}a - \frac{1}{2}b = \frac{-1}{8}$$

Now,

$$a + b = \frac{3}{4}$$

$$\Rightarrow \quad 4a + 4b = 3 \quad ...(iii)$$

and

$$\frac{a}{2} - \frac{b}{2} = \frac{-1}{8}$$

$$\Rightarrow \quad 4a - 4b = -1 \quad ...(iv)$$

Adding equation (iii) and (iv), we get

$$8a = 2$$

$$a = \frac{1}{2}$$

Putting the value of a in equation (iii), we get

$$\Rightarrow \quad 4\left(\frac{1}{2}\right) + 4b = 3$$

$$\Rightarrow \quad 4b = 3 - 2$$

$$\Rightarrow \quad b = \frac{-1}{2}$$

Now,

$$a = \frac{1}{2}$$

$$\Rightarrow \quad \frac{1}{3x+y} = \frac{1}{2} \quad \text{[From (i)]}$$

$$\Rightarrow \quad 3x + y = 2 \quad ...(v)$$

and

$$b = \frac{-1}{2}$$

$$\Rightarrow \quad \frac{1}{3x-y} = \frac{-1}{2} \quad \text{[From (ii)]}$$

$$\Rightarrow \quad 3x - y = -2 \quad ...(vi)$$

On adding equations (v) and (vi), we get

$$6x = 0$$

$$x = 0$$

Putting $x = 0$ in equation (v), we get

$$\Rightarrow \quad 3(0) + y = 2 \quad \text{[from (v)]}$$

$$\Rightarrow \quad y = 2.$$

Hence, $x = 0$ and $y = 2$. **Ans.**

Q. 12. Solve for x and y :

$$\frac{2}{3x+2y} + \frac{3}{3x-2y} = \frac{17}{5};$$

$$\frac{5}{3x+2y} + \frac{1}{3x-2y} = 2$$

Solution : Given pair of linear equations are,

$$\frac{2}{3x+2y} + \frac{3}{3x-2y} = \frac{17}{5};$$

and

$$\frac{5}{3x+2y} + \frac{1}{3x-2y} = 2$$

Let

$$\frac{1}{3x+2y} = a, \quad \frac{1}{3x-2y} = b$$

$$\Rightarrow \quad 2a + 3b = \frac{17}{5} \quad ...(i)$$

and $\quad 5a + b = 2 \quad ...(ii)$

On multiplying equation (ii) by 3, we get

$$15a + 3b = 6 \quad ...(iii)$$

Subtracting equation (iii) from equation (i), we get

$$-13a = \frac{17}{5} - 6$$

$$-13a = \frac{-13}{5} \Rightarrow a = \frac{1}{5}$$

Putting the value of a in equation (ii), we get

$$5\left(\frac{1}{5}\right) + b = 2$$

$$\Rightarrow \quad 1 + b = 2$$

$$\Rightarrow \quad b = 1$$

Now, $\quad a = \frac{1}{5}$

$$\Rightarrow \quad \frac{1}{3x+2y} = \frac{1}{5}$$

$$\Rightarrow \quad 3x + 2y = 5 \quad ...(iv)$$

and $\quad b = 1$

$$\Rightarrow \quad \frac{1}{3x-2y} = 1$$

$$\Rightarrow \quad 3x - 2y = 1 \quad ...(v)$$

Adding equation (iv) and (v), we get

$$6x = 6$$

$$\Rightarrow \quad x = 1$$

Putting the value of x in equation (iv), we get

$$3(1) + 2y = 5$$

$$\Rightarrow \quad 2y = 5 - 3$$

$$\Rightarrow \quad 2y = 2$$

$$\Rightarrow \quad y = 1$$

Hence, $x = 1$ and $y = 1$ **Ans.**

Chapter 4. Quadratic Equations

Q. 1. An equation has been given as $\frac{c}{x^2} + \frac{k}{x} = 1$.

Find the relation between c and k, if

(i) x has real values

(ii) x has no real values.

Solution : We have,

$$\frac{c}{x^2} + \frac{k}{x} = 1$$

Multiplying both the sides by x^2, we get

$$c + kx = x^2$$

$$\Rightarrow \quad x^2 - kx - c = 0$$

(i) For real values :

$$(-k)^2 - 4.1(-c) \geq 0 \Rightarrow k^2 + 4c \geq 0$$

$$\Rightarrow k^2 \geq -4c \Rightarrow -k^2 \leq 4c \Rightarrow c \geq \frac{-k^2}{4}$$

(ii) For non-real values :

$$(-k)^2 - 4.1(-c) \leq 0$$

$$\Rightarrow \quad k^2 + 4c \leq 0 \Rightarrow k^2 \leq -4c$$

$$\Rightarrow -k^2 \geq 4c \Rightarrow 4c \leq -k^2 \Rightarrow c \leq \frac{-k^2}{4}$$ **Ans.**

Q. 2. If α and β are roots of a quadratic equation such that $\alpha + \beta = 2$ and $\alpha^4 + \beta^4 = 272$, then find the equation.

Solution :
$$\alpha^4 + \beta^4 = (\alpha^2 + \beta^2)^2 - 2\alpha^2\beta^2$$
$$= \{(\alpha + \beta)^2 - 2\alpha\beta\}^2 - 2\alpha^2\beta^2$$
$$= (2^2 - 2\alpha\beta)^2 - 2\alpha^2\beta^2$$

Let $\quad \alpha\beta = t,$

Then $(4 - 2t)^2 - 2t^2 = 272$

$$\Rightarrow 16 - 16t + 4t^2 - 2t^2 = 272$$

$$\Rightarrow \quad 2t^2 - 16t - 256 = 0$$

$$\Rightarrow \quad t^2 - 8t - 128 = 0$$

$$\Rightarrow \quad t^2 - 16t + 8t - 128 = 0$$

$$\Rightarrow \quad t(t-16) + 8(t-16) = 0$$

$$\Rightarrow \quad (t-16)(t+8) = 0$$

$$\Rightarrow \quad t = 16 \text{ or } -8$$

For $\quad t = 16$

$$\alpha + \beta = 2 \text{ and } \alpha\beta = 16 \Rightarrow x^2 - 2x + 16 = 0$$

For $\quad t = -8$

$$\alpha + \beta = 2 \text{ and } \alpha\beta = -8 \Rightarrow x^2 - 2x - 8 = 0$$ **Ans.**

Q. 3. If p, q, r and s are real numbers, such that $pr = 2(q + s)$, then show that at least one of the two equations $x^2 + px + q = 0$ and $x^2 + rx + s = 0$ has real roots.

Solution : We have,

$$x^2 + px + q = 0 \quad ...(i)$$

and $\quad x^2 + rx + s = 0 \quad ...(ii)$

Let D_1 and D_2 be the discriminants of equation (i) and (ii) respectively. Then,

$$D_1 = p^2 - 4q \text{ and } D_2 = r^2 - 4s$$

$$\Rightarrow D_1 + D_2 = p^2 - 4q + r^2 - 4s = (p^2 + r^2) - 4(q + s)$$

$$\Rightarrow \quad D_1 + D_2 = p^2 + r^2 - 4\left(\frac{pr}{2}\right)$$

$$\left[\because pr = 2(q+s) \therefore q + s = \frac{pr}{2}\right]$$

$$\Rightarrow \quad D_1 + D_2 = p^2 + r^2 - 2pr = (p - r)^2 \geq 0$$

$$\left[\because (p-r)^2 \geq 0 \text{ for all real } p, r\right]$$

$\Rightarrow$ At least one of D_1 and D_2 is greater than or equal to zero.

$\Rightarrow$ At least one of the two equations has real roots.

Hence Proved.

Q. 4. Find the values of k for which the equation $x^2 + 5kx + 16 = 0$ has no real roots.

Solution : The given equation is $x^2 + 5kx + 16 = 0$.

Here, $a = 1$, $b = 5k$ and $c = 16$

$\therefore D = b^2 - 4ac = (5k)^2 - 4 \times 1 \times 16 = 25k^2 - 64$

The given equation will have no real roots, if

$$D < 0$$

$$\Rightarrow \quad 25k^2 - 64 < 0$$

$$\Rightarrow \quad 25\left(k^2 - \frac{64}{25}\right) < 0$$

$$\Rightarrow \quad k^2 - \frac{64}{25} < 0$$

$$\Rightarrow \quad K^2 < \frac{64}{25}$$

$$\Rightarrow \quad K^2 < \left(\frac{8}{9}\right)^2$$

$$\Rightarrow \quad K < -\frac{8}{5} \text{ and } K < +\frac{8}{5}$$

Thus, $$-\frac{8}{5} < k < \frac{8}{5}$$ **Ans.**

Q. 5. If the roots of the equation $x^2 + 2cx + ab = 0$ are real and unequal, prove that the equation

$$x^2 - 2(a+b)x + a^2 + b^2 + 2c^2 = 0$$

has no real roots.

Solution : The two equations are

$$x^2 + 2cx + ab = 0 \quad \text{...(i)}$$

and $$x^2 - 2(a+b)x + a^2 + b^2 + 2c^2 = 0 \quad \text{...(ii)}$$

Let D_1 and D_2 be the discriminants of equations (i) and (ii) respectively. Then,

$$D_1 = (2c)^2 - 4 \times 1 \times ab$$
$$= 4c^2 - 4ab = 4(c^2 - ab)$$
$$D_2 = \{-2(a+b)\}^2 - 4 \times 1 \times (a^2 + b^2 + 2c^2)$$
$$D_2 = 4(a+b)^2 - 4(a^2 + b^2 + 2c^2)$$
$$D_2 = 4\{a^2 + b^2 + 2ab - a^2 - b^2 - 2c^2\}$$
$$D_2 = (2ab - 2c^2)$$
$$= -8(c^2 - ab)$$

Since, the roots of equation (i) are real and unequal Therefore,

$$D_1 > 0$$
$$\Rightarrow \quad 4(c^2 - ab) > 0$$
$$\Rightarrow \quad c^2 - ab > 0$$
$$\Rightarrow \quad -8(c^2 - ab) < 0$$
$$\Rightarrow \quad D_2 < 0$$

i.e., roots of equation (ii) are not real.

Hence Proved.

Q. 6. Prove that the equation $x^2(a^2 + b^2) + 2x(ac + bd) + (c^2 + d^2) = 0$ has no real root, if $ad \neq bc$.

Solution : The discriminant of the given equation is given by

$$D = \{2(ac + bd)\}^2 - 4(a^2 + b^2)(c^2 + d^2)$$
$$D = 4(ac + bd)^2 - 4(a^2 + b^2)(c^2 + d^2)$$
$$D = 4[(ac + bd)^2 - (a^2 + b^2)(c^2 + d^2)]$$
$$D = 4[a^2c^2 + b^2d^2 + 2ac.\,bd - a^2c^2 - a^2d^2 - b^2c^2 - b^2d^2]$$
$$D = 4[2ac.\,bd - a^2d^2 - b^2c^2]$$
$$D = -4[a^2d^2 + b^2c^2 - 2ad.\,bc] = -4(ad - bc)^2$$

We have, $ad \neq bc$

$$ad - bc \neq 0 \Rightarrow (ad - bc)^2 > 0$$
$$\Rightarrow -4(ad - bc)^2 < 0 \Rightarrow D < 0$$

Hence, the given equation has no real roots.

Hence Proved.

Q. 7. Find the value of p if $\alpha^2 + \beta^2 - \alpha\beta = 3\frac{1}{4}$ where α and β are roots of quadratic equation $x^2 + px + 1 = 0$.

Solution : Given, $$\alpha^2 + \beta^2 - \alpha\beta = \frac{13}{4} \quad \text{...(i)}$$

If α and β are roots of the equation $x^2 + px + 1 = 0$ Then,

Sum of roots, $$\alpha + \beta = \frac{-b}{a} = -p$$

Product of roots, $$\alpha\beta = \frac{c}{a} = 1$$

Substitiuting $\alpha\beta = 1$ in equation (i), we get

$$\alpha^2 + \beta^2 - 1 = \frac{13}{4}$$
$$\Rightarrow \quad \alpha^2 + \beta^2 = \frac{13}{4} + 1 = \frac{17}{4}$$
$$\Rightarrow \quad (\alpha + \beta)^2 - 2\alpha\beta = \frac{17}{4}$$
$$\Rightarrow \quad (\alpha + \beta)^2 - 2 = \frac{17}{4} \quad [\because \alpha\beta = 1]$$
$$\Rightarrow \quad (\alpha + \beta)^2 = \frac{17}{4} + 2 = \frac{25}{4}$$
$$\therefore \quad \alpha + \beta = \sqrt{\frac{25}{4}} = \frac{5}{2} = -p$$

Thus, $$p = -\frac{5}{2}$$ **Ans.**

Q. 8. Out of a certain number of Saras birds, one-fourth the number are moving about in lotus plants, $\frac{1}{9}$th coupled (along) with $\frac{1}{4}$th as well as 7 times the square root of the number move on a hill, 56 birds remain in Vacula tree. What is the total number of birds?

Solution : Let the total number of birds be x.

$\therefore$ Number of birds moving in lotus plants = $\frac{x}{4}$

Number of birds moving on a hill

$$= \frac{x}{9} + \frac{x}{4} + 7\sqrt{x}$$

Number of birds in Vakula tree = 56.

Using the given informations, we have

$$\frac{x}{4} + \left(\frac{x}{9} + \frac{x}{4} + 7\sqrt{x}\right) + 56 = x$$

$$\Rightarrow \quad x - \frac{x}{4} - \frac{x}{9} - \frac{x}{4} - 7\sqrt{x} - 56 = 0$$

$$\Rightarrow \quad \frac{36x - 9x - 4x - 9x}{36} - 7\sqrt{x} - 56 = 0$$

$$\Rightarrow \quad \frac{14x}{36} - 7\sqrt{x} - 56 = 0$$

$$\Rightarrow \quad \frac{7x}{18} - 7\sqrt{x} - 56 = 0$$

$$\Rightarrow \quad \frac{x}{18} - \sqrt{x} - 8 = 0$$

$$\Rightarrow \quad x - 18\sqrt{x} - 144 = 0$$

Putting $\sqrt{x} = y$, we have

$$y^2 - 18y - 144 = 0$$

$$\Rightarrow \quad y^2 - 24y + 6y - 144 = 0$$

$$\Rightarrow \quad y(y-24) + 6(y-24) = 0$$

$$\Rightarrow \quad (y-24)(y+6) = 0$$

$$\Rightarrow \quad y = 24, -6$$

But $y \neq -6$, since $\sqrt{x} = y$ is positive.

$\therefore y = 24 \Rightarrow \sqrt{x} = 24 \Rightarrow x = 576$

Hence the total number of birds are 576. **Ans.**

Q. 9. O girl ! Out of a group of swans, $\frac{7}{2}$ times the square root of the number are playing on the shore of a tank. The two remaining ones are playing, with amorous fight, in the water. What is the total number of swans ?

Solution : Let the total number of swans be x. Then,

Number of swans playing on the shore of the tank

$$= \frac{7}{2}\sqrt{x}$$

It is given that there are two remaining swans.

$$\therefore \quad x = \frac{7}{2}\sqrt{x} + 2$$

$$x - \frac{7}{2}\sqrt{x} - 2 = 0$$

Putting $\sqrt{x} = y$, we have

$$y^2 - \frac{7}{2}y - 2 = 0$$

$$2y^2 - 7y - 4 = 0$$

$$2y^2 - 8y + y - 4 = 0$$

$$2y(y-4) + (y-4) = 0$$

$$(y-4)(2y+1) = 0$$

$$\Rightarrow \quad y = 4 \text{ or } y = -\frac{1}{2}$$

$$\Rightarrow \quad y = 4 \; [\because y = -\frac{1}{2} \text{ is not possible}]$$

$$\Rightarrow \quad x = y^2 = 4^2 = 16$$

Hence, the total number of swans is 16. **Ans.**

Q. 10. One-fourth of a herd of camels was seen in the forest. Twice the square root of the herd had gone to mountains and the remaining 15 camels were seen on the bank of a river. Find the total number of camels.

Solution : Let x be the total number of camels.

Then, number of camels in the forest = $\frac{x}{4}$

Number of camels on mountains = $2\sqrt{x}$

and number of camels on the bank of river = 15

Thus, total number of camels = $\frac{x}{4} + 2\sqrt{x} + 15$

Now, by hypothesis, we have

$$\frac{x}{4} + 2\sqrt{x} + 15 = x$$

$$\Rightarrow \quad x + 8\sqrt{x} + 60 = 4x$$

$$\Rightarrow \quad 3x - 8\sqrt{x} - 60 = 0$$

Let $\sqrt{x} = y$, then $x = y^2$

$$\Rightarrow \quad 3y^2 - 8y - 60 = 0$$

$$\Rightarrow \quad 3y^2 - 18y + 10y - 60 = 0$$

$$3y(y-6) + 10(y-6) = 0$$

$$\Rightarrow \quad (3y+10)(y-6) = 0$$

$$\Rightarrow \quad y = 6 \text{ or } y = -\frac{10}{3}$$

If, $\quad y = -\frac{10}{3}$

Then, $\quad x = \left(-\frac{10}{3}\right)^2 = \frac{100}{9} \qquad (\because x = y^2)$

But, the number of camels cannot be a fraction.

$\therefore \quad y = 6 \quad \Rightarrow \quad x = 6^2 = 36$

Hence, the number of camels = 36. **Ans.**

Q. 11. The denominator of a fraction is one more than twice the numerator. If the sum of the fraction and its reciprocal is $2\frac{16}{21}$, find the fraction.

Solution : Let the numerator of the fraction be x. Then,

$$\text{Denominator} = 2x + 1 \quad \text{[Given]}$$

$$\therefore \quad \text{Fraction} = \frac{x}{2x+1}$$

$\Rightarrow$ Reciprocal of the fraction $= \dfrac{2x+1}{x}$

It is given that the sum of the fraction and its reciprocal is $2\frac{16}{21}$.

$$\Rightarrow \quad \frac{x}{2x+1} + \frac{2x+1}{x} = 2\frac{16}{21}$$

$$\Rightarrow \quad \frac{x^2+(2x+1)^2}{x(2x+1)} = \frac{58}{21}$$

$$\Rightarrow \quad \frac{5x^2+4x+1}{2x^2+x} = \frac{58}{21}$$

$$\Rightarrow \quad 21(5x^2+4x+1) = 58(2x^2+x)$$

$$\Rightarrow \quad 105x^2 + 84x + 21 = 116x^2 + 58x$$

$$\Rightarrow \quad 11x^2 - 26x - 21 = 0.$$

$$\Rightarrow \quad 11x^2 - (33-7)x - 21 = 0$$

$$\Rightarrow \quad 11x^2 - 33x + 7x - 21 = 0$$

$$\Rightarrow \quad 11x(x-3) + 7(x-3) = 0$$

$$\Rightarrow \quad (11x+7)(x-3) = 0$$

$$\Rightarrow \quad x = \frac{-7}{11} \text{ or } x = 3$$

Since numerator of a fraction cannot be negative

$$\therefore \quad x = \frac{-7}{11} \text{ (Rejected)}$$

$$\therefore \quad x = 3$$

$$\text{Denominator} = 2x + 1 = 2(3) + 1 = 6 + 1 = 7$$

$\therefore$ Requried fraction $= \frac{3}{7}$. **Ans.**

Q. 12. Find three consecutive positive integers whose product is equal to sixteen times their sum.

Solution : Let integers be x, $x + 1$ and $x + 2$

According to the question,

$$x(x+1)(x+2) = 16(x+x+1+x+2)$$

$$\Rightarrow \quad (x^2+x)(x+2) = 16(3x+3)$$

$$\Rightarrow \quad x^3 + 2x^2 + x^2 + 2x = 48x + 48$$

$$\Rightarrow \quad x^3 + 3x^2 + 2x = 48x + 48$$

$$\Rightarrow x^3 + 3x^2 - 46x - 48 = 0$$

When $x = -1$, we have

L.H.S. $= (-1)^3 + 3(-1)^2 - 46 \times (-1) - 48 = -1 + 3 + 46 - 48 = 0 =$ R.H.S.

$\therefore$ $(x + 1)$ is a factor of above polynomial

For other factors

$$x+1\overline{)\,x^3 + 3x^2 - 46x - 48\,}(\,x^2 + 2x - 48$$

$$\begin{array}{l} x^3 + x^2 \\ -\quad - \\ \hline 2x^2 - 46x \\ 2x^2 + 2x \\ -\quad - \\ \hline -48x - 48 \\ -48x - 48 \\ +\quad + \\ \hline 0 \end{array}$$

$$\therefore \quad x^3 + 3x^2 - 46x - 48 = 0$$

$$\Rightarrow \quad (x+1)(x^2+2x-48) = 0$$

$$\Rightarrow \quad (x+1)(x^2+8x-6x-48) = 0$$

$$\Rightarrow (x+1)\{x(x+8) - 6(x+8)\} = 0$$

$$\Rightarrow \quad (x+1)(x+8)(x-6) = 0$$

$$\Rightarrow \quad x = -1, x = -8, x = 6.$$

Rejecting negative values

we get $x = 6$

$\therefore$ Positive integers are 6, 7 and 8. **Ans.**

Chapter 5. Arithmetic Progressions

Q. 1. If the pth term of an A.P. is $\frac{1}{q}$ and qth term $\frac{1}{p}$. Prove that the sum of the first pq term, is $\frac{1}{2}(pq+1)$.

Solution : Let a, d be the first term and the common difference, respectively.

It is given that

$$T_p = \frac{1}{q}$$

$$\Rightarrow \quad a + (p-1)d = \frac{1}{q} \quad \text{...(i)}$$

$$\text{and} \quad T_q = \frac{1}{p(1/c)}$$

$$\Rightarrow \quad a + (q-1)d = \frac{1}{p} \quad \text{...(ii)}$$

Subtracting equation (ii) from equation (i), we get

$$(q-1-p+1)\,d = \frac{1}{p} - \frac{1}{q}$$

$$\Rightarrow \quad (q-p)\,d = \frac{q-p}{pq}$$

$$\Rightarrow \quad d = \frac{1}{pq} \quad \text{...(iii)}$$

Put the value of d in equation (i), we get

$$a + \frac{p-1}{pq} = \frac{1}{q}$$

$$\Rightarrow \quad a = \frac{1}{q} - \frac{p-1}{pq}$$

$$= \frac{p-p+1}{pq} = \frac{1}{pq}$$

$$\therefore \quad a = \frac{1}{pq} \quad \text{...(iv)}$$

Now, we know $S_n = \frac{\pi}{2}[2a + (n-1)\,d\}$

$$\therefore \quad S_{pq} = \frac{pq}{2}\left[\frac{2}{pq} + (pq-1)\frac{1}{pq}\right]$$

$$= \frac{pq}{2}\left[\frac{2+pq-1}{pq}\right]$$

$$= \frac{1}{2}(pq+1)$$ **Hence Proved.**

Q. 2. Find the sum of the first 15 multiples of 8.

Solution : The first 15 multiples of 8 are

8, 16, 24,..... 120

Clearly, these numbers are in A.P. with first term, $a = 8$ and common difference, $d = 16 - 8 = 8$

$$\text{Thus,} \quad S_{15} = \frac{15}{2}[2\times 8 + (15-1)\times 8]$$

$$= \frac{15}{2}[16 + 14\times 8]$$

$$= \frac{15}{2}[16 + 112]$$

$$= \frac{15}{2}\times 128$$

$$= 15 \times 64 = 960.$$ **Ans.**

Q. 3. Find the sum of all two-digit natural numbers which when divided by 3 yield 1 as remainder.

Solution : Two-digit natural numbers which when divided by 3 yeild 1 as remainder are :

10, 13, 16, 19,.......... 97,

Clearly these numbers are in A.P. with $a = 10$, $d = 3$, $a_n = 97$

Now, $a_n = 97$

$$\Rightarrow \quad a + (n-1)d = 97$$

$$\Rightarrow \quad 10 + (n-1)3 = 97$$

$$\Rightarrow \quad (n-1) = \frac{87}{3} = 29$$

$$\Rightarrow \quad n = 30$$

$$\text{So,} \quad S_{30} = \frac{30}{2}[2\times 10 + (30-1)\times 3]$$

$$= \frac{30}{2}[20 + 29\times 3]$$

$$= 15[20 + 87]$$

$$= 15 \times 107$$

$$= 1605.$$ **Ans.**

Q. 4. A sum of ₹ 900 is to be used to give nine cash prizes to students of a school for their overall academic performance. If each prize is ₹ 10 less than its preceding prize, find the value of each of the prizes.

Solution : Since each prize is ₹ 20 less than its preceding prize, therfore, the values of the nine successive cash prizes will form an A.P.

Let the frist prize be ₹ a.

Then, successive prizes are ₹$(a - 10)$, ₹$(a - 20)$, ₹$(a - 30)$, ₹$(a - 40)$ .... ₹$(a - 80)$

Here, first term $= a$

Common difference, $d = (a - 10) - a = -10$

Number of terms, $n = 9$

and $S_9 = 900$

We know that, $S_n = \frac{n}{2}[2a + (n-1)\,d]$

$$\therefore \quad S_9 = \frac{9}{2}[2a + (9-1)(-10)]$$

$$\Rightarrow \quad 900 = \frac{9}{2}[2a + (8)(-10)]$$

$$\Rightarrow \quad 200 = 2a - 80$$

$$\Rightarrow \quad a = 140$$

Hence, value of first prize = ₹ 140

and successive prizes are ₹ 130, ₹ 120, ₹ 110, ₹ 100, ₹ 90, ₹ 80, ₹ 90, and ₹ 60. **Ans.**

Q. 5. If the mth term of an A.P. is $\frac{1}{n}$ and nth term is $\frac{1}{m}$, then show that its (mn)th term is 1.

Solution : Let a and d be the first term and the common difference, respectively of the given A.P. Then,

$$a_m = a + (m-1)d \quad \Rightarrow a + (m-1)d = \frac{1}{n} \quad ...(i)$$

and $$a_n = a + (n-1)d \Rightarrow a + (n-1)d = \frac{1}{m} \quad ...(ii)$$

Subtracting equation (ii) from equation (i), we get

$$(m-n)\,d = \frac{1}{n} - \frac{1}{m} \quad \Rightarrow (m-n)\,d = \frac{m-n}{mn}$$

$$\Rightarrow \quad d = \frac{1}{mn}$$

Putting $d = \frac{1}{mn}$ in equation (i), we get

$$a + (m-1)\frac{1}{mn} = \frac{1}{n} \Rightarrow a + \frac{m}{mn} - \frac{1}{mn} = \frac{1}{n}$$

$$\Rightarrow \quad a - \frac{1}{mn} = 0 \Rightarrow a = \frac{1}{mn}$$

$\therefore$ mnth term $= a + (mn-1)d$

$$= \frac{1}{mn} + (mn-1)\frac{1}{mn}$$

$$= \frac{1 + mn - 1}{mn} = 1$$

$\therefore$ (mn)th term = 1. **Hence Proved.**

Q. 6. If the sum of m terms of A.P. is the same as the sum of its n terms, show that the sum of its $(m+n)$ terms is zero.

Solution : Let a be the first term and d be the common difference of the given A.P. [Given]

Then, $S_m = S_n$

$$\Rightarrow \frac{m}{2}\{2a + (m-1)d\} = \frac{n}{2}\{2a + (n-1)d\}$$

$$\Rightarrow 2am + m(m-1)\,d = 2an + n\,(n-1)d$$

$$\Rightarrow 2a(m-n) + \{m\,(m-1) - n\,(n-1)\}\,d = 0$$

$$\Rightarrow 2a\,(m-n) + \{(m^2 - n^2) - (m-n)\}d = 0$$

$$\Rightarrow (m-n)\,\{2a + (m+n-1)\,d\} = 0$$

$$\Rightarrow 2a + (m+n-1)\,d = 0 \quad ...(i)$$

$$[\because m - n \neq 0]$$

Now, $$S_{m+n} = \frac{m+n}{2}\{2a + (m+n-1)d\}$$

$$\Rightarrow \quad S_{m+n} = \frac{m+n}{2} \times 0 = 0 \quad \text{[Using (i)]}$$

$\therefore$ Sum of its $(m+n)$ term is 0. **Hence Proved.**

Q. 7. The ratio of the sum of m and n terms of an A.P. is $m^2 : n^2$. Show that the ratio of the mth and nth terms is $(2m-1) : (2n-1)$.

Solution : Let a be the first term and d the common difference of the given A.P. Then,

$$S_m = \frac{m}{2}\{2a + (m-1)d\} \text{ and } S_n = \frac{n}{2}\{2a + (n-1)d\}$$

Given, $$\frac{S_m}{S_n} = \frac{m^2}{n^2}$$

$$\Rightarrow \quad \frac{\frac{m}{2}\{2a + (m-1)d\}}{\frac{n}{2}\{2a + (n-1)d\}} = \frac{m^2}{n^2}$$

$$\Rightarrow \quad \frac{2a + (m-1)d}{2a + (n-1)d} = \frac{m}{n}$$

$$\Rightarrow \{2a + (m-1)d\}\,n = \{2a + (n-1)\,d\}\,m$$

$$\Rightarrow \quad 2a\,(n-m) = d\,\{(n-1)\,m - (m-1)\,n\}$$

$$\Rightarrow \quad 2a\,(n-m) = d\,(n-m)$$

$$\Rightarrow \quad d = 2a$$

$$\therefore \quad \frac{a_m}{a_n} = \frac{a + (m-1)d}{a + (n-1)d}$$

$$= \frac{a + (m-1)2a}{a + (n-1)2a}$$

$$= \frac{2m-1}{2n-1}.$$ **Hence Proved.**

Q. 8. The sum of n, $2n$, $3n$ terms of an A.P. are S_1, S_2 and S_3 respectively. Prove that $S_3 = 3(S_2 - S_1)$.

Solution : Let a be the first term and d be the common difference of the A.P.

$$\therefore \quad S_1 = \frac{n}{2}[2a + (n-1)d] \quad ...(i)$$

$$S_2 = \frac{2n}{2}[2a + (2n-1)d] \quad ...(ii)$$

$$S_3 = \frac{3n}{2}[2a + (3n-1)d] \quad ...(iii)$$

$$\text{Now, } S_2 - S_1 = \frac{2n}{2}[2a + (2n-1)d] - \frac{n}{2}[2a + (n-1)d]$$

$$= \frac{n}{2}[2\{2a + (2n-1)d\}$$

$$- \{2a + (n-1)\,d\}]$$

$$= \frac{n}{2}[4a + 4nd - 2d - 2a - nd + d]$$

$$= \frac{n}{2}[2a + 3nd - d]$$

$$= \frac{n}{2}[2a + (3n-1)d]$$

$$\therefore \quad 3(S_2 - S_1) = \frac{3n}{2}[2a + (3n-1)d] = S_3 \quad \text{[Using (iii)]}$$

$\therefore$ $S_3 = 3(S_2 - S_1)$ **Hence Proved.**

Q. 9. If a^2, b^2, c^2 are in A.P., prove that

$$\frac{a}{b+c}, \frac{b}{c+a}, \frac{c}{a+b} \text{ are in A.P.}$$

Solution : Given, a^2, b^2, c^2 are in A.P.

$$\therefore \quad b^2 - a^2 = c^2 - b^2 \quad ...(i)$$

Now, $$\frac{b}{c+a} - \frac{a}{b+c} = \frac{b^2 + bc - ac - a^2}{(a+c)(b+c)}$$

$$= \frac{(b^2 - a^2) + c(b-a)}{(a+c)(b+c)}$$

$$= \frac{(b-a)(b+a+c)}{(a+b)(b+c)} \quad \text{...(ii)}$$

Also, $$\frac{c}{a+b} - \frac{b}{c+a} = \frac{c^2 + ac - ba - b^2}{(a+b)(c+a)}$$

$$= \frac{c^2 - b^2 + a(c-b)}{(a+b)(c+a)}$$

$$= \frac{(c-b)(c+b+a)}{(a+b)(a+c)}$$

$$= \frac{(c^2 - b^2)(c+b+a)}{(a+b)(b+c)(c+a)}$$

[Multiplying numerator and denominator by $b + c$]

$$= \frac{(b^2 - a^2)(a+b+c)}{(a+b)(b+c)(c+a)}$$

[Using (i)]

$$= \frac{(b-a)(a+b+c)}{(b+c)(c+a)} \quad \text{...(iii)}$$

From equations (ii) and (iii), we have

$$\frac{b}{c+a} - \frac{a}{b+c} = \frac{c}{a+b} - \frac{b}{c+a}$$

i.e., $\frac{a}{b+c}, \frac{b}{c+a}, \frac{c}{a+b}$ are in A.P. **Hence Proved.**

Q. 10. The sum of the third and seventh terms of an A.P. is 6 and their product is 8. Find the sum of first sixteen terms of the A.P.

Solution : Let a be the first term and d be the common difference of the A.P.

We have,

$$a_3 + a_7 - 6 \quad \text{...(i)}$$

and $$a_3 a_7 = 8 \quad \text{...(ii)}$$

From equation (i), we have

$\Rightarrow$ $$(a + 2d) + (a + 6d) = 6$$

$\Rightarrow$ $$2a + 8d = 6$$

$\Rightarrow$ $$a + 4d = 3$$

$\Rightarrow$ $$a = 3 - 4d \quad \text{...(iii)}$$

From equation (ii), we have

$$(a + 2d)(a + 6d) = 8$$

$\Rightarrow$ $$(3 - 4d + 2d)(3 - 4d + 6d) = 8$$

[Using (iii)]

$\Rightarrow$ $$(3 - 2d)(3 + 2d) = 8$$

$$\Rightarrow 9 - 4d^2 = 8 \Rightarrow 4d^2 = 1 \Rightarrow d^2 = \frac{1}{4} \Rightarrow d = \pm\frac{1}{2}$$

Case I : When $d = \frac{1}{2}$

Putting $d = \frac{1}{2}$ in equation (iii), we get

$$a = 3 - 4 \times \frac{1}{2} = 3 - 2 = 1$$

Now, we know

$$S_n = \frac{n}{2}[2a +)n - 1)d]$$

$$S_{16} = \frac{16}{2}\left[2 \times 1 + (16-1)\frac{1}{2}\right]$$

$$= \frac{16}{2}\left[2 + 15\left(\frac{1}{2}\right)\right]$$

$$= \frac{16}{2}\left[\frac{4+15}{2}\right]$$

$$= \frac{16}{2} \times \frac{19}{2}$$

$$= 4 \times 19$$

$$= 76$$

Case II : When $d = \frac{-1}{2}$

Putting $d = \frac{-1}{2}$ in equation (iii), we get

$$a - 3 - 4\left(\frac{-1}{2}\right) = 3 + 2 = 5$$

$\therefore$ $$S_{16} = \frac{16}{2}\left[2 \times 5 + (16-1)\left(\frac{-1}{2}\right)\right]$$

$$= \frac{16}{2}\left[10 + 15\left(\frac{-1}{2}\right)\right]$$

$$= \frac{16}{2}\left[\frac{20-15}{2}\right]$$

$$= \frac{16}{2} \times \frac{5}{2}$$

$$= 4 \times 5 = 20$$ **Ans.**

Q. 11. If in an A.P., the sum of *m* terms is equal to *n* and the sum of *n* terms is equal to *m*, then prove that the sum of (*m* + *n*) terms is – (*m* + *n*).

Solution : Let a be the first term and d be the common difference of the given A.P. Then,

$$S_m = n \quad \text{[Given]}$$

$$\Rightarrow \frac{m}{2}\{2a + (m-1)d\} = n$$

$$\Rightarrow 2am + m(m-1)d = 2n \quad \text{...(i)}$$

and $$S_n = m \quad \text{[Given]}$$

$$\Rightarrow \frac{n}{2}\{2a + (n-1)d\} = m$$

$$\Rightarrow 2an + n(n-1)d = 2m \quad \text{...(ii)}$$

Subtracting equation (ii) from equation (i), we get

$$2a(m-n) + \{m(m-1) - n(n-1)\}d = 2n - 2m$$

$$\Rightarrow 2a(m-n) + \{(m^2 - n^2) - (m-n)\}d = -2(m-n)$$

$$\Rightarrow \quad 2a + (m+n-1)d = -2 \quad ...(iii)$$

[On dividing both sides by $(m-n)$]

Now, $$S_{m+n} = \frac{m+n}{2}[2a + (m+n-1)d]$$

$$\Rightarrow \quad S_{m+n} = \frac{(m+n)}{2}(-2) \quad \text{[Using(iii)]}$$

$$\Rightarrow \quad S_{m+n} = -(m+n)$$ **Hence Proved.**

Q. 12. The sum of the first *p*, *q*, *r* terms of an A.P. are *a*, *b*, *c* respectively. Show that

$$\frac{a}{p}(q-r) + \frac{b}{q}(r-p) + \frac{c}{r}(p-q) = 0$$

Solution : Let A be the first term and D be the common difference of the given A.P. Then,

a = Sum of p terms

$$\Rightarrow \quad a = \frac{p}{2}\{2A + (p-1)D\}$$

$$\Rightarrow \quad \frac{2a}{p} = \{2A + (p-1)D\} \quad ...(i)$$

b = Sum of q terms

$$\Rightarrow \quad b = \frac{q}{2}\{2A + (q-1)D\} \quad ...(ii)$$

$$\Rightarrow \quad \frac{2b}{q} = \{2A + (q-1)D\}$$

and c = Sum of r terms

$$\Rightarrow \quad c = \frac{r}{2}\{2A + (r-1)D\} \quad ...(iii)$$

$$\Rightarrow \quad \frac{2c}{r} = \{2A + (r-1)D\}$$

Multiplying equations (i), (ii) and (iii) by $(q-r)$, $(r-p)$ and $(p-q)$ respectively and adding, we get

$$\frac{2a}{p}(q-r) + \frac{2b}{q}(r-p) + \frac{2c}{r}(p-q)$$
$$= 2A[q - r + r - p + p - q] + D[(p-1)(q-r) + (q-1) \cdot (r-p) + (r-1)(p-q)]$$

$$\Rightarrow \frac{a}{p}(q-r) + \frac{b}{q}(r-p) + \frac{c}{r}(p-q)$$
$$= D[pq - pr - q + r + qr - pq - r + p + pr - rq - p + q]$$

$$\Rightarrow \frac{a}{p}(q-r) + \frac{b}{q}(n-p) + \frac{c}{r}(p-q) = 0$$

Hence Proved.

Q. 13. If there are $(2n + 1)$ terms in A.P., then prove that the ratio of the sum of odd terms and the sum of even terms is $(n + 1) : n$.

Solution : Let a and d be the first term and common difference respectively of the given A.P.

Let a_k denote the kth term of the given A.P. Then,

$$a_k = a + (k-1)d$$

Let S_1 = Sum of odd terms

$$\Rightarrow \quad S_1 = a_1 + a_3 + a_5 + ... + a_{2n+1}$$

$$\Rightarrow \quad S_1 = \frac{n+1}{2}\{a_1 + a_{2n+1}\}$$

$$\Rightarrow \quad S_1 = \frac{n+1}{2}\{a + a + (2n+1-1)d\}$$

$$[\because a_{2n+1} = a + (2n+1-1)d]$$

$$\Rightarrow \quad S_1 = (n+1)(a + nd)$$

and, S_2 = Sum of even terms

$$\Rightarrow \quad S_2 = a_2 + a_4 + a_6 + ... + a_{2n}$$

$$\Rightarrow \quad S_2 = \frac{n}{2}[a_2 + a_{2n}]$$

$$\Rightarrow \quad S_2 = \frac{n}{2}[(a+d) + \{a + (2n-1)d\}]$$

$$[\because a_{2n} = a + (2n-1)d]$$

$$\Rightarrow \quad S_2 = n(a + nd)$$

$$\therefore S_1 : S_2 = (n+1)(a+nd) : n(a+nd) = (n+1) : n$$

Hence Proved.

Q. 14. Find the sum of the integers lying between 1 and 100 (both inclusive) and divisible by 3, 5 or 7.

Solution : Integers divisible by 3 from 1 to 100 are 3, 6, 9,.....99, *i.e.,* total 33 in number

Integers divisible by 5 are, 5, 10, 15,........100 *i.e.,* total 20 in number

Integers divisible by 7 are 7, 14,..., 98, *i.e.,* total 14 in number

Integers divisible by both 3 and 5 are 15, 30........90, i.e., total 6 in number

Integers divisible by both 3 and 7 are 21, 42, 63, and 84. *i.e.,* total 2 in number

and Integers divisible by both 5 and 7 are 35, 70 *i.e.,* total 2 in number

So, sum of numbers divisible by 3, 5 or 7 is

$$= \frac{33}{2}(3+99) + \frac{20}{2}(5+100) + \frac{14}{2}(7+98)$$
$$- \frac{6}{2}(15+90) - \frac{4}{2}(21+84) - \frac{2}{2}(35+70)$$

$$= \frac{33}{2} \times 102 + \frac{20}{2} \times 105 + \frac{14}{2} \times 105$$
$$- \frac{6}{2} \times 105 - \frac{4}{2} \times 105 - \frac{2}{2} \times 105$$

$= 33 \times 51 + 105\,(10 + 7 - 3 - 2 - 1)$

$= 1683 + 105 \times 11$

$= 1683 + 1155$

$= 2838$ **Ans.**

Q. 15. The digits of a positive integer having three digits are in A.P. and their sum is 15. If the number obtained by reversing the digits is 594 less than the original number, then find the number.

Solution : Let a and d be the first term and the common differnce of the A.P. Then,

Let $(a - d)$, a, $(a + d)$ be the digits of three-digit number.

Then, $a - d + a + a + d = 15$ [Given]

$\Rightarrow \quad 3a = 15$

$\Rightarrow \quad a = 5$ ...(i)

According to the question,

Original number $= (a - d) + 10a + 100\,(a + d)$

$= 111a + 99d$

$= 111 \times 5 + 99d$ [Using (i)]

$= 555 + 99d$

Number formed by reversing the digits

$= (a + d) + 10a + 100(a - d)$

$= 111a - 99d$ [Using (i)]

$= 111 \times 5 - 99d$

$= 555 - 99d$

$\therefore$ $(555 + 99d) - (555 - 99d) = 594$ [Given]

$\Rightarrow \quad 198d = 594$

$$\Rightarrow \quad d = \frac{594}{198} = 3$$

Thus, the digit in the unit's place is $5 - 3 = 2$, in the ten's place is 5 and in the hundred's place is $5 + 3 = 8$

Hence the number is 852. **Ans.**

Chapter 6. Coordinate Geometry

Q. 1. The line segment joining $A(2, 3)$ and $B(-3, 5)$ is extended through each end by a length equal to its original length. Find the coordinates of the new ends.

Solution : Let the line segment joining $A(2, 3)$ and $B(-3, 5)$ is extended through each end by a length equal to its original length AB as shown in figure.

P A B Q
(x_1, y_1) (2, 3) (–3, 5) (x_2, y_2)

Let $P(x_1, y_1)$ and $Q(x_2, y_2)$ be the new end points

Now, $PA = AB = BQ$

$\therefore$ A is the mid-point of PB.

$$\Rightarrow \quad 2 = \frac{x_1 + (-3)}{2} \text{ and } 3 = \frac{y_1 + 5}{2}$$

$\Rightarrow \quad 4 = x_1 - 3 \quad$ and $\quad 6 = y_1 + 5$

$\Rightarrow \quad x_1 = 7 \quad$ and $y_1 = 1$.

Thus, the coordinates of the point P are (7, 1).

Again, B is the mid-point of AQ

$$\therefore \quad \frac{2 + x_2}{2} = -3 \text{ and } \frac{3 + y_2}{2} = 5$$

$\Rightarrow \quad 2 + x_2 = -6$ and $3 + y_2 = 10$

$\Rightarrow \quad x_2 = -8$ and $\quad y_2 = 7$

$\therefore$ Therefore, the coordinates of Q are $(-8, 7)$.

Hence, the coordinates of the new ends are (7, 1) and $(-8, 7)$. **Ans.**

Q. 2. The line joining the points (2, 1) and (5, – 8) is trisected by the points P and Q. If the point P lies on the line $2x - y + k = 0$, find the value of k.

Solution : Let the line segment AB be trisected by the points P and Q. Therefore,

Case I : When $AP : PB = 1 : 2$

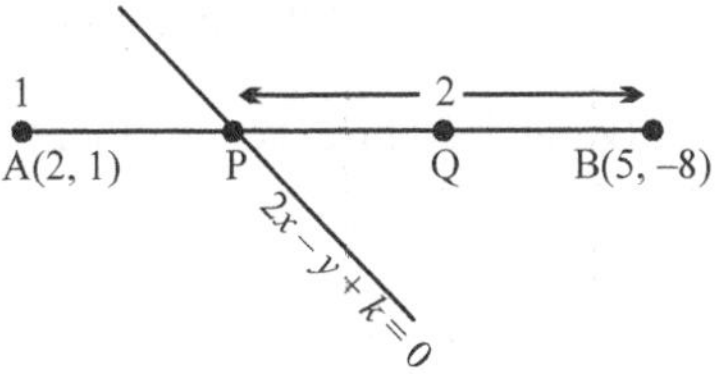

Then, coordinates of P are

$$\left\{\frac{1 \times 5 + 2 \times 2}{1+2}, \frac{1 \times (-8) + 1 \times 2}{1+2}\right\}$$

$\Rightarrow P\,(3, -2)$

Since the point $P(3, -2)$ lies on the line

$2x - y + k = 0$

$\therefore \quad 2 \times 3 - (-2) + k = 0$

$\Rightarrow \quad k = -8$

Case II : When $AP : PB = 2 : 1$.

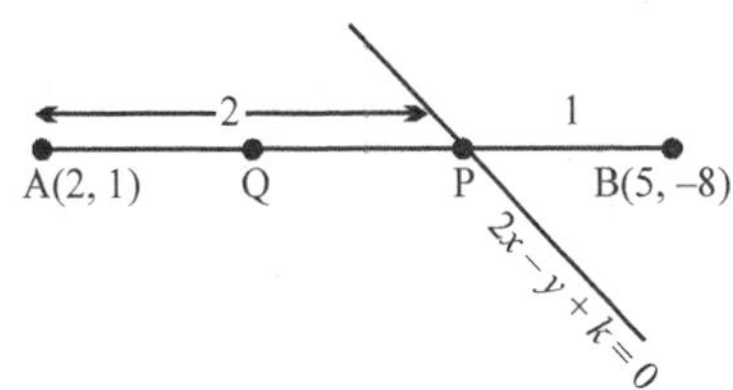

Then coordinates of point P are

$$\left\{\frac{2 \times 5 + 1 \times 2}{2+1}, \frac{2 \times (-8) + 1 \times 1}{2+1}\right\} = 4, -5$$

Since the point $P\,(4, -5)$ lies on the line

$2x - y + k = 0$

$\therefore \quad 2 \times 4 - (-5) + k = 0$

$\Rightarrow \quad k = -13.$ **Ans.**

Q. 3. Prove that the diagonals of a rectangle bisect each other and are equal.

Solution : Let $OACB$ be a rectangle such that OA is along X-axis and OB is along Y-axis. Let $OA = a$ and $OB = b$.

Then, the coordinates of A and B are $(a, 0)$ and $(0, b)$ respectively.

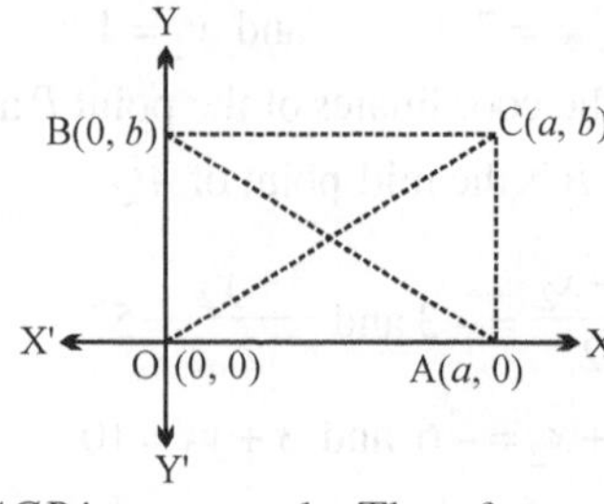

Since, $OACB$ is a rectangle. Therefore,

$AC = OB \Rightarrow AC = b$

Also, $OA = BC \Rightarrow BC = a$

So, the coordinates of C are (a, b).

The coordinates of the mid-points of OC are

$$\left(\frac{a+0}{2}, \frac{b+0}{2}\right) = \left(\frac{a}{2}, \frac{b}{2}\right)$$

Also, the coordinates of the mid-points of AB are

$$\left(\frac{a+0}{2}, \frac{b+0}{2}\right) = \left(\frac{a}{2}, \frac{b}{2}\right)$$

Clearly, coordinates of the mid-point of OC and AB are same.

Hence, OC and AB bisect each other.

Also, $OC = \sqrt{a^2 + b^2}$

and $BA = \sqrt{(a-0)^2 + (0-b)^2} = \sqrt{a^2 + b^2}$

$\therefore \quad OC = AB.$ **Hence Proved.**

Q. 4. In what ratio does the *Y*-axis divide the line segment joining the points *P*(– 4, 5) and *Q* (3, – 7)? Also, find the coordinates of the point of intersection.

Solution : Suppose Y-axis divides PQ in the ratio $k:1$. Then, the coordinates of the point of division are

$$R\left(\frac{3k-4}{k+1}, \frac{-7k+5}{k+1}\right)$$

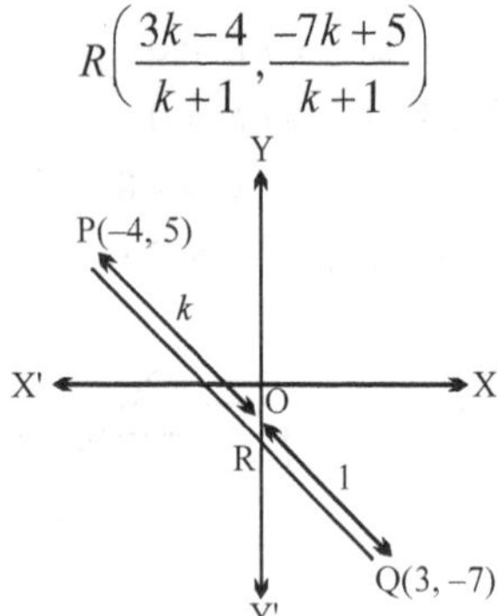

Since, R lies on Y-axis and x-coordinate of every point on Y-axis is zero.

$$\therefore \quad \frac{3k-4}{k+1} = 0$$

$$\Rightarrow \quad 3k - 4 = 0$$

$$\Rightarrow \quad k = \frac{4}{3}$$

Hence, the required ratio is $\frac{4}{3} : 1$ *i.e.,* $4 : 3$

Putting $k = 4/3$ in the coordinates of R, we find that its coordinates are $\left(0, \frac{-13}{7}\right)$. **Ans.**

Q. 5. Find the centre of a circle passing through the points (6, – 6), (3, – 7) and (3, 3).

Solution : Let $O(x, y)$ be the centre of circle. Given points are $A(6, -6)$, $B(3, -7)$ and $C(3, 3)$.

Then, $OA = \sqrt{(x-6)^2 + (y+6)^2}$;

$OB = \sqrt{(x-3)^2 + (y+7)^2}$

and $OC = \sqrt{(x-3)^2 + (y-3)^2}$

Since, each point on the circle is equidistant from centre.

$\therefore \quad OA = OB = OC = \text{Radius}$

$\Rightarrow \quad OA = OB$

$$\Rightarrow \sqrt{(x-6)^2 + (y+6)^2} = \sqrt{(x-3)^2 + (y+7)^2}$$

Squaring both sides, we get

$$(x-6)^2 + (y+6)^2 = (x-3)^2 + (y+7)^2$$

$$\Rightarrow x^2 - 12x + 36 + y^2 + 12y + 36 = x^2 - 6x + 9 + y^2 + 14y + 49$$

$$\Rightarrow \quad -6x - 2y = -14$$

$$\Rightarrow \quad 3x + y = 7 \quad \text{...(i)}$$

Similarly, $OB = OC$

$$\Rightarrow \sqrt{(x-3)^2 + (y+7)^2} = \sqrt{(x-3)^2 + (y-3)^2}$$

Squaring both sides, we get

$$(x-3)^2 + (y+7)^2 = (x-3)^2 + (y-3)^2$$

$$\Rightarrow \quad (y+7)^2 = (y-3)^2$$

$$\Rightarrow \quad y^2 + 14y + 49 = y^2 - 6y + 9$$

$$\Rightarrow \quad 20y = -40$$

$$\Rightarrow \quad y = -2$$

Putting $y = -2$ in equation (i), we get

$$3x - 2 = 7 \Rightarrow x = 3$$

Hence, the coordinates of the centre of circle are $(3, -2)$. **Ans.**

Q. 6. If the coordinates of the mid-points of the sides of a triangle are (1, 1), (2, – 3) and (3, 4). Find its centroid.

Solution : Let P (1, 1), $Q(2, -3)$ and $R(3, 4)$ be the mid-points of sides AB, BC and CA respectively, of ΔABC. Let $A(x_1, y_1)$, $B(x_2, y_2)$ and $C(x_3, y_3)$ be the vertices of triangle ABC.

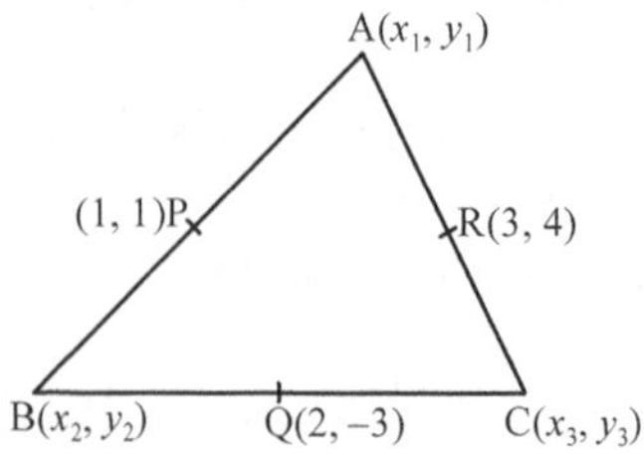

Then, P is the mid-point of AB

$$\Rightarrow \quad \frac{x_1+x_2}{2} = 1, \qquad \frac{y_1+y_2}{2} = 1$$

$$\Rightarrow \quad x_1+x_2 = 2 \text{ and } y_1+y_2 = 2 \qquad ...(i)$$

Q is the mid-point of BC

$$\Rightarrow \quad \frac{x_2+x_3}{2} = 2, \qquad \frac{y_2+y_3}{2} = -3$$

$$\Rightarrow \quad x_2+x_3 = 4 \text{ and } y_2+y_3 = -6 \qquad ...(ii)$$

R is the mid-point of AC

$$\Rightarrow \quad \frac{x_1+x_3}{2} = 3 \text{ and } \frac{y_1+y_3}{2} = 4$$

$$\Rightarrow \quad x_1+x_3 = 6 \text{ and } y_1+y_3 = 8 \qquad ...(iii)$$

On adding equations (i), (ii) and (iii), we get

$$x_1+x_2+x_2+x_3+x_1+x_3 = 2+4+6$$

and $$y_1+y_2+y_2+y_3+y_1+y_3 = 2-6+8$$

$$\Rightarrow \quad x_1+x_2+x_3 = 6 \text{ and } y_1+y_2+y_3 = 2 \qquad ...(iv)$$

The coordinates of the centroid of ΔABC is given as

$$\left(\frac{x_1+x_2+x_3}{3}, \frac{y_1+y_2+y_3}{3}\right) = \left(\frac{6}{3}, \frac{2}{3}\right) \quad \text{[Using (iv)]}$$

$$= \left(2, \frac{2}{3}\right)$$

Thus, $\left(2, \frac{2}{3}\right)$ is the centroid of ΔABC. **Ans.**

Q. 7. Determine the ratio in which the line $2x+y-4=0$ divides the line segment joining the points $A(2,-2)$ and $B(3, 7)$.

Solution : Let $P(x_1, y_1)$ be common point of both lines and divide the line segment joining $A(2, -2)$ and $B(3, 7)$ in the ratio $k : 1$.

Then, coordinates of P are

$$x_1 = \frac{3k+2}{k+1}$$

and $$y_1 = \frac{7k+1(-2)}{k+1} = \frac{7k-2}{k+1}$$

Since, point (x_1, y_1) lies on the line $2x + y = 4$

$$\therefore \quad 2\left(\frac{3k+2}{k+1}\right) + \left(\frac{7k-2}{k+1}\right) = 4$$

$$\Rightarrow \quad \frac{6k+4+7k-2}{k+1} = 4$$

$$\Rightarrow \quad 13k + 2 = 4k + 4$$

$$\Rightarrow \quad 9k = 2$$

$$\Rightarrow \quad k = \frac{2}{9}$$

So, required ratio is $\frac{2}{9} : 1$ or $2 : 9$. **Ans.**

Q. 8. Show that ΔABC with vertices $A(-2, 0)$, $B(2, 0)$ and $C(0, 2)$ is similar to ΔDEF with vertices $D(-4, 0)$ $E(4, 0)$ and $F(0, 4)$.

Solution : Given vertices of ΔABC and ΔDEF are

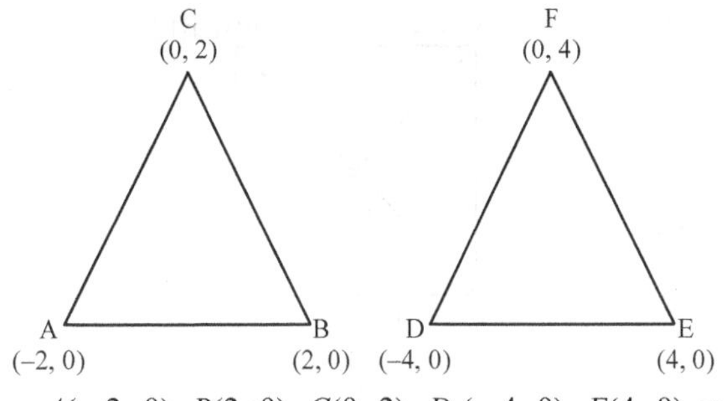

$A(-2, 0)$, $B(2, 0)$, $C(0, 2)$, $D(-4, 0)$, $E(4, 0)$ and $F(0, 4)$

$$AC = \sqrt{(0+2)^2+(2-0)^2}$$

$$= \sqrt{4+4} = 2\sqrt{2} \text{ units}$$

$$BC = \sqrt{(2-0)^2+(0-2)^2}$$

$$= \sqrt{4+4} = 2\sqrt{2} \text{ units}$$

$$AB = \sqrt{(-2-2)^2+(0-0)^2}$$

$$= \sqrt{(-4)^2+0} = 4 \text{ units}$$

$$FD = \sqrt{(0+4)^2+(0-4)^2}$$

$$= \sqrt{4^2+(-4)^2} = 4\sqrt{2} \text{ units}$$

$$FE = \sqrt{(4-0)^2+(0-4)^2}$$

$$= \sqrt{4^2+(-4)^2} = 4\sqrt{2} \text{ units}$$

$$DE = \sqrt{(-4-4)^2+(0-0)^2}$$

$$= \sqrt{(-8)^2} = 8 \text{ units}$$

Here, we see that sides of ΔDEF are twice the sides of a ΔABC.

Hence, both triangles are similar. **Hence Proved.**

Q. 9. The two opposite vertices of a square are (–1, 2) and (3, 2). Find the coordinates of the other two vertices.

Solution : Let the vertices of a square be A, B, C, D with $A = (-1, 2)$ and $C = (3, 2)$. Let $B = (x_1, y_1)$ and $D = (x_2, y_2)$

In a square, all sides are equal.

so, $AB = BC = CD = DA$

and both the diagonals are equal.

So, $AC = BD$.

Also, diagonals of a square bisect each other.

Let diagonals bisect each other at P. So, P is the mid-point of AC.

So, coordinates of P

$$= \left(\frac{3-1}{2}, \frac{2+2}{2}\right) = (1, 2)$$

A(–1, 2) B(x_1, y_1)

P (1, 2)

D(x_2, y_2) C(3, 2)

$$AB = BC$$

$$\Rightarrow \quad AB^2 = BC^2$$

$$\Rightarrow \quad (x_1+1)^2 + (y_1-2)^2 = (x_1-3)^2 + (y_1-2)^2$$

$$\Rightarrow \quad {x_1}^2 + 2x_1 + 1 = {x_1}^2 - 6x_1 + 9$$

$$\Rightarrow \quad 8x_1 = 8$$

$$\Rightarrow \quad x_1 = 1$$

Since P is also the mid-point of BD

$$\therefore \quad \frac{x_1+x_2}{2} = 1$$

$$\Rightarrow x_1 + x_2 = 2 \text{ or } 1 + x_2 = 2 \Rightarrow x_2 = 1$$

Now, $\quad AD^2 + DC^2 = AC^2$

$$(x_2+1)^2 + (y_2-2)^2 + (x_2-3)^2 + (y_2-2)^2 = (3+1)^2 + (2-2)^2$$

$$\Rightarrow 2(y_2-2)^2 + (1+1)^2 + (1-3)^2 = 4^2$$

$$\Rightarrow \quad 2(y_2-2)^2 + 4 + 4 = 16$$

$$\Rightarrow \quad 2(y_2-2)^2 = 8$$

$$\Rightarrow \quad (y_2-2)^2 = 4$$

$$\Rightarrow \quad y_2 - 2 = \pm 2$$

$$\Rightarrow \quad y_2 = 4 \text{ or } 0$$

Since, P is mid-point of BD

$$\Rightarrow \quad \frac{y_1+y_2}{2} = 2$$

$$\Rightarrow \quad y_1 + y_2 = 4$$

$$\Rightarrow \quad y_1 = 4$$

and if $y_2 = 4$, $y_1 = 0$

So, $\quad y_1 = 0$ and $y_2 = 4$ or $y_1 = 4$, $y_2 = 0$

Hence, coordinates of other two vertices are (1, 0) and (1, 4). **Ans.**

Chapter 7. Triangles

Q. 1. Prove that the ratio of corresponding sides of two similar triangles is the same as the ratio of their corresponding angle bisectors.

Solution : $\Delta ABC \sim \Delta PQR$ (Given)

$$\therefore \quad \frac{AB}{PQ} = \frac{BC}{QR} = \frac{CA}{RP} \quad ...(i)$$

Let AD and PS be the angle bisectors in ΔABC and ΔPQR respectively.

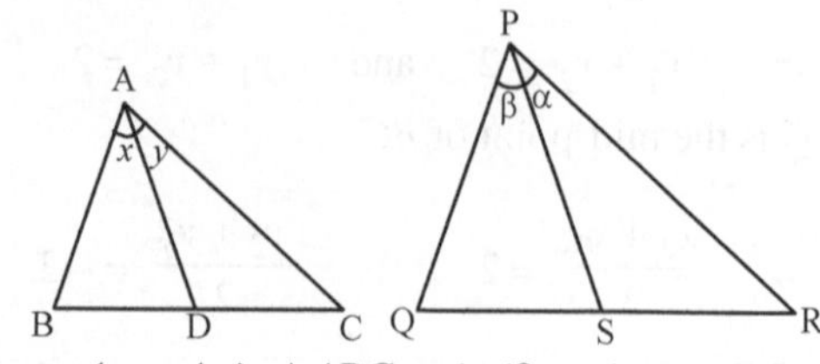

Thus, $\angle x = \angle y$ in ΔABC and $\angle\beta = \angle\alpha$ in ΔPQR

Now, $\angle B = \angle Q$ (corresponding angles of similar triangles) and $\angle A = \angle P$

$$\therefore \quad \frac{1}{2}\angle A = \frac{1}{2}\angle P \Rightarrow \angle x = \angle\beta \quad ...(i)$$

Now in ΔABD and ΔPQS,

$$\angle BAD = \angle QPS \quad \text{[From eq. (ii)]}$$

$$\angle B = \angle Q$$

$$\therefore \quad \Delta ABD \sim \Delta PQS \quad \text{[By AA Similarity]}$$

$$\therefore \quad \frac{AB}{PQ} = \frac{AD}{PS} \quad ...(iii)$$

From equations (i) and (iii),

$$\frac{AB}{PQ} = \frac{BC}{QR} = \frac{CA}{RP} = \frac{AD}{PS}$$ **Hence Proved.**

Q. 2. In the given figure, ABC is a right triangle, right angled at B AD and CE are the two medians drawn from A and C respectively. If $AC = 5$ cm and $AD = \frac{3\sqrt{5}}{2}$ cm, find the length of CE.

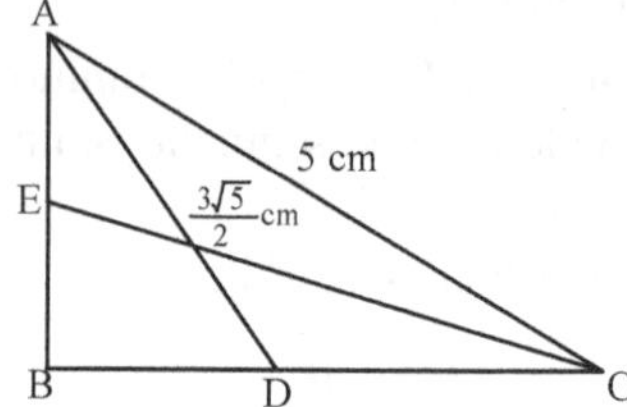

Solution : In ΔABC

By Pythagoras theorem,

$$AC^2 = AB^2 + BC^2 \quad ...(ii)$$

In right ΔABD, $AD^2 = AB^2 + BD^2$

$$\Rightarrow \quad AD^2 = AB^2 + \left(\frac{1}{2}BC\right)^2$$

$$\Rightarrow \quad AD^2 = AB^2 + \frac{1}{4}BC^2 \quad ...(ii)$$

Similarly, in ΔEBC,

$$EC^2 = BC^2 + \frac{1}{4}AB^2 \quad ...(iii)$$

Adding equation (ii) and (iii), we get

$$AD^2 + EC^2 = \frac{5}{4}AB^2 + \frac{5}{4}BC^2$$

$$\Rightarrow \quad AD^2 + EC^2 = \frac{5}{4}(AB^2 + BC^2)$$

$\Rightarrow \quad AD^2 + EC^2 = \frac{5}{4} AC^2$ [Using (i)]

$\Rightarrow \quad \left(\frac{3\sqrt{5}}{2}\right)^2 + EC^2 = \frac{5}{4}(5)^2$

$\Rightarrow \quad EC^2 = \frac{125}{4} - \frac{45}{4}$

$\Rightarrow \quad EC^2 = \frac{80}{4} = 20\text{ cm}$

$\Rightarrow \quad CE = 2\sqrt{5}\text{ cm.}$ **Ans.**

Q. 3. In an equilateral triangle with side a, prove that

(i) altitude = $\frac{\sqrt{3}}{2}a$ (ii) area = $\frac{\sqrt{3}}{4}a^2$

Solution : ΔABC is an equilateral with side a, then

$AB = AC = BC = a$

Draw $AD \perp BC$

Now, in ΔADB and ΔADC

$AB = AC,$

$\angle ADB = \angle ADC = 90°$

and $\angle B = \angle C = 60°$

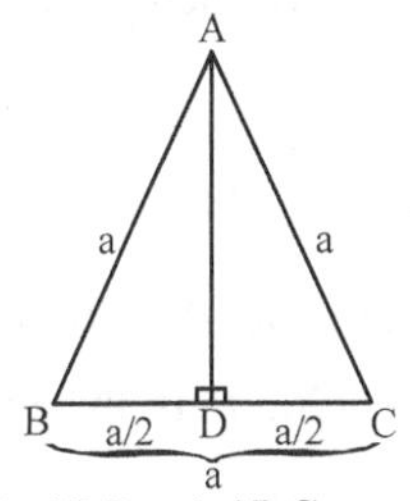

$\therefore \quad \Delta ADB \cong \Delta ADC$ (By ASA)

$\therefore \quad BD = DC = \frac{a}{2}$

(i) Now, in ΔADB

$AB^2 = AD^2 + BD^2$

[Using Pythagoras theorem]

$\Rightarrow \quad AD = \sqrt{AB^2 - BD^2}$

$= \sqrt{a^2 - \frac{a^2}{4}}$

$\Rightarrow \quad AD = \frac{\sqrt{3}}{2}a$ **Hence Proved.**

(ii) Now, area of ΔABC $= \frac{1}{2} \times BC \times AD$

$= \frac{1}{2} \times a \times \frac{\sqrt{3}}{2} a$

$= \frac{\sqrt{3}}{4} a^2$ sq. unit.

Hence Proved.

Q. 4. In Fig., $\Delta FEC \cong \Delta GDB$ and $\angle 1 = \angle 2$. Prove that $\Delta ADE \sim \Delta ABC$.

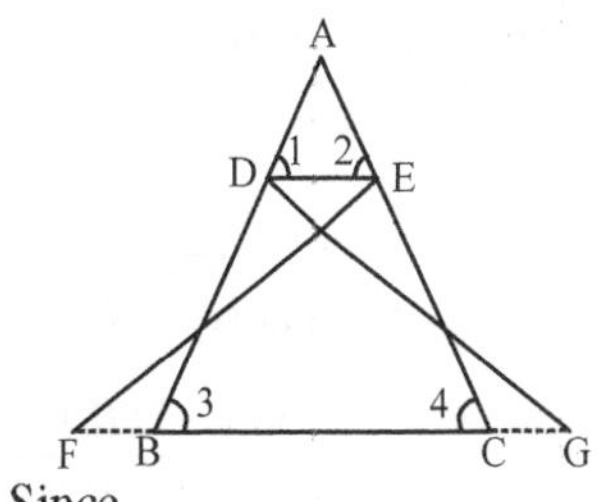

Solution : Since,

$\Delta FEC \cong \Delta GDB$

$\Rightarrow \quad EC = BD$ ...(i)

It is given that

$\angle 1 = \angle 2$

$\Rightarrow \quad AE = AD$ ...(ii)

[Sides opposite to equal angles are equal]

From equation (i) and (ii), we have

$\frac{AE}{EC} = \frac{AD}{BD}$

$\Rightarrow \quad DE \parallel BC$

[By the converse of basic proportionality theorem]

$\Rightarrow \quad \angle 1 = \angle 3$ and $\angle 2 = \angle 4$

[Corresponding angles]

Thus, in Δ's ADE and ABC, we have

$\angle A = \angle A$ [Common]

$\angle 1 = \angle 3$ [Proved above]

$\angle 2 = \angle 4$ [Proved above]

So, by AAA criterion of similarity, we have

$\Delta ADE \sim \Delta ABC.$ **Hence Proved.**

Q. 5. In ΔPQR, point M is on side PQ and S is on the side PR such that $QRSM$ is a trapezium. If $MS : QR = 3 : 5$, then find area (ΔPMS) : area (⏢ $QRSM$)

Solution : Given $QRSM$ is a trapezium,

$\therefore \quad QR \parallel MS$

$\Rightarrow \quad \frac{PM}{PQ} = \frac{PS}{PR}$

[Basic proportionality theorem]

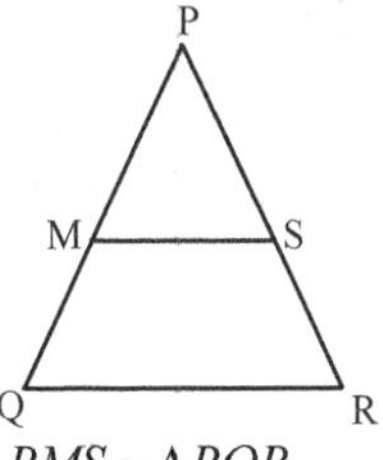

$\therefore \quad \Delta PMS \sim \Delta PQR$

$\Rightarrow \quad \frac{ar(\Delta PMS)}{ar(\Delta PQR)} = \frac{MS^2}{QR^2} = \frac{9}{25}$

$\Rightarrow \quad \frac{ar(\Delta PQR)}{ar(\Delta PMS)} = \frac{25}{9}$

$\Rightarrow \quad \frac{ar(\Delta PQR)}{ar(\Delta PMS)} - 1 = \frac{25}{9} - 1$

$$\Rightarrow \quad \frac{\text{ar}(\Delta PQR) - \text{ar}(\Delta PMS)}{\text{ar}(\Delta PMS)} = \frac{25-9}{9}$$

$$= \frac{\text{ar}(\text{quad. } QRSM)}{\text{ar}(\Delta PMS)} = \frac{16}{9}$$

$$= \frac{\text{ar}(\Delta PMS)}{\text{ar}(\text{quad. } QRSM)} = \frac{9}{16} \quad \textbf{Ans.}$$

Q. 6. Sides *AB* and *AC* and median *AD* of a triangle *ABC* are respectively proportional to sides *PQ* and *PR* and median *PM* of another triangle *PQR*. Show that $\Delta ABC \sim \Delta PQR$.

Solution : Given, in ΔABC and ΔPQR, *AD* and *PM* are their medians, respectively

such that $$\frac{AB}{PQ} = \frac{AD}{PM} = \frac{AC}{PR} \quad ...(i)$$

To prove : $\Delta ABC \sim \Delta PQR$

Construction : Produce *AD* to *E* such that $AD = DE$ and produce *PM* to *N* such that $PM = MN$.

Join *BE*, *CE*, *QN*, *RN*.

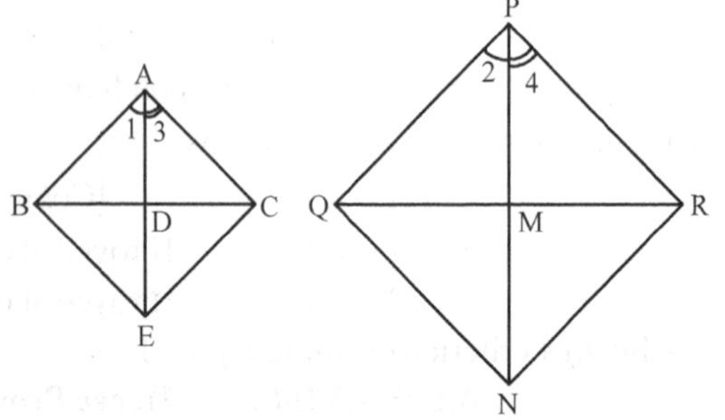

Proof : Quadrilaterals *ABEC* and *PQNR* are parallelogram because their diagonals bisect each other at *D* and *M* respectively.

$$\Rightarrow \quad BE = AC \text{ and } QN = PR$$

$$\Rightarrow \quad \frac{BE}{QN} = \frac{AC}{PR} \Rightarrow \frac{BE}{QN} = \frac{AB}{PQ} \quad \text{[From (i)]}$$

i.e., $$\frac{AB}{PQ} = \frac{BE}{QN} \quad ...(ii)$$

From equation (i),

$$\frac{AB}{PQ} = \frac{AD}{PM} = \frac{2AD}{2PM} = \frac{AE}{PN}$$

$$\Rightarrow \quad \frac{AB}{PQ} = \frac{AE}{PN} \quad ...(iii)$$

From equations (ii) and (iii),

$$\frac{AB}{PQ} = \frac{BE}{QN} = \frac{AE}{PN}$$

$$\Rightarrow \quad \Delta ABE \sim \Delta PQN$$

[*SSS* similarity criterion]

$$\Rightarrow \quad \angle 1 = \angle 2 \quad ...(iv)$$

Similarly, we can prove

$$\Delta ACE \sim \Delta PRN$$

$$\Rightarrow \quad \angle 3 = \angle 4 \quad ...(v)$$

Adding (iv) and (v), we get

$$\angle 1 + \angle 3 = \angle 2 + \angle 4$$

$$\Rightarrow \quad \angle A = \angle P$$

and $$\frac{AB}{PQ} = \frac{AC}{PR} \quad \text{[Given]}$$

$$\therefore \quad \Delta ABC \sim \Delta PQR$$

[By SAS criterion of similarity]

Hence Proved.

Q. 7. In fig., *P* is the mid-point of *BC* and *Q* is the mid-point of *AP*. If *BQ* when produced meets *AC* at *R*, prove that $RA = \frac{1}{3}CA$.

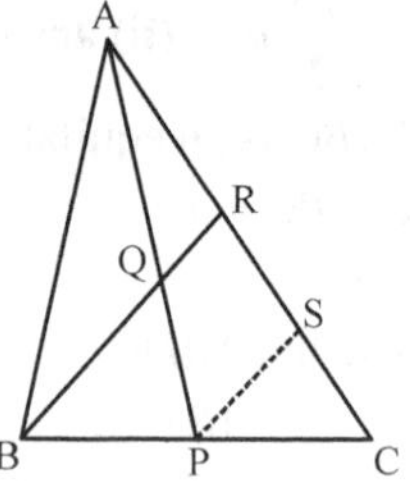

Solution : Given, In ΔABC, *P* is the mid-point of *BC*, *Q* is the mid-point of *AP* such that *BQ* produced meets *AC* at *R*.

To prove : $RA = \frac{1}{3}CA$

Construction : Draw $PS \parallel BR$, meeting *AC* at *S*.

Proof : In ΔBCR, *P* is the mid-point of *BC* and $PS \parallel BR$.

$\therefore$ *S* is the mid-point of *CR*.

$$\Rightarrow \quad CS = SR \quad ...(i)$$

In ΔAPS, *Q* is the mid-point of *AP* and $QR \parallel PS$.

$\therefore$ *R* is the mid-point of *AS*.

$$\Rightarrow \quad AR = RS \quad ...(ii)$$

From equation (i) and (ii), we get

$$AR = RS = SC$$

$$\Rightarrow \quad AC = AR + RS + SC = 3\,AR$$

$$\Rightarrow \quad AR = \frac{1}{3}AC = \frac{1}{3}CA.$$

Hence Proved.

Q. 8. In Fig., *ABC* and *DBC* are two triangles on the same base *BC*. If *AD* intersects *BC* at *O*, show that

$$\frac{\textbf{ar}(\Delta ABC)}{\textbf{ar}(\Delta DBC)} = \frac{AO}{DO}.$$

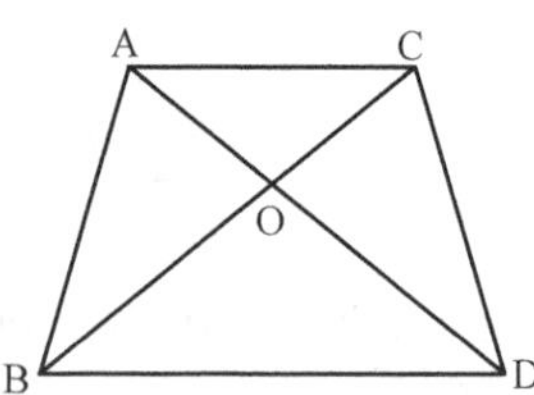

Solution : Given, two triangles ΔABC and ΔDBC which stand on the same base but on opposite sides of BC.

To prove : $\dfrac{ar(\Delta ABC)}{ar(\Delta DBC)} = \dfrac{AO}{DO}$

Construction : We draw $AE \perp BC$ and $DF \perp BC$.

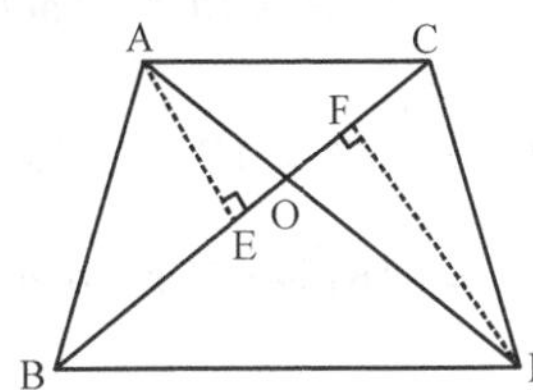

Proof : In ΔAOE and ΔDOF, we have

$$\angle AEO = \angle DFO = 90^\circ$$

$$\angle AOE = \angle DOF$$

[Vertically opposite angles]

$\therefore$ $\Delta AOE \sim \Delta DOF$

[By AA criterion of similarity]

$\Rightarrow$ $\dfrac{AE}{DF} = \dfrac{AO}{DO}$ [cpct] ...(i)

Now, $\dfrac{ar(\Delta ABC)}{ar(\Delta DBC)} = \dfrac{\frac{1}{2} \times BC \times AE}{\frac{1}{2} \times BC \times DF}$

$\Rightarrow$ $\dfrac{ar(\Delta ABC)}{ar(\Delta DBC)} = \dfrac{AE}{DF}$ ...(ii)

From (i) and (ii), we have

$\dfrac{ar(\Delta ABC)}{ar(\Delta DBC)} = \dfrac{AO}{DO}$ **Hence Proved.**

Q. 9. Two poles of height *a* metres and *b* metres are *p* metres apart. Prove that the height of the point of intersection of the lines joining the top of each pole to the foot of the opposite pole is given by $\dfrac{ab}{a+b}$ metres.

Solution : Let AB and CD be two poles of height a and b metres respectively such that the poles are p metres apart *i.e.,* $AC = p$ metres. Suppose the lines AD and BC meet at O such that $OL = h$ metres.

Let $CL = x$ and $LA = y$.

Then, $x + y = p$.

In ΔABC and ΔLOC, we have

$$\angle CAB = \angle CLO \quad \text{[Each 90°]}$$

$$\angle C = \angle C \quad \text{[Common]}$$

$\therefore$ $\Delta ABC \sim \Delta LOC$

[By AA criterion of similarity]

$\Rightarrow$ $\dfrac{CA}{CL} = \dfrac{AB}{LO}$ [cpct]

$\Rightarrow$ $\dfrac{p}{x} = \dfrac{a}{h}$

$\Rightarrow$ $x = \dfrac{ph}{a}$ ...(i)

In ΔALO and ΔACD, we have

$$\angle ALO = \angle ACD \quad \text{[Each 90°]}$$

$$\angle A = \angle A \quad \text{[Common]}$$

$\therefore$ $\Delta ALO \sim \Delta ACD$

[By AA criterion of similarity]

$\Rightarrow$ $\dfrac{AL}{AC} = \dfrac{OL}{DC}$ [cpct]

$\Rightarrow$ $\dfrac{y}{p} = \dfrac{h}{b}$

$\Rightarrow$ $y = \dfrac{ph}{b}$ ...(ii)

On adding equations (i) and (ii), we have

$$x + y = \frac{ph}{a} + \frac{ph}{b}$$

$\Rightarrow$ $p = ph\left(\dfrac{1}{a} + \dfrac{1}{b}\right)$ $[\because x + y = p]$

$\Rightarrow$ $1 = h\left(\dfrac{a+b}{ab}\right)$

$\Rightarrow$ $h = \dfrac{ab}{a+b}$ metres. **Hence Proved.**

Q. 10. In an equilateral triangle *ABC*, *D* is a point on side *BC* such that $BD = \dfrac{1}{3}BC$. Prove that $9AD^2 = 7AB^2$.

Solution : Given, an equilateral triangle ABC and D is a point on BC such that $BD = \dfrac{1}{3}BC$.

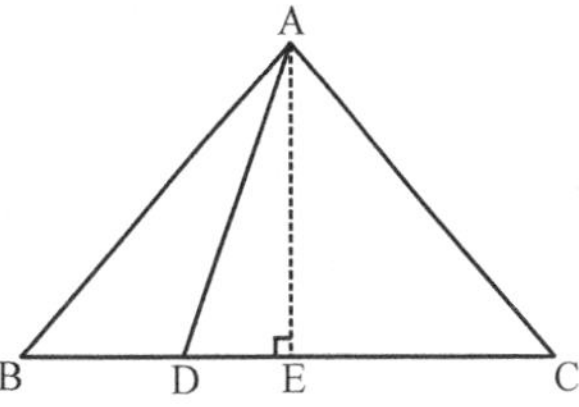

To Prove : $9AD^2 = 7AB^2$

Construction : Draw $AE \perp BC$. Join AD.

Proof : ΔABC is an equilateral triangle and $AE \perp BC$.

$\therefore$ $BE = EC$

Thus, we have

$$BD = \frac{1}{3}BC, \; DC = \frac{2}{3}BC$$

and $\quad BE = EC = \frac{1}{2} BC$

In ΔAEB

$$AE^2 + BE^2 = AB^2$$

[Using Pythagoras theorem]

$\Rightarrow \quad AE^2 = AB^2 - BE^2$

$\Rightarrow \quad AD^2 - DE^2 = AB^2 - BE^2$

[$\because$ In ΔAED, $AD^2 = AE^2 + DE^2$]

$$AD^2 = AB^2 - BE^2 + DE^2$$

$\Rightarrow \quad AD^2 = AB^2 - \left(\frac{1}{2}BC\right)^2 + (BE - BD)^2$

$\Rightarrow \quad AD^2 = AB^2 - \frac{1}{4}BC^2 + \left(\frac{1}{2}BC - \frac{1}{3}BC\right)^2$

$\Rightarrow \quad AD^2 = AB^2 - \frac{1}{4}BC^2 + \left(\frac{BC}{6}\right)^2$

$\Rightarrow \quad AD^2 = AB^2 - BC^2\left(\frac{1}{4} - \frac{1}{36}\right)$

$\Rightarrow \quad AD^2 = AB^2 - BC^2\left(\frac{8}{36}\right)$

$\Rightarrow \quad 9AD^2 = 9AB^2 - 2BC^2$

$\Rightarrow \quad 9AD^2 = 9AB^2 - 2AB^2 \quad$ [$\because AB = BC$]

$\Rightarrow \quad 9AD^2 = 7AB^2$. **Hence Proved.**

Q. 11. In a ΔABC, $\angle ABC = 45°$, prove that

$(AC)^2 = (AB)^2 + (BC)^2 - 4$ ar (ΔABC).

Solution : Given, in ΔABC

$$\angle B = 45°$$

Construction : Draw $AD \perp BC$

Proof : In ΔADB

$$\angle B + \angle BAD + \angle ADB = 180°$$

[Angle sum property of Δ]

$\Rightarrow \quad 45° + \angle BAD + 90° = 180°$

$\Rightarrow \quad 135° + \angle BAD = 180°$

$\Rightarrow \quad \angle BAD = 180° - 135°$

$= 45°$

i.e., $\quad \angle BAD = \angle B = 45°$

Thus, $\quad BD = AD$

Again,

In right ΔADC,

$$AC^2 = AD^2 + CD^2 \quad ...(i)$$

In right ΔADB, $\quad AB^2 = AD^2 + BD^2$

$\Rightarrow \quad AB^2 - BD^2 = AD^2$

Putting in equation (i), we get

$$AC^2 = AB^2 - BD^2 + CD^2$$

$$AC^2 = AB^2 + CD^2 - BD^2$$

$$= AB^2 + (CD - BD)(CD + BD)$$

$$= AB^2 + (CD - BD) BC$$

$$= AB^2 + (BC - BD - BD) . BC$$

$$= AB^2 + BC^2 - 2BC . BD$$

$$= AB^2 + BC^2 - 2. BC.AD$$

[Proved above]

$$= AB^2 + BC^2 - 4 \text{ ar } (\Delta ABC).$$

($\because$ ar $\Delta ABC = \frac{1}{2} BC \times AD$ $\therefore$ $BC \times AD = 2$ ar ΔABC)

Thus, $(AC)^2 = (AB)^2 + (BC)^2 - 4$ ar (ΔABC)

Hence Proved.

Q. 12. Prove that sum of the squares of the diagonals of a parallelogram is equal to sum of the squares of its sides.

Solution : We have, A parallelogram $ABCD$

To prove : $AC^2 + BD^2 = AB^2 + BC^2 + CD^2 + AD^2$

Construction : Draw $AM \perp CD$ and $BL \perp CD$ extended to L.

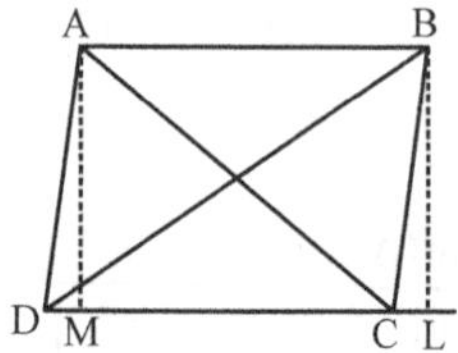

Proof : In ΔADM and ΔBLC,

$$AD = BC$$

[Opposite sides of parallelogram]

$$AM = BL$$

[$\perp$ distance between parallel sides]

$$\angle M = \angle L = 90°$$

$\Rightarrow \quad \Delta ADM \cong \Delta BLC \quad$ [RHS congruency]

$\Rightarrow \quad DM = CL \quad$ [cpct]

Now, in ΔAMD, $AD^2 = DM^2 + AM^2 \quad$ ...(i)

In ΔAMC, $\quad AC^2 = AM^2 + CM^2$

$= AM^2 + (CD - DM)^2 \quad$ ...(ii)

and in ΔBCL, $BC^2 = BL^2 + CL^2 \quad$ ...(iii)

In ΔBLD, $\quad BD^2 = BL^2 + DL^2 = BL^2 + (CD + CL)^2 \quad$ ...(iv)

Adding equations (ii) and (iv), we get

$$AC^2 + BD^2 = AM^2 + CD^2 + DM^2 - 2CD . DM + BL^2 + CD^2 + CL^2 + 2CD. CL$$

$$AC^2 + BD^2 = AM^2 + CD^2 + DM^2 + BL^2 + CD^2 + CL^2$$

$$= (AM^2 + DM^2) + (BL^2 + CL^2) + CD^2 + AB^2$$

[$\because AB = CD$; opposite sides of parallelogram]

$$= AD^2 + BC^2 + CD^2 + AB^2$$

$\Rightarrow \quad AC^2 + BD^2 = AD^2 + CB^2 + CD^2 + AB^2$

$\therefore$ Sum of squares of diagonals of a parallelogram is equal to the sum of squares of its sides. **Hence Proved.**

Chapter 8. Circles

Q. 1. The radii of two concentric circles are 13 cm and 8 cm. *PQ* is a diameter of the bigger circle. *QR* is a tangent to the smaller circle touching it at *R*. Find the length *PR*.

Solution : Let the line *QR* intersects the bigger circle at *S*.

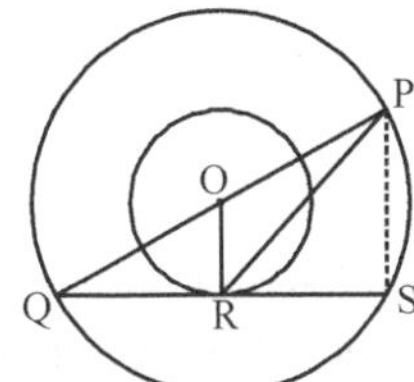

Join *PS*.

O is the mid-point of *PQ*.

[∵ *PQ* is a diameter of the bigger circle]

QR is a tangent to the smaller circle and *OR* is a radius through the point of contact *R*.

$\therefore \quad OR \perp QR \Rightarrow OR \perp QS$

Since, *OR* is perpendicular to a chord *QS* of the bigger circle.

$\therefore \quad QR = RS$

[∵ Perpendicular from the centre to a chord bisects the chord]

⇒ *R* is the mid-point of *QS*.

∴ In ΔQSP, *O* is the mid-point of *PQ* and *R* is the mid-point of *QS*.

$$\therefore \quad OR = \frac{1}{2}PS$$

[∵ segment joining the mid-points of any two sides of a triangle is half of the third side]

$\Rightarrow \quad PS = 2OR = 2 \times 8 \text{ cm} = 16 \text{ cm}$

In right ΔOQR,

$$OR^2 + QR^2 = OQ^2$$

$$\Rightarrow \quad 8^2 + QR^2 = 13^2$$

$$\Rightarrow \quad 64 + QR^2 = 169$$

$$\Rightarrow \quad QR^2 = 169 - 64 = 105$$

$$\Rightarrow \quad QR = \sqrt{105}$$

$$\therefore \quad RS = QR = \sqrt{105}$$

In ΔPRS

$$PR^2 = RS^2 + PS^2$$

$$= (\sqrt{105})^2 + 16^2$$

$\Rightarrow \quad PR = 19$ cm. **Ans.**

Q. 2. *A*, *B* and *C* are three points on a circle. The tangent at *C* meets *BA* produced at *T*. Given that $\angle ATC = 36°$ and that the $\angle ACT = 48°$, calculate the angle subtended by *AB* at the centre of the circle.

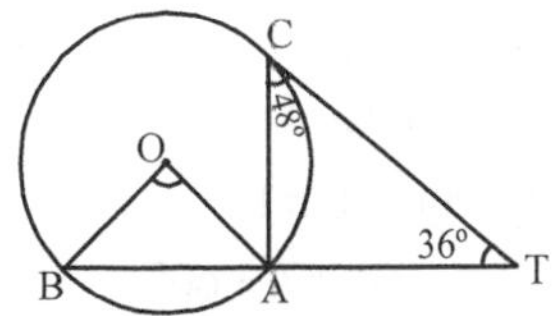

Solution : Let *O* be the centre of the circle. Join *OC*.

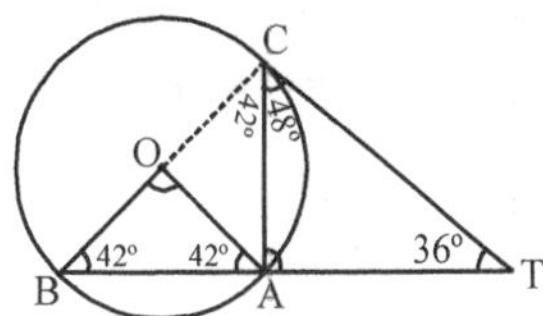

Since angle between the radius and the tangent is 90°,

$\therefore \quad \angle OCT = 90°$

$\Rightarrow \quad \angle OCA = \angle OCT - \angle ACT$

$= 90° - 48° = 42°$

and $\quad OA = OC = OB$ [Radii of the same circle]

$\Rightarrow \quad \angle OAC = \angle OCA = 42°$

[Angles opposite to equal sides are equal]

In ΔACT,

exterior $\angle CAB = \angle ACT + \angle CTA$

$= 48° + 36° = 84°$

$\therefore \quad \angle OAB = \angle CAB - \angle OAC$

$\Rightarrow \quad \angle OAB = 84° - 42° = 42°$

$\Rightarrow \quad \angle OBA = \angle OAB = 42° \quad$ [∵ $OA = OB =$ radii]

In isosceles ΔAOB,

$\angle OBA + \angle OAB + \angle AOB = 180°$

$\Rightarrow \quad 42° + 42° + \angle AOB = 180°$

$\Rightarrow \quad \angle AOB = 180° - 84° = 96°$

Hence, the angle subtended by *AB* at centre is 96°.

Ans.

Q. 3. In a circle *C*(*O*, *r*), *OP* is equal to diameter of the circle. *PA* and *PB* are two tangents. Prove that ΔABP is an equilateral triangle.

Solution : Let *OP* meet the circle at *Q* (fig).

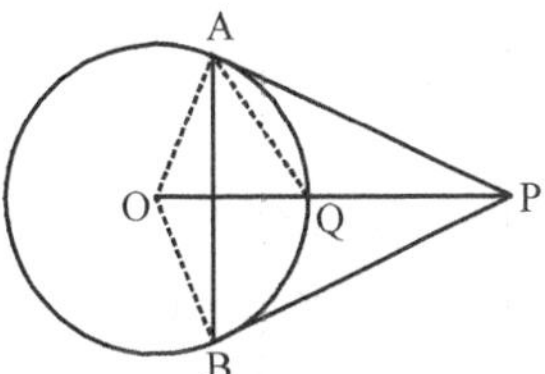

It is given that,

$OP =$ diameter $= 2r$

$\Rightarrow \quad OP = OQ + QP$

$\Rightarrow \quad 2r = r + QP$

$\Rightarrow \quad QP = r$

∴ *Q* is the mid-point of *OP*.

Now *AP* is tangent at *A* and *OA* is radius

$\Rightarrow \quad OA \perp AP$

$\therefore$ ΔOAP is right triangle with $A = 90°$ and Q is mid-point of hypotenuse OP.

$\therefore \quad AQ = OQ = OA$

$\therefore$ ΔOAQ is equilateral.

$\Rightarrow \quad \angle AOQ = 60°$

$\therefore \quad \angle APO = 180° - 90° - 60° = 30°$

$\therefore \quad \angle APB = 2\angle APO = 2 \times 30 = 60°$

Also, $\quad PA = PB$

$\Rightarrow \quad \angle PAB = \angle PBA$

Now in ΔAPB, we have

$\angle APB + \angle PAB + \angle PBA = 180°$

$\Rightarrow \quad 60° + 2\angle PAB = 180°$

$\Rightarrow \quad \angle PAB = 60°$

$\therefore \quad \angle PAB = \angle PBA = \angle APB = 60°$

$\therefore$ ΔAPB is equilateral. **Hence Proved.**

Q. 4. Prove that opposite sides of a quadrilateral circumscribing a circle subtend supplementary angles at the centre of the circle.

Solution : Let a circle with centre O touches the sides AB, BC, CD and DA of a quadrilateral $ABCD$ at the points P, Q, R and S respectively. Then, we have to prove that

$\angle AOB + \angle COD = 180°$ and $\angle AOD + \angle BOC = 180°$

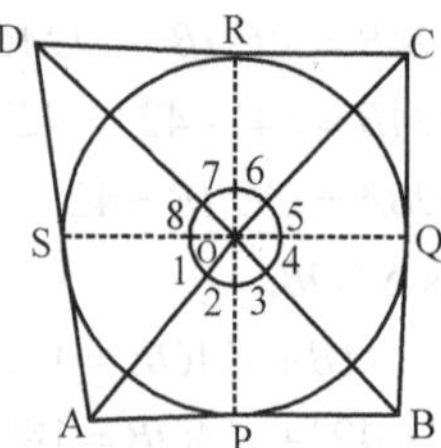

Now, Join OP, OQ, OR and OS.

Since the two tangents drawn from an external point to a circle subtend equal angles at the centre.

$\therefore \angle 1 = \angle 2, \angle 3 = \angle 4, \angle 5 = \angle 6$ and $\angle 7 = \angle 8$ ...(i)

Now, $\angle 1 + \angle 2 + \angle 3 + \angle 4 + \angle 5 + \angle 6 + \angle 7 + 8$
$= 360°$ ...(ii)

[Sum of all the angles subtended at a point is 360°]

$\Rightarrow 2(\angle 2 + \angle 3 + \angle 6 + \angle 7) = 360°$

[Using (i) and (ii)]

$\Rightarrow \quad (\angle 2 + \angle 3) + (\angle 6 + \angle 7) = 180°$

$\Rightarrow \quad \angle AOB + \angle COD = 180°$

Again, $2(\angle 1 + \angle 8 + \angle 4 + \angle 5) = 360°$

[Using (i) and (ii)]

$\therefore \quad (\angle 1 + \angle 8) + (\angle 4 + \angle 5) = 180°$

$\Rightarrow \quad \angle AOD + \angle BOC = 180°$

Hence Proved.

Q. 5. A triangle *ABC* is drawn to circumscribe a circle of radius 4 cm such that the segments *BD* and *DC* into which *BC* is divided by the point of contact *D* are of lengths 8 cm and 6 cm respecitvely. Find the lenght of sides *AB* and *AC*.★

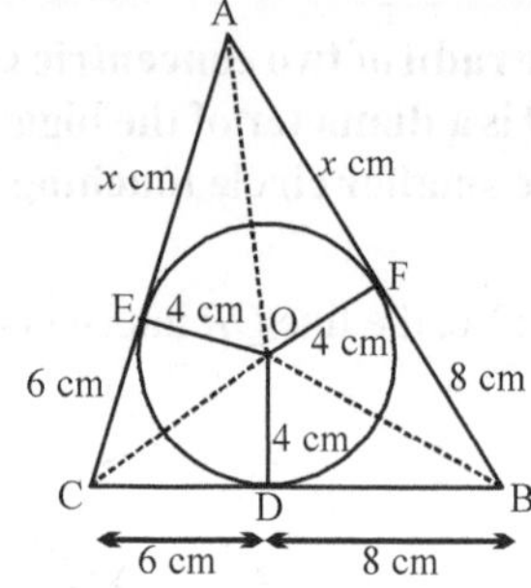

Solution : Let ΔABC be drawn to circumscribe a circle with centre O and radius 4 cm and circle touches the sides BC, CA and AB at D, E and F respectively.

It is given that $CD = 6$ cm and $BD = 8$ cm

$\therefore \quad BF = BD = 8$ cm and $CE = CD = 6$ cm

[Length of two tangents drawn from an external point to a circle are equal]

Now, let $\quad AF = AE = x$ cm

Then, $\quad AB = c = (x + 8)$ cm, $BC = a = 14$ cm, $CA = b = (x + 6)$ cm

$\therefore \quad 2s = (x + 8) + 14 + (x + 6)$

$\Rightarrow \quad 2s = 2x + 28 \quad$ or $\quad s = x + 14$

$\therefore \quad s - a = (x + 14) - 14 = x$

$s - b = (x + 14) - (x + 6) = 8$

$s - c = (x + 14) - (x + 8) = 6$

$\therefore$ Area of $\Delta ABC = \sqrt{s(s-a)(s-b)(s-c)}$

$= \sqrt{(x+14)(x)(8)(6)}$

$= \sqrt{48x(x+14)}$

Also, Area (ΔABC) = Area (ΔOBC) + Area (ΔOCA) + Area (ΔOAB)

$$= \frac{1}{2} \times BC \times OD + \frac{1}{2} \times CA \times OE + \frac{1}{2} \times AB \times OF$$

$$= \frac{1}{2} \times 14 \times 4 + \frac{1}{2} \times (x + 6) \times 4 + \frac{1}{2} \times (x + 8) \times 4$$

$= 28 + 2x + 12 + 12x + 16 = 4x + 56$

$\therefore \quad \sqrt{48x(x+14)} = 4x + 56$

$\Rightarrow \quad \sqrt{48x(x+14)} = 4(x + 14)$

Squaring both sides, we get

$48x(x + 14) = 16(x + 14)^2$

$\Rightarrow \quad 48x(x + 14) - 16(x + 14)^2 = 0$

$\Rightarrow \quad 16(x + 14)[3x - (x + 14)] = 0$

$\Rightarrow \quad 16(x + 14)(2x - 14) = 0$

Either $16(x + 14) = 0$ or $2x - 14 = 0$

$\Rightarrow \quad x = -14 \quad$ or $\quad 2x = 14$

$\Rightarrow \quad x = -14 \quad$ or $\quad x = 7$

★ are board exam questions from previous years

But x cannot be negative so $x \neq -14$

$\therefore \quad x = 7$ cm

Hence, $\quad AB = x + 8 = 7 + 8 = 15$ cm

$AC = x + 6 = 7 + 6 = 13$ cm. **Ans.**

Q. 6. In Fig., XY and $X'Y'$ are two parallel tangents to a circle with centre O and another tangent AB with point of contact C intersecting XY at A and $X'Y'$ at B. Prove that $\angle AOB = 90°$.★

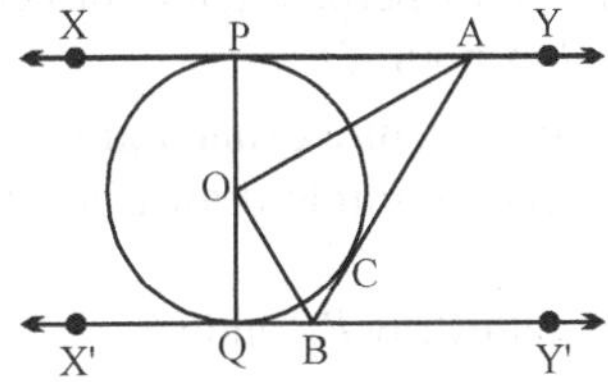

Solution : Join OC. In ΔAPO and ΔACO, we have

$AP = AC$

[Tangents drawn from external point A]

$AO = OA$ [Common]

$PO = OC$ [Radii of the same circle]

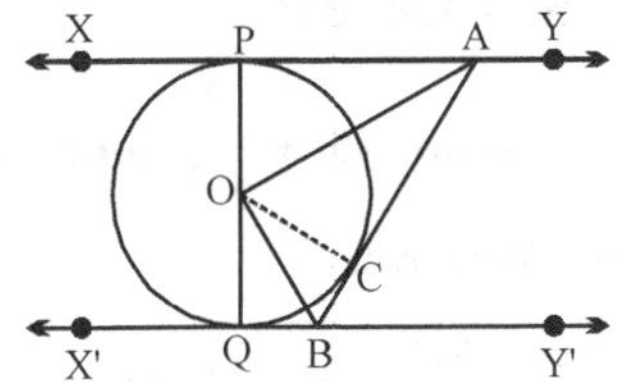

$\therefore \quad \Delta APO \cong \Delta ACO$

[By SSS criterion of congruence]

$\therefore \quad \angle PAO = \angle CAO$ [cpct]

$\Rightarrow \quad \angle PAC = 2\angle CAO$

Similarly, we can prove that

$\Delta OQB \cong \Delta OCB$

$\therefore \quad \angle QBO = \angle CBO$

$\Rightarrow \quad \angle CBQ = 2\angle CBO$

Now, $\angle PAC + \angle CBQ = 180°$

[Sum of interior angles on the same side of transveral is 180°]

$\Rightarrow 2\angle CAO + 2\angle CBO = 180°$

$\Rightarrow \quad \angle CAO + \angle CBO = 90°$

$\Rightarrow \quad 180° - \angle AOB = 90°$

$[\because \angle CAO + \angle CBO + \angle AOB = 180°]$

$\Rightarrow \quad 180° - 90° = \angle AOB$

$\Rightarrow \quad \angle AOB = 90°$. **Hence Proved.**

Chapter 9. Constructions

Q. 1. Draw a circle with the help of a bangle. Take a point outside the circle. Construct the pair of tangents from this point to the circle.

Solution :

Step I : Draw a circle with the help of a bangle.

★ are board exam questions from previous years

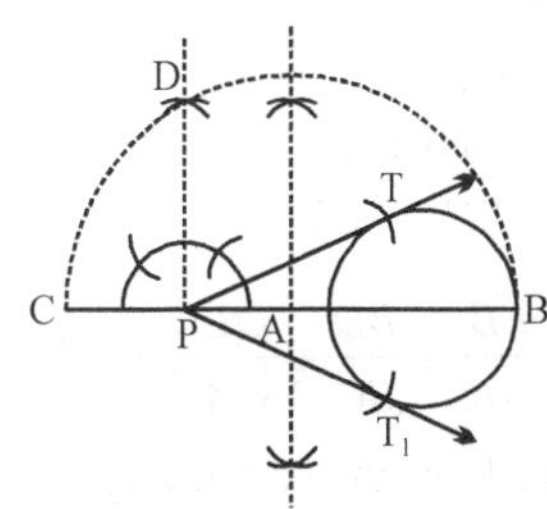

Step II : Let P be the external point from where the tangents are to be drawn to the given circle. Through P, draw a secant PAB to intersect the circle at A and B (say).

Step III : Produce AP to a point C, such that $AP = PC$, *i.e.*, P is the mid-point of AC.

Step IV : Draw a semi-circle with BC as diameter.

Step V : Draw $PD \perp CB$, intersecting the semi-circle at D.

Step VI : With P as centre and PD as radius, draw arcs to intersect the given circle at T and T_1.

Step VII : Join PT and PT_1.

Thus, PT and PT_1 are the required tangents.

Q. 2. Draw a ΔABC with side $BC = 7$ cm, $\angle B = 45°$, $\angle A = 105°$. Then, construct a triangle whose sides are $\frac{4}{3}$ times the corresponding sides of ΔABC.

Solution :

Step I : Construction a ΔABC in which $BC = 7$ cm, $\angle B = 45°, \angle C = 180° - (\angle A + \angle B)$

$= 180° - (105° + 45°) = 180° - 150° = 30°$.

Step II : Below BC, make an acute angle $\angle CBX$.

Step III : Along BX, mark off four arcs : B_1, B_2, B_3 and B_4 such that $BB_1 = B_1B_2 = B_2B_3 = B_3B_4$

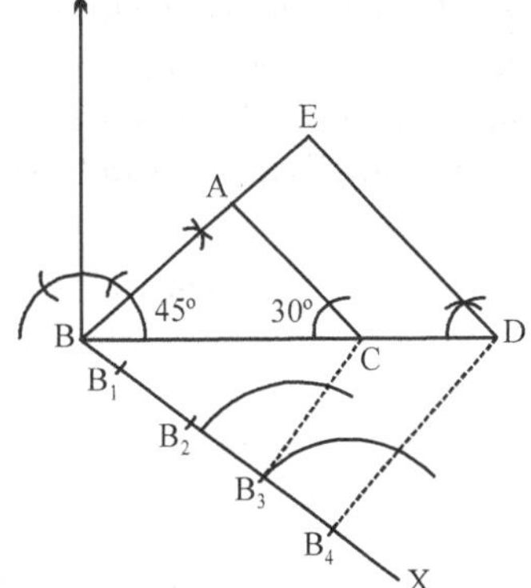

Step IV : Join B_3C.

Step V : From B_4, draw $B_4D \parallel B_3C$, meeting BC produced at D.

Step VI : From D, draw $ED \parallel AC$, meeting BA produced at E. Then

Thus, ΔEBD is the required triangle whose sides are $\frac{4}{3}$ times the corresponding sides of ΔABC.

Justification :

Since, $DE \parallel CA$

$\therefore \quad \Delta ABC \sim \Delta EBD$

and $\quad \dfrac{EB}{AB} = \dfrac{BD}{BC} = \dfrac{DE}{CA} = \dfrac{4}{3}$

Hence, we have the new triangle similar to the given triangle whose sides are equal to $\dfrac{4}{3}$ times the corresponding sides of ΔABC.

Q. 3. Draw a circle of radius 3 cm. Take two points *P* and *Q* on one of its extended diameter each at a distance of 8.5 cm from its centre. Draw tangents to the circle from the these two points *P* and *Q*.

Solution :

Step I : Draw a circle with O as centre and of radius 3 cm.

Step II : Draw a diameter ST and extend it to both the sides to P and Q such that $PO = OQ = 8.5$ cm.

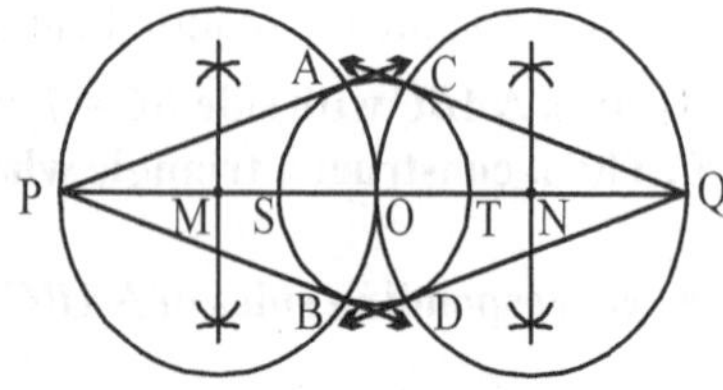

Step III : Draw perpendicular bisectors of PO and OQ. Let M and N be the mid-points of PO and OQ, respectively.

Step IV : Taking M as the centre and MP as radius, draw a circle with meets the circle drawn in step I at A and B. Similarly, taking N as centre and NQ as radius, draw a circle which meets the circle drawn in step I at C and D.

Step V : Join PA, PB, QC and QD.

Thus, PA, PB, QC and QD are the required tangents.

Q. 4. Draw two tangents from a point 9 cm away from the centre of a circle to the incircle of a triangle length of whose three sides are 7 cm, 6 cm and 5 cm.

Solution :

Step I : Draw a line segment BC of length 7 cm.

Step II : With centre B and radius 6 cm, draw an arc above BC.

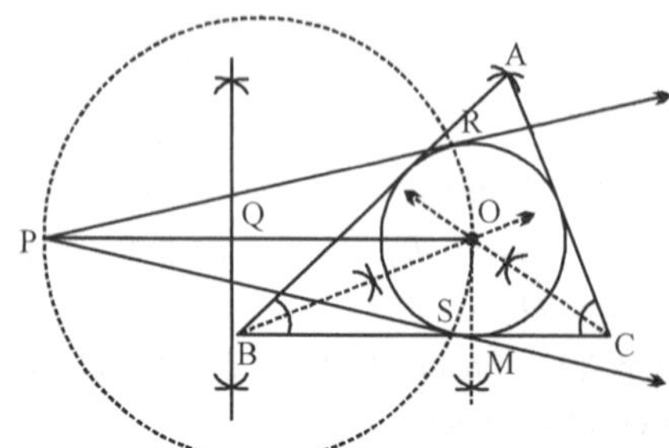

Step III : With centre C and radius 5 cm, draw an arc intersecting the arc drawn in step (II) at A.

Step IV : Join A to B and A to C.

Step V : Draw bisectors of $\angle ABC$ and $\angle ACB$, which intersects each other at point O.

Step VI : Draw perpendicular from O to BC which intersect BC at M.

Step VII : With centre O and radius OM, draw a circle which is incircle of ΔABC.

Step VIII : Draw a line segment OP through the centre O of length 9 cm.

Step IX : Draw perpendicular bisector of PO which intersects PO at point Q.

Step X : With centre Q and radius QO draw a circle which intersects the circle drawn in step (VII) at point R and S.

Step XI : Draw ray PR and PS.

Thus, PR and PS are the required tangents.

Chapter 10. Trigonometry

Q.1. Evaluate

$$\frac{\sec^2 54^\circ - \cot^2 36^\circ}{\operatorname{cosec}^2 57^\circ - \tan^2 33^\circ} + 2\sin^2 38^\circ \sec^2 52^\circ - \sin^2 45^\circ + \frac{2}{\sqrt{3}} \tan 17^\circ \tan 60^\circ \tan 73^\circ.$$

Solution : We have,

$$\frac{\sec^2 54^\circ - \cot^2 36^\circ}{\operatorname{cosec}^2 57^\circ - \tan^2 33^\circ} + 2\sin^2 38^\circ \sec^2 52^\circ - \sin^2 45^\circ + \frac{2}{\sqrt{3}} \tan 17^\circ \tan 60^\circ \tan 73^\circ$$

$$= \frac{\sec^2(90^\circ - 36^\circ) - \cot^2 36^\circ}{\operatorname{cosec}^2(90^\circ - 33^\circ) - \tan^2 33^\circ} + 2\sin^2 38^\circ \sec^2(90^\circ - 38^\circ) - \sin^2 45^\circ + \frac{2}{\sqrt{3}} \tan(90^\circ - 73^\circ) \tan 60^\circ \tan 73^\circ$$

$$= \frac{\operatorname{cosec}^2 36^\circ - \cot^2 36^\circ}{\sec^2 33^\circ - \tan^2 33^\circ} + 2\sin^2 38^\circ \operatorname{cosec}^2 38^\circ - \sin^2 45^\circ + \frac{2}{\sqrt{3}} \cot 73^\circ \tan 60^\circ \tan 73^\circ$$

$$= \frac{1}{1} + 2\sin^2 38^\circ \times \frac{1}{\sin^2 38^\circ} - \left(\frac{1}{\sqrt{2}}\right)^2 + \frac{2}{\sqrt{3}} \times \frac{1}{\tan 73^\circ} \times \sqrt{3} \times \tan 73^\circ$$

$$[\because \operatorname{cosec}^2\theta - \cot^2\theta = 1, \sec^2\theta - \tan^2\theta = 1]$$

$$= 1 + 2 - \frac{1}{2} + 2 = 5 - \frac{1}{2} = \frac{9}{2}.$$ **Ans.**

Q. 2. Evaluate :

(i) $\tan 35^\circ \tan 40^\circ \tan 45^\circ \tan 50^\circ \tan 55^\circ$

(ii) $\operatorname{cosec}(65^\circ + \theta) - \sec(25^\circ - \theta) - \tan(55^\circ - \theta) + \cot(35^\circ + \theta)$

Solution : (i) We have,

$\tan 35° \tan 40° \tan 45° \tan 50° \tan 55°$

$= \tan 35° \tan 40° \tan 45° \tan(90° - 40°) \tan (90° - 35°)$

$= \tan 35° \tan 40° \tan 45° \cot 40° \cot 35°$

$= (\tan 35°. \cot 35°) (\tan 40°. \cot 40°). \tan 45° = 1$

Ans.

(ii) We have,

$\text{cosec} (65° + \theta) - \sec (25° - \theta) - \tan (55° - \theta) + \cot (35° + \theta)$

$= \text{cosec}\{90° - (25° - \theta)\} - \sec(25° - \theta) - \tan (55° - \theta) + \cot \{90° - (55° - \theta)\}$

$= \sec (25° - \theta) - \sec(25° - \theta) - \tan (55° - \theta) + \tan (55° - \theta) = 0.$ **Ans.**

Q. 3. If $\cos \theta + \sqrt{3} \sin \theta = 2 \sin \theta$, show that $\sin \theta - \sqrt{3} \cos \theta = 2 \cos \theta$

Solution : Given, $\cos \theta + \sqrt{3} \sin \theta = 2 \sin \theta$

$\Rightarrow \cos \theta = 2 \sin \theta - \sqrt{3} \sin \theta = (2 - \sqrt{3}) \sin \theta$

Multiplying both sides by $2 + \sqrt{3}$, we get

$(2 + \sqrt{3}) \cos \theta = (2 + \sqrt{3}) (2 - \sqrt{3}) \sin \theta$

$\Rightarrow (2 + \sqrt{3}) \cos \theta = \{(2)^2 - (\sqrt{3})^2\} \sin \theta$ $[\because (a + b)(a - b) = a^2 - b^2]$

$\Rightarrow 2 \cos \theta + \sqrt{3} \cos \theta = (4 - 3) \sin \theta$

$\Rightarrow 2 \cos \theta + \sqrt{3} \cos \theta = \sin \theta$

$\Rightarrow \sin \theta - \sqrt{3} \cos \theta = 2 \cos \theta.$ **Hence Proved.**

Q. 4. If $a \cos \theta + b \sin \theta = m$ and $a \sin \theta - b \cos \theta = n$, prove that $a^2 + b^2 = m^2 + n^2$.

Solution : Given, $a \cos \theta + b \sin \theta = m$ ...(i)

and $a \sin \theta - b \cos \theta = n$ ...(ii)

Squaring and then adding equation (i) and (ii)

$(a^2 \cos^2 \theta + b^2 \sin^2 \theta + 2ab \sin \theta \cos \theta) + (a \sin^2 \theta + b^2 \cos^2 \theta - 2ab \sin \theta \cos \theta) = m^2 + n^2$

$\Rightarrow a^2 (\cos^2 \theta + \sin^2 \theta) + b^2 (\cos^2 \theta + \sin^2 \theta) = m^2 + n^2$

$\Rightarrow a^2 + b^2 = m^2 + n^2$ **Hence Proved.**

Q. 5. Show that $3 (\sin \theta - \cos \theta)^4 + 6 (\sin \theta + \cos \theta)^2 + 4 (\sin^6 \theta + \cos^6 \theta)$ is independent of θ.

Solution :

$3(\sin \theta - \cos \theta)^4 + 6 (\sin \theta + \cos \theta)^2 + 4 (\sin^6 \theta + \cos^6 \theta)$

$= 3 [(\sin \theta - \cos \theta)^2]^2 + 6 (\sin^2 \theta + \cos^2 \theta + 2 \sin \theta . \cos \theta) + 4 [(\sin^2 \theta + \cos^2 \theta)^3 - 3 \sin^2 \theta . \cos^2 \theta(\sin^2 \theta + \cos^2 \theta)]$

$= 3[\sin^2 \theta + \cos^2 \theta - 2 \sin \theta . \cos \theta]^2 + 6[1 + 2 \sin \theta . \cos \theta] + 4 [1 - 3 \sin^2 \theta. \cos^2 \theta]$

$= 3 (1 - 2 \sin \theta \cos \theta)^2 + 6 (1 + 2 \sin \theta. \cos \theta) + 4 (1 - 3 \sin^2 \theta. \cos^2 \theta)$

$= 3(1 - 4 \sin \theta. \cos \theta + 4 \sin^2 \theta. \cos^2 \theta) + 6(1 + 2 \sin \theta. \cos \theta) + 4 (1 - 3\sin^2 \theta. \cos^2 \theta)$

$= 3 - 12 \sin \theta. \cos \theta + 12 \sin^2 \theta. \cos^2 \theta + 6 + 12 \sin \theta. \cos \theta + 4 - 12 \sin^2 \theta. \cos^2 \theta$

$= 13$, which is independent of θ. **Hence Proved.**

Q. 6. Prove that : $\frac{\tan\theta}{1-\cot\theta} + \frac{\cot\theta}{1-\tan\theta} = 1 + \sec \theta \, \text{cosec}\, \theta = 1 + \tan \theta + \cot \theta$.

Solution : L.H.S. $= \frac{\tan\theta}{1-\cot\theta} + \frac{\cot\theta}{1-\tan\theta}$

$$= \frac{\frac{\sin\theta}{\cos\theta}}{1-\frac{\cos\theta}{\sin\theta}} + \frac{\frac{\cos\theta}{\sin\theta}}{1-\frac{\sin\theta}{\cos\theta}}$$

$$= \frac{\sin\theta \times \sin\theta}{\cos\theta(\sin\theta - \cos\theta)} + \frac{\cos\theta \times \cos\theta}{\sin\theta\,(\cos\theta - \sin\theta)}$$

$$= \frac{\sin^2\theta}{\cos\theta(\sin\theta - \cos\theta)} + \frac{\cos^2\theta}{\sin\theta\{-(\sin\theta - \cos\theta)\}}$$

$$= \frac{\sin^2\theta}{\cos\theta(\sin\theta - \cos\theta)} - \frac{\cos^2\theta}{\sin\theta(\sin\theta - \cos\theta)}$$

$$= \frac{\sin^3\theta - \cos^3\theta}{\cos\theta \sin\theta(\sin\theta - \cos\theta)}$$

$$= \frac{(\sin\theta - \cos\theta)(\sin^2\theta + \cos^2\theta + \sin\theta\cos\theta)}{\cos\theta\sin\theta(\sin\theta - \cos\theta)}$$

$$= \frac{1 + \sin\theta\cos\theta}{\sin\theta\cos\theta} \qquad [\because \sin^2\theta + \cos^2\theta = 1]$$

$$= \frac{1}{\sin\theta\cos\theta} + \frac{\sin\theta\cos\theta}{\sin\theta\cos\theta} = \frac{1}{\sin\theta}.\frac{1}{\cos\theta} + 1 \qquad ...(i)$$

$= \sec \theta \, \text{cosec}\, \theta + 1$ ...(ii)

For second part

Now from (i), we have

$$\text{L.H.S.} = \frac{1}{\sin\theta\cos\theta} + 1$$

$$= \frac{\sin^2\theta + \cos^2\theta}{\sin\theta\cos\theta} + 1$$

[Putting $1 = \sin^2 \theta + \cos^2 \theta$]

$$= \frac{\sin^2\theta}{\sin\theta\cos\theta} + \frac{\cos^2\theta}{\cos\theta\sin\theta} + 1$$

$$= \frac{\sin\theta}{\cos\theta} + \frac{\cos\theta}{\sin\theta} + 1 = \tan\theta + \cot\theta + 1 = \text{R.H.S.}$$

Hence Proved.

Q. 7. If $\tan A = n \tan B$ and $\sin A = m \sin B$, prove that

$$\cos^2 A = \frac{m^2 - 1}{n^2 - 1}$$

Solution : We have to find $\cos^2 A$ in terms of m and n. This means that the angle B is to be eliminated from the given relations.

Now, $\tan A = n \tan B$ [Given]

$$\Rightarrow \tan B = \frac{1}{n}\tan A$$

$$\because \cot B = \frac{1}{\tan B}$$

$$\Rightarrow \cot B = \frac{n}{\tan A}$$

and $\sin A = m \sin B$ [Given]

$$\Rightarrow \sin B = \frac{1}{m}\sin A$$

$$\because \operatorname{cosec} B = \frac{1}{\sin B}$$

$$\Rightarrow \operatorname{cosec} B = \frac{m}{\sin A}$$

Substituting the values of cot B and cosec B in $\operatorname{cosec}^2 B - \cot^2 B = 1$, we get

$$\frac{m^2}{\sin^2 A} - \frac{n^2}{\tan^2 A} = 1$$

$$\Rightarrow \frac{m^2}{\sin^2 A} - \frac{n^2\cos^2 A}{\sin^2 A} = 1$$

$$\Rightarrow \frac{m^2 - n^2\cos^2 A}{\sin^2 A} = 1$$

$$\Rightarrow m^2 - n^2\cos^2 A = \sin^2 A$$

$$\Rightarrow m^2 - n^2\cos^2 A = 1 - \cos^2 A$$

$$[\because \sin^2 A = 1 - \cos^2 A]$$

$$\Rightarrow m^2 - 1 = n^2\cos^2 A - \cos^2 A$$

$$\Rightarrow m^2 - 1 = (n^2 - 1)\cos^2 A$$

$$\Rightarrow \frac{m^2 - 1}{n^2 - 1} = \cos^2 A.$$ **Hence Proved.**

Q. 8. Prove the following identify, where the angle involved is acute angle for which the expressions are defined.

$$\frac{\cos A - \sin A + 1}{\cos A + \sin A - 1} = \operatorname{cosec} A + \cot A,$$

using the identity $\operatorname{cosec}^2 A = 1 + \cot^2 A$.

Solution : $$\text{L.H.S.} = \frac{\cos A - \sin A + 1}{\cos A + \sin A - 1}$$

Dividing Nr and Dr by $\sin A$,

$$= \frac{\frac{\cos A - \sin A + 1}{\sin A}}{\frac{\cos A + \sin A - 1}{\sin A}}$$

$$= \frac{\cot A - 1 + \operatorname{cosec} A}{\cot A + 1 - \operatorname{cosec} A}$$

$$= \frac{(\cot A + \operatorname{cosec} A) - (\operatorname{cosec}^2 A - \cot^2 A)}{\cot A - \operatorname{cosec} A + 1}$$

$$[\because \operatorname{cosec}^2 A - \cot^2 A = 1]$$

$$= \frac{(\cot A + \operatorname{cosec} A) - [(\operatorname{cosec} A + \cot A)(\operatorname{cosec} A - \cot A)]}{\cot A - \operatorname{cosec} A + 1}$$

$$= \frac{(\operatorname{cosec} A + \cot A)(1 - \operatorname{cosec} A + \cot A)}{(\cot A - \operatorname{cosec} A + 1)}$$

$= \operatorname{cosec} A + \cot A = \text{R.H.S.}$ **Hence Proved.**

Q. 9. If $x\sin^3\theta + y\cos^3\theta = \sin\theta\cos\theta$ and $x\sin\theta = y\cos\theta$, prove that $x^2 + y^2 = 1$.

Solution : We have,

$$x\sin^3\theta + y\cos^3\theta = \sin\theta\cos\theta$$

$$\Rightarrow (x\sin\theta)\sin^2\theta + (y\cos\theta)\cos^2\theta = \sin\theta\cos\theta$$

$$\Rightarrow x\sin\theta(\sin^2\theta) + (x\sin\theta)\cos^2\theta = \sin\theta\cos\theta$$

$$[\because x\sin\theta = y\cos\theta]$$

$$\Rightarrow x\sin\theta(\sin^2\theta + \cos^2\theta) = \sin\theta\cos\theta$$

$$\Rightarrow x\sin\theta = \sin\theta\cos\theta$$

$$[\because \sin^2\theta + \cos^2\theta = 1]$$

$$\Rightarrow x = \cos\theta$$

Now, we have

$$x\sin\theta = y\cos\theta$$

$$\Rightarrow \cos\theta\sin\theta = y\cos\theta \quad [\because x = \cos\theta]$$

$$\Rightarrow y = \sin\theta$$

Hence, $x^2 + y^2 = \cos^2\theta + \sin^2\theta = 1$. **Hence Proved.**

Q. 10. If $\tan\theta + \sin\theta = m$ and $\tan\theta - \sin\theta = n$, show that $(m^2 - n^2) = 4\sqrt{mn}$.

Solution : We have,

$$\tan\theta + \sin\theta = m$$

and $$\tan\theta - \sin\theta = n, \text{ then}$$

$$\text{L.H.S.} = (m^2 - n^2) = (\tan\theta + \sin\theta)^2 - (\tan\theta - \sin\theta)^2$$

$$= \tan^2\theta + \sin^2\theta + 2\tan\theta\sin\theta - \tan^2\theta - \sin^2\theta + 2\tan\theta\sin\theta$$

$$= 4\tan\theta\sin\theta = 4\sqrt{\tan^2\theta\sin^2\theta}$$

$$= 4\sqrt{\frac{\sin^2\theta}{\cos^2\theta}(1 - \cos^2\theta)} = 4\sqrt{\frac{\sin^2\theta}{\cos^2\theta} - \sin^2\theta}$$

$$= 4\sqrt{\tan^2\theta - \sin^2\theta}$$

$$= 4\sqrt{(\tan\theta - \sin\theta)(\tan\theta + \sin\theta)}$$

$= 4\sqrt{mn} = \text{R.H.S.}$ **Hence Proved.**

Q. 11. If $\sin\theta + \sin^2\theta = 1$, find the value of $\cos^{12}\theta + 3\cos^{10}\theta + 3\cos^8\theta + \cos^6\theta + 2\cos^4\theta + 2\cos^2\theta - 2$

Solution : We have,

$\sin\theta + \sin^2\theta = 1$

$\Rightarrow \quad \sin\theta = 1 - \sin^2\theta$

$\Rightarrow \quad \sin\theta = \cos^2\theta$

Now,

$\cos^{12}\theta + 3\cos^{10}\theta + 3\cos^8\theta + \cos^6\theta + 2\cos^4\theta + 2\cos^2\theta - 2$

$= (\cos^{12}\theta + 3\cos^{10}\theta + 3\cos^8\theta + \cos^6\theta) + 2(\cos^4\theta + \cos^2\theta - 1)$

$= (\cos^4\theta + \cos^2\theta)^3 + 2(\cos^4\theta + \cos^2\theta - 1)$

$= (\sin^2\theta + \cos^2\theta)^3 + 2(\sin^2\theta + \cos^2\theta - 1)$

$[\because \cos^2\theta = \sin\theta \ \therefore \ \cos^4\theta = \sin^2\theta]$

$= 1 + 2(1 - 1) = 1$ **Ans.**

Q. 12. If $a\sec\theta + b\tan\theta + c = 0$ and $p\sec\theta + q\tan\theta + r = 0$, prove that $(br - qc)^2 - (pc - ar)^2 = (aq - bp)^2$.

Solution : We have,

$$a\sec\theta + b\tan\theta + c = 0$$

$$p\sec\theta + q\tan\theta + r = 0$$

Solving two equations for sec θ and tan θ by the cross-multiplication, we get

$$\frac{\sec\theta}{br - qc} = \frac{\tan\theta}{cp - ar} = \frac{1}{aq - bp}$$

$$\Rightarrow \quad \sec\theta = \frac{br - cq}{aq - bp}$$

and $$\tan\theta = \frac{cp - ar}{aq - bp}$$

Now, we know

$$\sec^2\theta - \tan^2\theta = 1$$

$$\therefore \quad \left(\frac{br - cq}{aq - bp}\right)^2 - \left(\frac{cp - ar}{aq - bp}\right)^2 = 1$$

$$\Rightarrow \quad (br - cq)^2 - (cp - ar)^2 = (aq - bp)^2.$$

Hence Proved.

Q. 13. If $\sec\theta + \tan\theta = p$, obtain the values of sec θ, tan θ and sin θ in term of p.

Solution : We have,

$$\sec\theta + \tan\theta = p \qquad \text{...(i)}$$

We know, $\sec^2\theta - \tan^2\theta = 1$

$\Rightarrow \quad (\sec\theta + \tan\theta)(\sec\theta - \tan\theta) = 1$

$\Rightarrow \quad p(\sec\theta - \tan\theta) = 1$

$$\Rightarrow \quad \sec\theta - \tan\theta = \frac{1}{p} \qquad \text{...(ii)}$$

Adding equation (i) and (ii)

$$(\sec\theta + \tan\theta) + (\sec\theta - \tan\theta) = p + \frac{1}{p}$$

$$\Rightarrow \quad 2\sec\theta = p + \frac{1}{p}$$

$$\Rightarrow \quad 2\sec\theta = \frac{p^2 + 1}{p}$$

$$\Rightarrow \quad \sec\theta = \frac{1}{2}\left(\frac{p^2 + 1}{p}\right) \qquad \text{...(iii)}$$

Now, subtracting equation (i) and (ii)

$$(\sec\theta + \tan\theta) - (\sec\theta - \tan\theta) = p - \frac{1}{p}$$

$$2\tan\theta = p - \frac{1}{p}$$

$$2\tan\theta = \frac{p^2 - 1}{p}$$

$$\tan\theta = \frac{1}{2}\left(\frac{p^2 - 1}{p}\right) \qquad \text{...(iv)}$$

We know,

$$\tan\theta = \frac{\sin\theta}{\cos\theta}$$

or $$\tan\theta = \sin\theta \times \frac{1}{\cos\theta}$$

$$\Rightarrow \quad \tan\theta = \sin\theta \times \sec\theta$$

$$\Rightarrow \quad \sin\theta = \frac{\tan\theta}{\sec\theta}$$

$$= \frac{\frac{1}{2}\left(\frac{p^2 - 1}{p}\right)}{\frac{1}{2}\left(\frac{p^2 + 1}{p}\right)} = \frac{p^2 - 1}{p^2 + 1}$$ **Ans.**

Q. 14. If $\tan^2\theta = 1 - a^2$, prove that $\sec\theta + \tan^3\theta \operatorname{cosec}\theta = (2 - a^2)^{3/2}$

Solution : We have,

$\sec\theta + \tan^3\theta \operatorname{cosec}\theta$

$$= \sec\theta \left\{\frac{\sec\theta + \tan^3\theta \operatorname{cosec}\theta}{\sec\theta}\right\}$$

[Multiplying and dividing by sec θ]

$$= \sec\theta \left\{1 + \tan^3\theta . \frac{\cos\theta}{\sin\theta}\right\}$$

$$= \sec\theta \left\{1 + \tan^3\theta \times \cot\theta\right\}$$

$$= \sqrt{1 + \tan^2\theta}\left\{1 + \tan^2\theta\right\}$$

$$= \left\{1 + \tan^2\theta\right\}^{3/2}$$

$$= \left\{1 + (1 - a^2)\right\}^{3/2} \qquad [\because \tan^2\theta = 1 - a^2; \text{Given}]$$

$= (2 - a^2)^{3/2}$

Hence Proved.

Q. 15. If $a \cos\theta - b \sin\theta = c$, prove that

$$a \sin\theta + b \cos\theta = \pm\sqrt{a^2 + b^2 - c^2}.$$

Solution : We have, $a \cos\theta - b \sin\theta = c$

Squaring both sides

$$\Rightarrow (a \cos\theta - b \sin\theta)^2 = c^2$$

$$\Rightarrow a^2 \cos^2\theta + b^2 \sin^2\theta - 2ab \cos\theta.\sin\theta = c^2$$

$$\Rightarrow a^2(1 - \sin^2\theta) + b^2(1 - \cos^2\theta) - 2ab\cos\theta.\sin\theta = c^2$$

$$\Rightarrow a^2 - a^2\sin^2\theta + b^2 - b^2\cos^2\theta - 2ab\cos\theta.\sin\theta = c^2$$

$$\Rightarrow - a^2\sin^2\theta - b^2\sin^2\theta - 2ab\sin\theta.\cos\theta = -a^2 - b^2 + c^2$$

$$\Rightarrow -(a^2\sin^2\theta + b^2\sin^2\theta + 2ab\sin\theta.\cos\theta) = -(a^2 + b^2 - c^2)$$

$$\Rightarrow a^2\sin^2\theta + b^2\cos^2\theta + 2ab\cos\theta\sin\theta = a^2 + b^2 - c^2$$

$$\Rightarrow (a\sin\theta + b\cos\theta)^2 = a^2 + b^2 - c^2$$

$$\Rightarrow a\sin\theta + b\cos\theta = \pm\sqrt{a^2 + b^2 - c^2}.$$

Hence Proved.

Q. 16. If $\frac{x}{a}\cos\theta + \frac{y}{b}\sin\theta = 1$ and $\frac{x}{a}\sin\theta - \frac{y}{b}\cos\theta = 1$, prove that $\frac{x^2}{a^2} + \frac{y^2}{b^2} = 2$.

Solution : Given, $\frac{x}{a}\cos\theta + \frac{y}{b}\sin\theta = 1$

Squaring both sides, we get

$$\frac{x^2}{a^2}\cos^2\theta + \frac{y^2}{b^2}\sin^2\theta + \frac{2xy}{ab}\cos\theta\sin\theta = 1 \quad ...(i)$$

and $\frac{x}{a}\sin\theta - \frac{y}{b}\cos\theta = 1$

Squaring both sides, we get

$$\frac{x^2}{a^2}\sin^2\theta + \frac{y^2}{b^2}\cos^2\theta - \frac{2xy}{ab}\cos\theta\sin\theta = 1 \quad ...(ii)$$

Adding equations (i) and (ii), we get

$$\frac{x^2}{a^2}(\cos^2\theta + \sin^2\theta) + \frac{x^2}{a^2}(\sin^2\theta + \cos^2\theta) = 2$$

$$\Rightarrow \frac{x^2}{a^2} + \frac{y^2}{b^2} = 2$$

$[\because \sin^2\theta + \cos^2\theta = 1]$

Hence Proved.

Q. 17. If $x = a\cos^3\theta$, $y = b\sin^3\theta$, prove that

$$\left(\frac{x}{a}\right)^{2/3} + \left(\frac{y}{b}\right)^{2/3} = 1.$$

Solution : Given,

$x = a\cos^3\theta$ and $y = b\sin^3\theta$

$$\text{L.H.S.} = \left(\frac{x}{a}\right)^{2/3} + \left(\frac{y}{b}\right)^{2/3}$$

$$= \left(\frac{a\cos^3\theta}{a}\right)^{2/3} + \left(\frac{b\sin^3\theta}{b}\right)^{2/3}$$

$$= (\cos\theta)^{3\times 2/3} + (\sin\theta)^{3\times 2/3}$$

$$= \cos^2\theta + \sin^2\theta = 1 = \text{R.H.S.}$$

Hence Proved.

Q. 18. If $\text{cosec}\,\theta - \sin\theta = m$ and $\sec\theta - \cos\theta = n$, prove that $(m^2n)^{2/3} + (mn^2)^{2/3} = 1$.

Solution : Given, $\text{cosec}\,\theta - \sin\theta = m$

$$\Rightarrow \frac{1}{\sin\theta} - \sin\theta = m$$

$$\Rightarrow \frac{1 - \sin^2\theta}{\sin\theta} = m$$

$$\Rightarrow \frac{\cos^2\theta}{\sin\theta} = m$$

$[\because 1 - \sin^2\theta = \cos^2\theta]$ ...(i)

Also, $\sec\theta - \cos\theta = n$

$$\Rightarrow \frac{1}{\cos\theta} - \cos\theta = n$$

$$\Rightarrow \frac{1 - \cos^2\theta}{\cos\theta} = n$$

$$\Rightarrow \frac{\sin^2\theta}{\cos\theta} = n \quad [\because 1 - \cos^2\theta = \sin^2\theta]$$

...(ii)

Now, $\text{L.H.S.} = (m^2n)^{2/3} + (mn^2)^{2/3}$

$$= \left\{\left(\frac{\cos^2\theta}{\sin\theta}\right)^2 \frac{\sin^2\theta}{\cos\theta}\right\}^{2/3} + \left\{\frac{\cos^2\theta}{\sin\theta} \times \left(\frac{\sin^2\theta}{\cos\theta}\right)^2\right\}^{2/3}$$

$$= \left(\frac{\cos^4\theta.\sin^2\theta}{\sin^2\theta.\cos\theta}\right)^{2/3} + \left(\frac{\cos^2\theta}{\sin\theta} \times \frac{\sin^4\theta}{\cos^2\theta}\right)^{2/3}$$

$= (\cos^3\theta)^{2/3} + (\sin^3\theta)^{2/3} = \cos^2\theta + \sin^2\theta = 1 =$ R.H.S.

Hence Proved.

Chapter 11. Applications of Trigonometry

Q. 1. A boy on horizontal plane finds bird flying at a distance of 100 m from him at an elevation of 30°. A girl standing on the roof of 20 m high building, finds the angle of elevation of the same bird to be 45°. Both the boy and

the girl are on opposite sides of the bird. Find the distance of bird from the girl.

Solution : Let P, B and A be the positions of the bird, boy and girl respectively.

Given $PB = 100$ m, $AC = 20$ m

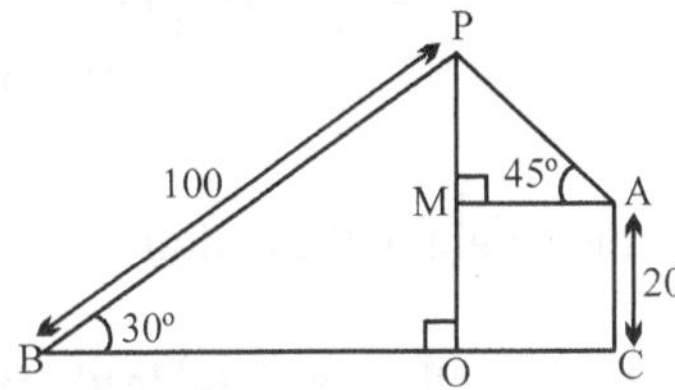

In right-angled ΔPBO, we have

$$\sin 30° = \frac{OP}{BP}$$

$$\Rightarrow \quad \frac{1}{2} = \frac{OP}{100}$$

$$\Rightarrow \quad OP = \frac{100}{2} = 50 \text{ m}$$

Now, $\quad OP = OM + MP$

$$\Rightarrow \quad 50 = 20 + MP \qquad [\because OM = AC = 20 \text{ m}]$$

$$\Rightarrow \quad MP = 30 \text{ m}$$

In ΔPMA,

$$\sin 45° = \frac{MP}{PA}$$

$$\Rightarrow \quad \frac{1}{\sqrt{2}} = \frac{30}{PA}$$

$$\Rightarrow \quad PA = 30\sqrt{2} \text{ m}$$

$$= 30 \times 1{\cdot}41 = 42{\cdot}3 \text{ m}$$

Hence, required distance is 42·3 m. **Ans.**

Q. 2. Two ships are sailing in the sea on the either side of the lighthouse. The angles of the depression of two ships as observed from the top of the lighthouse are 60° and 45° respectively. If the distance between the ships is $200\left(\frac{\sqrt{3}+1}{\sqrt{3}}\right)$ metre, find the height of the lighthouse.

Solution : Let height of the lighthouse, $AB = h$ m. Let C and D be the position of two ships, such that

$$CD = 200\left(\frac{\sqrt{3}+1}{\sqrt{3}}\right)$$

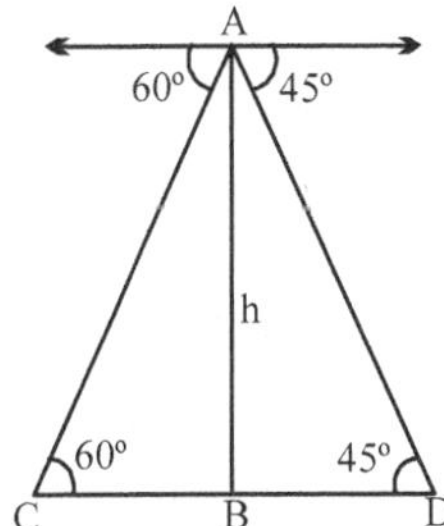

In ΔABC,

$$\tan 60° = \frac{AB}{CB}$$

$$\Rightarrow \quad \sqrt{3} = \frac{h}{CB}$$

$$\Rightarrow \quad CB = \frac{h}{\sqrt{3}} \qquad \text{...(i)}$$

In ΔABD,

$$\tan 45° = \frac{AB}{BD}$$

$$\Rightarrow \quad 1 = \frac{h}{BD}$$

$$h = BD \qquad \text{...(ii)}$$

Now, $\quad CD = CB + BD$

From equations (i) and (ii), we get

$$CD = \frac{h}{\sqrt{3}} + h$$

$$\Rightarrow \quad 200\left(\frac{\sqrt{3}+1}{\sqrt{3}}\right) = h\left(\frac{1}{\sqrt{3}}+1\right)$$

$$\Rightarrow \quad 200\left(1+\frac{1}{\sqrt{3}}\right) = h\left(\frac{1}{\sqrt{3}}+1\right)$$

$$\Rightarrow \quad h = 200 \text{ m}$$ **Ans.**

Q. 3. A man standing on the deck of a ship, which is 10 m above the water level observes the angle of elevation of the top of a hill as 60° and the angle of depression of the base of the hill as 30°. Calculate the distance of the hill from the ship and the height of the hill.

Solution : In Fig, let C represents the position of the man on the deck of the ship, A represents the top of hill and D its base.

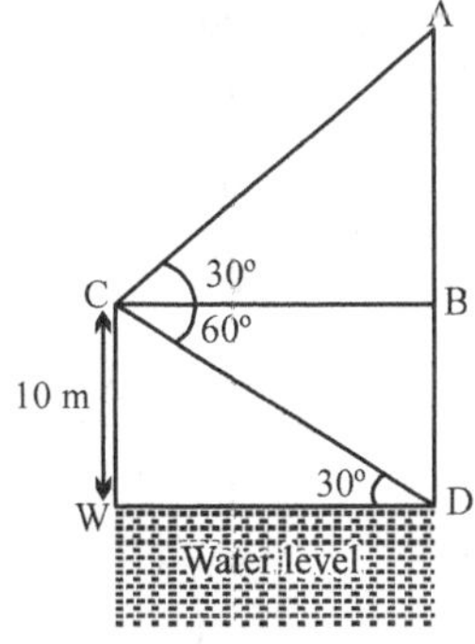

Now in right-angled ΔCWD,

$$\tan 30° = \frac{CW}{WD} \frac{10}{WD}$$

$$\Rightarrow \quad \frac{1}{\sqrt{3}} = \frac{10}{WD}$$

$$\Rightarrow \quad WD = 10\sqrt{3}$$

$$= 10 \times 1.73$$

$$= 17.3 \text{ m}$$

Also, in right-angled ΔABC,

$$\tan 60^\circ = \frac{AB}{BC} \text{ or } \frac{AB}{WD}$$

[From fig. $BC = WD$]

$$\Rightarrow \quad \sqrt{3} = \frac{AB}{10\sqrt{3}}$$

$$\Rightarrow \quad AB = 10\sqrt{3} \times \sqrt{3} = 30 \text{ m}$$

Now, $AD = AB + BD$

$= 30 \text{ m} + 10 \text{ m} = 40 \text{ m}.$

Therefore, the distance of the hill from the ship

$= 17.3$ m

and the height of the hill = 40 m **Ans.**

Q. 4. The angle of elevation of the top of a buliding from the foot of the tower is 30° and the angle of elevation of the top of the tower from the foot of the building is 60°. If the tower is 50 m high, find the length of the building.

Solution : Let AB be the building of height h m and CD be the tower of height 50 m. Then, we have

$$\angle ACB = 30^\circ$$

and $\angle DAC = 60^\circ$

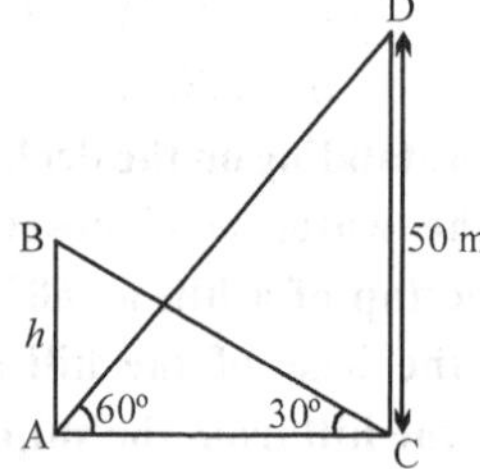

Now, in ΔACD, we have

$$\tan 60^\circ = \frac{DC}{AC}$$

$$\Rightarrow \quad \sqrt{3} = \frac{50}{AC}$$

$$\Rightarrow \quad AC = \frac{50}{\sqrt{3}}$$

$$= \frac{50}{\sqrt{3}} \times \frac{\sqrt{3}}{\sqrt{3}} = \frac{50\sqrt{3}}{3}$$

$$\Rightarrow \quad AC = \frac{50\sqrt{3}}{3} \quad \text{...(i)}$$

and in ΔABC, we have

$$\tan 30^\circ = \frac{AB}{AC}$$

$$\Rightarrow \quad \frac{1}{\sqrt{3}} = \frac{h}{AC}$$

$$\Rightarrow \quad AC = \sqrt{3}h$$

$$\Rightarrow \quad h = \frac{AC}{\sqrt{3}} = \frac{\frac{50\sqrt{3}}{3}}{\sqrt{3}}$$

$$= \frac{50}{3} = 16\frac{2}{3} \text{ m}$$

[From equation (i)]

Hence, the height of the building is $16\frac{2}{3}$ m. **Ans.**

Q. 5. Two poles of equal heights are standing opposite to each other on either side of the road which is 80 m wide. From a point P between them on the road, the angle of elevation of the top of a pole is 60° and the angle of depression from the top of the other pole of point P is 30°. Find the heights of the poles and the distance of the point P from the poles.★

Solution : Let AC is road of 80 m width. P is the point on road AC and height of poles AB and CD is h m.

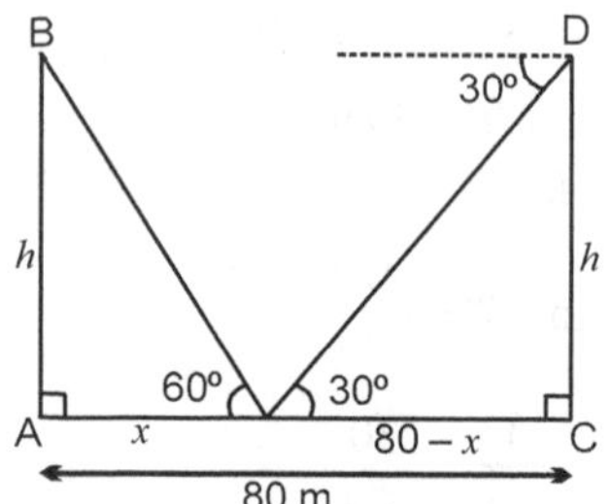

From right ΔPAB, we have

$$\frac{AB}{AP} = \tan 60^\circ = \sqrt{3}$$

$$\frac{h}{x} = \sqrt{3} \quad (\because AP = x)$$

$$h = \sqrt{3}x \quad \text{...(i)}$$

From right ΔDCP, we have

$$\frac{CD}{PC} = \tan 30^\circ = \frac{1}{\sqrt{3}}$$

$$\frac{h}{80 - x} = \frac{1}{\sqrt{3}}$$

$$\Rightarrow \quad h = \frac{80 - x}{\sqrt{3}} \quad \text{...(ii)}$$

Equating the values of h from equation (i) and (ii), we get

$$\Rightarrow \quad x\sqrt{3} = \frac{80 - x}{\sqrt{3}}$$

$$\Rightarrow \quad 3x = 80 - x$$

$$\Rightarrow \quad 4x = 80$$

$$\Rightarrow \quad x = 20 \text{ m}$$

★ are board exam questions from previous years

On putting $x = 20$ in equation (i), we get

$$h = \sqrt{3} \times 20$$
$$= 20\sqrt{3}$$
$$h = 20\sqrt{3} \text{ m}$$

Thus, height of poles is $20\sqrt{3}$ m and point P is at a distance of 20 m from left pole and $(80 - 20)$ *i.e.*, 60 m from right pole. **Ans.**

Q. 6. From a window (*h* metres high above the ground) of a house in a street, the angles of elevation and depression of the top and the foot of another house on the oppostie side of the street are θ and ϕ respectively. Show that the height of the opposite house is

$$\mathbf{h. (1 + \tan\theta \cot\phi)}$$

Solution : Let W be the window and AB be the house on the opposite side.

Then, WP is the width of the street (fig.)

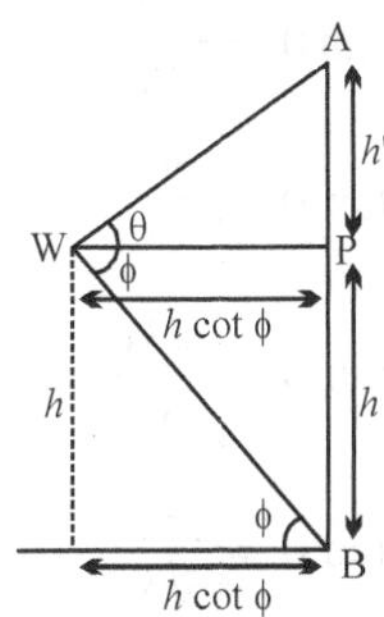

Let $AP = h'$ m

In ΔBPW, $\tan\phi = \dfrac{PB}{WP}$

$\Rightarrow$ $\tan\phi = \dfrac{h}{WP}$

$\Rightarrow$ $WP = h\cot\phi$ ...(i)

Now, in ΔAWP,

$$\tan\theta = \frac{AP}{WP} = \frac{h'}{WP}$$

$$h' = WP\tan\theta$$

$\Rightarrow$ $h' = h\cot\phi\tan\theta$ [From (i)]

$\therefore$ Height of house $= h' + h$

$= h\cot\phi\tan\theta + h$

$= h(1 + \tan\theta\cot\phi)$.

Hence Proved.

Q. 7. The angle of elevation of a jet plane from a point *A* on the ground is 60°. After a flight of 15 seconds, the angle of elevation changes to 30°. If the jet plane is flying at a constant height of $1500\sqrt{3}$ m, find the speed of the jet plane.

Solution : Let P and Q be the two positions of the plane and let A be the point of observation. Let ABC be the horizontal line through A. It is given that angles of elevation of the plane in two positions P and Q from a point A are 60° and 30°, respectively.

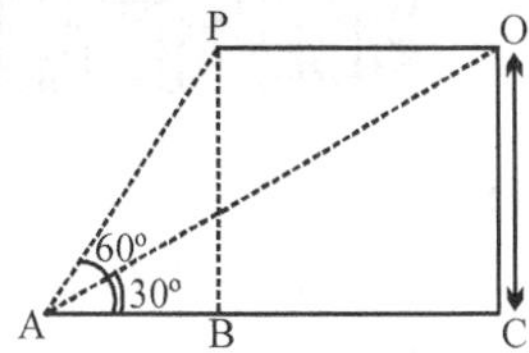

Then, $\angle PAB = 60^\circ$, $\angle QAB = 30^\circ$

It is also given that $PB = 1500\sqrt{3}$ metres

In $\angle ABP$, we have

$$\tan 60^\circ = \frac{BP}{AB} \text{ m}$$

$\Rightarrow$ $\sqrt{3} = \dfrac{1500\sqrt{3}}{AB}$

$\Rightarrow$ $AB = 1500$

In $\angle ACQ$, we have

$$\tan 30^\circ = \frac{CQ}{AC}$$

$\Rightarrow$ $\dfrac{1}{\sqrt{3}} = \dfrac{1500\sqrt{3}}{AC}$

$\Rightarrow$ $AC = 1500 \times 3 = 4500$ m

$\therefore$ $PQ = BC = AC - AB$

$= 4500 - 1500 = 3000$ m

Thus, the plane travels 3000 m in 15 seconds.

Hence, the speed of plane

$$= \frac{\text{Distance travelled}}{\text{Time taken}}$$

$$= \frac{3000}{15} = 200 \text{ m/s}$$

$$= 200 \times \frac{18}{5} \text{ km/h}$$

$= 720$ km/h. **Ans.**

Q. 8. If the angle of elevation of a cloud from a point *h* metres above a lake is α and the angle of depression of its reflection in the lake is β, prove that the height of the cloud is $\dfrac{h(\tan\beta + \tan\alpha)}{\tan\beta - \tan\alpha}$.

Solution : Let AB be the surface of the lake and let P be a point of observation such that $AP = h$ metres. Let C be the position of the cloud and C' be its reflection in the lake. Then, $CB = C'B$. Let PM be perpendicular from P on CB. Then, $\angle CPM = \alpha$ and $\angle MPC' = \beta$.

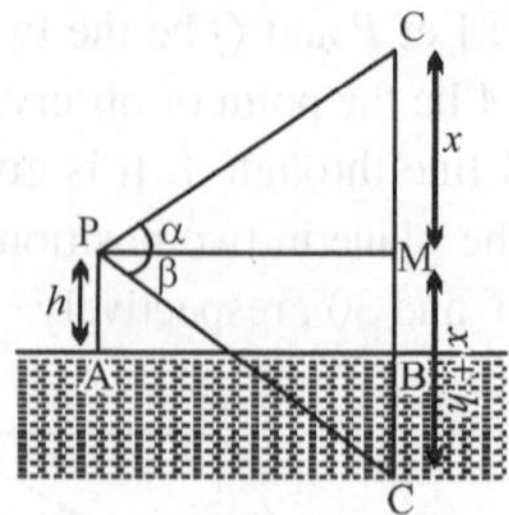

Let $CM = x$.

Then, $CB = CM + MB = CM + PA = x + h$.

In ΔCPM, we have

$$\tan \alpha = \frac{CM}{PM}$$

$$\Rightarrow \quad \tan \alpha = \frac{x}{AB} \qquad [\because PM = AB]$$

$$\Rightarrow \quad AB = x \cot \alpha \qquad ...(i)$$

In $\Delta PMC'$, we have

$$\tan \beta = \frac{C'M}{PM}$$

$$\Rightarrow \quad \tan \beta = \frac{x + 2h}{AB}$$

$$[\because C'M = C'B + BM = x + h + h]$$

$$\Rightarrow \quad AB = (x + 2h) \cot \beta \qquad ...(ii)$$

From equations (i) and (ii), we have

$$x \cot \alpha = (x + 2h) \cot \beta$$

$$\Rightarrow x(\cot \alpha - \cot \beta) = 2h \cot \beta$$

$$\Rightarrow x\left(\frac{1}{\tan\alpha} - \frac{1}{\tan\beta}\right) = \frac{2h}{\tan\beta}$$

$$\Rightarrow x\left(\frac{\tan\beta - \tan\alpha}{\tan\alpha\ \tan\beta}\right) = \frac{2h}{\tan\beta}$$

$$\Rightarrow \quad x = \frac{2h\tan\alpha}{\tan\beta - \tan\alpha}$$

Hence, the height CB of the cloud is given by

$$CB = x + h$$

$$\Rightarrow \quad CB = \frac{2h\tan\alpha}{\tan\beta - \tan\alpha} + h$$

$$\Rightarrow \quad CB = \frac{2h\tan\alpha + h\tan\beta - h\tan\alpha}{\tan\beta - \tan\alpha}$$

$$= \frac{h(\tan\alpha + \tan\beta)}{\tan\beta - \tan\alpha}$$ **Hence Proved.**

Q. 9. The shadow of a flagstaff is three times as long as the shadow of the flagstaff when the Sun rays meet the ground at an angle of 60°. Find the angle between the Sun rays and the ground at the time of longer shadow.

Solution : Let AB be the flagstaff and BC be the length of its shadow when the Sun rays meet the ground at an angle of 60°. Let θ be the angle between the Sun rays and the ground when the length of the shadow of the flagstaff is BD. Let h be the height of the flagstaff.

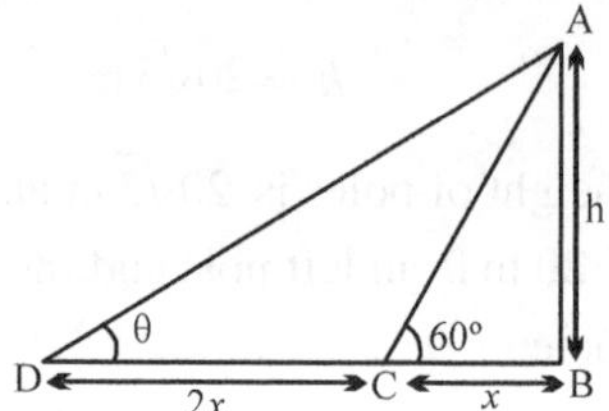

Let $BC = x$

Then, $BD = 3x$ and $CD = 2x$

In ΔACB, we have

$$\tan 60^\circ = \frac{AB}{BC}$$

$$\Rightarrow \quad \sqrt{3} = \frac{h}{x}$$

$$\Rightarrow \quad h = \sqrt{3}x \qquad ...(i)$$

In ΔADB, we have

$$\tan \theta = \frac{AB}{BD}$$

$$\Rightarrow \quad \tan \theta = \frac{h}{3x}$$

$$\Rightarrow \quad \tan \theta = \frac{\sqrt{3}x}{3x} \qquad [\text{From (i)}]$$

$$\Rightarrow \quad \tan \theta = \frac{1}{\sqrt{3}}$$

$$\Rightarrow \quad \tan \theta = \tan 30^\circ$$

$$\Rightarrow \quad \theta = 30^\circ$$ **Ans.**

Q. 10. From an aeroplane vertically above a straight horizontal plane, the angles of depression of two consecutive kilometre stones on the opposite sides of the aeroplane are found to be α and β. Show that the height of the aeroplane is $\frac{\tan\alpha . \tan\beta}{\tan\alpha + \tan\beta}$.

Solution : Let P be the position of plane, A and B be the positions of two stones one kilometre apart. Angles of depression of stones A and B are α and β respectively. Let $PC = h$ km.

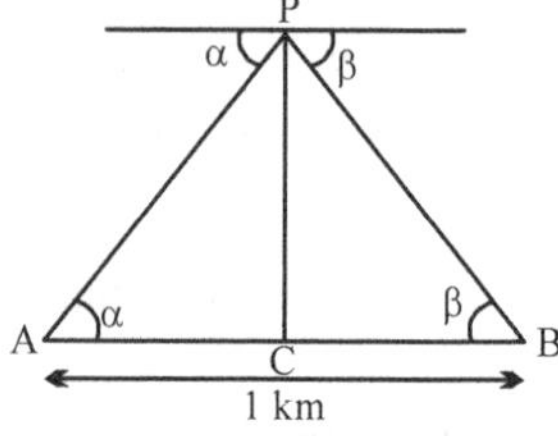

In right-angled triangle ACP, we have

$$\Rightarrow \quad \tan \alpha = \frac{PC}{AC} = \frac{h}{AC},$$

$$\Rightarrow \quad AC = \frac{h}{\tan\alpha} \qquad ...(i)$$

Now in right-angled triangle PCB, we have

$$\tan\beta = \frac{PC}{CB} = \frac{h}{CB}$$

$$\Rightarrow \quad CB = \frac{h}{\tan\beta} \quad ...(ii)$$

Adding equations (i) and (ii), we get

$$AC + CB = \frac{h}{\tan\alpha} + \frac{h}{\tan\beta}$$

$$= h\left(\frac{\tan\beta + \tan\alpha}{\tan\alpha . \tan\beta}\right)$$

As it is given that

$$AC + CB = 1$$

$$\therefore \quad 1 = h\left(\frac{\tan\beta + \tan\alpha}{\tan\alpha . \tan\beta}\right)$$

$$\Rightarrow \quad h = \frac{\tan\alpha . \tan\beta}{\tan\alpha + \tan\beta}$$ **Hence Proved.**

Q. 11. The angle of elevation of a cliff from a fixed point is θ. After going up a distance of k metres towards the top of the cliff at an angle of ϕ, it is found that the angle of elevation is α. Show that the height of the cliff is $\frac{k(\cos\phi - \sin\phi\cot\alpha)}{\cot\theta - \cot\alpha}$ metres.

Solution : Let AB be the cliff of height h m and O be the fixed point such that $\angle AOB = \theta$. Let $\angle AOC = \phi$ and $OC = k$ m. From C, draw CD and CE perpendiculars on AB and OA, respectively. Now, $\angle DCB = \alpha$.

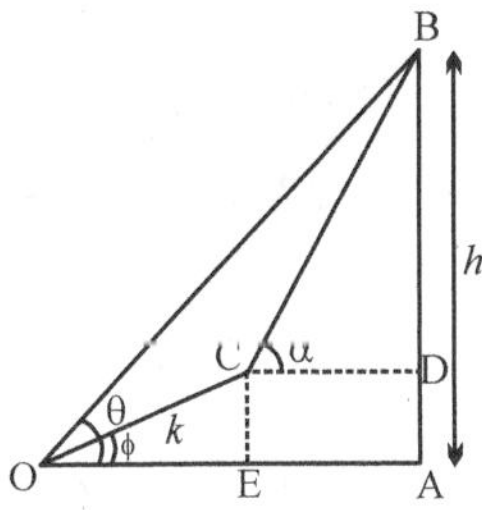

In ΔCOE, we have

$$\sin\phi = \frac{CE}{OC}$$

$$\Rightarrow \quad \sin\phi = \frac{CE}{k}$$

$$\Rightarrow \quad CE = k\sin\phi$$

$$\Rightarrow \quad AD = k\sin\phi \quad [\because CE = AD] \quad ...(i)$$

and

$$\cos\phi = \frac{OE}{OC}$$

$$\Rightarrow \quad \cos\phi = \frac{OE}{k}$$

$$\Rightarrow \quad OE = k\cos\phi \quad ...(ii)$$

In ΔAOB,

$$\tan\theta = \frac{AB}{OA}$$

$$\Rightarrow \quad \tan\theta = \frac{h}{OA}$$

$$\Rightarrow \quad OA = h\cot\theta \quad ...(iii)$$

$$\therefore \quad CD = EA = OA - OE$$

$$= h\cot\theta - k\cos\phi \quad ...(iv)$$

[Using (ii) and (iii)]

and

$$BD = AB - AD = AB - CE$$

$$= h - k\sin\phi \quad ...(v)$$

[Using (i)]

Now, in ΔBCD,

$$\tan\alpha = \frac{BD}{CD}$$

$$\Rightarrow \quad \tan\alpha = \frac{h - k\sin\phi}{h\cot\theta - k\cos\phi}$$

[Using (iv) and (v)]

$$\Rightarrow \quad \frac{1}{\cot\alpha} = \frac{h - k\sin\phi}{h\cot\theta - k\cos\phi}$$

$$\Rightarrow \quad h\cot\alpha - k\sin\phi\cot\alpha = h\cot\theta - k\cos\phi$$

$$\Rightarrow \quad h(\cot\theta - \cot\alpha) = k(\cos\phi - \sin\phi\cot\alpha)$$

$$\Rightarrow \quad h = \frac{k(\cos\phi - \sin\phi\cot\alpha)}{\cot\theta - \cot\alpha} \text{ m}$$

Hence Proved.

Q. 12. The angle of elevation of the top of a tower from a point A due south of the tower is α and from B due east of the tower is β. If $AB = d$, show that the height of the tower is $\frac{d}{\sqrt{\cot^2\alpha + \cot^2\beta}}$

Solution : Let OC be the tower of height h, A and B be two points due south and east of tower

Such that

$$\angle CAO = \alpha$$

and

$$\angle OBC = \beta$$

In ΔAOC, $\angle O = 90°$

$$\therefore \quad \cot\alpha = \frac{OA}{OC} = \frac{OA}{h}$$

$$\Rightarrow \quad OA = h\cot\alpha \quad ...(i)$$

In ΔBOC, $\angle O = 90°$

$$\therefore \quad \cot\beta = \frac{OB}{OC}\frac{OB}{h}$$

$$\Rightarrow \quad OB = h\cot\beta \quad ...(ii)$$

Now, in ΔAOB

$$AB^2 = OA^2 + OB^2$$

$$d^2 = h^2\cot^2\alpha + h^2\cot^2\beta$$

$$\Rightarrow \quad d = h\sqrt{\cot^2\alpha + \cot^2\beta}$$

$$\Rightarrow \quad h = \frac{d}{\sqrt{\cot^2\alpha + \cot^2\beta}}.$$

Hence Proved.

Q. 13. The elevation of a tower at a station A due north of it is α and at a station B due west of A is β. Prove that the height of the tower is $\dfrac{AB\sin\alpha\sin\beta}{\sqrt{\sin^2\alpha - \sin^2\beta}}$.

Solution : Let OP be the tower of height h and let A be a point due north of the tower OP and B be the point due west of A, such that $\angle OAP = \alpha$ and $\angle OBP = \beta$.

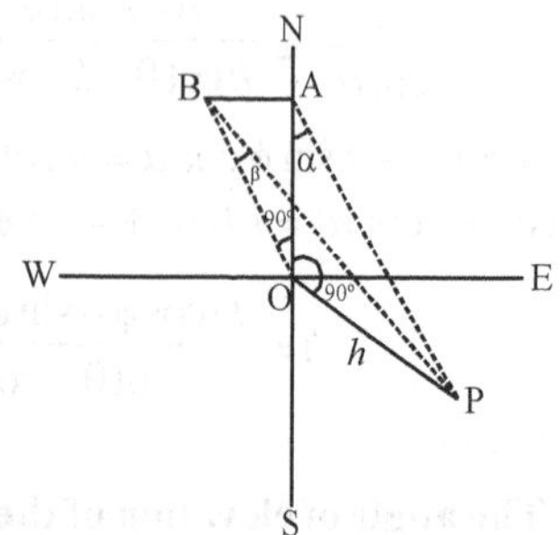

In right-angled triangle OAP and OBP, we have

$$\tan\alpha = \frac{h}{OA} \text{ and } \tan\beta = \frac{h}{OB}$$

$$\Rightarrow \quad OA = h\cot\alpha \text{ and } OB = h\cot\beta$$

In ΔOAB, we have

$$OB^2 = OA^2 + AB^2$$

$$\Rightarrow \quad AB^2 = OB^2 - OA^2$$

$$\Rightarrow \quad AB^2 = h^2\cot^2\beta - h^2\cot^2\alpha$$

$$\Rightarrow \quad AB^2 = h^2[\cot^2\beta - \cot^2\alpha]$$

$$\Rightarrow \quad AB^2 = h^2[(\operatorname{cosec}^2\beta - 1) - (\operatorname{cosec}^2\alpha - 1)]$$

$$\Rightarrow \quad AB^2 = h^2(\operatorname{cosec}^2\beta - \operatorname{cosec}^2\alpha)$$

$$\Rightarrow \quad AB^2 = h^2\left(\frac{\sin^2\alpha - \sin^2\beta}{\sin^2\alpha\sin^2\beta}\right)$$

$$\Rightarrow \quad h = \frac{AB\sin\alpha\sin\beta}{\sqrt{\sin^2\alpha - \sin^2\beta}}.$$

Hence Proved.

Q. 14. At the foot of a mountain the elevation of its summit is 45°, after ascending 1000 m towards the mountain up a slope of 30° inclination, the elevation is found to be 60°. Find the height of the mountain.

Solution : Let F be the foot and S be the summit of the mountain FOS. Then, $\angle OFS = 45°$ and therefore, $\angle OSF = 45°$. Consequently, $OF = OS = h$ km (say).

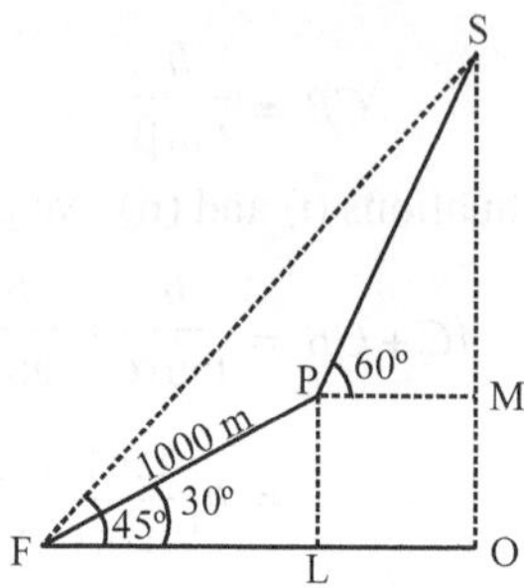

Let $FP = 1000$ m $= 1$ km be the slope so that $\angle OFP = 30°$.
Draw $PM \perp OS$ and $PL \perp OF$.
Join PS. It is given that $\angle MPS = 60°$.
In ΔFPL, we have

$$\sin 30° = \frac{PL}{PF}$$

$$\Rightarrow \quad PL = PF\sin 30° = 1\cdot\frac{1}{2}$$

$$= \frac{1}{2}\text{ km}$$

$$\therefore \quad OM = PL = \frac{1}{2}\text{ km}$$

$$\therefore \quad MS = OS - OM$$

$$= \left(h - \frac{1}{2}\right)\text{ km} \quad ...(i)$$

Also,

$$\cos 30° = \frac{FL}{PF}$$

$$\Rightarrow \quad FL = PF\cos 30°$$

$$= \left(1\times\frac{\sqrt{3}}{2}\right)\text{ km}$$

Now,

$$h = OS = OF = OL + LF$$

$$\Rightarrow \quad h = OL + \frac{\sqrt{3}}{2}$$

$$\Rightarrow \quad OL = \left(h - \frac{\sqrt{3}}{2}\right)\text{ km}$$

$$PM = \left(h - \frac{\sqrt{3}}{2}\right)\text{ km} \quad ...(ii)$$

$\therefore$ in ΔSPM, we have

$$\tan 60° = \frac{SM}{PM}$$

$$\Rightarrow \quad SM = PM\cdot\tan 60°$$

$$\Rightarrow \quad \left(h - \frac{1}{2}\right) = \left(h - \frac{\sqrt{3}}{2}\right)\sqrt{3}$$

[Using (i) and (ii)]

$$\Rightarrow \quad h - \frac{1}{2} = h\sqrt{3} - \frac{3}{2}$$

$$\Rightarrow \quad \sqrt{3}h - h = \frac{3}{2} - \frac{1}{2}$$

$$\Rightarrow \quad h(\sqrt{3} - 1) = 1$$

$$\Rightarrow \quad h = \frac{1}{\sqrt{3} - 1}$$

$$\Rightarrow \quad h = \frac{\sqrt{3}+1}{(\sqrt{3}-1)(\sqrt{3}+1)}$$

$$= \frac{\sqrt{3}+1}{2} = \frac{2.732}{2}$$

$$= 1.366 \text{ km}$$

Hence, the height of the mountain is 1.366 km.

Ans.

Q. 15. At a point on level ground, the angle of elevation of a vertical tower is found to be such that its tangent is 5/12. On walking 192 metres towards the tower, the tangent of the angle of elevation is 3/4. Find the height of the tower.

Solution : Let AB be the tower and let the angle of elevation of its top at C be α. Let D be the point at a distance of 192 metres from C such that the angle of elevation of the top of the tower at D be β.

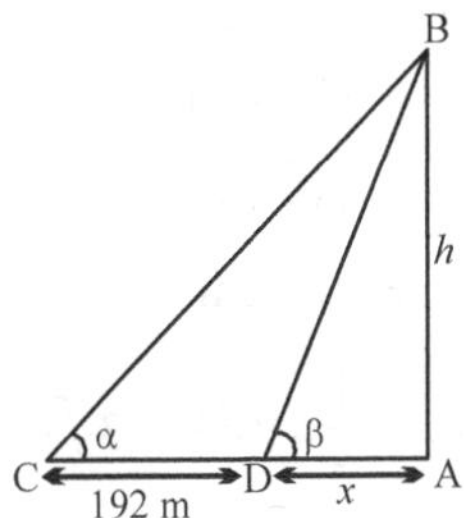

Let h be the height of the tower and $AD = x$.

It is given that

$$\tan\alpha = \frac{5}{12} \text{ and } \tan\beta = \frac{3}{4}$$

In ΔCAB, we have

$$\tan\alpha = \frac{AB}{AC}$$

$$\Rightarrow \quad \frac{5}{12} = \frac{h}{192 + x} \qquad ...(i)$$

In ΔDAB, we have

$$\tan\beta = \frac{AB}{AD}$$

$$\Rightarrow \quad \tan\beta = \frac{h}{x}$$

$$\Rightarrow \quad \frac{3}{4} = \frac{h}{x} \qquad ...(ii)$$

We have to find h. This means that we have to eliminate x from equations (i) and (ii).

From equation (ii), we have

$$3x = 4h \quad \Rightarrow x = \frac{4h}{3}$$

Substituting this value of x in equation (i), we get

$$\frac{5}{12} = \frac{h}{192 + 4h/3}$$

$$5\left(192 + \frac{4h}{3}\right) = 12h$$

$$5(576 + 4h) = 36h$$

$$2880 + 20h = 36h$$

$$16h = 2880$$

$$h = \frac{2880}{16} = 180$$

Hence, the height of the tower is 180 metres. **Ans.**

Chapter 12. Areas related to Circles

Q. 1. A circle of radius 'a' is divided into 6 equal sectors. An equilateral triangle is drawn on the chord of each sector to lie outside the circle. Find the area of the resulting figure.

Solution : Each chord of the circle with radius a, subtend an angle of 60° at the centre and make an equilateral triangle with the centre of circle as vertex. So, each side = a.

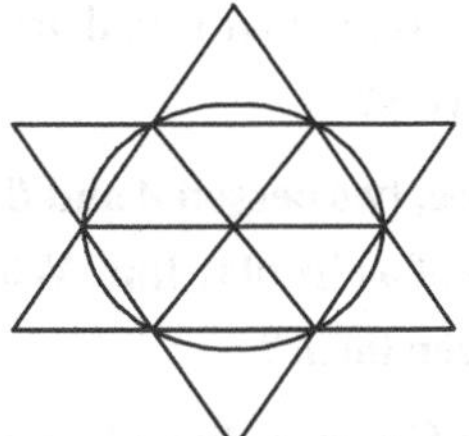

Since, on each chord an equilateral triangle is made, there are total 12 such triangles.

Area of one equilateral triangle = $\frac{\sqrt{3}}{4}a^2$.

Total area = $\frac{\sqrt{3}}{4}a^2 \times 12 = 3\sqrt{3}\,a^2$. **Ans.**

Q. 2. In the given figure, O is the centre of the circle with $AC = 24$ cm, $AB = 7$ cm and $\angle BOD = 90°$. Find the area of the shaded region. (Take $\pi = 3.14$)

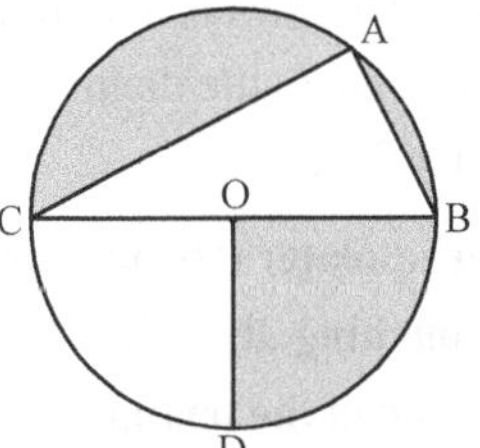

Solution : Here, BOC is a diameter of the circle.

$\therefore \angle BAC = 90°$ [Angle in a semi-circle is right angle]

In ΔBAC, by using Pythagoras theorem, we get

$$BC^2 = AB^2 + AC^2$$

$$BC^2 = (7)^2 + (24)^2$$

$\Rightarrow \quad BC = 25$ cm

$\therefore$ Radius of circle with centre $O = \frac{25}{2} = 12.5$ cm

Now, area of shaded region
= Area of circle – Area of quadrant COD – Area of ΔBAC

$$= \pi r^2 - \frac{\pi r^2}{4} - \frac{1}{2} \times AB \times AC$$

$$= \left(1 - \frac{1}{4}\right)\pi r^2 - \frac{1}{2} \times AC \times AC$$

$$= \frac{3}{4}\pi r^2 - \frac{1}{2} \times AB \times AC$$

$$= \frac{3}{4} \times 3.14 \times (12.5)^2 - \frac{1}{2} \times 7 \times 24$$

$= 367.97 - 84$
$= 283.97 \text{ cm}^2$

Hence, area of shaded region is 283.97 cm^2. **Ans.**

Q. 3. There is a circular park in front of Delhi Public School whose radius is 10 m. Around the park, one circular line is drawn, which is having radius 11 m.

(i) A person takes four round of a circular line. What distance he covers?

(ii) Suppose, two person *A* and *B* are standing on the opposite sides of a circular line. What is the maximum distance between them ?

Solution : Given, radius of circular park $r_1 = 10$ m and radius of circular line, $r_2 = 11$ m

(i) Now, perimeter of circular line $= 2\pi r_2$

$= 2 \times 3.14 \times 11 = 69.08$ m

As the person takes four round, then total distance covered by him

= $4 \times$ Perimeter of one circular line

$= 4 \times 69.08 = 276.32$ m **Ans.**

(ii) Since, the two persons are standing on the opposite of a circular line the maximum distance between A and B will be equal to the diameter of the circular park i.e., $2r = 2 \times 11 = 22$ m.

Q. 4. In an equilateral triangle of side 18 cm, a circle is inscribed touching its sides. Find the area of the remaining portion of the triangle.

Solution : Let ABC be an equilateral triangle of side 18 cm

$\Rightarrow \quad AB = BC = CA = 18$ cm

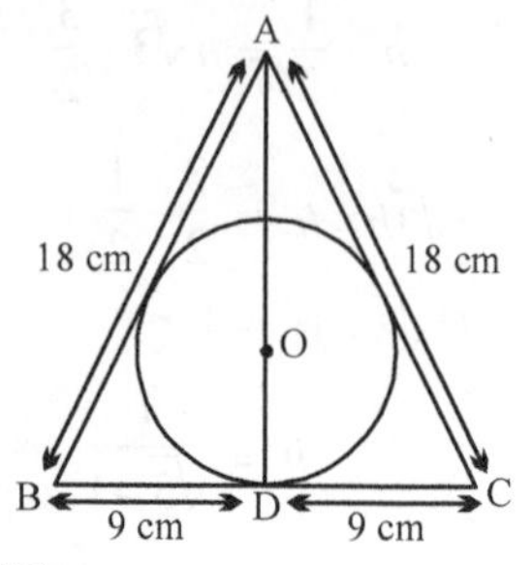

Draw $AD \perp BC$

$\Delta ADB \cong \Delta ADC$

[By R.H.S. criterion of congruence]

$\Rightarrow \quad BD = CD$ [cpct]

$\Rightarrow \quad BD = CD = \frac{1}{2} BC$

$= \frac{1}{2} \times 18 \text{ cm} = 9$ cm.

Since, the centre of the inscribed circle coincides with the centroid of the equilateral triangle.

$\therefore$ Radius of incircle, $OD = \frac{1}{3} AD$ [$\because AO : OD = 2 : 1$]

Now, in right ΔABD

$AB^2 = AD^2 + BD^2$ [By Pythagoras theorem]

$\Rightarrow \quad (18)^2 = AD^2 + 9^2$

$\Rightarrow \quad AD^2 = (18)^2 - 9^2$

$= 324 - 81 = 243$

$\Rightarrow \quad AD = +\sqrt{81 \times 3}$

$= 9\sqrt{3}$ cm

$\Rightarrow \quad OD = \frac{1}{3} \times 9\sqrt{3} = 3\sqrt{3}$

Area of incircle $= \pi r^2 = \frac{22}{7} \times (3\sqrt{3}) \text{ cm}^2$

$= \frac{22}{7} \times 27 \text{ cm}^2$

$= 84.86 \text{ cm}^2$

Area of equilateral ΔABC

$= \frac{\sqrt{3}}{4} (\text{side})^2$

$= \frac{\sqrt{3}}{4} \times (18)^2$

$= \sqrt{3} \times 81 \text{ cm}^2$

$= 1.732 \times 81$

$= 140.29 \text{ cm}^2$

Hence, area of the remaining portion of the triangle

$= (140.29 - 84.86) \text{ cm}^2$

$= 55.43 \text{ cm}^2$. **Ans.**

Q. 5. Two circles touch internally. The sum of their areas is 116 π cm² and distance between their centres is 6 cm. Find the radii of the circles.

Solution : Let R and r be the radii of the circles

Then, according to question,

$\Rightarrow \quad \pi R^2 + \pi r^2 = 116\pi$

$\Rightarrow \quad R^2 + r^2 = 116 \quad ...(i)$

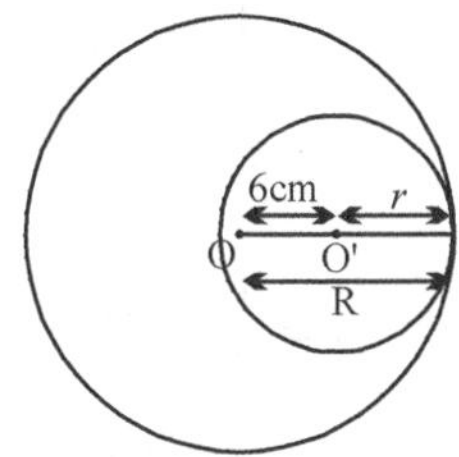

Also, distance between the centres = 6 cm

$\Rightarrow \quad OO' = 6$ cm

$\Rightarrow \quad R - r = 6 \quad ...(ii)$

Now, $(R + r)^2 + (R - r)^2 = 2(R^2 + r^2)$

Using the equations (i) and (ii), we get

$(R + r)^2 + (6)^2 = 2 \times 116$

$\Rightarrow \quad (R + r)^2 = 2 \times 116 - 36$

$= 196$

$\Rightarrow \quad R + r = 14 \quad ...(iii)$

Solving equations (ii) and (iii), we get $R = 10$ and $r = 4$.

Hence, radii of the given circles are 10 cm and 4 cm respectively. **Ans.**

Q. 6. A bicycle wheel makes 5000 revolutions in moving 11 km. Find the diameter of the wheel.

Solution : Let the radius of the wheel be r cm.

Distance covered by the wheel in one revolution

$$= \frac{\text{Distance moved}}{\text{Number of revolutions}}$$

$$= \frac{11}{5000} \text{ km}$$

$$= \frac{11}{5000} \times 1000 \times 100 \text{ cm}$$

$= 220$ cm

$\therefore$ Circumference of the wheel = 220 cm

$\Rightarrow \quad 2\pi r = 220$ cm

$\Rightarrow 2 \times \frac{22}{7} \times r = 220$

$\Rightarrow \quad r = \frac{220 \times 7}{2 \times 22}$

$\Rightarrow \quad r = 35$ cm

$\therefore$ Diameter $= 2r$ cm

$= (2 \times 35)$ cm

$= 70$ cm.

Hence, the diameter of the wheel is 70 cm. **Ans.**

Q. 7. Find the area of the shaded design of Fig., where *ABCD* is a square of side 10 cm and semicircles are drawn with each side of the square as diameter.

(Use π = 3.14)

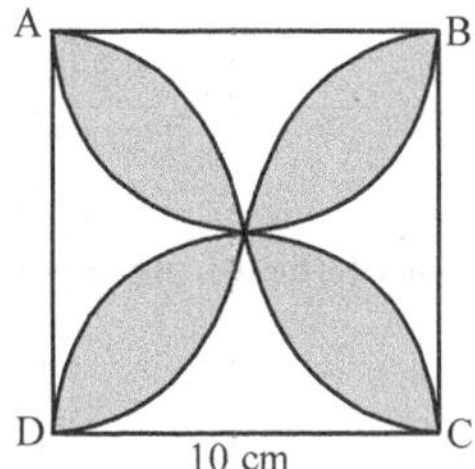

Solution : Let us mark the four unshaded regions as I, II, III, and IV, (see fig.)

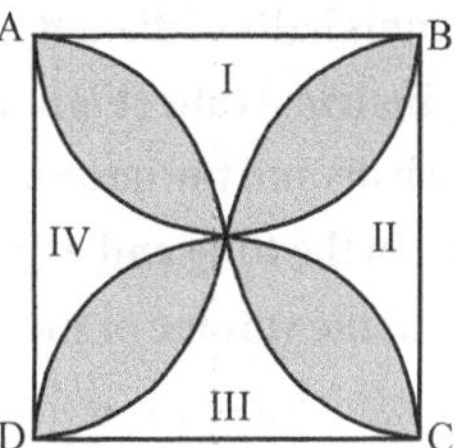

Area of I + Area of III

= Area of $ABCD$ – Areas of two semicircles of radius 5 cm each

$$= \left(10 \times 10 - 2 \times \frac{1}{2} \times \pi \times 5 \times 5\right) \text{cm}^2$$

$= (100 - 3.14 \times 25) \text{ cm}^2$

$= (100 - 78.5) \text{ cm}^2$

$= 21.5 \text{ cm}^2$

Similarly, Area of II + Area of IV = 21.5 cm²

So, Area of the shaded design

= Area of $ABCD$ – Area of (I + II + III + IV)

$= (100 - 2 \times 21.5) \text{ cm}^2$

$= (100 - 43) \text{ cm}^2$

$= 57 \text{ cm}^2$. **Ans.**

Q. 8. A copper wire, when bent in the form of a square, encloses an area of 484 cm². If the same wire is bent in the form of a circle, find the area enclosed by it.

Solution : We have, area of the square

$= a^2 = 484 \text{ cm}^2$

$\therefore \quad$ Side of the square $= \sqrt{484}$ cm = 22 cm

So, Perimeter of the square = 4 × side

$= (4 \times 22)$ cm

$= 88$ cm

Let r be the radius of the circle.

Then, according to question,

Circumference of the circle = Perimeter of the square

$\Rightarrow \qquad 2\pi r = 88$

$\Rightarrow \qquad 2 \times \frac{22}{7} \times r = 88$

$\Rightarrow \qquad r = \frac{88 \times 7}{2 \times 22}$

$\Rightarrow \qquad r = 14$ cm

$\therefore$ Area of the circle $= \pi r^2$

$= \left\{\frac{22}{7} \times (14)^2\right\}$ cm^2

$= 616$ cm^2. **Ans.**

Chapter 13. Surface Areas and Volumes

Q. 1. Six Tennis balls of diameter 62 mm are placed in cylindrical tube (Fig.). Find the volume of the six balls and the internal volume of unfilled space in the tube and express this as a percentage of the volume of the tube.

Solution : Diameter of the tennis balls = 62 mm.

$\because$ Radius of the balls and tube is half the diameter

$\therefore \qquad r = \frac{1}{2} \times 62 = 31$ mm

Volume of one ball $= \frac{4}{3}\pi r^3$

$= \frac{4}{3} \times \frac{22}{7} \times 31 \times 31 \times 31 = 124{,}838.476$ mm^3 = 124.838 cm^3

$\therefore$ Volume of 6 tennis balls $= 6 \times 124.838$

$= 749.028$ cm^3

Height of tube (h) is 6 times the diameter of the ball

$\therefore \qquad h = 6 \times 62 = 372$ mm

$\therefore$ Volume of the tube

$= \pi r^2 h = \frac{22}{7} \times 31 \times 31 \times 372$

$= 1123{,}546.29$ mm^3

$= 1123.54$ cm^3

$\therefore$ Volume of unfilled space (shaded area) in the tube

$= 1123.54 - 749.028 = 374.512$ cm^3

Space as a percentage of the volume of the tube

$= \frac{374.366}{1123.54} \times 100$

$= 33.33\%$

Hence, volume of unfilled space is 33.3% of tube.

Ans.

Q. 2. Find the volume and total surface area of a tumbler which is in the form of a frustum of a cone, if the diameter of the ends are 6.50 cm and 3.50 cm and the perpendicular height of the tumbler is 7.80 cm.

Solution : We have,

$R = \frac{1}{2} \times 6.50$

$= 3.25$ cm

$r = \frac{1}{2} \times 3.50$

$= 1.75$ cm

$h = 7.80$ cm.

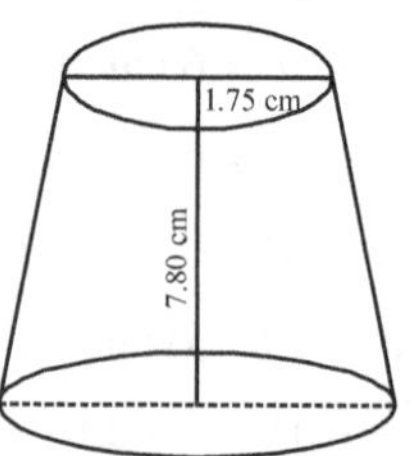

Volume of tumbler which is in the form of a frustum of a cone

$= \frac{\pi h}{3}[R^2 + r^2 + Rr]$

$= \frac{1}{3} \times \frac{22}{7} \times 7.80\,[(3.25)^2 + (1.75)^2 + 3.25 \times 1.75]$

$= \frac{1}{3} \times \frac{22}{7} \times 7.8\,[10.5625 + 3.0625 + 5.6875]$

$= \frac{1}{3} \times \frac{22}{7} \times 7.8\,(19.3125)$

$= \frac{22}{7} \times 2.6 \times 19.3125$

$= \frac{1104.675}{7}$ **Ans.**

$= 157.8$ cm^3

Total surface area of tumbler

$= \pi l(R + r) + \pi R^2 + \pi r^2$

Now $\qquad l^2 = h^2 + (R - r)^2$

$\Rightarrow \qquad l = \sqrt{(7.8)^2 + (3.25 - 1.75)^2}$

$= \sqrt{(7.8)^2 + (1.5)^2}$

$= \sqrt{60.84 + 2.25}$

$\Rightarrow \qquad l = \sqrt{63.09} = 7.94$ cm

$\therefore$ Total surface area of tumbler which is in the form of a frustum of cone

$$= \frac{22}{7} \times 7.94\,(3.25 + 1.75) + \frac{22}{7}(3.25)^2 + \frac{22}{7}(1.75)^2$$

$$= \frac{22}{7} \times 7.94 \times 5 + \frac{22}{7} \times 10.5625 + \frac{22}{7} \times 3.0625$$

$$= \frac{22}{7}(7.94 \times 5 + 10.5625 + 3.0625)$$

$$= \frac{22}{7}(39.7 + 10.5625 + 3.0625)$$

$$= \frac{22}{7} \times 53.325$$

$= 167.6\ \text{cm}^2$ **Ans.**

Q. 3. A sector of a circle of radius 6 cm has an angle of 120°. It is rolled up so that the two bounding radii are joined together to form a cone. Find volume of cone and T.S.A. of the cone.

Solution : On rolling up the sector, we get a cone whose slant height = radius of sector = 6 cm.

Circumference of the base of cone = Length of sector

$$\Rightarrow \quad 2\pi r = \frac{\theta}{360°}\, 2\pi r = \frac{120°}{360} \times 2 \times \pi \times 6$$

$$\Rightarrow \quad 2\pi r = 4\pi \text{ cm.}$$

$$\Rightarrow \quad r = 2 \text{ cm}$$

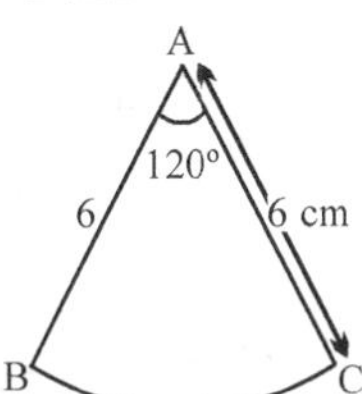

T.S.A. of cone = C.S.A. of cone + Area of base of cone

$$= \text{Area of sector} + \pi(r)^2$$

$$= \frac{\theta}{360°}\pi r^2 + \pi \times (2)^2$$

$$= \frac{120°}{360°} \times \pi \times (6)^2 + 4\pi$$

$$= 12\pi + 4\pi$$

$$= 16\pi \text{ cm}^2$$

Height of cone,

$$h = \sqrt{l^2 - r^2}$$

$$= \sqrt{6^2 - 2^2}$$

$$= \sqrt{36 - 4} = \sqrt{32}$$

$$= 4\sqrt{2} \text{ cm.}$$

Volume of cone $= \frac{1}{3}\pi r^2 h$

$$= \frac{1}{3}\pi(2)^2 \times (4\sqrt{2})$$

$$= \frac{16\pi\sqrt{2}}{3} \text{ cm}^3$$ **Ans.**

Q. 4. A well of diameter 3 m is dug 14 m deep. The earth taken out of it has been spread evenly all around it in the shape of a circular ring of width 4 m to form an embankment. Find the height of the embankment.

Solution : Here,

Diameter of the well = 3m

Radius of the well $= \frac{3}{2} = 1.5$ m

Depth of the well = 14 m

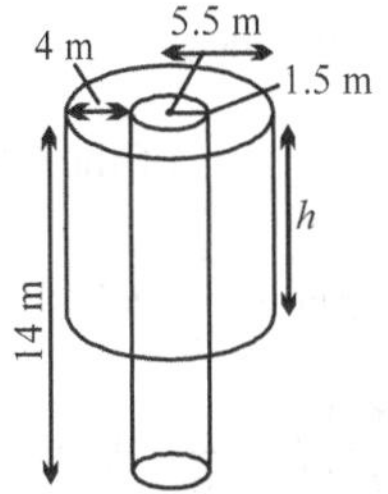

Width of the embankment = 4 m

∴ Radius of the embankment = 1.5 + 4 = 5.5 m

Let h be the height of the embankment.

∴ Volume of the embankment = Volume of the well (cylinder)

$$\Rightarrow \quad \pi[(5.5)^2 - (1.5)^2] \times h = \pi(1.5)^2 \times 14$$

$$\Rightarrow \quad (30.25 - 2.25) \times h = 2.25 \times 14$$

$$\Rightarrow \quad h = \frac{2.25 \times 14}{28}$$

$= 1.125$ m. **Ans.**

Q. 5. A hollow cone is cut by a plane parallel to the base and upper portion is removed. If the curved surface of the remainder is $\frac{8}{9}$ of the curved surface of the whole cone, find the ratio of the line segments into which the cone's altitude is divided by the plane.

Solution : In fig., let the smaller cone APQ has been cut off through the plane $PQ \parallel BC$. Let r and R be the radii of the smaller and larger cone and l and L be their slant heights, respectivley.

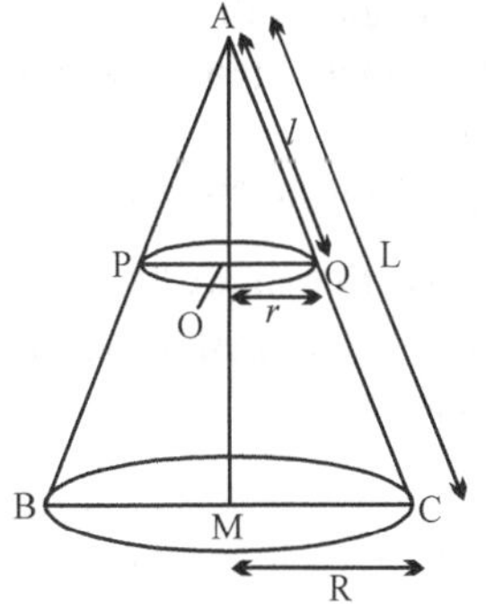

Here, in the adjoining figure

$$OQ = r,\ MC = R,\ AQ = l,\ AC = L.$$

Now, $\Delta AOQ \sim \Delta AMC$

$$\Rightarrow \quad \frac{OQ}{MC} = \frac{AQ}{AC} \quad \text{[cpct]}$$

$$\Rightarrow \quad \frac{r}{R} = \frac{l}{L}$$

$$\Rightarrow \quad r = \frac{Rl}{L} \quad \text{...(i)}$$

Since, curved surface area of the remainder = $\frac{8}{9}$ of the curved surface area of the whole cone, therefore, we get,

CSA of smaller cone

$$= \frac{1}{9} \text{ of the CSA of the whole cone}$$

$$\therefore \quad \pi r l = \frac{1}{9}\pi R L$$

$$\Rightarrow \quad \pi\left(\frac{Rl}{L}\right) l = \frac{1}{9}(\pi R L) \quad \text{[Using (i)]}$$

$$\Rightarrow \quad l^2 = \frac{L^2}{9}$$

$$\Rightarrow \quad l = \frac{L}{3}$$

$$\Rightarrow \quad \frac{l}{L} = \frac{1}{3}$$

Now, again in similar triangles AOQ and AMC, we have

$$\frac{AO}{AM} = \frac{AQ}{AC}$$

$$\Rightarrow \quad \frac{AO}{AM} = \frac{l}{L} = \frac{1}{3}$$

$$\Rightarrow \quad AO = \frac{AM}{3}$$

$$\Rightarrow \quad OM = AM - OA$$

$$= AM - \frac{AM}{3} = \frac{2}{3}AM$$

$$\therefore \quad \frac{AO}{OM} = \frac{AM/3}{2AM/3} = \frac{1}{2}.$$

Hence, the required ratio of their heights = 1 : 2.

Ans.

Q. 6. The height of the cone is 30 cm. A small cone is cut off at the top by a plane parallel to the base. If its volume be $\frac{1}{27}$ of the given cone, at what height above the base is the section made ?★

Solution : Let the small cone APQ be cut off at the top by the plane $PQ \parallel BC$.

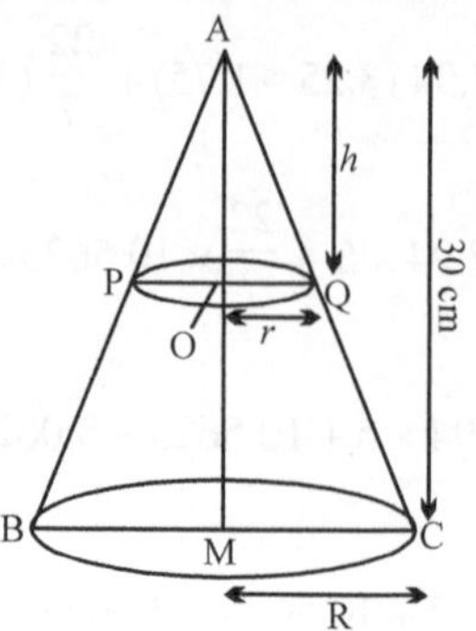

Let r and h be the radius and height of the smaller cone, respectively and also let the radius of larger cone = R.

Now, $\Delta AOQ \sim \Delta AMC$

$$\Rightarrow \quad \frac{AO}{AM} = \frac{OQ}{MC} \quad \text{[cpct]}$$

$$\Rightarrow \quad \frac{h}{30} = \frac{r}{R}$$

$$\therefore \quad r = \frac{hR}{30} \quad \text{...(i)}$$

Since, it is given that volume of the smaller cone

$$= \frac{1}{27}(\text{Volume of larger cone})$$

$$\frac{1}{3}\pi r^2 h = \frac{1}{27}\left(\frac{1}{3}\pi R^2 .30\right)$$

$$\Rightarrow \quad \frac{\pi}{3}\left(\frac{hR}{30}\right)^2 h = \frac{1}{27} \times \frac{1}{3}\pi R^2 . 30 \quad \text{[Using (i)]}$$

$$\Rightarrow \quad h^3 = \frac{30 \times 30 \times 30}{27} = 1000$$

$$\therefore \quad h = 10 \text{ cm.}$$

Hence, the smaller cone has been cut off at a height of (30 – 10) cm = 20 cm from the base. **Ans.**

Q. 7. A vessel is in the form of an inverted cone. Its height is 8 cm and the radius of its top, which is open, is 5 cm. It is filled with water up to the brim. When lead shots, each of which is a sphere of radius 0.5 cm are dropped into the vessel, one-fourth of the water flows out. Find the number of lead shots dropped in the vessel.

Solution : We have,

Height of conical vessel = h = 8 cm

and its radius = r = 5 cm

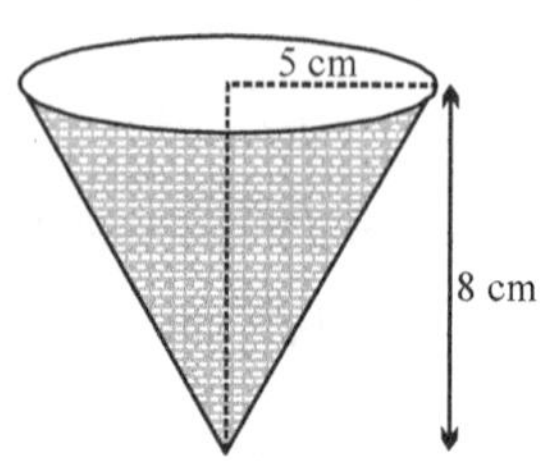

★ are board exam questions from previous years

Now,

Volume of cone = Volume of water in the cone

$$= \frac{1}{3}\pi r^2 h$$

$$= \frac{1}{3} \times \frac{22}{7} \times 5 \times 5 \times 8$$

$$= \frac{4,400}{21} \text{ cm}^3$$

Now, volume of water flows out

= Volume of lead shots

$$= \frac{1}{4} \times \text{Volume of water in the cone}$$

$$= \frac{1}{4} \times \frac{4,400}{21}$$

$$= \frac{1,100}{21} \text{ cm}^3$$

Now, radius of the lead shots $R = 0.5$ cm

$$= \frac{5}{10} \text{ cm} = \frac{1}{2} \text{ cm}$$

Volume of one spherical lead shot

$$= \frac{4}{3}\pi R^3$$

$$= \frac{4}{3} \times \frac{22}{7} \times \frac{1}{2} \times \frac{1}{2} \times \frac{1}{2}$$

$$= \frac{11}{21} \text{ cm}^3$$

$\therefore$ Number of lead shots dropped in the vessel

$$= \frac{\text{Volume of water flows out}}{\text{Volume of one lead shot}}$$

$$= \frac{\frac{1,100}{21}}{\frac{11}{21}}$$

$$= \frac{1,100}{21} \times \frac{21}{11} = 100. \quad \textbf{Ans.}$$

Q. 8. A right triangle with side 3 cm and 4 cm is revolved around its hypotenuse. Find the volume of double cone thus generated. (Use $\pi = 3.14$)

Solution : In the given Fig., ΔPQR is a right triangle, where $PQ = 3$ cm and $PR = 4$ cm

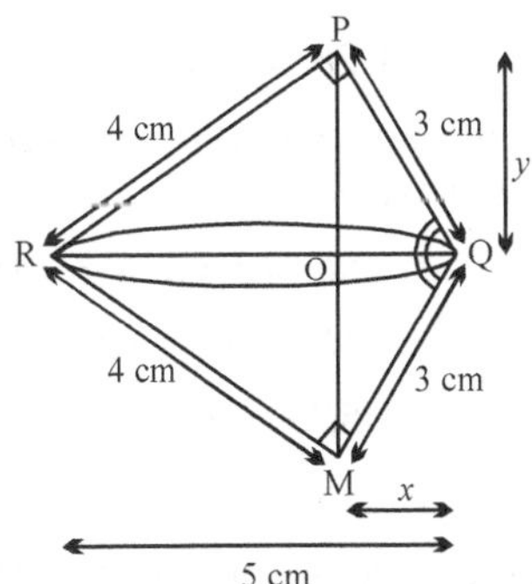

Now in ΔPQR

$$QR^2 = PQ^2 + PR^2$$

$$= 3^2 + 4^2 = 9 + 16 = 25$$

$$\Rightarrow \quad QR = 5 \text{ cm}$$

Let $OQ = x \Rightarrow OR = 5 - x$ and $OP = y$

Now, in right angled-triangle POQ, we have

$$PQ^2 = OQ^2 + OP^2$$

$$\Rightarrow \quad (3)^2 = x^2 + y^2$$

$$\Rightarrow \quad y^2 = 9 - x^2 \qquad ...(i)$$

Also from right angled triangle POR, we have

$$OP^2 + OR^2 = PR^2$$

$$\Rightarrow \quad y^2 + (5 - x)^2 = (4)^2$$

$$\Rightarrow \quad y^2 = 16 - (5 - x)^2 \qquad ...(ii)$$

From equation (i) and (ii), we get

$$9 - x^2 = 16 - (5 - x)^2$$

$$\Rightarrow \quad 9 - x^2 = 16 - (25 + x^2 - 10x)$$

$$\text{or} \quad 9 - x^2 = -9 - x^2 + 10x$$

$$\Rightarrow \quad 10x = 18$$

$$\Rightarrow \quad x = \frac{9}{5}$$

$$\because \quad OR = 5 - x = 5 - \frac{9}{5} = \frac{16}{5}$$

Now putting $x = \frac{9}{5}$ in (i), we get

$$y^2 = 9 - \left(\frac{9}{5}\right)^2$$

$$= 9 - \frac{81}{25} = \frac{144}{25}$$

$$\Rightarrow \quad y = \frac{12}{5}.$$

$$\therefore \quad OP = y = \frac{12}{5}$$

Now, for the cone PQM

$$\text{Radius, } OP = \frac{12}{5} \text{ cm}$$

$$\text{Height, } OQ = \frac{9}{5} \text{ cm}$$

$$\therefore \quad \text{Volume} = \frac{1}{3}\pi r^2 h$$

$$= \frac{1}{3}\pi\left(\frac{12}{5}\right)^2 \times \frac{9}{5}$$

$$= \frac{432\pi}{125} \text{ cm}^3$$

Also for the cone *PRM*, we have

$$\text{Radius, } OP = \frac{12}{5} \text{ cm,}$$

$$\text{Height, } OR = \frac{16}{5} \text{ cm}$$

$$\therefore \quad \text{Volume} = \frac{1}{3}\pi\left(\frac{12}{5}\right)^2 \times \frac{16}{5} \text{ cm}^3$$

$$= \frac{768\pi}{125} \text{ cm}^3$$

Hence, total volume, *i.e.,* volume of the double cone

$$= \left(\frac{432\pi}{125} + \frac{768\pi}{125}\right) \text{ cm}^3$$

$$= \frac{1200\pi}{125} \text{ cm}^3$$

$$= 9.6 \times 3.14 \text{ cm}^3$$

$$= 30.144 \text{ cm}^3.$$

Ans.

Q. 9. A right cylindrical container of radius 6 cm and height 15 cm is full of ice-cream, which has to be distributed to 10 children in equal cones having hemi-spherical shape on the top. If the height of the conical portion is four times its base radius, find the radius of the ice-cream cone.★

Solution : Let R and H be the radius and height of cylinder.

Given, R = 6 cm, H = 15 cm.

Volume of ice-cream in the cylinder = $\pi R^2 H$

$$= \pi \times 36 \times 15$$

$$= 540\,\pi \text{ cm}^3$$

Let the radius of cone be r cm

Height of the cone (h) = $4r$

Radius of hemispherical portion = r cm.

$\therefore$ Volume of ice-cream in cone

= Volume of cone + Volume of hemisphere

$$= \frac{1}{3}\pi r^2 h + \frac{2}{3}\pi r^3$$

r

r

h = 4r

$$= \frac{1}{3}\pi r^2 (h + 2r)$$

$$= \frac{1}{3}\pi r^2 (4r + 2r) \quad (\because h = 4r)$$

★ are board exam questions from previous years

$$= \frac{1}{3} \times \pi r^2 \times 6r$$

$$= 2\pi r^3$$

Number of ice cream cones distributed to the children = 10

$\Rightarrow$ 10 × Volume of ice-cream in each cone = Volume of ice-cream in cylindrical container

$$\Rightarrow \quad 10 \times 2\pi r^3 = 540\pi$$

$$\Rightarrow \quad 20r^3 = 540$$

$$\Rightarrow \quad r^3 = 27$$

$$r = 3$$

Thus, the radius of the ice-cream cone is 3 cm.

Ans.

Chapter 14. Statistics

Q. 1. The mean of $x_1, x_2, x_3 .... x_n$ is $\bar{x}$. If $(a - 2b)$ is added to each of the observations, show that the sum of the new observations = $\bar{x} + (a - 2b)$

Solution : Given, $\frac{x_1 + x_2 + x_3 + ... + x_n}{n} = \bar{x}$

$$\Rightarrow \quad x_1 + x_2 + x_3 + ... + x_n = n\bar{x} \quad ...(i)$$

Now, the sum of the new 'n' observations

$$= x_1 + (a - 2b) + x_2 + (a - 2b) + ... x_n + (a - 2b)$$

$$= x_1 + x_2 + ... + x_n + n(a - 2b)$$

$$= n\bar{x} + n(a - 2b) = n\{\bar{x} + (a - 2b)\} \quad \text{[From (i)]}$$

$$\text{New mean} = \frac{n\{\bar{x} + (a - 2b)\}}{n} = \bar{x} + (a - 2b).$$

Hence Proved.

Q. 2. Given $\sum_1^n (x_i - 3n) = 84$ and $\sum_1^n (x_i + 2n) =$ 144, find n and the mean.

Solution : $\sum_1^n (x_i - 3n) = x_1 + x_2 + ... + x_n - 3n = 84$

$$\therefore\Rightarrow \quad S - 3n = 84 \quad [\text{Let } S = x_1 + x_2 + ..x_n] \; ...(i)$$

and $\sum_1^n (x_i - 2n) = x_1 + x_2 + x_3 + ..x_n + 2n = 144$

$$\Rightarrow \quad S + 2n = 144 \quad ...(ii)$$

Multiplying equation (i) by 2 and equation (ii) by 3 and then adding, we get

$$5S = 2 \times 84 + 3 \times 114$$

$$\Rightarrow \quad 5S = 168 + 432$$

$$\Rightarrow \quad 5S = 600$$

$$\Rightarrow \quad S = 120$$

Putting the value of S in equation (i), we get

$$120 - 3n = 84$$
$$\Rightarrow \quad 3n = 120 - 84 = 36$$
$$\Rightarrow \quad n = \frac{36}{3} = 12$$
$$\text{Mean} = \frac{S}{n} = \frac{120}{12} = 10.$$ **Ans.**

Q. 3. The A.M. of n observations is M. If the sum of $n-4$ observations is a, then find the mean of remaining 4 observations.

Solution : Let the mean of the remaining 4 observations be $\bar{x}$.

Given, mean of n observations $= M$

$$\therefore \quad M = \frac{a + 4\bar{x}}{(n-4)+4}$$

where a = sum of $n-4$ observations

$$\Rightarrow \quad \bar{x} = \frac{nM - a}{4}$$

Hence, mean of remaining 4 observations $= \dfrac{nm - a}{4}$

Ans.

Q. 4. Prove that $\sum_{i=1}^{n}(x_i - \bar{x}) = 0$.

Solution : Let $x_1, x_2, x_3, \ldots\ldots, x_n$ be the set of n observations.

Then, $\bar{x} = \dfrac{x_1 + x_2 + x_3 + \ldots + x_n}{n}$

or $n\bar{x} = x_1 + x_2 + x_3 + \ldots\ldots + x_n$ ...(i)

The sum of the deviations of all the observations from their mean is given by

$$\sum_{i=1}^{n}(x_i - \bar{x}) = (x_1 - \bar{x}) + (x_2 - \bar{x}) + (x_3 - \bar{x}) + \ldots + (x_n - \bar{x})$$
$$= (x_1 + x_2 + x_3 + \ldots\ldots + x_n) - (\bar{x} + \bar{x} + \ldots n \text{ times})$$
$$= n\bar{x} - n\bar{x} = 0$$ [From (i)]

Hence Proved.

Q. 5. The mean of marks scored by 100 students was found to be 40. Later on, it was discovered that a score of 53 was misread as 83. Find the correct mean.

Solution : Here, $n = 100$, $\bar{x} = 40$

We know, $\bar{x} = \frac{1}{n}\left(\sum x_i\right) \Rightarrow 40 = \frac{1}{100}\left(\sum x_i\right)$

$\therefore$ Incorrect value of $\sum x_i = 4000$

Now, correct value of $\sum x_i = 4000 - 83 + 53 = 3970$

$$\therefore \quad \text{Correct mean} = \frac{\text{Correct value of } \sum x_i}{n}$$
$$= \frac{3970}{100} = 39.7$$

So, the correct mean is 39.7. **Ans.**

Q. 6. The mean of the following frequency table is 50. But the frequencies f_1 and f_2 in class 20 – 40 and 60 – 80, respectively are missing. Find the missing frequencies.

Classes	0 – 20	20 – 40	40 – 60	60 – 80	80 – 100	Total
Frequency	17	f_1	32	f_2	19	120

Solution : Let the assumed mean, $A = 50$ and $h = 20$.

Class Interval	Mid-Value (x_i)	Frequency (f_i)	$u_i = \frac{x_i - A}{h}$; h	$f_i u_i$
0 – 20	10	17	–2	–34
20 – 40	30	f_1	–1	$-f_1$
40 – 60	50	32	0	0
60 – 80	70	f_2	1	f_2
80 – 100	90	19	2	38
Total		$\Sigma f_i = 68 + f_1 + f_2$		$\Sigma f_i u_i = 4 - f_1 + f_2$

We have, $\quad \Sigma f_i = 120$ [Given]

$$\Rightarrow \quad 68 + f_1 + f_2 = 120$$
$$\Rightarrow \quad f_1 + f_2 = 52 \quad \ldots(i)$$

and $\quad$ Mean $= 50$

$$\Rightarrow \quad \bar{x} = A + \left(\frac{\Sigma f_i u_i}{\Sigma f_i}\right) \times h$$
$$\Rightarrow \quad 50 = 50 + \frac{4 - f_1 + f_2}{120} \times 20$$
$$\Rightarrow \quad 50 = 50 + \frac{4 - f_1 + f_2}{6}$$
$$\Rightarrow \quad 0 = \frac{4 - f_1 + f_2}{6}$$
$$\Rightarrow \quad f_1 - f_2 = 4 \quad \ldots(ii)$$

On adding equations (i) and (ii), we get

$$2f_1 = 56$$
$$\Rightarrow \quad f_1 = 28$$

Putting the value of f_1 in equation (i), we get

$$28 + f_2 = 52$$
$$\Rightarrow \quad f_2 = 24$$

Hence, the missing frequencies f_1 and f_2 are 28 and 24 respectively. **Ans.**

Q. 7. If the median of the following frequency distribution is 32·5. Find the values of f_1 and f_2.★

★ are board exam questions from previous years

Class	Frequency
0 – 10	f_1
10 – 20	5
20 – 30	9
30 – 40	12
40 – 50	f_2
50 – 60	3
60 – 70	2
Total	40

Solution : Median 32·5

Class	Frequency	Cumulative Frequency
0 – 10	f_1	f_1
10 – 20	5	f_1+5
20 – 30	9	f_1+14 $c.f.$
30 – 40	12	f_1+26
40 – 50	f_2	f_1+f_2+26
50 – 60	3	f_1+f_2+29
60 – 70	2	f_1+f_2+31

Total freqency = 40

$\therefore \quad f_1+f_2+31 = 40$

or $\quad f_1+f_2 = 9 \qquad$ ...(i)

Also $\quad \frac{n}{2} = \frac{40}{2} = 20$

Median = 32·5 (given)

which lies in the class interval (30 – 40)

$\therefore$ Median class = 30 – 40

$l = 30$

$f = 12, C.f. = f_1 + 14$

$h = 10$

So, $\quad \text{Median} = l + \left[\dfrac{\frac{n}{2} - C.f.}{f}\right] \times h$

$$32{\cdot}5 = 30 + \left[\frac{20-(f_1+14)}{12}\right] \times 10$$

$$32{\cdot}5 = 30 + \left(\frac{6-f_1}{6}\right) \times 5$$

or $\quad 2{\cdot}5 = \frac{5}{6}(6-f_1)$

or $\quad \frac{2{\cdot}5 \times 6}{5} = 6 - f_1$

or $\quad 6 - f_1 = 3, f_1 = 3$

From equation (i), we get

$f_2 = 6$

$\therefore \quad f_1 = 3, f_2 = 6 \qquad$ **Ans.**

Chapter 15. Probability

Q. 1. A jar contains only green, white and yellow marbles. The probability of selecting a green marble and white marble randomly from a jar is 1/4 and 1/3 respectively. If this jar contains 10 yellow marbles, what is the total number of marbles in the jar ?

Solution : Let the no. of green marbles $= x$

Let the no. of white marbles $= y$

$\therefore \quad$ Total no. of marbles $= x + y + 10$

Now, $P(\text{selecting a green marble}) = \dfrac{x}{x+y+10}$

But it is given that, $P(\text{green marble}) = \dfrac{1}{4}$

$\therefore \quad \dfrac{x}{x+y+10} = \dfrac{1}{4}$

$\Rightarrow \quad 4x = x + y + 10$

$\Rightarrow \quad 3x - y = 10 \qquad$ ...(i)

Similarly,

$P(\text{selecting a white marble}) = \dfrac{x}{x+y+10}$

$\therefore \quad \dfrac{y}{x+y+10} = \dfrac{1}{3}$

$3y = x + y + 10$

$\Rightarrow \quad 2y - x = 10 \qquad$ ...(ii)

On solving equations (i) and (ii), we get

$y = 8$ and $x = 6$.

$\therefore$ Total no. of marbles = 6 + 8 + 10 = 24. **Ans.**

Q. 2. Bag *A* contains 4 chips numbered 1, 3, 5 and 7 respectively. Bag *B* contains 3 chips numbered 2, 4 and 6 respectively. A chip is drawn at random from each bag.

(i) Draw a tree diagram to show all the possible outcomes.

(ii) Find the probability that the sum of the two chips drawn is (a) 7 (b) odd (c) even

Solution : (i) The diagram below is the required tree diagram.

Bag A	Bag B	Outcomes
1	2	(1, 2)
	4	(1, 4)
	6	(1, 6)
3	2	(3, 2)
	4	(3, 4)
	6	(3, 6)
5	2	(5, 2)
	4	(5, 4)
	6	(5, 6)
7	2	(7, 2)
	4	(7, 4)
	6	(7, 6)

There are 12 equally likely outcomes in the sample space.

(ii) (a) The favourables outcomes for a sum of 7 are (1, 6), (3, 4) and (5, 2)

$$\therefore \quad P(\text{the sum is 7}) = \frac{3}{12} = \frac{1}{4}$$ **Ans.**

(b) Since all the sums of the outcomes are odd,

$$P(\text{the sum is odd}) = \frac{12}{12} = 1$$ **Ans.**

(c) Since, there are no even sums,

$$\therefore P(\text{the sum is even}) = \frac{0}{12} = 0$$ **Ans.**

Q. 3. An unbiased die is rolled twice. Find the probability of getting (i) the sum of two numbers as a prime (ii) the sum of two numbers equal to 9.

Solution : (i) The sum of the two numbers lies between

2 and 12. So the required primes are 2. 3, 5, 7, 11.

No. of ways for getting 2 = {1, 1} = 1

No. of ways for getting 3 = {1, 2}, {2, 1} = 2

No. of ways of getting 5 = {(1, 4), (4, 1), (2, 3), (3, 2)} = 4

No. of ways of getting 7

= {(1, 6), (6, 1), (2, 5), (5, 2), (3, 4), (4, 3)}

= 6

No. of ways of getting 11 = (5, 6), (6, 5)

= 2

No. of favourable ways = 1 + 2 + 4 + 6 + 2 = 15

Total ways = 6 × 6 = 36

$$\text{Probability of the sum as a prime} = \frac{15}{36} = \frac{5}{12}$$ **Ans.**

(ii) No. of ways of getting a sum of 9

= {(3, 6), (6, 3), (4, 5), (5, 4)} = 4

$$\text{Probability of getting a sum of 9} = \frac{4}{36} = \frac{1}{9}$$ **Ans.**

Q. 4. Savita and Hamida are friends. What is the probability that both will have (i) different birthdays ? (ii) the same birthday ? (ignoring a leap year)

Solution : Out of the two friends, one girl, say, Savita's birthday can be any day of the year. Now, Hamida's birthday can also be any day of 365 days in the year. We assume that these 365 outcomes are equally likely.

(i) If Hamida's birthday is different from Savita's, the number of favourable outcome for her birthday is

365 – 1 = 364

So, P(Hamida's birthday is different from Savita's

$$\text{birthday}) = \frac{364}{365}$$ **Ans.**

(ii) P (Savita and Hamida have the same birthday)

= 1 – P (both have different birthdays)

$$= 1 - \frac{364}{365} = \frac{1}{365} \quad [\text{Using } P(\overline{E}) = 1 - P(E)]$$

Ans.

Q. 5. A die is thrown twice. What is the probability that

(i) 5 will not come up either time ?

(ii) 5 will come up at least once ?

Solution : Let E be the event that first throw shows 5

and F be the event that second throw shows 5.

$$\therefore \quad P(E) = \frac{1}{6} \text{ and } P(F) = \frac{1}{6}$$

Also, $P(\overline{E}) = 1 - P(E)$

$$= 1 - \frac{1}{6} = \frac{5}{6}$$

and $P(\overline{F}) = 1 - P(E)$

$$= 1 - \frac{1}{6} = \frac{5}{6}$$

(i) Probability that 5 will not come up either time

$$= P(\overline{E}) \times P(\overline{F})$$

$$= \frac{5}{6} \times \frac{5}{6} = \frac{25}{36}.$$ **Ans.**

(ii) Probability that 5 will come up at least once

$$= 1 - P(\overline{E}) \times (\overline{F})$$

$$= 1 - \frac{25}{36} = \frac{36 - 25}{36} = \frac{11}{36}.$$ **Ans.**

●●

Case Based Questions | Set 6 |

Chapter 1. Real numbers

Q. 1. Four friends John, Richard, Harry and Rozy start driving their car from the same starting point. They all travelled in different directions. The John travelled in East, Richard Davelled in west, harry travelled in North and Rozy in south. John, Richard, Harry and Rozy travelled with the speed of 30 km/hr, 40 km/hr, 50 km/hr and 60 km/hr respectively. They travelled for 1 hour.

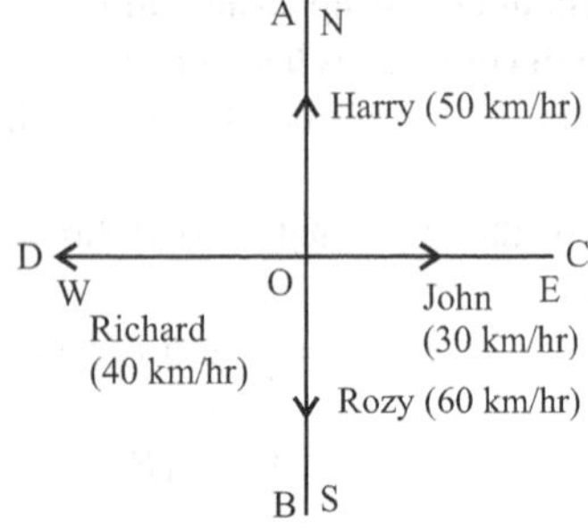

Answer the following questions :

(i) Find the distance between John and Harry.

(a) 35 km **(b)** $10\sqrt{34}$ km

(c) $138\sqrt{2}$ km **(d)** 37 km

Solution. (b) After 1 hour, distance covered by John is 30 km and by Harry is 50 km.

$\therefore$ Distance between Honey and John by pythagoras theorem $\sqrt{(50)^2+(30)^2}$

$= \sqrt{2500+900}$

$= \sqrt{3400} = 10\sqrt{34}$ km **Ans.**

(ii) What is the distance between John and Rozy after 2 hours?

(a) 123 km **(b)** 130 km

(c) 134 km **(d)** 138 km

Solution. (c) Distance of John after 2 hrs = 30 × 2 = 60 km

Distance of Rozy = 60 × 2 = 120 km

Distance between them as ΔBOC is a right-angled triangle

Then, $BC^2 = OB^2 + OC^2$

$= 60^2 + 120^2$

$= 3600 + 14400$

$= 18000$

$\therefore$ BC = 134.16 km = 134 km **Ans.**

(iii) What is the distance between John and Richard after 1 hour?

(a) 70 km **(b)** 110 km

(c) 90 km **(d)** 80 km

Solution. (a) Distance covered by John after 1 hr = 40 km

Distance covered by Richard after 1 hr = 30 km

$\therefore$ Distance between John and Richard after 1 hr = 30 + 40 = 70 km **Ans.**

(iv) Which of the following is not a pythagoras the plate?

(a) 7 cm, 24 m, 25 cm

(b) 3 cm, 8 cm, 6 cm

(c) 12 m, 16 cm, 20 cm

(d) 8 cm, 6 cm, 10 cm

Solution. (b) 3 cm, 8 cm, 6 cm

Since, $3^2 + 6^2 = 9 + 36$

$= 45 \neq 64$ **Ans.**

(v) If the length of diagonals of a rhombus are 16 cm and 12 cm. Then, the length of side of rhombus is ....

(a) 12 cm **(b)** 15 cm

(c) 18 cm **(d)** 10 cm

Solution. (d) Since, the diagonals of a rhombus bisect each other at right-angle.

$\text{side}^2 = 8^2 + 6^2$

$= 64 + 36 = 100$

side = 10 cm **Ans.**

Chapter 2. Polynomials

Q. 1. For the box to satisfy certain requirement its length must be three meter greater than the width and its height must be two meter less than the width.

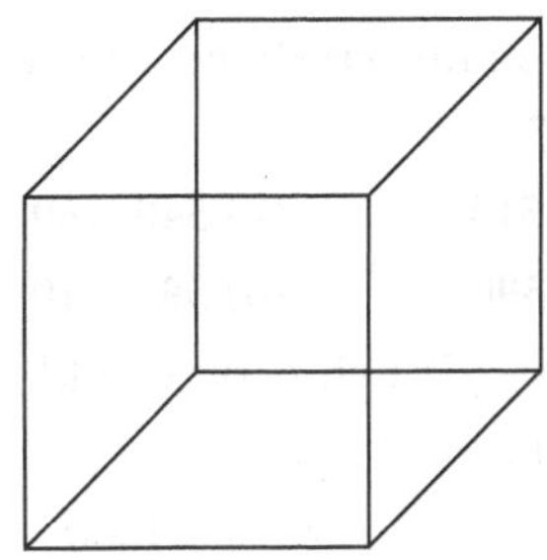

(i) If width is taken as x, which of the following polynomial represents volume of box?

(a) $x^2 - 5x - 6$ (b) $x^3 + x^2 - 6x$

(c) $x^3 - 6x^2 - 6x$ (d) $x^2 + x - 6$

Solution. (b) Ifwidth = x unit

$$\text{length} = (x + 3) \text{ unit}$$
$$\text{height} = (x - 2) \text{ unit}$$
$$\text{Volume of box} = \text{length} \times \text{breadth} \times \text{height}$$
$$= x \times (x + 3) \times (x - 2)$$
$$= (x^2 + 3x)(x - 2)$$
$$= x^3 + 3x^2 - 2x^2 - 6x$$
$$= x^3 + x^2 - 6x.$$

(ii) Which of the following polynomial represent the area of paper sheet used to make box?

(a) $x^2 - 5x - 6$ (b) $6x^2 + 4x + 2$

(c) $x^3 - 6x^2 - 6x$ (d) $6x^2 + 3x - 4$

Solution. (b) Paper area will be the total surface area of box

$$= 2[lb + bh + hl]$$
$$= 2[(x + 3)(x) + x(x - 2) + (x - 2)(x + 3)]$$
$$= 2[x^2 + 3x + x^2 - 2x + x^2 + x - 6]$$
$$= 2[3x^2 + 2x - 6] = 6x^2 + 4x - 12.$$

(iii) If it must have a volume of 18 m, what must be its length?

(a) 6 m (b) 3 m

(c) 4 m (d) 2 m

Solution. (a)

$$\begin{array}{r|l} & x^2 + 4x + 6 \\ \hline x - 3 & x^3 + x^2 - 6x - 18 \\ & x^3 - 3x^2 \\ & - \quad + \\ \hline & 4x^2 - 6x \\ & 4x \quad -12x \\ & - \quad + \\ \hline & 6x - 18 \\ & 6x - 18 \\ \hline & \times \end{array}$$

$$\text{Volume} = 18 \text{ m}$$
$$x^3 + x^2 - 6x = 18$$
$$x^3 + x^2 - 6x - 18 = 0$$

$(x - 3)$ is factor of above euation.

$$(x - 3)(x^2 + 4x + 6) = 0$$
$$(x - 3)(x^2 + 4x + 6) = 0$$
$$x = 3 \text{ or } x^2 + 4x + 6 = 0$$

$D \neq 0$, not real roots.

$\therefore$ Width = 3 m

Length = 3 + 3 = 6 m.

(iv) At a volume of 18 m, what must be its height?

(a) 1 m (b) 3 m

(c) 2 m (d) 4 m

Solution. (a) Height = 3 – 2 = 1 m.

(v) If box is made of a paper sheet which cores ₹ 100 per square metre, what is total the cost of per used?

(a) ₹ 5,400 (b) ₹ 10,800

(c) ₹ 2,700 (d) ₹ 3,400

Solution. (a) Cost = $[6(3)^2 + 4 \times 3 - 12] \times 100$

= ₹ 5,400.

Q. 2. A bag is playing with rope which moves as shown in the figure below:

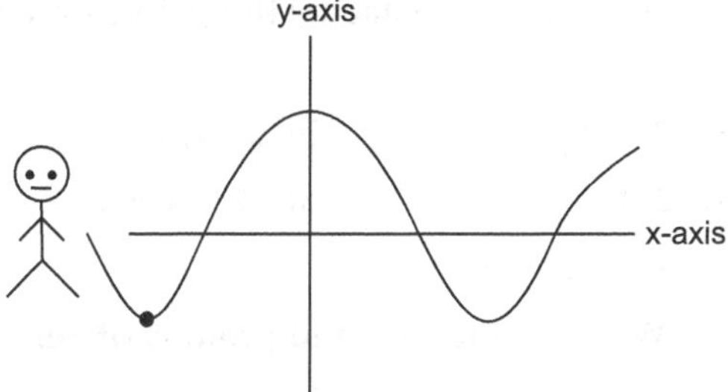

Now give the answers of following questions:

(i) The number of zeros of the above figure:

(a) 2 (b) – 2

(c) 3 (d) 1

Solution. (c) Figure cuts x-axis at 3 points, then no. of zeros are 3.

(ii) If above rope cuts x-axis at $-2, \frac{1}{2}, 1$, the polynomial $p(x)$ will be:

(a) $2x^3 + x^2 - 5x + 2$ (b) $2x^2 - x^2 - 5x + 2$

(c) $x^3 - 2x^2 - 5x + 2$ (d) $2x^3 - x^2 - 5x - 2$

Solution. (a) $P(x) = k[x^3 - (\text{sum of zeros})x^2 + (\text{sum of product of two zeros})x - \text{product of zeros}]$

Zeros are $-2, \frac{1}{2}, 1$

$$\alpha + \beta + \gamma = -2 + \frac{1}{2} + 1 = -1 + \frac{1}{2} = \frac{-1}{2}$$

$$\alpha\beta + \beta\gamma + \gamma\alpha = -2\times\frac{1}{2}+\frac{1}{2}\times 1 + -2\times 1$$

$$= -1+\frac{1}{2}-2 = \frac{1}{2}-3 = \frac{-5}{2}$$

$$\alpha\beta\gamma = -2\times\frac{1}{2}\times 1 = -1$$

$$= x^3+\frac{1}{2}x^2-\frac{5}{2}x+1$$

$$= \frac{1}{2}(2x^3+x^2-5x+2)$$

$$k = \frac{1}{2}.$$

(iii) If $P(x)$ is divded by $(x-1)$ then degree of quotient and remainder is:

(a) 2, 1 **(b) 3, 0**

(c) 2, 0 **(d) 3, 2**

Solution. (c) Degree of quotient = 2

Degree of remainder = 0.

(iv) If the boy moves the above rope with more force and it moves along with polynomial $P(x) = x^3 - 4x^2 + 5x - 2$, then what are the points on which it cuts x-axis:

(a) 2, 1, 1 **(b) 2, – 1, 1**

(c) 2, 1, 1 **(d) – 2, – 1, + 1**

Solution. (a) 2, 1, 1.

(v) What are the sum and product of zeros?

(a) 4, 2 **(b) – 4, 2**

(c) 5, 4 **(d) 4, – 2**

Solution. (a) $2+1+1=4$

$2\times 1\times 1 = 2.$

Chapter 3. Pair of Linear equations in two variables

Q. 1. A company manufactures two types of sanitizers Alpha and Beta. The cost of the small bottle of sanitizer is ₹ 10 and for beta sanitizer is ₹ 12. In the month of June, the company sold total 1000 bottles and makes a total sale of ₹ 10,820 seeing the great demand and short of supply, company decided to increase the price of both the sanitizer by ₹ 1. In the next month i.e., July, the company sold 2,500 bottles and total sales of ₹ 29,200.

Answer the following questions :

(i) How many sanitizers of each type was sold in June?

(a) 460, 510 **(b) 540, 460**

(c) 410, 590 **(d) 590, 410**

Solution. (d) Let, α type sold be x and β type sold be y.

Then, $x + y = 1000$ ...(i)

and $10x + 12y = 10,820$ ...(ii)

Now, multiply equation (i) by 10 and subtract the both, we get

$$10x + 10y = 10000$$
$$10x + 12y = 10,820$$
$$- \quad - \quad -$$
$$-2y = -820$$
$$y = 410$$

and $x = 1000 - 410 = 590$ **Ans.**

(ii) If the store sold 500 bottles of each type sanitizer in June, what would be their sales?

(a) ₹ 5500 **(b) ₹ 5600**

(c) ₹ 10,500 **(d) ₹ 11,000**

Solution. (d) $10 \times 5000 + 12 \times 50 = 5000 + 6000$

$= 11,000$ **Ans.**

(iii) How many bottles of each type were sold in the next month when rate was increased?

(a) 1200, 1300 **(b) 1300, 1200**

(c) 1550, 950 **(d) 1650, 800**

Solution. (d) $x + y = 2500$ ...(i)

$11x + 13y = 29,200$ ...(ii)

Multiply (i) by 11 and subtract

$$11x + 11y = 27500$$
$$11x + 13y = 29200$$
$$- \quad - \quad -$$
$$-2y = -1700$$
$$y = 850$$

Then, $x = 2500 - 850 = 1650$ **Ans.**

(iv) What percent of increased was found in sanitizer in July as compared to June?

(a) 160.55% **(b) 183.35%**

(c) 179.66% **(d) 100%**

Solution.

(c) In July sale of $\alpha = 1650$

In June sale of $\alpha = 590$

Then $\%\text{ increase} = \frac{1650-590}{590}\times 100$

$$= \frac{1060\times 10}{59} = 179.66\%$$

Ans.

(v) In July, if total of 1050 sanitizers of each type were sold, what would be the sale?

(a) ₹ 25,000 **(b) ₹ 25,200**

(c) ₹ 27,000 **(d) ₹ 28,500**

Solution.

(b) $11 \times 1050 + 13 \times 1050 = 11550 + 13650$

$= 25,200$ **Ans.**

Chapter 4. Quadratic Equations

Q. 1. A ball is thrown vertically upwards, from a roof top, which is 70 m above the ground. The ball once thrown, reach a maximum vertical height and then fall back again to the ground. The height of the ball from the ground at the time '*t*' is '*h*', which is given in the form of a quadratic equation represents as :

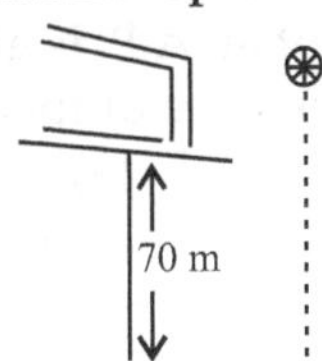

$$h = -16t^2 + 64t + 80$$

Answer the following questions :

(i) What is the height reached by the ball after 1 second?

(a) 128 m **(b) 130 m**

(c) 125 m **(d) 160**

Solution. (a) After 1 second, height of the ball is

$$P(t) = h = -16t^2 + 64\,t + 80$$
$$P(1) = h = -16(1)^2 + 64(1) + 80$$
$$= -16 + 144 = 129$$

(ii) Calculate the maximum height reached by the ball.

(a) 144 m **(b) 150 m**

(c) 124 m **(d) 164 m**

Solution. (a) Rearrange the given equation by completing the square method, we get

$$h = -16(t^2 - 4t - 5)$$
$$= -16[(t-2)^2 - 9]$$
$$= -16(t-2)^2 + 144$$

Height is maximum, when $t = 2$ sec then, $h = 144$ m

(iii) Find the roots of the given quadratic equation.

(a) 0, – 5 **(b) 0, 5**

(c) 5, 1 **(d) 5, – 1**

Solution. (d) Given,

$$P(t) = -16t^2 + 84t + 80 = 0$$

Then, $t^2 - 4t - 5 = 0$

$\Rightarrow$ $t^2 - 5t + 5 - 5 = 0$

$\Rightarrow$ $t(t-5) + 1(t-5) = 0$

$\Rightarrow$ $t = 5, -1$

(iv) What will be height of the ball at $t = 4$ sec?

(a) 30 **(b) 50**

(c) 80 **(d) 120**

Solution. (c) $P(t) = -16t^2 + 64t + 80$

$$P(4) = -16(4)^2 + 64 \times 4 + 80$$
$$= -256 + 256 + 80 = 80$$

(v) The shape of the graph of quadratic polynomial is a :

(a) straight line **(b) circle**

(c) parabola **(d) ellipse**

Solution. (c) Parabola.

Chapter 5. Arithmetic Progressions

Q. 1. 56 cards are stacked in the following manner. 14 cards are in the bottom row, 12 in the next row, 10 in the next to it and so on. Based on the observations, answer the following questions:

(i) If the above situation represented in an A.P., then what will be the common difference?

(a) 4 **(b) 2**

(c) – 2 **(d) – 4**

Solution. (c) 14, 12, 10, ...

$$a_1 = 14,\ a_2 = 12$$
$$d = a_2 - a_1 = 12 - 14 = -2.$$

(ii) The number of cards in the top row is:

(a) 4 **(b) 6**

(c) 1 **(d) 2**

Solution. (d) 14, 12, 10, 8, 6, 4, 2.

(iii) The total number of rows in which the cards are stacked:

(a) 7 **(b) 6**

(c) 8 **(d) 9**

Solution. (a) $a_n = a + (n-1)d$

$$2 = 14 + (n-1)(-2)$$
$$2 = 14 - 2n + 2$$

$$2 = 16 - 2n$$
$$2n = 14$$
$$n = 7.$$

(iv) If x, y, z are in A.P. then $\frac{x-y}{y-z}$ is equal to:

(a) $\frac{1}{2}$ **(b) 2**

(c) 1 **(d) 0**

Solution. (c) If x, y, z are in A.P. then

$$x - y = y - z$$
$$\frac{x-y}{y-z} = 1.$$

(v) The nth term of an A/P. is always a .... expression.

(a) Linear **(b) Quadratic**

(c) Cubic **(d) Bi-quadratic**

Solution. (a) Linear.

Q. 2. A man wants to participate in 400 m race. He can currently run that distance in 102 seconds. He wants to do in 62 seconds with each day of practice if takes him 4 seconds run.

(i) Which of the following terms are A.P. for the given situation?

(a) 102, 98, 94, 90 ... **(b) 102, 100, 98, 96 ...**

(c) 62, 60, 58, 56 ... **(d) 102, 101, 100, 99 ...**

Solution. (a) 102, 98, 94, 90 ...

(ii) What is the minimum number of days he needs to practice till his goal is achieved?

(a) 20 **(b) 11**

(c) 19 **(d) 16**

Solution. (b) $a = 102$

$$d = 98 - 102 = (-4)$$
$$a_n = 62$$
$$a + (n-1)d = 62$$
$$102 + (n-1)(-4) = 62$$
$$102 - 4n + 4 = 62$$
$$4n = 44$$
$$n = 11.$$

(iii) Which of the following term is not in A.P. of the above given situation?

(a) 97 **(b) 98**

(c) 94 **(d) 82**

Solution. (a) All terms of A.P. are even. So odd term is not in a.P.

(iv) If n^{th} term of an A.P. is given by $a_n = 2n + 3$ then common differences of the A.P. is:

(a) 2 **(b) 3**

(c) 5 **(d) 1**

Solution. (a) $a_n = 2n + 3$

$$a_1 = 2 + 3 = 5$$
$$a_2 = 4 + 3 = 7$$
$$d = 7 - 5 = 2.$$

(v) The value of n, for which $2x$, $x + 10$, $3x + 2$ are three consecutive terms of cm A.P.:

(a) 6 **(b) – 6**

(c) 18 **(d) – 18**

Solution. (a)$(x + 10) - 2x = (3x + 2) - (x + 10)$

$$10 - x = 2x - 8$$
$$18 = 3x$$
$$x = 6.$$

Chapter 6. Coordinate Geometry

Q. 1. Students of class X are on visit of Sansad Bhawan. Teacher assign them the activity to observe and take some pictures to analyses the seating arrangement between various MP and speaker based on coordinate geometry. The staff tour guide explained various facts related to Math's of Sansad Bhawan to the students, students were surprised when teacher ask them you need to apply coordinate geometry on the seating arrangement of MP's and speaker.

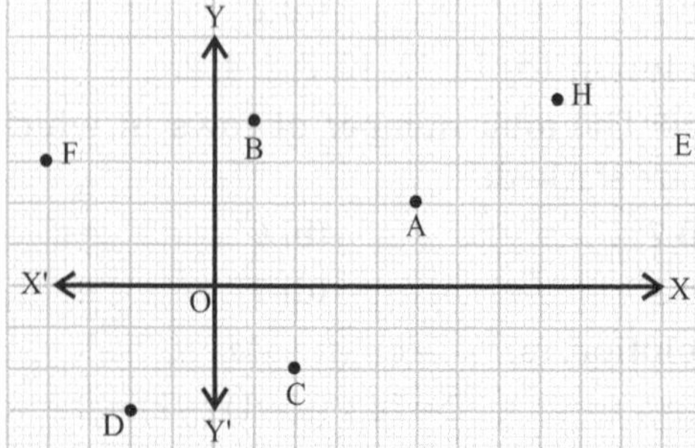

Calculate the following refer to the above image and graph.

Answer the following questions :

(i) Refer to the points *D* and *C*, find the distance between the points *C* and *D*, if the coordinates of *C* is (2, – 2) and *D* is (– 2, – 3)

(a) $\sqrt{17}$ **(b) 6.4**

(c) 5 **(d) 7**

Solution. (a) Distance between $C(2, -2)$ and $D(-2, -3)$ is

$$D = \sqrt{(-2-2)^2 + (-3+2)^2}$$

$$D = \sqrt{16+1}$$

$$D = \sqrt{17}$$ **Ans.**

(ii) Refer to the points *D*, *A* and *H*, condition for collinearity of points *DAH* satisfy by relation :

(a) *DA* = *AH* = *DH* **(b) *DA* + *AH* = *DH***

(c) *DA* + *AH* > *DH* **(d) *DA* + *AH* < *DH***

Solution. (b) *DA* + *AH* = *DH* **Ans.**

(iii) Refer to the points *B* and *C*, join *BC*. Mark a point *P* on *BC* and *P* divides the line segment *B*(1, 4) and *C*(2, – 2) such as $\frac{BP}{PC} = \frac{k}{1}$, if P lies on the line $x - y = -8$. Find the value of *k*.

(a) $\frac{-3}{2}$ **(b) $\frac{4}{5}$**

(c) $\frac{5}{12}$ **(d) $\frac{1}{2}$**

Solution. (c) Given that *P* divides the line segment joining *B*(1, 4) and *C*(2, – 2) in the ratio *k* : 1 so the coordinates of *P* are

$$\left(\frac{2k+1}{k+1}, \frac{-2k+4}{k+1}\right)$$

$P\left(\frac{2k+1}{k+1}, \frac{-2k+4}{k+1}\right)$ lies on the line segment

$$x - y = -8$$

$$\left(\frac{2k+1}{k+1}\right) - \left(\frac{-2k+4}{k+1}\right) = -8$$

$$2k + 1 + 2k - 4 = -8k - 8$$

$$12k = -5$$

$$k = \frac{5}{12}$$ **Ans.**

(iv) Refer to the points *F* and *E*, Find the mid-point of the line segment joining by the points *F*(– 4, 3) and *E*(12, 3).

(a) (4, 4) **(b) (3, – 4)**

(c) (– 4, 8) **(d) (4, 3)**

Solution. (d) Given point $F(-4, 3)$ and $E(12, 3)$.

Coordinates of mid-point are

$$\left(\frac{-4+12}{2}, \frac{3+3}{2}\right) \Rightarrow (4,3)$$ **Ans.**

(v) Refer to the point B and C, If B is, consider as origin then coordinates of C are :

(a) (1, 4) **(b) (1, – 6)**

(c) (– 4, – 6) **(d) (0, 0)**

Solution. (b) (1, – 6) **Ans.**

Chapter 7. Triangles

Q. 1. Scale factor: In 2 similar geometric figures, the ratio of their corresponding sides is called the scale factor. If you begin with the smaller figure, your scale factor will be less than one. If you begin with the larger figure, your scale factor will be greater than one.

(i) A model of a aeroplane (some what triangular in shape is made on the scale of 1:100. The model is 150 cm long, what is the length of the actual aeroplane?

(a) 170 m **(b) 150 m**

(c) 200 m **(d) 15 m**

Solution. (b) since, scale factor is $\frac{1}{100}$

Then, $$\frac{1}{100} = \frac{150}{x}$$

$\Rightarrow$ $$x = 15000 \text{ cm}$$

$$= 150 \text{ m}$$ **Ans.**

(ii) A 20 m long vertical pole casts a shadow 10 m long. At the same time, tower makes shadow 50 m long, the tower is..........long.

(a) 75 m **(b) 100 m**

(c) 105 m **(d) 120 m**

Solution.

(b)

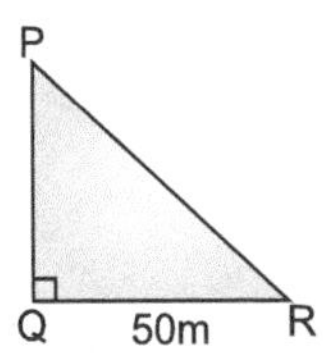

Here, AB is a pole and PQ is a tower.

Then, $\frac{AB}{BC} = \frac{PQ}{QR}$

$\Rightarrow$ $\frac{20}{10} = \frac{PQ}{50}$ $\Rightarrow$ PQ = 100 m. **Ans.**

(iii) Which of the following statement is incorrect:

(a) The ratio of perimeters of 2 similar triangles is the same as the ratio of their corresponding sides.

(b) If the areas of 2 similar triangles are equal, then they are congruent.

(c) The ratio of areas of 2 similar triangles is equal to ratio of sides.

(d) If ratio of corresponding sides is 5.8, then ratio of areas is 2.4.

Solution. (c) Since, ratio of area of two similar triangles is equal to ratio of square of their sides. **Ans.**

(iv) In ΔABC and ΔDEF, it is given that $\angle B = \angle E$, $\angle F = \angle C$ and AB = 3DE, then the two triangles are :

(a) Congruent but not similar

(b) Similar but not congruent

(c) Neither congruent nor similar

(d) Similar as well as congruent

Solution. (b) Their angles are equal and sides are proportional. **Ans.**

(v) The line segments joining the mid-points of the sides of a triangle form four triangles, each of which is:

(a) Congruent to the original triangle

(b) Similar to the original triangle

(c) An isosceles triangle

(d) An equilateral triangle.

Solution. (b) Since, their sides will be proportional. **Ans.**

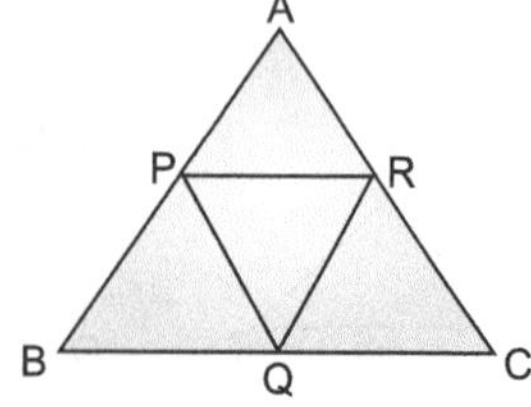

Chapter 8. Circles

Q. 1. A locality in the town has a beautiful triangular shaped park in the centre of their colony. It is a centre of attraction for all the residents of the colony. They come for walk and children use to play there Caretaker of park, want to build a circular portion inside a park, which touches all the sides of triangle, to be used as a special area for planting some beautiful flower trees.

The layout of the park is shown in the figure.

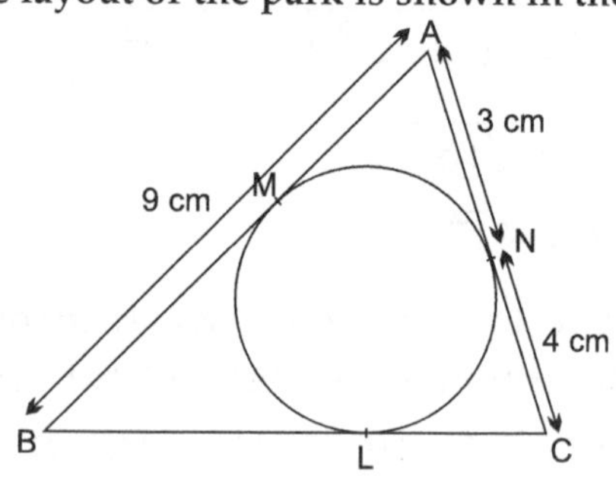

Now, answer the following questions:

(i) What is the length of MC?

(a) 3 cm **(b) 6 cm**

(c) 4 cm **(d) 7 cm**

Solution. (b) Here AN = AM

Since, AN and AM are tangents to the circle, so they will be equal in length

Then AN = AN = 3 cm

Then, MC = 9 – 3 = 6 cm **Ans.**

(ii) What is length of the portion LC?

(a) 3 cm **(b) 6 cm**

(c) 4 cm **(d) 7 cm**

Solution. (b) Since LC = MC

(tangent from a point)

$\therefore$ MC = LC = 6 cm **Ans.**

(iii) If the boundary of the triangular garden, need to be fenced, then calculate the length of boundary.

(a) 18 cm **(b) 27 cm**

(c) 26 cm **(d) 30 cm**

Solution. (c) Boundary = AB + AC + BC

= 4 + 3 + 9 + BL + LC

= 4 + 3 + 9 + 4 + 6

[$\because$ BM = BL]

= 26 cm **Ans.**

(iv) A line to a circle forms which angle with the radius, corresponding to the point of tangency?

(a) Obtuse angle **(b) Right angle**

(c) Acute angle **(d) Supplementary angle**

Solution. (b) Right angle **Ans.**

(v) If the radius radius of the incircle formed is 3 cm, find its boundary.

(a) 18 84 cm **(b) 15 81 cm**

(c) 10 05 cm **(d) 11 08 cm**

Solution. (a) Boundary of the circle is $2\pi r$

$= 2 \times 3{\cdot}14 \times 3 = 18{\cdot}84$ cm

Ans.

Chapter 10. Trignometry

Q. 1. There are in total six trigonometric ratios, namely sine (sin), cosine(cos), tangent (tan), cosecant (cosec), secant (sec) and cotangent (cot). The trigonometric functions cosecant, secant and cotangent are simly the reciprocals of the trigonometric functions sine, cosine and tangent for the angles of a triangle. The values of these trigonometric ratios gives a certain rational for some values of angle (, say α). Some such values for the angle of triangle are shown in the table below:

Angle-ratio	0°	30°	45°	60°	90°
sin θ	0	$\frac{1}{2}$	$\frac{1}{\sqrt{2}}$	$\frac{\sqrt{3}}{2}$	1
cos θ	1	$\frac{\sqrt{3}}{2}$	$\frac{1}{\sqrt{2}}$	$\frac{1}{2}$	0
tan θ	0	$\frac{1}{\sqrt{3}}$	1	$\sqrt{3}$	not defined
cosec θ	not defined	2	$\sqrt{2}$	$\frac{2}{\sqrt{3}}$	1
sec θ	1	$\frac{2}{\sqrt{3}}$	$\sqrt{2}$	2	not defined
cot θ	not defined	$\sqrt{3}$	1	$\frac{1}{\sqrt{3}}$	0

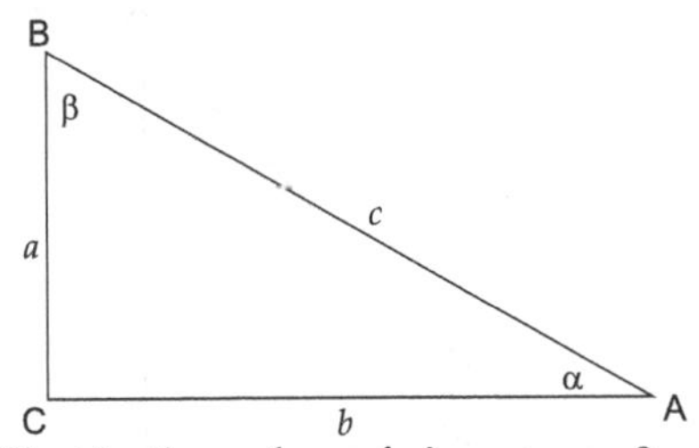

(i) What is the value of sin α + cos β, when the values of angle α and β are respectively 30° and 60°?

(a) 1 **(b) 1/2**

(c) 2 **(d) 0**

Solution.

(a) We have α = 30° and that of β = 60°

Thus, sin α + cos β = sin 30° + cos 60°

$= \frac{1}{2} + \frac{1}{2} = 1$ **Ans.**

(ii) Find the value of (sin² 30 – sin 0)/(cos² 90 – cos² 60).

(a) 0 **(b) –1**

(c) 1 **(d) 1/2**

Solution. (b) –1

Here $(\sin^2 30° - \sin 0°) = \left(\frac{1}{2}\right)^2 - 0 = \frac{1}{4}$ and $\cos^2 90° - \cos^2 60° = 0 - \frac{1}{4} = -\frac{1}{4}$ Thus, $(\sin^2 30° - \sin 0°)/(\cos^2 90° - \cos^2 60°) = \left(\frac{1}{4}\right)/\left(\frac{1}{-4}\right) = -1$ **Ans.**

(iii) If α = 90° and β = 60°, determine the value of sin² α + sin² β.

(a) 4/5 (b) 3/5

(c) 5/4 (d) 5/3

Solution. (c)

Here, $\sin^2 90° + \cos^2 60° = 1^? + \left(\frac{1}{2}\right)^2$

$= 1 + \frac{1}{4} = \frac{5}{4}$ **Ans.**

(iv) If both α and β = 60°, find the value of sin² α + cos² β.

(a) 1 (b) 2

(c) –1 (d) –2

Solution. (a)

Here, $\sin^2 60° + \cos^2 60° = \frac{1}{4} + \frac{3}{4} = 1$ **Ans.**

(v) If in the triangle ABC the lengths of sides AC and BC are in the ratio of 13:5, find the value of cosec β.

(a) 5/13 (b) 13/8
(c) 8/13 (d) 13/5

Solution. (d) From the table we can find that

$$\text{cosec}\ \beta = \frac{AC}{BC} = \frac{13k}{5k} = \frac{13}{5}$$ **Ans.**

Chapter 11. Applications of Trigonometry

Q. 1. A branch of mathematics, which relates an angle of a right-angled triangle to ratios of two side length. They are trignometric functions. it is widly used in measurement of heights, distances, science, engineering ring, video games and more. It is used in navigation, to find the height of mountain, height of tower etc.

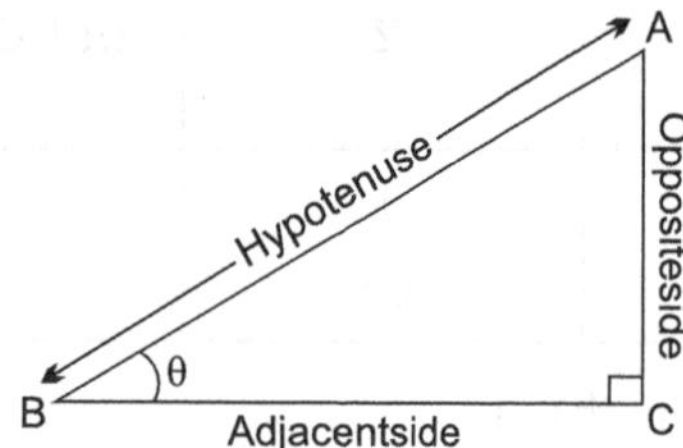

On the basis of this information, answer the following questions :

(i) Which trigonometric ratio relates the adjacent side and opposite side of the right-angled triangle?

(a) sin θ **(b) cos θ**
(c) tan θ **(d) sec θ**

Solution. (c)

$$\tan\theta = \frac{P}{B} = \frac{\text{Opposite side}}{\text{Adjacent side}}$$ **Ans.**

(ii) If the value of θ is 45°, then what is the value of sec 45°?

(a) 0 **(b) $\frac{2}{\sqrt{3}}$**
(c) $\sqrt{2}$ **(d) 1**

Solution. (c) $\sqrt{2}$ **Ans.**

(iii) Which among the following ratios, is not opposite of each other?

(a) sin θ and cos θ **(b) cos θ and sec θ**
(c) tan θ and cot θ **(d) cosec θ and sin θ**

Solution. (a)sin θ and cos θ

$$\because \quad \cos\theta = \frac{1}{\sec\theta}$$

$$\text{cosec}\ \theta = \frac{1}{\sin\theta}$$

$$\tan\theta = \frac{1}{\cot\theta}$$ **Ans.**

(iv) If the distance of a point from the base of a building is 15 m and angle of elevation to its top from that point is 60°. Then, what is the height of the building?

(a) 25 m **(b) 26 m**
(c) 31 m **(d) 18 m**

Solution. (b) Here, AB is building and C is the point of observation.

$$\tan 60° = \frac{AB}{BC}$$

$$\Rightarrow \quad \sqrt{3} = \frac{AB}{15}$$

$$\Rightarrow \quad AB = 15\sqrt{3}\ \text{m}$$

$$= 25.95\ \text{m}$$

$$\simeq 26\ \text{m}$$ **Ans.**

(v) Complete this identity : $\sec^2\theta - 1$ = ...

(a) $\cos^2\theta$ **(b) $\sin^2\theta$**
(c) $\tan^2\theta$ **(d) $\text{cosec}^2\theta$**

Solution. (c) $1 + \tan^2\theta = \sec^2\theta$ **Ans.**

Chapter 12. Areas Related to Circles

Q. 1. The given picture shows six circular portion of land each of radius 1 m, planted with different types of flowers are in an equilateral triangular garden as shown below. Area except the circular portion is covered with grass. The length of a side of an equilateral triangle is 7 m.

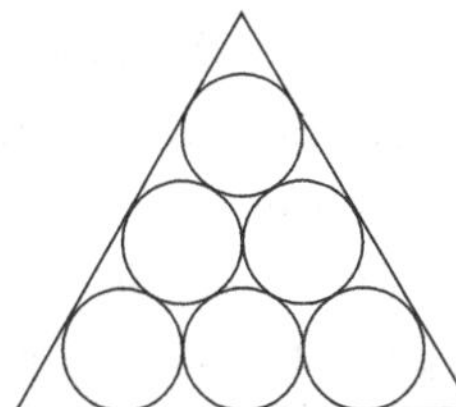

Answer the following questions :

(i) Calculate the area of garden, covered with the flowers.

(a) 19.50 m^2 **(b) 18.9 m^2**

(c) 17.6 m^2 **(d) 19.2 m^2**

Solution. (b) 18.9 m^2

Since, there are 6 circular portions which are covered with flowers.

Then area of garden covered with flowers = $6 \times \pi r^2$

$$= 6 \times \pi \times (1)^2$$

$$= 6 \times \frac{22}{7} = 18.86 \text{ m}^2$$

$$= 18.9 \text{ m}^2 \quad \textbf{Ans.}$$

(ii) Find the area of garden which is covered with grass.

(a) 5.25 m^2 **(b) 2 m^2**

(c) 1.25 m^2 **(d) 2.33 m^2**

Solution. (a) Area covered by grass = Area of equilateral triangle – Area of circles

$$= \frac{\sqrt{3}}{4} \times (\text{side})^2 - 6 \times \pi(1)^2$$

$$= \frac{\sqrt{3}}{4} \times 49 - 6\pi$$

$= 21.21 - 18.86$

$= 2.35$ m^2 **Ans.**

(iii) Find the cost of fencing the circular flower area at the cost of ₹ 7/m.

(a) ₹ 250 **(b) ₹ 264**

(c) ₹ 314 **(d) ₹ 350**

Solution. (b) Cost of fencing = Length need to be fenced × Rate of fencing

$$= 6 \times 2\pi r \times 7$$

$$= 2 \times \frac{22}{7} \times 1 \times 6 \times 7$$

$$= 44 \times 6 = ₹\ 264 \quad \textbf{Ans.}$$

(iv) If cost of plantation of flower in 1 sq m area is ₹ 140. Find the total cost of plantation of flowers.

(a) ₹ 1025 **(b) ₹ 1250**

(c) ₹ 2500 **(d) ₹ 2640**

Solution. (d) Area of flower reigon = 18.86 m^2

Cost of 1m^2 is ₹ 140.

Then, total cost = ₹ 140 × 18.86

= ₹ 2640.4

= ₹ 2640 **Ans.**

(v) If a boundary wall is need to be constructed along the boundary of the triangle, find the cost of making it, If cost of making boundary is ₹ 100/m.

(Including material and labour cost)

(a) ₹1200 **(b) ₹1000**

(c) ₹2100 **(d) ₹2200**

Solution. (c) Boundary = 7 + 7 + 7

= 21 m

∴ cost = 21 × 100

= ₹ 2100 **Ans.**

Chapter 13. Surface Areas and Volumes

Q. 1. Sides of a swimming pool is 122 m long and 120 m broad. Base of the swimming pool is same as its top layer of water. Depth of pool is 1.5 m. There are 10 water taps in the pool to fill water in it from each tap water is supplied at the rate of 55 m^3 minute.

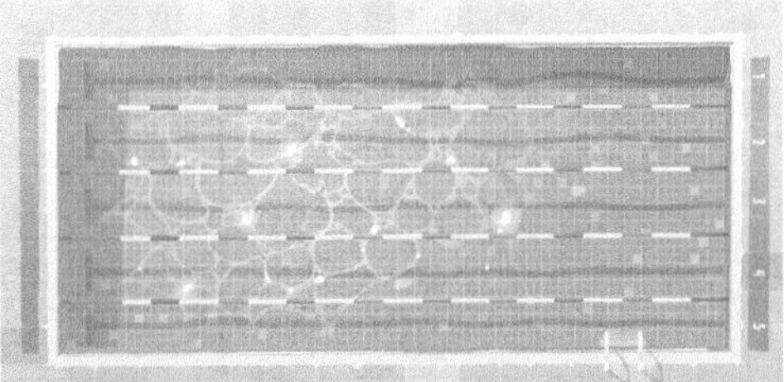

Now answer the following questions:

(i) Base of the pool is to be plastered at the rate of ₹ 25 per sq metre including material. Find the expense for palstering the base of pool after draining all the water.

(a) ₹ 3,66,000 **(b) ₹ 2,55,000**

(c) ₹ 2,75,000 **(d) ₹ 3,12,250**

Solution. (a) Area of base = $l \times b$

= 122 × 120

= 14640 m^2

Then cost of plastering it = ₹25 × 14,640

= ₹3,66,000 **Ans.**

(ii) What is the volume of water in pool if it is completely filled?(in litres)

(a) 2.196 10^4l **(b) 2.196 10^5l**

(c) 2.196 10^6l **(d) 2.196 10^7l**

Solution. (d) Since, the shape of the swimming pool is cuboid in shape

Then, its volume is = $l \times b \times h$

= 122 × 120 × 1.5

= 21,960 m^3

In litres = 21960 × 1000 l

[$\because$ 1m^3 = 1000 l]

= 2.196 × 10^7 l **Ans.**

(iii) How much time approximately it will take to fill the pool completely if all the water taps are opened to supply the water?

(a) 60 min **(b) 40 min**

(c) 30 min **(d) 50 min**

Solution. (d) There are 10 water taps. Each tap supply 55 m^3/min.

Then , in 1 minute, water supplied = 550 m^3

Then time taken = $\frac{21960}{550}$ min = 39.927 min $\simeq$ 40 min. **Ans.**

(iv) What is the surface area of the pool?

(a) 14,828 m^2 **(b) 15,366 m^2**

(c) 13,500 m^2 **(d) 15, 848 m^2**

Solution. (b) Surface area of pool = Area of 4 walls + Area of base

= $2h\,(l + b) + lb$

= 2 × 1.5 (122 + 120) + 122 × 120

= 3 × 242 + 14640

= 726 + 14640

= 15366 m^2 **Ans.**

(v) What is the formula for area of four walls of a cuboid shaped figure?

(a) $2(lb+bh+hl)$ **(b) $(l + b)\,h$**

(c) $2h\,(l + b)$ **(d) $(l + h)b$**

Solution. (c) Area of 4 walls = $2\,bh + 2\,lh$

= $2h\,(l + b)$ **Ans.**

Chapter 14. Statistics

Q. 1. The annual rainfall for 66 days is given in the following table :

Rainfall (in cm)	0–10	10–20	20–30	30–40	40–50	50–60
Number of days	22	10	8	15	5	6

Based on the record, answer the following questions:

(i) What is the mean rainfall of 66 days?

(a) 23·34 cm **(b) 33·23 cm** **(c) 20·34 cm** **(d) 23·43 cm**

Solution.

(a)

Class Interval	x_i	f_i	x_if_i	cf
0—10	5	22	110	22
10—20	15	10	150	32
20—30	25	8	200	40
30—40	35	15	525	55
40—50	45	5	225	60
50—60	55	6	330	66
Total		$\Sigma f_i = 66$	$\Sigma f_ix_i =$ 1540	

$$\text{Mean} = \frac{\Sigma f_i x_i}{\Sigma f_i} = \frac{1540}{66} = 23.34 \text{ cm.}$$ **Ans.**

(ii) For how many days rainfall is less than 30 cm?

(a) 8 (b) 18

(c) 40 (d) 22

Solution. (c) For rainfall less than 30 cm

Number of day = 22 + 10 + 8 = 40 **Ans.**

(iii) What is the lower limit of median class?

(a) 10 (b) 20

(c) 40 (d) 30

Solution. (b) As median class is 20–30 because

$$\frac{N}{2} = \frac{66}{2} = 33$$

33th is the middle value that lie in class 20-30. **Ans.**

(iv) What is the lower limit of modal class?

(a) 0 (b) 20

(c) 30 (d) 40

Solution. (a) Modal class is class with maximum frequency. Hence modal class is 0–10. **Ans.**

(v) For how many days it rained more than 40 cm?

(a) 5 (b) 6

(c) 11 (d) 26

Solution. (c) Rain full more than 40cm = 5 + 6 = 11. **Ans.**

Fill in the Blanks

Chapter 1. Real Numbers

Q. 1. HCF of 16 and 32 is ....

Solution. 16

Q. 2. The HCF of 36 and 54 is ....

Solution. 18

Q. 3. If HCF (336, 54) = 6, LCM (336, 54) = ...

Solution. 3025

Q. 4. If P is a prime number, then LCM of P^2 and P^3 is ....

Solution. P^3

Q. 5. The HCF of smallest prime number and the smallest composite number is ....

Solution. 2

Q. 6. HCF of 168 and 126 is ....

Solution. 42

Q. 7. HCF of 2 numbers is 113, and their LCM is 56952. If one number is 904, then the other number is ....

Solution. 7119

Q. 8. The decimal expansion of $\frac{185}{625}$ will terminate after .... places.

Solution. Three.

Q. 9. The decimal expansion of $\frac{3457}{2^2 \times 5}$ will terminate after .... places.

Solution. Two

Q. 10. The product of a non-zero rational number and an irrational number is ....

Solution. Irrational.

Q. 11. If two positive integers *a* and *b* can be expressed as $a = x^2y^3$ and $b = x^3y^2$, *x*, *y* are prime numbers, then the LCM of *a* and *b* is ....

Solution. x^3y^3

Q. 12. Prime factorisation of the denominator of any rational number having terminating decimal expansion is of the form $2^m \times$ ....

Solution. 5^n

Q. 13. If the HCF of 210 and 55 is expressible in the form $210 \times 5 + 55y$, then the value of *y* is ....

Solution. – 19

Q. 14. If a and b are co-prime numbers, then their HCF is ....

Solution. 1

Q. 15. Euclid's division lemma, state that for two positive integers *a* and *b*, there exist unique integers *q* and *r* such that $a = bq + r$, where *r* must satisfy ....

Solution. $0 \le r < b$

Q. 16. If *a* be a positive integer and *p* be a prime number such that *p* divides a^2, then *p* also divides.

Solution. divides *a*

Q. 17. For some integer *m*, every even integer is of the form ....

Solution. $2m$

Q. 18. If $a = 2^3 \times 3^2$, $b = 2 \times 3 \times 5$, $c = 3n \times 5$ and LCM $(a, b, c) = 2^3 \times 3^3 \times 5$, then *n* is ....

Solution. 3

Chapter 2. Polynomials

Q. 1. A quadratic polynomial, whose zeros are –3 and 4, is ....

Solution. $x^2 - x - 12$.

Q. 2. The zeroes of polynomial $x^3 - 9x$ are ....

Solution. 0, – 3, 3

Q. 3. The number of polynomials having zeros 2 and 3 is ....

Solution. one

Q. 4. If α and β are the zeroes of $x^2 + x + 1$, then the value of $\frac{1}{\alpha} + \frac{1}{\beta}$ is ....

Solution. – 1

Q. 5. If α and β are the zeroes of the polynomial $2x^2 - 5x + 1$, then the value of $\alpha^2 + \beta^2$ is ....

Solution. $\frac{21}{4}$

Q. 6. If α and β are the zeroes of the polynomial $5x^2 - 7x + 2$, then the sum of their reciprocals is ....

Solution. $\frac{7}{2}$

Q. 7. The graph of a polynomial of degree 3 intersects X-axis at most .... points.

Solution. 3

Q. 8. If α and β are the zeroes of $x^2 - 5x + 6$, then the value of $(\alpha + \beta)$ is ....

Solution. 5

Q. 9. The sum and product of the zeroes of the quadratic polynomial $4x^2 - 27x + 3k^2$ are equal. Then, the value of *k* is ....

Solution. ± 3

Q. 10. If the product of roots of the polynomial $x^2 - 7x + k - 4$ is 3, then the value of *k* is ....

Solution. 7

Q. 11. The ratio of the sum and product of the roots of the polynomial $7x^2 + 12x - 18$ is ....

Solution. $\frac{2}{3}$

Q. 12. If the sum of roots of the polynomial $4x^2 - 2x + (k - 4)$ is half of their product, then the value of k is ....

Solution. 8

Q. 13. If one root of the polynomial $x^2 - (k + 6) x + 2(2k - 1)$ is reciprocal of the other, then the value of k is ....

Solution. $\frac{3}{4}$

Q. 14. The graph of the polynomial $f(x) = x^2 - 9$ is a parabola which intersects X-axis at two points namely .... and ....

Solution. $-3, 3$

Q. 15. If the roots of the cubic polynomial $x^3 - 3x^2 + x + 1$ are $\alpha + \beta$, $\alpha - \beta$ and α, then the values of α and β are .... and .... respectively.

Solution. $1, \pm\sqrt{2}$

Q. 16. If one of the zeros of the cubic polynomial $ax^3 + bx^2 + cx + d$ is zero, then the product of other two zeros is ....

Solution. $\frac{c}{a}$

Q. 17. A quadratic polynomial is a polynomial of degree ....

Solution. 2

Q. 18. A real number α is a .... of a polynomial $f(x)$ if $f(\alpha) = 0$

Solution. zero.

Q. 19. If α and β are the zeroes of a quadratic polynomial $f(x)$. Then, the polynomial $f(x)$ is given by ....

Solution. $x^2 - (\alpha + \beta) x + \alpha\beta$

Q. 20. A quadratic polynomial, the sum of whose zeroes is 0 and one zero is 5 is ....

Solution. $x^2 - 25$.

Chapter 3. Pair of Linear Equations in Two Variables

Q. 1. The pair of linear equations $x - 3y + 7 = 0$ and $2x + ky - 3 = 0$ is inconsistent for $k =$ ....

Solution. -6

Q. 2. The value of c for which the system of equations $x + 2y - 3 = 0$ and $(c - 1)x + (c + 1) y - c - 3 = 0$ infinitely many solutions is ....

Solution. 3

Q. 3. If the pair of linear equations $2x + 3y = 11$ and $(m + n) x + (2m - n) y - 33 = 0$ has infinitely many solutions, then the values of m and n, are .... and .... respectively.

Solution. 5, 1

Q. 4. The pair of linear equations $2x + y - 5 = 0$ and $3x + 2y - 8 = 0$ has .... solution(s).

Solution. a unique.

Q. 5. If a pair of linear equations is inconsistent, then their graph lines will always be ....

Solution. Parallel.

Q. 6. If $2x - 3y = 7$ and $(a + b) x - (a + b - 3) y = 4a + b$ represent .... lines, then a and b satisfy the equation $a - 5b = 0$.

Solution. coincident

Q. 7. If $\frac{a_1}{a_2}$ .... $\frac{b_1}{b_2}$, then the system of equations $a_1x + b_1y + c_1 = 0$ and $a_2x + b_2y + c_2 = 0$ has a unique solution.

Solution. $\neq$

Q. 8. If $am \pm bl$, then the system of equations $ax + by = c$ and $lx + my = x$ has ....

Solution. a unique solution.

Q. 9. If a pair of linear equations in two variables is consistent then the lines represented by two equations are .... or ....

Solution. intersecting, coincident.

Q. 10. The sum of the digits of a two digit number is 9. If 27 is added to it, the digits of the number get reversed. The number is ....

Solution. 36.

Chapter 4. Quadratic Equations

Q. 1. The roots of the quadratic equation $2x^2 + 14x + 24 = 0$ are .... and ....

Solution. $-4, 3$.

Q. 2. The sum of the roots of the quadratic equation $3x^2 - 10x + 3 = 0$ is ....

Solution. $\frac{10}{3}$.

Q. 3. If 3 is a root of the equation $kx^2 - kx - 3 = 0$, then the value of k is ....

Solution. $\frac{1}{2}$

Q. 4. The value of k for which the quadratic equation $kx(x - 2) + 6 - 0$ has equal roots is ....

Solution. 6

Q. 5. If the zeros of the quadratic equation $x^2 + (a + 1) x + b + 1 = 0$ are 2 and -3, then the values of a and b, are .... and .... respectively.

Solution. $0, -7$

Q. 6. Discriminant of the quadratic equation $3\sqrt{3}x^2 + 10x + \sqrt{3} = 0$ is ....

Solution. 64.

Q. 7. The roots of the equation $\sqrt{2x + 9} + x = 13$ are .... and ....

Solution. 8, 20.

Q. 8. The nature of roots of the equation $9x^2 - 12x - 4 = 0$ is ....

Solution. Real and unequal.

Q. 9. The nature of roots of the equation $2x^2 + \sqrt{5}x - 1 = 0$ is ....

Solution. Real and unequal.

Q. 10. The nature of roots of the equation $4\sqrt{3}x^2 - 5x - 2\sqrt{3} = 0$ is real and ....

Solution. Unequal.

Q. 11. Constant term that must be added to the equation $2x^2 - 5x + 3 = 0$ to solve it by the method of completing the square is ....

Solution. $\frac{2.5}{16}$.

Q. 12. If $x = 1$ is a common root of $ax^2 + ax + 3 = 0$ and $x^2 + x + b = 0$, then the value of $a^2 - b^2$ is ....

Solution. $\frac{-7}{4}$.

Q. 13. If 2 is a root of the equation $x^2 + ax - 8 = 0$ and the quadratic equation $x_2 + 2ax + k = 0$ has equal roots, then the value of k is ....

Solution. – 12.

Q. 14. Equation $ax^2 + bx + c = 0$ represents a quadratic equation if and only if ....

Solution. $a \neq 0$.

Q. 15. If $ax^2 + bx + c = 0$ has equal roots, then the value of c is ....

Solution. $\frac{b^2}{4a}$.

Q. 16. If $x^2 + 4x + l$ is a perfect squate, then l is ....

Solution. 4.

Q. 17. The nature of the roots of the quadratic equation $ax^2 + bx + c = 0, a \neq 0$ is ...., if $(b^2 - 4ac) < 0$.

Solution. Imaginary

Chapter 5. Arithmetic Progressions

Q. 1. The sum of first 10 whole numbers is ....

Solution. 45

Q. 2. If $Sn = 2n^2 - 7n$, then the nth term is ....

Solution. $4n - 9$.

Q. 3. If $S_{n+1} = n^2 + 9n$, then the second term of the A.P. is ....

Solution. 10.

Q. 4. If the common difference of an A.P. is 3, then the value of $a_{37} - a_{15}$ is ....

Solution. 66.

Q. 5. If $\frac{4}{5}$, a, 2 are in A.P, then the value of a is ....

Solution. $\frac{7}{5}$.

Q. 6. The nth term of the A.P. 2, 5, 8, ... is ....

Solution. $3n - 1$

Q. 7. The next term of the A.P. $\sqrt{9}, \sqrt{36}, \sqrt{81}$, ... is ....

Solution. $\sqrt{144}$

Q. 8. The seventh term of the A.P. 19, 14, 9, ... is ....

Solution. – 11

Q. 9. If the seventh term of an arithmetic progression exceeds its tenth term by 3, then the common difference will be ....

Solution. – 1

Q. 10. .... term of the A.P. 3, 15, 27, 39 will be 132 more than its 54th term.

Solution. 65th

Chapter 6. Coordinate Geometry

Q. 1. .... is the point of intersection of the coordinate axis.

Solution. origin

Q. 2. In the second quadrant, for a point, the abscissa is .... and the ordinate is ....

Solution. negative, positive.

Q. 3. The distance between the points (x, y) and (0, 0) is .... Units.

Solution. $\sqrt{x^2 + y^2}$

Q. 4. The distance of the point (– 9, 5) from the origin is ....

Solution. $\sqrt{106}$ units.

Q. 5. The distance between the points (3, 4) and (– 5, 2) is ....

Solution. $2\sqrt{17}$ units.

Q. 6. If three points are collinear, then the area of triangle formed by them is ....

Solution. zero.

Q. 7. The ratio in which the line segment joining A(1, – 5) and B(– 4, 5) is divided by the X-axis is ....

Solution. 1 : 1.

Q. 8. If a point lies on Y-axis, then the abscissa of that point is ....

Solution. zero.

Chapter 7. Triangles

Q. 1. ABC and PQR are two equilateral triangles such that P is the mid-point of BC. The ratio of areas of the two triangles is ....

Solution. 4 : 1.

Q. 2. The perimeters of two similar triangles ABC and XYZ are 35 cm and 45 cm respectively. Then the ratio of the areas of the two triangles is ....

Solution. 49 : 81.

Q. 3. If the areas of two similar triangles are equal, then these triangles are ....

Solution. Congruent.

Q. 4. $\Delta ABC \sim \Delta PQR$ with $\frac{BC}{DF} = \frac{2}{3}$. Then, $\frac{ar(\Delta PQR)}{ar(\Delta ABC)} =$ ....

Solution. $\frac{9}{4}$.

Q. 5. If in ΔABC and ΔDEF, $\frac{AB}{DF} = \frac{AC}{DE}$, then they will be similar if $\angle A =$ ....

Solution. $\angle D$.

Q. 6. If $\Delta ABC \sim \Delta DEF$, $A = 30^\circ$ and $\angle C = 50^\circ$, then $\angle F =$ ....

Solution. 50°.

Q. 7. If the lengths of the diagonals of a rhombus are 16 cm and 12 cm, then the length of its sides is ....

Solution. 10 cm.

Q. 8. In ΔABC, D and E are mid-points of AC and BC respectively such that DE || AB. If $AD = 2x$, $BE = 2x - 1$, $CD = x + 1$ and $CE = x - 1$, then the value of x is ....

Solution. $x = \frac{1}{3}$.

Q. 9. In the figure given below, DE || BC and $AD = \frac{1}{2}BD$. If BC = 4.5 cm, find DE.

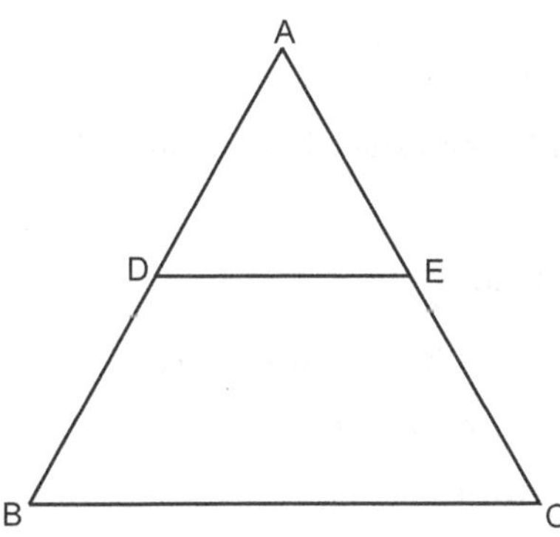

Solution. 1.5 cm

Q. 10. ABC is an isosceles triangle in which $\angle C = 90^\circ$ and AC = 4 cm. Then, $AB^2 =$ ....

Solution. 32 cm^2.

Q. 11. In figure, MN || BC and AM : MB = 1 : 2, then $\frac{ar(\Delta AMN)}{ar(\Delta ABC)} =$ ....

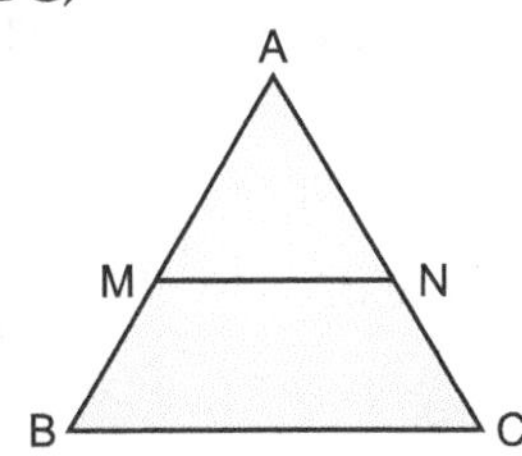

Solution : $\frac{1}{9}$

Q. 12. In ΔABC, $AB = 6\sqrt{3}$ cm, AC = 12 cm and BC = 6 cm, then $\angle B =$ ....

Solution : 90°

Q. 13. Two triangles are similar if their corresponding sides are ....

Solution. Proportional.

Q. 14. Given $\Delta ABC \sim \Delta PQR$, if $\frac{AB}{PQ} = \frac{1}{3}$, then $\frac{ar(\Delta ABC)}{ar(\Delta PQR)} =$ ....

Solution : $\frac{1}{9}$.

Q. 15. ABC is an equilateral triangle of side $2a$, then length of one of its altitude is ....

Solution : $\sqrt{3}\, a$

Q. 16. ABC and BDE are two equilateral triangles such that D is the mid-point of BC. Ratio of the areas of triangles ABC and BDE is ....

Solution : 4 : 1

Chapter 8. Circles

Q. 1. The lengths of two tangents drawn from an external point to a circle are ....

Solution. equal.

Q. 2. A circle may have .... parallel tangents.

Solution. two

Q. 3. The distance between two parallel tangents of a circle of radius 5cm is ....

Solution. 10 cm.

Q. 4. The perimeter of the ΔABC in the given figure is ....

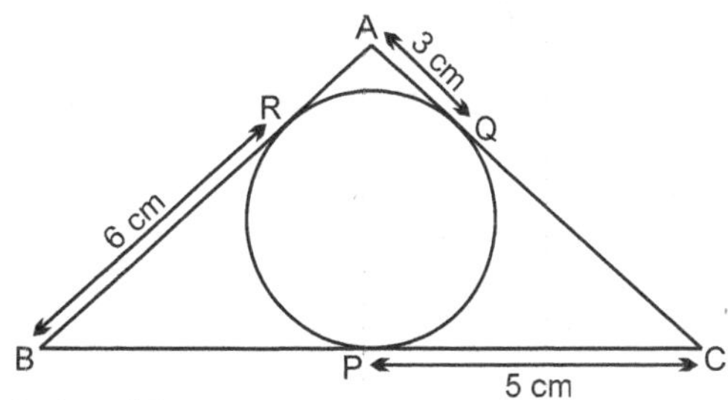

Solution. 28 cm.

Q. 5. If MP and MQ are two tangents to the circle with centre O, so that $\angle POQ = 120^\circ$, then $\angle PMQ$ is equal to ....

Solution. 60°.

Q. 6. The length of the tangent drawn from a point 10cm away from the centre of a circle of radius 6cm is ....

Solution. 8 cm.

Q. 7. If tangents PA and PB from a point P to a circle with centre O are inclined to each other at an angle of 100°, than ∠POA is equal to ....

Solution. 40°.

Q. 8. A line intersecting a circle in two points is called a ....

Solution. secant.

Q. 9. A tangent to a circle is a line that intersects the circle at .... point (s).

Solution. one

Q. 10. From a point Q, the length of the tangent to a circle is 12m and the distance of Q from the center is 13 cm. The radius of the circle is ....

Solution. 5 cm.

Q. 11. In given figure, the length PB = ... cm.

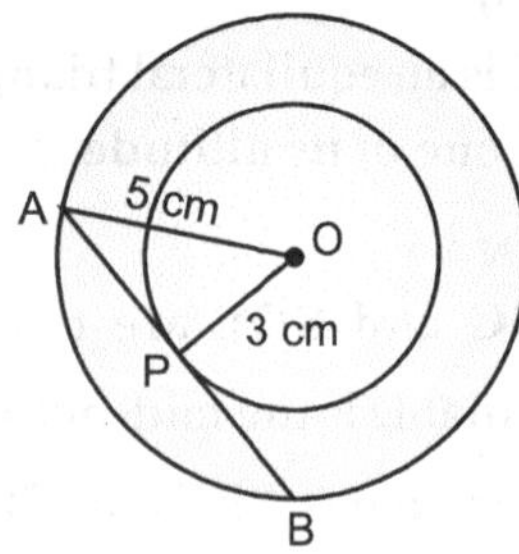

Solution : 4

Q. 12. In figure, ΔABC is circumscribing a circle, the length of BC is .... cm.

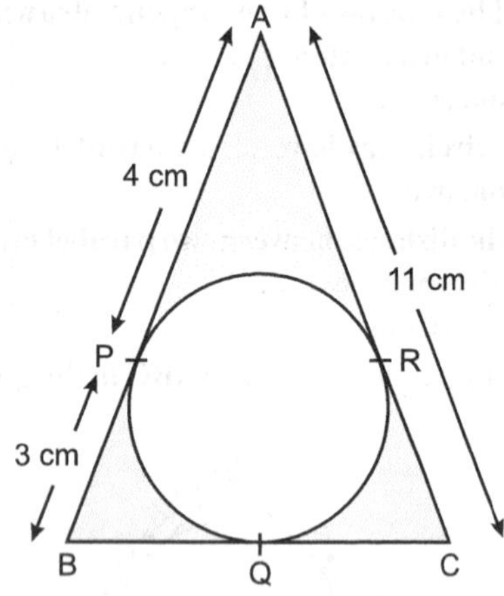

Solution : 10

Chapter 10. Trigonometry

Q. 1. Maximum value of $\frac{1}{\sec\theta}$ is ....

Solution. 1

Q. 2. If $3\cos\theta = 1$, then cosec θ is equal to ....

Solution. $\frac{3}{2\sqrt{2}}$.

Q. 3. If $\sin 5\theta = \cos 4\theta$, where 5θ and 4θ are acute angles, then the value of θ is ....

Solution. 10°

Q. 4. The value of $\sin^2 15° + \sin^2 75° - \tan^2 45°$ is ....

Solution. 0.

Q. 5. If A + B = 90° and $\tan A = \frac{3}{4}$, then the value of cot B is ....

Solution. $\frac{3}{4}$.

Q. 6. If $2\sin 2\theta = \sqrt{3}$, then the value of θ is ....

Solution. 30°.

Q. 7. The value of $\frac{\text{cosec } 65°}{\sec 25°}$ is ....

Solution. 1.

Q. 8. If ΔABC is right angled at C, then the value of cos (A + B) is ....

Solution. 0.

Q. 9. The value of $\sin^2 48° + \cos^2 48°$ is ....

Solution. 1.

Q. 10. $\frac{\sin 60°}{\cos 30°}$ = ...

Solution. 1

Q. 11. If tan θ + cot θ = 2, then the value of $\tan^2\theta + \cos^2\theta$ is ....

Solution. 2.

Q. 12. The value of θ for which $\sin 3\theta = \cos(\theta - 6°)$ is ....

Solution. 24°.

Q. 13. Two angle are said to be complementary, if their sum is ....

Solution. 90°.

Q. 14. The value of tan 5° tan 30° tan 85° is equal to ....

Solution. $\frac{4}{\sqrt{3}}$.

Q. 15. The value of (tan 1° tan 2° ..... tan 89°) is equal to ....

Solution : 1

Q. 16. The value of sin 23° cos 67° + cos 23° sin 67° is ....

Solution : 1

Q. 17. The value of $\frac{\tan 35°}{\cot 55°} + \frac{\cot 78°}{\tan 12°}$ is ....

Solution : 2

Q. 18. $\frac{\cos 80°}{\sin 10°}$ + cos 59° cosec 31° = ...

Solution : 2

Q. 19. The value of $\left(\sin^2\theta + \frac{1}{1+\tan^2\theta}\right)$ = ...

Solution : 1

Q. 20. The value of

$(1 + \tan^2 \theta)(1 - \sin \theta)(1 + \sin \theta) =$ ....

Solution : 1

Q. 21. $\left(\dfrac{\sin 35°}{\cos 55°}\right)^2 + \left(\dfrac{\cos 43°}{\sin 47°}\right)^2 - 2\cos 60° =$ ....

Solution : 1

Q. 22. $\dfrac{2\cos 67°}{\sin 23°} - \dfrac{\tan 40°}{\cot 50°} - \cos 0° =$ ....

Solution : 0

Chapter 11. Applications of Trigonometry

Q. 1. The height of the tower is 10m. Then the length of its shadow when sun's altitude is 45° is ....

Solution. 10m.

Q. 2. An observer, 1·5m tall, is 28·5m away from a 30m high tower, then the angle of elevation of the top of the tower from the eye of the observer is ....

Solution. 45°.

Q. 3. The length of shadow of a tower on the plane ground is $\sqrt{3}$ times the hight of a tower. The angle of elevation of sun is ....

Solution. 30°.

Q. 4. The angle of depression of a car parked on the road from the top of a 150m high tower is 30°, then the distance of the car from the tower (in meter) is ....

Solution. $150\sqrt{3}$.

Q. 5. If the height of the vertical pole is $\sqrt{3}$ times the length of its shadow on the ground, then the angle of elevation of the sun at that time is ....

Solution. 60°.

Q. 6. In figure, the angles of depressions from the observing positions O_1 and O_2 respectively of the object A are ...., ...

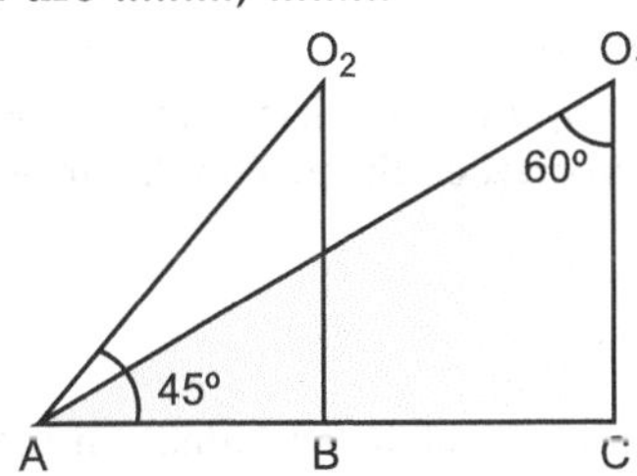

Solution : 30°, 45°

Q. 7. A ladder 10 m long reaches a window 8 m above the ground. The distance of the foot of the ladder from the base of the wall is .... m.

Solution : 6

Chapter 12. Areas related to Circles

Q. 1. Area of sector of circle is represented by the formula ....

Solution. $\dfrac{\theta}{360} \times \pi r^2$

Q. 2. Number of rounds that a wheel of diameter $\dfrac{7}{11}$ meter will make in moving a distance of 4 km is ....

Solution. 2000 rounds.

Q. 3. If the perimeter of a semi-circular protractor is 36 cm, then its diameter is ....

Solution. 14 cm.

Q. 4. A pendulum swings through an angle of 30° and describes an are 6·6 cm in length then the length of pendulum is ....

Solution. 12·6 cm.

Q. 5. If a chord of a circle of radius 14 cm makes an angle of 90° at the centre, then the area of major segment is ....

Solution. 560 cm^2.

Q. 6. If the radius of a circle is dimished by 10%, then its area is dimished by ....

Solution. 19 %.

Q. 7. The area of a square is scribed in a circle of diameter q cm is ....

Solution. $\dfrac{q^2}{2}$ cm^2.

Q. 8. If the area of a semi circular region is 308 sq cm, then its perimeter is ....

Solution. 72 cm.

Chapter 13. Surface Areas and Volumes

Q. 1. The formula of curved surface area of a frustum is ....

Solution. $\pi (r_1 + r_2) l$.

Q. 2. The radii of two cones are in the ratio 2 : 1 and their volumes are equal, then ratio of their heights is ....

Solution. 1 : 4.

Q. 3. A cylinder with done radius 7 cm and height 3 cm is melted to form a cone of height 9 cm. The radius of the cone is ....

Solution. 7 cm.

Q. 4. The volume of two spheres are in the ratio 64 : 27, then the ratio of their surface areas is ....

Solution. 16 : 9.

Q. 5. A solid sphere of radius r is melted and cost into the shape of a solid cone of height r, the redius of the base of the cone is ....

Solution. $2r$.

Q. 6. A solid sphere of radius 8 cm can be recasted into........... balls of radius 2 cm.

Solution. 64

Q. 7. If the length of each side of a cube is doubled, then its volume becomes........... times.

Solution. eight

Q. 8. A solid frustum is height 8 cm. It the radii of its lower and upper ends are 3 cm and 9 cm respectively, then its slant height is........... .

Solution. 10 cm.

Q. 9. A rectangular sheet of paper 40 cm × 22 cm, is rolled to form a hollow cylinder of height 40 cm. The radius of the cylinder, (in cm) is........... .

Solution. 3·5 cm.

Q. 10. The total surface area of a hemisphere of radius *r* is........... .

Solution. $3\pi r^2$.

Chapter 14. Statistics

Q. 1. Empirical relation between mean, median and mode is........... .

Solution. Mode = 3 Median –2 Mean.

Q. 2. Mode is the........... frequent value.

Solution. most.

Q. 3. Arithmetic mean of 1, 2, 3, ... *n*, is........... .

Solution. $\frac{n+1}{2}$.

Q. 4. .... is calculated using the formula :

$$l+\left(\frac{\frac{N}{2}-cf}{f}\right)\times h$$

Solution. Median

Q. 5. The mode of a frequency distribution can be determined graphically from........... .

Solution. Histogram.

Q. 6. The mean of first ten multiples of 2 is........... .

Solution. 11.

Q. 7. If the mode of the data : 16, 15, 17, 16, 15, x, 19, 17, 14 is 17, then the value of x is........... .

Solution. 17.

Q. 8. The arithmetic mean and mode of a data are 24 and 12 respectively, then its median is........... .

Solution. 20.

Q. 9. An ogive is used to determine........... .

Solution. Median.

Q. 10. If the mean of observations $x_1, x_2, x_3, ... x_n$ is $\bar{x}$, then the mean of $ax_1, ax_2, ax_3, ... ax_n$ is........... .

Solution. $a\bar{x}$.

Chapter 15. Probability

Q. 1. The probability of a sure event is........... .

Solution. 1.

Q. 2. The probability of 53 sundays in a leap year is........... .

Solution. $\frac{2}{7}$.

Q. 3. If $\overline{E}$ denote the negation of an event E, then the value of $P(E)+P(\overline{E})$ is........... .

Solution. 1.

Q. 4. One card is drawn at random from a well shuffled deck of 52 cards, then the probability of getting a black Jack is........... .

Solution. $\frac{1}{26}$.

Q. 5. The probability of getting a consonant from 'MATHEMATICS' is........... .

Solution. $\frac{7}{11}$.

Q. 6. The probability of any event lies between........... and........... .

Solution. 0, 1.

Q. 7. In a family of 3 children, the probability of having at most two girls is........... .

Solution. $\frac{7}{8}$.

Q. 8. A number x is chosen at random from the numbers – 4, – 3, – 2, – 1, 0, 1, 2, 3, 4 the probability that $|x| \le 3$ is........... .

Solution. $\frac{7}{9}$.

Q. 9. A bag contains used and 6 green balls. A ball is drawn at random, then the probability of drawing a red ball is........... .

Solution. $\frac{2}{5}$.

Q. 10. Probability of getting exactly 2 tails when two coins are tossed simultaneously is........... .

Solution. $\frac{1}{4}$.

Q. 11. The probability that it will rain tomorrow is 0.64, then the probability that it will not rain tomorrow is........... .

Solution. 0.36.

Q. 12. The probability of an impossible events is ....

Solution. 0.

Q. 13. A number is selected at random from first 50 natural number, then the probability of getting a perfect square is........... .

Solution. $\frac{7}{50}$.

Q. 14. The sum of the probability of all the elementary events of an experiment is........... .

Solution. 1.

Q. 15. In a single throw of a die, the probability of getting multiple of 2 is........... .

Solution. $\frac{1}{2}$.

●●

Passage Based Questions

Chapter 1. Real Numbers

Q. 1. In the small library, librarian sunita have three sets of books English, Hindi and Mathematics namely. She have to stack them in such a way that all the books are stored topic-wise and height of each stack should be same. There are 96 English books, 240 Hindi books and 336 Mathematics books.

Based on the above information, answer the following questions :

(i) How many books will be there in each stack?

(ii) Find the number of stacks of each English, Hindi and Mathematics books?

(iii) What is the total number of stacks formed?

Sol : (i) In order to arrange the books as required, we have to find the H.C.F of 96, 240 and 336.

$$96 = 2^5 \times 3$$
$$240 = 2^4 \times 3 \times 5$$
$$336 = 2^4 \times 3 \times 7$$

H.C.F. (96, 240, 336) = $2^4 \times 3 = 48$

Hence, there will be 48 books in each stack.

(ii) Number of stacks of English books

$$= \frac{\text{Number of English books}}{\text{Number of books in each stack}}$$

$$= \frac{96}{48} = 2$$

Number of stacks of Hindi books

$$= \frac{\text{Number of Hindi books}}{\text{Number of books in each stack}}$$

$$= \frac{240}{48} = 5$$

Number of stacks of Maths books

$$= \frac{\text{Number of Maths books}}{\text{Number of books in each stack}}$$

$$= \frac{336}{48} = 7.$$

(iii) Total number of stacks = 2 + 5 + 7 = 14.

Ans.

Q. 2. Suresh wants to renovate his house. He has to renovate ceiling of his room by putting square shape tiles on it. Ceiling of the room is 8 m 25 cm long and 6m 75 cm broad.

Based on the given information, answer the following questions :

(i) Find the dimensions of each tiles.

(ii) Find the number of tiles required for the project.

Sol : (i) Dimensions of tiles can be found by finding the H.C.F. of 8 m 25 cm and 6 m 75 cm

8 m 25 cm = 825 cm.

6 m 75 cm = 675 cm.

```
675 ) 825 ( 1
      675
      150 ) 675 ( 4
            600
             75 ) 150 ( 2
                  150
                   ×
```

∴ H.C.F. (825, 675) = 75.

So, side of each square tile is 75 cm.

(ii) Required number of titles = $\frac{\text{Area of ceiling}}{\text{Area of each tile}}$

$$= \frac{825 \times 675}{75 \times 75} = 99.$$

Q. 3. Sandhya on her very first day of her job, which was in a bank notice that there are six bells which keep on tolling at regular intervals. She noticed that toll of their intervals are 2, 4, 6, 8, 10, 12 minutes respectively. If all the six bells commence tolling together, at 10 AM, then answer the following questions :

(i) At what time will they again toll together?

(ii) How many times these bells will toll together during the working hours of Sandhya's job, if Sandhya works for 8 hours in a day ?

Sol : (i) In order to calculate the time interval at which bells toll together, we need to find out the LCM of 2, 4, 6, 8, 10, 12

By prime factorisation, we get

$$2 = 2,\ 4 = 2^2$$
$$6 = 2 \times 3,\ 8 = 2^3$$
$$10 = 2 \times 5,\ 12 = 2^2 \times 3$$

$\Rightarrow$ LCM (2, 4, 6, 8, 10, 12) = $2^3 \times 3 \times 5 = 120$ min.

So, after every 120 min. or 2 hrs. bell will toll together.

The time, these bells will again toll together

= 10 AM + 2 hours

= 12 noon

(ii) Number of times = $\frac{8}{2} = 4$

Chapter 2. Polynomials

Q. 1. Read the following passage and answer the questions that follows :

A teacher told 10 students to write a polynomial on the black board.

Students wrote

1. $x^2 + 2$ **6.** $x - 3$
2. $2x + 3$ **7.** $x^4 + x^2 + 1$
3. $x^3 + x^2 + 1$ **8.** $x^2 + 2x + 1$
4. $x^3 + 2x^2 + 1$ **9.** $2x^3 - x^2$
5. $x^2 - 2x - 1$ **10.** $x^4 - 1$

(i) How many students wrote cubic polynomial?

(ii) Divide the polynomial $(x^2 + 2x + 1)$ by $(x + 1)$.

Sol : (i) 3 students wrote cubic polynomial

(ii)
$$x+1)\,x^2+2x+1\,(x+1$$
$$\begin{array}{l} x^2 + x \\ -\;- \\ \hline x + 1 \\ x + 1 \\ -\;- \\ \hline \times \end{array}$$

$$\therefore \quad \frac{x^2+2x+1}{x+1} = x+1 \qquad \textbf{Ans.}$$

Q. 2. A teacher asked 10 of his students to write a polynomial in one variable on a paper and then to handover the paper. The following were the answers given by the students :★

$2x + 3$, $3x^2 + 7x + 2$, $4x^3 + 3x^2 + 2$, $x^3 + \sqrt{3x} + 7$, $7x + \sqrt{7}$, $5x^3 - 7x + 2$, $2x^2 + 3 - \frac{5}{x}$, $5x - \frac{1}{2}$, $ax^3 + bx^2 + cx + d$, $x + \frac{1}{x}$

Answer the following questions :

(i) How many of the above ten, are not polynomials?

(ii) How many of the above ten, are quadratic polynomials?

Solution : (i) 3

$$\left(x^2 + \sqrt{3x} + 7, 2x^2 + 3 - \frac{5}{x} \text{ and } x + \frac{1}{x}\right)$$

(ii) There is only one quadratic polynomial which is $3x^2 + 7x + 2$.

Chapter 3. Pair of Linear Equations in Two Variables

Q. 1. Ravi books a ola cab for his daily tour. Ola charges consist of fixed charges per day and the remaining charges depending upon the distance travelled in kilometres. If Ravi travels 110 km and he pays ₹ 1130. Next day he travels 200 km and pays ₹ 1850.

Based on above information, answer the following questions :

(i) What are the fixed charges and the running charges of the cab?

★ are board exam questions from previous years

(ii) What will he pay for his tour of 250 kms.

Sol : (i) Let the fixed charges for cab be ₹ x and running charges be ₹ y per km.

According to the given condition, we have

$$x + 110y = 1130 \qquad \text{...(i)}$$
$$x + 200y = 1850 \qquad \text{...(ii)}$$

Substracting equation (i) from (ii) we get

$$90y = 720$$
$$y = 8$$

Putting $y = 8$ in equation (i) we get,

$$x + 880 = 1130$$
$$x = 1130 - 880$$
$$x = 250$$

Hence, the fixed charges are ₹ 250 and running charges are ₹ 8 per km.

(ii) Total charges for travelling 250 km =

$$x + 250y = 250 + 2000$$
$$= 2,250$$

Hence, Ravi has to pay ₹ 2,250 for travelling 250 km.

Q. 2. Class IX consists of 2 sections A and B. From section A, 8 boys and 12 girls and from section B, 6 boys and 8 girls participated in a project. Section A can complete the project in 10 days and Section B can finish it in 14 days.

Based on the following information, answer the following questions :

(i) Find out in how many days only boys from both the section together can finish the project.

(ii) Also find out in how much time girls from both the section can finish the same project.

Sol : Suppose that one boy alone can finish the work in x days and one girl alone can finish it in y days. Then,

$$\text{One boy's one day's work} = \frac{1}{x}$$

$$\text{One girl's one day's work} = \frac{1}{y}$$

$$\therefore \quad \text{8 boy's one day's work} = \frac{8}{x}$$

$$\text{12 girl's one day's work} = \frac{12}{y}$$

Since, 8 boys and 12 girls can finish in 10 days

$$10\left(\frac{8}{x} + \frac{12}{y}\right) = 1$$

$$\frac{80}{x} + \frac{120}{y} = 1 \qquad \text{...(i)}$$

Again 6 boys and 8 girls can finish the work in 14 days.

$$\therefore \quad 14\left(\frac{6}{x} + \frac{8}{y}\right) = 1 \Rightarrow \frac{84}{x} + \frac{112}{y} = 1 \qquad \text{...(ii)}$$

putting $\frac{1}{x} = u$ and $\frac{1}{y} = v$ in equation (i) and (ii), we get

$$80u + 120v - 1 = 0 \quad ...(iii)$$
$$84u + 112v - 1 = 0 \quad ...(iv)$$

By using cross-multipication, we have

$$\frac{u}{-120+112} = \frac{-v}{-80+84} = \frac{1}{80\times112-120\times84}$$

$$\Rightarrow \frac{u}{-8} = \frac{v}{-4} = \frac{1}{-1120}$$

$$\Rightarrow u = \frac{8}{1120} = \frac{1}{140} \text{ and } v = \frac{-4}{-1120} = \frac{1}{280}$$

Now, $u = \frac{1}{140} \Rightarrow \frac{1}{x} = \frac{1}{140} \Rightarrow x = 140$

$v = \frac{1}{280} \Rightarrow \frac{1}{y} = \frac{1}{280} \Rightarrow y = 280.$

(i) One boy's one day's work = $\frac{1}{140}$

Total number of boys in both the section = 14

14 boy's one day's work = $\frac{14}{140} = \frac{1}{10}$

$\therefore$ 14 boys can finish the project in 10 days

(ii) One girl's one day's work = $\frac{1}{280}$

Total number of girls in both the section = 20

20 girl's one day's work = $\frac{20}{280} = \frac{1}{14}$

$\therefore$ 20 girls can finish the project in 14 days.

Q. 3. When Suresh asked Sanjay about his age, Sanjay answered it in Riddle. Sanjay said, "I am 3 times as old as my son, but in 12 years, I will be twice as old as my son". After reading the above information, answer the following questions :

(i) What is the present age of Sanjay and his son?

(ii) What will be their ages after 10 years?

Sol : (i) Let present ages of Sanjay be x years and has son be y years. Then,

$$x = 3y \quad ...(i)$$

12 years later, Sanjay's age will be $(x + 12)$ years and his son's age will be $(y + 12)$ years.

$$\therefore \quad x + 12 = 2(y + 12)$$
$$x + 12 = 2y + 24$$
$$\Rightarrow \quad x - 2y - 12 = 0 \quad ...(ii)$$

Putting $x = 3y$ in equation (ii), we get

$$3y - 2y - 12 = 0$$
$$\Rightarrow \quad y = 12$$

Putting $y = 12$ in equation (i), we get

$$x = 3 \times 12$$
$$= 36$$

i.e., Present age of Sanjay is 36 years and his son's age is 12 years.

(ii) After 10 years, Sanjay's age = 36 + 10 = 46 years. and his son's age = 12 + 10 = 22 years. **Ans.**

Chapter 4. Quadratic Equations

Q. 1. During the battle of Mahabharta, Arjun carried some arrows for fighting with Bheeshma. With half of the arrows, he cut down the arrows thrown by Bheeshma on him and with six other arrows he killed the rath driver of Bheeshma. With one arrow each, he knocked down respectively the rath, flag and the bow of Bheeshma. Finally, with one more than four times the square root of total arrows, he laid Bheeshma unconscious on an arrow bed. Based on above information answer the following questions :

(i) Find the total number of arrows Arjun had.

(ii) Find the number of arrows that Arjun used to lay unconsicious Bheeshma.

Sol : (i) Suppose Arjun had x arrows.

Number of arrows used to cut arrows of Bheeshma $= \frac{x}{2}$

Number of arrows used to kill the rath driver = 6

Number of arrows used to lay Bheeshma unconscious $= 4\sqrt{x} + 1$

According to question we have

$$\therefore \quad \frac{x}{2} + 6 + 3 + 4\sqrt{x} + 1 = x$$
$$\Rightarrow \quad x + 20 + 8\sqrt{x} = 2x$$
$$\Rightarrow \quad x = 20 + 8\sqrt{x}$$

Putting $x = y^2$, the above equation becomes

$$y^2 = 20 + 8y$$
$$\Rightarrow \quad y^2 - 8y - 20 = 0$$
$$\Rightarrow \quad y^2 - 10y + 2y - 20 = 0$$
$$\Rightarrow \quad (y - 10)(y + 2) = 0$$
$$\Rightarrow \quad y = 10 \text{ or } y = -2$$
$$\Rightarrow \quad y = 10 \quad [\because y \text{ cannot be negative}]$$
$$\Rightarrow \quad x = y^2 \Rightarrow x = 100$$

Hence, Arjun had 100 arrows.

(ii) Required no. of arrows= $4\sqrt{x} + 1$

$$= 4\sqrt{100} + 1$$
$$= 40 + 1 = 41. \quad \textbf{Ans.}$$

Q. 2. Abhay is planning a campaign. He investigates three possible routes.

- **If he takes route A, which is 600 km long, he expects to cover x km per day.**
- **Route B, whose distance is equal to the route A, has more difficult conditions and he would only expect to cover $(x - 5)$ km per day.**
- **Route C, which is 200 km longer than route A, has easier conditions and he would expect to cover $(x + 5)$ km per day.**

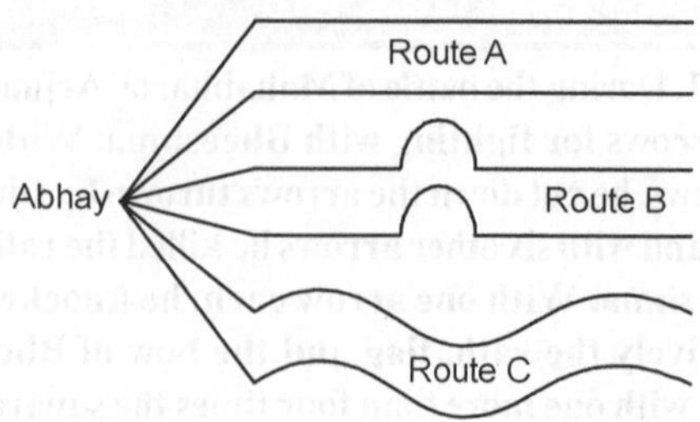

Based on the following figure and given information, answer the following questions :

(i) Abhay takes 20 days less, if he takes route C instead of route B. According to this statement, form an equation in x and reduce it to standard form.

(ii) Find the number of days taken by Abhay, if he choose route A.

Sol : (i) Number of days taken by Abhay, if he choose route C $= \frac{800}{x+5}$

Number of days taken by Abhay, if he choose route B

$$= \frac{600}{x-5}$$

$$\Rightarrow \quad \frac{800}{x+5} + 20 = \frac{600}{x-5}$$

$$\Rightarrow \quad \frac{800}{x+5} - \frac{600}{x-5} = -20$$

$$\Rightarrow \quad \frac{40(x-5) - 30(x+5)}{(x+5)(x-5)} = -1$$

$$\Rightarrow \quad \frac{40x - 200 - 30x - 150}{x^2 - 25} = -1$$

$$\Rightarrow \quad 10x - 350 = -x^2 + 25$$

$$\Rightarrow \quad x^2 + 10x - 375 = 0$$

(ii) On solving the equation $x^2 + 10x - 375 = 0$

We have,

$$x = \frac{-10 \pm \sqrt{100+1500}}{2}$$

$$= \frac{-5 \pm 40}{2}$$

$$= 17{\cdot}5 \text{ or } -22{\cdot}5 \quad (x \neq -22{\cdot}5)$$

So, the number of days, Abhay takes on route A

$$= \frac{600}{x}$$

$$= \frac{600}{17.5} \sim 34{\cdot}28.$$ **Ans.**

Q. 3. In the courtyard at the back of the house. Sunita prepared a small beautiful garden. The courtyard is of square shaped whose each side is 44 m. At the centre, she prepared a square flower bed leaving a gravel path all around it. One day a friend of her visited and praised the garden a lot. She also asked for the cost of laying the flower bed and gravelling the path, then sunita told her that the total cost of laying the flower bed and gravelling the path at ₹ 2.75 and ₹ 1.50 per square metre, respectively, is ₹ 4904. Using the above information find :

(i) The width of the path for gravelling.

(ii) The cost of laying the flower bed.

(iii) The cost of gravelling the path.

Sol. Let the width of the gravel path be x metres. Then, each side of the square flower bed is $(44 - 2x)$ metres.

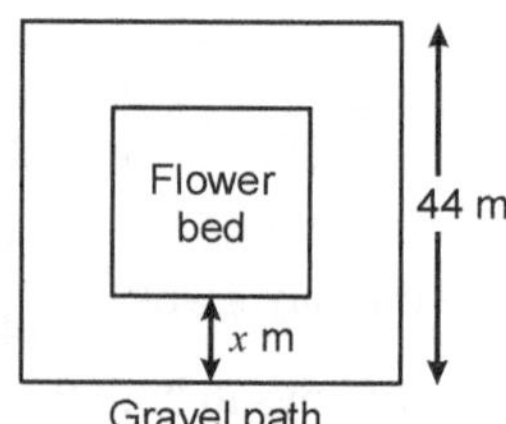

Now,

Area of the square field $= 44 \times 44 = 1936 \text{ m}^2$

Area of the flower bed $= (44 - 2x)^2 \text{ m}^2$

Area of gravel path

$= $ Area of square field – Area of flower bed

$= 1936 - (44 - 2x)^2$

$= 1936 - (1936 - 176x + 4x^2)$

$= (176x - 4x^2) \text{ m}^2$

Cost of laying the flower bed

$$= (44 - 2x)^2 \times \frac{275}{100}$$

$$= \frac{11}{4}(44 - 2x)^2$$

$$= 11(22 - x)^2$$

Cost of gravelling the path$= (176x - 4x^2)\frac{150}{100}$

$$= 6(44x - x^2)$$

It is given that the total cost of laying the flower bed and gravelling the path is ₹ 4904

$$\therefore \quad 11(22 - x)^2 + 6(44x - x^2) = 4904$$

$$\Rightarrow \quad 11[484 - 44x + x^2] + 264x - 6x^2 = 4904$$

$$\Rightarrow \quad 5x^2 - 220x + 5324 = 4904$$

$$\Rightarrow \quad 5x^2 - 220x + 420 = 0$$

$$\Rightarrow \quad x^2 - 44x + 84 = 0$$

$$\Rightarrow \quad x^2 - 42x - 2x + 84 = 0$$

$$\Rightarrow \quad x(x - 42) - 2(x - 42) = 0$$

$$\Rightarrow \quad (x - 2)(x - 42) = 0$$

$$\Rightarrow \quad x = 2 \text{ or } x = 42$$

By $x \neq 42$, as the side of the square is 44 m.

Therefore, $x = 2$

Hence, the width of the gravel path is 2.

(ii) Cost of laying flower bed = Area of flower bed $\times$ rate per m^2.

Area of flower bed $= [44 - 2 \times 2]^2$

$= (44 - 4)^2$

$= (40)^2 = 1600\ m^2$.

Hence, cost of laying flower bed= 1600×2.75

$= ₹\ 4400$. **Ans.**

(iii) Cost of gravelling the path = Area of path × Cost per m^2

Area of path = Area of square field – Area of flower bed

$= [(44)^2 - (40)^2] m^2$

$= [1936 - 1600] m^2$

$= 336\ m^2$

Total cost of gravelling the path $= 336 \times 1.50$

$= ₹\ 504$

Hence, the cost of gravelling the path is ₹ 504.

Ans.

Q. 4. Priyanka had her birthday next week. Some of her friends plan to arrange a surprise party for her. The budget for food was ₹ 2400. But at the end moment 4 of her friends refused to come in the party. Due to this, the cost of food for each friend went up by ₹ 50.

Based on the above information, answer the following questions :

(i) How many friends planned the party?

(ii) How many friends actually joined the party?

(iii) What was the final contribution each of them did?

Sol : Let the total number of friends be x.

$\therefore$ Share of each friend = ₹ $\frac{2400}{x}$

When 4 friends refused to join, then share of each friend = ₹ $\frac{2400}{x-4}$

According to the question,

$$\frac{2400}{x-4} - \frac{2400}{x} = 50$$

$$\Rightarrow \frac{2400x - 2400(x-4)}{x(x-4)} = 50$$

$$\Rightarrow \frac{2400x - 2400x + 9600}{x^2 - 4x} = 50$$

$$\Rightarrow \frac{9600}{x^2 - 4x} = 50$$

$$\Rightarrow x^2 - 4x = 192$$

$$\Rightarrow x^2 - 4x - 192 = 0$$

$$\Rightarrow x^2 - 16x + 12x - 192 = 0$$

$$\Rightarrow x(x - 16) + 12(x - 16) = 0$$

$$\Rightarrow (x - 16)(x + 12) = 0$$

$$x = 16 \text{ or } -12$$

[$x = -12$, rejected as number of friends cannot be negative]

(i) Total number of friends, who planned for the party were 16.

(ii) Number of friends, who actually joined the party were $16 - 4 = 12$.

(iii) Finaly contribution by each of them is ₹ $\frac{2400}{12}$ = ₹ 200. **Ans.**

Q. 5. An industry produces a certain number of toys in a day. On a particular day, the cost of production of each toy was 9 less than twice the number of toys produced on that day. The total cost of production on that day was ₹ 143.

Based on the given information, answer the following questions :

(i) Find the number of toys produced in the industry on that day.

(ii) What is the cost of each toy?

Sol : (i) Let the number of toys be n

According to the question,

$$\text{Cost of each toy} = 2n - 9$$

$$\text{Total cost of toys} = n(2n - 9)$$

But $\quad$ Total cost = ₹ 143

$$\therefore \quad n(2n - 9) = 143$$

$$\Rightarrow \quad 2n^2 - 9n - 143 = 0$$

$$\Rightarrow \quad n = \frac{-(-9) \pm \sqrt{(-9)^2 - 4 \times 2 \times (-143)}}{2 \times 2}$$

$$\Rightarrow \quad = \frac{9 \pm \sqrt{81 + 1144}}{4} = \frac{9 \pm \sqrt{1225}}{4} = \frac{9 \pm 35}{4}$$

$$= \frac{44}{4}, -\frac{26}{4}$$

$$\Rightarrow \quad n = 11, -\frac{26}{4}$$

Since, the number of toys cannot be negative or fraction,

$\therefore$ 11 toys were produced on that day.

(ii) Cost of each toy = ₹ $(2 \times 11) - 9)$ = ₹ 13

Q. 6. In a society, there is a big swimming pool. It has three pipes with uniform flow to fill the swimming pool. If first two pipes operate simultaneously then they fill the pool in same time during which the pool is filled by the third pipe alone. If second pipe is operated along it fills the pool five hour faster than the first pipe and four hours slower than the third pipe. Pool is closed for monthly maintenance. Based on the above information, answer the following questions :

(i) Find the time required by each pipe to fill the pool separately.

(ii) If all the three pipes are opened simultaneously then in how much time the pool be filled?

Sol : Let x be the number of hours required by the second pipe along to fill the pool. Then, the first pipe takes $(x + 5)$ hours, while the third pipe takes $(x - 4)$ hours to fill the pool. So, the parts of the pool filled by the first, second and third pipes in one hour are $\frac{1}{x+5}, \frac{1}{x}, \frac{1}{x-4}$ respectively.

Let the time taken by the first and second pipes to fill the pool simultaneously by t hours then, the third pipe also takes the same time to fill the pool.

$$\therefore \quad \left(\frac{1}{x+5}+\frac{1}{x}\right)t = \left(\frac{1}{x-4}\right)t$$

$$\Rightarrow \quad \frac{1}{x+5}+\frac{1}{x} = \frac{1}{x-4}$$

$$\Rightarrow \quad (2x+5)(x-4) = x^2+5x$$

$$\Rightarrow \quad x^2-8x-20 = 0$$

$$\Rightarrow \quad x^2-10x+2x-20 = 0$$

$$\Rightarrow \quad (x-10)(x+2) = 0$$

$$\Rightarrow \quad x = 10 \text{ or } x = -2$$

But time cannot be negative, So, $x = 10$.

Hence, the time required to fill the pool by first pipe is 15 hrs.

Second pipe is 10 hrs and third pipe is 6 hrs. **Ans.**

Chapter 5. Arithmetic Progressions

Q. 1. A man took a loan of ₹ 325,000 for his business. He repays it in instalments. In the first month he paid ₹ 2000, in the second month he paid ₹ 3500, in the third month he paid ₹ 5000, and so on.

Based on the given information, answer the following questions :

(i) How long will it take to clear the loan?

(ii) What amount he has to pay for his last instalment?

Sol : (i) Let, the number of months taken to clear the loan be n.

Clearly, the amounts form an A.P. with first term 2000 and the common difference = 3500 – 2000

= 1500

Sum of amount = 325000

$$\therefore \quad S_n = 325000$$

$$\Rightarrow \quad \frac{n}{2}[2\times 2000+(n-1)1500] = 325000$$

$$\Rightarrow \quad \frac{n}{2}[4000+1500n-1500] = 325000$$

$$\Rightarrow \quad \frac{n}{2}[2500+1500n] = 325000$$

$$\Rightarrow \quad n[3n+5] = \frac{325000\times 2}{500}$$

$$\Rightarrow \quad 3n^2+5n-1300 = 0$$

$$\Rightarrow \quad 3n^2+65n-60n-1300 = 0$$

$$\Rightarrow \quad n(3n+65)-20(3n+65) = 0$$

$$\Rightarrow \quad (3n+65)(n-20) = 0$$

$$\Rightarrow \quad n = 20 \text{ or } n = \frac{-65}{3}$$

$$\Rightarrow \quad n = 20 \qquad \left[\because n \neq \frac{-65}{3}\right]$$

So, it will take 20 months to clear the loan.

(ii) Amount of last instalment is

$$a_n = a+(n-1)d$$

$$a_{20} = 2000+(20-1)\times 1500$$

$$= 2000+19\times 1500$$

$$= 2000+25800$$

$$= 30500$$

So, the amount of last instalment ₹ 30500. **Ans.**

Q. 2. Ramkali would need ₹ 20,000 for admission fee and books etc., for her daughter admitted in a school after a year. She saved ₹ 500 in the first month of this year and increased her monthly saving by ₹ 200 every month.

Based on above information answer the following questions :

(i) After a year, how much money will she save?

(ii) How much short or more, she have in the end of a year?

Sol : Her savings in 12 months are 500, 700, 900, ...

Clearly, her savings form an A.P. with first term 500 and common difference 200.

Sum of amount after 1 year

$$= \frac{12}{2}[2\times 500+(12-1)200]$$

$$= 6[1000+2200]$$

$$= 6\times 3200 = 19200$$

After one year, she will save ₹ 19200.

(ii) She need ₹ 20,000 but she has ₹ 19200 at the end of year.

So, she is short with ₹ (20,000 – 19,200) i.e., ₹ 800. **Ans.**

Q. 3. 325 logs are stacked in the manner that there are 27 logs in the bottom row, 26 logs in the next row 25 logs in the row next to it and so on.

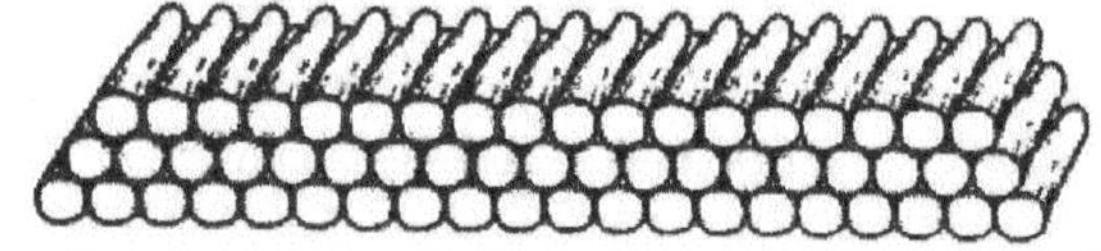

Base on the given information, answer the following questions :

(i) In how many rows are the 325 logs placed ?

(ii) How many logs are in the top row ?

(iii) which row is the middle row ?

Sol. (i) Clearly, logs stacked in each row form a sequence

27, 26, 25, 24,........

It is an A.P. with $a = 27$, $d = 26 - 27 = -1$

(i) Let there are n rows

$$S_n = \frac{n}{2}[2a+(n-1)d]$$

$$\Rightarrow \quad 325 = \frac{n}{2}[2\times 27+(n-1)(-1)]$$

$$\Rightarrow \quad 750 = n[55-n]$$

$$\Rightarrow \quad n^2-55n+750 = 0$$

$$\Rightarrow n^2-30n-25n+750 = 0$$

$\Rightarrow n(n-30)-25(n-3) = 0$

$\Rightarrow \quad (n-25)(n-30) = 0$

$\Rightarrow \quad n = 25$ or 30

For $n = 30$, we have

$$a_{30} = 27 + 29 \times (-1)$$
$$= 27 - 29 = -2$$

$\because$ Number of logs in the 30th row is – 2, which not possible

So, $\quad n = 25$

Thus there are, 25 rows.

(ii) Number of logs in the 25th row (top row) = a_{25}

$$\Rightarrow \quad a_{25} = a + 24d$$
$$= 27 + 24 \times (-1)$$
$$= 27 - 24 = 3 \qquad \textbf{Ans.}$$

(iii) Middle $= \left(\dfrac{25+1}{2}\right)^{th}$ row

$= 13$th row. **Ans.**

Q. 4. In a race, a bucket is placed at the starting point, which is 6 m apart from the first ball, and the other balls are placed 4 m apart in a straight line. There are 15 balls in the line. A competitor starts from the bucket, picks up the nearest ball, runs back with it, drops it in the bucket, runs back to pick up the next ball, runs to the bucket to drop it in and she continues in the same way untill all the balls are in the bucket.

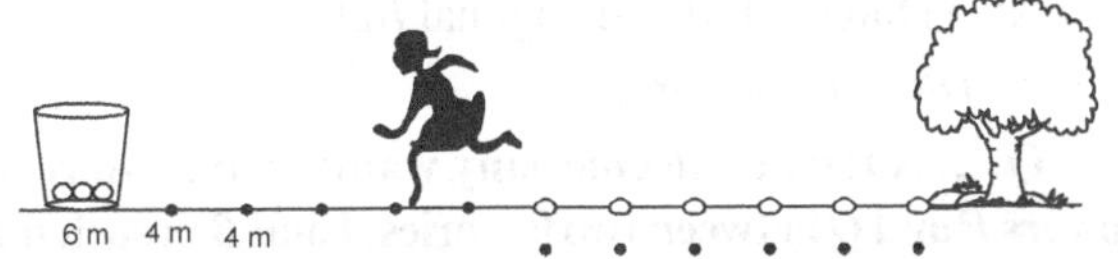

(i) Find the distance travelled by the competitor to pick up the seventh ball.

(ii) What is the total distance the competitor has to run?

(iii) If there are 20 balls in the line, then find the distance travelled by the competitor.

Sol : (i) Distance travelled to pick up first ball

$$= 2 \times 6 = 12 \text{ m}$$

Distance travelled to pick up second ball

$$= 2 \times (6 + 4) = 20 \text{ m}$$

Distance travelled to pick up third ball

$$= 2 \times (6 + 4 + 4) = 28 \text{ m}$$

The series is 12, 20, 28, ...

It forms an A.P.

Here, $n = 15$, $a = 12$ and $d = 20 - 12 = 8$

Distance travelled by competitor to pick up seventh ball

$$a_7 = a + (7-1)d$$
$$a_7 = 12 + 6 \times 8 = 12 + 48 = 60 \text{ m}$$

Ans.

(ii) We know, $\quad S_n = \dfrac{n}{2}[2a + (n-1)d]$

$$S_{15} = \frac{15}{2}[2 \times 12 + (15-1)\,8]$$
$$= \frac{15}{2}[24 + 14 \times 8]$$
$$= \frac{15}{2} \times 136$$
$$= 1020 \text{ m.}$$

Total distance travelled = 1020 m. **Ans.**

(iii) $\quad S_{20} = \dfrac{20}{2}[2 \times 12 + (20-1)8]$

$$= 10\,[24 + 19 \times 8]$$
$$= 10 \times 176$$
$$= 1760 \text{ m.} \qquad \textbf{Ans.}$$

Q. 5. The diagram shows a sequence of square wire frames. The lengths of the side of these frames are x cm, $(x + 2)$ cm, $(x + 4)$ cm ...... respectively. The sum of the areas of the first three squares is 440 sq. cm.

Based on the following information answer the following questions :

(i) Express the length of side of the nth frame in terms of x and n.

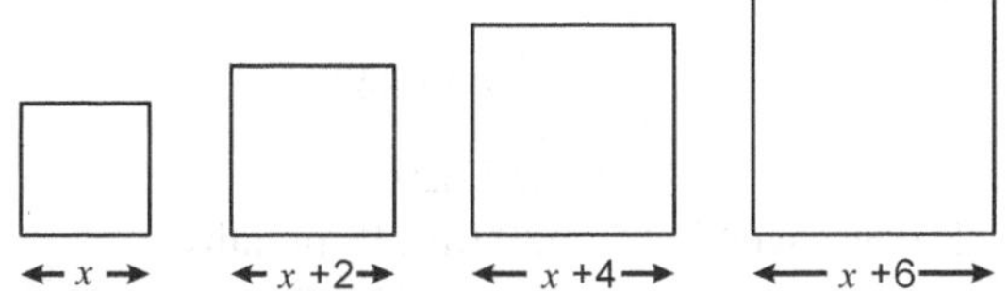

(ii) Find the value of x.

Sol : The length of the sides of the sequence of square frames are

$$x, x + 2, x + 4, ...$$

It is an A.P. with $a = x$, $d = 2$

(i) Side of the nth square = $a + (n-1)d$

$$= x + (n-1)(2)$$
$$= x + 2n - 2.$$

(ii) It is given that

$$x^2 + (x+2)^2 + (x+4)^2 = 440$$

$\Rightarrow \quad 3x^2 + 12x + 20 = 440$

$\Rightarrow \quad 3x^2 + 12x - 420 = 0$

$\Rightarrow \quad x^2 + 4x - 140 = 0$

$\Rightarrow \quad x^2 + 14x - 10x - 140 = 0$

$\Rightarrow \quad x(x + 14) - 10(+ 14) = 0$

$\Rightarrow \quad (x + 14)(x - 10) = 0$

$\Rightarrow \quad x = 10 \qquad (\because x \neq -14)$

Q. 6. Read the following passage and answer the questions that follows :

Ravi started to give home tuitions from the month of January. He charges ₹ 1000 from each student. He spends ₹ 1200 on rent, ₹ 2700 on his food, ₹ 980 on taxi fare and ₹ 1700 on electricity. In the first month he had 20 students and in each subsequent month, the number of students increases by 2.

(i) Do his savings form an arithmetic progression? If so, write the first three terms of this A.P.

(ii) How much did Ravi save till 11th month?

(iii) How much did he save in the month of July?

Sol : (i) Earnings from home tuition in the first month = ₹ (1000 × 20)

= ₹ 20,000.

Total expense = ₹ (1200 + 2700 + 980 + 1700)

= ₹ 6580

Ravi's savings in the 1st month

= (20,000 – 6580)

= ₹ 13,420

Ravi's savings in the 2nd month

= ₹ (13420 + 2000)

= ₹ 15,420

Ravi's savings in the 3rd month = (13,420 + 4000)

= 17,420

In each subsequent month, his savings increases by a fixed amount. Therefore, his savings forms an A.P.

The A.P. is 13420, 15420, 17420, ...

(ii) The amount, Ravi saved till 11 months will be the sum of 11 terms of this A.P.

Here, $a = 13420$ and $d = 2000$

$$S_{11} = \frac{11}{2}[2(13420 + 10(2000)]$$

$$= \frac{11}{2} \times 46,840$$

$$= 2,57,620.$$

So, he saved ₹ 2,57,620 till 11th months.

(iii) Ravi's saving in July will be 7th term of this A.P.

$$a_7 = 13420 + 6 \times 2000 \quad [\because a_n = a + (n-1)d]$$

$$= 13,420 + 12,000$$

$$= 25,420$$

So, he saved ₹ 25,420 in the month of July.

Chapter 6. Coordinate Geometry

Q. 1. In a colony's park there are three electric poles namely, *A*, *B* and *C* at the positions shown in the figure. Despite these three poles, some part of the park are still in dark. So, society decided to fix one more electric pole *D* in the park. Based on the following information, answer the following questions :

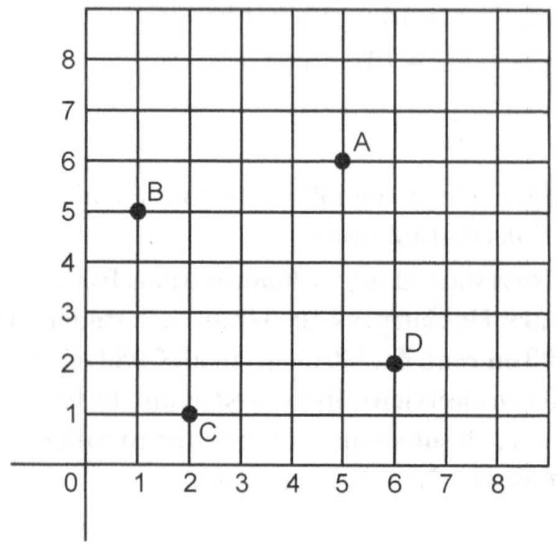

(i) Write the coordinates of *D*.

(ii) Find the distance of *D* from pole A and pole *C*.

(iii) Is *ABCD* a square?

Sol : (i) Coordinates of D are (6, 2)

(ii) $A(5, 6)$ and $D(6, 2)$

$$\text{Distance } AD = \sqrt{(6-5)^2 + (2-6)^2}$$

$$= \sqrt{1+16} = \sqrt{17} \text{ units}$$

$C(2, 1)$ and $D(6, 2)$

$$\text{Distance } CD = \sqrt{(6-2)^2 + (2-1)^2} = \sqrt{16+1}$$

$$= \sqrt{17} \text{ units.}$$

(iii) $A(5, 6)$, $B(1, 5)$, $C(2, 1)$ and $D(6, 2)$

$$AB = \sqrt{(1-5)^2 + (5-6)^2} = \sqrt{16+1}$$

$$= \sqrt{17} \text{ units}$$

$$BC = \sqrt{(2-1)^2 + (1-5)^2} = \sqrt{1+16}$$

$$= \sqrt{17} \text{ units}$$

$$CD = \sqrt{(6-2)^2 + (2-1)^2} = \sqrt{16+1} = \sqrt{17} \text{ units.}$$

$$DA = \sqrt{(5-6)^2 + (6-2)^2} = \sqrt{16+1} = \sqrt{17} \text{ units.}$$

$$AC = \sqrt{(2-5)^2 + (1-6)^2} = \sqrt{9+25} = \sqrt{34} \text{ units.}$$

$$BD = \sqrt{(6-1)^2 + (2-5)^2} = \sqrt{25+9} = \sqrt{34} \text{ units.}$$

$\because AB = BC = CD = AD$

Also Diagonal AC = Diagonal BD

$\therefore$ *ABCD* is a square. **Ans.**

Q. 2. A telephone company wants to fix two relay towers *P* and *Q* between two factories *A* and *B* in such that *AP* = *PQ* = *QB* points *P* and *Q* lies on the straight line joining A and B. Based on the given information, answer the following questions :

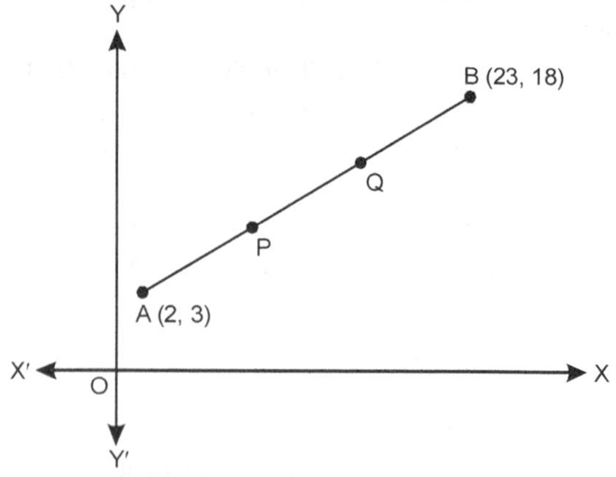

(i) Find the ratio in which *P* and *Q* divides *AB*.

(ii) Find the coordinates of *P*.

(iii) Find the coordinates of *Q*.

Sol : (i) Let the distance of point P from A be x.

Then, distance of point Q from P is also x and the distance of point B from Q is also x

$\therefore AQ = AP + PQ = 2x$ and $PB = PQ + QB = 2x$

$\Rightarrow \quad AP : PB = x : 2x = 1 : 2$

and $\quad AQ : QB = 2x : x = 2 : 1$

$\therefore$ The ratio in which P divides AB is 1 : 2 and the ratio in which Q divides AB is 2 : 1 **Ans.**

(ii) Coordinates of P

$$= \left(\frac{m_1x_2 + m_2x_1}{m_1 + m_2}, \frac{m_1y_2 + m_2y_1}{m_1 + m_2}\right)$$

$$= \left(\frac{1\times 23 + 2\times 2}{1+2}, \frac{1\times 18 + 3\times 2}{1+2}\right)$$

1 2
A P B
(2, 3) (23, 18)

$$= \left(\frac{23+4}{3}, \frac{18+6}{3}\right)$$

$$= \left(\frac{27}{3}, \frac{24}{3}\right)$$

$$= (9, 8)$$

(iii) Coordinates of Q

$$= \left(\frac{m_1x_2 + m_2x_1}{m_1 + m_2}, \frac{m_1y_2 + m_2y_1}{m_1 + m_2}\right)$$

2 1
A Q B
(2, 3) (23, 18)

$$= \left(\frac{2\times 23 + 2\times 1}{2+1}, \frac{18\times 2 + 3\times 1}{2+1}\right)$$

$$= \left(\frac{46+2}{3}, \frac{36+3}{3}\right)$$

$$= \left(\frac{48}{3}, \frac{39}{3}\right)$$

$$= (16, 13)$$ **Ans.**

Q. 3. Read the following passage and answer the questions that follows :

Sheenam starts walking from her house to office. Instead of going to the office directly, she goes to a bank first, from there to her daughter's school and then reaches the office. Assuming that all distances covered are in straight lines, answer the following questions.

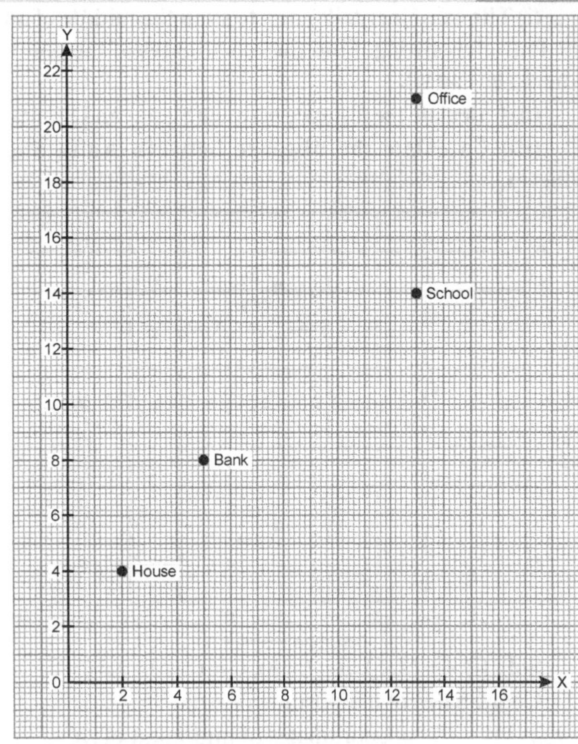

(i) What is the distance between her house and office?

(ii) Calculate the extra distance that is covered by sheenam in reaching her office.

Sol : (i) House = (2, 4), Bank = (5, 8), School = (13, 14) and Office = (13, 21)

$\therefore$ Distance between house and office

$$= \sqrt{(13-2)^2 + (21-4)^2}$$

$$= \sqrt{(11)^2 + (17)^2}$$

$$= \sqrt{121 + 289}$$

$$= \sqrt{410} \text{ units}$$

(ii) Distance covered from house to bank

$$= \sqrt{(5-2)^2 + (8-4)^2}$$

$$= \sqrt{9+16} = \sqrt{25} = 5 \text{ units}$$

Distance covered from bank to school

$$= \sqrt{(13-5)^2 + (14-8)^2}$$

$$= \sqrt{64+36} = \sqrt{100} = 10 \text{ units}$$

Distance covered from school to office

$$= \sqrt{(13-13)^2 + (21-14)^2}$$

$$= \sqrt{0+49} = \sqrt{49} = 7 \text{ units}$$

$\therefore$ Extra distance covered $= (5 + 10 + 7) - \sqrt{410}$

$$= (22 - \sqrt{410}) \text{ units.}$$

Q. 4. Read the following passage and answer the questions that follows :

In an examination room, four students Sunita, Anita, Vineeta and Punita are sitting at point *A*(2, 5), *B*(5, 8), *C*(8, 5) and *D*(5, 2) respectively. Then a new student Anju enters the room.

(i) The examiner tells Anju to sit in the middle of all the four girls sitting. Find the coordinates of the point, where she can sit.

(ii) Calculate the distance between Sunita and Punita.

(iii) Calculate the distance between Anita and Vineeta.

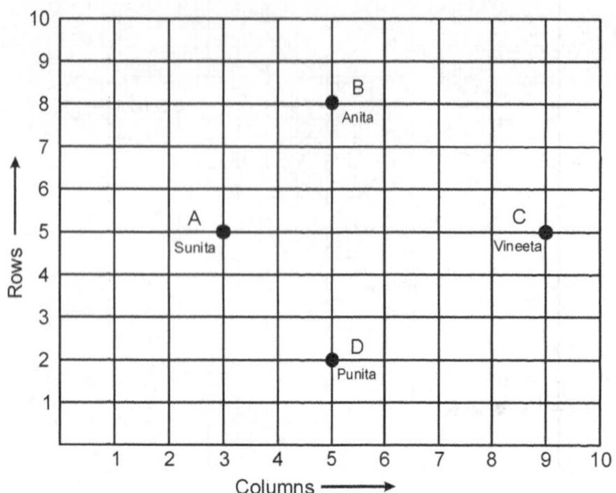

Sol : (i) $A = (2, 5), B = (5, 8), C = (8, 5), D = (5, 2)$

Coordinates of the position of Anju

= Coordinates of mid-point of AC

= Coordinates of mid-point of BD

$$\text{Mid-point of } AC = \left(\frac{2+8}{2}, \frac{5+5}{2}\right) = \left(\frac{10}{2}, \frac{10}{2}\right)$$

$$= (5, 5)$$

$$\text{Mid-point of } BD = \left(\frac{5+5}{2}, \frac{8+2}{2}\right) = \left(\frac{10}{2}, \frac{10}{2}\right)$$

$$= (5, 5)$$

So, the coordinates of the point, where Anju sits are (5, 5). **Ans.**

(ii) Distance between Sunita and Punita = AD

$$= \sqrt{(5-2)^2 + (2-5)^2}$$

$$= \sqrt{(3)^2 + (-3)^2}$$

$$= \sqrt{9+9}$$

$$= \sqrt{18} = 3\sqrt{2} \text{ units.}$$ **Ans.**

(iii) Distance between Anita and Vineeta = BC

$$= \sqrt{(8-5)^2 + (5-8)^2}$$

$$= \sqrt{(3)^2 + (-3)^2}$$

$$= \sqrt{9+9}$$

$$= \sqrt{18}$$

$$= 3\sqrt{2} \text{ units.}$$ **Ans.**

Q. 5. Three friends Aman, Raj and Deepak live in different cities. They decided to meet at a place in city Mr. Aman, Raj and Deepak lives in city *A*, city *R* and city *D* respectively. City *M*, city *A*, city *R* and city *D* are represented in the figure given below. Based on the given situation, answer the following questions.

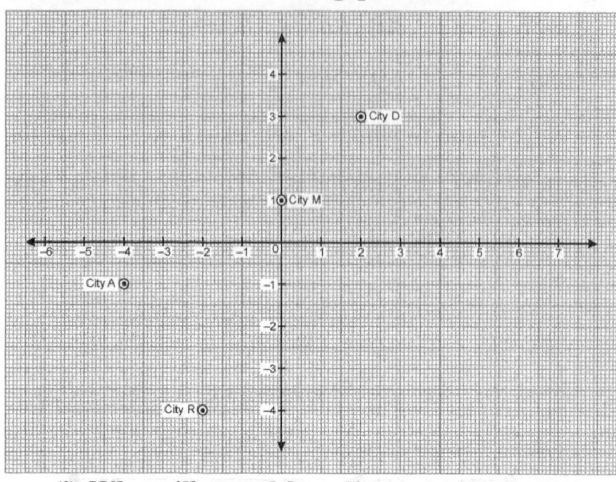

(i) Who will travel less distance, Aman, Raj or Deepak, to reach city *M*?

(ii) Find the distance between city *A* and city *D*.

(iii) Find the coordinates of city *P*, which is situated at point *P*, the mid-point of the line joining the points represented by city *R* and city *D*.

Sol : (i) $A(-4, -1), R(-2, -4), D(2, 3)$ and $M(0, 1)$

Distance travelled by Aman = $\sqrt{(0+4)^2 + (1+1)^2}$

$$= \sqrt{4^2 + 2^2} = \sqrt{16+4} = \sqrt{20} \text{ units}$$

Distance travelled by Raj = $\sqrt{(0+2)^2 + (1+4)^2}$

$$= \sqrt{2^2 + 5^2} = \sqrt{4+25} = \sqrt{29} \text{ units}$$

Distance travelled by Deepak

$$= \sqrt{(0-2)^2 + (1-3)^2} = \sqrt{4+4} = \sqrt{8} \text{ units}$$

∴ Deepak travels less distance. **Ans.**

(ii) Distance between city A and city D

$$= \sqrt{(2+4)^2 + (3+1)^2} = \sqrt{(6)^2 + (4)^2}$$

$$= \sqrt{36+16} = \sqrt{52} \text{ units}$$ **Ans.**

(iii) Coordinates of P are $\left(\frac{-2+2}{2}, \frac{-4+3}{2}\right)$

$$= \left(0, \frac{-1}{2}\right)$$ **Ans.**

Chapter 7. Triangles

Q. 1. The foot of a ladder is 12 m away from a wall and its top reaches a window 5 m above the ground. The ladder is shifted in such a way that its top touches the roof which is 6 m above the window. Based on the given information, answer the following questions :

(i) What is the length of the ladder?

(ii) How much the foot of ladder is shifted towards the wall, so that the top of ladder touches the roof?

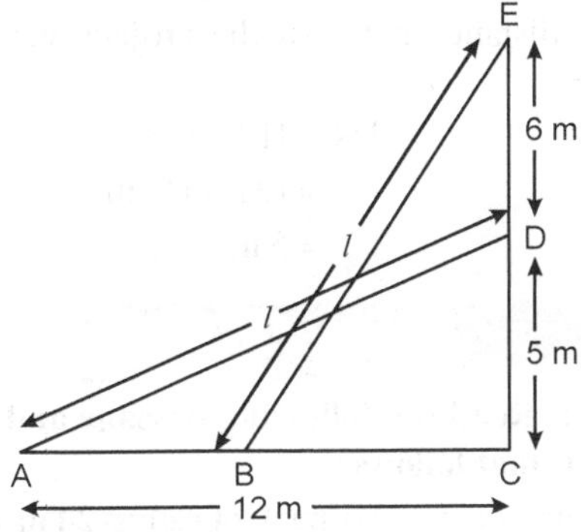

Sol : (i) In ΔACD,

AC = Distance of foot of ladder from wall = 12 m

CD = Height of window = 5 m.

Let, l be the length of ladder.

Applying pythagoras theorem in ΔACD, we get

$$l^2 = AC^2 + CD^2$$

$$\Rightarrow \quad l^2 = (12)^2 + (5)^2$$

$$\Rightarrow \quad l^2 = 144 + 25$$

$$\Rightarrow \quad l^2 = 169$$

$$\Rightarrow \quad l = 13 \text{ m.}$$

so, length of the ladder = 13 m. **Ans.**

(ii) Height of the roof = (5 + 6) m = 11 m.

length of the ladder = 13 m.

Now,

ΔBCE is a right angled triangle.

On applying pythagoras theorem, in ΔBCE, we get

$$BE^2 = EC^2 + BC^2$$

$$\Rightarrow \quad (13)^2 = (11)^2 + BC^2$$

$$\Rightarrow \quad 169 - 121 = BC^2$$

$$\Rightarrow \quad 48 = BC^2$$

$$\Rightarrow \quad BC = 4\sqrt{3} \text{ m.}$$

Required distance = AB

$$= AC - BC$$

$$= 12 - 4\sqrt{3}$$

$$= 4(3 - \sqrt{3}) \text{ m}$$ **Ans.**

Q. 2. An aeroplane leaves an airport and flies due North at a speed of 1000 km per hour. At the same time, another aeroplane leaves the same airport and flies due West at a speed of 1200 km per hour. Based on the following information, answer the following questions :

(i) How far apart will be the planes after $1\frac{1}{2}$ hour?

(ii) If the speed of first plane is 1600 km per hour, then find the distance between two planes after $1\frac{1}{2}$ hours?

(iii) If second plane flies in the South direction, then how far apart will be the planes after 1 hour?

Sol : (i) The distance travelled by first aeroplane in North direction = $1000 \times \frac{3}{2}$ km

AB = 1500 km

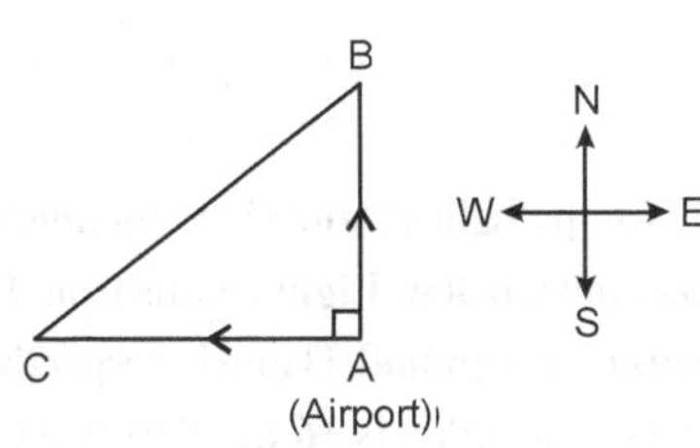

The distance travelled by second aeroplane in West direction

$$AC = 1200 \times \frac{3}{2} = 1800 \text{ km}$$

In ΔBAC, using pythagoras theorem

$$(BC)^2 = (1500)^2 + (1800)^2$$

$$\Rightarrow \quad (BC)^2 = 2250000 + 3240000$$

$$\Rightarrow \quad (BC)^2 = 5490000$$

$$\Rightarrow \quad BC = 300\sqrt{61} \text{ km}$$

Hence, the required distance = $300\sqrt{61}$ km

(ii) The distance travelled by first aeroplane

$$AB = 1600 \times \frac{3}{2} \text{ km}$$

$$AB = 2400 \text{ km}$$

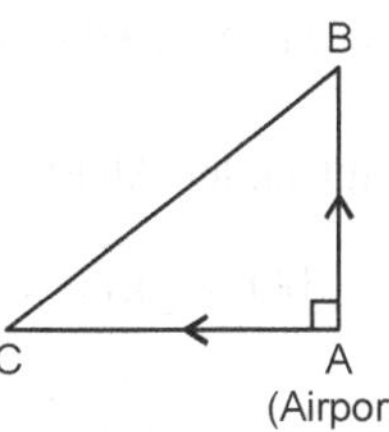

The distance travelled by second aeroplane, AC = 1800 km

In ΔBAC, using pythagoras theorem

$$(BC)^2 = (2400)^2 + (1800)^2$$

$$\Rightarrow \quad BC^2 = 5760000 + 3240000$$

$$\Rightarrow \quad BC = \sqrt{9000000}$$

$$BC = 3000 \text{ km}$$

Hence, the required distance is 3000 km.

(iii) Distance travelled by first aeroplane is 1 hour

$$AB = 1000 \text{ km}$$

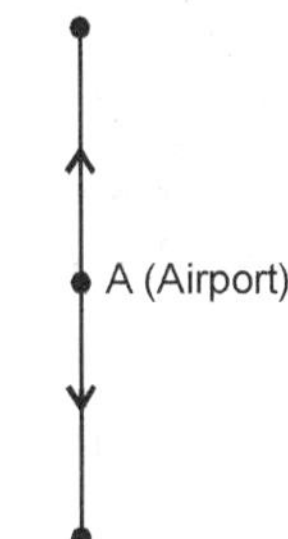

Distance travelled by second aeroplane in 1 hour

$$AC = 1200 \text{ km}$$

Hence, the required distance between two planes

$= AB + AC$

$= 1000 \text{ km} + 1200 \text{ km}$

$= 2200 \text{ km}$ **Ans.**

Q. 3. Two spotlights, P and Q are mounted on a vertical pole AB, as shown below. Light beams from P and Q shine on two points on the ground, H and K respectively. Distance between two spotlights is 16 m, distance between first shine point and vertical pole is 16 m, distance of spotlight *P* and *Q* to its shine points on ground H and K is 35 m and 20 m respectively,

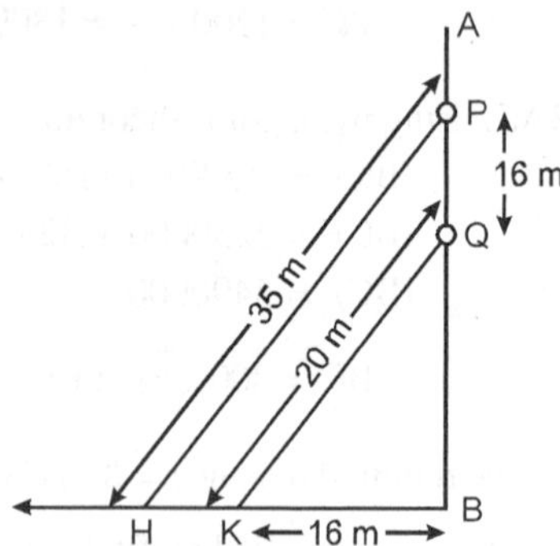

Based on the following information, answer the following questions :

(i) The height above the ground at which the spotlight *Q* is mounted.

(ii) The distance between the projections of the light beams.

Sol : (i) In right angled ΔKBQ

$$BQ = \sqrt{KQ^2 - KB^2}$$

$$= \sqrt{(20)^2 - (16)^2}$$

$$= \sqrt{400^2 - 256^2}$$

$$= \sqrt{144} = 12 \text{ m}$$

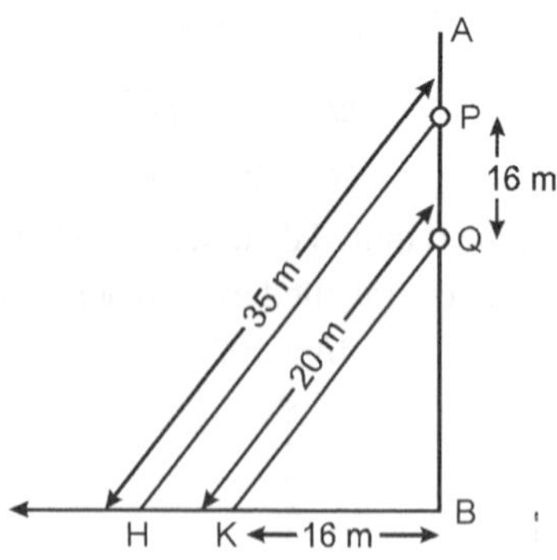

(ii) Using (i), BP = (12 + 16)m = 28 m.

In right angled, ΔHBP

$$HB = \sqrt{(HP)^2 - (PB)^2}$$

$$= \sqrt{(35)^2 - (28)^2}$$

$$= \sqrt{1225 - 784}$$

$$= \sqrt{441}$$

$$= 21 \text{ m.}$$

So, distance between the projections of the light beams =

$$HK = HB - KB$$

$$= (21 - 16) \text{ m}$$

$$= 5 \text{ m.}$$ **Ans.**

Chapter 8. Circles

Q. 1. Read the following passage and answer the questions that follows :

There is a circular park of radius 24 m and a pole at a distance of 26 m from the centre of the park, as shown in the figure. From the pole, there are two paths PR and PQ, which are tangential to the park.

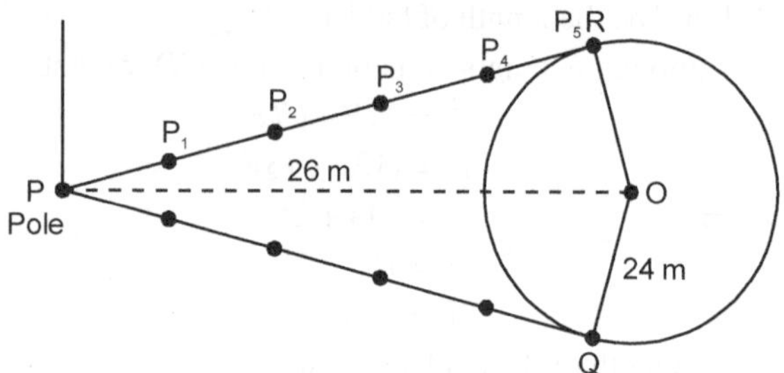

(i) Find the length of each path.

(ii) If five light poles are to be put long each tangential path at equal distances, find the distance between each consecutive light pole.

Sol : (i) In right-angled triangle *PRO*

We have,

$$PR = \sqrt{PO^2 - OR^2}$$

$$= \sqrt{(26)^2 - (24)^2}$$

$$= \sqrt{676 - 576}$$

$$= \sqrt{100}$$

$$= 10 \text{ m.}$$

As *PR* and *PQ* are tangents from same external point so, $PR = PQ = 10$ m. **Ans.**

(ii) As five poles are to be put along *PQ* and *PR* each. Let the distance between consecutive poles be *x*.

Hence,

$$5x = 10$$

$$\Rightarrow \quad x = 2 \text{ m.}$$

Each pole is at a distance of 2 m. **Ans.**

Chapter 9. Constructions

Q. 1. Read the following passage and answer the questions that follows :

A society has a right triangular flower bed whose sides are of length 3 m and 4 m (other than hypotenuse) The gardener has to enlarge the triangular flower bed whose sides are $\frac{5}{3}$ times that of earlier one.

(i) Construct the figure to help gardener in finding the length of new flower bed.

(ii) What will be the length of each side of enlarged flower bed?

Sol : (i) Take 1 m = 1 cm for construction.

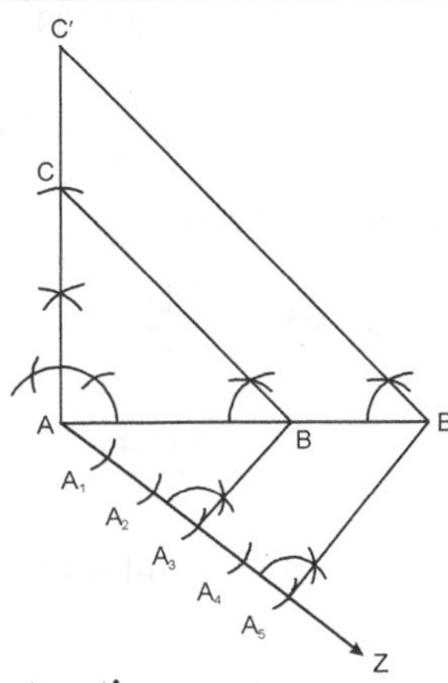

Steps of construction :

(1) Draw AB = 3 cm.

(2) At point A construct $\angle CAB = 90°$, with A as a centre and radius AC = 4 cm, draw an arc.

(3) Join BC to obtain ΔABC.

(4) Construct an acute $\angle BAZ$ at A on opposite side of vertex C of ΔABC.

(5) Mark-off five points A_1, A_2, A_3, A_4, A_5 on AZ such that $AA_1 = A_1A_2 = A_2A_3 = A_3A_4 = A_4A_5$.

(6) Join A_3 to B and draw a line through A_5B' parallel to A_3B intersecting the extended line segment AB to B′.

(7) Draw a line through B′ parallel to BC intersecting the extended line segement AC at C′.

(8) Triangle C′AB′ so obtained is the required triangle such that $\frac{C'A}{CA} = \frac{AB'}{AB} = \frac{B'C'}{BC} = \frac{5}{3}$.

(ii) $AB = 3$ cm

$\therefore \quad AB' = \frac{5}{3}AB = \frac{5}{3} \times 3 = 5$ cm.

$AC = 4$ cm

$\therefore \quad AC' = \frac{5}{3}AC = \frac{5}{3} \times 4 = \frac{20}{3} = 6.67$ cm

$BC = \sqrt{3^2 + 4^2}$

$= \sqrt{9+16} = \sqrt{25} = 5$

$\therefore \quad B'C' = \frac{5}{3}BC = \frac{5}{3} \times 5 = \frac{25}{3} = 8.33$ cm. **Ans.**

Chapter 10. Trigonometry

Q. 1. Read the passage and answer the questions that follows :

The students of a school are visiting India's tallest monument Qutub Minar. Students roam around the Minar and start discussing that height of Minar seems different as observed from the different points.

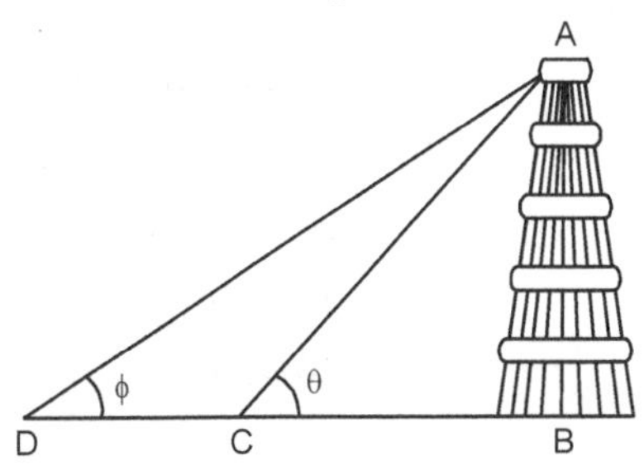

(i) If a student is standing at a distance of 75 m from the base of minar and the angle of elevation of the top of Minar is 45°, then find the height of Qutab Minar.

(ii) If two students, standing on same side, are looking at the Minar, at the same time, making angles of elevation of 30° and 60° respectively, then find the distance between them.

(iii) Evaluate : $\frac{\sin 70°}{\cos 20°} + \frac{2\cos 43°.\text{cosec } 47°}{\tan 10°.\tan 45°.\tan 80°}$

Sol : (i) Let the height of Minar be h.

$$\tan 45° = \frac{h}{75}$$

then, $$1 = \frac{h}{75}$$

$\Rightarrow$ $$h = 75$$

Hence, the height of minar is 75 m.

(ii) From the figure,

Distance between two students = DC

In ΔADB,

$$\tan 30° = \frac{75}{BD}$$

$\Rightarrow$ $$\frac{1}{\sqrt{3}} = \frac{75}{BD}$$

$\Rightarrow$ $$BD = 75\sqrt{3}$$

In ΔABC,

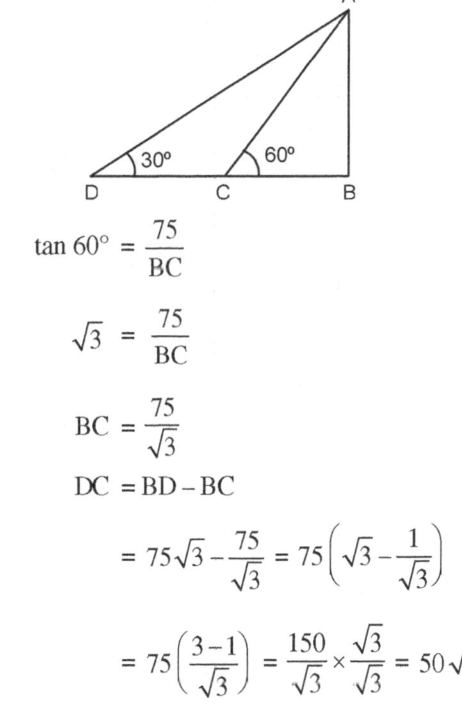

$$\tan 60° = \frac{75}{BC}$$

$$\sqrt{3} = \frac{75}{BC}$$

$\Rightarrow$ $$BC = \frac{75}{\sqrt{3}}$$

$\therefore$ $$DC = BD - BC$$

$$= 75\sqrt{3} - \frac{75}{\sqrt{3}} = 75\left(\sqrt{3} - \frac{1}{\sqrt{3}}\right)$$

$$= 75\left(\frac{3-1}{\sqrt{3}}\right) = \frac{150}{\sqrt{3}} \times \frac{\sqrt{3}}{\sqrt{3}} = 50\sqrt{3}$$

Hence, the distance between two students is $50\sqrt{3}$ m.

(iii) $\frac{\sin 70°}{\cos 20°} + \frac{2\cos 43°.\text{ cosec } 47°}{\tan 10°.\tan 45°.\tan 80°}$

$$= \frac{\sin 70°}{\cos(90° - 70°)} + \frac{2\cos(90° - 47°).\text{cosec } 47°}{\tan(90° - 80°).\tan 45°.\tan 80°}$$

$$= \frac{\sin 70°}{\sin 70°} + \frac{2 \sin 47° \operatorname{cosec} 47°}{\cot 80°.1. \tan 80°}$$

$$= 1 + \frac{2 \times 1}{1}$$

$= 3.$ **Ans.**

Q. 2. In an experiment, some masses are hung in equilibrium as shown in diagram. Distance between the two pullies A and B is 40 cm and ∠CAB = 45°. Based on given figure, answer the following questions:

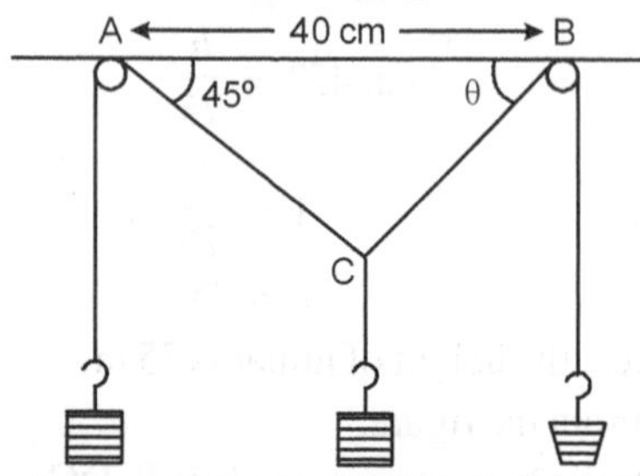

(i) In the figure, if sin θ = cos(3θ – 30°), where θ and 3θ – 30° are acute angles, then find the value of θ.

(ii) Using the value of θ obtained in (i), find the perpendicular distance of AB from C

(iii) Using the value of θ obtained in (i), find the length of AC and BC.

Sol : (i) $\sin \theta = \cos(3\theta - 30°)$

$\Rightarrow \cos(90° - \theta) = \cos(3\theta - 30°)$

$\Rightarrow 90° - \theta = 3\theta - 30°$

$\Rightarrow 4\theta = 120°$

$\Rightarrow \theta = 30°.$ **Ans.**

(ii) Draw CD ⊥ AB

In right angled ΔADC,

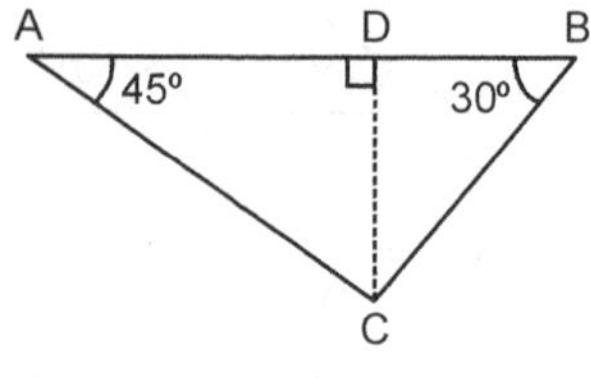

$$\cot 45° = \frac{AD}{CD}$$

$$\Rightarrow 1 = \frac{AD}{CD} \quad ...(i)$$

$\Rightarrow AD = CD$

In right angled ΔBDC

$$\cot 30° = \frac{BD}{CD}$$

$$\Rightarrow \sqrt{3} = \frac{BD}{CD}$$

$$\Rightarrow BD = \sqrt{3}\ CD \quad ...(ii)$$

Adding equations (i) and (ii), we get

$\Rightarrow AD + BD = CD + \sqrt{3}\ CD$

$\Rightarrow 40 = CD(1 + \sqrt{3})$

$\Rightarrow 40 = 2.732\ CD$

$$\Rightarrow CD = \frac{40}{2.732} = 14.64 \text{ cm}$$ **Ans.**

(iii) In right angled ΔADC,

$$\sin 45° = \frac{CD}{AC}$$

$$\Rightarrow \frac{1}{\sqrt{2}} = \frac{14.64}{AC}$$

$\Rightarrow AC = 1.414 \times 14.64 = 20.7$ cm. **Ans.**

In right angled ΔBDC,

$$\sin 30° = \frac{CD}{BC}$$

$$\Rightarrow \frac{1}{2} = \frac{14.64}{BC}$$

$\Rightarrow BC = 2 \times 14.64$

$BC = 29.28$ cm **Ans.**

Q. 3. In circus, artists show their different talents. One of these talents is climbing from the ground along a rope stretched from the top of a vertical pole and tied at the ground.

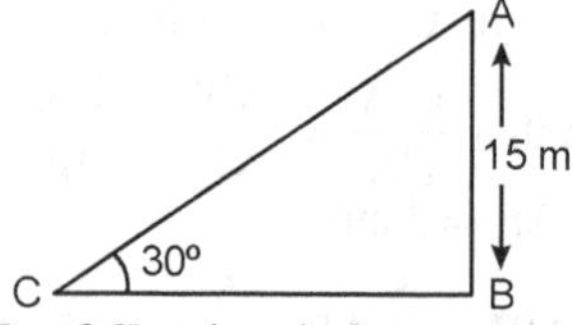

Based on the following information and figure, answer the following questions :

(i) What is length of the rope that artist climb to reach the top of the pole if height of the pole is 15 m and the rope makes an angle of 30° with the ground?

(ii) If tan 2θ = cot (θ – 18°) where, 2θ and (θ – 18°) are acute angles, then find the value of θ.

Sol : (i) In right angled ΔABC,

AB = 15 m, ∠ACB = 30°,

In ΔABC

$$\therefore \sin 30° = \frac{AB}{AC}$$

$$\Rightarrow \frac{1}{2} = \frac{15}{AC}$$

$\Rightarrow AC = 30$ m.

Hence, the length of pole is 30 m. **Ans.**

(ii) We have

$\tan 2\theta = \cot(\theta - 18°)$

$\Rightarrow \cot(90° - 2\theta) = \cot(\theta - 18°)$

$\Rightarrow 90° - 2\theta = \theta - 18°$

$\Rightarrow 3\theta = 108°$

$\Rightarrow \theta = 36°.$ **Ans.**

Q. 4. Fire in a building is reported on telephone to two fire stations P and Q, 20 kms apart from each other on a straight road. Station P observes that the fire is at an angle of 60° to the road and station Q observes that it is at an angle of 45° to the road.

Based on this information, answer the following questions :

(i) Which station should send its team?

(ii) How much will this team have to travel?

Sol : (i) Given, P and Q are 20 km apart from each other.

Let distance of A from station Q be x km.

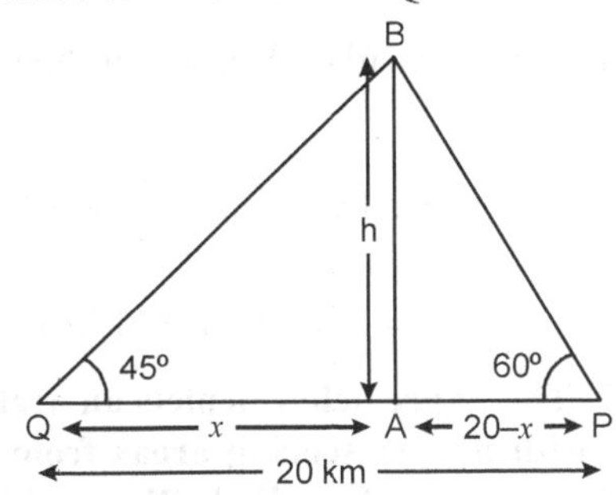

In ΔBAQ

$$\tan 45^\circ = \frac{AB}{AQ}$$

$$\Rightarrow \quad 1 = \frac{AB}{AQ}$$

$$\Rightarrow \quad 1 = \frac{h}{x}$$

$$\Rightarrow \quad h = x \qquad ...(i)$$

In ΔPAQ,

$$\tan 60^\circ = \frac{AB}{AP}$$

$$\Rightarrow \quad \sqrt{3} = \frac{h}{20-x}$$

$$\Rightarrow \quad h = (20-x)\sqrt{3} \qquad ...(ii)$$

From (i) and (ii), we get

$$x = (20-x)\sqrt{3}$$

$$\Rightarrow \quad x = 20\sqrt{3} - x\sqrt{3}$$

$$\Rightarrow \quad 20\sqrt{3} = x(1+\sqrt{3})$$

$$\Rightarrow \quad x = \frac{20\sqrt{3}}{1+\sqrt{3}} = \frac{20\sqrt{3}}{\sqrt{3}+1} \times \frac{\sqrt{3}-1}{\sqrt{3}-1}$$

$$= \frac{20\sqrt{3}(\sqrt{3}-1)}{3-1}$$

$$= 10(3-\sqrt{3})$$

$$= 10(3 - 1.732)$$

$$= 10 \times 1.268$$

$$= 12.68$$

Hence,

Distance of station Q from Building = 12.68 km

Distance of station P from Building = 20 – 12.68 = 7.32 km.

Therefore, station P should send its team.

(ii) Team P has to travel a distance of 7.32 km to reach the building.

Q. 5. A carpenter makes stools with two steps for electricians with a square top of side 0.5 m and at a height of 1.5 m above the ground. Each leg is inclined at an angle of 60° to the ground.

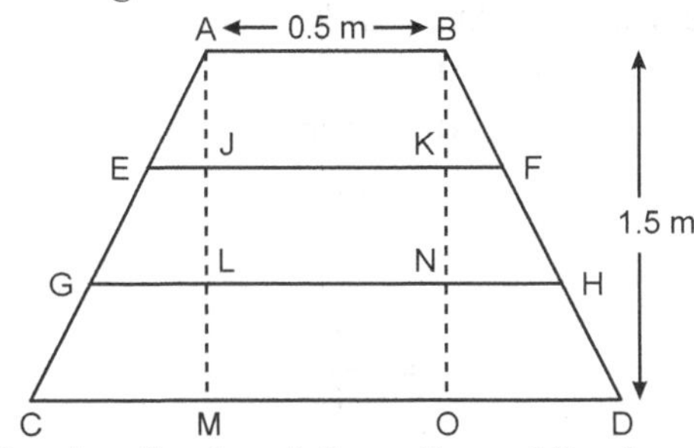

Based on the above information and the given figure, answer the following questions :

(i) Find the length of each leg.

(ii) Find the lengths of two steps EF and GH to be put at equal distances.

Sol : (i)

In ΔACM

∠C = 60°, AM = height of stool = 1.5 m.

$$\sin 60^\circ = \frac{AM}{AC}$$

$$\frac{\sqrt{3}}{2} = \frac{1.5}{AC}$$

$$\Rightarrow \quad AC = \frac{1.5 \times 2}{\sqrt{3}} = \frac{3}{\sqrt{3}} = \sqrt{3} = 1.732 \text{ m.}$$

Hence, the length of each leg is 1.732 m. **Ans.**

(ii) Lengths of steps are GH and EF and these steps are put at equal distance such that AJ = JL = LM

$$\therefore \quad AJ = JL = LM = \frac{1.5}{3} = 0.5 \text{ m}$$

$$AL = 0.5 + 0.5 = 1 \text{ m}$$

∵ CM and GL are parallel to each other

$$\angle G = \angle C = 60^\circ$$

Now consider ΔALG

$$\tan 60^\circ = \frac{AL}{GL}$$

$$\sqrt{3} = \frac{1}{GL}$$

$$GL = \frac{1}{\sqrt{3}} = \frac{1}{1.732} = 0.577$$

In ΔALG and ΔBNH

$\angle G = \angle H = 60°$
$\angle L = \angle M = 90°$
$AL = BN$
(since steps are at equal distance)

$\Rightarrow \Delta ALG \cong \Delta BNH$
(By ASA congruency rule)

$\therefore$ $GL = NH = 0.577$ (By CPCT)

Hence, the length of step GH = GL + LN + NH
$= 0.577 + 0.5 + 0.577$
$= 1.654$ m

Similarly, $\Delta AJE \cong \Delta BKF$

Now, consider ΔAEJ

$AJ = 0.5$ m

$$\tan 60° = \frac{AJ}{EJ}$$

$$\sqrt{3} = \frac{0.5}{EJ}$$

$$EJ = \frac{0.5}{1.732} = 0.288 \text{ m}$$

Now, $EF = EJ + JK + KF$
$= 0.288 + 0.5 + 0.288$
$= 1.076$ m.

$\therefore$ Lengths of steps EF and GH are 1.076 m and 1.654 m respectively. **Ans.**

Chapter 12. Areas related to Circles

Q. 1. Aman is cycling such that the wheels of the cycle are making 180 revolution per minute. If the diameter of the wheel is 70 cm.

Based on the given information, answer the following questions :

(i) Find the speed (in km/hr) with which the Aman is cycling.

(ii) Distance travelled by Aman in $\frac{1}{2}$ hour.

Sol : (i) Speed = Distance covered by the wheel in one hour.

Clearly, distance covered by wheel in one revolution is circumference of wheel.

$$\text{Radius of wheel} = \frac{70}{2} = 35 \text{ cm.}$$

$$\text{Circumference of wheel} = 2\pi r = 2 \times \frac{22}{7} \times 35 = 220 \text{ cm}$$

So,

Distance covered by wheel in one revolution = 220 cm

$\Rightarrow$ Distance covered by wheel in 180 revolutions
$= 220 \times 180 = 39600$ cm

$$= \frac{39600}{100} \text{ m}$$

$$= 396 \text{ m} = \frac{396}{1000} \text{ km}$$

$\Rightarrow$ Distance covered by wheel in one minute

$$= \frac{396}{1000} \text{ km}$$

$\Rightarrow$ Distance covered by wheel in one hour *i.e.*, 60 minutes

$$= \frac{396}{1000} \times 60 \text{ km} = 23.76 \text{ km}$$

Hence, the speed of the cycle is 23.76 km/hr. **Ans.**

(ii) Distance coverd by Aman is one hour = 23.76 km

So,

Distance covered by Aman in $\frac{1}{2}$ hour

$$= \frac{1}{2} \times 23.76 = 11.88 \text{ km}$$ **Ans.**

Q. 2. The figure below depicts an archery target marked with its five scoring areas from the centre outwards as Gold, Silver, Red, Blue and White. The diameter of the region representing Gold score is 21 cm and each of the other band is 10.5 cm wide.

Based on the above information answer the following quesitons.

(i) Find the area of Gold score band.

(ii) Find the area of Silver score band.

(iii) Find the ratio of areas of Gold and White score bands.

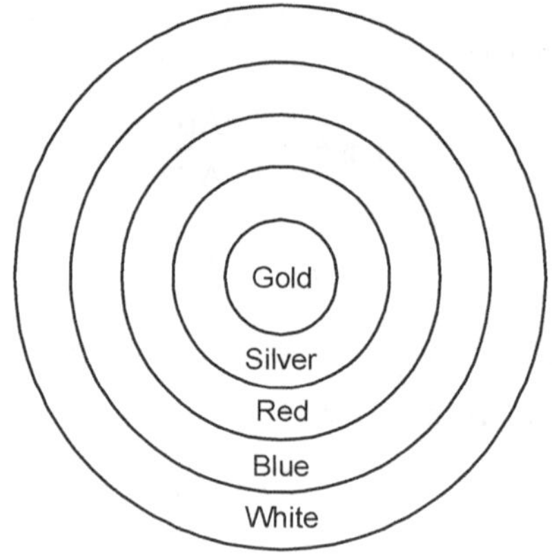

Sol : (i) Radius of Gold score band $r = \frac{21}{2}$ cm

Area Gold score band $= \pi r^2$

$$= \frac{22}{7} \times \left(\frac{21}{2}\right)^2$$

$$= \frac{22}{7} \times \frac{21}{2} \times \frac{21}{2}$$

$= 346.5 \text{ cm}^2$ **Ans.**

(ii) Outer radius of Silver score band = R

$$= \left(\frac{21}{2} + 10.5\right) \text{ cm}$$

$= 21$ cm.

Area of Silver score band = $\pi(R^2 - r^2)$

$$= \frac{22}{7}[(21)^2 - (10.5)^2]$$

$$= \frac{22}{7}[441 - 110.25]$$

$$= \frac{22}{7} \times 330.75$$

$$= 1039.5 \text{ cm}^2$$ **Ans.**

(iii) Outer radius of blue scoreband R_1

$$= 10.5 + 3 \times 10.5 = 42 \text{ cm}$$

Outer radius of White score band R_2

$$= [10.5 + 4 \times 10.5] = 10.5 + 42 = 52.5 \text{ cm}$$

$$R_2 = \frac{105}{2} \text{ cm}$$

Area of white score band = $\pi [R_2^2 - R_1^2]$

$$= \pi\left[\left(\frac{105}{2}\right)^2 - (42)^2\right]$$

$$= \pi\left(\frac{105}{2} - 42\right)\left(\frac{105}{2} + 42\right)$$

$$= \pi \times \frac{21}{2} \times \frac{189}{2} \text{ cm}^2$$

$$\therefore \text{ Required ratio } = \frac{\pi \times \frac{21}{2} \times \frac{21}{2}}{\pi \times \frac{21}{2} \times \frac{189}{2}}$$

$$= \frac{1}{9}$$

$\therefore$ The ratio of the areas of gold and white score band is 1 : 9. **Ans.**

Q. 3. Overgrazing occurs when plants are exposed to intensive grazing for extended periods of time, or without sufficient recovery periods. Overgrazing reduces the usefulness, productivity and biodiversity of the land and is one cause of desertification and erosion.

Based on the following figure related to overgrazing, answer the questions :

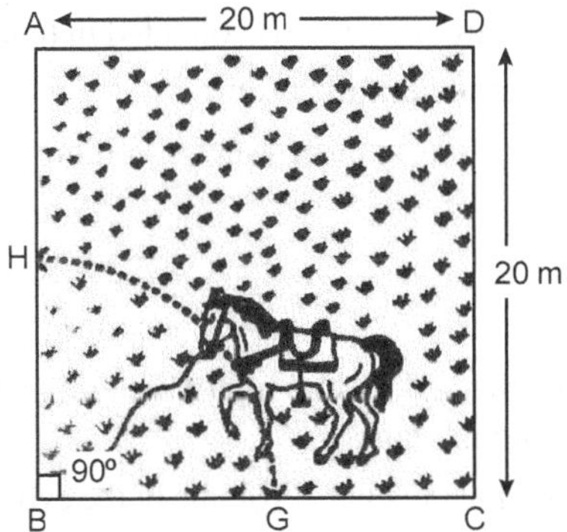

(i) If the length of a rope is 7 m, then find the area of that part of the field in which the horse can graze.

(ii) If the length of rope is 14 m, then find the area of the field in which the horse can graze.

(iii) Find the increase in the grazing area, if the rope is 14 m long instead of 7 m.

Sol : (i) Required Area $A_1 = \frac{\theta}{360°}\pi r^2$

$$= \frac{90°}{360°} \times \left(\frac{22}{7}\right) \times (7)^2$$

$$= \frac{1}{4} \times \frac{22}{7} \times 7 \times 7$$

$$= 38.5 \text{ m}^2$$

(ii) Required Area $A_2 = \frac{\theta}{360°}\pi r^2$

$$= \frac{90°}{360°} \times \frac{22}{7} \times 14 \times 14$$

$$= \frac{1}{4} \times \frac{22}{7} \times 14 \times 14$$

$$= 154 \text{ m}^2$$

(iii) Increase in grazing Area = $A_2 - A_1$

$$= (154 - 38.5) \text{ m}^2 = 115.5 \text{ m}^2$$ **Ans.**

Q. 4. Read the following passage and answer the questions that follows :

A wheel is made for a school project with silver wire in the form of circle as shown below. The diameter of the circle is 42 mm. The cost of each mm of silver wire is ₹ 30. The wire is also used in making 6 diameters which divide the circle into 12 equal sectors.

(i) How much silver wire will be required to make the wheel?

(ii) What is the area of each sector of the wheel?

(iii) What will be the cost of silver wire used to make the wheel?

Sol : (i) Total length of wire

= Length of 6 diameters + circumference of the circle

$$= \left[(6 \times 42) + \left(2 \times \frac{22}{7} \times \frac{42}{2}\right)\right] \text{ mm}$$

$$= (252 + 132) \text{ mm} = 384 \text{ mm.}$$

(ii) Area of each sector = $\frac{\theta}{360} \times \pi r^2$

$= \frac{30°}{360°} \times \frac{22}{7} \times \left(\frac{42}{2}\right)^2 \text{ mm}^2$

$= \frac{1}{12} \times \frac{22}{7} \times \frac{42}{2} \times \frac{42}{2} \text{ mm}^2$

$= 115.5 \text{ mm}^2$

(iii) Required amount= ₹ (384 × 30)

= ₹ 11520. **Ans.**

Q. 5. There is a semicircular grassy garden in a society with 4 m wide path running around it. The circumference of the plot is $81\frac{5}{7}$ m.

Based on the above information, answer the following questions :

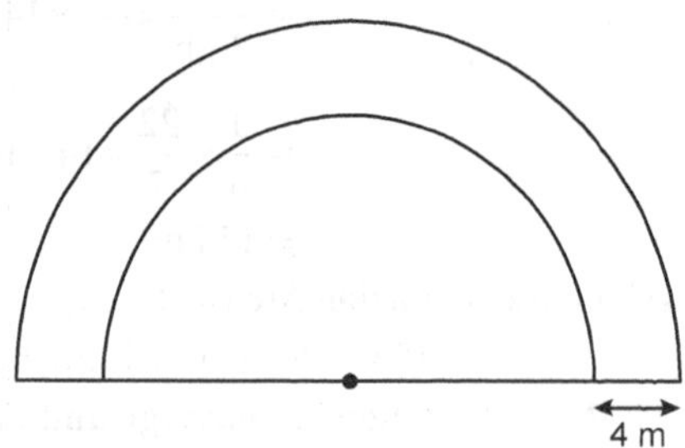

(i) Find the area of the path.

(ii) The cost of gravelling the path at the rate of ₹ 1.50 per square metre.

(iii) The cost of fencing the plot (alongwith path) all around at the rate of ₹ 35 per metre.

Sol. (i) Circumference of plot = $81\frac{5}{7}$ m $= \frac{572}{7}$ m

$\pi r = \frac{572}{7}$

$\frac{22}{7} \times r = \frac{572}{7}$

$\Rightarrow \quad r = 26$

So, radius of the plot = 26 m

Path is 4 m wide, so, radius of plot along with path $R = 26 + 4 = 30$ m.

$$\text{Area of path} = \frac{\pi(R^2 - r^2)}{2}$$

$$= \frac{\pi}{2}[(30)^2 - (26)^2]$$

$$= \frac{22}{7} \times \frac{1}{2}(30 + 26)(30 - 26)$$

$$= \frac{22}{7} \times \frac{1}{2} \times 56 \times 4$$

$= 352 \text{ m}^2$ **Ans.**

(ii) Cost of gravelling the path at rate ₹ 1.50 per m²

= Area of path × rate per m²

= ₹ 352 × 1.50

= ₹ 528. **Ans.**

(iii) Cost of fencing the plot is perimeter of plot × rate per meter.

Perimeter of plot $= \pi R + 2R$

$$= \frac{22}{7} \times 30 + 30 \times 2$$

$$= \frac{660}{7} + 60$$

$$= \frac{1080}{7} \text{ m}$$

Cost of fencing $= ₹ \frac{1080}{7} \times 3.5$

= ₹ 540 **Ans.**

Chapter 13. Surface Areas and Volumes

Q. 1. Vijay has rain water harvesting plant on his roof. After rain, all the water that is collected on the roof of 22 m × 20 m is drained into the cylindrical tank having diameter of base 2 m and height 3.5 m. It rained heavily last night and in the morning the tank is just full.

(i) How much water is collected in the tank (in litres)?

(ii) Find the rainfall in cm ?

Sol. (i) We have,

Radius of cylindrical tank $r = \frac{2}{2} = 1$ m

Height of cylindrical tank $h = 3.5$ m

Volume of cylindrical tank $= \pi r^2 h$

$$= \frac{22}{7} \times (1)^2 \times 3.5$$

$= 11 \text{ m}^3$

Since, $1 \text{ m}^3 = 1000\ l$

$11 \text{ m}^3 = 11000$ litres.

11000 litres of water is collected in tank. **Ans.**

(ii) Let the rainfall be x m,

Then, Volume of water = Volume of cuboid

$22 \text{ m} \times 20 \text{ m} \times x$ m.

$= (22 \times 20 \times x) \text{ m}^3$

$= 440x \text{ m}^3$

Since the tank is just full of water that drains out of the roof into the tank.

∴ Volume of water = Volume of the cylindrical vessel

$\Rightarrow \quad 22 \times 20 \times x = 11$

$$x = \frac{11}{440} = \frac{1}{40} \text{ m}$$

$$= \frac{100}{40} \text{ cm} = 2.5 \text{ cm}$$ **Ans.**

Q. 2. In a factory 120,000 pencils are manufactured daily. Pencils are cylindrical in shape, each of length 25 cm and the circumference of base is 1.5 cm. The cost of colouring curved surface of pencil is ₹ 0.05 per dm². Based on above information answer the following questions.

(i) How much area (in m²) of pencils is coloured daily ?

(ii) What is the cost of colouring all the pencils.

Sol. (i) Height of each pencil = 25 cm

Circumference of each pencil = $2\pi r = 1.5$ cm

Curved surface area of each pencil = $2\pi rh$

$\therefore \quad = 25 \times 1.5 \text{ cm}^2$

$= 37.5 \text{ cm}^2$

Total area of 120000 pencils = 37.5×120000

$= 4500000 \text{ cm}^2$

Area of pencils to be coloured daily

$$= \frac{4500000}{10000} \text{ m}^2$$

$= 450 \text{ m}^2$

(ii) Cost of colouring = 0.05 per dm²

Total area to be colour = 4500000 cm^2

$$= \frac{4500000}{100} \text{ dm}^2$$

$$\left[\because 1 \text{ cm}^2 = \frac{1}{100} \text{ dm}^2\right]$$

$= 45000 \text{ dm}^2$

Total cost of colouring all the pencils in a day

$= 45000 \times 0.05$

$= ₹ 2250$ **Ans.**

Q. 3. A wooden toy is in the shape of a right circular cylinder with hemisphere on one end and a cone on the other. The height and radius of cylinder part are 13 cm 5 cm respectively. The radii of the hemispherical and conical parts are same as that of the cylindrical part. Height of the conical part is 12 cm. The conical partion and the hemispherical portion is to be painted green and cylindrical portion is to be painted red.

Based on the given information, answer the following questions.

(i) How much area of the toy is painted with red colour ?

(ii) How much area of the toy is painted with green colour ?

(iii) Find the total surface area of the toy.

Sol. (i) Required area

= Curved surface area of cylinder

$= 2\pi rh$

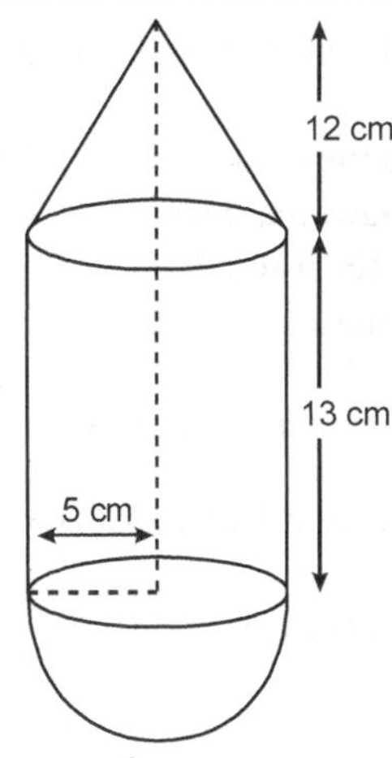

Here, $h = 13$ cm, $r = 5$ cm.

$$= 2 \times \frac{22}{7} \times 5 \times 13$$

$$= \frac{2860}{7} \text{ cm}^2$$

(ii) Required area = Curved surface area of cone + curved surface area of hemisphere.

$= \pi rl + 2\pi r^2$

Here, $r = 5$ cm, $h = 12$ cm

and $\quad l = \sqrt{r^2 + h^2} = \sqrt{5^2 + 12^2}$

$= \sqrt{25 + 144}$

$= 13$ cm

$$= \frac{22}{7} \times 5 \times 13 + 2 \times \frac{22}{7} \times 5 \times 5$$

$$= \frac{1430}{7} + \frac{1100}{7}$$

$$= \frac{2530}{7} \text{ cm}^2$$

(iii) Total surface area of toy = $\frac{2860}{7} + \frac{2530}{7}$

$$= \frac{5390}{7} = 770 \text{ cm}^2$$ **Ans.**

Q. 4. A Thirsty crows saw a spherical glass vessel with cylindrical neck 8 cm long and 2 cm in diameter. The diameter of spherical part is 8.5 cm. The vessel is half filled with water. But due to long neck of vessel crow was not able to drink water from it. He saw few spherical marbles of diameter 1 cm lying near by the vessel. He start dropping them inside the vessel one by one.

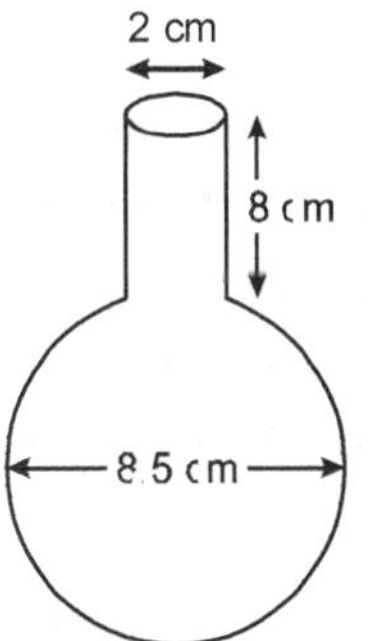

Based on the given information, answer the following questions :

(i) What is the total volume of the vessel ?

(ii) How many marbles crow has to drop inside the vessel to drink the water ?

Sol. (i) We have,

Radius of cylindrical neck $(r) = \frac{2}{2}\text{cm} = 1 \text{ cm}$

Length of cylindrical neck $(h) = 8$ cm

Radius of spherical part $(R) = \frac{8.5}{2} = 4.25$ cm

Volume of cylindrical part $= \pi r^2 h = \pi \times (1)^2 \times 8$
$= 8\pi \text{ cm}^3$

$$\text{Volume of spherical part} = \frac{4}{3}\pi R^3$$
$$= \frac{4}{3}\pi(4.25)^3$$
$$= 102.354\pi \text{ cm}^2$$

Total volume of vessel = Volume of cylindrical part + Volume of spherical
$= 8\pi + 102.354\pi \text{ cm}^2$
$= 110.354\pi \text{ cm}^3$

(ii) Since vessel is half filled = 54π

$$\therefore \text{Volume of water to be filled} = \frac{110.354}{2} \text{cm}^3$$
$$= 55.177\pi \text{ cm}^3$$

$$\text{Volume of each marble of radius } \frac{1}{2} \text{ cm} = \frac{4}{3}\pi r^3$$
$$= \frac{4}{3}\pi\left(\frac{1}{2}\right)^3 \text{cm}^3$$
$$= \frac{\pi}{6}\pi \text{ cm}^3$$

$$\text{No. of marbles} = \frac{\text{Volume of water to be filled}}{\text{Volume of each marble}}$$
$$= \frac{55.177\,\pi}{\frac{\pi}{6}}$$
$$= 331{\cdot}06$$
$$\approx 331$$

Crow has to drop 331 marbles to can drink water from it. **Ans.**

Q. 5. Atul has a container, open from the top, made up of a metal sheet in the form of frustum of a cone of height 16 cm with radii of its lower and upper ends as 8 cm and 20 cm respectively. He wants to fill it with milk. Based on the given information answer the following questions.

(i) Find the cost of milk which can completely fill the container at the rate of ₹ 15 per liter.

(ii) Find the cost of metal sheet to be used, if it costs ₹ 5 per 100 cm² (use $\pi = 3.14$)

Sol. (i) Let l be the slant height of container

Given, $r_1 = 20$ m, $r_2 = 8$ cm and $h = 16$ cm.

$$l = \sqrt{(r_1 - r_2)^2 + h^2}$$
$$= \sqrt{(20-8)^2 + 16^2}$$
$$= \sqrt{144 + 256} = \sqrt{400} = 20 \text{ cm.}$$

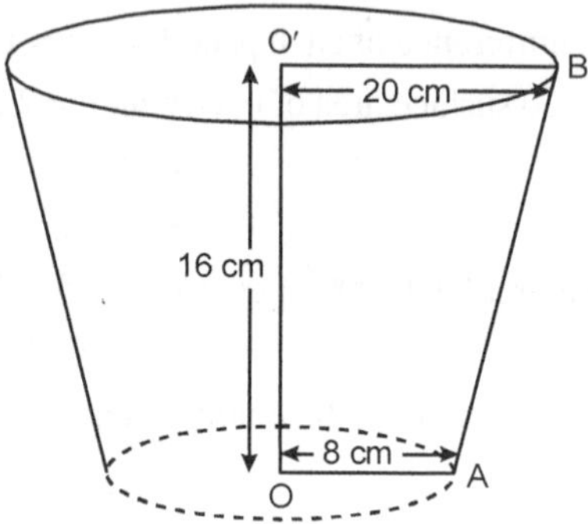

Let V be the volume of the container.

$$\therefore \quad V = \frac{\pi}{3}[r_1^2 + r_1^2 + r_1 r_2]h$$
$$= \frac{\pi}{3}[20^2 + 8^2 + 20 \times 8] \times 16 \text{ cm}^3$$
$$V = \frac{3.14 \times 624 \times 16}{3} \text{cm}^3$$
$$= 10449.92 \text{ cm}^3$$
$$= \frac{10449.92}{1000} l = 10.45 \text{ litres}$$

$\therefore$ cost of the milk at rate of ₹ 15 per/litre
= ₹ 10.45 × 15 = ₹ 156.75 **Ans.**

(ii) Let S be surface area of the container. Then,

$$S = \pi(r_1 + r_2)l + \pi r_2^2$$
$$= 3.14\,(20 + 8) \times 20 + 3.14 \times 8^2 \text{ cm}^2$$
$$= 3.14 \times [560 + 64] \text{ cm}^2$$
$$= 3.14 \times 624 \text{ cm}^2 = 1959.36 \text{ cm}^2$$

$$\therefore \text{cost of metal used} = \frac{₹\,(1959.36 + 5)}{100} = ₹\,97.97$$

Ans.

Chapter 14. Statistics

Q. 1. Research department in a company collected data about life time (in hours) of electrical components. The following table shows life time (in hours) of different electrical components :

Life time (in hours)	No. of components
0 - 20	10
20 - 40	35
40 - 60	52
60 - 80	61
80 - 100	38
100 - 120	29

Based on the given information, answer the following questions :

(i) Construct a 'less than type' ogive.

(ii) How many components have life more than 70 hours ?

Sol. (i)

Life time (in hours)	No. of components	Life time (less than)	Cumulative Frequency
0 – 20	10	less than 20	10
20 – 40	35	less than 40	45
40 – 60	52	less than 60	97
60 – 80	61	less than 80	158
80 – 100	38	less than 100	196
100 – 120	29	less than 120	225

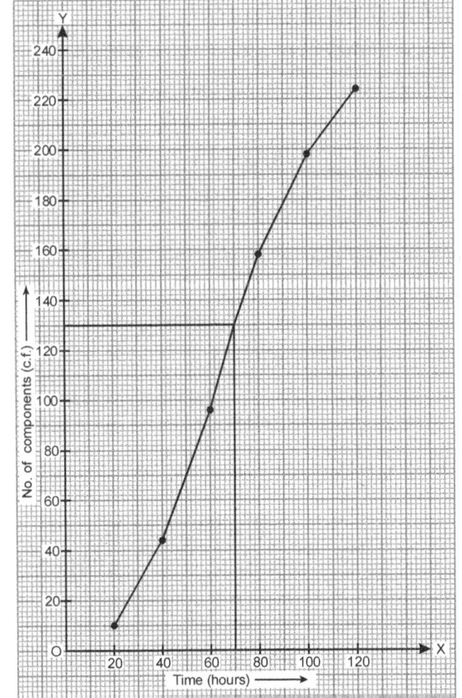

(ii) Number of components having life time more than 70 hours = (225 – 128) = 97. **Ans.**

Q. 2. A food officer has to find out the production yield of wheat in a village. For this, he surveyed 100 farms in that village. After collecting the data, he analyzed it and prepare report on the basis of production per hectare of wheat. Using this report he drew the following graph :

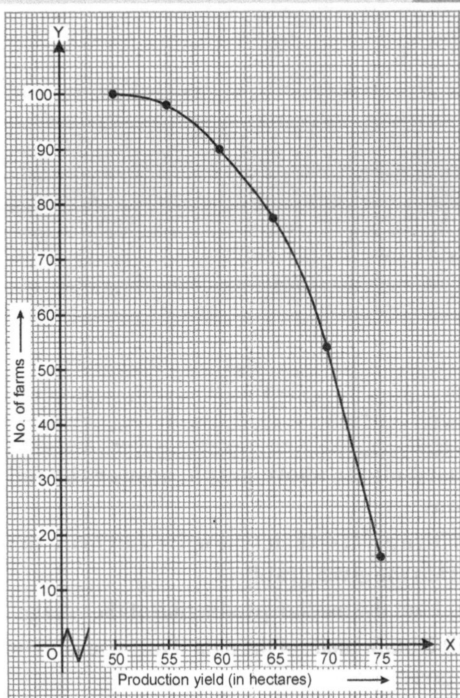

Based on the above graph, answer the following questions :

(i) Which type of ogive is given in the graph ?

(ii) Identify the median from the graph.

(iii) If mean production is 69.3 kg per hectare. Find out the mode.

Sol. (i) More than ogive is given in the graph.

(ii) From the graph median = 71 kg per hectare.

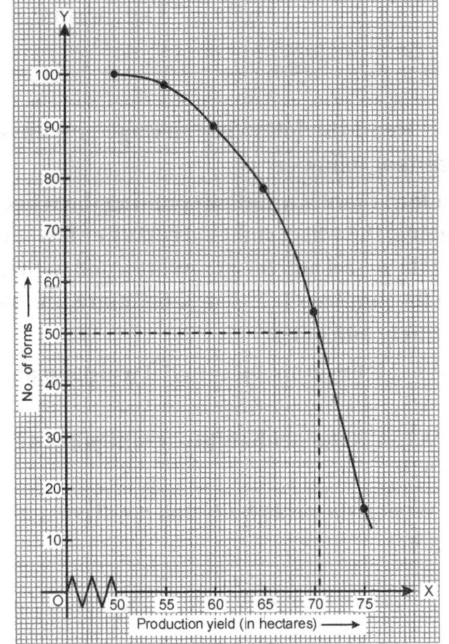

(iii) We know, Mode = 3 Median – 2 Mean

$= 3 \times 71 - 2 \times 69.3$

$= 213 - 138.6$

$= 74.4$ **Ans.**

Q. 3. A big agricultural farm of state prepare a report on the production yield per hectare of wheat of 100 farms of a village. Using this report he drew the following graph for a particular time period.

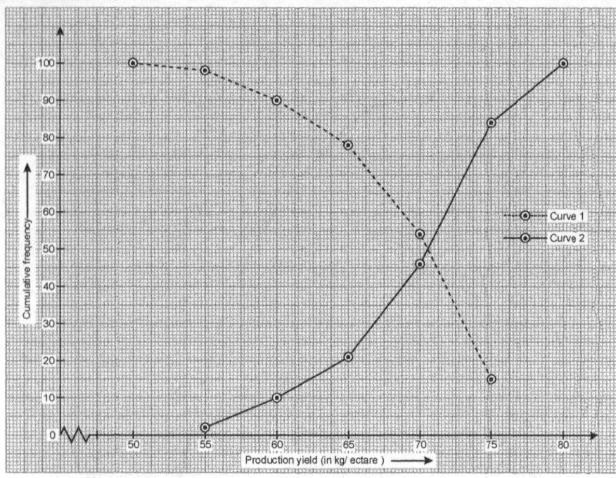

Based on the above graph, answer the following questions :

(i) Identify less than type ogive and more than type ogive from the given graph.

(ii) Find the median production of the village.

(iii) Obtain mode of the data if mean production is 70 kg/hectare.

Sol. (i) Curve 1 is 'more than' ogive and curve 2 is 'less than' ogive.

(ii) Median production = 71 kg/hectare.

(iii) $3 \text{ Median} = \text{Mode} + 2 \text{ Mean}$

$\Rightarrow$ $\text{Mode} = 3 \text{ Median} - 2 \text{ Mean}$

$\Rightarrow$ $\text{Mode} = 3 \times 71 - 21 \times 70$

$\Rightarrow$ $\text{Mode} = 213 - 140$

$\Rightarrow$ Mode = 73 kg/ hectare **Ans.**

Chapter 15. Probability

Q. 1. Read the following passage and answer the questions that follows :

In a fair, there is a stall of spinning the wheel game. In this game, the wheel comes to rest after few seconds of spinning the wheel and the arrow attached with it, points at one of the numbers 1, 2, 3, 4, 5, 6, 7 and 8 and these are equally likely outcomes.

(i) What is the probability that it will point at multiple of 2 ?

(ii) What is the probability it will point at a prime number ?

(iii) What is the probability that it will point at number less than 9 ?

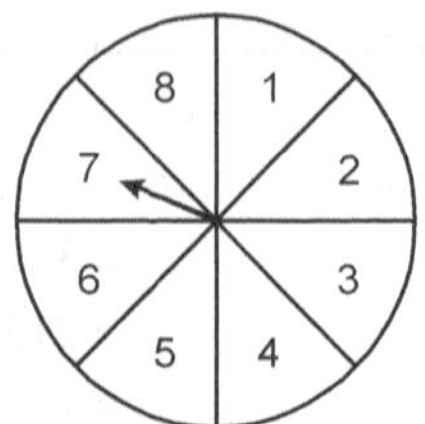

Sol. Total outcomes = {1, 2, 3, 4, 5, 6, 7, 8}

$\Rightarrow$ $n(s) = 8$

(i) Let A be the event that arrow will point at multiple of 2.

$\therefore$ $A = \{2, 4, 6, 8\}$

$\Rightarrow$ $n(A) = 4$

$\therefore$ $P(A) = \frac{4}{8} = \frac{1}{2}$ **Ans.**

(ii) Let B be the event that arrow will point at a prime number.

$\therefore$ $B = \{2, 3, 5, 7\}$

$\Rightarrow$ $n(B) = 4$

$\therefore$ $P(B) = \frac{4}{8} = \frac{1}{2}$ **Ans.**

(iii) Let C be the event that arrow will point at a number less than 9.

$\therefore$ $c = \{1, 2, 3, 4, 5, 6, 7, 8\}$

$\Rightarrow$ $n(c) = 8$

$\therefore$ $P(c) = \frac{8}{8} = 1$ **Ans.**

Q. 2. In a game, the entry fee is of ₹ 10. The game consists of tossing a coin 3 times. if it shows one or two heads, Pooja will get her entry fee back. If she throws 3 heads, She will get double the amount of entry fee, otherwise she will lose.

For tossing a coin three times, find the probability that she

(i) loses the entry fee, (ii) gets double entry fee

(iii) just gets her entry fee back.

Sol. When a coin is tossed three times, possible outcomes are

HHH, HHT, HTH, HTT, THH, THT, TTH, TTT

$\therefore$ Total numbers of elementary event = 8

(i) Let A be the event that she will lose the entry fee

$\therefore$ $A = \{TTT\}$

$n(A) = 1$

$P(A) = \frac{1}{8}$

(ii) Let B be the event that she will gets double the amount of entry fee.

$\therefore$ $B = \{HHH\}$

$n(B) = 1$

$P(B) = \frac{1}{8}$ **Ans.**

(iii) Let C be the event that she will just get her entry fee back.

$\therefore$ $C = [\text{HHT, HTH, HTT, THH, THT, TTH}]$

$n(C) = 6$

$P(C) = \frac{6}{8} = \frac{3}{4}$ **Ans.**

Q. 3. Read the following passage and answer the questions that follows :

Sunny has a piggy bank. he loves to collect coins. His piggy bank contains hundred (coins of ₹ 1, fifty coins of ₹ 2, twenty coins of ₹ 5 and ten coins ₹ 10. One day a coin dropped out from his piggy bank.

Based on the given information. Answer the following questions :

(i) What is the probability that the dropped coin will be a ₹ 2 coin ?

(ii) What is the probability that the dropped coin will not be ₹ 10 coin ?

Sol. Total number of coins = 100 + 50 + 20 + 10
= 180

∴ Total number of elementary events = 180.

(i) There are 50 coins of ₹ 2, out of which one coin can be chosen in 50 ways.

$$\therefore \quad P(A) = \frac{50}{180} = \frac{5}{18} \quad \textbf{Ans.}$$

(ii) Other than ₹ 10, there are 170 coins

∵ Probability that coin fallen out is not a 10 ₹ coins

$$= \frac{170}{180} = \frac{17}{18} \quad \textbf{Ans.}$$

Q. 4. Raju was playing ludo with his sister. Everytime, he throws a die, it always landed on the board and the ludo pieces get displaced from their positions. So, his sister scolded him to throw die on the circular part of the board only.

Based on the given situation and the following figure, answer the questions.

(i) If the diameter of the circular portion is 14 cm, what is the probability that die will land inside the circular portion ?

(ii) What is the probability of getting a prime number on a single throw of a die ?

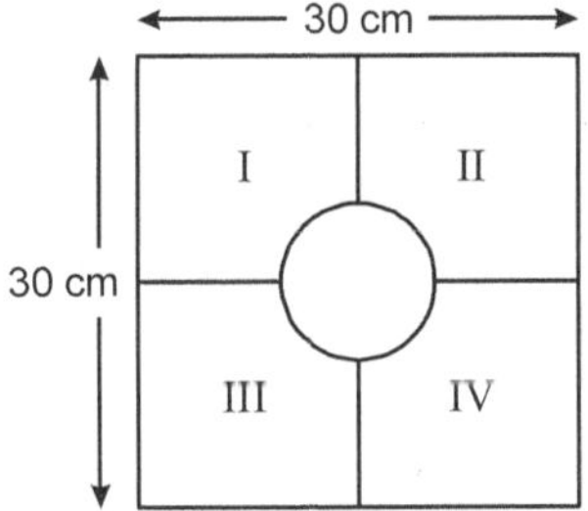

Sol. (i) Area of circular portion = πr^2

$$= \frac{22}{7} \times 7 \times 7 = 154 \text{ cm}^2$$

Area of board (square) = (side)^2
= 30 × 30 = 900 cm^2

$$\therefore \text{ P (die will land inside circular portion)} = \frac{154}{900}$$

$$= \frac{77}{450} \quad \textbf{Ans.}$$

(ii) Total number of outcomes = 6

And Prime numbers from 1 to 6 are {2, 3, 5}

⇒ Number of favourable outcomes = 3

$$\therefore \quad \text{P (Prime number)} = \frac{3}{6} = \frac{1}{2}. \quad \textbf{Ans.}$$

Q. 5. In a class there are 20 girls and 18 boys. The class teacher wants to choose one student for class monitor and she don't want to be partial. So, she writes the name of each student on a card and puts them into a basket and mixes thoroughly. A child is asked to pick one card from the basket.

Based on the given situation, answer the following questions :

(i) What is the probability, the name written on the card is the name of a girl ?

(ii) What is the probability, the name written on the card is the name of a boy ?

Sol. Total no. of students in the class = 20 + 18 = 38

(i) No. of girls in the class = 20

$$\text{Probability (the name is of a girl)} = \frac{20}{38} = \frac{10}{19}$$

Ans.

(ii) No. of boys in the class = 18.

$$\text{Probability (The name is of a boy)} = \frac{18}{38} = \frac{9}{19}$$

Ans.

Q. 6. A lot consists of 50 mobile phones of which 44 are good, 3 have only minor defects and 3 have major defects. Vanshika will buy a phone if it is good but the trader will only buy a mobile if it has no major defect. One phone is selected at random from she lot.

Based on the given information, answer the following questions :

(i) What is the probability that it is acceptable to vanshika ?

(ii) What is the probability that it is acceptable to trader ?

(iii) What is the probability that it is neither acceptable to trader nor to vanshika ?

Sol. Total number of phones in the lot = 50

(i) No. of mobile phones acceptable to Vanshika = 44.

Probability that mobile phone is acceptable to Vansika

$$= \frac{44}{50} = \frac{22}{25} \quad \textbf{Ans.}$$

(ii) No. of mobile phones acceptable to trader
= 44 + 3 = 47

Probability that mobile phone is acceptable to trader

$$= \frac{47}{50} \quad \textbf{Ans.}$$

(iii) No. of mobile phones not acceptable to both trader and Vanshika – 3

$$\text{Required Probability} = \frac{3}{50} \quad \textbf{Ans.}$$

Q. 7. A target game consist of three concentric circles of radii 3 cm, 5 cm and 9 cm respectively. A persson wins the game if dart hits on the shaded region, and gets a second chence if dart hits the inner most circle. And loses

the game if dart hits the outer most circle. Based on the given situation, answer the following questions :

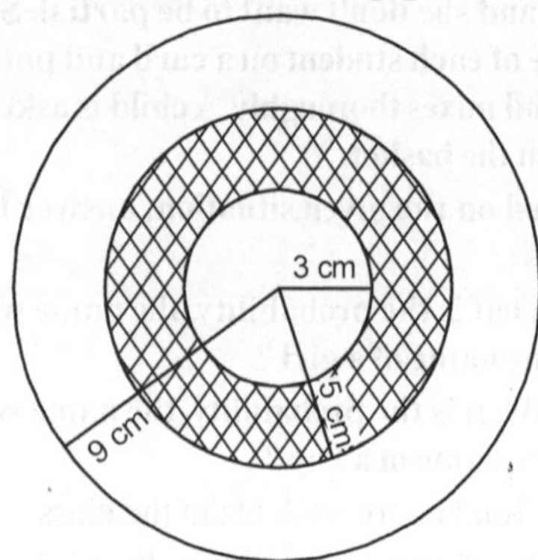

(i) What is the probability that a person will win the game ?

(ii) What is the probability that a person will get the second chance ?

(iii) What is the probability that a person will loose the game ?

Sol. Area of inner most circle with radius 3 cm

$$A_1 = \pi r^2 = \pi(3)^2 = 9\pi$$

Area of circle of with radius 5 cm

$$A_2 = \pi(5)^2 = 25\pi$$

Area of circle with radius 9 cm

$$A_3 = \pi(9)^2 = 81\pi$$

Area of shaded region $= A_2 - A_1$

$= 25\pi - 9\pi = 16\pi$

Area of outer ring $= A_3 - A_2$

$= 81\pi - 25\pi$

$= 56\pi$

(i) Probability of winning the game $= \dfrac{16\pi}{81\pi} = \dfrac{16}{81}$ **Ans.**

(ii) Probability of getting the second chance

$= \dfrac{9\pi}{81\pi} = \dfrac{1}{9}$ **Ans.**

(iii) Probability of lossing the game

$= \dfrac{56\pi}{81\pi} = \dfrac{56}{81}$ **Ans.**

●●

Test Your Knowledge

Chapter 1. Real Numbers

Q. 1. Define Euclid's Division Lemma.

Ans. According to Euclid's Division Lemma, if there are any two positive integers a and b, then their relationship is given by the relation $a = bq + r$, where $0 \leq r < b$ and r is the remainder whereas q is the quotient. Here a is the dividend and b is the divisor.

Q. 2. State the Fundamental Theorem of Arithmetic.

Ans. The Fundamental Theorem of Arithmetic states that every positive integer is a prime number or can be expressed as a product of prime numbers, which can be arranged in different orders. (Every composite number can be expressed as a product of primes and this factorisation is unique except for the order in which the prime factors occur).

Q. 3. Does the rational number $\frac{67}{2^3 5^5}$ has a terminating or non-terminating decimal representation ? If it is terminating then calculate after how many places of decimal will it do so ?

Ans. $\frac{67}{2^3 5^2}$ has a terminating decimal expansion. It terminates after 5 decimal places.

Q. 4. If $x + p$ is the G.C.D. of $x^2 + 7x - 18$ and $x^3 + 729$, then find the value of p.

Ans. 9.

Q. 5. Three iron rods of length 24 m, 94 m, 36 m have to be cut into poles of the same length. What is the greatest length possible ?

Ans. 2 m.

Q. 6. The traffic lights at three different crossing change after 36 s, 60 s and 96 s. If at 9 pm they changed together then when will they change together again ?

Ans. They will change together again after 24 minutes *i.e.*, at 9 : 24 pm.

Q. 7. A dealer has 60 l of blue paint, 84 l of violet paint and 132 l of white paint. What would be the capacity of the cans that he would to store all the three types of paint in equal quantities ? How many such cans will there be ?

Ans. The capacity of the cans should be 12 l each.

Number of such cans = 23.

Q. 8. In a class, there are 16 boys and 18 girls. Find the number of pens required to distribute them equally among the boys and the girls.

Ans. 144.

Q. 9. Prove the irrationality of the following numbers :

(i) $\sqrt{7} + \sqrt{11}$ (ii) $\sqrt{6} + 2\sqrt{3}$

(iii) $\sqrt{10} - \sqrt{5}$

Q. 10. Find the (i) H.C.F. of $2(x^3 - 4x^2 + 4x)$, $6(x^2 + x - 6)$ and $8(x^3 - 8)$ and (ii) L.C.M. of $3a^2 - 5ab - 12b^2$, $a^5 - 27a^2b^3$ and $9a^2 + 24ab + 16b^2$.

Ans. (i) $2(x - 2)$.

(ii) $a^2(a - 3b)(3a + 4b)^2(a^2 + 9b^2 + 3ab)$

Q. 11. Find the (i) H.C.F. of $8(x^2 - 4)$, $12(x^3 + 8)$ and $36(x^2 - 3x - 10)$ and (ii) L.C.M. of $(a^2 + 2a)^2$, $2a^3 - 2a$ and $2a^4 - 3a^3 - 14a^2$.

Ans. (i) $4(x + 2)$.

(ii) $a^2(a + 2)^2(2a - 1)(2a - 7)$.

Q. 12. Find the H.C.F. of (i) $x^4 - 2x^2 + 1$ and $x^4 - 2x^3 + 2x - 1$ and (ii) $4(x^2 - 7x + 12)$, $8(x^2 - 9)$ and $12(x^2 - 6x + 9)$.

Ans. (i) $(x - 1)^2(x + 1)$.

(ii) $4(x - 3)$.

Q. 13. Find the H.C.F. of $x^2 + 3x - 10$ and $x^3 - 8x$.

Ans. $(x - 2)$.

Chapter 2. Polynomials

Q. 1. If α and β are the zeros of the quadratic polynomial $p(x) = ax^2 + bx + c$, then evaluate :

(i) $\alpha - \beta$

(ii) $\alpha^2\beta + \alpha\beta^2$

(iii) $\alpha^4 + \beta^4$

(iv) $\frac{1}{a\alpha + b} + \frac{1}{a\beta + b}$

(v) $\frac{1}{\alpha} + \frac{1}{\beta} - 2\alpha\beta$

(vi) $a\left(\frac{\alpha^2}{\beta} + \frac{\beta^2}{\alpha}\right) + b\left(\frac{\alpha}{\beta} + \frac{\beta}{\alpha}\right)$

Ans. (i) $\frac{\sqrt{b^2 - 4ac}}{a}$

(ii) $\frac{-bc}{a^2}$

(iii) $\frac{(b^2-2ac)^2-2a^2c^2}{a^4}$

(iv) $\frac{b}{ac}$

(v) $-\left(\frac{b}{c}+\frac{2c}{a}\right)$

(vi) b

Q. 2. Find the remainder when

(i) $(3x^2 + 4x - 11)$ is divided by $(x - 1)$;

(ii) $(2x^3 - 3x + 2)$ is divided by $(x + 2)$;

(iii) $(x^5 - 4x^4 + 7x^3 - 11x - 13)$ is divisible by $(x - 2)$;

(iv) $(3x^4 + 5x^3 - 2x^2 + 4x + 6)$ is divisible by $(x - 1)$;

Ans. (i) – 4.

(ii) – 8.

(iii) – 11.

(iv) 16.

Q. 3. Show that

$$f(x) = (x^4 - 16x^3 + 86x^2 - 176x + 105)$$

is exactly divisible by $g(x) = (x - 3)$.

Q. 4. Show that $f(x) = (x^4 + 2x^3 - 13x^2 - 14x + 24)$ is exactly divisible by $g(x) = (x + 4)$.

Q. 5. Find the relation between a and b, if

$$f(x) = (4x^3 - 3x^2 + 2ax + b)$$

be exactly divisible by $(x + 2)$.

Ans. $-4a + b = 44$.

Q. 6. Find the relation between a and b, if

$$f(x) = (ax^5 + 3bx^3 + 8)$$

be exactly divisible by $(x - 2)$.

Ans. $4a + 3b = -1$.

Q. 7. Find the remainder when

(i) $(x^3 + 5x^2 + 3x + 2)$ is divided by $(x - 1)$

(ii) $(x^4 + 5x^3 + 4x^2 + 8x - 2)$ is divided by $(x + 2)$

(iii) $(2x^4 + 3x^3 - 9x^2 + 2x - 5)$ is divided by $(x + 3)$

(iv) $(3x^4 + 4x^3 + 7x^2 + 8x - 8)$ is divided by $(3x + 1)$

(v) $(3x^4 - x^3 + 2x^2 - 2x - 1)$ is divided by $(x - 3)$

Ans. (i) 11.

(ii) – 26.

(iii) – 11.

(iv) – 10.

(v) 227.

Q. 8. Express :

(i) $f(x) = 3x^3 - 4x^2 + 5x + 6$ as a polynomial of $x + 1$

(ii) $f(x) = x^4 - x^3 + 2x^2 - 3x + 1$ as a polynomial of $x - 3$

(iii) $f(x) = 3x^4 + 4x^3 + 7x^2 + 8x - 8$ as a polynomial of $x + 1$

(iv) $f(x) = 4x^5 - 6x^4 + 3x^3 - 5x + 2$ as a polynomial of $x + 2$.

Ans. (i) $3x^2(x + 1) - 7x(x + 1) + 12(x + 1) - 6$.

(ii) $x^3(x - 3) + 2x^2(x - 3) + 4x(x - 3) + 9(x - 3) + 28$.

(iii) $3x^3(x - 1) + 7x^2(x - 1) + 14x(x - 1) + 22(x - 1) + 14$.

(iv) $4x^5(x + 2) - 14x^4(x + 2) + 31x^3(x + 2) - 62x(x + 2) + 119(x + 2) - 236$.

Q. 9. (i) If $f(x) = x^4 - 3x^3 + 4x^2 - 5x - 9$ then show that

$$f(x + 2) = x^4 + 5x^3 + 10x^2 + 7x - 11.$$

(ii) If $f(x) = 2x^4 - x^3 - 2x^2 + 5x - 1$ then show that

$$f(x + 3) = 2x^4 + 23x^3 + 97x^2 + 182x + 131.$$

Q. 10. Find the quotient and remainder when $f(x) = x^5 - 5x^4 + 9x^3 - 6x^2 - 16x + 13$ is divided by $g(x) = x^2 - 3x + 2$.

Ans. Quotient $= x^3 - 2x^2 + x + 1$

Remainder $= -15x + 11$.

Q. 11. For $f(x) = (2x^3 + x^2 - 5x + 2)$, find the values for $f(x) = -1, 1, -2$ and prove that

$$f(1) = f(-2) \text{ and } 2f(-1) = f(2).$$

Ans. $f(-1) = 6, f(1) = 0, f(-2) = 0$

Q. 12. Obtain all zeros of the polynomial

$$f(x) = 2x^4 + x^3 - 14x^2 - 19x - 6$$

if two of its zeros are – 2 and – 1.

Ans. $-\frac{1}{2}, 3, -2, -1$.

Q. 13. Find the polynomial whose zeros are 1, – 2, 3 and – 4.

Ans. $f(x) = (x^4 + 2x^3 - 13x^2 - 14x + 24)$.

Q. 14. Find all the zeros of the polynomial $x^3 + 3x^2 - 2x - 6$, if two of its zeros are $-\sqrt{2}$ and $\sqrt{2}$.

Ans. $-\sqrt{2}, \sqrt{2}, -3$

Q. 15. Find the polynomial whose zeros are 2, – 3, 4 and – 1.

Ans. $f(x) = x^4 + 2x^3 - 13x^2 - 14x + 24$

Q. 16. In the equation $f(x) = (x^4 + 2x^3 - 13x^2 - 14x + 24)$, two zeros are 1 and – 2. Find the other zeros.

Ans. – 4 and 3.

Q. 17. Find the zeros of the following polynomials :

(i) $(x^4 - 9x^2 + 4x + 12) = 0$

(ii) $(x^4 - 6x^3 + 12x^2 - 10x + 3) = 0$.

Ans. (i) – 1, – 3, 2, 2.

(ii) 1, 1, 1, 3.

Q. 18. Find the zeros of the polynomial such that the sum of two of the zeros in each polynomial is 0 :

(i) $f(x) = x^3 - 5x^2 - 16x + 80$;

(ii) $f(x) = 4x^3 + 16x^2 - 9x - 36$.

Ans. (i) – 4, 4 and 5 (ii) $-\frac{3}{2}, \frac{3}{2}$ and – 4.

Q. 19. Find the zeros of x in $\frac{x+3}{x-3}+6\left(\frac{x-3}{x+3}\right)=5$.

Ans. 6 and 9.

Q. 20. Find the zeros of x in

$$\frac{x+1}{2}+\frac{2}{x+1}=\frac{x+1}{3}+\frac{3}{x+1}-\frac{5}{6}.$$

Ans. 0, – 7.

Chapter 3. Pair of Linear Equations in Two Variables

Q. 1. Romila went to a stationary stall and purchased 2 pencils and 3 erasers for ₹ 9. Her friend Sonali also bought 4 pencils and 6 erasers for ₹ 18. Represent this situation graphically. **[NCERT]**

Q. 2. Akhila went to a fair in her village. She wanted to ride on the giant wheel and play Hoopla. The number of times she played Hoopla is half the number of rides she had on the Giant Wheel. Each ride was priced at ₹ 3 and each attempt with the Hoopla cost her ₹ 4. If she spent a total of ₹ 20 at the fair then represent the situation graphically and also find the Ans. **[NCERT]**

Ans. 4, 2.

Q. 3. Aftab tells his daughter, "Seven year ago, I was seven times as old as you were then. In three years from now, I shall be three times as old as you will be." Represent the situation algebraically and graphically. Also, find their present ages. **[NCERT]**

Ans. Present age of Aftab is 56 and that of his daughter is 14.

Q. 4. Form the pair of linear equations and find their solution graphically, when 10 students of Class X participate in a mathematics quiz and the number of girls is 4 more than the number of boys. **[NCERT]**

Ans. The number of girls is 7 and the number of boys is 3.

Q. 5. 5 pencils and 7 pens together cost ₹ 50 whereas 7 pencils and 5 pens cost ₹ 46. Solving graphically, find the cost of 1 pencil and 1 pen. **[NCERT]**

Ans. Cost of 1 pencil is ₹ 3 and that of 1 pen is ₹ 5.

Q. 6. Champa bought skirts and trousers at a sale. The number of skirts purchased is two less than twice the number of trousers purchased. Also, the number of skirts purchased is four less than four times the number of trousers purchased. How many skirts and trousers did she buy ? Solve graphically. **[NCERT]**

Ans. The number of skirts bought is 0 and trousers bought is 1.

Q. 7. Given the linear equation $2x + 3y - 8 = 0$, write another equation in two variables such that the graphical representation of the pair so formed is (i) intersecting, (ii) parallel and (iii) coincident. **[NCERT]**

Ans. (i) $2x + 4y = 6$ (ii) $4x + 6y = 8$
(iii) $6x + 9y = 24$

Q. 8. The coach of a cricket team buys 7 bats and 6 balls for ₹ 3800. Later he buys 3 bats and 5 balls for ₹ 1750. Find the cost of each bat and ball. **[NCERT]**

Ans. 500, 50.

Q. 9. A lending library has a fixed change for the first three days and an additional charge for each day thereafter. Saritha paid ₹ 27 for a book kept for 7 days while Susy paid ₹ 21 for the book kept for five days. Find the fixed charge and the charge for each extra day. **[NCERT]**

Ans. 15, 3.

Q. 10. The sum of the numerator and denominator of a fraction is 12. If the denominator is increased by 3, the fraction becomes $\frac{1}{2}$· Find the fraction. **[NCERT]**

Ans. $\frac{5}{7}$·

Q. 11. The sum of digits of a two-digit number and the number obtained by reversing the order of digits is 66. If the two digits differ by 2, find the number. How many such numbers are there ? **[NCERT]**

Ans. There are two such numbers *i.e.*, 24 and 42.

Q. 12. The sum of the digits of a two-digit number is 9. Also, nine times the number is twice the number obtained by reversing the order of the digits. Find the number. **[NCERT]**

Ans. 18.

Q. 13. Five years ago, A was thrice as old as B and 10 years later, A shall be twice as old as B. What are the present ages of A and B ? **[NCERT]**

Ans. A is 50 years old and B is 20 years old.

Q. 14. A fraction becomes $\frac{9}{11}$ if 2 is added to both the numerator and the denominator and $\frac{5}{6}$ if 3 is added to both the numerator and the denominator. Find the fraction. **[NCERT]**

Ans. $\frac{7}{9}$·

Q. 15. A fraction becomes 1 if 2 is added to the numerator and $\frac{1}{2}$ if 1 is added to only the denominator. Find the fraction. **[NCERT]**

Ans. $\frac{3}{5}$·

Q. 16. The ages of Ani and Biju differ by 3 years. Ani's father, Dharam is twice as old as Ani and Biju is twice as old as his sister Cathy. The ages of Cathy and Dharam differ by 30 years. Find the ages of Ani and Biju. [NCERT]

Ans. Case 1 : When Ani is older than Bijju. Age of Ani is 19 years and that of Bijju is 16 years.

Case 2 : When Ani is younger than Bijju. Age of Ani is 16 years and that of Bijju is 19 years.

Q. 17. A man travels 600 km partly by train and partly by car. If he covers 120 km by train and 480 km by car, it takes him 8 hours. But if he travels 200 km by train and 400 km by car, he takes 20 minutes longer. Find the speed of the train and the car. [NCERT]

Ans. The speed of the train and car are 60 km/hr and 80 km/hr respectively.

Q. 18. A train covered a distance at a uniform speed. If it would have been faster by 10 km/hr, the time taken would be 2 hours lesser than the scheduled time. If it would have been slower by 10 km/hr, the time taken would have been 3 hours more than the scheduled time. Find the distance covered by the train. [NCERT]

Ans. 600 km.

Q. 19. Places *A* and *B* are 100 km apart on the highway. One car starts from *A* and another from *B* at the same time. If they travel in the same direction, they meet in 5 hours. If they travel in opposite directions, they meet in 1 hours. Find the speed of each car.★ [NCERT]

Ans. The speeds of *A* and *B* are 60 km/hr and 40 km/hr respectively.

Q. 20. Ritu can row 20 km downstream in 2 hours and 4 km upstream in 2 hours. Find her speed of rowing in still water and the speed of the current. [NCERT]

Ans. Ritu's speed in still water is 6 km/hr and the speed of the current is 4 km/hr.

Q. 21. The taxi charges in a city comprise of a fixed charge together with the charge for the distance covered. For a journey of 10 km, the charge paid is ₹ 75 and for a journey of 15 km, the charge paid is ₹ 110. What will a person pay for travelling a distance of 25 km ? [NCERT]

Ans. ₹ 180.

Q. 22. The area of a rectangle gets reduced by 9 square units if its length is reduced by 5 units and the breadth is increased by 3 units. If the length is increased by 3 units and the breadth by 2 units, the area is increased by 67 square units. Find the length and breadth of the rectangle. [NCERT]

Ans. 17 units, 9 units.

★ **are board exam questions from previous years**

Q. 23. The ratio of incomes of two persons is 9 : 7 and the ratio of their expenditures is 4 : 3. If each of them saves ₹ 200 per month, find their monthly incomes. [NCERT]

Ans. *A*'s monthly income is ₹ 1800 and *B*'s monthly income is ₹ 1400.

Q. 24. A part of the monthly hostel charges is fixed and the remaining depends on the number of days food has been taken in the mess. *A* takes food for 20 days, he pays ₹ 1000 as total charges and *B* takes food for 26 days and pays ₹ 1180 as charges. Find the fixed charge and the cost of the food per day. [NCERT]

Ans. The fixed charge is ₹ 400 and the cost of food per day is ₹ 30.

Q. 25. The larger of two supplementary angles exceeds the smaller by 18°. Find them. [NCERT]

Ans. The larger angle is 99° and the smaller is 81°.

Q. 26. 2 women and 5 men can together finish a job in 4 days, while 3 women and 6 men can finish it in 3 days. Find the time taken by 1 woman alone and 1 man alone to finish the job. [NCERT]

Ans. 1 woman will take 18 days and 1 man will take 36 days to complete the same task individually.

Q. 27. Meena went to the bank and withdrew ₹ 2000. The cashier paid in ₹ 50 and ₹ 100 notes. If Meena got 25 notes in all, how many note of each type did she receive ? [NCERT]

Ans. The numbers of ₹ 50 and ₹ 100 notes are 10 and 15 respectively.

Q. 28. Yash scored 40 marks in a test getting 3 marks for every right answer and losing 1 mark for every wrong answer. Had 4 marks been awarded for each correct answer and 2 marks deducted for each incorrect answer, then he would have scored 50 marks. How many questions were there in the test ? [NCERT]

Ans. 20.

Q. 29. Solve for x and y : $\frac{x}{6}-\frac{y}{3}=\frac{x}{12}-\frac{2y}{3}=4$.

Ans. $x = 16$ and $y = -4$.

Q. 30. Solve for x and y :

$$\frac{1}{x-1}+\frac{1}{y-2}=3 \text{ and } \frac{2}{x-2}+\frac{3}{y-2}=5.$$

Ans. $x=\frac{5}{4}$ and $y = 1$.

Q. 31. Solve for x and y :

$$\frac{2}{x}+\frac{3}{y}=18 \text{ and } \frac{5}{x}+\frac{4}{y}=31.$$

Ans. $x=\frac{1}{3}$ and $y=\frac{1}{4}$.

Q. 32. Solve for x and y :

$$\frac{2}{x}+\frac{3}{y}=2 \text{ and } \frac{5}{x}+\frac{10}{y}=5\frac{5}{6}.$$

Ans. $x = 2$ and $y = 3$.

Q. 33. Solve for x and y :

$$\frac{x}{a}+\frac{y}{b}=2 \text{ and } ax-by=a^2-b^2.$$

Ans. $x = a$ and $y = b$.

Q. 34. Solve for x and y :

$$\frac{2}{x}+\frac{3}{y}=1 \text{ and } \frac{3}{x}+\frac{2}{y}=1.$$

Ans. $x = 5$ and $y = 5$.

Chapter 4. Quadratic Equations

Q. 1. In a flight of 6000 km, an aircraft was slowed down due to bad weather. The average speed for the trip was reduced by 400 km/hr and the time of the flight was increased by 30 minutes. Find the original duration of the flight.

Ans. $2\frac{1}{2}$ hrs.

Q. 2. The area of a rectangular plot is 528 m^2. The length of the plot (in metres) is one more than twice the breadth. Formulate the quadratic equation to determine the length and breadth of the plot. **[NCERT]**

Ans. $2x^2 + x - 528 = 0$.

Q. 3. Rohan's mother is 26 years older than him. The product of their ages 3 years from now will be 360. Formulate the quadratic equation to find their ages. **[NCERT]**

Ans. $x^2 + 32x - 273 = 0$.

Q. 4. A train travels a distance of 480 km at uniform speed. If the speed had been 8 km/hr less, it would have taken 3 hours more to cover the same distance. Formulate the quadratic equation in terms of the speed of the train. **[NCERT]**

Ans. $x^2 - 8x - 1280 = 0$.

Q. 5. The product of two consecutive positive integers is 306. Form the quadratic equation to find the integers, if x denotes the smaller integer. **[NCERT]**

Ans. $x^2 + x - 306 = 0$.

Q. 6. A cottage industry produces a certain number of toys in a day. The cost of production of each toy (in rupees) was found to be the product of the numbers of toys produced per day and 55 minus the number of toys produced in a day. On a particular day, the total cost of production was ₹ 750. If x denotes the number of toys produced that day, form the quadratic equation to find x. **[NCERT]**

Ans. $x^2 - 55x + 750 = 0$.

Q. 7. The height of a right-triangle is 7 cm less than the base. If the hypotenuse is 13 cm form the quadratic equation to find the base of the triangle. **[NCERT]**

Ans. $x^2 - 7x - 60 = 0$.

Q. 8. Solve : $\frac{1}{x-2}+\frac{2}{x-1}=\frac{6}{x}$. **[NCERT]**

Ans. $x = 3$ and $\frac{4}{3}$.

Q. 9. Solve : $9x^2 - 15x + 6 = 0$. **[NCERT]**

Ans. $x = 1$ and $\frac{2}{3}$.

Q. 10. Solve : $5x^2 - 6x - 2 = 0$. **[NCERT]**

Ans. $x = \frac{3 \pm \sqrt{19}}{5}$.

Q. 11. Solve : $4x^2 + 3x + 5 = 0$. **[NCERT]**

Ans. $\frac{-3 \pm \sqrt{-71}}{8}$.

Q. 12. Solve: $2x^2 - 2\sqrt{2}x + 1 = 0$. **[NCERT]**

Ans. $x = \frac{1}{\sqrt{2}}, \frac{1}{\sqrt{2}}$.

Q. 13. Solve : $3x^2 - 5x + 2 = 0$. **[NCERT]**

Ans. $x = 1$ and $\frac{2}{3}$.

Q. 14. Solve : $x + \frac{1}{x} = 3$, where $x \neq 0$. **[NCERT]**

Ans. $x = \frac{3 \pm \sqrt{5}}{2}$.

Q. 15. Determine the roots of the equation

$2x^2 - 3x + 5 = 0$. **[NCERT]**

Ans. $x = \frac{3 \pm \sqrt{-31}}{4}$.

Q. 16. Determine the roots of the equation

$2x^2 - 6x + 3 = 0$. **[NCERT]**

Ans. $x = \frac{3 \pm \sqrt{6}}{2}$.

Q. 17. Determine the roots of the equation

$3x^2 - 4\sqrt{3}x + 4 - 0$. **[NCERT]**

Ans. $x = \frac{2}{\sqrt{3}}, \frac{2}{\sqrt{3}}$.

Q. 18. Find the value of k for which the equation $2x^2 + kx + 3 = 0$ has real and equal roots. **[NCERT]**

Ans. $k = \pm 2\sqrt{6}$.

Q. 19. Find the value of k for which the equation $kx(x-2)+6=0$ has real and equal roots. **[NCERT]**

Ans. $k = 6$.

Q. 20. The difference of the squares of two numbers is 180. If the square of the smaller number is 8 times the larger number, then find the two numbers. **[NCERT]**

Ans. The two numbers are 18 and 12 or 18 and – 12.

Q. 21. An express train takes 1 hour less than a passenger train to travel 132 km between Mysore and Bangalore. If the average speed of the express train is 11 km/hr more than the passenger train, find the speeds of the two trains. **[NCERT]**

Ans. The speed of the passenger train is 33 km/hr and the speed of the express train is 44 km / hr.

Q. 22. The sum of the ages of two friends is 20 years. Four years ago, the product of their ages was 48. Find the ages if the situation is possible. **[NCERT]**

Ans. The situation is not possible.

Q. 23. The sum of the reciprocals of Rehman's ages 3 years ago and 5 years hence is $\frac{1}{3}$. Find his present age. **[NCERT]**

Ans. 7 years.

Q. 24. A pole has to be erected at a point on the boundary of a circular park of diameter 13 m in such a way that the difference of its distances from two diametrically opposite fixed gates A and B on the boundary is 7 metres. If it is possible to do so, at what distance from the gates should the pole be erected ? **[NCERT]**

Ans. 5 metres and 12 metres.

Q. 25. Is it possible to design a rectangular mango grove whose length is twice its breadth and area is 800 m^2 ? If so, find the length and the breadth. **[NCERT]**

Ans. Breadth = 20 m and length = 40 m.

Q. 26. Is it possible to design a rectangular park of perimeter 80 m and area 400 m^2 ? If so, find the length and the breadth. **[NCERT]**

Ans. The dimensions are 20 m and 20 m.

Q. 27. Two water taps together take $9\frac{3}{8}$ hours to fill a tank. If the tap with the larger diameter takes 10 hours lesser than the tap with the smaller diameter, then find the time in which each tap can separately fill the tap. **[NCERT]**

Ans. The smaller tap takes 25 hours to fill the tank and the larger one takes 15 hours to do so.

Q. 28. In a class test, the sum of the marks obtained by Shefali in Mathematics and English was 30. Had she secured 2 more marks in Mathematics and 3 less in English then the product of the marks in both the tests would have been 210. Find the marks obtained by her in the two subjects separately. **[NCERT]**

Ans. The marks obtained in Mathematics are either 13 or 12 and in English are 17 or 18.

Chapter 5. Arithmetic Progressions

Q. 1. The nth term of an A.P. is $6n+2$. Find the common difference. **[NCERT]**

Ans. 6.

Q. 2. A sum of ₹ 1000 is invested at 8% S.I. per annum. Calculate the rate of interest at the end of 1, 2, 3, ... years. Is the sequence of the interests an A.P. ? Find the interest at the end of 30 years. **[NCERT]**

Ans. The interest after 1 year = ₹ 80

The interest after 2 years = ₹ 160

The interest after 3 years = ₹ 240

Yes, the sequence of the interests is an A.P.

The interest after 30 years = ₹ 2400.

Q. 3. In a flower bed, there are 23 rose plants in the first row, 21 in the second row, 19 in the third row and so on. There are five plants in the last row. How many rows are there in the flower bed ? **[NCERT]**

Ans. 10.

Q. 4. A person started working in 1995 at a salary of ₹ 5000 per month with a yearly increment of ₹ 200. In which year did his salary reach ₹ 7000 per month ? **[NCERT]**

Ans. In the year 2005.

Q. 5. A housewife saved ₹ 5 in the first week of the year and thereafter increased her weekly savings by ₹ 1.75. After how many weeks will her weekly savings be ₹ 20.75 ? **[NCERT]**

Ans. 10 weeks.

Q. 6. Find the 12th term from the end of the following arithmetic progression : 3, 8, 13, ..., 253. **[NCERT]**

Ans. 198.

Q. 7. The sum of the 4th and 8th terms of an A.P. is 24 and the sum of the 6th and 10th term is 34. Find the first term and the common difference of the A.P. **[NCERT]**

Ans. – 0.5, 2.5.

Q. 8. Find the A.P. whose third term is 16 and the seventh term exceeds its fifth term by 12. **[NCERT]**

Ans. 4, 10, 16, 22, ...

Q. 9. Two A.P.s have the same common difference. The difference between their 100th terms is 100. What is the difference between their 1000th terms ? **[NCERT]**

Ans. 100.

Q. 10. A manufacturer of TV sets produced 600 units in the 3rd year and 700 units in the 7th year. Assuming that the production increases uniformly by a fixed number every year, find the production in (i) the first year, (ii) the 10th year and (iii) the total production in 7 years. **[NCERT]**

Ans. (i) 550, (ii) 775, (iii) 4375.

Q. 11. The contract of a construction job specifies a penalty for delay of completion beyond a certain date as follows : ₹ 200 for the first day, ₹ 250 for the second day, ₹ 300 for the third day and so on. How much does a delay of 30 days cost the contractor ? **[NCERT]**

Ans. ₹ 27,750.

Q. 12. A sum of ₹ 280 is to be used to award four prizes. If each prize after the first is ₹ 20 less than its preceding prize, find the value of each of the prizes. **[NCERT]**

Ans. The first prize is ₹100, second prize is ₹ 80, third prize is ₹ 60 and fourth prize is ₹ 40.

Q. 13. Find the total number of trees planted by the students of a school if there are three sections in class and if each class plants trees according to the number of classes (*e.g.*, Class 1 plants 1 tree, Class 2 plants 2 trees, ... Class 12 plants 12 trees). **[NCERT]**

Ans. 234.

Q. 14. 200 logs are placed in the following manner : 20 logs in the bottom row, 19 in the next, 18 in the next and so on. How many rows would there be ? Also find the number of the logs in the top now. **[NCERT]**

Ans. The total number of rows is 16 and the number of logs in the last row is 5.

Q. 15. The houses of a row are numbered consecutively from 1 to 49. Show that there is a house x such that sum of the number of houses preceding it is equal to the sum of numbers following it. Consequently, find x. **[NCERT]**

Ans. 35.

Q. 16. A football gallery is 50 m wide and each step is 0.5 m in length. If the height of each step increases by 0.25 m from the first, find the volume of concrete required to make a gallery of 15 steps. **[NCERT]**

Ans. 11250 m^3.

★ are board exam questions from previous years

Q. 17. How many terms of the A.P. 9, 17, 25, ... must be taken so that their sum is 636 ? **[NCERT]**

Ans. 12.

Q. 18. Find the sum of the first 15 terms of the series where $t_n = 9 - 5n$. **[NCERT]**

Ans. – 465.

Q. 19. Find the sum of the first 51 terms of an A.P. whose 2nd and 3rd terms are 14 and 18 respectively. **[NCERT]**

Ans. 5610.

Q. 20. The first term of an A.P. is 5, the last term is 45 and the sum is 400. Find the number of terms and the common difference. **[NCERT]**

Ans. 16, $\frac{8}{3}$.

Q. 21. Find the nth term and the last term of an A.P. whose first term is 2, the common difference is 8 and the sum of all the terms is 90. **[NCERT]**

Ans. $8n - 6$, 34.

Q. 22. If the sum of the first n terms of an A.P. is $4n - n^2$, find the first term, second term and sum of the first two terms. **[NCERT]**

Ans. 3, 1, 4.

Q. 23. The first term of an A.P. is 17 and the last term is 350. If the common difference is 9, how many terms are there and what is their sum ? **[NCERT]**

Ans. 38, 6973.

Q. 24. The sum of n terms of an A.P. is $5n - n^2$. Find nth term of this A.P. ★

Ans. $6 - 2n$.

Q. 25. The sum of first n terms of an A.P. is $3n^2 + 4n$. Find the 25th term of this A.P.

Ans. 157.

Chapter 6. Coordinate Geometry

Q. 1. Find the point on the X-axis which is equidistant from the points $A(2, -5)$ and $B(-2, 9)$.★ **[NCERT]**

Ans. (– 7, 0).

Q. 2. Find the point on the Y-axis which is equidistant from the points (6, 5) and (– 4, 3). **[NCERT]**

Ans. (0, 9).

Q. 3. Do the points $A(3, 2)$, $B(-2, -3)$ and $C(2, 3)$ form a triangle ? If so, name the type of triangle formed.

Ans. The triangle formed is a right-angled triangle.

Q. 4. Show that the points $A(1, 7)$, $B(4, 2)$, $C(-1, -1)$ and $D(-4, 4)$ are the vertices of a square. **[NCERT]**

Q. 5. If the point $Q(0, 1)$ is equidistant from the points $P(5, -3)$ and $R(x, 6)$, find the values of x. Also find the distances QR and PR. **[NCERT]**

Ans. $x = \pm 4$, $QR = \sqrt{41}$, $PR = \sqrt{82}$ or $9\sqrt{2}$ units.

Q. 6. Find the values of y for which the distance between the points $P(2, -3)$ and $Q(10, y)$ is 10 units. **[NCERT]**

Ans. 3 or -9.

Q. 7. Name the quadrilateral formed by the following points $A(4, 5)$, $B(7, 6)$, $C(4, 3)$ and $D(1, 2)$. **[NCERT]**

Ans. Parallelogram.

Q. 8. Find the equation of the perpendicular bisector of the line segment joining the points (7, 1) and (3, 5). **[NCERT]**

Ans. $x - y = 2$.

Q. 9. Find the relation between x and y such that the point (x, y) is equidistant from the points (3, 6) and $(-3, 4)$. **[NCERT]**

Ans. $3x + y = 5$.

Q. 10. Find the coordinates of the points of trisection of the line segment joining the points $A(2, -2)$ and $B(-7, 4)$. **[NCERT]**

Ans. $(-1, 0)$ and $(-4, 2)$

Q. 11. If A and B are two points having coordinates $(-2, -2)$ and $(2, -4)$ respectively, find the coordinates of P such that $AP = \frac{3}{7}AB$.★★ **[NCERT]**

Ans. $\left(-\frac{2}{7}, -\frac{20}{7}\right)$.

Q. 12. Find the coordinates of the points which divide the line segment joining $A(-2, 2)$ and $B(2, 8)$ into four equal parts. **[NCERT]**

Ans. $\left(-1, \frac{7}{2}\right)$, $(0, 5)$, $\left(1, \frac{13}{2}\right)$.

Q. 13. $ABCD$ is a rectangle where $A(-1, -1)$, $B(-1, 4)$, $C(5, 4)$ and $D(5, -1)$ are the vertices. P, Q, R, S are the mid-points of sides AB, BC, CD and AD respectively. Is the quadrilateral $PQRS$ a square or a rectangle or a rhombus ? Justify your answer. **[NCERT]**

Ans. Rhombus because all the sides of quadrilateral are equal and diagonals are not equal to each other.

Q. 14. Find the ratio in which the point $P(-4, 6)$ divides the line segment joining the points $A(-6, 10)$ and $B(3, -8)$. **[NCERT]**

Ans. 2 : 7.

Q. 15. Find the coordinates of a point A where AB is a diameter of a circle whose centre is $(2, -3)$ and B is $(1, 4)$. **[NCERT]**

Ans. $(3, -10)$.

★ are board exam questions from previous years
★★ are frequently asked board exam questions

Q. 16. Find the area of the triangle formed by joining the mid-points of the sides of the triangle whose vertices are $(0, -1)$, $(2, 1)$ and $(0, 3)$. Find the ratio of the area of a triangle formed to the area of the given triangle. **[NCERT]**

Ans. 1 sq. units, 1 : 4.

Q. 17. The vertices of ΔABC are $A = (4, 6)$, $B = (1, 5)$ and $C = (7, 2)$. A line is drawn to intersect sides AB and AC at D and E respectively such that $\frac{AD}{AB} = \frac{AE}{AC} = \frac{1}{4}$. Calculate the area of ΔADE and compare it with the area of ΔABC. **[NCERT]**

Ans. $\frac{15}{32}$ sq. units, 1 : 16.

Q. 18. Find the area of the quadrilateral $ABCD$ whose vertices are $A(-4, -2)$, $B(-3, -5)$, $C(3, -2)$ and $D(2, 3)$. **[NCERT]**

Ans. 28 sq. units.

Q. 19. If the distance between $(x, 3)$ and $(1, 7)$ is 5. Calculate the value of x.

Ans. 4, -2.

Q. 20. For what value of p will the points (22, 2), $\left(2, -\frac{1}{2}\right)$ and $\left(-\frac{p}{2}, 2\right)$ will be collinear ?

Ans. -44.

Q. 21. Determine the mid-point of the line segment joining the points $(4, -2)$ and $(7, 5)$.

Ans. $\left(\frac{11}{2}, \frac{3}{2}\right)$.

Chapter 7. Triangles

Q. 1. In the given figure, $EF \parallel AB \parallel DC$. Prove that : $\frac{AE}{ED} = \frac{BF}{FC}$. **[NCERT]**

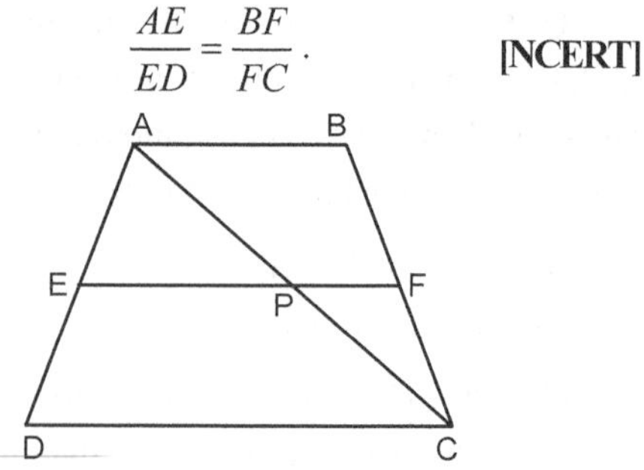

Q. 2. In the given figure, $\frac{PS}{SQ} = \frac{PT}{TR}$ and $\angle PST = \angle PRQ$. Prove that ΔPQR is an isosceles triangles. **[NCERT]**

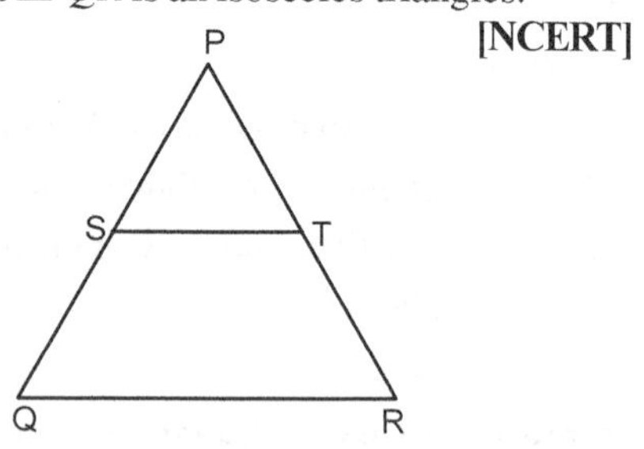

Q. 3. Prove that the diagonals of a trapezium divide each other proportionally. **[NCERT]**

Q. 4. Prove that the line joining the mid-points of two sides of a triangle is parallel to the third side.

Q. 5. Prove that if the diagonals of a quadrilateral divide each other proportionally then it is a trapezium. **[NCERT]**

Q. 6. Prove that two equiangular triangles are similar. **[NCERT]**

Q. 7. If in two triangles, one pair of corresponding sides is proportional and the included angles are equal, then show that the two triangles are similar. **[NCERT]**

Q. 8. If the corresponding sides of two triangles are proportional then show that they are similar. **[NCERT]**

Q. 9. In the given figure, if $\Delta EDC \sim \Delta EBA$, $\angle BEC = 115°$ and $\angle EDC = 70°$, find $\angle DEC$, $\angle DCE$, $\angle EAB$, $\angle AEB$ and $\angle EBA$. **[NCERT]**

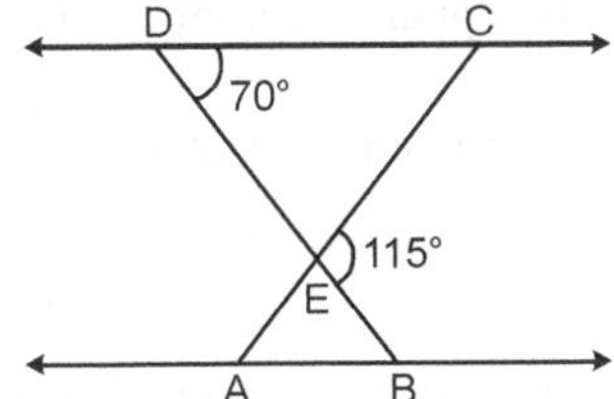

Ans. $\angle DEC = 65°$, $\angle DCE = 45°$, $\angle EAB = 45°$, $\angle AEB = 65°$ and $\angle EBA = 70°$.

Q. 10. In the given figure, if $\Delta POS \sim \Delta ROQ$, prove that $PS \parallel QR$. **[NCERT]**

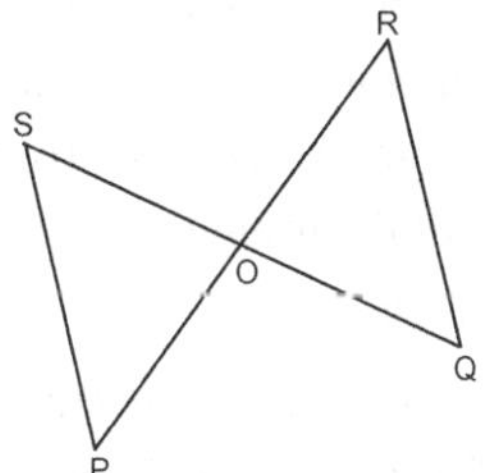

Q. 11. In the given figures, if CD and GH (D and H lie on AB and FE) are the bisectors of $\angle ACB$ and $\angle EGF$ respectively and $\Delta ABC \sim \Delta FEG$, prove that :

(i) $\Delta DCA \sim \Delta HGF$, (ii) $\dfrac{CD}{GH} = \dfrac{AC}{FG}$,

(iii) $\Delta DCB \sim \Delta HGE$. **[NCERT]**

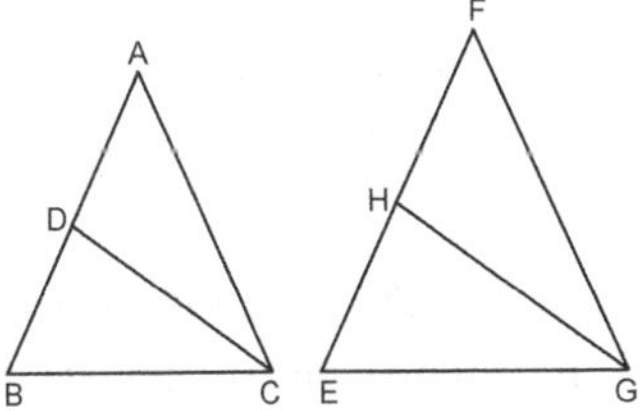

Q. 12. In the figures given below, CD and GH are respectively the medians of ΔABC and ΔFEG. If $\Delta ABC \sim \Delta FEG$, prove that :

(i) $\Delta ADC \sim \Delta FGH$, (ii) $\dfrac{CD}{GH} = \dfrac{AC}{FG}$,

(iii) $\Delta CDB \sim \Delta GHE$. **[NCERT]**

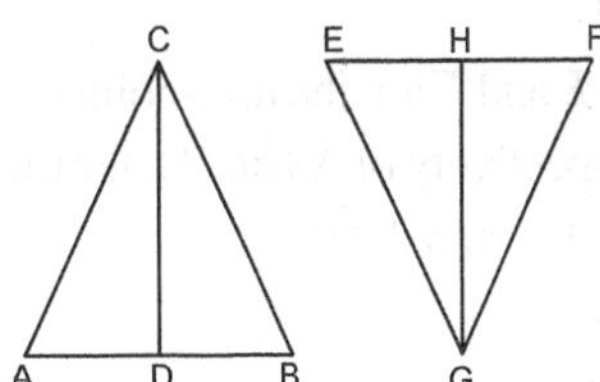

Q. 13. E is a point on side AD produced on a parallelogram $ABCD$ and BE bisects CD at F. Prove that $\Delta ABE \sim \Delta CFB$. **[NCERT]**

Q. 14. In the given figure, ABC is a right-angled triangle, where $\angle B = 90°$ and D is the foot of the perpendicular drawn from B on AC. If DM is perpendicular to BC and DN to AB, prove that : (i) $DM^2 = DN \times MC$ and (ii) $DN^2 = DM \times AN$. **[NCERT]**

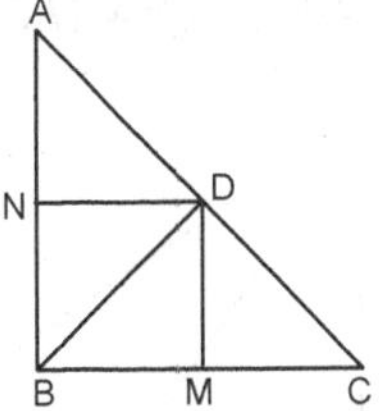

Q. 15. In the figure given below, E is a point on side CB produced in the isosceles ΔABC, where $AB = AC$. If AD is perpendicular to BC and EF is perpendicular to AC, prove that $\Delta ABD \sim \Delta ECF$.★ **[NCERT]**

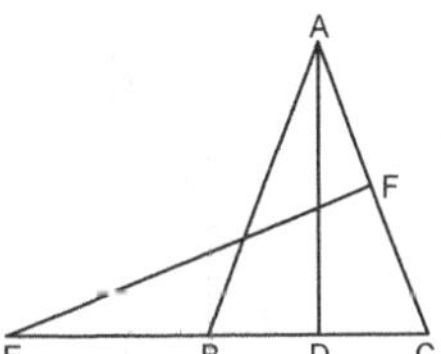

Q. 16. In the figure given below, $\dfrac{OA}{OC} = \dfrac{OD}{OB}$.

Prove that : $\angle A = \angle C$ and $\angle B = \angle D$. **[NCERT]**

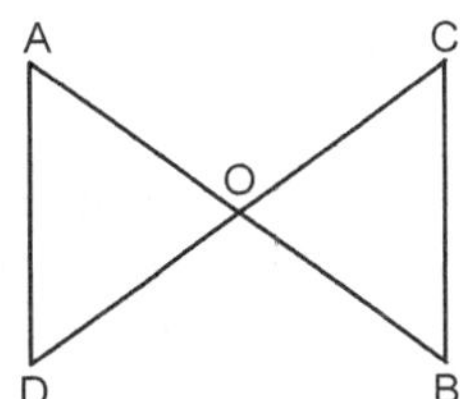

Q. 17. A vertical stick of length 6 m casts a shadow 4 m long on the ground at the same time a tower casts a shadow 28 m long. Find the height of the tower. **[NCERT]**

Ans. 42 m.

Q. 18. Prove that if the areas of two similar triangles are equal then they are congruent. **[NCERT]**

★ are board exam questions from previous years

Q. 19. $ABCD$ is a trapezium, in which $AB \parallel CD$ and $AB = 2CD$. Determine the ratio of the areas of ΔAOB and ΔCOD. **[NCERT]**

Ans. 4 : 1.

Q. 20. D, E and F are the mid-points of the sides BC, CA and AB respectively of ΔABC. Determine the ratios of the areas of ΔDEF and ΔABC. **[NCERT]**

Ans. 1 : 4.

Q. 21. D and E are points on the sides AB and AC respectively of ΔABC such that $DE \parallel BC$ and divides ΔABC into two parts, equal in area. Find $\frac{BD}{AB}$. **[NCERT]**

Ans. $\frac{\sqrt{2}-1}{\sqrt{2}}$.

Q. 22. The given figure represents ΔABC, obtuse-angled at B. If AD is perpendicular to CB, prove that

$$AC^2 = AB^2 + BC^2 + 2BC \times BD.$$ **[NCERT]**

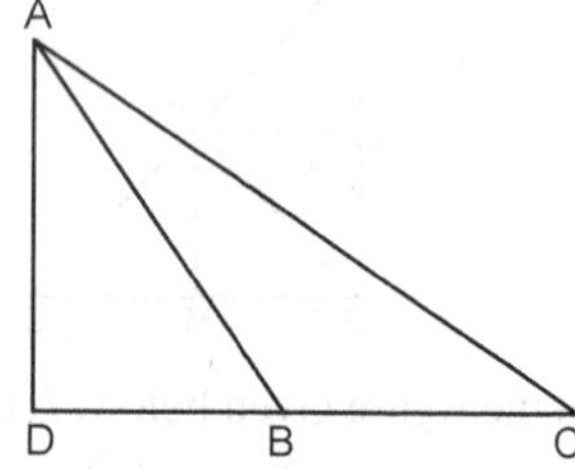

Q. 23. The given figure represents ΔABC, acute-angled at B. If AD is perpendicular to CB, prove that

$$AC^2 = AB^2 - BC^2 + 2BC \times BD.$$ **[NCERT]**

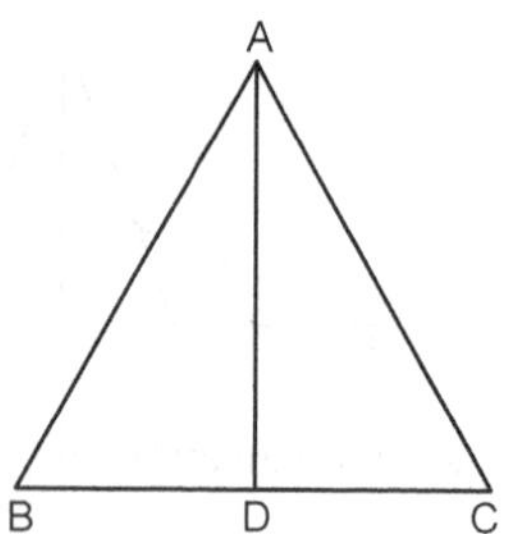

Q. 24. ABC is an isosceles triangle, right-angled at C. Prove that $AB^2 = 2AC^2$. **[NCERT]**

Q. 25. From a point O in the interior of ΔABC perpendicular OD, OE and OF are drawn to the sides BC, CA and AB respectively. Prove that :

(i) $AF^2 + BD^2 + CE^2$

$$= OA^2 + OB^2 + OC^2 - OD^2 - OE^2 - OF^2$$

(ii) $AF^2 + BD^2 + CE^2 = BF^2 + CD^2 + AE^2$. **[NCERT]**

Q. 26. A point O in the interior of a rectangle $ABCD$ is joined with each of the vertices A, B, C and D. Prove that $OB^2 + OD^2 = OC^2 + OA^2$. **[NCERT]**

★ **are board exam questions from previous years**

Q. 27. $ABCD$ is a rhombus. Prove that

$$AB^2 + BC^2 + CD^2 + DA^2 = AC^2 + BD^2.$$ **[NCERT]**

Q. 28. In an equilateral ΔABC, the side BC is trisected at D. Prove that $9AD^2 = 7AB^2$. **[NCERT]**

Q. 29. In ΔPQR, QM is perpendicular to PR and $PR^2 - PQ^2 = QR^2$. Prove that $QM^2 = PM \times MR$. **[NCERT]**

Q. 30. Prove that the sum of the squares of the diagonals of a parallelogram is equal to the sum of the squares of its sides. **[NCERT]**

Q. 31. A guy wire attached to a vertical pole of height 18 m is 24 m long hand has a stake attached to the other end. How far from the base of the pole should the stake be driven so that the wire will be taut ? **[NCERT]**

Ans. $6\sqrt{7}$ m.

Q. 32. An aeroplane leaves an airport and flies due north at speed of 1000 km/hr. At the same time, another plane leaves due west at a speed of 1200 km/hr. How far apart will the two planes be after 1 hour and 30 minute ? **[NCERT]**

Ans. 2343 km (approximately)

Chapter 8. Circles

Q. 1. A tangent PQ at a point P of a circle of radius 5 cm meets a line through the centre O at a point Q so that $OQ = 13$ cm. Find the length of PQ. **[NCERT]**

Ans. 12 cm

Q. 2. The radius of the incircle of a triangle is 8 cm and the segments into which one side is divided by the point of contact are 12 cm and 16 cm. Determine the other two sides of the triangle.

Ans. 26 cm, 30 cm.

Q. 3. Two tangents TP and TQ are drawn to a circle with centre O from an external point T. Prove that :★

$$\angle PTQ = 2\angle OPQ.$$ **[NCERT]**

Q. 4. PQ is a chord of length 8 cm of a circle of radius 5 cm. The tangents P and Q intersect at a point T. Find the length of TP. **[NCERT]**

Ans. 6.67 cm.

Q. 5. Two concentric circles are of radii 10 cm and 6 cm. Find the length of the chord of the larger circle which touches the smaller circle. **[NCERT]**

Ans. 16 cm.

Q. 6. From an external point P, tangents PA and PB are drawn to a circle with centre O. If CD is the tangent to the circle at a point E and $PA = 14$ cm, find the perimeter of ΔPCD.

Ans. 28 cm.

Q. 7. In Fig., there are two concentric circles with centre O of radii 5 cm and 3 cm. From an external point P, tangents PA and PB are drawn to these circle. If $AP = 12$ cm, find the length of BP.

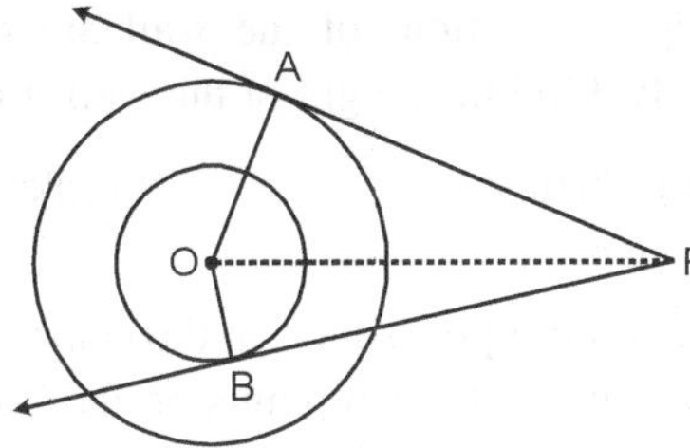

Ans. $4\sqrt{10}$ cm.

Q. 8. In Fig., AB is a chord of length 16 cm of a circle of radius 10 cm. The tangents at A and B intersect at a point P. Find the length of PA.

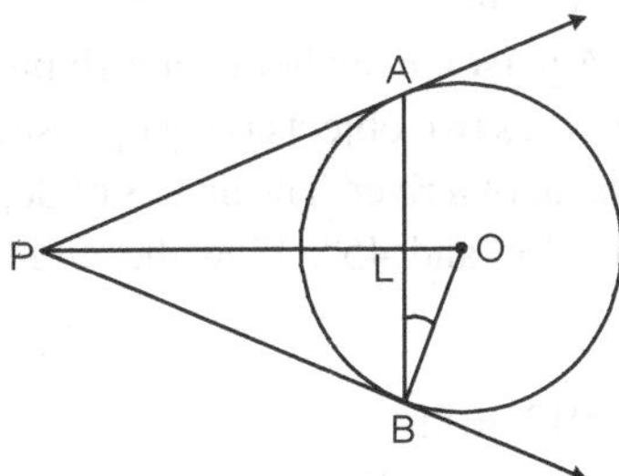

Ans. $\frac{40}{3}$ cm.

Q. 9. In Fig., PA and PB are tangents from an external point P to a circle with centre O. LN touches the circle at M. Prove that : $PL + LM = PN + MN$.

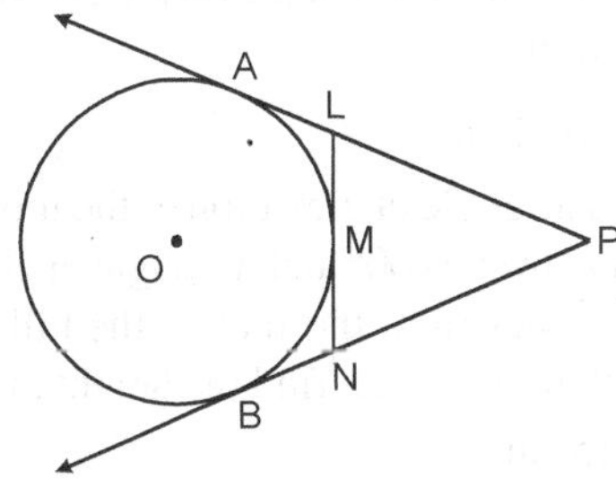

Chapter 10. Trigonometry

Q. 1. In a right triangle ABC, right-angled at C, if $\tan A = 1$, then verify that $2 \sin A \cos A = 1$. **[NCERT]**

Q. 2. In ΔABC right – angled at C, $AB = 29$ units, $BC = 21$ units and $\angle ABC = \theta$. Determine the values of (i) $\cos^2 \theta + \sin^2 \theta$, and (ii) $\cos^2 \theta - \sin^2 \theta$. **[NCERT]**

Ans. (i) 1, (ii) $\frac{41}{841}$

Q. 3. In ΔOPQ, right-angled at P, $OP = 7$ cm, $OQ - PQ = 1$ cm. Determine the values of $\sin Q$ and $\cos Q$. **[NCERT]**

Ans. $\sin Q = \frac{7}{25}$, $\cos Q = \frac{24}{25}$

Q. 4. In ΔPQR, right-angled at Q, $PQ = 5$ cm, $PR + QR = 25$ cm. Determine the values of $\sin P$, $\cos P$ and $\tan P$. **[NCERT]**

Ans. $\sin P = \frac{12}{13}$, $\cos P = \frac{5}{13}$ and $\tan P = \frac{12}{5}$.

Q. 5. In ΔABC, right-angled at B, if $\tan A = \frac{1}{\sqrt{3}}$, find the value of (i) $\sin A \cos C + \cos A \sin C$, and (ii) $\cos A \cos C - \sin A \sin C$. **[NCERT]**

Ans. (i) $\sin A \cos C + \cos A \sin C = 1$

(ii) $\cos A \cos C - \sin A \cos C = 0$.

Q. 6. If $\angle B$ and $\angle Q$ are acute angles such that $\sin B = \sin Q$, then prove $\angle B = \angle Q$. **[NCERT]**

Q. 7. In a ΔABC, right-angled at B, $AB = 24$ cm, $BC = 7$ cm. Determine (i) $\sin A$, $\cos A$ and (ii) $\sin C$, $\cos C$. **[NCERT]**

Ans. (i) $\sin A = \frac{7}{25}$ and $\cos A = \frac{24}{25}$

(ii) $\sin C = \frac{24}{25}$ and $\cos C = \frac{7}{25}$

Q. 8. Given $15 \cot A = 8$. Find $\sin A$ and $\sec A$. **[NCERT]**

Ans. $\sin A = \frac{15}{17}$, $\sec A = \frac{17}{8}$.

Q. 9. If $\cot \theta = \frac{7}{8}$ evaluate (i) $\frac{(1+\sin\theta)(1-\sin\theta)}{(1+\cos\theta)(1-\cos\theta)}$ and (ii) $\cot^2 \theta$. **[NCERT]**

Ans. (i) $\frac{49}{64}$, (ii) $\frac{49}{64}$.

Q. 10. If $3 \cot A = 4$, check whether

$\frac{1-\tan^2 A}{1+\tan^2 A} = \cos^2 A - \sin^2 A$ or not. **[NCERT]**

Ans. Both are equal to $\frac{7}{25}$

Q. 11. If $\angle A$ and $\angle B$ are acute angles such that $\cos A = \cos B$, then show that $\angle A = \angle B$. **[NCERT]**

Q. 12. If $\tan(A - B) = \frac{1}{\sqrt{3}}$ and $\tan(A + B) = \sqrt{3}$, $0° < A + B \le 90°$ and $A > B$, find A and B. **[NCERT]**

Ans. $A = 45°$ and $B = 15°$.

Q. 13. If $\sin(A - B) = \frac{1}{2}$ and $\cos(A + B) = \frac{1}{2}$, $0° < A + B \le 90°$ and $A > B$, find A and B. **[NCERT]**

Ans. $A = 45°$ and $B = 15°$.

Q. 14. In ΔPQR, right-angled at Q, $PQ = 3$ cm and $PR = 6$ cm. Determine $\angle P$ and $\angle R$. **[NCERT]**

Ans. $\angle P = 60°$ and $\angle R = 30°$.

Q. 15. Prove that : $\frac{\sec A - 1}{\sec A + 1} = \frac{1 - \cos A}{1 + \cos A}$.

Q. 16. Prove that :
$(\text{cosec } A - \sin A)(\sec A - \cos A)\sec^2 A = \tan A$.

Q. 17. If $\tan A = \cot B$, prove that $A + B = 90°$. **[NCERT]**

Q. 18. If $\tan 2A = \cot (A - 18°)$, here $2A < 90°$, find the value of A. **[NCERT]**

Ans. 36°.

Q. 19. If $\sec 4A = \text{cosec } (A - 20°)$, where $4A$ is an acute angle, find the value of A.★ **[NCERT]**

Ans. 22°.

Q. 20. If $\sin 3A = \cos (A - 26°)$, where $3A$ is an acute angle, find the value of A. **[NCERT]**

Ans. 29°.

Q. 21. If A, B and C are the interior angles of a triangle, prove that :

(i) $\tan\left(\frac{C + A}{2}\right) = \cot \frac{B}{2}$,

(ii) $\sin\left(\frac{B + C}{2}\right) = \cos \frac{A}{2}$. **[NCERT]**

Chapter 11. Applications of Trigonometry

Q. 1. A ramp for unloading a moving truck, has an angle of elevation of 30°. If the top of the ramp is 0.9 m above the ground level, then find the length of the ramp.

Ans. 1·8 m.

Q. 2. A girl standing on a light-house built on a cliff near the seashore, observes two boats due East of the light-house. The angles of depression of the two boats are 30° and 60°. The distance between two boats is 300 m. Find the distance of the top of the light-house from the sea level. (Boats and foot of the light-house are in a straight line).

Ans. $150\sqrt{3}$ m.

Q. 3. Two crows A and B are sitting at a height of 15 m and 10 m in two different trees vertically opposite to each other. They view a vadai (an eatable) on the ground at an angle of depression 45° and 60° respectively. They start at the same time and fly at the same speed along the shortest path to pick up the vadai. Which bird will succeed in it ?

[Hint : Foot of two trees and vadai (an eatable) are in a straight line]

Ans. Crow B.

Q. 4. A vertical wall and a tower are on the ground. As seen from the top of the tower, the angles of depression of the top and bottom of the wall are 45° and 60° respectively. Find the height of the wall if the height of the tower is 90 m. (Take $\sqrt{3} = 1.732$)

Ans. 38.04 m.

Q. 5. A lamp-post stands at the centre of a circular park. Let P and Q be two points on the boundary such that PQ subtends an angle 90° at the foot of the lamp-post and the angle of elevation of the top of the lamp from P is 30°. If $PQ = 30$ m, then find the height of the lamp-post.

Ans. $5\sqrt{6}$ m.

Q. 6. A person in an helicopter flying at a height of 700 m, observes two objects lying opposite to each other on either bank of a river. The angles of depression of the objects are 30° and 45°. Find the width of the river. (Take $\sqrt{3} = 1.732$)

Ans. 1912.40 m.

Q. 7. A person X standing on a horizontal plane, observes a bird flying at a distance of 100 m from him at an angle of elevation of 30°. Another person Y standing on the roof of a 20 m high building, observes the bird at the same time at an angle of elevation of 45°. If X and Y are on the opposite sides of the bird, then find the distance of the bird from Y.

Ans. $30\sqrt{2}$ m.

Q. 8. The angle of elevation of the top of a hill from the foot of a tower is 60° and the angle of elevation of the top of the tower from the foot of the hill is 30°. If the tower is 50 m high, then find the height of the hill.

Ans. 150 m.

Q. 9. Suppose two insects A and B can hear each other up to a range of 2 m. The insect A is on the ground 1 m away from a wall and sees her friend B on the wall, about to be eaten by a spider. If A sounds a warning to B and if the angle of elevation of B from A is 30°, will the spider have a meal or not ? (Assume that B escapes if she hears A calling)

Ans. No.

Q. 10. A simple pendulum of length 40 cm subtends 60° at the vertex in one full oscillation. What will be the shortest distance between the initial position and the final position of the bob ? (between the extreme ends)

Ans. 40 cm.

Q. 11. A straight highway leads to the foot of a tower. A man standing on the top of the tower spots a van at an angle of depression of 30°. The van is approaching the

★ are board exam questions from previous years

tower with a uniform speed. After 6 minutes, the angle of depression of the van is found to be 60°. How many more minutes will it take for the van to reach the tower ?

Ans. 3 minutes.

Q. 12. The angles of elevation of an artificial earth satellite measured from two earth stations, situated on the same side of the satellite are found to be 30° and 60°. The two earth stations and the satellite are in the same vertical plane. If the distance between the earth stations is 4000 km, find the distance between the satellite and earth. (Use $\sqrt{3} = 1.732$)

Ans. 3464 km.

Q. 13. The angle of elevation of an aeroplane from a point *A* on the ground is 60°. After a flight of 15 seconds horizontally, the angle of elevation changes to 30°. If the aeroplane is flying at a speed of 200 m/s, then find the constant height at which the aeroplane is flying.

Ans. $1500\sqrt{3}$ m.

Q. 14. A student sitting in a classroom sees a picture on the black board at a height of 1.5 m from the horizontal level of sight. The angle of elevation of the picture is 30°. As the picture is not clear to him, he moves straight towards the black board and sees the picture at an angle of elevation of 45°. Find the distance moved by the student.

Ans. 1.098 m.

Q. 15. A boy standing on the ground, spots a balloon moving with the wind in a horizontal line at a constant height. The angle of elevation of the balloon from the boy at an instant is 60°. After 2 minutes, from the same point of observation, the angle of elevation reduces to 30°. If the speed of wind is $29\sqrt{3}$ m/min, then find the height of the balloon from the ground level.

Ans. 87 m.

Q. 16. The angle of elevation of a hovering helicopter as seen from a point 45 m above a lake is 30° and the angle of depression of its reflection in the lake, as seen from the same point and at the same time is 60°. Find the distance of the helicopter from the surface of the lake.

Ans. 90 m.

Q. 17. A boy is standing at some distance from a 30 m tall building and his eye level from the ground is 1.5 m. The angle of elevation from his eyes to the top of the building increases from 30° to 60° as he walks towards the building. Find the distance he walked towards the building.

Ans. $19\sqrt{3}$ m.

Chapter 12. Areas related to Circles

Q. 1. The cost of fencing a circular field at the rate of ₹ 24 per metre is ₹ 5280. The field is to be ploughed at the rate of ₹ 0.50 per m^2. Find the cost of ploughing the field. $\left[\text{Take } \pi = \frac{22}{7}\right]$ **[NCERT]**

Ans. ₹ 1925.

★ **are board exam questions from previous years**

Q. 2. The radius of two circles are 8 cm and 6 cm respectively. Find the radius of the circle having its area equal to the sum of the areas of the two circles. **[NCERT]**

Ans. 10 cm.

Q. 3. The radii of two circles are 19 cm and 9 cm respectively. Find the radius and the area of the circle which has its circumference equal to the sum of the circumference of the two circles. **[NCERT]**

Ans. 28 cm, 2464 cm^2.

Q. 4. A car has two wipers which do not overlap. Each wiper has a blade of length 25 cm sweeping through an angle of 115°. Find the total area cleaned in each sweep of the blades. **[NCERT]**

Ans. 1254.96 cm^2.

Q. 5. To warn ships about underwater rocks, a lighthouse spread a red-coloured light over a sector of 80° angle to a distance of 16.5 km. Find the area of the sea over which the ships are warned. [Use $\pi = 3.14$] **[NCERT]**

Ans. 189.97 km^2.

Q. 6. An umbrella has 8 ribs. Assuming the umbrella to be a flat circle of radius 45 cm, find the area between two consecutive ribs of the umbrella. **[NCERT]**

Ans. 795.53 cm^2.

Q. 7. A circular brooch is made of silver wire and consists of 5 diameters (35 mm each) forming of equal dimensions. Calculate the total length of silver wire required and the area of each sector that is formed. **[NCERT]**

Ans. 285 mm, 96.25 mm^2.

Q. 8. In a circle of radius 21 cm, an arc subtends an angle of 60° at the centre. Find : (i) the length of the arc, (ii) area of the sector formed by the arc, (iii) area of the segment formed by the corresponding chord of the arc. **[NCERT]**

Ans. (i) 22 cm, (ii) 231 cm^2, (iii) 40.04 cm^2.

Q. 9. A chord *PQ* of length 12 cm subtends an arc of 120° at the centre of a circle. Find the area of the minor segment cut off by the chord *PQ*. **[NCERT]**

Ans. 88.44 cm^2.

Q. 10. *AB* and *CD* respectively are the arcs of two concentric circles of radii 21 cm and 7 cm and centre *O*. If $\angle AOB = 30°$, find the area of the shaded region. **[NCERT]**

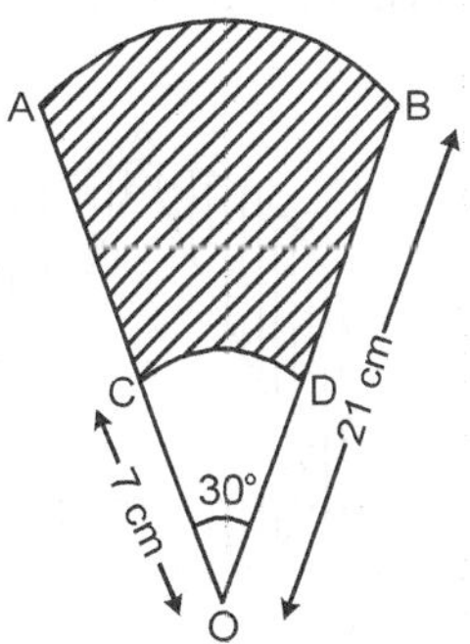

Ans. 102.67 cm^2.

Q. 11. Find the area of the shaded region in the given figure if *ABCD* is a square of side 14 cm and *APD* and *BPC* are semi-ircles. **[NCERT]**

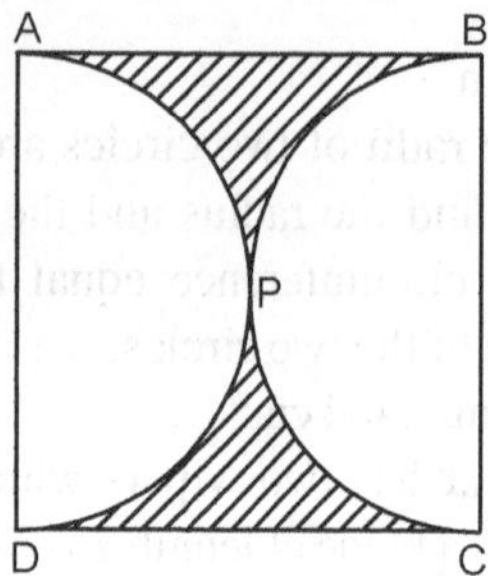

Ans. 42 cm^2.

Q. 12. Find the area of the shaded region in the given figure, where a circular arc of radius 6 cm has been drawn with the vertex *O* of an equilateral triangle *OAB* of side 12 cm as centre. **[NCERT]**

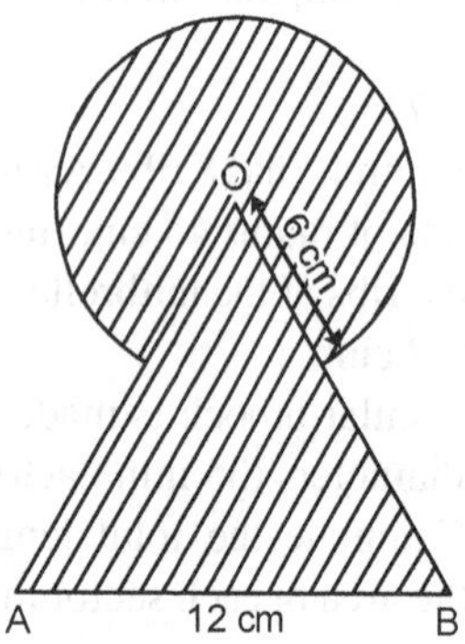

Ans. 156.6 cm^2.

Q. 13. A circle is inscribed in an equilateral triangle *ABC* of side 12 cm, touching its sides. Find the radius of the inscribed circle and the area of the shaded part.

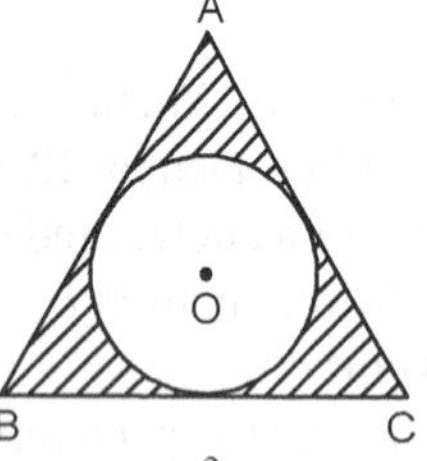

Ans. $2\sqrt{3}$ cm, 24.638 cm^2.

Q. 14. In Fig., an equilateral triangle *ABC* of side 6 cm has been inscribed in a circle. Find the area of the shaded region. (Take $\pi = 3.14$)

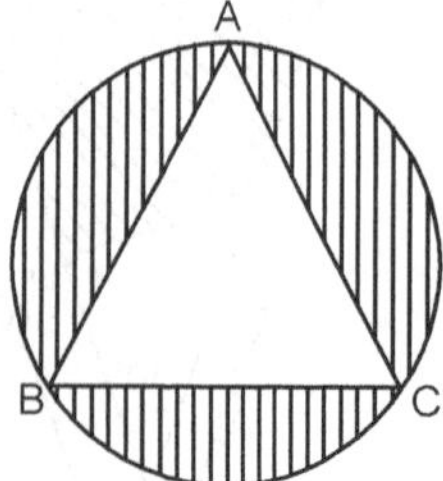

Ans. 22.126 cm^2.

Q. 15. A circular field has a perimeter of 650 m. A square plot having its vertices on the circumference of the field is marked in the field. Calculate the area of the square plot.

Ans. 21387 m^2.

Q. 16. In Fig., the square *ABCD* is divided into five equal parts, all having same area. The central part is circular and the lines *AE, GC, BF* and *HD* lie along the diagonals *AC* and *BD* of the square. If *AB* = 22 cm, find :

(i) the circumference of the central part, (ii) the perimeter of the part *ABEF*.

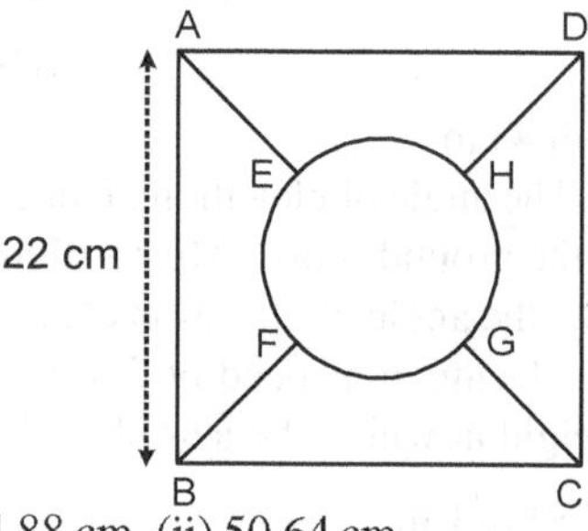

Ans. (i) 34.88 cm, (ii) 50.64 cm.

Q. 17. In Fig., *ABCD* is a rectangle having *AB* = 20 cm and *BC* = 14 cm. Two sectors of 180° have been cut off. Calculate :

(i) the area of the shaded region.

(ii) the length of the boundary of the shaded region.

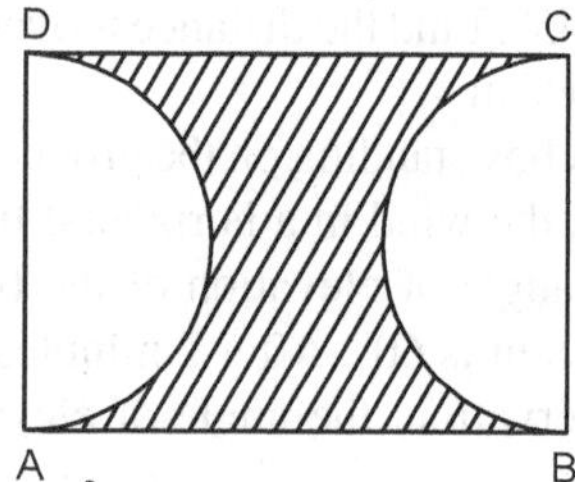

Ans. (i) 126 cm^2, (ii) 84 cm.

Q. 18. The diameter of a coin is 1 cm. If four such coins be placed on a table so that rim of each touches that of the other two, find the area of the shaded region. (Take $\pi = 3.1416$)

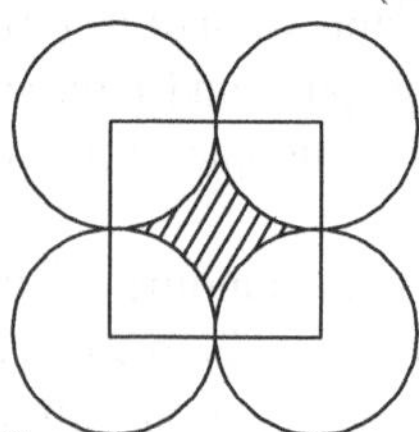

Ans. 0.2146 cm^2.

Q. 19. In Fig., a circle is inscribed in a quadrilateral *ABCD* in which $\angle B = 90°$. If *AD* = 23 cm, *AB* = 29 cm and *DS* = 5 cm, find the radius *r* of the circle.

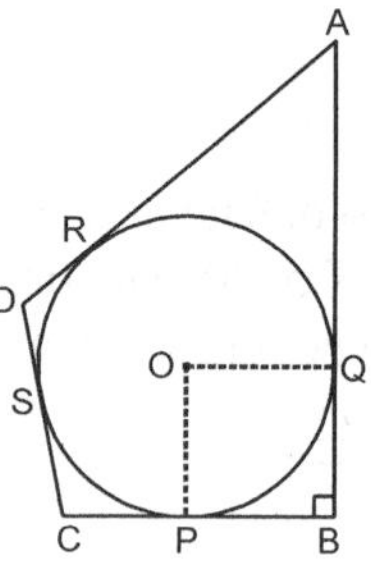

Ans. 11 cm.

Q. 20. A circular pend is of diameter 17.5 m. It is surrounded by a 2 m wide path. Find the cost of constructing the path at the rate of ₹ 25 per square metre.★ (Use $\pi = 3.14$)

Ans. ₹ 3061.50.

Chapter 13. Surface Areas and Volumes

Q. 1. If the diameter of a sphere is 14 cm, then what is its curved surface area ?

Ans. 616 cm^2

Q. 2. The radius and height of a right-circular cone are 12 cm and 9 cm respectively. What is the curved surface area of that cone ?

Ans. 565.71 cm^2

Q. 3. The volume of a right-circular cylinder is 352 cm^3 and the height is 7 cm. Find the radius of the base.

Ans. 4 cm

Q. 4. If the radii of a sphere and a right-circular cylinder are 3 cm each and if their volumes are equal as well then find the height of the cylinder :

Ans. 4 cm

Q. 5. The curved surface area and the volume of a right circular cylinder are numerically equal. Find the radius of the cylinder.

Ans. 2 cm

Q. 6. The volumes of a sphere and cylinder are equal. The diameter of the base of the cylinder is equal to the diameter of the sphere. Find the ratio between the radius of the base and the height of the cylinder.

Ans. 3 : 4

Q. 7. 77 m^2 of canvas is required to make a conical tent of slant height 7 m. Find the area of the base.

Ans. $\frac{77}{2}$ m^2.

Q. 8. The volume of the right-circular cone of height 24 cm is 1232 cm^3. Find the lateral surface area of the cone.

Ans. 550 cm^2.

Q. 9. Two cubes each of volume 64 cm^3 are joined end to end to form a solid. Find the surface area and volume of the resulting cuboid. **[NCERT]**

Ans. 160 cm^2, 128 cm^3.

Q. 10. A metallic sphere of radius 4.2 cm is melted and recast into the shape of a cylinder of radius 6 cm. Find the height of the cylinder. **[NCERT]**

Ans. 2.74 cm.

Q. 11. Metallic spheres of radii 6 cm, 8 cm and 10 cm respectively are melted to form a single solid sphere. Find the radius of the resulting sphere. **[NCERT]**

Ans. 12 cm.

★ are board exam questions from previous years

Q. 12. A 20 m deep well with diameter 7 m is dug and the earth from digging is evenly spread out to form a platform of 22 m by 14 m. Find the height of the platform. **[NCERT]**

Ans. 2.5 m.

Q. 13. A copper wire, 3 mm in diameter is wound about a cylinder whose length is 1.2 m and diameter 10 cm, so as to cover the curved surface of the cylinder. Find the length and mass of the wire, assuming the density of the copper wire to be 8.88 g/cm. **[NCERT]**

Ans. 125.6 m, 111 kg 532 g.

Q. 14. Selvi's house has an overhead tank in the shape of a cylinder. It is filled up by pumping water from an underground tank that is cuboid in shape. The dimensions of the cuboid are 1.57 m × 1.44 m × 0.95 m. The radius of the overhead tank is 60 cm and its height is 95 cm. Find the height of the water-level in the underground tank after the overhead tank has been filled up completely. Compare the capacities of both the tanks. (Use $\pi = 3.14$) **[NCERT]**

Ans. 47.5 cm, 2 : 1.

Q. 15. A copper rod of diameter 1 cm and length 8 cm is drawn into a wire of length 18 cm of uniform thickness. Find the thickness of the wire. **[NCERT]**

Ans. 0.67 cm.

Q. 16. How many coins 1.75 cm in diameter and 2 mm thickness must be melted to form a cuboid of dimensions 5.5 cm × 10 cm × 3.5 cm ? **[NCERT]**

Ans. 400.

Q. 17. Water in a canal, 6 m deep and 1.5 m wide is flowing at a speed of 10 km/hr. How much area will it irrigate in 30 minutes if 8 cm of standing water is needed for irrigation ? **[NCERT]**

Ans. 562500 m^2.

Q. 18. A farmer connects a pipe of internal diameter 20 cm from a canal into a cylindrical tank which is 10 m in diameter and 2 m deep. If the water flows through the pipe at the rate of 3 km/hr in how much time will the tank be completely filled ? **[NCERT]**

Ans. 1 hour and 40 minutes.

Q. 19. A solid toy is in the shape of a hemisphere surmounted by a right circular cone. Height of the cone is 2 cm and the diameter of the base is 4 cm. If a right circular cylinder circumscribes the solid, find how much more space it will circumscribe. **[NCERT]**

Ans. 8π cm^3.

Q. 20. A wooden toy rocket is in the shape of a cone mounted on a cylinder. The height of the entire rocket is 26 cm while the height of the conical part is 6 cm. The base of the conical portion has a diameter of 5 cm while the diameter of the cylinder is 3 cm. If the conical portion is to be painted orange and the cylindrical portion yellow, find the area of the rocket painted with each of these colours. [Use $\pi = 3.14$] **[NCERT]**

Ans. 195.47 cm^2.

Q. 21. A wooden toy was made from the rest of the solid cylinder after scooping out a hemisphere of same radius from each of it. If the height of the cylinder is 10 cm and its base radius is 3.5 cm, find the total surface area. $\left[\text{Use } \pi = \frac{22}{7}\right]$ **[NCERT]**

Ans. 374 cm^2.

Q. 22. Mayank made a birdhouse in the shape of a cylinder with a hemispherical depression of same radius at one end of it. If the height of the cylinder is 1.45 m and its base radius is 30 cm, find the total surface area. $\left[\text{Use } \pi = \frac{22}{7}\right]$. **[NCERT]**

Ans. 3.3 m^2.

Q. 23. A tent is in the form of a right-circular cylinder of base diameter 4 m and height 2.1 m surmounted by a right circular cone of the same base radius and slant height 2.8 cm. Find the area of the canvas used and the cost of canvas at ₹ 500 per square metre. **[NCERT]**

Ans. ₹ 22000.

Q. 24. From a solid cylinder of height 2.4 cm and diameter 1.4 cm, a conical cavity of the same height is hollowed out. Find the total surface area of the remaining solid. $\left[\text{Use } \pi = \frac{22}{7}\right]$. **[NCERT]**

Ans. 17.6 cm^2.

Q. 25. A solid is in the shape of a cone mounted on a hemisphere, the radius of each of them being 1 cm and the total height of the cone equal to its radius. Find the volume of the solid in terms of π. **[NCERT]**

Ans. π cm^3

Q. 26. A solid toy is in the form of a right circular cylinder with two conical shapes at both ends. Their common diameter is 3 cm and the heights of the conical portions are 2 cm each and the total height is 12 cm. Find the volume of the given toy. $\left[\text{Use } \pi = \frac{22}{7}\right]$. **[NCERT]**

Ans. 66 cm^3.

Q. 27. A pen stand consists of a wooden cuboid with 4 conical depressions. The dimensions of the cuboid are 15 cm by 10 cm by 3.5 cm, the diameter of each depression is 1 cm and the depth of each depression is 1.4 cm. Find the volume of wood in the stand. $\left[\text{Use } \pi = \frac{22}{7}\right]$. **[NCERT]**

Ans. 523.53 cm^3.

Q. 28. A vessel is in the form of an inverted cone. Its height is 8 cm and the radius of its top, which is open is 5 cm. It is filled with water up to the brim. When lead shots each of which is a sphere of radius 0.5 cm are dropped in the vessel, one-fourth of the water flows out. Find the number of lead shots dropped in the vessel. **[NCERT]**

Ans. 100.

Q. 29. A solid consists of a hemisphere surmounted by a right circular cone placed upright in a cylinder full of water. Height of the cylinder is 180 m and the radius of the base is 60 cm. If a right circular cylinder circumscribes the solid, find how much more space it will circumscribe. **[NCERT]**

Ans. 1.13 m^3.

Q. 30. A solid iron pole consists of a cylinder of height 220 cm and base diameter 24 cm that is surmounted by another cylinder of height 60 cm and radius 8 cm. Find the mass of the pole given that the density of the pole is 8 g/cm^3. [Use $\pi = 3.14$] **[NCERT]**

Ans. 892.26 kg.

Q. 31. A spherical glass vessel has a cylindrical neck 8 cm long and 2 cm in diameter. The diameter of the spherical part is 8.5 cm. Find the quantity of water it can hold. [Use $\pi = 3.14$] **[NCERT]**

Ans. 346.51 cm^3.

Q. 32. For his birthday, Rasheed got a playing top which is shaped like a cone surmounted by a hemisphere. The entire top is 5 cm long and the diameter of the top is 3.5 cm. Find the surface area of the entire top. $\left[\text{Use } \pi = \frac{22}{7}\right]$ **[NCERT]**

Ans. 39.55 cm^2.

Q. 33. A hemispherical depression is cut out from one face of a cubical block such that the diameter of the hemisphere is equal to the edge of the cube, where a is the edge of cube. Find the surface area of the remaining solid. **[NCERT]**

Ans. $\left[\frac{24a^2 + \pi a^2}{4}\right]$.

Q. 34. A solid is made up of a cube surmounted by a hemisphere. If the edges of the cube equal to 7 cm each, what is the greatest possible diameter of the hemisphere? Also find the total surface area. $\left[\text{Use } \pi = \frac{22}{7}\right]$. **[NCERT]**

Ans. 7 cm, 332.5 cm^2.

Q. 35. A capsule is in the shape of a cylinder with hemispherical ends. If the whole length of the capsule is 14 mm and the diameter of the capsule is 5 mm, find its surface area. $\left[\text{Use } \pi = \frac{22}{7}\right]$. **[NCERT]**

Ans. 220 mm^2.

Q. 36. A factory shed is in the shape of a cuboid surmounted by a cylinder cut up laterally (length-wise). If the base of the shed is 7 m by 15 m and height of the cuboidal portion is 8 m, find the volume of the shed. If machinery inside the shed occupies total space of 300 m^3 and each of the 20 workers occupy a spade of 0.08 m^3 how much air is in the shed ? **[NCERT]**

Ans. 1128.75 m^3, 827.15 m^3.

Q. 37. A cistern, internally measuring 150 cm × 120 cm × 110 cm has 129600 cm^3 of water in it. Porous bricks are placed in the water until the cistern is full of the brim. Each brick absorbs one-seventeenth of its own volume of water. How many bricks can be placed without the water overflowing and each brick being 22.5 cm × 7.5 cm × 6.5 cm ? **[NCERT]**

Ans. 1793.

Q. 38. A tin oil funnel is in the shape of a frustum over a cylinder. If the cylindrical portion is 10 cm long, the diameter of the cylinder and lower end of the frustum is the same, the total height of the funnel is 22 cm, the diameter of the lower end is 8 cm and the diameter of the upper end of the funnel is 18 cm, find the total area of the tin required to make the funnel. **[NCERT]**

Ans. 782.57 cm^2.

Q. 39. The lower and upper diameters of a frustum are 30 cm and 35 cm respectively. The height of the frustum is 14 cm. Find its mass if the density of the material is 1.2 g/cm^3. $\left[\text{Use } \pi = \frac{22}{7}\right]$. **[NCERT]**

Ans. 13 kg 970 g.

Q. 40. Find the area of the clothing material required to make a cap shaped like a frustum the radius of whose closed side is 4 cm, of the open side is 10 cm and the slant height of the frustum is 15 cm. **[NCERT]**

Ans. 710.29 cm^2.

Chapter 14. Statistics

Q. 1. A survey was conducted by a group of students as a part of their environmental awareness programme, in which they collected the following data regarding the number of plants in 20 houses of a locality. Find the mean number of plants per house. **[NCERT]**

Number of Plants	Number of Houses
0 – 2	1
2 – 4	2
4 – 6	1
6 – 8	5
8 – 10	6
10 – 12	2
12 – 14	3

Ans. 8 plants.

Q. 2. Thirty women were examined in a hospital by a doctor and the number of heart-beats per minute were recorded and surmmarised as follows. Find the mean heart-beats per minute. **[NCERT]**

Heart-beats per minute	Number of Women
65 – 68	2
68 – 71	4
71 – 74	3
74 – 77	8
77 – 80	7
80 – 83	4
83 – 86	2

Ans. 75.9 heart-beats per minute.

Q. 3. The following distribution shows the daily pocket allowance given to the children of a multi-storey building. The average pocket allowance is ₹ 18. Find out the missing frequency. **[NCERT]**

Class Interval	Frequency
11 – 13	7
13 – 15	6
15 – 17	9
17 – 19	13
19 – 21	f
21 23	5
23 – 25	4

Ans. 20.

Q. 4. In the retail market, fruit vendors were selling mangoes kept in packing boxes. These boxes contained varying number of mangoes. The following was their distribution according to the number of boxes.

No. of Mangoes	No. of Boxes
50 – 52	15
53 – 55	110
56 – 58	135
59 – 61	115
62 – 64	25

Find the mean number of mangoes kept in a packing box. **[NCERT]**

Ans. 57.

Q. 5. The table below shows the daily expenditure on food of 25 households in a locality. Find the Mean expenditure. **[NCERT]**

Expenditure	Number of Households
100 – 150	4
150 – 200	5
200 – 250	12
250 – 300	2
300 – 350	2

Ans. ₹ 211.

Q. 6. To find out the concentration of SO_2 in air, (in parts per million *i.e.,* ppm), data was collected from 30 localities in a certain city. Calculate the mean concentration of SO_2 in the air. **[NCERT]**

SO_2 in air	Frequency
0.00 – 004	4
0.04 – 0.08	9
0.08 – 0.12	9
0.12 – 0.16	2
0.16 – 0.20	4
0.20 – 0.24	2

Ans. 0.099 ppm.

Q. 7. A class teacher has the following absentee record of 40 students for the whole term. Find the mean number of days a student remained absent. **[NCERT]**

Number of Days	Number of Students
0 – 6	11
6 – 10	10
10 – 14	7
14 – 20	4
20 – 28	4
28 – 38	3
38 – 40	1

Ans. 12.475 days.

Q. 8. The following table gives the literacy rate (in percentage) of 35 cities. Find the mean literacy rate. **[NCERT]**

Literacy %	Cities
45 – 55	3
55 – 65	10
65 – 75	11
75 – 85	8
85 – 95	3

Ans. 69.43%.

Q. 9. The following is the distribution of the marks obtained by 74 students in a class test. Find the mean of the distribution.

Marks	Students
4 – 6	5
6 – 12	6
12 – 15	15
15 – 27	12
27 – 32	15
32 – 40	21

Ans. $19.77 \approx 20$.

Q. 10. Find the mean of the following distribution :

Class	Frequency
24 – 36	15
36 – 42	15
42 – 47	51
47 – 52	21
52 – 70	18

Ans. 52.

Q. 11. If the median of the distribution given below is 28.5 and $\Sigma f_i = 60$, find the value of x and y. **[NCERT]**

Class Interval	Frequency
0 – 10	5
10 – 20	x
20 – 30	20
30 – 40	15
40 – 50	y
50 – 60	5

Ans. $x = 8, y = 7$.

Q. 12. A survey was conducted related to the heights (in cm) of 51 girls of class X and the following data was obtained :

Heights	Number of Girls
135 – 140	4
140 – 145	7
145 – 150	18
150 – 155	11
155 – 160	6
160 – 165	5

Determine the median height.

Ans. 149.03 cm.

Q. 13. A life insurance agent found the following data for distribution of ages of 100 policy holders. Calculate the median age, if policies are only given to persons in the age group of 18 to 60 years. **[NCERT]**

Age (in years)	Number of Policy-holders
Below 20	2
Below 25	6
Below 30	24
Below 35	45
Below 40	78
Below 45	89
Below 50	92
Below 55	98
Below 60	100

Ans. 35.76 years.

Q. 14. The lengths of 40 leaves of a plant are measured correct to the nearest millimetre and the data obtained is represented in the following table. Find the median length of the leaves. **[NCERT]**

Length (in mm)	Number of Leaves
118 – 126	3
127 – 135	5
136 – 144	9
145 – 153	12
154 – 162	5
163 – 171	4
172 – 180	2

Ans. 146.75 mm.

Q. 15. The following table gives the distribution of the life time (in hours) of 400 neon lamps. Find the median life time of a lamp **[NCERT]**

Life time (in hours)	Number of Lamps
1500 – 2000	14
2000 – 2500	56
2500 – 3000	60
3000 – 3500	86
3500 – 4000	74
4000 – 4500	62
4500 – 5000	48

Ans. 3406.98 hours.

Q. 16. The following distribution gives the weight (in kg) of 30 students in a class. Find the median weight of the students. **[NCERT]**

Weight (in kg)	Number of Students
40 – 45	2
45 – 50	3
50 – 55	8
55 – 60	6
60 – 65	6
65 – 70	3
70 – 75	2

Ans. 56.67 kg.

Q. 17. The following data gives the distribution of the total household expenditure (in rupees) of manual workers in a city :

Expenditure (in Rupees)	Frequency
1000 – 1500	24
1500 – 2000	40
2000 – 2500	33
2500 – 3000	28
3000 – 3500	30
3500 – 4000	22
4000 – 4500	16
4500 – 5000	7

Find the mode of the average expenditure of the families. **[NCERT]**

Ans. ₹ 1847.83.

Q. 18. The following table shows the ages of the patients admitted to a hospital during a year.

Ages (in years)	Number of Patients
5 – 15	6
15 – 25	11
25 – 35	21
35 – 45	23
45 – 55	14
55 – 65	5

Find the Mean and Mode of the data given above. **[NCERT]**

Ans. Mean = 35.375
Mode = 36.8

Q. 19. The following distribution provides information about the observed lifetime (in hours) of 225 electrical components. Determine the modal lifetime of the components. **[NCERT]**

Lifetime (in years)	Number of Components
0 – 20	10
20 – 40	35
40 – 60	52
60 – 80	61
80 – 100	38
100 – 120	29

Ans. 65.625 hours.

Q. 20. The following distribution gives the state-wise teacher-student ratio in higher-secondary schools of India. Find the mean and mode from the data. **[NCERT]**

Number of Students/Teachers	Number of States/U.T.s
15 – 20	3
20 – 25	8
25 – 30	9
30 – 35	10
35 – 40	3
40 – 45	0
45 – 50	0
50 – 55	2

Ans. Mean = 29.2

Mode = 30.6

Q. 21. The following data shows the number of runs scored in ODI cricket by some of the top batsmen of the world :

Runs Scored	Number of Batsmen
3000 – 4000	4
4000 – 5000	18
5000 – 6000	9
6000 – 7000	7
7000 – 8000	6
8000 – 9000	3
9000 – 10000	1
10000 – 11000	1

Find the mode of the data. **[NCERT]**

Ans. 4608.7 runs.

Q. 22. A student noted the number of cars passing through a spot on a road for 100 periods, each of 3 minutes and summarized it in the table given below. Find the mode. **[NCERT]**

Number of Cars	Frequency
0 – 10	7
10 – 20	14
20 – 30	13
30 – 40	12
40 – 50	20
50 – 60	11
60 – 70	15
70 – 80	8

Ans. 44.71 cars $\approx$ 45 cars.

Q. 23. Find the median of the following data after drawing the 'more than' and 'less than' ogive :

Class	Frequency
0 – 5	10
5 – 10	16
10 – 15	30
15 – 20	42
20 – 25	50

Ans. 13.5.

Q. 24. Find the median of the following data after drawing the 'more than' and 'less than' ogive :

Temperature (in °c)	0 – 10	10 – 20	20 – 30
Number of Days	18	10	2

Ans. 15°.

Q. 25. Find the median of the following data after drawing the 'more than' and 'less than' ogive :

Class	Frequency
140 – 160	15
160 – 180	29
180 – 200	8
200 – 220	12
220 – 240	14

Ans. 19.

Chapter 15. Probability

Q. 1. A bag contains 5 red balls and some blue balls. If the probability of drawing a blue ball from the bag is double that of a red ball, find the number of blue balls in the bag. **[NCERT]**

Ans. 10.

Q. 2. A bag contains 12 balls out of which x are white. If one ball is drawn at random what is the probability that it will be white ? If six more white balls are put in the bag the probability of drawing a white ball doubles. Find x. **[NCERT]**

Ans. $\frac{1}{4}$, 3.

Q. 3. A child has a block in the shape of a cube with a letter written on each face (A, B, C, D, E, A).The cube is thrown once. What is the probability of getting (i) an A and (ii) a D ? **[NCERT]**

Ans. (i) $\frac{1}{3}$, (ii) $\frac{1}{6}$.

Q. 4. There are 40 students in Class X of a school of whom 25 are girls and 15 are boys. The class teacher, in order to select a class representative, writes down the names of all the students on same-sized pieces of cardboard and puts them in a box. Before drawing out a name, she stirs it well. What is the probability of the name being that of a (i) girl's and (ii) a boy's ? **[NCERT]**

Ans. (i) $\frac{5}{8}$, (ii) $\frac{3}{8}$.

Q. 5. A carton consists of 100 shirts of which, 88 are good, 8 have minor defects and 4 have major defects. Jimmy, a trader will accept only the good shirts but Sujatha another trader will reject only the shirts with major defects. One shirt is drawn at random from the carton. What is the probability that it is acceptable to (i) Jimmy and (ii) to Sujatha ?

Ans. (i) 0.88 (ii) 0.96.

Q. 6. Gopi buys a fish from a shop for his aquarium. The shopkeeper picks one at random from a tank containing 8 female fish and 5 male fish. What is the probability that the fish taken out is a male fish ? [NCERT]

Ans. $\frac{5}{13}$.

Q. 7. A jar contains 24 marbles of which some are green and the rest are blue. If a marble is drawn at random from the jar, the probability that it is green is $\frac{2}{3}$. Find the number of blue marbles in the jar. [NCERT]

Ans. 8.

Q. 8. A die is thrown. Find the probability of getting (i) a prime number, (ii) an even prime number, (iii) a number greater than 5 and (iv) a number lying between 2 and 6. [NCERT]

Ans. (i) $\frac{1}{2}$ (ii) $\frac{1}{6}$ (iii) $\frac{1}{6}$ (iv) $\frac{1}{2}$

Q. 9. One card is drawn from a well-shuffled deck of 52 cards. Find the probability of getting (i) a red king, (ii) a face card, (iii) a red face card, (iv) a black queen, (v) a jack of hearts, and (vi) a spade. [NCERT]

Ans. (i) $\frac{1}{26}$ (ii) $\frac{3}{13}$ (iii) $\frac{3}{26}$ (iv) $\frac{1}{26}$ (v) $\frac{1}{52}$ (vi) $\frac{1}{4}$.

Q. 10. A bag contains 3 red and 5 black balls. A ball is drawn at random from the bag. What is the probability that the ball is (i) red, (ii) black ? [NCERT]

Ans. (i) $\frac{3}{8}$ (ii) $\frac{5}{8}$.

Q. 11. Two customers are visiting a particular shop in the same week (Monday to Saturday). Each is equally likely to visit the shop on any day as on another. What is the probability that both will visit the shop on (i) the same day, (ii) on different days and (iii) on consecutive days ? [NCERT]

Ans. (i) $\frac{1}{6}$ (ii) $\frac{5}{6}$ (iii) $\frac{5}{36}$.

Q. 12. It is given that in a group of 3 students, the probability of 2 students not having the same birthday is 0.992. What is the probability that the 2 students have the same birthday ? [NCERT]

Ans. 0.008.

Q. 13. (i) A lot of 20 electric bulbs contains 4 defective bulbs. One bulb is taken out at random from the box. What is the probability that the bulb drawn is defective ? (ii) Suppose the bulb is not defective and not replaced and another bulb is drawn from the lot, what is the probability that the second bulb is also not defective ? [NCERT]

Ans. (i) $\frac{1}{5}$ (ii) $\frac{15}{19}$.

Q. 14. A box contains 90 discs that are numbered from 1 to 90. If a disc is drawn at a random from the box, find the probability that it bears (i) a two-digit number, (ii) a perfect square number and (iii) a number divisible by 5. [NCERT]

Ans. (i) $\frac{9}{10}$ (ii) $\frac{1}{10}$ (iii) $\frac{1}{5}$.

Q. 15. A lot consists of 144 ball pens of which 20 are defective and others are good. Nuri will buy a pen if it is good but will not buy one if it is defective. The shopkeeper draws one pen at a random and gives it to her. Find the probability of Nuri's buying and not buying it. [NCERT]

Ans. $\frac{31}{36}, \frac{5}{36}$.

Q. 16. A lot consists of 144 pens of which 12 are defective and others are good. One pen is drawn out at random. Find the probability of the pen being a good one. [NCERT]

Ans. $\frac{19}{18}$

Q. 17. An unbiased die is thrown. What is the probability of getting (i) a multiple of 3, (ii) an even number or a multiple of 3 and (iii) an odd number ?★★ [NCERT]

Ans. (i) $\frac{1}{3}$ (ii) $\frac{2}{3}$ (iii) $\frac{1}{2}$.

Q. 18. A box contains 4 green, 8 white and 5 red marbles. If a marble is drawn out at random from the box, what is the probability that it will be (i) red, (ii) white, and (iii) not green marble ? [NCERT]

Ans. (i) $\frac{5}{17}$ (ii) $\frac{8}{17}$ (iii) $\frac{13}{17}$.

Q. 19. What is the probability of having 53 sundays in a leap year ?

Ans. $\frac{2}{7}$.

Q. 20. Two dice are thrown together. What is the probability of having one number being twice the other ?

Ans. $\frac{1}{12}$.

Q. 21. Two dice are thrown together. What is probability of having a difference of 3 between the two numbers ?

Ans. $\frac{1}{6}$.

Q. 22. Two dice are thrown together. What is the probability of having a product of 6 between the two numbers ?

Ans. $\frac{1}{9}$.

Q. 23. A card is drawn from a deck of 52 cards. What is the probability of drawing a red king or a black jack ?

Ans. $\frac{1}{13}$.

Q. 24. From a deck of 52 cards the face cards are removed and they are replaced by two jokers and two blank cards. What is the probability of drawing a joker or a blank card ?

Ans. $\frac{1}{11}$.

Q. 25. For a game, the entry fee is ₹ 5. The game consists of tossing a coin three times. If one or two heads show up then Shweta gets back her entry free. If she tosses three heads, she receives double the entry fee or else she loses. After tossing the coin thrice, find the probabilities that she (i) loses the entry fee, (ii) gets double the entry fee and (iii) gets her entry free back.

Ans. (i) $\frac{1}{8}$ (ii) $\frac{1}{8}$ (iii) $\frac{3}{4}$.

Q. 26. A die has six faces marked as 0, 1, 1, 1, 6, 6. Two such dices are thrown together and the total score is recorded. (i) How many different scores are possible ? (ii) What is the probability of getting a total of 7 ?

Ans. (i) 6 (ii) $\frac{1}{3}$.

Q. 27. A bag contains white, black and red balls only. A ball is drawn at random from the bag. The probability of getting a white ball is $\frac{3}{10}$ and that of a black ball is $\frac{2}{5}$. Find the probability of getting a red ball if the bag contains 20 black balls. Also find the total number of balls.

Ans. $\frac{3}{10}$, 50.

Q. 28. A bag contains 24 balls of which x are red, $2x$ are white and $3x$ are blue. A ball is selected at random. What is the probability that the ball drawn is (i) not white, (ii) blue ?

Ans. (i) $\frac{2}{3}$ (ii) $\frac{1}{2}$.

Q. 29. The probability of guessing the correct answer to a certain test is $\frac{P}{12}$. If the probability of not guessing the correct answer to this question is $\frac{1}{3}$. Find the value of P.

Ans. 8.

Q. 30. A number is selected at random from the numbers 3, 5, 5, 7, 7, 7, 9, 9, 9, 9. Find the probability that the selected number is their average.

Ans. $\frac{3}{10}$.

Q. 31. If a number x is chosen from the sequence 1, 2, 3 and a number y is chosen from the sequence 1, 4, 9, find the probability that $xy = 10$.

Ans. 0.

Q. 32. If 65% of the population has black eyes, 25% have brown eyes and the remaining have blue eyes, what is the probability that a person selected at random has (i) blue eyes, (ii) brown or black eyes, (iii) blue or black eyes and (iv) neither blue nor brown eyes ?

Ans. (i) $\frac{1}{10}$ (ii) $\frac{9}{10}$ (iii) $\frac{3}{4}$ (iv) $\frac{13}{20}$.

Q. 33. If a number x is chosen from the sequence 1, 2, 3 and a number y is chosen from the sequence 1, 4, 9, find the probability that xy will be less than 9 ?

Ans. $\frac{5}{9}$.

Q. 34. A jar contains 54 marbles each of which is blue, green or white. The probability of selecting a blue marble at random is $\frac{1}{3}$ and the probability of selecting a green marble at random is $\frac{4}{9}$. How many white marbles does the jar contain ?

Ans. 12.

Q. 35. In a game of musical chair, the person playing the music has been asked to stop the music within 2 minutes. What is the probability that she would stop the music within the first half minute ?

Ans. $\frac{1}{4}$.

Q. 36. What is the probability that the month of june will have 5 Mondays in a (i) Leap Year and (ii) non-Leap Year ?

Ans. (i) $\frac{2}{7}$, (ii) $\frac{2}{7}$.

Q. 37. What is the probability that the month of February will have 5 wednesdays in a (i) Leap Year and (ii) non-Leap Year ?

Ans. (i) $\frac{1}{7}$ (ii) 0.

Q. 38. A number x is chosen from the numbers – 4, – 3 – 2, – 1, 0, 1, 2, 3, 4. Find the probability that $x < 3$.

Ans. $\frac{7}{9}$.

●●

Topper's Answers - 2019

Section – A

Q. 1. Two concentric circles of radii a and b $(a > b)$ are given. Find the length of the chord of the larger circle which touches the smaller circle.

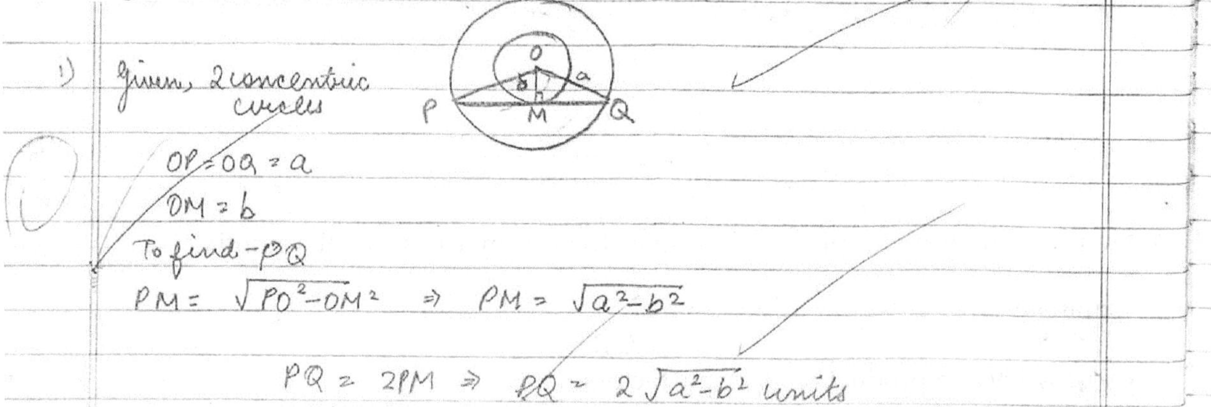

Q. 2. In Figure 1, PS = 3 cm, QS = 4 cm, $\angle PRQ = \theta$, $\angle PSQ = 90°$, $PQ \perp RQ$ and RQ = 9 cm. Evaluate $\tan \theta$.

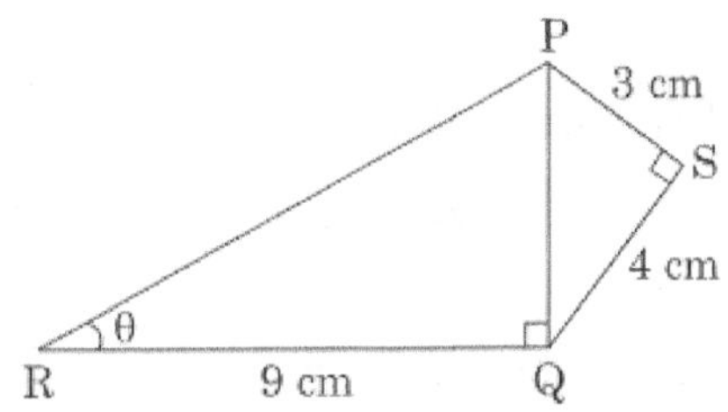

Figure - 1

OR

If $\tan \alpha = \frac{5}{12}$, find the value of $\sec \alpha$.

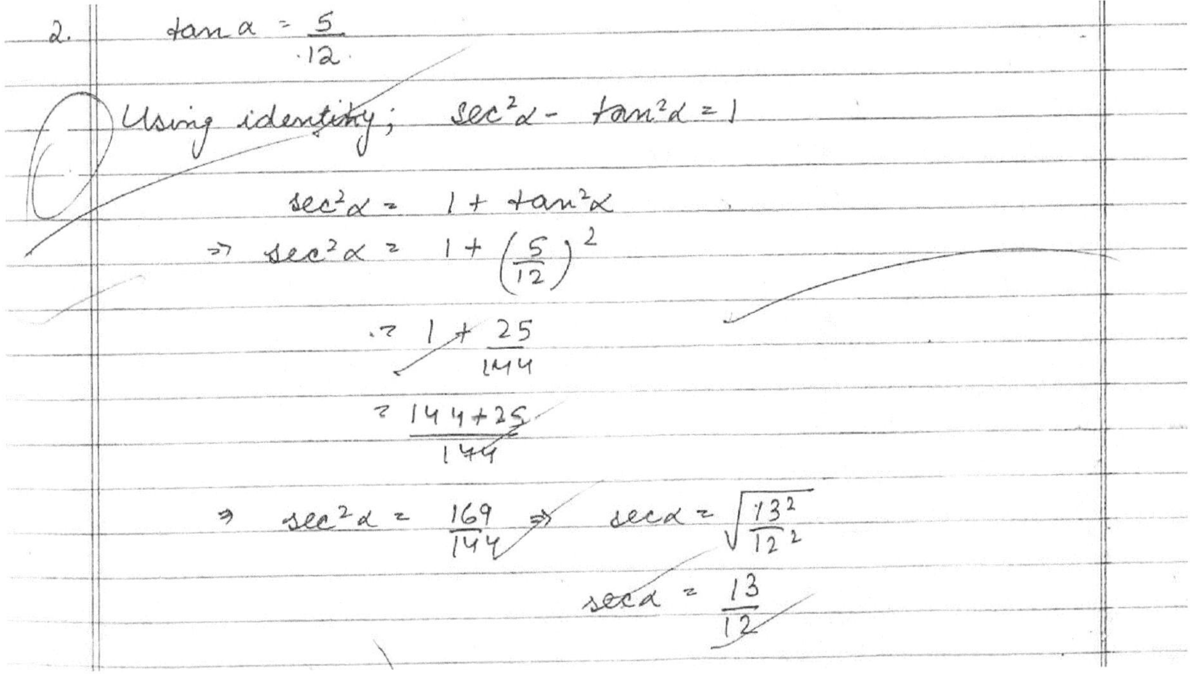

Q. 3. Write the discriminant of the quadratic equation $(x + 5)^2 = 4\,(5x - 3)$.

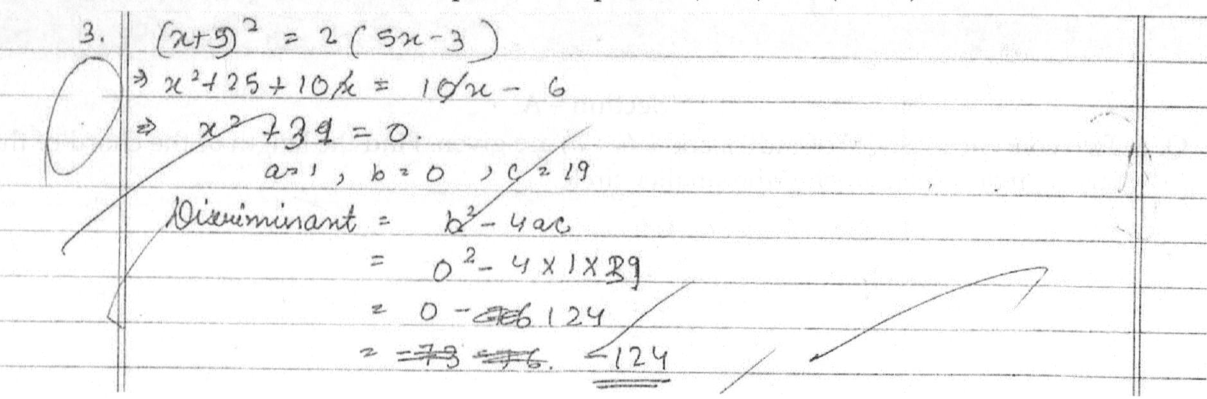

3. $(x+5)^2 = 2(5x-3)$

$\Rightarrow x^2+25+10x = 10x - 6$

$\Rightarrow x^2+29 = 0.$

$a=1, \; b=0, \; c=29$

Discriminant $= b^2-4ac$

$= 0^2-4\times1\times29$

$= 0-116$

$= -116$. -124

Q. 4. Find after how many places of decimal the decimal form of the number $\dfrac{27}{2^3.5^4.3^2}$ will terminate.

OR

Express 429 as a product of its prime factors.

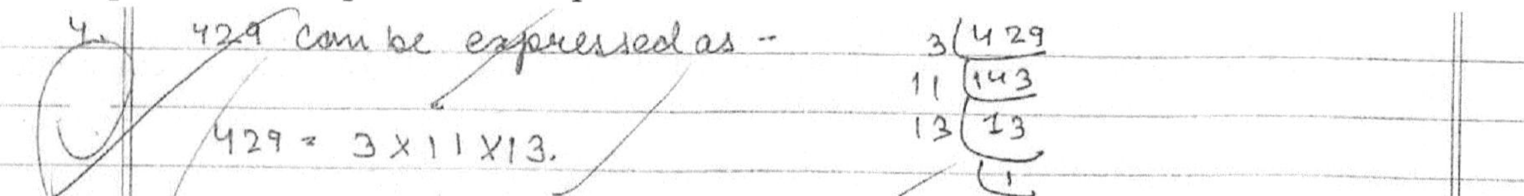

4. 429 can be expressed as -

429 = 3 × 11 × 13.

3 | 429
11 | 143
13 | 13
 | 1

Q. 5. Find the sum of the first 10 multiples of 6.

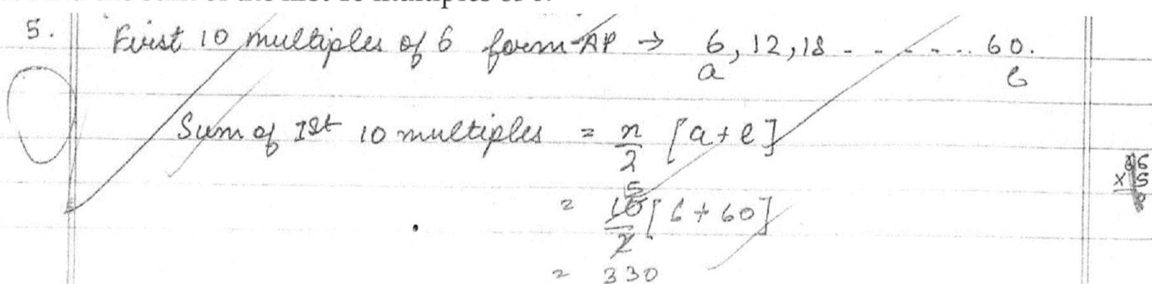

5. First 10 multiples of 6 form AP → 6, 12, 18 - - - - - - 60.
a ... l

Sum of 1st 10 multiples $= \frac{n}{2}[a+l]$

$= \frac{10}{2}[6+60]$

$= 330$

Q. 6. Find the value(s) of *m*, if the distance between the points A(5, –3) and B(13, *m*) is 10 units.

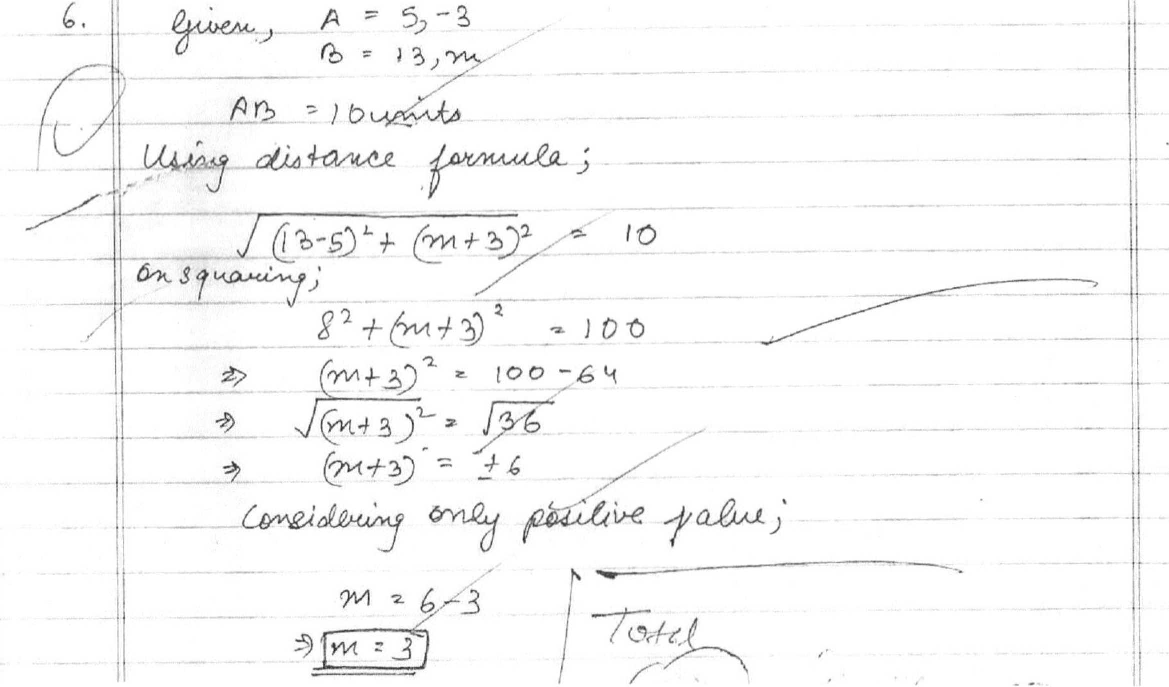

6. Given, A = 5, −3
B = 13, m

AB = 10 units

Using distance formula;

$\sqrt{(13-5)^2+(m+3)^2} = 10$

On squaring;

$8^2+(m+3)^2 = 100$

$\Rightarrow (m+3)^2 = 100-64$

$\Rightarrow \sqrt{(m+3)^2} = \sqrt{36}$

$\Rightarrow (m+3) = \pm 6$

Considering only positive value;

$m = 6-3$

$\Rightarrow \boxed{m = 3}$

Total

Section - B

Q. 7. A die is thrown once. Find the probability of getting (i) a composite number, (ii) a prime number.

7. Event : Dice is thrown.
Outcomes : 1, 2, 3, 4, 5, 6 (6 outcomes)
Favourable events :-

i) Composite number = 4, 6
Probability of getting a composite no. = $\frac{\text{no. of favourable outcomes}}{\text{Total outcomes}}$
$= \frac{2}{6} = \frac{1}{3}$

ii) prime no. : 2, 3, 5.
Probability = $\frac{\text{no. of outcomes favourable to the event}}{\text{Total possible outcomes}}$
or $\left(\frac{\text{no. of prime nos.}}{\text{Total outcomes}}\right)$
$= \frac{3}{6} = \frac{1}{2}$

Q. 8. Cards numbered 7 to 40 were put in a box. Poonam selects a card at random. What is the probability that Poonam selects a card which is a multiple of 7 ?

8. Event : cards numbered from 7-40 are chosen.
Total possible outcomes : (7, 8, 9 - - 40) = 34 cards.

For Favourable event : cards multiple of 7.
Favourable outcomes : (7, 14, 21, 28, 35) 5 cards

Probability of selecting a cards multiple of 7 = $\frac{\text{Favourable outcomes}}{\text{Total no. of outcomes}}$
$= \frac{5}{34}$

Q. 9. Points A(3, 1), B(5, 1), C(a, b) and D(4, 3) are vertices of a parallelogram ABCD. Find the values of a and b.

OR

Points P and Q trisect the line segment joining the points A(– 2, 0) and B(0, 8) such that P is near to A. Find the coordinates of points P and Q.

9. Points A, B, C, D are vertices of a parallelogram.

D(4, 3), C(a, b), O(x, y), B(5, 1), A(3, 1)

We know that
diagonals of a parallelogram bisect each other.

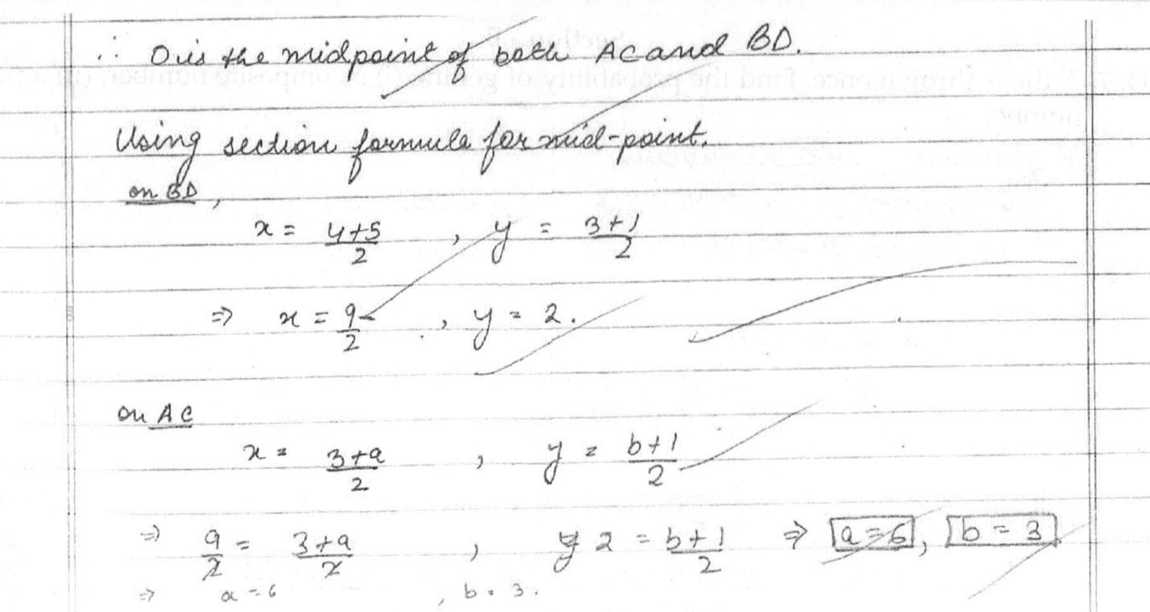

∴ O is the midpoint of both AC and BD.

Using section formula for mid-point.

on BD,

$x = \frac{4+5}{2}$, $y = \frac{3+1}{2}$

$\Rightarrow x = \frac{9}{2}$, $y = 2$.

on AC

$x = \frac{3+a}{2}$, $y = \frac{b+1}{2}$

$\Rightarrow \frac{9}{2} = \frac{3+a}{2}$, $2 = \frac{b+1}{2}$ $\Rightarrow$ $a = 6$, $b = 3$

$\Rightarrow a = 6$, $b = 3$.

Q. 10. Solve the following pair of linear equations :

$$3x - 5y = 4$$
$$2y + 7 = 9x$$

10. Given -

$3x - 5y = 4$ ——(1)

$9x - 2y = 7$ ——(2)

To find - 'x' and 'y'

Multiplying (1) × 3, and (2) × 1 and subtracting (2) from (1), we get :

$(3x - 5y) \times 3 - (9x - 2y) \times 1 = 4 \times 3 - 7 \times 1$

$\Rightarrow 9x - 15y - 9x + 2y = 12 - 7$

$\Rightarrow -13y = 5$ $\Rightarrow$ $y = \frac{-5}{13}$

Then, putting $y = \frac{-5}{13}$ in (1);

$3x = 4 + 5y$

$\Rightarrow 3x = 4 + 5 \times \frac{-5}{13}$

$\Rightarrow 3x = \frac{52 - 25}{13}$ $\Rightarrow x = \frac{27}{13 \times 3}$

$\Rightarrow x = \frac{9}{13}$

Q. 11. If HCF of 65 and 117 is expressible in the form 65*n* – 117, then find the value of *n*.

OR

On a morning walk, three persons step out together and their steps measure 30 cm, 36 cm and 40 cm respectively. What is the minimum distance each should walk so that each can cover the same distance in complete steps ?

11. Using Euclid's Division Lemma (states that $a = bq + r$, $0 \le r < b$) we can find the HCF of 65 and 117.

$117 = 65 \times 1 + 52$
$65 = 52 \times 1 + 13$
$52 = 13 \times 4 + 0$

$\therefore$ HCF of 65, 117 = 13

But,
$65n - 117 = 13$
$\Rightarrow 65n = 13 + 117$
$\Rightarrow n = \frac{130}{65}$
$\Rightarrow n = 2$

Q. 12. In a quadratic equation $kx^2 - 6x - 1 = 0$ determine the value of k for which the equation does not have real roots.

12) given, quadratic equation $\Rightarrow kx^2 - 6x - 1 = 0$.
where $a = k$, $b = -6$, $c = -1$.

For no real roots (for imaginary roots), discriminant must be less than 0.

That is, $D < 0$
$\Rightarrow b^2 - 4ac < 0$
$\Rightarrow (-6)^2 - 4 \times k \times -1 < 0$
$\Rightarrow 36 + 4k < 0$
$\Rightarrow 4k < -36$ $\Rightarrow k < -9$
$\therefore$ k should be less than −9. ($k = -10, -11 \ldots$)

Section – C

Q. 13. A, B and C are interior angles of a triangle ABC. Show that

(i) $\sin\left(\frac{B+C}{2}\right) = \cos\frac{A}{2}$

(ii) If $\angle A = 90°$, then find the value of $\tan\left(\frac{B+C}{2}\right)$

OR

If the $(A + B) = 1$ and $\tan(A - B) = \frac{1}{\sqrt{3}}$, $0° < A + B < 90°$, $A > B$, then find the values of A and B.

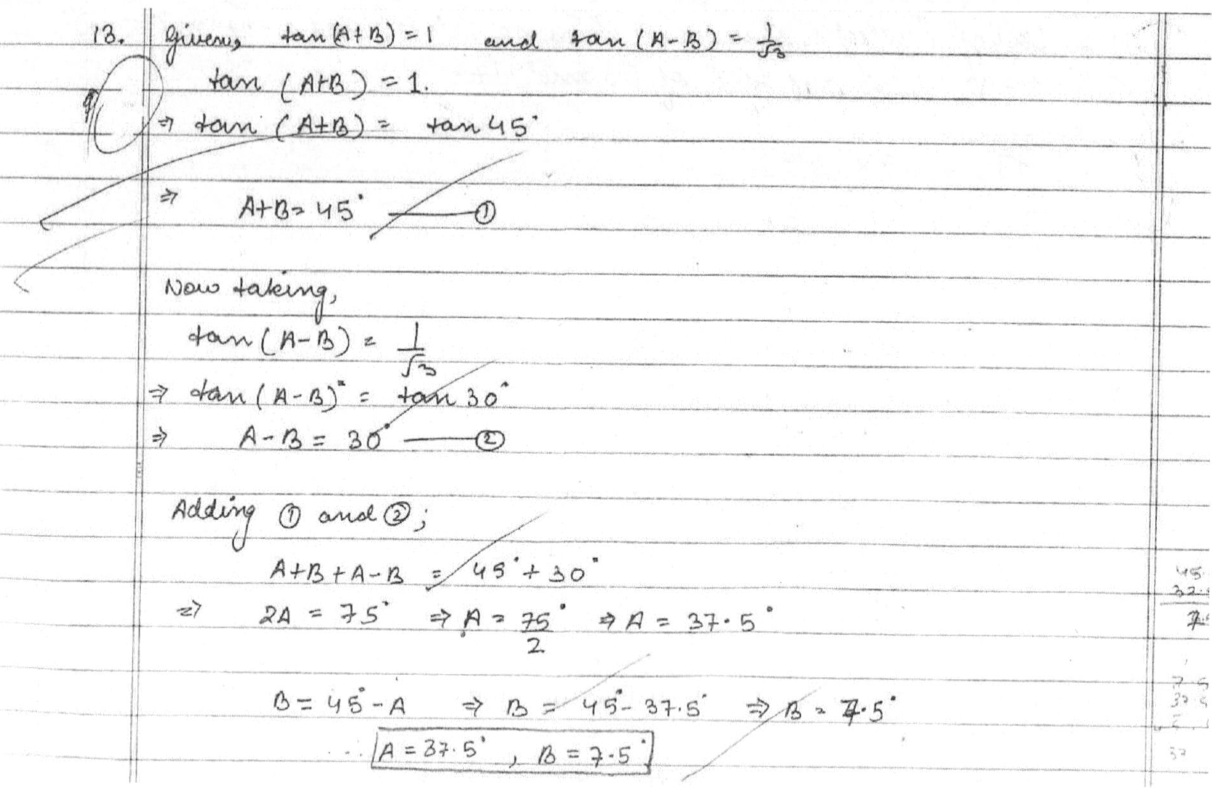

13. Given, $\tan(A+B) = 1$ and $\tan(A-B) = \frac{1}{\sqrt{3}}$

$\tan(A+B) = 1.$

$\Rightarrow \tan(A+B) = \tan 45^\circ$

$\Rightarrow A+B = 45^\circ$ ——①

Now taking,

$\tan(A-B) = \frac{1}{\sqrt{3}}$

$\Rightarrow \tan(A-B) = \tan 30^\circ$

$\Rightarrow A-B = 30^\circ$ ——②

Adding ① and ②;

$A+B+A-B = 45^\circ + 30^\circ$

$\Rightarrow 2A = 75^\circ \Rightarrow A = \frac{75^\circ}{2} \Rightarrow A = 37.5^\circ$

$B = 45^\circ - A \Rightarrow B = 45^\circ - 37.5^\circ \Rightarrow B = 7.5^\circ$

$\therefore \boxed{A = 37.5^\circ,\ B = 7.5^\circ}$

Q. 14. In Figure 2, PQ is a chord of length 8 cm of a circle of radius 5 cm. The tangents at P and Q intersect at a point T. Find the length TP.

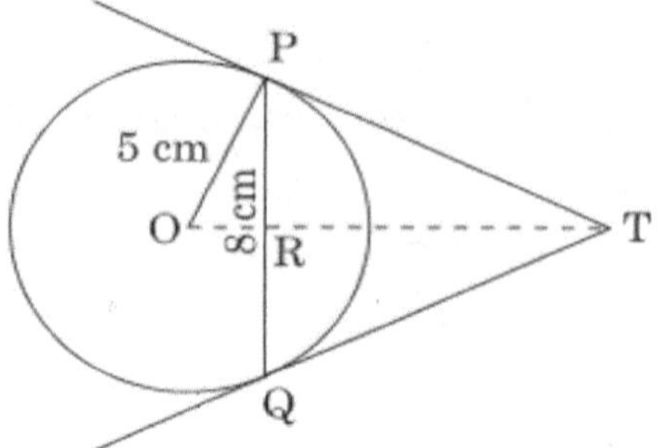

Figure - 2

OR

Prove that opposite sides of a quadrilateral circumscribing a circle subtend supplementary angles at the centre of the circle.

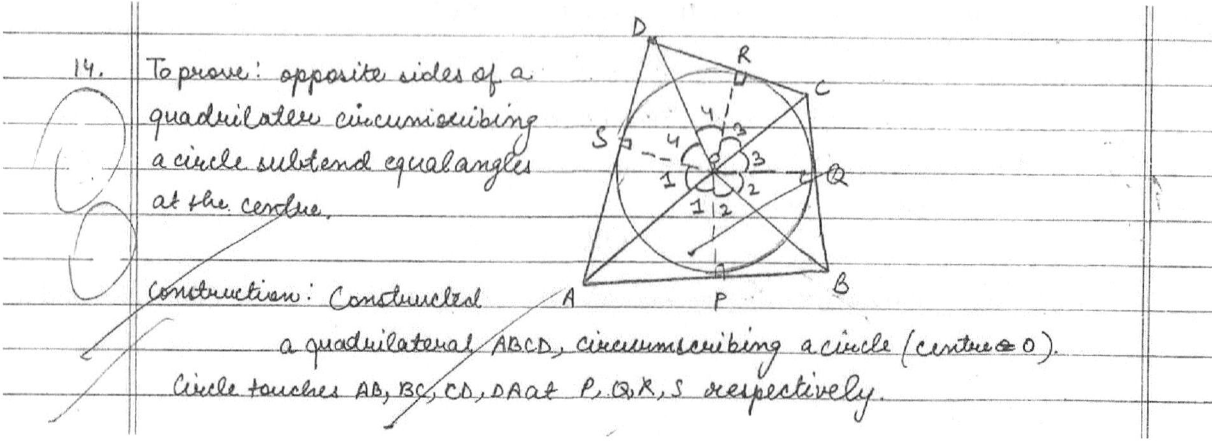

14. To prove: opposite sides of a quadrilateral circumscribing a circle subtend equal angles at the centre.

Construction: Constructed a quadrilateral ABCD, circumscribing a circle (centre O). Circle touches AB, BC, CD, DA at P, Q, R, S respectively.

To prove : ∠AOB + ∠COD = 180°
Or ∠AOD + ∠BOC = 180°

We know, that tangents from same exterior point subtend equal angle at the centre of circle with radius.

∴ ∠AOP = ∠AOS = ∠1 (say)

Similarly, ∠BOP = ∠BOQ = ∠2

∠COQ = ∠COR = ∠3

∠DOR = ∠ ∠DOS = ∠4.

∴ ∠AOP + ∠BOP + ∠BOQ + ∠COQ + ∠COR + ∠DOR + ∠DOS + ∠AOS = 360° [Complete angle around a point]

⇒ 2∠1 + 2∠2 + 2∠3 + 2∠4 = 360°

⇒ ∠1 + ∠2 + ∠3 + ∠4 = 180°

⇒ (∠1 + ∠2) + (∠3 + ∠4) = 180°.

⇒ ∠AOB + ∠COD = 180°

Or (∠1 + ∠4) + (∠2 + ∠3) = 180°

⇒ ∠AOD + ∠BOC = 180°.

Hence, proved !

Q. 15. A class teacher has the following absentee record of 40 students of a class for the whole term. Find the mean number of days a student was absent.

Number of days :	0 – 6	6 – 12	12 – 18	18 – 24	24 – 30	30 – 36	36 – 42
Number of Students :	10	11	7	4	4	3	1

15. Calculating mean using step deviation method.

No. of days (Class interval)	No. of students (f_i)	x_i (Upper + lower limit / 2)	$u_i = \frac{x_i - A}{h}$	$f_i \times u_i$
0 - 6	10	3	$\frac{3-21}{6} = -3$	10×-3 = -30.
6 - 12	11	9	$\frac{9-21}{6} = -2$	11×-2 = -22
12 - 18	7	15	$\frac{15-21}{6} = -1$	7×-1 = -7
18 - 24	4	A (21)	$\frac{21-21}{6} = 0$	4×0 = 0
24 - 30	4	27	$\frac{27-21}{6} = 1$	4×1 = 4
30 - 36	3	33	$\frac{33-21}{6} = 2$	3×2 = 6
36 - 42	1	39	$\frac{39-21}{6} = 3$	1×3 = 3
Total :	$\Sigma f_i = 40$			$\Sigma f_i u_i = -46$

Class size (h) = 6 − 0 = 6

Assumed mean (A) = 21

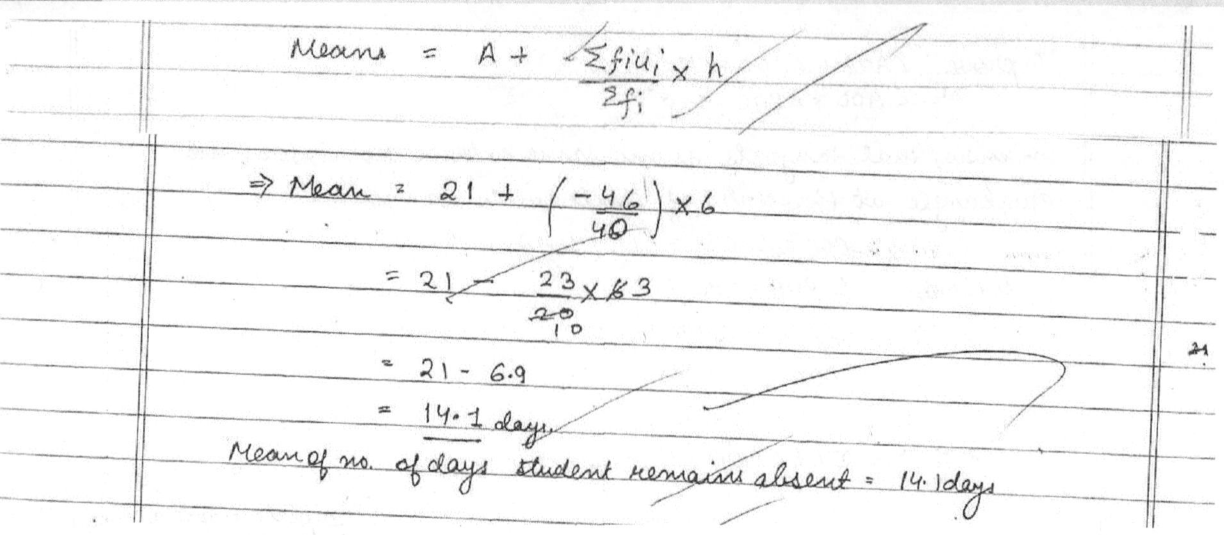

Means = A + $\frac{\Sigma f_i u_i}{\Sigma f_i} \times h$

⇒ Mean = $21 + \left(\frac{-46}{40}\right) \times 6$

$= 21 - \frac{23}{20} \times 6$

= 21 − 6.9

= 14.1 days

Mean of no. of days student remains absent = 14.1 days

Q. 16. A car has two wipers which do not overlap. Each wiper has a blade of length 21 cm sweeping through an angle 120°. Find the total area cleaned at each sweep by the blades. (Take $\pi = \frac{22}{7}$)

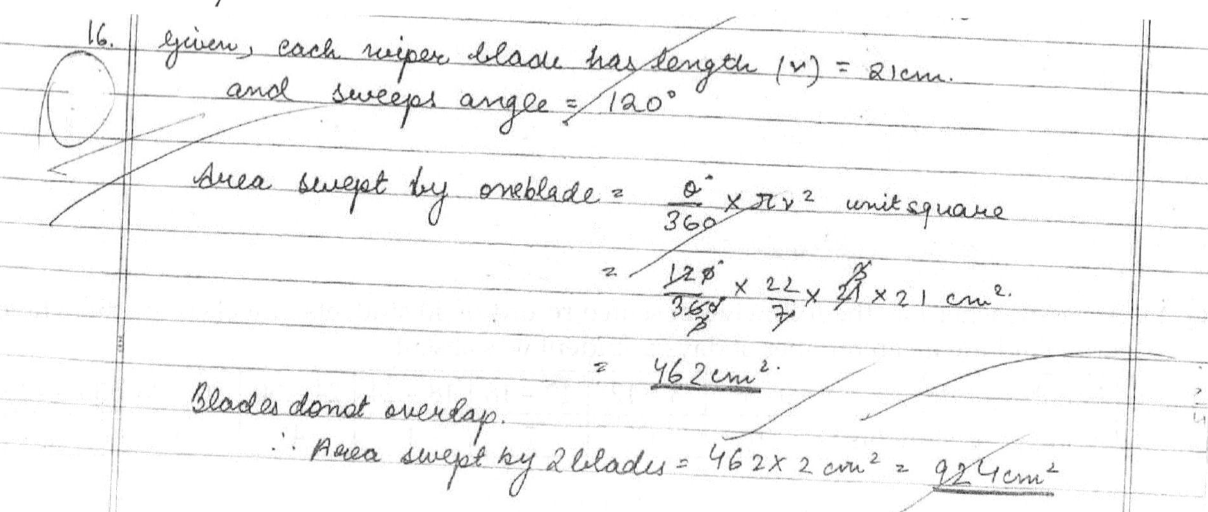

16. given, each wiper blade has length (r) = 21 cm.
and sweeps angle = 120°

Area swept by one blade = $\frac{\theta}{360} \times \pi r^2$ unit square

$= \frac{120}{360} \times \frac{22}{7} \times 21 \times 21 \text{ cm}^2$

= 462 cm²

Blades donot overlap.

∴ Area swept by 2 blades = 462 × 2 cm² = 924 cm²

Q. 17. The perpendicular from A on side BC of a ΔABC meets BC at D such that DB = 3CD. Prove that $2AB^2 = 2AC^2 + BC^2$.

OR

AD and PM are medians of triangles ABC and PQR respectively where ΔABC ~ ΔPQR. Prove that $\frac{AB}{PQ} = \frac{AD}{PM}$.

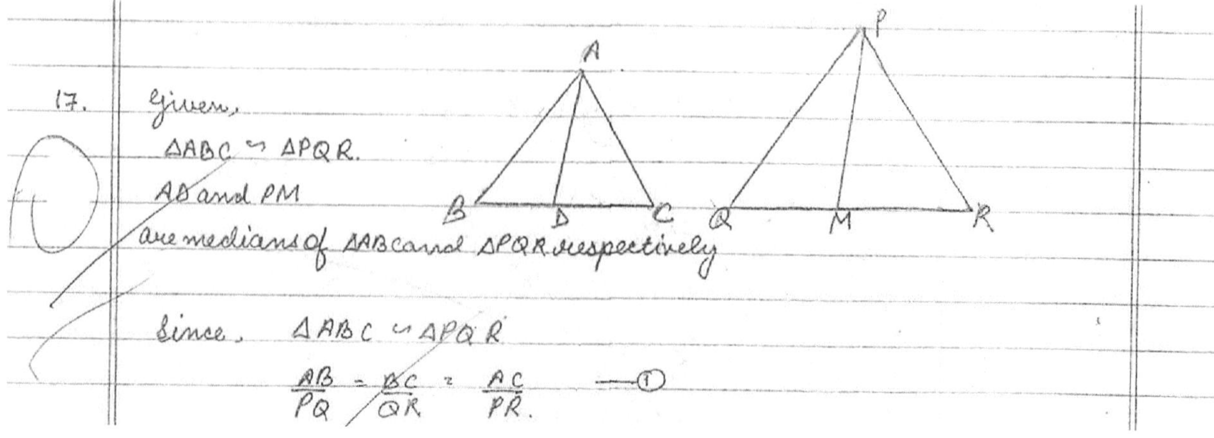

17. given,
ΔABC ~ ΔPQR.
AD and PM are medians of ΔABC and ΔPQR respectively

Since, ΔABC ~ ΔPQR

$\frac{AB}{PQ} = \frac{BC}{QR} = \frac{AC}{PR}$ —(1)

D is the midpoint of BC (AD is median)
M is the midpoint of QR (PM is median)

$\therefore$ BC = 2BD
QR = 2QM } (2)

$\therefore \frac{AB}{PQ} = \frac{BC}{QR}$ [from (1)]

$\Rightarrow \frac{AB}{PQ} = \frac{2BD}{2QM}$ [from (2)]

$\Rightarrow \frac{AB}{PQ} = \frac{BD}{QM} \Rightarrow \therefore \Delta ABD \sim \Delta PQM$ — (a)

That is, $\frac{AB}{PQ} = \frac{BD}{QM} = \frac{AD}{PM}$

Similarly, $\frac{BC}{QR} = \frac{AC}{PR}$ [from (1)]

$\Rightarrow \frac{2BD}{2QM} = \frac{AC}{PR}$

$\Rightarrow \frac{AC}{PR} = \frac{BD}{QM}$.

$\therefore \Delta ACD \sim \Delta PRM$ — (b) That is, $\frac{AC}{PR} = \frac{AD}{PM} = \frac{CD}{RM}$

From both (a) and (b); we get that.

$\frac{AB}{PQ} = \frac{AD}{PM}$

Hence proved!

Q. 18. Check whether $g(x)$ is a factor of $p(x)$ by dividing polynomial $p(x)$ by polynomial $g(x)$, where $p(x) = x^5 - 4x^3 + x^2 + 3x + 1$, $g(x) = x^3 - 3x + 1$

18. Given: $p(x) = x^5 - 4x^3 + x^2 + 3x + 1$.
$g(x) = x^3 - 3x + 1$

To check: if $g(x)$ is a factor of $p(x)$ or not.

Method: simply divide.

$\Rightarrow$

$x^3 - 3x + 1 \overline{) x^5 - 4x^3 + x^2 + 3x + 1} (x^2 - 1$
$x^5 - 3x^3 + x^2$
$- \quad + \quad -$
$-x^3 + 3x + 1$
$-x^3 + 3x - 1$
$+ \quad - \quad +$
2

We get remainder = 2

Therefore, $p(x)$ is not completely divisible by $g(x)$

$g(x)$ is not a factor of $p(x)$

Q. 19. Prove that $\sqrt{3}$ is an irrational number.

OR

Find the largest number which on dividing 1251, 9377 and 15628 leaves remainders 1, 2 and 3 respectively.

19. Let us assume, if possible, that $\sqrt{3}$ is rational

Then, $\sqrt{3}$ can be expressed as $\frac{p}{q}$ where ($q \neq 0$) and p, q are coprimes [HCF(p, q) = 1]

$\therefore \sqrt{3} = \frac{p}{q}$ [$p, q \in Z$; HCF(p, q) = 1]

On squaring both sides,

$3 = \frac{p^2}{q^2}$

$\Rightarrow p^2 = 3q^2$. —— ①

3 divides p^2

$\therefore$ 3 divides p.

Then, p can be written as;

$p = 3a$ for some integer 'a'.

On squaring,

$p^2 = 9a^2$

Put $p^2 = 3q^2$ from ①

$\Rightarrow 3q^2 = 9a^2$

$\Rightarrow q^2 = 3a^2$.

3 divides q^2

$\therefore$ 3 divides q.

$\because$ 3 divides both p and q, 3 is a common factor of p and q.

But, p and q are co-primes.

Therefore, our assumption is wrong

$\therefore \sqrt{3}$ is irrational.

from Section D (4 marks)

Q. 20. Find the area of triangle ABC with A(1, – 4) and the mid-points of sides through A being (2, – 1) and (0, – 1).

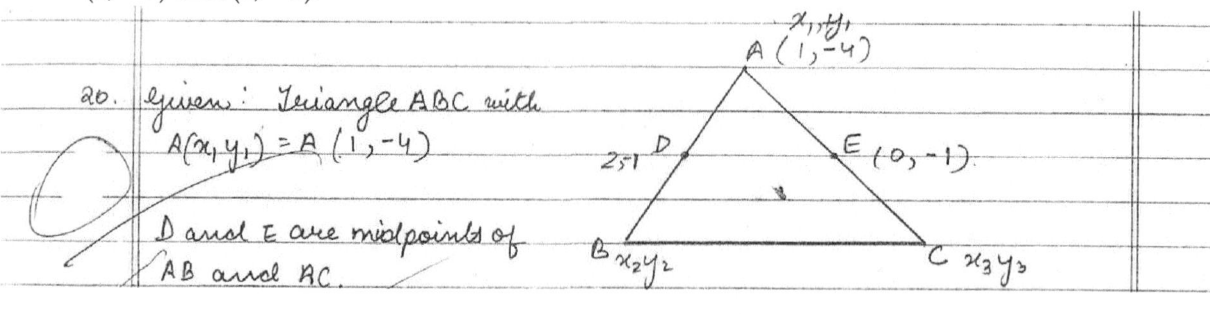

20. Given: Triangle ABC with

$A(x_1, y_1) = A(1, -4)$

D and E are midpoints of AB and AC.

Let coordinates of B be (x_2, y_2) and that of C be (x_3, y_3).
using section formula for mid-point;

$$\frac{1+x_2}{2} = 2 \quad , \quad \frac{-4+y_2}{2} = -1$$

$$\Rightarrow x_2 = 3 \quad , \quad y_2 = 2$$

$$(x_2, y_2) = (3, 2)$$

Similarly, $\frac{1+x_3}{2} = 0$, $\frac{-4+y_3}{2} = -1$

$$\Rightarrow x_3 = -1 \quad , \quad y_3 = 2.$$

$x_3, y_3 = -1, 2$

Area of triangle $= \frac{1}{2}|x_1(y_2-y_3) + x_2(y_3-y_1) + x_3(y_1-y_2)|$ unit square

$= \frac{1}{2}|1(2-2) + 3(2+4) + (-1)(-4-2)|$ unit square

$= \frac{1}{2}|1\times 0 + 3\times 6 + (-1\times -6)|$ sq. units

$= \frac{1}{2}|18+6|$ sq. units $= 12$ sq. units

$\therefore$ Area of $\Delta = 12$ sq. units

Q. 21. Two numbers are in the ratio 5 : 6. If 7 is subtracted from each of the number, the ratio becomes 4 : 5. Find the numbers.

21. Let the required two numbers be $5x$ and $6x$.
Given, if 7 is subtracted from both nos, ratio becomes 4:5.
New. nos. = $(5x-7)$ and $(6x-7)$.
According to the question;

$$\frac{5x-7}{6x-7} = \frac{4}{5} \quad \text{—(1)}$$

Solving (1);

$$5(5x-7) = 4(6x-7)$$

$$\Rightarrow 25x - 35 = 24x - 28$$

$$\Rightarrow x = 35-28$$

$$\Rightarrow x = 7$$

$\therefore$ The required nos. are : $5x = 35$
$6x = 42$

Q. 22. Water is flowing at the rate of 2.52 km/h through a cylindrical pipe into a cylindrical tank, the radius of whose base is 40 cm. If the increase in the level of water in the tank, in half an hour is 3.15 m, find the internal diameter of the pipe.

22. Let the radius of cylindrical pipe be 'r' metres

cylindrical pipe

cylindrical tank

R = 40cm.

Given -

Radius (R) of cylindrical tank = 40 cm $= \frac{2}{5}m$

Height of tank filled = 3.15 m

Time taken = $\frac{1}{2}h$ = 30 minutes = 30×60 s.

Rate of flow of water via pipe = 2.52 km/h $= \frac{252}{100} \times \frac{5}{18}$ m/s. = 0.7 m/s.

To find - internal diameter of pipe (2r).

Solution:

Volume of water passed through pipe in $\frac{1}{2}$ hour = $\pi r^2 \times h$ unit cube

$= \pi r^2 \times$ rate of flow × time

$= \pi r^2 \times 0.7 \times 30 \times 60 \; m^3$.

Volume of water in tank in $\frac{1}{2}$ hour $= \pi R^2 \times h$

$= \pi\left(\frac{2}{5}\right)^2 \times 3.15 \; m^3$

But, volume of water passed through pipe = Volume of water collected in tank

$\therefore \pi r^2 \times \frac{7}{10} \times 30 \times 60 = \pi\left(\frac{2}{5}\right)^2 \times \frac{315}{100}$

$\Rightarrow r^2 = \frac{4}{5 \times 8} \times \frac{315}{5100} \times \frac{1}{7} \times \frac{1}{3} \times \frac{1}{60}$

$\Rightarrow r^2 = \frac{1}{2500} \Rightarrow r = \sqrt{\frac{1}{50^2}} \Rightarrow r = \pm\frac{1}{50}$

Radius is always positive, so $r = -\frac{1}{50}$ can be ignored.

$\Rightarrow r = \frac{1}{50} m$.

$\Rightarrow r = \frac{1}{50} \times 100$ cm $\Rightarrow r = 2$ cm

Internal diameter of pipe = d = 2r = 4cm

Or 0.04m.

Section – D

Q. 23. In Figure 3, a decorative block is shown which is made of two solids, a cube and a hemisphere. The base of the block is a cube with edge 6 cm and the hemisphere fixed on the top has a diameter of 4.2 cm. Find

(a) the total surface area of the block.

(b) the volume of the block formed. (Take $\pi = \frac{22}{7}$)

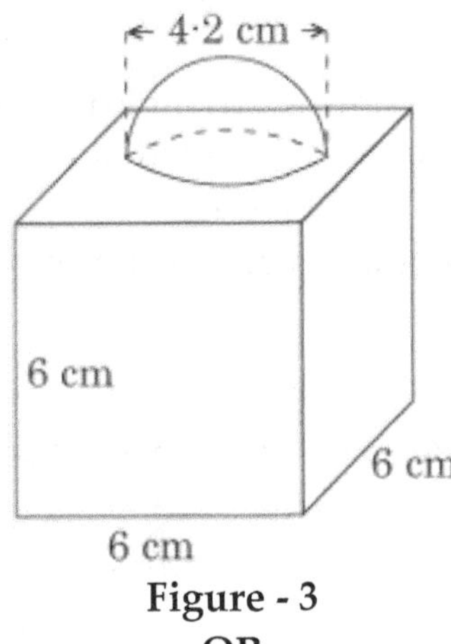

Figure - 3

OR

A bucket open at the top is in the form of a frustum of a cone with a capacity of 12308.8 cm^3. The radii of the top and bottom circular ends are 20 cm and 12 cm respectively. Find the height of the bucket and the area of metal sheet used in making the bucket. (Use π = 3.14)

Section - D

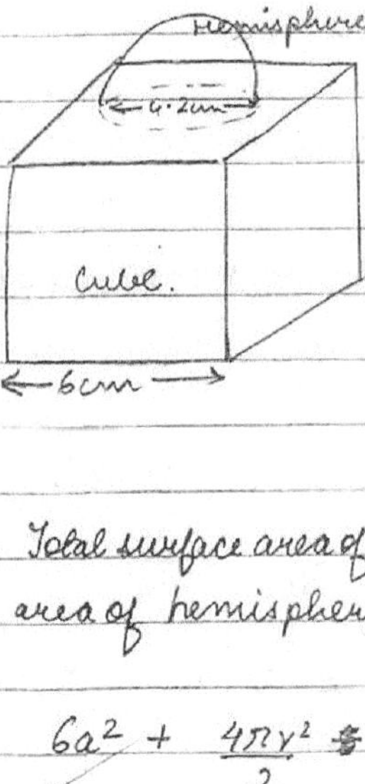

23. Radius of the hemisphere (r) $= \frac{d}{2}$

$= \frac{4 \cdot 2 \text{ cm}}{2}$

$= 2 \cdot 1 \text{ cm}$

Hemisphere, 4·2cm, Cube, ← 6cm →

Side of cube (a) = 6cm.

a) Total surface area of block = Total surface area of cube + Curved surface area of hemisphere - area enclosed by base of hemisphere

$= 6a^2 + \frac{4\pi r^2}{2} - \pi r^2$

$= 6a^2 + \pi r^2$

$= 6 \times 6^2 + \frac{22}{7} \times (2 \cdot 1)^2 \text{ cm}^2$

$= 216 + \frac{22}{7} \times 2 \cdot 1 \times 2 \cdot 1 \text{ cm}^2$

$= [216 + 13 \cdot 86] \text{ cm}^2$

$= 229 \cdot 86 \text{ cm}^2$

b) Volume of block formed = volume of cube + volume of hemisphere

$= a^3 + \frac{2}{3}\pi r^3$

$$= 6^3 + \frac{2}{3} \times \frac{22}{7} \times 2.1^3 \text{ cm}^3$$

$$= 216 + \frac{2}{3} \times \frac{22}{7} \times \frac{21 \times 21 \times 21}{1000} \text{ cm}^3$$

$$= 216 + 19.404 \text{ cm}^3$$

$$= 235.404 \text{ cm}^3.$$

Q. 24. If a line is drawn parallel to one side of a triangle to intersect the other two sides in distinct points, prove that the other two sides are divided in the same ratio.

OR

Prove that in a right triangle, the square of the hypotenuse is equal to the sum of the squares of the other two sides.

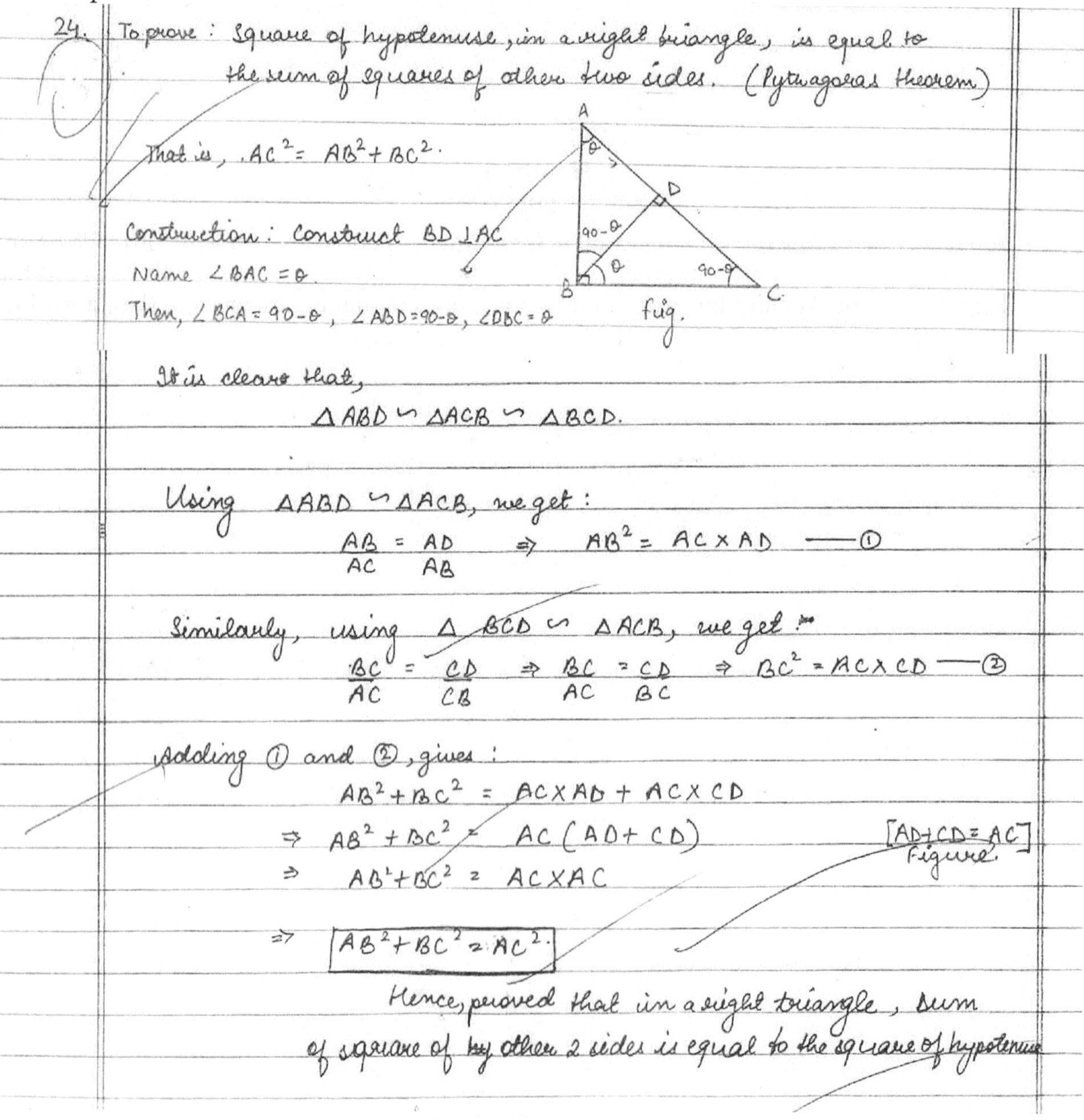

24. To prove : Square of hypotenuse, in a right triangle, is equal to the sum of squares of other two sides. (Pythagoras theorem)

That is, $AC^2 = AB^2 + BC^2$.

Construction : Construct $BD \perp AC$

Name $\angle BAC = \theta$.

Then, $\angle BCA = 90 - \theta$, $\angle ABD = 90 - \theta$, $\angle DBC = \theta$

It is clear that,

$\Delta ABD \sim \Delta ACB \sim \Delta BCD$.

Using $\Delta ABD \sim \Delta ACB$, we get :

$\frac{AB}{AC} = \frac{AD}{AB} \Rightarrow AB^2 = AC \times AD$ —— ①

Similarly, using $\Delta BCD \sim \Delta ACB$, we get :

$\frac{BC}{AC} = \frac{CD}{CB} \Rightarrow \frac{BC}{AC} = \frac{CD}{BC} \Rightarrow BC^2 = AC \times CD$ —— ②

Adding ① and ②, gives :

$AB^2 + BC^2 = AC \times AD + AC \times CD$

$\Rightarrow AB^2 + BC^2 = AC(AD + CD)$ [AD + CD = AC Figure.]

$\Rightarrow AB^2 + BC^2 = AC \times AC$

$\Rightarrow \boxed{AB^2 + BC^2 = AC^2}$.

Hence, proved that in a right triangle, sum of square of the other 2 sides is equal to the square of hypotenuse

Q. 25. Change the following distribution to a 'more than type' distribution. Hence draw the 'more than type' ogive for this distribution.

Class interval :	20 – 30	30 – 40	40 – 50	50 – 60	60 – 70	70 – 80	80 – 90
Frequency :	10	8	12	24	6	25	15

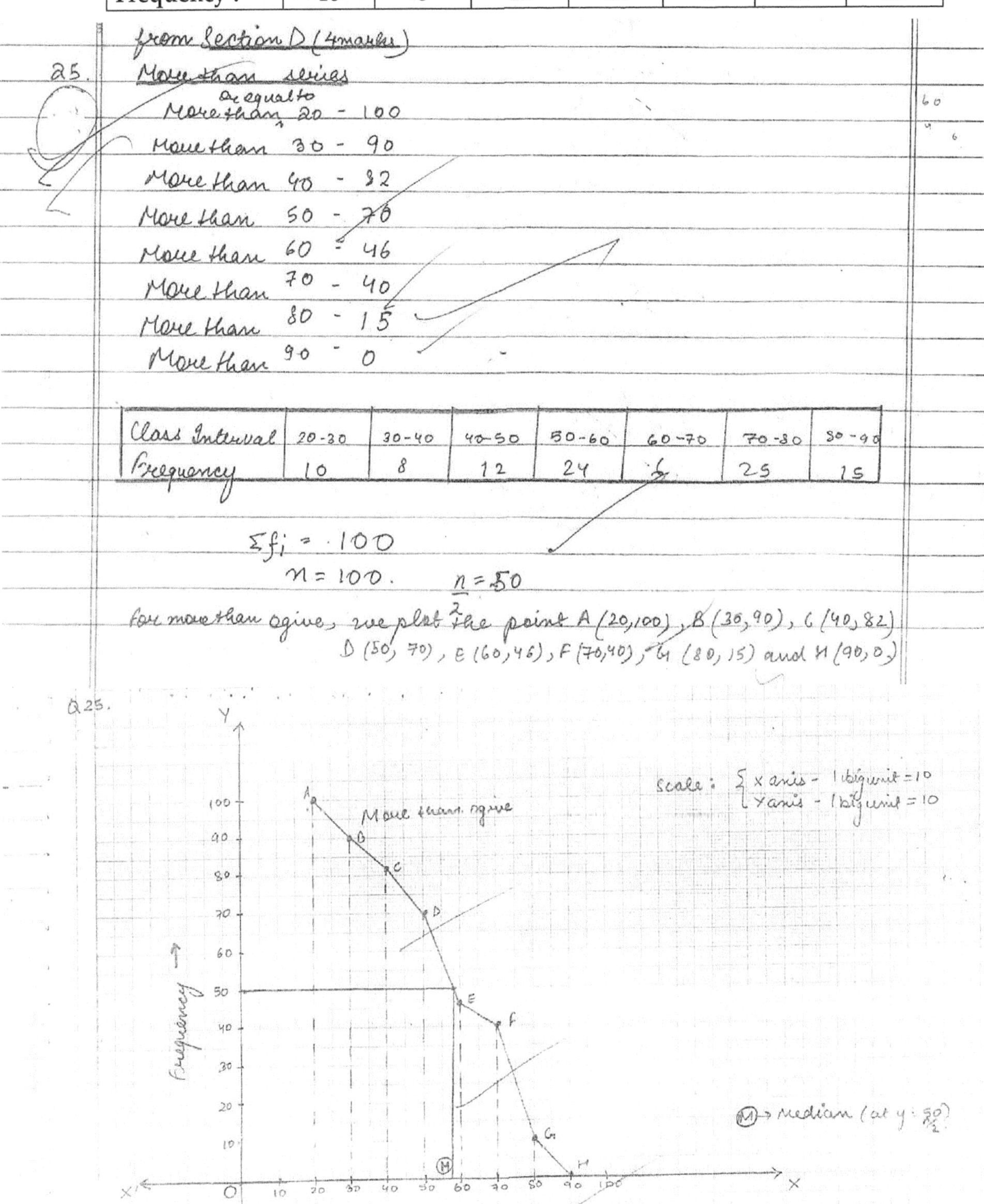

from Section D (4 marks)

25. More than series

More than or equal to 20 - 100

More than 30 - 90

More than 40 - 82

More than 50 - 70

More than 60 - 46

More than 70 - 40

More than 80 - 15

More than 90 - 0

Class Interval	20-30	30-40	40-50	50-60	60-70	70-80	80-90
Frequency	10	8	12	24	6	25	15

$\Sigma f_i = 100$

$n = 100$. $\frac{n}{2} = 50$

For more than ogive, we plot the point A (20, 100), B (30, 90), C (40, 82), D (50, 70), E (60, 46), F (70, 40), G (80, 15) and H (90, 0)

Q 25.

More than ogive.

Q. 26. The shadow of a tower standing on a level ground is found to be 40 m longer when the Sun's altitude is 30° than when it was 60°. Find the height of the tower. (Given $\sqrt{3} = 1.732$)

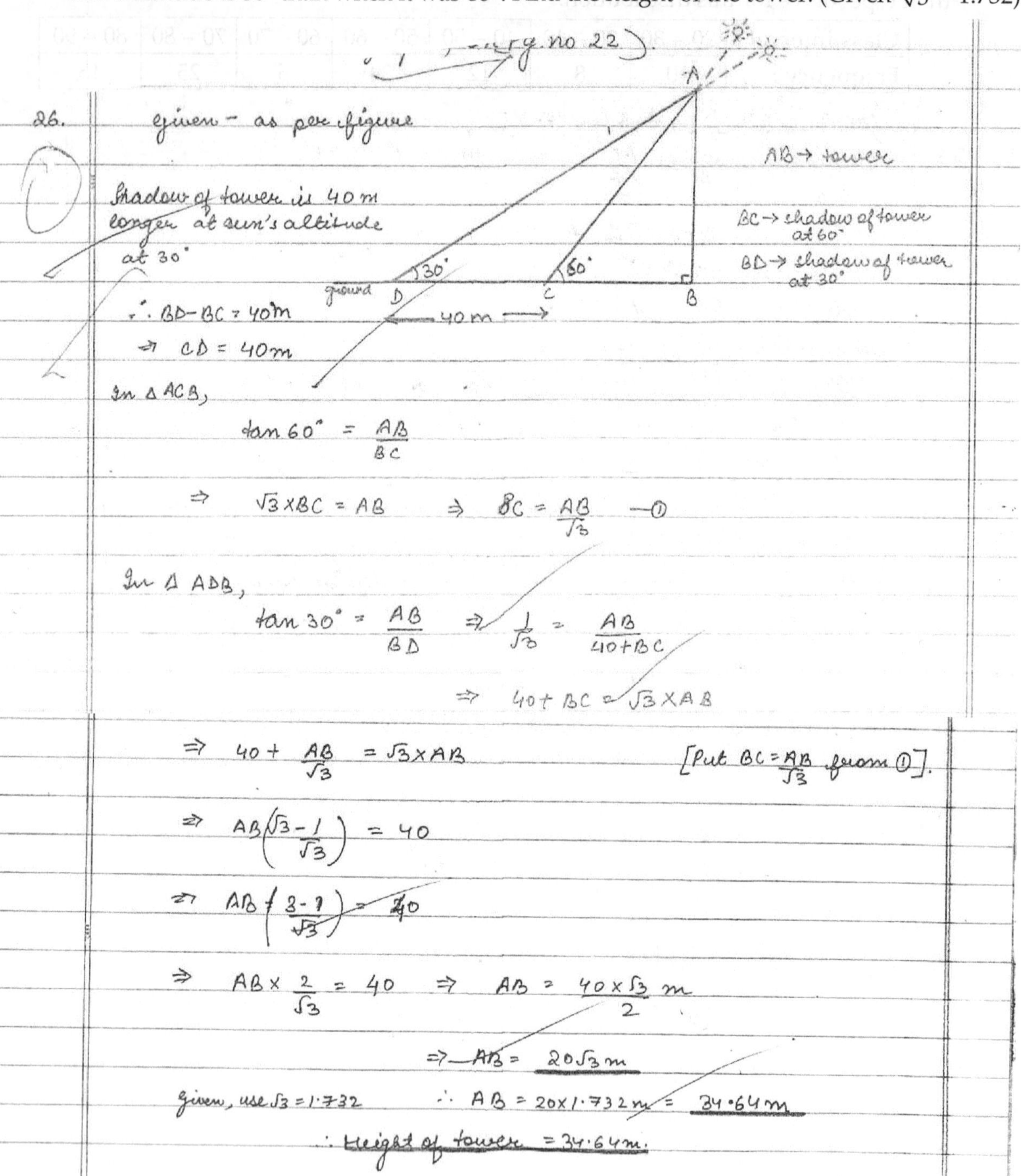

26. given – as per figure

AB → tower

BC → shadow of tower at 60°

BD → shadow of tower at 30°

Shadow of tower is 40 m longer at sun's altitude at 30°

$\therefore BD - BC = 40\text{m}$

$\Rightarrow CD = 40\text{m}$

In $\triangle ACB$,

$\tan 60° = \frac{AB}{BC}$

$\Rightarrow \sqrt{3} \times BC = AB \Rightarrow BC = \frac{AB}{\sqrt{3}}$ —①

In $\triangle ADB$,

$\tan 30° = \frac{AB}{BD} \Rightarrow \frac{1}{\sqrt{3}} = \frac{AB}{40 + BC}$

$\Rightarrow 40 + BC = \sqrt{3} \times AB$

$\Rightarrow 40 + \frac{AB}{\sqrt{3}} = \sqrt{3} \times AB$ [Put $BC = \frac{AB}{\sqrt{3}}$ from ①].

$\Rightarrow AB\left(\sqrt{3} - \frac{1}{\sqrt{3}}\right) = 40$

$\Rightarrow AB\left(\frac{3-1}{\sqrt{3}}\right) = 40$

$\Rightarrow AB \times \frac{2}{\sqrt{3}} = 40 \Rightarrow AB = \frac{40 \times \sqrt{3}}{2}\text{ m}$

$\Rightarrow AB = 20\sqrt{3}\text{ m}$

given, use $\sqrt{3} = 1.732$ $\therefore AB = 20 \times 1.732\text{ m} = 34.64\text{ m}$

$\therefore$ Height of tower $= 34.64\text{ m}$.

Q. 27. If m times the mth term of an Arithmetic Progression is equal to n times its nth term and $m \neq n$, show that the $(m + n)^{th}$ term of the A.P. is zero.

OR

The sum of the first three numbers in an Arithmetic Progression is 18. If the product of the first and the third term is 5 times the common difference, find the three numbers.

27. Let the first term of given A.P. be 'a'
And, the common difference be 'd'.
and a_p dentotes p^{th} term.

Given; $m(a_m) = n(a_n)$ $[m \neq n]$.

To show: $a_{(m+n)} = 0$

$m(a_m) = n(a_n)$

$\Rightarrow m[a+(m-1)d] = n[a+(n-1)d]$

$\Rightarrow am + md(m-1) = an + nd(n-1)$

$\Rightarrow am - an = nd(n-1) - md(m-1)$

$\Rightarrow a(m-n) = d[n(n-1) - m(m-1)]$

$\Rightarrow a(m-n) = d[n^2 - n - m^2 + m]$

$\Rightarrow a(m-n) = d[n^2 - m^2 + m - n]$

$\Rightarrow a(m-n) = d[(m+n)(n-m) + (m-n)]$

$\Rightarrow a(m-n) = d(m-n)[-(m+n) + 1]$

$\Rightarrow a - d[-(m+n)+1] = 0$

$\Rightarrow \boxed{a + (m+n-1)d = 0}$

$\Rightarrow \boxed{a_{m+n} = 0}$

$\therefore a_{m+n} = 0$

Hence, proved!

Q. 28. A shopkeeper buys some book for ks 80. If he had bought 4 more books for the same amount, each book have cost ₹ 1 less. Find the total number of books he bought.

28. Let the no. of books bought by the shopkeeper be 'n'.
Total money spent = Rs 80
$\therefore$ Cost of each book = Rs $\frac{80}{n}$

Now, given: He buys 4 more books, no. of books bought = $n+4$
(for same amount)

New cost of each book = Rs $\frac{80}{n+4}$

Given, new cost of each book is Rs. 1 less than earlier.

$\therefore \frac{80}{n} - \frac{80}{n+4} = 1$

$\Rightarrow 80\left(\frac{1}{n} - \frac{1}{n+4}\right) = 1$

$\Rightarrow \dfrac{n+4-n}{n(n+4)} = \dfrac{1}{80} \Rightarrow 4 \times 80 = n(n+4)$

$\Rightarrow n^2 + 4n - 320 = 0$

Using quadratic formula; $\Rightarrow n = \dfrac{-4 \pm \sqrt{16 + 4 \times 320}}{2}$

~~$\Rightarrow n = \dfrac{-4 \pm \sqrt{676}}{2} = \dfrac{-4 \pm 26}{2} \Rightarrow n = \dfrac{-30}{2}$ or $\dfrac{22}{2}$~~

$\Rightarrow n = \dfrac{-4 \pm \sqrt{1296}}{2}$

$\Rightarrow n = \dfrac{-4 \pm 36}{2}$

$\Rightarrow n = \dfrac{-40}{2}$ or $\dfrac{32}{2} \Rightarrow n = -20$ or 16

Since, no. of books is a whole no., it cannot be negative

$n = -20$ can be ignored.

$\therefore$ $\boxed{n = 16}$

No. of books bought by the shopkeeper = 16.

Q. 29. Draw a circle of radius 4 cm. Construct a pair of tangents from a point at distance of 6 cm from the centre of circle.

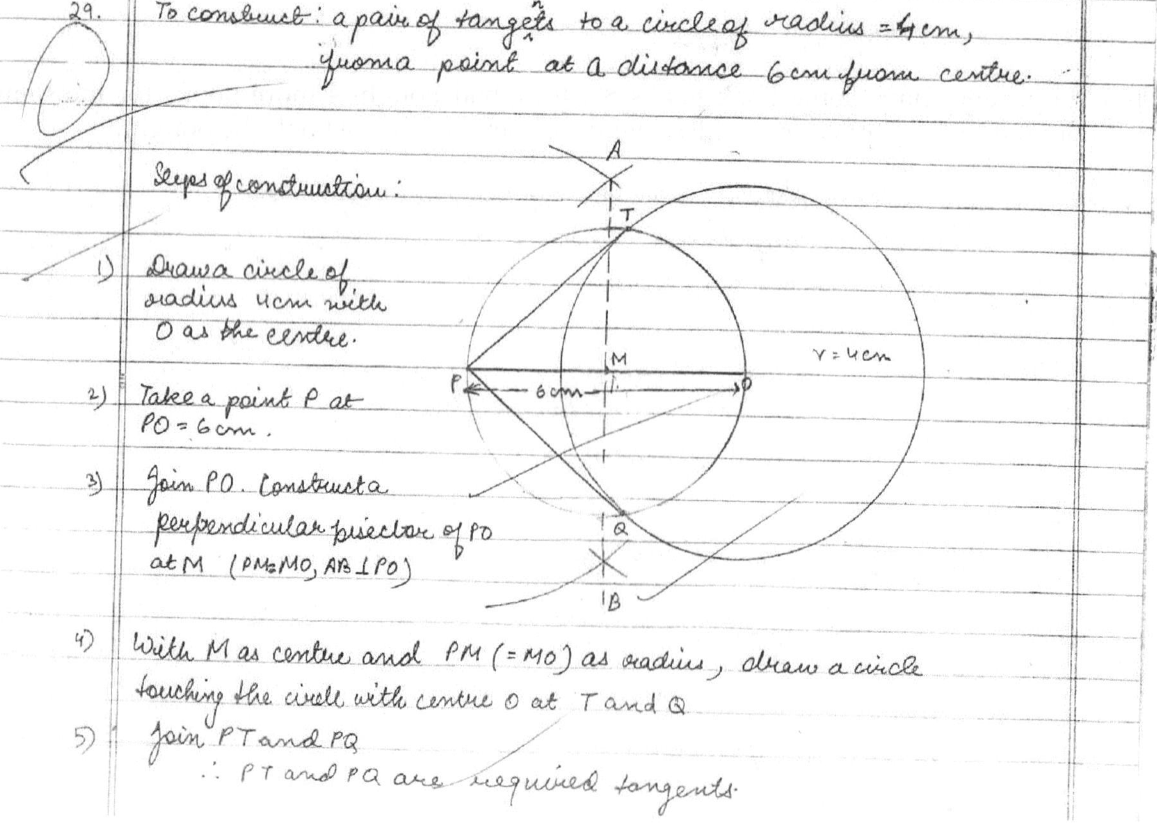

Q. 30. Prove that $\dfrac{1}{1+\sin^2\theta}+\dfrac{1}{1+\cos^2\theta}+\dfrac{1}{1+\sec^2\theta}+\dfrac{1}{1+\text{cosec}^2\theta}=2$

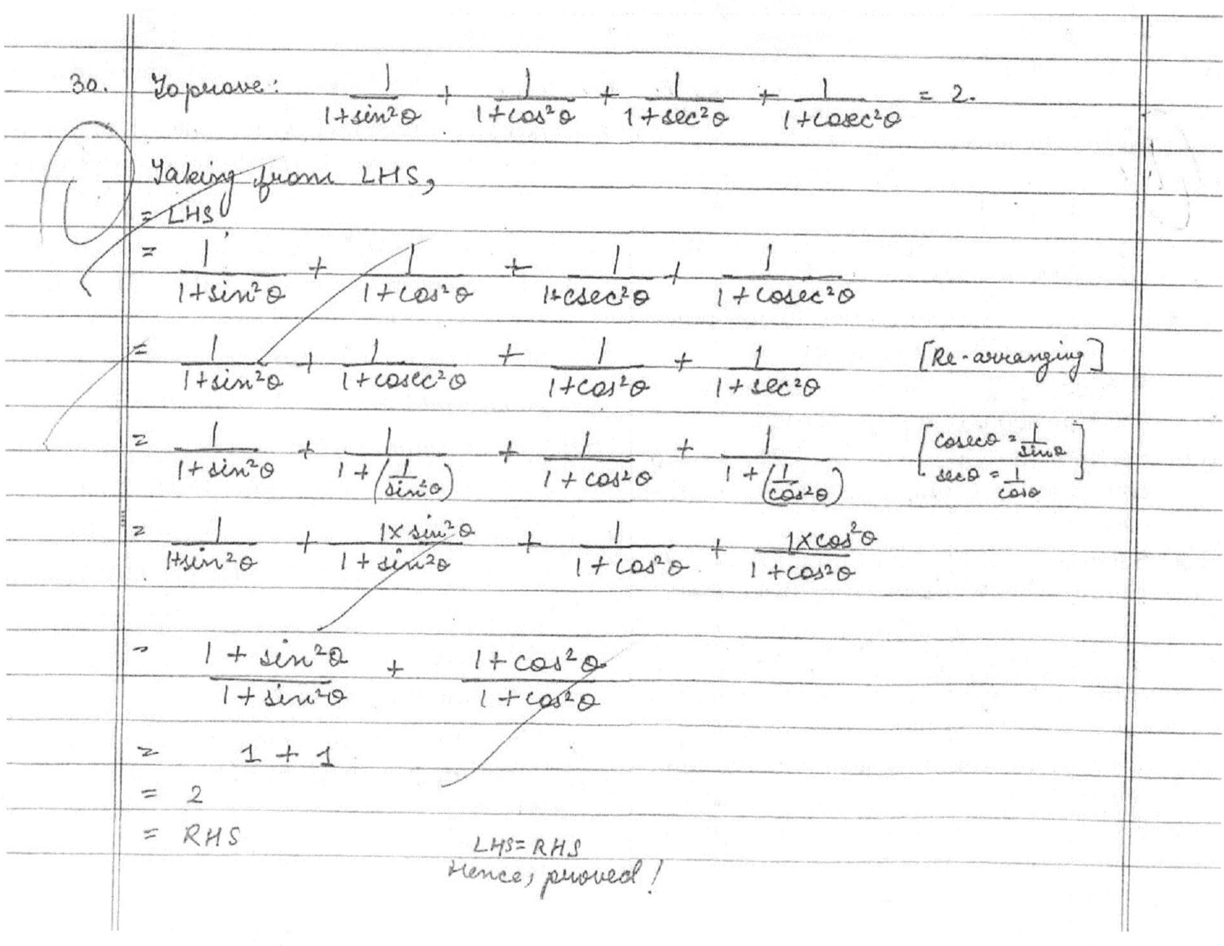

30. To prove: $\dfrac{1}{1+\sin^2\theta}+\dfrac{1}{1+\cos^2\theta}+\dfrac{1}{1+\sec^2\theta}+\dfrac{1}{1+\text{cosec}^2\theta}=2.$

Taking from LHS,

= LHS

$=\dfrac{1}{1+\sin^2\theta}+\dfrac{1}{1+\cos^2\theta}+\dfrac{1}{1+\sec^2\theta}+\dfrac{1}{1+\text{cosec}^2\theta}$

$=\dfrac{1}{1+\sin^2\theta}+\dfrac{1}{1+\text{cosec}^2\theta}+\dfrac{1}{1+\cos^2\theta}+\dfrac{1}{1+\sec^2\theta}$ [Re-arranging]

$=\dfrac{1}{1+\sin^2\theta}+\dfrac{1}{1+\left(\frac{1}{\sin^2\theta}\right)}+\dfrac{1}{1+\cos^2\theta}+\dfrac{1}{1+\left(\frac{1}{\cos^2\theta}\right)}$ $\left[\text{cosec}\,\theta=\frac{1}{\sin\theta},\ \sec\theta=\frac{1}{\cos\theta}\right]$

$=\dfrac{1}{1+\sin^2\theta}+\dfrac{1\times\sin^2\theta}{1+\sin^2\theta}+\dfrac{1}{1+\cos^2\theta}+\dfrac{1\times\cos^2\theta}{1+\cos^2\theta}$

$=\dfrac{1+\sin^2\theta}{1+\sin^2\theta}+\dfrac{1+\cos^2\theta}{1+\cos^2\theta}$

$=1+1$

$=2$

= RHS

LHS = RHS

Hence, proved!

MATHEMATICS (STANDARD) 2020

Outside Delhi [Set-I]

Time allowed : 3 hours

Maximum Marks : 80

General Instructions :

Read the following instructions very carefully and strictly follow them :

(i) This question paper comprises **four** sections A, B, C and D.
This question paper carries 40 questions. All questions are compulsory.

(ii) **Section A**—Question no. **1** to **20** comprises of **20** questions of **one** mark each.

(iii) **Section B**—Question no. **21** to **26** comprises of **6** questions of **two** marks each.

(iv) **Section C**—Question no. **27** to **34** comprises of **8** questions of **three** marks each.

(v) **Section D**—Question no. **35** to **40** comprises of **6** questions of **four** marks each.

(vi) There is no overall choice in the question paper. However, an internal choice has been provided in **2** questions of **one** mark, **2** questions of **two** marks, **3** questions of **three** marks and **3** questions of **four** marks. You have to attempt only **one** of the choices in such questions.

(vii) In addition to this, separate instructions are given with each section and question, wherever necessary.

(viii) Use of calculators is **not** permitted.

SECTION – A

Q. Nos. 1 to 10 are multiple choice questions of 1 mark each. Select the correct option.

1. The sum of exponents of prime factors in the prime-factorisation of 196 is **[1]**

(a) 3 (b) 4 (c) 5 (d) 2

Sol. Prime factors of 196 $= 2^2 \times 7^2$

So, sum of exponents $= 2 + 2 = 4$

$\therefore$ Option (b) is correct. **Ans.**

2. Euclid's division Lemma states that for two positive integers *a* and *b*, there exists unique integer *q* and *r* satisfying $a = bq + r$, and **[1]**

(a) $0 < r < b$ (b) $0 < r \leq b$

(c) $0 \leq r < b$ (d) $0 \leq r \leq b$

Sol. Euclid's division Lemma states that for two positive integers *a* and *b*, there exist unique integers *q* and *r* satisfying $a = bq + r$, and $0 \leq r < b$.

$\therefore$ Option (c) is correct. **Ans.**

3. The zeroes of the polynomial $x^2 - 3x - m(m + 3)$ are **[1]**

(a) $m, m + 3$ (b) $-m, m + 3$

(c) $m, -(m + 3)$ (d) $-m, -(m + 3)$

Sol. $x^2 - 3x - m(m + 3)$

$$= x^2 - (m + 3 - m)x - m(m + 3)$$
$$= x^2 - (m + 3)x + mx - m(m + 3)$$
$$= x\{x - (m + 3)\} + m\{x - (m + 3)\}$$
$$= (x + m)\{x - (m + 3)\}$$

$\therefore$ Its zeroes are $-m, (m + 3)$.

$\therefore$ Option (b) is correct. **Ans.**

4. The value of *k* for which the system of linear equations $x + 2y = 3, 5x + ky + 7 = 0$ is inconsistent is **[1]**

(a) $-\frac{14}{3}$ (b) $\frac{2}{5}$

(c) 5 (d) 10

Sol. For the system of linear equations to be incosistent, we have

$$\frac{a_1}{a_2} = \frac{b_1}{b_2} \neq \frac{c_1}{c_2}$$

Here $a_1 = 1, b_1 = 2, c_1 = -3$ and $a_2 = 5, b_2 = k, c_2 = 7$

So, $$\frac{1}{5} = \frac{2}{k} \neq \frac{-3}{7}$$

$\Rightarrow$ $k = 10$

$\therefore$ Option (d) is correct. **Ans.**

5. The roots of the quadratic equation $x^2 - 0.04 = 0$ are **[1]**

(a) ± 0.2 (b) ± 0.02

(c) 0.4 (d) 2

Sol. Given : $x^2 - 0.04 = 0$

$\Rightarrow$ $x^2 - (0.2)^2 = 0$

$\Rightarrow$ $(x + 0.2)(x - 0.2) = 0$

$\Rightarrow$ $x = -0.2, 0.2$

$\therefore$ Option (a) is correct. **Ans.**

6. The common difference of the A.P. $\frac{1}{p}, \frac{1-p}{p}, \frac{1-2p}{p}$, ... is **[1]**

(a) 1 **(b) $\frac{1}{p}$** **(c) – 1** **(d) $-\frac{1}{p}$**

Sol. We know,

Common difference, $d = a_2 - a_1$

$$= \frac{1-p}{p} - \frac{1}{p}$$

$$= \frac{1-p-1}{p}$$

$$= -\frac{p}{p}$$

$$= -1$$

∴ Option (c) is correct. **Ans.**

7. The nth term of the A.P. a, $3a$, $5a$, ... is **[1]**

(a) na **(b) $(2n - 1)a$**

(c) $(2n + 1)a$ **(d) $2na$**

Sol. In the A.P., $a = a$ and $d = 3a - a = 2a$

We know, n^{th} term of an A.P. is given by,

$$a_n = a + (n - 1)d$$

$$\Rightarrow a_n = a + (n - 1)2a$$

$$\Rightarrow a_n = a + 2an - 2a$$

$$\Rightarrow a_n = 2an - a$$

$$\Rightarrow a_n = a(2n - 1)$$

∴ Option (b) is correct. **Ans.**

8. The point P on x-axis equidistant from the points A(– 1, 0) and B(5, 0) is **[1]**

(a) (2, 0) **(b) (0, 2)**

(c) (3, 0) **(d) (2, 2)**

Sol. Let P(h, 0) be the required point on x-axis.

So, according to the question

$$PA = PB$$

or $$(PA)^2 = (PB)^2$$

$$\Rightarrow (h + 1)^2 + (0 - 0)^2 = (h - 5)^2 + (0 - 0)^2$$

$$\Rightarrow (h + 1)^2 = (h - 5)^2$$

$$\Rightarrow h^2 + 2h + 1 = h^2 - 10h + 25$$

$$\Rightarrow 12h = 24$$

$$\Rightarrow h = 2$$

So, the required point is (2, 0)

∴ Option (a) is correct. **Ans.**

9. The co-ordinates of the point which is reflection of point (– 3, 5) in x-axis are **[1]**

(a) (3, 5) **(b) (3, – 5)**

(c) (– 3, – 5) **(d) (– 3, 5)**

Sol. The co-ordinates of the point which is reflection of point (– 3, 5) in x-axis is (– 3, – 5).

∴ Option (c) is correct. **Ans.**

10. If the point P(6, 2) divides the line segment joining A(6, 5) and B(4, y) in the ratio 3 : 1, then the value of y is **[1]**

(a) 4 **(b) 3**

(c) 2 **(d) 1**

Sol. Using section formula,

$$y\text{-coordinate of P} = \frac{3 \times y + 1 \times 5}{3 + 1}$$

$$\Rightarrow 2 = \frac{3y + 5}{4}$$

$$\Rightarrow 8 = 3y + 5$$

$$\Rightarrow 3y = 3$$

$$\Rightarrow y = 1$$

∴ Option (d) is corrrect. **Ans.**

In Q. Nos. 11 to 15, fill in the blanks. Each question is of 1 marks.

11. In Fig. 1, MN || BC and AM : MB = 1 : 2, then $\frac{ar(\Delta AMN)}{ar(\Delta ABC)}$ = ... **[1]**

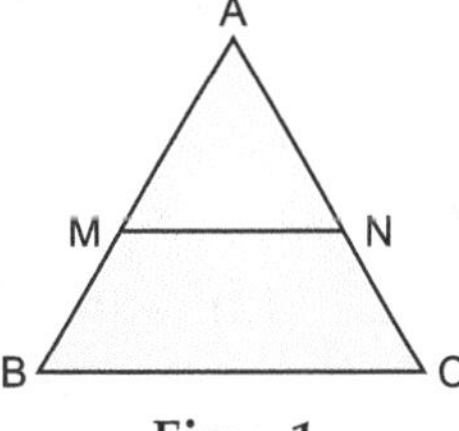

Fig. - 1

Sol. $\frac{1}{9}$

Explanation :

Given : MN || BC and AM : MB = 1 : 2

In ΔAMN and ΔABC,

$\angle MAN = \angle BAC$ [Common angle]

$\angle AMN = \angle ABC$ [Corresponding angles]

∴ $\Delta AMN \sim \Delta ABC$

$$\therefore \frac{ar(\Delta AMN)}{ar(\Delta ABC)} = \left(\frac{AM}{AB}\right)^2 = \left(\frac{AM}{AM + MB}\right)^2$$

$$= \left(\frac{1}{1+2}\right)^2 = \frac{1}{9}$$ **Ans.**

12. In given Fig. 2, the length PB = ... cm. **[1]**

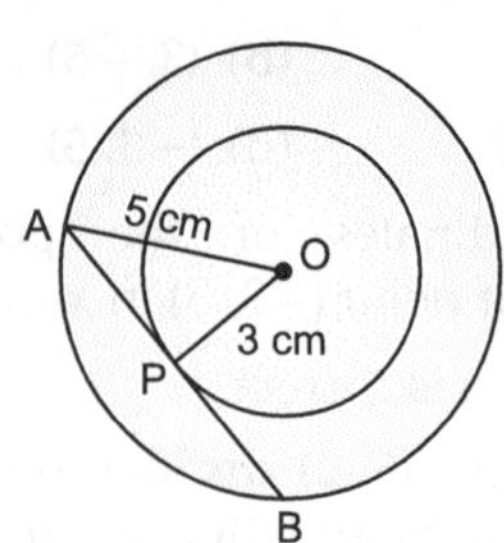

Fig. - 2

Sol. 4

Explanation :

We have, $OP \perp AB$

$\therefore$ P is the mid-point of AB.

$\Rightarrow$ $AP = PB$

Now, in ΔOAP

$$OA^2 = OP^2 + AP^2$$

[Pythagoras theorem]

$\Rightarrow$ $(5)^2 = (3)^2 + AP^2$

$\Rightarrow$ $25 = 9 + AP^2$

$\Rightarrow$ $AP^2 = 16$

$\Rightarrow$ $AP = 4$

$\Rightarrow$ $PB = 4$ cm **Ans.**

13. In ΔABC, AB = $6\sqrt{3}$ cm, AC = 12 cm and BC = 6 cm, then $\angle B$ = ... **[1]**

OR

Two triangles are similar if their corresponding sides are ....

Solution : 90°

Explanation :

$\because$ $(12)^2 = (6\sqrt{3})^2 + (6)^2$

$\Rightarrow$ $AC^2 = AB^2 + BC^2$

$\therefore$ By the converse of Pythagoras theorem,

$\therefore$ $\angle B = 90°$. **Ans.**

OR

Proportional.

14. The value of (tan 1° tan 2° ..... tan 89°) is equal to .... **[1]**

Solution : 1

Explanation :

tan 1° tan 2° .... tan 89°

= tan 1° tan 2° ..... tan 45° ... tan 88° tan 89°

= tan 1° tan 2° .... tan 45° ... tan (90° – 2°) tan (90° – 1°)

= tan 1° tan 2° .... tan 45° ..... cot 2° cot 1°

[$\because$ tan (90° – θ) = cot θ]

$$= \tan 1° \tan 2° ..... \tan 45° ... \frac{1}{\tan 2°} \frac{1}{\tan 1°}$$

= tan 45°

= 1 **Ans.**

15. In Fig. 3, the angles of depressions from the observing positions O_1 and O_2 respectively of the object A are ...., ... **[1]**

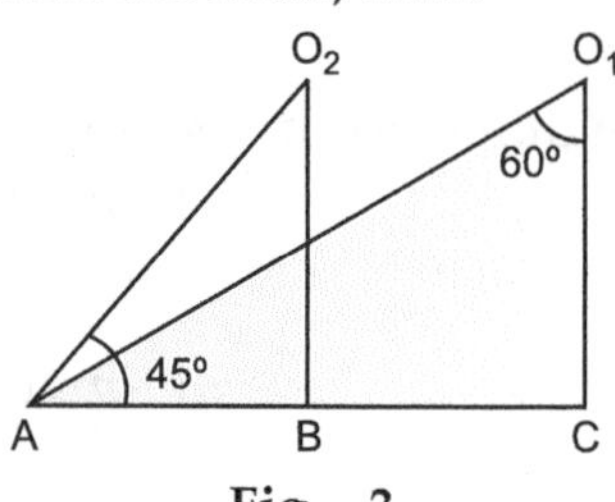

Fig. - 3

Sol. 30°, 45°

Explanation :

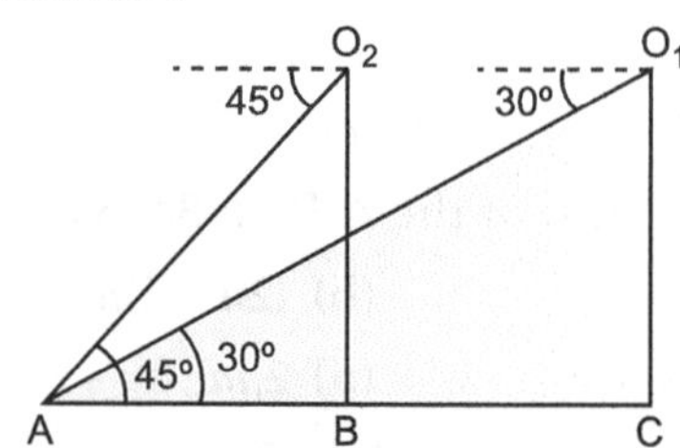

Q. Nos. 16 to 20 are short answer type questions of 1 mark each.

16. If sin A + sin² A = 1, then find the value of the expression (cos² A + cos⁴ A). **[1]**

Sol. Given : $\sin A + \sin^2 A = 1$

$\Rightarrow$ $\sin A = 1 - \sin^2 A$

$\Rightarrow$ $\sin A = \cos^2 A$

[$\because \sin^2 A + \cos^2 A = 1$]

Squaring both sides, we get

$$\sin^2 A = \cos^4 A$$

$\Rightarrow$ $1 - \cos^2 A = \cos^4 A$

$\Rightarrow$ $\cos^2 A + \cos^4 A = 1$ **Ans.**

17. In fig. 4 is a sector of circle of radius 10.5 cm. Find the perimeter of the sector. $\left(\text{Take } \pi = \frac{22}{7}\right)$ **[1]**

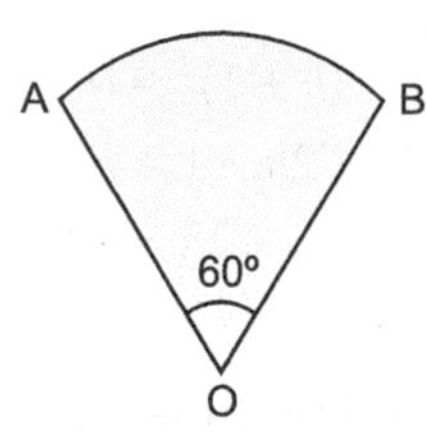

Fig. - 4

Sol. Given : OA = OB = 10.5 cm

and $\angle AOB = 60°$

Perimeter of the sector = $OA + OB + \widehat{AB}$

$$= 10.5 + 10.5 + \frac{60°}{360°} \times 2 \times \frac{22}{7} \times 10.5$$

$$\left[\because l = \frac{\theta}{360°} \times 2\pi r\right]$$

$$= 21 + 11 = 32 \text{ cm}$$ **Ans.**

18. If a number x is chosen at random from the numbers – 3, – 2, – 1, 0, 1, 2, 3, then find the probability of $x^2 < 4$. [1]

OR

What is the probability that a randomly taken leap year has 52 Sundays?

Sol. Total outcomes = {– 3, – 2, – 1, 0, 1, 2, 3}

⇒ Number of total outcomes, $n(S) = 7$

Let A be the event of chosing a number whose square is less than 4.

$\therefore$ A = {– 1, 0, 1}

⇒ $n(A) = 3$

$\therefore$ $P(x^2 < 4) = \frac{n(A)}{n(S)}$

$= \frac{3}{7}$ **Ans.**

OR

A leap year has 366 days *i.e.* 52 × 7 + 2 days

So, we have 52 weeks and 2 days

⇒ 52 weeks will have 52 sundays

And, these 2 days can be the following pair of days: (Sunday, Monday), (Monday, Tuesday), (Thursday, Wednesday), (Wednesday, Thursday), (Thursday, Friday), (Friday, Saturday), (Saturday, Sunday).

$\therefore$ Out of 7 possible pairs, only 2 pairs have Sunday and 5 pairs do not have Sunday.

$\therefore$ Probability that a randomly taken leap year has 52 Sundays = $\frac{5}{7}$. **Ans.**

19. Find the class-marks of the classes 10 – 25 and 35 – 55. [1]

Sol. We know,

$$\text{Class mark} = \frac{\text{Lower limit} + \text{Upper limit}}{2}$$

$\therefore$ Class mark of 10 – 25 = $\frac{10+25}{2}$

$= \frac{35}{2}$

$= 17.5$

and class mark of 35 – 55 = $\frac{35+55}{2}$

$= \frac{90}{2}$

$= 45$ **Ans.**

20. A die is thrown once. What is the probability of getting a prime number? [1]

Sol. When a die is thrown once,

Total number of possible outcomes = 6

Also, prime numbers from 1 to 6 = {2, 3, 5}

⇒ Number of favourable outcomes = 3

$\therefore$ P(prime number) = $\frac{3}{6} = \frac{1}{2}$ **Ans.**

SECTION – B

Q. Nos. 21 to 26 carry 2 marks each.

21. A teacher asked 10 of his students to write a polynomial in one variable on a paper and then to handover the paper. The following were the answers given by the students : [2]

$2x + 3$, $3x^2 + 7x + 2$, $4x^3 + 3x^2 + 2$, $x^3 + \sqrt{3x} + 7$, $7x + \sqrt{7}$, $5x^3 - 7x + 2$, $2x^2 + 3 - \frac{5}{x}$, $5x - \frac{1}{2}$,

$ax^3 + bx^2 + cx + d$, $x + \frac{1}{x}$

Answer the following questions :

(i) How many of the above ten, are not polynomials?

(ii) How many of the above ten, are quadratic polynomials?

Sol. (i) 3

$$\left(x^2 + \sqrt{3x} + 7, 2x^2 + 3 - \frac{5}{x} \text{ and } x + \frac{1}{x}\right)$$

(ii) There is only one quadratic polynomial which is $3x^2 + 7x + 2$.

22. In fig. 5, ABC and DBC are two triangles on the same base BC. If AD intersects BC at O, show that $\frac{ar(\Delta ABC)}{ar(\Delta DBC)} = \frac{AO}{DO}$ [2]

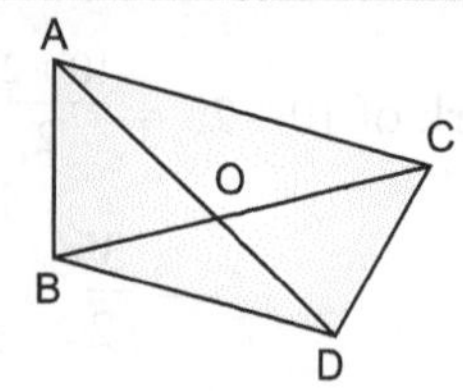

Fig. - 5

OR

In fig. 6, if AD $\perp$ BC, then prove that

$$AB^2 + CD^2 = BD^2 + AC^2 \quad \textbf{[2]}$$

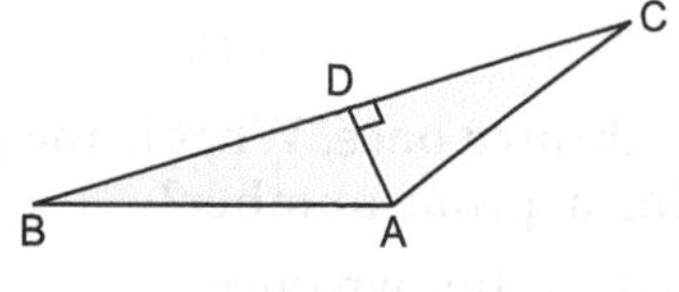

Fig - 6

Sol. Given : ΔABC and ΔDBC are on the same base BC and AD intersect BC at O.

To prove : $\dfrac{ar(\Delta ABC)}{ar(\Delta DBC)} = \dfrac{AO}{DO}$.

Construction : Draw AL $\perp$ BC and DM $\perp$ BC.

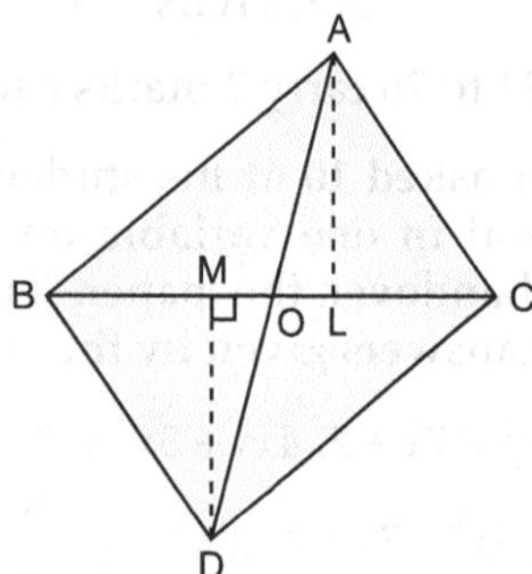

Proof : In ΔALO and ΔDMO, we have

$$\angle ALO = \angle DMO = 90°$$

and $\quad \angle AOL = \angle DOM$

[Vertically opposite angles]

$\therefore \quad \angle ALO \sim \angle DMO$

[AA-similarity]

$\therefore \quad \dfrac{AL}{DM} = \dfrac{AO}{DO}$

[corresponding parts of similar triangles] ...(i)

Now, $\dfrac{ar(\Delta ABC)}{ar(\Delta DBC)} = \dfrac{\frac{1}{2} \times BC \times AL}{\frac{1}{2} \times BC \times DM} = \dfrac{AL}{DM} = \dfrac{AO}{DO}$

[using (i)]

$\Rightarrow \quad \dfrac{ar(\Delta ABC)}{ar(\Delta DBC)} = \dfrac{AO}{DO}$. **Hence Proved.**

OR

Given : ΔABC in which AD $\perp$ BC.

To prove : $AB^2 + CD^2 = BD^2 + AC^2$

Proof : From right ΔADB, we have

$$AB^2 = AD^2 + BD^2$$

[By Pythagoras theorem]

$\Rightarrow \quad AB^2 - BD^2 = AD^2$. ...(i)

From right ΔADC, we have

$$AC^2 = AD^2 + CD^2$$

$\Rightarrow \quad AC^2 - CD^2 = AD^2$ ...(ii)

From (i) and (ii), we get

$$AB^2 - BD^2 = AC^2 - CD^2$$

$\Rightarrow \quad AB^2 + CD^2 = BD^2 + AC^2$

Hence Proved.

23. Prove that $1 + \dfrac{\cot^2 \alpha}{1 + \operatorname{cosec} \alpha} = \operatorname{cosec} \alpha$ [2]

OR

Show that $\tan^4 \theta + \tan^2 \theta = \sec^4 \theta - \sec^2 \theta$ [2]

Sol. Consider, L.H.S. $= 1 + \dfrac{\cot^2 \alpha}{1 + \operatorname{cosec} \alpha}$

$$= 1 + \frac{\operatorname{cosec}^2 \alpha - 1}{1 + \operatorname{cosec} \alpha}$$

$[\because \cot^2 \alpha = \operatorname{cosec}^2 \alpha - 1]$

$$= 1 + \frac{(\operatorname{cosec} \alpha + 1)(\operatorname{cosec} \alpha - 1)}{1 + \operatorname{cosec} \alpha}$$

$$= 1 + \operatorname{cosec} \alpha - 1$$

$$= \operatorname{cosec} \alpha = \text{R.H.S.}$$

Hence Proved.

OR

Consider,

$$\text{L.H.S.} = \tan^4 \theta + \tan^2 \theta$$

$$= \tan^2 \theta (\tan^2 \theta + 1)$$

$$= (\sec^2 \theta - 1) \sec^2 \theta$$

$[\because \sec^2 \theta - \tan^2 \theta = 1]$

$$= \sec^4 \theta - \sec^2 \theta = \text{R.H.S.}$$

Hence Proved.

24. The volume of a right circular cylinder with its height equal to the radius is $25\frac{1}{7}$ cm^3. Find the height of the cylinder. $\left(\text{Use } \pi = \frac{22}{7}\right)$

[2]

Sol. Given :

$$\text{Height of cylinder} = \text{Radius of cylinder}$$

$$\Rightarrow \quad h = r$$

And, Volume of cylinder $= 25\frac{1}{7}\text{ cm}^3 = \frac{176}{7}\text{ cm}^3$

We know, Volume of cylinder $= \pi r^2 h$

$$= \pi h^3 \qquad [\because h = r]$$

$$\Rightarrow \quad \frac{176}{7} = \frac{22}{7}h^3$$

$$\Rightarrow \quad h^3 = 8$$

$$\Rightarrow \quad h = 2.$$

Hence, the height of the cylinder is 2 cm. **Ans.**

25. A child has a die whose six faces show the letters as shown below : [2]

A	B	C	D	E	A

The die is thrown once. What is the probability of getting (i) A, (ii) D?

Solution : Total outcomes = {A, B, C, D, E, A}

(i) Probability of getting A $= \frac{2}{6} = \frac{1}{3}$ **Ans.**

(ii) Probability of getting D $= \frac{1}{6}$ **Ans.**

26. Compute the mode for the following frequency distribution : [2]

Size of items (in cm)	Frequency
0 – 4	5
4 – 8	7
8 – 12	9
12 – 16	17
16 – 20	12
20 – 24	10
24 – 28	6

Sol. Here, the maximum frequency is 17 which lies in the class interval 12 – 16.

$\therefore$ Modal class = 12 – 16

So, $l = 12, f_1 = 17, f_0 = 9, f_2 = 12$ and $h = 4$

We know, $\quad \text{Mode} = l + \left(\frac{f_1 - f_0}{2f_1 - f_0 - f_2}\right) \times h$

$$= 12 + \left(\frac{17-9}{34-9-12}\right) \times 4$$

$$= 12 + \left(\frac{8}{34-21}\right) \times 4$$

$$= 12 + \left(\frac{8}{13}\right) \times 4$$

$$= 12 + \frac{32}{13}$$

$$= 12 + 2.46$$

$$= 14.46$$ **Ans.**

SECTION – C

Q. Nos. 27 to 34 carry 3 marks each.

27. If $2x + y = 23$ and $4x - y = 19$, find the value of $(5y - 2x)$ and $\left(\frac{y}{x} - 2\right)$. [3]

OR

Solve for x : $\frac{1}{x+4} - \frac{1}{x+7} = \frac{11}{30}$, $x \neq -4, 7$.

Sol. Given : $\quad 2x + y = 23 \quad$ ...(i)

and $\quad 4x - y = 19 \quad$ ...(ii)

From equation (i), we have

$$y = 23 - 2x$$

Putting this value in equation (ii), we get

$$4x - (23 - 2x) = 19$$

$$\Rightarrow \quad 4x - 23 + 2x = 19$$

$$\Rightarrow \quad 6x = 19 + 23$$

$$\Rightarrow \quad 6x = 42$$

$$x = 7$$

So, $\quad y = 23 - 2 \times 7 = 23 - 14 = 9$

So, $\quad 5y - 2x = 5 \times 9 - 2 \times 7$

$$= 45 - 14 = 31$$

And $\quad \frac{y}{x} - 2 = \frac{9}{7} - 2 = \frac{9-14}{7}$

$$= -\frac{5}{7}$$ **Ans.**

OR

Given : $\quad \frac{1}{x+4} - \frac{1}{x+7} = \frac{11}{30}$

$$\Rightarrow \quad \frac{x+7-x-4}{(x+4)(x+7)} = \frac{11}{30}$$

$$\Rightarrow \quad \frac{3}{x^2+11x+28} = \frac{11}{30}$$

$$\Rightarrow \quad 11x^2 + 121x + 218 - 0$$

$$\Rightarrow \quad x = \frac{-121 \pm \sqrt{(121)^2 - 4 \times 11 \times 218}}{2 \times 11}$$

$$\Rightarrow \quad x = \frac{-121 \pm \sqrt{5049}}{22}$$

28. Show that the sum of all terms of an A.P. whose first term is a, the second term is b and the last term is c is equal to $\frac{(a+c)(b+c-2a)}{2(b-a)}$ **[3]**

OR

Solve the equation :

$1 + 4 + 7 + 10 + ... + x = 287$.

Sol. Given : First term of A.P. = a

Second term of A.P. = b

and, Last term of A.P. = c

So, Common difference, $d = b - a$

We know,

$$l = a + (n-1)d$$

$$\Rightarrow \quad c = a + (n-1)(b-a)$$

$$\Rightarrow \quad c - a = (n-1)(b-a)$$

$$\Rightarrow \quad n - 1 = \frac{c-a}{b-a}$$

$$\Rightarrow \quad n = \frac{c-a}{b-a} + 1$$

$$\Rightarrow \quad n = \frac{b+c-2a}{b-a}$$

Now, $$S_n = \frac{n}{2}(a+l)$$

$$\Rightarrow \quad S_n = \frac{(b+c-2a)(a+c)}{2(b-a)}$$

Hence Proved.

OR

Given : $1 + 4 + 7 + 10 + ... + x = 287$

In this A.P., $a = 1$, $d = 3$, $S = 287$ and $l = x$

We know that,

$$l = a + (n-1)d$$

$$\Rightarrow \quad x = 1 + (n-1)3$$

$$\Rightarrow \quad x = 1 + 3n - 3$$

$$\Rightarrow \quad x = 3n - 2$$

$$\Rightarrow \quad n = \frac{x+2}{3}$$

Now, $$S_n = \frac{n}{2}(a+l)$$

$$\Rightarrow \quad 287 = \frac{(x+2)(1+x)}{6}$$

$$\Rightarrow \quad x^2 + 3x + 2 = 1722$$

$$x^2 + 3x - 1720 = 0$$

$$x^2 + 43x - 40x - 1720 = 0$$

$$x(x+43) - 40(x+43) = 0$$

$$x = -43 \text{ or } x = 40$$

$\therefore$ $x = -43$ is not possible because it is increasing A.P.

$\therefore$ $x = 40$. **Ans.**

29. In a flight of 600 km, an aircraft was slowed down due to bad weather. The average speed of the trip was reduced by 200 km/hr and the time of flight increased by 30 minutes. Find the duration of flight. **[3]**

Sol. Let the original speed of the aircraft be x km/hr.

So, time taken to cover 6000 km = $\frac{600}{x}$ hours

Now, Reduced speed = $(x - 200)$km/hr.

So, time taken to cover 600 km at reduced speed

$$= \frac{600}{(x-200)} \text{ hours}$$

$\therefore$ According to the question,

$$\Rightarrow \quad \frac{600}{x-200} - \frac{600}{x} = \frac{1}{2}$$

$$\Rightarrow \quad \frac{1}{x-200} - \frac{1}{x} = \frac{1}{1200}$$

$$\Rightarrow \quad \frac{x-(x-200)}{x(x-200)} = \frac{1}{1200}$$

$$\Rightarrow \quad \frac{200}{x^2-200x} = \frac{1}{1200}$$

$$\Rightarrow \quad x^2 - 200x = 240000$$

$$\Rightarrow \quad x^2 - 200x - 240000 = 0$$

$$\Rightarrow \quad x^2 - 600x + 400x - 240000 = 0$$

$$\Rightarrow \quad x(x-600) + 400(x-600) = 0$$

$$\Rightarrow \quad (x-600)(x+400) = 0$$

$$\Rightarrow \quad x - 600 = 0 \text{ or } x + 400 = 0$$

$$\Rightarrow \quad x = 600 \text{ or } x = -400$$

$$\Rightarrow \quad x = 600$$

[$\because$ speed cannot be negative]

$\therefore$ Original speed of aircraft = 600 km/hr

$\therefore$ Original duration of flight = $\frac{600}{600}$ = 1 hour

And, Increased duration of flight

$$= \left(1 + \frac{1}{2}\right) \text{ hours}$$

= 1.5 hours. **Ans.**

30. If the mid-point of the line segment joining the points A(3, 4) and B(k, 6) is P(x, y) and $x + y - 10 = 0$. Find the value of k. **[3]**

OR

Find the area of triangle ABC with A(1, – 4) and the mid-points of sides through A being (2, – 1) and (0, – 1).

Sol. Coordinates of the mid-point of A(3, 4) and B(k, 6) $= \left\{\frac{3+k}{2}, \frac{4+6}{2}\right\}$

$$= \left(\frac{3+k}{2}, 5\right)$$

$\because$ P(x, y) is the mid-point of AB

$$\therefore \quad P(x, y) = \left(\frac{3+k}{2}, 5\right)$$

On comparing, we get

$$x = \frac{3+k}{2} \text{ and } y = 5$$

Now, $\quad x + y - 10 = 0$

$$\Rightarrow \quad \frac{3+k}{2} + 5 - 10 = 0$$

$$\Rightarrow \quad \frac{3+k}{2} - 5 = 0$$

$$\Rightarrow \quad 3 + k - 10 = 0$$

$$\Rightarrow \quad k - 7 = 0$$

$$\Rightarrow \quad k = 7$$ **Ans.**

OR

Let the points M and N be the mid-points of AB and AC, respectively.

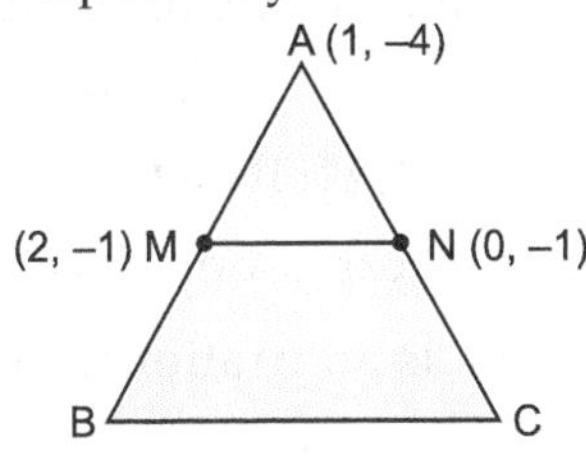

Now, $\quad ar(\Delta AMN) = \frac{1}{2} |1(-1+1) + 2(-1+4) + 0(-4+1)|$

$$= \frac{1}{2} |1(0) + 2(3) + 0|$$

$$= \frac{1}{2} |6| = 3 \text{ sq. units}$$

$\because \quad \Delta AMN \sim \Delta ABC$

$$\therefore \quad \frac{ar(\Delta AMN)}{ar(\Delta ABC)} = \frac{AM^2}{AB^2}$$

$$= \left(\frac{AM}{AB}\right)^2$$

$$= \left(\frac{1}{2}\right)^2 = \frac{1}{4}$$

$$\Rightarrow \quad \frac{3}{ar(\Delta ABC)} = \frac{1}{4}$$

$$\Rightarrow \quad ar(\Delta ABC) = 12 \text{ sq. units.}$$ **Ans.**

31. In Fig. 7, if ΔABC ~ ΔDEF and their sides of lengths (in cm) are marked along them, then find the lengths of sides of each triangle. **[3]**

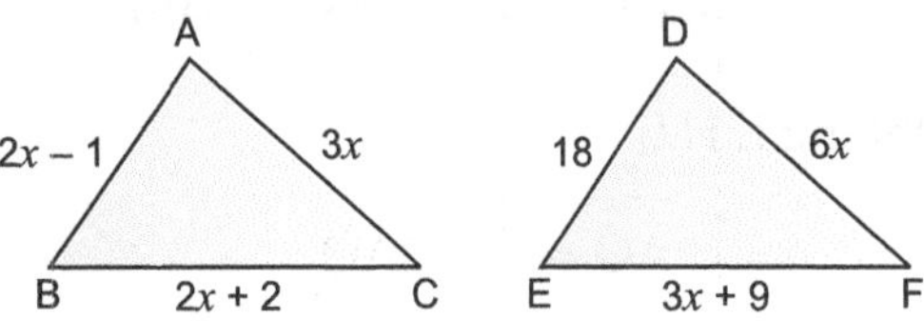

Figure – 7

Sol. Given : $\quad \Delta ABC \sim \Delta DEF$

$$\therefore \quad \frac{AB}{DE} = \frac{AC}{DF}$$

[Corresponding parts of similar triangles]

$$\Rightarrow \quad \frac{2x-1}{18} = \frac{3x}{6x}$$

$$\Rightarrow \quad \frac{2x-1}{18} = \frac{1}{2}$$

$$\Rightarrow \quad 4x - 2 = 18$$

$$\Rightarrow \quad 4x = 20$$

$$\Rightarrow \quad x = 5$$

Now, lengths of sides of triangle ABC are,

$$AB = 2x - 1 = 9 \text{ cm}$$

$$BC = 2x + 2 = 12 \text{ cm}$$

$$AC = 3x = 15 \text{ cm}$$

And, lengths of sides of triangle DEF are,

$$DE = 18 \text{ cm}$$

$$EF = 3x + 9 = 24 \text{ cm}$$

$$DF = 6x = 30 \text{ cm}$$ **Ans.**

32. If a circle touches the side BC of a triangle ABC at P and extended sides AB and AC at Q and R, respectively, prove that

$$\mathbf{AQ = \frac{1}{2}(BC + CA + AB)}$$ **[3]**

Sol. We know that the lengths of tangents drawn from an external point to a circle are equal.

$\therefore$ AQ = AR ...(i) [Tangents from A]

BP = BQ ...(ii) [Tangents from B]

CP = CR ...(iii) [Tangents from C]

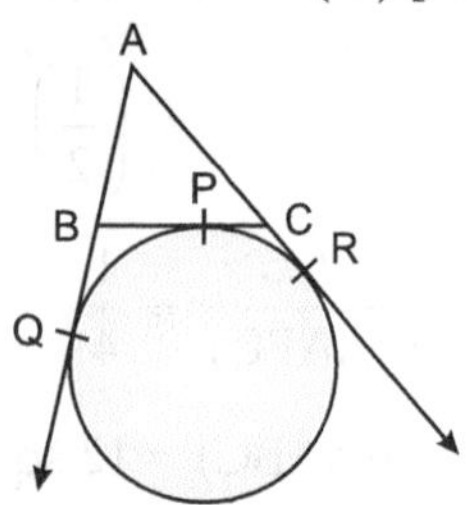

We have to show that

$$AQ = \frac{1}{2}(BC + CA + AB)$$

Taking L.H.S.

$$AQ = \frac{1}{2} \times 2AQ$$

$$\Rightarrow \quad AQ = \frac{1}{2} \times (AQ + AQ)$$

$$\Rightarrow \quad AQ = \frac{1}{2} \times (AQ + AR)$$

[From equation (i)]

$$\Rightarrow \quad AQ = \frac{1}{2} \times (AB + BQ + AC + CR)$$

$$\Rightarrow \quad AQ = \frac{1}{2} \times (AB + BP + AC + CP)$$

[Using equations (ii) and (iii)]

$$\Rightarrow \quad AQ = \frac{1}{2} \times \{AB + (BP + CP) + AC\}$$

$$\Rightarrow \quad AQ = \frac{1}{2} \times (AB + BC + AC)$$

Hence Proved.

33. If $\sin\theta + \cos\theta = \sqrt{2}$, prove that $\tan\theta + \cot\theta = 2$. **[3]**

Sol. Given : $\sin\theta + \cos\theta = \sqrt{2}$

Squaring both sides,

$$(\sin\theta + \cos\theta)^2 = (\sqrt{2})^2$$

$$\Rightarrow \sin^2\theta + \cos^2\theta + 2\sin\theta\cos\theta = 2$$

$$1 + 2\sin\theta\cos\theta = 2$$

$[\because \sin^2\theta + \cos^2\theta = 1]$

$$\Rightarrow \quad 2\sin\theta\cos\theta = 1$$

$$\Rightarrow \quad \sin\theta\cos\theta = \frac{1}{2} \quad ...(i)$$

We have to prove that $\tan\theta + \cot\theta = 2$

Taking L.H.S.,

$$\tan\theta + \cot\theta = \frac{\sin\theta}{\cos\theta} + \frac{\cos\theta}{\sin\theta}$$

$$= \frac{\sin^2\theta + \cos^2\theta}{\sin\theta\cos\theta}$$

$$\frac{1}{\sin\theta\cos\theta} = \frac{1}{1/2} = 2 = \text{R.H.S. [from eq (i)]}$$

Hence Proved.

34. The area of a circular play ground is 22176 cm^2. Find the cost of fencing this ground at the rate of ₹ 50 per metre. **[3]**

Sol. Given : Area of play ground = 22176 cm^2

Let the radius of play ground be r cm.

$\therefore$ Area of play ground = πr^2

$$\Rightarrow \quad 22176 = \frac{22}{7} \times r^2$$

$$\Rightarrow \quad r^2 = \frac{22176 \times 7}{22}$$

$$\Rightarrow \quad r^2 = 7056$$

$$\Rightarrow \quad r = \sqrt{7056} = 84 \text{ cm}$$

$$= 0.84 \text{ m}$$

Now, circumference of play ground = $2\pi r$

$$= 2 \times \frac{22}{7} \times 0.84 = 5.28 \text{ m}$$

$\because$ Rate of fencing = ₹ 50 per metre

$\therefore$ Total of fencing the circular ground = 50 × 5.28

= ₹ 264 **Ans.**

SECTION – D

Q. Nos. 35 to 40 carry 4 marks each.

35. Prove that $\sqrt{5}$ is an irrational number. **[4]**

Sol. Let us assume, to the contrary that $\sqrt{5}$ is a rational number.

Then, there exists co-prime positive integers a and b such that $b \neq 0$.

Then, $$\sqrt{5} = \frac{a}{b}$$

Squaring both sides, we get

$$(\sqrt{5})^2 = \left(\frac{a}{b}\right)^2$$

$$\Rightarrow \quad 5 = \frac{a^2}{b^2}$$

$$\Rightarrow \quad 5b^2 = a^2 \quad \text{...(i)}$$

$\Rightarrow$ 5 divides a^2

$\Rightarrow$ 5 divides a

$\Rightarrow a = 5k$ for some integer k

Substituting the value of a in equation (i),

$$5b^2 = (5k)^2$$

$$\Rightarrow \quad 5b^2 = 25k^2$$

$$\Rightarrow \quad b^2 = 5k^2$$

$\Rightarrow$ 5 divides b^2

$\Rightarrow$ 5 divides b

$\Rightarrow a$ and b have 5 as a common factor.

But this contradicts the fact that a and b are co-prime integers.

$\therefore$ Our assumption that $\sqrt{5}$ is rational is incorrect.

$\therefore$ We conclude that $\sqrt{5}$ is an irrational number.

Hence Proved.

36. It can take 12 hours to fill a swimming pool using two pipes. If the pipe of larger diameter is used for four hours and the pipe of smaller diameter for 9 hours, only half of the pool can be filled. How long would it take for each pipe to fill the pool separately? [4]

Sol. Let the time taken by pipe of larger diameter be x hours and the time taken by the pipe of smaller diameter by y hours.

So, amount of water filled by pipe of larger diameter in 1 hr $= \frac{1}{x}$

and amount of water filled by pipe of smaller diameter in 1 hr $= \frac{1}{y}$

$$\therefore \quad \frac{1}{x} + \frac{1}{y} = \frac{1}{12} \quad \text{...(i)}$$

Now, amount of water filled by pipe of larger diameter in 4 hr $= \frac{4}{x}$

and amount of water filled by pipe of smaller diameter in 9 hr $= \frac{9}{y}$

$$\therefore \quad \frac{4}{x} + \frac{9}{y} = \frac{1}{2} \quad \text{...(ii)}$$

Consider $\frac{1}{x} = a$ and $\frac{1}{y} = b$

So, by equation (i)

$$a + b = \frac{1}{12}$$

$$12a + 12b = 1 \quad \text{...(iii)}$$

and, by equation (ii)

$$4a + 9b = \frac{1}{2}$$

$$\Rightarrow \quad 8a + 18b = 1 \quad \text{...(iv)}$$

Multiplying equation (iii) by 2 and equation (iv) by 3 and then subtracting them, we get

$$\begin{array}{r} 24a + 24b = 2 \\ 24a + 54b = 3 \\ - \quad - \quad - \\ \hline -30b = -1 \end{array}$$

$$\Rightarrow \quad b = \frac{1}{30}$$

Putting the value of b in equation (iii), we get

$$12a + 12\left(\frac{1}{30}\right) = 1$$

$$\Rightarrow \quad 12a + \frac{2}{5} = 1$$

$$\Rightarrow \quad 12a = 1 - \frac{2}{5}$$

$$\Rightarrow \quad 12a = \frac{3}{5}$$

$$\Rightarrow \quad a = \frac{1}{20}$$

$$\because \quad a = \frac{1}{x} \Rightarrow x = 20 \text{ hr}$$

$$\text{and} \quad b = \frac{1}{y} \Rightarrow y = 30 \text{ hr}$$

Hence, the larger diameter pipe would take 20 hours and the smaller diameter pipe would take 30 hours to fill the pool separately. **Ans.**

37. Draw a circle of radius 2 cm with centre O and take a point P outside the circle such that OP = 6.5 cm. From P, draw two tangents to the circle. [4]

OR

Construct a triangle with sides 5 cm, 6 cm and 7 cm and then construct another triangle

whose sides are $\frac{3}{4}$ times the corresponding sides of the first triangle.

Solution : Steps of construction :

(i) Draw a circle with centre O and radius 2 cm.

(ii) Take a point P outside the circle at a distance of 6.5 cm from O and join OP.

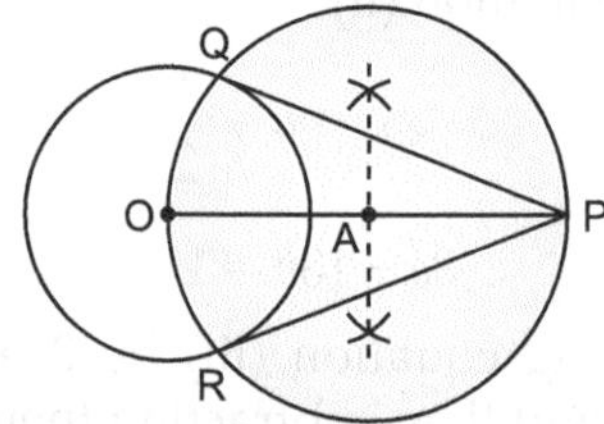

(iii) Draw perpendicular bisector of OP to get its mid-point A.

(iv) Taking A as centre and OA (or AP) as radius, draw a circle which intersects the circle of radius 2 cm at points Q and R.

(v) Join PQ and PR.

Thus, PQ and PR are the required tangents.

OR

Steps of construction :

(i) Draw a line segment BC = 7 cm.

(ii) Taking B as centre, draw an arc of radius 5 cm.

(iii) Taking C as centre, draw another arc of radius 6 cm. Let the two arcs intersects each other at point A.

(iv) Join AB and AC. Thus, ΔABC is obtained.

(v) Draw a ray BX at an acute angle with side BC.

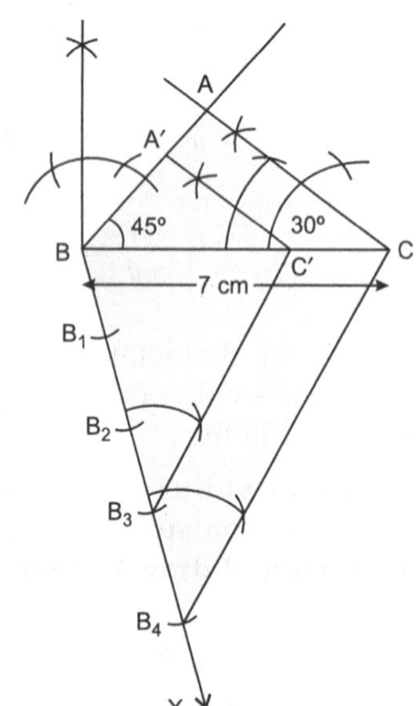

(vi) Locate 4 point B_1, B_2, B_3 and B_4 on BX such that $BB_1 = B_1B_2 = B_2B_3 = B_3B_4$.

(vii) Join B_4C and draw a line through B_3, parallel to B_4C, to intersect BC at C′.

(viii) Draw a line C′A′, parallel to CA, to intersect AB at A′.

Thus, ΔA'BC′ is the required triangle whose sides are $\frac{3}{4}$ times the corresponding sides of ΔABC.

38. From a point on the ground, the angles of elevation of the bottom and the top of a tower fixed at the top of a 20 m high building are 45° and 60° respectively. Find the height of the tower. **[4]**

Sol. Let the building be AB, tower be CA and the point of observation on the ground be P.

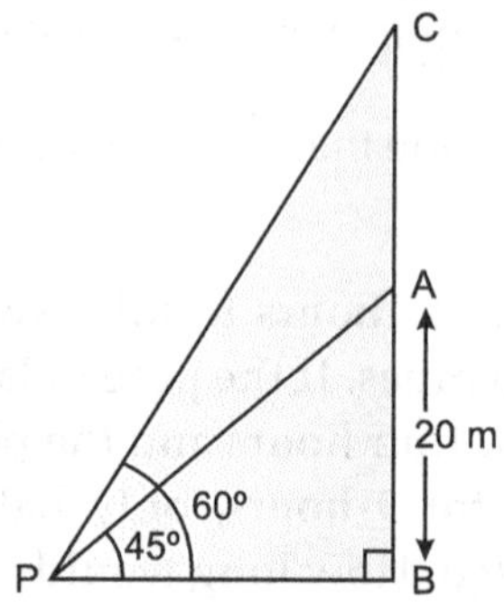

$\therefore$ AB = 20 m, ∠APB = 45° and ∠CPB = 60°

Now, in right triangle ABP,

$$\tan 45^\circ = \frac{AB}{PB}$$

$$\Rightarrow \quad 1 = \frac{20}{PB}$$

$$\Rightarrow \quad PB = 20 \text{ m}$$

Similarly, in right triangle CBP

$$\tan 60^\circ = \frac{BC}{PB}$$

$$\Rightarrow \quad \sqrt{3} = \frac{BC}{20}$$

$$\Rightarrow \quad BC = 20\sqrt{3} \text{ m}$$

Now, $\quad AC = BC - AB = 20\sqrt{3} - 20$

$$= 20(\sqrt{3} - 1) \text{ m}$$

Hence, the height of the tower is $20(\sqrt{3}-1)$ m.

Ans.

39. Find the area of the shaded region in fig. 8, if PQ = 24 cm, PR = 7 cm and O is the centre of the circle. [4]

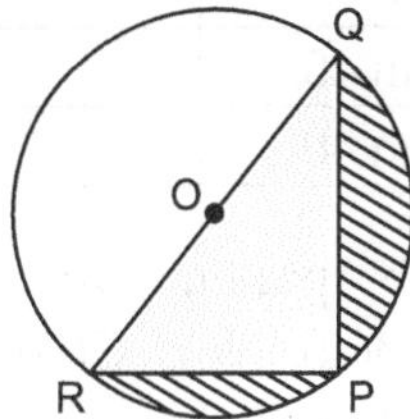

Fig - 8

OR

Find the curved surface area of the frustum of a cone, the diameters of whose circular ends are 20 m and 6 m and its height is 24 m.

Sol. Given : PQ = 24 cm, PR = 7 cm.

We have, $\angle RPQ = 90°$ [$\because$ Angle in a semicircle is 90°]

$\therefore$ In right ΔRPQ, we have

$$RQ^2 = PR^2 + PQ^2$$ [Pythagoras theorem]

$$= (7)^2 + (24)^2$$
$$= 49 + 576$$
$$\Rightarrow RQ^2 = 625$$
$$\Rightarrow RQ = \sqrt{625} = 25 \text{ cm}$$
$$\therefore \text{Radius of circle} = \frac{1}{2}\times 25 \text{ cm} = \frac{25}{2} \text{ cm}$$

Now, Area of shaded region

= Area of semicircle with radius $\frac{25}{2}$ cm – Area of ΔPQR

$$= \frac{1}{2}\pi\left(\frac{25}{2}\right)^2 - \frac{1}{2}\times 7\times 24$$
$$= \left(\frac{1}{2}\times\frac{22}{7}\times\frac{25}{2}\times\frac{25}{2}\right) - (7\times 12)$$
$$= \left(\frac{11\times 25\times 25}{28}\right) - 84$$
$$= \frac{6875}{28} - 84$$
$$= (245.54 - 84) \text{ cm}^2$$
$$= 161.54 \text{ cm}^2.$$ **Ans.**

OR

Given : Diameter of circular ends are 20 m and 6 m.

And, Height = 24 m

$\therefore$ R = 10 m, r = 3 m and h = 24 m

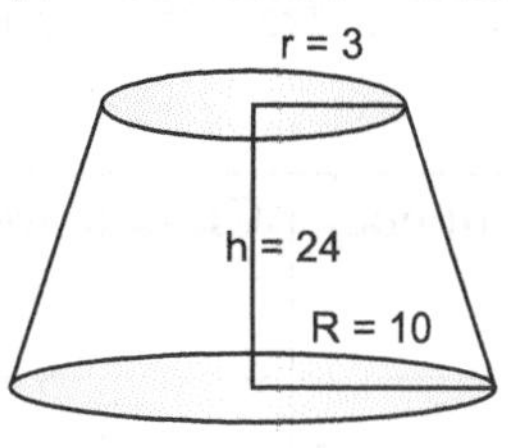

We know, $l^2 = h^2 + (R - r)^2$

$$\Rightarrow l^2 = (24)^2 + (10 - 3)^2$$
$$\Rightarrow l^2 = (24)^2 + (7)^2$$
$$\Rightarrow l^2 = 576 + 49$$
$$\Rightarrow l^2 = 625$$
$$\Rightarrow l = 25$$

Now, curved surface area of the frustum of a cone

$$= \pi(R + r)l$$
$$= \frac{22}{7}\times(10 + 3)\times 25$$
$$= \frac{22}{7}\times 13\times 25$$
$$= \frac{7250}{7}$$
$$= 1021.43 \text{ m}^2$$ **Ans.**

40. The mean of the following frequency distribution is 18. The frequency f in the class interval 19 – 21 is missing. Determine f. [4]

Class Interval	Frequency
11 – 13	3
13 – 15	6
15 – 17	9
17 – 19	13
19 – 21	f
21 – 23	5
23 – 25	4

OR

The following table gives production yield per hectare of wheat of 100 farms of a village:

Production Yield	No. of farms
40 – 45	4
45 – 50	6
50 – 55	16
55 – 60	20
60 – 65	30
65 – 70	24

Change the distribution of a 'more than' type distribution and draw its ogive.

Sol.

Class Interval	Frequency f_i	Class-mark x_i	f_ix_i
11 – 13	3	12	36
13 – 15	6	14	84
15 – 17	9	16	144
17 – 19	13	18	234
19 – 21	f	20	$20f$
21 – 23	5	22	110
23 – 25	4	24	96
	$\Sigma f_i = 40 + f$		$\Sigma f_ix_i = 704 + 20f$

We know, $\text{Mean} = \frac{\Sigma f_i x_i}{\Sigma f_i}$

$\Rightarrow \quad 18 = \frac{704 + 20f}{40 + f}$

$\Rightarrow \quad 720 + 18f = 704 + 20f$

$\Rightarrow \quad 720 - 704 = 20f - 18f$

$\Rightarrow \quad 16 = 2f$

$\Rightarrow \quad f = 8$

So, missing frequency of class inteval 19 – 21 is 8.

Ans.

OR

More than type distribution :

Production yield	No. of farms
More than 65	24
More than 60	54
More than 55	74
More than 50	90
More than 45	96
More than 40	100

On a graph paper, we plot the points A(40, 100), B(45, 96), C(50, 90), D(55, 74), E(60, 54) and F(65, 24).

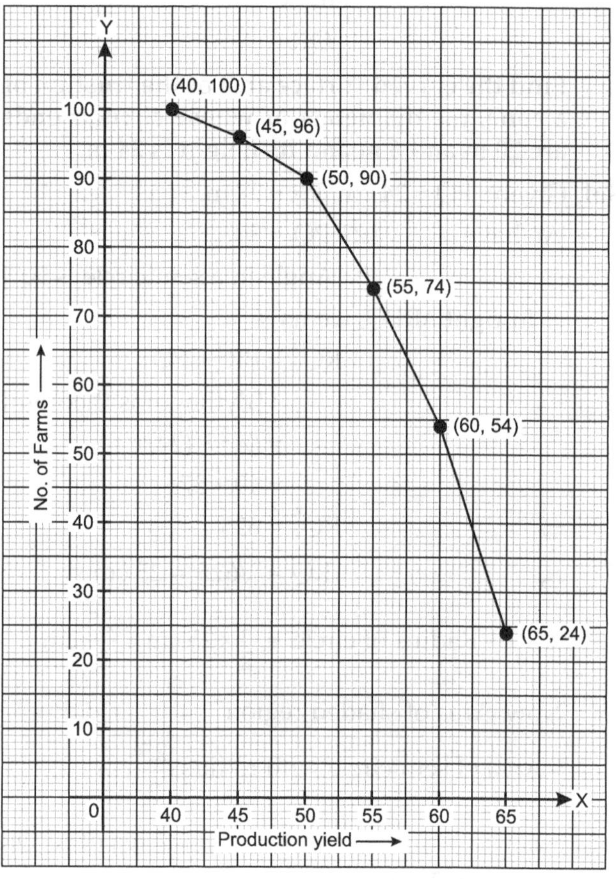

Outside Delhi [Set-II]

Time allowed : 3 hours **Maximum Marks : 80**

Note : Except for the following questions, all the remaining questions have been asked in previous set.

SECTION–A

15. The value of sin 23° cos 67° + cos 23° sin 67° is .... **[1]**

Sol. 1

Explanation : $\sin 23° \cos 67° + \cos 23° \sin 67°$

$= \sin 23° \cos (90° - 23°) + \cos 23° \sin (90° - 23°)$

$= \sin 23°. \sin 23° + \cos 23°. \cos 23°$

$\left[\because \cos(90° - \theta) = \sin\theta \text{ and } \sin(90° - \theta) = \cos\theta\right]$

$= \sin^2 23° + \cos^2 23° \quad [\because \sin^2\theta + \cos^2\theta = 1]$

$= 1$ **Ans.**

19. If tan A = cot B, then find the value of (A + B). **[1]**

Sol. Given : $\tan A = \cot B$

$\Rightarrow \quad \tan A = \tan (90° - B)$

$[\because \tan (90° - \theta) = \cot \theta]$

On comparing both sides, we get

$$A = 90° - B$$

$$\Rightarrow \quad A + B = 90°. \qquad \textbf{Ans.}$$

20. Find the class marks of the classes 15 – 35 and 45 – 60. [1]

Sol. We know,

$$\text{Class Mark} = \frac{\text{Lower limit} + \text{Upper limit}}{2}$$

$$\therefore \quad \text{Class mark of } 15 - 35 = \frac{15+35}{2} = \frac{50}{2} = 25$$

$$\text{and, class mark of } 45 - 60 = \frac{45+60}{2} = \frac{105}{2} = 52.5$$

Ans.

SECTION–B

25. A child has a die whose six faces show the letters as shown below : [2]

A	A	B	C	C	C

The die is thrown once. What is the probability of getting (i) A, (ii) C ?

Sol. Total outcomes = {A, A, B, C, C, C}

(i) Probability of getting A = $\frac{2}{6} = \frac{1}{3}$ **Ans.**

(ii) Probability of getting C = $\frac{3}{6} = \frac{1}{2}$ **Ans.**

26. A solid is in the shape of a cone mounted on a hemisphere of same base radius. If the curved surface areas of the hemispherical part and the conical part are equal, then find the ratio of the radius and the height of the conical part. [2]

Sol. Let the radii of conical part and the hemispherical part be r, slant height of conical part be l and the height of conical part be h.

Now, according to the question,

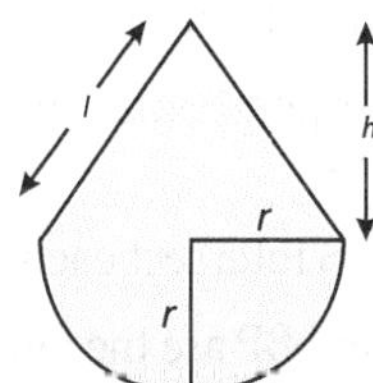

C.S.A of hemispherical part = C.S.A. of conical part

$$2\pi r^2 = \pi r l$$

Squaring both sides,

$$\Rightarrow \quad 4r^2 = l^2$$

$$\Rightarrow \quad 4r^2 = r^2 + h^2 \qquad [\because l^2 = r^2 + h^2]$$

$$\Rightarrow \quad 3r^2 = h^2$$

$$\Rightarrow \quad \frac{r^2}{h^2} = \frac{1}{3}$$

$$\Rightarrow \quad \frac{r}{h} = \frac{1}{\sqrt{3}}$$

$$\Rightarrow \quad r : h = 1 : \sqrt{3} \qquad \textbf{Ans.}$$

SECTION–C

32. If in an A.P., the sum of first m terms is n and the sum of its first n terms is m, then prove that the sum of its first $(m + n)$ terms is – $(m + n)$. [3]

OR

Find the sum of all 11 terms of an A.P. whose middle term is 30.

Sol. Given : Sum of first m terms of an A.P. is n.

$$\therefore \quad \frac{m}{2}[2a + (m-1)d] = n$$

$$\Rightarrow \quad [2am + m(m-1)d] = 2n \qquad ...(i)$$

Also, sum of first n terms of an A.P. is m.

$$\therefore \quad \frac{n}{2}[2a + (n-1)d] = m$$

$$\Rightarrow \quad [2an + n(n-1)d] = 2m \qquad ...(ii)$$

Subtracting (ii) from (i),

$$\Rightarrow [2am + m(m-1)d] - [2an + n(n-1)d] = 2n - 2m$$

$$\Rightarrow 2a\,(m-n) + [(m^2 + n^2) - (m-n)]d = 2(n-m)$$

$$\Rightarrow 2a\,(m-n) + [(m+n)\,(m-n) - (m-n)]d = -2(m-n)$$

$$\Rightarrow (m-n)\,[2a + \{(m+n) - 1\}d] = -2\,(m-n)$$

$$\Rightarrow \quad 2a + (m+n-1)d = -2$$

$$\Rightarrow \quad \frac{m+n}{2}[2a + (m+n-1)d] = -2 \times \frac{m+n}{2}$$

$$\left[\text{Multiplying } \frac{m+n}{2} \text{ both sides}\right]$$

$$\Rightarrow \quad S_{m+n} = -(m+n) \qquad \textbf{Hence Proved.}$$

OR

Total number of terms = 11 *i.e.*, odd

$$\therefore \quad \text{Middle term} = \left(\frac{11+1}{2}\right)^{\text{th}} \text{term}$$

$$= 6^{\text{th}} \text{ term}$$

$$\Rightarrow \quad a_6 = 30$$

Let a be the first term and d be the common difference.

Then, $\quad a + 5d = 30 \qquad ...(i)$

Now, $S_{11} = \frac{11}{2}[2a+(11-1)d]$

$S_{11} = \frac{11}{2}[2a+10d]$

$S_{11} = 11\,[a+5d]$

$S_{11} = 11 \times 30$ [Using (i)]

$= 330$ **Ans.**

33. A fast train takes 3 hours less than a slow train for a journey of 600 km. If the speed of the slow train is 10 km/h less than that of the fast train, find the speed of each train. [3]

Sol. Let the speed of the slow train be x km/hr.

$\therefore$ Speed of fast train $= (x + 10)$ km/hr

We know, $\text{Time} = \frac{\text{Distance}}{\text{Speed}}$

$\therefore$ Time taken by slow train to cover 600 km $= \frac{600}{x}$

Time taken by fast rain to cover 600 km $= \frac{600}{x+10}$

Given,

Time taken by slow train – Time taken by fast train $= 3$ hours

$\Rightarrow \frac{600}{x} - \frac{600}{x+10} = 3$

$\Rightarrow \frac{600x + 6000 - 600x}{x(x+10)} = 3$

$\Rightarrow \frac{6000}{x(x+10)} = 3$

$\Rightarrow \frac{2000}{x(x+10)} = 1$

$\Rightarrow x^2 + 10x - 2000 = 0$

$\Rightarrow x^2 + (50-40)x - 2000 = 0$

$\Rightarrow x^2 + 50x - 40x - 2000 = 0$

$\Rightarrow x(x+50) - 40\,(x+50) = 0$

$\Rightarrow (x-40)\,(x+50) = 0$

$\Rightarrow x = 40$ or $x = -50$ (Rejected, as speed cannot be ngative)

$\therefore$ Speed of slow train $= 40$ km/hr

And, Speed of fast train $= 50$ km/hr **Ans.**

34. If $1 + \sin^2\theta = 3\sin\theta\cos\theta$, prove that $\tan\theta = 1$ or $\frac{1}{2}$. [3]

Sol. Given : $1 + \sin^2\theta = 3\sin\theta\cos\theta$

Dividing both sides by $\cos^2\theta$, we get

$\sec^2\theta + \tan^2\theta = 3\tan\theta$

$\Rightarrow (\tan^2\theta + 1) + \tan^2\theta = 3\tan\theta$

$[\because \sec^2\theta = \tan^2\theta + 1]$

$\Rightarrow 1 + 2\tan^2\theta = 3\tan\theta$

$\Rightarrow 2\tan^2\theta - 3\tan\theta + 1 = 0$

$\Rightarrow 2\tan^2\theta - 2\tan\theta - \tan\theta + 1 = 0$

$\Rightarrow 2\tan\theta\,(\tan\theta - 1) - 1\,(\tan\theta - 1) = 0$

$\Rightarrow (\tan\theta - 1)\,(2\tan\theta - 1) = 0$

$\Rightarrow \tan\theta = 1$ or $\frac{1}{2}$

Hence Proved.

SECTION–D

39. Draw two tangents to a circle or radius 4 cm, which are inclinded to each other at an angle of 60°. [4]

OR

Construct a triangle ABC with sides 3 cm, 4 cm and 5 cm. Now, construct another triangle whose sides are $\frac{4}{5}$ times the corresponding sides of ΔABC.

Sol. Steps of construction :

(i) Draw a circle with O as centre and radius = 4 cm.

(ii) Take a point A on the circumference of the circle and join OA.

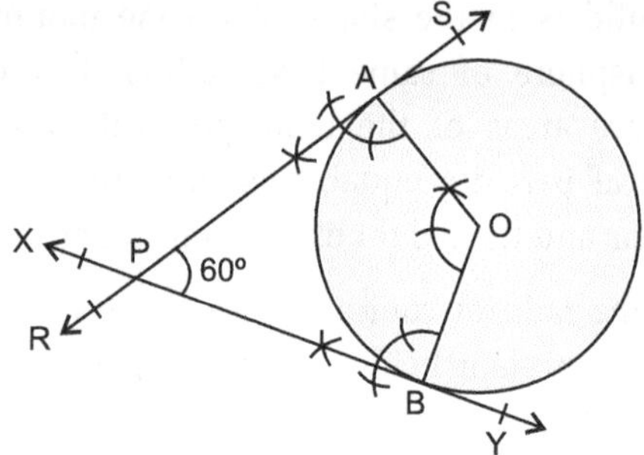

(iii) Construct $\angle$AOB = 120° such that point B is on circumference of the circle.

(iv) Draw RS perpendicular to OA and XY perpendicular to OB.

Let XY and RS intersect each other at P.

Hence, AP and BP are the tangents inclined at 60°.

OR

Steps of construction :

(i) Draw a line segment BC = 5 cm.

(ii) Taking B as centre, draw an arc of radius 3 cm.

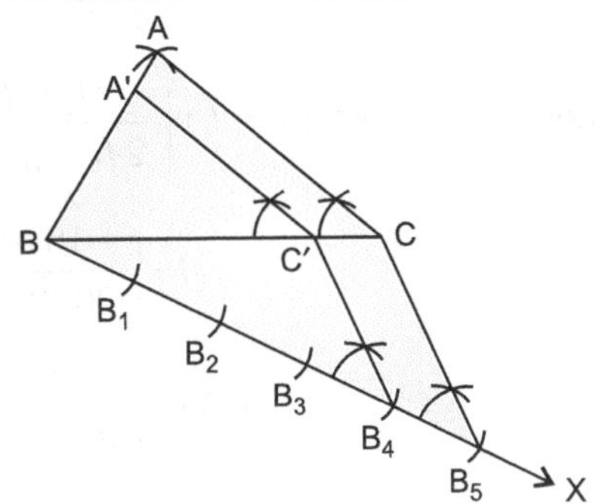

(iii) Taking C as centre, draw an another arc of radius 4 cm. Let the two arcs intersect each other at point A.

(iv) Join AB and AC. Thus, ΔABC is obtained.

(v) Draw a ray BX at an acute angle with side BC.

(vi) Locate 5 points B_1, B_2, B_3, B_4 and B_5 on BX such that $BB_1 = B_1B_2 = B_2B_3 = B_3B_4 = B_4B_5$.

(vii) Join B_5C and draw a line through B_4, parallel to B_5C, to intersect BC at C′.

(viii) Draw a line C′A′, parallel to CA, to intersect BA that A′.

Thus, ΔA′BC′ is the required triangle whose sides are $\frac{4}{5}$ times the cooresponding sides of ΔABC.

40. The angle of elevation of the top of a building from the foot of a tower is 30° and the angle of elevation of the top of a tower from the foot of the building is 60°. If the tower is 50 m high, then find the height of the building. **[4]**

Sol. Let AB be the building and CD be the tower.

$\therefore$ CD = 50 m $\angle ACB = 30°$ and $\angle DBC = 60°$.

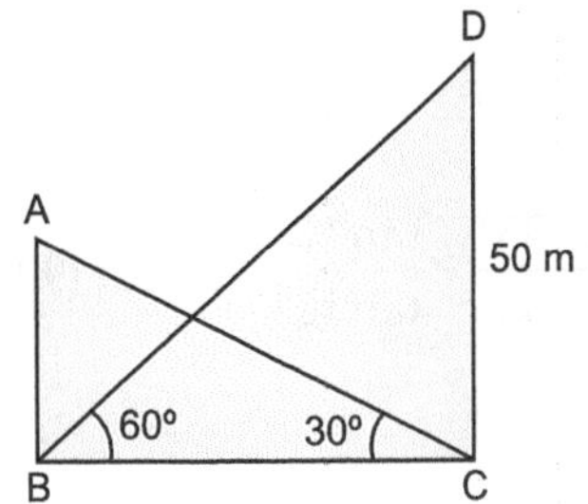

Now, in a right angled triangle DBC,

$$\tan 60° = \frac{DC}{BC}$$

$$\Rightarrow \quad \sqrt{3} = \frac{50}{BC}$$

$$\Rightarrow \quad BC = \frac{50}{\sqrt{3}}$$

And, in a right angled triangle ABC

$$\tan 30° = \frac{AB}{BC}$$

$$\Rightarrow \quad \frac{1}{\sqrt{3}} = \frac{AB}{50/\sqrt{3}}$$

$$\Rightarrow \quad \frac{1}{\sqrt{3}} = AB\times\frac{\sqrt{3}}{50}$$

$$\Rightarrow \quad AB = \frac{1}{\sqrt{3}}\times\frac{50}{\sqrt{3}} = \frac{50}{3}$$

$$\Rightarrow \quad AB = 16.67\text{m}$$

Hence, the height of the building is 16.67 m. **Ans.**

Outside Delhi [Set-III]

Note : Except for the following questions, all the remaining questions have been asked in previous sets.

SECTION–A

15. The value of sin 32° cos 58° + cos 32° sin 58° is .... **[1]**

OR

The value of $\frac{\tan 35°}{\cot 55°} + \frac{\cot 78°}{\tan 12°}$ is ....

Sol. 1

Explanation : sin 32° cos 58° + cos 32° sin 58°

$= \sin 32° \cos (90° - 32°) + \cos 32° \sin (90° - 32°)$

$= \sin 32° \sin 32° + \cos 32° \cos 32°$

$$\begin{bmatrix} \because \quad \cos(90° - \theta) = \sin\theta \\ \text{and } \sin(90° - \theta) = \cos\theta \end{bmatrix}$$

$= \sin^2 32° + \cos^2 32°$ $\quad [\because \sin^2\theta + \cos^2\theta = 1]$

$= 1.$ **Ans.**

OR

2

Explanation : $\frac{\tan 35°}{\cot 55°} + \frac{\cot 78°}{\tan 12°}$

$$= \frac{\tan 35°}{\cot(90° - 35°)} + \frac{\cot(90° - 12°)}{\tan 12°}$$

$$= \frac{\tan 35°}{\tan 35°} + \frac{\tan 12°}{\tan 12°} \quad [\because \cot(90° - \theta) = \tan\theta]$$

$= 1 + 1 = 2$ **Ans.**

19. Find the area of the sector of a circle of radius 6 cm whose central angle is 30°. (Take $\pi = 3.14$) **[1]**

Sol. Given : Radius of sector (r) = 6 cm and central angle (θ) = 30°.

$$\text{Area of sector of a circle} = \frac{\theta}{360°} \times \pi r^2$$

$$= \frac{30°}{360°} \times 3.14 \times 6^2$$

$$= 3 \times 3.14 = 9.42 \text{ cm}^2$$

Ans.

20. Find the class marks of the classes 20 – 50 and 35 – 60. **[1]**

Sol. We know,

$$\text{Class mark} = \frac{\text{Lower limit} + \text{Upper limit}}{2}$$

$\therefore$ Class mark of 20 – 50

$$= \frac{20+50}{2} = \frac{70}{2} = 35$$

and, class mark of 35 – 60 $= \frac{35+60}{2} = \frac{95}{2} = 47.5$

Ans.

SECTION–B

25. Find the mode of the following frequency distribution : **[2]**

Class	Frequency
15 – 20	3
20 – 25	8
25 – 30	9
30 – 35	10
35 – 40	3
40 – 45	2

Sol. Here, the maximum frequency is 10 which lies in the class interval 30 – 35.

$\therefore$ Modal class = 30 – 35.

So, $l = 30, f_1 = 10, f_0 = 9, f_2 = 3$ and $h = 5$.

We know, $$\text{Mode} = l + \left(\frac{f_1 - f_0}{2f_1 - f_0 - f_2}\right) \times h$$

$$= 30 + \left(\frac{10-9}{20-9-3}\right) \times 5$$

$$= 30 + \left(\frac{1}{20-12}\right) \times 5$$

$$= 30 + \left(\frac{1}{8}\right) \times 5$$

$$= 30 + \frac{5}{8}$$

$$= 30 + 0.625$$

$= 30.625$ **Ans.**

26. From a solid right circular cylinder of height 14 cm and base radius 6 cm, a right circular cone of same height and same base radius is removed. Find the volume of the remaining solid. **[2]**

Sol. Given : Height of cylinder (h) = 14 cm

And, Base radius of cylinder (r) = 6 cm

Radius and height of right circular cone is same as that of the cylinder.

$\therefore$ Height of cone = 14 cm

and radius of cone = 6 cm

Now, Volume of remaining solid

= Volume of cylinder – Volume of cone

$$= \pi r^2 h - \frac{1}{3}\pi r^2 h$$

$$= \frac{2}{3}\pi r^2 h$$

$$= \frac{2}{3} \times \frac{22}{7} \times 6^2 \times 14$$

$= 1056 \text{ cm}^3$ **Ans.**

SECTION–C

32. Which term of the A.P. 20, $19\frac{1}{4}, 18\frac{1}{2}, 17\frac{3}{4}, \ldots$ is the first negative term. **[3]**

OR

Find the middle term of the A.P. 7, 13, 19, ... 247.

Sol. Given A.P. is $20, 19\frac{1}{4}, 18\frac{1}{2}, 17\frac{3}{4}, ...$

$\therefore$ First term $(a) = 20$

and common difference (d)

$$= 19\frac{1}{4} - 20$$

$$= \frac{77}{4} - 20 = \frac{77-80}{4}$$

$$= -\frac{3}{4}$$

Let n^{th} term be the first negative term

So, $a_n < 0$

$\Rightarrow \quad a + (n-1)d < 0$

$\Rightarrow \quad 20 + (n-1)\left(-\frac{3}{4}\right) < 0$

$\Rightarrow \quad 20 - \frac{3n}{4} + \frac{3}{4} < 0$

$\Rightarrow \quad \frac{83}{4} - \frac{3n}{4} < 0$

$\Rightarrow \quad \frac{3n}{4} > \frac{83}{4}$

$\Rightarrow \quad 3n > 83$

$\Rightarrow \quad n > \frac{83}{3}$

$\Rightarrow \quad n > 27\frac{2}{3}$

So, $n = 28$

Hence, 28th term is the first negative term. **Ans.**

OR

Given A.P. is 7, 13, 19, ..., 247.

$\therefore a = 7$ and $d = 13 - 7 = 6$

Let n be the number of terms in the given A.P.

So, $a_n = a + (n-1)d$

$\Rightarrow \quad 247 = 7 + (n-1)6$

$\Rightarrow \quad 240 = (n-1)6$

$\Rightarrow \quad n - 1 = 40$

$\Rightarrow \quad n = 41$

So, number of terms are odd.

Hence, middle term $= \frac{n+1}{2} = \frac{41+1}{2} = 21$

$\therefore$ Middle term $= a_{21}$

$$= a + (21-1)d$$

$$= 7 + 20 \times 6$$

$$= 7 + 120$$

$$= 127$$ **Ans.**

33. Water in a canal, 6 m wide and 1.5 m deep, is flowing with a speed of 10 km/h. How much area will it irrigate in 30 minutes, if 8 cm standing water is required? [3]

Sol. Canal is in the shape of cuboid with breadth = 6 m and height = 1.5 m.

Also, speed of water = 10 km/hr

$\Rightarrow$ Length of water flowing in 1 hour = 10 km

$\Rightarrow$ Length of canal in $\frac{1}{2}$ hours (30 minutes)

= 5 km = 5000 m

Volume of water flow in canal

= Length×Breadth×Height

$= 5000 \times 6 \times 1.5 \text{ m}^3$

Now, Volume of water in canal

= Volume of area irrigated

$\Rightarrow$ Volume of water in canal

= Area irrigated × Height

$\Rightarrow 5000 \times 6 \times 1.5 \text{ m}^3$ = Area irrigated $\times \frac{8}{100}$ m

$\Rightarrow$ Area irrigated $= \frac{5000 \times 6 \times 1.5 \times 100}{8} \text{ m}^2$

$= 562500 \text{ m}^2$ **Ans.**

34. Show that :

$$\frac{\cos^2(45^\circ + \theta) + \cos^2(45^\circ - \theta)}{\tan(60^\circ + \theta)\tan(30^\circ - \theta)} = 1.$$ **[3]**

Sol. L.H.S. $= \frac{\cos^2(45^\circ + \theta) + \cos^2(45^\circ - \theta)}{\tan(60^\circ + \theta)\tan(30^\circ - \theta)}$

$$= \frac{\cos^2\{90^\circ - (45^\circ - \theta)\} + \cos^2(45^\circ - \theta)}{\tan\{90^\circ - (30^\circ - \theta)\}\tan(30^\circ - \theta)}$$

$$= \frac{\sin^2(45^\circ - \theta) + \cos^2(45 - \theta)}{\cot(30^\circ - \theta)\tan(30^\circ - \theta)}$$

$\left[\because \begin{array}{l}\cos(90^\circ - \theta) = \sin\theta \\ \tan(90^\circ - \theta) = \cot\theta\end{array}\right]$

$= 1 \quad \left[\because \begin{array}{l}\sin^2\theta + \cos^2\theta = 1 \\ \cot\theta.\tan\theta = 1\end{array}\right]$

= R.H.S. **Hence Proved.**

SECTION–D

39. Draw a circle of radius 3.5 cm. From a point P, 6 cm from its centre, draw two tangents to the circle. [4]

OR

Construct a ΔABC with AB = 6 cm, BC = 5 cm and ∠B = 60°. Now construct another triangle whose sides are $\frac{2}{3}$ times the corresponding sides of ΔABC.

Sol. Steps of construction :

(i) Draw a circle with centre O and radius 3.5 cm.

(ii) Take a point P outside the circle at a distance 6 cm from centre O and join OP.

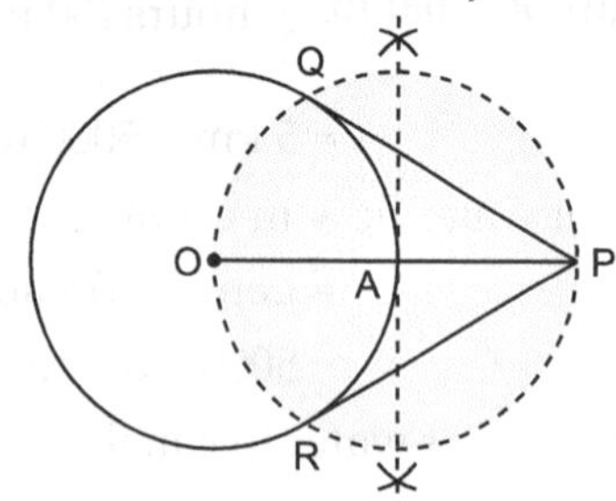

(iii) Draw perpendicular bisector of OP to get its mid-point A.

(iv) Taking A as centre and OA (or AP) as radius, draw a circle which intersects the circle of radius 3.5 cm at points Q and R.

(v) Join PQ and PR.

Thus, PQ and PR are the required tangents.

OR

Steps of construction :

(i) Draw a line segment BC = 5 cm.

(ii) At point B, construct ∠MBC = 60°.

(iii) Taking B as centre, draw an arc of radius 6 cm, which intersect ray BM at point A.

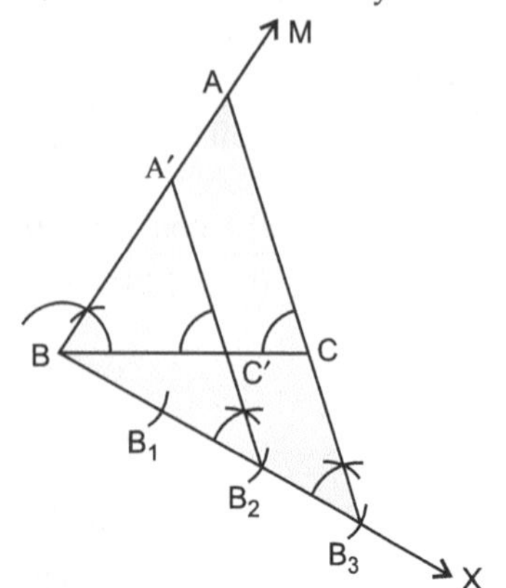

(iv) Join AC.

Thus, ΔABC is obtained.

(v) Draw a ray BX at an acute angle with side BC.

(vi) Locate 3 points B_1, B_2 and B_3 on BX such that $BB_1 = B_1B_2 = B_2B_3$.

(vii) Join B_3C and draw a line throuhg B_2, parallel to B_3C, to intersect BC at C′.

Draw a line C′A′, parallel to CA, to intersect BA at A′.

Thus, ΔA′BC′ is the required triangle, whose sides are $\frac{2}{3}$ times the corresponding sides of ΔABC.

40. A solid is in the shape of a hemisphere surmounted by a cone. If the radius of hemisphere and base radius of cone is 7 cm and height of cone is 3.5 cm, find the volume of the solid. $\left(\text{Take } \pi = \frac{22}{7}\right)$ [4]

Sol. Given : Radii of hemisphere and cone (r) = 7 cm

And, Height of the cone (h) = 3.5 cm = $\frac{7}{2}$cm

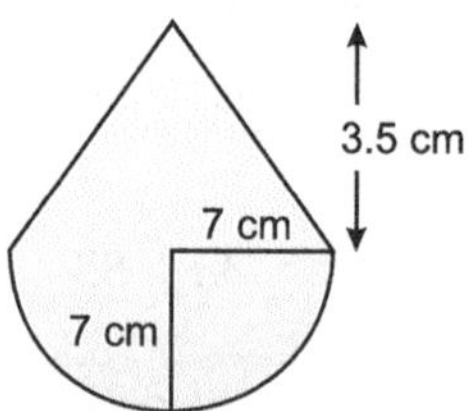

Now, Volume of solid

= Volume of hemisphere + Volume of cone

$$= \frac{2}{3}\pi r^3 + \frac{1}{3}\pi r^2 h = \frac{1}{3}\pi r^2(2r + h)$$

$$= \frac{1}{3} \times \frac{22}{7} \times 7 \times 7\left(2 \times 7 + \frac{7}{2}\right)$$

$$= \frac{154}{3} \times \frac{35}{2} = \frac{77 \times 35}{3}$$

= 898.33 cm^3 **Ans.**

MATHEMATICS (STANDARD) 2020
Delhi [Set-I]

Time allowed : 3 hours Maximum Marks : 80

Note : Except for the following questions, all the remaining questions have been asked in outside delhi sets

SECTION – A

Q. Nos. 1 to 10 are multiple choice type questions of 1 mark each. Select the correct option.

1. If one of the zeroes of the quadratic polynomial $x^2 + 3x + k$ is 2, then the value of k is [1]

(a) 10 (b) – 10
(c) – 7 (d) – 2

Sol. Let $p(x) = x^2 + 3x + k$

$\because x = 2$ is a zero of $p(x)$

$$\therefore \quad p(2) = 0$$
$$\Rightarrow \quad (2)^2 + 3(2) + k = 0$$
$$\Rightarrow \quad 4 + 6 + k = 0$$
$$\Rightarrow \quad 10 + k = 0$$
$$\Rightarrow \quad k = -10$$

$\therefore$ Option (b) is correct. **Ans.**

2. The total number of factors of a prime number is [1]

(a) 1 (b) 0
(c) 2 (d) 3

Sol. As we know that prime numbers are the numbers with two factors, 1 and the number itself.

$\therefore$ Total number of factors of a prime number will be 2.

$\therefore$ Option (c) is correct. **Ans.**

3. The quadratic polynomial, the sum of whose zeroes is – 5 and their product is 6, is [1]

(a) $x^2 + 5x + 6$ (b) $x^2 - 5x + 6$
(c) $x^2 - 5x - 6$ (d) $-x^2 + 5x + 6$

Sol. Let the polynomial be

$$p(x) - k\,[x^2 - (\text{Sum of zeroes})\,x + \text{Product of zeroes}]$$

Given, Sum of zeroes = – 5

Product of zeroes = 6

$\therefore$ Required polynomial is

$$p(x) = k[x^2 - (-5)x + (6)]$$
$$= k[x^2 + 5x + 6]$$

Let $k = 1$

$$\therefore \quad p(x) = x^2 + 5x + 6$$

$\therefore$ Option (a) is correct. **Ans.**

4. The value of k for which the system of equations $x + y - 4 = 0$ and $2x + ky = 3$, has no solution, is [1]

(a) – 2 (b) $\neq 2$
(c) 3 (d) 2

Sol. Given,

$$x + y - 4 = 0$$
$$2x + ky - 3 = 0$$
$$a_1 = 1, b_1 = 1, c_1 = -4$$
$$a_2 = 2, b_2 = k, c_2 = -3$$

For no solution, the condition is

$$\frac{a_1}{a_2} = \frac{b_1}{b_2} \neq \frac{c_1}{c_2}$$
$$\Rightarrow \quad \frac{1}{2} = \frac{1}{k} \neq \frac{-4}{-3}$$
$$\Rightarrow \quad \frac{1}{2} = \frac{1}{k} \quad \text{and} \quad \frac{1}{k} \neq \frac{4}{3}$$
$$\Rightarrow \quad k = 2 \quad \text{and} \quad k \neq \frac{3}{4}$$
$$\Rightarrow \quad k = 2$$

$\therefore$ Option (d) is correct. **Ans.**

5. The HCF and the LCM of 12, 21, 15 respectively are [1]

(a) 3, 140 (b) 12, 420
(c) 3, 420 (d) 420, 3

Sol.

2	12
2	6
3	3
	1

3	21
7	7
	1

3	15
5	5
	1

$\therefore$ Prime factors of 12, 21 and 15 are

$$12 = 2 \times 2 \times 3 = 2^2 \times 3^1$$
$$21 = 3 \times 7 = 3^1 \times 7^1$$
$$15 = 3 \times 5 = 3^1 \times 5^1$$

$\Rightarrow$ HCF (12, 21, 15) $= 3^1 = 3$

$\Rightarrow$ LCM (12, 21, 15) $= 2^2 \times 3^1 \times 5^1 \times 7^1 = 420$

$\therefore$ Option (c) is correct. **Ans.**

6. The value of x for which $2x$, $(x + 10)$ and $(3x + 2)$ are the three consecutive terms of an A.P. is [1]

(a) 6 **(b) – 6**

(c) 18 **(d) – 18**

Sol. Let $a_1 = 2x$, $a_2 = (x + 10)$, $a_3 = (3x + 2)$

We know, $a_2 - a_1 = a_3 - a_2$

$$(x + 10) - 2x = (3x + 2) - (x + 10)$$

$$\Rightarrow \quad 3x = 18$$

$$\Rightarrow \quad x = \frac{18}{3} = 6$$

$$\Rightarrow \quad x = 6$$

$\therefore$ Option (a) is correct. **Ans.**

7. The first term of an A.P. is p and the common difference is q, then its 10th term is [1]

(a) $q + 9p$ **(b) $p - 9q$**

(c) $p + 9q$ **(d) $2p + 9q$**

Sol. We know that

$$a_n = a + (n - 1)d$$

Here $a = p$, $d = q$, $n = 10$

$$\therefore \quad 10^{th} \text{ term}, a_{10} = p + (10 - 1)(q)$$

$$\Rightarrow \quad a_{10} = p + 9q$$

$\therefore$ Option (c) is correct. **Ans.**

8. The distance between the points

$(a \cos\theta + b \sin\theta, 0)$ and $(0, a \sin\theta - b \cos\theta)$, is [1]

(a) $a^2 + b^2$ **(b) $a^2 - b^2$**

(c) $\sqrt{a^2 + b^2}$ **(d) $\sqrt{a^2 - b^2}$**

Sol. The distance formula,

$$d = \sqrt{(x_2 - x_1)^2 + (y_2 - y_1)^2}$$

Here $x_1 = a \cos\theta + b \sin\theta$, $y_1 = 0$

$x_2 = 0$, $y_2 = a \sin\theta - b \cos\theta$

$$\therefore d = \sqrt{[0 - (a\cos\theta + b\sin\theta)]^2 + [(a\sin\theta - b\cos\theta) - 0]^2}$$

$$= \sqrt{(a\cos\theta + b\sin\theta)^2 + (a\sin\theta - b\cos\theta)^2}$$

$$= \sqrt{\begin{aligned}&a^2\cos^2\theta + b^2\sin^2\theta + 2ab\sin\theta\cos\theta + a^2\sin^2\theta\\ &+ b^2\cos^2\theta - 2ab\sin\theta\cos\theta\end{aligned}}$$

$$= \sqrt{a^2(\sin^2\theta + \cos^2\theta) + b^2(\sin^2\theta + \cos^2\theta)}$$

$$= \sqrt{a^2 \times 1 + b^2 \times 1} \qquad \{\because \sin^2\theta + \cos^2\theta = 1\}$$

$$= \sqrt{a^2 + b^2}$$

$\therefore$ Option (c) is correct. **Ans.**

9. If the point P(k, 0) divides the line segment joining the points A(2, – 2) and B(– 7, 4) in the ratio 1 : 2, then the value of k is [1]

(a) 1 **(b) 2**

(c) – 2 **(d) – 1**

Sol.

(m) (n)
1 : 2
A (2, –2) (x_1, y_1) P (k, 0) (x, y) B (–7, 4) (x_2, y_2)

By section formula,

$$x = \frac{mx_2 + nx_1}{m + n}, \; y = \frac{my_2 + ny_1}{m + n}$$

Here $x = k$, $m = 1$, $n = 2$, $x_1 = 2$, $x_2 = -7$

$\therefore$ we get,

$$k = \frac{(1)(-7) + (2)(2)}{1 + 2}$$

$$\Rightarrow \quad k = \frac{-7 + 4}{3}$$

$$\Rightarrow \quad k = \frac{-3}{3}$$

$$\Rightarrow \quad k = -1$$

$\therefore$ Option (d) is correct. **Ans.**

10. The value of p, for which the points A(3, 1), B(5, p) and C(7, – 5) are collinear, is [1]

(a) – 2 **(b) 2**

(c) – 1 **(d) 1**

Sol. Given,

$$A(3, 1) \Rightarrow x_1 = 3, y_1 = 1$$

$$B(5, p) \Rightarrow x_2 = 5, y_2 = p$$

$$C(7, -5) \Rightarrow x_3 = 7, y_3 = -5.$$

If A, B and C are collinear points,

then, ar(ΔABC) = 0

$$\Rightarrow \frac{1}{2}[x_1(y_2 - y_3) + x_2(y_3 - y_1) + x_3(y_1 - y_2)] = 0$$

$$\Rightarrow \quad \frac{1}{2}[3(p - (-5)) + 5(-5 - 1) + 7(1 - p)] = 0$$

$$\Rightarrow \quad \frac{1}{2}[3p + 15 - 25 - 5 + 7 - 7p] = 0$$

$$\Rightarrow \quad \frac{1}{2}[-4p - 8] = 0$$

$\Rightarrow \quad -2p - 4 = 0$

$\Rightarrow \quad p = -2$

$\therefore$ Option (a) is correct. **Ans.**

In Q. Nos. 11 to 15, fill in the blanks. Each question is of 1 mark.

11. In Fig 1, ΔABC is circumscribing a circle, the length of BC is .... cm. **[1]**

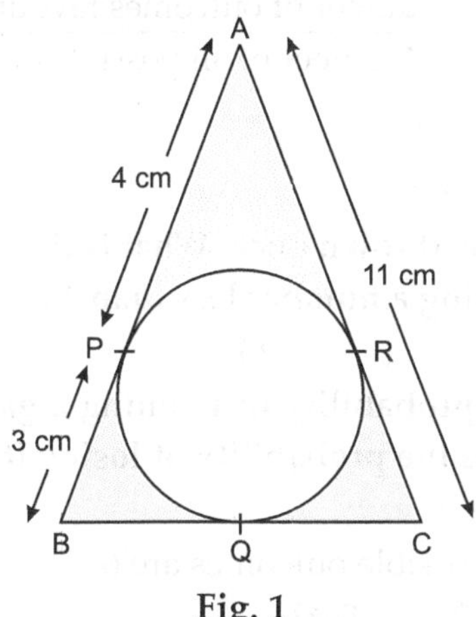

Fig. 1

Sol. 10

Explanation : We know that

The lengths of tangents drawn from an external point to a circle are equal.

$\therefore \quad BP = BQ, \quad AP = AR$

$BQ = 3 \text{ cm}, \quad AR = 4 \text{ cm}$

So, $\quad AC = AR + CR$

$11 = 4 + CR$

$CR = 7 \text{ cm}$

Now, $\quad CQ = CR$

$\Rightarrow \quad CQ = 7 \text{ cm}$

So, $\quad BC = BQ + CQ$

$= 3 + 7$

$= 10$

$BC = 10 \text{ cm.}$ **Ans.**

12. Given $\Delta ABC \sim \Delta PQR$, if $\frac{AB}{PQ} = \frac{1}{3}$, then $\frac{ar(\Delta ABC)}{ar(\Delta PQR)} =$ .... **[1]**

Sol. $\frac{1}{9}$.

Explanation : As $\Delta ABC \sim \Delta PQR$

$$\therefore \quad \frac{ar(\Delta ABC)}{ar(\Delta PQR)} = \left(\frac{AB}{PQ}\right)^2 = \left(\frac{1}{3}\right)^2 = \frac{1}{9}$$

13. ABC is an equilateral triangle of side $2a$, then length of one of its altitude is .... **[1]**

Sol. $\sqrt{3}\,a$

Explanation : We know that

For an equilateral triangle.

$$\text{Length of altitude} = \frac{\sqrt{3}}{2} \times \text{side}$$

$$= \frac{\sqrt{3}}{2} \times 2a = \sqrt{3}a$$ **Ans.**

14. $\frac{\cos 80^\circ}{\sin 10^\circ} + \cos 59^\circ \operatorname{cosec} 31^\circ =$ .... **[1]**

Sol. 2

Explanation : $\frac{\cos 80^\circ}{\sin 10^\circ} + \cos 59^\circ \operatorname{cosec} 31^\circ$

$$= \frac{\sin(90 - 80)^\circ}{\sin 10^\circ} + \cos 59^\circ \sec(90 - 31)^\circ$$

$$\left[\begin{array}{l} \because \cos\theta = \sin(90^\circ - \theta) \\ \operatorname{cosec}\theta = \sec(90^\circ - \theta) \end{array}\right]$$

$$= \frac{\sin 10^\circ}{\sin 10^\circ} + \cos 59^\circ . \sec 59^\circ$$

$$= 1 + \cos 59^\circ \times \frac{1}{\cos 59^\circ} \qquad \left[\because \sec\theta = \frac{1}{\cos\theta}\right]$$

$= 1 + 1 = 2$ **Ans.**

15. The value of $\left(\sin^2\theta + \frac{1}{1 + \tan^2\theta}\right) =$ .... **[1]**

OR

The value of

$(1 + \tan^2\theta)(1 - \sin\theta)(1 + \sin\theta) =$ ....

Sol. 1

Explanation : $\sin^2\theta + \frac{1}{1 + \tan^2\theta}$

$$= \sin^2\theta + \frac{1}{\sec^2\theta}$$

$$\{\because \sec^2\theta = 1 + \tan^2\theta\}$$

$$= \sin^2\theta + \left(\frac{1}{\sec\theta}\right)^2$$

$$= \sin^2\theta + \cos^2\theta \quad \left\{\because \cos\theta = \frac{1}{\sec\theta}\right\}$$

$= 1 \quad \{\because \sin^2\theta + \cos^2\theta = 1\}$ **Ans.**

OR

Sol. 1

Explanation : $(1 + \tan^2\theta)(1 - \sin\theta)(1 + \sin\theta)$

$= (1 + \tan^2\theta)[(1)^2 - (\sin\theta)^2]$

$= (\sec^2\theta)(1-\sin^2\theta)$ $\{\because \sec^2\theta = 1+\tan^2\theta\}$

$= \dfrac{1}{\cos^2\theta}\times\cos^2\theta$ $[\because \cos^2\theta = 1-\sin^2\theta]$

$= 1$ **Ans.**

Q. Nos. 16 to 20 are short answer type questions of 1 mark each.

16. The ratio of the length of a vertical rod and the length of its shadow is $1:\sqrt{3}$. Find the angle of elevation of the sun at that moment ? [1]

Sol. Let the angle of elevation be θ.

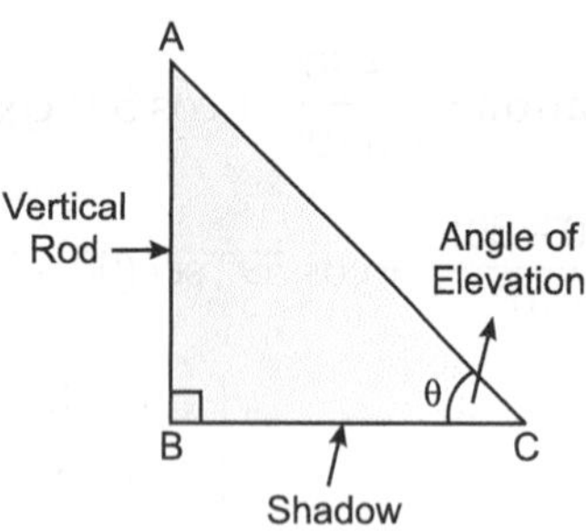

We know that,

$$\tan\theta = \frac{\text{Perpendicular}}{\text{Base}}$$

$$\Rightarrow \quad \tan\theta = \frac{AB}{BC}$$

$$= \frac{1}{\sqrt{3}}$$

$$= \tan 30°$$

$$\therefore \quad \tan\theta = \tan 30°$$

$$\Rightarrow \quad \theta = 30°$$ **Ans.**

17. Two cones have their heights in the ratio 1 : 3 and radii in the ratio 3 : 1. What is the ratio of their volumes? [1]

Sol. We have, $\dfrac{h_1}{h_2} = \dfrac{1}{3}$ and $\dfrac{r_1}{r_2} = \dfrac{3}{1}$

Now, Ratio of volumes $= \dfrac{V_1}{V_2}$

$$= \frac{\frac{1}{3}\pi r_1^2 h_1}{\frac{1}{3}\pi r_2^2 h_2}$$

$$= \left(\frac{r_1}{r_2}\right)^2\left(\frac{h_1}{h_2}\right)$$

$$= \left(\frac{3}{1}\right)^2\times\left(\frac{1}{3}\right)$$

$$= 9\times\frac{1}{3} = \frac{3}{1}$$

∴ The ratio of their volumes is 3 : 1. **Ans.**

18. A letter of English alphabet is chosen at random. What is the probability that the chosen letter is a consonant? [1]

Sol. We know,

Total number of alphabets = 26

Number of consonants = 21

$$P(E) = \frac{\text{Number of outcomes favourable to E}}{\text{Number of all possible outcomes}}$$

$$= \frac{21}{26}$$ **Ans.**

19. A die is thrown once. What is the probability of getting a number less than 3 ? [1]

OR

If the probability of winning a game is 0.07, what is the probability of losing it ?

Sol. We know,

Total possible outcomes are 6

i.e., {1, 2, 3, 4, 5, 6}

Favourable outcome are 2

i.e., {1, 2}

∴ P (getting number less than 3)

$$= \frac{\text{Number of favourable outcomes}}{\text{Number of all possible outcomes}}$$

$$= \frac{2}{6} = \frac{1}{3}$$ **Ans.**

OR

We have,

Probability of winning a game *i.e.,* P(W) = 0.07

Probability of losing a game *i.e.,* $P(\overline{W})$

$$= 1 - P(W)$$
$$= 1 - 0.07$$
$$= 0.93$$ **Ans.**

20. If the mean of the first *n* natural number is 15, then find *n*. [1]

Sol. First *n* natural numbers are 1, 2, 3, ... *n*.

$$\therefore \quad \text{Sum of numbers} = \frac{n(n+1)}{2}$$

Given, Mean = 15

$$\Rightarrow \quad \frac{\text{Sum of numbers}}{n} = 15$$

$$\Rightarrow \quad \frac{n(n+1)}{2}\times\frac{1}{n} = 15$$

$$\Rightarrow \quad n+1 = 30$$

$$n = 29$$ **Ans.**

SECTION — B

Q. No. 21 to 26 carry 2 marks each.

21. Show that $(a-b)^2$, (a^2+b^2) and $(a+b)^2$ are in AP. **[2]**

Sol. Let $a_1 = (a-b)^2$, $a_2 = (a^2+b^2)$, $a_3 = (a+b)^2$

$$\therefore \quad a_2 - a_1 = (a^2+b^2) - (a-b)^2$$
$$= a^2 + b^2 - (a^2 + b^2 - 2ab)$$
$$= a^2 + b^2 - a^2 - b^2 + 2ab$$
$$= 2ab$$
$$a_3 - a_2 = (a+b)^2 - (a^2+b^2)$$
$$= a^2 + b^2 + 2ab - a^2 - b^2$$
$$= 2ab$$
$$\therefore \quad a_2 - a_1 = a_3 - a_2$$

Hence $(a-b)^2$, (a^2+b^2) and $(a+b)^2$ are in A.P.

Hence Proved.

22. In Fig. 2, DE || AC and DC || AP. Prove that $\frac{BE}{EC} = \frac{BC}{CP}$ **[2]**

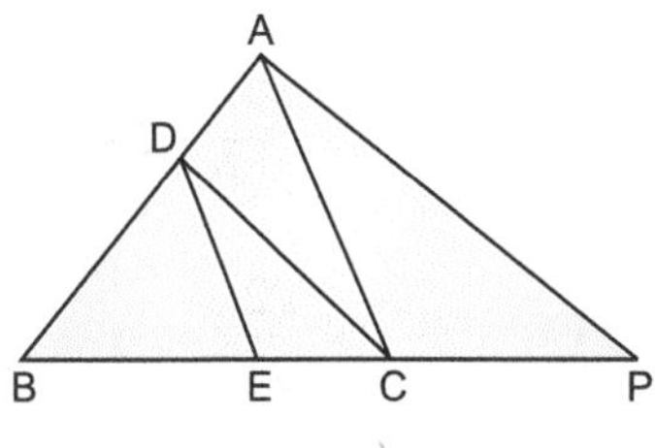

Fig. 2

OR

In Fig. 3, two tangents TP and TQ are drawn to a circle with centre O from an external point T. Prove that ∠PTQ = 2 ∠OPQ.

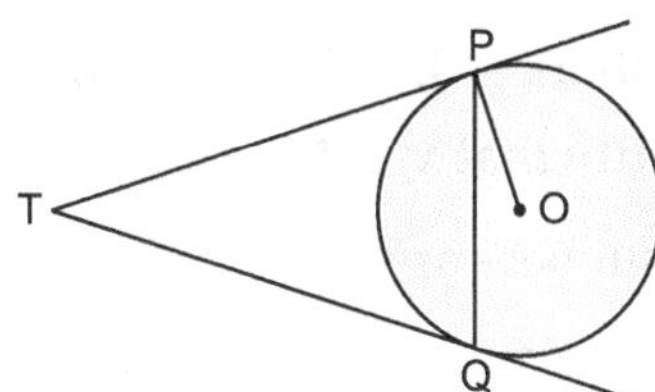

Fig. 3

Sol. Given : DE || AC and DC || AP.

To Prove : $\frac{BE}{EC} = \frac{BC}{CP}$

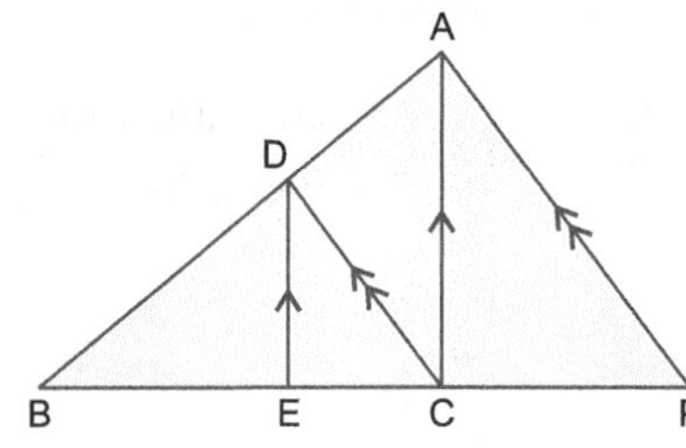

Proof : In ΔABC,

DE || AC

$$\therefore \quad \frac{BE}{EC} = \frac{BD}{AD} \quad ...(i)$$

(By Basic Proportionality Theorem)

Similarly, In ΔABP,

DC || AP

$$\therefore \quad \frac{BC}{CP} = \frac{BD}{AD} \quad ...(ii)$$

From (i) and (ii), we get

$$\frac{BE}{EC} = \frac{BC}{CP}$$ **Hence Proved.**

OR

Given : A circle with centre O, TP and TQ are two tangents from external point T.

To Prove : ∠PTQ = 2∠OPQ.

Proof : Let ∠PTQ = α

In ΔPTQ, TP = TQ

[∵ The lengths of the two tangents from an external point to a circle are equal]

∴ ∠TPQ = ∠TQP

[∵ Angles opposite to equal sides are equal]

So, ∠PTQ + ∠TPQ + ∠TQP = 180°

[Angle sum property of a triangle]

$$\Rightarrow \alpha + \angle TPQ + \angle TPQ = 180°$$
$$\Rightarrow 2\angle TPQ = 180° - \alpha$$
$$\Rightarrow \angle TPQ = \frac{1}{2}(180° - \alpha)$$
$$\Rightarrow \angle TPQ = 90° - \frac{\alpha}{2}$$

Now, OP ⊥ TP

{∵ The tangent to a circle is perpendicular to the radius through the point of contact}

$$\therefore \angle OPT = 90°$$
$$\Rightarrow \angle OPQ + \angle TPQ = 90°$$
$$\Rightarrow 90° - \frac{\alpha}{2} + \angle OPQ = 90°$$
$$\Rightarrow \angle OPQ = 90° - 90° + \frac{\alpha}{2}$$
$$\Rightarrow \angle OPQ = \frac{\alpha}{2}$$

$$\Rightarrow \quad \angle OPQ = \frac{\angle PTQ}{2}$$

or $$\angle PTQ = 2\angle OPQ$$

Hence Proved.

23. The rod AC of a TV disc antenna is fixed at right angles to the wall AB and a rod CD is supporting the disc as shown in Fig. 4. If AC = 1.5 m long and CD = 3 m, find (i) tan θ (ii) sec θ + cosec θ

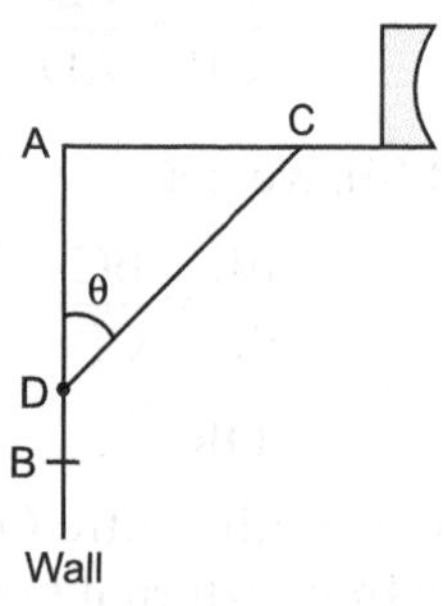

Fig. 4

Sol. Given : AC = 1.5 m and CD = 3 m

In ΔADC,

$$\sin\theta = \frac{\text{Perpendicular}}{\text{Hypotenuse}}$$

$$= \frac{AC}{CD}$$

$$= \frac{1.5}{3}$$

$$= \frac{1}{2}$$

$$\therefore \quad \sin\theta = \frac{1}{2}$$

$$\Rightarrow \quad \sin\theta = \sin 30°$$

$$\Rightarrow \quad \theta = 30°$$

(i) $$\tan\theta = \tan 30°$$

$$= \frac{1}{\sqrt{3}}$$ **Ans.**

(ii) $$\sec\theta + \text{cosec}\,\theta = \sec 30° + \text{cosec}\, 30°$$

$$\Rightarrow \quad = \frac{2}{\sqrt{3}} + \frac{2}{1}$$

$$= \frac{2+2\sqrt{3}}{\sqrt{3}}$$ **Ans.**

24. A piece of wire 22 cm long is bent into the form of an arc of a circle subtending an angle of 60° at its centre. Find the radius of the circle. $\left[\text{Use } \pi = \frac{22}{7}\right]$ **[2]**

Sol. Let the radius of the arc of the circle be r cm

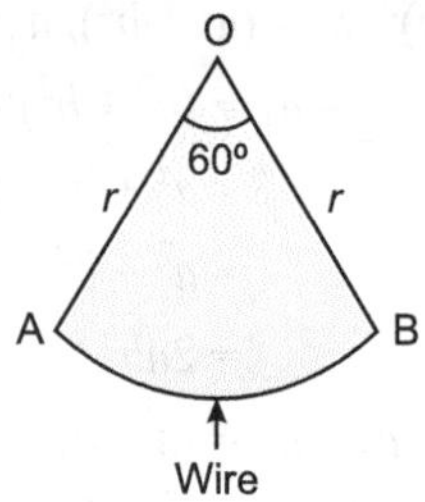

We have

Length of wire (l) = 22 cm

and, θ = 60°

We know,

$$\text{Length of arc/ wire} = \frac{\theta}{360} \times 2\pi r$$

$$= \frac{60}{360} \times 2 \times \frac{22}{7} \times r$$

$$= \frac{22}{21} r$$

$$\therefore \quad \frac{22}{21} r = 22$$

$$\Rightarrow \quad r = 21 \text{ cm}$$ **Ans.**

25. If a number x is chosen at random from the numbers – 3, – 2, – 1, 0, 1, 2, 3. What is probability that $x^2 \leq 4$? **[2]**

Sol. Given numbers are

– 3, – 2, – 1, 0, 1, 2, 3

∴ Total numbers = 7

If $x^2 \leq 4$, then favourable numbers are

0, 1, 2, – 1, – 2

∴ Favourable outcomes = 5

$$\therefore \quad P(x^2 \leq 4) = \frac{\text{Favourable outcomes}}{\text{Total possible outcomes}}$$

$$= \frac{5}{7}$$

26. Find the mean of the following distribution : **[2]**

Class	Frequency
3 – 5	5
5 – 7	10
7 – 9	10
9 – 11	7
11 – 13	8

OR

Find the mode of the following data :

Class	Frequency
0 – 20	6
20 – 40	8
40 – 60	10
60 – 80	12
80 – 100	6
100 – 120	5
120 – 140	3

Sol.

Class	Frequency (f_i)	Class Mark (x_i)	Product ($f_i \times x_i$)
3 – 5	5	4	20
5 – 7	10	6	60
7 – 9	10	8	80
9 – 11	7	10	70
11 – 13	8	12	96
	$\sum_{i=1}^{n} f_i = 40$		$\sum_{i=1}^{n} f_i x_i = 326$

$$\text{Mean } (\overline{x}) = \frac{\sum_{i=1}^{n} f_i x_i}{\sum_{i=1}^{n} f_i}$$

$$= \frac{326}{40} = 8.15$$ **Ans**

OR

Class	Frequency
0 – 20	6
20 – 40	8
40 – 60	10
60 – 80	12
80 – 100	6
100 – 120	5
120 – 140	3

Maximum frequency = 12

$\therefore$ Modal class = 60 – 80

So, $l = 60, f_1 = 12, f_0 = 10, f_2 = 6, h = 20.$

We know that,

$$\text{Mode} = l + \left(\frac{f_1 - f_0}{2f_1 - f_0 - f_2}\right) \times h$$

$$\therefore \quad \text{Mode} = 60 + \left(\frac{12-10}{2(12)-10-6}\right) \times 20$$

$$= 60 + \frac{2}{24-16} \times 20$$

$$= 60 + \frac{2}{8} \times 20$$

$$= 60 + 5 = 65$$ **Ans.**

SECTION – C

Q. Nos. 27 to 34 carry 3 marks each.

27. Find a quadratic polynomial whose zeroes are reciprocals of the zeroes of the polynomial $f(x) = ax^2 + bx + c, a \neq 0, c \neq 0$. **[3]**

OR

Divide the polynomial $f(x) = 3x^2 - x^3 - 3x + 5$ by the polynomial $g(x) = x - 1 - x^2$ and verify the division algorithm.

Sol. Given, $f(x) = ax^2 + bx + c, a \neq 0, c \neq 0.$

Let α and β be the zeroes of $f(x)$

$\therefore \alpha + \beta = -\frac{b}{a}$ and $\alpha\beta = \frac{c}{a}$

Now let the zeroes of required polynomial be α' and β'.

So, $\alpha' = \frac{1}{\alpha}, \beta' = \frac{1}{\beta}$

Now, sum of zeroes (S) = $\alpha' + \beta'$

$$= \frac{1}{\alpha} + \frac{1}{\beta}$$

$$= \frac{\beta + \alpha}{\alpha\beta}$$

$$= \frac{-b/a}{c/a}$$

$$= -\frac{b}{c}$$

Product of zeroes (P) $= \alpha' \times \beta'$

$$= \frac{1}{\alpha} \times \frac{1}{\beta}$$

$$= \frac{1}{\alpha\beta}$$

$$= \frac{1}{c/a}$$

$$= \frac{a}{c}$$

$\therefore$ Required polynomial

$$g(x) = k[x^2 - Sx + P]$$

$$= k\left[x^2 - \left(-\frac{b}{c}\right)x + \frac{a}{c}\right]$$

$$= k\left[x^2 + \frac{b}{c}x + \frac{a}{c}\right]$$

$$= k\left[\frac{cx^2 + bx + a}{c}\right]$$

$\because$ k/c = any constant

$\therefore$ $g(x) = cx^2 + bx + a$ is the required polynomial.

Ans.

OR

Given, $f(x) = 3x^2 - x^3 - 3x + 5$

$$= -x^3 + 3x^2 - 3x + 5$$

and, $g(x) = x - 1 - x^2$

$$= -x^2 + x - 1$$

Dividing $f(x)$ by $g(x)$

$$-x^2 + x - 1 \overline{) -x^3 + 3x^2 - 3x + 5} (x - 2$$

$$-x^3 + x^2 \quad - x$$

$$(+) \quad (-) \quad (+)$$

$$2x^2 - 2x + 5$$

$$2x^2 - 2x + 2$$

$$(-) \quad (+) \quad (-)$$

$$3$$

$\therefore$ $q(x) = (x - 2)$, $r(x) = 3$

Now,

Dividend = Divisor × Quotient + Remainder.

$$f(x) = g(x) \times q(x) + r(x)$$

$$= (-x^2 + x - 1)(x - 2) + 3$$

$$= -x^2(x - 2) + x(x - 2) - 1(x - 2) + 3$$

$$= -x^3 + 2x^2 + x^2 - 2x - x + 2 + 3$$

$$= -x^3 + 3x^2 - 3x + 5$$

$= f(x)$ **Hence Verified.**

28. Determine graphically the coordinates of the vertices of a triangle, the equations of whose sides are given by $2y - x = 8$, $5y - x = 14$ and $y - 2x = 1$. **[3]**

OR

If 4 is zero of the cubic polynomial $x^3 - 3x^2 - 10x + 24$, find its other two zeroes.

Sol. Given the equations,

$2y - x = 8$

x	0	-8	2
y	4	0	5

$5y - x = 14$

x	1	6	-4
y	3	4	2

$y - 2x = 1$

x	1	0	2
y	3	1	5

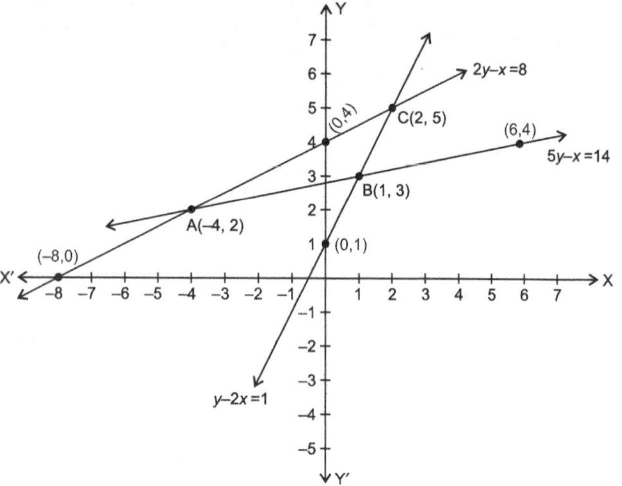

Therefore, the coordinates of the vertices of the triangle are A(– 4, 2), B(1, 3) and C(2, 5).

Ans.

OR

Let $p(x) = x^3 - 3x^2 - 10x + 24$

$\because$ $x = 4$ is a zero of $p(x)$.

$\Rightarrow$ $(x - 4)$ is a factor of $p(x)$

Dividing $p(x)$ by $(x-4)$,

$$x-4\,)\,\overline{x^3-3x^2-10x+24}\,(\,x^2+x-6$$
$$x^3-4x^2$$
$$(-)\quad(+)$$
$$\overline{\quad x^2-10x}$$
$$x^2-4x$$
$$(-)\quad(+)$$
$$\overline{\quad -6x+24}$$
$$-6x+24$$
$$(+)\quad(-)$$
$$\overline{\quad\times\quad}$$

Now, to find other zeroes of $p(x)$,

Put $\quad x^2+x-6=0$

$\Rightarrow \quad x^2+3x-2x-6=0$

$\Rightarrow \quad x(x+3)-2(x+3)=0$

$\Rightarrow \quad (x+3)(x-2)=0$

$\Rightarrow \quad x=-3, 2$

$\therefore$ -3 and 2 are the other two zeroes of the given polynomial. **Ans.**

30. Find the area of triangle PQR formed by the points P(– 5, 7), Q(– 4, – 5) and R(4, 5). **[3]**

OR

If the point C(– 1, 2) divides internally the line segment joining A(2, 5) and B(x, y) in the ratio 3 : 4, find the coordinates of B.

Sol. Given P(– 5, 7), Q(– 4, – 5), R(4, 5)

$\therefore \quad x_1=-5, x_2=-4, x_3=4$

$\quad y_1=7, y_2=-5, y_3=5$

Area of ΔPQR

$$=\frac{1}{2}\,|x_1(y_2-y_3)+x_2(y_3-y_1)+x_3(y_1-y_2)|$$

$$=\frac{1}{2}\,|(-5)(-5-5)+(-4)(5-7)+4(7-(-5))|$$

$$=\frac{1}{2}\,|(-5)(-10)-4(-2)+4(7+5)|$$

$$=\frac{1}{2}\,|50+8+48|$$

$$=\frac{1}{2}\times 106=53 \text{ sq. units.}$$ **Ans.**

OR

3 : 4

A(2, 5) (x_1, y_1) —— C(–1, 2) (x_a, y_a) —— B(x, y) (x_2, y_2)

By section formula,

$$x_a=\frac{mx_2+nx_1}{m+n}, \quad y_a=\frac{my_2+ny_1}{m+n}$$

$$\Rightarrow \quad -1=\frac{3(x)+4(2)}{3+4}, \quad 2=\frac{3(y)+4(5)}{3+4}$$

$$\Rightarrow \quad -1=\frac{3x+8}{7}, \quad 2=\frac{3y+20}{7}$$

$$\Rightarrow \quad -7=3x+8, \quad 14=3y+20$$

$$\Rightarrow \quad 3x=-15, \quad y=\frac{-6}{3}$$

$$\Rightarrow \quad x=-5, \quad y=-2$$

$\therefore$ (– 5, – 2) are the coordinates of the point B. **Ans.**

31. In Fig. 5, $\angle$D = $\angle$E and $\frac{AD}{DB}=\frac{AE}{EC}$, prove that BAC is an isosceles triangle. **[3]**

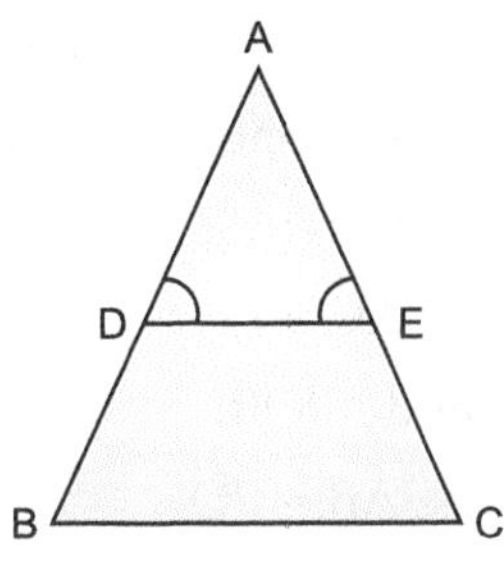

Fig. 5

Sol. Given : $\quad \angle D=\angle E$

and, $\quad \frac{AD}{DB}=\frac{AE}{EC}$

To prove : ΔBAC is an isosceles triangle

Proof : In ΔABC, $\frac{AD}{DB}=\frac{AE}{EC}$ (given)

$\Rightarrow \quad DE \parallel BC$

{By converse of Basic Proportionality theorem}

$\therefore \quad \angle ADE=\angle ABC$ ...(i)

and $\quad \angle AED=\angle ACB$ ...(ii)

But $\quad \angle ADE=\angle AED$ (Given) ...(iii)

$\therefore \quad \angle ABC=\angle ACB$

(From eq. (i), (ii) and (ii))

$\Rightarrow \quad AB=AC$

$\therefore$ ΔABC is an isosceles triangle as two of its sides are equal. **Hence Proved.**

32. In a triangle, if square of one side is equal to the sum of the squares of the other two sides, then prove that the angle opposite of the first side is a right angle. [3]

Sol. Given : ΔABC in which

$$AC^2 = AB^2 + BC^2$$

To prove : $\angle B = 90°$

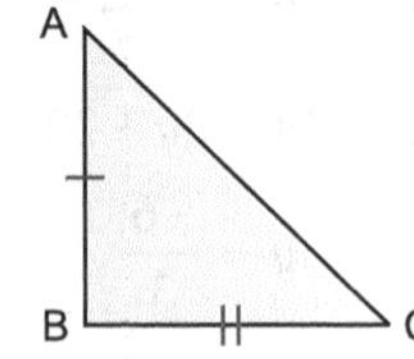

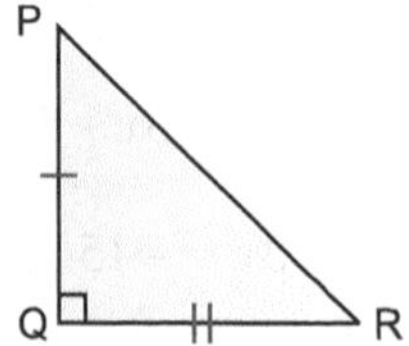

Construction : Draw ΔPQR in which $PQ = AB$, $QR = BC$ and $\angle Q = 90°$.

Proof : In ΔABC,

$\Rightarrow \quad AC^2 = AB^2 + BC^2$ (Given)

$\Rightarrow \quad AC^2 = PQ^2 + QR^2$ (Given) ...(i)

Now, $\quad PR^2 = PQ^2 + PR^2$ ...(ii)

(By Pythagoras Theorem)

From equations (i) and (ii), we get

$$AC^2 = PR^2$$

$\Rightarrow \quad AC = PR$

$\therefore$ In ΔABC and ΔPQR,

$$AB = PQ$$
$$BC = QR$$
$$AC = PR$$

$\therefore \quad \Delta ABC \cong \Delta PQR$ [By SSS rule]

$\Rightarrow \quad \angle ABC = \angle PQR$ (C.P.C.T.)

$\therefore \quad \angle ABC = 90°$ **Hence Proved.**

33. If $\sin\theta + \cos\theta = \sqrt{3}$, then prove that

$$\tan\theta + \cot\theta = 1$$ **[3]**

Sol. Given : $\quad \sin\theta + \cos\theta = \sqrt{3}$

To prove : $\tan\theta + \cot\theta = 1$

Proof : We have $\sin\theta + \cos\theta = \sqrt{3}$

Squaring both the sides, we get

$$(\sin\theta + \cos\theta)^2 = (\sqrt{3})^2$$

$\Rightarrow \sin^2\theta + \cos^2\theta + 2\sin\theta\cos\theta = 3$

$\Rightarrow \quad 1 + 2\sin\theta\cos\theta = 3$

$[\because \sin^2\theta + \cos^2\theta = 1]$

$\Rightarrow \quad \sin\theta\cos\theta = \dfrac{2}{2}$

$\Rightarrow \quad \sin\theta\cos\theta = 1$ ...(i)

Now, L.H.S. $= \tan\theta + \cot\theta$

$$= \frac{\sin\theta}{\cos\theta} + \frac{\cos\theta}{\sin\theta}$$

$$\left[\because \tan\theta = \frac{\sin\theta}{\cos\theta}; \cot\theta = \frac{\cos\theta}{\sin\theta}\right]$$

$$= \frac{\sin^2\theta + \cos^2\theta}{\sin\theta\cos\theta}$$

$$= \frac{1}{\sin\theta\cos\theta} \quad [\because \sin^2\theta + \cos^2\theta = 1]$$

$$= \frac{1}{1} = 1 = \text{R.H.S.} \quad \text{[Using eq. (i)]}$$

Hence Proved.

34. A cone of base radius 4 cm is divided into two parts by drawing a plane through the mid-point of its height and parallel to its base. Compare the volume of the two parts. [3]

Sol. $r_1 = 4$ cm, $r_2 = ?$, $h_1 = 2h$ cm, $h_2 = h$ cm.

Now, $\quad \Delta ABE \sim \Delta ACD$

(By AA Similarity Criterion)

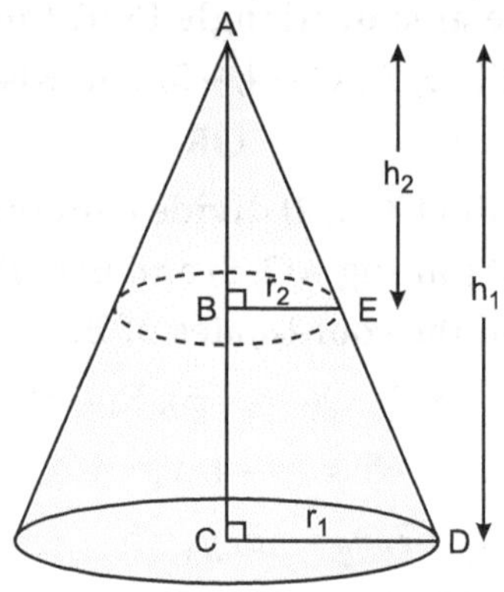

$$\therefore \quad \frac{AB}{AC} = \frac{AE}{AD} = \frac{BE}{CD}$$

{$\because$ Corresponding sides of similar triangle are proportional}

$$\Rightarrow \quad \frac{AB}{AC} = \frac{BE}{CD}$$

$$\Rightarrow \quad \frac{h_2}{h_1} = \frac{r_2}{r_1}$$

$$\Rightarrow \quad \frac{h}{2h} = \frac{r_2}{4}$$

$$\Rightarrow \quad r_2 = \frac{4 \times 1}{2} = 2$$

$$r_2 = 2 \text{ cm}$$

Ratio of volume of the two parts

$$= \frac{\text{Volume of the small cone}}{\text{Volume of the frustum of the cone}}$$

$$= \frac{V_2}{V_1} = \frac{\frac{1}{3}\pi r_2^2 h_2}{\frac{1}{3}\pi h[r_1^2 + r_2^2 + r_1 r_2]}$$

$$= \frac{r_2^2 h_2}{h(r_1^2 + r_2^2 + r_1 r_2)}$$

$$= \frac{(2)^2 \times h}{h(4^2 + 2^2 + 4 \times 2)} = \frac{4}{28} = \frac{1}{7} = 1:7$$

$\therefore$ 1 : 7 is the required ratio of the volumes of the two parts. **Ans.**

SECTION – D

Q. Nos. 35 to 40 carry 4 marks each.

35. Show that the square of any positive integer cannot be of the form $(5q + 2)$ or $(5q + 3)$ for any integer q. **[4]**

OR

Prove that one of every three consecutive positive integers is divisible by 3.

Sol. Let a be the given positive integer.

To prove : a^2 cannot be of the from $(5q + 2)$ or $(5q + 3)$ for any integer q.

Proof : By Euclid's division lemma,

$a = bm + r, 0 \le r < b$

Let $b = 5$

$\therefore r = 0, 1, 2, 3$ and 4.

$\Rightarrow a = 5m, 5m + 1, 5m + 2, 5m + 3$ and $5m + 4$.

Case I : $a = 5m$

$a^2 = (5m)^2$
$= 25m^2$
$= 5(5m)^2$
$= 5(q)$
$= 5q$, where $q = 5m^2$

Case II : $a = 5m + 1$

$a^2 = (5m + 1)^2$
$= 25m^2 + 1 + 2(5m)(1)$
$= 25m^2 + 10m + 1$
$= 5(5m^2 + 2m) + 1$
$= 5q + 1$, where $q = 5m^2 + 2m$

Case III : $a = 5m + 2$

$a^2 = (5m + 2)^2$
$= 25m^2 + 4 + 2(5m)(2)$
$= 25m^2 + 20m + 4$
$= 5(5m^2 + 4m) + 4$
$= 5q + 4$, where $q = 5m^2 + 4m$

Case IV : $a = 5m + 3$

$a^2 = (5m + 3)^2$
$= 25m^2 + 9 + 2(5m)(3)$
$= 25m^2 + 30m + 9$
$= 5(5m^2 + 6m) + 5 + 4$
$= 5(5m^2 + 6m + 1) + 4$
$= 5q + 4$, where $q = 5m^2 + 6m + 1$

Case V : $a = (5m + 4)$

$a^2 = (5m + 4)$
$= 25m^2 + 16 + 2(5m)(4)$
$= 25m^2 + 40m + 16$
$= 25m^2 + 40m + 15 + 1$
$= 5(5m^2 + 8m + 3) + 1$
$= 5q + 1$, where $q = 5m^2 + 8m + 3$.

$\therefore a^2$ is of the form $5q$, $5q + 1$, $5q + 4$ but not of the form $5q + 2$ or $5q + 3$ for any integer q.

Hence Proved.

OR

Let n, $n + 1$ and $n + 2$ are the three consecutive positive integers.

By Euclid's division lemma, we have,

$$n = mq + r, 0 \le r < m$$

Let $m = 3$

$\therefore r = 0$, 1 and 2

$\therefore n = 3q$, $3q + 1$ and $3q + 2$

Case I : $n = 3q$

So, $n + 1 = 3q + 1$

$n + 2 = 3q + 2$

In this case, only $n = 3q$ is divisible by 3 but $n + 1$ and $n + 2$ are not divisible by 3.

Case II : $n = 3q + 1$

So, $n + 1 = (3q + 1) + 1 = 3q + 2$

$n + 2 = (3q + 1) + 2 = 3q + 3 = 3(q + 1)$

In this case, only $n + 2 = 3(q + 1)$ is divisible by 3 but n and $n + 1$ are not divisible by 3.

Case III : $n = 3q + 2$

So, $n + 1 = (3q + 2) + 1 = 3q + 3 = 3(q + 1)$

$n + 2 = (3q + 2) + 2 = 3q + 4$

In this case, only $n + 1 = 3(q + 1)$ is divisible by 3 but n and $n + 2$ are not divisible by 3

Hence, one of every three consecutive positive integers is divisible by 3. **Hence Proved.**

36. The sum of four consecutive numbers in A.P. is 32 and the ratio of the product of the first and last terms to the product of two middle terms is 7 : 15. Find the number. [4]

Sol. Let the four consecutive terms of the A.P. be $a - 3d, a - d, a + d, a + 3d$.

According to the question,

$$(a - 3d) + (a - d) + (a + d) + (a + 3d) = 32$$

$$\Rightarrow \quad 4a - 3d + 3d - d + d = 32$$

$$\Rightarrow \quad 4a = 32$$

$$\Rightarrow \quad a = 8$$

Also, $$\frac{(a-3d)(a+3d)}{(a-d)(a+d)} = \frac{7}{15}$$

$$\Rightarrow \quad \frac{(a)^2-(3d)^2}{(a)^2-(d)^2} = \frac{7}{15}$$

$$\Rightarrow \quad \frac{(8)^2-9d^2}{(8)^2-d^2} = \frac{7}{15} \qquad [\because a = 8]$$

$$\Rightarrow \quad \frac{64-9d^2}{64-d^2} = \frac{7}{15}$$

$$\Rightarrow \quad 15(64 - 9d^2) = 7(64 - d^2)$$

$$\Rightarrow \quad 960 - 135d^2 = 448 - 7d^2$$

$$\Rightarrow \quad -128d^2 = -512$$

$$\Rightarrow \quad d^2 = \frac{-512}{-128}$$

$$d = \pm 2$$

For $d = 2$, four terms of A.P. are,

$$a - 3d = 8 - 3(2) = 2$$
$$a - d = 8 - 2 = 6$$
$$a + d = 8 + 2 = 10$$
$$a + 3d = 8 + 3(2) = 14$$

For $d = -2$, four terms are,

$$a - 3d = 8 - 3(-2) = 14$$
$$a - d = 8 - (-2) = 10$$
$$a + d = 8 + (-2) = 6$$
$$a + 3d = 8 + 3(-2) = 2$$

Thus, the four terms of A.P. series are 2, 6, 10, 14 or 14, 10, 6, 2. **Ans.**

37. Draw a line segment AB of length 7 cm. Taking A as centre, draw a circle of radius 3 cm and taking B as centre, draw another circle of radius 2 cm. Construct tangents to each circle from the centre of the other circle. [4]

Sol. Steps of construction :

(i) Draw AB = 7 cm.

(ii) With centres A and B and radii $r_1 = 3$ cm and $r_2 = 2$ cm respectively, draw two circles.

(iii) Draw perpendicular bisector of AB. Let M is the mid-point of AB.

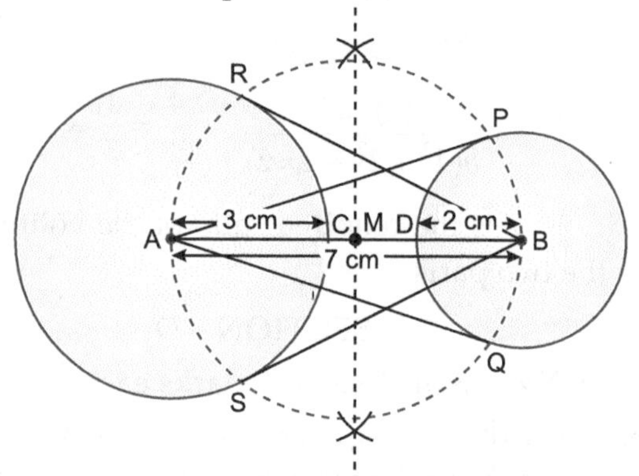

(iv) With M as centre and radius MA or MB, draw a circle cutting the pervious circles at P, Q, R and S.

(v) Join AP, AQ, BR and BS.

Then AP, AQ, BR and BS are the required tangents.

38. A vertical tower stands on a horizontal plane and is surmounted by a vertical flag-staff of height 6 m. At a point on the plane, the angle of elevation of the bottom and top of the flag-staff are 30° and 45° respectively. Find the height of the tower.(Take $\sqrt{3}$ = 1.73) [4]

Sol. Let the height of the tower AB = h.
and height of the flag-staff AD = 6 m.

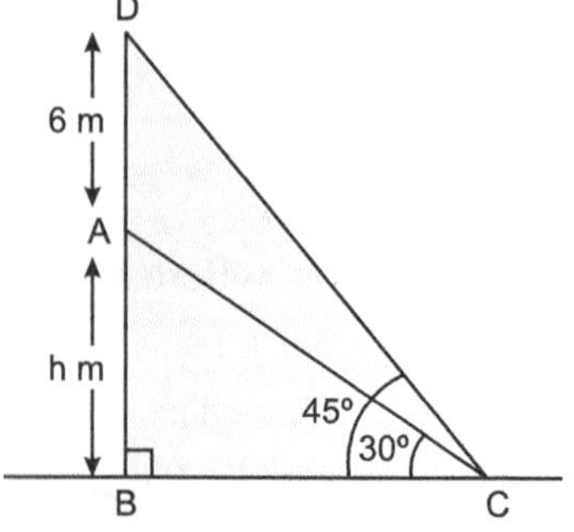

Now, In ΔABC,

$$\tan 30° = \frac{AB}{BC}$$

$$\Rightarrow \quad \frac{1}{\sqrt{3}} = \frac{h}{BC}$$

$\therefore \quad BC = h\sqrt{3}$ m ...(i)

In ΔDBC,

$$\tan 45° = \frac{DB}{BC}$$

$$\Rightarrow \quad 1 = \frac{(6+h)}{BC}$$

$\Rightarrow \quad BC = (6+h)$m ...(ii)

From (i) and (ii), we get

$$h\sqrt{3} = (6+h)$$

$$h\sqrt{3} - h = 6$$

$$h(\sqrt{3}-1) = 6$$

$$h = \frac{6}{\sqrt{3}-1}$$

$$= \frac{6}{\sqrt{3}-1} \times \frac{\sqrt{3}+1}{\sqrt{3}+1} = \frac{6(\sqrt{3}+1)}{(\sqrt{3})^2-(1)^2}$$

$$= \frac{6(\sqrt{3}+1)}{3-1} = \frac{6(\sqrt{3}+1)}{2}$$

$$= 3(\sqrt{3}+1)$$

$$= 3\,(1.73+1)$$

$$= 3(2.73)$$

$$= 8.19 \text{ m}$$

∴ 8.19 m is the height of the tower. **Ans.**

39. A bucket in the form of a frustum of a cone of height 30 cm with radii of its lower and upper ends as 10 cm and 20 cm, respectively. Find the capacity of the bucket. Also find the cost of milk which can completely fill the bucket at the rate of ₹ 40 per litre. $\left(\text{Use } \pi = \frac{22}{7}\right)$ **[4]**

Sol. Here $h = 30$ cm, $r_1 = 10$ cm and $r_2 = 20$ cm

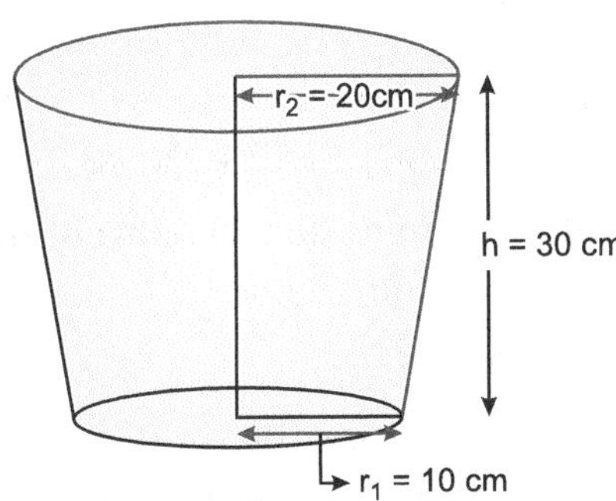

The capacity of the bucket

= Volume of the frustum of cone

$$= \frac{1}{3}\pi h[r_1^2 + r_2^2 + r_1r_2]$$

$$= \frac{1}{3} \times \frac{22}{7} \times 30 \times [10^2 + 20^2 + 10 \times 20]$$

$$= \frac{1}{3} \times \frac{22}{7} \times 30 \times [100 + 400 + 200]$$

$$= \frac{1}{3} \times \frac{22}{7} \times 30 \times 700$$

$$= 22 \times 10 \times 100 \text{ cm}^3$$

$$= 22000 \text{ cm}^3$$

$$= \frac{22000}{1000}\, l \qquad [\because 1000 \text{ cm}^3 = 1l]$$

$$= 22\, l$$

∴ Total cost of milk at ₹ 40/litre = ₹ 22 × 40

= ₹ 880

∴ ₹ 880 is the cost of milk and 22 l is the capacity of the bucket. **Ans.**

40. The median of the following data is 525. Find the values of x and y, if total frequency is 100 : **[4]**

Sol.

Class	Frequency	c.f.
0 – 100	2	2
100 – 200	5	7
200 – 300	x	$7 + x$
300 – 400	12	$19 + x$
400 – 500	17	$36 + x$
500 – 600	20	$56 + x$
600 – 700	y	$56 + x + y$
700 – 800	9	$65 + x + y$
800 – 900	7	$72 + x + y$
900 – 1000	4	$76 + x + y$
	$\Sigma f_i = 76 + x + y$	

∵ Median = 525

∴ Median class = 500 – 600

$\therefore l = 500, h = 100, c.f. = 36 + x, f = 20, \frac{n}{2} = 50$

We know that

$$\text{Median} = l + \left[\frac{\frac{n}{2} - c.f.}{f}\right] \times h$$

$$\Rightarrow \quad 525 = 500 + \left[\frac{50 - (36 + x)}{20}\right] \times 100$$

$$\Rightarrow \quad 525 = 500 + \left[\frac{50 - 36 - x}{20}\right] \times 100$$

$$\Rightarrow \quad 525 - 500 = \left(\frac{14 - x}{1}\right) \times 5$$

$$\Rightarrow \quad 25 = 70 - 5x$$

$$\Rightarrow \quad 5x = 45$$

$$\Rightarrow \quad x = 9$$

Also, $x + y + 76 = 100$

$$\Rightarrow \quad 9 + y + 76 = 100$$

$$\Rightarrow \quad y + 85 = 100$$

$$\Rightarrow \quad y = 15$$

$\therefore x = 9$ and $y = 15$. **Ans.**

Delhi [Set-II]

Note : Except for the following questions, all the remaining questions have been asked in previous set.

SECTION – A

14. $\left(\dfrac{\sin 35^\circ}{\cos 55^\circ}\right)^2 + \left(\dfrac{\cos 43^\circ}{\sin 47^\circ}\right)^2 - 2\cos 60^\circ = \ldots\ldots\ldots\ldots$ **[1]**

Sol. 1

Explanation : $\left(\dfrac{\sin 35^\circ}{\cos 55^\circ}\right)^2 + \left(\dfrac{\cos 43^\circ}{\sin 47^\circ}\right)^2 - 2\cos 60^\circ$

$$= \left[\frac{\cos(90 - 35)^\circ}{\cos 55^\circ}\right]^2 + \left[\frac{\sin(90 - 43)^\circ}{\sin 47^\circ}\right]^2 - 2\cos 60^\circ$$

$$\left[\because \begin{array}{l} \cos(90^\circ - \theta) = \sin\theta \\ \text{and, } \sin(90^\circ - \theta) = \cos\theta \end{array}\right]$$

$$= \left(\frac{\cos 55^\circ}{\cos 55^\circ}\right)^2 + \left(\frac{\sin 47^\circ}{\sin 47^\circ}\right)^2 - 2\cos 60^\circ$$

$$= (1)^2 + (1)^2 - 2 \times \frac{1}{2} \qquad \left(\because \cos 60^\circ = \frac{1}{2}\right)$$

$$= 1 + 1 - 1$$

$$= 1$$

15. ABC and BDE are two equilateral triangles such that D is the mid-point of BC. Ratio of the areas of triangles ABC and BDE is .... **[1]**

Sol. 4 : 1

Explanation :

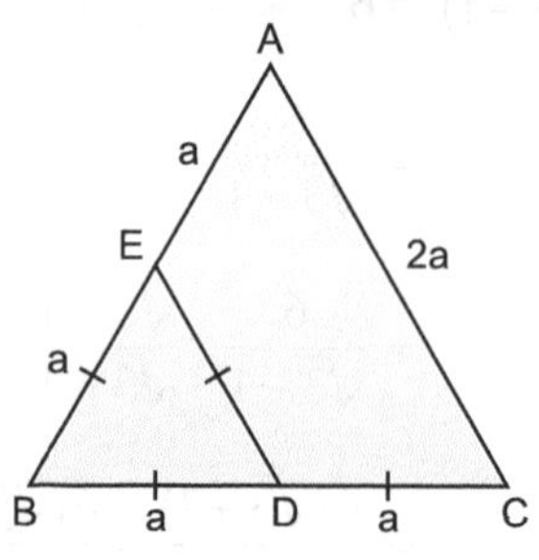

Figure – 1

$\Delta ABC \sim \Delta BDE$ ($\because$ both are equilateral)

$$\therefore \frac{ar\,(\Delta ABC)}{ar\,(\Delta BDE)} = \left(\frac{2a}{a}\right)^2$$

$$= \frac{4}{1} = 4 : 1$$

20. A die is thrown once. What is the probability of getting an even prime number? **[1]**

Sol. We know,

All possible outcomes = {1, 2, 3, 4, 5, 6}

Favourable outcomes = {2}

P(even prime number)

$$= \frac{\text{No. of favourable outcomes}}{\text{No. of all possible outcomes}}$$

$$= \frac{1}{6}$$

SECTION – B

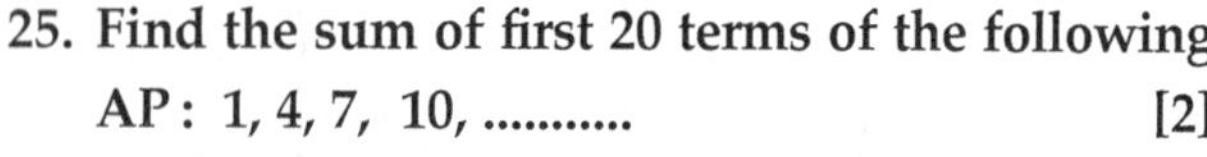

25. Find the sum of first 20 terms of the following AP : 1, 4, 7, 10, ... **[2]**

Sol. Given, A.P. is

1, 4, 7, 10, ...

$\therefore a = 1, d = 4 - 1 = 3$ and $n = 20$

We know that,

$$S_n = \frac{n}{2}[2a + (n-1)d]$$

$$\Rightarrow \quad S_{20} = \frac{20}{2}[2(1) + (20-1) \times 3]$$

$$\Rightarrow \quad S_{20} = 10 \times [2 + 57]$$

$$= 10 \times 59 = 590 \qquad \textbf{Ans.}$$

26. The perimeter of a sector of a circle of radius 5.2 cm is 16.4 cm. Find the area of the sector.[2]

Sol. We have,

Radius (r) = 5.2 cm.

Perimeter of sector (P) = 16.4 cm

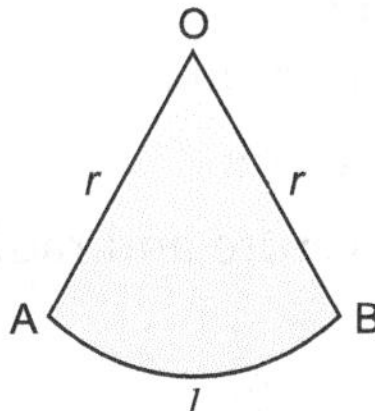

We know that,

$$P = l + 2r$$

$$\Rightarrow \quad 16.4 = l + 2(5.2)$$

$$\Rightarrow \quad 16.4 = l + 10.4$$

$$\Rightarrow \quad l = 16.4 - 10.4$$

$$l = 6 \text{ cm.}$$

Now, Area of sector $= \frac{1}{2}lr$

$$= \frac{1}{2} \times 6 \times 5.2$$

$$= 3 \times 5.2$$

$$= 15.6 \text{ cm}^2 \qquad \textbf{Ans.}$$

SECTION – C

32. A train covers a distance of 480 km at a uniform speed. If the speed had been 8 km/h less, then it would have taken 3 hours more to cover the same distance. Find the original speed of the train. [3]

Sol. Let the original speed of the train

$= x$ km/h

Total distance to be covered = 480 km.

According to the question,

$$\frac{480}{(x-8)} - \frac{480}{x} = 3$$

$$\Rightarrow \quad \frac{480x - 480(x-8)}{x(x-8)} = 3$$

$$\Rightarrow \quad \frac{480x - 480x + 3840}{x^2 - 8x} = 3$$

$$\Rightarrow \quad 3840 = 3(x^2 - 8x)$$

$$\Rightarrow \quad 3x^2 - 24x - 3840 = 0$$

$$\Rightarrow \quad x^2 - 8x - 1280 = 0$$

$$\Rightarrow \quad x^2 - 40x + 32x - 1280 = 0$$

$$\Rightarrow \quad (x - 40)(x + 32) = 0$$

$$\Rightarrow \quad (x - 40) = 0, \text{ or } (x + 32) = 0$$

$\Rightarrow x = 40$ or $x = -32$ (Neglected)

$\therefore$ Original speed of the train is 40 km/h. **Ans.**

33. Prove that the parallelogram circumscribing a circle is a rhombus. [3]

Sol. Given : A parallelogram ABCD circumscribing a circle.

To prove : ABCD is a rhombus.

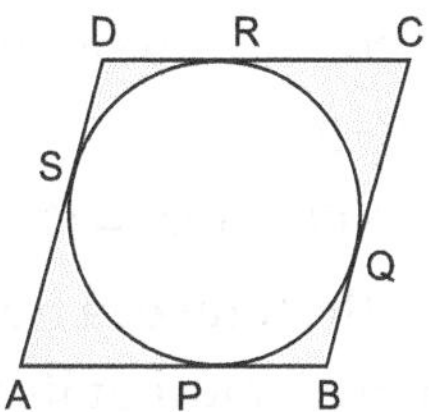

Proof : We know that, the length of two tangents from an external point to a circle are equal.

$\therefore$ for point A, AP = AS ...(i)

for point B, BP = BQ ...(ii)

for point C, CR = CQ ...(iii)

and for point D, DR = DS ...(iv)

On adding equations (i), (ii), (iii) and (iv), we get

$\Rightarrow$ AP + BP + CR + DR = AS + BQ + CQ + DS

$\Rightarrow$ (AP + BP) + (CR + DR)

= (AS + DS) + (BQ + CQ)

$\Rightarrow$ AB + CD = AD + BC

$\Rightarrow$ AB + AB = AD + AD

{$\because$ opposite sides of a parallelogram are equal *i.e.*, AB = CD, AD = BC}

$\Rightarrow$ 2AB = 2AD

$\Rightarrow$ AB = AD

$\therefore$ ABCD is a parallelogram with adjacent sides AB = AD.

Hence, it is a rhombus. **Hence Proved.**

34. Prove that : $2(\sin^6\theta + \cos^6\theta) - 3(\sin^4\theta + \cos^4\theta) + 1 = 0$. **[3]**

Sol. To prove : $2(\sin^6\theta + \cos^6\theta) - 3(\sin^4\theta + \cos^4\theta) + 1 = 0$

L.H.S. $= 2(\sin^6\theta + \cos^6\theta) - 3(\sin^4\theta + \cos^4\theta) + 1$

$= 2[(\sin^2\theta)^3 + (\cos^2\theta)^3] - 3[(\sin^2\theta)^2 + (\cos^2\theta)^2] + 1$

$= 2[(\sin^2\theta + \cos^2\theta)^3 - 3\sin^2\theta\cos^2\theta(\sin^2\theta + \cos^2\theta)] - 3[(\sin^2\theta + \cos^2\theta)^2 - 2\sin^2\theta\cos^2\theta] + 1$

$$\left\{\begin{array}{l}\because a^3 + b^3 = (a+b)^3 - 3ab(a+b) \\ \quad a^2 + b^2 = (a+b)^2 - 2ab\end{array}\right\}$$

$= 2[(1)^3 - 3\sin^2\theta\cos^2\theta(1)] - 3[(1)^2 - 2\sin^2\theta\cos^2\theta] + 1$

$\{\because \sin^2\theta + \cos^2\theta = 1\}$

$= 2[1 - 3\sin^2\theta\cos^2\theta] - 3[1 - 2\sin^2\theta\cos^2\theta] + 1$

$= 2 - 6\sin^2\theta\cos^2 - 3 + 6\sin^2\theta\cos^2\theta + 1$

$= 2 - 3 + 1 = 0 =$ R.H.S. **Hence Proved.**

SECTION – D

39. A bucket is in the form of a frustum of a cone of height 16 cm with radii of its lower and upper circular ends as 8 cm and 20 cm respectively. Find the cost of milk which can completely fill the bucket, at the rate of ₹ 40 per litre. (Use π = 3.14) [4]

Sol. Here,

$h = 16$ cm, $r_1 = 8$ cm, $r_2 = 20$ cm

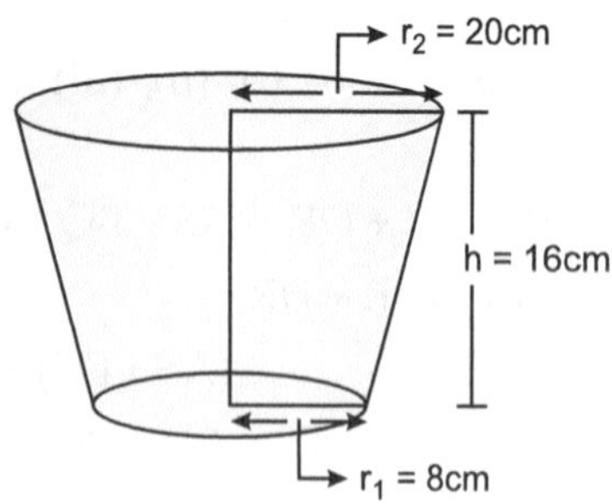

Volume of milk in the bucket = Volme of frustum of cone

$$= \frac{1}{3}\pi h[r_1^2 + r_2^2 + r_1r_2]$$

$$= \frac{1}{3} \times 3.14 \times 16 \times [8^2 + 20^2 + 8 \times 20]$$

$$= \frac{1}{3} \times 3.14 \times 16 \times [64 + 400 + 160]$$

$$= \frac{1}{3} \times 3.14 \times 16 \times 624$$

$= 3.14 \times 16 \times 208 \text{ cm}^3$

$= 10449.92 \text{ cm}^3$

$$= \frac{10449.92}{1000}\, l = 10.4492\, l \approx 10.45\, l$$

Cost of milk at ₹ 40 per litre = 10.45 × 40 = ₹ 418 **Ans.**

40. Construct a triangle with sides 4 cm, 5 cm and 6 cm. Then construct another triangle whose sides are $\frac{2}{3}$ times the corresponding sides of the first triangle **[4]**

Sol. Steps of construction :

(i) Draw BC = 6 cm

(ii) With B as centre and radius 4 cm, draw an arc.

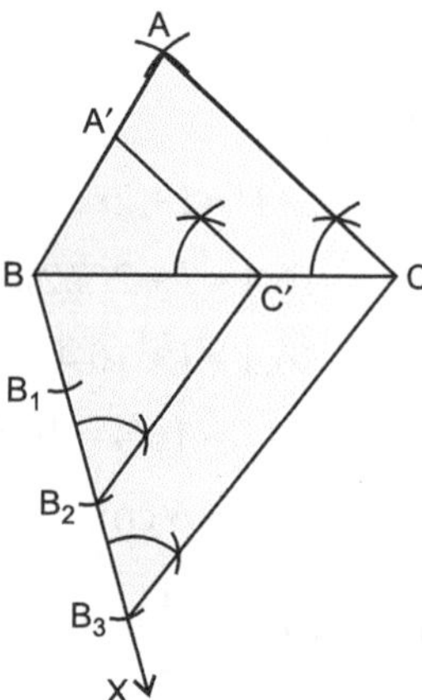

(iii) With C as centre and radius 5 cm, draw another arc intersecting the previous arc at A.

(iv) Join AB and AC to get ΔABC.

(v) Draw a ray BX making an acute angle with BC.

(vi) With any convenient radius draw 3 arcs on BX *i.e.*, B_1, B_2 and B_3 such that $BB_1 = B_1B_2 = B_2B_3$.

(vii) Join B_3C.

(viii) Draw $B_2C' \parallel B_3C$ making $\angle BB_3C = \angle BB_2C'$ meeting BC at C′.

(ix) Draw $C'A' \parallel CA$ making $\angle ACB = \angle A'C'B$ metting AB at A′.

Then ΔA′BC′ is the required triangle.

●●

Delhi [Set-III]

Note : Except for the following questions, all the remaining questions have been asked in previous set.

SECTION – A

14. A ladder 10 m long reaches a window 8 m above the ground. The distance of the foot of the ladder from the base of the wall is .... m. **[1]**

Sol. 6

Explanation : In ΔABC,

$AC = 10$ m (length of ladder)

$AB = 8$ m (height of window)

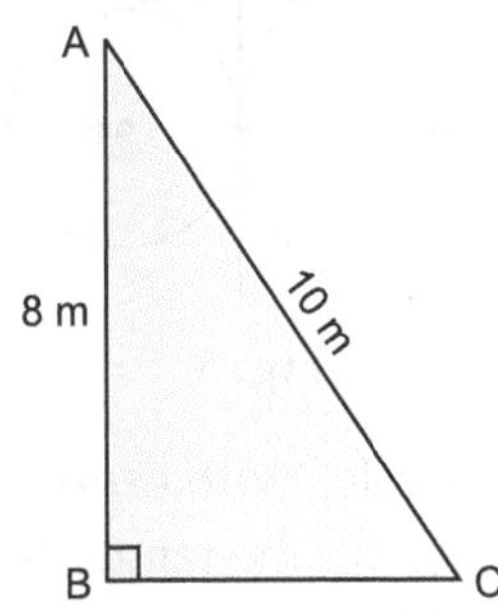

Now $AC^2 = AB^2 + BC^2$ (Pythagoras theorem)

$\Rightarrow \quad (10)^2 = (8)^2 + BC^2$

$\Rightarrow \quad 100 - 64 = BC^2$

$\Rightarrow \quad BC^2 = 36$

$\Rightarrow \quad BC = 6$ m

15. $\dfrac{2\cos 67^\circ}{\sin 23^\circ} - \dfrac{\tan 40^\circ}{\cot 50^\circ} - \cos 0^\circ =$ **................** **[1]**

Sol. 0

Explanation : $\dfrac{2\cos 67^\circ}{\sin 23^\circ} - \dfrac{\tan 40^\circ}{\cot 50^\circ} - \cos 0^\circ$

$$= \frac{2\cos(90-23)^\circ}{\sin 23^\circ} - \frac{\tan(90-50)^\circ}{\cot 50^\circ} - \cos 0^\circ$$

$$= \frac{2\sin 23^\circ}{\sin 23^\circ} - \frac{\cot 50^\circ}{\cot 50^\circ} - \cos 0^\circ$$

$$\left\{\begin{array}{l} \because \cos(90^\circ - \theta) = \sin\theta \\ \tan(90^\circ - \theta) = \cot\theta \end{array}\right\}$$

$= 2(1) - 1 - 1$ $\{\cos 0^\circ = 1\}$

$= 2 - 2 = 0$

20. A pair of dice is thrown once. What is the probability of getting a doublet ? **[1]**

Sol. We know,

Total number of possible outcomes = 36

Favourable outcomes = {(1, 1), (2, 2), (3, 3), (4, 4), (5, 5), (6, 6)}

$\Rightarrow$ No. of favourable outcomes = 6

$\therefore$ P(getting a doublet)

$$= \frac{\text{No. of favourable outcomes}}{\text{No. of total possible outcomes}}$$

$$= \frac{6}{36}$$

$$= \frac{1}{6}$$ **Ans.**

SECTION – B

25. The minute hand of a clock is 12 cm long. Find the area of the face of the clock described by the minute hand in 35 minutes. **[2]**

Sol. Given, $r = 12$ cm

Angle swept by minute hand in 35 minutes

$= 35 \times 6^\circ$ ($\because$ 1 minute = 6°)

$\Rightarrow \quad \theta = 210^\circ$

Area swept by minute hand

= Area of sector

$$= \frac{\theta}{360^\circ} \times \pi r^2$$

$$= \frac{210}{360} \times \frac{22}{7} \times 12 \times 12$$

$= 22 \times 12$ cm^2

$= 264$ cm^2 **Ans.**

26. The sum of the first 7 terms of an A.P. is 63 and that of its next 7 terms is 161. Find the A.P. **[2]**

Sol. For the given A.P.,

$S_7 = 63$ and $S_{14} - S_7 = 161$

Now we know,

$$S_n = \frac{n}{2}[2a + (n-1)d]$$

$$\therefore \quad S_7 = \frac{7}{2}[2a + (7-1)d]$$

$$\Rightarrow \quad 63 = \frac{7}{2}[2a + 6d]$$

$$\Rightarrow \quad a + 3d = 9 \quad \text{...(i)}$$

Also, $S_{14} - S_7 = 161$

$$\Rightarrow \quad S_{14} = S_7 + 161$$

$$\Rightarrow \quad \frac{14}{2}[2a + (14-1)d] = 63 + 161$$

$$\Rightarrow \quad 2a + 13d = \frac{224}{7}$$

$$\Rightarrow \quad 2a + 13d = 32$$

From equation (i) we get $a = 9 - 3d$

Putting this value of a in equation (ii), we get

$$2(9 - 3d) + 13d = 32$$

$$\Rightarrow \quad 18 - 6d + 13d = 32$$

$$\Rightarrow \quad 7d = 32 - 18$$

$$\Rightarrow \quad 7d = 14$$

$$\Rightarrow \quad d = 2$$

Putting d in equation (i)

$$\Rightarrow \quad a = 9 - 3(2)$$

$$\Rightarrow \quad a = 3$$

$$a + d = 3 + 2 = 5$$

$$a + 2d = 3 + 2(2) = 7$$

$\therefore$ A.P. is 3, 5, 7 ...... **Ans.**

SECTION – C

32. A man can row a boat downstream 20 km in 2 hours and upstream 4 km in 2 hours. Find his speed of rowing in still water. Also find the speed of the stream. [3]

Sol. Let speed of the stream be x km/hr

speed of the boat in still water be y km/hr.

Speed of the boat upstream = $(y - x)$ km/hr

speed of the boat downstream = $(y + x)$ km/hr

For downstream motion :

$$\frac{20}{y+x} = 2$$

or $\quad y + x = 10 \quad$ ...(i)

For upstream motion :

$$\frac{4}{y-x} = 2$$

or $\quad y - x = 2 \quad$ ...(ii)

On adding equation (i) and (ii), we get

$$2y = 12$$

$$\Rightarrow \quad y = 6$$

Putting $y = 6$ in equation (i), we get

$$6 + x = 10$$

$$x = 4$$

$\therefore$ Speed of the stream = 4 km/hr

Speed of the boat in still water = 6 km/hr

Ans.

33. In given Fig. 5, two circles touch each other at the point C. Prove that the common tangent to the circles at C, bisects the common tangent at P and Q. [3]

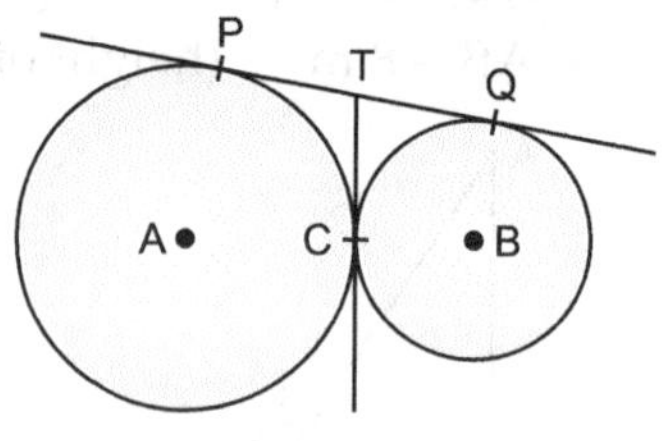

Fig. 5

Sol. Given : Two circles with centres A and B. Also, PT, TQ and TC are the tangents.

To Prove : CT bisects PQ *i.e.*, PT = QT

Proof : We know that,

The length of two tangents drawn to a circle from an external point are equal.

$\therefore$ For point T and circle with centre A,

$$TP = TC \quad \text{...(i)}$$

For point T and circle with centre B,

$$TQ = TC \quad \text{...(ii)}$$

From (i) and (ii),

$$TP = TQ$$

$\therefore$ T is mid-point of PQ

Hence, TC bisects PQ **Hence Proved.**

34. Prove that : $\frac{\cot\theta + \text{cosec}\,\theta - 1}{\cot\theta - \text{cosec}\,\theta + 1} = \frac{1 + \cos\theta}{\sin\theta}$ [3]

Sol. $\frac{\cot\theta + \text{cosec}\,\theta - 1}{\cot\theta - \text{cosec}\,\theta + 1}$

$$= \frac{\cot\theta + \text{cosec}\,\theta - (\text{cosec}^2\theta - \cot^2\theta)}{(\cot\theta - \text{cosec}\,\theta + 1)}$$

$[\because \text{cosec}^2\theta - \cot^2\theta = 1]$

$$= \frac{(\text{cosec}\,\theta + \cot\theta) - (\text{cosec}\,\theta + \cot\theta)(\text{cosec}\,\theta - \cot\theta)}{(\cot\theta - \text{cosec}\,\theta + 1)}$$

$$= \frac{(\text{cosec}\,\theta + \cot\theta)[1 - (\text{cosec}\,\theta - \cot\theta)]}{(\cot\theta - \text{cosec}\,\theta + 1)}$$

$$= \frac{(\text{cosec}\,\theta + \cot\theta)[1 - \text{cosec}\,\theta + \cot\theta]}{(1 - \text{cosec}\,\theta + \cot\theta)}$$

$$= \text{cosec}\,\theta + \cot\theta$$

$$= \frac{1}{\sin\theta} + \frac{\cos\theta}{\sin\theta}$$

$$= \frac{1+\cos\theta}{\sin\theta}$$

Hence Proved.

SECTION – D

39. Draw a ΔABC with BC = 7 cm, $\angle B = 45°$ and $\angle A = 105°$. Then construct another triangle whose sides are $\frac{3}{4}$ times the corresponding sides of ΔABC. **[4]**

Sol. In ΔABC,

$$\angle A + \angle B + \angle C = 180° \quad \text{(Angles sum property)}$$

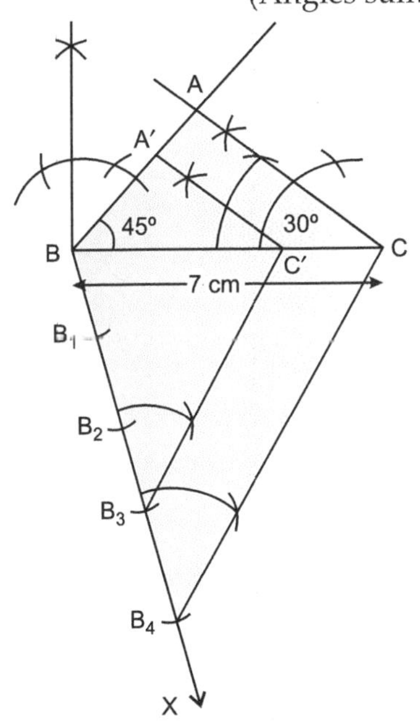

$$\Rightarrow \quad 105° + 45° + \angle C = 180°$$

$$\Rightarrow \quad 150° + \angle C = 180°$$

$$\Rightarrow \quad \angle C = 30°$$

Steps of construction :

(i) Draw BC = 7 cm

(ii) At point B, construct $\angle B = 45°$.

(iii) At point C, construct $\angle C = 30°$. Thus ΔABC is obtained.

(iv) Draw a ray BX making an acute angle with BC

(v) With any convenient radius draw 4 arcs on BX *i.e.*, B_1, B_2, B_3 and B_4 such that $BB_1 = B_1B_2 = B_2B_3 = B_3B_4$.

(vi) Join B_4C.

(vii) Draw $B_3C' \parallel B_4C$ making $\angle BB_4C = \angle BB_3C'$ meeting BC at C'.

(viii) Draw $C'A' \parallel CA$ making $\angle ACB = \angle A'C'B$ meeting BA at A'.

Then, $\Delta A'BC'$ is the required triangle.

40. From the top of a 7 m high building the angle of elevation of the top of a tower is 60° and the angle of depression of its foot is 45°. Determine the height of the tower. **[4]**

Sol. Let AB be the building and CD be the tower.

$$AB = 7 \text{ m}$$

Also, let $\quad DC = h$ m.

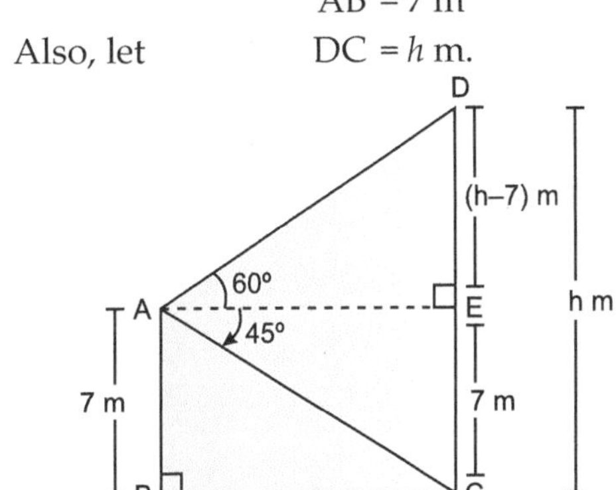

Now, in ΔAEC, we have

$$\tan 45° = \frac{EC}{AE}$$

$$\Rightarrow \quad 1 = \frac{7}{AE}$$

$$\Rightarrow \quad AE = 7$$

In ΔADE, we have

$$\tan 60° = \frac{DE}{AE}$$

$$\Rightarrow \quad \sqrt{3} = \frac{h-7}{7}$$

$$\Rightarrow \quad 7\sqrt{3} = h - 7$$

$$\Rightarrow \quad h = 7\sqrt{3} + 7$$

$$\Rightarrow \quad h = 7(\sqrt{3}+1)$$

$\therefore$ The height of the tower is $7(\sqrt{3}+1)$ meters.

Ans.